15.00
D1552279

Dynamics
for Engineers

Springer
New York
Berlin
Heidelberg
Barcelona
Budapest
Hong Kong
London
Milan
Paris
Santa Clara
Singapore
Tokyo

B.B. Muvdi
A.W. Al-Khafaji
J.W. McNabb
Bradley University

Dynamics for Engineers

With 1118 Illustrations

Springer

Bichara B. Muvdi
Amir W. Al-Khafaji
J.W. McNabb
Civil Engineering and Construction
Bradley University
Peoria, IL 61625
USA

The original illustration depicting an object impacting the Earth (Chapter 16) courtesy of William K. Hartmann of the Planetary Science Institute, Tucson, AZ.

Library of Congress Cataloging-in-Publication Data
Muvdi, B.B.
Dynamics for engineers / B.B. Muvdi, A.W. Al-Khafaji, J.W. McNabb.
p. cm.
Includes bibliographical references and index.
ISBN 0-387-94798-1 (hardcover: alk. paper)
1. Dynamics. 2. Mechanics, Applied. 3. Kinematics. I. Al-Khafaji, Amir Wadi. II. Title.
TA352.M88 1996
620.1'04—dc20 96-13159

Printed on acid-free paper.

© 1997 Springer-Verlag New York, Inc.
All rights reserved. This work may not be translated or copied in whole or in part without the written permission of the publisher (Springer-Verlag New York, Inc., 175 Fifth Avenue, New York, NY 10010, USA), except for brief excerpts in connection with reviews or scholarly analysis. Use in connection with any form of information storage and retrieval, electronic adaptation, computer software, or by similar or dissimilar methodology now known or hereafter developed is forbidden.
The use of general descriptive names, trade names, trademarks, etc., in this publication, even if the former are not especially identified, is not to be taken as a sign that such names, as understood by the Trade Marks and Merchandise Marks Act, may accordingly be used freely by anyone.

Production managed by Lesley Poliner; manufacturing supervised by Jeffrey Taub.
Typeset by Asco Trade Typesetting Ltd., Hong Kong.
Printed and bound by R.R. Donnelley and Sons, Harrisonburg, VA.
Printed in the United States of America.

9 8 7 6 5 4 3 2 1

ISBN 0-387-94798-1 Springer-Verlag New York Berlin Heidelberg SPIN 10539336

To my wife, Gladys, children, B. Charles, Diane, Katherine, Patti, and George, and grandchildren, Valerie, Christopher, and Richard, as well as my close friends, for their patience and continued support and encouragement during the preparation of the manuscripts.

B.B. Muvdi

To my children, Ali, Laith, and Elise, for keeping me young and happy.

A.W. Al-Khafaji

To Ada Spring McNabb, our children, their spouses, and young writers: Amy Chloe, Nicholas, and Phoebe.

J.W. McNabb

Preface

"Mechanics is one of the branches of physics in which the number of principles is at once very few and very rich in useful consequences. On the other hand, there are few sciences which have required so much thought—the conquest of a few axioms has taken more than 2000 years."—Rene Dugas, *A History of Mechanics*

Introductory courses in *engineering mechanics* (statics and dynamics) are generally found very early in engineering curricula. As such, they should provide the student with a thorough background in the basic fundamentals that form the foundation for subsequent work in engineering analysis and design. Consequently, our primary goal in writing *Statics for Engineers* and *Dynamics for Engineers* has been to develop the fundamental principles of *engineering mechanics* in a manner that the student can readily comprehend. With this comprehension, the student thus acquires the tools that would enable him/her to think through the solution of many types of engineering problems using logic and sound judgment based upon fundamental principles.

Approach

We have made every effort to present the material in a concise but clear manner. Each subject is presented in one or more sections followed by one or more examples, the solutions for which are presented in a detailed fashion with frequent reference to the basic underlying principles. A set of problems is provided for use in homework assignments. Great care was taken in the selection of these problems to ensure that all of the basic fundamentals discussed in the preceding section(s) have been given adequate coverage. The problems in a given set are organized so that the set begins with the simplest and ends with the most difficult ones. A sufficient number of problems is provided in each set to allow the use of both books for several semesters without having to assign the same problem more than once. Also, each and every one of the twenty three chapters in both books contains a set of review problems which, in general, are a little more challenging than the homework set of problems. This feature allows the teacher the freedom to choose homework assignments from either or from both sets. The two books have a total number of problems in excess of 2,600. All of the examples and problems were selected to reflect realistic situations. However, because these two books were written for beginning courses in statics and dynamics, the principal objective of these examples and problems remains to demonstrate the subject matter and to illustrate how the fundamental principles of mechanics may be used in the solution of practical problems.

Math Background

The prerequisite mathematical background needed for mastery of the material in both books consists primarily of high school courses in algebra and trigonometry and a beginning course in differential and integral calculus. We have made occasional use of vector analysis in the development of some concepts and basic principles and in the solution of some problems. The needed background in vector operations, however, is introduced and developed as needed throughout the books, particularly, in Chapters 3, 5, and 13. It should be emphasized, however, *that the vector approach is used only when it is judged to offer distinct advantages over the scalar method.* Such is the case, for example, in the solution of three-dimensional problems.

Free-Body Diagram

The very important concept of the free-body diagram is introduced in Chapters 2 and 4 but is used extensively throughout. It is our firm belief that the free-body diagram greatly enhances the understanding of the fundamental principles of mechanics. Thus, the free-body diagram is used not only in the solution of statics problems but also in the solution of dynamics problems whether Newton's second law, the energy method, or the impulse-momentum technique is used in their solution.

Organization

We have organized the material in both books to enable us to present the simple concepts before embarking on more difficult ones. Thus, the treatment of the mechanics of particles is dealt with before considering the mechanics of rigid bodies. Also, two-dimensional mechanics is presented separately, and before, three-dimensional mechanics. This approach allows the instructor to focus on simple concepts early in the semester and to postpone the more complex concepts until a later date when the student has had a chance to develop some maturity in the principles of mechanics. Furthermore, the separation between two- and three-dimensional mechanics in both *Statics for Engineers* and *Dynamics for Engineers*, provides the instructor flexibility in selecting topics to teach during a given semester.

Nontraditional Topics

In addition to the traditional topics found in existing books, we have included several new topics that we felt may be of interest to some teachers. These include: Axial Force and Torque Diagrams, General Theorem for Cables, a brief treatment of the Six Fundamental Machines and the use of Lagrange's Equations in the formulation of the equations of motion. It should be stated, however, that among the topics covered (both traditional and new) are some that are judged to be *not* essential for an understanding of the basic concepts of mechanics of rigid bodies. These topics are identified by asterisks in the Table of Contents.

Units

In view of the fact that the international system of units, referred to as SI (System International), is now beginning to gain acceptance in this country, it was decided to use it in this book. However, it is realized that a complete transition from the U.S. Customary to SI units will be a slow and costly process that may last as long as 20 years and possibly longer. Some factors that will play a significant role in slowing down the transformation process are the existing literature of engineering research and development, plans, and calculations, as well as structures and production machinery, that have been conceived and built using largely the U.S. Customary system of units. Thus, the decision was made to use both systems of units in this book. Approximately one-half of the examples and one-half of the problems are stated in terms of the U.S. Customary system whereas the remainder are given in terms of the emerging SI system of units.

Special Features

In addition to the features described under APPROACH, these two textbooks contain some special features that may be summarized as follows:

1. Many of the example and homework problems are designed to obtain *general symbolic solutions* that allow the student to view engineering problems from a broad point of view before assigning specific numerical data.
2. Each of the twenty-three chapters in both books is prefaced by a carefully written vignette designed to motivate the student prior to undertaking the study of the chapter.
3. The two books contain over 350 examples carefully worked out in sufficient detail to make the solutions easily understood by the student.
4. The two textbooks are characterized by the extensive use of free-body diagrams as well as impulse-momentia and inertia-force diagrams. Each of these diagrams is accompanied by a right-handed coordinate system that establishes the sign convention being used. With a few exceptions, three-dimensional, right-handed, x-y-z coordinate systems are shown with the x axis coming out of the paper. However, all of the two-dimensional, right-handed, x-y coordinate systems are shown with the x axis pointing to the right.
5. Each of the two volumes, *Statics for Engineers* and *Dynamics for Engineers* have a companion *Solutions Manual.*

In addition to providing complete solutions to all of the homework problems, each *Solutions Manual* contains suggested outlines for courses in statics and in dynamics.

Appendices

Six Appendices containing information useful in the solution of many problems have been included at the end of each of the two books. Appendix A contains information about a selected set of areas, Appendix B, information about a selected set of masses, Appendix C, useful mathematical relations, Appendix D, selected derivatives, Appendix E, selected integrals, and Appendix F, information about supports and connections.

Acknowledgments

We acknowledge with much gratitude the assistance we received in the typing of the manuscript by Ms. Sharon McBride, Ms. Janet Maclean, and Wilma Al-Khafaji. We also acknowledge with thanks the help given by Dr. Farzad Shahbodaghlou and Dr. Akthem Al-Manaseer in the typing of the Solutions Manuals. The authors are grateful to the many colleagues and students who have contributed significantly and often indirectly to their understanding of statics and dynamics. Contributions by many individuals are given credit by reference to their published work and by quotations. The source of photographs is indicated in each case. These books have been written on the proposition *that good judgment comes from experience and that experience comes from poor judgment*. We certainly feel that the books are a real contribution to our profession, but we have miscalculated the enormous sacrifices they required. The quality of these books has been and will continue to be judged by our students and colleagues whose comments and suggestions have contributed greatly to the successful completion of the final manuscript. We would not have been able to complete this project without the help and support of our families. We apologize for any omissions.

The authors appreciate the efforts of the reviewers, who, by their criticisms and helpful comments, have encouraged us in the preparation and completion of the manuscript.

Our thanks to our editor Mr. Thomas von Foerster. All the people at Springer-Verlag who were involved with the production of this book deserve special acknowledgment for their dedication and hard work.

B.B.M.
A.W.A.
J.W.M.

Contents: Dynamics for Engineers

23 Vibrations 769

Appendices

Answers 872

Index 895

Contents: Statics for Engineers

13

Kinematics of Particles

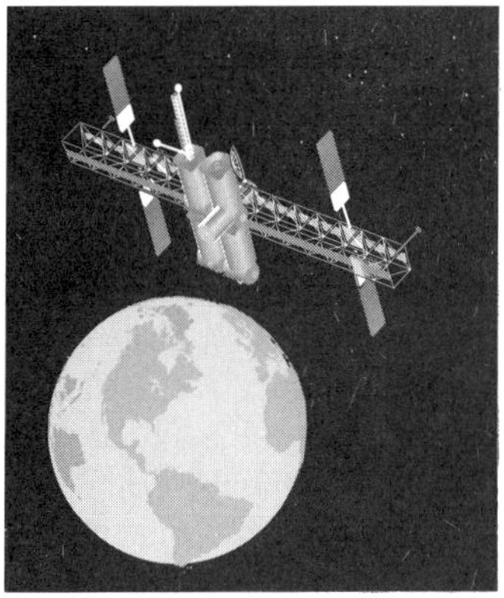

As you begin the study of dynamics, you should be filled with excitement, interest, and a craving to discover new principles, conquer new frontiers, and develop new insights. This is a necessary step to establishing a solid foundation for your future development as an engineer. At times you may question the benefits of some topics and their real engineering value. This is not an unexpected reaction for a beginning engineering student whose sense of the profession is yet to be developed.

As we write this book, NASA is busy developing plans for a space station that will serve as a base from which future spacecraft will be launched to Mars and other planets. Using only existing materials and technology, permanent habitats in space could be established before the end of this century. Such colonies would help alleviate the problems of overcrowding, pollution, drought, and energy shortages on Earth. The contemplation of such a future is ancient and has been a subject of scientific speculation for several decades. Among the surprises to develop during the recent past was the realization that space possesses unique qualities of definite advantage to future colonists. Sunlight is available in a continuous stream to provide power 24 hours per day, to grow crops throughout the year, and to provide light and warmth constantly. Furthermore, the lack of gravity makes it easier to move materials between points with energy needed to change speed and direction, rather than to maintain them. This is precisely what Newton's laws state.

The scientific contributions made over the centuries by giants, such as Isaac Newton, form the foundation upon which dynamics principles are based. Undoubtedly, you have already been exposed to the wonders of dynamics through the gigantic scientific marvels, such as the space shuttle. Additionally, you must have been exposed to the serious challenges facing mankind from pollution on Earth to the depletion of the ozone layer in space. The topics that follow illustrate the need for competent engineers who will be able to tackle, in space, the challenges facing Earth. Just remember that you may be the first astronaut who will land on Mars or the scientist who will make the next great discovery.

RECTILINEAR MOTION

13.1 Position, Velocity and Acceleration

Theoretical research in astrophysics together with experimental data on X-ray emissions from the binary star Cygnus X-1 strongly suggests that very massive dying stars collapse to form *black holes*. This has revised our thinking about whether it is reasonable to assume that a mass may be located at a point. With this in mind, we define a particle as a *point mass* and imagine that a mass may be located at a mathematical point in space. What we term *a particle* depends upon our judgment as engineers or scientists who create mathematical models. If we wish to study the motion of a baseball as it moves through the atmosphere, we would usually consider it to be a particle. If we were interested in forces delivered by the bat to the ball, then, initially, we may model the ball as an elastic sphere. Space mechanics suggests more practical examples. A space craft traveling in the solar system is readily modeled as a particle. If it is designed to land on another planet, a more elaborate model is required to study effects of landing forces on the craft. The mathematical model we adopt for a given body depends upon the nature of the study we wish to conduct. A knowledge of the fundamentals of dynamics, experience, and judgment are all involved in the successful creation of mathematical models. Whenever experimental results are available for comparison with our model predictions, we are able to decide that our model is satisfactory or needs modification.

The term *particle kinematics* refers to the study of the geometry of motion of a particle without concern for the forces causing that motion. The geometry of particle motion is completely defined by specifying its *position*, its *velocity*, and its *acceleration*.

Position

The *position* of a particle at a given instant of time refers to its location relative to some reference point usually taken as the origin of a coordinate system. Thus, if a particle moves along a straight line path, we refer to the motion as one-dimensional or rectilinear. Consider a particle P, at any time t, which moves along the s axis as shown in Figure 13.1(a). The $\mathbf{r} = s\boldsymbol{\lambda}$ vector referred to as the position vector of the particle, enables us to determine the position of the particle P, at any time t, because s is a function of time. Note that s is the magnitude of $\mathbf{r}$, and $\boldsymbol{\lambda}$ is a unit vector along the positive s axis. In general, we select an origin O arbitrarily on the straight line path and indicate a positive sense for the s coordinate as shown. When $s = 0$, the particle is at the origin O, and, when s is positive, the particle lies to the right of origin. The function $s(t)$ provides us with *timetable* information on the motion of the particle. If we specify the time, say $t = t_1$, then the value of the function at this time $s(t_1)$ gives us the position of the particle P on its path.

Displacement

As the particle moves along its straight-line path, its position changes with time. By definition, the displacement $\Delta\mathbf{r}$ of the particle during a

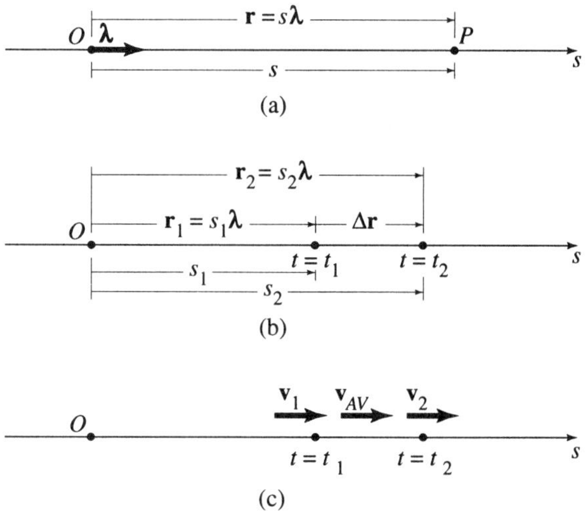

FIGURE 13.1.

time interval Δt is defined as the change in its position during this time interval. Thus, as shown in Figure 13.1(b),

$$\Delta \mathbf{r} = \mathbf{r}_2 - \mathbf{r}_1, \tag{13.1a}$$

and

$$\Delta s = s_2 - s_1. \tag{13.1b}$$

Average and Instantaneous Velocities

The *average velocity* of the particle during the time interval ($t_1 \leq t \leq t_2$) is defined as

$$\mathbf{v}_{AV} = \frac{\Delta \mathbf{r}}{\Delta t} \tag{13.2a}$$

and

$$v_{AV} = \frac{\Delta s}{\Delta t}. \tag{13.2b}$$

Equations (13.2) assume that the change in displacement occurs at a constant rate during the time interval t. A positive sign for $\mathbf{v}_{AV}$ indicates that this vector is directed in the positive sense along the s axis. A negative sign for $\mathbf{v}_{AV}$ would indicate that this vector is directed in the negative sense along the s axis. A positive $\mathbf{v}_{AV}$ is shown in Figure 13.1(c) during the time interval $\Delta t = t_2 - t_1$.

To define the *instantaneous velocity* $\mathbf{v}$ of a particle P on its straight-line path, we let the time interval approach zero in the limit in Eqs. (13.2). Thus,

$$\mathbf{v} = \lim_{\Delta t \to 0} \frac{\Delta \mathbf{r}}{\Delta t} = \frac{d\mathbf{r}}{dt} \tag{13.3a}$$

and

$$v = \lim_{\Delta t \to 0} \frac{\Delta s}{\Delta t} = \frac{ds}{dt} \tag{13.3b}$$

where v is known as the *instantaneous speed* of the particle. A positive instantaneous velocity is directed along the positive s axis. Instantaneous velocities v_1 and v_2 at $t = t_1$ and $t = t_2$, respectively, are indicated in Figure 13.1(c). Because the displacement of a particle, moving along a straight line path, lies along the path and since time is a scalar quantity, we observe from Eqs. (13.3) that the velocity vector is also directed along the straight-line path of the particle. Both velocity and speed have the dimensions of length divided by time and are measured in units, such as m/s or ft/s.

Average and Instantaneous Accelerations

In our study of kinematics of straight-line motion, in addition to position, displacement, and velocity, we need to learn how the velocity varies with time which requires definitions for *average* and *instantaneous accelerations.* We define average acceleration by

$$\mathbf{a}_{AV} = \frac{\Delta \mathbf{v}}{\Delta t} \tag{13.4a}$$

and

$$a_{AV} = \frac{\Delta v}{\Delta t}. \tag{13.4b}$$

Reference to Figure 13.2 will enhance our understanding of this definition. A particle is initially in some position along its path defined by $\mathbf{r} = \mathbf{r}_1$ at $t = t_1$. At some later time $t = t_2$, it is in a new position defined by $\mathbf{r} = \mathbf{r}_2$. Initially, it has a velocity $\mathbf{v}_1$ and, finally, a velocity $\mathbf{v}_2$. The velocity has changed from an instantaneous value $\mathbf{v}_1$ to an instantaneous value $\mathbf{v}_2$ during the time interval, $\Delta t = t_2 - t_1$. In Eqs. (13.4), the velocity change is assumed to occur at a constant rate over the time interval $\Delta t = t_2 - t_1$. A positive $\mathbf{a}_{AV}$ for this time interval is shown in Figure 13.2(b).

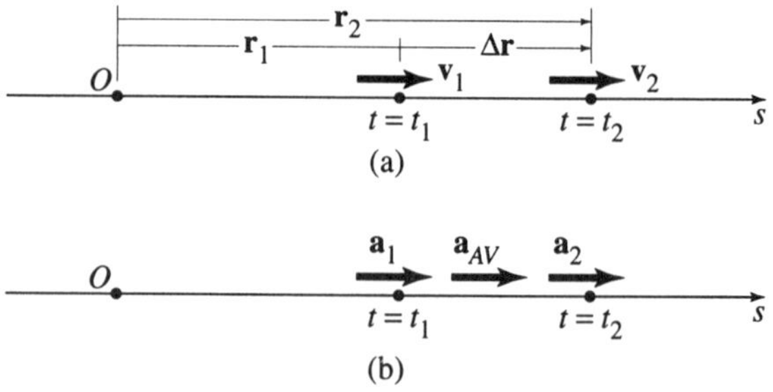

FIGURE 13.2.

The instantaneous acceleration is defined by letting the time interval Δt approach zero in the limit in Eqs. (13.4) to yield

$$\mathbf{a} = \lim_{\Delta t \to 0} \frac{\Delta \mathbf{t}}{\Delta t} = \frac{d\mathbf{v}}{dt} \tag{13.5a}$$

and

$$a = \lim_{\Delta t \to 0} \frac{\Delta v}{\Delta t} = \frac{dv}{dt}. \tag{13.5b}$$

Because the particle travels a straight-line path, the velocity vector does not change direction and, thus, the acceleration vector is directed along the path. A positive sign for **a** indicates that the vector is directed in the positive sense along the path. A negative acceleration is termed a *deceleration*. Instantaneous accelerations $\mathbf{a}_1$ and $\mathbf{a}_2$ at $t = t_1$ and $t = t_2$, respectively, are shown in Figure 13.2(b). From Eqs. (13.4) and (13.5), the unit of acceleration is length divided by time squared. Thus, such units as m/s^2 and ft/s^2 are common. Graphical interpretations of Eqs. (13.3b) and (13.5b) are given in Figure 13.3.

Substitution of Eq. (13.3b) in Eq. (13.5b) provides an alternate equation for the magnitude of the acceleration. Thus,

$$a = \frac{d}{dt}\left(\frac{ds}{dt}\right) = \frac{d^2 s}{dt^2}. \tag{13.6}$$

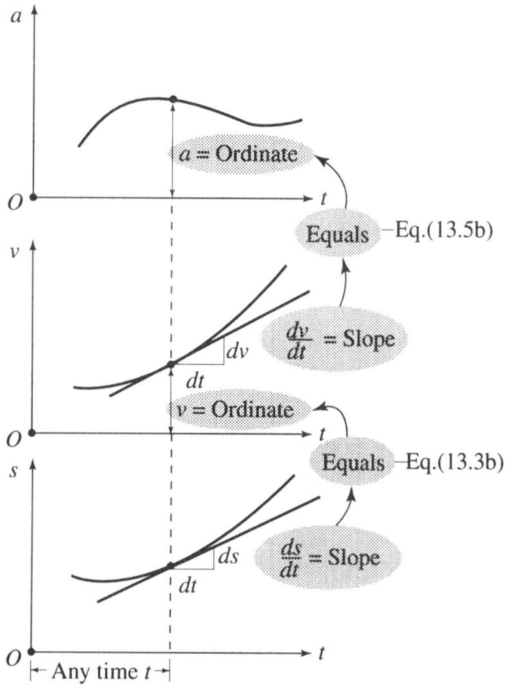

FIGURE 13.3.

Another alternate and very useful equation for the magnitude of the acceleration is obtained by multiplying Eq. (13.5b) on the right side by the unit ratio *ds*/*ds*. Thus,

$$a = \frac{dv}{dt}\left(\frac{ds}{ds}\right).$$

This may be written as

$$a = \left(\frac{ds}{dt}\right)\left(\frac{dv}{ds}\right)$$

from which

$$a = v\frac{dv}{ds}. \tag{13.7}$$

When a particle executes rectilinear motion that repeats itself over a period of time, it is said to have *vibratory* motion. Such is the case, for example, if the position of the particle is defined by a sine function. A detailed study of vibratory motion is given in Chapter 23. For our present needs, however, Figure 13.4 provides an illustration of such motion and defines the terms *amplitude* and *period* of vibration. The term amplitude represents the maximum displacement experienced by the particle during its vibratory (oscillatory) motion. The period represents the time it takes for the particle to execute one complete cycle of vibration.

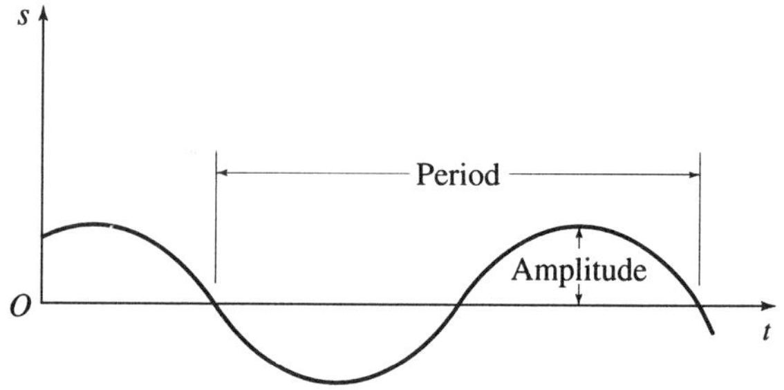

FIGURE 13.4.

The following examples illustrate some of the concepts discussed above.

■ Example 13.1

A remote-controlled vehicle moves along a straight-line path, such that its position is given by

$$s = t^3 + t^2 + t + 10.$$

This function of time is known as the position function and s is expressed in meters whereas t is given in seconds. Determine (a) the velocity at $t = 1$ s, (b) the acceleration at $t = 1$ s, (c) the displacement during the time interval from $t = 0$ s to $t = 3$ s, (d) the average velocity during this same time interval, and (e) the average acceleration during this same time interval.

Solution

Refer to Figure E13.1 as you study this solution. It depicts the answers to each part.

(a) Using Eq. (13.3b),

$$v = \frac{ds}{dt} = 3t^2 + 2t + 1$$

$$v|_{t=1} = 3(1)^2 + 2(1) + 1 = 6.00 \text{ m/s.} \qquad \text{ANS.}$$

(b) From Eq. (13.5b), the acceleration is given by

$$a = \frac{dv}{dt} = 6t + 2$$

$$a|_{t=1} = 6(1) + 2 = 8.00 \text{ m/s}^2. \qquad \text{ANS.}$$

(c) The net displacement after three seconds is calculated as follows:

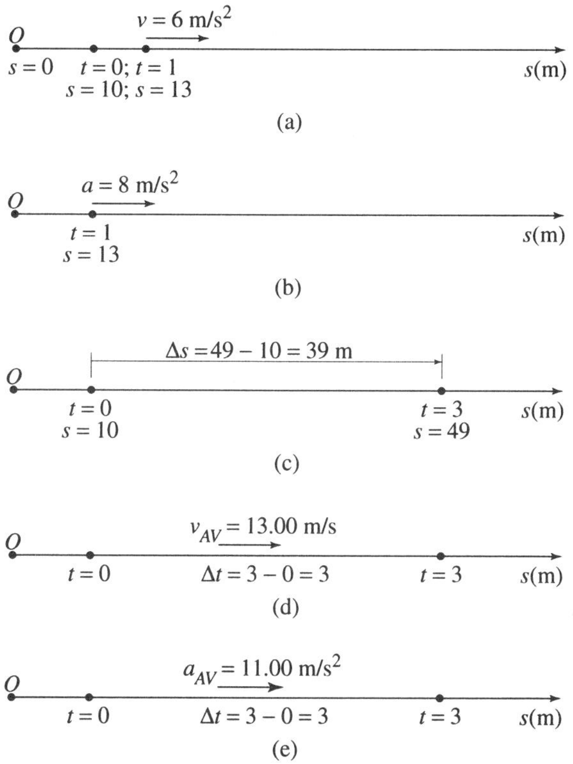

FIGURE E13.1.

$$\Delta s = s|_{t=3} - s|_{t=0}$$
$$s|_{t=3} = (3)^3 + (3)^2 + 3 + 10 = 49 \text{ m}$$
$$s|_{t=0} = (0)^3 + (0)^2 + 0 + 10 = 10 \text{ m}$$
$$\Delta s = 49 - 10 = 39.0 \text{ m.} \qquad \text{ANS.}$$

(d) The average velocity during the first three seconds is calculated using Eq. (13.2b). Thus,

$$v_{AV} = \frac{\Delta s}{\Delta t} = \frac{39.0}{3} = 13.00 \text{ m/s.} \qquad \text{ANS.}$$

(e) The average acceleration during the first three seconds is given by Eq. (13.4b). Thus,

$$a_{AV} = \frac{\Delta v}{\Delta t} = \frac{v|_{t=3} - v|_{t=0}}{\Delta t} = \frac{34 - 1}{3} = 11.00 \text{ m/s}^2. \qquad \text{ANS.}$$

■ Example 13.2

A motorized toy car moves along a straight-line path as shown in Figure E13.2(a) such that, for a short period of time, $s = t^3 - 5t^2 + 2t + 8$ where s is expressed in ft and t in s. Determine (a) the velocity as a function of time, (b) the times at which the velocity is zero, and (c) the acceleration as a function of time.

Solution

(a) Using Eq. (13.3b), the velocity becomes

$$v = \frac{ds}{dt} = 3t^2 - 10t + 2. \qquad \text{ANS.}$$

A sketch of this function is shown in Figure 13.2(b) for $0 \le t < 4$.

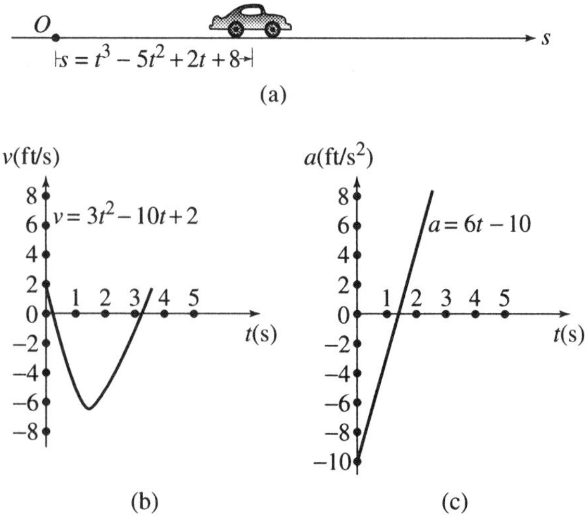

FIGURE E13.2.

(b) Equate v to zero, and solve the resulting quadratic equation for the two values of t.

$$3t^2 - 10t + 2 = 0,$$

$$t = \frac{10 \pm \sqrt{(-10)^2 - 4(3)(2)}}{2(3)} = \frac{10 \pm \sqrt{76}}{6},$$

$$t = 0.214 \text{ s}, 3.12 \text{ s} \qquad \text{ANS.}$$

At these times, the car stops (i.e., $v = 0$) and *reverses the direction of its motion.*

(c) Using Eq. (13.5b), the acceleration is given as

$$a = \frac{dv}{dt} = 6t - 10. \qquad \text{ANS.}$$

A sketch of this function is shown in Figure E13.2(c) for $0 \leq t < 3$.

Example 13.3

Refer to Example 13.2 and determine (a) the displacement during the interval from $t = 0$ to $t = 4$ s, and (b) the distance traveled during the interval from $t = 0$ to $t = 4$ s.

Solution

(a) From Eq. (13.1b), $\Delta s = s_2 - s_1$. Here, s_1 is the value of s at $t = 0$ or the initial position of the particle. Thus,

$$s_1 = f(0) = 8 \text{ ft}.$$

The value s_2 is the position at $t = 4$ s or the final position of the particle. Thus,

$$s_2 = f(4) = 4^3 - 4(4)^2 + 2(4) + 8 = 0,$$

$$\Delta s = s_2 - s_1 = 0 - 8 = -8.00 \text{ ft}. \qquad \text{ANS.}$$

Refer to Figure E13.3(a) to understand the significance of the negative sign.

(b) The displacement during a time interval need not be equal to the distance traveled during the same time interval. An important question to ask is: Did the particle come to rest ($v = 0$) during the interval of interest? In this case, the answer is *yes* and note that the time values for $v = 0$ are $t = 0.214$ s and $t = 3.12$ s from part (b) in Example 13.2. Because the interval contains both $t = 0.214$ s and $t = 3.12$, we are interested in both values. Where is the particle when $v = 0$?

$$s(0.214) = (0.214)^3 - 5(0.214)^2 + 2(0.214) + 8 = 8.209 \text{ ft}.$$

$$s(3.12) = (3.12)^3 - 5(3.12)^2 + 2(3.12) + 8 = -4.061 \text{ ft}.$$

We summarize the position information of interest

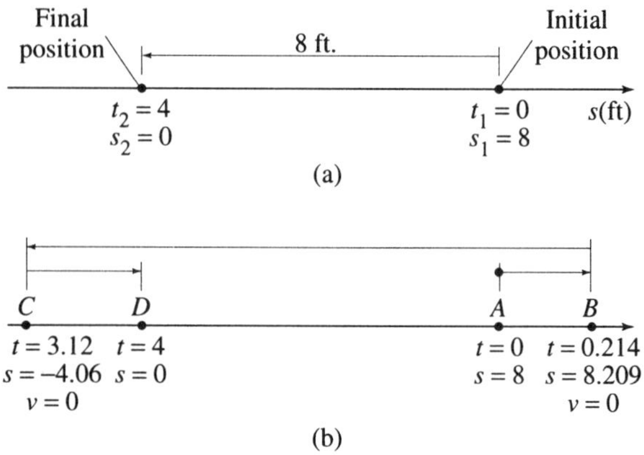

FIGURE E13.3.

$$s(0) = 8.00 \text{ ft},$$

$$s(0.214) = 8.209 \text{ ft},$$

$$s(3.12) = -4.061 \text{ ft},$$

and

$$s(4) = 0.0 \text{ ft}.$$

Reviewing these four values, we describe the particle motion as follows: At $t = 0$, the particle is 8 ft to the right of the origin, at $t = 0.214$ s, it has moved 8.209 ft to the right of the origin and stopped. Then, it begins to move to the left (i.e., v is negative) until, at 3.12 s, it lies at -4.061 ft or 4.061 ft to the left of the origin. Then, it begins to move to the right (i.e., v is positive) until, at $t = 4$, it arrives at the origin (i.e., $s = 0$). The total distance traveled (DT) is

$$\text{DT} = 0.209 + 8.209 + 2(4.061) = 16.54 \text{ ft}. \qquad \text{ANS.}$$

This result is depicted in Figure E13.3(b).

■ Example 13.4

A small block of mass m is attached to a vertical spring of spring constant k and can slide freely in the smooth vertical guide shown in Figure E13.4(a). It is set in motion and vibrates about the equilibrium position (E.P.) such that $s = 20 \sin 5t$ where s in millimeters and t is in seconds. A sketch of this function is shown in Figure E13.4(b). Determine (a) the velocity of the block as a function of time, (b) its acceleration as a function of time, and (c) its position, velocity, and acceleration when $t = 3$ s. Note that the argument $5t$ of the sine function is expressed in radians.

Solution

(a) The velocity of the block as a function of time is obtained by differentiating the given displacement function. Thus,

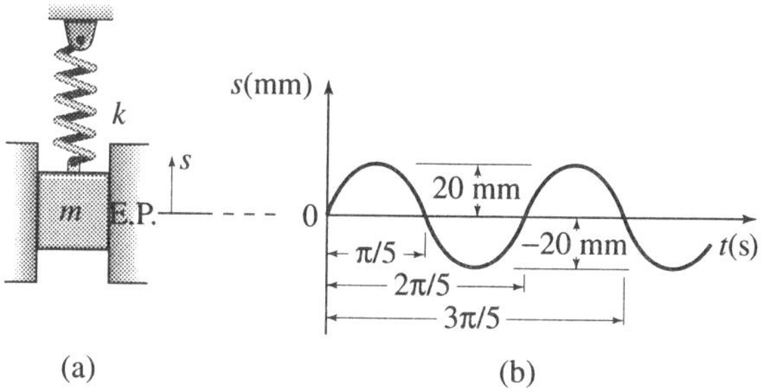

FIGURE E13.4.

$$v = \frac{ds}{dt} = (100 \cos 5t) \text{ mm/s}. \qquad \text{ANS.}$$

(b) The acceleration of the block as a function of time is obtained by differentiating the velocity function. Thus,

$$a = \frac{dv}{dt} = (-500 \sin 5t) \text{ mm/s}^2. \qquad \text{ANS.}$$

(c) For $t = 3$ s,

$$s = 20 \sin 15 = 13.01 \text{ mm},$$
$$v = 100 \cos 15 = -76.0 \text{ mm/s}, \qquad \text{ANS.}$$

where the negative sign implies that the velocity vector is directed downward.

$$a = -500 \sin 15 = -325 \text{ mm/s}^2. \qquad \text{ANS.}$$

Again, the negative sign means that the acceleration vector is directed downward.

■

Problems

Provide sketches showing answers physically in each problem. Assume each particle moves along a horizontal s axis which is directed positively to the right.

13.1 A particle moves along a straight line such that $s = 2t^3$. Determine the velocity and acceleration as functions of time. If t is given in seconds and s is given in meters, state the velocity and acceleration units.

13.2 A vehicle moves along a straight, level roadway as shown in Figure P13.2 such that, for a short period of time, $s = 4t^2 + 12t + 30$ where t is given in seconds and s is given in feet. Determine

FIGURE P13.2.

the velocity and acceleration as functions of time. State the units of v and a.

13.3 For a period of time, the position function for the straight-line motion of a cyclist is given by $s = t^2 - 4t + 6$ where t is given in seconds and s is given in meters. Determine the velocity as a function of time, and find the time at which the velocity is zero.

13.4 Given $s = 2t^2 - 8t + 12$ where t is measured in seconds and s is measured in feet, determine (a) the position of the particle when $t = 0$, (b) the position of the particle when $t = 4$ s, (c) the displacement of the particle from $t = 0$ to $t = 4$ s, (d) the time at which the velocity equals zero, and (e) the total distance traveled during the time interval from $t = 0$ to $t = 4$ s.

13.5 Given $s = t^3 - 6t^2 + 11t - 6$ 6 where t is measured in seconds and s is measured in meters, determine for this rectilinear particle motion (a) the position of the particle when $t = 0$ s, (b) the position of the particle when $t = 2$ s, (c) the displacement of the particle from $t = 0$ s to $t = 2$ s, (d) the times at which the velocity equals zero, and (e) the total distance traveled during the time interval defined in part (c).

13.6 Refer to problem 13.5, and use the same position function to find (a) the position of the particle when $t = 1$ s, (b) the position of the particle when $t = 3$ s, (c) the displacement of the particle during the time interval from $t = 1$ s to $t = 3$ s, (d) the times at which the velocity equals zero, and (e) the total distance traveled during the time interval defined in part (c).

13.7 For a short time interval, the position function for rectilinear motion of a collar in a mechanism is given by $s = t^2 - 16t + 48$ where t is expressed in seconds and s is expressed in feet. Determine (a) the times at which the particle is at the origin for s, (b) the time at which the velocity is zero, (c) the displacement of the particle from $t = 4$ s to $t = 12$ s, and (d) total distance traveled during the time interval defined in part (c).

13.8 A particle oscillates along a straight line such that its position at any time t is given by $s = 10 \sin 4t$ where t is given in seconds and s is given in inches. Determine (a) the velocity as a function of time, (b) the acceleration as a function of time, (c) the amplitude of the motion, and (d) the position, velocity and acceleration of the particle when $t = 0.20$ s. Note that the argument $10t$ of the sine function is given in radians.

13.9 The oscillatory motion of a particle on a straight line is given by $s = 200 \sin 10t$ where t is given in seconds and s is given in millimeters. Determine (a) the velocity as a function of time, (b) the acceleration as a function of time, (c) the amplitude of this vibratory motion, and (d) the position, velocity, and acceleration of the particle when $t = 0.25$ s. Note that the argument $10t$ of the sine function is given in radians.

13.10 In Figure P13.10, the vibratory motion of a small body along a vertical straight-line path is given by $s = 0.5 \cos 2t$ where t is given in seconds and s is given in feet. Determine (a) the amplitude of this motion, (b) the velocity as a function of time, (c) the acceleration as a function of time, and (d) the position, velocity, and acceleration of the particle when $t = 1$ s. Note that the argument $2t$ of the cosine function is given in radians.

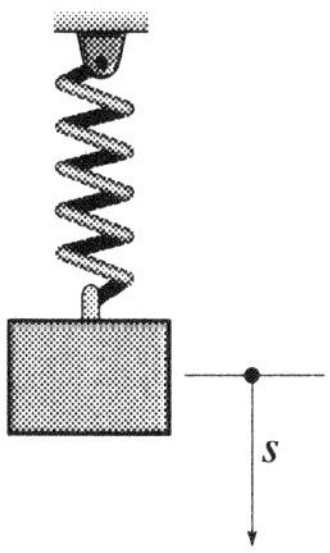

FIGURE P13.10.

13.11 For the time interval $(0 \leq t \leq 20)$, a particle moves along a straight line according to the position function $s = 10 \sinh 0.2t$ where t is given in seconds and s is given in inches. The argument of the hyperbolic sine function $0.2t$ is expressed in radians. Determine (a) the initial velocity of the particle, (b) the initial acceleration of the particle (c) the position of the particle when $t = 20$ s, (d) the velocity of the particle when $t = 10$ s, and (e) briefly explain why this motion is not oscillatory.

13.12 For the time interval $(0 \leq t \leq 10)$, a particle moves along a straight line according to the position function $s = 5 \cosh 0.3t$ where t is given in seconds and s is given in feet. The argument of the hyperbolic cosine function $0.3t$ is expressed in radians. Determine (a) the position of the particle when $t = 0$, (b) the initial velocity of the particle, (c) the initial acceleration of the particle, (d) the velocity of the particle when $t = 5$ s, and (e) the displacement of the particle during the time interval from $t = 0$ to $t = 1$ s.

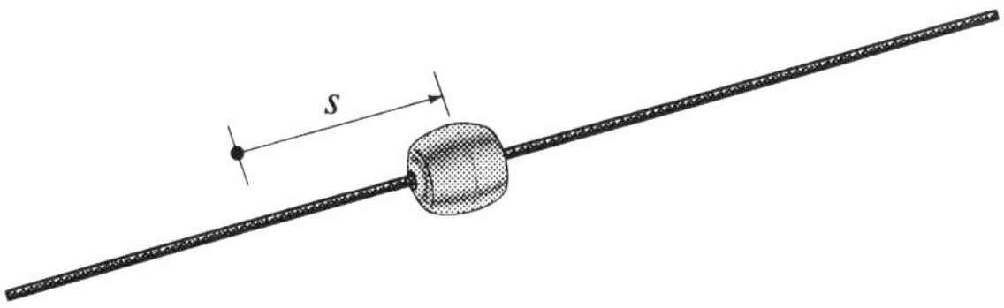

FIGURE P13.13.

13.13 In Figure P13.13, the position function for the rectilinear motion of a bead on a wire is given by $s = t^3 - 4t^2 + 4t$ where t is measured in seconds and s is measured in meters. Determine (a) the times at which the bead is at the origin for s, (b) the velocity of the bead, (c) the times at which the velocity equals zero, (d) the displacement from $t = 0$ to $t = 2$ s, and (e) the total distance traveled during the time interval of part (d).

13.14 A particle moves along a straight line such that $s = -t^3 + 8t^2 - 16t$ where t is measured in seconds and s is measured in feet. Determine (a) a plot for s vs. t showing the times at which the particle is at the origin for s, (b) the velocity of the particle, (c) the times at which the velocity equals zero, (d) the displacement from $t = 0$ to $t = 4$ s, and (e) the total distance traveled during the time interval of part (d) above.

13.15 The rectilinear motion of a remote-controlled vehicle proceeds according to the relationship $s = \frac{1}{24}v^2 + 4$ where v is given in meters per second and s is given in meters. Determine (a) the acceleration of the particle and (b) the position function which expresses s as a function of time, knowing that $s = 4$ m when $t = 0$ s.

13.16 The straight-line motion of a particle takes place as specified by the relationship $s = \frac{1}{8}v^2 - 2$ where v is given in ft/s and s is given in ft. Determine (a) the acceleration of the particle and (b) the position function which expresses s as a function of time, knowing that $s = 0$ when $t = 1$ s.

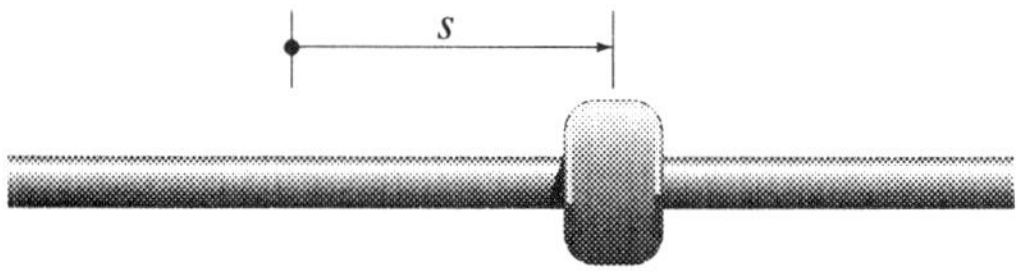

FIGURE P13.17.

13.17 In Figure P13.17, a collar moves along a rod such that $s = \frac{1}{3}t^3 + \frac{1}{2}t^2 + 5$ where s is expressed in feet and t is given in seconds. For the collar, determine (a) the velocity at $t = 1$ s, (b) the acceleration at $t = 1$ s, (c) the displacement during the time interval, from $t = 0$ s to $t = 2$ s, (d) the average velocity during this same interval, and (e) the average acceleration during this same time interval.

13.18 A particle moves along a straight line path such that $s = 2t^3 - 6t^2 + 5t - 6$ where s is expressed in meters and t is given in seconds. For the particle, determine (a) the velocity as a function of time, (b) the times at which the velocity is zero, (c) the acceleration as a function of time, (d) the displacement during the interval from $t = 0$ s to $t = 1$ s, and (e) the distance traveled during the interval from $t = 1$ s to $t = 2$ s.

13.19 A particle starts from rest when $t = 0$ and $s = -8$ and moves along a straight-line path according to the function $s = \frac{1}{16}v^2 - 8$ where v is given in m/s and s is given in m. For the particle, determine (a) the acceleration, (b) the position function which expresses s as a function of time, and (c) the velocity as a function of time.

13.20 The position function for the straight-line motion of a particle is given by $s = 2t^2 - 8t + 12$ where t is given in seconds and s is given in feet. Determine (a) the velocity as a function of time and find the time at which the velocity is zero, (b) the acceleration of the particle, and (c) the distance traveled by the particle during the time interval from $t = 0$ to $t = 3$ s.

13.21 Given $s = 4t^3 - 5t^2 + 10t - 4$ where t is measured in seconds and s is measured in meters, for this straight line motion, find (a) the position of the particle when $t = 0$ s, (b) the position of the particle when $t = 1$ s, (c) the displacement of the particle during the time interval from $t = 0$ to $t = 1$ s, and (d) the time at which the acceleration equals zero.

13.22 A mass attached to a spring oscillator moves along a straight line, as shown in Figure P13.22, according to the position function $s = 20 \cos 2t$ where s is expressed in feet and t is expressed in

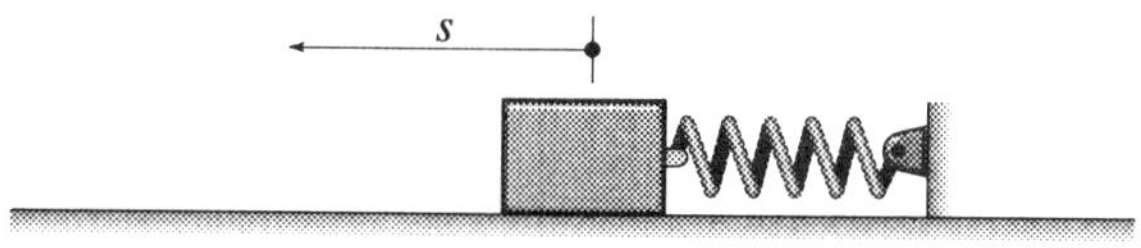

FIGURE P13.22.

seconds. Determine (a) the velocity as a function of time, (b) the acceleration as a function of time, and (c) the position, velocity, and acceleration of the particle when $t = 0.1$ s. Note that the argument of the cosine function is given in radians.

13.23 The oscillatory motion of a particle on a straight line is given by $s = 2 \sin 5t$ where t is in seconds and s is in feet. Determine (a) the velocity as a function of time, (b) the acceleration as a function of time, and (c) the amplitude of this vibratory motion, and (d) the position, velocity and acceleration of this particle when $t = 0.2$ s. Note that the argument of the sine function is expressed in radians.

13.24 The vibratory motion of a particle along a straight-line path is given by $s = 1.2 \sin 4t$ where t is in seconds and s is in meters. Determine (a) the amplitude of this motion, (b) the velocity as a function of time, (c) the acceleration as a function of time, and (d) the position, velocity, and acceleration of the particle when $t = 0.5$ s. Note that the argument of the sine function is given in radians.

13.25 For the time interval ($0 \leq t \leq 10$ s), a toy car moves along a straight line according to the position function $s = 5 \sinh 0.1t$ where t is in seconds and s is in inches. The argument of the hyperbolic sine function is expressed in radians. Determine (a) the initial velocity of the particle, (b) the initial acceleration of the particle, (c) the position of the particle when $t = 10$ s, (d) the velocity of the particle when $t = 10$ s, and (e) briefly explain why this motion is not oscillatory.

13.26 For the time interval ($0 < t < 5$ s), a toy train moves along a straight line according to the position function $s = 4 \cosh 0.2t$ where t is in seconds and s is in feet. The argument of the hyperbolic cosine is given in radians. Determine (a) the position of the particle when $t = 0$, (b) the initial velocity of the particle, (c) the initial acceleration of the particle, (d) the velocity of the particle when $t = 2$ s, and (e) the displacement of the particle during the time interval from $t = 0$ to $t = 2$ s.

13.27 The position function for the rectilinear motion of a particle is given by $s = t^3 - 6t^2 + 11t - 6$ where t is measured in seconds and s in inches. Determine (a) the position of the particle for $t = 0$, $t = 1$, $t = 2$ and $t = 3$ s, (b) the velocity

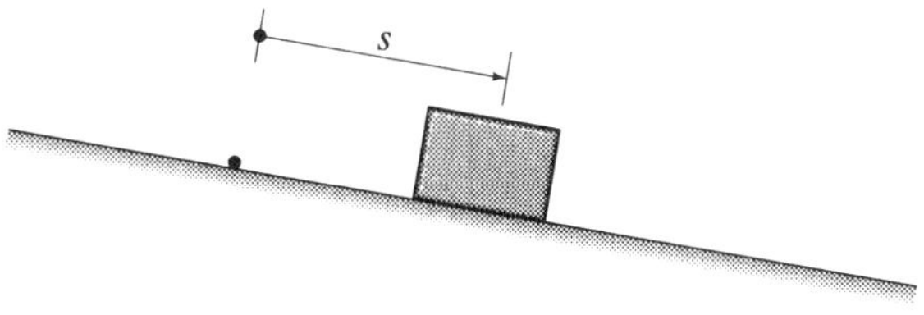

FIGURE P13.28.

of the particle at any time, (c) the times at which the velocity equals zero, and (d) the total distance traveled during the time interval from $t = 0$ to $t = 2$ s.

13.28 In Figure P13.28, a block moves along an inclined plane such that $s = t^3 - 7t^2 + 12t$ where t is measured in seconds and s in feet. Determine (a) a plot for s vs. t showing the times at which the block is at the origin for s, (b) the velocity of the block, (c) the times at which the velocity equals zero, and (d) the position, velocity, and acceleration of the block at $t = 1$ s.

13.29 The rectilinear motion of a particle follows the following function relating v to s, $v^2 = s + 1$, where v is in inches per second and s is in inches. Determine (a) the acceleration of the particle and (b) the function expressing the position s as a function of time, knowing that $s = 0$ when $t = 0$.

13.2 Integral Analysis of Rectilinear Motion

The differential definitions developed in Section 13.1 for velocity and acceleration are useful if we know the mathematical relationship between position and time. In most cases, the available information obtained from observation and measurement relates the acceleration a to one or more of the variables position s velocity v and time t. Using integration along with the available information, we may develop other needed relationship among the variables of the motion. *If the available information cannot be accurately expressed in equation form, the needed integration would have to be performed numerically.* Three cases of practical interest are discussed below:

A. Given $a = f(t)$, develop the v vs. t and s vs. t relationships.
From Eq. (13.5b), we conclude that

$$dv = a\,dt.$$

Substituting the given $a = f(t)$ function and integrating between limits yields

$$\int_{v_0}^{v} dv = \int_0^t f(t)\,dt$$

from which

$$v = v_0 + \int_0^t f(t)\,dt. \tag{a}$$

Equation (a) expresses the velocity as a function of time. Note that the integral in Eq. (a) represents the area under the a-t curve. From Eq. (13.3b), we obtain

$$ds = v\,dt.$$

Integrating between limits we obtain

$$\int_{s_0}^{s} ds = \int_0^t v\,dt$$

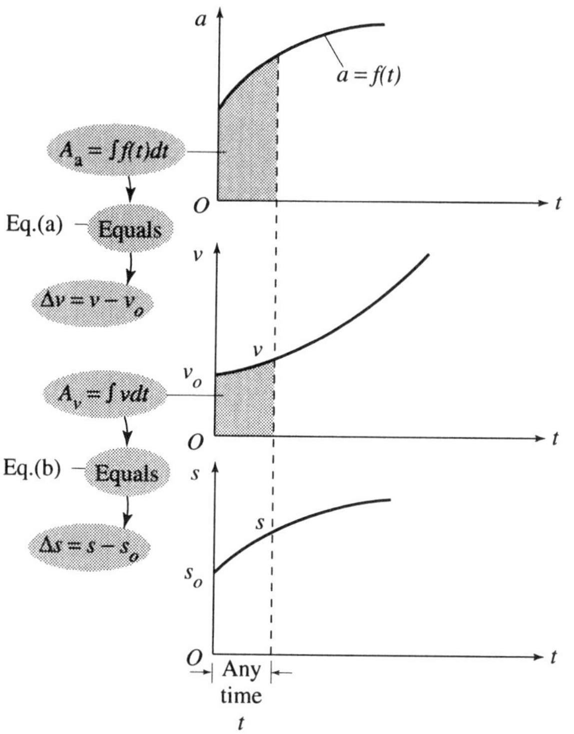

FIGURE 13.5.

from which

$$s = s_0 + \int_0^t v\,dt. \tag{b}$$

Graphical interpretations of Eqs. (a) and (b) are shown in Figure 13.5. The integration in Eq. (b) is performed after substituting for v from Eq. (a) to express the position as a function of time. If the integral in Eq. (b) is too complex for a closed-form solution, a numerical integration is used. Note that the integral in Eq. (b) represents the area under to v-t curve.

B. Given $a = f(s)$, develop the v vs. s and s vs. t relationships.

From Eq. (13.7), we conclude that

$$v\,dv = a\,ds.$$

Substituting the given $a = f(s)$ function and integrating between limits yields

$$\int_{v_0}^{v} v\,dv = \int_{s_0}^{s} f(s)\,ds$$

from which

$$v^2 = v_0^2 + 2\int_{s_0}^{s} f(s)\,ds. \tag{c}$$

Equation (c) expresses the velocity as a function of position. Note that the integral in Eq. (c) represents the area under the *a*-*s* curve. From Eq. (13.3b), we obtain

$$dt = \frac{dt}{v}.$$

Integrating between limits we obtain

$$\int_0^t dt = \int_{s_0}^{s} \frac{ds}{v}$$

from which

$$t = \int_{s_0}^{s} \frac{ds}{v}. \tag{d}$$

The integration in Eq. (d) is performed after substituting for v from Eq. (c) to express the position as a function of time. Again, numerical integration may become necessary.

C. Given $a = f(v)$, develop the v vs. t and s vs. t relationships. Substituting the given function $a = f(v)$ in Eq. (13.5b),

$$f(v) = \frac{dv}{dt}.$$

Rearranging terms and integrating between limits,

$$\int_0^t dt = \int_{v_0}^{v} \frac{dv}{f(v)}$$

from which

$$t = \int_{v_0}^{v} \frac{dv}{f(v)}. \tag{e}$$

Equation (e) expresses the velocity as a function of time. Note that numerical integration may be needed. From Eq. (13.3b), we conclude that

$$ds = v\,dt.$$

Integrating between limits, we obtain

$$\int_{s_0}^{s} ds = \int_0^t v\,dt$$

from which

$$s = s_0 + \int_0^t v\,dt. \tag{f}$$

The integration (closed form or numerical) in Eq. (f) is performed after substituting for v from Eq. (e) to express the position as a function of time.

Alternatively, we may begin with Eq. (13.7), and substitute the given function $a = f(v)$ to obtain

$$f(v) = v\frac{dv}{ds}.$$

Rearranging terms and integrating between limits, we obtain

$$\int_{s_0}^{s} ds = \int_{v_0}^{v} \frac{v\,dv}{f(v)}$$

from which

$$s = s_0 + \int_{v_0}^{v} \frac{v\,dt}{f(v)}. \tag{g}$$

Equation (g) expresses the velocity as a function of position. This expression may now be substituted in Eq. (b) to obtain the relationship between position and time. Note that both of the above operations may require numerical integration.

The following examples illustrate some of the concepts discussed above.

■ Example 13.5

The acceleration of a particle moving along a straight-line path is given by $a = 6t - 20$ where a is expressed in m/s^2 and t in s. If the initial conditions are $s = -18$ m and $v = 27$ m/s when $t = 0$, determine (a) the velocity as a function of time, (b) the position as a function of time, and (c) sketch the a-t, v-t, and s-t curves and provide a graphical interpretation of the relationships that exist among the three curves.

Solution

(a) Using Eq. (13.5b),

$$a = \frac{dv}{dt} = 6t - 20.$$

Separating variables and integrating with the initial conditions ($t = 0$, $v = 27$ m/s) as lower limits and general upper limits, we obtain

$$\int_{27}^{v} dv = \int_{0}^{t} (6t - 20)\,dt$$

$$v|_{27}^{v} = 3t^2 - 20t|_0^t$$

$$v = 3t^2 - 20t + 27. \qquad \text{ANS.}$$

(b) From Eq. (13.3b),

$$v = \frac{ds}{dt} = 3t^2 - 20t + 27.$$

Separating variables and integrating with the given initial conditions ($t = 0$, $s = -18$ m) and general upper limits, we obtain

$$\int_{-18}^{s} ds = \int_{0}^{t} (3t^2 - 20t + 27)\,dt$$

$$s + 18 = t^3 - 10t^2 + 27t.$$

$$s = t^3 - 10t^2 + 27t - 18. \qquad \text{ANS.}$$

(c) The sketches of the motion curves are shown in Figure E13.5 as plotted from the above a vs. t, v vs. t, and s vs. t functions.

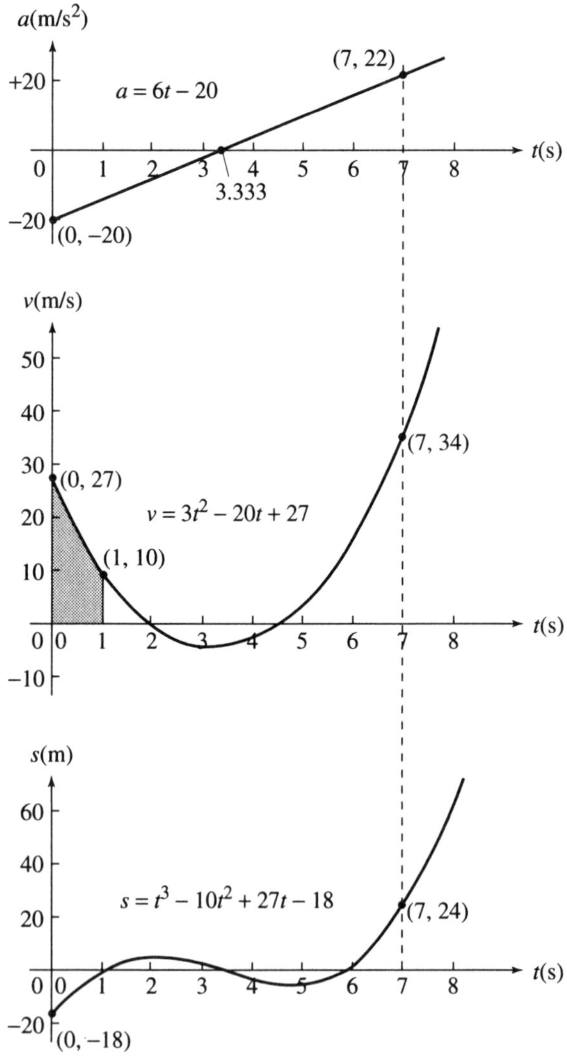

FIGURE E13.5.

Graphically, we may say that the velocity v at any time t can be obtained from the relationship

$$v = \text{Initial velocity} + \text{Area under } a\text{-}t \text{ curve from } t = 0 \text{ to the chosen value of } t.$$

For example,

$$v_{t=7} = 27 + \tfrac{1}{2}(-20)(3.333) + \tfrac{1}{2}(22)(7 - 3.333) = 34.0 \text{ m/s}.$$

Also, the position s at any time t can be obtained from the relationship

$$s = \text{initial displacement} + \text{Area under } v\text{-}t \text{ curve from } t = 0 \text{ to the chosen value of } t.$$

Application of the relationship for s is not easily accomplished in this case because the v-t curve does not consist of straight line segments. However, one may choose small time increments and use a numerical procedure to determine approximate values of the position s corresponding to the chosen time increments. Note that the smaller the time increments, the more accurate will be the values of the computed displacement. If, for example, we wanted to find s at $t = 1$ s, we would approximate the v-t curve in the interval $t = 0$ and $t = 1$ s by a straight line segment and approximate the area under the v-t curve by a trapezoid, as shown shaded in Figure E13.5. Thus,

$$s_{t=1} = -18 + \tfrac{1}{2}(27 + 10)(1) = 0.5 \text{ m}.$$

The exact value of s for $t = 1$ obtained from the s vs. t function is zero. The difference between the exact and the approximate value can be reduced by using a smaller time increment.

■ Example 13.6

A proposal would have a fighter plane landing on a carrier and subjected to the decelerations given by the a-t graph shown in Figure E13.6(a). If the fighter plane is to touch the carrier deck with a velocity of 120 ft/s at $t = 0$, determine the time t_f when the plane comes to a complete stop. How far does it travel on the deck?

Solution

From the given a-t graph,

$$a = -30 \text{ ft/s}^2, \quad \text{for } 0 < t < 1 \text{ s},$$

$$a = -60 \text{ ft/s}^2, \quad \text{for } 1 < t < t_f.$$

Using Eq. (13.5b) for the time interval $0 < t < 1$ s, we obtain

$$a = -30 = \frac{dv}{dt}.$$

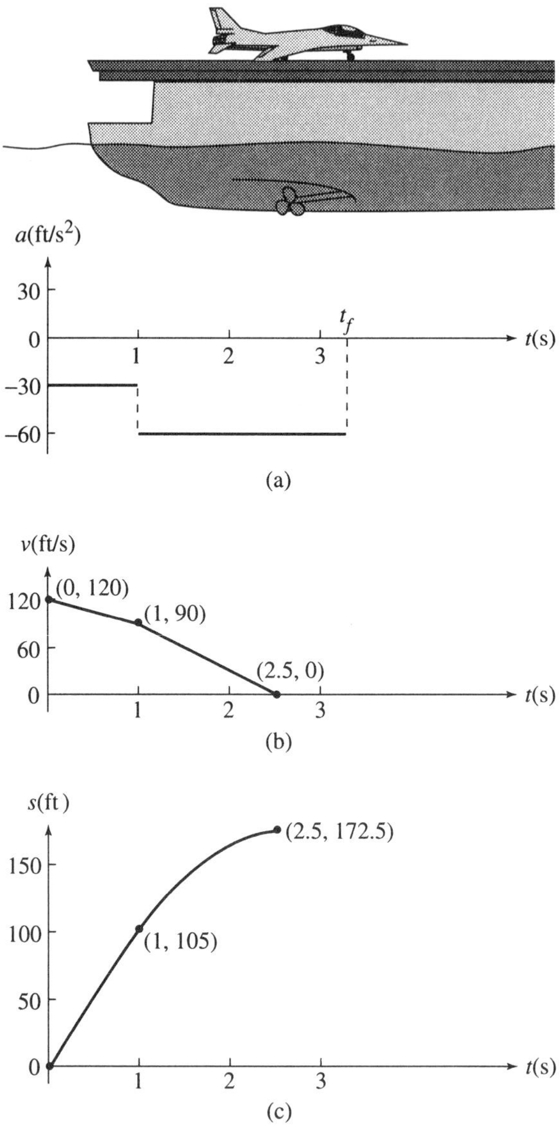

FIGURE E13.6.

Rearranging terms and integrating using the initial conditions ($t = 0$, $v = 120$ ft/s) for lower limits and general upper limits,

$$\int_{120}^{v} dv = -30 \int_{0}^{t} dt$$

which leads to

$$v = 120 - 30t, \quad 0 < t < 1 \text{ s}.$$

At $t = 1$ s, $v = 90$ ft/s.

Now, using Eq. (13.5b) for the time interval $1 < t < t_f$, we conclude that

$$a = -60 = \frac{dv}{dt}.$$

Rearranging terms and integrating using the initial conditions ($t = 1$ s, $v = 90$ ft/s) for lower limits and general upper limits,

$$\int_{90}^{v} dv = -60 \int_{1}^{t} dt$$

which leads to

$$v = 150 - 60t, \quad 1 < t < t_f.$$

When the fighter plane comes to a full stop, its velocity is zero and $t = t_f$. Thus,

$$v = 0 = 150 - 60\, t_f$$

from which

$$t_f = 2.50 \text{ s.} \qquad \text{ANS.}$$

The v-t functions are plotted in Figure E13.6(b).

Using Eq. (13.3b) for the time interval $0 < t < 1$ s, we obtain

$$v = 120 - 30t = \frac{ds}{dt}.$$

Rearranging terms and integrating using the initial conditions ($t = 0$, $s = 0$) for lower limits and general upper limits,

$$\int_{0}^{s} ds = \int_{0}^{t} (120 - 30t)\, dt$$

from which

$$s = 120t - 15t^2, \quad 0 < t < 1 \text{ s.}$$

At $t = 1$ s, $s = 105$ ft.

Now, using Eq. (13.3b) for the time interval $1 < t < t_f$, we obtain

$$v = 150 - 60t = \frac{ds}{dt}.$$

Rearranging terms and integrating using the initial conditions ($t = 1$ s, $s = 105$ ft) for lower limits and general upper limits,

$$\int_{105}^{s} ds = \int_{1}^{t} (150 - 60t)\, dt$$

from which

$$s = 150t - 30t^2 - 15, \quad 1 < t < t_f.$$

The s-t function is plotted in Figure E13.6(c).

The stopping distance s_f is obtained by letting $t = t_f = 2.50$ in the second s-t function to get

$$s_f = 172.5 \text{ ft.} \qquad \text{ANS.}$$

Example 13.7

The acceleration of a particle moving along a straight-line path is given by the relationship $a = 4$ s where a is in ft/s^2 and s is in ft. It is known that, when $t = 0$, $s = 0$ and $v = 2$ ft/s. Develop (a) the v vs. s function and (b) the s vs. t function.

Solution

(a) Because the acceleration is given in terms of the position, we use Eq. (13.7). Thus,

$$a = 4s = v\frac{dv}{ds}.$$

Separating variables and integrating, we obtain

$$4\int_0^s ds = \int_2^v v\,dv$$

from which

$$2s^2 = \left.\frac{v^2}{2}\right|_2^v = \frac{v^2 - 4}{2}.$$

Rearranging terms, we obtain

$$v^2 = 4(s^2 + 1)$$

from which

$$v = 2\sqrt{s^2 + 1}. \qquad \text{ANS.}$$

(b) Now that we have expressed the velocity as a function of displacement, we use Eq. (13.3b) to obtain

$$v = 2\sqrt{s^2 + 1} = \frac{ds}{dt}.$$

Separating variables and integrating, we obtain

$$2\int_0^t dt = \int_0^s \frac{ds}{\sqrt{s^2 + 1}}$$

from which

$$2t = \ln(s + \sqrt{s^2 + 1}).$$

This relationship between s and t may be expressed in the form

$$s + \sqrt{s^2 + 1} = e^{2t}. \qquad \text{ANS.}$$

Example 13.8

A train in a railroad station is brought to a complete stop by engaging a stationary bumper attached to springs. This action provides a deceleration to the train that may be approximated by $a = -v/2$ where a is in m/s^2 and v is in m/s. Determine (a) the speed of the train v as a function of position after the engagement assuming that $v = 10$ m/s for $s = 0$ and (b) the speed of the train after the engagement as a function of time assuming that $v = 10$ m/s for $t = 0$.

Solution

(a) The acceleration is expressed in terms of displacement by the relationship $a = v\dfrac{dv}{ds}$. Thus,

$$a = v\frac{dv}{ds} = -\frac{1}{2}v.$$

Therefore,

$$dv = -\frac{1}{2}ds.$$

Integrating using given initial conditions ($v = 10$ m/s, $s = 0$) for lower limits and general upper limits,

$$\int_{10}^{v} dv = -\frac{1}{2}\int_{0}^{s} ds,$$

$$v - 10 = -\frac{1}{2}s,$$

$$v = \left(10 - \frac{1}{2}s\right) \text{ m/s.} \qquad \text{ANS.}$$

(b) The acceleration is expressed in terms of time by the relationship $a = \dfrac{dv}{dt}$. Therefore,

$$a = \frac{dv}{dt} = -\frac{1}{2}v$$

$$\frac{dv}{v} = -\frac{1}{2}dt.$$

Integrating using the given initial conditions ($v = 10$ m/s, $t = 0$) for lower limits and general upper limits we obtain,

$$\int_{10}^{v} \frac{dv}{v} = -\frac{1}{2}\int_{0}^{t} dt,$$

$$\ln v \Big|_{10}^{v} = -\frac{1}{2}t \Big|_{0}^{t},$$

$$\ln\left(\frac{v}{10}\right) = -\frac{1}{2}t,$$

$$v = (10e^{-t/2})\ \text{m/s}. \qquad \text{ANS.}$$

■

Problems

13.30 A bead moves along a taut wire as shown in Figure P13.30 such that its acceleration in m/s^2 is given as a function of time in s by $a = 6t - 14$. Initially, when $t = 0$ s, the particle is 8 m to the left of the origin and moving 14 m/s to the right. Determine (a) the velocity as a function of time, (b) the position as a function of time, and (c) the displacement during the time interval from $t = 4$ s to $t = 8$ s.

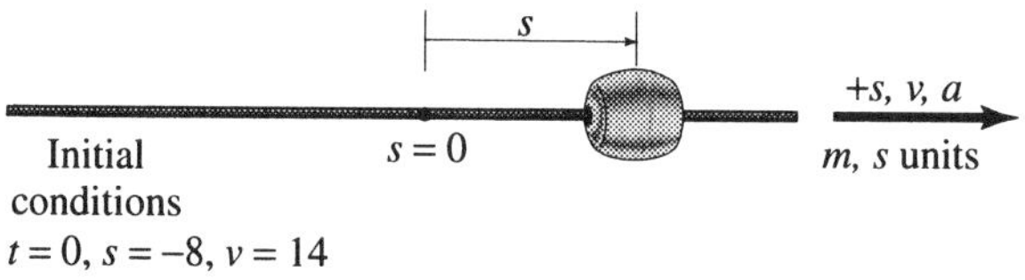

FIGURE P13.30.

13.31 The velocity function $v = t^2 - 25t + 150$ is specified for the rectilinear motion of a particle where v is expressed in ft/s and t is expressed in s. The particle is at the origin when the clock is started. Determine (a) the acceleration as a function of time, (b) the position as a function of time, (c) the times for which the velocity equals zero, (d) the displacement from $t = 0$ to $t = 5$ s, and (e) the distance traveled during the time interval from $t = 0$ to $t = 8$ s.

13.32 A vehicle starts from rest at the origin at $t = 0$ and moves such that $a = 10 - 5t$ where a is measured in m/s^2 and t in s. Find the acceleration, velocity, and position of the particle when $t = 2$ s.

13.33 The straight line motion of a small block shown in Figure P13.33 is such that $a = Ak^2 \sinh kt$ where a is expressed in m/s^2 and t in s. The block is at the origin when the clock is started. When $t = 0$, $v = 10$ m/s and $t = 1$ s, $v = 20$ m/s. Determine (a) the velocity as a function of time and (b) the position as a function of time.

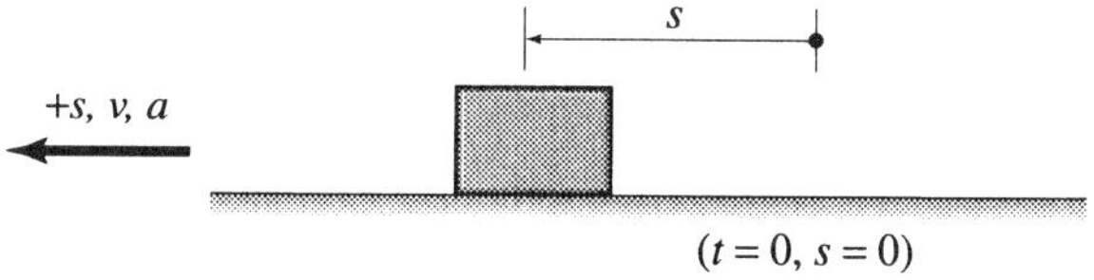

FIGURE P13.33.

13.34 Rectilinear particle motion occurs such that the acceleration is given by $a = A \sinh kt$ where a is expressed in ft/s^2 and t in s. When $t = 0$, $v = 100$ ft/s and when $t = 5$ s, $v = 200$ ft/s. The ratio of A to k is 50, and $s = 0$ at $t = 0$. Determine (a) the numerical values for A and k, (b) the position, velocity, and acceleration of the particle when $t = 2$ s and (c) the units of A and k.

13.35 A particle starts from rest at the origin when $t = 0$ and moves according to $a = t^3 - 6t^2 + 11t - 6$ where a is expressed in m/s^2 and t in s. Find (a) the velocity as a function of time, (b) the position as a function of time, and (c) sketch the motion curves (i.e., a-t, v-t and s-t curves) for the time interval from $t = 0$ to $t = 4$ s.

13.36 The acceleration of a small block, which oscillates along the inclined plane shown in Figure P13.36, is given by $a = -\pi^2 A \sin \pi t$ where a is in inch/s^2 and t is in s. When $t = 0$, then, $v = \pi A$ in./s and $s = 0$. The constant A is the amplitude of the vibration. Determine (a) the velocity as a function of time, (b) the position as a function of time, and (c) sketch the motion curves (i.e., a-t, v-t and s-t curves) for the time interval from $t = 0$ to $t = 2$ s.

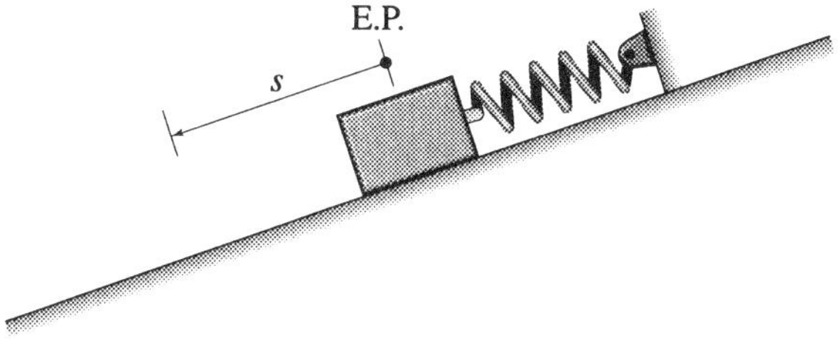

FIGURE P13.36.

13.37 In Figure P13.37 the vibratory motion of a small body along a vertical straight line is defined by $a = -A\pi^2 \cos \pi t$ where a is in m/s^2 and t is in s. When $t = 0$, $s = 1$ m, and $v = 0$, determine (a) the velocity as a function of time, (b) the position as a function of time, and (c) let $A = 1.00$, and sketch the motion curves (i.e., a-t, v-t and s-t curves) for the time interval from $t = 0$ to $t = 2$ s.

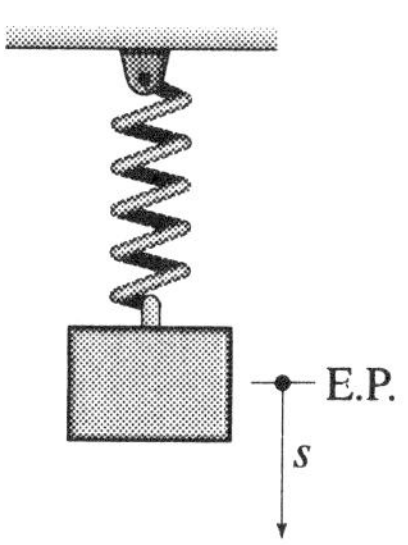

FIGURE P13.37.

13.38 A particle is at rest at the origin when $t = 0$. It moves such that $a = mt + b$ where m and b are constants and a is measured in inch/s^2 and t in s. Find (a) the velocity as a function of time, (b) the position as a function of time, and (c) what special case arises for $m = 0$?

13.39 The motion of a particle along a straight line is defined by $a = A + B \sin \pi t$ where A and B are constants. The acceleration is measured in m/s^2 and the time in s. When $t = 0$, $s = 0$ and $v = -1/\pi$, determine (a) the velocity as a function of time and specialize it for $A = 2$ and $B = 1$, and (b) the position as a function of time and specialize it for $A = 2$ and $B = 1$.

13.40 The velocity function $v = At^2 + Bt + C$ is specified for the rectilinear motion of a particle where v is expressed in m/s and t in s. When $t = 0$, $s = 0$, $v = 1$ m/s; $t = 0$, $a = 2$ m/s^2 and at $t = 1$, $v = 6$ m/s. Determine (a) the constants A, B, C. Be sure to state their units, (b) the acceleration-time function, and (c) the position-time function.

13.41 Refer to the acceleration-time plot shown in Figure P13.41 for a vehicle which moves along a straight-line path. It has an initial velocity of 10 ft/s. Plot the velocity-time curve for this vehicle for the 10 s interval shown. What is the velocity of the vehicle when $t = 3$ s and when $t = 10$ s?

13.42 Refer to the velocity-time plot shown in Figure P13.42 for a vehicle which moves

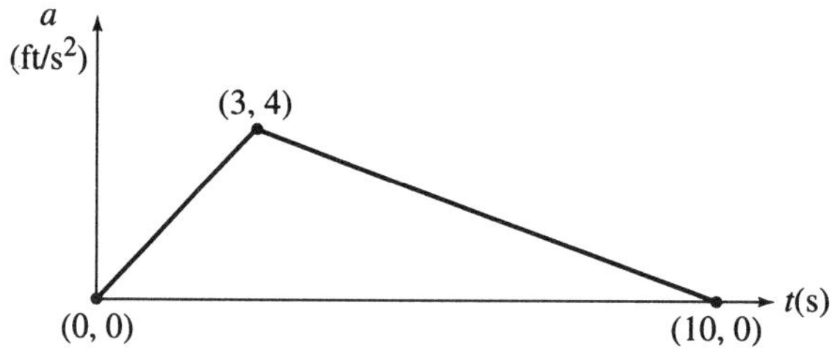

FIGURE P13.41.

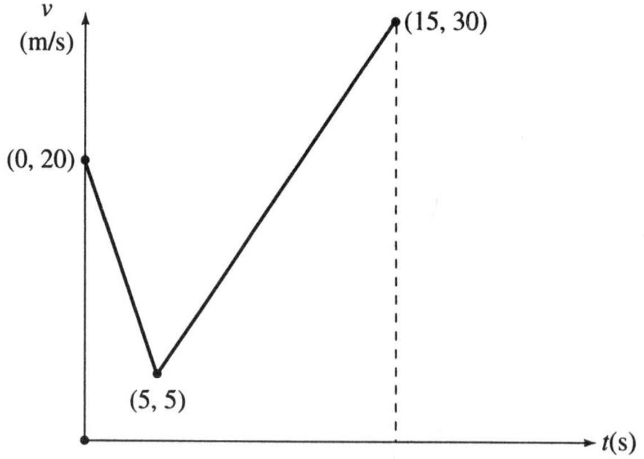

FIGURE P13.42.

along a straight-line path. When $t = 0$, the vehicle is at the origin for the position function. Plot the acceleration-time curve and the position-time curve for this vehicle for the 15 s interval shown. What is the position of this vehicle when $t = 5$ s and when $t = 15$ s?

13.43 The rectilinear motion of a particle proceeds such that $a = 6t - 16$ where t is in seconds and a is in m/s^2. When $t = 0$, the particle is 10 meters to the left of the origin and moving 17 m/s to the right. Determine (a) the velocity as a function of time, (b) the position as a function of time, and (c) the displacement during the time interval from $t = 0$ to $t = 2$ s.

13.44 The velocity as a function of time is given by $v = t^2 - 9t + 18$ where v is in ft/s and t in s. The particle moves along a straight line and is at the origin when $t = 0$. Determine (a) the acceleration as a function of time, (b) the position as a function of time, (c) the times for which the velocity equals zero, (d) the displacement during the time interval $t = 0$ to $t = 4$ s, and (e) the distance traveled during the time interval $t = 0$ to $t = 4$ s.

13.45 The straight-line motion of a car is in accord with the (a-t) function shown in Figure P13.45. During ($0 < t \leq 4$), $a = -2$ m/s^2. Write the acceleration as a function of time for the interval ($4 \leq t \leq 10$). Construct the (v-t) plot for ($0 < t \leq 10$), and write the corresponding functions. The particle is at rest at $t = 0$.

13.46 Refer to the velocity vs. time plot of a toy train shown in Figure P13.46. Construct the corresponding (a-t) and (s-t) plots for this straight-line motion of a particle. The particle is 2 ft to the right of the origin when $t = 0$.

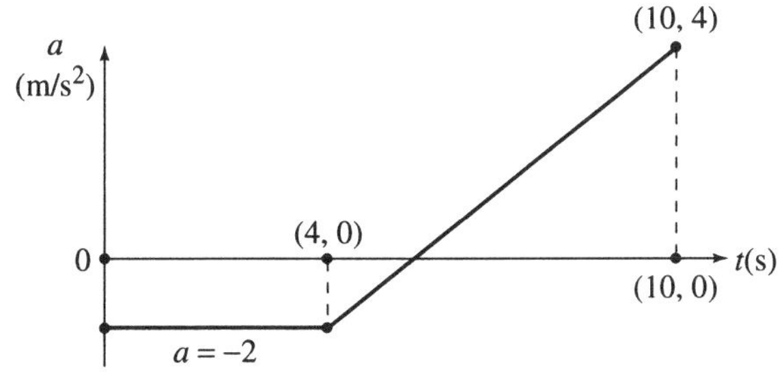

FIGURE P13.45.

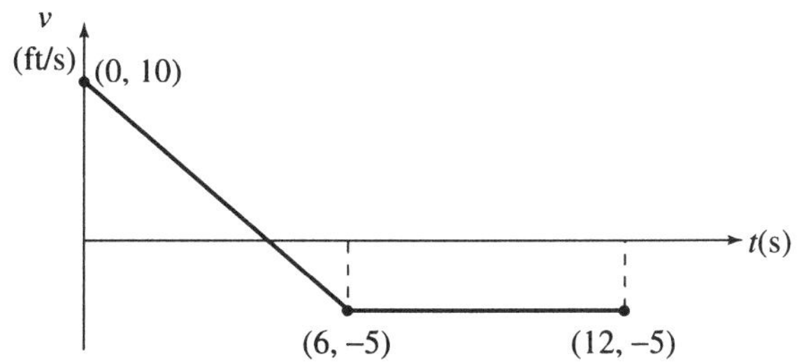

FIGURE P13.46.

13.47 The acceleration as a function of time is given for the rectilinear motion of a particle by $a = 3.5 - t^2$ where a is in m/s^2 and t is in s. This particle is at rest at the origin when $t = 0$. Determine the position, velocity, and acceleration of this particle when $t = 6$ s.

13.48 A particle starts from rest at the origin when $t = 0$ and moves according to $a = t^3 - 15t^2 + 54t - 40$ where a is expressed in in./s^2 and t in s. Find (a) the velocity as a function of time and (b) the position as a function of time.

13.49 The acceleration of a particle, which oscillates along a straight line, is given by $a = -4 \sin 2t$ where a is in ft/s and t is in s. When $t = 0$, $v = 4$ ft/s and $s = 2$ ft. Determine (a) the velocity as a function of time, (b) the position as a function of time, and (c) sketch the (a-t), (v-t) and (s-t) curves for t between zero and π s.

13.50 The (a-t) curve for one cycle of the oscillatory motion of a spring-mass system is shown in Figure P13.50. The particle moves on a straight line and starts from rest at the origin when $t = 0$. Sketch the (v-t) and (s-t) curves for this motion. Show the ordinates for $t = 0$, $\pi/2$ and π s.

13.51 The (v-t) curve for the rectilinear motion of a motorcyclist is shown in Figure P13.51. Sketch the corresponding (a-t) and (s-t) curves. Label the ordinates for $t = 0$, 2 and 4 s. The particle starts from rest at the origin.

13.52 A car moves on a straight line such that its velocity varies with time as shown in Figure P13.52. It starts from rest at $t = 0$ and is 10 ft to the right of the origin for s at this time. Sketch the (a-t) and (s-t) curves for this motion. Show the ordinates for $t = 0$, 4, 8 and 12 s.

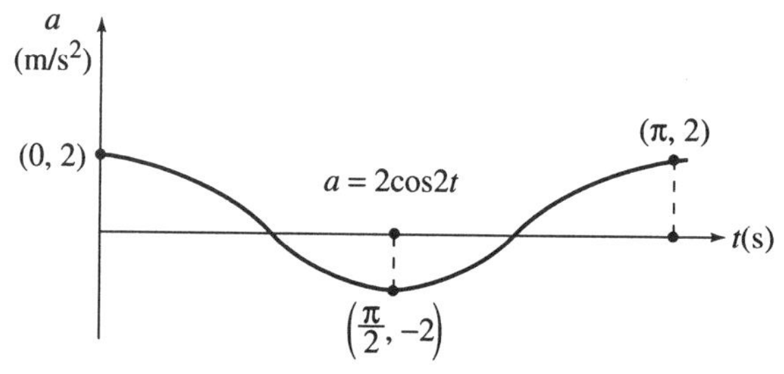

FIGURE P13.50.

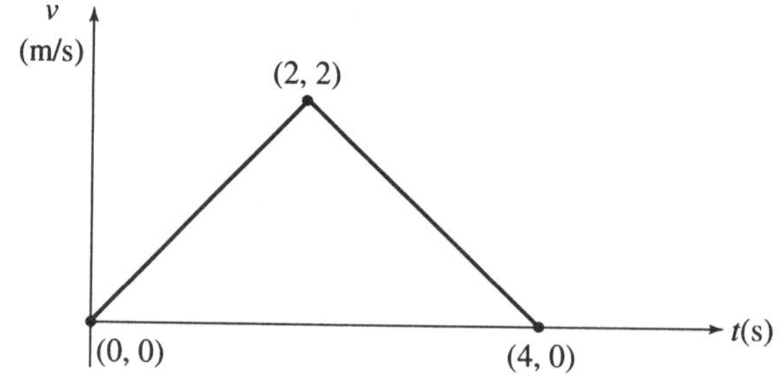

FIGURE P13.51.

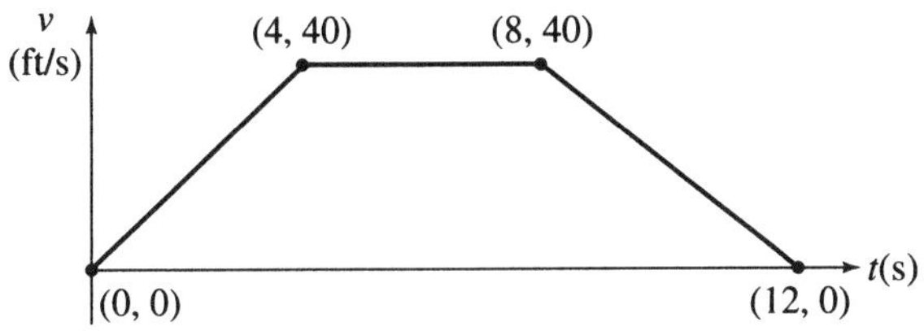

FIGURE P13.52.

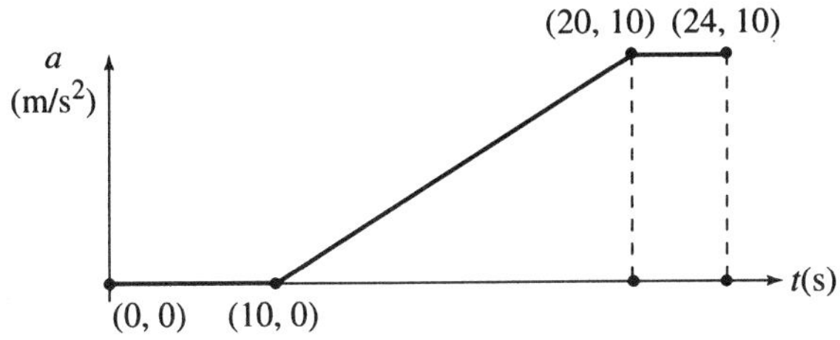

FIGURE P13.53.

13.53 The $(a\text{-}t)$ curve for the rectilinear motion of a train is shown in Figure P13.53. At $t = 0$, this particle has a velocity of 10 m/s directed to the right, and it is located 5 m to the right of the origin for s. Sketch the $(v\text{-}t)$ and $(s\text{-}t)$ curves for this motion. Show the ordinates for $t = 0$, 10, 20 and 24 s.

13.54 A particle is moving with a velocity of 48 in/s at $t = 0$ and is located at the origin for s. Its acceleration is given by $a = 2t - 16$ where a is measured in in./s^2 and t in s. Determine the velocity v and the position s as functions of time t.

13.55 For a period of 4 s, a particle executes rectilinear motion such that $a = 3t^2$ where a is expressed in m/s^2 and t in s. When $t = 0$, $v = 2$ m/s and $s = 4$ m. Determine the velocity and position as functions of time. Find the position, velocity, and acceleration when $t = 4$ s.

13.56 In Figure P13.56, the small block moves on a table between vertical guides. The acceleration of this block, as it moves on a straight line is expressed by $a = -s$ with initial conditions $t = 0$, $s = 0$ and $t = 0$, $v = 1$. Units are (ft·s). Show that the motion is oscillatory. Determine v, s, and a as functions of time.

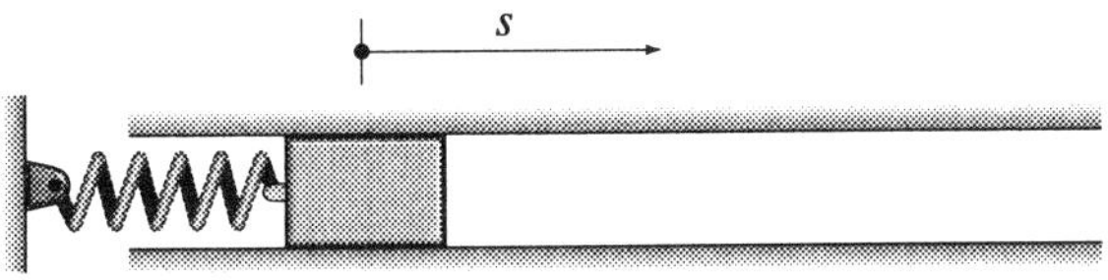

Plan view

FIGURE P13.56.

13.57 A particle moves along a straight line such that $a = -kv$. When $s = 0$, $v = v_0$. Determine v as a function of s. If $v_0 = 4$ m/s and if $v = 2$ m/s when $s = 2$ m, determine k. What are the units of k?

13.58 The acceleration a of a plunger moving through oil may be approximated by $a = -kv$ where v is its velocity. When $t = 0$, $s = 0$ and when $t = 0$, $v = v_0$. Determine s and v as a function of t. If $v_0 = 8$ ft/s and if $v = 4$ ft/s when $s = 4$ ft, determine k. What are the units of k?

13.59 The motion of a particle is specified by stating that $a = 2$ s and the initial conditions $t = 0$, $s = 4$ m and $t = 0$, $v = 12$ m/s. Determine (a) v as a function of s and (b) t as a function of s.

13.60 The rectilinear motion of a particle is specified by $a = -ks^{-2}$. When $s = 2$ m, $v = 4$ m/s and when $s = 10$ m, $v = 0$. Determine numerical values for k and s as a function of v. What are the units of k? Sketch this function.

13.61 A small collar moves along the inclined rod shown in Figure P13.61. Its acceleration is given as a function of its position coordinate by $a = 100 - 5s^2$ where a is in m/s^2 and s is in m. When $s = 0$, $v = 0$. Determine (a) the velocity v as a function of s and (b) the velocity when $s = 3$ m.

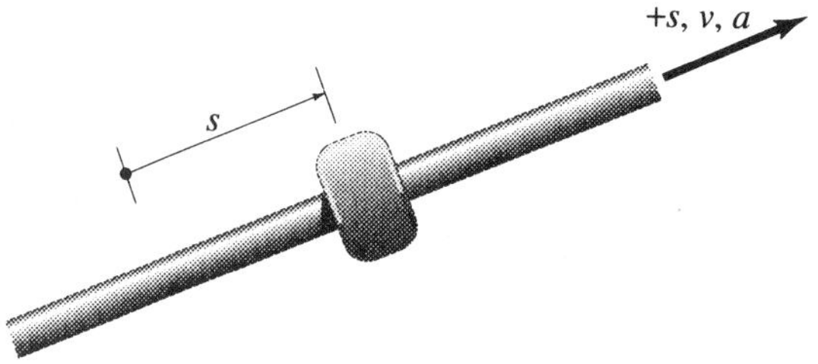

FIGURE P13.61.

13.62 The acceleration for the straight-line motion of a particle is given by $a = 20 - 2s^2$ where a is in inch/s^2 and s is in inches. When $s = 0$, $v = 0$. Determine (a) the velocity v as a function of s and (b) the velocity when $s = 2$ in.

13.63 The acceleration of a particle in rectilinear motion is given by $a = kv$ with the initial conditions $t = 0$, $s = s_0$; $s = s_0$, $v = v_0$. Units are (m·s). Assume $k = 1.00$ 1/s, then, determine (a) the velocity v as a function of the position coordinate s and (b) the position coordinate s as a function of time t.

13.64 The acceleration of a particle which moves along a straight-line path is given by $a = 6s^2 + 2$ where s is in meters and a is in m/s^2. When $s = 0$, $v = 0$. Determine (a) the velocity v as a function of s, and (b) the velocity when $s = 1$ m.

13.65 A boat moves across a lake at a velocity $v_0 = 10$ m/s when the motor is turned off. The fluid resistance to its motion may be approximated by the function $a = -0.2v$ where a is in m/s^2 and v in m/s. Determine (a) the velocity as a function of s, (b) the distance traveled by the boat before coming to rest, (c) the time required, theoretically, for the boat to come to rest, and (d) the time required for the boat to move 49 m after the motor is turned off.

13.66 Atmospheric resistance to the motion of a locomotive may be approximated by $a = -kv$, where a is in ft/s^2 and v in ft/s. The locomotive is traveling along a straight, level track at 100 ft/s when its diesel engine stops. If it comes to rest after moving 50 ft, determine the value of the constant k and state its units. Show that the theoretical time required for the locomotive to come to rest is infinite, and, then, calculate the time required for it to travel the first 49.995 ft after the diesel engine stops.

13.67 In Figure P13.67, the straight-line motion of a magnet is governed by $a = -100\ s^{-2}$, where s is in inches and a is in inch/s^2. If $v = 1.414$ in/s when $s = 100$ in., determine v as a function of s. What is the velocity of this magnet when $s = 50$ in?

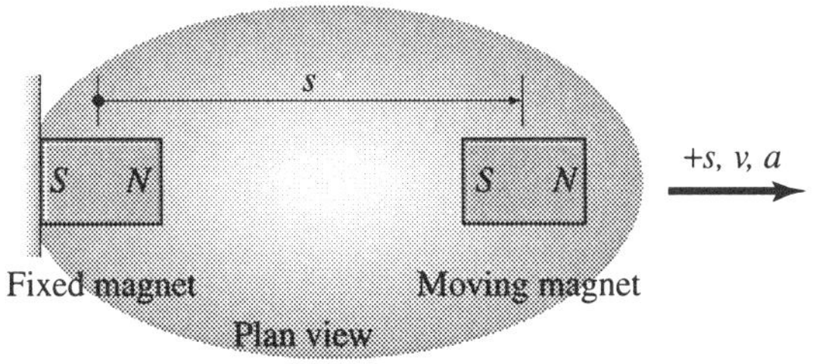

FIGURE P13.67.

13.68 The rectilinear motion of a particle is governed by $a = s^2 + 4s$ where a is in m/s^2 and s is in meters. When $s = 1$ m, $v = 1$ m/s. Determine v as a function of s. Explain why $s = 0$ is inadmissible for this motion. Are there other inadmissible values of s?

13.4 Rectilinear Motion at Constant Acceleration

Consider the special case of motion of a particle at a constant acceleration a_c. Using Eq. (13.5b), $a_c = \dfrac{dv}{dt}$. Multiplying by dt and utilizing the initial condition of $t = 0$, $v = v_0$ for lower limits and general upper limits of v and t, we integrate to obtain,

$$\int_{v_0}^{v} dv = \int_{0}^{t} a_c\, dt.$$

Because a_c is constant, we may take it outside the integral sign before integrating to yield

$$v = v_0 + a_c t. \qquad (13.8)$$

Using Eqs. (13.3b and (13.8), $v = \dfrac{ds}{dt} = v_0 + a_c t$. Multiplying by t and utilizing the initial condition $t = 0$, $s = s_0$ for lower limits and general upper limits of s and t, we integrate to obtain

$$\int_{s_0}^{s} ds = \int_0^t (v_0 + a_c t)\, dt.$$

Because v_0 and a_c are constants, we may take them outside the integral sign before integrating to obtain

$$s = s_0 + v_0 t + \tfrac{1}{2} a_c t^2. \tag{13.9}$$

The motion curves for this special case are shown in Figure 13.6.

An equation which does not contain the time explicitly and expresses v as a function of s for constant acceleration will be developed. Using Eq. (13.7), $a = v\dfrac{dv}{ds} = a_c$. Multiplying by ds and using the initial condition $v = v_0$, $s = s_0$ for lower limits and general upper limits of v and s, we integrate to obtain

$$\int_{v_0}^{v} v\, dv = \int_{s_0}^{s} a_c\, ds.$$

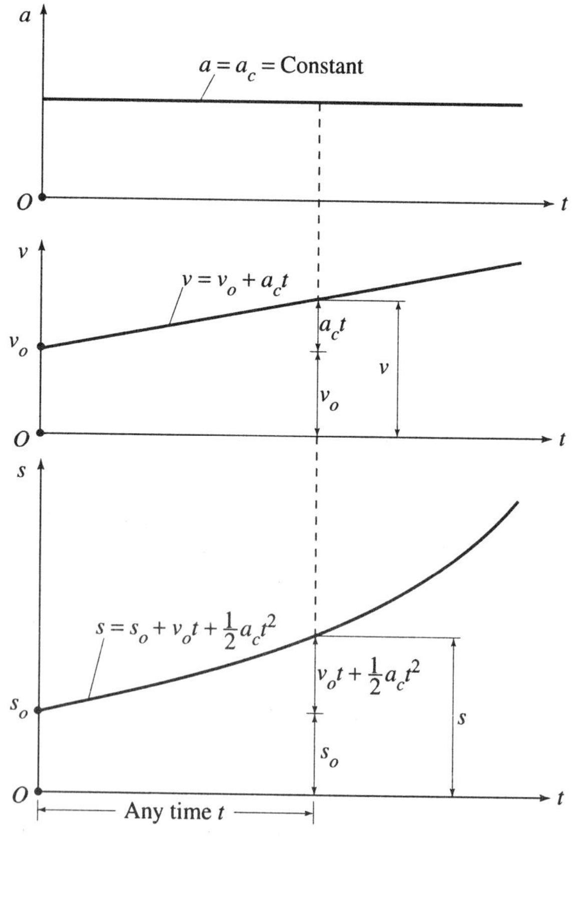

FIGURE 13.6.

Because a_c is constant, we may take it outside the integral sign before integrating to yield

$$v^2 = v_0^2 + 2a_c(s - s_0). \tag{13.10}$$

Example 13.9

A cyclist moves at a constant acceleration of 4 ft/s^2 along a straight-line path with the positive sense directed to the right. The initial conditions are $t = 0$, $v_0 = -20$ ft/s and , $s_0 = 5$ ft. Determine (a) the position of the cyclist at $t = 8$ s, (b) the velocity of the cyclist at $t = 8$ s, (c) the displacement of the cyclist during the time interval from $t = 0$ s to $t = 8$ s, and (d) the distance traveled during the time interval from $t = 0$ s to $t = 8$ s.

Solution

(a) From Eq. (13.9),

$$s = s_0 + v_0t + \tfrac{1}{2}a_ct^2 = 5 + (-20)(8) + \tfrac{1}{2}(4)(8)^2 = -27.0 \text{ ft.} \qquad \text{ANS.}$$

(b) From Eq. (13.8),

$$v = v_0 + a_ct = -20 + 4(8) = 12.0 \text{ ft/s.} \qquad \text{ANS.}$$

(c) From part (a) $s|_{t=8} = -27$ ft. Also from the given initial conditions, $s|_{t=0} = s_0 = 5$ ft. Therefore, since $\Delta s = s|_{t=8} - s|_{t=0}$, we obtain,

$$\Delta s = -27 - 5 = -32.0 \text{ ft.} \qquad \text{ANS.}$$

(d) Does the cyclist reverse direction of motion during the time interval of interest? We answer this question by equating the velocity to zero and solving for the time. Thus,

$$v = v_0 + a_ct = -20 + 4t = 0,$$

$$t = 5 \text{ s.}$$

Thus, when $t = 5$ s, the cyclist comes to rest. His location at this time is determined by

$$s = s_0 + v_0t + \tfrac{1}{2}a_ct^2 = 5 + (-20)(5) + \tfrac{1}{2}(4)(5)^2 = -45 \text{ ft.}$$

Let us tabulate the time, position and velocity information of interest to find the total distance traveled (DT).

t(s)	s(ft)	v(ft/s)
0	5	−20
5	−45	0
8	−27	12

$$\text{DT} = 5 + 45 + (45 - 27) = 68.0 \text{ ft.} \qquad \text{ANS.}$$

■ Example 13.10

A young man tosses a baseball upward while standing on the ground next to a building as shown in Figure E13.10. The baseball rises vertically upward, reaches its maximum height h_{max} and on its way down, hits the top of the building 2 s after release. The baseball, which is released with a velocity of 40 ft/s at a height of 5 ft above the ground, is subjected to the constant acceleration of gravity $g = 32.2$ ft/s^2. Determine (a) the maximum height h_{max} and (b) the height of the building h.

Solution

(a) When the baseball reaches its maximum height, its velocity is reduced to zero because of its constant deceleration $a_c = -g = -32.2$ ft/s^2. Thus, using Eq. (13.10),

$$v^2 = v_0^2 + 2a_c(s - s_0).$$

Therefore,

$$0 = (40)^2 + 2(-32.2)(h_{max} - 5),$$

$$h_{max} = 29.8 \text{ ft.} \qquad \text{ANS.}$$

(b) We now use Eq. (13.8) to determine the time it takes the baseball to reach its maximum height. Thus,

$$v = v_0 + a_c t,$$

$$0 = 40 - 32.2t,$$

$$t = 1.242 \text{ s.}$$

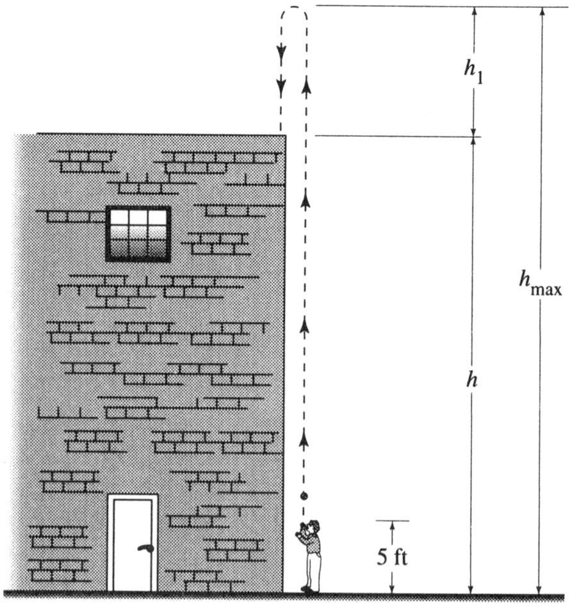

FIGURE E13.10.

Therefore, the time t_1 it takes the baseball to travel the height h_1 from its maximum height to the top of the building is $t_1 = 2 - 1.242 = 0.758$ s. Thus, using Eq. (13.9) with $s_0 = v_0 = 0$,

$$s = \tfrac{1}{2}a_c t^2$$

$$h_1 = \tfrac{1}{2}(32.2)(0.758)^2 = 9.25 \text{ ft.} \qquad \text{ANS.}$$

Note that whereas a_c is negative (deceleration) when the baseball is moving upward (against gravity), a_c is positive (acceleration) when it is moving downward (with gravity). Therefore,

$$h = h_{max} - h_1 = 20.6 \text{ ft.} \qquad \text{ANS.}$$

■

Problems

13.69 An automobile is traveling along a straight, level roadway, as shown in Figure P13.69, at a speed of 55 mph. Determine its stopping distance if the deceleration is constant and equal to 15.0 ft/s. How much time is required for this vehicle to stop?

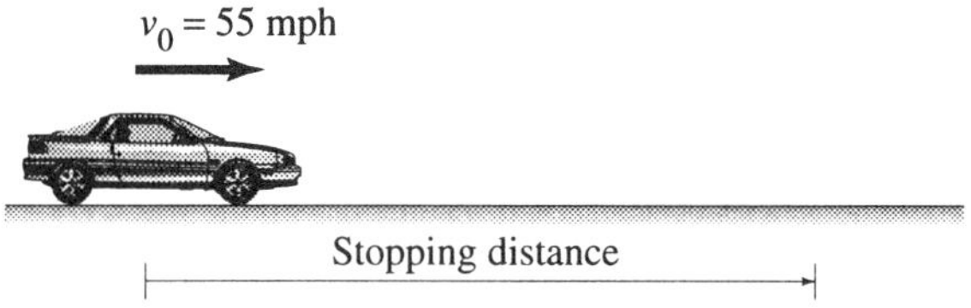

FIGURE P13.69.

13.70 An automobile is traveling along a straight level roadway at a speed of 35 km/hr. Determine its stopping distance and time if the deceleration is constant and equal to 6.0 m/s^2.

13.71 When $t = 0$, a motorcycle starts from rest at a distance of 80 ft to the right of an origin for s which is measured positively to the right. If the acceleration is constant and equal to 10 ft/s^2, determine (a) the velocity of the motorcycle when $t = 4$ s and (b) the displacement during the time interval from $t = 0$ to $t = 4$ s. Sketch the (a-t), (v-t), and (s-t) curves.

13.72 When $t = 0$, a car starts from rest at a distance 25 m to the left of an origin for s which is measured positively to the right. If the acceleration is constant and equal to 4 m/s^2, determine (a) the velocity of the car when $t = 3$ s and (b) the displacement during the time interval from $t = 0$ to $t = 3$ s. Sketch the (a-t), (v-t), and (s-t) curves.

13.73 The velocity of a vehicle is reduced from 50 ft/s to 30 ft/s as it moves 100 ft along a straight-line path. Determine the constant deceleration of the vehicle.

13.74 The velocity of a vehicle is increased from 30 m/s to 40 m/s as it moves 45 m along a straight-line path. Determine the constant acceleration of the vehicle.

13.75 A high speed train is traveling along a straight, level roadbed at a speed of 240 km/hr as shown in Figure P13.75. Determine its stopping distance if the deceleration is constant and equal to 7.0 m/s^2. How much time elapsed during which the brakes were applied to stop this train?

FIGURE P13.75.

13.76 A supersonic jet aircraft, shown in Figure P13.76, must reduce its velocity from 1000 mph to 600 mph in 5 seconds. Determine the constant deceleration required. Express your answer in ft/s^2. What distance did this jet plane travel in straight, level flight during this deceleration phase?

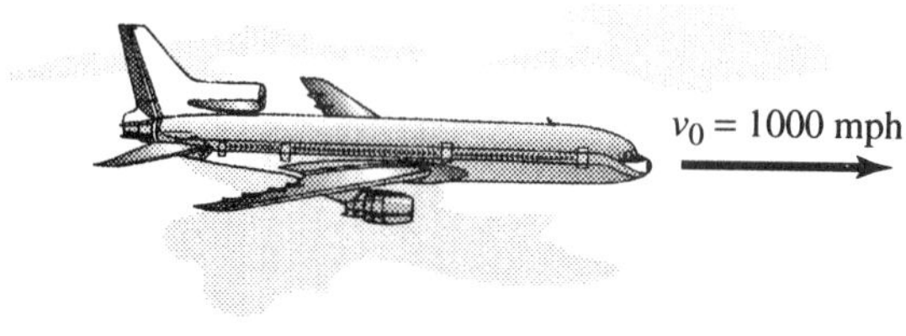

FIGURE P13.76.

13.77 An automobile's velocity is increased from 45 ft/s to 90 ft/s as it moves 120 ft over a straight, level highway. Determine its constant acceleration. How long did this accelerating phase last?

13.78 A toy car moves with constant acceleration of 2 m/s^2 along a straight-line path. The positive sense is directed to the right for this motion. Initial conditions are $t = 0$, $s = -10$ m; $t = 0$, $v = -4$ m/s. Express s and v as functions of time. Determine the position and velocity of the toy car when $t = 2$ s.

13.79 A racing car is moving along a straight stretch of race track shown in Figure P13.79. Its velocity is increased from 120 mph to 125 mph as it moves 60 ft. Determine its constant acceleration.

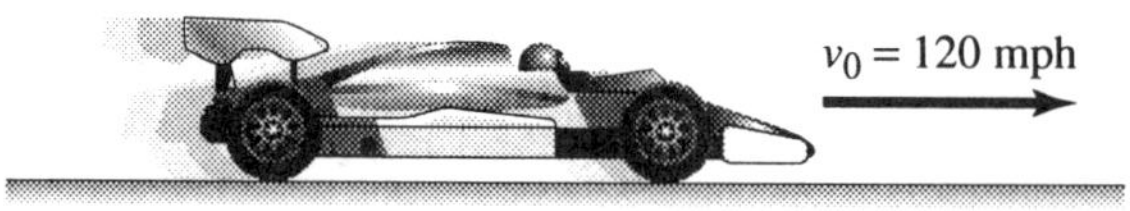

FIGURE P13.79.

13.80 A skater moving along a straight line on the surface of a frozen lake is decelerated at 0.1 m/s^2. If she comes to rest after moving 50 m, what was her initial velocity? How long did she move across the ice?

13.81 If a pitcher's fast ball can be thrown at 95 mph and we neglect air resistance, how long does it take for the ball to go from the pitching mound to home plate if the distance is 60.5 ft? If a very good catcher can catch the ball and throw it to the second baseman in 2.80 s, what average speed would a runner need to exceed in order to be safe at second base? The distance from first to second base is 90 ft, and assume that the runner has an 8 ft *leadoff* when the pitcher releases the ball.

13.82 When $t = 1$ s, a particle is moving to the right at 5 ft/s and is located 10 ft to the left of the origin for the linear coordinate s. If the constant acceleration of the particle is 4 ft/s^2, determine (a) the velocity of the particle when $t = 5$ s and (b) the displacement of the particle during the time interval from $t = 1$ s to $t = 4$ s. The positive sense for s is to the right.

13.5 Relative Motion of Two Particles

Independent Motion

In certain physical situations, we encounter several particles that move along the same straight line independently of one another. Under these conditions, it is possible to consider the motion of two such particles and relate their positions, velocities, and accelerations. Thus, refer to Figure 13.7(a), where two particles A and B are positioned on a straight line path by s_A and s_B which are both functions of time. These absolute position coordinates are given with respect to point O, the origin for s, where $s = 0$, and the positive sense is directed to the right. If we chose to state the relative position of A with respect to B, then, we imagine an origin at B and give $s_{A/B}$ which is also a function of time. This relative position coordinate is directed positively to the right. Note that the absolute positions of A and B given by s_A and s_B are seen by an observer at point O while the relative position $s_{A/B}$ is seen by an observer at B.

From the geometry in Figure 13.7(a),

$$\begin{aligned} OA &= OB + BA \\ s_A &= s_B + s_{A/B}. \end{aligned} \tag{13.11}$$

If we differentiate Eq. (13.11) term by term with respect to time, we obtain

$$\frac{ds_A}{dt} = \frac{ds_B}{dt} + \frac{ds_{A/B}}{dt}.$$

By definition, $\dfrac{ds_A}{dt} = v_A$ and $\dfrac{ds_B}{dt} = v_B$. We define $\dfrac{ds_{A/B}}{dt}$ as $v_{A/B}$, the relative velocity of A with respect to B. An observer moving with B would see particle A moving with the relative velocity $v_{A/B}$. Thus,

$$v_A = v_B + v_{A/B}. \tag{13.12}$$

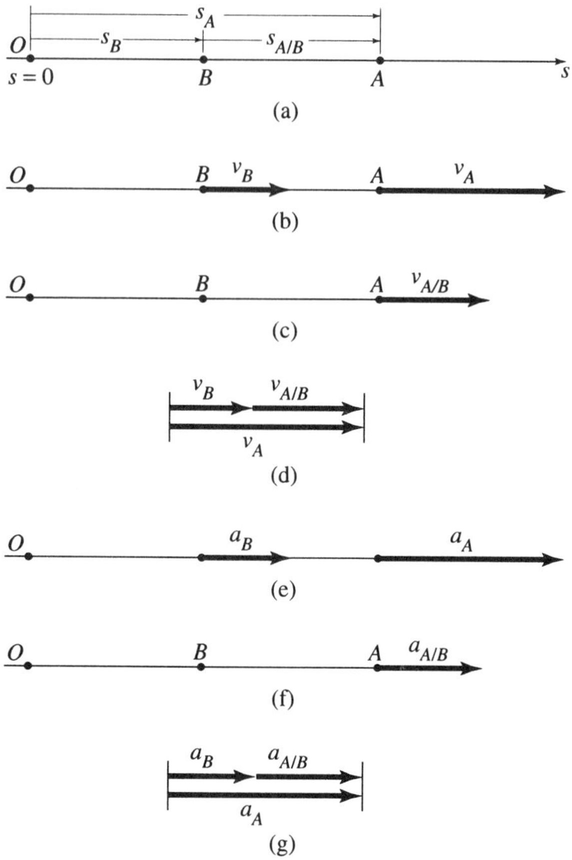

FIGURE 13.7.

The absolute velocities v_A and v_B, the relative velocity $v_{A/B}$, and their velocity vector addition by Eq. (13.12) are shown in Figures 13.7(b), (c), and (d), respectively. Physically, this equation states that, if we add the absolute velocity v_B of B to the relative velocity $v_{A/B}$ of A with respect to B, the result is the absolute velocity v_A of A.

If we differentiate Eq. (13.12) term by term with respect to time, we obtain

$$\frac{dv_A}{dt} = \frac{dv_B}{dt} + \frac{dv_{A/B}}{dt}.$$

By definition $\frac{dv_A}{dt} = a_A$ and $\frac{dv_B}{dt} = a_B$. We define $\frac{dv_{A/B}}{dt}$ as $a_{A/B}$, the relative acceleration of A with respect to B. An observer moving with B would see particle A accelerating at the relative acceleration $a_{A/B}$. Thus,

$$a_A = a_B + a_{A/B}. \tag{13.13}$$

The absolute accelerations a_A and a_B, the relative acceleration $a_{A/B}$, and their acceleration vector addition by Eq. (13.13) are shown in Figures 13.7 (e), (f), and (g) respectively. Physically, this equation states that if, we add the absolute acceleration a_B of B to the relative acceleration $a_{A/B}$ of A with respect to B, the result is the absolute acceleration a_A of A.

Dependent Motion

Bodies are sometimes constrained to move so that their displacements, velocities, and accelerations are not independent of each other. In Figure 13.8, for example, the motions of blocks A and B depend upon each other. At any time t, s_A and s_B position blocks A and B while d_1, d_2, d_3, and r are constant. The cable passing over the pulleys is assumed to be inextensible, which means that its length remains constant. To position block A, we choose a horizontal datum through the fixed centers of pulley C. The position of block A is measured positively downward. Block B is positioned with respect to the fixed center of pulley D. If we let L be the total length of the cable passing over the pulleys, it follows that

$$L = 2(s_A - d_2) + d_1 + 2\pi r + \frac{\pi}{2} r + s_B$$

or, because L, d_1, d_2 and r are constants, we obtain

$$2s_A + s_B = \text{constant.}$$

Differentiating this equation term by term with respect to time gives

$$2\frac{ds_A}{dt} + \frac{ds_B}{dt} = 0.$$

Since $v_A = \dfrac{ds_A}{dt}$ and $v_B = \dfrac{ds_B}{dt}$, we conclude that

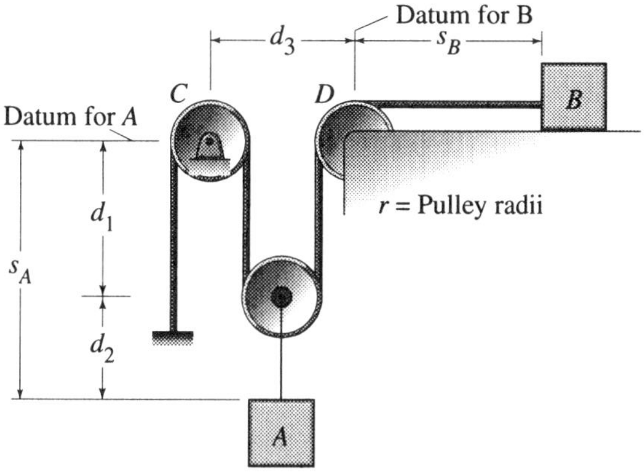

FIGURE 13.8.

$$v_B = -2v_A.$$

This equation means that, if v_A is known, then, v_B is determined. If we differentiate again with respect to time,

$$\frac{dv_B}{dt} = -2\frac{dv_A}{dt}.$$

Because $a_B = \dfrac{dv_B}{dt}$ and $a_A = \dfrac{dv_A}{dt}$, it follows that

$$a_B = -2a_A.$$

The accelerations of A and B are related to each other in the same fashion as their the velocities. It is also true that position changes of A and B are related in the same way. Thus,

$$\Delta s_B = -2\Delta s_A.$$

Refer to Figure 13.9 as another example. Choose fixed point P on the lower horizontal plane and position block A with coordinate s_A and block B with coordinate s_B. The positive sense for the position of both blocks A and B is directed to the right. Block C is positioned by s_C directed positively downward from the fixed center of the lower pulley. The quantities d_1, d_2, and r are constant. If we let L be the total length of cable of the system

$$L = (s_B - d_1 - s_A) + d_2 - d_1 - s_A)$$

and, since L, d_1, d_2, and r are constants, we conclude that

$$s_B + s_C - 2s_A = \text{constant}.$$

Differentiation of this equation with respect to time gives

$$\frac{ds_B}{dt} + \frac{ds_C}{dt} - 2\frac{ds_A}{dt} = 0.$$

In terms of velocities,

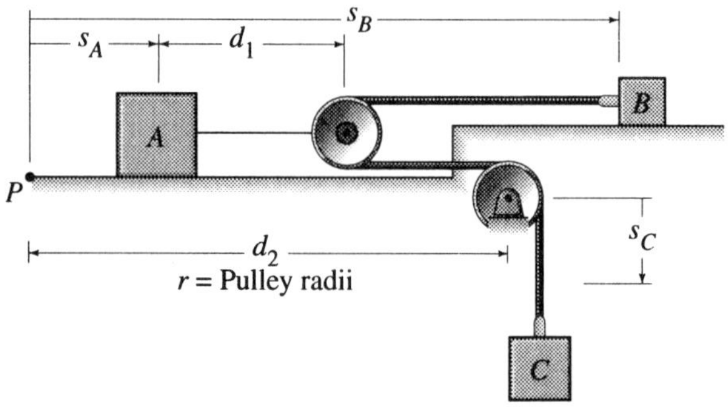

FIGURE 13.9.

$$2v_A = v_B + v_C.$$

Differentiation of this velocity equation with respect to time gives

$$2\frac{dv_A}{dt} = \frac{dv_B}{dt} + \frac{dv_C}{dt}.$$

In terms of accelerations,

$$2a_A = a_B + a_C.$$

Of course, these equations relate the *magnitudes* of the velocities and accelerations of the blocks without regard to their directions. In Figure 13.8, the velocity and acceleration of block A are directed vertically and the velocity and acceleration of block B are directed horizontally. In Figure 13.9, both A and B move along horizontal planes and block C moves vertically.

The following examples illustrate some of the above concepts.

Example 13.11

At the instant shown in Figure E13.11, boat A is traveling along the positive x axis at a speed of 30 mph which is decreasing at the rate of 60 mi/h^2. At the same instant, boat B is traveling along the negative x' axis at a speed of 20 mph which is increasing at the rate of 40 mi/h^2. Determine the relative velocity and relative acceleration of boat A with respect to boat B. What are the velocity and acceleration of boat B relative to boat A?

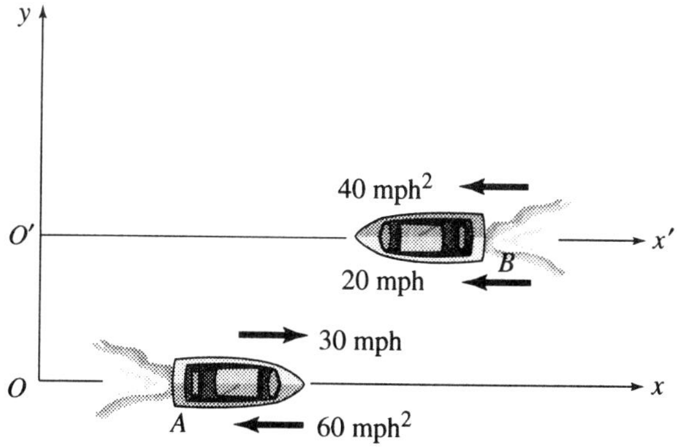

FIGURE E13.11.

Solution

By Eq. (13.12)

$$v_A = v_B + v_{A/B}$$

from which

$$v_{A/B} = v_A - v_B = 30 - (-20) = 50.0 \text{ mph.} \qquad \text{ANS.}$$

By Eq. (13.13)

$$a_A = a_B + a_{A/B}$$

from which

$$a_{A/B} = a_A - a_B = -60 - (-40) = -20.0 \text{ mi/hr}^2. \qquad \text{ANS.}$$

The velocity of boat B relative to boat A is the negative of that found for $v_{A/B}$. This may be shown by using Eq. (13.12) in the form

$$v_B = v_A + v_{B/A}$$

from which

$$v_{B/A} = v_B - v_A = -20 - 30 = -50.0 \text{ mph}. \qquad \text{ANS.}$$

The acceleration of boat B relative to boat A is the negative of that obtained from $a_{A/B}$. This may be shown by using Eq. (13.13) in the form

$$a_B = a_A + a_{B/A}$$

from which

$$a_{B/A} = a_B - a_A = -40 - (-60) = 20.0 \text{ mi/hr}^2. \qquad \text{ANS.}$$

■ Example 13.12

Refer to Figure E13.12 and determine the velocity and acceleration of block A if block B is moving downward at the rate of 10 ft/s and is being decelerated at the rate of 2 ft/s^2. The connecting cable is assumed to be inextensible. The radius of each pulley is r.

Solution

Establish a fixed reference level or datum through the fixed center of the upper pulley with the positive sense directed downward. Because the cable is inextensible, its length will remain constant with the passage of time. Let the total length of cable be L. Therefore,

$$L = s_A + 2(s_B - d) + 2\pi r + b$$

or, because L, d, r, and b are constants,

$$s_A + 2s_B = \text{constant}.$$

Differentiating this equation term by term with respect to time and recalling that the time derivative of a constant is zero,

$$\frac{ds_A}{dt} + 2\frac{ds_B}{dt} = 0.$$

Recalling the definition of the velocity of a particle,

$$v_A + 2v_B = 0 \quad \text{or} \quad v_A = -2v_B.$$

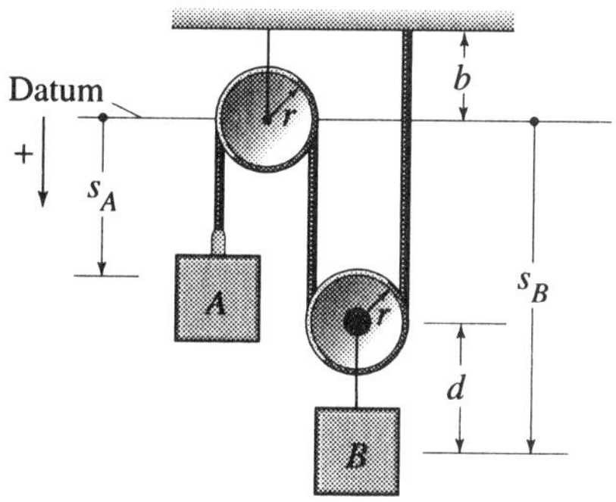

FIGURE E13.12.

Therefore,

$$v_A = -2(10) = -20.0 \text{ ft/s.} \qquad \text{ANS.}$$

The negative sign means that the velocity of A is directed upward, opposite to the positive downward sense.

Differentiating the velocity relationship $v_A = -2v_B$ with respect to time yields

$$\frac{dv_A}{dt} = -2\frac{dv_B}{dt}.$$

Recalling the definition of the acceleration of a particle,

$$a_A = -2a_B.$$

Because $a_B = -2 \text{ ft/s}^2$,

$$a_A = -2(-2) = 4.00 \text{ ft/s}^2. \qquad \text{ANS.}$$

The positive sign means that block A is being accelerated downward in the positive sense.

■ Example 13.3

In Figure E13.13, block A is moving down the inclined plane with a velocity of 10 ft/s and is being decelerated at 4 ft/s^2. Determine the velocity and acceleration of block B at this instant. The cable is inextensible and each pulley has a radius r. The positive sense for body A is down the plane and the positive sense for body B is downward as shown in the figure.

Solution

The total length of cable of this system does not change with time. Choosing a datum as shown in the figure and letting the length of cable be L,

$$L = s_A + \left(\beta + \frac{\pi}{2}\right)r + 2(s_B - d) + \pi r + d + b,$$

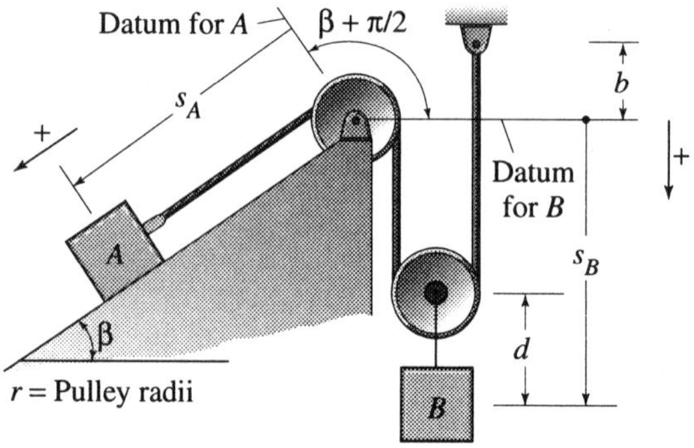

FIGURE E13.13.

and, because L, r, d, and b are constants, we conclude that

$$s_A + 2s_B = \text{constant.}$$

Differentiating this equation with respect to time and recalling that the time derivative of a constant vanishes, we obtain

$$\frac{ds_A}{dt} + 2\frac{ds_B}{dt} = 0.$$

By definition, $v_A = \dfrac{ds_A}{dt}$ and $v_B = \dfrac{ds_B}{dt}$. Thus,

$$v_B = -\frac{1}{2}v_A.$$

Because $v_A = 10$ ft/s,

$$v_B = -\frac{1}{2}(10) = -5.00 \text{ ft/s.} \qquad \text{ANS.}$$

The negative sign means that block B is moving upward. Differentiation of the velocity equation with respect to time gives

$$\frac{dv_B}{dt} = -\frac{1}{2}\frac{dv_A}{dt}.$$

By definition, $a_B = \dfrac{dv_B}{dt}$ and $a_A = \dfrac{dv_A}{dt}$. Thus, $a_A = -\dfrac{1}{2}a_A$. Because $a_A = -4$ ft/s^2,

$$a_A = -\frac{1}{2}(-4) = 2.00 \text{ ft/s}^2. \qquad \text{ANS.}$$

The positive sign means that block B has an acceleration directed downward.

■

Problems

13.83 Toy car A lies 10 ft to the left of an origin and toy car B lies 5 ft to the right of this same origin. Car A is moving to the left at 20 ft/s and car B is moving to the right at 15 ft/s. At this same instant, A has a negative acceleration of 2 ft/s^2 and B has a positive acceleration of 3.5 ft/s^2. As shown in Figure P13.83 the positive sense is directed to the right on a horizontal straight line along which A and B are moving. Determine (a) the position of A relative to B, (b) the position of B relative to A, (c) the relative velocity of A with respect to B, and (d) the relative acceleration of A with respect to B.

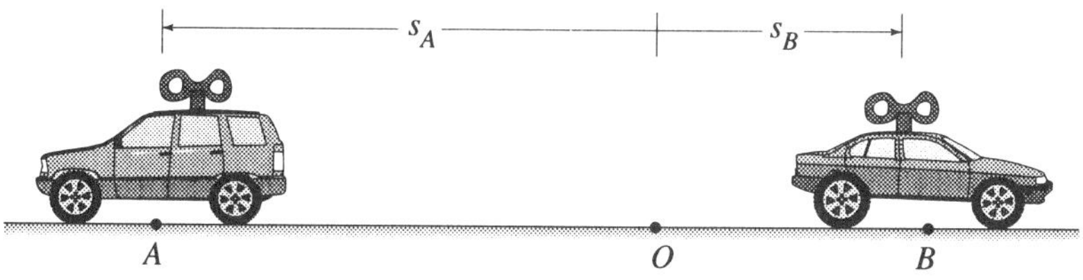

FIGURE P13.83.

13.84 The motion of particles A and B takes place along an inclined plane sloping at 30° with the horizontal and a positive sense directed up and to the right along this plane. Particle A lies at the origin and B lies 5 m along the inclined plane, up and to the right of the origin. If A is moving at 4 m/s positively and B is moving at 3 m/s negatively, determine the velocity of A with respect to B. If A has a positive acceleration of 2 m/s^2 and B has a negative acceleration of 1.5 m/s^2, determine the acceleration of A with respect to B.

13.85 Two adjacent elevators, A and B, shown in Figure P13.85 have instantaneous velocities $v_A = 15$ ft/s and $v_B = -20$ ft/s, where the positive sense is directed upward. Determine the relative velocity of elevator A with respect to elevator B. If elevator A has an upward acceleration of 1.5 ft/s^2 and elevator B has a downward acceleration of 5.0 ft/s^2, determine the relative acceleration of elevator A with respect to elevator B.

13.86 Two aircraft are flying straight, level paths at a safe spacing. If aircraft A is flying at 500 mph due north and aircraft B is flying at 400 mph due south, what is the relative velocity of A with respect to B? What is the relative velocity of B with respect to A? In each case, carefully note the location of an observer.

13.87 A ship moves due east at 20 knots while another ship moves due west at 25 knots as shown in Figure P13.87. Determine the velocity of the first ship with respect to an observer on the second ship.

13.88 Refer to Figure P13.88. If A has a downward velocity of 2.50 m/s and an upward acceleration of 0.25 m/s^2, determine the velocity and acceleration of B.

13.89 Solve problem 13.88, if A has an upward velocity of 0.40 m/s and a downward acceleration of 0.10 m/s^2.

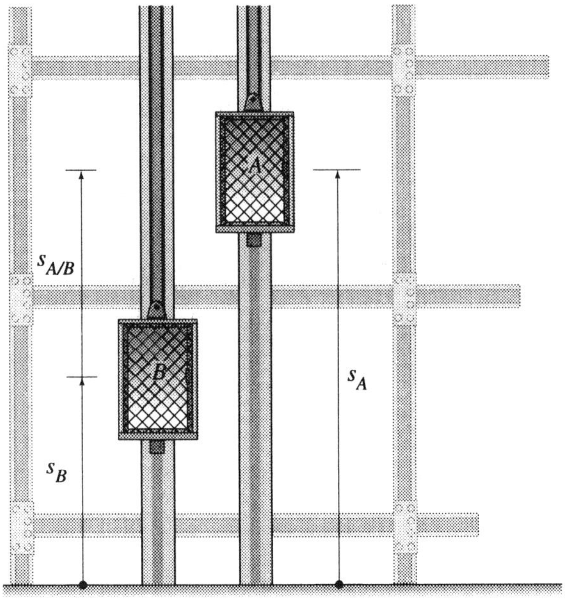

FIGURE P13.85.

FIGURE P13.87.

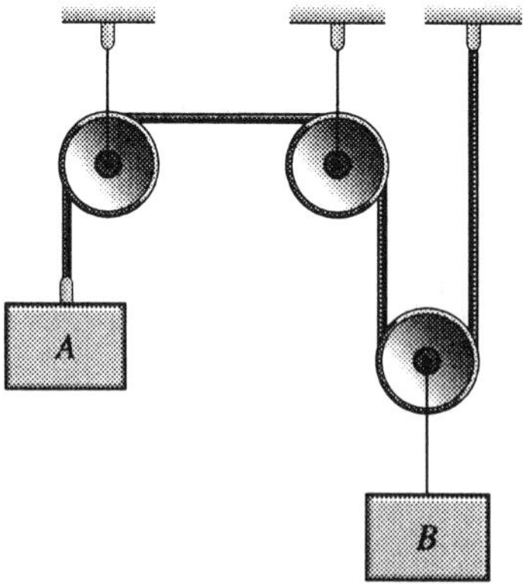

FIGURE P13.88.

13.90 In Figure P13.90, block A is moving down the inclined plane with a velocity of 4.50 ft/sec and is accelerated up this plane at 2.00 ft/s^2. Determine the velocity and acceleration of block B.

13.91 Refer to Figure P13.91 and determine the velocities and accelerations of block B and the center of pulley C given that A is moving upward at 1.5 ft/s and is accelerated downward at 0.4 ft/s^2. Also find the relative velocity and acceleration of B with respect to A.

13.92 Automobiles A and B are traveling east along a straight, level roadway. If A has a velocity of 80 ft/s and a deceleration of 4 ft/s^2 and B has a velocity of 60 ft/s and

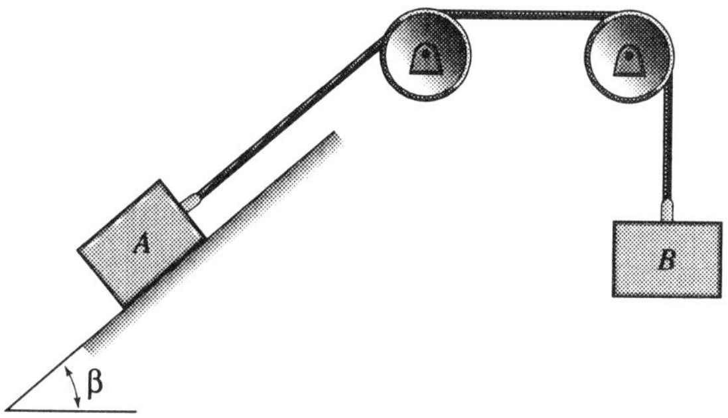

FIGURE P13.90.

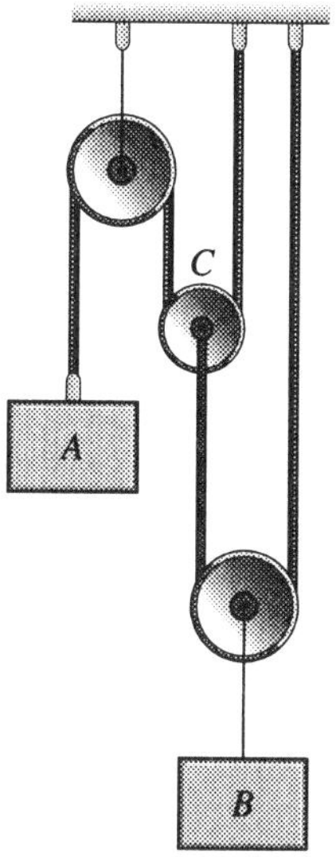

FIGURE P13.91.

an acceleration of 2 ft/s^2, determine the relative velocity and acceleration of B with respect to A.

13.93 Automobiles A and B are moving along a grade of +5%. Automobile A is at the foot of the grade which is the origin and B is 200 ft up the grade. If A is moving at 75 ft/s and is accelerating at 5 ft/s^2 and B is moving at 60 ft/s and is decelerating at 3 ft/s^2, determine the relative velocity and acceleration of B with respect to A.

13.94 In Figure P13.94, block B is moving upward with a velocity of 10 ft/s and is being decelerated at the rate of 4 ft/s^2. Determine the corresponding velocity and acceleration of block A.

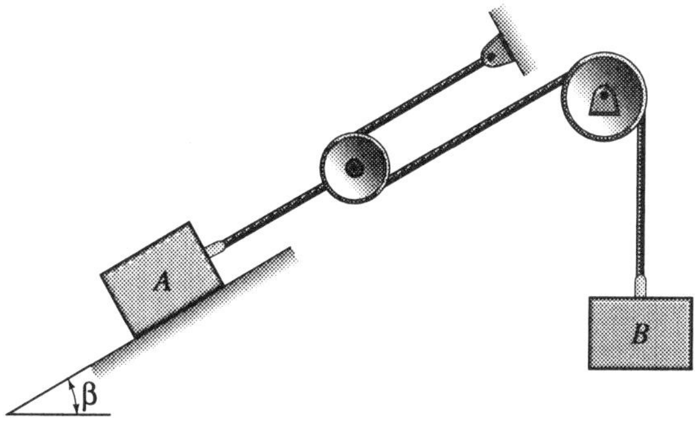

FIGURE P13.94.

CURVILINEAR MOTION

13.6 Position, Velocity, and Acceleration

Position: *Curvilinear motion* of a particle occurs when the particle moves along a path different from a straight line. Consider, for example, the case of a particle moving along a curved path in a plane, say the *x*-*y* plane, as indicated in Figure 13.10(a). An origin is chosen arbitrarily at a point Q on the path and the curvilinear coordinate s is measured along the arc length of the path. Consider the particle at the position denoted by P_1 at any time t. The position vector $\mathbf{r}$ from the fixed origin O to point P_1 varies in magnitude and direction with time t. During a time interval Δt, the particle moves along the path from point P_1 to point P_2. The displacement vector from P_1 to P_2 is the vector $\Delta\mathbf{r}$ which also varies in magnitude and direction with time.

Velocity: We define the instantaneous velocity of the particle as

$$\mathbf{v} = \lim_{\Delta t \to 0} \frac{\Delta\mathbf{r}}{\Delta t} = \frac{d\mathbf{r}}{dt}. \tag{13.14}$$

As P_2 approaches P_1 along the path (as the time interval Δt approaches zero in the limit), the change in the position vector $\Delta\mathbf{r}$ takes on the direction of the tangent to the path at P_1. Referring to the definition of the velocity vector v and knowing that Δt is a scalar, we conclude that the velocity vector $\mathbf{v}_1$ also takes on the direction of the tangent to the path at P_1 and $\mathbf{v}_2$, the direction of the tangent to the path at P_2 as indicated.

As P_2 approaches P_1 along the path, the time interval Δt approaches zero in the limit, and the magnitude of the change in the position vector $\Delta\mathbf{r}$ (i.e., Δr) is equal to the change in the variable s measured along the arc length of the path, (i.e., Δs). In equation form,

$$\Delta r = \Delta s \quad (\text{as } \Delta t \to 0).$$

The time rate of change of the magnitude of the position vector is termed the speed of the particle. Because the time interval Δt is a scalar, we are dealing with a scalar in defining speed. Thus,

$$v = \lim_{\Delta t \to 0} \frac{\Delta r}{\Delta t} = \lim_{\Delta t \to 0} \frac{\Delta s}{\Delta t},$$

$$v = \frac{ds}{dt}. \tag{13.15}$$

The particle instantaneous speed v is the time rate of change of the distance s measured along the arc length of the path. Although similar at first glance, Equations (13.14) and (13.15) are indeed very different when carefully examined. Equation (13.14) is a *vector* differential equation for the instantaneous velocity of a particle whereas Eq. (13.15) is a *scalar* differential equation for the instantaneous speed of a particle.

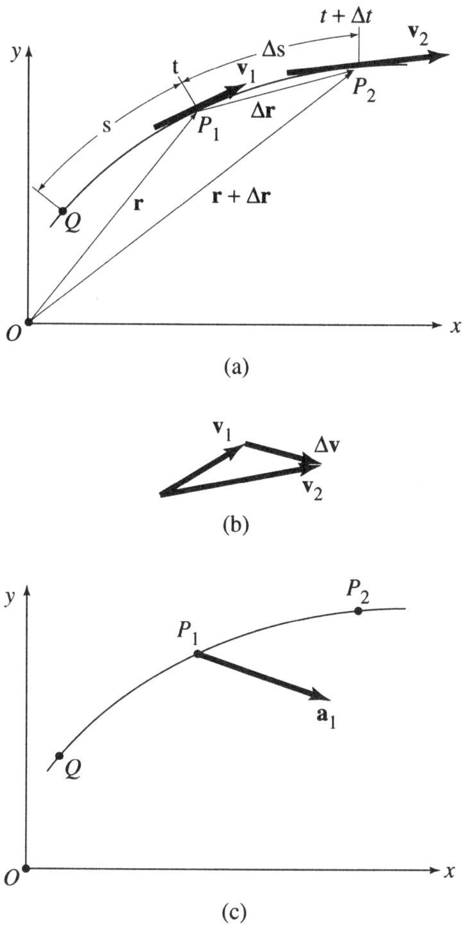

FIGURE 13.10.

To develop further our understanding of the difference between velocity and speed, let us consider the motion of a jogger along a circular path. The number of strides taken by the jogger per second is a measure of his speed. This speed is obviously a scalar quantity because it does not take into consideration the direction in which the jogger is moving, but only the number of strides per second. However, as the jogger moves along the circular path, his direction (measured along the circular path) is constantly changing and, therefore, his velocity is also changing. Hence, although the speed of the jogger is constant, his velocity is not.

When the particle is at P_1, its velocity is $\mathbf{v}_1$, and, when it is at P_2, its velocity is $\mathbf{v}_2 = \mathbf{v}_1 + \Delta\mathbf{v}$. Both the magnitude and direction of the particle's velocity vector have changed as it moved from position P_1 to position P_2. If these velocity vectors $\mathbf{v}_1$ and $\mathbf{v}_2$ are drawn from a common origin, as shown in Figure 13.10(b), we observe that $\Delta\mathbf{v}$ is the vector which must be added to $\mathbf{v}_1$ to obtain $\mathbf{v}_2$. This vector $\Delta\mathbf{v}$ reflects both magnitude and directional changes of the velocity vector in a time increment Δt as the particle moves from P_1 to P_2 along its path.

Acceleration

We define the instantaneous acceleration of the particle as

$$\mathbf{a} = \lim_{\Delta t \to 0} \frac{\Delta \mathbf{v}}{\Delta t} = \frac{d\mathbf{v}}{dt}. \tag{13.16}$$

The time rate of change of the velocity vector equals the acceleration vector. If we imagine P_2 approaching P_1 along the path, as the time interval Δt approaches zero in the limit, then, the acceleration vector **a** has a direction which is the same as $\Delta\mathbf{v}$, because the time increment Δt is a scalar. As shown in Figure 13.10(c), in general, the acceleration vector **a** has no particular direction with respect to the path of the particle.

It is to be noted that the definitions of the velocity **v** given by Eq. (13.14), the speed v given by Eq. (13.15), and the acceleration of **a** given by Eq. (13.16) are all independent of the choice of a coordinate system. The position vector **r** is measured from a fixed origin at O and the curvilinear coordinate s is measured from an arbitrary fixed point Q on the particle's path. The arc length s is an intrinsic property of the path and is independent of the coordinate system chosen to study the motion of a given particle. In Sections 13.7, 13.8 and 13.9, three different coordinate systems are discussed for studying the curvilinear motion of a particle.

13.7 Curvilinear Motion—Rectangular Coordinates

Let us express the position vector **r** to point P_1 on the path of the particle in terms of x and y components indicated in Figure 13.11(a). Thus,

$$\mathbf{r}(t) = x(t)\mathbf{i} + y(t)\mathbf{j}.$$

where (t) emphasizes that r, x, and y are functions of time t. For convenience, let us omit (t) but keep in mind that **r** is a vector function of time and x and y are scalar functions of time. Therefore,

$$\mathbf{r} = x\mathbf{i} + y\mathbf{j}. \tag{13.17}$$

Differentiating this equation with respect to time and noting that **i** and **j** are unit vectors which do not change in magnitude or direction with time

$$\frac{d\mathbf{r}}{dt} = \frac{dx}{dt}\mathbf{i} + \frac{dy}{dt}\mathbf{j}.$$

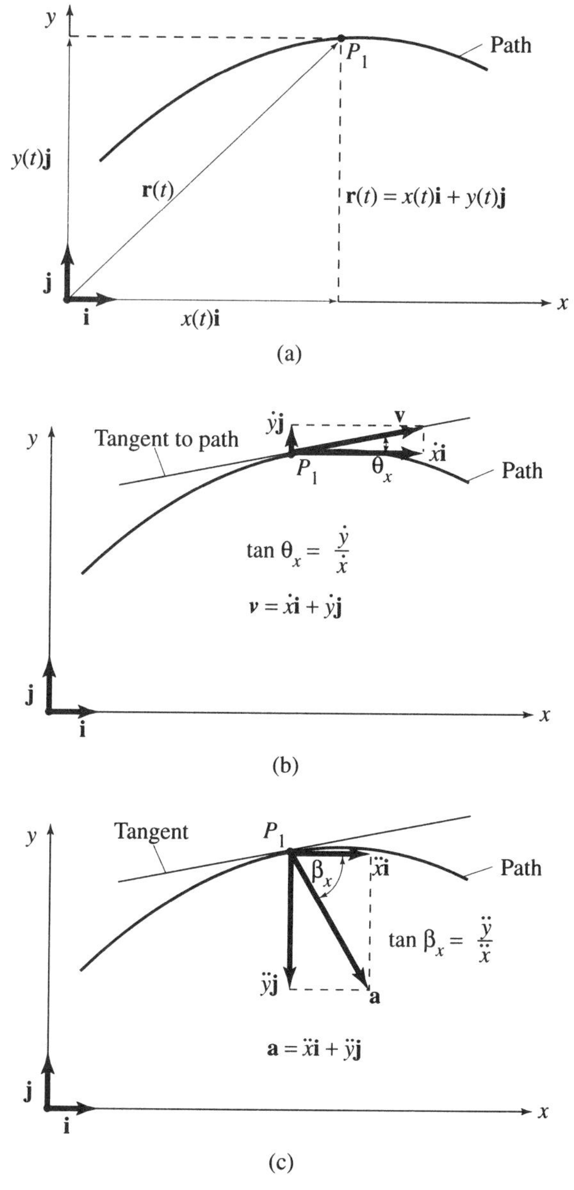

FIGURE 13.11.

By Eq. (13.14), $\mathbf{v} = \frac{d\mathbf{r}}{dt}$ and, adopting the dot notation devised by Isaac Newton, $\dot{x} = \frac{dx}{dt}$ and $\dot{y} = \frac{dy}{dt}$. Thus,

$$\mathbf{v} = \dot{x}\mathbf{i} + \dot{y}\mathbf{j} \tag{13.18}$$

(we say *x dot* for $\dot{x}$ and *y dot* for $\dot{y}$). The total velocity vector $\mathbf{v}$ and its vector components $\dot{x}\mathbf{i}$ and $\dot{y}\mathbf{j}$ are shown in Figure 13.11(b). We refer to $\dot{x}$ and $\dot{y}$ as the scalar components of the velocity vector $\mathbf{v}$. The Pythagorean theorem will be used to enable us to find the magnitude and direction of the velocity vector $\mathbf{v}$ once we know the scalar components $\dot{x}$ and $\dot{y}$. The speed $\mathbf{v}$ is given by

$$v = \sqrt{\dot{x}^2 + \dot{y}^2} \tag{13.19a}$$

and the direction θ_x by

$$\theta_x = \tan^{-1}\left(\frac{\dot{y}}{\dot{x}}\right). \tag{13.19b}$$

Because the slope of a tangent to a curve in the x-y plane is given by $\tan\theta_x = \dfrac{dy}{dx}$ it follows that

$$\tan\theta_x = \frac{\dot{y}}{\dot{x}} = \frac{dy}{dx}$$

which proves once again that the velocity vector for a particle is always directed along the tangent to the path at each point during its motion.

To write an equation for the acceleration of a particle in rectangular coordinates, we differentiate Eq. (13.18) with respect to time. Thus,

$$\mathbf{a} = \frac{d\mathbf{v}}{dt} = \ddot{x}\mathbf{i} + \ddot{y}\mathbf{j} \tag{13.20}$$

where $\ddot{x} = \dfrac{d^2x}{dt^2}$ and $\ddot{y} = \dfrac{d^2y}{dt^2}$. This equation for the $\mathbf{a}$ vector is represented graphically in Figure 13.11(c). To determine the magnitude and direction of this acceleration vector, we apply the Pythagorean theorem as done for the velocity vector. Therefore,

$$a = \sqrt{\ddot{x}^2 + \ddot{y}^2}, \tag{13.21a}$$

$$\beta_x = \tan^{-1}\left(\frac{\ddot{y}}{\ddot{x}}\right). \tag{13.21b}$$

Although we are able to determine the direction of the acceleration vector from this equation, the acceleration vector, in general, has no particular direction with respect to a curved path. In contrast, the velocity vector for a particle is always directed tangent to the path.

The equations developed above for the two-dimensional motion of a particle may be extended to three-dimensional motion by simply adding a component of velocity and acceleration corresponding to the third axis, say the z axis. Thus, Eqs. (13.18) and (13.20) become, respectively,

$$\mathbf{v} = \dot{x}\mathbf{i} + \dot{y}\mathbf{j} + \dot{z}\mathbf{k} \tag{13.22}$$

and

$$\mathbf{a} = \ddot{x}\mathbf{i} + \ddot{y}\mathbf{j} + \ddot{z}\mathbf{k}. \tag{13.23}$$

The problems following this section may be classified broadly in two categories:

1. *The x and y scalar components of the position vector **r** are known functions of time. Differentiation will be required to determine the velocity and acceleration vectors. Elimination of the parameter t from x(t) and y(t) will provide the equation of the path. An illustration of this type of problem is given in Example 13.14.*
2. *The x and y scalar components of the acceleration vector **a** are given as functions of time together with appropriate initial conditions. Integrations will be required to determine the velocity and position vectors. Again, eliminating the parameter t from x(t) and y(t) will provide the equation of the path. An illustration of this type of motion is provided by Example 13.15.*

A very good physical example of the second category, mentioned above, is provided by the motion of a projectile, which is conveniently described in terms of rectangular components. If we assume that air resistance is negligibly small, the component of the acceleration in the horizontal x direction is zero ($\ddot{x} = 0$). Thus, the only acceleration to which the projectile is subjected is that due to gravity. Thus, although $\ddot{x} = 0$, $\ddot{y} = -g$ which may be taken as a constant if we assume that the motion occurs close to the surface of Earth. The equations that govern the motion of a projectile under these conditions are developed in Example 13.16, and a physical problem using these equations is given in Example 13.17.

■ Example 13.14

The parametric equations for the x-y motion of a particle are given as $x = 2t$ and $y = 4(1 + t^2)$ where x and y are measured in m and time t in s. Determine (a) the equation of the path, (b) the magnitude and direction of the velocity vector when $t = 1$ s, and (c) the magnitude and direction of the acceleration vector when $t = 1$ s.

Solution

(a) Eliminate the time t from the given x and y equations to determine the equation of the path. Thus $x = 2t$ leads to $t = x/2$. Substituting for t in terms of x in $y = 4(1 + t^2)$ gives

$$y = 4\left[1 + \left(\frac{x}{2}\right)^2\right] = 4 + x^2. \qquad \text{ANS.}$$

This is the equation of a parabola shown in Figure E13.14.

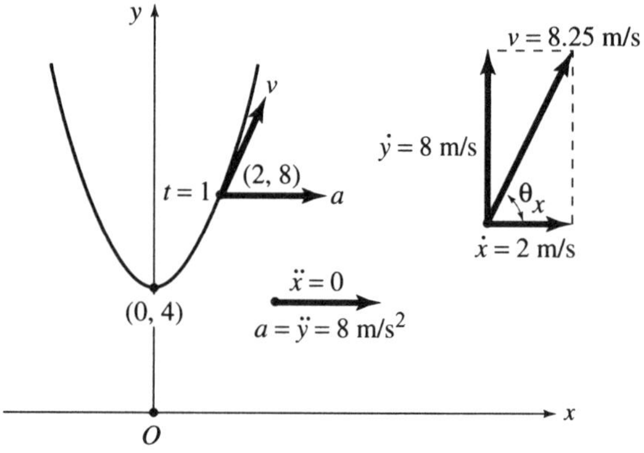

FIGURE E13.14.

(b) Differentiating the given functions with respect to time to determine the scalar velocity components,

$$\dot{x} = 2 \text{ m/s} \quad \text{and} \quad \dot{y} = 8t \text{ m/s}$$

When $t = 1$ s, $\dot{x} = 2$ m/s, and $\dot{y} = 8$ m/s. Therefore,

$$v = \sqrt{\dot{x}^2 + \dot{y}^2} = \sqrt{2^2 + 8^2} = 8.25 \text{ m/s}, \qquad \text{ANS.}$$

and

$$\theta_x = \tan^{-1}\left(\frac{\dot{y}}{\dot{x}}\right) = \tan^{-1}\left(\frac{8}{2}\right) = 76.0°. \qquad \text{ANS.}$$

The resulting vector is shown in Figure E13.14.

(c) Because $\dot{x} =$ constant, then, $\ddot{x} = 0$. Also,

$$\dot{y} = 8t \Rightarrow \ddot{y} = 8 \text{ m/s}^2.$$

Thus, the scalar components of the acceleration do not vary with time and are constant at $\ddot{x} = 0$ and $\ddot{y} = 8$ m/s^2. Thus,

$$a = \sqrt{\ddot{x}^2 + \ddot{y}^2} = \sqrt{0^2 + 8^2} = 8 \text{ m/s}^2, \qquad \text{ANS.}$$

and

$$\beta_x = \tan^{-1}\left(\frac{0}{2}\right) = 0° \qquad \text{ANS.}$$

The acceleration vector is also shown in Figure E13.14. It does not change in magnitude or direction with time.

■ Example 13.15

The motion of a toy train in a horizontal x-y plane is such that $\ddot{x} = -16\cos 2t$ and $\ddot{y} = -16\sin 2t$ where x and y are measured in feet and t in seconds. If the toy train starts when $t = 0$ so that $x = 4$ ft, $y = 0$,

$\dot{x} = 0$, and $\dot{y} = 8$ ft/s, determine (a) the equation of the path followed by the train, (b) the velocity of the train for $t = 0.5$ s, and (c) the acceleration of the train for $t = 0.5$ s.

Solution

(a) Starting with the given components of acceleration,

$$\ddot{x} = \frac{d\dot{x}}{dt} = -16\cos 2t \quad \text{and} \quad \ddot{y} = \frac{d\dot{y}}{dt} = -16\sin 2t. \tag{a}$$

Integrating using the given initial conditions for lower limits and general upper limits,

$$\int_0^{\dot{x}} d\dot{x} = \int_0^t -16\cos 2t\,dt \quad \text{and} \quad \int_0^{\dot{y}} d\dot{y} = \int_0^t -16\sin 2t\,dt.$$

These equations lead to

$$\dot{x} = \frac{dx}{dt} = -8\sin 2t \quad \text{and} \quad \dot{y} = \frac{dy}{dt} = -16\sin 2t. \tag{b}$$

A second integration using the given initial conditions for lower limits and general upper limits yields

$$\int_0^{x} dx = \int_0^t -8\sin 2t\,dt \quad \text{and} \quad \int_0^{y} dy = \int_0^t 8\cos 2t\,dt.$$

$$x = 4\cos 2t, \quad \text{and} \quad y = 4\sin 2t.$$

Squaring both sides of the two equations above and adding them, we conclude that

$$x^2 + y^2 = 16 \qquad \text{ANS.}$$

which is the equation of a circle with a radius of 4 ft centered at the origin of the x-y coordinate system. This circular path is shown in Figure E13.15.

(b) The x and y components of the train's velocity are given by Eqs. (b). Thus, for $t = 0.5$ s,

$$\dot{x} = -8\sin[2(0.5)] = -6.73 \text{ ft/s}, \quad \text{and} \quad \dot{y} = 8\cos[2(0.5)] = 4.32 \text{ ft/s}.$$

Thus,

$$v = \sqrt{\dot{x}^2 + \dot{y}^2} = 8.00 \text{ ft/s}. \qquad \text{ANS.}$$

When $t = 0.5$ s, the toy train is at point B in Figure E13.15 where the angle $\alpha = 2t = 2(0.5) = 1$ rad $\doteq 57.3°$ as shown. The angle θ_y defining the direction of the velocity from the y axis is given by

$$\theta_y = \tan^{-1}\left(\frac{\dot{x}}{\dot{y}}\right) = \tan^{-1}\left(\frac{-6.73}{4.32}\right) = 57.3°$$

which again shows that the velocity is tangent to the path.

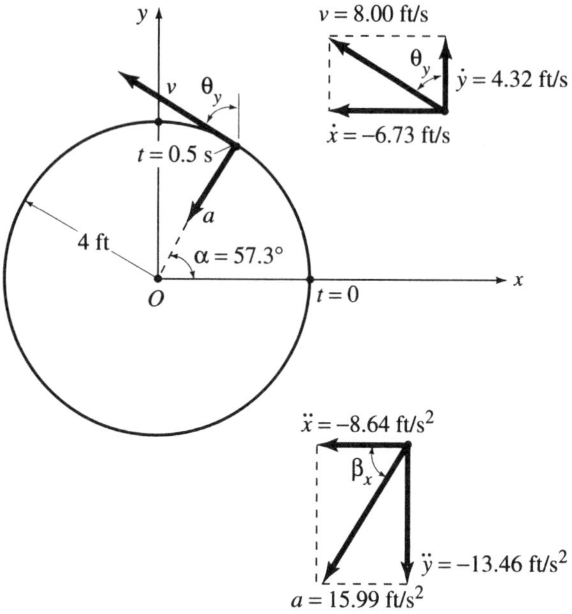

FIGURE E13.15.

(c) The x and y components of the train's acceleration are given by Eqs. (a). Thus, for $t = 0.5$ s,

$$\ddot{x} = -16\cos[2(0.5)] = -8.64 \text{ ft/s}^2, \quad \text{and}$$

$$\ddot{y} = -16\sin[2(0.5)] = -13.46 \text{ ft/s}^2.$$

Therefore,

$$a = \sqrt{\ddot{x}^2 + \ddot{y}^2} = 15.99 \text{ ft/s}^2. \qquad \text{ANS.}$$

As shown in Figure E13.15, the angle β_x defining the direction of the acceleration from the x axis is given by

$$\beta_x = \tan^{-1}\left(\frac{\ddot{x}}{\ddot{y}}\right) = \tan^{-1}\left(\frac{-13.46}{-8.64}\right) = 57.3^\circ$$

The value obtained for β_x shows that the acceleration is perpendicular to the path's tangent. In other words, the acceleration is directed towards the center of the circular path.

■ Example 13.16

Neglecting air resistance and assuming a constant acceleration due to gravity g, consider the general motion of a projectile. In Figure E13.16 a projectile is launched from the point whose coordinates are (x_0, y_0). Starting with the known acceleration components $\ddot{x} = 0$ and $\ddot{y} = -g$, develop general equations for the velocity components $\dot{x}$ and $\dot{y}$ and for

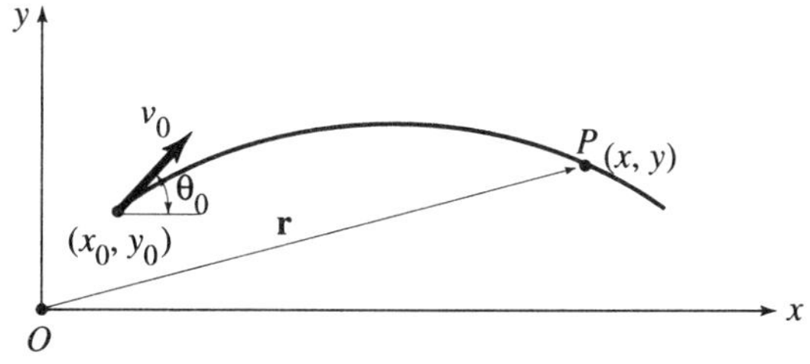

FIGURE E13.16.

the position coordinates x and y. State these quantities as functions of time.

Solution

(a) Neglecting air resistance implies that the acceleration in the x direction $\ddot{x}$ will vanish. Because the projectile is close to the Earth's surface, the acceleration in the y direction $\ddot{y}$ equals $-g$. Galileo was first to observe that the projectile's motion may be broken down into separate motions in the horizontal and vertical directions. It is convenient to imagine that we are studying the motion of the *shadows* of the projectile along the x and y axes.

Horizontal Motion

Because $a_x = \ddot{x} = 0$ it follows that $v_x = \dot{x} =$ constant. When the projectile is fired from the point (x_0, y_0), it has a horizontal velocity of $v_0 \cos\theta_0$ which remains constant. Thus,

$$v_x = \dot{x} = \frac{dx}{dt} = v_0 \cos\theta_0. \qquad \text{ANS.}$$

Multiplying by dt and integrating from $t = 0$, $x = x_0$ to general upper limits yields

$$\int_{x_0}^{x} dx = v_0 \cos\theta_0 \int_0^t dt$$

$$x = x_0 + v_0 t \cos\theta_0 \qquad \text{ANS.}$$

This equation expresses the x coordinate of the projectile as a function of time, provided we know the initial conditions, the elevation angle θ_0, the muzzle or launch velocity v_0, and the abscissa x_0 of the firing or launch position.

Vertical Motion

In the y direction, $\ddot{y} = d\dot{y}/dt = -g =$ constant. Multiply by dt and integrate from $t = 0$, $v_y = \dot{y} = v_0 \sin\theta_0$ to general upper limits. Thus,

$$\int_{v_0 \sin\theta_0}^{v_y} \ddot{y}\, dt = -g \int_0^t dt,$$

$$v_y = \dot{y} = \frac{dy}{dt} = v_0 \sin\theta_0 - gt. \qquad \text{ANS.}$$

This equation expresses the y component of the velocity of the projectile as a function of t, provided we know θ_0, v_0 and g. Of course, $g = 32.2\ \text{ft/s}^2$ in U.S. Customary units and $g = 9.81\ \text{m/s}^2$ in SI units. Multiply this equation by dt and integrate from $t = 0$, $y = y_0$ to general upper limits to obtain

$$\int_{y_0}^{y} dy = v_0 \sin\theta_0 \int_0^t dt - g \int_0^t t\,dt,$$

$$y = y_0 + (v_0 \sin\theta_0)t - \tfrac{1}{2}gt^2.$$

This equation expresses the y coordinate of the projectile as a function of time, provided we know the initial conditions given by θ_0, v_0, and y_0.

The position vector $\mathbf{r}$ is given by

$$\mathbf{r} = (x_0 + v_0 \cos\theta_0 t)\mathbf{i} + (y_0 + v_0 \sin\theta_0 t - \tfrac{1}{2}gt^2)\mathbf{j}$$

This is a compact way to state the position of the projectile at any time t during its flight. The reader should differentiate this position vector to obtain the velocity and acceleration vectors for the projectile at any time t. The component velocities and accelerations will check those stated above.

■ Example 13.17

A person holds a hose as shown in Figure E13.17 at an angle of 30° to the horizontal. If he wishes to have the water strike an opening in the wall at B, 14 ft above the ground, determine the distance x from the wall at which she should position the nozzle. Hint: Refer to Example 13.16.

Solution

Motion in the y Direction

From Example 13.16,

$$y = y_0 + (v_0 \sin\theta_0)t - \tfrac{1}{2}gt^2$$

$$14 = 4 + (60 \sin 30°)t - \tfrac{1}{2}(32.2)t^2.$$

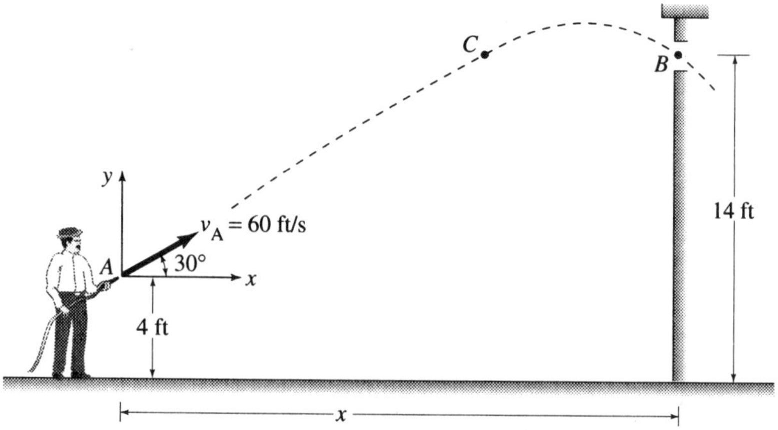

FIGURE E13.17.

Simplifying yields

$$t^2 - 1.863t + 0.621 = 0.$$

Therefore, using the quadratic formula,

$$t = 1.428 \text{ s} \quad \text{or} \quad t = 0.435 \text{ s}.$$

Motion in the *x* Direction

From Example 13.16,

$$x = x_0 + v_0 t \cos\theta,$$
$$= 0 + 60\cos 30°t.$$

For $t = 1.428$ s,

$$x = (60\cos 30°)(1.428) = 74.2 \text{ ft.} \qquad \text{ANS.}$$

For $t = 0.435$ s,

$$x = (60\cos 30°)(0.435) = 22.6 \text{ ft.} \qquad \text{ANS.}$$

Note that both answers are physically acceptable. The first answer ($t = 1.428$ s, $x = 74.2$ ft.) corresponds to point B after the jet of water has reached its maximum height and is on the way down. The second answer ($t = 0.435$ s, $x = 22.6$ ft) corresponds to point C which is located at a height of 14 ft on the projectile path of the water prior to reaching its maximum height.

■

Problems

13.95 An automobile moves in the x-y plane such that $x = t$ and $y = 2t$ (m and s units apply). Determine the equation of the path and the velocity and acceleration vectors.

13.96 Parametric equations of the x-y motion of a particle are given by $y = t$ and $x = t^2 + 1$ where the parameter is the time t. Units for x and y are m, and the time is measured in s. Provided $y \geq 0$, determine (a) the path equation and the scalar components of the velocity vector, (b) the components of the velocity vector and the magnitude and direction of the velocity vector when $t = 2$ s, and (c) the magnitude and direction of the acceleration vector at any time t.

13.97 Parametric equations of the x-y motion of a toy car are given by $x = 2t^2$ and $y = 4(1 + t^4)$ where the parameter is the time t. Units are ft and s. Determine (a) the equation of the path and sketch it, (b) the velocity vector when $t = 1$ s, and (c) the acceleration vector when $t = 1$ s.

13.98 A particle moves on a path in the x-y plane such that $\ddot{x} = -a\omega^2 \sin\omega t$ and $\ddot{y} = -a\omega^2 \cos\omega t$ where a is a constant measured in ft., ω is a constant measured in rad/s, and t is the time measured in s. Initial conditions of the motion are $t = 0$, $x = 0$, $y = a$, $\dot{x} = a\omega$, and $\dot{y} = 0$. Determine the scalar components of velocity, the x and y coordinates of the particle at any time t, and the equation of the path.

13.99 A particle moves on a path in the x-y plane such that $\ddot{x} = -a\omega^2 \sin \omega t$ and $\ddot{y} = -a\omega^2 \cos \omega t$. Initial conditions of the motion are $t = 0$, $x = c$, $y = d + a$, $\dot{x} = a\omega$, and $\dot{y} = 0$. Constants a, c, and d are measured in m, the constant ω is measured in rad/s, and the time t is given in s. Determine the velocity vector and the position vector at any time t, and find the equation of the path.

13.100 Planar motion of a particle takes place such that the scalar acceleration components are given by $\ddot{x} = 2$ and $\ddot{y} = 12t^2$ where the units are in. and s. The initial conditions are $t = 0$, $x = c$, $y = d$, $\dot{x} = 0$, $\dot{y} = 0$. Constants c and d are measured in inches. Determine the velocity and position vectors as functions of time. Find the equation of the path by eliminating the parameter t. Sketch the path and the position and velocity vectors.

13.101 Curvilinear motion takes place in the x-y plane such that the scalar acceleration components are given by $\ddot{x} = 12t^2$ and $\ddot{y} = 56t^6$. Units are m and s. Initial conditions of the motion are $t = 0$, $x = c$, $y = d$, $\dot{x} = 0$, $\dot{y} = 0$. Constants c and d are expressed in m. Determine the velocity and position vectors of the particle when $t = 1$ s if $c = -0.5$ m and $d = 2$ m. Sketch the path and the required vectors.

13.102 The position vector from a fixed origin to a particle moving along a curvilinear path is given by $\mathbf{r} = (c + t^2)\mathbf{i} + (d + t^6)\mathbf{j}$. Units are ft and s, and the constants c and d are given in ft. Determine (a) the path and its equation, (b) the velocity and acceleration vectors, and (c), for $t = 1$ s, $c = 1$ ft, $d = -2$ ft, sketch the path, and show the above vectors on your sketch.

13.103 A particle moving along a curvilinear path in the x-y plane is positioned by the vector $\mathbf{r} = a \sin \omega t \mathbf{i} + b \cos \omega t \mathbf{j}$, where the constants a and b are given in m, the constant ω is given in rad/s, and time is given in s. Determine (a) the path and its equation, (b) the velocity vector, and (c) the acceleration vector.

13.104 Acceleration components for the motion of a particle in the x-y plane are given by $\ddot{x} = -a\omega^2 \sin \omega t$ and $\ddot{y} = -b\omega^2 \cos \omega t$. Initial conditions of the motion are $t = 0$, $x = 0$, $y = b$, $\dot{x} = a\omega$, $\dot{y} = 0$. Constants a and b are measured in ft., the constant ω is measured in rad/s, and time is measured in s. Determine the velocity vector and the position vector at any time t, and find the equation of the path. Sketch the path and the velocity and position vectors. Why is this a periodic motion?

13.105 The straight-line path of a toy train is given by $y = 6 + 4x$ where $x = 4t$. Units are in. and s. Find y as a function of t and the vectors $\mathbf{r}$, $\mathbf{v}$ and $\mathbf{a}$. Is the train accelerating?

13.106 The position of a cannon ball shown in Figure P13.106 may be determined

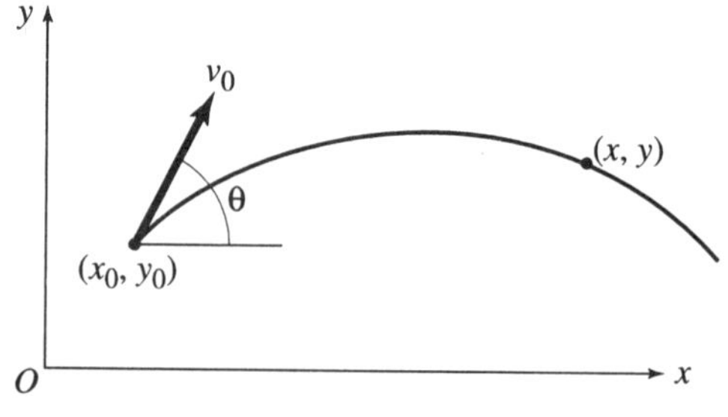

FIGURE P13.106.

from the equations $x = x_0 + (v_0 \sin\theta)t$ and $y = y_0 + (v_0 \cos\theta)t - \frac{1}{2}gt^2$, where (x_0, y_0) denotes the initial position of the projectile when it is fired at an angle θ to the horizontal and v_0 is its initial velocity. The ball's range and its altitude during its motion are such that the acceleration due to gravity g may be assumed constant. Atmospheric resistance is neglected. (a) Differentiate these equations for x and y with respect to time to write equations for the velocity components in the x and y directions. (b) Differentiate the velocity components equations to obtain the acceleration components in the x and y directions.

13.107 A baseball is hit as indicated in Figure P13.107. You are to decide which trajectory is the correct one. If trajectory 1 is correct, where does the ball hit the ground, assuming a fielder is unable to catch it? If trajectory 2 is correct, where will the ball hit the wall? If trajectory 3 is correct, by what distance will the ball clear the wall? The origin is selected at the point where the ball is hit. Break the motion into x and y components. The velocity in the x direction remains constant if we neglect air resistance. The constant acceleration in the y direction is $-g = -32.2$ ft/s^2.

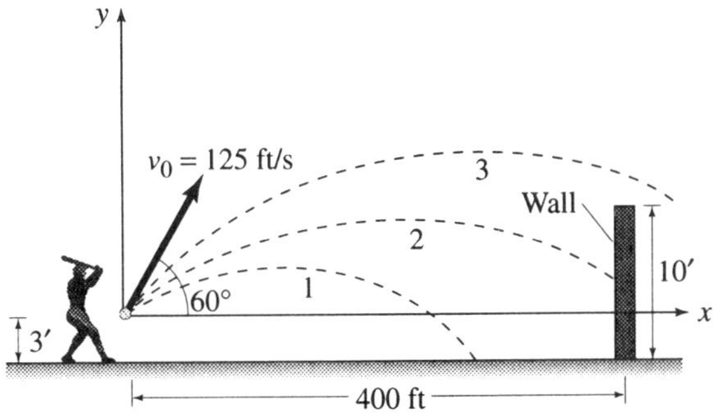

FIGURE P13.107.

13.108 A projectile is fired from the origin with an initial velocity of 400 ft/s at an angle of 30° to the horizontal. (a) Write the equation of its path, (b) Determine the range of this projectile, and (c) Find the maximum height above the horizontal reached by the projectile.

13.109 A projectile is fired from the point whose coordinates are $x_0 = 200$ ft, $y_0 = 400$ ft. It has an initial velocity of 500 ft/s and makes an initial angle of 40° with the positive x axis. (a) Express x and y as functions of time. (b) Determine the x coordinate on the path corresponding to $y = -100$ ft.

13.110 A projectile is launched from the origin with an initial velocity of 300 m/s at an angle of θ to the horizontal. It strikes a point 2000 m to the right of the origin on a horizontal plane. Determine the two possible values for θ, and sketch the trajectories.

13.111 A person is shot from a cannon at a circus as shown in Figure P13.111. For trajectory 1, find θ_1, and, for trajectory 2, find θ_2. Assume $v_1 = v_2 = 200$ ft/s,

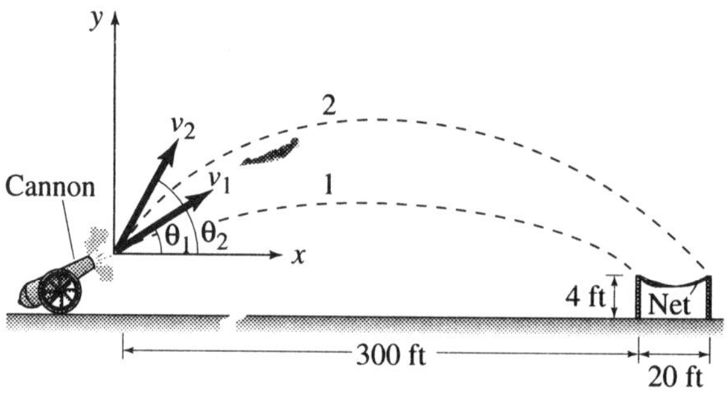

FIGURE P13.111.

and choose θ values corresponding to smaller times. Break the motion into x and y components. The velocity in the x direction remains constant if we neglect air resistance. The acceleration in the y direction is $-g = -32.2 \text{ ft/s}^2$. What is the range of values of θ so that the person hits the safety net?

13.8 Curvilinear Motion—Tangential and Normal Coordinates

In many types of problems dealing with the motion of a particle along a curved path, it is convenient to express the acceleration in terms of two components, one along the tangent to the path, known as the *tangential component*, and the second along the inward normal to the path, known as the *normal component*. For this purpose, we define two new unit vectors $\boldsymbol{\lambda}_n$ and $\boldsymbol{\lambda}_t$, along the tangent and along the inward normal to the path of motion, respectively.

Consider a particle moving along the curved path shown in Figure 13.12. At a given instant, the particle is at point P which serves as the origin of a t-n coordinate system that moves with the particle. The axis t and the unit vector $\boldsymbol{\lambda}_t$ coincide with the tangent to the path at P and

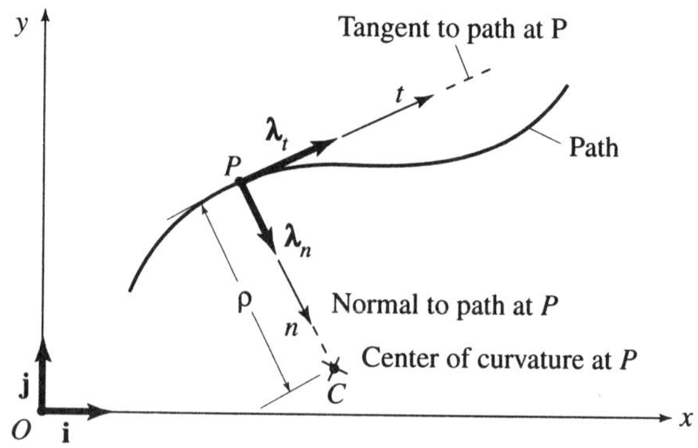

FIGURE 13.12.

are pointed in the direction of motion. The axis n and the unit vector λ_n coincide with the normal to the path at P and are pointed along the inward normal towards the *center of curvature* C of the path at P. Note that the radius of curvature of the path is denoted by ρ in Figure 13.12. Note also the unit vectors **i** and **j** which are fixed not only in magnitude but also in direction along the fixed x and y axes, respectively. In contrast, whereas the unit vectors λ_t and λ_n are fixed in magnitude, they continuously change direction as the particle moves along the path.

Now consider a particle moving along a curved path as shown in Figure 13.13(a). At time t the particle is at P and its velocity, which is along the tangent to the path at P, is $\mathbf{v}$. At $t_1 = t + \Delta t$, the particle is at P_1, after having moved a distance Δs along the path, and its velocity $\mathbf{v}_1$ is directed along the tangent at P_1. If both of these velocity vectors are placed at the same origin as shown in Figure 13.13(b), we conclude that the change in velocity $\Delta\mathbf{v} = \mathbf{v}_1 - \mathbf{v}$ may be decomposed into the components $\Delta\mathbf{v}_n$ and $\Delta\mathbf{v}_t$ as shown. Thus,

$$\Delta\mathbf{v} = \Delta\mathbf{v}_t + \Delta\mathbf{v}_n.$$

In the limit as $\Delta t \to 0$ and P_1 approaches P, $\Delta\mathbf{v}_t$ and $\Delta\mathbf{v}_n$ will be directed, respectively, along the tangent and inward normal to the path at P.

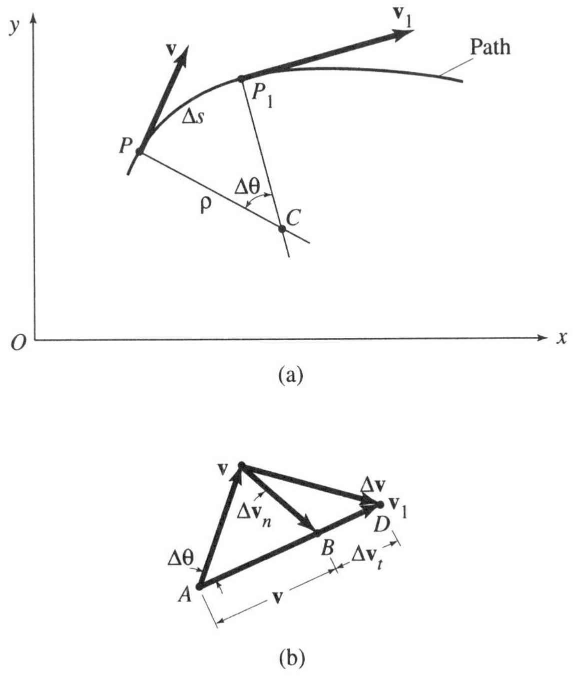

FIGURE 13.13.

If the above equation for $\Delta\mathbf{v}$ is divided by Δt and the limit is taken as $\Delta t \to 0$,

$$\lim_{\Delta t \to 0} \frac{\Delta \mathbf{v}}{\Delta t} = \lim_{\Delta t \to 0} \left(\frac{\Delta \mathbf{v}}{\Delta t}\right)_{\mathrm{t}} + \lim_{\Delta t \to 0} \left(\frac{\Delta \mathbf{v}}{\Delta t}\right)_{\mathrm{n}}$$

which may be written as

$$\frac{d\mathbf{v}}{dt} = \left(\frac{d\mathbf{v}}{dt}\right)_{\mathrm{t}} + \left(\frac{d\mathbf{v}}{dt}\right)_{\mathrm{n}}.$$

Because, by definition, $\frac{d\mathbf{v}}{dt}$ is the acceleration of the particle, the above relationship may expressed in the form

$$\mathbf{a} = \mathbf{a}_{\mathrm{t}} + \mathbf{a}_{\mathrm{n}}. \tag{13.24}$$

Equation (13.24) states that the acceleration of a particle moving along a curved path may be decomposed into the components $\mathbf{a}_{\mathrm{t}}$ along the tangent to the path and $\mathbf{a}_{\mathrm{n}}$ along the inward normal to the path. Using the unit vectors $\boldsymbol{\lambda}_{\mathrm{t}}$ and $\boldsymbol{\lambda}_{\mathrm{n}}$, the two components of the acceleration may expressed as

$$\mathbf{a}_{\mathrm{t}} = a_{\mathrm{t}}\boldsymbol{\lambda}_{\mathrm{t}} \tag{13.25a}$$

and

$$\mathbf{a}_{\mathrm{n}} = a_{\mathrm{n}}\boldsymbol{\lambda}_{\mathrm{n}} \tag{13.25b}$$

where a_{t} and a_{n} are, the magnitudes of the vectors $\mathbf{a}_{\mathrm{t}}$ and $\mathbf{a}_{\mathrm{n}}$, respectively.

Let us now determine the magnitudes a_{t} and a_{n}. From line ABD in Figure 13.13(b), we conclude that $\Delta\mathbf{v}_{\mathrm{t}} = \mathbf{v}_1 - \mathbf{v}$. Because all of these vectors lie along the same straight line, we may deal only with magnitudes and write $\Delta v_{\mathrm{t}} = v_1 - v$. Because $v_1 - v$ represents the change in magnitude of the velocity (i.e., change in speed) Δv, we conclude that

$$\Delta v_{\mathrm{t}} = \Delta v$$

Dividing this equation by Δt and taking the limit as $\Delta t \to 0$,

$$\lim_{\Delta t \to 0} \left(\frac{\Delta v}{\Delta t}\right)_{\mathrm{t}} = \lim_{\Delta t \to 0} \frac{\Delta v}{\Delta t}.$$

Therefore,

$$\left(\frac{dv}{dt}\right)_{\mathrm{t}} = \frac{dv}{dt}.$$

Because $\left(\frac{dv}{dt}\right)_{\mathrm{t}} = a_{\mathrm{t}}$, we conclude that

$$a_{\mathrm{t}} = \frac{dv}{dt} \tag{13.26}$$

which states that the scalar component of acceleration a_t represents the *rate of change in the speed of the particle.*

The magnitude of $\mathbf{a}_n$ may also be found by reference to Figure 13.13(b). Because $\Delta\theta$ is a very small angle we may approximate the arc length by the chord length and obtain

$$\Delta v_n = v\,\Delta\theta.$$

If this equation is divided by Δt, then, taking the limit as $\Delta t \to 0$,

$$\lim_{\Delta t \to 0}\left(\frac{\Delta v}{\Delta t}\right)_n = \lim_{\Delta t \to 0} v\left(\frac{\Delta\theta}{\Delta t}\right)$$

which becomes

$$\left(\frac{dv}{dt}\right)_n = v\frac{d\theta}{dt}$$

Because $\left(\dfrac{dv}{dt}\right)_n = a_n$ and $d\theta = \dfrac{ds}{\rho}$ where ρ is the radius of curvature of the path, we conclude that

$$a_n = \frac{v}{\rho}\frac{ds}{dt} = \frac{v^2}{\rho}. \tag{13.27}$$

Equation (13.27) states that the scalar component of acceleration a_n is obtained by squaring the speed of the particle and dividing it by the radius of curvature at the point of interest. It represents the *change in the direction of the velocity* of the particle and is always pointed toward the center of curvature of the path. In summary,

$$\mathbf{a} = \left(\frac{dv}{dt}\right)\boldsymbol{\lambda}_t + \left(\frac{v^2}{\rho}\right)\boldsymbol{\lambda}_n \tag{13.28}$$

where the radius of curvature ρ is given by

$$\rho = \left| \pm\frac{\left[1 + \left(\dfrac{dy}{dx}\right)^2\right]^{3/2}}{\left(\dfrac{d^2y}{dx^2}\right)} \right|. \tag{13.29}$$

Equation (13.28) is represented graphically in Figure 13.14. The magnitude of the resultant acceleration may be found from the relationship

$$a = \sqrt{a_t^2 + a_n^2} \tag{13.30a}$$

The direction of the acceleration may be specified by finding the angle θ_n it makes with the inward normal. Thus.

$$\theta_n = \tan^{-1}\left(\frac{a_t}{a_n}\right). \tag{13.30b}$$

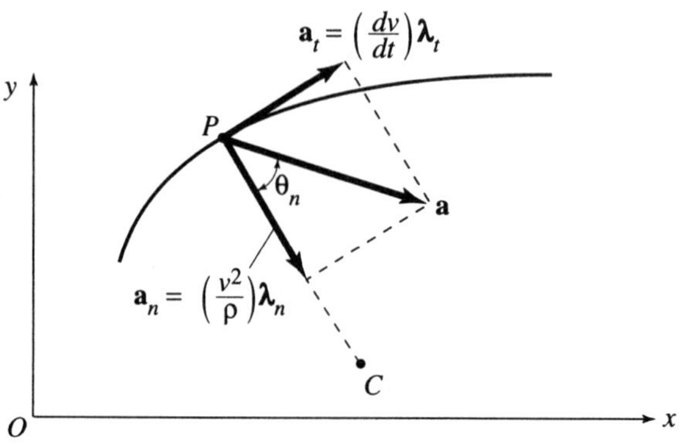

FIGURE 13.14.

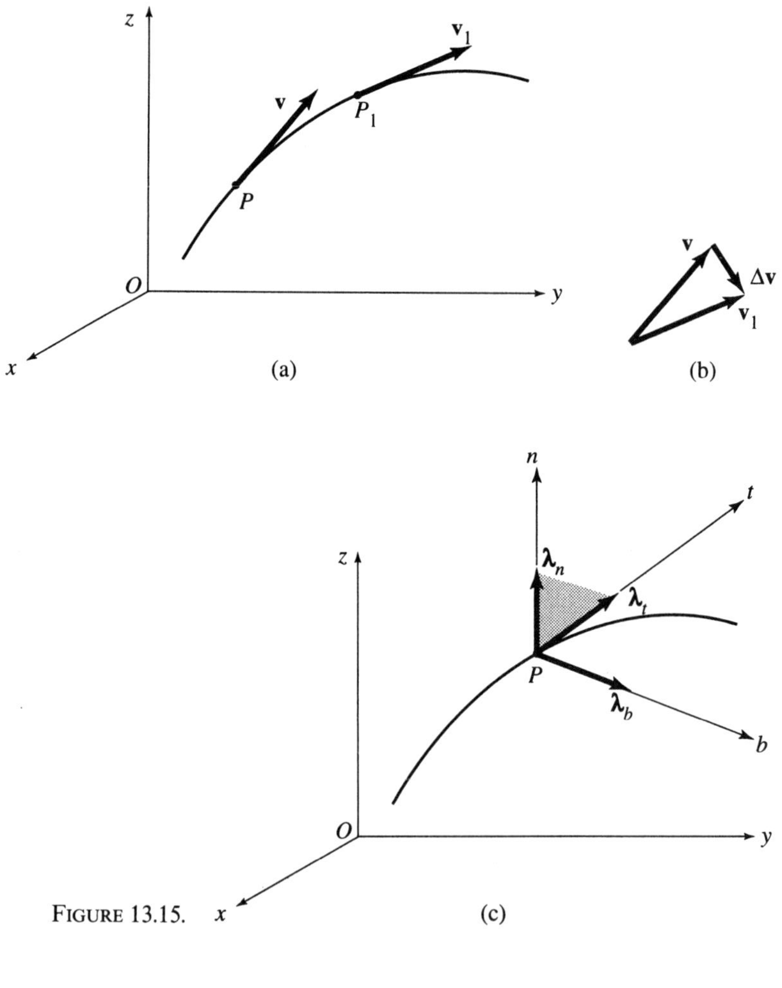

FIGURE 13.15.

The equations developed above for two-dimensional motion of a particle are also valid for three-dimensional motion. Because there are infinitely many lines that may be drawn normal to the tangent at any point on a three-dimensional curved path, we need to be careful in our definition of the axes t and n.

Consider the three-dimensional path shown in Figure 13.15(a). At time t, the particle is at P and its velocity is $\mathbf{v}$ as shown. At time $t_1 = t + \Delta t$, the particle is at P_1 and its velocity is $\mathbf{v}_1$ such that $\mathbf{v}_1 = \mathbf{v} + \Delta\mathbf{v}$ as shown by the velocity diagram of Figure 13.15(b). This velocity diagram defines a plane in space. Let us now imagine that we construct another plane at P parallel to the plane just defined. As $\Delta t \to 0$ and P_1 approaches P, the resulting plane at P is referred to as the *osculating** plane and it contains the t and n axes. This n axis is known as the *principal normal* at P. As shown in Figure 13.15(c), an axis b is constructed perpendicular to the osculating plane at P. This axis, referred to as the *binormal* axis, has a positive sense determined by the relationship $\boldsymbol{\lambda}_b = \boldsymbol{\lambda}_t \times \boldsymbol{\lambda}_n$. Thus, once the t and n axes are properly determined at any point P, the acceleration of the particle is contained in the osculating plane defined by these two axes. Therefore, the acceleration of the particle at any point P has no component along the binormal axis b and, as expressed in Eq. (13.28), may be decomposed into tangential and normal components.

The following examples illustrate some of the concepts in this section.

Example 13.18

A highway engineer needs to find the least radius R of an exit circular ramp as shown schematically in Figure E13.18. His computations are to be based on a constant exit ramp speed of 75 km/h and that the passengers in the cars are subjected to accelerations not to exceed 1.1 m/s^2. What should be the least value of the radius R?

Solution

Since the exit ramp speed is to be constant, there is no tangential component of acceleration. Thus, the only acceleration acting on the car passengers is the normal component. Hence, using Eq. (13.27) we obtain

$$\rho = \frac{v^2}{a_n}$$

where

$$v = 75 \text{ km/h} = \frac{75 \times 10^3}{3600} = 20.83 \text{ m/s}$$

* An illustration of the osculating plane is given in Example 13.22.

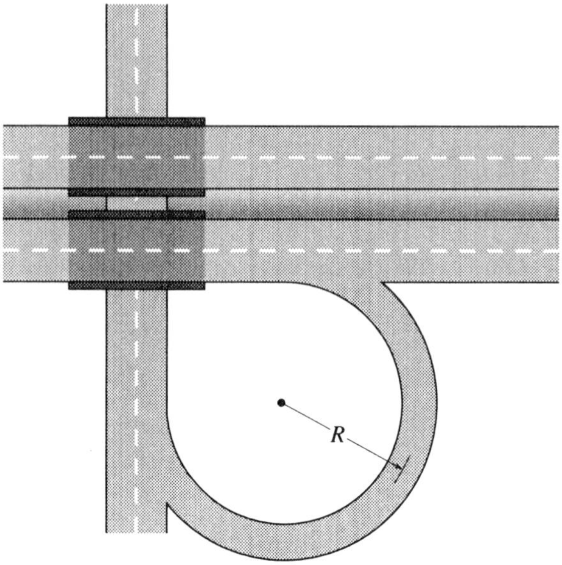

FIGURE E13.18.

and

$$a_n = 1.1 \text{ m/s}^2.$$

Thus,

$$R_{min} = \frac{(20.833)^2}{1.1} = 395.0 \text{ m.} \qquad \text{ANS.}$$

Example 13.19

The pilot of a small plane flying along a horizontal circular path as shown in Figure E13.19(a) increases his speed at a constant rate from 100 mph to 150 mph in 3 s. If the radius of the circular path is 1800 ft, determine the pilot's acceleration at the instant he reaches the speed of 150 mph. Sketch a vector that represents the pilots' acceleration.

Solution

To determine the acceleration of the pilot, we need to find the tangential and normal components of his acceleration at the instant he reaches the speed of 150 mph. Thus, by Eq. (13.8),

$$v = v_0 + (a_t)t$$

from which

$$a_t = \frac{v - v_0}{t}$$

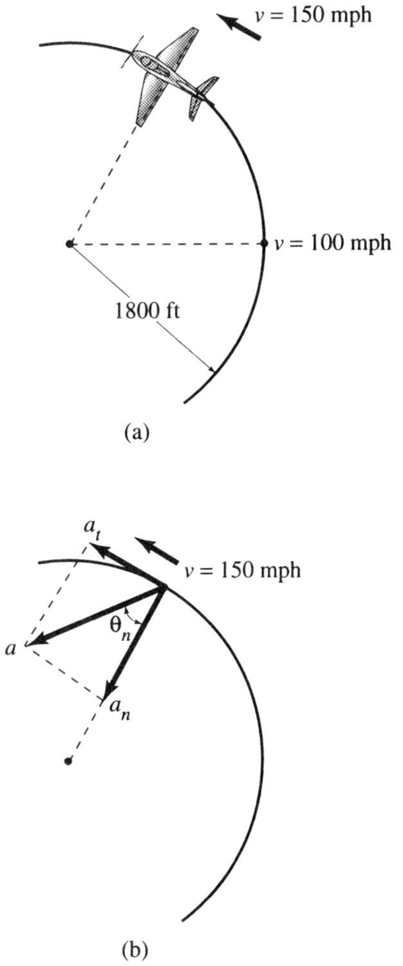

FIGURE E13.19.

where

$$v = 150 \text{ mph} = 220 \text{ ft/s},$$

$$v_0 = 100 \text{ mph} = 146.7 \text{ ft/s},$$

and

$$t = 3 \text{ s}.$$

Thus,

$$a_t = \frac{220 - 146.7}{3} = 24.4 \text{ ft/s}^2.$$

The normal component of acceleration is obtained from Eq. (13.27). Thus,

$$a_n = \frac{v^2}{\rho} = \frac{(220)^2}{1800} = 26.9 \text{ ft/s}^2.$$

Therefore, because $a = \sqrt{a_t^2 + a_n^2}$,

$$a = \sqrt{(24.4)^2 + (26.9)^2} = 36.3 \text{ ft/s}^2, \qquad \text{ANS.}$$

$$\theta_n = \tan^{-1}\left(\frac{a_t}{a_n}\right) = \tan^{-1}\left(\frac{24.4}{26.9}\right) = 42.2^\circ. \qquad \text{ANS.}$$

The acceleration **a** of the pilot, when he reaches the speed of 150 mph, is shown in Figure E13.19(b).

Example 13.20

The motion of a skier moving down a ski slope may be approximated by the equation $y = \frac{1}{12}x^2$ as shown in Figure E13.20(a). At point A on the slope, the skier's speed is clocked at 10 m/s and the change in speed is measured at 5 m/s^2. Determine the magnitude and direction of the skier's acceleration at this point.

Solution

The tangential component of acceleration is equal to the change in speed. Thus,

$$a_t = 5 \text{ m/s}^2.$$

The normal component of acceleration is given by

$$a_n = \frac{v^2}{\rho}$$

where $v = 10$ m/s and

$$\rho = \left| \frac{\left[1 + \left(\dfrac{dy}{dt}\right)^2\right]^{3/2}}{\pm\left(\dfrac{d^2y}{dx^2}\right)} \right|.$$

Because $y = \frac{1}{12}x^2$,

$$\frac{dy}{dx} = \frac{1}{6}x \quad \text{and} \quad \frac{d^2y}{dx^2} = \frac{1}{6}.$$

Therefore, at $x = 12$ m,

$$\frac{dy}{dx} = 2 \quad \text{and} \quad \frac{d^2y}{dx^2} = \frac{1}{6}.$$

Thus,

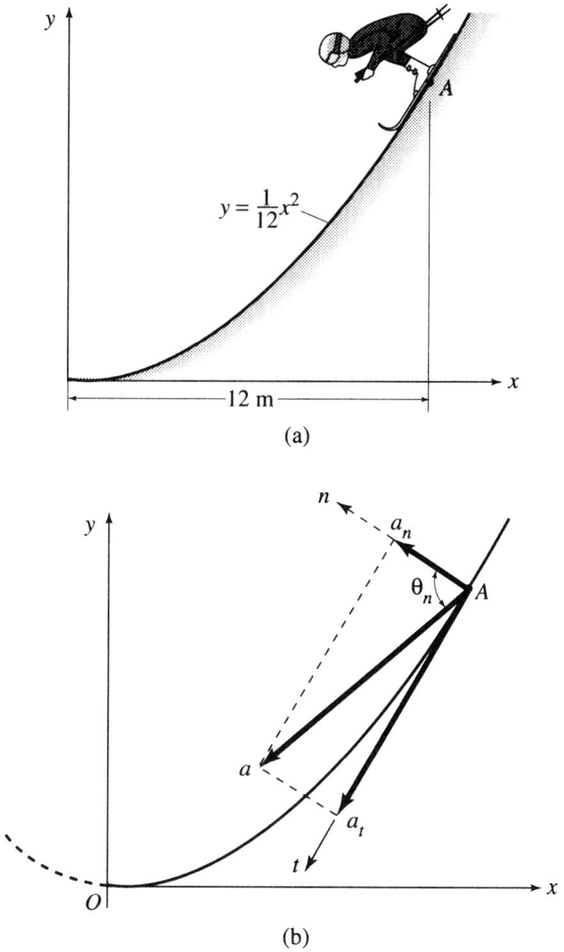

FIGURE E13.20.

$$\rho = \frac{[1 + 2^2]^{3/2}}{(\frac{1}{6})} = 67.1 \text{ m}$$

and

$$a_n = \frac{(10)^2}{67.1} = 1.490 \text{ m/s}^2.$$

Therefore,

$$a = \sqrt{a_t^2 + a_n^2} = 5.22 \text{ m/s}^2. \qquad \text{ANS.}$$

The direction of acceleration at point A is shown in Figure 13.20(b), given by the angle θ_n measured from the inward normal at this point, where

$$\theta_n = \tan^{-1}\left(\frac{a_t}{a_n}\right) = \tan^{-1}\left(\frac{5}{1.490}\right) = 73.4^\circ \qquad \text{ANS.}$$

Example 13.21

A cyclist moves in the horizontal plane along a circular path with a radius of 50 ft. For a short time interval, the cyclist's position along the path measured in feet is given by $s = 2t^2$ where t is in seconds. Assume that when $t = 0$, $s = 0$ and determine the cyclist's acceleration when $t = 5$ s. Show the acceleration on a sketch of the path.

Solution

To find the cyclist's acceleration, we need to determine the tangential and normal components, a_t and a_n, respectively. The normal component of the acceleration requires knowledge of the velocity which may be obtained by differentiating the given position function. Thus,

$$v = \frac{ds}{dt} = 4t$$

for $t = 5$ s,

$$v = 20 \text{ ft/s},$$

and

$$a_n = \frac{v^2}{\rho} = \frac{(20)^2}{50} = 8 \text{ ft/s}^2.$$

The tangential component of acceleration becomes

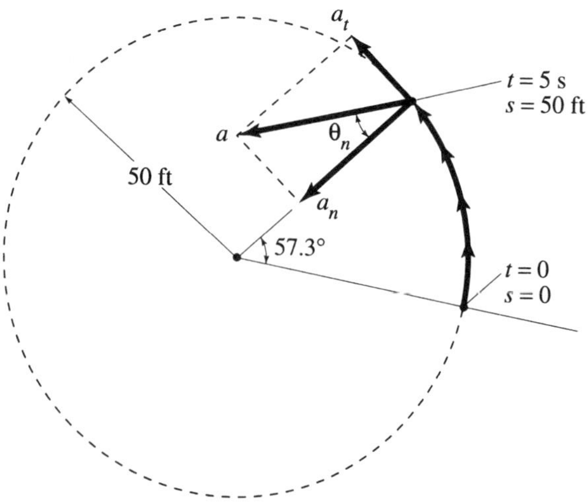

FIGURE E13.21.

$$a_t = \frac{dv}{dt} = 4 \text{ ft/s}^2.$$

The distance s traveled along the circular path during 5 s becomes

$$s = 2(5^2) = 50 \text{ ft}.$$

The position of the cyclist at $t = 5$ s is thus established as shown in Figure E13.21. At this position, the acceleration of the cyclist becomes

$$a = \sqrt{a_t^2 + a_n^2} = 8.94 \text{ ft/s}^2. \qquad \text{ANS.}$$

This acceleration, along with its two components is shown in Figure E13.21, where the angle θ_n that it makes with the inward normal to the path is

$$\theta_n = \tan^{-1}\left(\frac{a_t}{a_n}\right) = 26.6°. \qquad \text{ANS.}$$

■

Problems

13.112 A toy train moves along a straight-line path whose equation is $y = 2x + 4$. It intercepts the y axis when $t = 0$ and moves such that $s = t^2 - 10t + 16$ where s is in m and t in s. Direct a unit tangent vector up to the right along the line as positive, and, on this basis, determine the velocity and acceleration vectors of the train. What is the value of s when $t = 0$?

13.113 In Figure P13.113, a circular path is traversed by a particle such that $s = t^3 - 16t^2 + 60t$ where units are inches and seconds. The path center lies at the origin and the particle begins moving at $t = 0$ in a counterclockwise (ccw) sense around the 10 in. radius circle from its initial position on the positive x axis. Determine the velocity and the normal and tangential components of acceleration of the particle when $t = 1$ s. Show the position of the particle and these vectors on sketches.

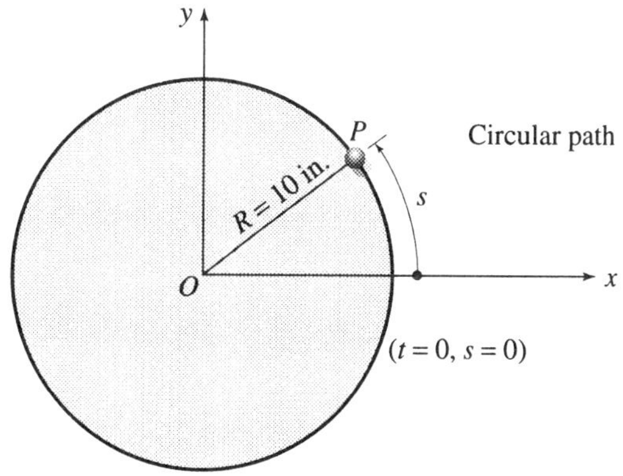

FIGURE P13.113.

13.114 A toy car moves along an elliptic path shown in Figure P13.114 and given by the equation $\frac{x^2}{16} + \frac{y^2}{9} = 1$. When the car crosses the positive y axis, it is moving 10 ft/s leftward and this speed is increasing at the rate of 20 ft/s^2 at

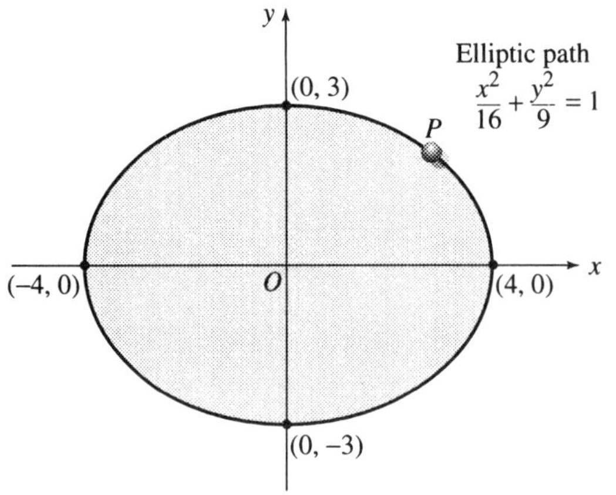

FIGURE P13.114.

that instant. Determine the total acceleration of the car at this instant and show this vector on a sketch of the path.

13.115 In Figure P13.115 a particle moves along a hyperbolic path given by the equation $\frac{x^2}{36}-\frac{y^2}{25}=1$. Consider the particle moving along the branch in the first and fourth quadrants. When the particle crosses the positive x axis, it is moving upward with a speed of 2 m/s, and this speed is increasing at the rate of 1 m/s^2 at that instant. Determine the total acceleration of the particle at that instant and show this vector on a sketch of the path.

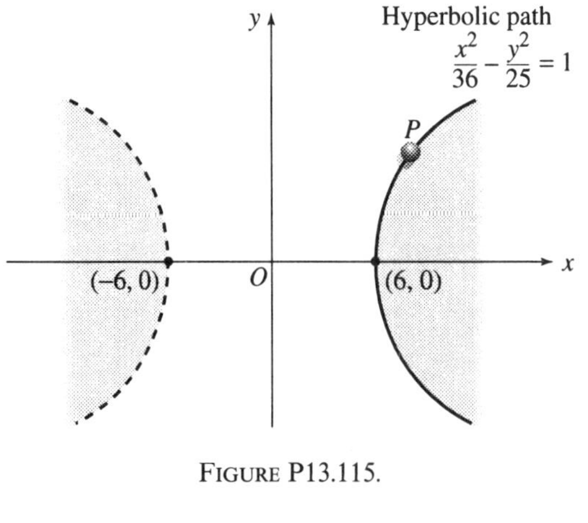

FIGURE P13.115.

13.116 A particle moves along a path in the x-y plane such that $x = t$ ($t \geq 0$) and $y = -t^2$ where x and y are in inches and t is in seconds. Determine (a) the path of the particle and sketch it, (b) for $t = 1$ s, the velocity and acceleration vectors of the particle, and (c), for $t = 1$ s, the tangential and normal components of the acceleration. Show a sketch of these components on the path of the particle.

13.117 A particle moves along a straight-line path whose equation is $y = 20 - x$. It intercepts the y axis when $t = 0$ and moves such that $s = t^2 - 8t$ where s is the distance measured down to the right along the line given in inches, time is given in seconds. Direct a unit tangent vector positively down to the right along the line, and determine the velocity and acceleration vectors of the particle.

13.118 The path of a particle in the x-y plane is given parametically by $x = -4t^2$, $y = 2t$ ($t \geq 0$) where x and y are in meters and t is in seconds. Determine (a) the path of the particle and sketch it, (b) for $t = 0.5$ s, the velocity and acceleration vectors of the particle, and (c), for $t = 0.5$ s, the tangential and normal components of the acceleration. Show a sketch of these components on the path of the particle.

13.119 A circular path is traversed by a particle as shown in Figure P13.119 such that $s = \frac{1}{3}t^3 - \frac{3}{2}t^2 + 2t$ where units are ft and s. The path center lies at the origin and the particle begins moving at $t = 0$ in a cw sense around the 2 ft. radius circle from its initial position on the positive x axis. Determine the velocity and the normal and tangential components of acceleration of the particle when $t = 2$ s. Show the position

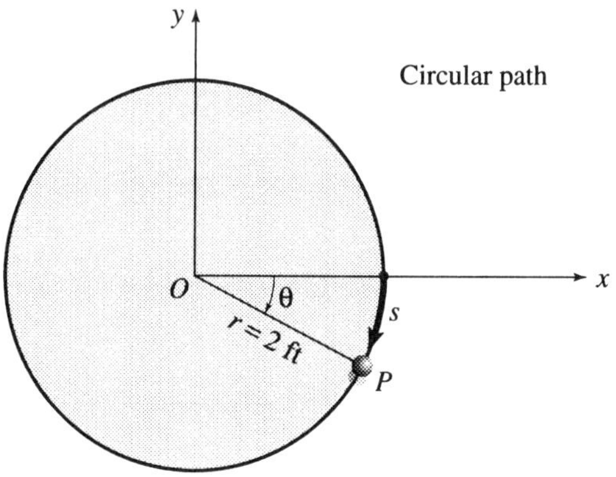

FIGURE P13.119.

of the particle and these vectors on a sketch.

13.120 A toy train moves along the path given by $y = \sinh x$. When $x = 2$ ft, the particle has a speed of 16 ft/s which is increasing at the rate of 20 ft/s^2. The train is moving up to the right at this instant. Locate the train on a sketch of the path. Determine the total acceleration of the train when $x = 2$ ft. Sketch this acceleration vector along with its normal and tangential components.

13.121 The path of a vehicle is given by $y = \cosh x$. When $x = -2$ m, the car is moving down to the right with a speed of 20m/s which is decreasing at the rate of 24 m/s^2. Locate the car on a sketch of the path. Determine the total acceleration of the car when $x = -2$ m. Sketch this acceleration vector along with its normal and tangential components.

13.122 A roller coaster is depicted in Figure P13.122. At point A, the radius of curvature of the path has a magnitude of 80 ft. A coaster car arrives at this point with a speed of 60 ft/s which is increasing at the rate of 20 ft/s^2 at the instant. Determine the instantaneous acceleration of the car in magnitude and direction at the instant when it is at point A, the low point on the track.

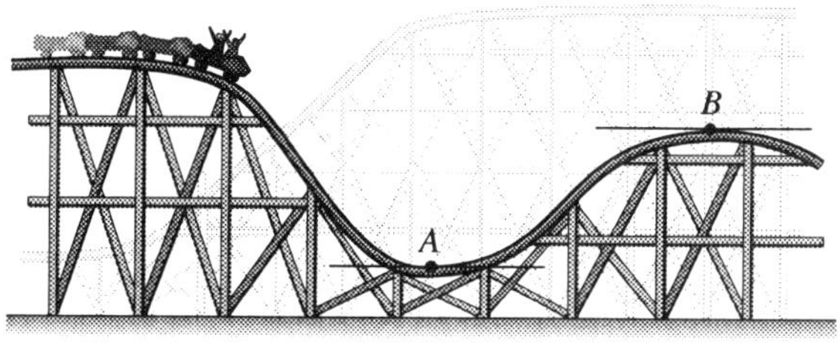

Tangents to track are horizontal at A and B.
Roller coaster

FIGURE P13.122.

13.123 Refer to the roller coaster shown in Figure P13.122. At point B, the track has a radius of curvature of 25 m. A coaster car arrives at this point with a speed of 18 m/s which is decreasing at the rate of 6 m/s^2 at the instant. Determine the instantaneous acceleration of the car in magnitude and direction at the instant it arrives at point B.

13.124 A carnival swing is depicted in Figure P13.124. Cables of length L are attached to a shaft which rotates about a vertical axis and swings so that P moves along a circular path in a hori-

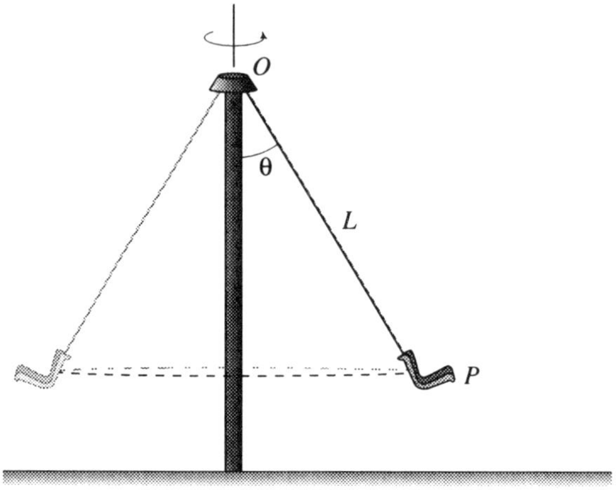

Carnival swing

FIGURE P13.124.

zontal plane. Given that the speed of P is constant at 30.5 ft/s, $\theta = 30°$, and $L = 100$ ft, determine the acceleration **a** of the seat P. Show **a** on a sketch of the circular path.

13.125 Refer to the carnival swing shown in Figure P13.124 Swings, such as P, travel a circular path in a horizontal plane. Determine the acceleration **a** of the seat P if its constant speed is 47.7 ft/s, $\theta = 45°$, and $L = 100$ ft. Show **a** on a sketch of the circular path.

13.126 A particle moves along a sinusoidal path given by $y = \sin x$, with $x = t$. Show that the acceleration vector has a magnitude equal to the ordinate of the path.

13.127 A curvilinear path in the x-y plane is defined parametically by $y = \cos t$ and $x = t$. (a) Determine general equations for the velocity and acceleration vectors at any time t. (b) For $t = \pi$ s, determine the velocity and acceleration vectors.

13.9 Curvilinear Motion—Cylindrical Coordinates

Three-Dimensional Motion

In certain types of the three-dimensional motion of a particle, it is convenient to use cylindrical coordinates to position points such as P of Figure 13.16, in three-dimensional space. We use three coordinates r, θ, z where r is measured outward from an origin at O in the radial direction, θ is measured from the positive x axis to the radius r in a ccw sense as viewed from above, and z is measured positively in the upward vertical direction. The positive sense of the angle θ establishes the positive sense of what is termed the *transverse direction*, perpendicular to the vertical plane determined by the radial and vertical directions. It is informative to refer to Figure 13.16 and imagine the geometric shapes generated by holding each of the variables r, θ, and z constant, in turn. First, consider r held constant, and imagine z and θ varying to

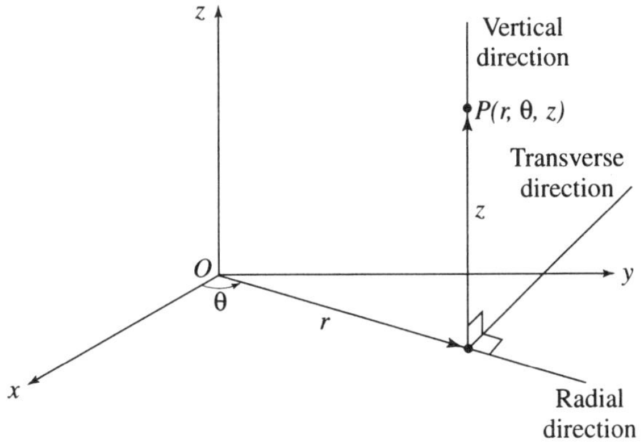

FIGURE 13.16.

obtain a right circular cylinder whose axis is the z axis and whose depth extends from $-\infty$ to $+\infty$. Next, consider θ held constant and imagine r and z varying to obtain a vertical plane of indefinite extent which makes an angle of θ with the x-z plane. Finally, consider z held constant and imagine r and θ varying to obtain a horizontal plane of indefinite extent which is parallel to the x-y plane.

We will now define three mutually perpendicular unit vectors, as shown in Figure 13.17(a), which move along the path with the particle. The unit vector $\boldsymbol{\lambda}_z$ does not change in magnitude or direction with time because it is always directed along the fixed positive z axis. However, a change in θ will produce changes in the directions of the unit vectors $\boldsymbol{\lambda}_r$ and $\boldsymbol{\lambda}_\theta$, and we must account for these changes when we consider time derivatives.

As before, we begin by writing the position vector $\mathbf{r}$ as a function of time in terms of the unit vectors. Thus, as shown in Figure 13.17(b),

$$\mathbf{r} = r\boldsymbol{\lambda}_r + 0\boldsymbol{\lambda}_\theta + z\boldsymbol{\lambda}_z. \tag{13.31}$$

We write $0\boldsymbol{\lambda}_\theta$ to emphasize that $\mathbf{r}$ does not have a component in the θ direction because the vector $\mathbf{r}$ lies in a vertical plane determined by $\boldsymbol{\lambda}_r$ and $\boldsymbol{\lambda}_z$. Recalling the definition of instantaneous velocity $\mathbf{v} = \dfrac{d\mathbf{r}}{dt}$,

$$\mathbf{v} = \frac{d\mathbf{r}}{dt} = r\frac{d\boldsymbol{\lambda}_r}{dt} + \frac{dr}{dt}\boldsymbol{\lambda}_r + z\frac{d\boldsymbol{\lambda}_z}{dt} + \frac{dz}{dt}\boldsymbol{\lambda}_z.$$

Using the dot notation for time derivatives and noting that $\dfrac{d\boldsymbol{\lambda}_z}{dt} = \mathbf{0}$ from the foregoing discussion,

$$\mathbf{v} = r\frac{d\boldsymbol{\lambda}_r}{dt} + \dot{r}\boldsymbol{\lambda}_r + \dot{z}\boldsymbol{\lambda}_z. \tag{13.32}$$

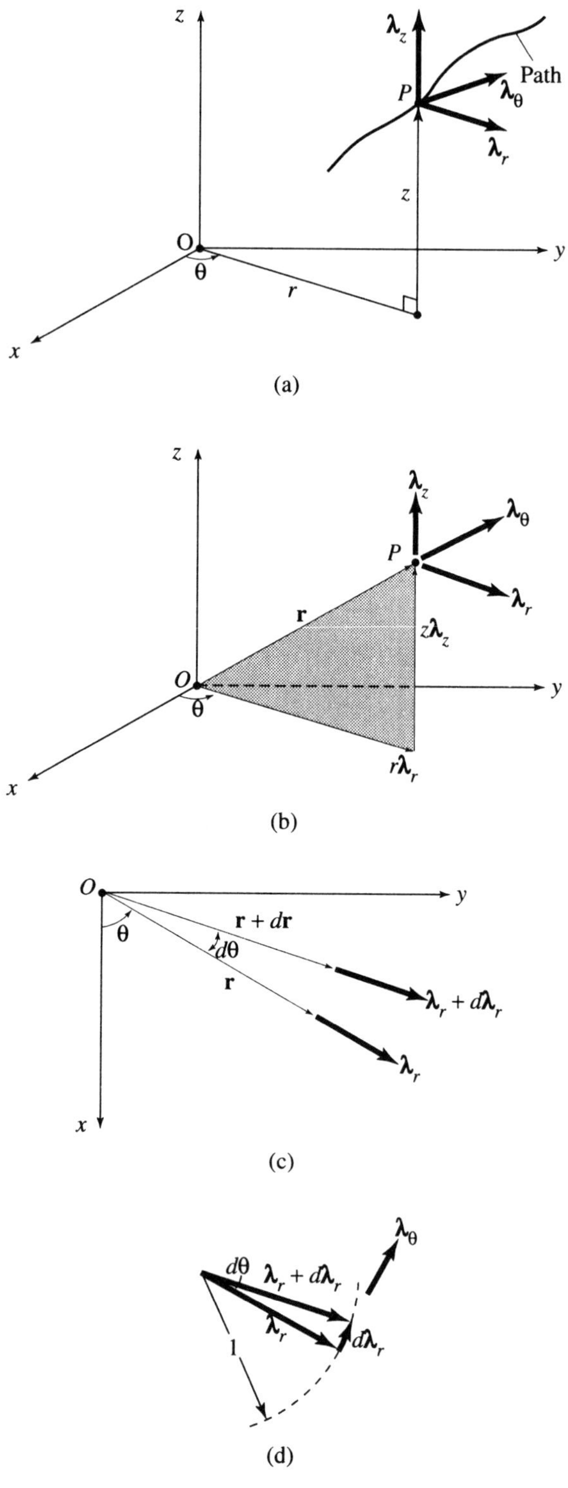
z
λz
Path
P
λθ
λr
z
O
y
θ
r
x
(a)
z
λz
λθ
P
λr
r
zλz
O
y
θ
rλr
x
(b)
O
y
r + dr
θ
dθ
r
λr + dλr
λr
x
(c)
λθ
dθ
λr + dλr
λr
dλr
1
(d)

FIGURE 13.17.

We would be able to determine the velocity $\mathbf{v}$ from this equation provided we knew an expression for the vector derivative $\dfrac{d\boldsymbol{\lambda}_r}{dt}$.

In the development of an expression for $\dfrac{d\boldsymbol{\lambda}_r}{dt}$, we first utilize the chain rule to write

$$\frac{d\boldsymbol{\lambda}_r}{dt} = \frac{d\boldsymbol{\lambda}_r}{d\theta}\frac{d\theta}{dt} = \dot{\theta}\frac{d\boldsymbol{\lambda}_r}{d\theta}. \tag{13.33a}$$

Because $\dot{\theta}$ is a scalar derivative, which we can determine if we know θ as a function of time, we now focus on determining an expression for the vector derivative, $\dfrac{d\boldsymbol{\lambda}_r}{d\theta}$. During a differential time dt, the angle θ will change by a differential amount $d\theta$, as shown in Figure 13.17(c). The unit vector $\boldsymbol{\lambda}_r$ will change in direction, but not in magnitude, during this change $d\theta$. In Figure 13.17(d), the unit vectors are drawn from a common origin containing an angle $d\theta$ which subtends an arc length $d\lambda_r$ on a circle of unit radius. We may, thus, write

$$|d\lambda_r| = (1)\,d\theta$$

from which

$$\frac{d\lambda_r}{d\theta} = 1.$$

Thus, the magnitude of $\dfrac{d\boldsymbol{\lambda}_r}{d\theta}$ is unity. The direction of $\dfrac{d\boldsymbol{\lambda}_r}{d\theta}$ is that of $d\boldsymbol{\lambda}_r$ because $d\theta$ is a scalar. Because $d\theta$ is a differential angle, the direction of the vector $d\boldsymbol{\lambda}_r$ is that of $\boldsymbol{\lambda}_\theta$ as shown in Figure 13.17(d). A vector $\dfrac{d\boldsymbol{\lambda}_r}{d\theta}$ with a magnitude of unity and a direction of $\boldsymbol{\lambda}_\theta$ enables us to write

$$\frac{d\boldsymbol{\lambda}_r}{d\theta} = \boldsymbol{\lambda}_\theta. \tag{13.33b}$$

Substitute (13.33b) in (13.33a) to give

$$\frac{d\boldsymbol{\lambda}_r}{dt} = \dot{\theta}\boldsymbol{\lambda}_\theta. \tag{13.33c}$$

Substituting Eq. (13.33c) in Eq. (13.32), we obtain

$$\mathbf{v} = \dot{r}\boldsymbol{\lambda}_r + r\dot{\theta}\boldsymbol{\lambda}_\theta + \dot{z}\boldsymbol{\lambda}_z. \tag{13.34}$$

The three vector components of this velocity vector are shown in Figure 13.18 consistent with the positive senses of the unit vectors. The

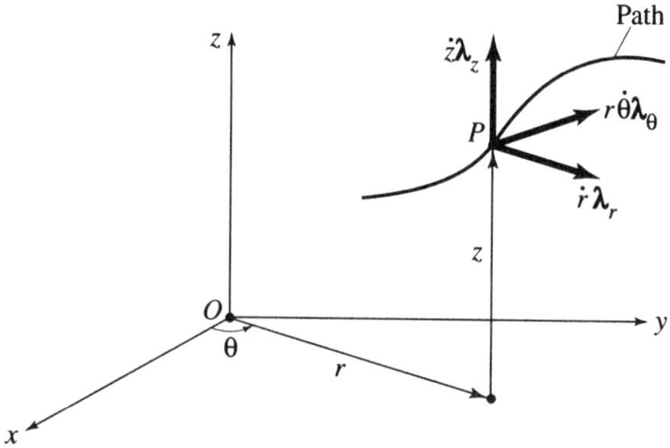

FIGURE 13.18.

magnitude of the velocity vector, termed the speed, is given by the square root of the sum of the squares of the mutually perpendicular scalar components. Mathematically, this becomes

$$v = \sqrt{\dot{r}^2 + (r\dot{\theta})^2 + \dot{z}^2} \tag{13.35}$$

The directional cosines of the angles between the velocity vector and each of the unit vectors in the radial, transverse, and vertical directions are given by

$$\cos\alpha_r = \frac{\dot{r}}{v}, \quad \cos\alpha_t = \frac{r\dot{\theta}}{v}, \quad \text{and} \quad \cos\alpha_z = \frac{\dot{z}}{v}. \tag{13.36}$$

The acceleration vector **a** equals the time rate of change of the velocity vector **v**. Thus, differentiation of Eq. (13.34) yields

$$\mathbf{a} = \frac{d\mathbf{v}}{dt} = \dot{r}\frac{d\boldsymbol{\lambda}_r}{dt} + \ddot{r}\boldsymbol{\lambda}_r + r\dot{\theta}\frac{d\boldsymbol{\lambda}_\theta}{dt} + r\ddot{\theta}\boldsymbol{\lambda}_\theta + \dot{r}\dot{\theta}\boldsymbol{\lambda}_\theta + \dot{z}\frac{d\boldsymbol{\lambda}_z}{dt} + \ddot{z}\boldsymbol{\lambda}_z. \tag{13.37}$$

Because $\boldsymbol{\lambda}_z$ does not change in magnitude or direction, $\frac{d\boldsymbol{\lambda}_z}{dt}$ vanishes. Recall that, by Eq. (13.33c), $\frac{d\boldsymbol{\lambda}_r}{dt} = \dot{\theta}\boldsymbol{\lambda}_\theta$. In an analogous fashion, it can be shown that $\frac{d\boldsymbol{\lambda}_\theta}{dt} = -\dot{\theta}\boldsymbol{\lambda}_r$. Substituting for these vector derivatives in Eq. (13.37) and rearranging terms, we obtain

$$\mathbf{a} = (\ddot{r} - r\dot{\theta}^2)\boldsymbol{\lambda}_r + (r\ddot{\theta} + 2\dot{r}\dot{\theta})\boldsymbol{\lambda}_\theta + \ddot{z}\boldsymbol{\lambda}_z. \tag{13.38}$$

The three vector components of the acceleration vector are shown in Figure 13.19 consistent with the positive senses of the unit vectors. The magnitude of the acceleration vector is given by the square root of the

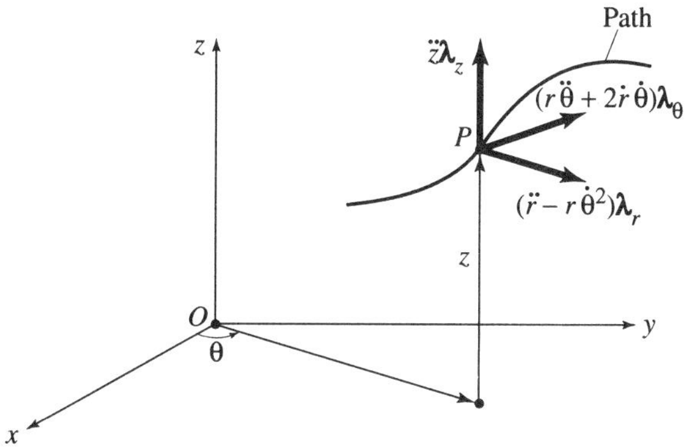

FIGURE 13.19.

sum of the squares of the mutually perpendicular scalar components. Mathematically, this becomes

$$a = \sqrt{(\ddot{r} - r\dot{\theta}^2)^2 + (r\ddot{\theta} + 2\dot{r}\dot{\theta})^2 + \ddot{z}^2}. \qquad (13.39)$$

The directional cosines of the angles between the acceleration vector and each of the unit vectors in the radial, transverse, and vertical directions are given by

$$\cos\beta_r = \frac{\ddot{r} - r\dot{\theta}^2}{a}, \quad \cos\beta_\theta = \frac{r\ddot{\theta} + 2\dot{r}\dot{\theta}}{a} \quad \text{and} \quad \cos\beta_z = \frac{\ddot{z}}{a}. \qquad (13.40)$$

The following example illustrates some of the concepts discussed in this Section.

■ Example 13.22

A helical ramp of radius $r = 100$ ft, shown schematically in Figure E13.22(a) is used by vehicles in a parking deck. A motorcyclist travels down the ramp such that $\theta = 0.3t$ rad and $z = 15t$ where t is in seconds and z is in feet. Determine (a) the equation of the ramp expressing z in terms of θ and (b) the velocity and acceleration of the motorcyclist for any position along the helical ramp.

Solution

(a) Solve $\theta = 0.3t$ for t and substitute in the equation $z = 15t$ to obtain

$$z = 50\theta. \qquad \text{ANS.}$$

(b) The velocity of the motorcyclist is found by using Eq. (13.34) which requires that we find $\dot{r}$, $\dot{\theta}$, and $\dot{z}$. Because $r =$ constant, then,

$$\dot{r} = 0$$

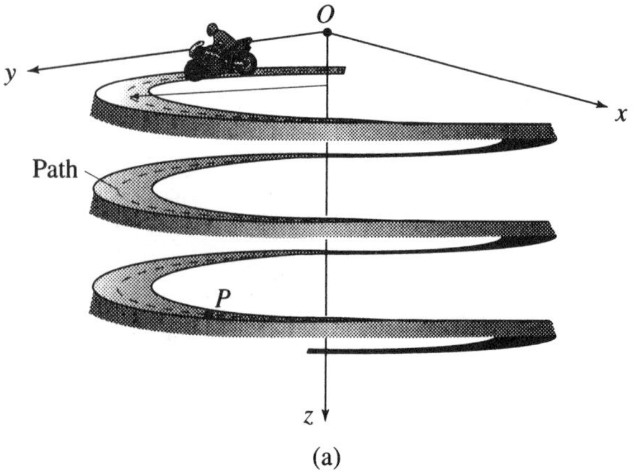

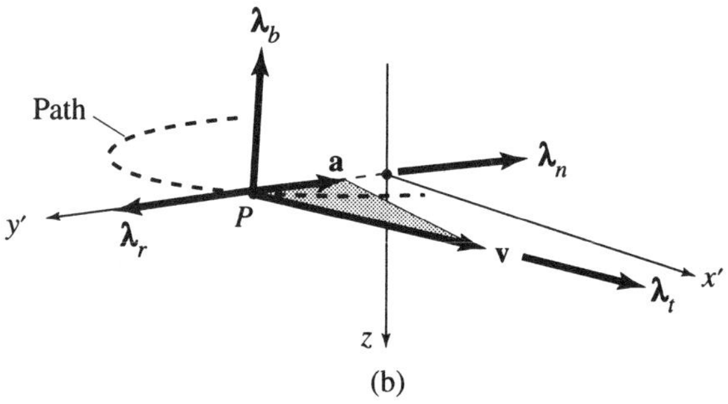

FIGURE E13.22.

Also, from the relationships $\theta = 0.3t$ and $z = 15t$ we obtain

$$\dot{\theta} = 0.3 \text{ rad/s}$$

and

$$\dot{z} = 15 \text{ ft/s}.$$

Therefore, by Eq. (13.34), we obtain

$$\mathbf{v} = 100(0.3)\boldsymbol{\lambda}_\theta + 15\boldsymbol{\lambda}_z = (30\boldsymbol{\lambda}_\theta + 15\boldsymbol{\lambda}_z) \text{ ft/s}. \qquad \text{ANS.}$$

The acceleration is obtained by using Eq. (13.38) which, in addition to the above derivatives, requires that we find $\ddot{r}$, $\ddot{\theta}$, and $\ddot{z}$. Thus, because $\dot{r}$, $\dot{\theta}$, and $\dot{z}$ are constants, $\ddot{r}$, $\ddot{\theta}$, and $\ddot{z}$ are all equal to zero. Therefore, by Eq. (13.38),

$$\mathbf{a} = (0 - 100[0.3]^2)\boldsymbol{\lambda}_r + (0 + 0)\boldsymbol{\lambda}_\theta + 0\boldsymbol{\lambda}_z = (-9\boldsymbol{\lambda}_r) \text{ ft/s}^2.$$

The velocity and acceleration vectors of the motorcyclist are shown at any point P in Figure E13.22(b). Note that the velocity is tangent to the path along which the unit vector is $\boldsymbol{\lambda}_t$. The acceleration points in the negative r direction and, therefore, is pointed toward center of curvature of the path (i.e., along the unit vector $\boldsymbol{\lambda}_n$). The osculating plane is defined by the vectors $\boldsymbol{\lambda}_t$ and $\boldsymbol{\lambda}_n$ as shown shaded in the diagram. The binormal axis is perpendicular to the osculating plane along which the unit vector is $\boldsymbol{\lambda}_b$ where, as stated earlier, $\boldsymbol{\lambda}_b = \boldsymbol{\lambda}_t \times \boldsymbol{\lambda}_n$.

■

Two-Dimensional Motion

Plane *polar coordinates* are a special case of cylindrical coordinates. The equation $z =$ constant defines a one parameter family of planes. For each new value of the constant, a plane parallel to the x-y plane is defined. For convenience, let us assign to the constant a value zero, and note that, when $z = 0$, the motion lies in the x-y plane. Substitution of $z = 0$, $\dot{z} = 0$, and $\ddot{z} = 0$ in Eqs. (13.31), (13.34), and (13.38), respectively, yields the following vectors for curvilinear motion, expressed in polar coordinates, along a path in the x-y plane. For the position vector at any time t,

$$\mathbf{r} = r\boldsymbol{\lambda}_r. \tag{13.41}$$

For the velocity vector at any time t,

$$\mathbf{v} = \dot{r}\boldsymbol{\lambda}_r + r\dot{\theta}\boldsymbol{\lambda}_\theta = v_r\boldsymbol{\lambda}_r + v_\theta\boldsymbol{\lambda}_\theta \tag{13.42a}$$

where $v_r = \dot{r}$, $v_\theta = r\dot{\theta}$, and and the speed is given by

$$v = \sqrt{v_r^2 + v_\theta^2}. \tag{13.42b}$$

For the acceleration vector at anytime t,

$$\mathbf{a} = (\ddot{r} - r\dot{\theta}^2)\boldsymbol{\lambda}_r + (r\ddot{\theta} + 2\dot{r}\dot{\theta})\boldsymbol{\lambda}_\theta = a_r\boldsymbol{\lambda}_r + a_\theta\boldsymbol{\lambda}_\theta \tag{13.43a}$$

where $a_r = \ddot{r} - r\dot{\theta}^2$, $a_\theta = r\ddot{\theta} + 2\dot{r}\dot{\theta}$, and the magnitude of the acceleration is obtained from

$$a = \sqrt{a_r^2 + a_\theta^2}. \tag{13.43b}$$

These vectors are shown in Figure 13.20. Plane polar coordinate equations are most useful when the path is given or may be expressed in plane polar form either as $r = f(\theta)$ or, parametically, as a function of the parameter time, such as $r = g(t)$ and $\theta = h(t)$.

The following examples illustrate the use of polar coordinates in the solution of problems.

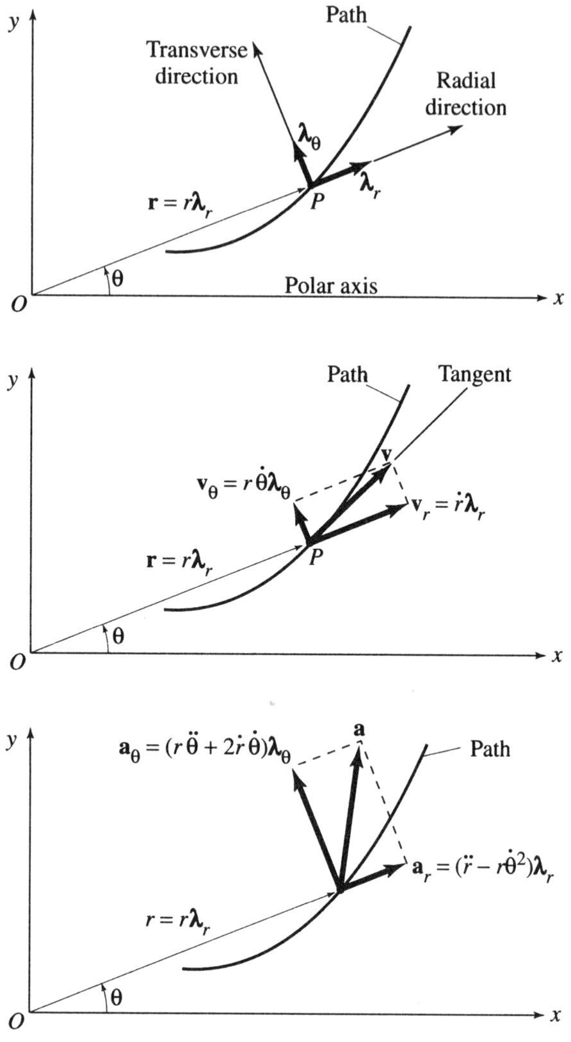

FIGURE 13.20.

■ Example 13.23

The slotted member AB shown in Figure E13.23(a) rotates about the hinge at A such that $\theta = 0.2t^{3/2}$ where θ is in radians and t is in seconds. The pin P slides freely in the slot according to the relationship $r = 0.5 + 0.1t^2$ where r is in feet. When $t = 5$ s, determine the position, velocity, and acceleration of thc pin.

Solution

The position of the pin, when $t = 5$ s, is found by determining r and θ. Thus, substituting for t in the given expressions

$$\theta = 0.2(5)^{3/2} = 2.236 \text{ rad} = 128.1^\circ \quad \text{and} \qquad \text{ANS.}$$

$$r = 0.5 + 0.1(5)^2 = 3.00 \text{ ft.} \qquad \text{ANS.}$$

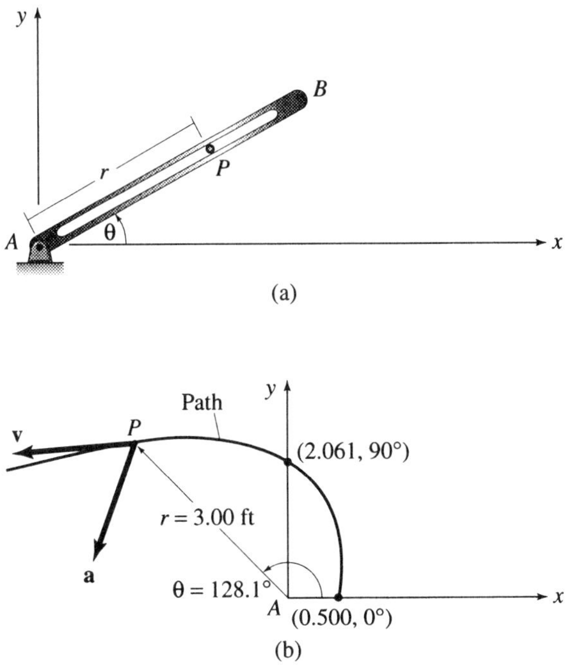

FIGURE E13.23.

The velocity of the pin is determined by, first, differentiating the given expressions with respect to time. Thus,

$$\dot{\theta} = 0.3t^{1/2} \quad \text{and} \quad \dot{r} = 0.2t.$$

For $t = 5$ s,

$$\dot{\theta} = 0.671 \text{ rad/s}, \quad \text{and} \quad \dot{r} = 1.000 \text{ ft/s}.$$

Thus, using Eq. (13.42a),

$$\mathbf{v} = 1.000\boldsymbol{\lambda}_r + (3.00)(0.671)\boldsymbol{\lambda}_\theta = (1.000\boldsymbol{\lambda}_r + 2.013\boldsymbol{\lambda}_\theta) \text{ ft/s.} \quad \text{ANS.}$$

The acceleration of the pin is found from Eq. (13.43) in which, at $t =$ 5 s,

$$\ddot{r} = 0.2,$$

$$r\dot{\theta}^2 = (3.00)(0.671)^2 = 1.350,$$

$$\ddot{\theta} = \frac{0.5}{t^{1/2}} = 0.0671,$$

$$r\ddot{\theta} = (3.00)(0.0671) = 0.201, \quad \text{and}$$

$$2\dot{r}\dot{\theta} = 2(1.000)(0.671) = 1.342.$$

Therefore, because $\mathbf{a} = (\ddot{r} - r\dot{\theta}^2)\boldsymbol{\lambda}_r + (r\ddot{\theta} + 2\dot{r}\dot{\theta})\boldsymbol{\lambda}_\theta$,

$$a = (-1.150\boldsymbol{\lambda}_r + 1.543\boldsymbol{\lambda}_\theta)\ \text{ft/s}^2. \qquad \text{ANS.}$$

The position, velocity and acceleration of the pin, when $t = 5$ s, are shown in Figure E13.23(b) along with a sketch of the path followed by the pin. The equation of this path is obtained by eliminating the variable t from the given functions which yields

$$r = 0.5 + 0.1(5\theta)^{4/3} = 0.5 + 0.855\theta^{4/3}.$$

■ Example 13.24

A radar station tracks the flight of a jet plane flying along a straight-line path as shown in Figure E13.24(a). The flight path is directly above the radar station at a height $h = 2000$ ft. When the angle $\theta = 45°$, the plane has a speed of 750 ft/s which is increasing at 400 ft/s². Determine, at the instant $\theta = 45°$, the angular velocity $\dot{\theta}$ and the angular acceleration $\ddot{\theta}$ of the radar station.

Solution

From the given geometry, we conclude that

$$r = \frac{h}{\sin\theta}$$

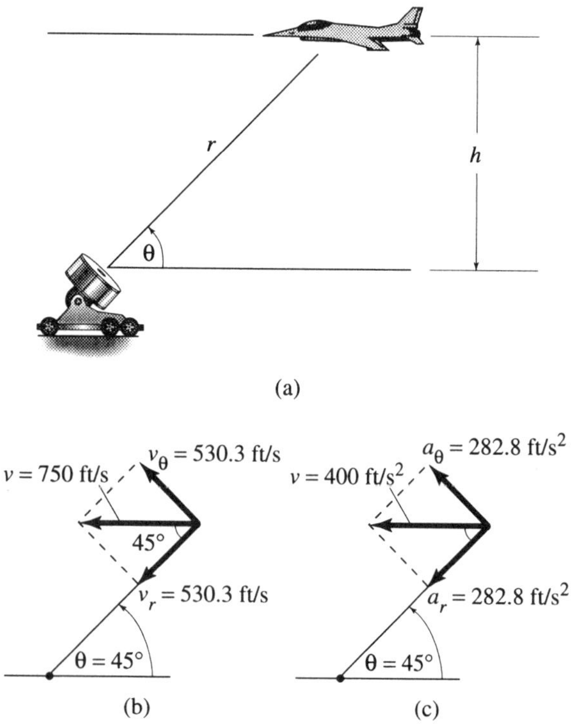

FIGURE E13.24.

Differentiating this expression with respect to time, we obtain

$$\dot{r} = -h\dot{\theta}\left(\frac{\cos\theta}{\sin^2\theta}\right)$$

and

$$\ddot{r} = -h\left[\ddot{\theta}\left(\frac{\cos\theta}{\sin^2\theta}\right)\right] - \dot{\theta}^2\left[\left(\frac{1+\cos^2\theta}{\sin^3\theta}\right)\right].$$

The speed of the plane is given by Eq. (13.42b) where, at $\theta = 45^\circ$,

$$v_r = \dot{r} = -2000\dot{\theta}\left(\frac{\cos 45^\circ}{\sin^2 45^\circ}\right) = -2828.4\dot{\theta}$$

$$v_\theta = r\dot{\theta} = \frac{2000}{\sin 45^\circ}\dot{\theta} = 2828.4\dot{\theta}$$

Thus, by Eq. (13.42b),

$$v = 750 = \sqrt{(-2828.4\dot{\theta})^2 + (2828.4\dot{\theta})^2}.$$

Solving for $\dot{\theta}$, we obtain

$$\dot{\theta} = 0.1875 \text{ rad/s}. \qquad \text{ANS.}$$

Therefore $v_r = -530.3$ ft/s and $v_\theta = 530.3$ ft/s. These velocity components are shown in Figure E13.24(b). The magnitude of the acceleration is given by Eq. (13.43b) where, at $\theta = 45^\circ$

$$a_r = \ddot{r} - r\dot{\theta}^2 = -2000\left[\ddot{\theta}\left(\frac{\cos 45^\circ}{\sin^2 45^\circ}\right) - (0.1833)^2\left(\frac{1+\cos^2 45^\circ}{\sin^3 45^\circ}\right)\right]$$

$$- 2828.4(0.1833)^2$$

$$= -2828.4\ddot{\theta} + 190.1.$$

$$a_\theta = (r\ddot{\theta} + 2\dot{r}\dot{\theta}) = 2828.4\ddot{\theta} + 2(-2828.4)(0.1833)^2$$

$$= 2828.4\ddot{\theta} - 190.1.$$

Thus, by Eq. (13.43b) we obtain

$$a = 400 = \sqrt{(-2828.4\ddot{\theta} + 190.1)^2 + (2828.4\ddot{\theta} - 190.1)^2}$$

which reduces to

$$\ddot{\theta}^2 - 0.1344\ddot{\theta} - 0.00548 = 0.$$

The solution of the above quadratic equation yields

$$\ddot{\theta} = 0.1672 \quad \text{or} \quad -0.0328.$$

The negative answer is discarded on physical grounds, and, therefore,

$$\ddot{\theta} = 0.1672 \text{ rad/s}^2. \qquad \text{ANS.}$$

Thus, $a_r = -282.8$ ft/s^2, and $a_\theta = 282.8$ ft/s^2. These acceleration components are shown in Figure E13.24(c).

■

Problems

13.128 Consider the following special-case motion using cylindrical coordinates. Let $r = 10$ m, $\theta = t^3 + t^2 + 1$, and $z = 5$ m. Use m and s units. What is the path? Sketch it. Determine the velocity and the normal and tangential components of acceleration when $t = 1$ s. Show these vectors on a sketch.

13.129 Consider the following special-case motion using cylindrical coordinates. Let $r = t^2 + 4t + 1$, $\theta = \pi/4$, and $z = 0$. Use in. and s units. What is the path? Sketch it. Determine the velocity and acceleration when $t = 1$ s. Show these vectors on a sketch.

13.130 Consider the following special-case motion using cylindrical coordinates. Let $r = 2$ ft, $\theta = \pi/4$ and $z = 4 \sin t$. Use ft and s units. What is the path? Sketch it. Determine the velocity and acceleration when $t = 1$ s. Show these vectors on a sketch.

13.131 A particle travels a circular path of radius $b = 10$ ft as shown in Figure P13.131 such that $s = t^2 + t$ where s is measured in ft along the circumference and t is measured in seconds. (a) Use tangential and normal unit vectors in order to find the velocity vector and the acceleration vector when $t = 4$ s. (b) Determine the same two vectors (i.e., **v** and **a**) using radial and transverse unit vectors for $t = 4$ s. (c) Sketch the position of the particle on its path when $t = 4$ s, and show **v** and **a** at this same instant on a sketch. When $t = 0$, $s = 0$ and $\theta = 0$. The angle θ is positive ccw.

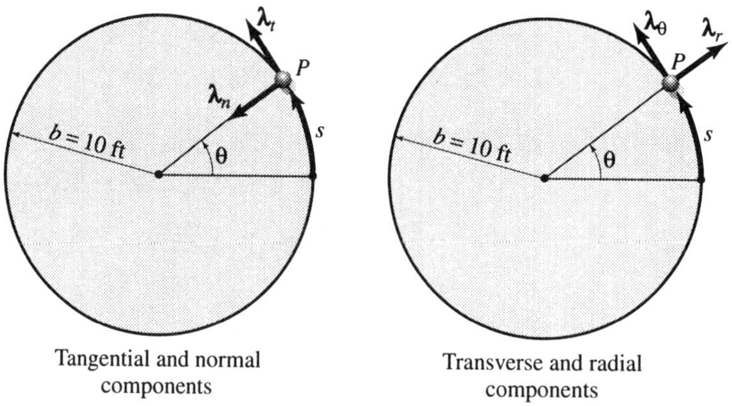

FIGURE P13.131.

13.132 A circular path of radius $b = 2$ m is traversed such that $s = t^3 + t^2 + t$ where s is in meters and t is in seconds. For $t = 1$ s, determine the velocity vector $\mathbf{v}$ and the acceleration vector $\mathbf{a}$ using (a) radial and transverse unit vectors and (b) tangential and normal unit vectors. Draw a sketch showing the position of the particle on its path when $t = 1$ s, and show $\mathbf{v}$ and $\mathbf{a}$ at the same instant on a sketch. When $t = 0$, $s = 0$ and $\theta = 0$. The angle θ is positive ccw.

13.133 A particle moves on a hyperbolic spiral given by $r\theta = a$, $\theta > 0$ and $\theta = \omega t$ where ω and a are constants. Note that, as $r \to \infty$, $\theta \to 0$. Let $\omega = 2$ rad/s and $a = 4$ m. Determine, for $t = 1$ s, the velocity and acceleration vectors, and show them on sketches. The angle θ is positive ccw.

13.134 A particle moves on a hyperbolic spiral given by $r\theta = -a$, $\theta < 0$, and $\theta = \omega t$ where ω and a are constants. Note that as $r \to \infty$, $\theta \to 0$. Let $\omega = 4$ rad/s and $a = 4$ m. For $t = 0.25$ s, determine the velocity and acceleration vectors.

13.135 Show that the equation of a vertical straight-line path of Figure P13.135 may be written in polar form as $r\cos\omega t = b$ where $\theta = \omega t$ and ω and b are constants. For $b = 1$ ft, $\omega = 2$ rad/s, and $t = 0.5$ s, determine the magnitude and direction of the velocity vector using (a) transverse and radial components and (b) a coordinate s measured vertically along the straight path with $s = 0$ when $t = 0$. Hint: $s = b\tan\omega t$ where $\theta = \omega t$.

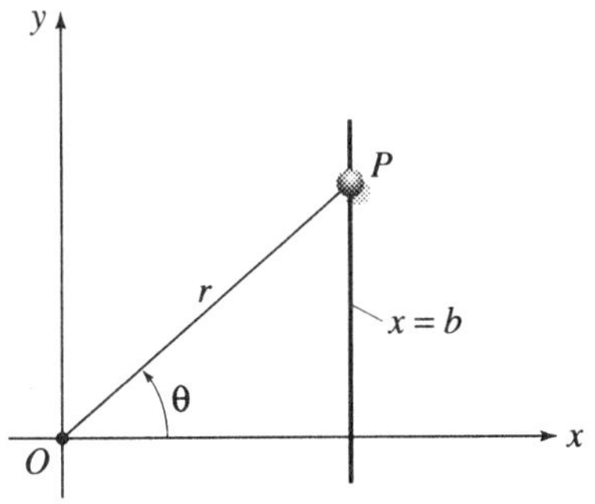

FIGURE P13.135.

13.136 Refer to Figure P13.136 and show that the equation of the horizontal straight-line path may be written in polar form as $r\sin\omega t = b$ where $\theta = \omega t$ and ω and b are constants. For $b = 2$ m, $\omega = 4$ rad/s, and $t = 0.25$ s, determine the magnitude and direction of the velocity vector using (a) transverse and radial components and (b) a coordinate s measured along the straight path. Hint: $s = b\,\text{cotan}\,\theta$ where $\theta = \omega t$.

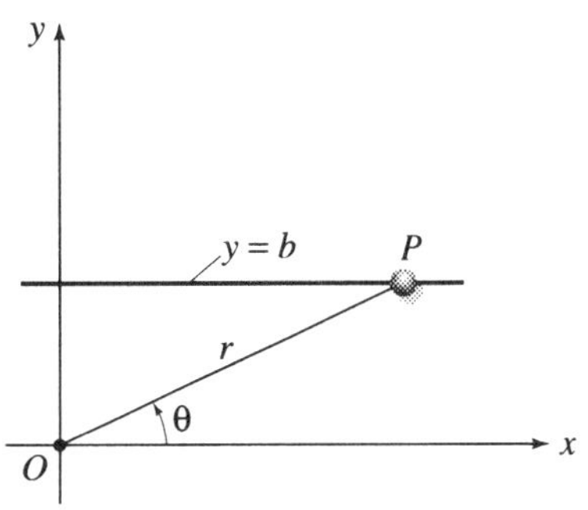

FIGURE P13.136.

13.137 Refer to Problem 13.135 and determine the magnitude and direction of the acceleration vector for the particle at $t = 0.5$ s, using radial and transverse components.

13.138 Refer to Problem 13.136 and determine the magnitude and direction of the acceleration vector for the particle at $t = 0.25$ s, using radial and transverse components.

13.139 Figure P13.139 shows a path known as a lemniscate. A toy train is constrained

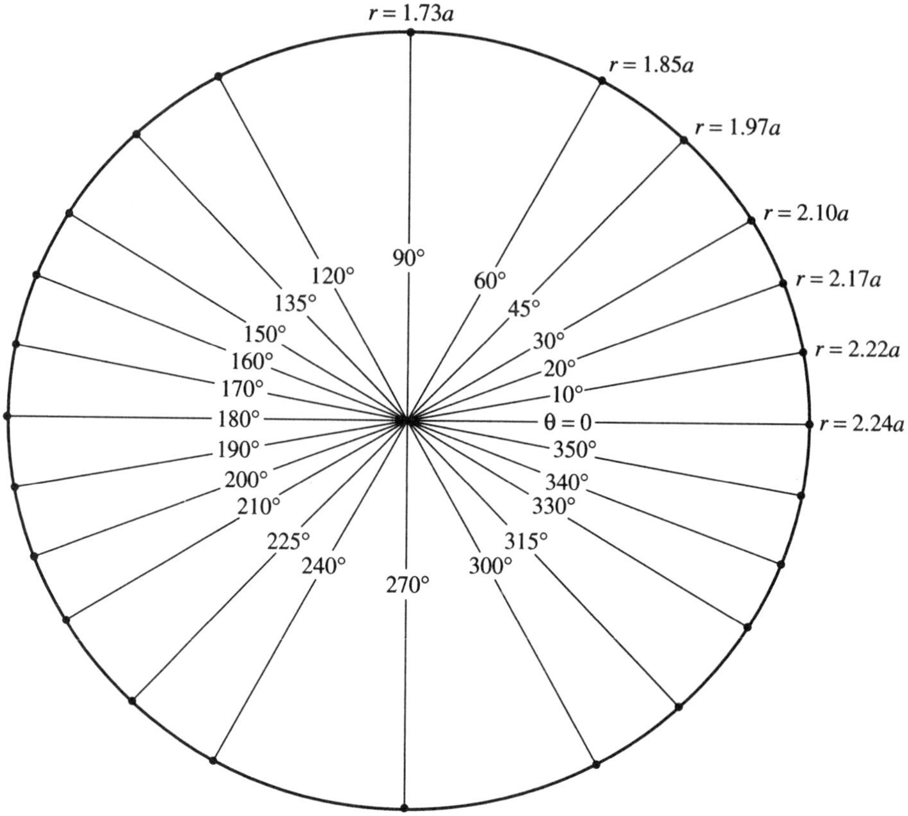

FIGURE P13.139.

to move on this path such that the parametric equations of the path are $r^4 - 2a^2r^2 \cos 2\theta - 15a^4 = 0$ where $\theta = t^2$. If $a = 3$ ft, determine, for $t = 0.5$ s, (a) the position vector and (b) the velocity vector using transverse and radial unit vectors which move with the train.

13.140 An elliptic path is given parametrically as $r = \dfrac{100}{(1 - 0.5 \cos \theta)}$ and $\theta = \omega t$ where ω is a constant equal to 2 rad/s. For $t = 0$, determine the velocity and acceleration of a particle moving on this path, using radial and transverse unit vectors which move with the particle. Sketch these vectors showing their relationship to the path (units are m and s).

13.141 A hyperbolic path is given parametrically as $r = \dfrac{200}{(1 - 1.5 \cos \theta)}$ and $\theta = \omega t$ where ω is a constant equal to 4 rad/s. For $t = 0.25$ s, determine the velocity and acceleration of a particle moving on this path, using radial and transverse unit vectors which movc with the particle (units are ft and s).

13.142 A logarithmic spiral path is defined by $r = be^{\theta}$ where $\theta = \omega t$. For $b = 1$ ft, $\omega = 2$ rad/s, and $t = 0.5$ s, determine (a) the position vector and (b) the velocity vector, using radial and transverse unit vectors.

13.143 Determine the speed and the magnitude of the acceleration vector of a particle moving on the logarithmic spiral path defined by $r = be^{\theta}$ where $\theta = \omega t$. Use radial and transverse components, and assume b and ω are constants.

13.144 A particle travels on the surface of a right elliptic cylinder along what could be called an elliptic helix. Its parametric equations are $r = \dfrac{50e}{(1 - e\cos\theta)}$, $\theta = \omega t$ and $z = v_1 t$, (provided $e < 1$) where ω is a constant in rad/s and v_1 is a constant in/s. Use cylindrical coordinates to write an equation for the speed v of this particle.

13.145 Refer to Problem 13.144, and write an equation for the magnitude of the acceleration of this particle.

13.146 Refer to Problem 13.144, and write an equation for the velocity vector.

13.147 Refer to Problem 13.144, and write an equation for the acceleration vector.

13.148 A particle travels on the surface of a right lemniscate cylinder along what could be termed a lemniscate helix. Its parametric equations are $r^4 - 2a^2r^2\cos 2\theta - 15a^4 = 0$, $\theta = \omega t$, and $z = v_1 t$ where a, ω, and v_1 are constants. Use cylindrical coordinates to write an equation for the velocity of this particle.

13.10 Relative Motion—Translating Frame of Reference

In Figure 13.21 we will be dealing with two reference frames, X-Y-Z which is fixed and x-y-z which is translating with respect to X-Y-Z. Physically we may think of these axes as three mutually perpendicular slender rods which are welded together to form a rigid body. When we say that x-y-z translates with respect to X-Y-Z, we mean that, for the times of interest during the motion, the x axis remains parallel to the X axis, the y axis remains parallel to the Y axis, and the z axis remains parallel to the Z axis. Of course, there is no frame of reference which is truly fixed in space because the billions of stars of each galaxy and the billions of galaxies of the universe have all been in motion since the *big bang* or the instant of creation. Nevertheless, we find it convenient and useful to refer to certain reference frames as *fixed*. In engineering, we often attach a frame of reference to Earth and refer to it as fixed. Position vectors $\mathbf{r}_A$ and $\mathbf{r}_B$ for particles A and B shown in Figure

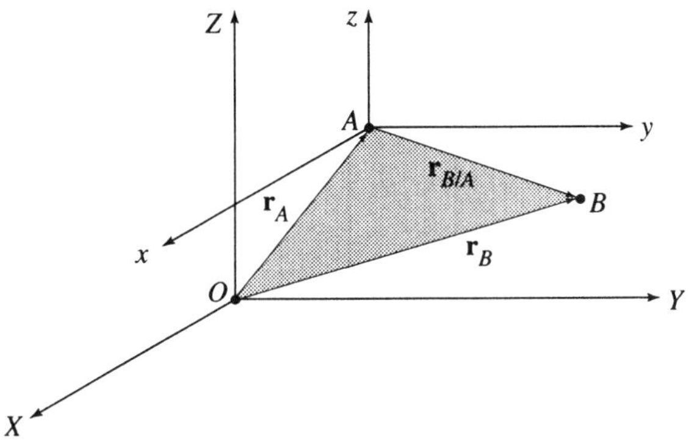

FIGURE 13.21.

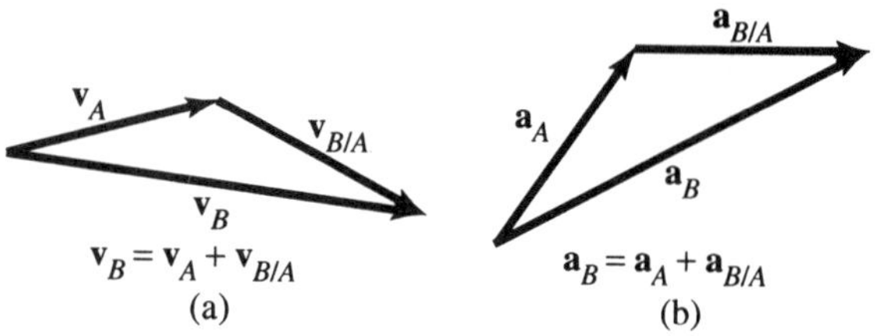

FIGURE 13.22.

13.21 are referred to as absolute position vectors because they originate at the origin O of a fixed reference frame. The position of any particle B with respect to the origin A of the x-y-z frame, $\mathbf{r}_{B/A}$, is referred to as the relative position vector of particle B with respect to particle A which is the origin of the moving frame x-y-z. The three position vectors $\mathbf{r}_A$, $\mathbf{r}_B$, and $\mathbf{r}_{B/A}$ determine a plane. Using the triangle law for vector addition, as shown in Figure 13.21, for any time t,

$$\mathbf{r}_B = \mathbf{r}_A + \mathbf{r}_{B/A}. \tag{13.44}$$

If we differentiate this equation, term by term, with respect to time, then we obtain the relative velocity equation. Thus,

$$\mathbf{v}_B = \mathbf{v}_A + \mathbf{v}_{B/A} \tag{13.45}$$

where $\mathbf{v}_{B/A}$ is the relative velocity of B with respect to A and represents the velocity of particle B as measured by an observer stationed at particle A, the origin of the translating x-y-z coordinate system. Next, differentiate this velocity equation with respect to time to obtain

$$\mathbf{a}_B = \mathbf{a}_A + \mathbf{a}_{B/A} \tag{13.46}$$

where $\mathbf{a}_{B/A}$ is the relative acceleration of B with respect to A and represents the acceleration of particle B as measured by an observer stationed at particle A, the origin of the translating x-y-z coordinate system. Equations (13.45) and (13.46) are shown pictorially in Figure 13.22(a) and (b), respectively.

The following examples illustrate some of the concepts discussed in this Section.

■ Example 13.25

Passenger plane A travels due east at a constant speed of 660 km/h as shown in Figure E13.25(a). At the instant plane A passes the origin of the x-y coordinate system, passenger plane B is 20 km north of the origin, traveling due north at the same elevation as plane A at a speed of 540 km/h which is increasing at the constant rate of 14,400 km/h^2. Determine the position, velocity, and acceleration of plane A as perceived by a passenger in plane B two minutes after plane A passes the origin of the x-y coordinate system.

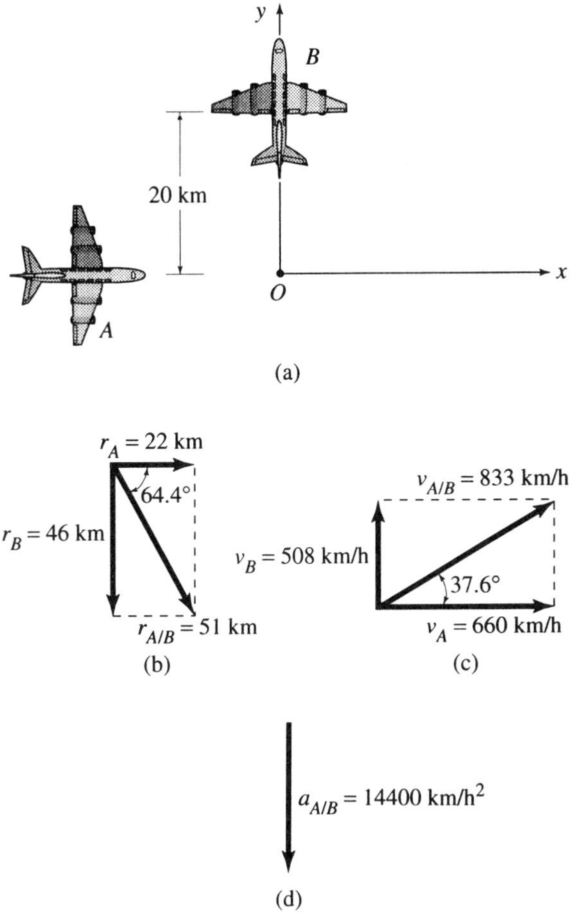

FIGURE E13.25.

Solution

Two minutes after plane A passes point O, the positions of the two planes r_A and r_B are found by using Eq. (13.9). Thus,

$$r_A = (r_A)_O + v_A t + \frac{1}{2}(a_A)_C t^2$$

$$= 0 + \left(\frac{660}{60}\right)(2) + 0 = 22 \text{ km},$$

and

$$r_B = (r_B)_O + v_B t + \frac{1}{2}(a_B)_C t^2$$

$$= 20 + \left(\frac{540}{60}\right)(2) + \frac{1}{2}\left(\frac{1440}{60^2}\right)(2^2) = 46 \text{ km}.$$

Therefore, $\mathbf{r}_A = (22\mathbf{i})$ km and $\mathbf{r}_B = (46\mathbf{j})$ km, and, by Eq. (13.44), we obtain

$$\mathbf{r}_{A/B} = \mathbf{r}_A - \mathbf{r}_B = (22\mathbf{i} - 46\mathbf{j}) \text{ km}.$$

Thus,

$$r_{A/B} = \sqrt{(22)^2 + (46)^2} = 51.0 \text{ km}. \qquad \text{ANS.}$$

This is shown in Figure E13.25(b). Thus, to a passenger in plane B, plane A is at a distance of 51 km along a line making an angle of 64.4° cw with the horizontal.

Two minutes after plane A passes point O, the speeds of the two planes, v_A and v_B, are found using Eq. (13.8). Thus,

$$v_A = (v_A)_O + (a_A)_C t = 660 + 0 = 660 \text{ km/h},$$

and

$$v_B = (v_B)_O + (a_B)_C t = 500 + \left(\frac{14400}{60^2}\right)(2) = 508 \text{ km/h}.$$

Therefore, $\mathbf{v}_A = (600\mathbf{i})$ km/h and $\mathbf{v}_B = (508\mathbf{j})$ km/h, and, by Eq. (13.45),

$$\mathbf{v}_{A/B} = \mathbf{v}_A - \mathbf{v}_B = (600\mathbf{i} - 508\mathbf{j}) \text{ km/h}.$$

Thus,

$$v_{A/B} = \sqrt{660^2 + 508^2} = 833 \text{ km/h} \qquad \text{ANS.}$$

as shown in Figure E13.25(c). Thus, to a passenger in plane B, plane A moves at a speed of 833 km/h in a direction 37.6° ccw from the horizontal. Finally, we know that the acceleration of plane A is zero and that of plane B is 14400 km/h^2. Thus, $\mathbf{a}_A = \mathbf{0}$ and $\mathbf{a}_A = 14400\mathbf{j}$, and by Eq. (13.46),

$$\mathbf{a}_{A/B} = \mathbf{a}_A - \mathbf{a}_B = (\mathbf{0} - 14400\mathbf{j}) \text{ km/h}^2$$

where

$$a_{A/B} = 14400 \text{ km/h}^2 \qquad \text{ANS.}$$

as shown in Figure E13.25(d). Therefore, to a passenger in plane B, plane A is accelerating due south at the rate of 14400 km/h^2.

■ Example 13.26

Car A travels at a constant speed of 70 ft/s while car B travels at a constant speed v_B as shown in Figure E13.26(a). Both cars pass point O at the same instant, and 5 s later, car B appears to an observer in car A to be moving at 127.5 ft/s in a direction that makes an angle of 12.97° with the negative x axis. Determine (a) the constant speed v_B of car B and the angle θ defining its path and (b) the position of car B, 5 s after point O, as measured by the observer in car A.

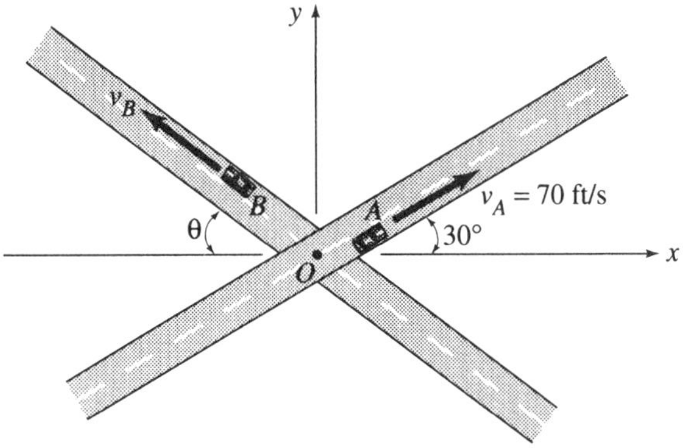

(a)

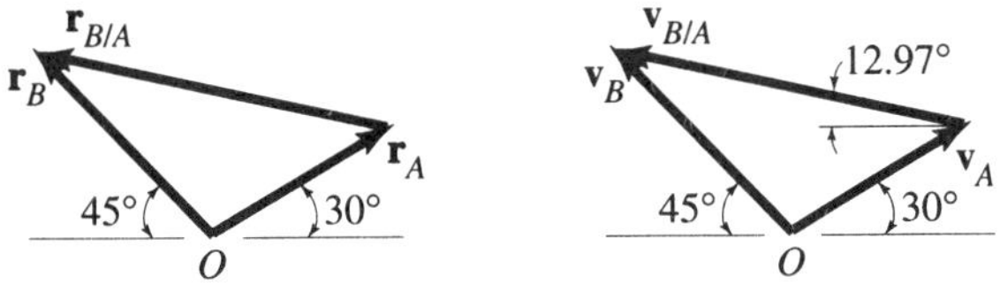

FIGURE E13.26.

Solution

(a) From the information given, we conclude that

$$\mathbf{v}_{\mathrm{B/A}} = -(127.5\cos 12.97^\circ)\mathbf{i} + (127.5\sin 12.97^\circ)\mathbf{j}$$
$$= (-124.2\mathbf{i} + 28.6\mathbf{j})\ \text{ft/s}.$$

Also,

$$\mathbf{v}_{\mathrm{A}} = (70\cos 30^\circ)\mathbf{i} + (70\sin 30^\circ)\mathbf{j} = (60.6\mathbf{i} + 35.0\mathbf{j})\ \text{ft/s}.$$

Using Eq. (13.45),

$$\mathbf{v}_{\mathrm{B}} = \mathbf{v}_{\mathrm{A}} + \mathbf{v}_{\mathrm{B/A}} = (60.6 - 124.2)\mathbf{i} + (35.0 + 28.6)\mathbf{j}$$
$$= (-63.6\mathbf{i} + 63.6\mathbf{j})\ \text{ft/s}.$$

Therefore,

$$v_{\mathrm{B/A}} = \sqrt{(63.6)^2 + (63.6)^2} = 89.9\ \text{ft/s}$$ 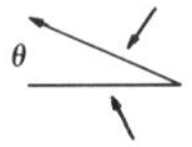

ANS.

$$\theta = \tan^{-1}\left(\frac{63.6}{63.6}\right) = 45^\circ$$

The vector diagram showing the relation among $\mathbf{v}_{\mathrm{A}}$, $\mathbf{v}_{\mathrm{B}}$, and $\mathbf{v}_{\mathrm{B/A}}$ is shown in Figure E13.26(b).

(b) The positions of cars A and B, r_{A} and r_{B}, respectively, 5 s after point O, are given by Eq. (13.9). Thus,

$$r_{\mathrm{A}} = (r_{\mathrm{B}})_0 + v_{\mathrm{A}}t + \tfrac{1}{2}(a_{\mathrm{A}})_{\mathrm{C}}t^2 = 0 + 70(5) + 0 = 350\ \text{ft}.$$

Hence,

$$\mathbf{r}_A = (350 \cos 30°)\mathbf{i} + (350 \sin 30°)\mathbf{j} = (303.1\mathbf{i} + 175.0\mathbf{j}) \text{ ft.}$$

Also

$$r_B = (r_B)_0 + v_B t + \tfrac{1}{2}(a_B)_C t^2 = 0 + 89.9(5) + 0 = 449.5 \text{ ft.}$$

Hence,

$$\mathbf{r}_B = (-449.5 \cos 45°)\mathbf{i} + (449.5 \sin 45°)\mathbf{j} = (-317.8\mathbf{i} + 317.8\mathbf{j}) \text{ ft.}$$

Using Eq. (13.44),

$$\begin{aligned} \mathbf{r}_{B/A} &= \mathbf{r}_B - \mathbf{r}_A \\ &= (-317.8 - 303.1)\mathbf{i} + (317.8 - 175.0)\mathbf{j} \\ &= (-620.9\mathbf{i} + 142.8\mathbf{j}) \text{ ft.} \end{aligned} \qquad \text{ANS.}$$

The vector diagram showing the relation among $\mathbf{r}_A$, $\mathbf{r}_B$ and $\mathbf{r}_{B/A}$ is shown in Figure E13.26(c).

■ Example 13.27

Toy car A moves along a cardioid-shaped path while toy car B moves along a circular-shaped path as shown in Figure E13.27(a). When $t = 0$, both cars are on the x axis. The path equations are given parametrically by

$$r_A = 2(1 - \cos 2t) \text{ m}, \quad \theta_A = 2t \text{ rad}$$

and

$$r_B = 2 \text{ m}, \quad \theta_A = t \text{ rad}$$

where t is measured in seconds. Find $\mathbf{r}_{B/A}$ and $\mathbf{v}_{B/A}$ for $t = 2$ s.

Solution

Let us first examine the motion of car A.

$$r_A = 2(1 - \cos 2t) \quad \text{and} \quad \theta_A = 2t$$

and

$$\dot{r}_A = 4 \sin 2t \quad \text{and} \quad \dot{\theta}_A = 2.$$

Now,

$$\mathbf{r}_A = r_A \boldsymbol{\lambda}_{r_A} = 2(1 - \cos 2t)\boldsymbol{\lambda}_{r_A}.$$

By Eq. (13.42a),

$$\begin{aligned} \mathbf{v}_A &= \dot{r}_A \boldsymbol{\lambda}_{r_A} + r_A \dot{\theta}_A \boldsymbol{\lambda}_{\theta_A} \\ &= (4 \sin 2t)\boldsymbol{\lambda}_{r_A} + 4(1 - \cos 2t)\boldsymbol{\lambda}_{\theta_A}. \end{aligned}$$

Next, we examine the motion of car B.

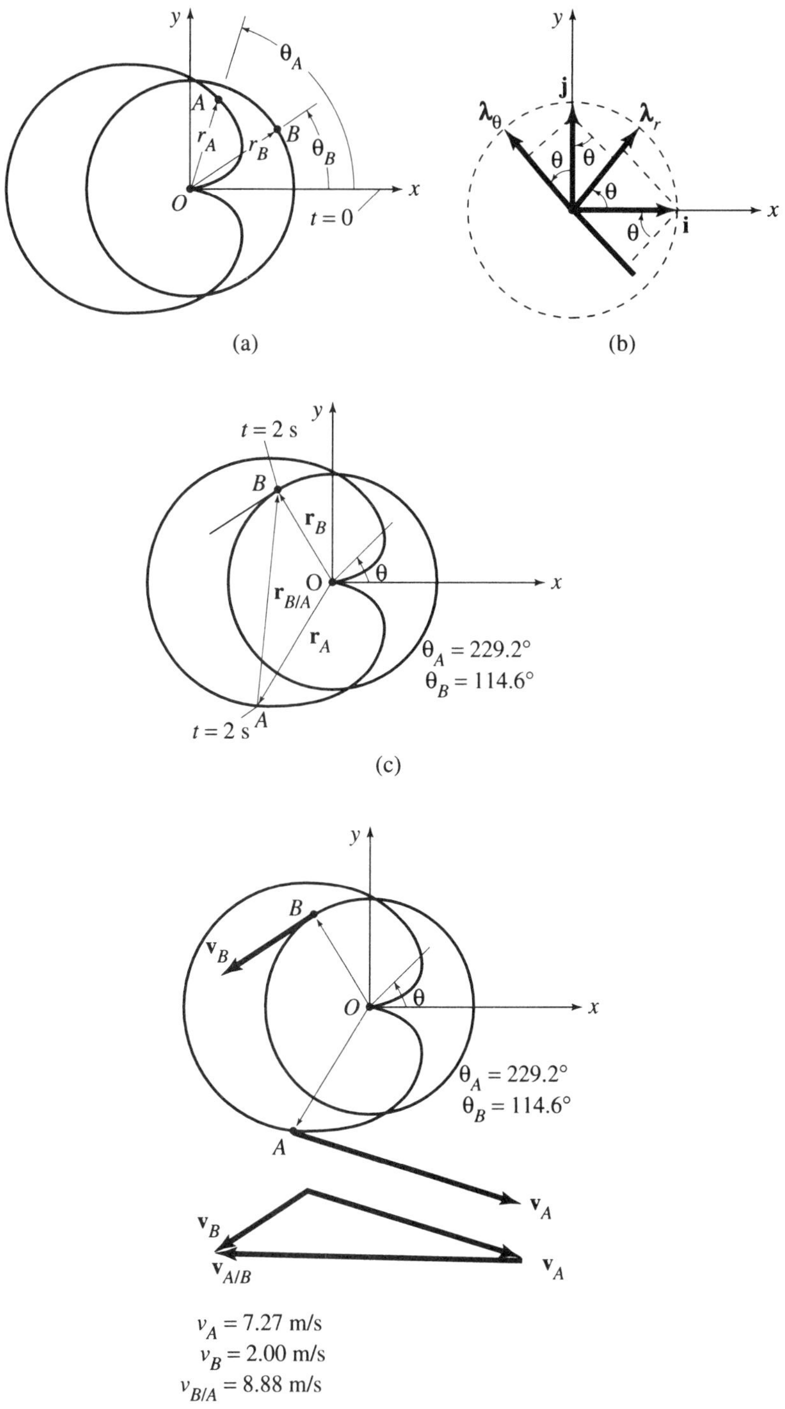

FIGURE E13.27.

$$r_B = 2 \quad \text{and} \quad \theta_A = t,$$

$$\dot{r}_B = 0 \quad \text{and} \quad \dot{\theta}_B = 1,$$

and

$$\mathbf{r}_B = r_B \boldsymbol{\lambda}_{r_B} = 2\boldsymbol{\lambda}_{r_B}.$$

By Eq. (13.42a),

$$\begin{aligned}\mathbf{v}_B &= \dot{r}_B \boldsymbol{\lambda}_{r_B} + r_B \dot{\theta}_B \boldsymbol{\lambda}_{\theta_B} \\ &= 0\boldsymbol{\lambda}_{r_B} + 2\boldsymbol{\lambda}_{\theta_B}.\end{aligned}$$

Because $\boldsymbol{\lambda}_{r_A}$, $\boldsymbol{\lambda}_{\theta_A}$, $\boldsymbol{\lambda}_{r_B}$, and $\boldsymbol{\lambda}_{\theta_B}$ move at different rates, it is necessary to express them in terms of the fixed **i** and **j** unit vectors to simplify the interpretation of the relative position $\mathbf{r}_{B/A}$ and the relative velocity $\mathbf{v}_{B/A}$. Reference to Figure E13.27(b) allows us to conclude that

$$\boldsymbol{\lambda}_r = (\cos\theta)\mathbf{i} + (\sin\theta)\mathbf{j}$$

and

$$\boldsymbol{\lambda}_\theta = (-\sin\theta)\mathbf{i} + (\cos\theta)\mathbf{j}.$$

Thus, for car A, because $\theta_A = 2t$,

$$\left.\begin{aligned}\boldsymbol{\lambda}_{r_A} &= (\cos 2t)\mathbf{i} + (\sin 2t)\mathbf{j} \\ \boldsymbol{\lambda}_{\theta_A} &= (-\sin 2t)\mathbf{i} + (\cos 2t)\mathbf{j}\end{aligned}\right\}. \tag{a}$$

Similarly for car B, because $\theta_B = t$,

$$\left.\begin{aligned}\boldsymbol{\lambda}_{r_B} &= (\cos t)\mathbf{i} + (\sin t)\mathbf{j} \\ \boldsymbol{\lambda}_{\theta_B} &= (-\sin t)\mathbf{i} + (\cos t)\mathbf{j}\end{aligned}\right\}. \tag{b}$$

Substituting Eqs. (a) and (b) in the above expressions and after simplification,

$$\begin{aligned}\mathbf{r}_A &= [2(1-\cos 2t)\cos 2t]\mathbf{i} + [2(1-\cos 2t)\sin 2t]\mathbf{j}, \\ \mathbf{r}_B &= (2\cos t)\mathbf{i} + (2\sin t)\mathbf{j}, \\ \mathbf{v}_A &= [4\sin 2t\cos 2t - 4(1-\cos 2t)\sin 2t]\mathbf{i} \\ &\quad + [4\sin^2 2t + 4(1-\cos 2t)\cos 2t]\mathbf{j},\end{aligned}$$

and

$$\mathbf{v}_B = (-2\sin t)\mathbf{i} + (2\cos t)\mathbf{j}.$$

For $t = 2$ s, these expressions yield

$$\mathbf{r}_A = (-2.16\mathbf{i} - 2.50\mathbf{j})\ \text{m}$$

and

$$\mathbf{r}_B = (-0.832\mathbf{i} + 1.819\mathbf{j})\ \text{m}.$$

Therefore, by Eq. (13.44)

$$\mathbf{r}_{B/A} = \mathbf{r}_B - \mathbf{r}_A = (1.328\mathbf{i} + 4.32\mathbf{j})\ \text{m}, \qquad \text{ANS.}$$

$$r_{B/A} = \sqrt{(1.328)^2 + (4.32)^2} = 4.52\ \text{m},$$

$$\mathbf{v}_A = (6.98\mathbf{i} - 2.03\mathbf{j})\ \text{m/s}, \qquad \text{ANS.}$$

$$v_A = \sqrt{(6.98)^2 + (2.03)^2} = 7.27\ \text{m/s},$$

$$\mathbf{v}_B = (-1.819\mathbf{i} - 0.832\mathbf{j})\ \text{m/s}, \qquad \text{ANS.}$$

and

$$v_B = \sqrt{(1.819)^2 + (0.832)^2} = 2.00\ \text{m/s}.$$

Thus, by Eq. (13.45),

$$\mathbf{v}_{B/A} = (-8.80\mathbf{i} + 1.198\mathbf{j})\ \text{m/s} \qquad \text{ANS.}$$

and

$$v_{B/A} = \sqrt{(-8.80)^2 + (1.198)^2} = 8.88\ \text{m/s}.$$

The relationships among $\mathbf{r}_A$, $\mathbf{r}_B$, and $\mathbf{r}_{B/A}$ and among $\mathbf{v}_A$, $\mathbf{v}_B$, and $\mathbf{v}_{B/A}$ are shown in Figure E13.27(c) and (d) respectively.

■

Problems

13.149 People are riding the two Ferris wheels as shown in Figure P13.149. Those at A have a constant speed of 30 ft/s and those at B have a constant speed of 40 ft/s. Determine the relative velocity $v_{B/A}$ and the relative acceleration $a_{B/A}$ at the instant depicted. Ignore the *rocking* motion of the chairs induced by the occupants.

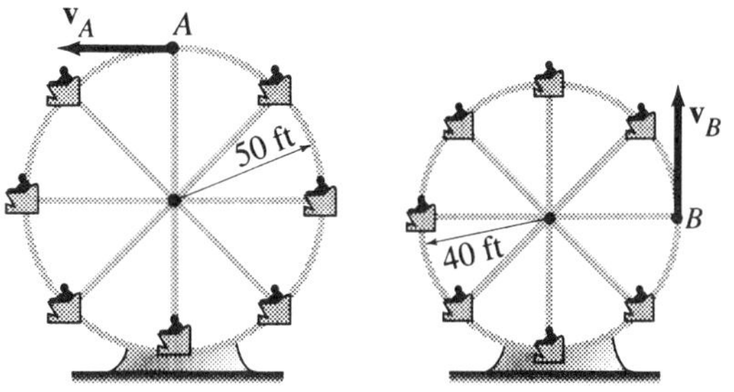

FIGURE P13.149.

13.150 Two aircraft are flying at a constant altitude of 10,000 ft as shown in Figure P13.150. Aircraft A is flying N30° W at a speed of 900 mph which is being reduced at the rate of 6,000 mph^2 while aircraft B is flying S at a speed of 1,200 mph which is being reduced at the rate of 8,000 mph^2. Determine the relative velocity of B with respect to A and the relative acceleration of B with respect to A.

13.151 As shown in Figure P13.151, car A travels along the x axis at a constant speed of 50 mph while car B moves along a straight path inclined at 60°

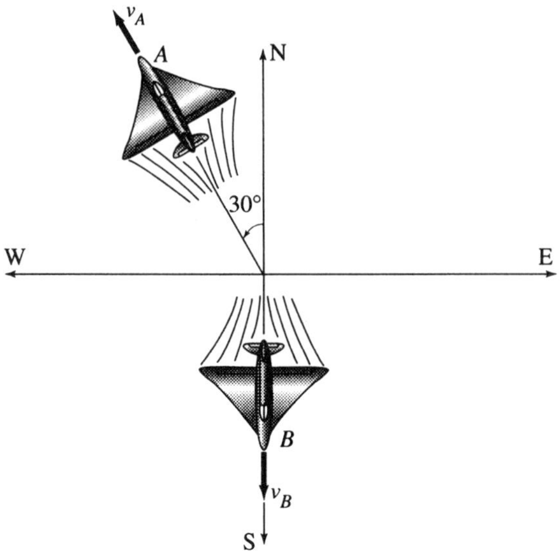

FIGURE P13.150.

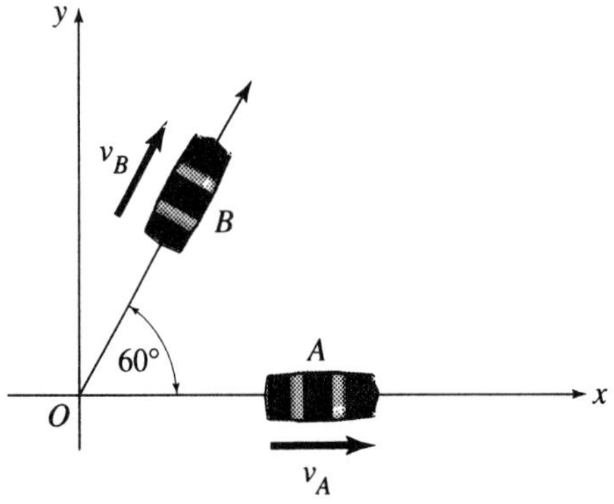

FIGURE P13.151.

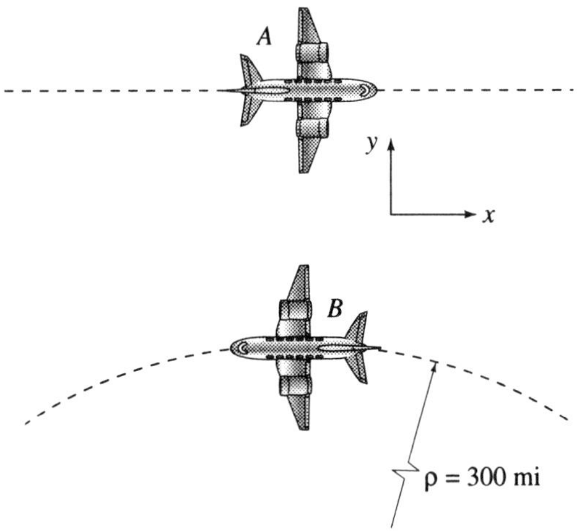

FIGURE P13.152.

ccw with the x axis, at a speed of 30 mph which is increasing at the rate 7200 mi/h^2. Determine the velocity and acceleration of car B relative to car A.

13.152 Planes A and B are flying at the same elevation. At the instant shown in Figure P13.152, plane A flies parallel to the x axis at a speed of 500 mph which is decreasing at the rate 600 mi/h^2, and plane B flies at a constant speed of 600 mph along a circular path with a radius of curvature of 300 miles. Determine the velocity and acceleration of plane A as measured by a passenger in plane B.

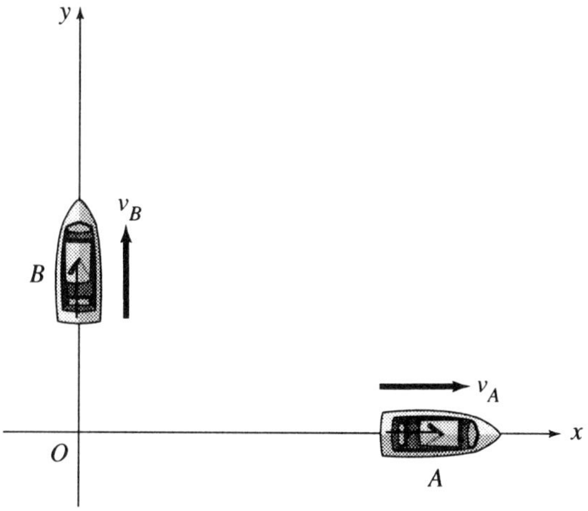

FIGURE P13.153.

13.153 Boat A moves along the x axis while boat B moves along the y axis as shown in Figure P13.153. Boat A has a constant speed of 10 m/s while boat B has a constant speed of 7 m/s. Boat A passes point O, 10 s before boat B. Determine, 5 s after boat B passes point O, (a) the velocity of A as measured by an observer in B and (b) the position of A as measured by an observer in B.

13.154 Skater A moves with a constant speed along a straight path $\theta°$ ccw from the x axis while skater B moves with a constant speed of 10 ft/s along the x axis, as shown in Figure P13.154. Skater A arrives at point O, 5 s before skater

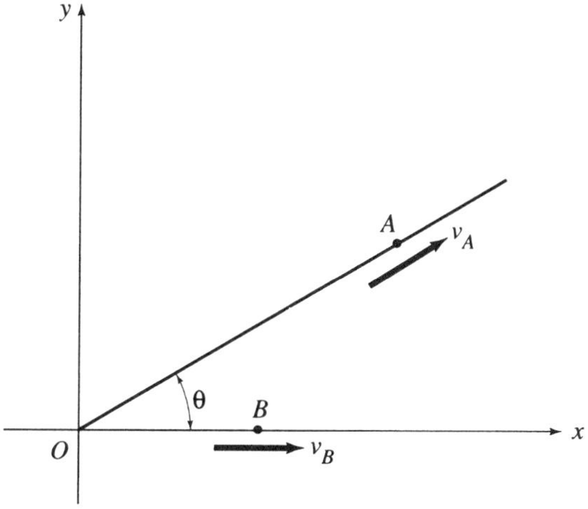

FIGURE P13.154.

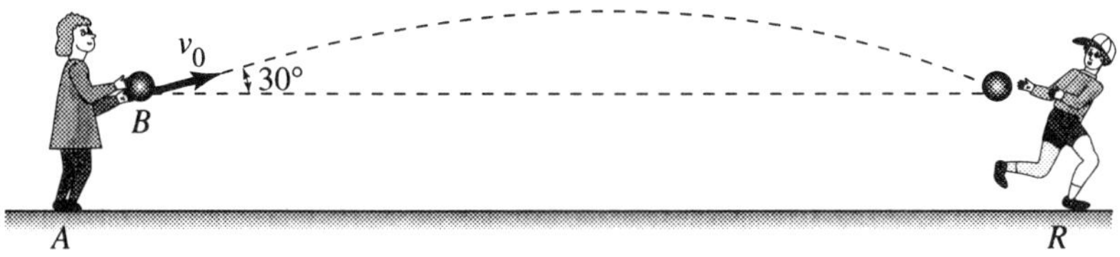

FIGURE P13.155.

B, and 10 s later, skater A appears to skater B to be moving at 13.74 ft/s in a direction that makes and angle of 73.68° with the x axis. Determine (a) the speed and the angle θ of skater A and (b) the position of skater A as measured by skater B 10 s after skater B passes point O.

13.155 A girl A throws a baseball B at a speed $v_0 = 40$ ft/s to a friend who is running at a constant speed, as shown in Figure P13.155. The ball is released when the runner R is at a distance of 20 ft from the point where the ball was released. If the runner catches the ball at the same elevation as it was released, determine the constant speed of the runner. Find also the velocity of the ball relative to the runner at the instant it is caught.

13.156 Refer to Example 13.27, and solve it after changing r_A, r_B, θ_A and θ_B as follows:

$$r_A = 4(1 - \cos 3t) \text{ m}; \quad \theta_A = 3t \text{ rad};$$
$$r_B = 4 \text{ m}; \quad \theta_A = 2t \text{ rad}.$$

13.157 Refer to Example 13.27 and solve it after changing r_A, r_B, θ_A and θ_B as follows:

$$r_A = 4(1 - \cos t) \text{ m}; \quad \theta_A = t \text{ rad};$$
$$r_B = 1 \text{ m}; \quad \theta_B = t \text{ rad}.$$

Review Problems

13.158 A car moves along a straight-line path such that, for a short period of time, its position is given by $s = 2t^2 - t + 3$ where s is in feet and t is in seconds. Determine the position, velocity, and acceleration when $t = 10$ s.

13.159 A bullet is fired horizontally into the vertical wall of a container of liquid as shown in Figure P13.159. It penetrates the wall, enters the fluid with a speed of 200 m/s, and continues its horizontal straight-line path through the fluid as shown. The fluid retards the motion of the bullet by applying a deceleration given by $a = -0.2v^2$ m/s², where v is in m/s. Find the speed and position of the bullet 0.3 s after it enters the fluid.

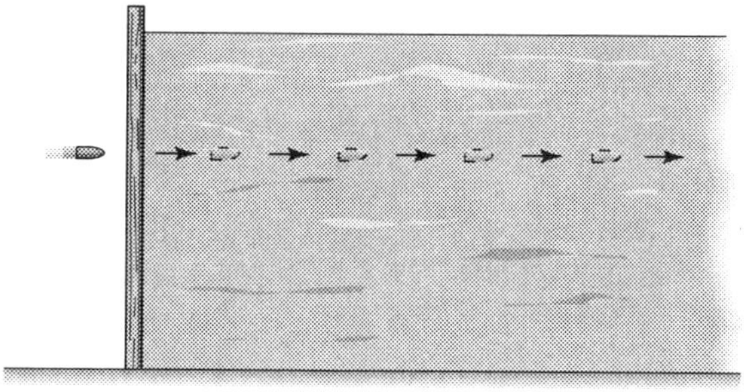

FIGURE P13.159.

13.160 A locomotive starts from rest at $s = 0$ when $t = 0$ and moves along a straight-line path. For a limited time interval, its position is given by $s = \left(\frac{1}{30}\right)v^2 + 2$ where v is in ft/s and s is in feet. Determine (a) the acceleration of the locomotive recalling that, by Eq. (13.11), $a = v\frac{dv}{ds}$ and (b) the velocity as a function of time.

13.161 The acceleration of an airplane flying along a straight-line path right after takeoff is approximated, for a limited period of time, by the expression $a = 8t + 1$ where a is in m/s^2 and t is in seconds. When $t = 0$, $s = 200$ m and $v = 60$ m/s. Determine (a) the velocity as a function of time and (b) the position as a function of time. (c) Sketch the a-t, v-t, and s-t curves for the interval $0 \leq t \leq 4$ s.

13.162 A small stone is thrown vertically into a tank containing a fluid and enters the fluid with a vertical downward speed of 50 ft/s. The fluid imparts a deceleration given by $a = -0.5v^2$ where a is in ft/s^2 and v is in ft/s. Determine (a) the velocity as a function of time and (b) the position as a function of time. (c) Sketch a-t, v-t, and s-t curves for the interval $0 \leq t \leq 5$ s.

13.163 The recoil mechanism of a gun consists of a piston, attached to the end of the gun barrel, moving in a cylinder attached to springs. When the gun recoils, the springs are compressed, causing the gun barrel to have a deceleration that may be approximated by $a = -2v$ where a is in m/s^2 and v is the speed in m/s. Determine (a) the recoil speed as a function of the position s assuming that $v = 100$ m/s for $s = 0$ and (b) the recoil speed as a function of time assuming that $v = 100$ m/s for $t = 0$.

13.164 A motorcyclist starts from rest and moves along a straight-line section of road at a constant acceleration a_1 and attains a speed of 50 mph after moving a distance of 300 ft. At this point, the acceleration is changed to another constant value a_2 such that the motorcyclist attains a speed of 70 mph after having moved an additional distance of 200 ft. Determine (a) the accelerations a_1 and a_2 for each part of the motion and (b) the average speed and average acceleration over the entire distance of 500 ft.

13.165 A stone A is thrown vertically upward from a height of 2 m above the ground with a speed of 10 m/s. One half second later, stone B is thrown vertically upward from a height of 2 m above the ground with a speed v_B. If both stones reach the maximum altitude reached by stone A at the same time, find v_B. Note that the stones move under the influence of the constant acceleration of gravity g.

13.166 A car moving at a constant speed of 65 mph is brought to a complete stop in a distance of 300 ft by applying the brakes, thus, creating a constant deceleration. Determine this constant deceleration and the time it took to come to a complete stop from the moment the brakes were applied.

13.167 An electric motor is designed to operate at 1800 rpm. If it is assumed that it comes up to operating speed at a constant angular acceleration in 0.5 s, determine (a) this constant angular acceleration and (b) the number of revolutions the motor shaft makes before reaching operating speed.

13.168 A race car, moving along a straight-line section of a race track, has an acceleration that, for a short period of time, is given by $a = t^2 - 14t + 40$ where a is in ft/s^2 and t in seconds. When $t = 0$, $s = 20$ ft and $v = 10$ ft/s. Determine (a) the velocity as a function of time and (b) the position as a function of time.

13.169 In Figure P13.169, block B is moving down with a speed of 6 m/s and is decelerating at 3 m/s^2. Determine the speed and acceleration of block A.

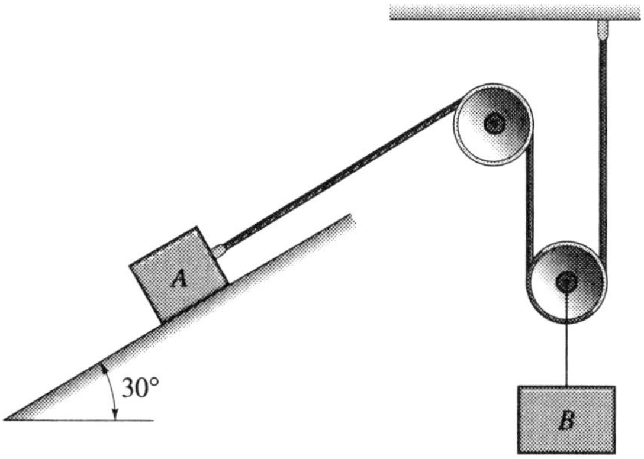

FIGURE P13.169.

13.170 A remote-controlled toy car moves in the x-y plane such that its scalar acceleration components are $\ddot{x} = 3$ ft/s^2, $\ddot{y} = 6t^2$ ft/s^2 where t is in seconds. The initial conditions for the motion are known to be $t = 0$, $x = 5$ ft, $y = 3$ ft, and $\dot{x} = \dot{y} = 0$. Find the velocity and position vectors as functions of time. Also find the equation of the car's path in the x-y plane.

13.171 A plane flying at an altitude $h = 3$ km at an angle of 15° with the horizontal as shown in Figure P13.171, releases a bomb. If, at the instant of release, the plane has a speed of 400 km/h, determine the distance x_1 at which the

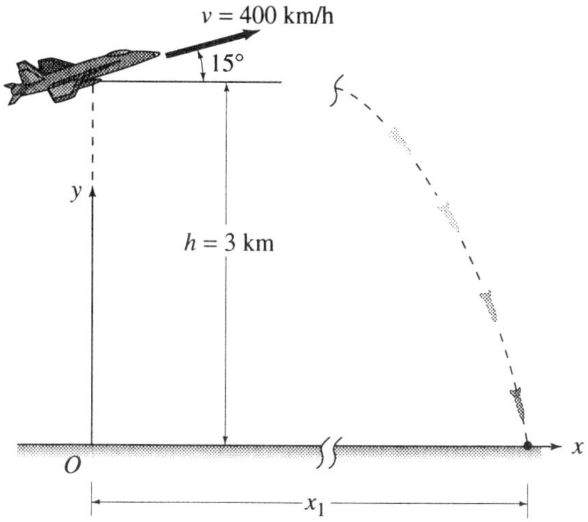

FIGURE P13.171.

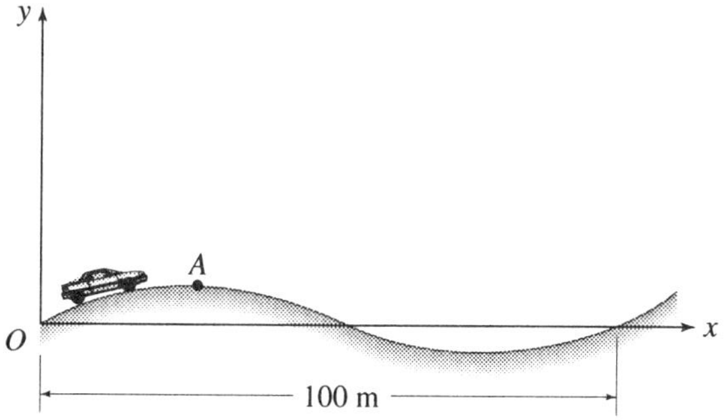

FIGURE P13.173.

bomb hits the ground. Note (see Example 13.17) that motion of the bomb is such that its scalar acceleration components are $\ddot{x} = 0$ and $\ddot{y} = -g$, where g is the acceleration due to gravity, equal to 9.81 m/s^2. Also find the velocity with which the bomb hits the ground.

13.172 A communications satellite circles the Earth in a circular orbit of radius $\rho =$ 26,250 mi measured from the center of Earth. The speed of the satellite is constant at 6,870 mi/h. For any position of the satellite, determine the resultant acceleration that acts on it.

13.173 A car travels a section of road that may be approximated by the relationship $y = 2\sin\left(\frac{\pi x}{50}\right)$ as shown in Figure P13.173 where both x and y are in meters. The car arrives at point A with a speed of 80 km/h which is increasing at the rate of 10 m/s^2. Determine the resultant acceleration of the car at point A giving both its magnitude and direction.

13.174 A train starts from rest on a horizontal section of tracks of circular shape with a radius of 700 ft and accelerates at a constant rate of 3 ft/s^2. Determine the time required for the train to acquire a total resultant acceleration of 5 ft/s^2. What is the distance traveled during this time period?

13.175 Arm OA oscillates about the vertical axis forcing pin C to slide freely around the circular groove as shown in Figure P13.175. For a certain period of time during its motion, arm OA rotates such that $\theta = 1.5t$ rad where t is in seconds. Determine the magnitude and direction of the resultant acceleration of the pin for any position defined by the angle θ during the period of interest.

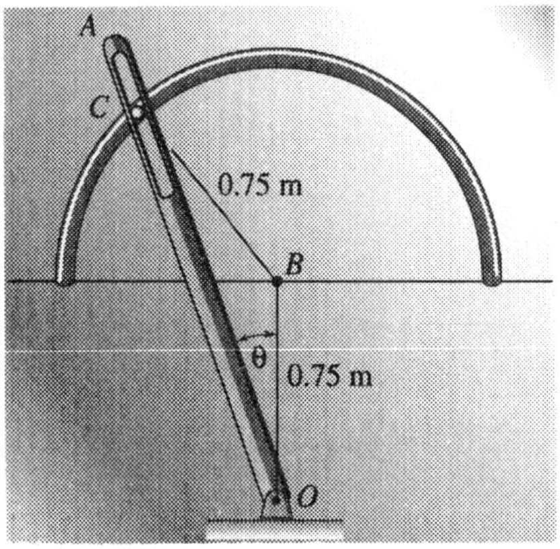

FIGURE P13.175.

13.176 A race car travels at a constant speed of 120 mph along a horizontal curved road given by $v = 1000 \sin\theta$ where v is in feet and θ is in radians, as shown in Figure P13.176. Determine the resultant acceleration of the car for any position defined by the angle θ.

13.177 Refer to Figure P13.177. Boat A moves parallel to the x axis at a speed of 40 km/h. At the moment that it crosses the y axis at point C, boat B starts from rest at point D and proceeds along a path, which is 30° ccw from the horizontal, at a constant acceleration of 2 m/s^2. Determine the position, velocity, and acceleration of boat A relative to boat B ten seconds after boat A crosses the y axis at point C.

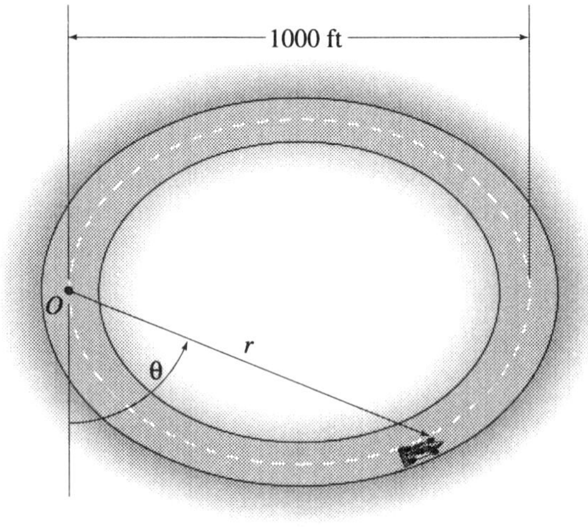

FIGURE P13.176.

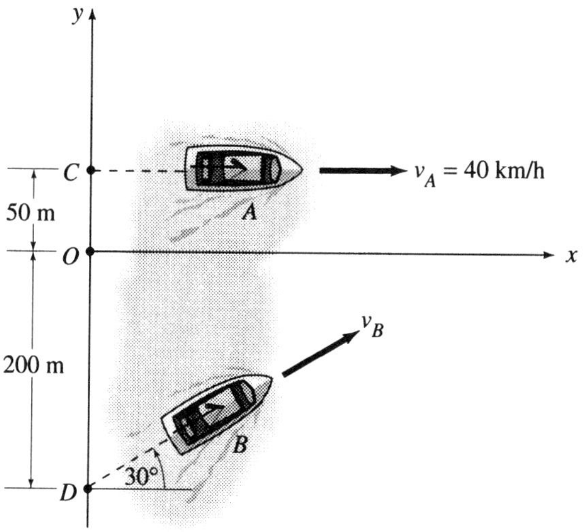

FIGURE P13.177.

14

Particle Kinetics: Force and Acceleration

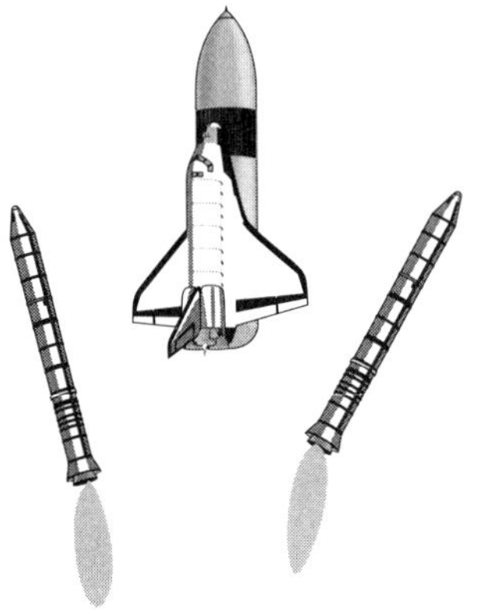

Newton's laws of motion are valid here on Earth as well as in space. The study of dynamics involves the geometry of motion and the forces involved.

The design of machines to achieve maximum performance is one of humanity's oldest engineering challenges. The design of sailing vessels and some understanding of the wind's propulsive force go far back in history. The first sailing ship may have been designed by the Egyptians some 5000 years ago. The Egyptians learned about sail forces, as they affected steering, and placed the single mast well forward for sailing down the Nile before a constant southerly breeze. For use on the Mediterranean, the mast was located nearly amidships. Such vessels were followed by Homeric Greek ships which became more specialized and larger.

Although the physical and applied sciences have experienced tremendous leaps over the ages, the essence of engineering problems, as they pertain to forces and motion geometry, have not changed. Presently, engineering systems from sailing ships to spacecraft are designed more efficiently and precisely than ever before. This is made possible because of the cumulative knowledge and experience gained over the years. Insofar as it is a study of dynamic systems, one needs theoretical, observational, and experimental findings. This is precisely what Newton's laws are based on. It is remarkable that Isaac Newton's laws of motion were found to be accurate despite the fact that he had no access to computers.

In this chapter, we learn about concepts pertaining to forces and motion. We will also learn that the logical standards and methods of reasoning, which have been pursued for over two thousand years, are evolutionary rather than stationary. Despite the rigor of some concepts, we should not overlook the real importance of the subject. The basic principles covered have made it possible for us to secure a remarkable grip on our universe.

14.1 Newton's Laws of Motion

In Chapter 13, a study was made of the *kinematics* of particle motion. We recall that *kinematics* deals only with the geometry of motion and ignores completely the forces that cause this motion. In this chapter, we deal with the *kinetics* of motion which is concerned with the study of the relationships that exist between the motion of a particle and the forces that cause this motion.

Newton's three laws of motion and his law of gravitational attraction were introduced in Chapter 1. As stated there, these laws are basic to the study of both statics and dynamics. Because of this significance, Newton's three laws of motion and his law of gravitation are summarized below:

Newton's First Law: *A particle will remain at rest or move with constant velocity unless it is subjected to an unbalanced force.*

Newton's Second Law: *A particle of mass m, subjected to a nonzero resultant force, will be accelerated in the direction of the force, and the magnitude of this acceleration will be proportional to the magnitude of the resultant force.*

Newton's Third Law: *If bodies A and B are in contact, then the force exerted by body B on body A is equal and opposite in sense to the force exerted by body A on body B.*

Newton's Law of Gravitation: *The gravitational force between two particles varies directly with the product of their masses and inversely with the square of the distance between them. In equation form, this law may be represented by Eq. (1.2) which is repeated here for convenience. Thus,*

$$F = G\frac{m_1 m_2}{r^2} \tag{14.1}$$

where F is the gravitational attractive force, G is the universal constant equal to $6.673 \times 10^{-11}\ \text{m}^3/\text{kg}\cdot\text{s}^2 = 3.438 \times 10^{-8}\ \text{ft}^4/\text{lb}^2$, m_1 and m_2 are the masses of the two particles, and r is the distance between them.

Newton's second law of motion forms the foundation for the study of the kinetics of a particle. Because a particle occupies no space, i.e., it is a point in space, all of the forces that act on it must necessarily be concurrent. Thus, the resultant of any system of forces acting on a particle is a force which may be obtained by forming the vector sum of all of the forces in the force system (i.e., $\mathbf{R} = \sum \mathbf{F}$). Therefore, Newton's second law may be stated mathematically by

$$\sum \mathbf{F} = m\mathbf{a} \tag{14.2a}$$

where $\sum \mathbf{F}$ represents the resultant $\mathbf{R}$ of all of the forces acting on the particle, $\mathbf{a}$ is the acceleration vector of the particle, and m is a scalar constant of proportionality between the resultant vector force $\sum \mathbf{F}$ and the resultant vector acceleration $\mathbf{a}$. Because Eq. (14.2a) is a vector equa-

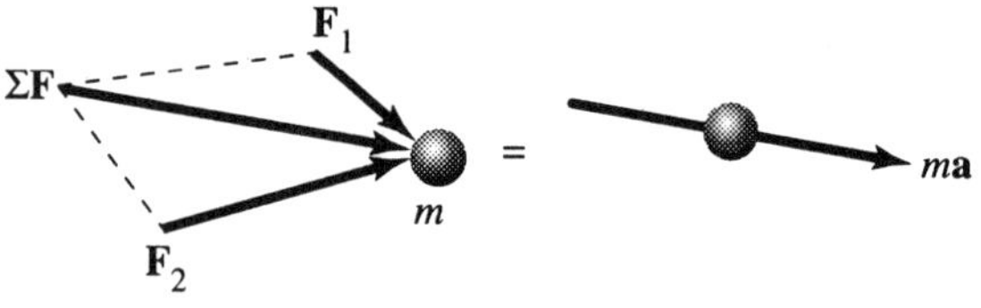

FIGURE 14.1.

tion and m is a positive scalar quantity, it follows that the acceleration vector $\mathbf{a}$ is in the same direction as the resultant vector force $\sum \mathbf{F}$, as shown in Figure 14.1 which provides a graphical representation of Newton's second law. Also, the scalar constant of proportionality m represents a property of the particle known as its *mass*. This property of the particle, *mass*, may be considered a measure of the resistance of the particle to a change in its velocity. As discussed in Chapter 1, the mass m of a body is a fundamental quantity that remains constant anywhere in the universe. It is related to the weight W and the acceleration of gravity g by Newton's second law, namely, $m = W/g$, where both W and g change depending upon the position of the body in the universe. For example, if a body is taken from the surface of Earth to the surface of another planet, both the weight W and the acceleration g change so that the mass m remains unchanged.

The reason that the SI system of units is regarded as preferable to the U.S. Customary system, particularly for space mechanics, is that mass is considered fundamental in the former system and force, including weight, is considered fundamental in the latter system. It is more convenient to regard a constant physical quantity, such as the mass m, as fundamental compared to the quantity force (weight) which changes from location to location in the universe. However, the U.S. Customary system presents little difficulty in practice, provided we know the acceleration of gravity g that applies locally.

Newton's second law as expressed in Eq. (14.2a) is the fundamental equation of motion underlying the classical study of dynamics. The methods of work-energy and impulse-momentum, which will be studied in subsequent chapters, although representing different formulations of mechanics problems, are both derivable from the fundamental equation of motion. It should be noted that the acceleration $\mathbf{a}$ in Eq. (14.2a) is an absolute quantity measured relative to a *fixed* coordinate system known as a *Newtonian or inertial coordinate system.* The term *fixed* implies that the coordinate system neither rotates nor translates, or translates with zero acceleration. Because the distant stars *may be assumed* fixed in position, a coordinate system, whose origin is fixed to these stars or moving at constant velocity with respect to them and which does not rotate with respect to them, is a Newtonian or an inertial coordinate system. However, strictly speaking, a coordinate system on the surface of Earth is not a Newtonian or inertial coordinate system even though it may be fixed against translation and rotation with respect to Earth. This is so because Earth is not fixed in space but moves relative to the fixed stars and rotates about its own polar axis. Nevertheless, the acceleration created by the motions of the earth are sufficiently small and may be neglected in most cases dealing with engineering applications.

It is possible to rewrite Newton's second law of motion in the form

$$\sum \mathbf{F} - m\mathbf{a} = \mathbf{0} \tag{14.2b}$$

In Eq. (14.2b), the vector $m\mathbf{a}$ is a fictitious force known as the *inertial force vector*, and is treated as though it were a real force, balancing the effect of the applied resultant force $\sum \mathbf{F}$, to produce a condition that has come to be known as *dynamic equilibrium*. This condition of dynamic equilibrium is illustrated for a particle in Figure 14.2. Note that the inertial force vector $m\mathbf{a}$ is shown in its opposite sense (i.e., $-m\mathbf{a}$), thus, opposing the applied resultant force $\sum \mathbf{F}$ to produce the fictitious condition of dynamic equilibrium. When expressed in the form of Eq. (14.2b), Newton's second law of motion is known as *D'Alembert's principle* after the French mathematician Jean D'Alembert who was first to propose it. It should be emphasized, however, that this brief discussion of D'Alembert's principle is given here more for its historical than for its utilitarian significance. All of the examples in this chapter are formulated on the basis of Newton's equation of motion expressed in Eq. (14.2a), not on the basis of D'Alembert's principle expressed in Eq. (14.2b).

As in the case of Statics, the solution of problems in Dynamics is greatly facilitated by the use of correct free-body diagrams. Thus, the student is strongly urged to review the procedure discussed in Chapters 2 and 3 for the construction of the free-body diagram for a particle. In the case of dynamics problems, it is useful to show both sides of Newton's equation of motion on a diagram. One part of such a diagram shows the free-body diagram indicating all of forces acting on the particle and the second part, the inertial force vector $m\mathbf{a}$. This procedure is illustrated in Figure 14.1 and is the one used for the solution of dynamics problems in this book. Note that, although the particle in Figure 14.1 is subjected to only two forces $\mathbf{F}_1$ and $\mathbf{F}_2$, whose resultant

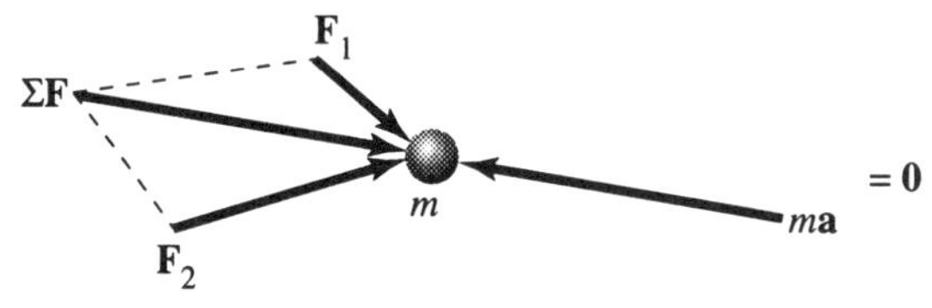

FIGURE 14.2.

is $\sum \mathbf{F}$, the procedure would be the same if the particle were subjected to any number of forces.

When solving dynamics problems, it is convenient to express Newton's second law of motion in terms of components referred to some coordinate system. Two such coordinate systems commonly used in the solution of problems in dynamics are the rectangular coordinate system and the cylindrical coordinate system. In addition to components of force and acceleration in terms of these two coordinate systems, Newton's second law of motion is also discussed in terms of normal and tangential components. Which of these three types of components is most suitable in a given case depends upon the specific problem at hand. The formulations of dynamics problems on the basis of these three types of acceleration and force components are discussed in Sections 14.3, 14.4 and 14.5.

14.2 Newton's Second Law Applied to a System of Particles

Now, consider a system of n particles contained in the space within a closed envelope as shown in Figure 14.3(a). In general, anyone of the n particles is subjected to the action of external and internal forces. For example, the ith particle, located by the position vector $\mathbf{r}_i$ from the origin of the inertial x-y-z coordinate system, whose mass is m_i, is subjected to externally applied forces, whose resultant is $\sum \mathbf{F}_i$, and to internally applied forces, whose resultant is $\sum \mathbf{f}_i$. The external forces represented by $\sum \mathbf{F}_i$ are the action on the ith particle of other contacting bodies and/or the action of externally applied gravitational, magnetic, or electric forces. The internally applied forces, represented by $\sum \mathbf{f}_i$, are the forces of action and reaction exerted on the ith particle by all of its neighboring particles in the envelope.

The free-body diagram of the ith particle is shown in Figure 14.3(b). Application of Newton's second law of motion to this particle yields

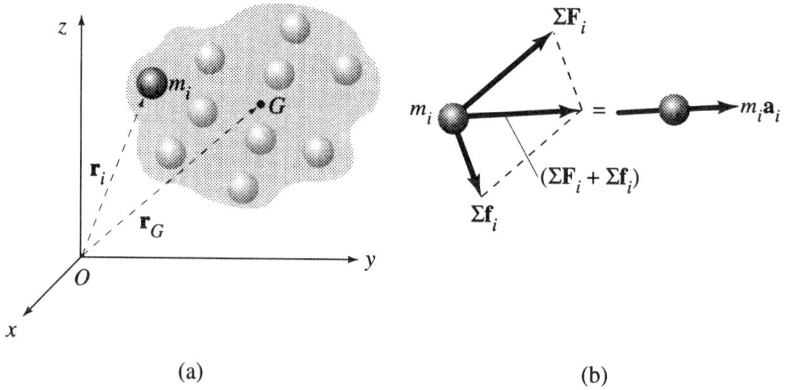

FIGURE 14.3.

$$\sum \mathbf{F}_i + \sum \mathbf{f}_i = m_i \mathbf{a}_i \tag{14.3}$$

where $\mathbf{a}_i$ is the acceleration of the ith particle. In the same manner, an expression similar to Eq. (14.3) may be written for each one of the n particles in the system and, if all of these expressions are added vectorially, the result would be

$$\sum \mathbf{F} + \sum \mathbf{f} = \sum m_i \mathbf{a}_i \tag{14.4}$$

where the quantity $\sum \mathbf{F}$ represents the vector sum of *all* external forces acting on the system of particles and the quantity $\sum \mathbf{f}$ represents the vector sum of *all* internal forces of action and reaction acting on *all* particles of the system. Because, by Newton's third law, these internal forces of action and reaction occur in pairs which are equal in magnitude, collinear and opposite in sense, the quantity $\sum \mathbf{f}$ in Eq. (14.4) vanishes and the resulting relationship becomes

$$\sum \mathbf{F} = \sum m_i \mathbf{a}_i. \tag{14.5}$$

Now, refer to Figure 14.3(a) and let G be the center of mass for the system of particles. The position vector $\mathbf{r}_G$ is determined using the definition of the center of mass expressed in scalar form by Eqs. (10.7). Expressing this definition in vector form,

$$\mathbf{r}_G = \frac{\sum m_i \mathbf{r}_i}{\sum m_i} \tag{14.6a}$$

from which

$$(\textstyle\sum m_i)\mathbf{r}_G = \sum m_i \mathbf{r}_i. \tag{14.6b}$$

Because $\sum m_i = m$, we rewrite Eq. (14.6b) in the form

$$m\mathbf{r}_G = \sum m_i \mathbf{r}_i. \tag{14.6c}$$

If Eq. (14.6c) is differentiated twice with respect to time, the result is

$$m\frac{d^2\mathbf{r}_G}{dt^2} = \sum m_i \frac{d^2\mathbf{r}}{dt^2}$$

or

$$m\mathbf{a}_G = \sum m_i \mathbf{a}_i \tag{14.7}$$

where $\mathbf{a}_G = d^2\mathbf{r}_G/dt^2$ is the acceleration of the mass center of the system of n particles and, as stated earlier, $\mathbf{a}_i = d^2\mathbf{r}_i/dt^2$ is the acceleration of the ith particle. Note that, in differentiating Eq. (14.6), the assumption was made that the mass m is a constant and does not vary with time. However, there are cases, such as rockets and missiles, in which the mass of the system changes with time. Such cases are beyond the scope of this chapter but will be discussed in Chapter 16.

If the term $\sum m_i \mathbf{a}_i$ in Eq. (14.5) is replaced by its equivalent from Eq. (14.7), the result is

$$\sum \mathbf{F} = m\mathbf{a}_G. \tag{14.8}$$

Equation (14.8) expresses the *principle of motion of the mass center* of a system of particles. This principle states that the resultant $\sum \mathbf{F}$ of all externally applied forces acting on a system of particles, is equal to the total mass of the system m multiplied by the acceleration of its mass center $\mathbf{a}_G$. Note that there are no restrictions made as to how the n particles are attached to each other within the containing envelope, and, therefore, Eq. (14.8) may be applied to a system of particles representing a rigid, deformable, fluid or gaseous body. Note also that the resultant external force $\sum \mathbf{F}$ may or may not pass through the mass center G of the system of particles. If this system of particles represents a rigid body, it will experience translation only if the resultant force passes through the mass center. On the other hand, if the resultant force does not pass through the mass center, a moment will be created about the mass center and the rigid body will experience rotation in addition to translation. These concepts will be developed in depth in connection with rigid bodies discussed in later chapters of this book.

14.3 Newton's Second Law in Rectangular Components

Equation (14.1), expressing Newton's second law of motion, is a vector equation, and each of its two terms may be expressed in terms of its rectangular x, y, and z components. Thus, the resultant force $\sum \mathbf{F}$ may be written in the form

$$\sum \mathbf{F} = (\sum F_x)\mathbf{i} + (\sum F_y)\mathbf{j} + (\sum F_z)\mathbf{k}$$

and the resultant acceleration, $\mathbf{a}$, in the form

$$\mathbf{a} = (a_x)\mathbf{i} + (a_y)\mathbf{j} + (a_z)\mathbf{k}.$$

It follows, therefore, that Eq. (14.1) may be reexpressed in the form

$$(\sum F_x)\mathbf{i} + (\sum F_y)\mathbf{j} + (\sum F_z)\mathbf{k} = m[(a_x)\mathbf{i} + (a_y)\mathbf{j} + (a_z)\mathbf{k}]. \tag{14.9}$$

In Eq. (14.9), the quantities $\sum F_x$ and $\mathbf{a}_x$ represent, respectively, the sum of *all* force components and the component of acceleration in the x direction. The same is true for the quantities $\sum F_y$ and a_y and the quantities $\sum F_z$ and a_z in the y and z directions, respectively. A graphical representation of Eq. (14.9) is shown in Figure 14.4, where the magnitude of the resultant force $\mathbf{R} = \sum \mathbf{F}$ and that of the resultant acceleration $\mathbf{a}$ are given by

$$R = \sqrt{\sum F_x^2 + \sum F_y^2 + \sum F_z^2} \tag{4.10}$$

and

$$a = \sqrt{a_x^2 + a_y^2 + a_z^2}. \tag{4.11}$$

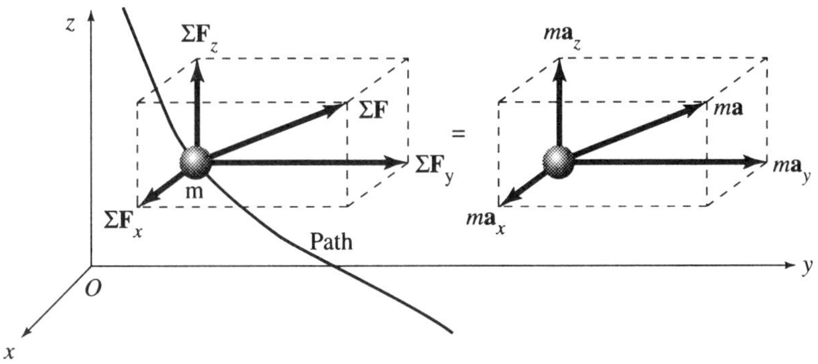

FIGURE 14.4.

It is obvious that, if Eq. (14.9) is to be satisfied, the coefficients of the **i** terms on both sides of the equal sign must be equal to each other and, similarly, for the **j** and the **k** terms. Thus, the following three scalar equations of motion are obtained

$$\left.\begin{aligned} \sum F_x &= ma_x, \\ \sum F_y &= ma_y, \\ \sum F_z &= ma_z. \end{aligned}\right\} \qquad (4.12)$$

Equations (14.12) are independent of each other. For example, if a particle is constrained to move along the x axis (rectilinear translation), then, only the first of Eqs. (14.12) is applicable while $\sum F_y = \sum F_z = 0$ because the accelerations of the particle in the y and z directions are both zero. As a second example, if a particle is constrained to move in the x-y plane, only the first two of Eqs. (14.12) are applicable while $\sum F_z = 0$ because the acceleration in the z direction is zero.

Because $a_x = d^2x/dt^2 = \ddot{x}$, $a_y = d^2y/dt^2 = \ddot{y}$, and $a_z = d^2z/dt^2 = \ddot{z}$, Eqs. (14.12) may be written in the form of differential equations as

$$\left.\begin{aligned} \sum F_x &= m\ddot{x}, \\ \sum F_y &= m\ddot{y}, \\ \sum F_z &= m\ddot{z}. \end{aligned}\right\} \qquad (4.13)$$

In the form of Eqs. (14.13), Newton's second law of motion may be used in the solution of problems in which the applied forces are given as functions of time. In such cases, Eqs. (14.13) would be set up and integrated once, to obtain the velocities as functions of time, and a second time to obtain the displacements as functions of time.

When using Eqs. (14.12) or (14.13), it is essential that a coordinate system be established to have a frame of reference from which the motion may be measured. It is very convenient in solving problems to assume that the unknown quantities (forces and/or accelerations) lie in the positive direction of the coordinate axes. A positive answer would, then, indicate a correct assumption and a negative answer, an incorrect one. The following examples illustrate some of the concepts developed in this section.

■ Example 14.1

A crate weighing 100 lb is placed on the flat surface of a cart that weighs 50 lb as shown in Figure E14.1(a). The cart is being pulled up the 20° inclined plane by a force $P = 80$ lb. If the crate is not to slip on the cart, determine the minimum coefficient of friction between the crate and the cart. Deal with the crate and cart as though they were particles. Ignore rolling friction of the cart's wheels.

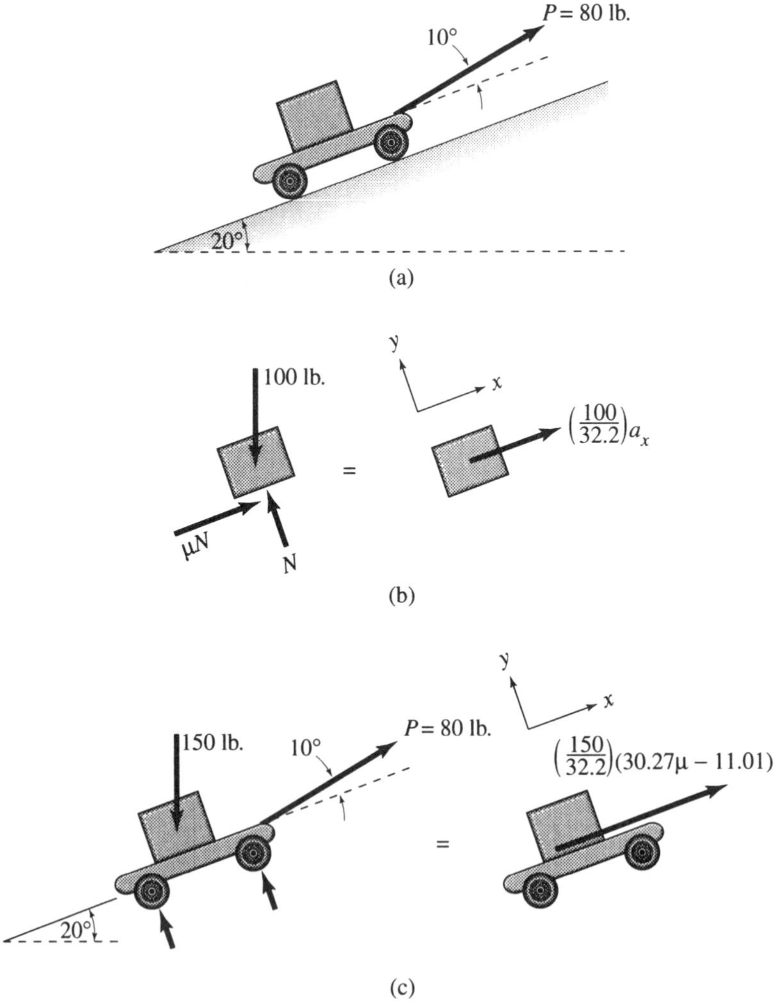

FIGURE E14.1.

Solution

The free-body diagram of the crate and its inertia force vector are shown in Figure E14.1(b) for the condition when the crate *is on the verge of slipping* with respect to the cart. Using the coordinate system shown in Figure E14.1(b),

$$\sum F_y = ma_y = 0$$

$$N - 100\cos 20^\circ = 0$$

$$N = 93.97\ \text{lb} \tag{a}$$

$$\sum F_x = ma_x$$

$$\mu N - 100\sin 20^\circ = (100/32.2)a_x \tag{b}$$

Substituting from Eq. (a) and solving Eq. (b) for a_x,

$$a_x = 30.27\mu - 11.01 \tag{c}$$

where a_x represents the common acceleration of the cart and the crate.

Now, consider now the entire system consisting of the cart and the crate. The free-body diagram and the inertial force vector for this entire system are shown in Figure E14.1(c). Thus,

$$\sum F_x = ma_x,$$

$$80\cos 10^\circ - 150\sin 20^\circ = 4.66(30.27\mu - 11.01),$$

$$27.48 = 4.66(30.27\mu - 11.01). \tag{e}$$

The solution of Eq. (e) for μ yields

$$\mu = 0.559. \qquad \text{ANS.}$$

■ Example 14.2

An elevator is designed to reach operating speed with a constant acceleration a. A person of weight W stands on a bathroom scale inside the elevator. Determine an expression for the scale reading when the elevator is accelerating (a) upward and (b) downward. What are the scale readings after the elevator reaches the constant operating speed while moving upward and downward.

Solution

(a) The free-body diagram of the person and his inertia force vector are shown in Figure E14.2(a) for the case where the elevator is accelerating upward. Also shown is a convenient coordinate system. The quantity Q represents the force between the person and the bathroom scale (i.e., the scale reading). Thus,

$$\sum F_y = ma_y$$

$$Q - W = (W/g)a,$$

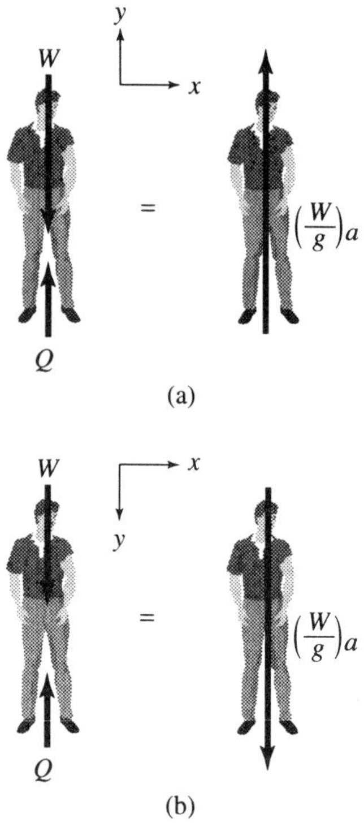

FIGURE E14.2.

from which

$$Q = W\left(1 + \frac{a}{g}\right). \qquad \text{(a) ANS.}$$

Thus, the apparent weight of the person inside an elevator that is accelerating upward is more than his real weight. The increase in the apparent weight depends upon the ratio a/g where a is the acceleration of the elevator and g is the acceleration of gravity.

(b) The free-body diagram and the inertial force vector are shown in Figure E14.2(b) for the case where the elevator is accelerating downward. Again, a convenient coordinate system is chosen and the force Q represents the bathroom scale reading. Thus,

$$\sum F_y = ma_y$$
$$W - Q = (W/g)a$$

from which

$$Q = W\left(1 - \frac{a}{g}\right) \qquad \text{(b) ANS.}$$

It is clear from this equation that the apparent weight of the person inside an elevator that is accelerating downward is less than his true weight. The decrease in the apparent weight is again a function of a/g.

When the elevator reaches its constant operating speed, its acceleration a vanishes and, whether it is moving up or down, Eqs. (a) and (b) both show that the scale reading Q becomes equal to the real weight W.

Example 14.3

The frictional force on the skin of a motor boat created by the surrounding water is given by the relation $F = kv$ where k is a constant and v is the speed of the boat. The boat has a weight W and is moving at a constant speed v_0 when the motor is suddenly turned off. (a) In terms of W, k, g, v and v_0, develop an expression for the distance traveled by the boat after the motor is turned off. (b) For the case where $k = 2$ lb·s/ft, $v_0 = 35$ ft/s, $W = 1200$ lb, determine the distance traveled corresponding to (i) $v = 3.5$ ft/s and (ii) $v = 0.035$ ft/s.

Solution

(a) The free-body diagram of the boat and its inertial force vector are shown in Figure E14.3 immediately after the motor is turned off. Thus

$$\sum F_x = ma_x,$$

$$-kv = \frac{W}{g}a_x,$$

from which

$$a_x = -\frac{kg}{W}v. \tag{a}$$

Because $v\,dv = a_x\,dx$, it follows that

$$v\,dv = -\frac{kg}{W}v\,dx$$

and

$$dv = -\frac{kg}{W}dx. \tag{b}$$

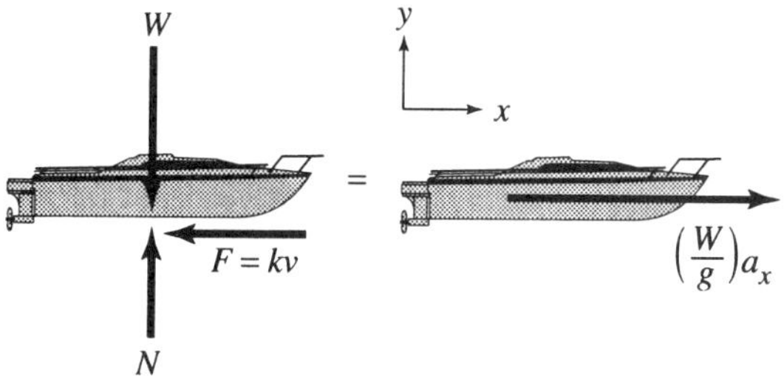

FIGURE E14.3.

Integrate Eq. (b) using the lower limits $v = v_0$ and $x = 0$ and general upper limits. Thus,

$$\int_{v_0}^{v} dv = -\frac{kg}{W}\int_0^x dx$$

from which

$$v - v_0 = -\frac{kg}{W}x$$

and

$$x = (v - v_0)\frac{W}{kg}. \qquad \text{(c) ANS.}$$

(b) For the given numerical data, Eq. (c) yields

Part (i):

$$x = \left[\frac{1200}{2(32.2)}\right](35 - 3.5) = 587 \text{ ft}; \qquad \text{ANS.}$$

Part (ii):

$$x = \left[\frac{1200}{2(32.2)}\right](35 - 0.035) = 652 \text{ ft}. \qquad \text{ANS.}$$

■ Example 14.4

Refer to the solution of Example 14.3. (a) In terms of W, k, g, v, and v_0, develop an expression for the time t needed for the boat to reach any speed v after the motor is turned off. (b) For the case where $k = 2$ lb·s/ft, $v_0 = 35$ ft/s, $W = 1200$ lb, determine the time t needed to reach (i) $v = 3.5$ ft/s and (ii) $v = 0.035$ ft/s.

Solution

Because $a_x = dv/dt$, it follows from Eq. (a) of Example 14.3 that

$$\frac{dv}{dt} = -\left(\frac{kg}{W}\right)v$$

from which

$$\frac{dv}{v} = -\left(\frac{kg}{W}\right)dt.$$

Integrating using the lower limits $v = v_0$ and $t = 0$ and general upper limits yields

$$\int_{v_0}^{v}\frac{dv}{v} = -\frac{kg}{W}\int_0^t dt.$$

Thus,

$$\ln\left(\frac{v}{v_0}\right) = -\left(\frac{kg}{W}\right)t$$

Solving for t gives

$$t = -\left(\frac{W}{kg}\right)\ln\left(\frac{v}{v_0}\right). \qquad \text{(a)} \quad \text{ANS.}$$

(b) For the given numerical values, Eq. (a) yields

Part (i):

$$t = \left(\frac{1200}{2 \times 32.2}\right)\ln\left(\frac{3.5}{35}\right) = 42.9 \text{ s}; \qquad \text{ANS.}$$

Part (ii):

$$t = -\left(\frac{1200}{2 \times 32.2}\right)\ln\left(\frac{0.035}{35}\right) = 128.7 \text{ s}. \qquad \text{ANS.}$$

■ Example 14.5

Blocks A and B, connected by a cable that passes over a smooth pulley, slide on inclined planes as shown in Figure E14.5(a). Block A has a mass of 50 kg and block B a mass of 20 kg. The coefficients of kinetic friction (μ) between the blocks and the corresponding planes are shown in the diagram. If the system is released from rest, determine (a) its acceleration and the tension in the cable and (b) the speed of the system 3 s after release.

Solution

(a) The free-body diagram and the corresponding inertial force vectors of blocks A and B are shown in Figures E14.5(b) and E14.5(c), respectively. Also, a convenient x-y coordinate system has been established in each case. Note that the symbol T has been used for the tension in the cable. Application of Newton's second law of motion yields,

for block A,

$$\sum F_y = ma_y,$$

$$N_A - 490.5\cos 60^\circ = 0,$$

$$N_A = 245.25 \text{ N},$$

$$\sum F_x = ma_x,$$

$$490.5\sin 60^\circ - 0.2(245.25) - T = 50a_x,$$

$$375.735 - T = 50a_x; \qquad \text{(a)}$$

for block B,

$$\sum F_y = ma_y = 0,$$

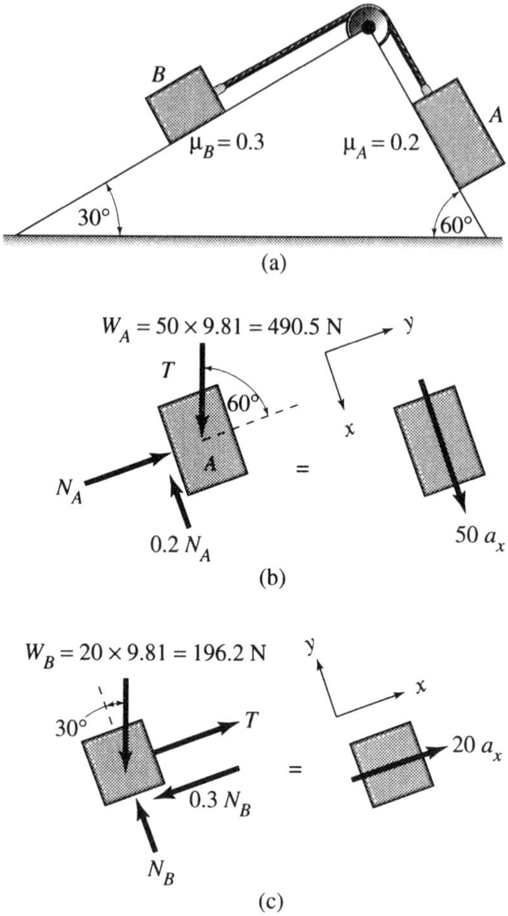

FIGURE E14.5.

$$N_B - 196.2\cos 30^\circ = 0,$$

$$N_B = 169.914\ \text{N},$$

$$\sum F_x = ma_x,$$

$$T - 0.3(169.914) - 196.2\sin 30^\circ = 20a_x,$$

$$T - 149.074 = 20a_x. \qquad \text{(b)}$$

A simultaneous solution of Eqs. (a) and (b) yields

$$a_x = 3.24\ \text{m/s}^2 \qquad \text{ANS.}$$

and

$$T = 214\ \text{N}. \qquad \text{ANS.}$$

(b) Because the acceleration a_x is constant, Eq. (13.19) may be used with $v_0 = 0$ (the system starts from rest). Thus,

$$v = v_0 + a_x t$$

$$v = 0 + 3.24(3) = 9.72 \text{ m/s}. \qquad \text{ANS.}$$

■ Example 14.6

A block of weight W, which may be considered as a particle, is given an initial velocity v up the inclined plane as shown in Figure E14.6(a). If the coefficient of kinetic friction is m determine, in terms of v, μ and θ, the distance traveled by the block up the inclined plane. Express this distance in dimensionless form and discuss its implications.

Solution

The free-body diagram and the inertial force vector of the block are shown in Figure E14.6(b), along with a coordinate system in which the positive x direction has been chosen in the direction of the initial velocity. Application of Newton's second law of motion yields

$$\sum F_y = ma_y = 0,$$

$$N - W\cos\theta = 0, \tag{a}$$

$$\sum F_x = ma_x,$$

$$-\mu N - W\sin\theta = (W/g)a_x. \tag{b}$$

where g is the acceleration of gravity. The simultaneous solution of Eqs. (a) and (b) leads to

$$a_x = -g(\mu\cos\theta + \sin\theta) \tag{c}$$

where the minus sign signifies a deceleration instead of an acceleration as assumed in the solution. Note that, because the forces acting on the block are constant, the acceleration a_x is also constant and, consequently, Eq. (13.21), developed in Chapter 13 for motion under constant acceleration, is applicable. Thus,

$$v^2 = v_0^2 + 2a_c(x - x_0),$$

$$0 = v^2 - 2g(\mu\cos\theta + \sin\theta)x.$$

Therefore,

$$x = \left(\frac{v^2}{2g}\right)\left[\frac{1}{\mu\cos\theta + \sin\theta}\right] \tag{d}$$

Equation (d) may be expressed in dimensionless form by dividing both sides by the quantity $v^2/2g$. Thus,

$$\frac{x}{(v^2/2g)} = \frac{1}{\mu\cos\theta + \sin\theta}. \tag{e}$$

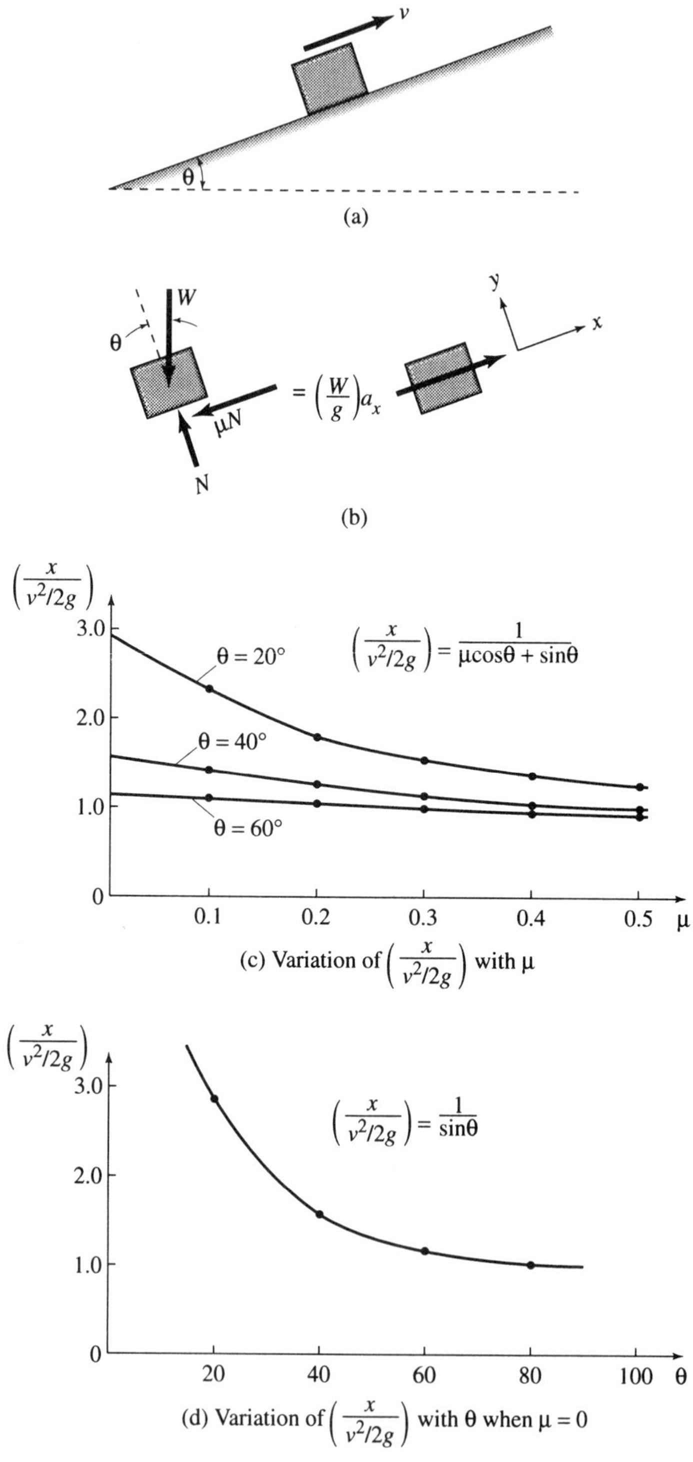

v
θ
(a)
W
θ
μN
N
= (W/g) a_x
y
x
(b)
(x / (v²/2g))
3.0
2.0
1.0
0
θ = 20°
θ = 40°
θ = 60°
(x / (v²/2g)) = 1 / (μcosθ + sinθ)
0.1
0.2
0.3
0.4
0.5
μ
(c) Variation of (x / (v²/2g)) with μ
(x / (v²/2g))
3.0
2.0
1.0
0
(x / (v²/2g)) = 1 / sinθ
20
40
60
80
100
θ
(d) Variation of (x / (v²/2g)) with θ when μ = 0

FIGURE E14.6.

A plot of the dimensionless quantity $x/(v^2/2g)$ as a function of μ for three different values of θ is shown in Figure E14.6(c). It is evident from this plot that, regardless of the value of θ, the dimensionless quantity $x/(v^2/2g)$ decreases as the value of μ increases. Physically, this statement means that the rougher the surface, the less will be the distance traveled by the block up the inclined plane for given values of the initial velocity v and the inclination of the plane θ. Note also that the decrease in the dimensionless quantity $x/(v^2/2g)$ becomes more rapid as the value of θ decreases. In other words, the dimensionless quantity $x/(v^2/2g)$ is very sensitive to μ for small values of θ but not so sensitive for larger values of θ.

It should also be observed that the distance traveled by the block is independent of its weight. Furthermore, if we consider the case of a frictionless plane (i.e., $\mu = 0$), the dimensionless quantity from Eq. (e) becomes

$$\frac{x}{(v^2/2g)} = \frac{1}{\sin\theta}. \tag{f}$$

Equation (f), plotted in Figure E14.6(d), shows that, for a given initial velocity and for $\mu = 0$, the dimensionless distance decreases as the inclination of the plane becomes larger. Note also, that, for $\theta = 0$, the dimensionless quantity $x/(v^2/2g)$ becomes infinite which means that, on a frictionless horizontal plane, the block, with any initial velocity, will travel indefinitely.

■

Problems

14.1 A block of weight $W = 300$ lb resting on a horizontal plane is subjected suddenly to a force $P = 100$ lb as shown in Figure P14.1. If $\theta = 30°$ and the coefficient of kinetic friction between the block and the plane is $\mu = 0.20$, determine the velocity of the block after it has moved from rest a distance $s = 20$ ft.

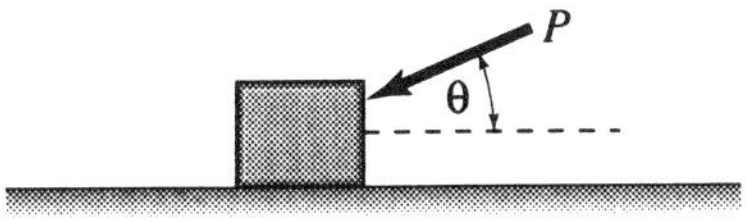

FIGURE P14.1.

14.2 A block of weight W resting on a horizontal plane is subjected suddenly to a force P as shown in Figure P14.1. If the coefficient of kinetic friction between the block and the plane is μ, show that the time t required for the block (assumed to start from rest at $t = 0$) to reach a velocity v may be expressed as

$$t = \frac{vW}{\mu g\left[W + P\left(\sin\theta - \dfrac{1}{\mu}\cos\theta\right)\right]}.$$

14.3 A block of mass $m = 75$ kg is given an initial velocity $v_0 = 2$ m/s down an in-

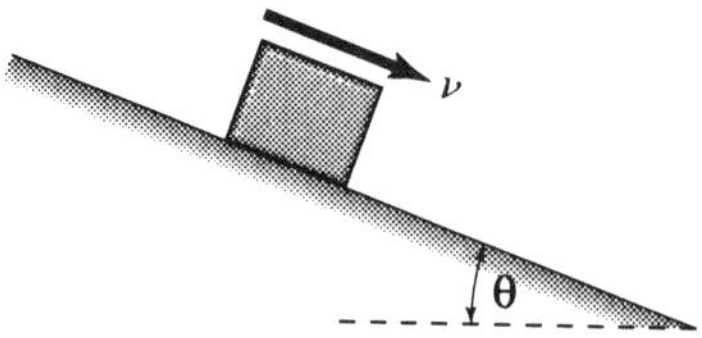

FIGURE P14.3.

clined plane as shown in Figure P14.3. If the coefficient of kinetic friction between the block and the plane is $\mu = 0.80$ and $\theta = 35^\circ$, determine the distance s along the inclined plane through which the block moves before coming to rest.

14.4 A block of mass m is given an initial velocity v_0 down an inclined plane as shown in Figure P14.3. If the coefficient of kinetic friction between the block and the plane is μ, determine, in terms of v, g, μ, and θ, an expression for the time t required for the block to come to a complete stop. Discuss the limitations of this expression.

14.5 A pickup truck, traveling on a horizontal road with a speed of $v = 40$ mph, carries a load $W = 1500$ lb as shown in Figure P14.5. The brakes are applied suddenly and the truck comes to a full stop in a distance $s = 200$ ft. Determine the minimum coefficient of friction between the load and the truck bed if the load is not to move during the stopping operations.

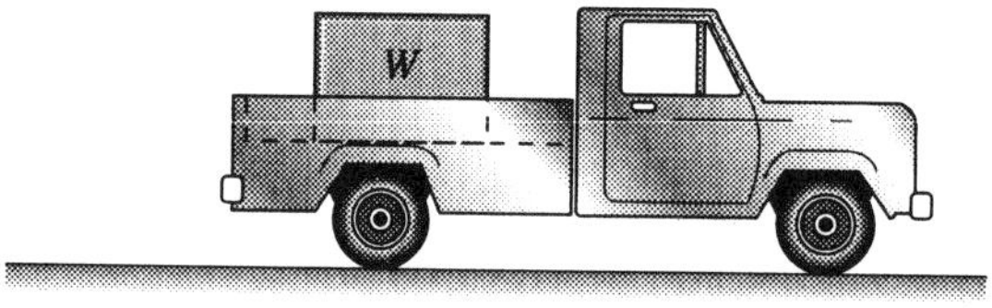

FIGURE P14.5.

14.6 A pickup truck, traveling on a horizontal road with a speed v, carries a load W as shown in Figure P14.5. The coefficient of friction between the load and the truck bed is μ. In terms of v, μ, and g, determine the smallest stopping distance s if the load is not to slide on the truck bed.

14.7 A collar of mass $m = 10$ kg is acted upon by a horizontal force $P = 100$ N as shown in Figure P14.7. If the collar starts from rest and the coefficient of kinetic friction between it and the rod is $\mu = 0.2$, determine (a) the acceleration of the collar, (b) the distance moved by the collar after 5 s, and (c) the velocity of the collar after it has moved a distance of 2 m along the rod.

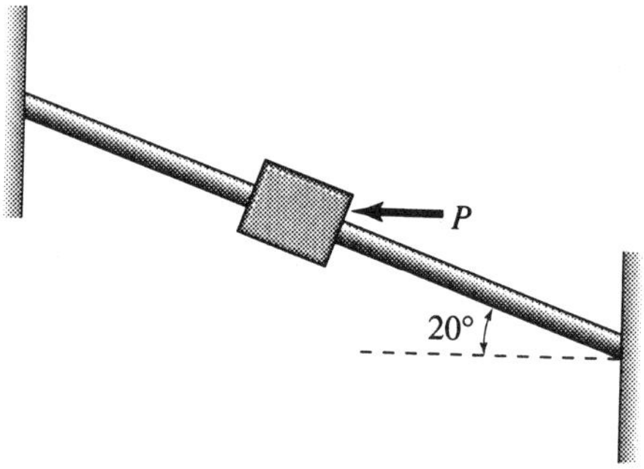

FIGURE P14.7.

14.8 Rework Problem 14.7 for the case when $P = 0$.

14.9 The system shown in Figure P14.9 is released from rest. The cable ABC is

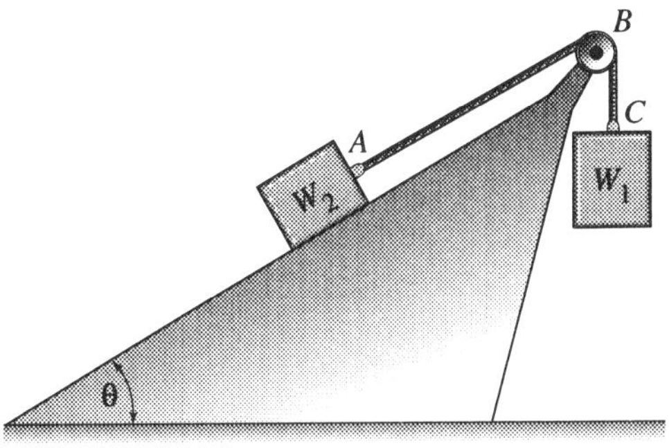

FIGURE P14.9.

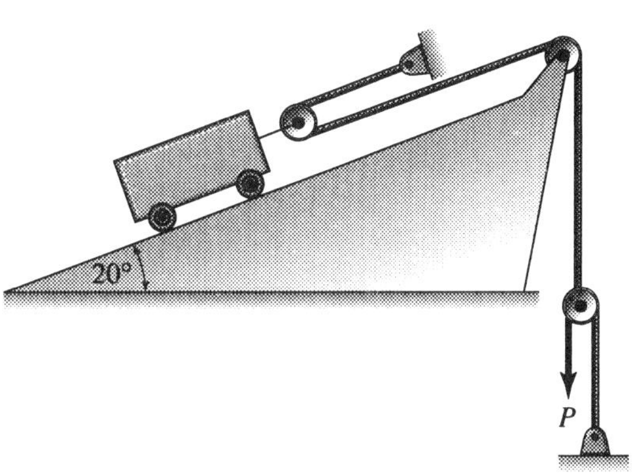

FIGURE P14.11.

flexible but inextensible, and the pulley at B may be assumed weightless and frictionless. If W_2 moves up the inclined plane, determine its magnitude if the system reaches a velocity $v = 15$ ft/s after 5 s has elapsed, given $W_1 = 100$ lb, $\theta = 30^\circ$ and the coefficient of kinetic friction between W_2 and the inclined plane $\mu = 0.25$.

14.10 The system shown in Figure P14.9 is released from rest. The cable ABC is flexible but inextensible and the pulley at B may be assumed weightless and frictionless. If W_2 moves up the inclined plane and the coefficient of kinetic friction between it and the inclined plane is μ, determine, symbolically, in terms of W_1, W_2, μ, and θ, (a) the acceleration of the system and (b) the tension in the cable.

14.11 The pulley system shown in Figure P14.11 is used to pull the 500 kg cart up the 20° inclined plane. If the cart is to start from rest and travel for 5 s, determine the needed constant force P. What is the speed of the cart at the end of 5 s? Assume weightless and frictionless pulleys, and ignore friction and rolling resistance at the cart's wheels.

14.12 A weight $W = 20$ lb is suspended from a cable which passes over a frictionless pulley attached to the roof of an elevator as shown in Figure P14.12. The other end of the cable is attached to a spring (spring constant $k = 100$ lb/in.) which, in turn, is fastened to the floor of the elevator. Determine the total deformation of the spring (including the equilibrium deformation) when the elevator is accelerating (a) upward at a constant acceleration $a = 5$ ft/s^2 and (b) downward at a constant acceleration $a = 5$

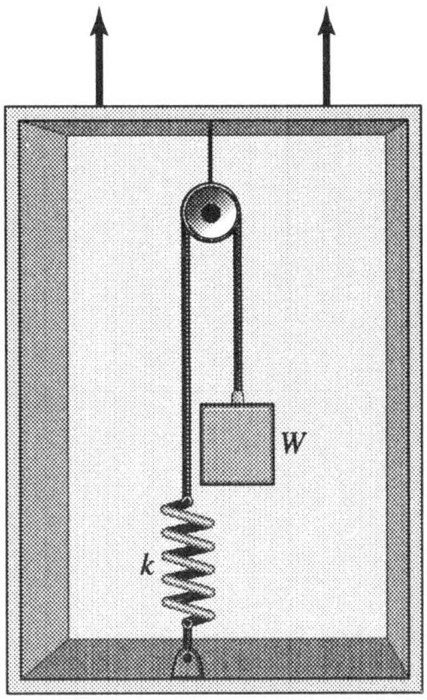

FIGURE P14.12.

ft/s^2. Assume the pulley and cable to be weightless.

14.13 A weight W is suspended from a cable which passes over a frictionless pulley attached to the roof of an elevator as shown in Figure P14.12. The other end of the cable is attached to a spring (spring constant k) which, in turn, is fastened to the floor of the elevator. Let the total deformation of the spring (including the equilibrium deformation) be δ. Develop expressions, in terms of W, k, δ, and g, for (a) the upward acceleration of the elevator and (b) the downward acceleration of the elevator.

14.14 A body of weight W is suspended from the roof of an elevator by two identical springs (spring constant $k = 10$ kN/m) as shown in Figure P14.14. When the elevator moves up with an acceleration $a = 1.5$ m/s^2, the total spring deformations (including the equilibrium deformation) are found to be $\delta = 0.02$ m each. Determine the magnitude of the weight W. Assume that the change in the 25° angles due to the spring deformations is negligible.

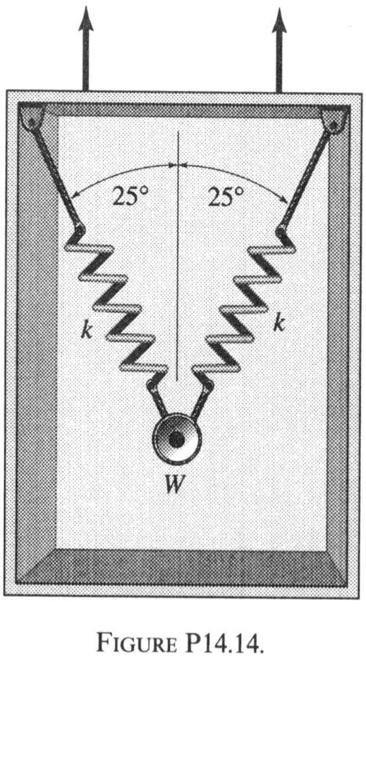

FIGURE P14.14.

14.15 A cylinder of weight $W = 75$ lb is placed against the back of a cart as shown in Figure P14.15. The cart is accelerated up the inclined plane with an acceleration $a = 7$ ft/s^2. If $\theta = 15°$ determine the forces F_A and F_B exerted on the cylinder at contact points A and B, respectively.

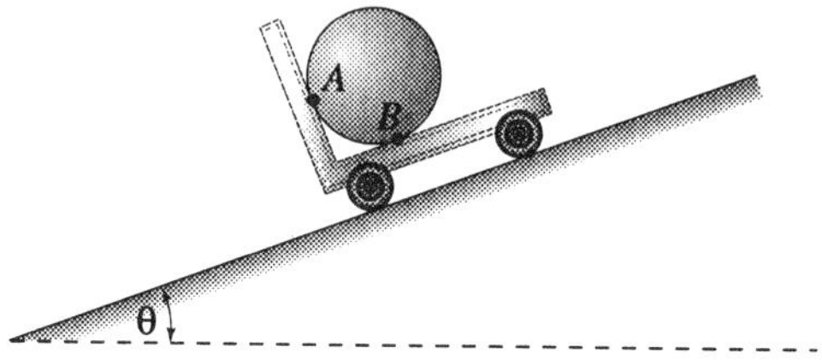

(a)

FIGURE P14.15.

14.16 A cylinder of weight W is placed against the back of a cart as shown in Figure P14.15. The cart is accelerated down the inclined plane with an acceleration a. Develop an expression for the acceleration a in terms of the angle θ for which the cylinder will separate from the cart at point A. What is the numerical value of a for $\theta = 20°$.

14.17 A workman hoists himself and his cage by means of the pulley-rope system shown in Figure P14.17. Assume that the pulley-rope system is weightless and frictionless. If the mass of the workman and his cage $m = 105$ kg., determine their acceleration if the workman pulls the rope with a constant force $P = 260$ N. If the workman starts from rest, how long must he maintain a constant pull of 260 N to reach a speed of 1 m/s?

14.18 A workman lowers himself and his cage by means of the pulley-rope system shown in Figure P14.17. Assume that

FIGURE P14.17.

the pulley-rope system is weightless and frictionless. If the mass of the workman and his cage is m and the workman maintains a constant pull P on the rope, develop a symbolic expression relating the acceleration a to the constant pull P and the mass m. Also, if the cage starts from rest, develop a symbolic expression for its speed after it has moved a distance h downward. What is the maximum value of P in terms of g and m if the cage is to accelerate downward?

14.19 The passenger train shown in Figure P14.19 is moving at a constant speed on level tracks when the brakes are applied suddenly to produce a constant deceleration $a = 3$ ft/s^2. Assume that each of the two cars and the engine develops a braking force equal to F. If $W_1 = 50{,}000$ lb and $W_2 = 60{,}000$ lb, determine (a) the magnitude of the braking force F and (b) the coupling force between the engine and the first car.

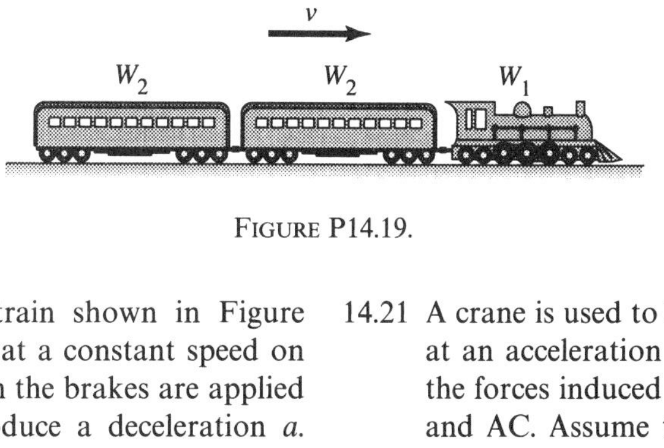

FIGURE P14.19.

14.20 The passenger train shown in Figure 14.19 is moving at a constant speed on level tracks when the brakes are applied suddenly to produce a deceleration a. Assume that each of the two cars and the engine develop a braking force F. Determine, symbolically, in terms of W_1, W_2, a, and g, (a) an expression for the braking force F and (b) an expression for the coupling force between the first and second cars. If $W_1 = 300$ kN, $W_2 = 400$ kN and $a = 1.5$ m/s^2, find a numerical value for this coupling force.

14.21 A crane is used to hoist a 4,000-lb beam at an acceleration of 4 ft/s^2. Determine the forces induced in truss members AB and AC. Assume the pulley at A to be weightless and frictionless.

14.22 The coefficient of kinetic friction μ between the pavement and the tires of a car of mass m was obtained experimentally by bringing up the speed of the car to a certain value v, applying the brakes, and allowing the car to come to a complete stop. If the length of the skid marks is denoted by s, develop an expression

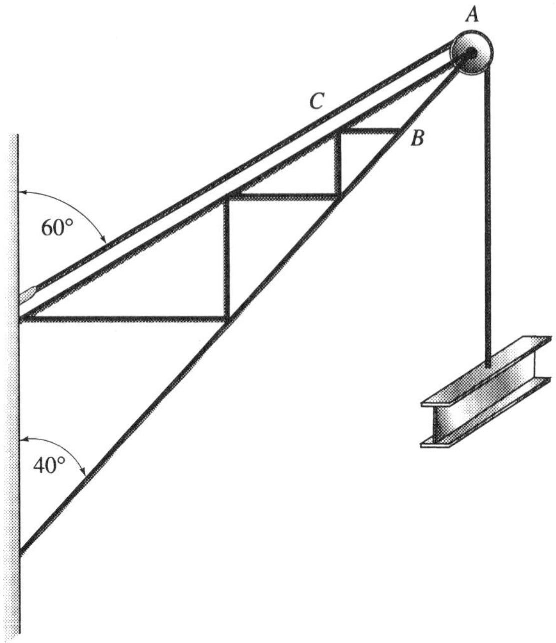

FIGURE P14.21.

for μ in terms of v, g, and s. What is the numerical value of the coefficient of kinetic friction if $m = 1{,}600$ kg, $v = 100$ km/h and $s = 80$ m?

14.23 A 100-lb package is released from rest at time $t = 0$ on a 30° inclined plane as shown in Figure P14.23. The coefficient of friction between the package and the plane is 0.3. When $t = 5$ s, a force $F = (5 + 2t^3)$ lb, where t is in seconds, is applied to the package as shown. Determine (a) the velocity of the package and the distance it has traveled when F is applied and (b) the additional time that elapses before the package is brought to a full stop. *Hint: Use the relationship $a = dv/dt$, and solve the resulting expression for time t by trial and error.*

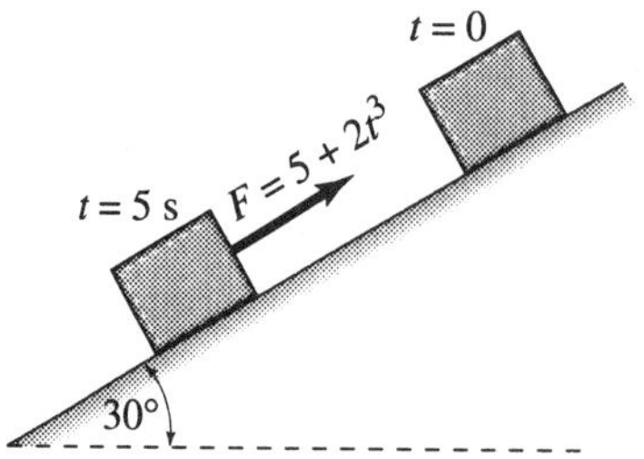

FIGURE P14.23.

14.24 A motorcyclist wants to jump across a river from point A to point B, which is at a horizontal distance X and a vertical distance Y from point A, as shown in Figure P14.24. To do so, he has to get a running start by accelerating up the inclined plane and leaving point A with a minimum speed v. Consider the motorcyclist to be a particle, neglect air resistance, and determine this minimum speed symbolically in terms of X, Y and θ.

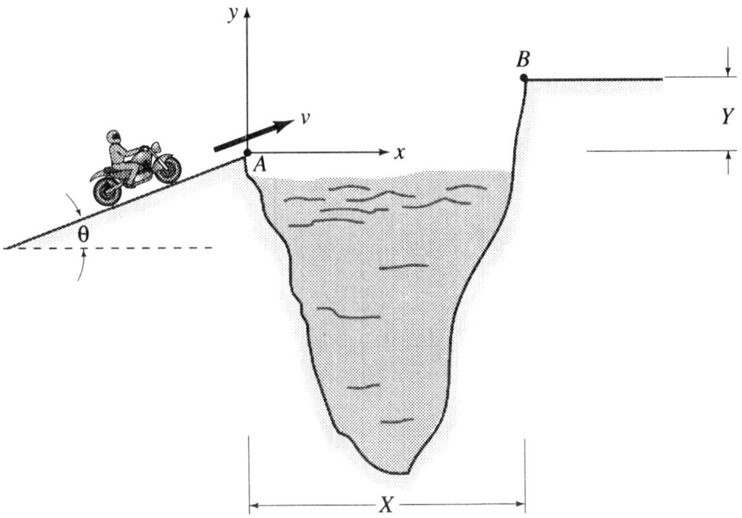

FIGURE P14.24.

14.25 An experimental 2,000-lb missile is fired vertically upward as shown in Figure P14.25. At a height of 4,000 ft above ground level, the propelling rockets fail when the missile has a velocity of 800 ft/s and the missile continues on a vertical free-flight motion. If the air resistance is approximated by the relation $R = 0.1v$ lb where v is the speed of the missile in ft/s, determine the additional height h to which the missile rises.

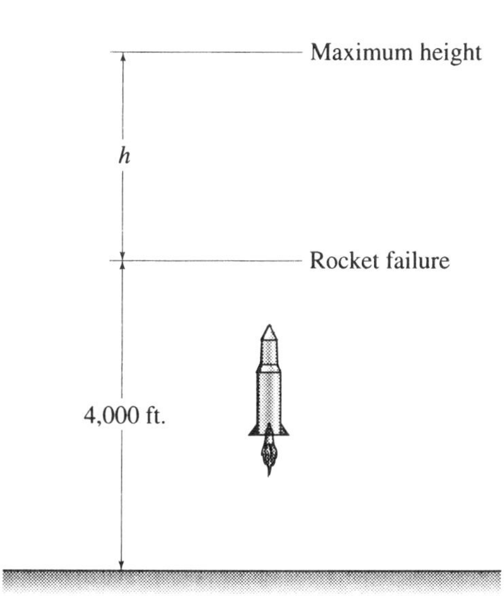

FIGURE P14.25.

14.26 Refer to Problem 14.25 and determine the speed with which the missile hits the ground. *Hint: Use the height* $h = 9{,}680$ *ft found in Problem 14.25, the relation* $v\,dv = a\,ds$, *and solve the resulting velocity equation by trial and error.*

14.27 A rock of weight W, dropped above the surface of the water in a tank, hits the water with a downward speed v_1 as shown in Figure P14.27. The resistance of the water to the motion of the rock is approximated by the relation $R = Cv$ where C is a constant and v is the speed of the rock. If the velocity with which the rock hits the bottom of the tank is v_2, determine the depth h of the tank in terms of v_1, v_2, C, g, and W.

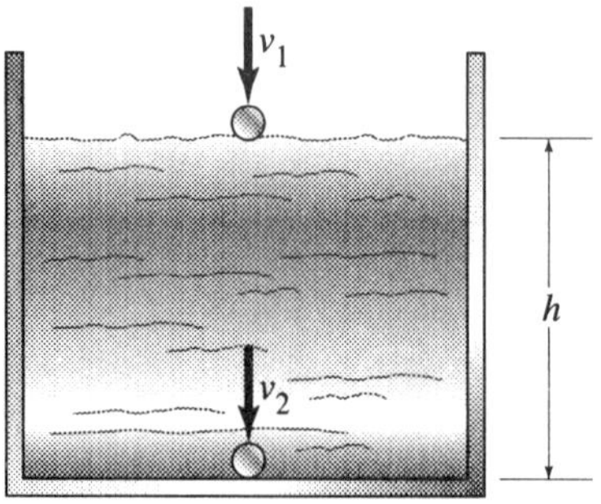

FIGURE P14.27.

14.28 A small experimental fighter plane of mass $m = 6{,}000$ kg uses a parachute to reduce the stopping distance s after landing as shown in Figure P14.28. If the plane touches ground at a speed of 250 km/h, and, if resistance to motion, including air drag, is approximated by the relation $R = (10000 + 0.5v^2)$ N where v is the speed of the plane in m/s, determine the stopping distance s.

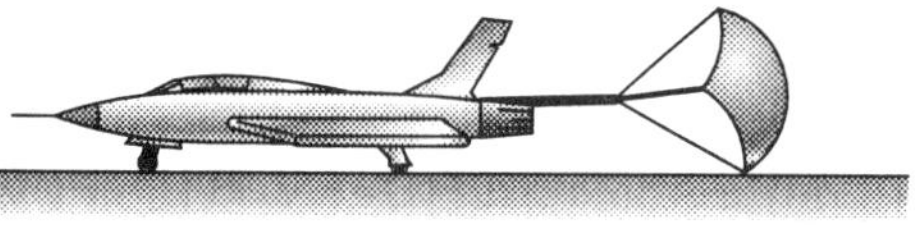

FIGURE P14.28.

14.29 A block of weight $W = 10$ lb slides in a straight line along a horizontal plane under the influence of a force $F = (2 + t^2)$ lb where t is the time in seconds. The coefficient of kinetic friction between the block and the plane is $\mu = 0.3$. If the block starts with an initial velocity $v_0 = 5$ ft/s, determine its final velocity after 4 s.

14.30 The slider shown in Figure P14.30 moves in the frictionless horizontal slot under the influence of the spring whose spring constant is $k = 15$ N/m. The mass of the slider $m = 6$ kg, and the unstretched length of the spring $L_u = 0.5$ m. Determine the velocity of the slider at point B if it is released from rest at point A.

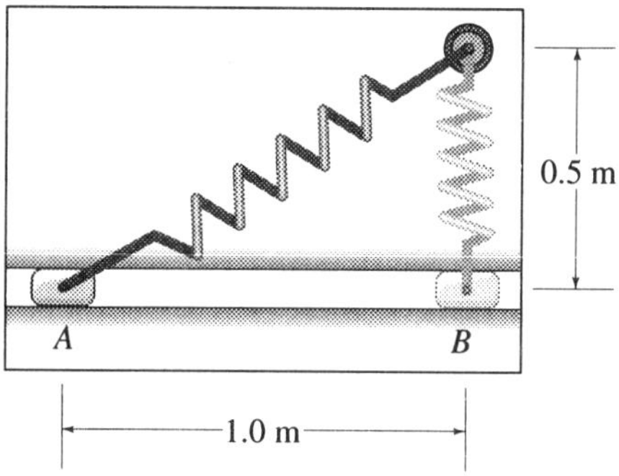

FIGURE P14.30.

14.31 Block A, which weighs 100 lb, is placed on block B which weighs 300 lb. The system is placed on a 20° inclined plane and attached to a pulley arrangement as shown in Figure P14.31. The coefficient of kinetic friction between blocks A and B is 0.25 and that between block B and

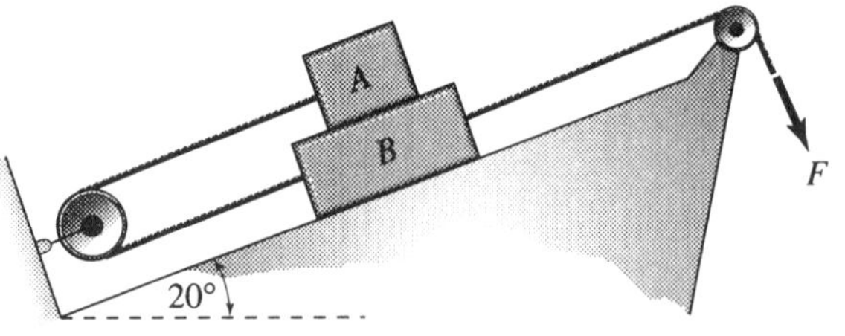

FIGURE P14.31.

the inclined plane is 0.15. Determine the magnitude of the force F needed to produce an acceleration of block B equal to 5 ft/s^2. What would be the acceleration of block A? Assume weightless and frictionless pulleys.

14.32 Consider the system shown in Figure P14.32 in which $m_A = 50$ kg, $m_B = 40$ kg, and $m_C = 40$ kg. The coefficient of kinetic friction between the inclined plane and block B is $\mu = 0.2$. Determine the acceleration of block A and the tension in cables AB and BC after the system is released from rest.

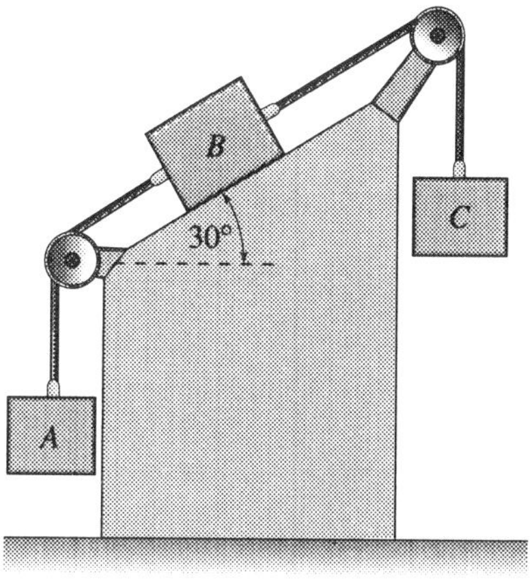

FIGURE P14.32.

14.4 Newton's Second Law in Normal and Tangential Components

In certain cases dealing with the motion of a particle of mass m along a curved path, it is convenient to express Newton's second law of motion in terms of normal and tangential components. Consider, for example, a particle of mass m moving along a curved path as shown in Figure 14.5. For any position of the particle along its path, a rectangular t-n-b coordinate system moving with the particle may be established. The coordinate t is referred to as the tangential coordinate and is obviously along the tangent to the path and pointed in the direction of motion. The coordinate n is known as the principal normal coordinate and is always directed toward the center of curvature of the path, point C. As

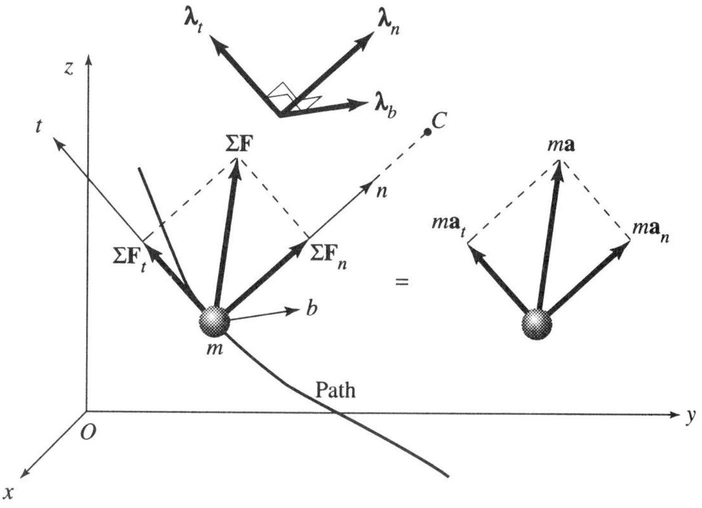

FIGURE 14.5.

discussed in Section 13.8, the coordinate b is called the binormal coordinate and is perpendicular to the *osculating* plane formed by t and n. Note that $\boldsymbol{\lambda}_b = \boldsymbol{\lambda}_t \times \boldsymbol{\lambda}_n$. Note also that, because the particle is constrained to move along the path, no acceleration occurs in the b direction and, therefore, $\sum \mathbf{F}_b = m\mathbf{a}_b = \mathbf{0}$. Thus, as seen from Figure 14.5, the resultant force $\sum \mathbf{F}$ may be expressed in terms of only two components in the form

$$\sum \mathbf{F} = (\sum F_n)\boldsymbol{\lambda}_n + (\sum F_t)\boldsymbol{\lambda}_t$$

where $\sum F_n$ and $\sum F_t$ represent, respectively, the sum of *all* forces along the normal and tangential directions, and $\boldsymbol{\lambda}_n$ and $\boldsymbol{\lambda}_t$ are unit vectors along these two directions, respectively. Note that, unlike the **i**, **j** and **k** unit vectors, $\boldsymbol{\lambda}_n$ and $\boldsymbol{\lambda}_t$ move with the particle and assume different orientations during the motion. Also, the resultant acceleration **a** may be expressed in terms of its two components in the form

$$\mathbf{a} = (a_n)\boldsymbol{\lambda}_n + (a_t)\boldsymbol{\lambda}_t.$$

Therefore, Newton's second law of motion may be written as

$$(\sum F_n)\boldsymbol{\lambda}_n + (\sum F_t)\boldsymbol{\lambda}_t = m[(a_n)\boldsymbol{\lambda}_n + (a_t)\boldsymbol{\lambda}_t]. \tag{14.14}$$

A graphical representation of Eq. (14.14) is shown in Figure 14.5 in which the diagram on the left side of the equal sign depicts the free-body diagram of the particle (the left-hand side of Eq. (14.14), and the diagram on the right side of the equal sign shows the inertial force vector (the right-hand side of Eq. (14.14)). Note that the magnitude of the resultant force, $\mathbf{R} = \sum \mathbf{F}$ and that of the resultant acceleration **a** are given by

$$R = \sqrt{[\sum F_n]^2 + [\sum F_t]^2} \tag{14.15}$$

and

$$a = \sqrt{[a_n]^2 + [a_t]^2}. \tag{14.16}$$

It is obvious that, if Eq. (14.14) is to be satisfied, the coefficients of the $\boldsymbol{\lambda}_n$ terms on both sides of the equal sign must be equal to each other and, similarly, for the $\boldsymbol{\lambda}_t$ terms. Thus, the following two scalar equations of motion are obtained:

$$\sum F_n = ma_n,$$

and

$$\sum F_t = ma_t. \tag{14.17}$$

As discussed in Chapter 13, the normal component of the acceleration a_n may be expressed in the following form:

$$a_n = \rho\omega^2 = \frac{v^2}{\rho}$$

where ρ is the radius of curvature of the path, v the linear speed, and ω the angular speed of the line connecting the particle to the center of curvature of the path. The specific form used would depend upon the requirements of a given problem. Also, as was discussed in Chapter 13, the normal component of the acceleration a_n is always pointed toward the center of curvature of the path. It follows, then, by the first of Eqs. 14.17, that the force $\sum F_n$ must also be pointed toward the center of curvature of the path as shown in Figure 14.5. In the particular case when the particle is moving along a *circular* path, the force $\sum F_n$ is called the *centripetal* force. Also, the reversed inertial force vector ma_n, which, by Newton's second law of motion, is equal and opposite to $\sum F_n$, is given the name *centrifugal* force.

The tangential component of acceleration a_t represents the time rate of change in the speed of the particle and is directed along the path. This component of acceleration may be stated in two different forms as follows:

$$a_t = \frac{dv}{dt} = \frac{d^2 s}{dt^2} = \rho\alpha$$

where $\alpha = a_t/\rho$ is the angular acceleration of the line connecting the particle to the path's center of curvature. The specific form used would depend upon the requirements of a given problem. We should note that, by the second of Eqs. (14.17), if the force $\sum F_t$ acts opposite to the direction of motion, then, the quantity a_t is a negative acceleration or a deceleration.

Occasionally, it becomes necessary to find the radius of curvature for a given path described by the relation $y = f(x)$. This is possible by using the following expression which may be found in any analytic geometry textbook:

$$\rho = \left| \pm \frac{\left[1 + \left(\frac{dy}{dx}\right)^2\right]^{3/2}}{\left(\frac{d^2 y}{dx^2}\right)} \right|.$$

The following examples illustrate the use of some of the above concepts.

Example 14.7

One end of a 20-ft string is attached to a model airplane and the other to a vertical pole as shown in Figure E14.7(a). Starting with an initial velocity of 15 ft/s, the airplane is made to fly at a constant acceleration $a_t = 0.5$ ft/s^2 along a circular path that lies in a horizontal plane. The string has a breaking strength of 10 lb, and the weight of the model airplane is 1.5 lb. Assume negligible air resistance and determine (a) the time it would take for the string to break (b) the force that propels the plane and (c) the magnitude of the lift force that keeps the model airplane in level flight.

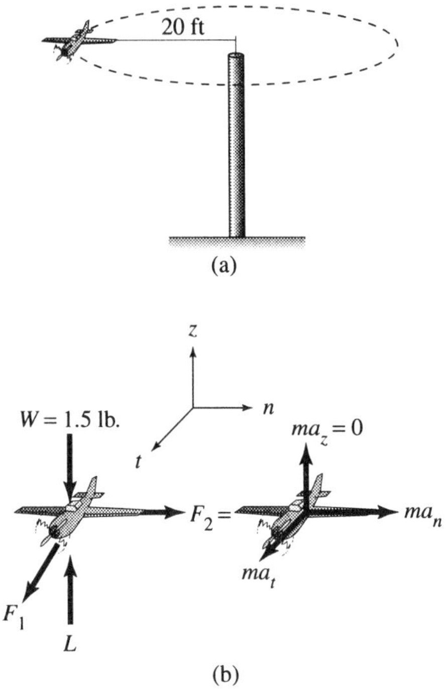

FIGURE E14.7.

Solution

The free-body diagram of the model airplane and the inertial force vector components in the normal and tangential directions are shown in Figure E14.7(b). The force L represents the lift that keeps the model airplane in level flight, the force F_1 is the thrust that propels it, and the force F_2 is the tension in the string that constrains the plane to fly in a circular path. Thus

(a)

$$\sum F_n = ma_n,$$

$$F_2 = (1.5/32.2)(v^2/20). \tag{a}$$

Setting $F_2 = 10$ lb, Eq. (a) may be used to find the speed v of the plane at which the string breaks. Thus,

$$v = 65.5 \text{ ft/s}.$$

Because the acceleration is constant,

$$v = v_0 + at,$$

$$65.5 = 15 + 0.5t$$

from which

$$t = 101.0 \text{ s}. \qquad \text{ANS.}$$

(b)

$$\sum F_t = ma_t,$$

$$F_1 = (1.5/32.2)(0.5),$$

$$F_1 = 0.023 \text{ lb.} \qquad \text{ANS.}$$

(c)

$$\sum F_z = ma_z = 0,$$

$$L - 1.5 = 0,$$

$$L = 1.5 \text{ lb.} \qquad \text{ANS.}$$

Example 14.8

A stunt car of mass m moves along a vertical loop with a constant speed v. The vertical loop is circular with radius R as shown in Figure E14.8(a). Neglect friction and air resistance, and, if the stunt car is to stay in contact with the loop all around its circumference, determine (a) the minimum constant speed v and (b) the force between the stunt car and the loop at point A. Express answers in terms of R, m and g.

Solution

(a) The minimum speed is determined such that the stunt car is just about to separate from the loop at its highest point. Thus, for minimum speed, the normal force N between the stunt car and the loop at point B is zero. The free-body diagram of the stunt car at point B and its inertial force vectors are shown in Figure E14.8(b). Note that because the speed is constant, $a_t = 0$ and the corresponding force F_t is also zero. Thus,

$$\sum F_n = ma_n,$$

$$mg = m(v^2 R).$$

Therefore,

$$v = \sqrt{gR}. \qquad \text{(a)} \quad \text{ANS.}$$

(b) The free-body diagram of the stunt car at point A and its inertial force vectors are shown in Figure E14.8(c). Therefore,

$$\sum F_n = ma_n,$$

$$N - mg = m(v^2/R)$$

from which

$$N = m(g + v^2/R). \qquad \text{(b)}$$

Substituting the value of v from Eq. (a) in Eq. (b) yields

$$N = 2mg. \qquad \text{ANS.}$$

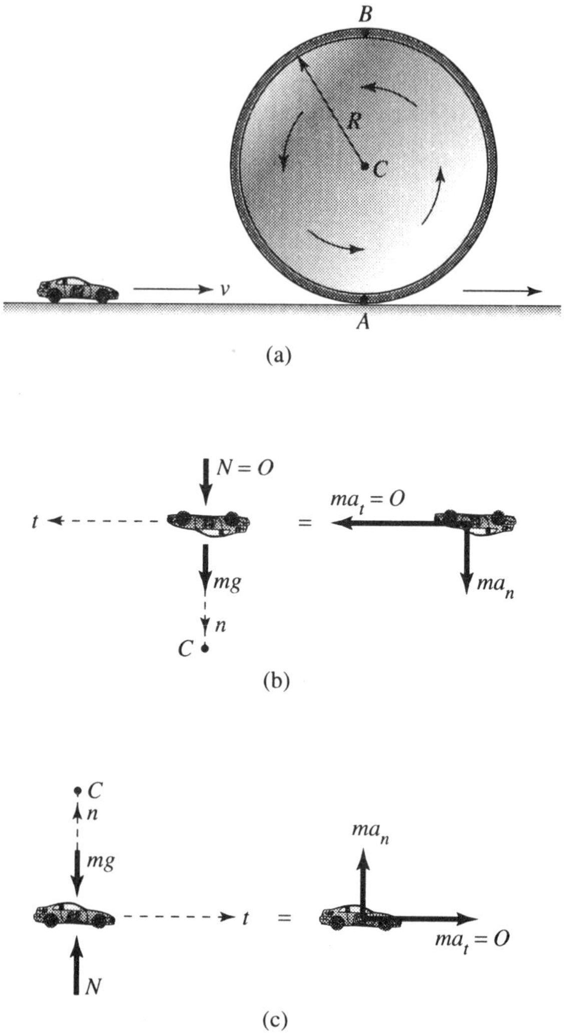

FIGURE E14.8. (c)

■ Example 14.9

The driver of a car weighing 750 N maintains a constant speed as he drives down dip ABC of the highway shown in Figure E14.9(a). The radius of curvature of the dip in the neighborhood of point B is 500 m. (a) What is this constant speed if the apparent weight of the driver at point B is 1000 N? (b) If he continues at the same constant speed over hump CDE and his apparent weight at point D is 400 N, determine the radius of curvature of the hump in the neighborhood of point D.

Solution

(a) The free-body diagram of the driver at point B and the inertial force vectors are shown in Figure E14.9(b). The normal force N repre-

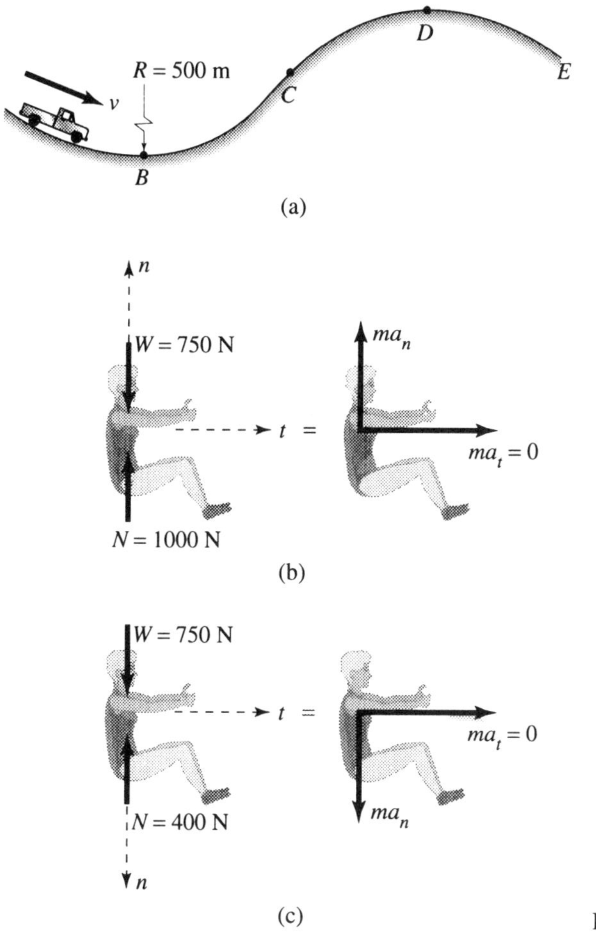

FIGURE E14.9.

sents the reaction between the driver and the car and is, therefore, his apparent weight. Also, because the speed is constant, $a_t = 0$. Thus,

$$\sum F_n = ma_n,$$

$$100 - 750 = \frac{750}{9.81}\left(\frac{v^2}{500}\right)$$

from which

$$v = 40.4 \text{ m/s} = 145.4 \text{ km/h}. \qquad \text{ANS.}$$

(b) The free-body diagram of the driver at point D and in inertial force vectors are shown in Figure E14.9(c). Thus,

$$\sum F_n = ma_n,$$

$$750 - 400 = \frac{750}{9.81}\left(\frac{40.4^2}{\rho}\right)$$

from which

$$\rho = 357 \text{ m.} \qquad \text{ANS.}$$

■ Example 14.10

Pendulum CB is held in position by cord AB as shown in Figure E14.10(a). The weight of the bob B is W and the length of weightless cord CB is L. If cord AB is suddenly cut, determine the speed v of the bob and the tension T in cord CB for any angle α, as the pendulum swings about its pivot at C. Express your answers in terms of α, α_0, L, g, and W. Form the ratio T/W and plot vs. α for values of α between the limits $\alpha_0 = 20°$ and $90°$.

Solution

The free-body diagram of the bob and the corresponding inertial force vectors are shown in Figure E14.10(b) for any position defined by the angle α. The force T represents the tension in cord CB. Thus,

$$\sum F_n = ma_n,$$

$$T - W \sin \alpha = \frac{W}{g}\left(\frac{v^2}{L}\right)$$

from which

$$T = W\left[\sin \alpha + \frac{v^2}{gL}\right] \tag{a}$$

where v is the instantaneous speed of the bob as it swings about its pivot at C. Also,

$$\sum F_t = ma_t,$$

$$W \cos \alpha = \frac{W}{g} a_t$$

from which

$$a_t = g \cos \alpha. \tag{b}$$

Now, $v\,dv = a_t\,ds$, and because $ds = L\,d\alpha$, it follows that

$$v\,dv = a_t L\,d\alpha. \tag{c}$$

Substitute Eq. (b) in Eq. (c) to obtain

$$v\,dv = gL \cos \alpha\,d\alpha. \tag{d}$$

Integrating,

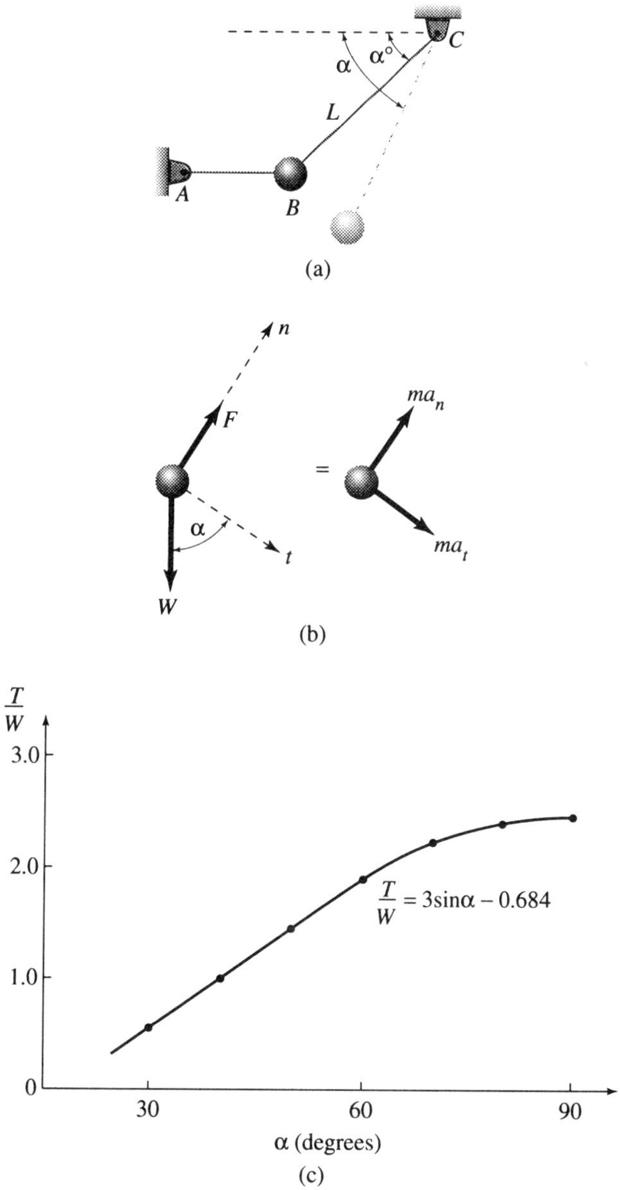

FIGURE E14.10.

$$\int_0^v v\,dv = gL\int_{\alpha_0}^{\alpha}\cos\alpha\,d\alpha$$

from which

$$v = \sqrt{2gL(\sin\alpha - \sin\alpha_0)}. \qquad \text{(e)} \quad \text{ANS.}$$

Substituting Eq. (e) in Eq. (a),

$$T = W(3\sin\alpha - 2\sin\alpha_0). \qquad \text{(f)} \quad \text{ANS.}$$

Note that the angle α_0 represents the initial position of the pendulum. From Eq. (f),

$$T/W = 3\sin\alpha - 2\sin\alpha_0. \tag{g}$$

If $\alpha_0 = 20^\circ$ is substituted in Eq. (g), we obtain

$$T/W = 3\sin\alpha - 0.684. \tag{h}$$

Equation (h) is plotted in Figure E14.10(c) for values $20^\circ \leq \alpha \leq 90^\circ$. Note the rapid increase in the dimensionless ratio T/W from a value of 0.342 at $\alpha = 20^\circ$ to a value of 2.32 at $\alpha = 90^\circ$.

Example 14.11

A block of weight $W = 10$ lb is placed on plank AB which is pivoted at point A as shown in Figure E14.11(a). The coefficient of friction between the block and the plank is $\mu = 0.1$. The block is attached to a string whose breaking strength is $T = 3$ lb. If the plank starts from rest, when $\theta = 0$, rotates in a vertical plane in a ccw manner with a constant angular acceleration α, and if the string breaks when $\theta = 75^\circ$, determine the magnitude of α.

Solution

The free-body diagram of the block for any arbitrary angle between 0° and 75° is shown in Figure E14.11(b) along with the corresponding inertial force vectors. Thus,

$$\sum F_n = ma_n,$$

$$T + \mu N + W\sin\theta = (W/g)R\omega^2. \tag{a}$$

Also, because $a_t = R\alpha$,

$$\sum F_t = ma_t,$$

$$N - W\cos\theta = (W/g)R\alpha$$

from which

$$N = W\cos\theta + (W/g)R\alpha. \tag{b}$$

Substitute Eq. (b) in Eq. (a) to obtain

$$T + W(\mu\cos\theta + \sin\theta) = (W/g)R(\omega^2 - \mu\alpha). \tag{c}$$

Because the angular acceleration is constant, it follows that $\omega^2 = \omega_0^2 + 2\alpha\theta$ where $\omega_0 = 0$. Thus,

$$\omega^2 = 2\alpha\theta \tag{d}$$

Substituting Eq. (d) in Eq. (c),

$$T + W(\mu\cos\theta + \sin\theta) = (W/g)R(2\alpha\theta - \mu\alpha)$$

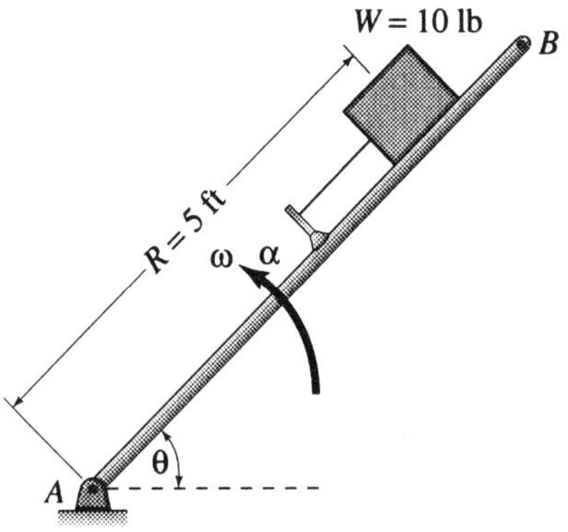

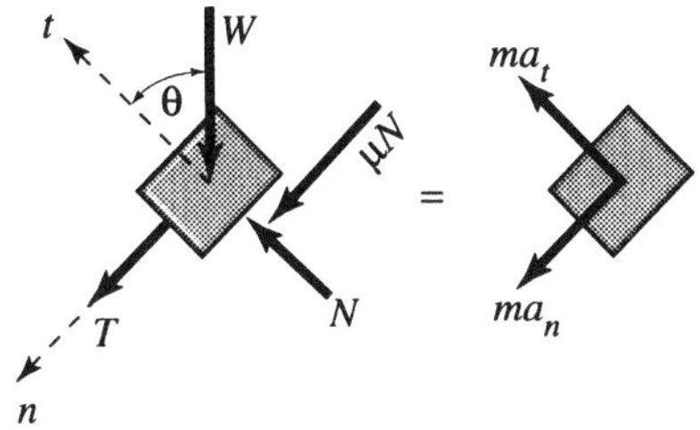

FIGURE E14.11.

which, when solved for α, yields

$$\alpha = \frac{g[T + W(\mu\cos\theta + \sin\theta)]}{WR(2\theta - \mu)}. \tag{e}$$

If the given numerical values are substituted in Eq. (e),

$$\alpha = 3.30 \text{ rad/s}^2. \qquad \text{ANS.}$$

■

Problems

14.33 A person loosely attaches one end of a string of length $L = 18$ in. to his index finger and the other end to a small stone of weight $W = 2$ oz. He imparts a constant angular speed $\omega = 6$ rad/s to the string-stone system to make it rotate in a vertical plane describing a circular path. Determine the force that the string exerts on his finger when the stone is (a) at its highest point and (b) at its lowest point.

14.34 Refer to Problem 14.33 and determine the constant angular speed ω that must be maintained if the finger is to experience no force from the string when the stone is at its highest point.

14.35 A bob B of mass m is attached to a cord BA of length L as shown in Figure P14.35 and made to rotate in a horizontal plane at constant speed v. (a) Determine an expression for this constant speed v in terms of g, L, and θ. (b) If $L = 1.5$ m and $\theta = 60°$, find the constant angular speed at which the cord-bob system must rotate about the vertical axis through the pivot at A.

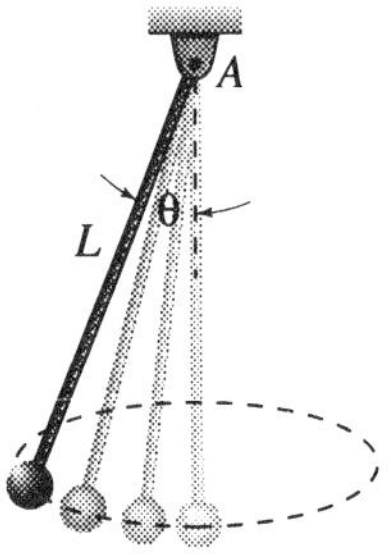

FIGURE P14.35.

14.36 A block of weight $W = 12$ lb is placed on board AB which can rotate freely about the hinge at B in a vertical plane as shown in Figure P14.36. The coefficient of friction between the block and the board is μ. The board starts from rest, when $\theta = 0$, rotating in a cw direction at a constant angular acceleration $\alpha = 4$ rad/s^2. If the block is on the verge of motion with respect to the board when θ reaches a value of 60°, determine the coefficient of friction.

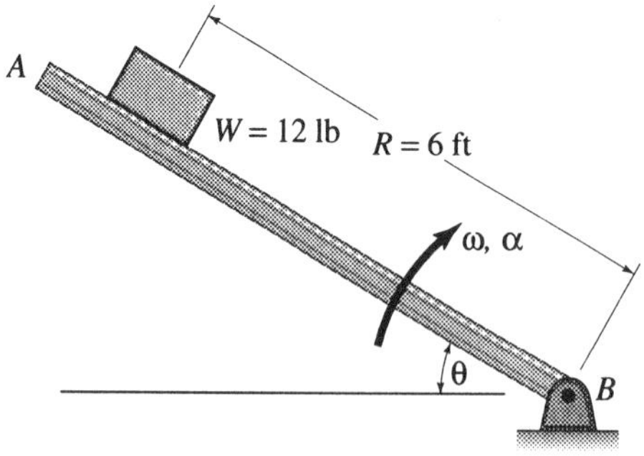

FIGURE P14.36.

14.37 A locomotive engine weighing 20,000 lb is traveling along the horizontal, curved tracks AB shown in Figure P14.37. At point A where the radius of curvature is 1,200 ft, it has a constant speed of 60 mph. The engineer applies the brakes and slows down the locomotive so that, when it reaches point B, where the radius of curvature is 300 ft, it has a constant speed of 40 mph. Determine the horizontal force exerted by the tracks on the wheels of the locomotive at (a) point A and (b) point B.

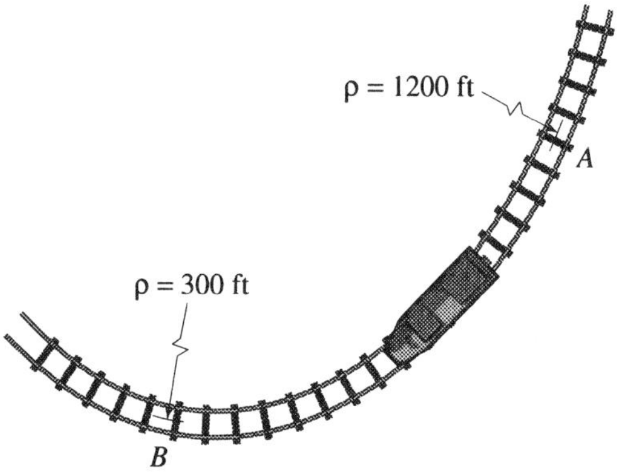

FIGURE P14.37.

14.38 The pendulum shown in Figure P14.38 consists of a bob of weight W and a rigid rod of length L. It is released from rest in the position when $\theta = 0$. (a) Develop symbolic expressions, in terms of W, L, g, and θ for the speed v of the bob and the tension T in the rod. (b) If $W =$ 20 N, $L = 2$ m and $\theta = 45°$, find numerical values for v and T. (c) What are the

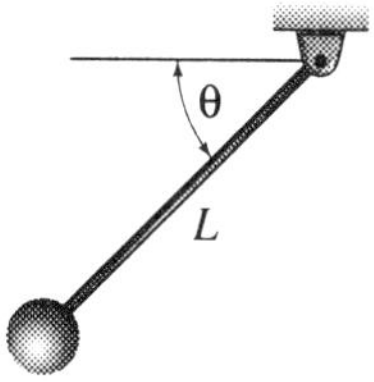

FIGURE P14.38.

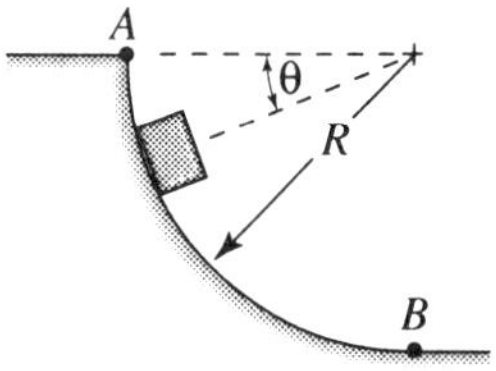

FIGURE P14.39.

values of θ ($0^\circ \le \theta \le 90^\circ$) for which v and T assume their maximum magnitudes. What are these maximum magnitudes?

14.39 A small package of weight W is released from rest at point A down the smooth circular surface AB whose radius is R as shown in Figure P14.39. (a) Develop symbolic expressions, in terms of W, R, g, and θ, for the speed v of the package and for the force N between it and the smooth surface. (b) If $W = 5$ lb, $R = 10$ ft and $\theta = 30^\circ$, find v and N. (c) What are the values of θ ($0^\circ \le \theta \le 90^\circ$) for which v and N attain their maximum magnitudes. What are these maximum magnitudes?

14.40 A stunt driver is to drive his car on the inside surface of a vertical cylindrical structure of radius R as shown in Figure P14.40. (a) Develop a symbolic expression for the minimum radius R in terms of the coefficient of friction μ between the tires and the wall and the constant speed of the car v. (b) What is the value of R if $\mu = 0.5$ and $v = 70$ km/h.

14.41 A stunt car of total weight W passes point A at the bottom of a frictionless circular path with a speed v, as shown in Figure P14.41. If the engine is turned off

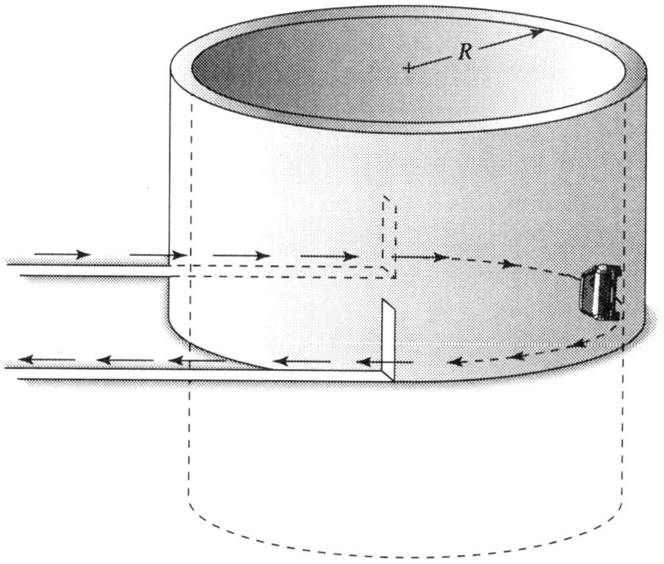

FIGURE P14.40.

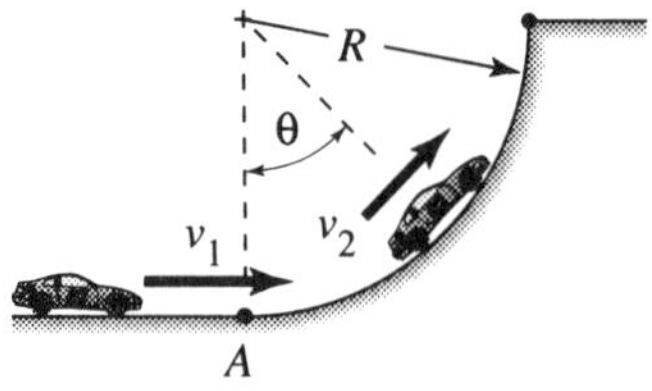

FIGURE P14.41.

at point A, derive symbolic relations for the angle θ and for the normal force N between the car and the path, for any speed v_2 along the circular path. Express your answers in terms of W, R, v_1, v_2, and g. What are the values of θ and N when $v_2 = 0$. Let $W = 3{,}000$ lb, $R = 200$ ft, and $v_1 = 75$ mph.

14.42 A person skis down the smooth slope AB which may be approximated by the equation $y = 0.05x^2$, as shown in Figure P14.42. If he starts from rest at point A, determine (a) his speed at point B and (b) the force between him and the snow at point B. The mass of the person and his equipment $m = 80$ kg. *Hint: Integrate the expression* $v\,dv = a\,ds$ *to find the speed at B.*

14.43 A stunt pilot of weight $W = 160$ lb flies his plane at a constant speed $v = 120$ mph in a vertical circular loop as shown in Figure P14.43. Determine (a) the ra-

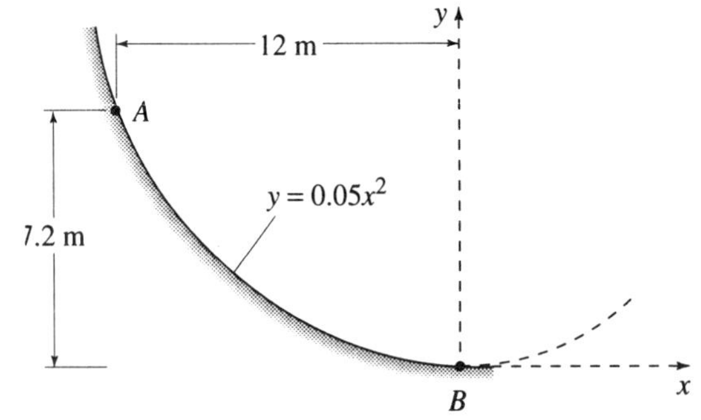

FIGURE P14.42.

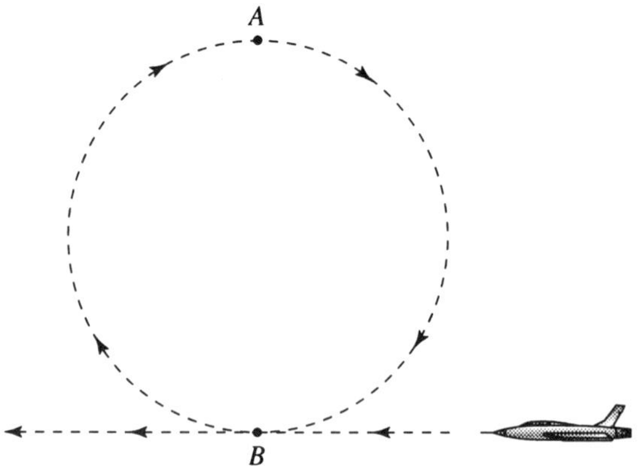

FIGURE P14.43.

dius of the circular loop if he experiences weightlessness (i.e., an apparent weight of zero) at the top of the loop (point A) and (b) his apparent weight at the bottom of the loop (point B).

14.44 A small body of weight $W = 10$ N is suspended by two wires AC and BC as shown in Figure P14.44. Determine the force in wire AC (a) immediately after wire BC is cut and (b) when it reaches the vertical position (i.e., when the body is directly under support A).

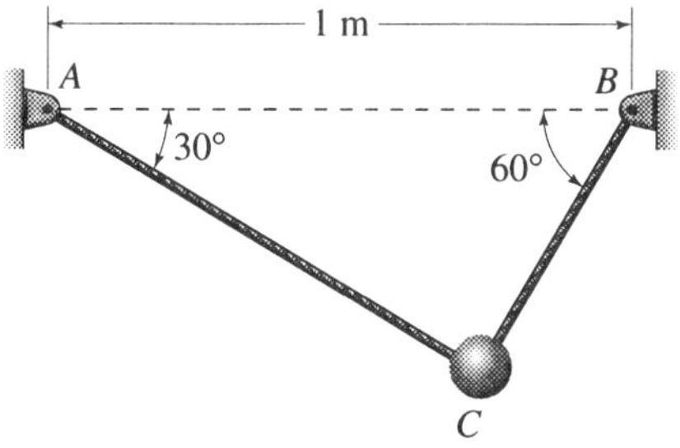

FIGURE P14.44.

14.45 A small body (particle) of weight $W = 10$ N is attached to two wires AC and BC, as shown in Figure P14.44. This particle is given a circular motion forcing it to rotate about a horizontal axis AB. If, at the bottom of the swing (the position shown), the force in wire BC is known to be 15 N, determine, for this position, the force in wire AC and the linear speed of the body.

14.46 Figure P14.46 represents, schematically, the mechanism for an amusement ride in a carnival. The mechanism rotates about vertical axis AB at a constant angular speed ω. (a) Develop a symbolic expression for ω in terms of θ, L, and g where θ and L are defined in Figure P14.46. (b) If $L = 5$ ft, find a numerical value for the angular speed ω in order

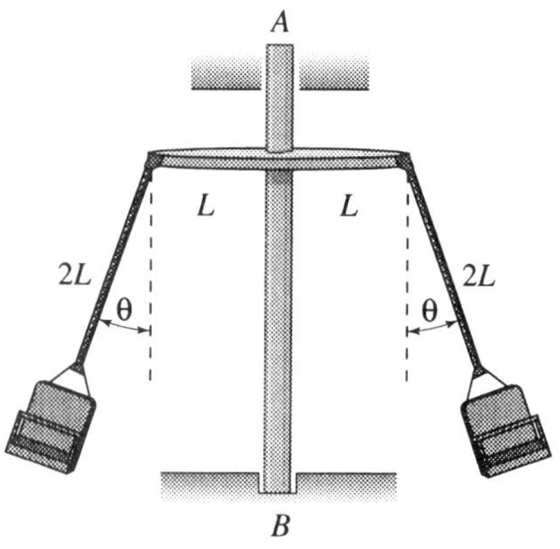

FIGURE P14.46.

for the angle θ to have values of 30° and 60°.

14.47 A racetrack is banked at an angle θ to the horizontal and has a radius of curvature R measured in a horizontal plane as shown in Figure P14.47. The coefficient of friction between the tires of the race car and the pavement is μ. (a) Develop a symbolic expression, in terms of R, θ, μ, and g, for the maximum speed of the car v so that it does not skid on the pavement away from the track's center of curvature. (b) Specialize this expression for the case when $\mu = 0$. Find, for this case, the needed bank angle θ if $v = 100$ mph and $R = 1000$ ft.

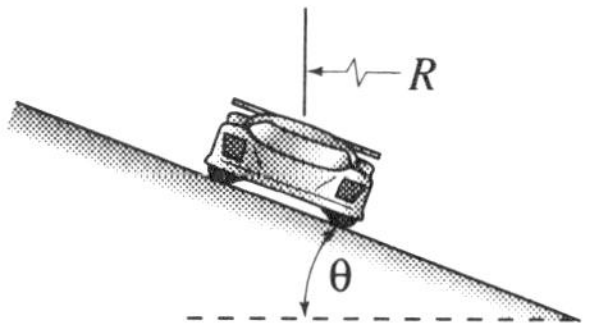

FIGURE P14.47.

14.48 Refer to Problem 14.47. Specialize the expression developed in part (a) for the case where $v = 150$ km/h, $R = 250$ m,

$\theta = 25°$, and determine the minimum coefficient of friction μ needed between the car tires and the pavement.

14.49 The conical container shown in Figure P14.49 rotates about its geometric vertical axis at a constant angular speed $\omega = 60$ rpm. A retaining wall keeps objects from 'flying' out during rotation. Spheres of weight $W = 10$ lb each contact the container at point A and the retaining wall at B. Determine the reactive forces at A and B for the case where $R = 3$ ft and $\theta = 50°$.

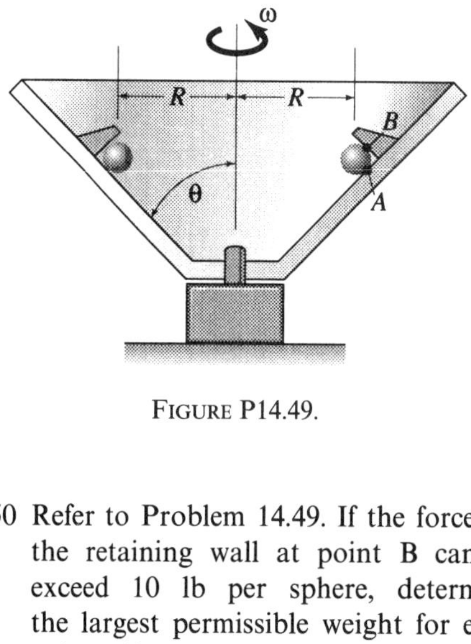

FIGURE P14.49.

14.50 Refer to Problem 14.49. If the force on the retaining wall at point B cannot exceed 10 lb per sphere, determine the largest permissible weight for each sphere if $\omega = 120$ rpm, $R = 2$ ft and $\theta = 40°$. What is the reaction experienced by the container under these conditions?

14.51 Arm AB rotates freely in a horizontal plane about a vertical axis at B as shown in Figure P14.51. A block of mass m is placed on the arm at a distance R from the axis of rotation. The block is restrained from movement in the radial direction by a string and by frictional forces. The breaking strength of the string is T_0, and the coefficient of friction is μ. Starting from rest, the arm is rotated at a constant angular acceleration α. Develop an expression, in terms of m, R, T_0, and α for the time t required for the block to overcome friction, break the string, and *fly* off the arm radially. Determine a numerical value for the time t for the case when $m = 10$ kg, $R = 0.75$ m, $T_0 = 200$ N, $\alpha = 2$ rad/s^2, and $\mu = 0.25$. In solving the problem, assume that friction is sufficiently large to keep the block from sliding off the arm tangentially. Check this assumption.

14.52 A communications satellite has a circular orbit around Earth and travels at a constant speed $v = 25{,}000$ km/h. If the mass of the earth is known to be $m = 5.976 \times 10^{24}$ kg, determine the radius of the circular orbit measured from the center of Earth. Recall that Newton's law of gravitation provides the force of attraction between any two bodies.

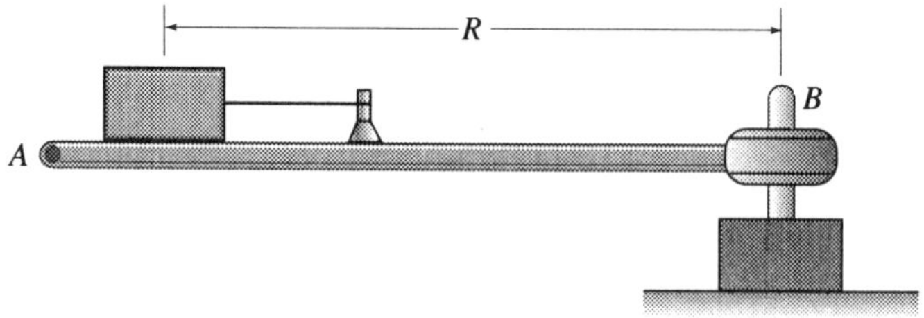

FIGURE P14.51.

14.53 An automobile travels along a road shaped in the form of a circle in the vertical plane as shown in Figure P14.53. The speed of the automobile is 40 mph at point A where it begins to accelerate until it reaches point B which is directly below the center of curvature of the road. The reaction between the driver and his seat at point B is 1.5 times that at point A. If $R = 200$ ft and $\theta = 50^\circ$, determine the speed of the automobile at point B.

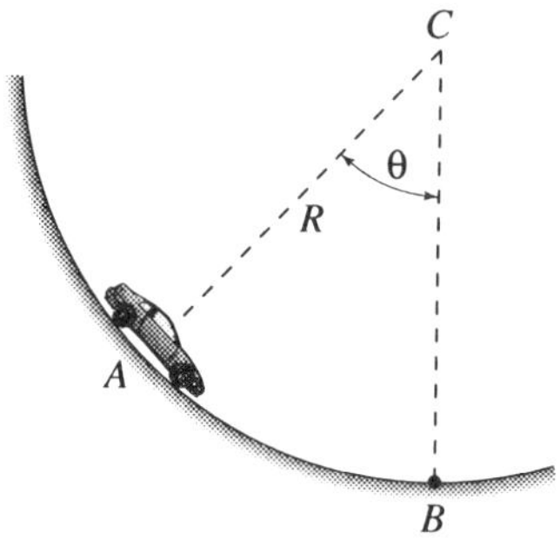

FIGURE P14.53.

14.54 A small object is released from rest at point A, slides down the frictionless circular path AB of radius R_1, and continues on the frictionless circular path BC of radius R_2, leaving it as a projectile at point D, as shown in Figure P14.54. Show that the angle α defining the location of point D is given by the relationship $\cos\alpha = \frac{2}{3}\left(1 + \frac{R_1}{R_2}\right)$. *Hint: First, determine the speed v_B at point B, and use it as the initial velocity for motion on path BC.*

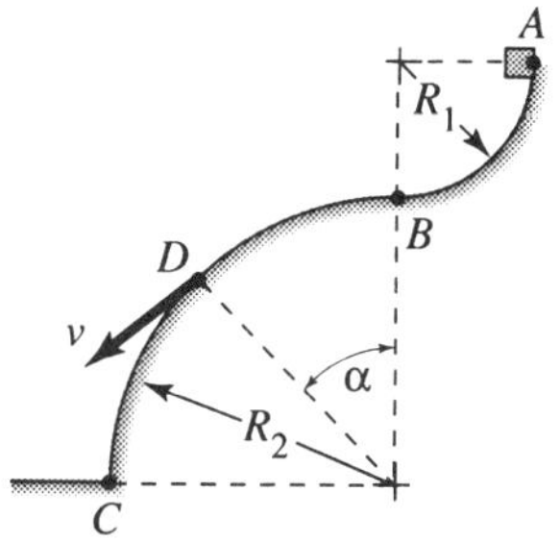

FIGURE P14.54.

14.55 Refer to Problem 14.54. (a) Determine the ratio R_1/R_2 so that the body slides down path AB and leaves the path as a projectile at point B. (b) If the body is released with a very small (negligible) speed from point B instead of point A, determine the angle α defining the location of point D at which it would leave the circular path.

14.56 A small body of mass m is held on the inside surface of a rotating conical container by centrifugal action as shown in Figure P14.56. If $\mu = 0.2$ is the coefficient of friction, $R = 0.5$ m, and $\theta = 50^\circ$, determine the value of the maximum rotational speed ω_{max} at which the body begins to slide up the inside surface of the container.

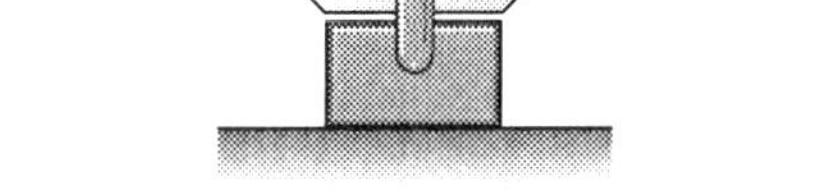

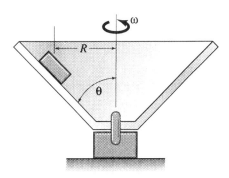

FIGURE P14.56.

14.57 Refer to Problem 14.56 and determine the value of the minimum rotational speed ω_{min} at which the body begins to slide down the inside surface of the container.

14.58 A conveyor system, used to transport small packages from one location to another in a factory, is shown in Figure P14.58. The conveyor belt and the packages move at a constant speed v and the packages begin to slip off the belt at point A where $\theta = 20°$. If the coefficient of friction between the belt and the packages is $\mu = 0.50$ and $r = 1.5$ ft, determine the constant speed of the belt v.

14.59 A section of a roller-coaster tracks is shown in Figure P14.59 and may be assumed to lie entirely in the vertical plane. Throughout this section of tracks, the cars maintain a constant speed v. (a) Determine the maximum permissible constant speed if the cars are not to leave the tracks at the top of the curve. (b) For the speed determined in part (a), find the normal force exerted on the tracks by each car if the mass of the car and its passengers $m = 200$ kg.

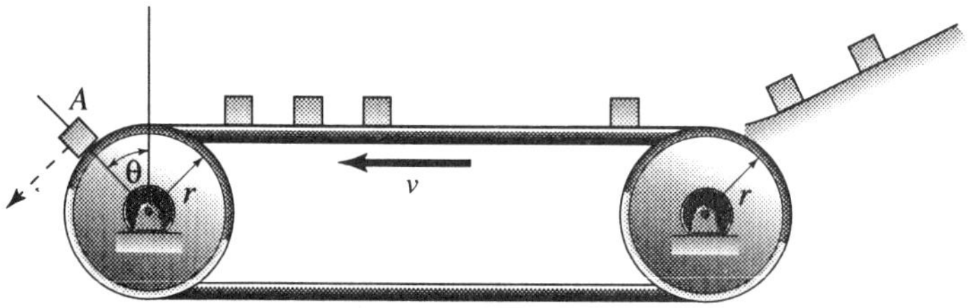

FIGURE P14.58.

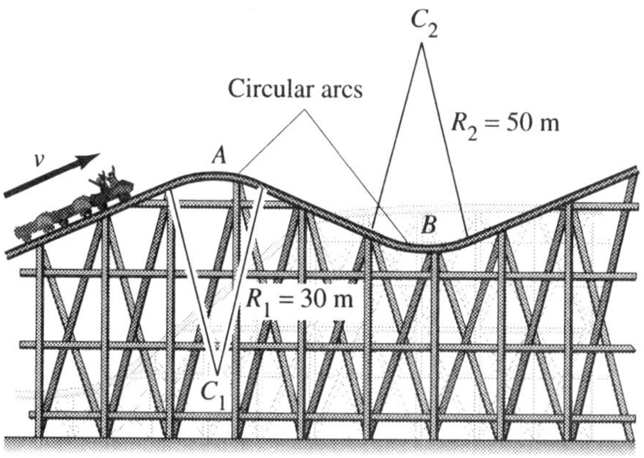

FIGURE P14.59.

14.60 A small collar may slide on rod AB which is a part of a mechanism that rotates about vertical axis AC as shown in Figure P14.60. The coefficient of friction between the collar and rod AB $\mu = 0.5$, and the inclination of the rod to the axis of rotation is θ. The mechanism rotates at a constant angular speed $\omega = 14.5$ rad/s. Find the needed value of the angle θ if the collar is not to slide with respect to rod AB if $L = 1.00$ m.

14.61 A small collar of weight $W = 2.0$ lb is

attached to a linear spring of spring constant k and may slide on the smooth rod AB which is a part of a mechanism that rotates about vertical axis AC as shown in Figure P14.61. The mechanism rotates at a constant angular speed $\omega = 20$ rad/s. If $\theta = 50°$, the unstretched length of the spring $L_u = 0.5$ ft, and the position of the collar when the mechanism is rotating is defined by $L = 2.0$ ft, determine the spring constant k.

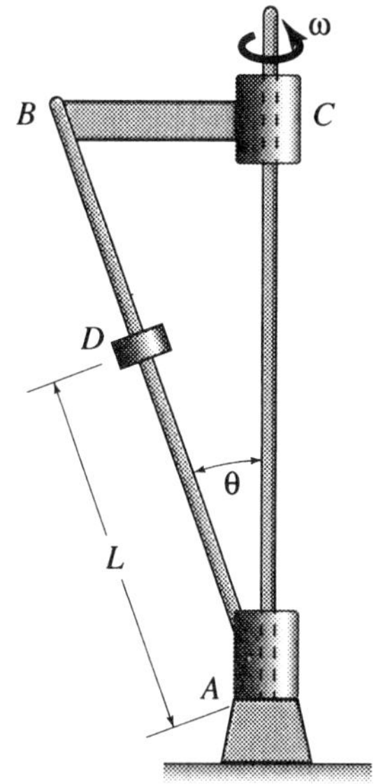

FIGURE P14.60.

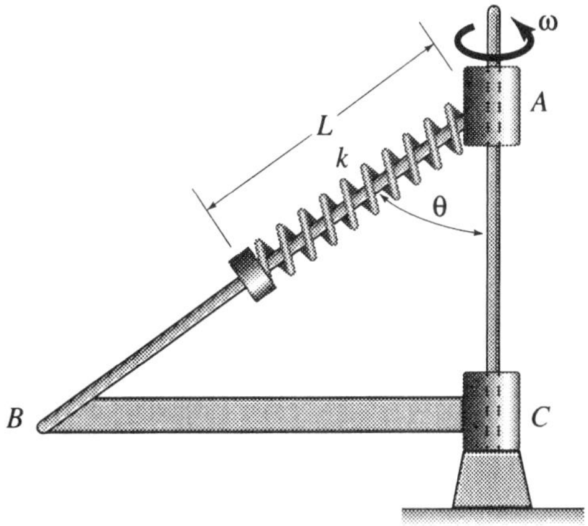

FIGURE P14.61.

14.5 Newton's Second Law of Motion in Cylindrical Components

In many cases dealing with the motion of a particle along a curvilinear path, its acceleration and the forces acting on it may be conveniently expressed in terms of cylindrical components. Thus, consider the case of a particle of mass m moving along a curved path as shown in Figure 14.6. The position of the particle at any instant is defined by the coordinates r, θ, and z. The resultant force $\sum \mathbf{F}$ acting on the particle may be expressed in terms of the three components $\sum \mathbf{F}_r$, $\sum \mathbf{F}_\theta$, and $\sum \mathbf{F}_z$ in the form

$$\sum \mathbf{F} = (\sum F_r)\boldsymbol{\lambda}_r + (\sum F_\theta)\boldsymbol{\lambda}_\theta + (\sum F_z)\boldsymbol{\lambda}_z$$

where $\sum F_r$, $\sum F_\theta$, and $\sum F_z$ represent, respectively, the sum of *all* forces in the radial, transverse, and vertical directions and $\boldsymbol{\lambda}_r$, $\boldsymbol{\lambda}_\theta$, and $\boldsymbol{\lambda}_z$ are unit vectors along these directions. Also, the resultant acceleration $\mathbf{a}$ may be expressed in terms of the three components $\mathbf{a}_r$, $\mathbf{a}_\theta$ and $\mathbf{a}_z$ in the form

$$\mathbf{a} = (a_r)\boldsymbol{\lambda}_r + (a_\theta)\boldsymbol{\lambda}_\theta + (a_z)\boldsymbol{\lambda}_z.$$

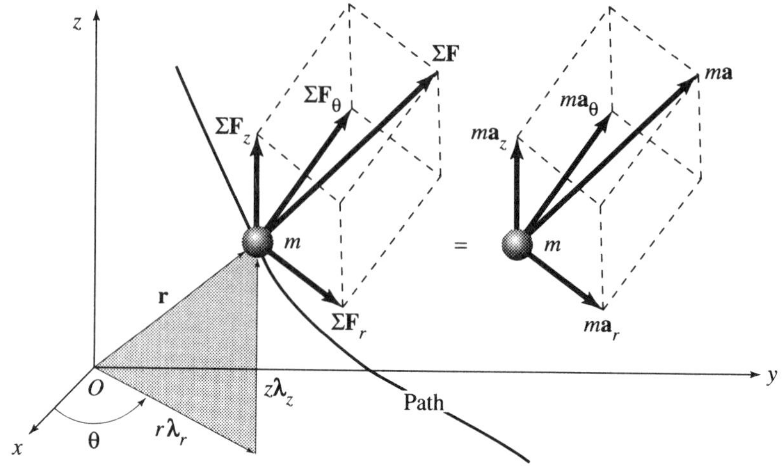

FIGURE 14.6.

Thus, Newton's second law of motion may be written as

$$(\sum F_r)\boldsymbol{\lambda}_r + (\sum F_\theta)\boldsymbol{\lambda}_\theta + (\sum F_z)\boldsymbol{\lambda}_z = m[(a_r)\boldsymbol{\lambda}_r + (a_\theta)\boldsymbol{\lambda}_\theta + (a_z)\boldsymbol{\lambda}_z]. \quad (14.19)$$

A graphical representation of Eq. (14.19) is shown in Figure 14.6 in which the diagram on the left side of the equal sign depicts the free-body diagram of the particle (the left-hand side of Eq. (14.19)) and the diagram on the right side of the equal sign depicts the inertial force vector (the right-hand side of Eq. (14.19)). Note that the magnitude of the resultant force $\mathbf{R} = \sum F$ and that of the resultant acceleration $\mathbf{a}$ are given by

$$R = \sqrt{[\sum F_r]^2 + [\sum F_\theta]^2 + [\sum F_z]^2} \quad (14.20)$$

and

$$a = \sqrt{[\sum a_r]^2 + [\sum a_\theta]^2 + [\sum a_z]^2}. \quad (14.21)$$

In order for Eq. (14.19) to be satisfied, the coefficients of the $\boldsymbol{\lambda}_r$ terms on both sides of the equation must be equal to each other and, similarly, for the $\boldsymbol{\lambda}_\theta$ and $\boldsymbol{\lambda}_z$ terms. Thus, the following three scalar equations of motion are obtained

$$\left.\begin{aligned} \sum F_r &= ma_r, \\ \sum F_\theta &= ma_\theta, \\ \sum F_z &= ma_z. \end{aligned}\right\} \quad (14.22)$$

where, as discussed in Chapter 13,

$$\left.\begin{aligned} a_r &= \ddot{r} - r\dot{\theta}^2 \\ a_\theta &= r\ddot{\theta} + 2\dot{r}\dot{\theta} \\ a_z &= \ddot{z} \end{aligned}\right\}. \quad (14.23)$$

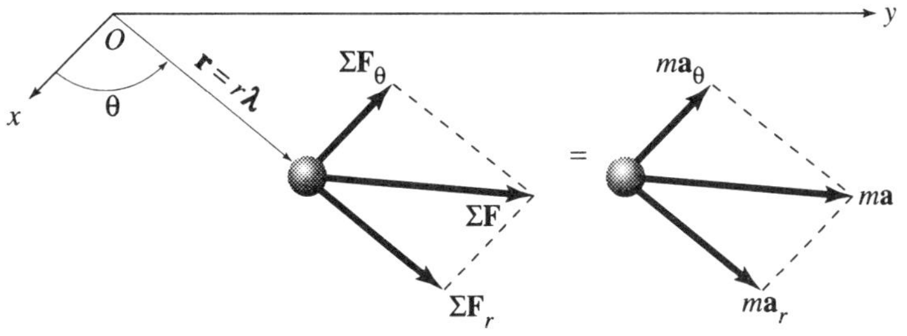

FIGURE 14.7.

A special two-dimensional case of Eqs. (14.22) deals with the motion of a particle in a plane, say the x-y plane, in which case the variable z is set equal to a constant. Thus, $a_z = \ddot{z} = 0$ and the third of Eqs. (14.22) reduces to $\sum F_z = ma_z = 0$ which refers to a condition of static equilibrium in the z direction. The remaining two equations define fully the motion of the particle in the plane in which case the motion is said to be described in terms of *polar coordinates* for which the equations of motion are

$$\left.\begin{aligned}\sum F_r &= ma_r\\ \sum F_\theta &= ma_\theta\end{aligned}\right\}. \tag{14.24}$$

These equations of motion are depicted graphically in the x-y plane of Figure 14.7.

Note that, when dealing with polar coordinates, the direction θ is sometimes referred to as the *transverse* direction. Note also that, for the very special case of a planar circular path centered at the origin of the coordinate system, the radial and transverse directions coincide, respectively, with the normal and tangential directions of the path. It follows, therefore, that Eqs. (14.17) represent a special case of Eqs. (14.24).

The following examples illustrate some of the above concepts.

Example 14.12

The three-dimensional motion of a particle is defined by the coordinates $r = 5t - t^2$, $\theta = 7t$ and $z = 3\sin \pi t$ where t is time in seconds, θ is in radians, and r and z are in inches. If the weight of the particle is 50 lb, determine the force components, F_r, F_θ, and F_z that act on the particle when (a) $t = 5$ seconds and (b) $t = 2.5$ seconds.

Solution

The first and second time derivatives of r, θ, and z are determined in terms of t and their values used to find the acceleration components a_r, a_θ, and a_z. Thus,

$$\begin{aligned}\dot{r} &= 5 - 2t; & \ddot{r} &= -2;\\ \dot{\theta} &= 7; & \ddot{\theta} &= 0;\\ \dot{z} &= 3\pi\cos \pi t; & \ddot{z} &= -3\pi^2 \sin \pi t.\end{aligned}$$

Therefore,

$$a_r = \ddot{r} - r\dot{\theta}^2 = -2 - (5 - 2t))(7)^2 = -247 + 98t,$$

$$a_\theta = r\ddot{\theta} + 2\dot{r}\dot{\theta} = (5t - t^2))(0) + 2(5 - 2t)(7) = 10 - 14t,$$

$$a_z = \ddot{z} = -3\pi^2 \sin \pi t.$$

(a) For $t = 5$ s,

$$a_r = 243.0 \text{ in./s}^2 = 20.3 \text{ ft/s}^2,$$

$$a_\theta = -60.0 \text{ in./s}_2 = -5.0 \text{ ft/s}^2,$$

$$a_z = 0.$$

The free-body diagram of the particle for any time t and the corresponding inertial force components are shown in Figure E14.12. Application of Newton's second law of motion yields

$$\sum F_r = ma_r,$$

$$F_r = (W/g)a_r,$$

$$F_r = \frac{50}{32.2}(20.3) = 31.4 \text{ lb.} \qquad \text{ANS.}$$

$$\sum F_\theta = ma_\theta,$$

$$F_\theta = (W/g)a_\theta,$$

$$F_\theta = \frac{50}{32.2}(-5.0) = -7.8 \text{ lb.} \qquad \text{ANS.}$$

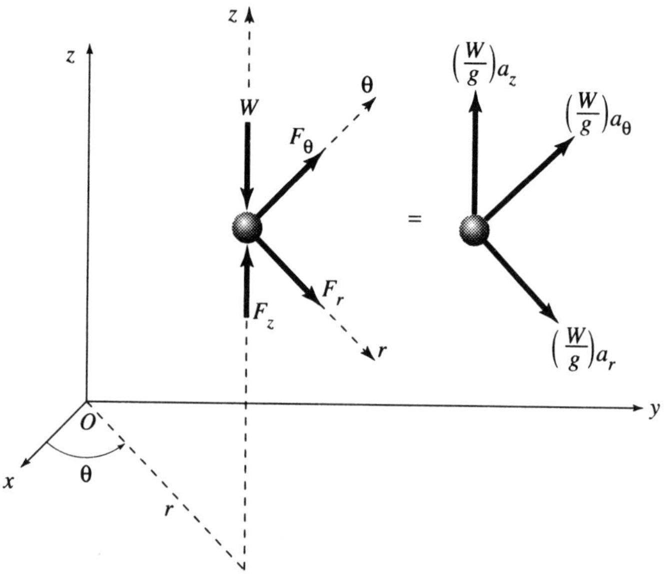

FIGURE E14.12.

$$\sum F_z = ma_z,$$
$$F_z - W = (W/g)a_z,$$
$$= (50/32.2)(0).$$

Therefore,

$$F_z = W = 50.0 \text{ lb.} \qquad \text{ANS.}$$

(b) For $t = 2.5$ s

$$a_r = -2.0 \text{ in./s}^2 = -0.2 \text{ ft/s}^2,$$
$$a_\theta = -25.0 \text{ in./s}^2 = -2.1 \text{ ft/s}^2,$$
$$a_z = 29.6 \text{ in./s}^2 = 2.5 \text{ ft/s}^2.$$

Thus,

$$\sum F_r = ma_r,$$
$$F_r = (W/g)a_r,$$
$$F_r = \frac{50}{32.2}(-0.20) = -0.300 \text{ lb.} \qquad \text{ANS.}$$
$$\sum F_\theta = ma_\theta,$$
$$F_\theta = (W/g)a_\theta,$$
$$F_\theta = \frac{50}{32.2}(-2.1) = -3.30 \text{ lb.} \qquad \text{ANS.}$$
$$\sum F_z = ma_z,$$
$$F_z - W = (W/g)a_z.$$

Solving for F_z gives

$$F_z = W(1 + a_z/g),$$
$$F_z = 50\left(1 + \frac{2.5}{32.2}\right) = 53.9 \text{ lb.} \qquad \text{ANS.}$$

Example 14.13

A skier of weight W slides down a small mountain side along a spiral path which may be approximated by the system shown in Figure E14.13(a). The spiral path is properly banked to keep the skier from 'flying off' as a projectile. His angular position is expressed by the equation $\theta = 1.2t$ where t is in seconds and θ in radians. The skier moves a distance q, measured along the z axis, every time he makes one complete revolution. In terms of t, D, q, W, and H, determine the components F_r, F_θ, and F_z of the force exerted by the path on the skier

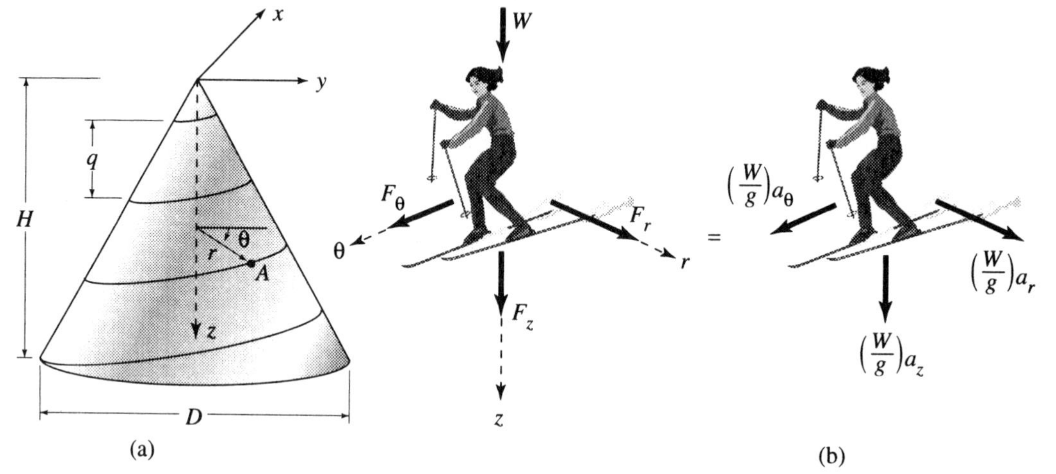

FIGURE E14.13.

for any position along the spiral path. Find these force components for the case where $W = 750$ N, $q = 20$ m, $D = 300$ m, $H = 400$ m, and $t = 10$ s.

Solution

The three cylindrical coordinates of the skier for any arbitrary position such as A on the spiral path expressed in terms of the time t are

$$\theta = 1.2t,$$

$$z = \left(\frac{q}{2\pi}\right)\theta,$$

$$z = 0.191qt,$$

$$r = \left(\frac{D}{2H}\right)z,$$

$$r = 0.096\left(\frac{D}{H}\right)qt.$$

The time derivatives of θ, z, and r are given below. Note that these derivatives are independent of the time t and, consequently, they have the same values for any position along the spiral path.

$$\dot{\theta} = 1.2; \qquad \ddot{\theta} = 0;$$

$$\dot{z} = 0.19q; \qquad \ddot{z} = 0;$$

$$\dot{r} = 0.096\left(\frac{D}{H}\right)q; \quad \ddot{r} = 0.$$

The acceleration components may now be determined as

$$\begin{aligned} a_\theta &= r\ddot{\theta} + 2\dot{r}\dot{\theta} \\ &= 0 + 2q\left(\frac{0.096D}{H}\right)(1.2) \\ &= 0.230\frac{Dq}{H}, \end{aligned}$$

$$a_z = \ddot{z} = 0,$$

$$\begin{aligned} a_r &= \ddot{r} - r\dot{\theta}^2 \\ &= 0 - \left(\frac{0.096D}{H}\right)(qt)(1.2)^2 \\ &= -0.138\frac{Dqt}{H}. \end{aligned}$$

The free-body diagram of the skier for the arbitrary position A as well as the inertial force vector components are shown in Figure E14.13(b). Thus,

$$\sum F_r = ma_r,$$

$$F_r = \frac{W}{g}\left(-0.138\frac{Dqt}{H}\right) = -0.138\left(\frac{WDqt}{gH}\right). \qquad \text{ANS.}$$

$$\sum F_\theta = ma_\theta,$$

$$F_\theta = \frac{W}{g}\left(0.230\frac{Dq}{H}\right) = 0.230\left(\frac{WDq}{gH}\right). \qquad \text{ANS.}$$

$$\sum F_z = ma_z,$$

$$W + F_z = 0,$$

$$F_z = -W. \qquad \text{ANS.}$$

The negative signs on F_z and F_r imply that these force components are opposite those shown in Figure E14.13(b). Thus, for the data provided,

$$\begin{aligned} F_r &= 0.138(750 \times 300 \times 20 \times 10/9.81 \times 400), \\ &= 1583\ \text{N}. \end{aligned} \qquad \text{ANS.}$$

$$\begin{aligned} F_\theta &= 0.230(750 \times 300 \times 20/9.81 \times 400), \\ &= 264\ \text{N}. \end{aligned} \qquad \text{ANS.}$$

$$F_z = 750\ \text{N}. \qquad \text{ANS.}$$

Example 14.14

A pin P of weight $W = 5$ lb is constrained to move along a circular path in the vertical plane while sliding freely in the slot of the rotating arm AB as shown in Figure E14.14(a). Arm AB rotates freely about the hinge at A such that $\theta = \pi t$ where t is the time in seconds and θ represents the angular position of the arm in radians. The diameter of the circular path is $D = 20$ in. Determine the forces exerted by the arm and by the circular path on the pin when (a) $\theta = 30°$ and (b) $\theta = 120°$.

Solution

From the geometry shown in Figure E14.14(a), the radial distance r from point A to pin P for any angle θ may be written in the form $r = D \sin \theta$. The time derivatives of r and those of θ are determined as follows:

$$\dot{r} = (D \cos \theta)\dot{\theta}; \quad \ddot{r} = (D \cos \theta)\ddot{\theta} - (D \sin \theta)\dot{\theta}^2;$$

$$\dot{\theta} = \pi; \qquad \ddot{\theta} = 0.$$

Therefore,

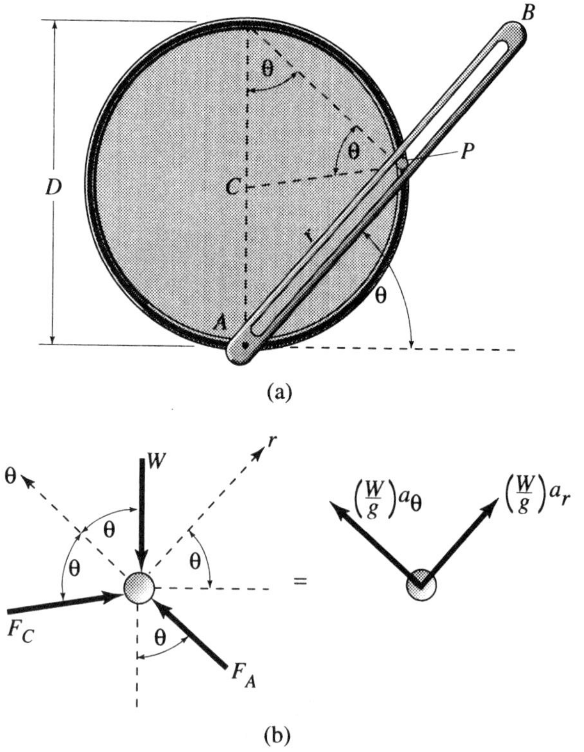

FIGURE E14.14.

$$a_r = \ddot{r} - r\dot{\theta}^2$$
$$= -2\pi^2 D \sin\theta,$$
$$a_\theta = r\ddot{\theta} + 2\dot{r}\dot{\theta}$$
$$= 2\pi^2 D \cos\theta.$$

The free-body diagram of pin P for any position defined by the angle θ is shown in Figure E14.14(b) along with the r and θ inertial force vectors. The forces F_A and F_C represent, respectively, the action of the arm and that of the circular path on pin P. Note that the force F_C has a direction from C to P in Figure E14.14(a) which makes the angle θ from the θ axis. Thus,

$$\sum F_r = ma_r,$$
$$F_C \cos(90 - \theta) - W\cos(90 - \theta) = (W/g)(-2\pi^2 D \sin\theta).$$

Because $\cos(90 - \theta) = \sin\theta$, it follows that

$$F_C = W[1 - (2\pi^2 D/g)]. \tag{a}$$

Clearly, the force exerted by the circular path is a constant regardless of the angle θ. Also

$$\sum F_\theta = ma_\theta,$$
$$F_A - W\cos\theta - F_C\cos\theta = (W/g)(2\pi^2 D \cos\theta).$$

Substituting for F_C from Eq. (a) and solving for F_A,

$$F_A = 2W\cos\theta. \tag{b}$$

Therefore, the force exerted by the rotating arm on pin P is a function of the angle θ.

(a) For $\theta = 30°$

$$F_C = 5\left[1 - \frac{2\pi^2(20)}{12(32.2)}\right] = -0.1085 \text{ lb} \qquad \text{ANS.}$$

where the negative sign indicates that F_C is directed toward the center of the circular path and not away from it, as assumed.

$$F_A = 2(5)\cos 30° = 8.66 \text{ lb.} \qquad \text{ANS.}$$

(b) For $\theta = 120°$,

$$F_C = -0.1085 \text{ lb as before,} \qquad \text{ANS.}$$

$$F_A = 2(5)\cos 120° = -5.00 \text{ lb} \qquad \text{ANS.}$$

where the negative sign implies that the force exerted by the arm on the pin is in the negative transverse direction and not as assumed in the free-body diagram.

■

Problems

14.62 Solve Problem 14.33 (p. 145) using polar coordinates.

14.63 Solve Problem 14.35 (p. 146) using polar coordinates.

14.64 Solve Problem 14.36 (p. 146) using polar coordinates.

14.65 Solve Problem 14.39 (p. 147) using polar coordinates.

14.66 Solve Problem 14.41 (p. 147) using polar coordinates.

14.67 Solve Problem 14.43 (p. 148) using polar coordinates.

14.68 The three dimensional motion of a particle is defined by the coordinates $r = 2t^2 - t^3$, $\theta = 3t^2$, and $z = 2\cos \pi t$ where t is time in seconds, θ is in radians, and r and z are in inches. If the weight of the particle is 40 lb, determine the force components, F_r, F_θ, and F_z that act on the particle when (a) $t = 1.5$ s and (b) $t = 4$ s.

14.69 A particle moves in space such that its coordinates are given by $r = 0.1t^2 - t$, $\theta = 3t$, and $z = 0.01t^3$ where t is time in

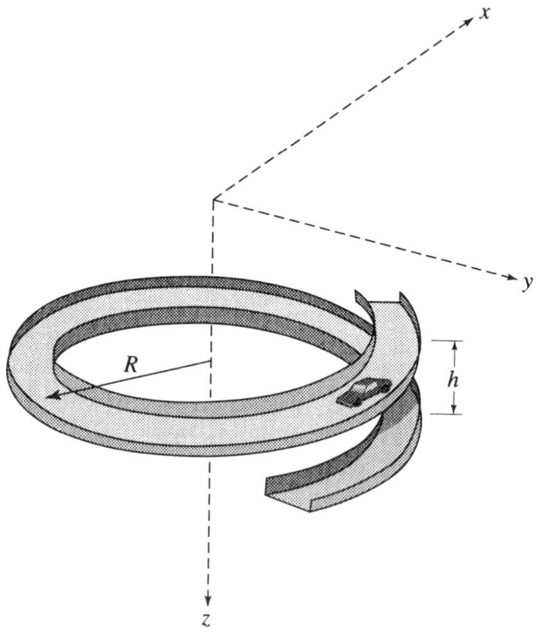

FIGURE P14.70.

seconds, θ is in radians, and r and z are in meters. If the particle has a mass of 4 kg, determine the force components, F_r, F_θ, and F_z that act on the particle when (a) $t = 3$ s and (b) $t = 8$ s.

14.70 An automobile of weight $W = 4000$ lb moves down the exit ramp of a parking deck which is a cylindrical spiral of radius $R = 100$ ft, a section of which is shown in Figure P14.70. The pitch (the vertical distance traveled for every complete revolution) is $h = 12$ ft. The automobile starts when $\theta = 0°$ and moves down such that $\theta = 0.2t$ where t is the time in seconds and θ is in radians. Consider the automobile when $t = 10$ s, and determine the force components F_r, F_θ, and F_z that it exerts on the ramp.

14.71 A particle of weight $W = 2$ lb is constrained to move such that its position can be defined by the coordinates $r = 10\sin^2\theta$, $\theta = 2t$, and $z = 13 + 2\sin\theta$ where r and z are in inches and θ in radians. Determine the force components F_r, F_θ, and F_z of the constraining force that act on the particle when $t = 20$ s.

14.72 A toy car of mass m moves along a spiral track on the inside surface of a conical depression as shown schematically in Figure P14.72. The angular position of the car is approximated by the equation $\theta = 0.1t^2$ where t is the time in seconds and θ is in radians. The pitch of the spiral track is h in meters (i.e., the vertical distance measured along the z axis for every complete revolution). In terms of t, D, h, m, and H, determine the components F_r, F_θ, and F_z of the force exerted by the track on the car for any position of the car.

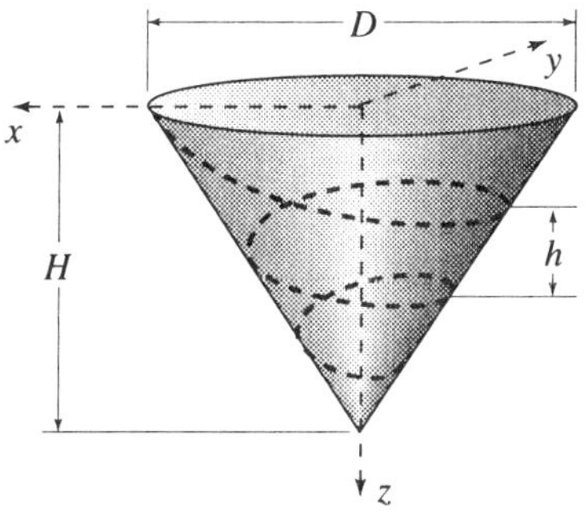

FIGURE P14.72.

14.73 Rod AB rotates in a vertical plane about a horizontal axis at A as shown in Figure P14.73. The coefficient of kinetic friction between the collar C of weight W and the rod is μ. In the position shown, $r = 5$ ft, $\dot{r} = 4$ ft/s, $\ddot{r} = 2\text{ft/s}^2$, $\theta = 20°$, $\dot{\theta} = 2$ rad/s, and $\ddot{\theta} = 0$. Determine the value of μ.

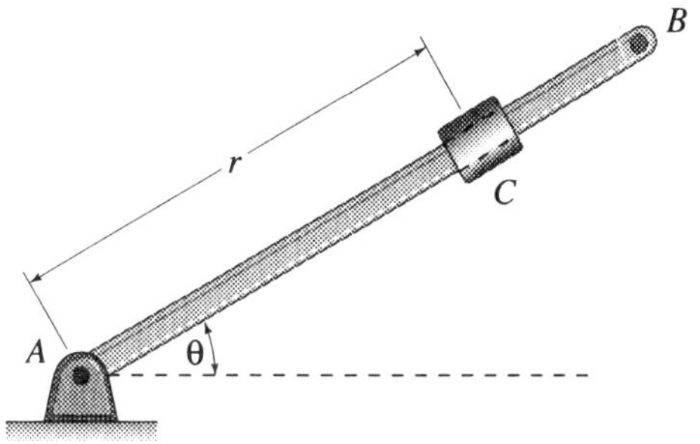

FIGURE P14.73.

14.74 A car of mass m moves at a constant speed v_0 along the horizontal curved road defined by $r = b\sin\theta$ as shown in Figure P14.74. In terms of v_0, b and m, determine the magnitude of the resultant horizontal force that the pavement exerts on the wheels to keep the car on the road for any position defined by the angle θ.

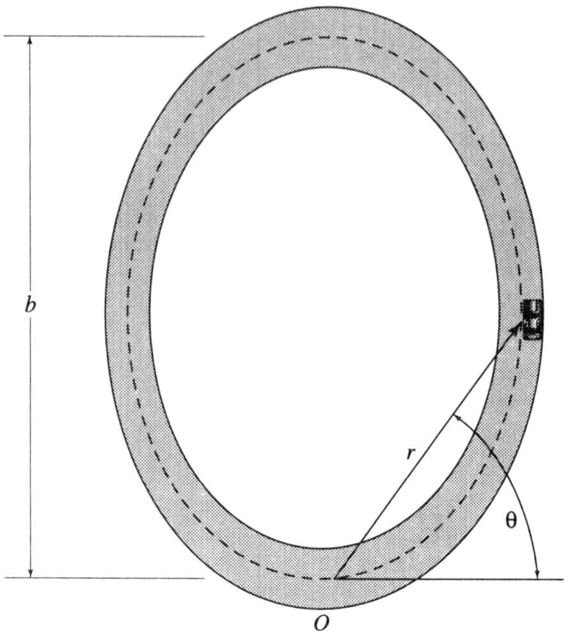

FIGURE P14.74.

14.75 A particle of mass m is constrained to move at a constant angular speed $\dot{\theta} = \omega_0$ along a horizontal curve given by $r = b\sin^2\theta$. (a) In terms of b, ω_0, m, and θ, determine the components F_r and F_θ of the horizontal force needed to keep the particle on the curve. (b) Find the values of the angle θ for which F_r and F_θ take on their maximum magnitudes. What are these maximum magnitudes?

14.76 Rod AB in Figure P14.76 rotates about a vertical axis at A. A small collar C of mass $m = 4$ kg can slide along rod AB but is constrained to move along a horizontal curve defined by $r = 5\theta$ and $\theta = 0.5t$ where r is in meters, θ is in radians, and t is in seconds. The particle starts from rest when $t = 0$. Determine r, t, and θ $(0 \leq \theta \leq \pi)$ for which the radial component of the force exerted by the curve on the particle is $F_r = -12.5$ N. Compute the transverse component of the force F_θ at this position of the particle.

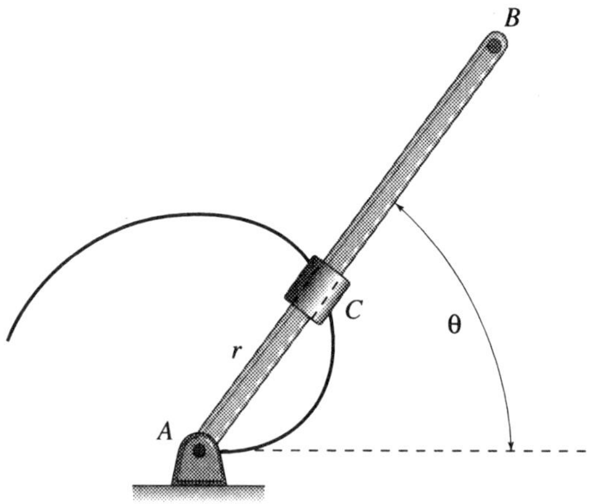

FIGURE P14.76.

14.77 A small motorized pulley P is used to apply the necessary force to slide collar C of mass m along the fixed rod AB as shown in Figure P14.77. Both the pulley and rod AB lie in the same horizontal plane. If the pulley rotates at a constant angular speed ω_0 and its radius is R,

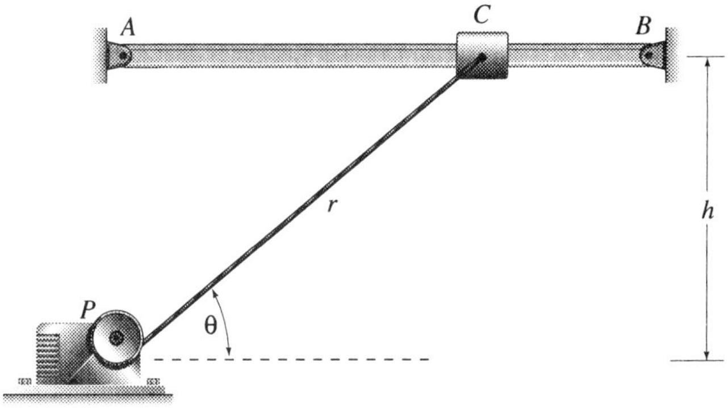

FIGURE P14.77.

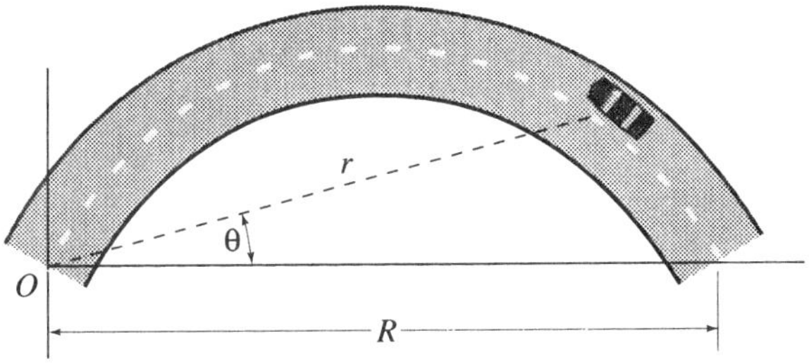

FIGURE P14.78.

determine the force in the cable for any position defined by the angle θ. Express your answer in terms of m, R, ω_0, h, θ, and μ, where μ is the coefficient of kinetic friction between the collar and the rod.

14.78 A car of weight W travels at constant speed v_0 along a horizontal curved road defined by $r = R\cos\theta$, as shown in Figure P14.78. Determine the magnitude of the resultant horizontal force that the road exerts on the car for any position defined by θ.

14.79 Arm OA rotates in a vertical plane forcing peg B of weight W to slide along a vertical slot as shown in Figure P14.79. Assume frictionless conditions, and let $\theta = 2t^2$ where t is the time in seconds and θ is in radians. Let F_A be the normal force exerted on the peg by arm OA and F_S the normal force exerted on it by the slot. Compute values of F_A/W and F_S/W at the instant when $t = 3$ s.

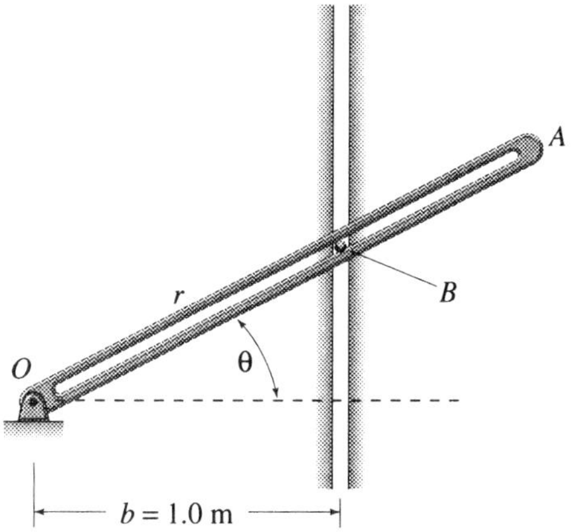

FIGURE P14.79.

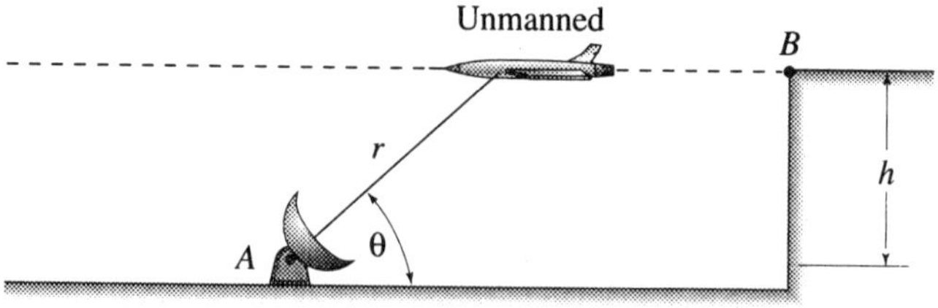

FIGURE P14.80.

14.80 A radar station positioned at point A is used to track the horizontal flight of an unmanned aircraft which is fired from point B, as shown in Figure P14.80, where the angular position of the radar is θ. The weight of the aircraft is W and may be assumed constant for the duration of the tracking process. Develop expressions for the velocity v and for the propelling thrust T of the aircraft in terms of W, h, θ, $\dot{\theta}$., and $\ddot{\theta}$.

14.6 Motion of a Particle under a Central Force

There are physical problems when a body is subjected to a force directed toward a fixed point in space. The body is said to be experiencing motion under a *central force.*

Consider, for example, a particle of mass m moving in space under the central force **F** which is always directed toward the fixed point O as shown in Figure 14.8(a). It is assumed that **F** is the only force acting on

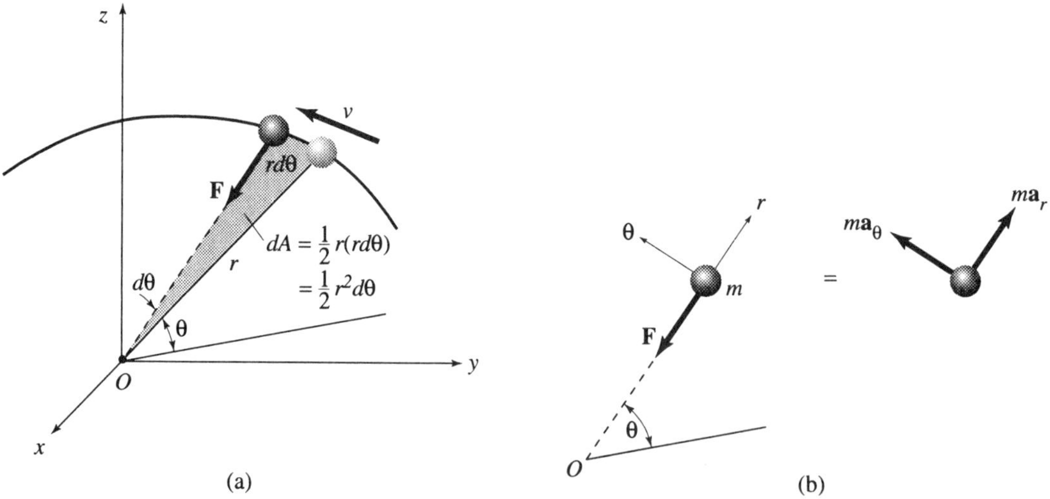

FIGURE 14.8.

the particle or that other forces that may be acting are insignificant compared to **F**. An artificial satellite orbiting Earth represents a very good example of the motion of a particle under the influence of a central force. In this case, the central force is the gravitational force of attraction experienced by the satellite which is always directed toward the center of Earth.

The motion of the particle of Figure 14.8(a) may be described by using the polar coordinates r and θ. The free-body diagram of the particle and the corresponding inertial force vectors in terms of these polar coordinates are shown in Figure 14.8(b). Therefore, using Newton's second law of motion in terms of polar coordinates (i.e., $\sum F_r = ma_r$ and $\sum F_\theta = ma_\theta$),

$$-F = m(\ddot{r} - r\dot{\theta}^2) \tag{14.25}$$

and

$$0 = m(r\ddot{\theta} + 2\dot{r}\dot{\theta}) \tag{14.26}$$

where $\dot{r} = dr/dt$, $\ddot{r} = d^2r/dt^2$, $\dot{\theta} = d\theta/dt$ and $\ddot{\theta} = d^2\theta/dt^2$. Thus, Eq. (14.26) may be written in the form

$$\frac{d(r^2\dot{\theta})}{dt} = 0$$

from which

$$r^2\dot{\theta} = K \tag{14.27}$$

and

$$\dot{\theta} = \left(\frac{1}{r^2}\right)K = Kq^2 \tag{14.28}$$

where K is a constant of integration and q was substituted for the quantity $1/r$ in Eq. (14.28). Now,

$$\dot{r} = \left(\frac{dr}{dt}\right)\left(\frac{d\theta}{dt}\right) = \left(\frac{dr}{d\theta}\right)\dot{\theta}. \tag{14.29}$$

Substituting Eq. (14.28) in Eq. (14.29),

$$\dot{r} = Kq^2\left(\frac{dr}{dq}\right). \tag{14.30}$$

Differentiation of Eq. (14.30) with respect to time yields

$$\begin{aligned} \ddot{r} &= K\left(\frac{d}{dt}\right)\left(q^2\frac{dr}{d\theta}\right), \\ \ddot{r} &= K\left(\frac{d}{d\theta}\right)\left(q^2\frac{dr}{d\theta}\right)\dot{\theta}, \\ \ddot{r} &= (Kq)^2\left(\frac{d}{d\theta}\right)\left(q^2\frac{dr}{d\theta}\right). \end{aligned} \tag{14.31}$$

Because $r = 1/q$, it follows that

$$\frac{dr}{d\theta} = -\left(\frac{1}{q^2}\right)\frac{dq}{d\theta}. \tag{14.32}$$

If Eq. (14.32) is substituted in Eq. (14.31),

$$\ddot{r} = -(Kq)^2\frac{d^2q}{d\theta^2}. \tag{14.33}$$

Recalling that $r = 1/q$ and substituting Eq. (14.28) and (14.33) in Eq. (14.25) after simplification,

$$\frac{d^2q}{d\theta^2} + q = \frac{F}{mK^2q^2}. \tag{14.34}$$

Equation (14.34) is the differential relationship that yields the trajectory of a particle moving under the influence of a central force **F**. This differential relationship is used in Section 14.7 to solve problems dealing with the flight of artificial Earth satellites and other space vehicles.

Now, return to Figure 14.8 and observe that the differential area defined by the radius r, as it moves through the differential angle $d\theta$, is

$$dA = \frac{r}{2}(r\,d\theta) = \frac{r^2}{2}\,d\theta. \tag{14.35}$$

Dividing both sides of Eq. (14.35) by dt,

$$\frac{dA}{dt} = \frac{r^2}{2}\left(\frac{d\theta}{dt}\right) = \frac{1}{2}r^2\dot{\theta} \tag{14.36}$$

Substituting $r^2\dot{\theta} = K$ from Eq. (14.27) in Eq. (14.36) yields

$$dA/dt = K/2 = \text{constant}. \tag{14.37}$$

The quantity dA/dt is known as the *areal* velocity of the particle, and Eq. (14.37) shows that if a particle moves under the influence of a central force, its areal velocity is constant.

14.7 Space Mechanics

14.7.1 Governing Equation

As stated earlier, the free-flight motion of artificial satellites and other spacecraft in the vicinity of Earth or in the vicinity of other planets, such as Venus and Mars, represents a very good example of motion under a central force for which the governing differential relationship is given by Eq. (14.34). In such a case, the central force **F** is the gravitational attraction between the spacecraft and the planet given by Newton's Law of Gravitation, $F = G(m_1m_2/r^2)$, discussed in Chapter 1 and repeated earlier in this chapter. It is assumed here that other forces acting on the spacecraft or satellite, such as the gravitational attraction of other planets, are negligibly small as long as the satellite remains in the neighborhood of its own planet. Recall that, in Newton's law of gravitation, G is a universal constant whose value is known on the basis of experimental evidence and r is the distance between the two particles of masses m_1 and m_2. To specialize Newton's law of gravitation to the motion of a spacecraft, we let $m_1 = \mathrm{M}$, the mass of the planet, and $m_2 = m$, the mass of the spacecraft, and write the force of gravitation in the form

$$F = \frac{GMm}{r^2} = gR^2 mq^2 \tag{14.38}$$

where $1/r$ was replaced by q and by Eq. (1.5), the product GM was replaced by the product gR^2. In the particular case of Earth, $g = 32.2\ \mathrm{ft/s^2}$ or $9.81\ \mathrm{m/s^2}$, and $R = 3960$ mi or 6370 km. Substituting Eq. (14.38) in the differential relationship given in Eq. (14.34),

$$\frac{d^2 q}{d\theta^2} + q = \frac{gR^2}{K^2}. \tag{14.39}$$

Equation (14.39) is a second-order, nonhomogeneous differential equation, the general solution for which is the sum of the particular solution $q = gR^2/K^2$ and the complementary (or homogeneous) solution which may be written in the form $q = C\cos\theta$ if $\theta = 0$ at the *initial axis* (i.e., polar axis) of the polar coordinate system. Therefore, the general solution of Eq. (14.39) becomes

$$q = \frac{1}{r} = \frac{gR^2}{K^2} + C\cos\theta \tag{14.40}$$

which may be written in the form

$$r = \frac{K^2}{(gR^2 + K^2 C\cos\theta)},$$

$$r = \left(\frac{K^2}{gR^2}\right)\frac{1}{\left(1 + \dfrac{K^2 C}{gR^2}\cos\theta\right)},$$

$$r = \left(\frac{K^2}{gR^2}\right)\frac{1}{(1 + e\cos\theta)}, \tag{14.41}$$

where

$$e = K^2 C/gR^2. \tag{14.42}$$

As will be shown later, the values of the constants K and C are determined from knowledge of the position and speed of the space vehicle at the beginning of its free flight.

14.7.2 Conic Sections

Note that Eq. (14.41), which defines the free-flight trajectory of a spacecraft, represents a *conic section*, and the quantity e, defined by Eq. (14.42), is known as the *eccentricity* of the conic section. The value of

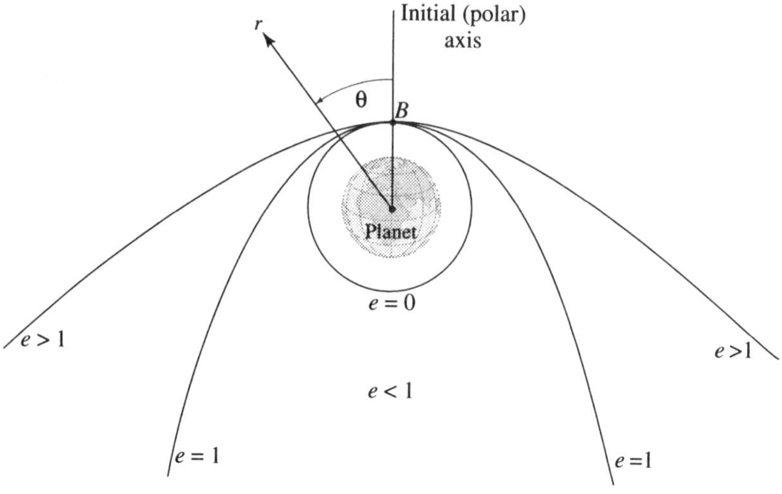

FIGURE 14.9.

the eccentricity e determines the specific type of conic section that exists in a given situation. The four types of conic sections, depicted in Figure 14.9 and discussed below, are of interest in the study of space mechanics. Note that, in Figure 14.9, point O, the center of the planet, is the origin of the polar coordinate system and coincides with the *focus* of the conic section. *The initial (polar) axis of the polar coordinate system is assumed in the vertical direction, and it is an axis of symmetry of the conic section. In other words, the initial axis is directed from the focus to the nearest point on the conic section.* Point B, at some height above the surface of the planet, represents the end of the powered phase of the flight and the beginning of the free flight of the spacecraft. It is assumed, for our present purposes, that the powered flight is designed to produce a velocity of the vehicle parallel to the surface of the planet at the beginning of its free flight. The case in which this condition is not met is best treated using the principle of conservation of angular momentum discussed in Chapter 16.

The Circular Trajectory: Equation (14.41) defines a circular path if $e = 0$ because the radius vector becomes a constant given by the relationship $r = K^2/gR^2$.

The Elliptic Trajectory: If $e < 1$, the denominator of Eq. (14.41) is always different from zero because the quantity $e \cos\theta$ is always greater than -1. Thus, the radius vector, although not constant, is always finite and Eq. (14.41) describes an elliptic orbit. Obviously, the circle with $e = 0$ is but a special case of the ellipse.

The Parabolic Trajectory: If $e = 1$, the denominator of Eq. (14.41) vanishes for $\theta = \pi$ or $\theta = -\pi$. Thus, for these values of θ, the radius vector becomes infinite and Eq. (14.41) defines a parabolic trajectory.

The Hyperbolic Trajectory: The radius vector becomes infinite when the denominator of Eq. (14.41) vanishes. This condition exists if $\cos\theta = -1/e$. Thus, if we assume that $e > 1$, then, the quantity $\cos\theta$ has a negative value resulting in two values of θ, one positive and one negative, for which the radius vector becomes infinite. Therefore, if $e > 1$, Eq. (14.41) defines a hyperbola for the free-flight path of a space vehicle.

14.7.3 Determination of K and C

The constants K and C in Eqs. (14.41) and (14.42) are determined from the initial conditions of the vehicle's free flight. Thus, let us assume that the space vehicle begins its free flight at point B which has an elevation defined by the coordinate r_B measured from the center of the planet as shown in Figure 14.10. Also, as stated earlier, we are going to assume that the velocity of the space vehicle at the beginning of its free flight is parallel to the surface of the planet with a magnitude of v_B. Thus, because the velocity at point B has no radial component, it follows that $v_B = r_B\dot{\theta}$ and, by Eq. (14.27) for $r = r_B$, we obtain

$$K = r^2\dot{\theta} = r_B v_B. \tag{14.43}$$

Letting $\theta = 0$ and $r = r_B$ in Eq. (14.41) yields

$$r_B = \frac{K^2}{gR^2(1+e)}. \tag{14.44}$$

Substituting for e from Eq. (14.42) and solving for the constant C, we obtain

$$C = \frac{1}{r_B} - \frac{gR^2}{K^2}. \tag{14.45}$$

The constants K and C are termed *orbital constants* for a given spacecraft in free flight. They can be changed only if energy is added, for example, by firing rockets to change the orbit of the spacecraft.

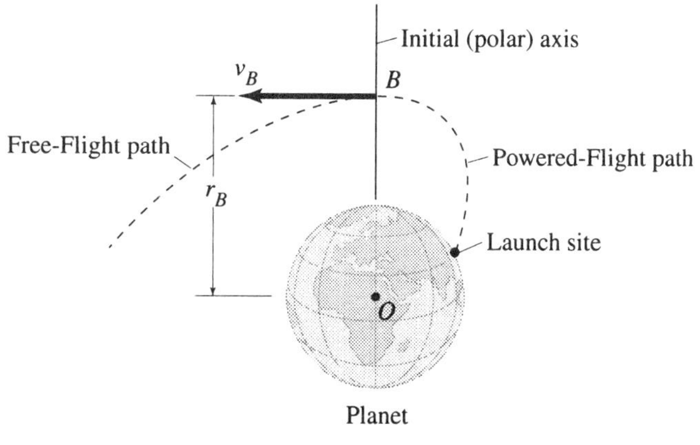

FIGURE 14.10.

14.7.4 Initial Velocities

As stated earlier, the spacecraft is assumed to have an initial velocity $\mathbf{v}_B$ parallel to the surface of the planet at the beginning of its free-flight trajectory (i.e., at the end of its powered flight) the magnitude of which is compatible with the desired trajectory. The case in which v_B is *not* parallel to the surface of the planet is conveniently treated using the principle of conservation of angular momentum and is discussed in Chapter 16.

When the initial velocity $\mathbf{v}_B$ is parallel to the surface of the planet, its magnitude, for each of the four basic free-flight paths, is developed as follows:

The Circular Trajectory: A circular trajectory requires that $e = 0$ which leads to $C = 0$ by Eq. (14.42). Thus, using Eq. (14.45) with $C = 0$ yields

$$v_B = v_{CIR} = \sqrt{\frac{gR^2}{r_B}} \tag{14.46}$$

where v_{CIR} represents the speed needed at point B for a circular orbit. Thus Eq. (14.46) gives the magnitude of the initial velocity v_B in order for the space vehicle to achieve a circular, free-flight trajectory. Such a trajectory is shown in Figure 14.11 and labeled $v_B = v_{CIR}$.

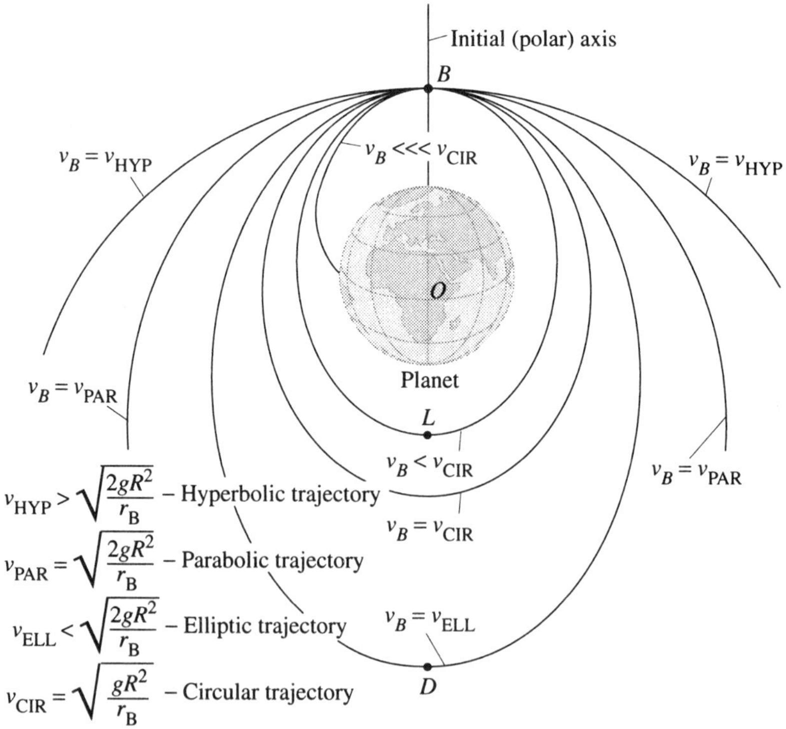

FIGURE 14.11.

The Elliptic Trajectory: As stated earlier, the elliptic curve is characterized by $e < 1$, which, by Eq. (14.42), requires that $C < gR^2/K^2$. If this condition is used in Eq. (14.45),

$$v_B = v_{ELL} < \sqrt{\frac{2gR^2}{r_B}} \tag{14.47}$$

where v_{ELL} is the speed at B for an elliptic orbit. Thus, Eq. (14.47) provides the magnitude of the initial velocity v_B in order for the spacecraft to travel along an elliptic, free-flight path as shown in Figure 14.11. By comparing Eqs. (14.46) and (14.47), we again observe that the condition leading to a circular trajectory is a special case of that for an elliptic trajectory. The orbit of most artificial satellites is elliptical. Such an orbit is shown in Figure 14.11 and denoted by $v_B = v_{ELL}$. Point B, at the beginning of free flight, is closest to the surface of the earth and is known as the perigee. Point D on the elliptic orbit, farthest away from the surface of the planet, is known as the *apogee*. Note that, for values of $v_B < v_{CIR}$, point B at the beginning of the free flight becomes the apogee and point L becomes the perigee . Such a path is shown in Figure 14.11 and identified by $v_B < v_{CIR}$. However, for values of $v_B <<< v_{CIR}$, point L falls inside the surface of the planet and the vehicle, obviously, will not go into orbit but returns to the planet as depicted in Figure 14.11 by the path labeled $v_B <<< v_{CIR}$. Ballistic missiles represent an example of space vehicles which are designed not to go into orbit but to hit specific targets on the surface of the earth.

The Parabolic Trajectory: In the case of the parabolic path, $e = 1$. Therefore, it follows from Eq. (14.42) that $C = gR^2/K^2$. If this value of C is substituted in Eq. (14.45)

$$v_B = v_{PAR} = \sqrt{\frac{2gR^2}{r_B}} = v_{ESC} \tag{14.48}$$

where v_{PAR} symbolizes the speed necessary for a parabolic trajectory and is known as the *escape velocity* v_{ESC} because it represents the least speed needed for the spacecraft to escape the influence of the planet's gravity. Therefore, Eq. (14.48) gives the magnitude of the initial velocity v_B in order for the space vehicle to achieve a parabolic, free-flight trajectory as shown in Figure 14.11 by the path identified by $v_B = v_{PAR}$.

The Hyperbolic Trajectory: The hyperbolic trajectory is characterized by the condition $e > 1$ which, by Eq. (14.42), reduces to $C > gR^2/K^2$. When this value of C is used in Eq. (14.45), it leads to

$$v_B = v_{HYP} > \sqrt{\frac{2gR^2}{r_B}} \tag{14.49}$$

where v_{HYP} represents the speed necessary for a hyperbolic trajectory. Thus, Eq. (14.49) gives the magnitude of the initial velocity v_B in order for the spacecraft to follow a hyperbolic, free-flight path as shown in Figure 14.11 by the trajectory denoted by $v_B = v_{HYP}$.

14.7.5 Determination of Orbital Period

Planets in our solar system follow elliptic orbits with the sun located at one of the two foci. Also, most artificial satellites are placed in elliptic or circular orbits. Obviously, all such orbits are repetitive. For example, an artificial satellite designed to execute an elliptic orbit around Earth as shown in Figure 14.12, will complete the entire orbit in a time interval τ known as the *orbital period* of the satellite. Provided

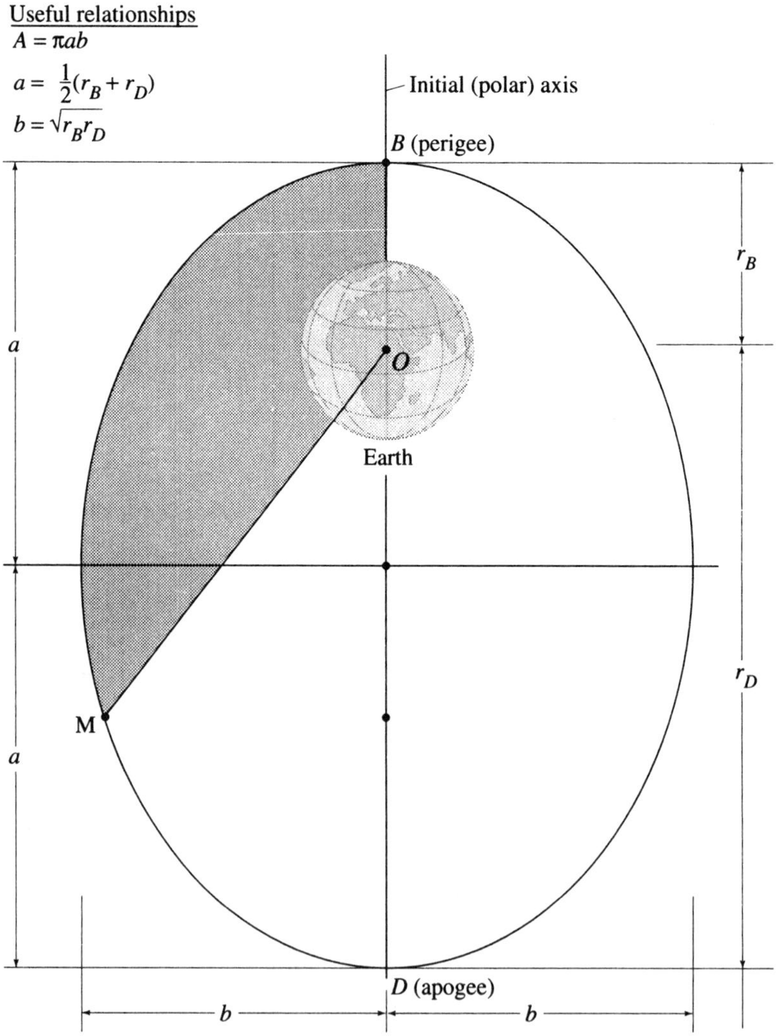

FIGURE 14.12.

that atmospheric resistance is not a factor, the satellite will execute the same orbit indefinitely in the same orbital period.

To determine the time t required for a satellite or planet to travel a certain distance along its orbital path, we use Eq. (14.37) from which $dt = (2/K)\,dA$. If this relation is integrated,

$$t = (2/K)A \tag{14.50}$$

where t is the time required for the radius vector to cover the area A while the particle (satellite or planet) moves along its trajectory. Thus, for example, the time t needed to travel from point B to point M in Figure 14.12 may be found from Eq. (14.50) if we substitute for A the area contained inside the ellipse between the radius vectors OB and OM shown hatched in the sketch. Also, to determine the orbital period τ for the elliptic orbit of Figure 14.12, we need to find the time t required to complete the entire orbit, from point B all around to the apogee at D and back again to the starting point at B. In this case, the quantity A in Eq. (14.50) is the area of the entire ellipse which is given by the product πab. Thus,

$$\tau_{\text{ELL}} = 2\pi ab/K \tag{14.51}$$

where τ_{ELL} is the orbital period for an elliptic orbit and where K is defined by Eq. (14.43). The semiaxes a and b are determined from the radius vectors r_{B} and r_{D} that position the points closest and farthest, respectively, from the center of the planet which coincide with one of the two foci of the ellipse. It can be shown from analytic geometry that $a = (r_{\text{B}} + r_{\text{D}})/2$ and $b = \sqrt{r_{\text{B}} r_{\text{D}}}$.

Equation (14.51) may be specialized for the case of a circular orbit by setting $b = a$. This yields

$$\tau_{\text{CIR}} = 2\pi a^2/K \tag{14.52}$$

where τ_{CIR} is the orbital period and a the radius for a circular orbit.

14.7.6 Kepler's Laws

Before Isaac Newton (1642–1727) presented his laws of motion and his law of gravitation, Johanns Kepler (1571–1630) introduced his three laws of planetary motion as follows:

1. Planets describe elliptic orbits around the Sun which is located at one of the two foci.
2. The areal velocity of a planet is constant.
3. The orbital period of a planet about the Sun is directly proportional to the semimajor axis of its orbit raised to the 3/2 power.

It is interesting to note that Kepler developed his laws of planetary motion solely on the basis of observations of the motion of planets around the Sun and without the benefit of Newton's laws of motion or his law of gravitation. Despite this obstacle, Kepler was able to de-

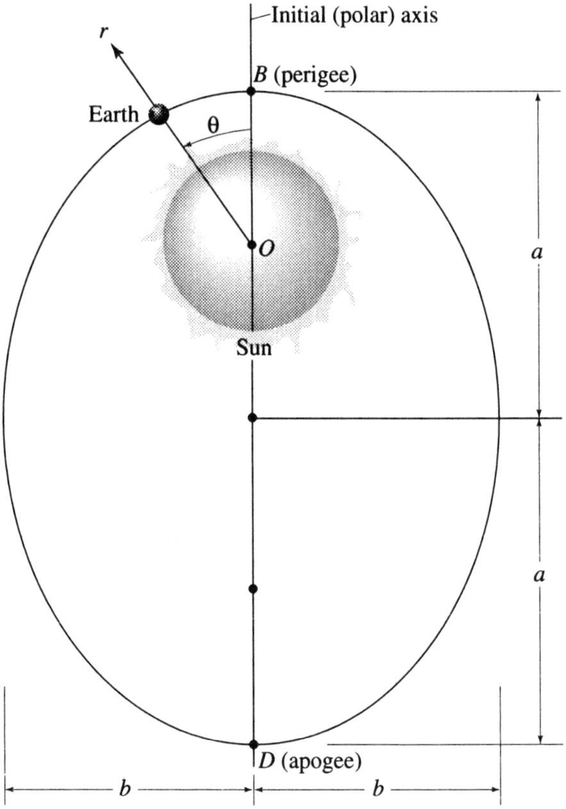

FIGURE 14.13.

scribe the orbits of planets very accurately. The correctness of the first two laws has already been demonstrated and the third law may be also derived on the basis of the conclusions reached earlier. Thus, consider the elliptic orbit shown in Figure 14.13 which is described by a planet, such as Earth, as it moves around the Sun. By Eq. (14.41), at the perigee point B, $\theta = 0$ and

$$r_{\mathrm{B}} = \frac{K^2}{gR^2(1+e)}$$

from which

$$e = \frac{(K^2/gR^2)}{r_{\mathrm{B}}} - 1. \tag{14.53a}$$

Also, at the apogee point D, $\theta = \pi$, and

$$r_{\mathrm{D}} = \frac{K^2}{gR^2(1-e)}$$

from which

$$-e = \frac{K^2}{gR^2 r_D} - 1. \tag{14.53b}$$

If Eqs. (14.53a) and (14.53b) are added and the resulting relationship solved for gR^2/K^2,

$$\frac{gR^2}{K^2} = \frac{1}{2}\left(\frac{r_B + r_D}{r_B r_D}\right) \tag{14.53c}$$

where $r_B + r_D = 2a$ and $r_B r_D = b^2$. Therefore,

$$gR^2/K^2 = a/b^2 \tag{14.53d}$$

from which

$$K = Rb\sqrt{\frac{g}{a}} \tag{14.53e}$$

Substituting Eq. (14.53e) in Eq. (14.51) yields

$$\tau_{ELL} = \frac{2\pi}{R\sqrt{g}} a^{3/2} \tag{14.54}$$

which is a mathematical expression of Kepler's third law. Note that R and g in Eq. (14.54) refer, respectively, to the radius and the acceleration of gravity of the Sun. Note also that R and g are, respectively, the radius and the acceleration of gravity of the Sun in any of the equations developed in this Section, if these equations are used to investigate the motion of a given planet around the Sun.

The following examples illustrate some of the concepts discussed above.

■ Example 14.15

A satellite was launched from point A and was programmed to reach an altitude of 300 mi at the end of the powered flight at point B with a velocity of 20,000 mi/h parallel to surface of Earth, as shown in Figure E14.15, (a) Show that the trajectory of the satellite is elliptic as shown in Figure E14.15, (b) Find the maximum altitude and the velocity of the satellite at this altitude and (c) Find the flight time from point B to point M which is at one end of the minor axis of the elliptic path.

Solution

(a) The eccentricity e is determined by Eq. (14.42) where

$$\begin{aligned} K = r_B v_B &= (3{,}960 + 300)(5{,}280)(20{,}000)(5{,}280/3{,}600) \\ &= 6.598 \times 10^{11}\ \text{ft}^2/\text{s}, \\ C = 1/r_B - gR^2/K^2 &= 1/[(3{,}960 + 300)(5{,}280)] \\ &\quad - (32.2)(3{,}960 \times 5{,}280)^2/(6.598 \times 10^{11})^2 \\ &= 1.212 \times 10^{-8}\ \text{ft}^{-1}. \end{aligned}$$

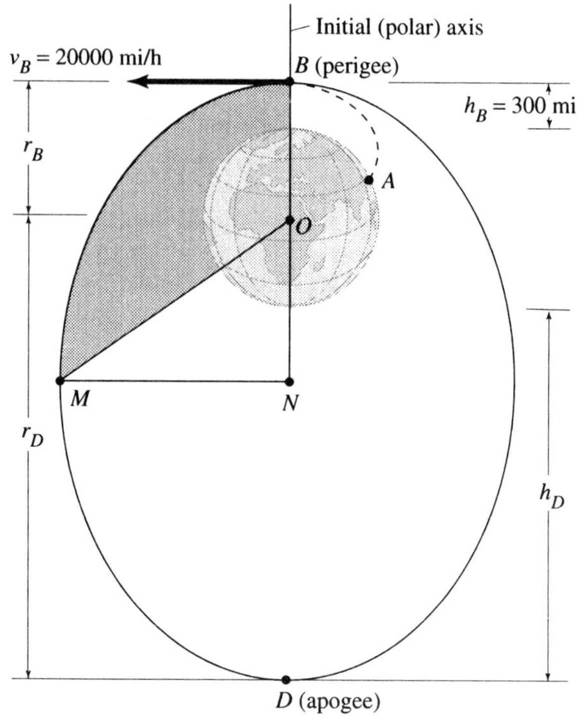

FIGURE E14.15.

Therefore,

$$e = K^2C/gR^2 = (6.598 \times 10^{11})^2(1.212 \times 10^{-8})/(32.2)(3{,}960 \times 5{,}280)^2$$
$$= 0.375.$$

Because $0 < e < 1$, the trajectory is elliptic. ANS.

(b) The maximum radius vector at point D (the apogee) is determined from Eq. (14.41) by setting $\theta = \pi$. Thus,

$$r_D = (K^2/gR^2)/(1 + e\cos\pi)$$
$$= [(6.598 \times 10^{11})^2/(32.2)(3{,}960 \times 5{,}280)^2]/(1 - 0.375),$$
$$= 4.948 \times 10^7 \text{ ft},$$
$$= 9{,}371 \text{ mi}.$$

The maximum altitude h_D at the apogee, therefore, is

$$h_D = r_D - R = 9{,}371 - 3{,}960 = 5{,}410 \text{ mi}. \qquad \text{ANS.}$$

Also, because the velocity has no radial component at D, $K = r_D v_D$, and it follows that

$$v_D = K/r_D = 6.598 \times 10^{11}/4.948 \times 10^7 = 1.333 \times 10^4 \text{ ft/s}$$
$$= 9{,}090 \text{ mi/h.} \qquad \text{ANS.}$$

(c) Let us first determine the semimajor and semiminor axes as follows:

$$a = (r_B + r_D)/2 = [(3{,}960 + 300) + 9{,}371] \times 5{,}280/2 = 3.599 \times 10^7 \text{ ft},$$
$$b = \sqrt{r_B r_D} = \sqrt{(3{,}960 + 300)(5{,}280)^2} = 3.336 \times 10^7 \text{ ft}.$$

The area A inside the ellipse, contained between the radius vectors OB and OM, is found by subtracting the area of triangle OMN from one fourth the area of the ellipse. Thus,

$$A = \frac{\pi}{4}ab - \frac{1}{2}(a - r_B)b$$
$$= \frac{\pi}{4}(3.599)(3.336) \times 10^{14} - \frac{3.336}{2}(3.599 - 2.249) \times 10^{14}$$
$$= 7.178 \times 10^{14} \text{ ft}^2$$

where $r_B = (3{,}960 + 300)(5{,}280) = 2.249 \times 10^7$ ft was substituted in the above equation. Thus, by Eq. (14.50),

$$\tau = \frac{2}{6.598 \times 10^{11}}(7.178 \times 10^{14}) = 2.176 \times 10^3 \text{ s} = 36.3 \text{ min.} \qquad \text{ANS.}$$

Example 14.16

An unmanned spacecraft is executing an elliptic orbit around Earth, as shown in Figure E14.16, before it is injected into an escape trajectory to a distant planet. The altitudes of the spacecraft at the perigee and apogee are, respectively, $h_B = 2{,}000$ km and $h_D = 80{,}000$ km. To inject the spacecraft into an escape trajectory, an increase in its speed is necessary. (a) Determine the minimum increase in speed at the perigee, point B, to initiate the escape trajectory. (b) What is the orbital period of the spacecraft prior to the initiation of the escape trajectory?

Solution

(a) At point B, $r = r_B$, $\theta = 0$, and Eq. (14.41) yields

$$r_B = \frac{K^2}{gR^2(1 + e)}. \qquad \text{(a)}$$

At point D, $r = r_D$, $\theta = \pi$, and Eq. (14.41) yields

$$r_D = \frac{K^2}{gR^2(1 - e)}. \qquad \text{(b)}$$

Elimination of the eccentricity e between Eqs. (a) and (b) leads to

$$K^2 = \frac{2gR^2 r_B r_D}{(r_B + r_D)} \qquad \text{(c)}$$

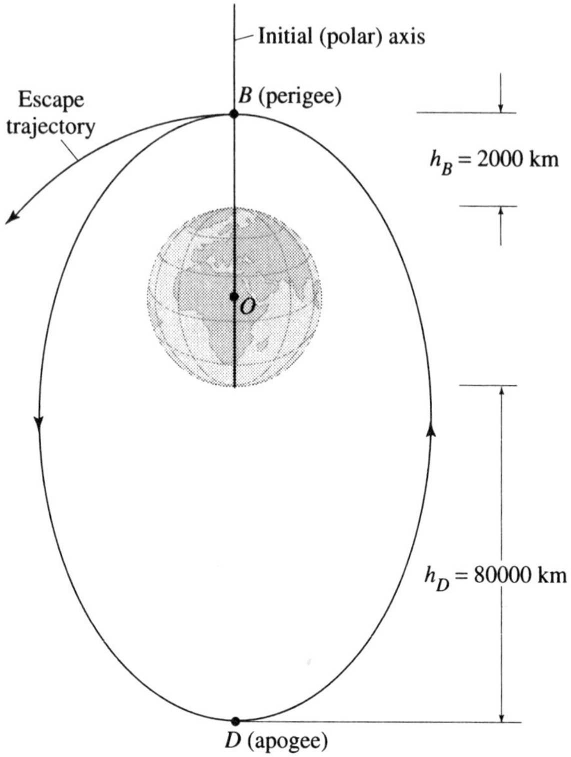

FIGURE E14.16.

where

$$r_B = 6{,}370 + 2{,}000 = 8{,}370 \text{ km},$$

and

$$r_D = 6{,}370 + 80{,}000 = 86{,}370 \text{ km}.$$

At point B, $K = r_B v_B$. Substituting this value of K in Eq. (c) and solving for v_B,

$$v_B = \sqrt{\frac{2gR^2 r_D}{r_B(r_B + r_D)}} = 9{,}312 \text{ m/s}.$$

The escape velocity at point B is given by Eq. (14.48). Thus,

$$v_{ESC} = \sqrt{\frac{2gR^2}{r_B}} = \sqrt{\frac{2(9.81)(6{,}370 \times 10^3)^2}{8{,}370 \times 10^3}} = 9{,}753 \text{ m/s}.$$

Therefore, the change in speed Δv_B required at the perigee point B, is

$$\Delta v_B = 9{,}753 - 9{,}312 = 441 \text{ m/s}. \qquad \text{ANS.}$$

(b) The orbital period for the elliptic orbit is given by Eq. (14.51), $\tau_{ELL} = 2\pi ab/K$, where

$$a = (r_B + r_D)/2 = (8{,}370 + 86{,}370)/2 = 47{,}370 \text{ km},$$
$$b = \sqrt{r_B r_D} = \sqrt{(8{,}370)(86{,}370)} = 26{,}887 \text{ km},$$
$$K = r_B v_B = (8{,}370 \times 10^3)(9{,}312) = 7.794 \times 10^{10} \text{ m}^2/\text{s}.$$

Therefore,

$$\tau_{ELL} = \frac{2\pi(47{,}370 \times 10^3)(26{,}887 \times 10^3)}{7.794 \times 10^{10}}$$
$$= 1.027 \times 10^5 \text{ s} = 28.5 \text{ h} \qquad \text{ANS.}$$

■

Problems

Use the tabular data in the solution of the following problems. Note that these values were determined on the basis of $gR^2 = GM$ for each body

	g		R	
Star/Planet, ... etc.	ft/s²	m/s²	mi	km
Sun	894.0	272.37	435,200	700,000
Mercury	12.4	3.78	1,520	2,440
Venus	28.2	8.60	3,760	6,050
Earth	32.2	9.81	3,960	6,370
Moon	5.3	1.61	1,080	1,740
Mars	12.2	3.72	2,110	3,400

14.81 A satellite is to be launched so that it will begin its free flight at an altitude of 200 mi with a velocity which is parallel to the surface of Earth. Determine the needed initial velocities and the corresponding eccentricities for each of the following orbits: (a) circular, (b) elliptic, (c) parabolic, and (d) hyperbolic.

14.82 A spacecraft is orbiting the Moon at a constant altitude of 300 km. Determine (a) the orbital period and (b) the minimum increase in speed necessary to escape the Moon's gravitational attraction.

14.83 A satellite is launched at an altitude of 400 mi above the surface of Mars with an initial velocity of 10,200 mi/h parallel to the surface of the planet. Determine (a) the eccentricity of the satellite's trajectory, (b) the maximum altitude reached and the velocity at this altitude, and (c) the orbital period.

14.84 Communications satellites are placed into a *geostationary* circular orbit around Earth so that they appear stationary to an Earth observer. This means that the orbital period of such satellites must be a day (approximately 24 h). Using a period of 24 h, determine the altitude and the corresponding speed of communications satellites.

14.85 A space station describes an elliptic orbit around Earth as shown in Figure

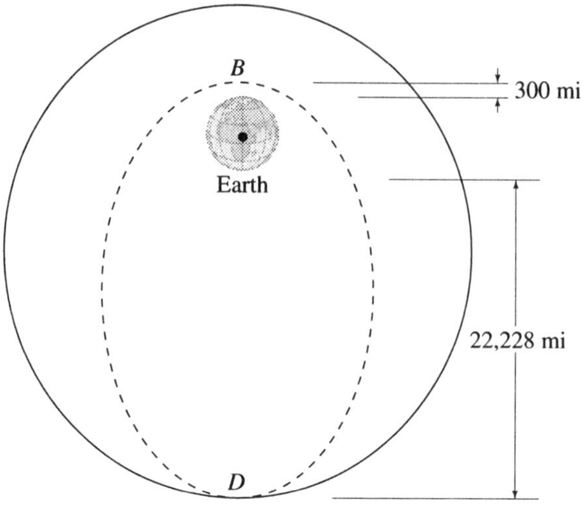

FIGURE P14.85.

P14.85. It is to inject a communications satellite into a geostationary orbit (see Problem 14.84) when it reaches its apogee, point D. Determine (a) the speed of the satellite relative to the space station at point D and (b) the orbital period of the space station and that of the communications satellite.

14.86 An unmanned probe describes an elliptic orbit about the planet Venus such that its altitude at the perigee, $h_B = 500$ km and at the apogee $h_D = 30{,}000$ km. Find the minimum increase in speed needed at the perigee to place the probe on an escape trajectory.

14.87 Refer to Problem 14.86 and determine (a) the orbital period for the elliptic trajectory and (b) the increase in speed needed at the perigee to place the probe on a hyperbolic trajectory with $e = 1.30$.

14.88 A space probe to Mercury is injected, first, into a circular orbit of altitude $h =$ 3,000 mi as shown in Figure P14.88. At point D, the speed is reduced to place the probe into an elliptic orbit that would bring it to a touchdown on Mercury at point B where point B is the perigee and point D is the apogee of the elliptic trajectory. Find (a) the change in speed at D to accomplish this mission and (b) the time it takes the probe to travel the distance from D to B.

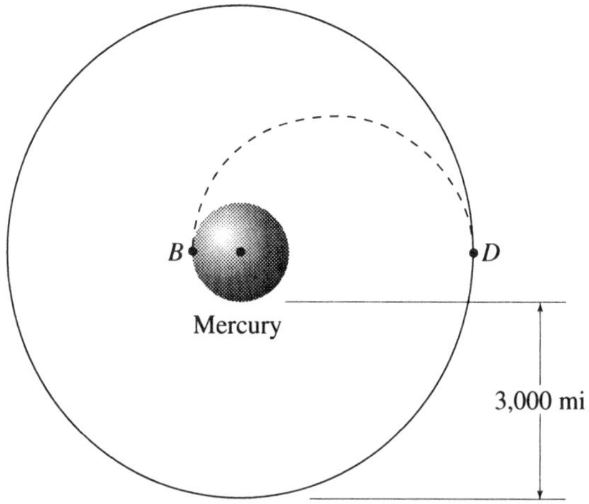

FIGURE P14.88.

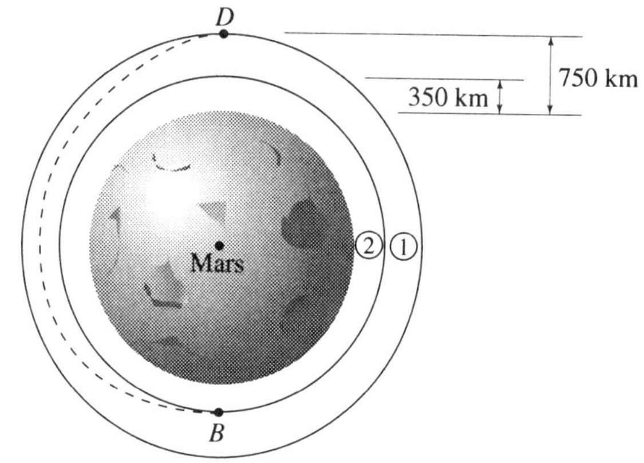

FIGURE P14.89.

14.89 A spacecraft describes circular orbit (1) around the planet Mars at a constant altitude $h_1 = 750$ km as shown in Figure P14.89. It is necessary to place the spacecraft in circular orbit (2) which has a constant altitude above the surface of the planet $h_2 = 350$ km. To accomplish this mission, the spacecraft is injected into an elliptic orbit DB as shown. (a) Find the change in speed at D necessary to change from circular orbit (1) to elliptic orbit DB. (b) Compute the speed change needed at B to place the spacecraft in circular orbit (2).

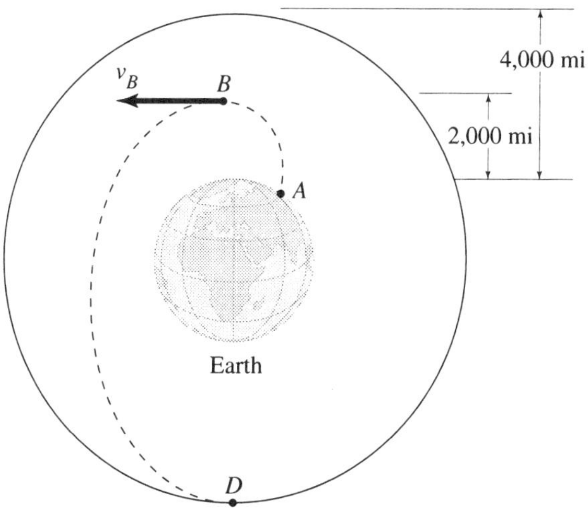

FIGURE P14.90.

14.90 A space station describes a circular orbit around Earth at an altitude of 4,000 mi as shown in Figure P14.90. A spacecraft is to be launched from point A and is programmed to reach point B at an altitude of 2,000 mi with a velocity v_B parallel to the surface of Earth where point B is the beginning of its free flight. (a) Determine the value of v_B so that the spacecraft reaches the orbit of the space station at point D with a velocity tangent to it. (b) What adjustment in velocity must be made at D so that the spacecraft travels in the same orbit as the space station?

14.91 Refer to Problem 14.90 and determine the angle θ defining the position of the space station when the spacecraft is at B so that they reach point D simultaneously.

14.92 A spacecraft executes an elliptic orbit around the planet Venus as shown in Figure P14.92. Determine (a) the eccentricity of the orbit, (b) the period of the orbit, and (c) the change in speed at B needed to achieve a circular orbit of constant altitude $h = 5{,}000$ km.

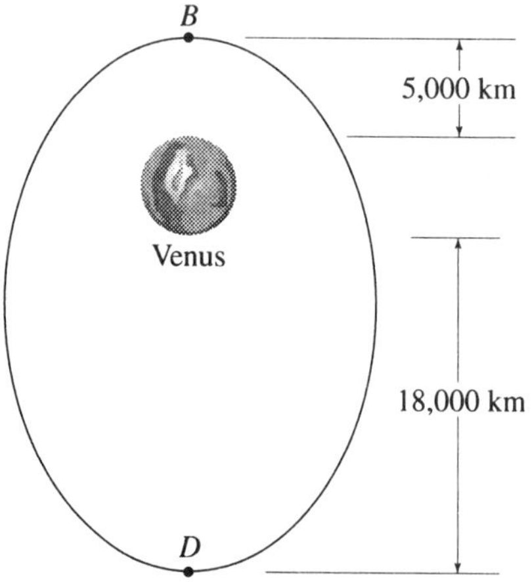

FIGURE P14.92.

14.93 Refer to Problem 14.92 and determine (a) the minimum increase in speed at B to achieve the escape velocity and (b) the increase in speed at B to achieve a hyperbolic trajectory with $e = 1.4$.

14.94 A proposed manned mission to Mars would place a command module in a circular orbit around the planet at an altitude of 500 mi as shown in Figure P14.94. A manned exploration module would be injected into an elliptic orbit a point D. If the exploration module is to land at point A, defined by the angle $\phi = 50°$, determine the speed it must have at point D which is the beginning of its free-flight trajectory. Note that point D is the apogee, and point B is the perigee of the elliptic orbit of the exploration module.

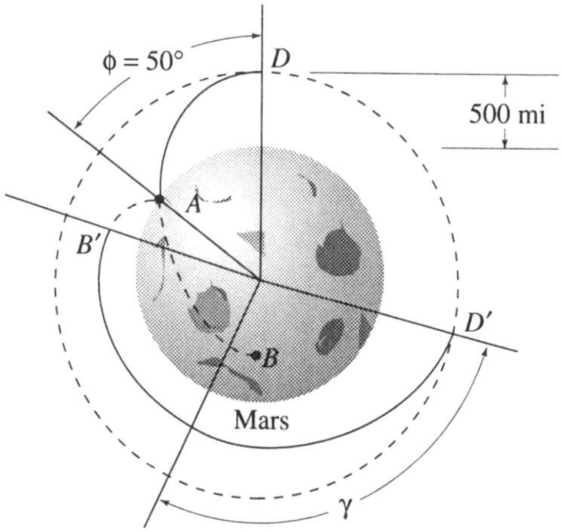

FIGURE P14.94.

14.95 Refer to Figure 14.94. After completing their mission on Mars, the astronauts would rocket their exploration module to point B′ at an altitude of 30 mi and begin their free-flight elliptic trajectory at that point with a velocity v'_B parallel to the surface of the planet. If the exploration module is to rendezvous with the command module at point D′ with a velocity tangent to the command module's circular orbit, determine (a) the velocity v_B' and (b) the speed adjustment needed at D′ to allow docking with the command module.

14.96 Refer to Problem 14.95, and determine the position of the command module, measured by the angle γ in Figure P14.94, when the exploration module begins its free flight at point B′.

14.97 The planet Mars describes an elliptic orbit around the Sun such that its semimajor axis $a = 249.1 \times 10^6$ km and its semiminor axis $b = 206.7 \times 10^6$ km. Develop the equation of the elliptic path in polar coordinates.

14.98 Earth describes an elliptic orbit around the Sun such that its semimajor axis $a = 94.51 \times 10^6$ mi and its semiminor axis $b = 94.50 \times 10^6$ mi. Determine (a) the speed of Earth at its perigee and at its apogee and (b) the eccentricity of its orbit.

Review Problems

14.99 A package of mass $m = 100$ kg is initially at rest and is attached to a rope which passes over a frictionless and massless pulley as shown in Figure P14.99. A constant force $P = 1200$ N is applied to the rope as shown. Determine (a) the speed of the package at the instant it has moved a distance of 5 m upward and (b) the time it takes to move this distance.

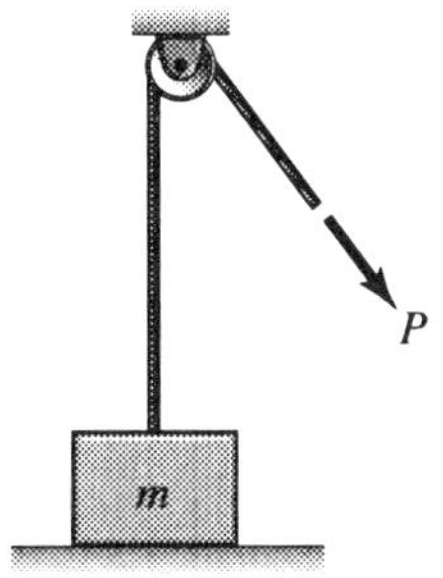

FIGURE P14.99.

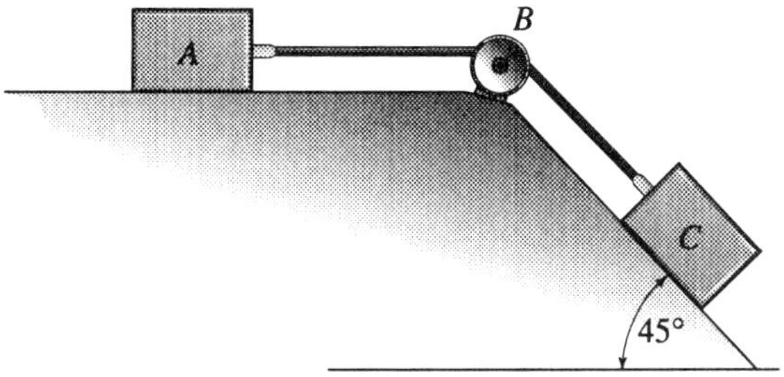

FIGURE P14.100.

14.100 The system shown in Figure P14.100 is released from rest. The cable ABC is flexible but inextensible, and the pulley at B may be assumed weightless and frictionless. If $W_A = 20$ lb and $W_C = 50$ lb and the coefficient of kinetic friction between the blocks and their planes is $\mu = 0.10$, determine the acceleration of the system and the tension in the cable.

14.101 A block of mass $m = 50$ kg is placed on the frictionless bed of a cart and kept in position by means of string AB as shown in Figure P14.101. If the string has a breaking strength of 500 N, determine the maximum acceleration up the inclined plane that may be imparted to the cart without breaking the string.

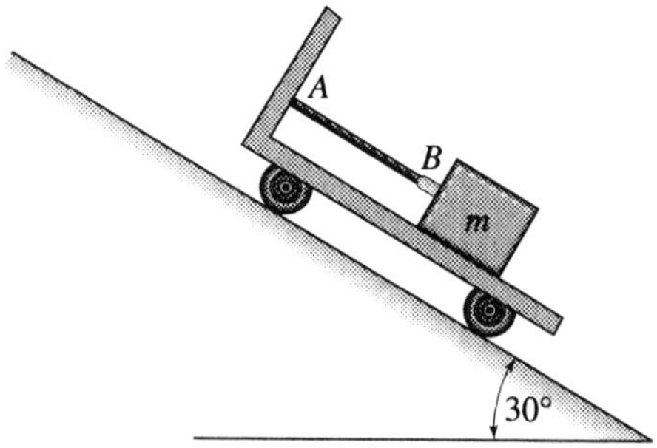

FIGURE P14.101.

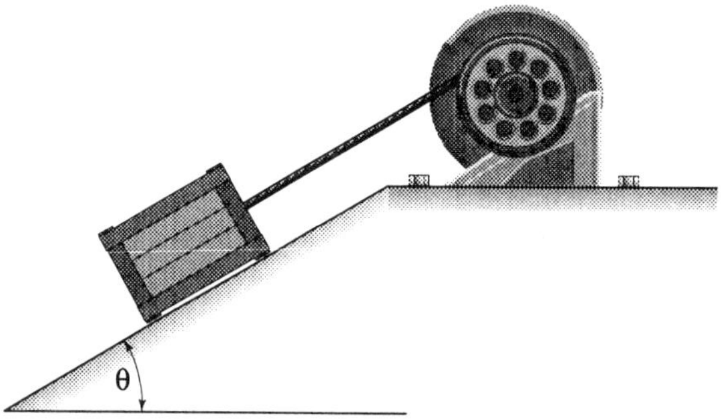

FIGURE P14.102.

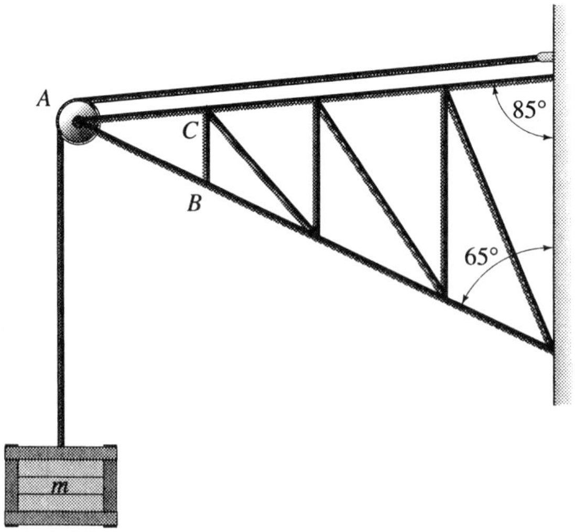

FIGURE P14.103.

14.102 A motorized pulley system is used to move a crate of weight W up a rough inclined plane (coefficient of kinetic friction μ) as shown in Figure P14.102. If the acceleration of the crate is to be limited to 0.3 g, develop, in terms of W, μ and θ, an expression for the force F that the motor exerts on the cable. Assume that the pulley is weightless and frictionless.

14.103 The crane shown schematically in Figure P14.103 is used to lift packages at an acceleration of 2 m/s^2. If member AC of the crane has a maximum strength of 10,000 N, determine the largest mass m that may be lifted by this crane and the force developed in member AB.

14.104 A cyclist approaches point O at a speed v_0 as shown in Figure P14.104. He stops peddling at this point, continues traveling along the vertical frictionless parabolic path, and comes to a complete stop at point B, 10 ft above point O. If the cyclist and his bicycle together weigh 200 lb, determine the speed he had at point O. What is the normal force that the parabolic path exerts on the bicycle at points A and B.

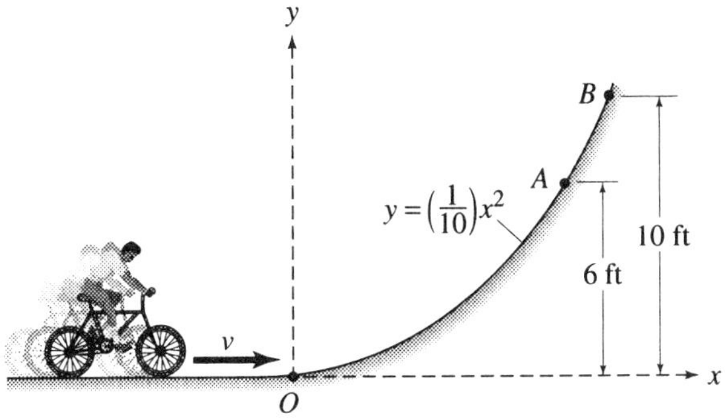

FIGURE P14.104.

14.105 A pilot of weight W flies his plane at a constant speed v in a vertical elliptic loop as shown in Figure P14.105. Find (a) the constant speed at which he is flying if his apparent weight at the top of the loop (point A) is $\frac{2}{5}W$ and (b) his apparent weight at the bottom of the loop (point B).

14.106 A circular highway curve is banked at an angle of 20° as shown in Figure P14.106. Determine the maximum allowable speed of a car on this curve if no frictional forces along the inclined plane are to be experienced by the tires. The radius of curvature of the curve is 150 m.

14.107 A car weighing 3000 lb travels a section of a road that may be approximated by the equation $y = 5\sin\left(\frac{\pi x}{100}\right)$ where both x and y are in ft as shown in Figure P14.107. If the driver maintains a constant speed of 60 mph, determine the normal force exerted by the road on the car at (a) the highest point A and (b) the lowest point B.

14.108 A satellite of mass m_s is to circle the planet Mars in a circular orbit of ra-

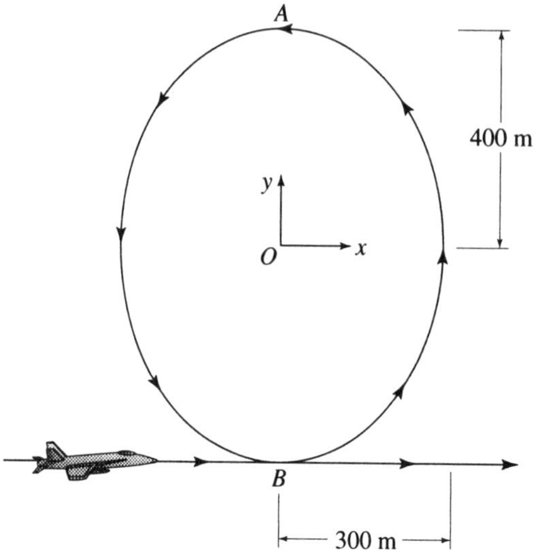

FIGURE P14.105.

dius $R = 3{,}000$ mi measured from the center of the planet. If the mass of Mars is estimated to be $m_m = 4.404 \times 10^{22}$ slug, determine the speed at which the satellite must travel. Hint: Newton's law of gravitation ($G = 3.438 \times 10^{-8}$ $ft^4/lb \cdot s^2$) provides the force of attraction between the planet and the satellite.

14.109 A car of mass $m = 1{,}000$ kg travels along a curved highway that lies in a horizontal plane as shown in Figure P14.109. At point A, the radius of curvature $\rho_A = 350$ m and the car is traveling at a constant speed of 100 km/hr. The driver applies the brakes so that, when he reaches point B, the car has a constant speed of 50 km/hr, and the

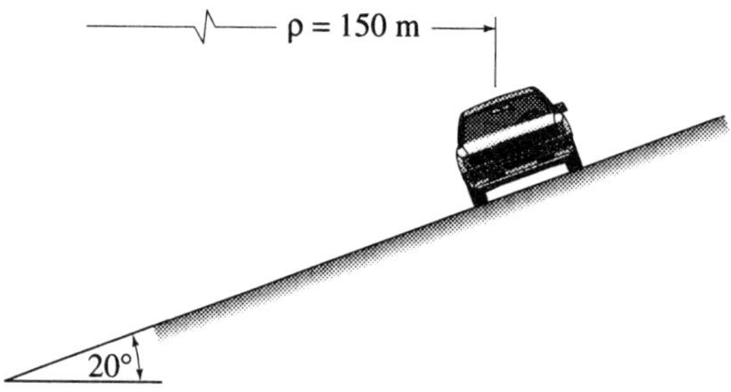

FIGURE P14.106.

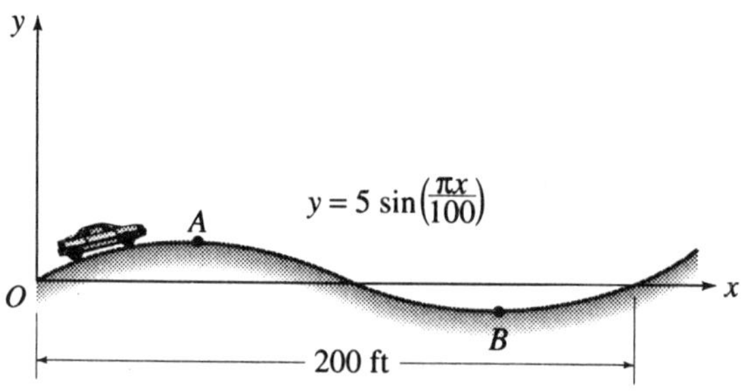

FIGURE P14.107.

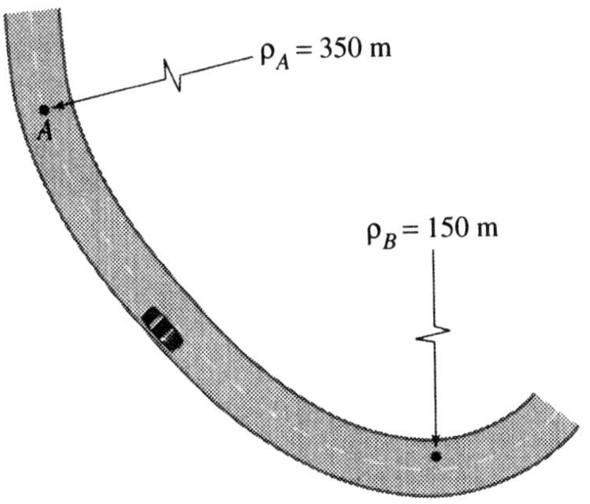

FIGURE P14.109.

corresponding radius of curvature $\rho_B = 150$ m. Find the magnitude of the horizontal force of friction that keeps the car on the pavement at (a) point A and (b) point B.

14.110 Arm OA rotates in a horizontal plane forcing peg B of weight $W = 2$ lb to slide along a stationary horizontal slot as shown in Figure P14.110. Assume frictionless conditions, and let $\theta = 0.5t$ where t is the time in seconds and θ is in radians. Determine the forces exerted on the peg by the rotating arm and by the stationary slot when $t = 1$ s.

14.111 A locomotive of weight W travels along a horizontal section of tracks that is semicircular in shape with a radius R as shown in Figure P14.111. Express the position of the locomotive in polar coordinates (i.e., $r = f(\theta)$. If the locomotive moves along the tracks such that $\theta = \omega =$ constant, determine the magnitude of the resultant horizontal force that the tracks exert on the locomotive for the position defined by the angle (a) $\theta = 45°$ and (b) $\theta = 90°$.

14.112 A spacecraft describes an elliptic orbit around Earth as shown in Figure P14.112. What change of speed is

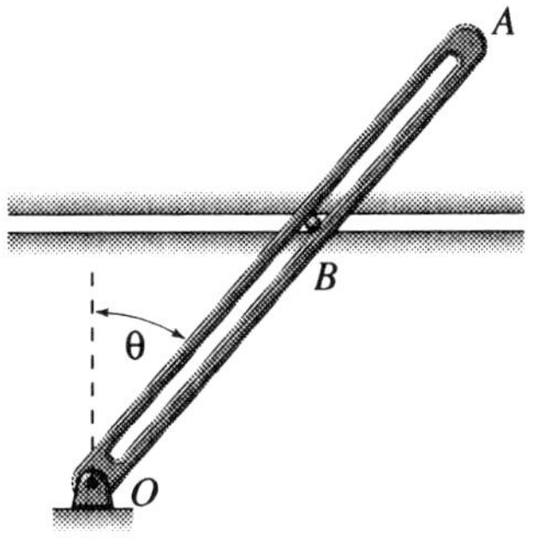

FIGURE P14.110.

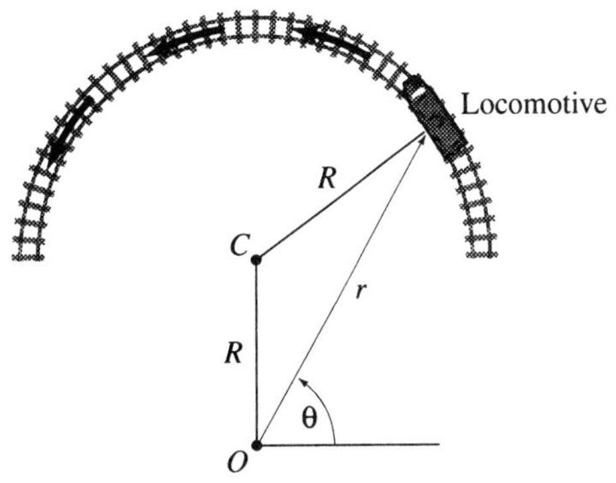

FIGURE P14.111.

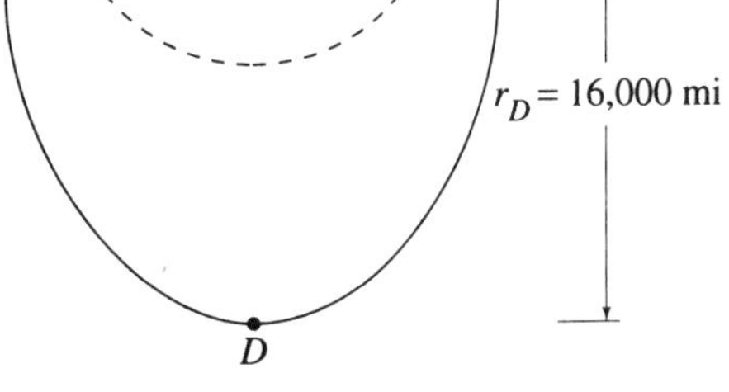

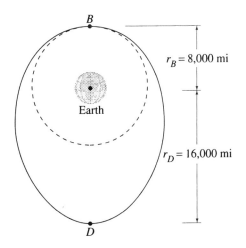

FIGURE P14.112.

needed at point B to change into a circular orbit having a radius of 8,000 miles?

14.113 A spacecraft describes circular orbit 1 around the planet Mercury at a constant altitude of 700 km as shown in Figure P14.113. It is necessary to place the spacecraft in circular orbit 2 which has a constant altitude of 1,500 km. To accomplish this mission, the spacecraft is injected into an elliptic orbit BD as shown. (a) Find the change in speed at B necessary to change from circular orbit 1 to elliptic orbit BD. (b) Find the change in speed at D needed to place the spacecraft in circular orbit 2.

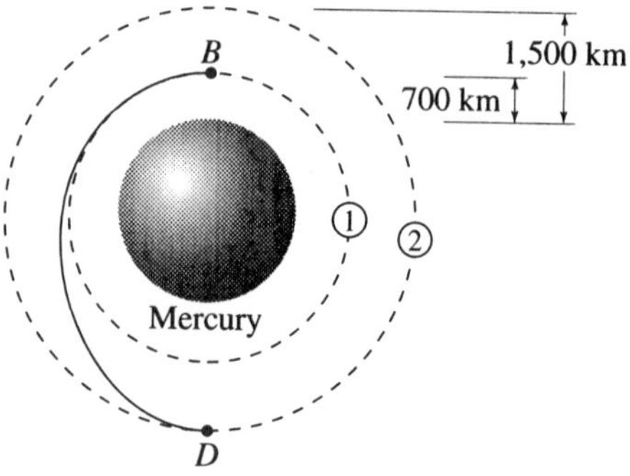

FIGURE P14.113.

15

Particle Kinetics: Energy

The elegance of work-energy techniques for solving problems in dynamics can be matched only by the beauty of systems they help us analyze.

As a youngster, you may have visited an amusement park where you were captivated by the variety of entertainment, diversions, rides, and contests. You may have since wondered about the engineering marvels ranging from the stupefying Ninga challenge roller coaster to the pleasurable Ferris wheel. As an adult, you have undoubtedly been exposed to the technological wonders of space exploration and to the stunning accuracy of modern weapons systems. The analysis and design of such systems of peace and combat require experimentation, innovative approaches, unique experience, and much insight. The examination of many such kinetic problems is often made easier using energy principles.

As a student, you should know that there are different solution techniques for analyzing the same problem. Some solution techniques are easier to implement than others. For you to succeed in practice, you need to develop special engineering skills, common sense, and that special gift we call instinct. *Remember that good engineering judgment can be developed from experience and that experience comes by learning from your mistakes. As Louis Pasteur once said, "Chance favors the prepared mind." Therefore, you can add to your reservoir of talent through initiative and independence when solving problems.*

As an engineer, you may be called upon to tackle problems dealing with systems that are varied in their application and complex in their use. Consequently, you should be adequately prepared to select the appropriate solution technique to achieve the desired results. Irrespective of the problems being solved, you are expected to design and build systems that are both safe and economical. In this chapter, you will learn about the versatility of adapting different solution techniques for efficiently solving complex engineering problems. The energy principles are introduced for a single particle and systems of particles. Their use in the solution of complex engineering problems is demonstrated through several practical examples.

15.1 Definition of Work

The concepts discussed in this section were developed in detail in Chapter 12. Some of these concepts are summarized here for convenience.

The force **F** acts on a particle P as it moves along the path from $s = s_1$ to $s = s_2$ as indicated in Figure 15.1(a). By definition, (see Chapter 12) the differential work dU is given by the dot product of the force **F** and the differential displacement vector, $d\mathbf{r}$. Thus,

$$dU = \mathbf{F} \cdot d\mathbf{r}$$

Because dU is defined as the dot product of two vectors it is a scalar quantity. Imagine that the particle moves from position P to P_1 through a differential displacement ds measured along the path. As P_1 approaches P along the path, the magnitude of $d\mathbf{r}$ equals ds and the direction of $d\mathbf{r}$ becomes the direction of a unit tangent vector $\boldsymbol{\lambda}_t$, as shown in Figure 15.1(b). Therefore,

$$d\mathbf{r} = ds\,\boldsymbol{\lambda}_t.$$

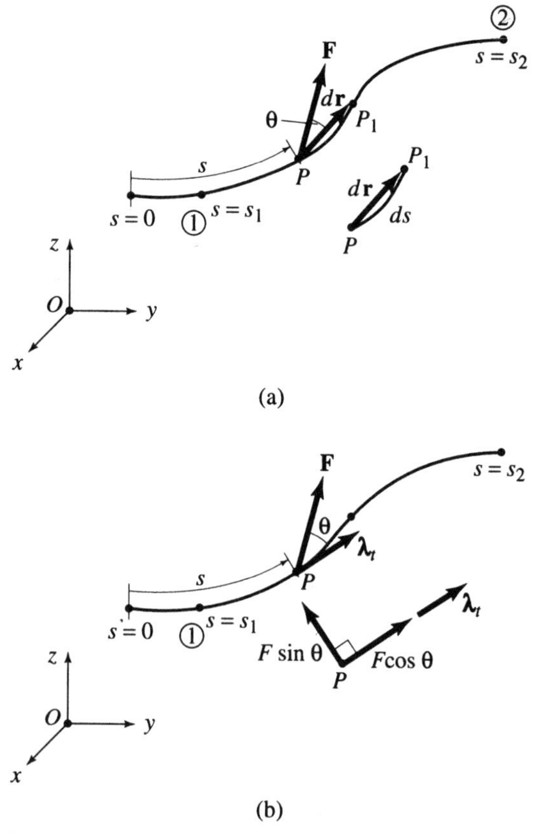

FIGURE 15.1.

Substitute for $d\mathbf{r}$ in the equation for dU to obtain

$$dU = \mathbf{F} \cdot \boldsymbol{\lambda}_t \, ds. \tag{15.1}$$

To find the work done as P moves along the path from position 1 to position 2 we integrate. Thus,

$$\int_{s_1}^{s_2} dU = U(s_2) - U(s_1) = U_{1 \to 2} = \int_{s_1}^{s_2} \mathbf{F} \cdot \boldsymbol{\lambda}_t \, ds \tag{15.2}$$

where we adopt the notation $U_{1 \to 2}$ to denote the work done by **F** as P moves from the initial point, $s = s_1$, to the final point, $s = s_2$. Work is defined as the line integral of the dot product of the force **F** and the differential displacement vector ds $\boldsymbol{\lambda}_t$. The force **F** is evaluated from point to point along the path since, in general, the magnitude and direction of the force are functions of the coordinates of its point of application.

Work of a Constant Force During Straight Line Displacement

Consider a force $\mathbf{F} = \mathrm{F} \cos\theta \mathbf{i} + \mathrm{F} \sin\theta \mathbf{j}$ which is constant in magnitude and direction acting in the x-y plane and which does work as it acts during a displacement directed along the horizontal x axis as shown in Figure 15.2(a). Thus, by Eq. (15.2),

$$dU = \mathbf{F} \cdot \boldsymbol{\lambda}_t \, ds.$$

In this case, $\boldsymbol{\lambda}_t = \mathbf{i}$ and $ds = dx$. Therefore,

$$dU = (F \cos\theta \mathbf{i} + F \sin\theta \mathbf{j}) \cdot \mathbf{i} \, dx.$$

Recalling that $\mathbf{i} \cdot \mathbf{i} = 1$ and $\mathbf{j} \cdot \mathbf{i} = 0$ leads to

$$dU = F \cos\theta \, ds.$$

Integrating from the initial point 1 to the final point 2,

$$U_{1 \to 2} = \int_{x_1}^{x_2} F \cos\theta \, dx.$$

Because F is constant, this equation leads to

$$U_{1 \to 2} = F \cos\theta (x_2 - x_1). \tag{15.3}$$

Note that $F \cos\theta$ is constant during the integration on x.

We may interpret Eq. (15.3) in one of two ways. Thus, we may think of the work U_{1-2} as either the product of the scalar force component in the direction of the displacement $(F \cos\theta)$ multiplied by the magnitude of the linear displacement $x_2 - x_1$ (see Fig. 15.2(b)), or as the product of the magnitude of the force F multiplied by the scalar displacement component measured in the direction of the force $(x_2 - x_1)\cos\theta$ (see Fig. 15.2(c)). Of course the x axis need not be oriented horizontally, and the result is a general one for a constant force displaced along any straight-line path.

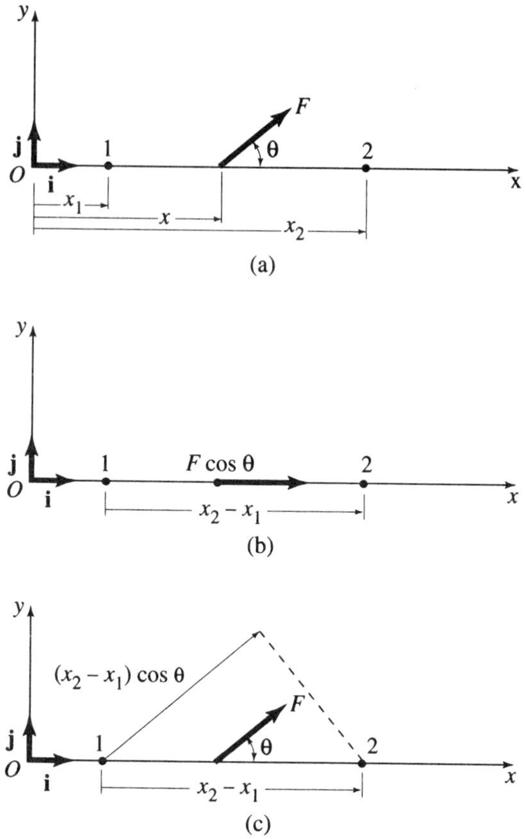

FIGURE 15.2.

Work of the Force of Gravity with g Assumed Constant

Refer to Figure 15.3(a) and imagine P to be any point on a path connecting points 1 and 2. We wish to develop an equation for the work done by the force of gravity as displacement of a particle of weight **W** occurs along this path from position 1 to position 2. The force of gravity (i.e., the weight **W**) may be written as $\mathbf{W} = -W\mathbf{j}$ where $W = mg$ is a constant since we regard the mass m and the acceleration due to gravity g remaining constant during the displacement. This approximation is very reasonable provided $(y_1 - y_2)$ does not take on large values. In other words, for earthbound engineering problems, the approximation is very satisfactory but for earth satellite motion it would not be acceptable. The differential work dU is given by

$$dU = -W\mathbf{j}\cdot d\mathbf{r} = -mg\mathbf{j}\cdot d\mathbf{r}.$$

The position vector **r** to any point on the path is expressed by

$$\mathbf{r} = x\mathbf{i} + y\mathbf{j}$$

where x and y are the coordinates of any position of the particle P on the path. Differentiating this equation yields

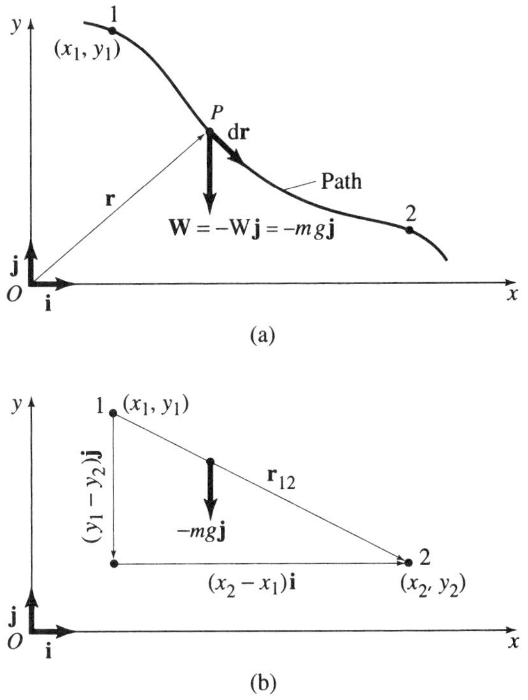

FIGURE 15.3.

$$\mathbf{dr} = dx\,\mathbf{i} + dy\,\mathbf{j}.$$

Substitute for **dr** into the equation for dU to obtain

$$dU = -mg\mathbf{j}\cdot(dx\,\mathbf{i} + dy\,\mathbf{j}) = -mg\,dy.$$

Integrating this equation from the initial point 1 to the final point 2 leads to

$$\int_1^2 dU = U_{1\to 2} = -mg\int_{y_1}^{y_2} dy$$

$$U_{1\to 2} = -mg(y_2 - y_1) = mg(y_1 - y_2) \qquad (15.4)$$

Equation (15.4) expresses the work of the force of gravity when both m and g are assumed constant. We observe that this result is independent of the equation of the path which connects point 1 to point 2 and depends only on the difference in elevation of points 1 and 2, that is $(y_1 - y_2)$. We also note from Eq. (15.4) that U_{1-2} is independent of the horizontal distance between points 1 and 2, that is, $(x_2 - x_1)$.

Work Done by a Constant Couple

The couple C shown in Figure 15.4(a) is represented by a ccw curved arrow. It has a constant magnitude, and, if we assume it always acts in the same plane, then, by the right hand rule, it would be represented by a vector perpendicular to that plane and would not change in direction

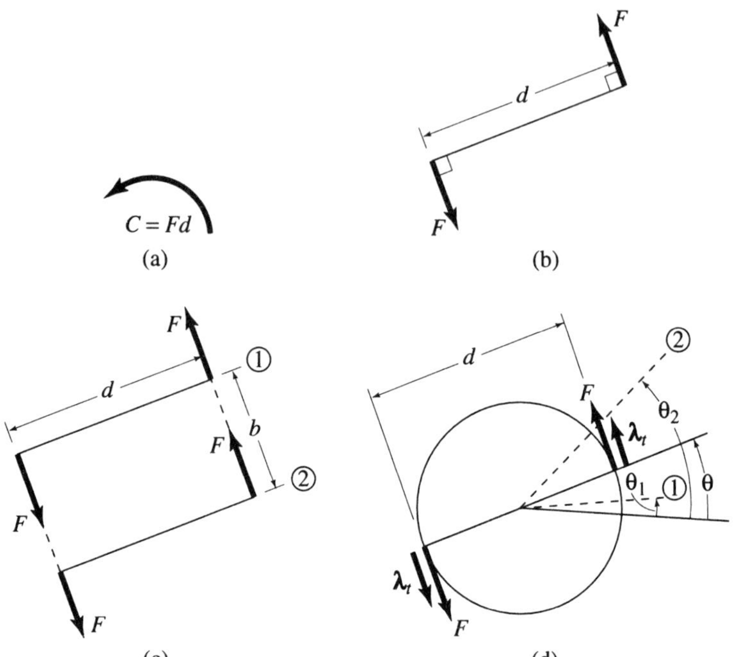

FIGURE 15.4.

or sense. We may also represent the couple $C = Fd$, as shown in Figure 15.4(b), using the fundamental definition of a couple.

First, consider the work done by a couple as it is translated from position 1 to position 2, as indicated in Figure 15.4(c), through a translatory displacement b. The force F on the right does positive work Fb and the force F on the left does negative work $-Fb$ which leads us to conclude that the work done by a couple in translation vanishes.

Next, inscribe a circle of diameter d which is tangent to the action lines of the forces **F** which comprise the couple as shown in Figure 15.4(d). Imagine the couple being rotated from an initial position $\theta = \theta_1$ to a final position $\theta = \theta_2$, and compute the work done by the rotated couple. Note that **F** always acts tangent to the circle or in the same direction as a unit tangent vector $\boldsymbol{\lambda}_t$ which enables us to represent **F** as $F\boldsymbol{\lambda}_t$. Using Eq. (15.1) and considering the fact that both forces perform work,

$$dU = 2\boldsymbol{\lambda}_t F \cdot \boldsymbol{\lambda}_t \, ds.$$

But $\boldsymbol{\lambda}_t \cdot \boldsymbol{\lambda}_t = 1$ which yields $dU = 2F\,ds$. For this circular path, $ds = \frac{d}{2} d\theta$. Substitute for ds to obtain

$$dU = Fd(d\theta).$$

Integrate from the initial to the final position to obtain

$$\int_1^2 dU = \int_{\theta_1}^{\theta_2} Fd\,d\theta$$

$$U_{1\to 2} = Fd(\theta_2 - \theta_1) = C(\theta_2 - \theta_1) \tag{15.5}$$

Equation (15.5) shows that the work done by a constant couple equals the product of the couple and the angular displacement $(\theta_2 - \theta_1)$ through which the couple rotates. This displacement is measured in radians which is a dimensionless quantity. Therefore, we conclude that the couple and its work have the same dimensions, that is, the product of a length unit and a force unit.

Work Done in Stretching or Compressing Linearly Elastic Springs

As discussed in Chapter 2 and 12, a linearly elastic spring is a spring which requires a force directly proportional to its deformation to stretch or compress it and, upon removal of the force, the spring returns to its undeformed state. The constant of proportionality which relates the spring force to its deformation is termed the spring constant k which is measured in force units per unit length, such as N/m or lb/in. Figure 15.5(a) shows the force-deformation line $F = ks$ for a linearly

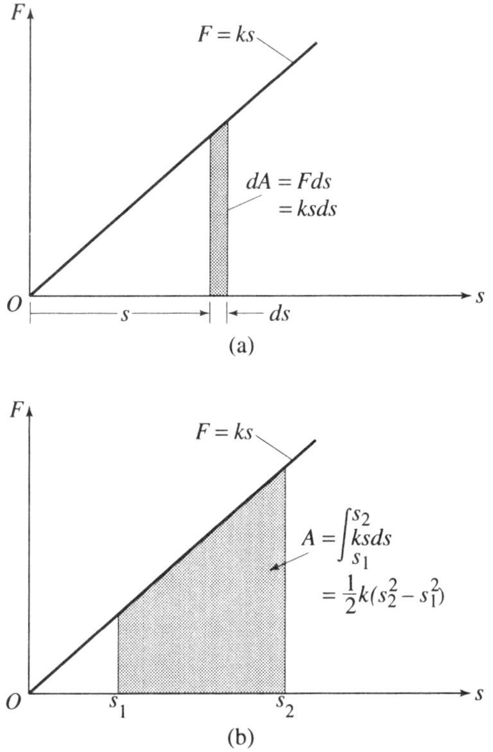

FIGURE 15.5.

elastic spring. If we double the deformation (either tensile or compressive) of a linear spring, then, the force required is also doubled, which is a convenient way to recall what we mean by linear behavior. Stretching or compressing an elastic spring is reversible which means that $F = ks$ is the same for loading or unloading the spring. When the force in a spring is zero, its length is termed its unstretched length L_u. The spring *deformation* s is found from the relationship

$$s = L - L_u \tag{15.6}$$

As stated earlier, the spring force F and its deformation s are related by

$$F = ks. \tag{15.7}$$

This function plots as a sloping straight line through the origin of an $F - s$ set of axes, as shown in Figure 15.5(a). The slope of this straight line equals the spring constant k.

To determine *the work done on a linearly elastic spring* we consider the force $F = ks$ corresponding to a deformation s and imagine a differential increase ds in deformation of the spring as shown in Figure 15.5(a). During this differential deformation, which is the displacement of the applied force F, we may consider F constant and write

$$dU = \mathbf{F} \cdot d\mathbf{r} = F\,ds = ks\,ds.$$

Integrating from s_1 to s_2, we determine the work done on the spring. Thus,

$$\int_1^2 dU = U_{1\to 2} = \int_{s_1}^{s_2} ks\,ds$$

$$U_{1\to 2} = \frac{k}{2}(s_2^2 - s_1^2). \tag{15.8a}$$

However, the work done by the spring on an attached body is given by

$$U_{1\to 2} = -\frac{k}{2}(s_2^2 - s_1^2) \tag{15.8a}$$

Equation (15.8a) shows that the *work done on the spring* equals the area under the $F - s$ curve as shown in Figure 15.5(b). If we ignore losses, this external work done on the spring by the applied force which varies from ks_1 to ks_2 is stored in the spring as energy. Because s_2 and s_1 are both squared in Eq. (15.8), it is clear that they will both be positive terms and we conclude that the work done on the spring is independent of whether we stretch it or compress it, provided we assume the spring does not buckle laterally or is prevented from doing so. Thus, as was discussed in Chapter 12, if we consider a spring attached to a rigid body, *the spring performs negative work on the body, and the body performs positive work on the spring when the system is displaced from the equilibrium position. Also, the spring performs positive work on the rigid body and the body performs negative work on the spring when the system returns toward the equilibrium position.*

Forces Which Do No Work

It is convenient to discuss and classify those forces which do no work. Refer to Figure 15.6(a) which shows a force **N** which is always normal to the path of motion and recall that $dU = \mathbf{N} \cdot \boldsymbol{\lambda}_t \, ds$. Since **N** and $\boldsymbol{\lambda}_t$ are perpendicular vectors, we conclude that their dot product vanishes

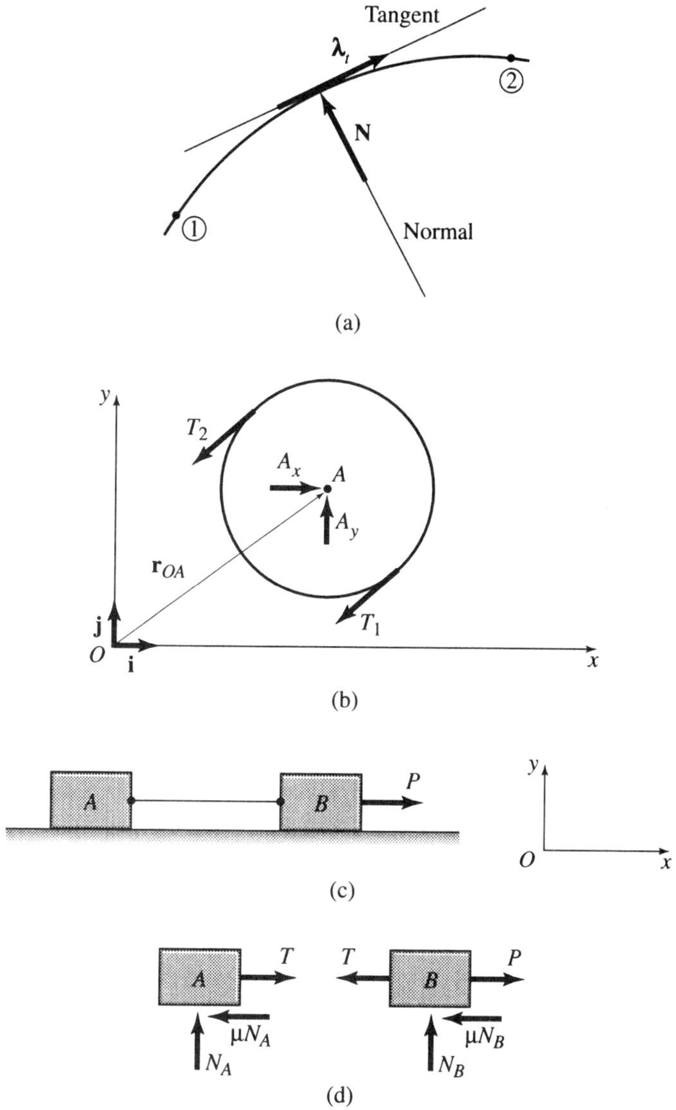

FIGURE 15.6.

and, therefore, dU is zero. Because $dU = 0$ and remains zero for all positions along the path from point 1 to point 2, we conclude that $U_{1\to 2} = 0$ *for forces which act normal to the path of motion.*

Forces which act at a fixed point are not displaced and, thus, do no work. Consider the reaction components A_x and A_y acting at point A, which represent the shaft of the pulley shown in Figure 15.6(b). Since A remains fixed during the rotation of the pulley, A_x and A_y do not displace, and the work done by the reaction components at A vanishes.

The adjectives *internal* and *external* are relative when they modify the word *force*. Depending upon the system considered, a force may be internal for one free body but external for another free body. Refer to Figure 15.6(c), and consider the system consisting of blocks A and B where the two blocks are connected by a rigid strut. An external force P is applied, which is large enough to move the system through a horizontal distance x_0 to the right on the smooth horizontal plane. It can be shown that the work done by the internal tension $\mathbf{T}$ in the inextensible strut on the combined system will vanish. In order to show this, we construct individual free-body diagrams of the two parts of the system, as shown in Figure 15.6(d), and compute the work done by the external tension $\mathbf{T}$ in the strut. Because $\mathbf{T}$ and x_0 are both along the positive x direction,

$$\text{Work on Body A} = \int_0^{x_0} (T\mathbf{i})\cdot(dx\,\mathbf{i}) = \int_0^{x_0} T\,dx$$

and

$$\text{Work on Body B} = \int_0^{x_0} -(T\mathbf{i})\cdot(dx\,\mathbf{i}) = -\int_0^{x_0} T\,dx.$$

If we add these results algebraically, the total work done by the tension $\mathbf{T}$ vanishes. *Thus, we conclude that forces internal to a rigid body do no work.* A convenient way to remember that forces do no work on a system for which they are internal is to recall that, by Newton's Third Law, internal forces occur in equal and opposite pairs, and, if one force of the pair does positive work, then, the equal and opposite force with which it is paired will do an equal amount of negative work such that the net amount of work done is zero. If the strut were deformable, then, work could be done on it internally and energy stored in it just as energy is stored in a spring.

The following examples illustrate some of the above concepts.

■ Example 15.1

For each situation depicted in Figure E15.1, determine the work done by the force P. Other forces may act on the body and do work on it, but we are only concerned with the work done by $\mathbf{P}$. In each case, 1

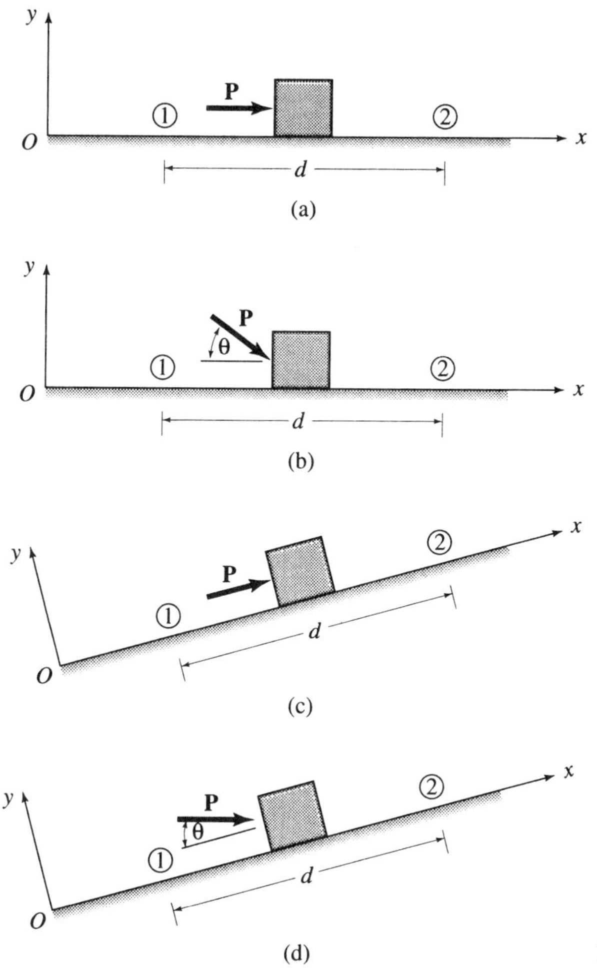

FIGURE E15.1.

denotes the initial position and 2 the final position of the body. The force P remains constant in magnitude and direction throughout the displacement from 1 to 2.

Solution

Refer to the development of Eq. (15.3) and use it as the basis of our solution. Recall that work is the scalar or dot product of the force and displacement vectors.

(a) Using Eq. (15.3) and noting that $\theta = 0$ and $\cos\theta = 1$,

$$U_{1\to 2} = P(1)(x_2 - x_1) = P(x_2 - x_1) = Pd. \qquad \text{ANS.}$$

(b) Making use of Eq. (15.3) and noting that $P\cos\theta$ and the displacement $(x_2 - x_1)$ are both directed in the same sense,

$$U_{1\to 2} = P\cos\theta(x_2 - x_1) = Pd\cos\theta. \qquad \text{ANS.}$$

(c) We note that the sense of the force and displacement agree because both are directed up the inclined plane. Thus, using Eq. (15.3) with $\theta = 0$,

$$U_{1\to 2} = P(1)(x_2 - x_1) = Pd. \qquad \text{ANS.}$$

(d) By Eq. (15.3),

$$U_{1\to 2} = P\cos\theta(x_2 - x_1) = Pd\cos\theta. \qquad \text{ANS.}$$

Observe that, in each of the cases above, the force did positive work on the body. If we were to interchange the initial and final positions in the figure, then, the force would do negative work on the body. In this latter case, the force would oppose the motion of the body.

Example 15.2

Assume that the 50-N force shown in Figure E15.2 remains constant in magnitude and direction. Determine (a) the work done by all forces acting on the block of mass m = 10 kg as it moves from initial position 1 to final position 2 and (b) the work done by all forces acting on the block if we interchange positions 1 and 2, i.e., the block moves down the plane. In both cases, assume that $\mu_k = 0.2$ and that, in the initial

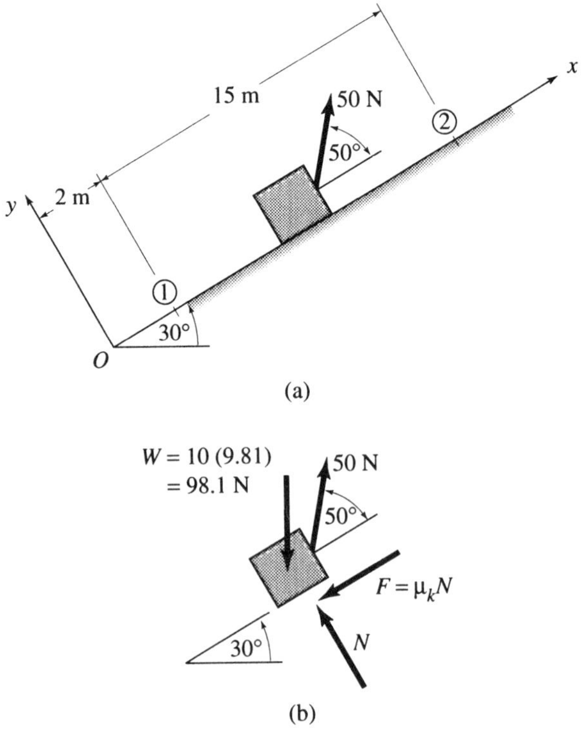

FIGURE E15.2.

position, the block has enough energy of motion to assure that it will reach the final position regardless of the sign of the work done.

Solution

(a) The free body diagram of the block shown in Figure E15.2(b) applies for motion up the inclined plane since the friction force opposes motion, i.e., it acts down the inclined plane. Consider each force separately and note that it remains constant during the displacement.

Work of weight W
By Eq. (15.4),

$$\begin{aligned} U_{1\to2} &= mg(y_1 - y_2) \\ &= 10(9.81)(2\sin 30^\circ - 17\sin 30^\circ) \\ &= -735.75\ \text{N}\cdot\text{m}. \end{aligned}$$

Work of 50-N force
By Eq. (15.3)

$$\begin{aligned} U_{1\to2} &= F\cos\theta(x_2 - x_1) \\ &= 50\cos 30^\circ(17 - 2) \\ &= 482.09\ \text{N}\cdot\text{m}. \end{aligned}$$

Work of friction force *F*

Although the normal force N does no work, we need to find it to determine the friction force F which does work on the block. Thus,

$$\sum F_n = 0:\ N - 98.1\cos 30^\circ + 50\sin 50^\circ = 0$$

$$N = 46.65\ \text{N}.$$

Therefore,

$$F = \mu_k N = 0.20\,(46.65) = 9.33\ \text{N}.$$

The work done by friction is negative because the block is displaced up the plane but the friction force acts down the plane. We account for this by using $\theta = 180^\circ$ in Eq. (15.3). Thus,

$$U_{1\to2} = F\cos\theta(x_2 - x_1) = 9.33\cos 180^\circ(17 - 2) = -139.95\ \text{N}\cdot\text{m}$$

Work is a scalar quantity, and we simply add the foregoing values algebraically to determine the total work done by all of the forces. Thus,

$$U_{1\to2} = -735.75 + 482.09 - 139.95 = -394\ \text{N}\cdot\text{m}. \qquad \text{ANS.}$$

The negative sign means that the body will have its speed reduced as it moves from 1 to 2.

(b) Because the block moves down the plane, the friction force in the free-body diagram of Figure 15.2(b) will be reversed in sense. Therefore, the work performed by it during the downward motion of the block will still be negative. However, the quantities of work performed by W and by the 50-N force will be the same in magnitude but opposite in sense to those found in part (a). Thus, the total work performed by all forces acting on the block during its downward motion will be

$$U_{1\to 2} = 735.75 - 482.09 - 139.95 = 113.7 \text{ N}\cdot\text{m}. \qquad \text{ANS.}$$

■ Example 15.3

Refer to arm OP shown in Figure E15.3(a) which rotates in a vertical plane about an axis at O. Four positions of interest are indicated. Points O and Q are fixed, and the arm of negligible weight has enough energy to move around the circle through the four positions indicated. The arm which is hinged at point O has a particle weighing 100 lb attached at its center, point G. A constant couple $C = 400$ lb·ft acts on the arm. Attached to the rod at point P is a spring with an unstretched length of $L_u = 5$ ft. and a spring constant $k = 150$ lb/ft. The hinge at O may be assumed to be frictionless. Determine the total work done on the rod as it moves from position 1 to (a) position 2, (b) position 3, and (c) position 4.

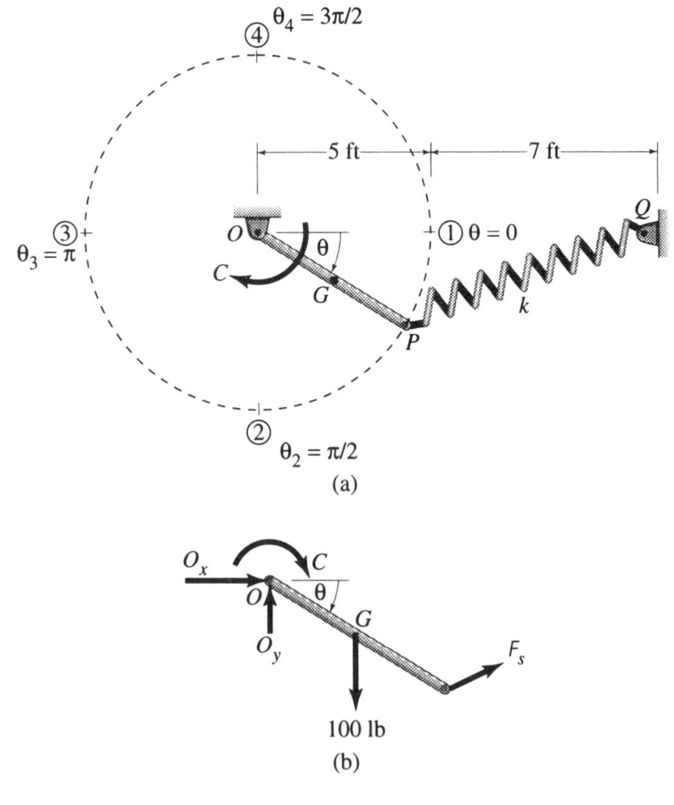

FIGURE E15.3.

Solution

The free-body diagram of the rotating arm OP is shown in Figure E15.3(b) for any position defined by the angle θ. The fixed reactions at O, O_x and O_y, do no work on the rod because they are not displaced. Because the hinge at O is frictionless, there is no couple associated with friction. The total work U_T performed on the arm is the algebraic sum of the work done by the weight W, the couple C, and the spring force F_S. Thus,

$$U_T = U_W + U_C + U_S$$

where U_W is given by Eq. (15.4), U_C by Eq. (15.5), and U_s by Eq. (15.8) after changing the sign because we are now dealing with the work of the spring on the arm.

(a) For position 1, $\theta = 0$,

$$L_1 = 7 \text{ ft}, \quad \text{and} \quad s_1 = L - L_u = 7 - 5 = 2 \text{ ft}.$$

For position 2, $\theta = \pi/2$,

$$L_2 = \sqrt{12^2 + 5^2} = 13 \text{ ft}, \quad \text{and} \quad s_2 = L_2 - L_u = 13 - 5 = 8 \text{ ft}.$$

Therefore,

$$U_T = W(y_1 - y_2) + C(\theta_1 - \theta_2) - \frac{1}{2}k(s_2^2 - s_1^2)$$

$$= 100[0 - (-2.5)] + 400\left(\frac{\pi}{2} - 0\right) - \frac{1}{2}(150)(8^2 - 2^2)$$

$$= -3620 \text{ ft·lb.} \qquad \text{ANS.}$$

(b) For position 3, $\theta = \pi$,

$$L_3 = 5 + 5 + 7 = 17 \text{ ft}, \quad \text{and} \quad s_3 = L_3 - L_u = 17 - 5 = 12 \text{ ft}.$$

Thus,

$$U_T = W(y_1 - y_3) + C(\theta_3 - \theta_1) - \tfrac{1}{2}k(s_3^2 - s_1^2)$$
$$= 100[0 - 0] + 400(\pi - 0) - \tfrac{1}{2}(150)(12^2 - 2^2) = -9240 \text{ ft·lb.} \qquad \text{ANS.}$$

(c) For position 4, $\theta = 3\pi/2$,

$$L_4 = \sqrt{12^2 + 5^2} = 13 \text{ ft}, \quad \text{and} \quad s_4 = L_4 - L_u = 13 - 5 = 8 \text{ ft}.$$

Therefore,

$$U_T = W(y_1 - y_4) + C(\theta_4 - \theta_1) - \frac{1}{2}k(s_4^2 - s_1^2)$$

$$= 100[0 - (2.5)] + 400\left[3\left(\frac{\pi}{2}\right) - 0\right] - \frac{1}{2}(150)(8^2 - 2^2)$$

$$= -2870 \text{ ft·lb.} \qquad \text{ANS.}$$

We observe that all three answers are negative which means that the system is being *slowed down* or decelerated by these working forces. If the system were to move in a horizontal plane, rather than in a vertical plane, then, the work of gravity would vanish because point G would not change elevation.

■

Problems

Assume that the bodies have enough kinetic energy in their initial positions to assure that they will move to their final positions regardless of the total work done on them.

15.1 A block weighing 200 lb moves along a horizontal plane as shown in Figure P15.1. The constant force P has a magnitude of 50 lb and the coefficient of kinetic friction between the block and plane is 0.2. Determine the total work done on the block as it moves through a displacement of 10 ft directed to the right.

15.2 A block weighing 600 N moves along a horizontal plane as shown in Figure P15.2. The force P has a magnitude of 100 N. If the coefficient of kinetic friction between the block and plane is 0.1, determine the total work done on the block as it moves through a displacement of 4 m to the right.

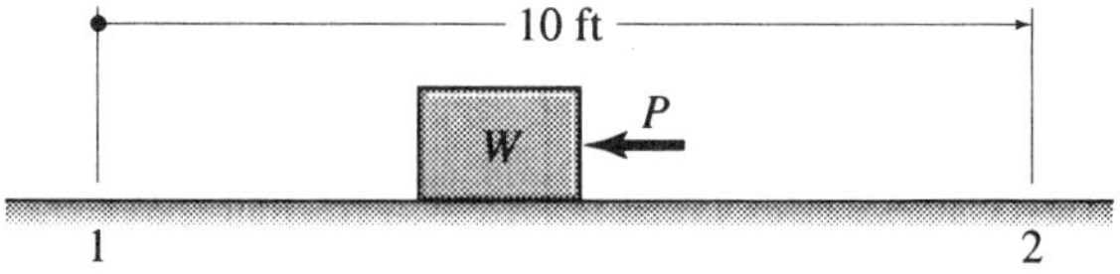

FIGURE P15.1.

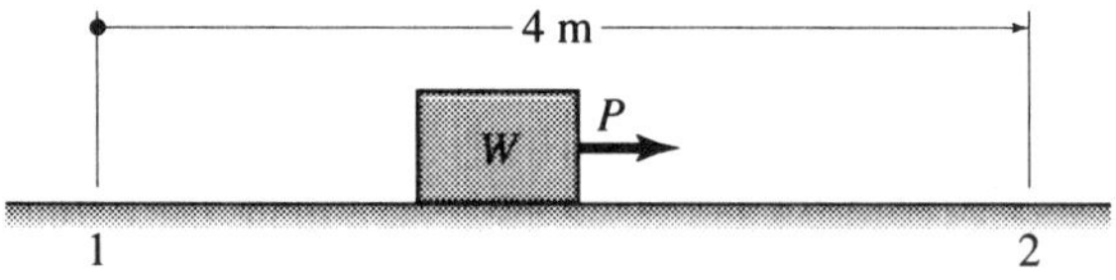

FIGURE P15.2.

15.3 A block weighing 500 lb moves 30 ft down an inclined plane as shown in Figure P15.3. If the coefficient of kinetic friction between the block and plane is 0.1, determine the total work done on the block as it moves 30 ft down the plane.

15.4 Determine the total work done on the block as it moves 6 m up the inclined plane shown in Figure P15.4. The block

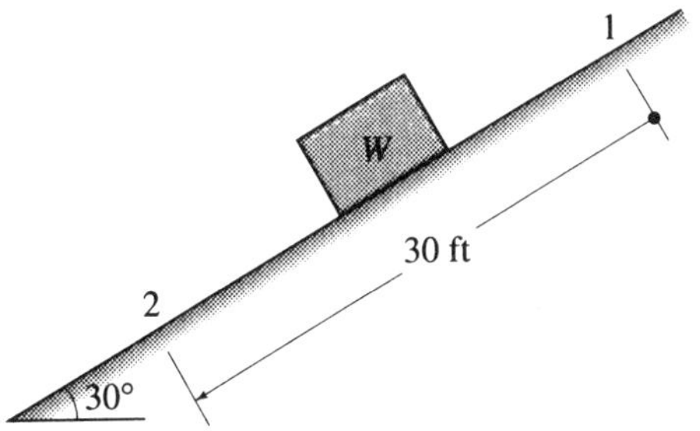

FIGURE P15.3.

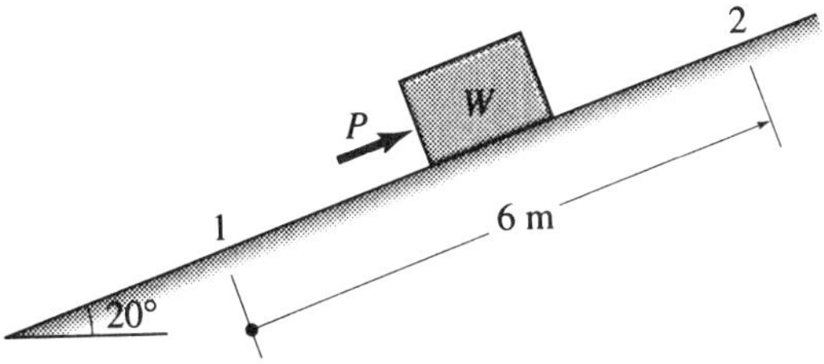

FIGURE P15.4.

has a mass of 120 kg, the constant force P has a magnitude of 600 N, and the coefficient of kinetic friction for the block and plane is 0.1.

15.5 A block with a mass of 10 slugs moves horizontally in frictionless guides, as shown in Figure P15.5. A spring ($k = 50$ lb/in) is attached to the block as indicated. The unstretched length of the spring is 1.00 ft. Determine the total work done on the block as it moves 4 ft horizontally to the left.

15.6 Determine the total work done on the block shown in Figure P15.6 as it moves 6 m up the inclined plane. The spring attached to the block has a constant $k = 10{,}000$ N/m and an unstretched length of 2 m. The block has a mass of 120 kg and the coefficient of kinetic friction between the block and plane is 0.1.

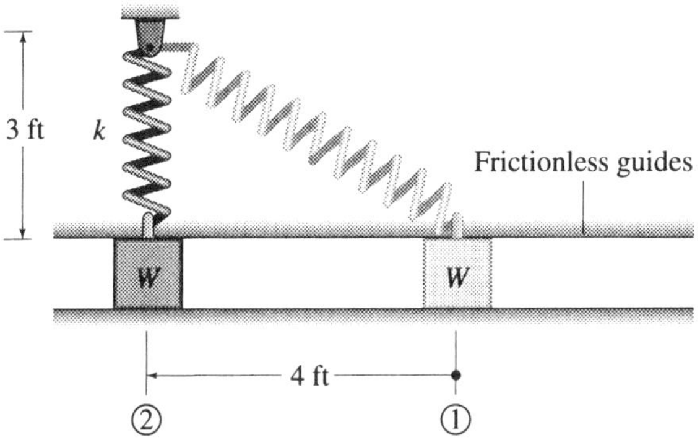

FIGURE P15.5.

15.7 Determine the total work done on bar OA shown in Figure P15.7 as it rotates from vertical position 1 to horizontal position 2 through a 90° ccw angle. The bar weighs 150 lb and the spring constant $k = 600$ lb/ft. The unstretched length of the spring is 1.5 ft and end B of the spring remains fixed as the bar rotates.

15.8 Rod AB weighs 800 N and is subjected to a constant couple C as indicated in Figure P15.8. Determine the couple C such that the total work done on the rod is $+1500$ N·m as it rotates from position 1 to position 2. Assume the hinge at A to be frictionless.

15.9 Two weights W are connected by a spring which is unstretched as they lie

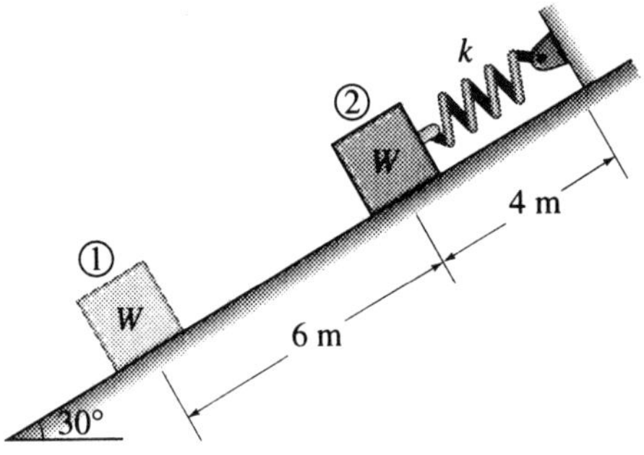

FIGURE P15.6.

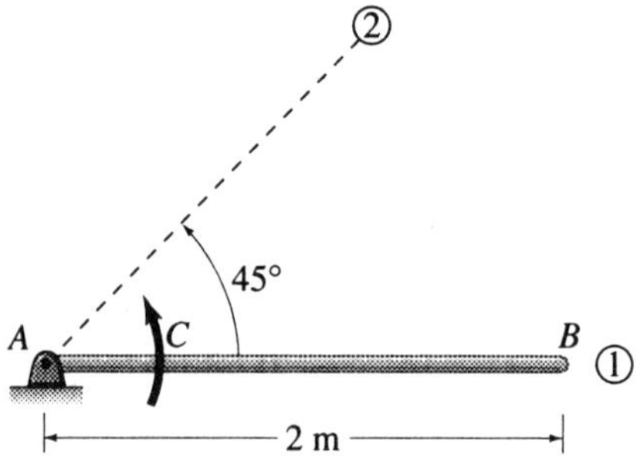

FIGURE P15.8.

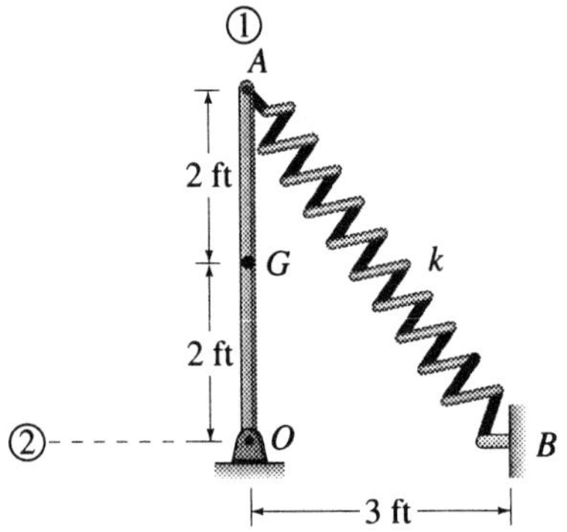

FIGURE P15.7.

initially at rest on a horizontal table, as shown in Figure P15.9. The spring is carefully draped over a fixed peg P which is indicated as position 2. Develop an equation for the work done on the spring in terms of W and k. Ignore the size of peg P in your analysis.

15.10 In Figure P15.10, two very light rods, AB and DE, whose weights may be ignored, are attached to a spring which is unstretched in position 1. Couples C are applied to the rods and the rods are rotated to position 2 as indicated. Determine an expression for the total work done on the two rods. All connecting pins are frictionless.

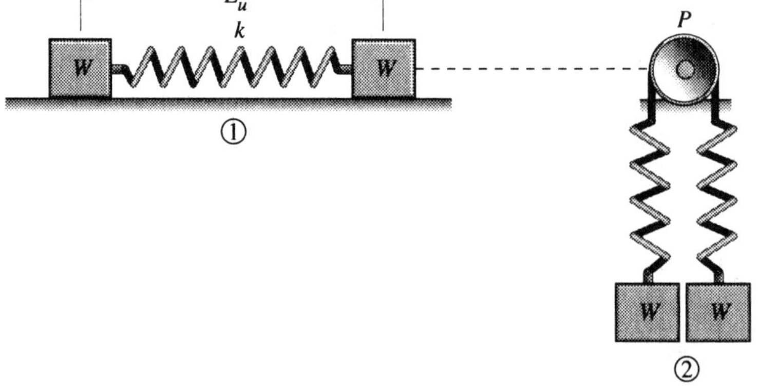

FIGURE P15.9.

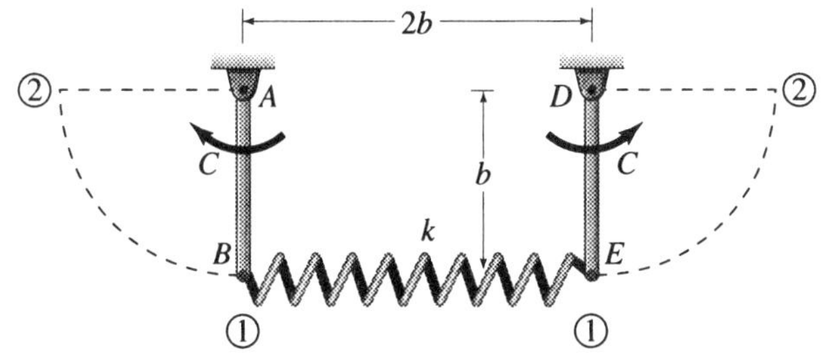

FIGURE P15.10.

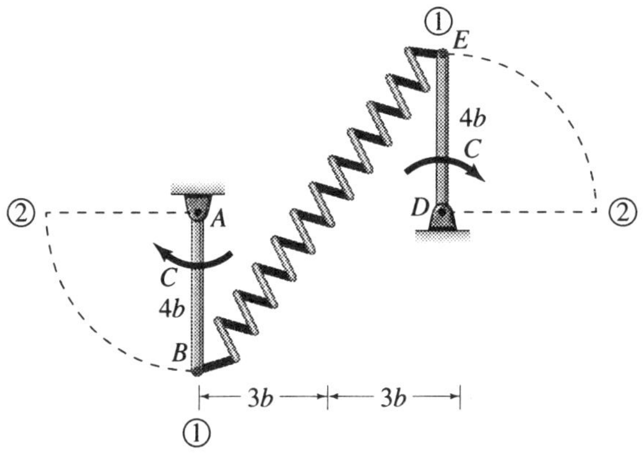

FIGURE P15.11.

15.11 As shown in Figure P15.11, two very light rods, AB and DE, whose weights may be ignored, are attached to a spring which is unstretched in position 1. Couples C are applied to the rods and the rods are rotated to position 2 as indicated. Determine an expression for the total work done on the two rods.

15.12 Solve Problem 15.10 if each of the rods is uniform and has a weight equal to W. All other information given is unchanged.

15.13 Solve Problem 15.11 if each of the rods is uniform and has a weight equal to W. All other information given is unchanged.

15.14 A very light rod AB supports a weight W at B and is attached to a spring which is fixed at D as shown in Figure P15.14. A constant couple C is applied to the rod, and it rotates through an angle θ to a new position such that B travels a circular arc to B′. If the total work done on the rod vanishes, determine C as a function of θ, b, W, and k. The spring has an unstretched length of b.

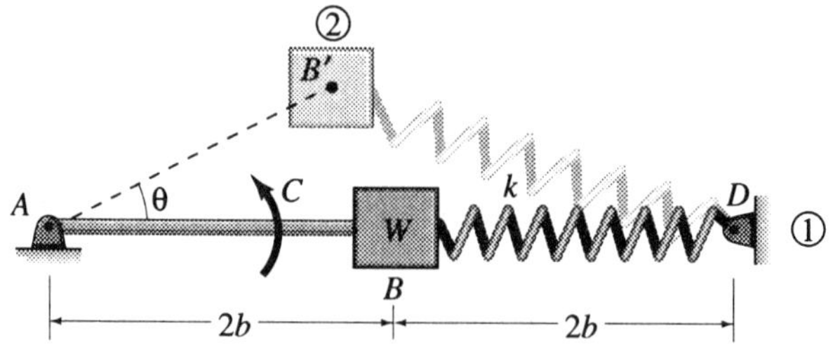

FIGURE P15.14.

15.2 Kinetic Energy—The Energy of Motion

Quantities used in science and engineering must be precisely defined. The criteria for defining a given quantity are that it be useful in enabling the investigator to find solutions and that it leads to time saving in thinking and discussing ideas and in recording results. Kinetic energy is an example of such a quantity. The kinetic energy of a particle is defined as

$$T = \tfrac{1}{2}m\mathbf{v}\cdot\mathbf{v} = \tfrac{1}{2}mv^2 \qquad (15.9)$$

where the symbol T represents the kinetic energy, m the mass, and v the speed of the particle. Note that the kinetic energy, by definition, is a scalar quantity because it involves a scalar multiplier of $1/2m$ and the dot or scalar product of the velocity vector $\mathbf{v}$ with itself. This scalar quantity is evaluated instantaneously because, in general, the speed v is a function of time.

15.3 Work-Energy Principle for a Single Particle

The work-energy principle, to be derived, relates scalar quantities arising from the dot or scalar products of vectors. Refer to the particle P of mass m shown at any time t along its path in space as depicted in Figure 15.7(a). The differential work dU during a differential displacement $d\mathbf{r}$ of the particle was defined earlier and is repeated here for convenience. Thus,

$$dU = \sum \mathbf{F}\cdot d\mathbf{r} = \mathbf{R}\cdot d\mathbf{r} = (R\cos\theta)\,ds.$$

From Figure 15.7(b) we note that $R_t = R\cos\theta$ is the tangential component of the resultant force R acting on the particle. As stated in Section 15.1, the normal component $R_n = R\sin\theta$ performs no work as the particle moves along its path. By Newton's second law, the tangential component equals the product of the mass of the particle and the tangential component of acceleration. Thus, applying Newton's second law to the motion of the particle along the tangent to the path

$$R_t = ma_t.$$

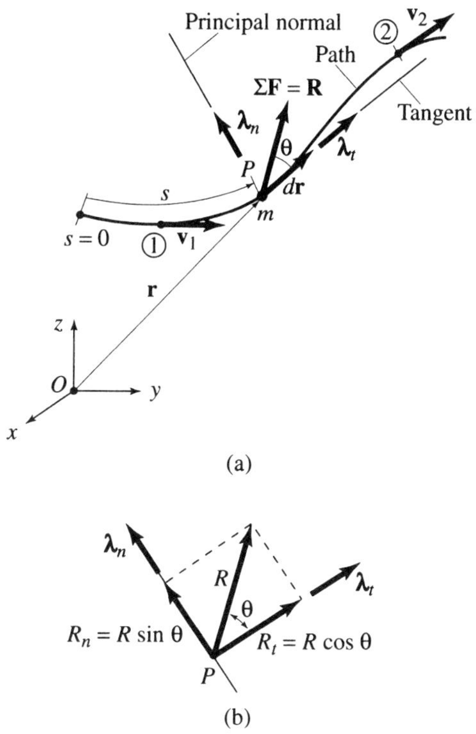

FIGURE 15.7.

Because $dU = R_t\,ds$, by definition, we conclude that

$$dU = ma_t\,ds.$$

Using the relationship $a_t = v\dfrac{dv}{ds}$ from Eq. (13.7), the above equation yields

$$dU = mv\frac{dv}{ds}ds = mv\,dv.$$

The particle moves from initial position 1 to final position 2 over the path as indicated in Figure 15.7(a). Integrating,

$$\int_1^2 dU = m\int_{v_1}^{v_2} v\,dv = m\frac{v^2}{2}\bigg|_{v_1}^{v_2}$$

$$U_{1\to 2} = \tfrac{1}{2}mv_2^2 - \tfrac{1}{2}mv_1^2 \tag{15.10}$$

We recognize $U_{1\to 2}$ as the work done on the particle by the sum of all the forces which act upon it as it moves along the path from initial position 1 to final position 2 and the terms on the right hand side as kinetic energies T_2 and T_1. Therefore,

$$U_{1\to 2} = T_2 - T_1 \tag{15.11}$$

This single scalar equation states that the work done on the particle as it moves along its path from the initial to the final position equals the change in kinetic energy of the particle $T_2 - T_1$. Equation (15.11) is known as the *principle of work-energy*. It is well to emphasize that the kinetic energy T is evaluated when the particle is in the initial and final positions, i.e., instantaneously, and that the work $U_{1\to 2}$ is done on the particle by the forces which act on it as it moves along its path from position 1 to position 2. In general, the work $U_{1\to 2}$ done on the particle will depend upon the path.

The following examples illustrate the use of the principle of work-energy in the solution of problems.

Example 15.4

The block shown in Figure E15.4(a) has a mass of 10 kg. In position 1 it is moving up the inclined plane with a speed of 8 m/s. (a) If the coefficient of static friction is 0.4 and that of kinetic friction is 0.1, find the distance s the block will move along the plane before coming to rest. (b) Determine the speed of the block when it returns to position 1.

Solution

(a) For motion up the plane, the free body diagram is shown in Figure E15.4(b). Since the block does not accelerate perpendicular to the plane,

$$\sum F_y = 0: N - mg\cos 30^\circ = 0$$

$$N = 10(9.81)(0.866) = 84.95 \text{ N}.$$

The friction force is given by

$$F = \mu_k N = 0.1(84.95) = 8.5 \text{ N}.$$

The force of gravity W and the friction force are the forces that do work. These forces remain constant during the motion. Because velocities and displacement are involved in the problem statement, the work-energy principle is an appropriate choice. Thus,

$$U_{1\to 2} = T_2 - T_1$$

where $T_2 = 0$. Therefore,

$$-(98.1)\, s\,(\sin 30^\circ) - 8.50\text{ s} = 0 - \tfrac{1}{2}(10)(8)^2$$

from which

$$s = 5.56 \text{ m}. \qquad \text{ANS.}$$

(b) For motion down the plane, the friction force acts up as shown in Figure E15.4(c). Thus,

$$U_{2\to 1} = T_1 - T_2$$

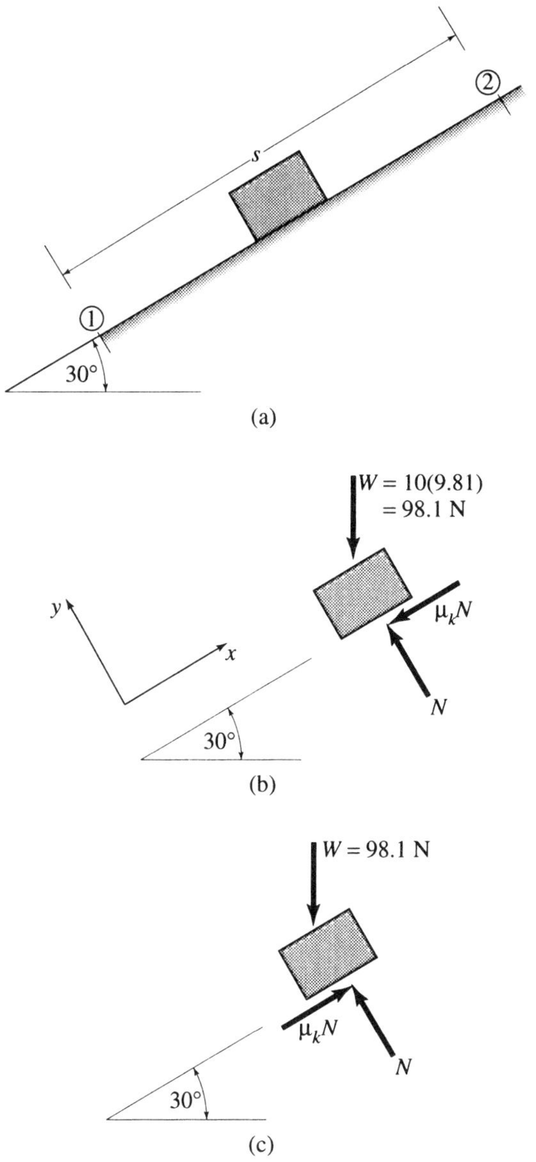

FIGURE E15.4.

where, once again, $T_2 = 0$. Therefore,

$$-98.1(5.56)(\sin 30°) - 8.5(5.56) = \tfrac{1}{2}(10)v_1^2 - 0$$

from which

$$v_1 = 6.72 \text{ m/s}. \qquad \text{ANS.}$$

The block began its journey up the plane with a speed of 8 m/s but has returned to the same position with a lower speed of 6.72 m/s. This loss of kinetic energy is accounted for by the friction force which dissipated mechanical energy by increasing the temperature of the plane and the block.

■ Example 15.5

The 4-lb collar C, attached to a spring as shown in Figure E15.5(a) slides along the frictionless circular rod which lies in a vertical plane.

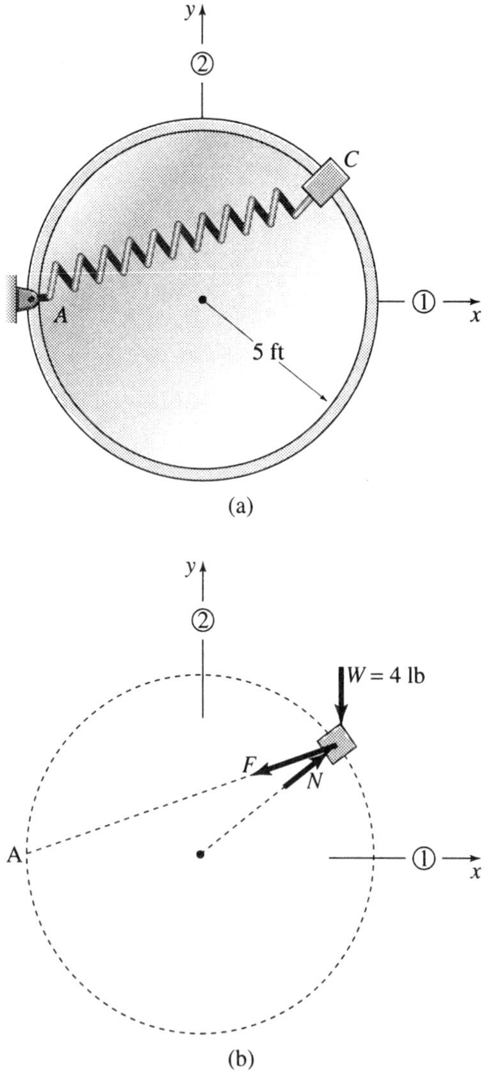

FIGURE E15.5.

The spring has an unstretched length $L_u = 6$ ft and, when in position 1, the collar has a speed of 8 ft/s. If the spring constant is $k = 10$ lb/ft, determine the speed of the collar when it reaches position 2.

Solution

The free-body diagram of the collar for any position between 1 and 2 is shown in Figure E15.5(b). Note that the spring force F is directed from the collar to point A where the spring is attached. As the collar moves from position 1 to position 2, the work done by the spring force on the collar is given by

$$(U_{1\to 2})_S = \tfrac{1}{2}k(s_1^2 - s_2^2)$$

where

$$s_1 = L - L_u = 10 - 6 = 4 \text{ ft}$$

and

$$s_2 = L - L_u = \sqrt{5^2 + 5^2} - 6 = 1.07 \text{ ft}.$$

Therefore,

$$(U_{1\to 2})_S = \tfrac{1}{2}(10)(4^2 - 1.07^2) = 74.28 \text{ ft·lb}.$$

Also, the normal force N between the smooth rod and the collar does no work as it is always perpendicular to the direction of motion. The work performed by the weight of the collar is given by

$$(U_{1\to 2})_W = mg(y_1 - y_2) = 4(-5) = -20 \text{ ft·lb}.$$

The kinetic energy in position 1 is

$$T_1 = \frac{1}{2}mv_1^2 = \frac{1}{2}\left(\frac{4}{32.2}\right)(8^2) = 3.975 \text{ lb·ft}.$$

The kinetic energy in position 2 is

$$T_2 = \frac{1}{2}mv_2^2 = \frac{1}{2}\left(\frac{4}{32.2}\right)v_2^2.$$

Thus, the work-energy equation, $U_{1\to 2} = T_2 - T_1$, becomes

$$74.28 - 20 = 0.0621v_2^2 - 3.975$$

from which

$$v_2 = 30.6 \text{ ft/s}. \qquad \text{ANS.}$$

■

Problems

15.15 The variable force shown acting on the block in Figure P15.15 is given by $F(s) = 4s^2 + 2s + 9$ where s is measured in ft and F is given in lb. If the block weights 150 lb and starts from rest, determine its velocity after it has moved 10 ft to the right. What is the initial acceleration of the block? Neglect friction and air resistance. Assume that $s = 0$ at $t = 0$.

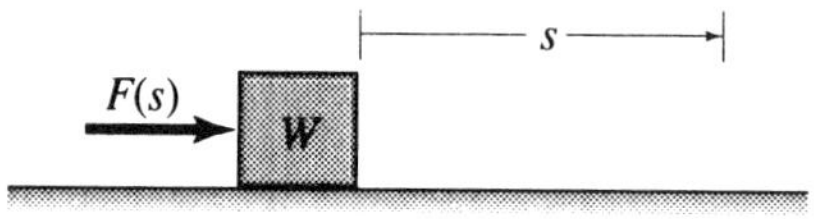

FIGURE P15.15.

15.16 A variable force acts parallel to the inclined plane shown in Figure P15.16 and is given by $F(s) = 30s + 200$ where s is measured in m and F is given in N. If the block weighs 200 N and starts from rest, determine its velocity after it has moved 5 m up the plane. What is the initial acceleration of the block? Neglect friction and air resistance. Assume that $s = 0$ at $t = 0$.

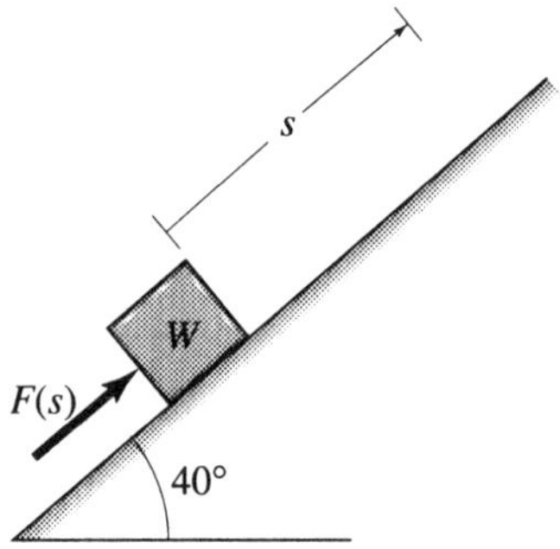

FIGURE P15.16.

15.17 A mass m is free to slide on a horizontal frictionless plane and is attached to a spring which is fixed at point O as shown in Figure P15.17. If the mass is released from rest in position 1, determine its velocity when it reached position 2 in terms of k, m, and b. In position 3, the spring is unstretched.

15.18 In Figure P15.18, the spring is horizontal in position 1 and perpendicular to the inclined plane in position 2. The mass m has a hole drilled in it which enables it to move along the rod without friction. If this mass is released from rest in position 1, determine its velocity when it reaches position 2 as a function

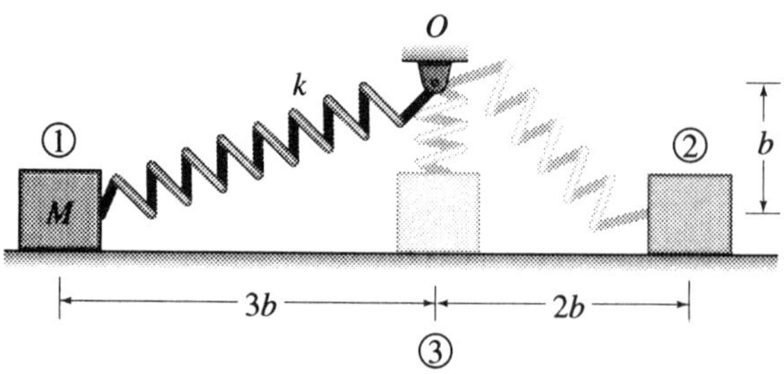

FIGURE P15.17.

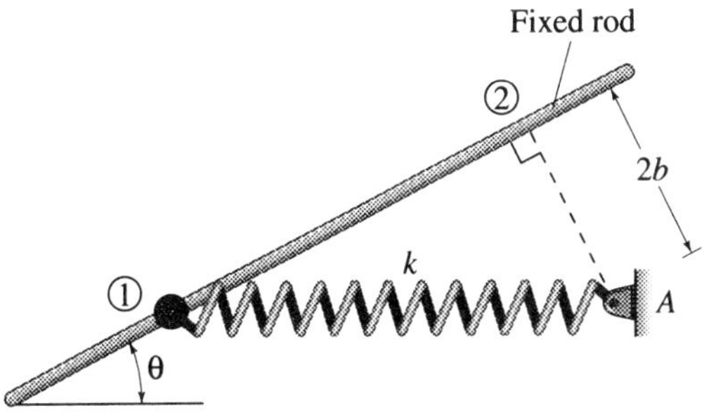

FIGURE P15.18.

of m, k, θ, and b. The unstretched length of the spring is b.

15.19 Refer to Problem 15.18. Let $\theta = 30°$, and determine the acceleration of the mass when it is in position 1 in terms of k, b, and g. Also, determine its acceleration when it is in position 2.

15.20 A block of mass m is placed on an inclined plane as shown in Figure P15.20 with an initial velocity v_1. If the kinetic coefficient of friction is μ determine, v_2 as a function of v_1, θ, m, μ, d, and g. What condition must be satisfied among these quantities which will assure that $v_2 \geq 0$?

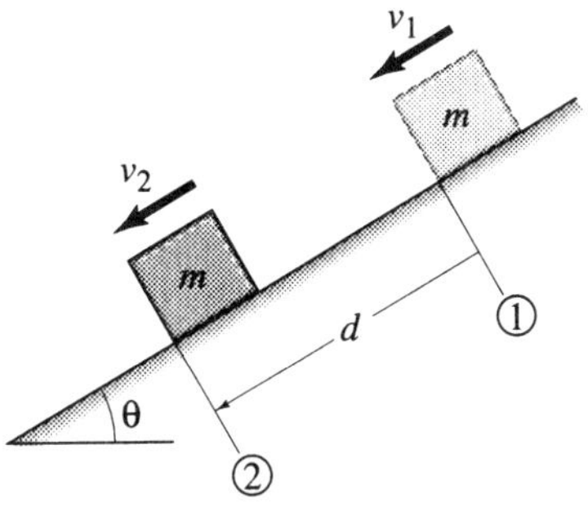

FIGURE P15.20.

15.21 Rod ABC shown in Figure P15.21 has negligible weight. When the system is in position 1, the mass m has a velocity $v_1 = 15$ ft/s directed horizontally to the left. The spring is attached to the rod at B and fixed point D. Let $m = 0.31$ slug, $b = 2$ ft, and $k = 5$ lb/ft. Then, determine the velocity v_2 for $\theta = 45°$.

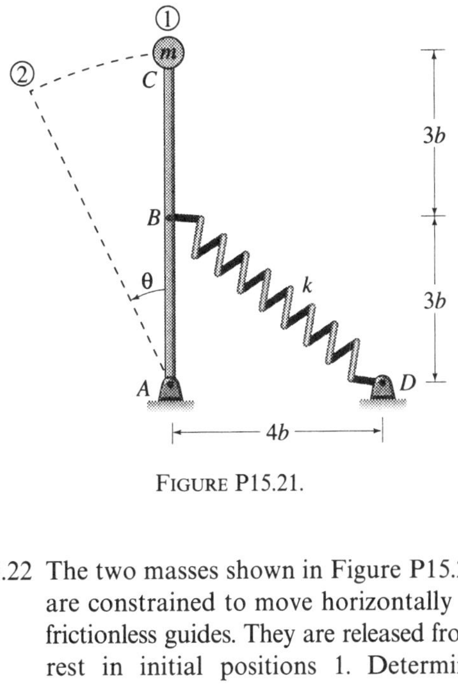

FIGURE P15.21.

15.22 The two masses shown in Figure P15.22 are constrained to move horizontally in frictionless guides. They are released from rest in initial positions 1. Determine

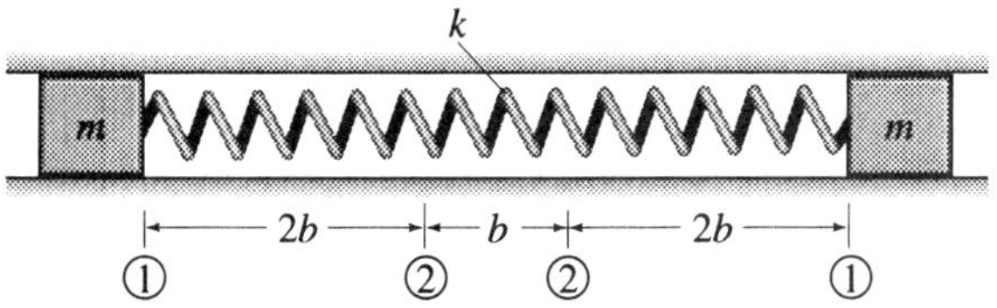

FIGURE P15.22.

their relocities in final positions denoted by '2' as a function of k, b, and m. The unstretched length of the spring is b.

15.23 Mass m is attached to a very light rod which rotates about A as shown in Figure P15.23. It is released from rest in position 1. Determine the speed of m in position 2 as a function of r, θ, and g. What is the corresponding speed of m when it arrives at position 3? Specialize your results for $\theta = 90°$. Ignore friction at the pin A and air resistance.

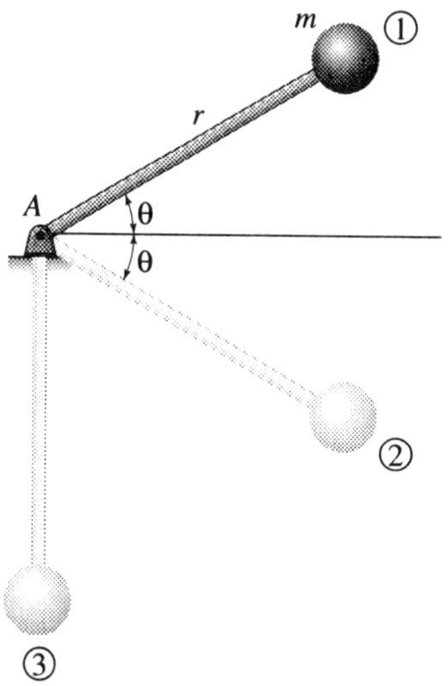

FIGURE P15.23.

15.24 Determine the speed v_1 of mass m, shown in Figure P15.24, such that, when the rod has rotated cw through 90°, the final speed of mass m will be zero. The unstretched length of the spring is $3b$. Express v_1 as a function of m, k, g and b. What relationship must hold among these quantities to assure a solution? Neglect the rod mass, air resistance and pin friction at A.

15.25 Mass m slides along a frictionless vertical rod as shown in Figure P15.25. It is released from rest in position 1 and the spring has an unstretched length of b. Determine the speed of the mass when it arrives at position 2. Express your answer in terms of k, m, g, and y.

15.26 As shown in Figure P15.26, the mass m is held against a spring which has a deformation δ and then released in position 1. The spring is not attached to the mass. Determine the velocity of the mass when it arrives at position 2 as a function of k, δ, m, g, L, and θ. Find a numerical value for this velocity corresponding to $k = 600$ lb./ft., $\delta = 1$ ft., $m = 0.5$ slug, $g = 32.2$ ft./s^2, $L = 6$ ft. and $\theta = 30°$. Next, the particle becomes a projectile launched from position 2,

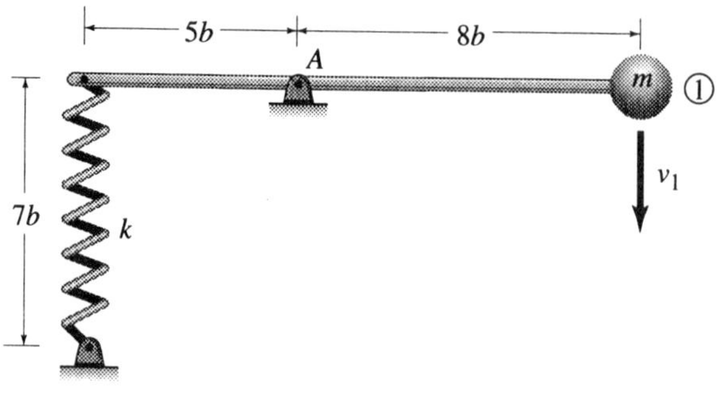

FIGURE P15.24.

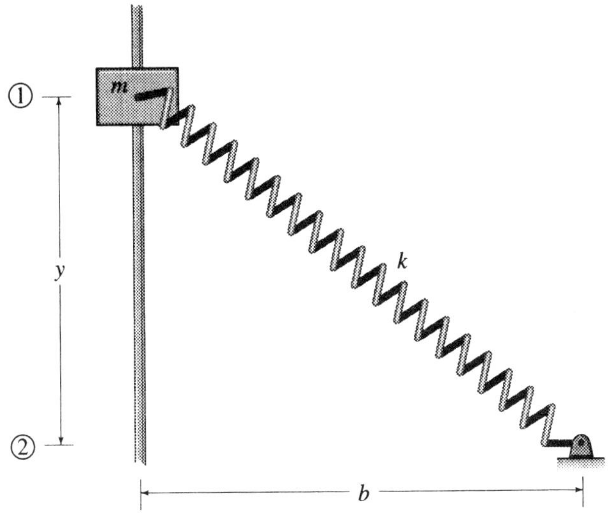

FIGURE P15.25.

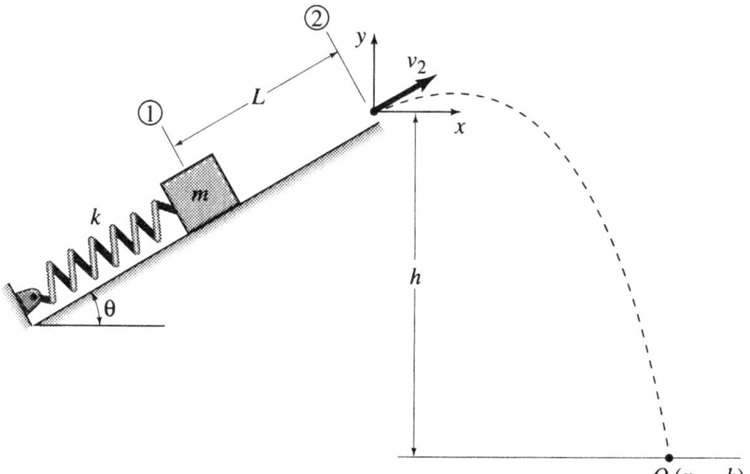

FIGURE P15.26.

where an origin to study this motion is shown in the figure. Determine the coordinate x_1 of point Q where the projectile strikes the horizontal plane if h is given a numerical value of 30 ft. Ignore friction and air resistance.

15.4 Work-Energy Principle for a System of Particles

In this section we restrict our discussion to a system of particles which are connected by inextensible cables or struts. In Chapter 19, the work energy principle is discussed for systems of connected rigid bodies.

Equation (15.11), $U_{1\to 2} = T_2 - T_1$, is the work energy principle for a single particle. Work and energy are both scalars, and, for a system of particles, we introduce summations in order to write the appropriate work energy principle. Thus,

$$\sum U_{1\to 2} = \sum T_2 - \sum T_1. \tag{15.12}$$

The term, $\sum U_{1\to 2}$ in Eq. (15.12), represents the total work done by all of the external working forces of the system. The term $\sum T_2$ represents the final kinetic energy of all of the particles of the system and the term $\sum T_1$ represents the initial kinetic energy of all the particles of the system.

A free-body diagram of the entire system should be constructed with all external forces shown. Those forces which do no work in the system need not be included in the work summation. Because the connecting cables or struts are inextensible, energy cannot be stored in these components, and forces in them are internal to the system and can do no external work on the system.

Equation (15.12) is a single scalar equation which can be solved for a single scalar unknown. This means that the speeds of the various particles of a given system must be expressed in terms of the speed of one of the particles, and the displacements of the working forces acting on the system of particles must be expressed in terms of the displacement of one of the working forces. This is accomplished by considering the geometric constraints governing the motion of the various particles in the system.

Examples 15.6 and 15.7 illustrate some of the concepts discussed above.

■ Example 15.6

Refer to the system shown in Figure E15.6(a). Block A has a mass of 4 kg, and block B has a mass of 6 kg. The pulley C is of negligible mass and is frictionless. The connecting cable is inextensible and of negligible mass. The kinetic coefficient of friction between block A and the horizontal plane is $\mu = 0.1$. (a) If this system is released from rest, determine the velocities of blocks A and B after block A has moved 2 meters to the right. (b) Determine the magnitude of the tension in the connecting cable.

Solution

(a) Refer to Figure E15.6(b) which shows the free-body diagram of the entire system consisting of blocks A and B, connected by the cable which passes over the pulley C. Before writing the work-energy equation for this system, we note the nonworking forces as follows:

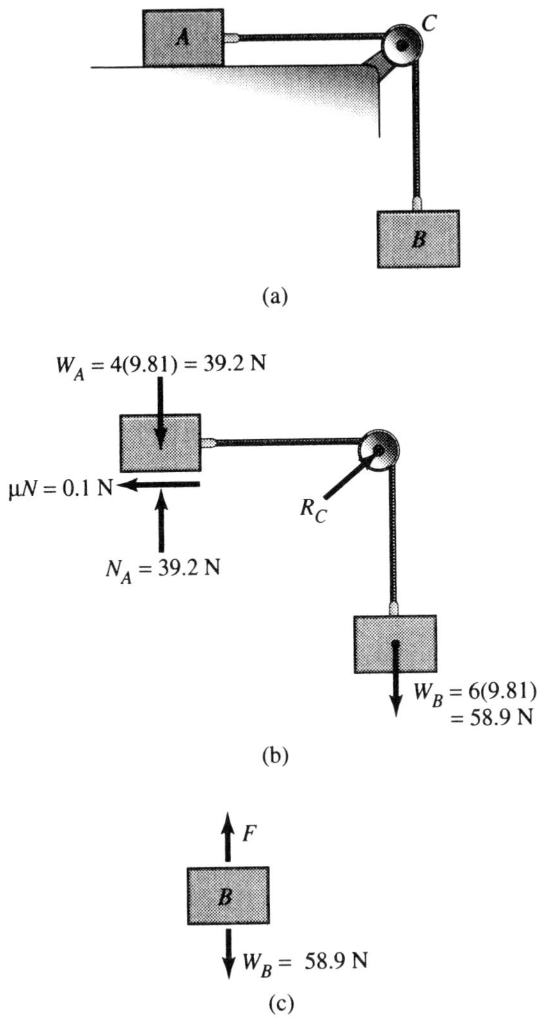

FIGURE E15.6.

(i) The weight W_A and the normal force N_A. These forces act perpendicular to the horizontal displacement of body A and, therefore, do no work on this system.
(ii) The pulley reaction R_C is not displaced during motion of the system and, therefore, does no work on it.
(iii) Internal cord forces, not shown in Figure E15.6(b), occur in equal and opposite pairs and do no work on the system.

Eq. (15.12) states that

$$\sum U_{1\to 2} = \sum T_2 - \sum T_1.$$

Because both masses, A and B, are initially at rest, the initial kinetic energy of the system $\sum T_1$ vanishes. The working forces are the friction force $\mu_k N_A$ and the weight of block B, W_B. Note that the normal N_A equals W_A because block A is not accelerated vertically. Thus,

$$\sum U_{1\to 2} = -\mu N_A(2) + W_B(2) = -0.1(39.2)(2) + 58.9(2) = 110 \text{ N·m}.$$

Because the blocks are connected by an inextensible cable, they move equal distances in equal times and, therefore, have the same speed, i.e., $v_A = v_B = v$. The final kinetic energy of the system is given by

$$\sum T_2 = \tfrac{1}{2}m_A v^2 + \tfrac{1}{2}m_B v^2 = \tfrac{1}{2}(4)v^2 + \tfrac{1}{2}(6)v^2 = 5v^2.$$

Substitution in Eq. (15.12) yields

$$110 = 5v^2 - 0.$$

Thus, $v = 4.69$ m/s, and

$$v_A = 4.69 \text{ m/s}\uparrow, \quad v_B = 4.69 \text{ m/s}\downarrow. \qquad \text{ANS.}$$

(b) Consider the free-body diagram of block B as shown in Figure E15.6(c), and write the work-energy equation to determine the cable tension F. Note that the tension F is internal for the system shown in Figure E15.6(b), but external for block B as shown in Figure E15.6(c). Therefore, since F is constant,

$$U_{1\to 2} = T_2 - T_1,$$
$$58.9(2) - F(2) = \tfrac{1}{2}(6)(4.69)^2$$

from which

$$F = 25.9N. \qquad \text{ANS.}$$

Example 15.7

Refer to the system shown in Figure E15.7(a). Block A has a mass of 1 slug, and block B has a mass of 10 slug. The system starts from rest, and, after block B moves downward 1 ft, it has a downward velocity of 3 ft/s. Determine the spring constant k consistent with this motion. The cord is inextensible, and the system is frictionless. Initially the spring is unstretched. Assume the two pulleys to be weightless and frictionless.

Solution

The length L of the cord connecting the two blocks remains constant. Thus,

$$L = d_1 - s_A + \left(\frac{90^\circ + 30^\circ}{180^\circ}\pi\right)r + 2(s_B - d_2) + \pi r + d_2$$

or

$$-s_A + 2s_B = \text{constant}.$$

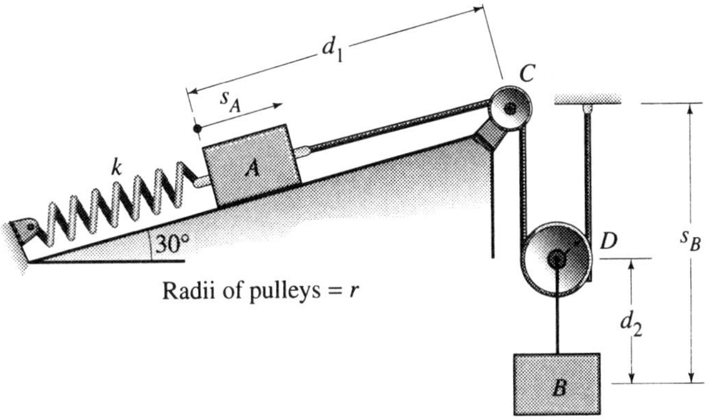

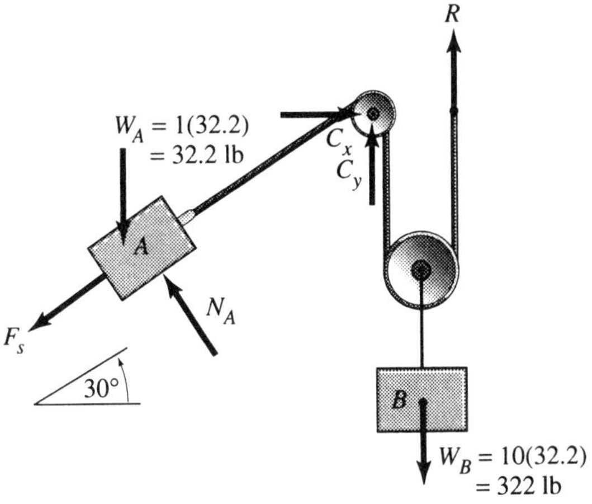

FIGURE E15.7.

Therefore,

$$-\frac{ds_A}{dt} + 2\frac{ds_B}{dt} = 0$$

and

$$v_A = 2v_B.$$

At any instant during this motion, the speed of block A is twice the speed of block B. Because the final speed of block B is 3 ft/s, the corresponding final speed of block A is 6 ft/s. Also, since block B moves downward a distance of 1 ft, block A moves upward a distance of 2 ft along the inclined plane.

Refer to the free-body diagram of the entire system shown in Figure E.15.7(b) and note that the normal force N_A acting on block A does no work on the system because it always acts perpendicular to the displacement of block A. The upper pulley reactions C_x and C_y do no work on the system because they are not displaced during the motion. The vertical support reaction R does no work on the system because it acts at a fixed point and is not displaced during the motion.

The work energy equation for the system is given by

$$U_{1\to 2} = \sum T_2 - \sum T_1$$

where

$$\sum U_{1\to 2} = \tfrac{1}{2}k(2)^2 - 32.2(2\sin 30°) + 322(1) = -2k + 289.8 \text{ (ft·lb)},$$

$$\sum T_1 = 0,$$

and

$$\sum T_2 = \tfrac{1}{2}(1)(6)^2 + \tfrac{1}{2}(10)(3)^2 = 63.0 \text{ ft·lb}.$$

Therefore,

$$-2k + 289.8 = 63.0 - 0,$$

$$k = 113.4 \text{ lb/ft}. \qquad \text{ANS.}$$

■

Problems

Assume all cables and struts are inextensible and massless and that all pulleys are frictionless and massless.

15.27 Refer to Figure P15.27, and neglect friction between block B and the horizontal plane and at the pulley. Block A has a mass of 2 kg, and block B has a mass of 3 kg. Initially block A is moving downward at a rate of 4 m/s. Determine the velocity of each block after block A has moved 2 m downward.

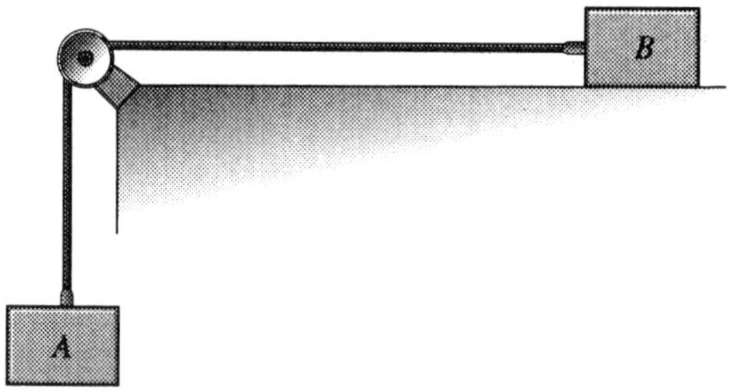

FIGURE P15.27.

15.28 Refer to Figure P15.27 and neglect friction between block B and the horizontal plane and at the pulley. Block A has a mass of 4 slugs and block B has a mass of 2 slugs. Initially block B is moving to the right at a rate of 6 ft/s. Determine how far block B moves before the system comes to rest.

15.29 Very light struts connect the three masses shown in Figure P15.29. Initially the system is moving to the left at a speed of 5 ft/s when the variable force $F = 3s^2 - 2$ is applied. F is expressed in pounds when s is given in feet. Determine the distance s which the system moves before coming to rest. The masses are $m_A = 2$ slug, $m_B = 3$ slug, and $m_C = 4$ slug. The system is frictionless.

15.30 Solve Problem 15.29 for $F = 2s - 4$. All other conditions remain unchanged.

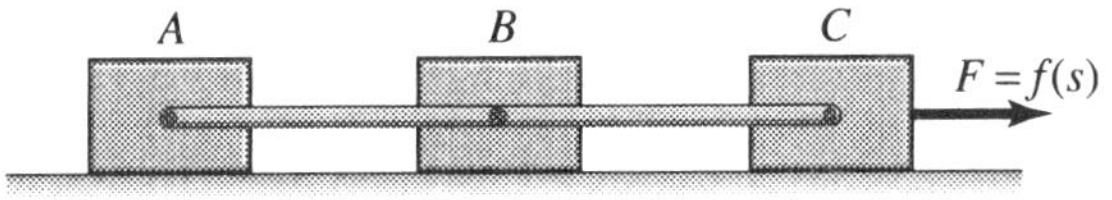

FIGURE P15.29.

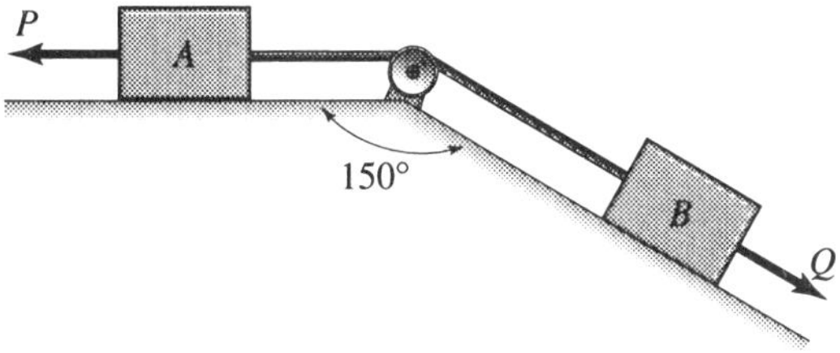

FIGURE P15.31.

15.31 Block B of Figure P15.31 is moving 2 m/s down the inclined plane when $P = 40$ N and Q = 200 N are applied. What is the velocity of each block after block A has moved 1.00 m to the right? Given masses are $m_A = 3.50$ kg and $m_B = 6.00$ kg. The system is frictionless.

15.32 Solve Problem 15.31 for $P = 60$ N and $Q = 160$ N. All other conditions remain unchanged.

15.33 Refer to Figure P15.31 and let $P = 4Q$. These forces are applied when block B is moving down the incline at 10 ft/s. Determine Q such that the system comes to rest after block A has moved 6 ft to the right. The system is frictionless, and the masses are $m_A = 2$ slug and $m_B = 1$ slug.

15.34 In Figure P15.34, block B has a negligible velocity directed to the left when the light strut AB is vertical. Block A is constrained to move vertically, and block B is constrained to more horizontally. The system is frictionless. Determine the velocity of each block after block B has moved 5 ft to the left, given that $m_B = 2m_A$.

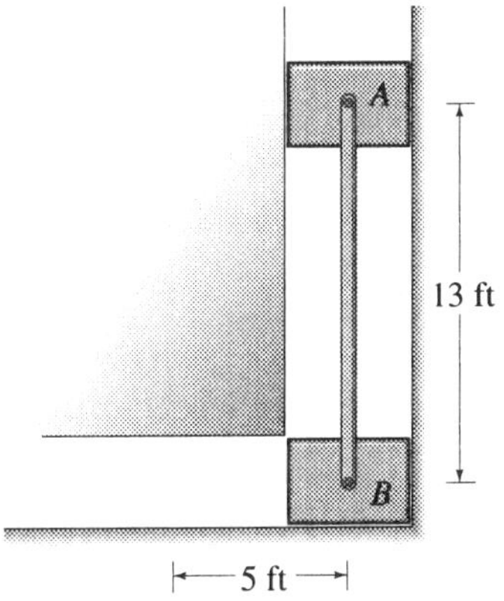

FIGURE P15.34.

15.35 Solve Problem 15.34 for the case where m_B is very small compared to m_A. Review and explain your result if A were to move downward vertically through a distance h, rather than the specific distance of Problem 15.34.

15.36 The system shown in Figure P15.36 is frictionless. Block A is constrained to move vertically, and block B is constrained to move horizontally. In the position shown, block B has a negligible velocity directed to the right, and the spring is unstretched. Given that $k = 2 m_A g/L$, express s as a function of L such that the system comes to rest after the spring is compressed by an amount s.

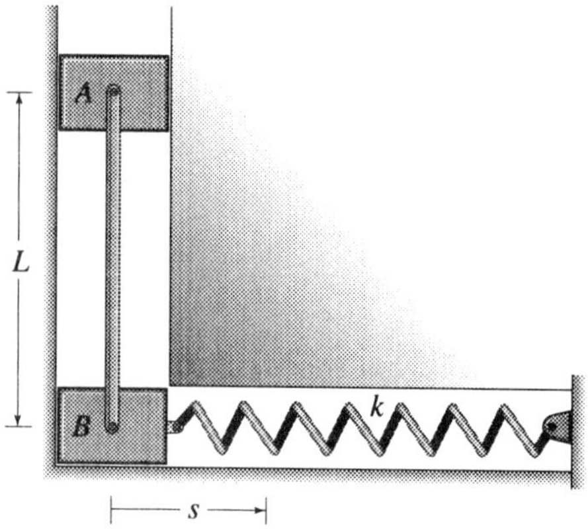

FIGURE P15.36.

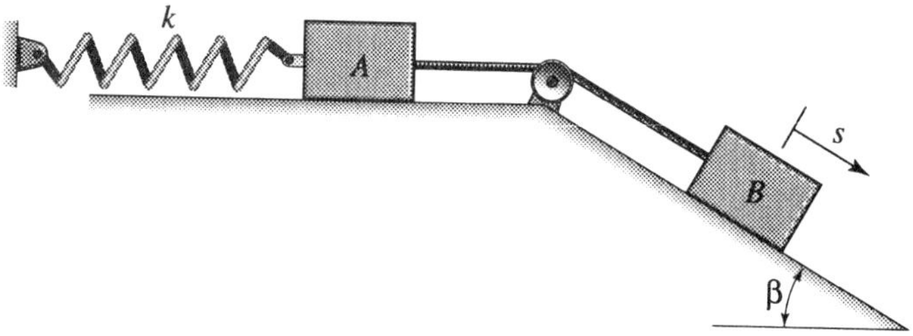

FIGURE P15.37.

15.37 The frictionless system shown in Figure P15.37 is initially at rest when the spring is unstretched. After block B moves down the inclined plane a distance s, the system again comes to rest. Express s as a function of m_B, g, b, and k. Discuss the special cases of $\beta = 0°$ and $\beta = 90°$.

15.38 The frictionless system shown in Figure P15.38 is moving such that block A has a upward velocity of 10 ft/s and block B has a rightward velocity of equal magnitude. At this instant the spring is unstretched, and $Q = 30s^2$ is applied to the right. Q is given in pounds when s is stated in feet. Determine the velocities of these masses for $s = 3.00$ ft. Given $m_A = 2$ slug, $m_B = 4$ slug, and $k = 50$ lb/ft.

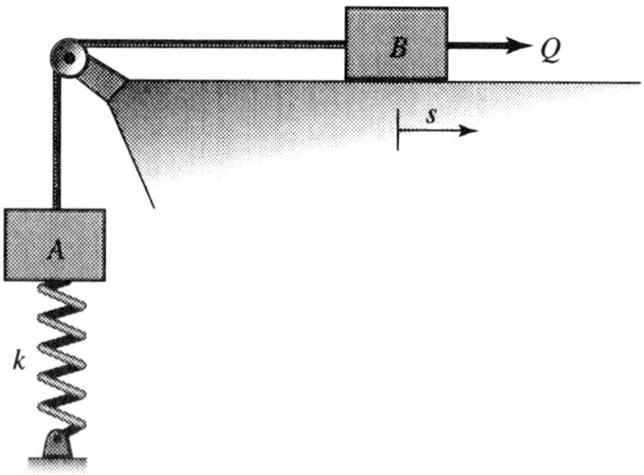

FIGURE P15.38.

15.5 Work of Conservative Forces

The work of a force was defined in Section 12.3 by Eq. (12.6) which is repeated here for convenience.

$$U_{1\to 2} = \int_{s_1}^{s_2} (F\cos\beta)\,ds. \tag{12.6}$$

In this equation, s is the distance measured along the path between position s_1 and s_2 and $F\cos\beta$ represents the components of the force **F** along the tangent to the path. As stated previously, integration of Eq. (12.6) is possible only if (a) the force $F\cos\beta$ is constant or (b) we can relate the quantities F and β to the path s. In either case, it is seen that, in general, the work $U_{1\to 2}$ depends upon the path followed by the point of application of the force. There are special cases of forces, however, for which the work $U_{1\to 2}$ is independent of the specific path followed and depends only on the initial and final positions expressed by the symbols s_1 and s_2. Such forces are known as *conservative* forces. Also, a mechanical system acted upon by only conservative forces is known as a *conservative system.* Thus, a conservative force would perform the same amount of work in moving from position s_1 to position s_2 whether path A or path B is followed as shown in Figure 15.8. One consequence of this fact is that a conservative force moving through a closed loop from s_1 to s_2 and back to s_1, *using any path,* would perform *no* work. On the other hand, a nonconservative force performs work that depends upon the specific path followed. Therefore, the work performed by a nonconservative force in moving from position s_1 to position s_2 would have one value for path A and an entirely different value for path B, and, of course, its work for a closed loop would *not* be zero.

A mathematical proof of the above concepts is given below.

In Chapter 12 (see Eq. 12.22) we concluded that $U_{1\to 2} = -V$ where

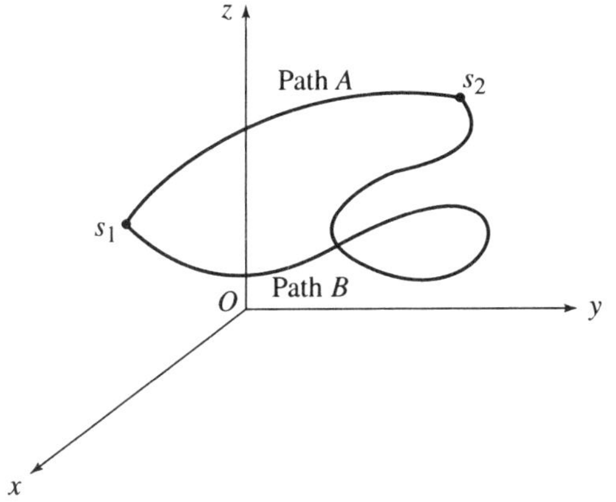

FIGURE 15.8.

V is defined as the potential energy of a given mechanical system. In differential form, this relation becomes $dU = -dV$, where

$$dU = \mathbf{F}\cdot d\mathbf{r} = (F_x\mathbf{i} + F_y\mathbf{j} + F_z\mathbf{k})\cdot(dx\,\mathbf{i} + dy\,\mathbf{j} + dz\,\mathbf{k})$$
$$= F_x\,dx + F_y\,dy + F_z\,dz$$

and

$$dV = \frac{\partial V}{\partial x}dx + \frac{\partial V}{\partial y}dy + \frac{\partial V}{\partial z}dz.$$

It follows from $dU = -dV$ that

$$F_x\,dx + F_y\,dy + F_z\,dz = -\left(\frac{\partial V}{\partial x}dx + \frac{\partial V}{\partial y}dy + \frac{\partial V}{\partial z}dz\right).$$

Thus, we conclude that $F_x = -\frac{\partial V}{\partial x}$, $F_y = -\frac{\partial V}{\partial y}$, and $F_z = -\frac{\partial V}{\partial z}$, and, consequently,

$$\mathbf{F} = -\left(\frac{\partial V}{\partial x}\mathbf{i} + \frac{\partial V}{\partial y}\mathbf{j} + \frac{\partial V}{\partial z}\mathbf{k}\right),$$

$$\mathbf{F} = -\left(\frac{\partial}{\partial x}\mathbf{i} + \frac{\partial}{\partial y}\mathbf{j} + \frac{\partial}{\partial z}\mathbf{k}\right)V.$$

The quantity inside the brackets is known as the *gradient* operator and is given the symbol ∇. Thus,

$$\nabla = \frac{\partial}{\partial x}\mathbf{i} + \frac{\partial}{\partial y}\mathbf{j} + \frac{\partial}{\partial z}\mathbf{k}, \qquad (15.13)$$

and

$$\mathbf{F} = -\nabla V. \qquad (15.14)$$

Thus, a conservative force $\mathbf{F}$ is one that is defined as the negative gradient of the scalar potential function V.

Now, because $dU = \mathbf{F}\cdot d\mathbf{r}$, it follows that

$$U_{1\to 2} = \int_1^2 \mathbf{F}\cdot d\mathbf{r}$$
$$= -\int_1^2 \left(\frac{\partial V}{\partial x}\mathbf{i} + \frac{\partial V}{\partial y}\mathbf{j} + \frac{\partial V}{\partial z}\mathbf{k}\right)\cdot(dx\,\mathbf{i} + dy\,\mathbf{j} + dz\,\mathbf{k})$$
$$= -\int_1^2 \frac{\partial V}{\partial x}dx + \frac{\partial V}{\partial y}dy + \frac{\partial V}{\partial z}dz.$$

Since $dV = \frac{\partial V}{\partial x}dx + \frac{\partial V}{\partial y}dy + \frac{\partial V}{\partial z}dz$, we conclude that

$$U_{1\to 2} = -\int_1^2 dV = V_1 - V_2 \tag{15.15}$$

where V_1 and V_2 are the values of the scalar potential function V at positions 1 and 2, respectively. Therefore, Eq. (15.15) shows that, if the force **F** is the gradient of the scalar function V, the work done by it is a function *only* of its initial and final positions, and, consequently, the force is conservative.

Many examples of conservative forces may be cited, but we will limit our discussion only to two such forces that have been discussed earlier. It was shown in Chapter 12 and again earlier in this Chapter that the work of the force of gravity (i.e., $W = mg$ with $g =$ constant), given by Eq. (15.4), and that by the force of a spring ($F = ks$), given by Eq. (15.8), are independent of the path followed by the forces and depend only on the initial and final positions. Thus, we conclude that the force of gravity and the force of a spring are both conservative.

The *curl of a vector* is defined as the vector product of the gradient operator ∇ and the vector in question. Thus, the curl of a force $\mathbf{F} = F_x\mathbf{i} + F_y\mathbf{j} + F_z\mathbf{k}$ is given by

$$\begin{aligned}\nabla \times \mathbf{F} &= \begin{bmatrix} \mathbf{i} & \mathbf{j} & \mathbf{k} \\ \frac{\partial}{\partial x} & \frac{\partial}{\partial y} & \frac{\partial}{\partial z} \\ F_x & F_y & F_z \end{bmatrix} \\ &= \left(\frac{\partial F_z}{\partial y} - \frac{\partial F_y}{\partial z}\right)\mathbf{i} + \left(\frac{\partial F_z}{\partial x} - \frac{\partial F_x}{\partial z}\right)\mathbf{j} + \left(\frac{\partial F_y}{\partial x} - \frac{\partial F_x}{\partial y}\right)\mathbf{k}.\end{aligned} \tag{15.16}$$

It can be shown that a necessary and sufficient condition for a force **F** to be conservative is that $\nabla \times \mathbf{F} = \mathbf{0}$. It follows, therefore, from Eq. (15.16) that for a force **F** to be conservative, the following conditions must be met:

$$\left.\begin{aligned} \frac{\partial F_z}{\partial y} &= \frac{\partial F_y}{\partial z} \\ \frac{\partial F_z}{\partial x} &= \frac{\partial F_x}{\partial z} \\ \frac{\partial F_y}{\partial x} &= \frac{\partial F_x}{\partial y} \end{aligned}\right\}. \tag{15.17}$$

The following examples illustrate some of the above concepts.

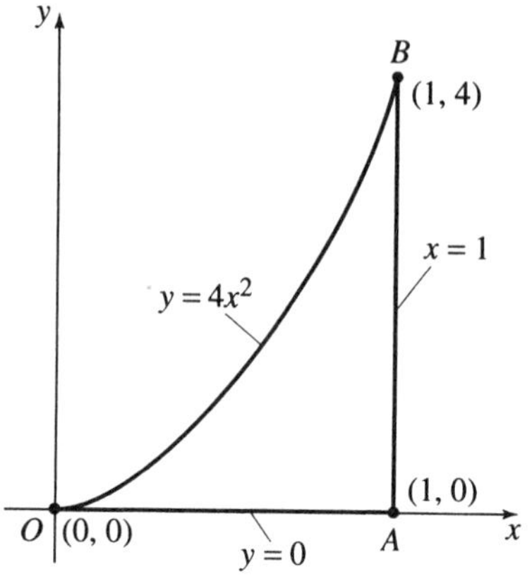

FIGURE E15.8.

Example 15.8

Refer to Figure E15.8 for path information, and determine the work done for the following cases using SI units. (a) the nonconservative force $\mathbf{F} = 2y\mathbf{i} + 4x\mathbf{j}$ along the parabolic path OB given by $y = 4x^2$, (b) the nonconservative force of part (a) along the path OA followed by AB, (c) the conservative force $\mathbf{F} = 2x\mathbf{i} + 4y\mathbf{j}$ along the parabolic path OB.

Solution

(a) A position vector to any point in the x-y plane is given by

$$\mathbf{r} = x\mathbf{i} + y\mathbf{i}$$

$$\mathbf{dr} = dx\,\mathbf{i} + dy\,\mathbf{j}$$

$$U_{\text{OB}} = \int \mathbf{F} \cdot \mathbf{dr} = \int 2y\,dx + \int 4x\,dy.$$

The path equation for the parabola is $y = 4x^2$. Differentiation of this equation gives $dy = 8x\,dx$. Substituting in the work integrals and introducing appropriate limits leads to

$$U_{\text{OB}} = \int_0^1 2(4x^2)\,dx + \int_0^1 4x(8x\,dx)$$

$$= \frac{8x^3}{3}\bigg|_0^1 + \frac{32x^3}{3}\bigg|_0^1 = \frac{8}{3} + \frac{32}{3} = 13.33 \text{ N·m.} \qquad \text{ANS.}$$

(b) The basic integrals are the same, but the path equations are for OA and AB. Along OA, $y = 0$, and $dy = 0$. Therefore,

$$U_{OA} = \int 2y\,dx + \int 4x\,dy = 0.$$

Along AB, $x = 1$, and $dx = 0$. Hence,

$$U_{AB} = \int 2y\,dx + \int 4x\,dy = 0 + \int_0^4 4x\,dy = 16.00\ \text{N}\cdot\text{m}.$$

Therefore,

$$U_{OB} = U_{OA} + U_{AB} = 16.00\ \text{N}\cdot\text{m}. \qquad \text{ANS.}$$

We note that U_{OB} is different along the two different paths. If the force **F** were conservative, then, the work done would be the same regardless of the path chosen.

(c) Forming the work integrals for the given conservative force, we obtain

$$U_{OB} = \int \mathbf{F}\cdot d\mathbf{r} = \int 2x\,dx + \int 4y\,dy.$$

We observe that these integrals may be evaluated simply by introducing the appropriate limits. The path equation is not required because the force is conservative, and the work done is independent of the path. Thus,

$$U_{OB} = \int_0^1 2x\,dx + \int_0^4 4y\,dy = \left.\frac{2x^2}{2}\right|_0^1 + \left.\frac{4y^2}{2}\right|_0^4 = 33.0\ \text{N}\cdot\text{m}.$$

ANS.

Example 15.9

Use Eq. (15.17) to show that

(a) the force $\mathbf{F} = 5x\mathbf{i} - 3y\mathbf{j} + 2z\mathbf{k}$ is conservative, and
(b) the force $\mathbf{F} = 5x\mathbf{i} + (x + 4z)\mathbf{j} + (4y + z)\mathbf{k}$ is nonconservative.

Solution

(a) Because $\mathbf{F} = 5x\mathbf{i} - 3y\mathbf{j} + 2z\mathbf{k}$, then,

$$F_x = 5x, \qquad \frac{\partial F_x}{\partial y} = 0, \qquad \frac{\partial F_x}{\partial z} = 0$$

$$F_y = -3y, \qquad \frac{\partial F_y}{\partial x} = 0, \qquad \frac{\partial F_y}{\partial z} = 0$$

$$F_z = 2z, \qquad \frac{\partial F_z}{\partial x} = 0, \qquad \frac{\partial F_z}{\partial y} = 0.$$

Therefore, all of the conditions in Eq. (15.17) are met, and the force is conservative. ANS.

(b) Since $\mathbf{F} = 5x\mathbf{i} + (x + 4z)\mathbf{j} + (4y + z)\mathbf{k}$, then,

$$F_x = 5x, \qquad \frac{\partial F_x}{\partial y} = 0, \qquad \frac{\partial F_x}{\partial z} = 0,$$

$$F_y = x + 4z, \qquad \frac{\partial F_y}{\partial x} = 1, \qquad \frac{\partial F_y}{\partial z} = 4,$$

$$F_z = 4y + z, \qquad \frac{\partial F_z}{\partial x} = 0, \qquad \frac{\partial F_z}{\partial y} = 4.$$

The first and second conditions in Eq. (15.17) are satisfied. However, the third condition is not and the force is nonconservative. ANS.

■

15.6 Gravitational and Elastic Potential Energy

The concept of potential energy was discussed in detail in Chapter 12. However, for convenience, some of the significant aspects of that discussion are repeated here. The term *potential energy* of a mechanical system is used to signify the capacity or potential of a system to do work. For example, the weight W sitting on the table at height h above the floor, as shown in Figure 15.9(a), has the ability to perform a certain amount of work relative to the floor. In other words, if the weight W is released from the table and allowed to drop, it would be able to do Wh units of work assuming the floor is used as a reference or datum plane. Consider also the spring-weight system shown in Figure 15.9(b) where the weight W slides freely on a frictionless surface. In the displaced position shown, the system has a certain amount of energy stored in it as a consequence of the spring deformation. If the system is released and allowed to return to the equilibrium position, it has the capacity to do $\frac{1}{2}ks^2$ units of work on the weight W where s is the displacement of the spring.

Gravitational Potential Energy V_g

In Section 15.5, we concluded that the work of a conservative force is a function only of its initial and final positions and is independent of the path it follows. The work performed by the weight of a body is given by Eq. (15.4) which states that the work is a function only of the difference in elevation from the initial to the final position of the body. Also, if the elevation of the weight W is increased, the work is negative,

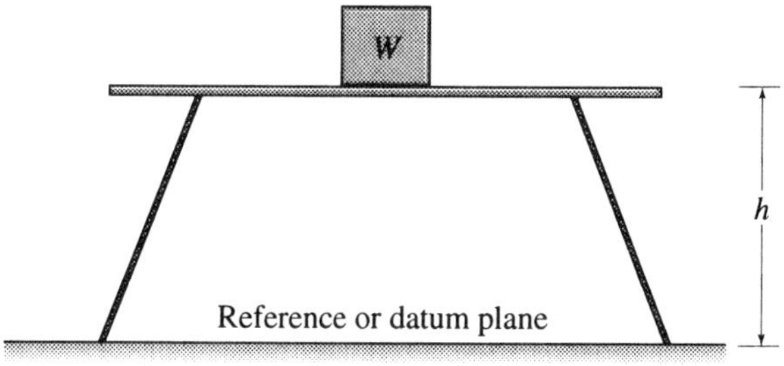

(b)

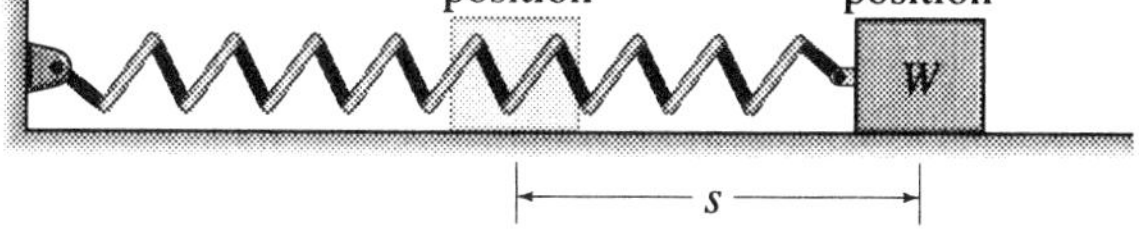

(b)

FIGURE 15.9.

but its capacity to do work becomes greater, and we say that its *potential energy due to gravity* V_g increases. On the other hand, if the elevation of the weight W is decreased, the work is positive *but* its potential energy V_g decreases. Therefore, we may state that the change in potential energy ΔV_g is equal to the *negative* of the work performed as the body changes elevation from one level to another. Thus, using Eq. (15.4),

$$\Delta V_g = -U_{1\to 2} = -W(y_1 - y_2) = Wy_2 - Wy_1$$
$$= (V_g)_2 - (V_g)_1 \tag{15.18}$$

where elevation 2 is lower than elevation 1. Equation (15.18) may be written in the following form:

$$(V_g)_2 = \Delta V_g + (V_g)_1 \tag{15.19}$$

which states that the gravitational potential energy of the body in the final elevation is equal to that in the initial elevation plus the change ΔV_g. However, as was shown in Chapter 12, only changes in potential energy are significant, and, therefore, the potential energy in the final elevation may be measured from *any* convenient reference or datum plane. Thus, the potential energy of a body due to gravity may be expressed as

$$V_g = -U_{1\to 2} = Wy. \tag{15.20}$$

where y is the height of the body measured from any arbitrary position.

Elastic Potential Energy V_e

As expressed in Eq. (15.8), the work done by a linearly elastic member, such as a spring, on an attached body is a function *only* of its initial and final positions. Also, the *work of the spring force on an attached body is negative if it is displaced away from its equilibrium position,* but its capacity to do work increases, and we say that its potential energy due to elastic behavior V_e increases. On the other hand, the *work of the spring force on an attached body is positive if it is allowed to return towards its equilibrium position, but* its potential energy V_e decreases. Thus, we may state that the change in potential energy is equal to the negative of the work done by the spring on an attached body. Therefore, by Eq. (15.8b),

$$\Delta V_e = -U_{1\to 2} = \tfrac{1}{2}k(s_2^2 - s_1^2) = \tfrac{1}{2}ks_2^2 - ks_1^2 = (V_e)_2 - (V_e)_1 \qquad (15.21)$$

If follows, therefore, that

$$(V_e)_2 = \Delta V_e + (V_e)_1 \qquad (15.22)$$

which states that the elastic potential energy of an elastic system, such as a spring, in its final position is equal to that in the initial position plus the change ΔV_e. However, the elastic potential energy $(V_e)_1$ is zero because it represents the potential energy of the elastic system in its equilibrium position. Thus, the elastic potential energy $(V_e)_2$ becomes equal to the change ΔV_e *which must be measured from the equilibrium position of the elastic system.* Therefore, in the case of a spring, for example,

$$V_e = -U_{1\to 2} = \tfrac{1}{2}ks^2 \qquad (15.23)$$

where s is the deformation of the spring measured from its equilibrium position.

Consequently, if a given mechanical system possesses gravitational potential energy V_g and elastic potential energy V_e, its total potential energy V will be given by the algebraic sum of V_g and V_e. Thus,

$$V = V_g + V_e \qquad (15.24)$$

where V would be a function of one or more variables depending upon the degrees of freedom of the mechanical system under consideration and is known as the *potential energy function.* Also, because the potential energy is equal to the negative of the work done, it follows that

$$U_{1\to 2} = -V \qquad (15.25)$$

where $U_{1\to 2}$ is the work performed by conservative forces. As a matter of fact, Eq. (15.25) is valid only for the case of conservative systems.

15.7 Principle of Conservation of Mechanical Energy

The *principle of conservation of mechanical energy* is more restrictive than the work-energy principle discussed in Section 15.3. Only when conservative forces perform work on a system is it valid to apply the conservation of energy principle developed here.

To derive the principle of conservation of mechanical energy, we use Eq. (15.25). In differential form Eq. (15.25) may be written as $dU = -dV$. Integration between initial position 1 and final position 2 yields

$$\int_1^2 dU = -\int_1^2 dV$$

from which

$$U_{1\to 2} = -(V_2 - V_1).$$

Substituting $U_{1\to 2} = T_2 - T_1$ from Eq. (15.11) yields

$$T_2 - T_1 = -(V_2 - V_1)$$

from which

$$T_1 + V_1 = T_2 + V_2. \tag{15.26}$$

This equation expresses the principle of conservation of mechanical energy. It states that the initial kinetic energy added to the initial potential energy (total initial mechanical energy) equals the final kinetic energy added to the final potential energy (total final mechanical energy).

Equation (15.26) applies so long as the working forces are conservative. If a nonconservative force, such as a friction force, acts on the system, then, the work-energy principle of Eq. (15.11) should be applied.

In Section 15.4, the work-energy principle was extended to a system of particles. In a similar fashion, Eq. (15.26) will be extended to a system of particles by the introduction of summation signs to obtain

$$\sum T_1 + \sum V_1 = \sum T_2 + \sum V_2. \tag{15.27}$$

Equation (15.27) represents the principle of conservation of mechanical energy for a system of particles. To avoid the storage of energy in the members that connect the various particles, we restrict its application to systems of particles which are connected by inextensible cables or struts. The subscripts 1 and 2 refer to initial and final positions of the system. The summations of kinetic energy extend over the total number of particles of the system. The summations for the potential energy $V = V_g + V_e$ extend over the total number of external gravitational forces (or weights) of the system and over the total number of linearly elastic spring forces which act on the system.

The following examples illustrate some of the concepts discussed above.

■ Example 15.10

A collar of mass m travels along a circular rod which lies in a vertical plane as shown in Figure E15.10. The rod is frictionless, and the spring is unstretched when the collar is in position 1 where its velocity is $v_1 = 5\text{m/s}$ directed to the right. If $k = 50$ N/m, $r = 2$ m, and $m = 10$ kg, find (a) the velocity of the collar when it reaches position 2. (b) What is the velocity of the collar when it reaches position 3?

Solution

Because the system is frictionless, the principle of conservation of mechanical energy applies because only gravity and spring forces work on the collar. As shown in Figure E15.10, a datum through position 1 is used for gravitational potential energy.

(a) By Eq. (15.26),

$$T_1 + V_1 = T_2 + V_2$$

where

$$T_1 = \tfrac{1}{2}mv_1^2 = \tfrac{1}{2}(10)(5^2) = 125 \text{ N}\cdot\text{m},$$

$$V_1 = V_{g1} + V_{e1} = 0 + 0 = 0 \quad \text{(by Eq. (15.24))},$$

$$T_2 = \tfrac{1}{2}mv_2^2 = \tfrac{1}{2}(10)v_2^2 = 5v_2^2 \text{ N}\cdot\text{m},$$

and

$$V_2 = V_{g2} + V_{e2} = -mgy_2 + \tfrac{1}{2}ks_2^2 \quad \text{(by Eq. (15.24))},$$
$$= -10(9.81)(2) + \tfrac{1}{2}(50)(\sqrt{6^2 + 2^2} - 4)^2 = -61.1 \text{ N}\cdot\text{m}.$$

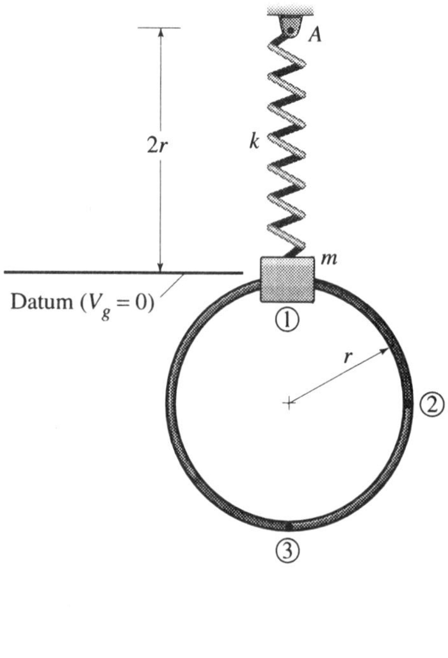

FIGURE E15.10.

Therefore,

$$125 + 0 = 5v_2^2 - 61.1$$

from which

$$v_2 = 6.10 \text{ m/s}. \qquad \text{ANS.}$$

(b) As in part (a),

$$T_1 = 125 \text{ N·m}, \; V_1 = 0,$$
$$T_3 = \tfrac{1}{2}mv_3^2 = \tfrac{1}{2}(10)v_3^2 = 5v_3^2 \text{ N·m},$$

and

$$V_3 = V_{g3} + V_{e3} = -mgy_3 + \tfrac{1}{2}ks_3^2$$
$$= -10(9.81)(4) + \tfrac{1}{2}(50)(8.4)^2 = 7.60 \text{ N·m}.$$

Thus,

$$125 + 0 = 5v_3^2 + 7.60$$

and

$$v_3 = 4.85 \text{ m/s}. \qquad \text{ANS.}$$

Example 15.11

A block of weight $W = 100$ lb is attached to a spring ($k = 1000$ lb/ft) and is free to move along the frictionless inclined plane shown in Figure E15.11. The block is released from rest in position 1. When the block is in position 2, the spring is unstretched. (a) What will be its velocity in position 2? (b) What distance s will the block move down the plane until it again comes to rest in position 3?

Solution

(a) Choose a horizontal datum for zero gravitational potential energy through position 2 of the block, and write the law of conservation of

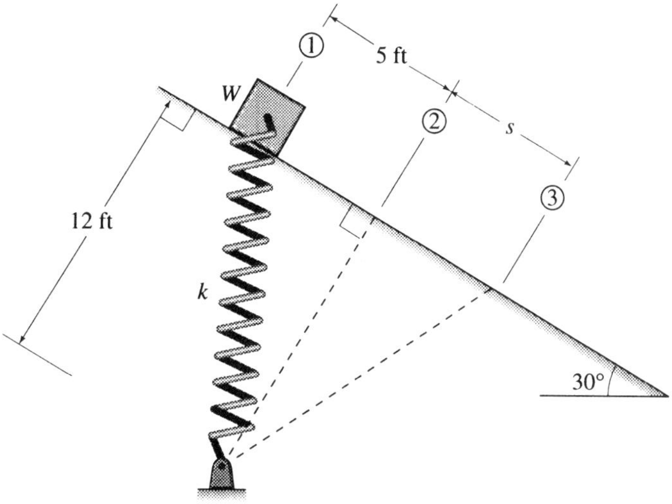

FIGURE E15.11.

mechanical energy. Thus, by Eq. (15.26),

$$T_1 + V_1 = T_2 + V_2.$$

By Eq. (15.24) and applying Eqs. (15.20) and (15.23), this relationship may be expressed in the forms

$$T_1 + (V_{g1} + V_{e1}) = T_2 + (V_{g2} + V_{e2}),$$

$$T_1 + (mgh_1 + \tfrac{1}{2}ks_1^2) = T_2 + (mgh_2 + \tfrac{1}{2}ks_2^2)$$

$$0 + \left[100(5\sin 30^\circ) + \frac{1}{2}(1000)(\sqrt{12^2 + 5^2} - 12)^2\right]$$

$$= \frac{1}{2}\left(\frac{100}{32.2}\right)v_2^2 + (0 + 0)$$

$$v_2 = 22.0 \text{ ft/s} \quad 30^\circ$$ ANS.

(b) Choose a horizontal datum for zero gravitational potential energy through position 3 of the block, and write the law of conservation of mechanical energy. Thus, by Eq. (15.26),

$$T_2 + V_2 = T_3 + V_3.$$

By Eq. (15.24) and applying Eqs. (15.20) and (15.23), this relationship may be expressed in the form

$$T_2 + (V_{g2} + V_{e2}) = T_3 + (V_{g3} + V_{e3}),$$

$$T_2 + (mgh_2 + \tfrac{1}{2}ks_2^2) = T_3 + (mgh_3 + \tfrac{1}{2}ks_3^2)$$

$$\frac{1}{2}\left(\frac{100}{32.2}\right)(22.0)^2 + [100(s\sin 30^\circ) + 0]$$

$$= 0 + \left[0 + \frac{1}{2}(1000)(\sqrt{12^2 + s^2} - 12)^2\right]$$

Simplifying yields

$$(\sqrt{12^2 + s^2} - 12)^2 - 0.100s - 1.503 = 0.$$

A trial-and-error solution yields

$$s = 6.08 \text{ ft.}$$ ANS.

■ Example 15.12

Refer to the frictionless system shown in Figure E15.12 which is connected by massless and inextensible cables. The masses are $m_A = 2$ slug and $m_B = 6$ slug. The upper spring ($k_1 = 1000$ lb/ft) is initially stretched by $d_A = 0.5$ ft, and the lower spring ($k_2 = 500$ lb/ft) is initially compressed by $d_B = 0.25$ ft. The lower spring is supported laterally to prevent it from buckling. Initially, m_A is moving at 40 ft/s to the right and moves to the right through a horizontal distance of 1.5 ft. Determine the final velocities of the two masses.

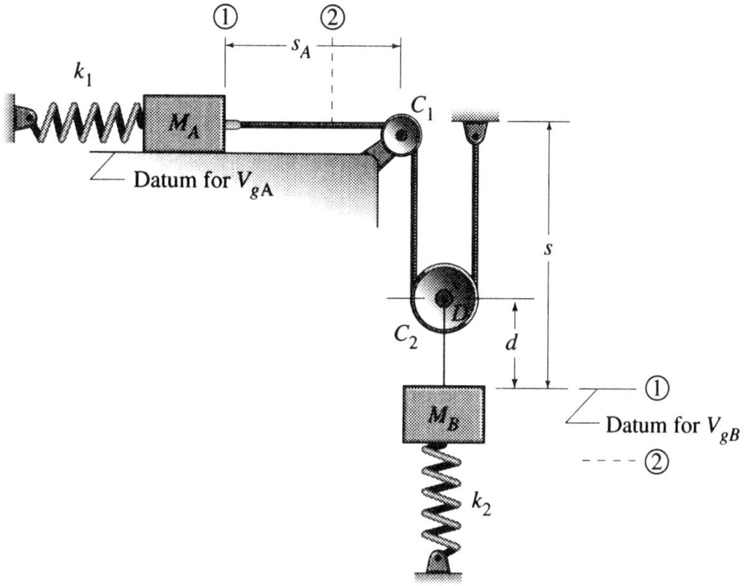

FIGURE E15.12.

Solution

Because the length of the connecting cable L is constant, it follows that

$$s_A + c_1 + 2(s_A - d) + c_2 = L.$$

Therefore, because c_1, c_2, d, and L are constants, the above relationship may be written as

$$s_A + 2s_B = \text{constant}$$

Thus,

$$|\Delta s_A| = 2|\Delta s_B| \tag{a}$$

Because $|\Delta s_A| = 1.5$ ft, $|\Delta s_B| = 0.75$ ft. Also, if we divide Eq. (a) by Δt and let $\Delta t \to 0$,

$$|v_A| = 2|v_B|. \tag{b}$$

Because initially $v_A = 40$ ft/s→, it follows that initially $v_B = 20$ ft/s ↓.

By Eq. (15.27),

$$\sum T_1 + \sum V_1 = \sum T_2 + \sum V_2.$$

By Eq. (15.24) and applying Eqs. (15.20) and (15.23), this relationship may be written in the form

$$\sum T_1 + \left(\sum V_{g1} + \sum V_{e1}\right) = \sum T_2 + \left(\sum V_{g2} + \sum V_{e2}\right),$$
$$\sum T_1 + \left(\sum mgh_1 + \sum \tfrac{1}{2}ks_1^2\right) = \sum T_2 + \left(\sum mgh_2 + \sum \tfrac{1}{2}ks_2^2\right)$$

where 1 and 2 refer to initial and final positions. Thus,

$$\tfrac{1}{2}(20)(40)^2 + \tfrac{1}{2}(6)(20)^2 + [0 + 0 + \tfrac{1}{2}(1000)(0.5^2) + \tfrac{1}{2}(500)(0.25^2)]$$
$$= \tfrac{1}{2}(2)v_{A2}^2 + \tfrac{1}{2}(6)v_{B2}^2 + [0 - 6(32.2)(0.75)$$
$$+ \tfrac{1}{2}(1000)(0.5 + 1.50)^2 + \tfrac{1}{2}(500)(0.25 + 0.75)^2].$$

Substituting $v_{B2} = v_{A2}/2$ from Eq. (b) and solving,

$$v_{A2} = 21.9 \text{ ft/s} \rightarrow, \qquad \text{ANS.}$$

and

$$v_{B2} = 10.93 \text{ ft/s} \downarrow. \qquad \text{ANS.}$$

■

Problems

15.39 Path information is shown in Figure P15.39. Determine the work done for the following cases. Forces are expressed in pounds and distances are given in feet. (a) The nonconservative force $\mathbf{F} = 20y\mathbf{i} + 40x\mathbf{j}$ along the path OB. (b) The nonconservative force of part (a) along the path OA followed by AB.

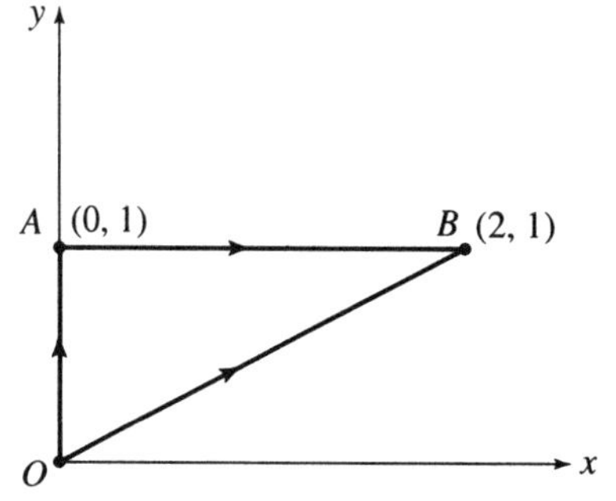

FIGURE P15.39.

15.40 Solve Problem 15.39 for the conservative force $\mathbf{F} = 20y\mathbf{i} + 40x\mathbf{j}$.

15.41 Consider the closed path OAO in the directions indicated in Figure P15.41. Determine the work done around this closed path by the force $\mathbf{F} = x^2\mathbf{i} + y^2\mathbf{j}$. Units are in pounds and inches.

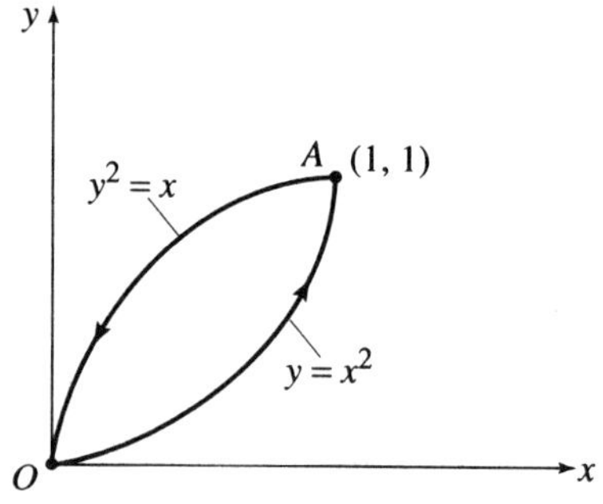

FIGURE P15.41.

15.42 Solve Problem 15.41 for the force $\mathbf{F} = 10(x + y)\mathbf{i} + 10(x - y)\mathbf{j}$.

15.43 Determine whether the force field $\mathbf{F} = 4xz\mathbf{i} + (x^2 + y)\mathbf{j} + (4z - x^2)\mathbf{k}$ is conservative.

15.44 Given the conservative force $\mathbf{F} = (4x\mathbf{i} + 8y\mathbf{j})$ N, consider the work done as this force is displaced from the origin to the point (4,4) with lengths given in meters.

The work done will be independent of the path you choose. Units are in Newtons and meters.

15.45 A person using an exerciser is shown in Figure P15.45. When the fabric is horizontal the person, who weighs 160 lb, is moving 25 ft/s. Determine the spring constant k for each of the twenty springs if the person is brought to rest when the maximum deformation of the springs is as shown in Figure P15.45(b). Neglect any energy losses, and assume that the springs are the only deformable parts of the system. The unstretched length of each spring is 6 in. Assume that the mass center of the person moves along a vertical line above the center of the exerciser and ignore curvature of the fabric beneath the person's shoes.

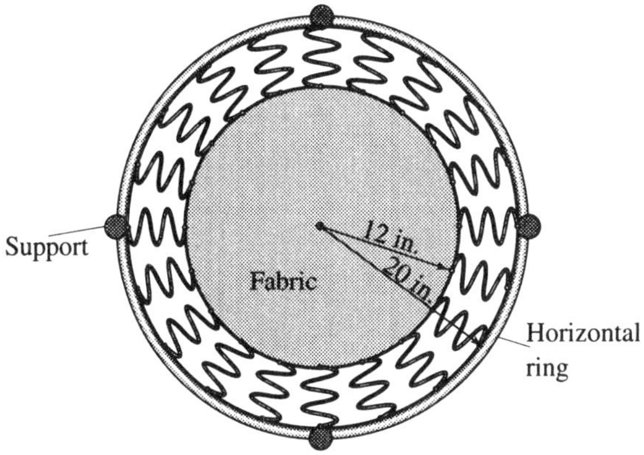

(a) Plan view with springs horizontal

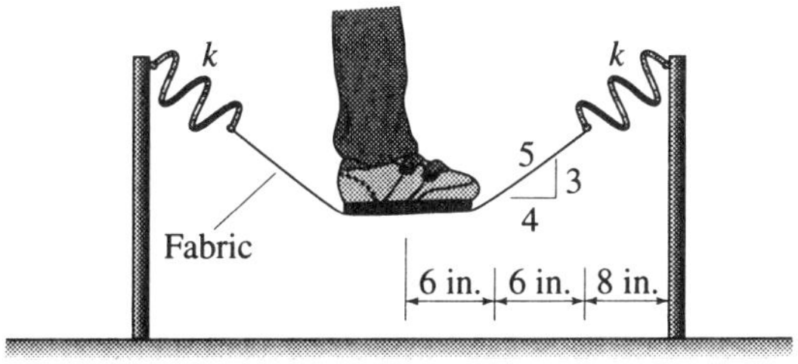

(b) Person on exerciser. Springs deformed the maximum amount.

FIGURE P15.45.

15.46 A weight $W = 400$ N is held in position 1 as shown in Figure P15.46. It is held such that the spring is compressed 0.3 m. The spring constant $k = 8000$ N/m and the weight is not attached to the spring. Determine the distance b which the weight will move up the inclined plane before coming to rest again if

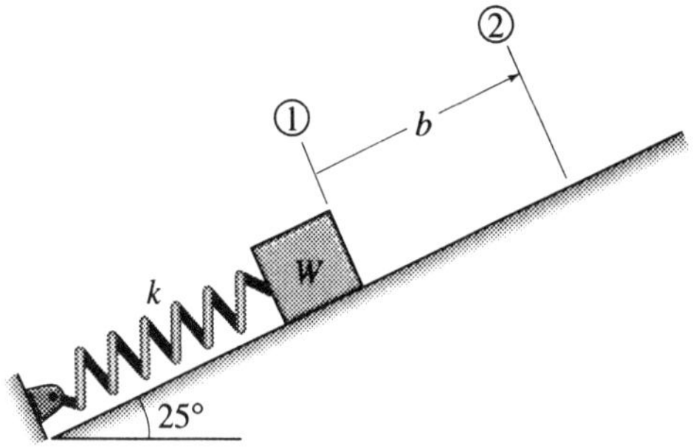

FIGURE P15.46.

it is released from rest in position 1. Neglect friction and air resistance in your analysis.

15.47 Refer to Figure P15.47 and determine the spring constant k as a function of m and h if the weight is released from rest in position 1 and arrives at position 2 with zero speed. Neglect track friction. The spring is unstretched in position 1.

15.48 In Figure P15.47, the mass $m = 2$ slug, $h = 9$ ft, and $k = 15$ lb/ft for the relatively flexible spring. In position 1 the weight is moving at a speed of 5 ft/s. If the unstretched length of the spring is 1 ft, determine the speed of the weight when it reaches position 2. Neglect track friction.

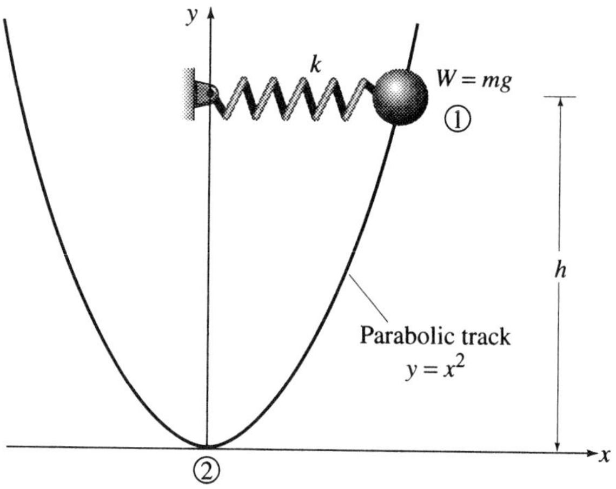

FIGURE P15.47.

15.49 A block weighing 50 lb moves down an inclined plane as shown in Figure P15.49. It is released from rest, moves $s = 10$ ft along the plane, and strikes a spring for which $k = 200$ lb/ft. Determine the maximum compression x of the spring if we neglect friction and air resistance.

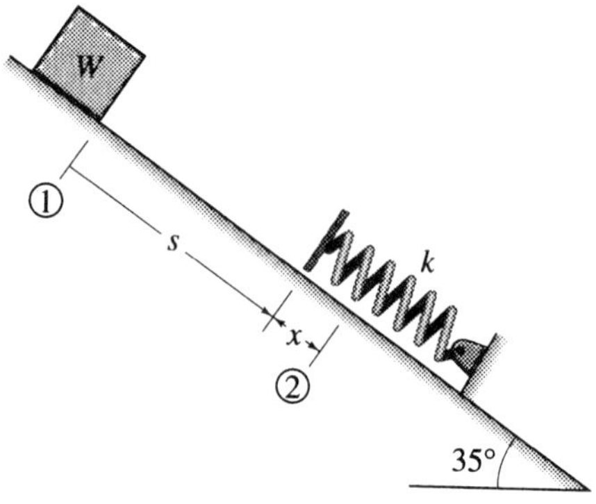

FIGURE P15.49.

15.50 A block weighing 200 N moves down an inclined plane as shown in Figure P15.49. It is released from rest, moves $s = 1.50$ m along the plane, and strikes a spring for which $k = 2500$ N/m. Determine the maximum compression x of the spring if we neglect friction and air resistance.

15.51 A weight $W = 25$ lb is held in position 1 of Figure P15.51 such that the spring is compressed 6 in. Knowing that the spring constant $k = 400$ lb/ft, determine the maximum height h to which the weight will rise if it is released from rest. Find the acceleration immediately before release of the weight which is not attached to the spring. Neglect air resistance.

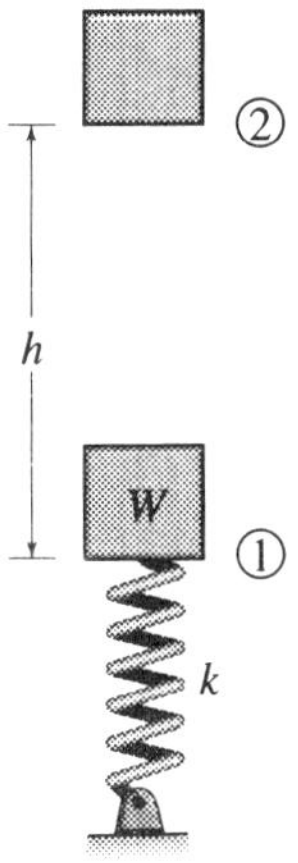

FIGURE P15.51.

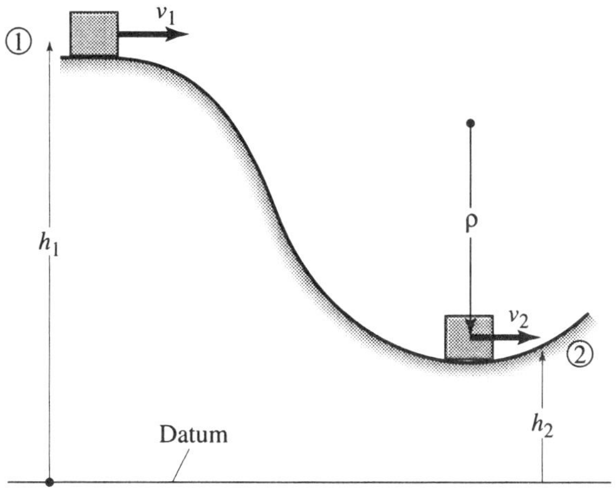

FIGURE P15.53.

15.52 A block weighing 120 N is held in position 1 of Figure P15.51 such that the spring is compressed 0.15 m. Knowing that the spring constant $k = 5000$ N/m, determine the maximum height h to which the weight will rise if it is released from rest. Find the acceleration immediately before release of the weight which is not attached to the spring. Neglect air resistance.

15.53 A block of mass m travels the frictionless track shown in Figure P15.53. At positions 1 and 2, it has horizontal velocities v_1 and v_2. Determine v_2 as a function of v_1, h_1, and h_2. Express the normal force acting on the block when

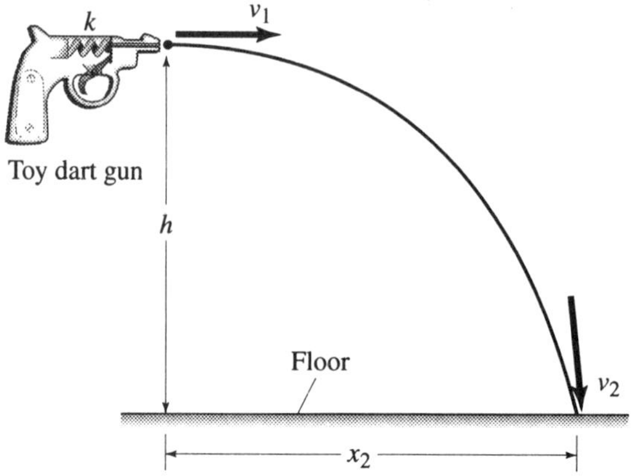

FIGURE P15.54.

it arrives at position 2 as a function of m, v_2, and the radius of curvature ρ at position 2.

15.54 Energy is stored in the spring of the toy dart gun shown in Figure P15.54. Pulling the trigger releases this energy and imparts it to the dart. If we ignore frictional losses and sound generated when the toy gun is fired, then, the energy of the spring is all transmitted to the dart. Express the horizontally directed v_1 as a function of the spring constant k, the dart mass m, and the initial spring compression s_1. The spring is unstretched after the dart is released. Use the conservation of energy principle to find the speed of the dart just before it strikes the floor as a function of v_1 and the height h at which the toy gun is held. Then, consider the kinematics of the dart motion and break it into an x and y motion. If we ignore air resistance, then, the x motion will be at a constant velocity v_1 and the y motion will be at a constant acceleration of g downward. Express the time of travel of the dart in terms of g and h, and determine x_2 in terms of g, h, k, m, and s_1.

15.55 The particle of weight W is released from rest in position 1 as shown in Figure P15.55 and moves for a distance d along the inclined plane until it strikes the spring and compresses it a maximum amount s. Determine s as a function of W, k, d, and β. Ignore friction between the plane and the particle and losses during impact.

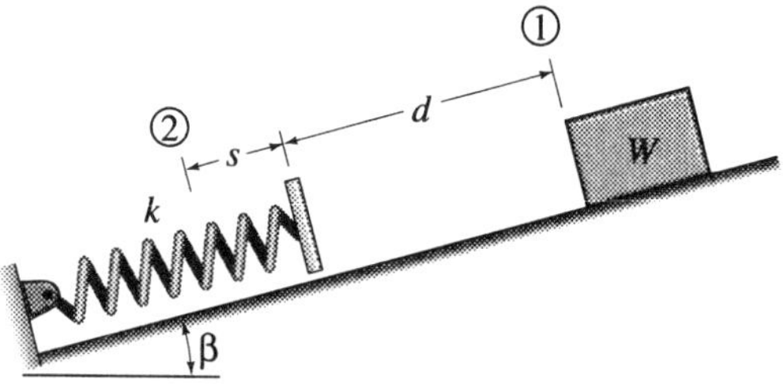

FIGURE P15.55.

15.56 A rod of negligible weight supports the mass m as shown in Figure P15.56. In position 1, the mass is given a very small velocity to the right. When the mass has moved to position 2, it again has a zero velocity. The spring is attached to the lower end of the rod and has an unstretched length of c. Determine the dimension b so that the kinetic energy will vanish in positions 1 and 2. Let $m = 2$ slug, $c = 1.5$ ft, and $k = 100$ lb/ft. The system moves in a vertical plane.

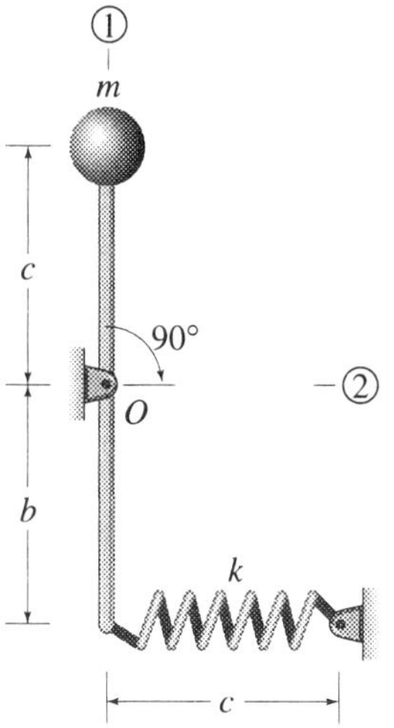

FIGURE P15.56.

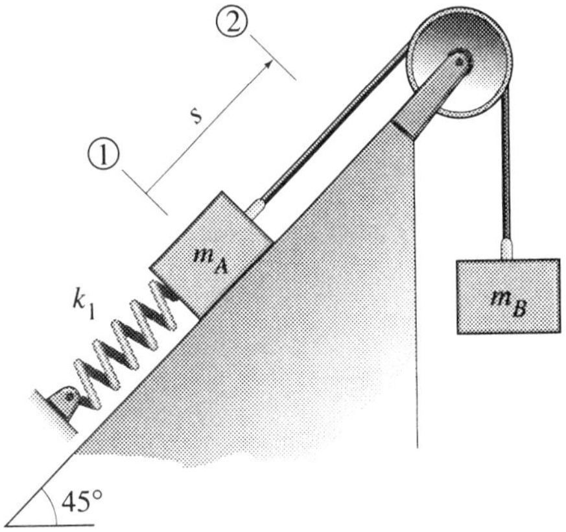

FIGURE P15.57.

15.57 Use the conservation of energy principle for the system depicted in Figure P15.57. The system is released from rest when the spring is unstretched, and mass m_1 moves a distance $s = 1$ ft along the frictionless inclined plane. Determine the corresponding final velocities of these masses, given that $m_1 = 1$ slug, $m_2 = 10$ slug, and $k_1 = 400$ lb/ft.

15.58 The system shown in Figure P15.58 is frictionless. It is released from rest, and mass $m_2 = 4$ slug moves downward a distance of 2 ft. Determine the velocities of the masses after this downward movement, given that $m_1 = 2$ slug and $k_1 =$ 40 lb/ft. The cable is inextensible and the springs are unstretched initially.

15.59 Refer to Figure P15.59 which depicts a frictionless system. Initially the spring is unstretched and the system is at rest. Determine the final velocity of mass m_3 after it moves a distance $s = 1$ ft down the inclined plane. Given $m_1 = 2$ slug, $m_2 = 3$ slug, $m_3 = 6$ slug, and $k_1 = 20$ lb/ft.

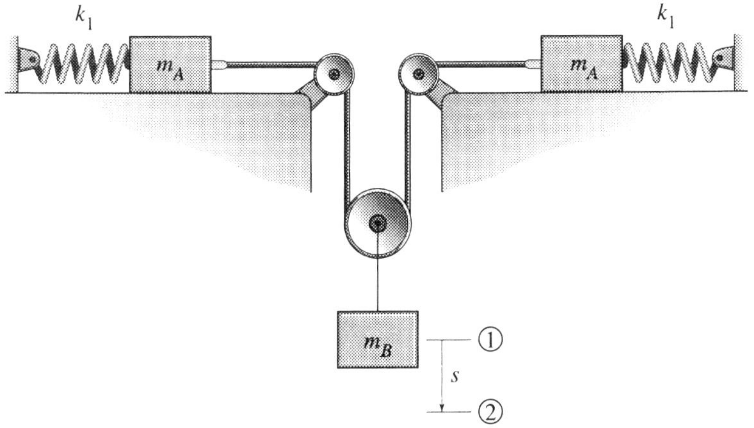

FIGURE P15.58.

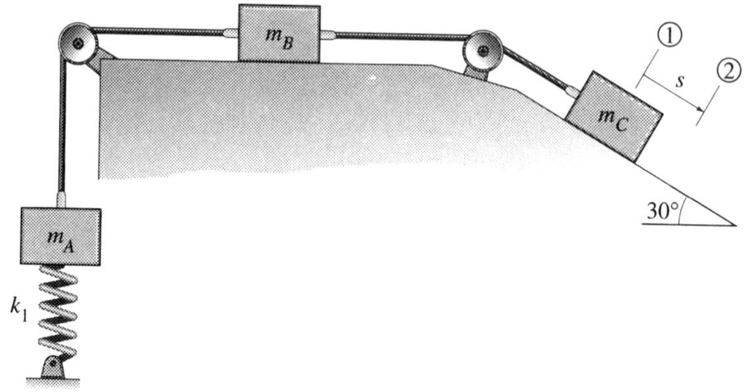

FIGURE P15.59.

Review Problems

15.60 The block of weight $W = 30$ lb shown in Figure P15.60 moves from position 1 to position 2 on a rough horizontal plane. The coefficient of kinetic friction between the block and the plane is $\mu = 0.15$. The spring has a spring constant $k = 50$ lb/ft and an unstretched length $L_u = 2$ ft. Determine the work done by all forces acting on the block as it moves from position 1 to position 2.

15.61 The homogeneous rod of weight $W = 50$ N rotates from position 1 to position 2 as shown in Figure P15.61. The spring has a spring constant $k = 20$ N/m and

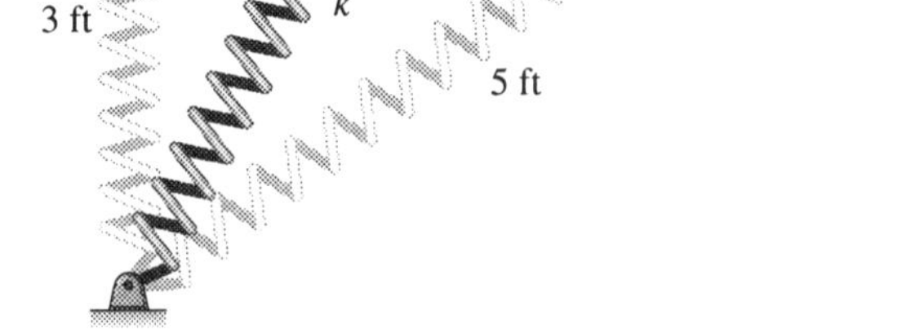

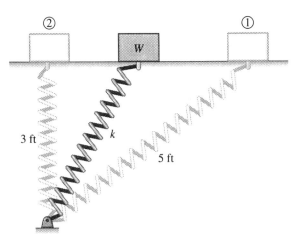

FIGURE P15.60.

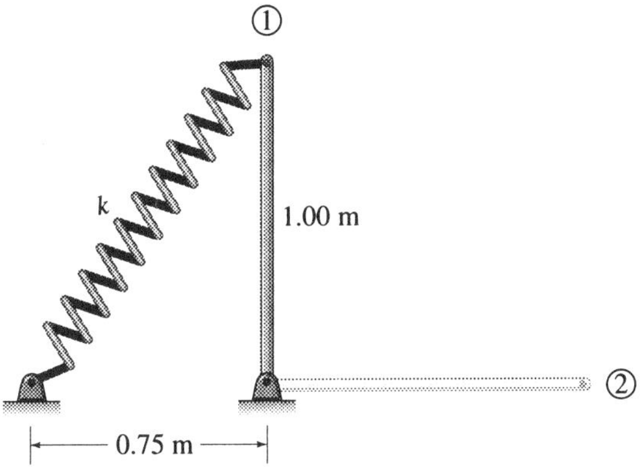

FIGURE P15.61.

an unstretched length $L_u = 0.8$ m. If the friction at the hinge at A provides a constant resisting couple of 10 N·m during the rod's rotation, determine the work done by all the forces and couples acting on the rod as it rotates from position 1 to position 2.

15.62 A block weighing 30 lb is released from rest in the position shown in Figure P15.62. After sliding a distance of 10 ft down the inclined plane, its speed is measured at 5 ft/s. What is the kinetic coefficient of friction between the block and the inclined plane?

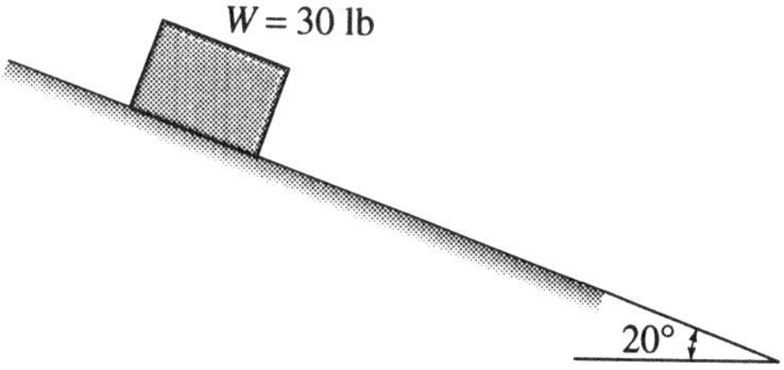

FIGURE P15.62.

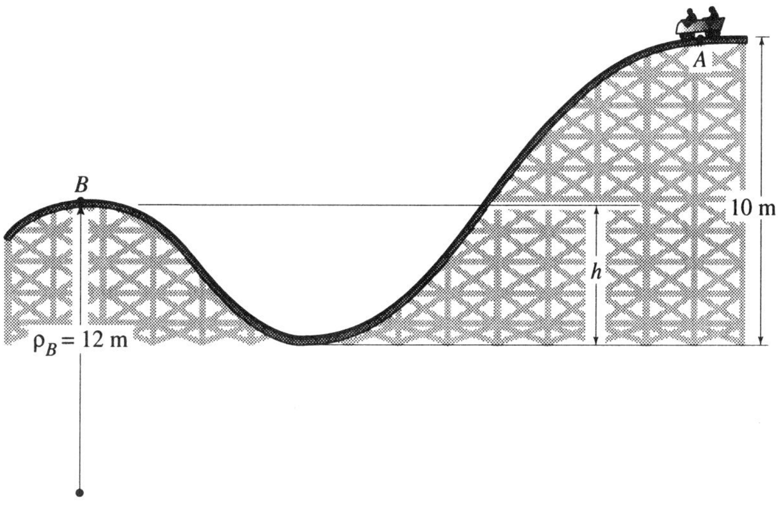

FIGURE P15.63.

15.63 A roller-coaster car and its occupant in an amusement park has a mass of 600 kg. It has a speed of 5 m/s at point A and rolls along the frictionless tracks to point B where the radius of curvature is $\rho_B = 12$ m, as shown in Figure P15.63. Determine the minimum height h, so that the car does not leave the track at point B.

15.64 The system shown in Figure P15.64 is proposed for transporting packages from one level to another 10 ft below. It consists of a conveyor belt, an inclined plane AB, and a spring arrangement. A given package on the belt reaches point A with a speed of 4 ft/s, slides down the plane, and compresses the spring (initially compressed 2 in. by strings as shown) an additional amount of 6 in. until it reaches point B with zero speed.

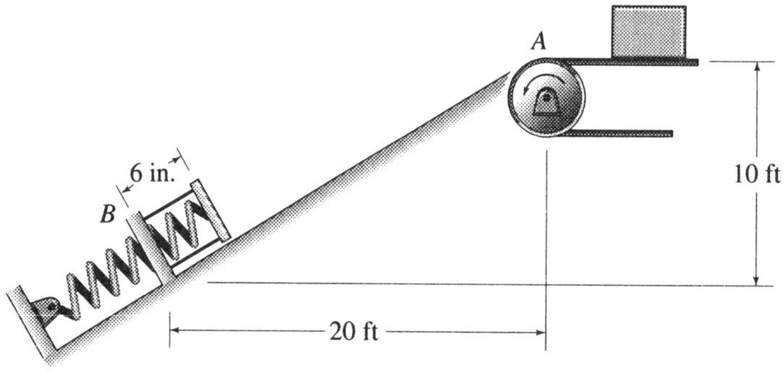

FIGURE P15.64.

The package is then removed by a handler at point B to make room for the next. If the weight of the average package is 30 lb and the coefficient of kinetic friction is 0.10, determine the needed spring constant.

15.65 The pulley system shown in Figure P15.65 is used to move block A along the horizontal plane. Let $m_A = 20$ kg, $m_B = 80$ kg, and the kinetic coefficient of friction $\mu = 0.15$. If the system is released from rest, determine the velocities of blocks A and B after block B moves down a distance of 2 m. Assume small frictionless pulleys.

15.66 A block of weight $W = 15$ lb is held at rest against an unattached spring as shown in Figure P15.66. In the position shown, the spring has an initial compression of 4 in. The spring constant

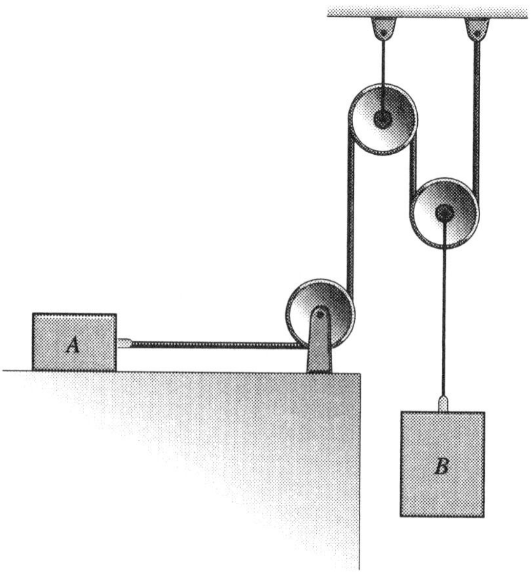

FIGURE P15.65.

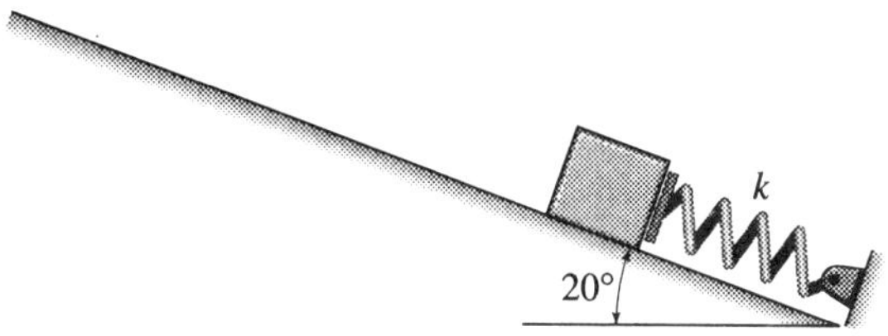

FIGURE P15.66.

$k = 20$ lb/in. and the inclined plane may be assumed frictionless. If the block is released from the position shown, determine the distance along the plane that it will travel before coming to rest.

15.67 A boy starts from rest at A and rides his sled down the ice-covered path, as shown in Figure P15.67. If the boy has a mass $m = 50$ kg and his apparent weight at B is twice his actual weight, determine the radius of curvature of the ice path at B. Assume the ice path to be frictionless.

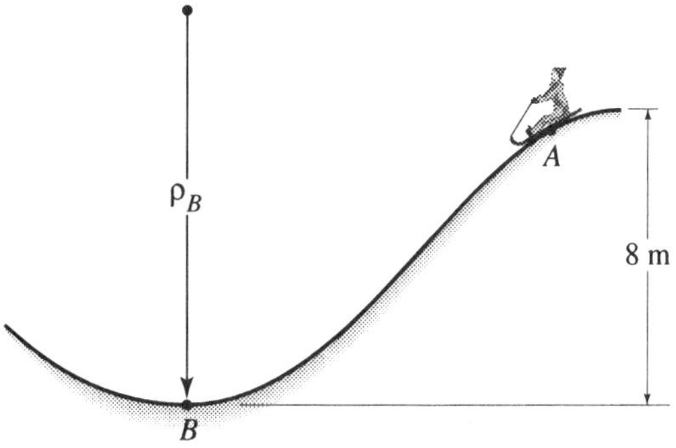

FIGURE P15.67.

15.68 Collar A of weight W = 7.5 lb slides on frictionless rod BC and is attached to two identical springs as shown in Figure P15.68. Let $k = 3$ lb/in. and the unstretched length of the springs $L_u = 10$ in. If the collar is given a velocity of 10 ft/s to the right when $x = 0$, determine the maximum distance x that the collar will travel on rod BC before coming to rest and reversing direction.

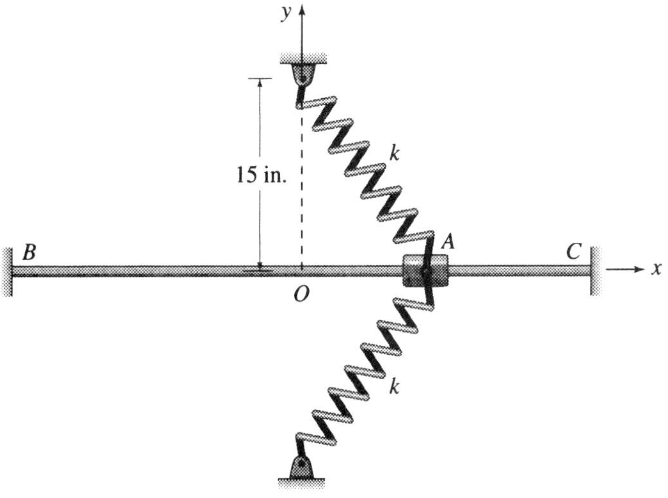

FIGURE P15.68.

16

Kinetics of Particles: Impulse-Momentum

This is an artistic rendering of an object impacting the Earth. Illustration by William K. Hartmann

Since the beginning of time, the principles of impact and momentum have influenced our daily lives here on Earth and in the depth of space. There is strong evidence that the universe exploded from an infinitely dense primal seed, possibly 8 to 18 billion years ago, and that it will continue to expand forever. In these respects, this event inaugurated an eternal potential for collisions between debris and planet Earth. In fact, it is now believed that the collision of a large comet with Earth is the reason for the extinction of the dinosaurs from the face of our planet.

Engineers have always sought to understand the forces and the physical laws affecting the inner structure of systems from the small atom to the extraordinary galaxies. The most powerful conservation principle of quantum theory relates to the mass, energy, and momentum of particles. Physicists observed that mass and energy are always conserved in chemical reactions until 1905 when Einstein proposed that the energy of a system is equal its mass times the square of the velocity of light ($E = mc^2$). The implication is that, in a particle collision or an annihilation, mass and energy can be transformed into each other or exchanged among particles, but not lost.

More recently, scientists have discovered that the classical laws Newton introduced to describe the behavior of our macroscopic universe fail to account for the data gathered in subatomic experiments. One may question, therefore, whether the presumably finite structure of matter is based on an endlessly layered quantum world whose innermost secrets, like the most distant objects in the universe, constantly retreat from the inquiring human mind. Perhaps, there is a symmetry in our universe between its largest and smallest dimensions that intertwines the infinite and the finite.

In this chapter, you will learn about the macroscopic world in which we live. You will learn how Newton's laws of motion can be extended to the

solution of problems affecting our daily lives. The principles of conservation of energy and momentum will be discussed as they relate to object collisions, fluid flow, space mechanics, and other relevant engineering applications. It is hoped that you may be inspired to become the scientist who will form the unifying theory for the subatomic and the macroscopic worlds.

16.1 Definition of Linear Momentum and Linear Impulse

Momentum: Linear momentum **L** of a particle is a vector quantity which is defined as the mass m multiplied by the velocity vector **v** for a particle at a given instant. Mathematically,

$$\mathbf{L} = m\mathbf{v}. \tag{16.1}$$

Because the mass m of a particle is a positive scalar quantity, the linear momentum vector **L** has the same direction and sense as the velocity vector **v**. The magnitude of the linear momentum vector is equal to the mass multiplied by the particle speed v. Thus,

$$\mathrm{L} = \mathrm{mv} \tag{16.2}$$

Although Eqs. (16.1) and (16.2) have a similar appearance they are quite different. Equation (16.1) is a vector equation defining the linear momentum vector **L** which has both magnitude and direction whereas Eq. (16.2) is a scalar equation which defines the magnitude L of the linear momentum vector.

In Figure 16.1, a particle of mass m is shown at any time t as it moves along its path. It has a velocity **v** which changes with time, in general, in both magnitude and direction. Because **L** gets its vector character from **v**, then, the linear momentum **L** changes with time, in general, in both magnitude and direction. Because **v** is directed along the tangent to the path, then, **L** is also a tangential vector.

If we express the velocity **v** in rectangular cartesian coordinates,

$$\begin{aligned}\mathbf{v} &= v_x\mathbf{i} + v_y\mathbf{j} + v_z\mathbf{k}\\ &= \dot{x}\mathbf{i} + \dot{y}\mathbf{j} + \dot{z}\mathbf{k}.\end{aligned}$$

Multiplication of this vector by the scalar mass m gives the linear momentum vector. Thus,

$$\mathbf{L} = m\dot{x}\mathbf{i} + m\dot{y}\mathbf{j} + m\dot{z}\mathbf{k} = L_x\mathbf{i} + L_y\mathbf{j} + L_z\mathbf{k}. \tag{16.3}$$

This vector and its components are shown in Figure 16.2.

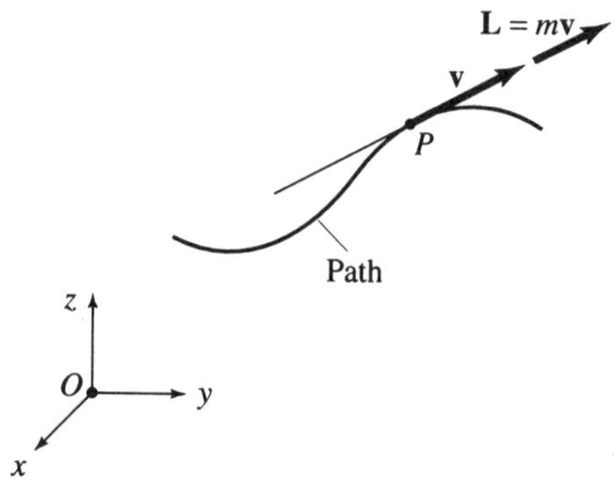

FIGURE 16.1.

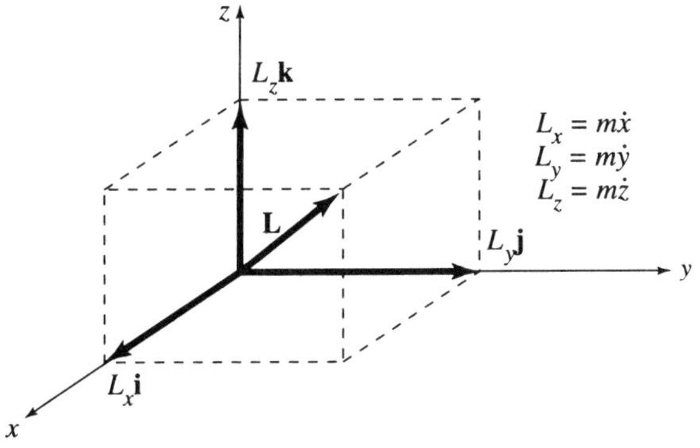

FIGURE 16.2.

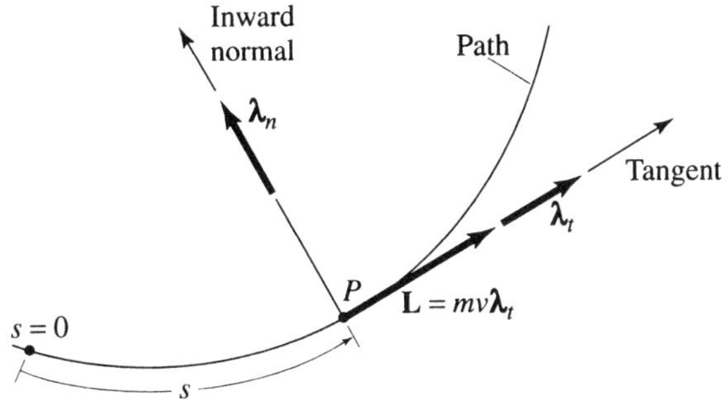

FIGURE 16.3.

In Sec. (13.7), the velocity vector $\mathbf{v}$ in tangential and normal components is given by $\mathbf{v} = v\boldsymbol{\lambda}_t$ where $v = \dfrac{ds}{dt}$ and $\boldsymbol{\lambda}_t$ is a unit tangent vector. If we multiply this equation by the scalar m, we obtain the linear momentum vector $\mathbf{L}$ suitable for use with tangential and normal components as shown in Figure 16.3. Therefore,

$$\mathbf{L} = mv\boldsymbol{\lambda}_t = m\frac{ds}{dt}\boldsymbol{\lambda}_t. \tag{16.4}$$

We note that the unit tangent vector $\boldsymbol{\lambda}_t$ provides the direction for $\mathbf{L}$, and the product $mv = m\dfrac{ds}{dt}$ provides its magnitude, because the mass m, the speed v, and the time derivative of the distance measured along the path $\dfrac{ds}{dt}$ are all scalars.

In Section 13.8 the velocity vector $\mathbf{v}$ in transverse and radial components is given by

$$\mathbf{v} = \dot{r}\boldsymbol{\lambda}_r + r\dot{\theta}\boldsymbol{\lambda}_\theta. \tag{16.5}$$

This form is useful when the path is conveniently expressed in plane polar coordinates. When the path is expressed parametrically with time as the parameter, then, we know r and θ as functions of t which enables us to write $\mathbf{v}$ as a function of time t by finding the derivatives $\dot{r}$ and $\dot{\theta}$ for substitution in Eq. (16.5). Multiplication of Eq. (16.5) by the mass m enables us to write the linear momentum vector. Thus,

$$\mathbf{L} = m\mathbf{v} = m\dot{r}\boldsymbol{\lambda}_r + mr\dot{\theta}\boldsymbol{\lambda}_\theta. \tag{16.6}$$

This equation is represented pictorially in Figure 16.4.

Following the same procedure, expressing the velocity vector in cylindrical coordinates,

$$\mathbf{L} = m\mathbf{v} = m\dot{r}\boldsymbol{\lambda}_r + mr\dot{\theta}\boldsymbol{\lambda}_\theta + m\dot{z}\boldsymbol{\lambda}_z. \tag{16.7}$$

This equation is represented pictorially in Figure 16.5.

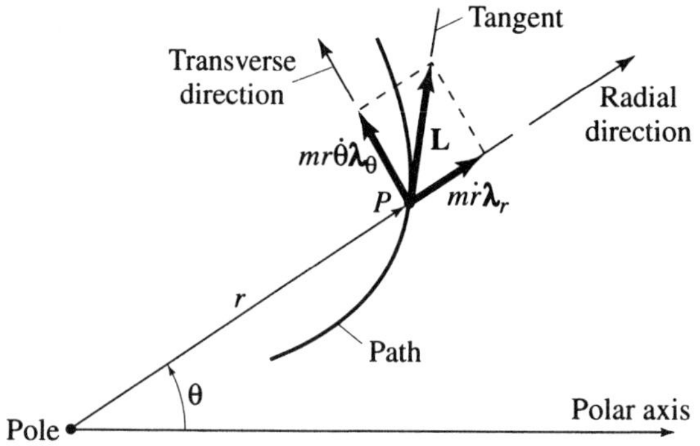

FIGURE 16.4.

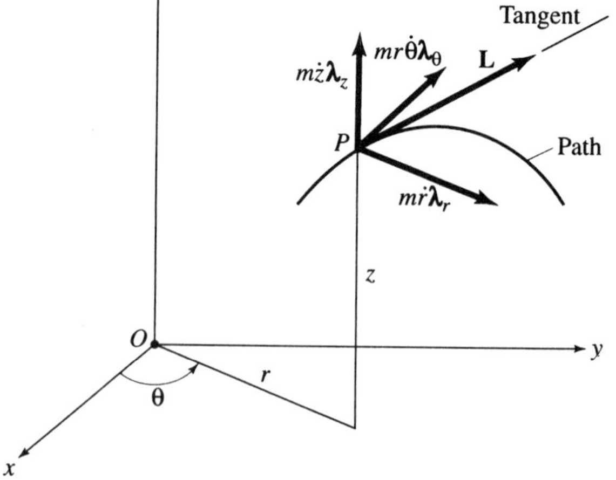

FIGURE 16.5.

The linear momentum has dimensions of mass multiplied by velocity. However, using Newton's second law $\mathbf{F} = m\mathbf{a}$, this dimension may be reduced to that of force multplied by time. Typical units of linear momentum are lb·s and N·s.

Impulse: The linear impulse **I** of a force is a vector quantity. To define this vector quantity, we begin with a definition of the differential linear impulse given by

$$\mathbf{dI} = \mathbf{F}\,dt. \tag{16.8}$$

Now we integrate this equation from t_1 to t_2 which provides a definition of the vector quantity termed *linear impulse*. Thus,

$$\mathbf{I} = \int_{t_1}^{t_2} \mathbf{F}\,dt \tag{16.9}$$

Because both **I** and **F** are vectors, we may write **I** in rectangular cartesian coordinates. Thus,

$$\mathbf{I} = I_x\mathbf{i} + I_y\mathbf{j} + I_z\mathbf{k} = \mathbf{i}\int_{t_1}^{t_2} F_x\,dt + \mathbf{j}\int_{t_1}^{t_2} F_y\,dt + \mathbf{k}\int_{t_1}^{t_2} F_z\,dt \tag{16.10}$$

Equating scalar multipliers on both sides of this equation,

$$I_x = \int_{t_1}^{t_2} F_x\,dt, \quad I_y = \int_{t_1}^{t_2} F_y\,dt, \quad \text{and} \quad I_z = \int_{t_1}^{t_2} F_z\,dt. \tag{16.11}$$

Equation (16.11) shows that the three scalar components of the linear impulse vector **I** are obtained by integrating each of the force components (F_x, F_y, F_z) over the time interval of interest (i.e., from t_1 to t_2). A pictorial representation of the linear impulse vector components is shown in Figure 16.6 for rectangular cartesian coordinates. The scalar components of the linear impulse vector may be interpreted as areas under their respective force-time curves as shown in Figure 16.7.

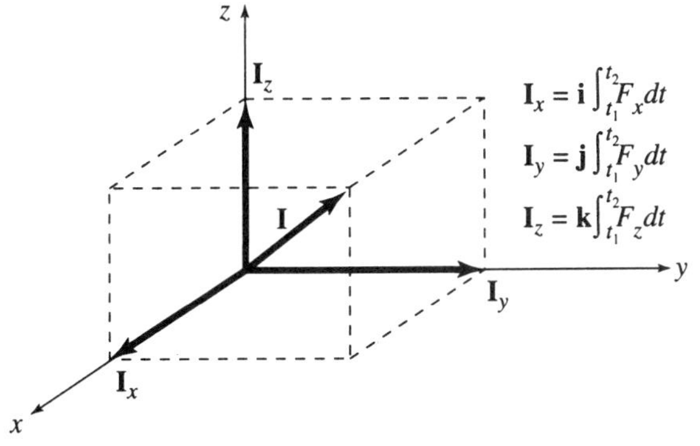

FIGURE 16.6.

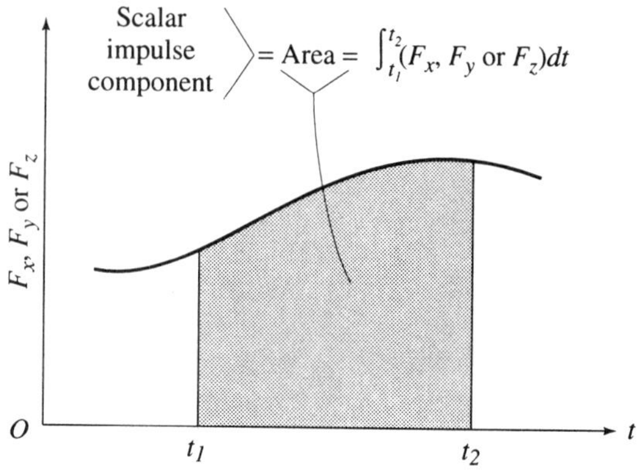

FIGURE 16.7.

Considering the definition of the linear impulse of a force, we conclude that it has the dimension of force multiplied by time which is the same as the dimension of the linear momentum. Thus, typical units of the linear impulse are lb·s and N·s.

The following examples illustrate the determination of the linear impulse and linear momentum for a particle under specific conditions.

Example 16.1

A 3000 lb automobile travels a parabolic vertical curve whose equation is given by $y = 100 - 0.1111(t - 30)^2$ and $x = 40\,(t - 30)$ where x and y are measured in ft and t is s. The path and three points on it are shown in Figure E16.1(a). Find the linear momentum of the automobile when $t = 30$ s and $t = 60$ s.

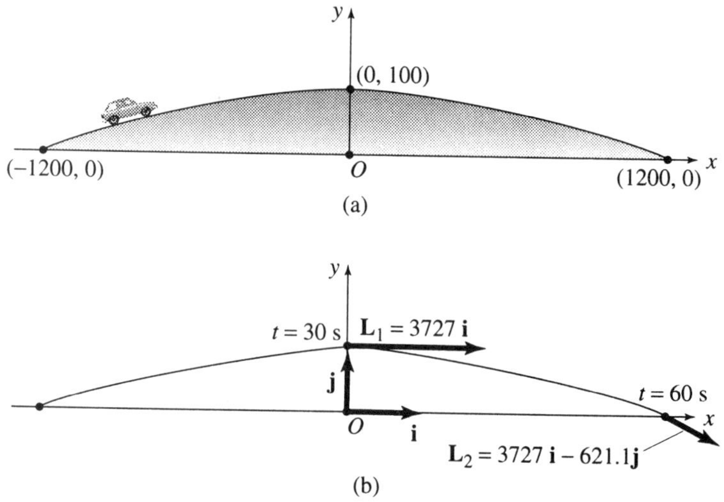

FIGURE E16.1.

Solution

To determine the linear momentum, we need to find the velocity components $\dot{x}$ and $\dot{y}$ by differentiation of x and y with respect to time.

$$x = 40\,(t - 30), \quad \dot{x} = 40 \text{ ft/s},$$

$$y = 100 - 0.1111(t - 30)^2,$$

$$\dot{y} = -0.2222\,(t - 30) \text{ ft/s}.$$

Evaluating at $t = 30$ and $t = 60$,

$$t = 30: \quad \dot{x} = 40 \text{ ft/s}, \quad \dot{y} = 0,$$

$$t = 60: \quad \dot{x} = 40 \text{ ft/s}, \quad \dot{y} = -6.666 \text{ ft/s}.$$

The mass of the automobile is $3000/32.2 = 93.17$ slug. The linear momentum vector is given by the two-dimensional form of Eq. (16.3). Thus,

$$\mathbf{L} = m\mathbf{v} = m\dot{x} + m\dot{y},$$

$$t = 30: \quad \mathbf{L}_1 = 93.17(40)\mathbf{i} = 3730\mathbf{i} \text{ lb·s}, \qquad \text{ANS.}$$

$$t = 60: \quad \mathbf{L}_2 = 93.17(40)\mathbf{i} + 93.17(-6.666)\mathbf{j}$$

$$= (3730\mathbf{i} - 621\mathbf{j}) \text{ lb·s}. \qquad \text{ANS.}$$

These vectors are shown in Figure E16.1(b). Note that both vectors are tangent to the parabolic path.

■ Example 16.2

A force which varies in magnitude and direction is given by $\mathbf{F} = 2t\mathbf{i} + 2\sin t\mathbf{j} + 3t^2\mathbf{k}$ where t is in s and F is in kN. (a) Determine the impulse $\mathbf{I}$ exerted by this force during the time interval from $t = 1$ to $t = 3$ s. (b) Find the magnitude of this vector and the angles it makes with each of the coordinate axes. (c) Numerically approximate I_y, the impulse component in the y direction by using 4 trapezoids to determine the area under the $F_y - t$ curve.

Solution

(a) By Eq. (16.10)

$$\mathbf{I} = \int \mathbf{F}\,dt = \mathbf{i}\int_1^3 2t\,dt + \mathbf{j}\int_1^3 2\sin t\,dt + \mathbf{k}\int_1^3 3t^2\,dt$$

from which

$$\mathbf{I} = [8.00\mathbf{i} + 3.06\mathbf{j} + 26.0\mathbf{k}] \text{ kN·s}. \qquad \text{ANS.}$$

(b) The magnitude of $\mathbf{I}$ is given by

$$I = \sqrt{8.00^2 + 3.06^2 + 26.0^2} = 27.4 \text{ kN·s} \qquad \text{ANS.}$$

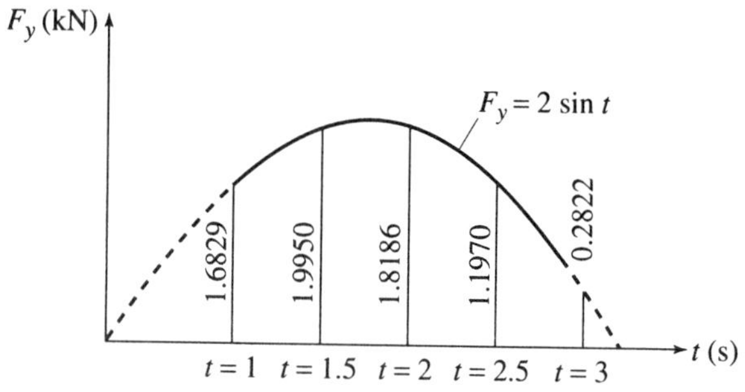

FIGURE E16.2.

$$\cos\theta_x = \frac{8.00}{27.4} = 0.292 \Rightarrow \theta_x = \cos^{-1}(0.292) = 73.0^\circ \qquad \text{ANS.}$$

$$\cos\theta_y = \frac{3.06}{27.4} = 0.112 \Rightarrow \theta_y = \cos^{-1}(0.112) = 83.6^\circ \qquad \text{ANS.}$$

$$\cos\theta_z = \frac{26.0}{27.4} = 0.949 \Rightarrow \theta_z = \cos^{-1}(0.949) = 18.4^\circ \qquad \text{ANS.}$$

(c) Refer to Figure E16.2 where a plot of the $(F_y - t)$ curve is shown with ordinates required for the numerical integration. The area of a single trapezoid is equal to the product of the average height and base. In each case, the base of the trapezoids is 0.5 s and the heights are the ordinates shown in Figure E16.2. An approximate value of I_y is given by the sum of the areas of the 4 trapezoids. Thus,

$$\begin{aligned} I_y &= \tfrac{1}{2}(0.5)(1.6829 + 1.9950) + \tfrac{1}{2}(0.5)(1.9950 + 1.8186) \\ &\quad + \tfrac{1}{2}(0.5)(1.8186 + 1.1970) + \tfrac{1}{2}(0.5)(1.1970 + 0.2822) \\ &= 3.00 \text{ kN}\cdot\text{s}. \end{aligned} \qquad \text{ANS.}$$

This value is about 2% lower than the value of 3.06 obtained by integration. An increase in the number of trapezoids would decrease this 2% error.

Comments

Numerical integration was not required in this case, but engineers often encounter the following two situations where numerical integration using digital computers is the best approach:

1. Ordinates to the force-time curve are known at discrete points usually from experiments. In this case, we either work directly with the measured ordinates or curve-fit the data. Numerical integration or integration of the fitted functions is employed.

2. The functional form of the force is known but the integrand is either very difficult or impossible to integrate. In such cases we employ numerical methods such as the trapezoidal rule, Simpson rule, etc.

■

Problems

Problems 16.1 to 16.8 deal with linear momentum.
Problems 16.9 to 16.16 deal with linear impulse.

16.1 A particle of mass $m = 2$ slugs travels the circular path shown in Figure P16.1 such that $s = \pi(t^2 + 8t)$ where t is measured in seconds and s is measured in feet. Determine the linear momentum of this particle when it is in position 1 and in position 2.

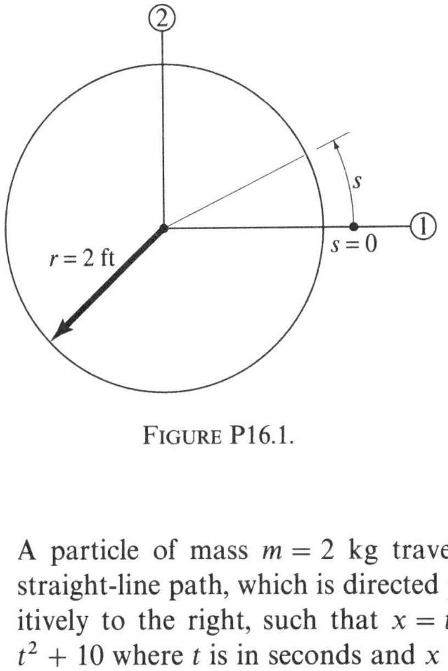

FIGURE P16.1.

16.2 A particle of mass $m = 2$ kg travels a straight-line path, which is directed positively to the right, such that $x = t^3 + t^2 + 10$ where t is in seconds and x is in meters. Determine the linear momentum of this particle when $t = 1$ s and when $t = 2$ s.

16.3 The parametric equations of the path of a particle are given by $x = 2t$ and $y = 4t^2 - 8t + 8$ where t is in seconds and x and y are in feet. If the particle has a mass of one slug, determine (a) the linear momentum of the particle for $t = 1$ s, (b) the path of the particle, and (c) the acceleration of the particle when $t = 1$ s.

16.4 In cylindrical coordinates, the parametric equations of the path of a particle are given in m-s units by $r = 2$, $\theta = 2t$, and $z = 4t$. Determine the velocity **v** and acceleration **a** of the particle at any time t. If the particle has a mass of 1 kg, determine its linear momentum.

16.5 In plane polar coordinates, the parametric equations of a particle path are given in m-s units by $r = \dfrac{2}{\cos 4t}$ and $\theta = 4t$. Determine the path of the particle and its polar velocity vector when $t = 0.25$ s. If the particle has a mass of 2 kg, determine its linear momentum vector at $t = 0.25$ s. Sketch the path and this vector at $t = 0.25$ s.

16.6 In plane polar coordinates, the parametric equations of a particle path are given in ft-s units by $r = \dfrac{10}{\cos 6t}$ and $\theta = 4t$. Determine the path of the particle and its velocity vector using transverse and radial components at $t = 0.1$ s. If the particle has a mass of 1 slug, determine its linear momentum at this same time.

16.7 A particle of mass 0.5 kg moves on a path in the x-y plane such that $x = \cos \omega t$ and $y = 0.6 \sin \omega t$ in m-s units. For $\omega = 2$ rad/s and $t = 0.2$ s, determine the linear momentum of this particle.

16.8 A particle of 2 kg mass moves along the path in the first quadrant defined by $x = a \cosh \omega t$ and $y = b \sinh \omega t$. For $a = 2$ m, $b = 3$ m, $\omega = 2$ rad/s, and $t = 0.5$ s, determine the linear momentum of this particle.

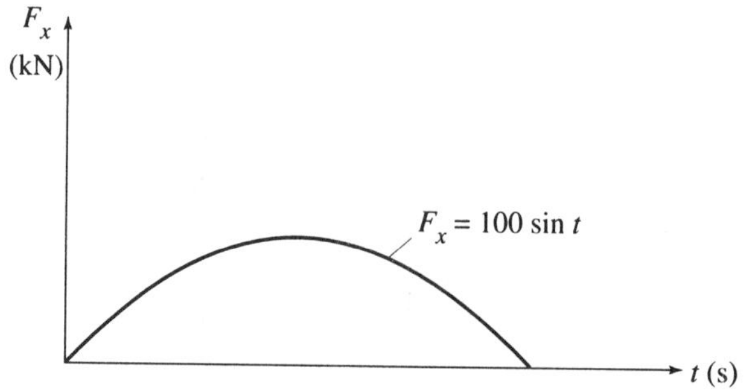

FIGURE P16.9.

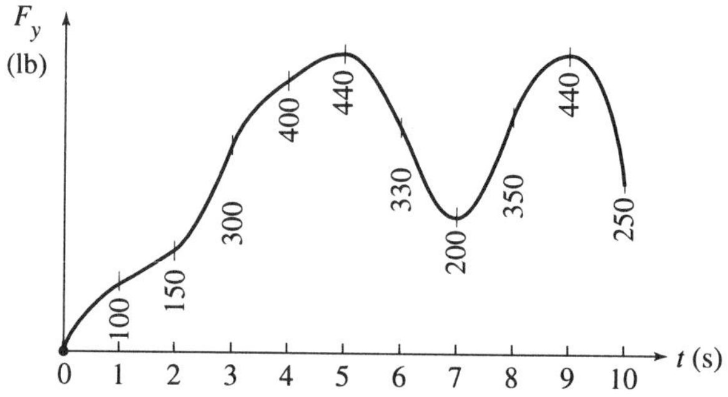

FIGURE P16.10.

16.9 The variation of a force component F_x with time is sinusoidal as shown in Figure P16.9. For the time interval from $t = 0$ to $t = \pi$, determine the impulse exerted by this force component.

16.10 Use the trapezoidal rule for numerical integration to determine the impulse exerted in 10 s by the force which varies with time as shown in Figure P16.10. Let $\Delta t = 1$ s.

16.11 A force measured in kN varies with time measured in s as shown in Figure P16.11. Use the trapezoidal rule for numerical integration to determine the impulse exerted by the force in 10 seconds. Let $\Delta t = 1$ s.

16.12 A force varies with time such that $F_x = 100 \sinh 4t$ where the force is in lb and time t is in s. Determine the impulse exerted by this force from $t = 0$ to $t = 1$ s.

16.13 A force varies with time such that $F_y = 10 \cosh 2t$ where the force is in kN and time t is in s. Determine the impulse exerted by this force from $t = 0$ to $t = 0.5$ s.

16.14 The force exerted on a particle is given by $\mathbf{F} = 4t\mathbf{i} + 6t^2\mathbf{j} + 10\mathbf{k}$ where F is in lb and t is in s. Determine the impulse exerted by this force from $t = 0$ to $t = 2$ s. What is the magnitude of this impulse vector? What angle does this vector make with the x axis?

16.15 A force varies with time such that $F_z = t^2 + 2$ where F_z is in N and t is in s. (a) Determine the impulse exerted by

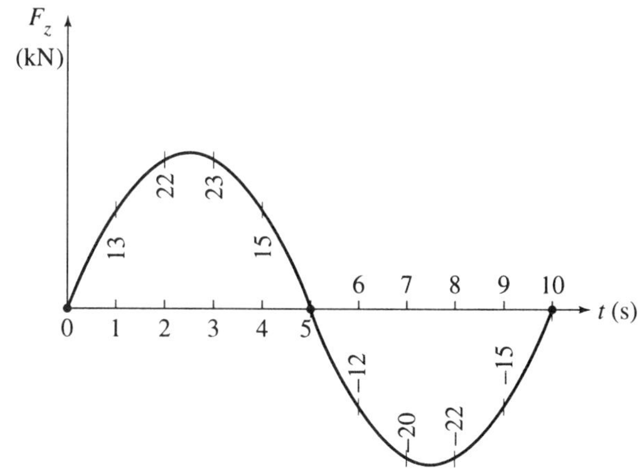

FIGURE P16.11.

this force from $t = 0$ to $t = 4$ s. (b) Use the trapezoidal rule to find an approximate value for the impulse during the same time interval. Let $\Delta t = 1$ s. (c) What is the percentage error in the approximation of part (b)

16.16 Repeat Problem 16.5 for the case where $F_x = 4t^3 + 6$ in which F_x is in lb and t is in s.

16.2 The Principle of Linear Impulse and Momentum

Newton's second law for the motion of a particle is given by

$$\mathbf{R} = \sum \mathbf{F} = m\mathbf{a}$$

where $\mathbf{R} = \sum \mathbf{F}$ represents the resultant of all the forces acting on the particle. To derive the principle of linear impulse and momentum, we use the relationship $\mathbf{a} = \dfrac{d\mathbf{v}}{dt}$ and substitute to obtain

$$\mathbf{R} = m\frac{d\mathbf{v}}{dt}$$

Multiplying by dt and integrating yields

$$\int_{t_1}^{t_2} \mathbf{R}\,dt = m\int_{v_1}^{v_2} d\mathbf{v} = m\mathbf{v}_2 - m\mathbf{v}_1$$

where, as defined in Section 16.1, the quantity $\int_{t_1}^{t_2} \mathbf{R}\,dt$ is the impulse $\mathbf{I}$ and the quantities $m\mathbf{v}_1$ and $m\mathbf{v}_2$ represent, respectively, the linear momenta $\mathbf{L}_1$ and $\mathbf{L}_2$ corresponding to t_1 and t_2. This equation expresses the principle of linear impulse and momentum and it may be written as

$$\mathbf{I} = \mathbf{L}_2 - \mathbf{L}_1. \tag{16.12}$$

Equation (16.12) states that the change in the linear momentum of a

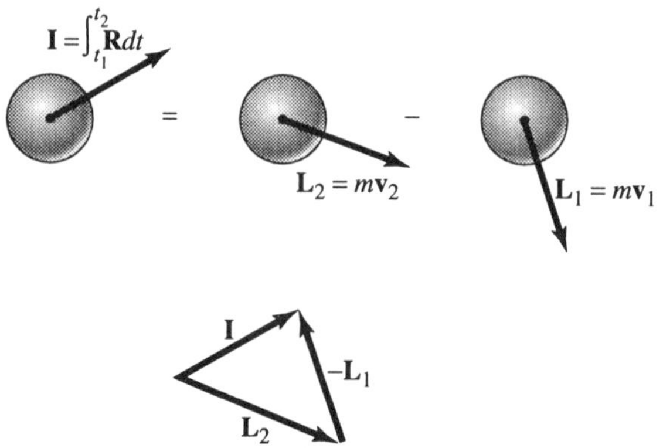

FIGURE 16.8. Eq. (16.13): $\mathbf{I} = \mathbf{L}_2 - \mathbf{L}_1$

particle, $\mathbf{L}_2 - \mathbf{L}_1$, is equal to the linear impulse of the resultant force delivered over the time interval from t_1 to t_2. It is important to note that the initial and final momenta are evaluated instantaneously whereas the impulse is an integrated effect extending over the time interval from t_1 to t_2. Equation (16.12) is represented pictorially by the impulse-momenta diagrams shown in Figure 16.8. It is strongly recommended that such diagrams be constructed before attempting a solution by the principle of linear impulse and momentum.

Expressing the three vectors in Eq. (16.12) in terms of rectangular components, we conclude that this vector equation may be written as three scalar equations. Thus,

$$\left.\begin{aligned} I_x &= (L_2 - L_1)_x \\ I_y &= (L_2 - L_1)_y \\ I_z &= (L_2 - L_1)_z \end{aligned}\right\}. \tag{16.13}$$

The following examples illustrate the use of the principle of linear impulse and momentum in the solution of problems.

■ Example 16.3

A block of mass $m = 2$ kg is moving to the right with a speed of 4 m/s on a frictionless horizontal plane as shown in Figure E16.3(a) when a force $F = (3t^2 + 10)$ N, where t is in seconds, is applied as shown. Determine the velocity of the block after the force F has acted for a period of 2 s.

Solution

The impulse-momenta diagrams are shown in Figure E16.3(b) along with a convenient coordinate system. Note that because the normal N is equal in magnitude to the weight W of the block, their impulses during the 2 s period will cancel each other and, therefore, there is no

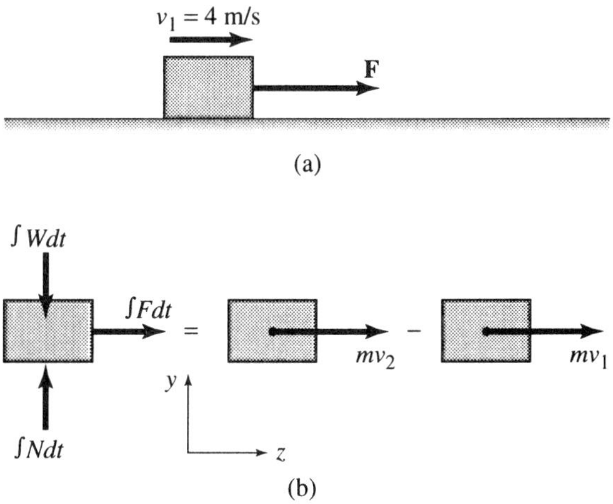

FIGURE E16.3.

change in momentum in the y direction. However, the impulse of the applied force F produces a change in the momentum in the x direction. Thus,

$$I_x = (L_2 - L_1)_x,$$

$$\int_0^2 (3t^2 + 10)\,dt = mv_2 - mv_1,$$

$$t^3 + 10t\big|_0^2 = 2(v_2 - 4),$$

$$28 = 2(v_2 - 4).$$

from which

$$v_2 = 18.00 \text{ m/s} \rightarrow. \qquad \text{ANS.}$$

■ Example 16.4

A car of weight $W = 3220$ lb travels in a circular horizontal path of radius $r = 200$ ft as shown in Figure E16.4(a). The distance s measured along the path is given by $s = r\omega t$ where $\omega = 0.33$ rad/s. Determine the impulse exerted by the pavement on the car as it moves from position 1 $(t_1 = 0)$ to position 2 $\left(t_2 = \dfrac{\pi}{2\omega} \text{ second}\right)$ by (a) the linear impulse-momentum principle and by (b) direct integration of the force-time function.

Solution

(a) The impulse-momenta diagrams are shown in Figure E16.4(b) for any position of the car defined by the angle $\theta = \omega t$. Using Eq. (16.12),

$$\mathbf{I} = \mathbf{L}_2 - \mathbf{L}_1 = -mv_2\mathbf{i} - mv_1\mathbf{j}$$

$$= -m(v_2\mathbf{i} + v_1\mathbf{j}). \qquad \text{(a)}$$

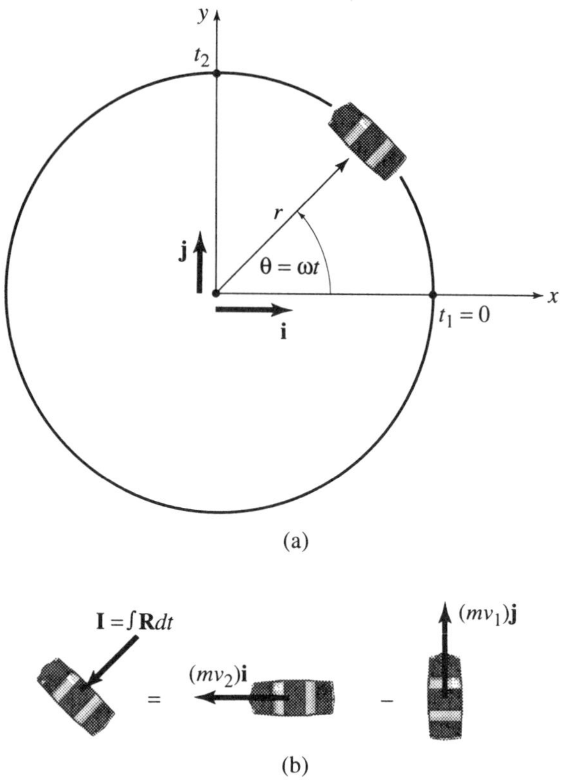

FIGURE E16.4.

In general,

$$v = \frac{ds}{dt} = \frac{d}{dt}(r\omega t) = r\omega = 200(0.33) = 66 \text{ ft/s}.$$

Because r and ω are both constants, the speed is also a constant. Thus,

$$v_1 = v_2 = 66 \text{ ft/s}$$

and, by Eq. (a),

$$\mathbf{I} = -\left(\frac{3220}{32.2}\right)(66)(\mathbf{i} + \mathbf{j}) = -6600\,(\mathbf{i} + \mathbf{j}) \text{ lb·s.} \qquad \text{ANS.}$$

(b) To find the resultant pavement force acting on the car, we first determine its acceleration by differentiating the position vector twice with respect to time. Thus, because $\mathbf{r} = (r\cos\omega t)i + (r\sin\omega t)\mathbf{j}$,

$$\frac{\mathbf{dr}}{dt} = (-r\omega\sin\omega t)\mathbf{i} + (r\omega\cos\omega t)\mathbf{j},$$

$$\frac{\mathbf{d}^2\mathbf{r}}{dt^2} = (-r\omega^2\cos\omega t)\mathbf{i} - (r\omega^2\sin\omega t)\mathbf{j}.$$

Because $\mathbf{R} = m\mathbf{a}$,

$$\mathbf{R} = -mr\omega^2 \cos \omega t\, \mathbf{i} - mr\omega^2 \sin \omega t\, \mathbf{j}.$$

Because the car moves at constant speed, $a_t = 0$ and $a_n = \dfrac{v^2}{\rho} = r\omega^2$. Thus, the total acceleration of the car for any position is $a = a_n = r\omega^2$, directed toward the center of the circular path. Also, the magnitude of the resultant force $\mathbf{R}$ that the pavement exerts on the car is $R = ma = mr\omega^2$ also directed toward the center of the circular path.

$$\mathbf{I} = \int_{t_1}^{t_2} \mathbf{R}\, dt = \int_0^{\pi/2\omega} (-mr\omega^2 \cos \omega t \mathbf{i} - mr\omega^2 \sin \omega t \mathbf{j})\, dt$$

$$= -mr\omega\mathbf{i}[\sin \omega t]\big|_0^{\pi/2\omega} - mr\omega\mathbf{j}[-\cos \omega t]\big|_0^{\pi/2\omega}$$

which, after simplification, yields

$$\mathbf{I} = -mr\omega(\mathbf{i} + \mathbf{j}) = -6600(\mathbf{i} + \mathbf{j})\,\text{lb·s}. \qquad \text{ANS.}$$

This result agrees with that of part (a). Note that $\mathbf{R}$ changed in direction throughout the time interval. We were able to handle this directional variation because we projected this resultant force in the direction of two unit vectors ($\mathbf{i}$ and $\mathbf{j}$) which did not vary in direction or magnitude.

Example 16.5

A block of mass $m = 13$ kg is being pushed up a plane as shown in Figure E16.5(a). Initially, it is moving up the inclined plane at a velocity $v_1 = 0.5$ m/s at time $t_1 = 1$ s. A worker applies a force $F(t) = (20 + 12t^2)$ N parallel to the plane as shown. The coefficient of kinetic friction for the box and plane is $\mu = 0.15$. Determine the velocity v_2 at time $t_2 = 4$ s.

Solution

The linear impulse-momentum principle is expressed with vectors and it is convenient to represent these vector quantities as shown in Figure E16.5(b), where, on the left of the equal sign, we have the *impulse diagram* and, on the right, we have the two *momenta diagrams*. Choose an x axis parallel to the plane and a y axis perpendicular to the plane and write the first two of Eqs. (16.13). The scalar unknowns in these equations are the normal force N and the final velocity $\mathbf{v}_2$. Note that both $\mathbf{L}_1$ and $\mathbf{L}_2$ are directed parallel to the plane because the block moves along this plane.

In the y direction,

$$I_y = (L_2 - L_1)_y = 0,$$

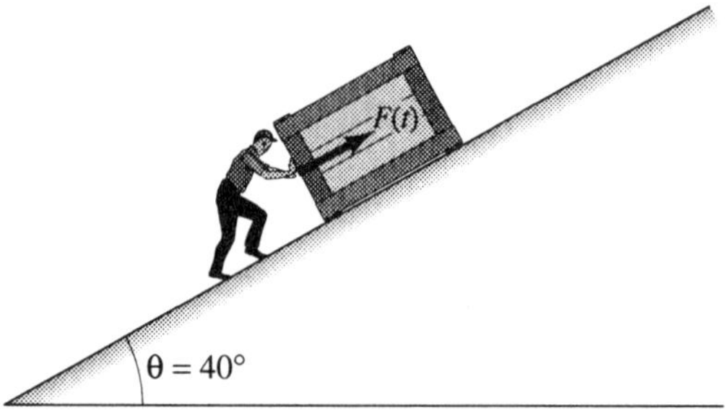

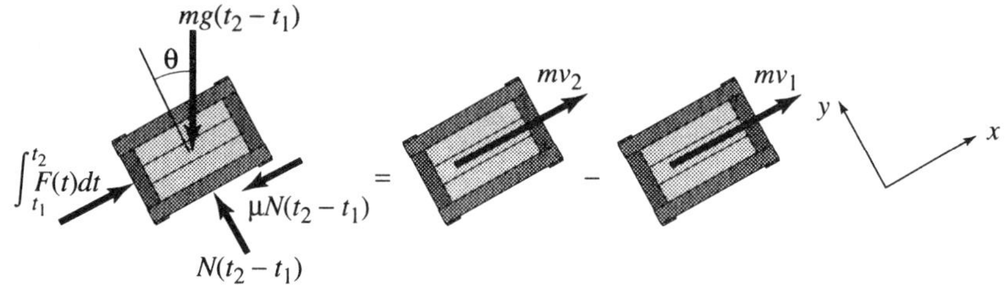

FIGURE E16.5.

$$N(t_2 - t_1) - mg(t_2 - t_1)\cos\theta = 0,$$

$$N(4 - 1) - 13(9.81)(4 - 1)\cos 40^\circ = 0,$$

$$N = 97.7\ \text{N}.$$

In the x direction,

$$I_x = (L_2 - L_1)_x,$$

$$\int_{t_1}^{t_2} F(t)\,dt - \mu N(t_2 - t_1) - mg(t_2 - t_1)\sin\theta = mv_2 - mv_1,$$

$$\int_1^4 (20 + 12t^2)\,dt - 0.15(97.7)(4 - 1) - 13(9.81)(4 - 1)\sin 40^\circ$$

$$= 13v_2 - 13(0.5)$$

Solve for v_2 to obtain

$$v_2 = 2.20\ \text{m/s} \quad \angle 40^\circ .$$ ANS.

■

Problems

Use impulse and momenta diagrams in the solution of the following problems.

16.17 A mass of 1 kg moves along the positive x axis on a horizontal plane with a velocity of 10 m/s. If it is to be brought to rest in 10 s, what constant force F_x must be applied to it? What is the acceleration of this particle?

16.18 A mass of 2 slug travels along the positive y axis on a horizontal plane with a velocity of 20 ft/s. If it is to be brought to rest in 15 s, what constant force F_y must be applied to it? What is the acceleration of this particle?

16.19 Refer to Example 16.5. Let $F(t) = 6t$, and solve the stated problem. $F(t)$ is expressed in N and t in s. All other conditions and numerical values stated in Example 16.5 remain unchanged.

16.20 A variable force expressed in N as a function of time in s is given by $Q(t) = 3t^2$. This force acts horizontally to the right on a mass m which moves along a smooth horizontal plane. Initial and final velocities are v_1 and v_2, respectively. (a) Determine v_2 at any time t_2 as a function of m, v_1, t_1, and t_2. (b) Find v_2 if $m = 1$ kg, $v_1 = 1$ m/s, $t_1 = 2$ s, and $t_2 = 6$ s.

16.21 A variable force expressed in lb as a function of time in s is given by $Q(t) = 4t^3 + 10t$. This force acts horizontally to the right on a mass m which moves along a smooth horizontal plane. Initial and final velocities are v_1 and v_2, respectively. (a) Determine v_2 at any time t_2 as a function of m, v_1, t_1, and t_2. (b) Find v_2 if $m = 2$ slug, $v_1 = 10$ ft/s, $t_1 = 4$ s, and $t_2 = 7$ s.

16.22 Refer to Figure P16.22 where $P(t) = 3t^2 + 5$, t is in seconds, and P is in pounds. The block weighs 32.2 lb and the coefficient of friction is 0.2. If the initial velocity is 2 ft/s to the right, determine the velocity 4 seconds later.

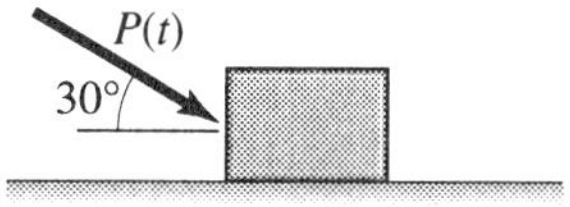

FIGURE P16.22.

16.23 A mass of 4 slug moves from point 1 to point 2 in a time interval $t_2 - t_1$ measured in s as shown in Figure P16.23. Determine the impulse **I** exerted on the mass during this time interval. Assume frictionless conditions.

16.24 Refer to the connected system shown in Figure P16.24 at $t = t_1$. Determine the tension in the cable and the final velocity v_2 at time t_2 as a function of m_A, m_B, v_1, t_1, t_2, and g. Neglect air resistance

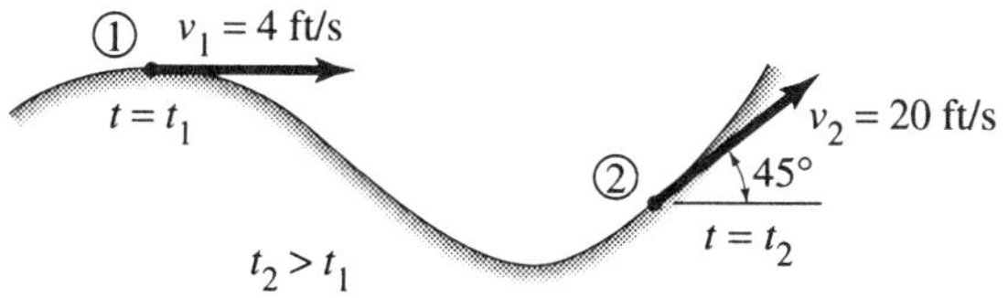

FIGURE P16.23.

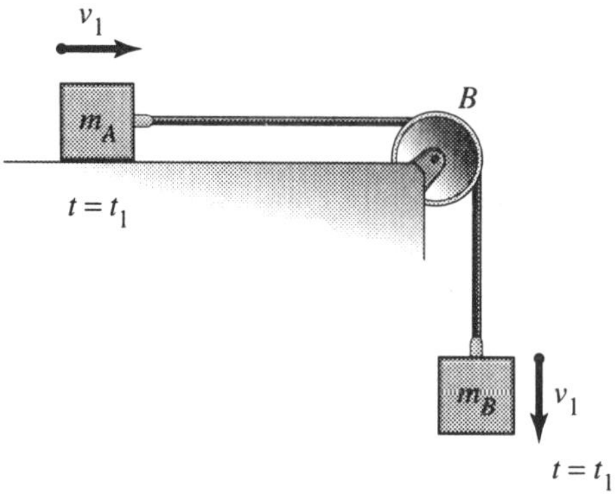

FIGURE P16.24.

and friction between block A and the horizontal plane.

16.25 Refer to the connected system shown in Figure P16.25 at $t = t_1$. Determine the tension in the cable and the final velocity v_2 at time t_2 as functions of m_A, m_B, v_1, t_1, t_2, θ, and g. Neglect air resistance and friction between block A and the inclined plane.

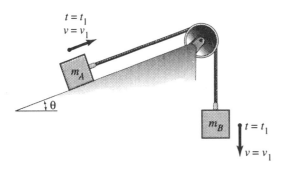

FIGURE P16.25.

16.26 A block of weight 40 lb rests on a horizontal plane for which the static coefficient of friction is 0.2 and the kinetic coefficient of friction is 0.1. A force $Q = t + 2$ acts to the right on the block where Q is in pounds and t is in seconds. (a) How much time elapses before the block begins to move to the right? (b) Once the block begins to move, how much additional time elapses before it reaches a velocity of 10 ft/s to the right?

16.27 A block which weighs 400 N rests on a horizontal plane for which the static coefficient of 0.3 and the kinetic coefficient of friction is 0.2. A force $Q = 20t + 60$ acts to the right on the block where Q is in Newtons and t is in seconds. (a) How much time elapses before the block begins to move to the right? (b) Once the block begins to move, how much additional time elapses before it reaches a velocity of 10 m/s to the right?

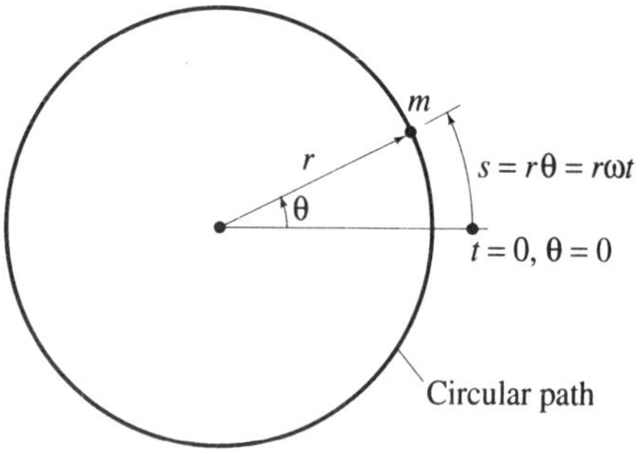

FIGURE P16.28.

16.28 A mass of 2 kg travels, with constant speed, the circular path shown in Figure P16.28. It starts at $\theta = 0$ when $t = 0$ as shown. Determine the impulse exerted on this particle during the time interval from $t_1 = \pi/\omega$ to $t_2 = 5\pi/2\omega$ (a) by computing the difference between the final and initial momenta and (b) by determining the force from Newton's second law and evaluating the defining integral for impulse. Use $\omega = 4$ rad/s and $r = 3$m.

16.29 A mass m travels a path in the x-y plane given parametrically by $y = b + 4t^2$ and $x = 2t$. Determine the path of the particle and a general equation for the impulse vector exerted on the particle from $t = 0$ to $t = t$ by (a) computing the difference between the final and initial momenta and (b) determining the force from Newton's second law and integrating the defining integral for impulse.

16.30 A mass m travels a path in the x-y plane given parametrically by $x = 3t$ and $y = 9t^2$. Determine the path of the particle and a general equation for the impulse vector exerted on the particle from $t = 0$ to $t = t$ by (a) computing the difference between the final and initial momenta and (b) determining the force from Newton's second law and integrating the defining integral for impulse.

16.3 System of Particles—Principle of Linear Impulse and Momentum

A system of particles, depicted at any time t, is shown in Figure 16.9. Equation (14.5) ($\sum \mathbf{F} = \sum m_i \mathbf{a}_i$) states that the vector sum of all the external forces acting on the system of particles at a given instant equals the vector sum of the inertial forces for the system at the same instant. The summations extend over the n particles of the system and the subscript i represents the ith particle in the system. Substitution of $\mathbf{a} = d\mathbf{v}_i/dt$ enables us to write

$$\sum \mathbf{F} = \sum m_i \frac{d\mathbf{v}_i}{dt}.$$

Multiplying through by the scalar dt yields

$$\sum \mathbf{F}\,dt = \sum m_i\,d\mathbf{v}_i.$$

Integration of this equation from the initial time t_1 when $\mathbf{v}_i = \mathbf{v}_{i1}$ to the final time t_2 when $\mathbf{v}_i = \mathbf{v}_{i2}$ gives

$$\sum \int_{t_1}^{t_2} \mathbf{F}\,dt = \sum m_i \mathbf{v}_{i2} - \sum m_i \mathbf{v}_{i1}. \tag{16.14}$$

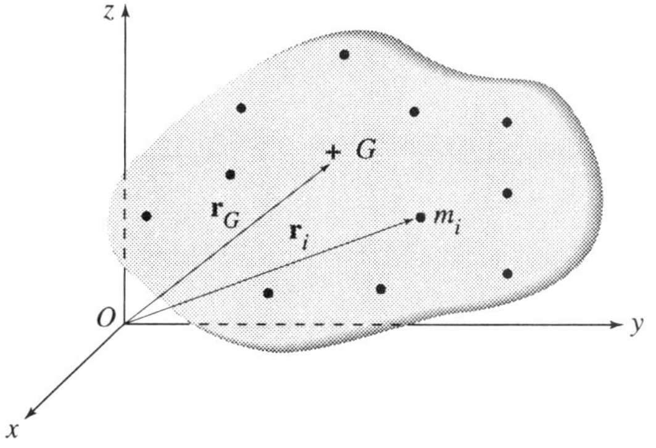

FIGURE 16.9.

This equation states that the vector sum of the linear impulses of the external forces acting on the system of n particles during the time interval from t_1 to t_2 equals the change of linear momenta of the system. Internal forces do not appear in the linear impulse summation because, by Newton's third law, they occur in equal and opposite collinear pairs and, thus, cancel out.

The mass center G of any system of particles is located by $\mathbf{r}_G$ which, by definition, is given by

$$m\mathbf{r}_G = \sum m_i \mathbf{r}_i$$

where the total mass m of the system is given by the arithmetic sum $\sum m_i$. Differentiation of this equation with respect to time gives

$$m\mathbf{v}_G = \sum m_i \mathbf{v}_i.$$

This equation states that the total linear momentum of the system at any instant is equal to the product of the total mass of the system and the velocity of its mass center G. This enables us to write Eq. (16.14) in the form

$$\sum \int_{t_1}^{t_2} \mathbf{F}\, dt = m\mathbf{v}_{G2} - m\mathbf{v}_{G1}. \tag{16.15}$$

Equation (16.15) states that the vector sum of the linear impulses of the external forces acting on the system of n particles is equal to the vector difference between the final and initial momenta of the system. Because a rigid body is an infinite collection of particles, Eq. (16.15) applies also to rigid bodies. Once again, the pictorial representation of the linear impulse and the linear momenta terms of this equation (see Figure 16.8) will be employed in solving problems. Care must always be taken to clearly define the system for which equations are written.

16.4 Conservation of Linear Momentum

Conservation of linear momentum is a special case of the principle of linear impulse and momentum in which the impulse can be shown to vanish. Letting $\mathbf{I} = \mathbf{0}$ in Eq. (16.12), we obtain

$$\mathbf{L}_2 - \mathbf{L}_1 = \mathbf{0}$$

from which

$$\mathbf{L}_1 = \mathbf{L}_2. \tag{16.16a}$$

Equation (16.16a) states that if the sum of all impulses exterted on a particle by external forces vanish (i.e., $\mathbf{I} = \mathbf{0}$), then, the initial linear momentum vector of the particle equals its final linear momentum vector. This conservation equation means that the particle's linear momentum vector remains constant in magnitude and does not change direction with time. If we assume the particle's mass to be constant during the motion, we conclude that its velocity will also remain constant (i.e., $v_1 = v_2$).

Equation (16.16a) may be expressed in terms of its three rectangular x, y, and z scalar components. Thus,

$$\left.\begin{aligned} L_{1x} &= L_{2x} \\ L_{1y} &= L_{2y} \\ L_{1z} &= L_{2z} \end{aligned}\right\}. \tag{16.16b}$$

Equation (16.16b) could also be obtained directly from Eq. (16.13) by setting $I_x = I_y = I_z = 0$.

For a system of particles, consider Eq. (16.15), and let the sum of the impulses vanish to obtain

$$m\mathbf{v}_{G2} - m\mathbf{v}_{G1} = \mathbf{0}$$

from which

$$m\mathbf{v}_{G1} = m\mathbf{v}_{G2}. \tag{16.17}$$

If the mass m of the system of particles remains constant, Eq. (16.17) reduces to

$$\mathbf{v}_{G1} = \mathbf{v}_{G2}. \tag{16.18}$$

Equation (16.17) expresses the conservation of momentum for a system of particles. This conservation equation states that if no net external forces (i.e., $\sum \int_{t_1}^{t_2} \mathbf{F}\, dt = \mathbf{0}$), then the product of the total mass of the system and the velocity of its mass center G remains constant in magnitude and does not change direction with time. Equation (16.18) expresses the fact that the velocity of the mass center of the system remains constant in magnitude and does not change direction with time if its mass remains unchanged.

In general, it will be preferable to solve problems by the use of the impulse momentum equations (16.12) and (16.15) and to show clearly that the impulse terms $\sum \int_{t_1}^{t_2} \mathbf{F}\, dt$ vanish. If one starts with the conservation equations (16.16) and (16.17) or (16.18), then, the assumption has been made that the impulse terms vanish and this may or may not be true for the particle or system under consideration.

16.5 Impulsive Forces and Impact

When two bodies strike each other for a very short duration, very large forces known as *impulsive forces* are developed between the two bodies. By definition, then, an impulsive force is one that reaches a relatively large magnitude in a relatively small period of time. The two striking bodies are said to have experienced *impact*. In dealing with the impact of bodies that may be idealized as particles, we can identify two types of impact as follows:

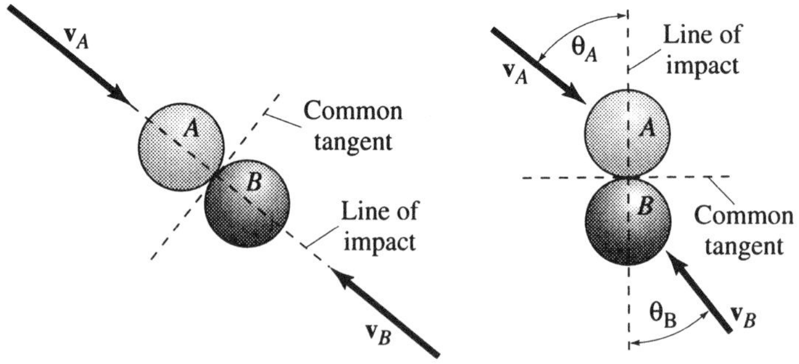

FIGURE 16.10. (a) Central impact (b) Oblique impact

Central impact between two particles takes place when the velocities of the two particles lie along the line connecting their mass centers which is known as the *line of impact*. Central impact between two particles A and B is illustrated in Figure 16.10(a). Note that the line of impact is perpendicular to the common tangent between particles A and B.

Oblique impact between two particles occurs when either or both particles have velocities which do not lie along the line of impact. Oblique impact between two particles A and B is shown in Figure 16.10(b).

Analysis of Central Impact

Consider the two particles A and B shown in Figure 16.11(a) moving along the same straight frictionless path which lies in a horizontal plane. This path is labeled in Figure 16.11(a) as the x axis. Particle A has a mass m_A and is moving at a velocity v_{A1}. Particle B has a mass m_B and moves at a velocity v_{B1} such that $v_{B1} < v_{A1}$. Thus, particle A eventually overtakes particle B and impacts it as shown in Figure 16.11(b). The process of impact consists of a period of *deformation* and a period of *restitution*. During deformation, the two particles experience a certain amount of deformation at the end of which the two particles have exactly the same velocity v. Following deformation, a period of restitution occurs at the end of which the two particles either regain their original shape or remain permanently deformed. Which of the two possibilities actually takes place depends upon the materials of the two particles and upon the magnitudes of the impulsive forces. At the end of the period of restitution, particle A has a velocity v_{A2} and particle B a velocity v_{B2}, both along the x axis as shown in Figure 16.11(c). It is useful and interesting to determine the two unknown quantities v_{A2} and v_{B2}. Two equations are, therefore, needed to accomplish this purpose.

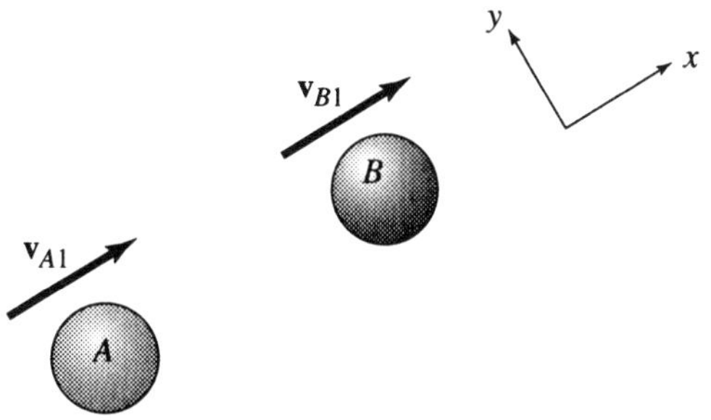

(a) Prior to impact

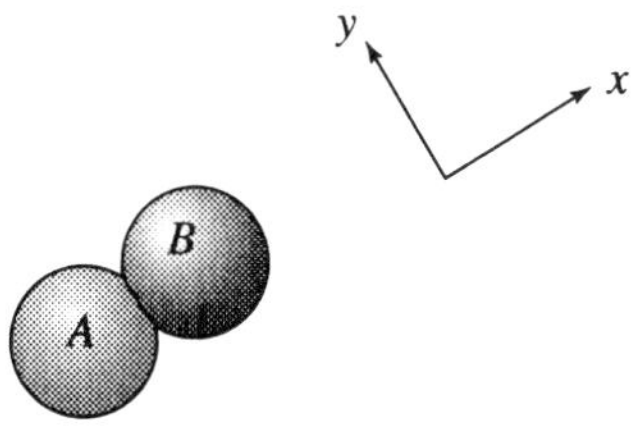

(b) During impact

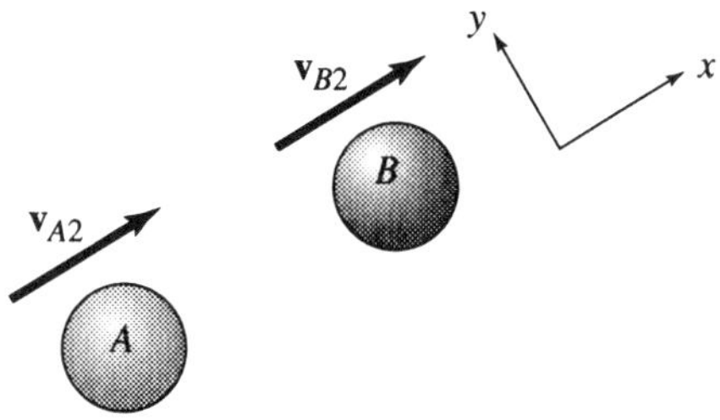

(c) After impact

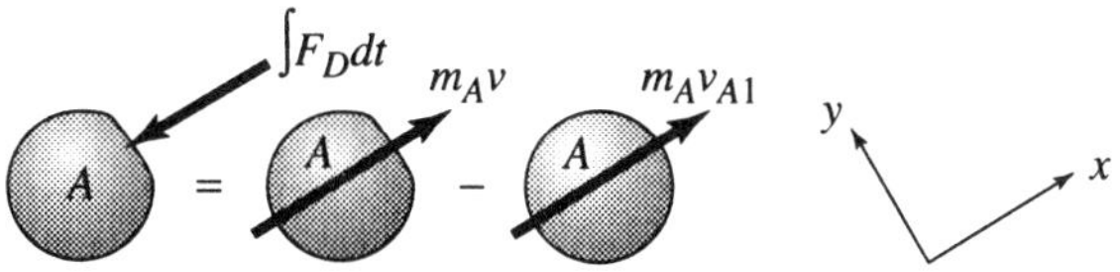

(d) Impulse-Momentum during deformation for body A.

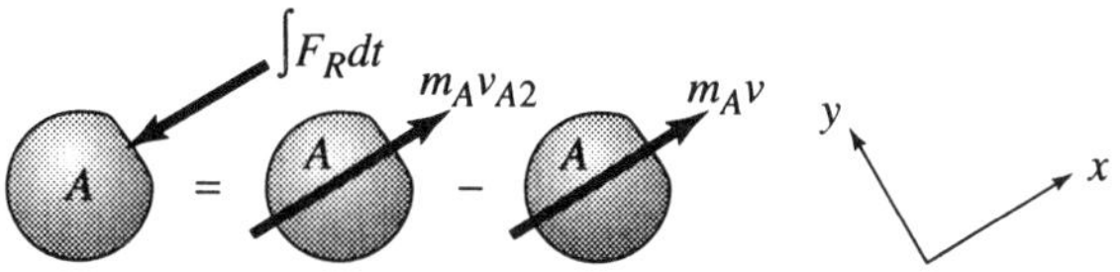

(e) Impulse-Momentum during restitution for body A.

FIGURE 16.11.

The first equation is obtained by considering the system of both particles during impact. During this short period of time, the linear momentum of the system is conserved because the internal impulses cancel out. Thus,

$$m_A v_{A1} + m_B v_{B1} = m_A v_{A2} + m_B v_{B2} \tag{16.19}$$

The second necessary equation follows from consideration of the periods of deformation and restitution. Applying the principle of impulse and momentum to particle A during the period of deformation (see Fig. 16.11(d)),

$$-\int F_D\,dt = m_A v - m_A v_{A1}$$

where $\int F_D\,dt$ represents the impulse of the only impulsive force F_D that particle B exerts on particle A during the period of deformation. Also, applying the principle of impulse and momentum to particle A during the period of restitution (see Fig. 16.11(e)),

$$-\int F_R\,dt = m_A v_{A2} - m_A v$$

where $\int F_R\,dt$ represents the impulse of the only impulsive force F_R that particle B exerts on particle A during the period of restitution. By definition, the *coefficient of restitution* e is the ratio of the impulse during restitution to that during deformation. Thus, for particle A,

$$e = \frac{\int F_R\,dt}{\int F_D\,dt} = \frac{v - v_{A2}}{v_{A1} - v}.$$

Similarly, for particle B, we can show that

$$e = \frac{\int F_R\,dt}{\int F_D\,dt} = \frac{v_{B2} - v}{v - v_{B1}}.$$

Eliminating the quantity v from the two equations above for e yields

$$e = \frac{v_{B2} - v_{A2}}{v_{A1} - v_{B1}}. \tag{16.20}$$

Because $(v_{B2} - v_{A2})$ represents the relative velocity after impact and $(v_{A1} - v_{B1})$ the relative velocity before impact, Eq. (16.20) shows that the coefficient of restitution is the ratio of the relative velocity after impact to the relative velocity before impact.

Analysis of Oblique Impact

As stated earlier, oblique impact occurs when either or both particles have velocities which do not lie along the line of impact. Consider, for example, the oblique impact between particles A and B depicted in Figure 16.12. If it is assumed that, prior to impact, the quantities v_{A1},

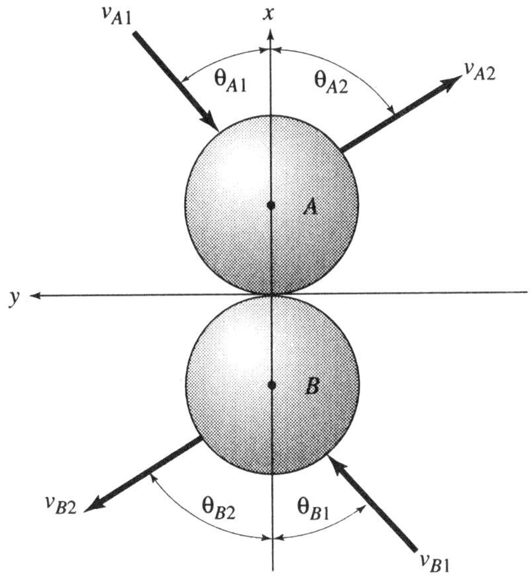

FIGURE 16.12.

θ_{A1}, v_{B1} and θ_{B1} are known, then, after impact we have four unknown quantities, namely, v_{A2}, θ_{A2}, v_{B2}, θ_{B2}. These four unknown quantities require four independent equations for their determination. Thus, referring to Figure 16.12, if we assume frictionless conditions, we conclude that the momentum of each of the two particles is conserved in the y direction, thus providing two of the needed equations among the unknown quantities. A third relationship is obtained by considering the system of both particles and concluding that the component of its total momentum in the x direction is conserved. Finally, the fourth relationship follows from application of Eq. (16.20) in the x direction. Example 16.10 illustrates application of the above steps in the solution of a problem dealing with oblique impact.

Two special values of the coefficient of restitution e are of considerable interest. If $e = 0$, the impact is termed *perfectly plastic*, and we note, from the defining equation for e, that this implies $v_{B2} = v_{A2}$. This means that the two particles move off together after impact with a common velocity. In other words, after a perfectly plastic impact, the particles *stick together*. If $e = 1$, the impact is termed *perfectly elastic*. This means that the relative velocity $(v_{B2} - v_{A2})$ after impact equals the relative velocity $(v_{A1} - v_{B1})$ before impact. It is readily shown that the kinetic energy of the colliding particles is conserved during a perfectly elastic impact, provided we neglect relatively small impulses delivered by friction forces during the small time of impact. For values of e less than unity, there is a loss of energy which is dissipated in sound or in heating the impacting particles.

A pictorial representation of the linear impulse-momentum equation is an excellent way to relate the physical and mathematical aspects of impact analysis. Diagrams showing the two colliding particles before, during, and after impact, together with a stated sign convention, is an excellent way to start solving impact problems. Prior to beginning the solution, it is good practice to count the unknowns and list the available equations. Before writing a given equation, care should be taken to state the system for which it is to be written. In impact problems, the isolated system will consist of either both of the colliding particles or one of the two particles. The needed relations will be either conservation of momentum equations or impulse-momentum equations for given directions, as well as the defining equation for the coefficient of restitution. Once the necessary equations have been formulated, they are solved for the unknowns. Careful interpretation of the resulting signs of the answers is required to find the direction of motion of the particles. A check by back substitution should be made to see that these answers satisfy all of the original equations. An intuitive appraisal of the results is often of great value in detecting errors.

The following examples illustrate some of the above concepts.

■ Example 16.6

The system shown in Figure E16.6(a) consists of a tugboat A weighing 50 tons connected by a cable to a loaded barge B weighing 400 tons. Initially the tugboat is moving to the right with a velocity $v_{A1} = 25$ ft/s,

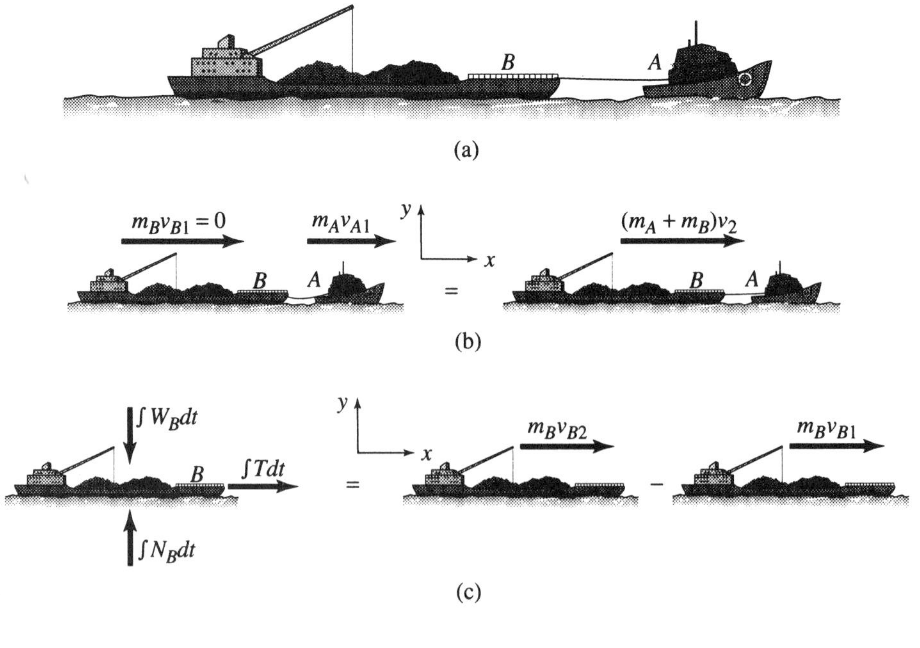

FIGURE E16.6.

the barge is at rest, and the cable is slack. Determine (a) the velocity of the barge v_{B2} after the cable becomes taut and (b) the impulse exerted on the barge by the taut cable. Neglect air and water resisting forces.

Solution

(a) Consider the system to be the tugboat and the barge as shown in Figure E16.6(b). Since air and water resisting forces are ignored, the momentum of the system in the x direction is conserved. Thus,

$$m_B(0) + m_A(25) = m_B v_{B2} + m_A v_{A2}$$

Because the tugboat and barge move together after the cable becomes taut, $v_{A2} = v_{B2} = v_2$. Thus,

$$\frac{50(2000)}{32.2}(25) = \left[\frac{400(2000)}{32.2} + \frac{50(2000)}{32.3}\right] v_2,$$

$$v_2 = v_{A2} = v_{B2} = 2.78 \text{ ft/s} \rightarrow. \qquad \text{ANS.}$$

Note that the conversion factors 2000 lb/ton and 32.2 ft/s^2 cancel from this conservation equation, but they will not cancel from an impulse-momentum equation.

(b) Consider the system to be the barge B, and note that the cable tension T is now an external force. Write the impulse-momentum equation in the x direction by referring to Figure E16.6(c). Thus,

$$\int_{t_1}^{t_2} T\,dt = m_B v_{B2} - m_B(0),$$

$$I = \int_{t_1}^{t_2} T\,dt = \frac{400(2000)}{32.3}(2.78) = 69100 \text{ lb·s} \rightarrow. \qquad \text{ANS.}$$

■ Example 16.7

As shown in Figure E16.7(a), a package P weighing 15 lb is projected with a velocity $v_{P1} = 6$ ft/s onto a hand cart C weighing 75 lb. The cart is at rest before the package strikes it. Neglect rolling resistance of the wheels and impulses of the weights of the cart and package during the small time of impact. Determine (a) the velocity of the cart after the package comes to rest on it, (b) the impulse exerted on the package by the cart, and (c) the percentage of initial kinetic energy lost in the impact.

Solution

(a) The cart and package are taken as the system. Once the package comes to rest on the cart they both have a common velocity $v_{P2} = v_{C2} = v_2$, as shown in Figure E16.7(b). Because the only external impulse is delivered in the y direction, momentum is conserved in the x direction. Thus,

$$m_P v_{P1} \cos 20^\circ + m_C v_{C1} = (m_P + m_C) v_2,$$

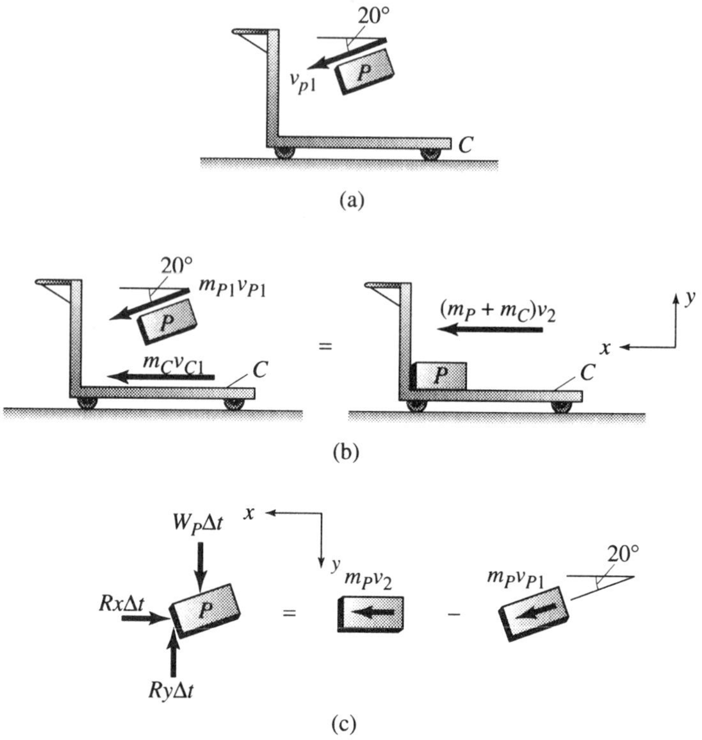

FIGURE E16.7.

$$\frac{15}{32.2}(6)\cos 20^\circ + 0 = \left(\frac{15 + 75}{32.2}\right)v_2,$$

$$v_{P2} = v_{C2} = v_2 = 0.940 \text{ ft/s} \leftarrow. \qquad \text{ANS.}$$

(b) Refer to Figure E16.7(c), and write the impulse-momentum equation for the horizontal and vertical directions for the package. Momentum is not conserved in either of these directions because forces are exerted by the cart on the package. Since the impulse of the weight is to be neglected, by Eq. (16.13),

$$I_x = (L_2 - L_1)_x\!: \; -R_x\Delta t = \frac{15}{32.2}(0.940) - \frac{15}{32.2}(6)\cos 20^\circ$$

$$= 2.19 \text{ lb·s} \rightarrow. \qquad \text{ANS.}$$

$$I_y = (L_2 - L_1)_y\!: \; -R_y\Delta t = 0 - \frac{15}{32.2}(6)\cos 20^\circ$$

$$= 0.956 \text{ lb·s} \uparrow. \qquad \text{ANS.}$$

If these impulses are added vectorially, we obtain the resultant impulse exerted by the cart on the package. Thus,

$$\mathrm{I} = 2.39 \text{ lb·s} \quad \angle\, 23.6^\circ. \qquad \text{ANS.}$$

(c) The loss of kinetic energy is given by

$$\begin{aligned}\Delta T &= T_1 - T_2,\\ &= \frac{1}{2}m_P v_{P1}^2 - \frac{1}{2}(m_P + m_C)v_{C2}^2\\ &= \frac{1}{2}\left(\frac{15}{32.2}\right)(6)^2 - \frac{1}{2}\left(\frac{15+75}{32.2}\right)(0.940)^2 = 7.15 \text{ ft·lb}.\end{aligned}$$

The initial kinetic energy is

$$T_1 = \frac{1}{2}m_P v_{P1}^2 = \frac{1}{2}\left(\frac{15}{32.2}\right)(6)^2 = 8.39 \text{ ft·lb}.$$

$$\% \text{ of initial kinetic energy lost} = \frac{7.15}{8.29}(100) = 85.2\% \qquad \text{ANS.}$$

■ Example 16.8

Two small spheres A and B approach each other as shown in Figure E16.8(a). Sphere A weighs 4 ounces and sphere B weighs 2 ounces. Just before impact, A is moving at 5 ft/s directed to the right and B is moving at 6 ft/s directed to the left. The impact is central and the coefficient of restitution e has an experimentally determined value of 0.6. Neglect frictional impulses exerted on the spheres during impact, and determine (a) the velocities of spheres A and B just after impact and (b) the loss of energy during impact.

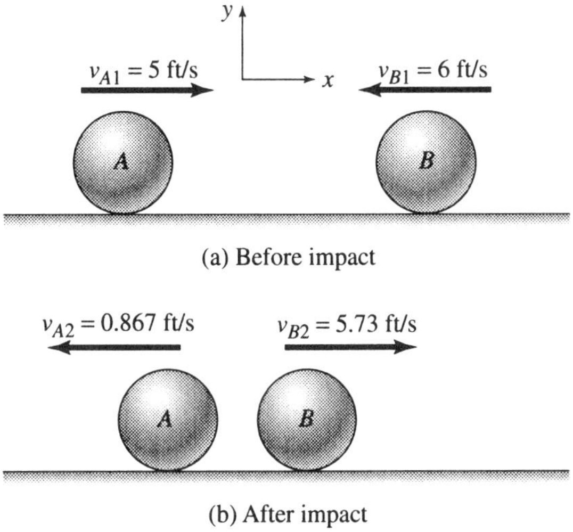

FIGURE E16.8.

Solution

(a) Because frictional impulses are negligible, we conclude that, during impact, the linear momentum in the x direction is conserved for the system consisting of both A and B.

$$L_{1x} = L_{2x}.$$

Therefore,

$$m_A v_{A1} + m_B v_{B1} = m_A v_{A2} + m_B v_{B2}$$

where, as indicated in Figure E16.8(a), positive x is assumed to the right. Substituting values,

$$\frac{4/16}{g}(5) + \frac{2/16}{g}(-6) = \frac{4/16}{g} v_{A2} + \frac{2/16}{g} v_{B2}$$

from which

$$2v_{A2} + v_{B2} = 4. \tag{a}$$

The equation for the coefficient of restitution e is

$$e = \frac{v_{B2} - v_{A2}}{v_{A1} - v_{B1}}$$

Therefore,

$$0.60 = \frac{v_{B2} - v_{A2}}{5 - (-6)}$$

from which

$$v_{B2} - v_{A2} = 6.6 \tag{b}$$

Solving Eqs. (a) and (b),

$$v_{A2} = -0.867 \text{ ft/s} \qquad \text{ANS.}$$

where the negative sign means that A moves to the left after impact, and

$$v_{B2} = 5.73 \text{ ft/s} \qquad \text{ANS.}$$

where the positive sign means that B moves to the right after impact. Figure E16.8(b) shows the actual velocities of spheres A and B after impact.

(b) The loss of energy during impact is given by

$$\Delta T = T_1 - T_2,$$

$$\Delta T = \tfrac{1}{2} m_A v_{A1}^2 + \tfrac{1}{2} m_B v_{B1}^2 - \tfrac{1}{2} m_A v_{A2}^2 - \tfrac{1}{2} m_B v_{B2}^2$$

Substitute numerical values to give

$$\Delta T = \left[\frac{1}{2}\left(\frac{4/16}{32.2}\right)(5)^2 + \frac{1}{2}\left(\frac{2/16}{32.2}\right)(-6)^2\right] - \frac{1}{2}\left(\frac{4/16}{32.2}\right)(-0.867)^2 - \frac{1}{2}\left(\frac{2/16}{32.2}\right)(5.73)^2$$

$$= 0.100 \text{ ft}\cdot\text{lb.} \qquad \text{ANS.}$$

This kinetic energy was dissipated in the form of sound and heat during the short time of impact.

■ Example 16.9

Refer to Example 16.8 and determine (a) the impulse exerted by sphere B on sphere A and (b) the time average value of the force exerted by sphere B on sphere A if the duration of the impact is 0.02 s.

Solution

(a) To determine the impulse exerted by sphere B on sphere A, consider the free-body diagram of sphere A as shown in Figure E16.9(a). The impulse exerted by sphere B on sphere A is now external. We write the linear impulse-momentum equation in the x direction by referring to the figure. Thus,

$$-\int R\,dt = m_A v_{A2} - m_A v_{A1}$$

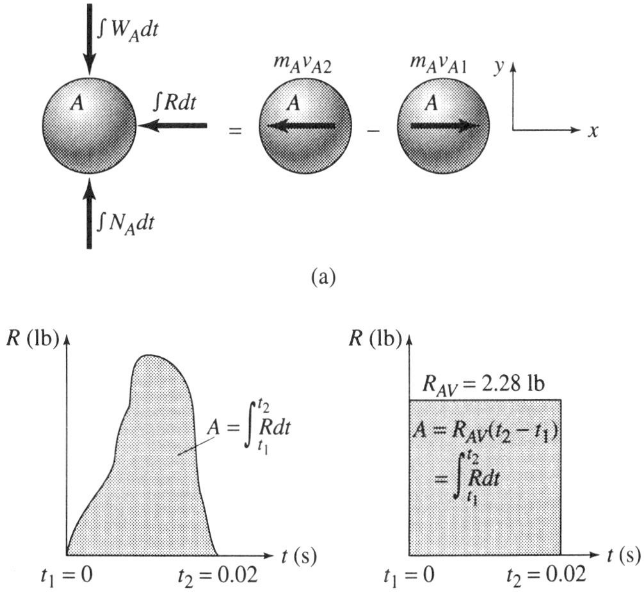

(b)

FIGURE E16.9.

where from Example 16.8, $v_{A1} = 5$ ft/s and $v_{A2} = -0.867$ ft/s. Thus,

$$-\int R\,dt = \frac{4/16}{32.2}(-0.867) + \frac{4/16}{32.2}(5)$$

from which

$$\int R\,dt = 0.0456 \text{ lb·s.} \qquad \text{ANS.}$$

Although we have been able to determine the total impulse, we are not able to determine the impulsive force R as a function of time.

(b) Assuming we know the duration of the impact to be 0.02 s from observation, we are able to find the time average value of the force, R_{AV}. This concept of a time average value for an impulsive force is illustrated pictorially in Figure E16.9(b). In this case, $t_2 - t_1 = 0.02$s. Thus,

$$R_{AV} = \frac{1}{0.02}(0.0456) = 2.28 \text{ lb.} \qquad \text{ANS.}$$

Although we do not think of this as a large force, when compared to the weights of the spheres, it is relatively large. Thus,

$$\frac{R_{AV}}{W_A} = \frac{2.28}{4/16} = 9.12, \qquad \frac{R_{AV}}{W_B} = \frac{2.28}{2/16} = 18.24.$$

Example 16.10

Two balls of mass m_A and m_B collide in oblique impact as shown in Figure E16.10(a) just before colliding. (a) Let $m_A = 2$ kg, $v_{A1} = 3$ m/s, $\theta_{A1} = 30°$, $m_B = 4$ kg, $v_{B1} = 5$ m/s, $\theta_{B1} = 60°$ and $e = 0.8$ and find values for v_{A2}, θ_{A2}, v_{B2} and θ_{B2}. (b) Determine the loss of energy during the impact and express this as a percentage of the kinetic energy of the system before impact. Assume frictionless conditions.

Solution

(a) Select an x axis directed along the line of impact of the colliding balls as shown in Figure E16.10(a). The y axis is directed along the common tangent.

Consider the system consisting of both balls, neglect frictional impulses, and write the conservation of momentum equation in the x direction. Thus,

$$m_A v_{A1} \cos\theta_{A1} - m_B v_{B1} \cos\theta_{B1} = m_A v_{A2} \cos\theta_{A2} + m_B v_{B2} \cos\theta_{B2},$$

$$2(3)\cos 30° - 4(5)\cos 60° = 2(v_{A2})_x + 4(v_{B2})_x,$$

$$(v_{A2})_x + 2(v_{B2})_x = -2.40. \qquad \text{(a)}$$

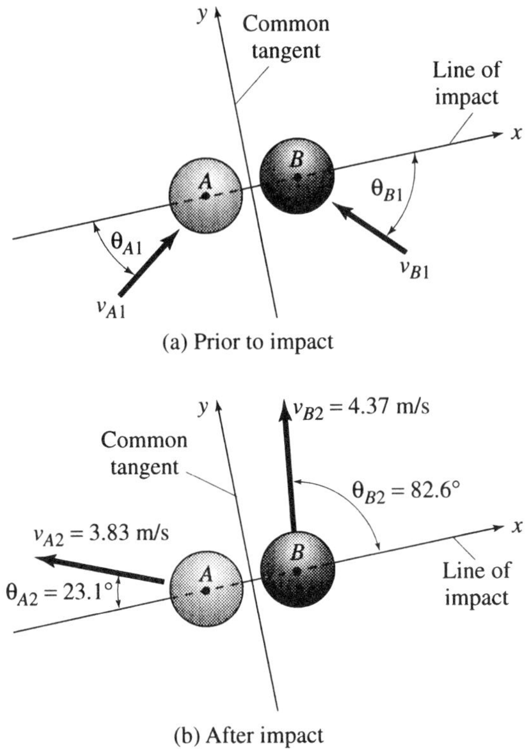

FIGURE E16.10.

The impulsive forces are delivered in the x direction and, once again, for the system consisting of both balls, we write the defining equation for the coefficient of restitution. Thus,

$$e = \frac{(v_{B2})_x - (v_{A2})_x}{(v_{A1})_x - (v_{B1})_x}$$

$$0.8 = \frac{(v_{B2})_x - (v_{A2})_x}{2.60 - (-2.50)}$$

$$(v_{B2})_x - (v_{A2})_x = 4.08. \qquad \text{(b)}$$

Solve (a) and (b) simultaneously to obtain

$$(v_{A2})_x = 3.52 \text{ m/s},$$

$$(v_{B2})_x = 0.560 \text{ m/s}.$$

No impulsive forces are applied in the y direction and, thus, for each ball, separately, we write a conservation of momentum equation in this direction. Thus, for ball A,

$$m_A v_{A1} \sin\theta_{A1} = m_A v_{A2} \sin\theta_{A2},$$

$$2(3)\sin 30^\circ = 2(v_{A2})_y,$$

$$(v_{A2})_y = 1.50 \text{ m/s}.$$

For ball B,

$$m_B v_{B1} \sin\theta_{B1} = m_B v_{B2} \sin\theta_{B2},$$

$$4(5)\sin 60^\circ = 4(v_{B2})_y,$$

$$(v_{B2})_y = 4.33 \text{ m/s}.$$

Therefore, because we now have the components of the velocities for both balls after impact, we are able to determine the magnitudes and directions of their velocities after impact. Thus, for ball A,

$$v_{A2} = \sqrt{(v_{A2})_x^2 + (v_{A2})_y^2} = \sqrt{(3.52)^2 + (1.50)^2} = 3.83 \text{ m/s}, \quad \text{ANS.}$$

$$\theta_{A2} = \tan^{-1}\frac{(v_{A2})_y}{(v_{A2})_x} = \tan^{-1}\left(\frac{1.50}{3.52}\right) = 23.1^\circ \quad \text{ANS.}$$

And for ball B,

$$v_{B2} = \sqrt{(v_{B2})_x^2 + (v_{B2})_y^2} = \sqrt{(0.560)^2 + (4.33)^2} = 4.37 \text{ m/s}, \quad \text{ANS.}$$

$$\theta_{B2} = \tan^{-1}\frac{(v_{B2})_y}{(v_{B2})_x} = \tan^{-1}\left(\frac{4.33}{0.560}\right) = 82.6^\circ. \quad \text{ANS.}$$

All of this information is depicted in Figure E16.10(b).

(c) The energy loss during impact equals the initial kinetic energy minus the final kinetic energy. Thus,

$$\Delta T = T_1 - T_2$$

$$\Delta T = [\tfrac{1}{2}(2)(3^2) + \tfrac{1}{2}(4)(5^2)] - [\tfrac{1}{2}(2)(3.83^2) + \tfrac{1}{2}(4)(4.37^2)]$$

$$= 6.14 \text{ N·m}. \quad \text{ANS.}$$

This energy is dissipated as sound and heat.

$$\% \text{ kinetic energy loss} = \frac{\Delta T}{T_1}(100) = \frac{6.14}{59}(100) = 10.41\% \quad \text{ANS.}$$

About one tenth of the original kinetic energy is lost during the impact.

■

Problems

16.31 In Figure P16.31 block A of mass M rests on a frictionless horizontal plane and is initially at rest. Block B of mass m is projected as shown with $v_{B1} = v_0$ such that it impacts block A and they move off together. Neglect impulses associated with body weights and determine (a) the common velocity of these blocks after impact, (b) the impulse components exerted on block B, and (c) the percentage of initial kinetic energy lost in the impact. Express your answers in terms of M, m, v_0 and θ.

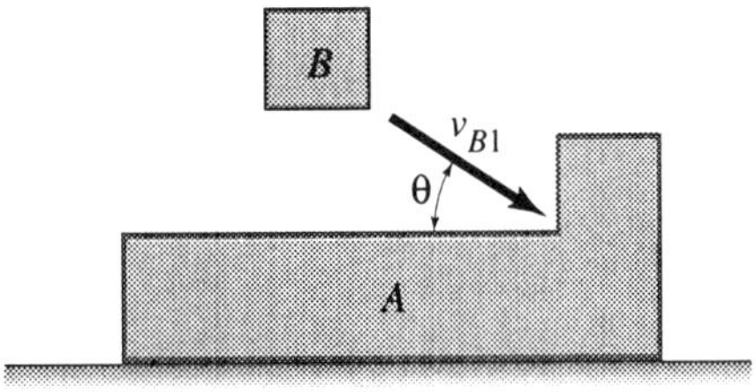

FIGURE P16.31.

16.32 Solve Problem 16.31 for the following inputs $M = 10$ kg, $m = 1$ kg, $v_0 = 1.25$ m/s and $\theta = 50°$.

16.33 Two boats, of mass M_A and M_B, are connected by a slack rope as shown in Figure P16.33. Boat A is initially moving to the left with a velocity $v_{A1} = v_0$ and boat B is at rest. Neglect frictional effects of the water on the boats. Determine (a) the common velocity of both boats after the rope becomes taut and (b) the impulse exerted on boat B by boat A. Express your answers in terms of M_A, M_B and v_0.

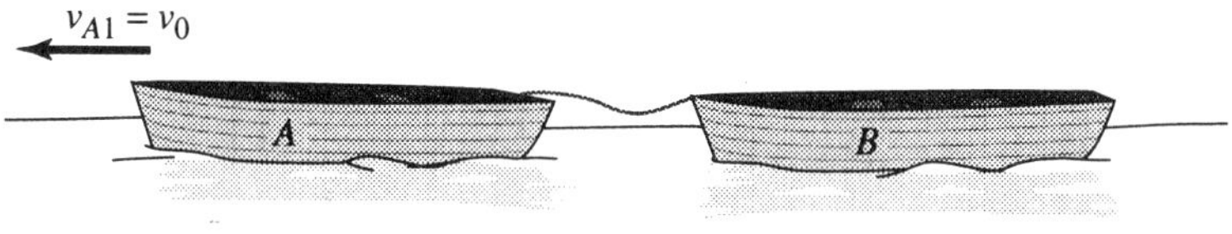

FIGURE P16.33.

16.34 Solve Problem 16.33 if boat B is initially moving to the right with a velocity $v_{B1} = 0.2\, v_0$.

16.35 Swimmer A weighs 130 lb and swimmer C weighs 175 lb. Swimmer A can dive to the left with a relative velocity with respect to the boat B of $v_{A/B} = 10$ ft/s while C can dive to the right with $v_{C/B} = 15$ ft/s. If the boat weighs 395 lb and is initially at rest, determine the final velocity of the boat if (a) swimmers A and C dive simultaneously, (b) swimmer A dives before swimmer C, and (c) swimmer C dives before swimmer A.

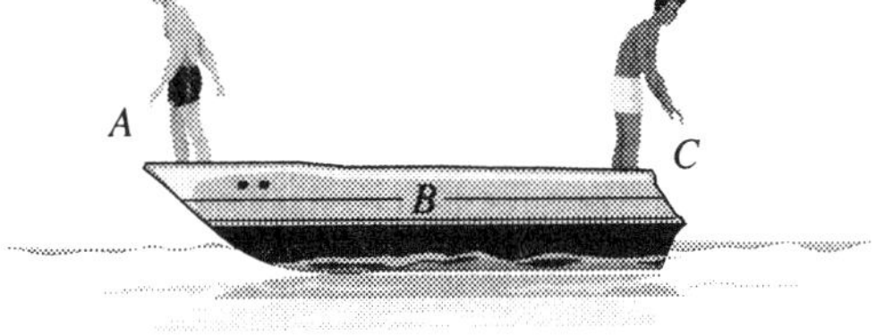

FIGURE P16.35.

16.36 An American Civil War cannon of mass M_c is depicted in Figure P16.36. It can fire a shell of mass m_s with a velocity

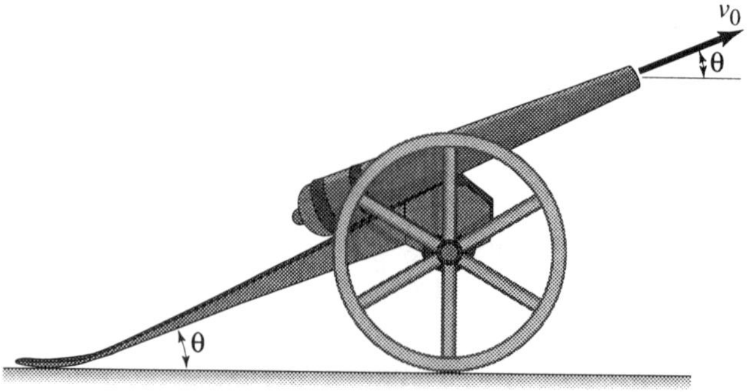

FIGURE P16.36.

v_0 at an angle θ with respect to the horizontal. Neglect horizontal resisting forces acting on the cannon and, in terms of M_c, m_s, θ, and v_0 determine (a) the velocity of the cannon after it is fired. and (b) the vertical impulse exerted on the cannon by the horizontal surface.

16.37 Solve Problem 16.36 for the following inputs $M_c = 1500$ kg, $m_s = 15$ kg, $\theta = 24°$, and $v_0 = 500$ m/s.

16.38 A car C of mass m_C collides with a loaded truck of mass m_T as shown in Figure P16.38. Before impact the car moves to the right with a velocity $v_{C1} = v_0$ and the truck moves to the left with a velocity $v_{T1} = 0.5\ v_0$. Neglect frictional impulses during the small time of impact and assume the vehicles move as a single body after impact. (a) Determine their common velocity v_f after the collision in terms of m_C, m_T and v_0. (b) Prepare a dimensionless plot of v_f/v_0 as a function of m_T/m_C. Let m_T/m_C take on values of 2, 4, 6, 8 and 10.

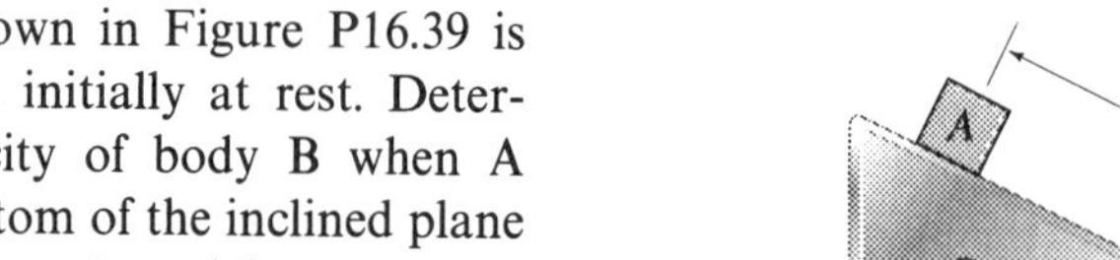

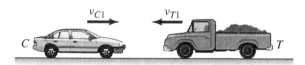

FIGURE P16.38.

16.39 The system shown in Figure P16.39 is frictionless and initially at rest. Determine the velocity of body B when A reaches the bottom of the inclined plane in terms of m_A, m_B, L, and θ.

16.40 Solve Problem 16.39 for $m_A = 2$ slug, $m_B = 10$ slug, $L = 4$ ft, and $\theta = 25°$.

16.41 In Figure P16.41, box car A is at rest and flat car B is moving to the left at $v_{B1} = v_0$ just prior to impact. Neglect frictional impulses and determine (a) the

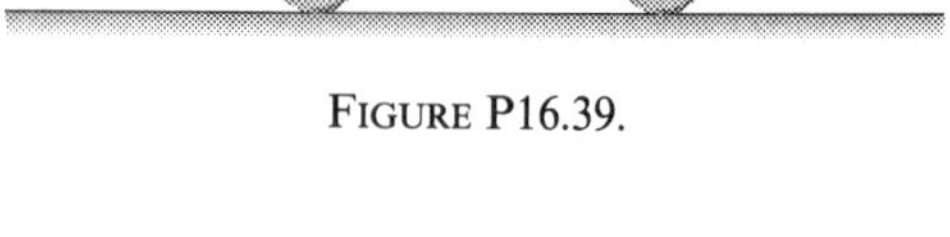

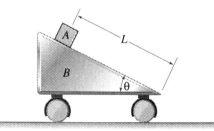

FIGURE P16.39.

FIGURE P16.41.

final velocity of the coupled cars in terms of m_A, m_B and v_0 and, (b) the time average coupling force acting on the flat car in terms of m_A, m_B, θ, and t_c where t_c is the coupling time.

16.42 Solve Problem 16.41 given that $m_A = 2500$ slug, $m_B = 1500$ slug, $v_0 = 5$ ft/s and $t_c = 0.08$ s.

16.43 A system of 10 particles has a total mass of 200 kg and a mass center located at $\mathbf{r}_G = 4t\mathbf{i} + t^2\mathbf{j} - 8t\mathbf{k}$ where t is in seconds and $\mathbf{r}_G$ in meters. Determine the resultant momentum vector for this system at $t = 1$ s. Also find the angles that this momentum vector makes with the three coordinate axes.

16.44 At a given instant, the mass center of a system of particles is moving with a velocity $\mathbf{v}_G = (4\mathbf{i} + 6\mathbf{j} - 8\mathbf{k})$ ft/s. The system of particles has a combined mass of 30 slug. If a force $\mathbf{F} = (5t\mathbf{i} - 3t^2\mathbf{j} + 2\mathbf{k})$ lb is applied to the mass center, determine its velocity after the force $\mathbf{F}$ has acted for a period of 5 s.

16.45 Refer to Example 16.8, and determine the quantities required in parts (a) and (b) for the following inputs: Sphere A weighs 50 N, and sphere B weighs 80 N. Just before impact, A is moving 3 m/s directed to the left, and B is moving 6 m/s directed to the left. The coefficient of restitution is 0.4.

16.46 Refer to Example 16.8, and determine the quantities required in parts (a) and (b) for the following inputs: Sphere A weighs 2 lb, and sphere B weighs 3 lb. Just before impact, A is moving 6 ft/s directed to the right, and B is moving 4 ft/s directed to the right. Use $e = 0.3$.

16.47 Use the following numerical inputs for Example 16.10: $m_A = 3$ kg, $v_{A1} = 2$ m/s, $\theta_{A1} = 0°$, $m_B = 5$ kg, $v_{B1} = 6$ m/s, $\theta_{B1} = 90°$ and $e = 0.5$. Determine the velocities of each of the masses after impact. How much energy was lost during the impact?

16.48 Refer to Example 16.10 and use the following inputs: $m_A = 0.2$ slug, $v_{A1} = 2.5$ ft/s, $\theta_{A1} = 45°$, $m_B = 0.1$ slug, $v_{B1} = 3.5$ ft/s, $\theta_{B1} = 75°$, and $e = 0.25$. Determine the velocities of each of the two masses after impact. How much energy was lost during impact.

16.49 A small mass A is released from rest at a distance $h = 0.4$ m above the relatively massive floor B as shown in Figure P16.49. It rebounds to a height $h' = 0.3$ m. Determine the coefficient of restitution e.

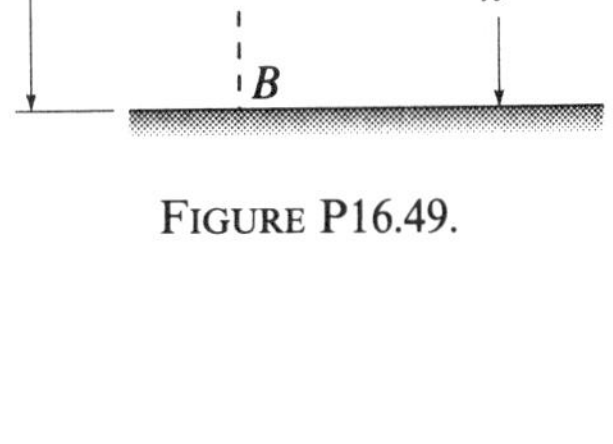

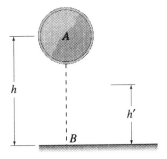

FIGURE P16.49.

16.50 A plan view of a small mass A striking a very massive frictionless wall is shown in Figure P16.50. The speed of mass A before impact is $v_A = 12$ ft/s. The angle of incidence is $\theta_1 = 40°$ and the angle of rebound is $\theta_2 = 65°$. Determine the coefficient of restitution e.

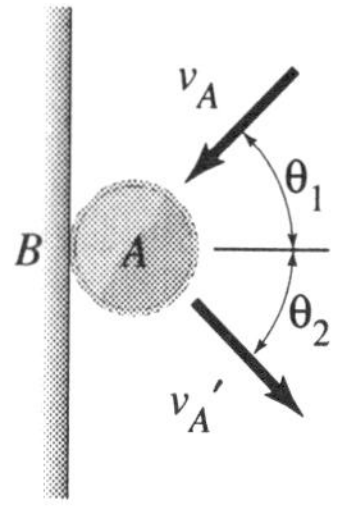

FIGURE P16.50.

16.51 Two disks collide in oblique impact as shown in Figure P16.51. Before impact, their velocities are $v_A = 5$ m/s and $v_B = 4$ m/s. Disk A weighs 40 N and disk B weighs 60 N. The coefficient of restitution is $e = 0.45$. Determine the velocity of each disk after impact.

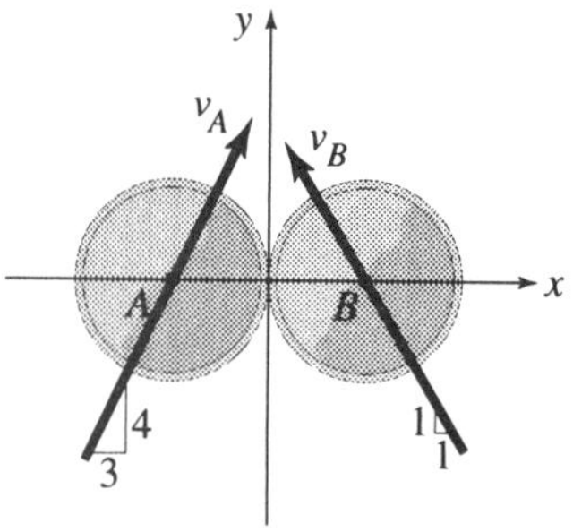

FIGURE P16.51.

16.52 In Figure P16.52, mass A is twice as large as mass B. Both bodies are moving to the right with $v_A = 4$ m/s and $v_B = 2$ m/s. An impact occurs when A overtakes B. Determine the velocity of each mass after impact for (a) $e = 0$, (b) $e = 0.5$, and (c) $e = 1.0$.

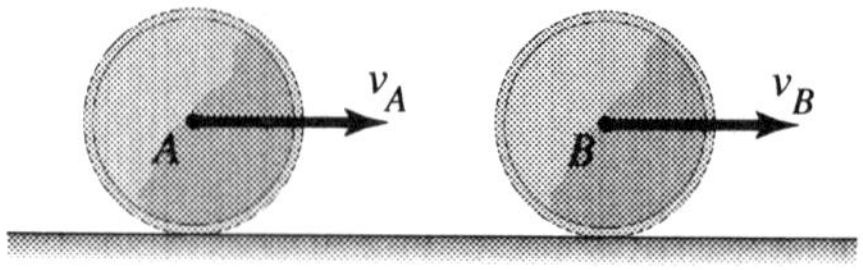

FIGURE P16.52.

16.53 A small sphere A is given a relatively small horizontal speed at a distance $h_1 = 1.00$ m above the relatively massive floor B as shown in Figure P16.53. If the coefficient of restitution $e = 0.25$, determine the maximum height h_2 after the first bounce and the maximum height h_3 after the second bounce. Neglect air resistance.

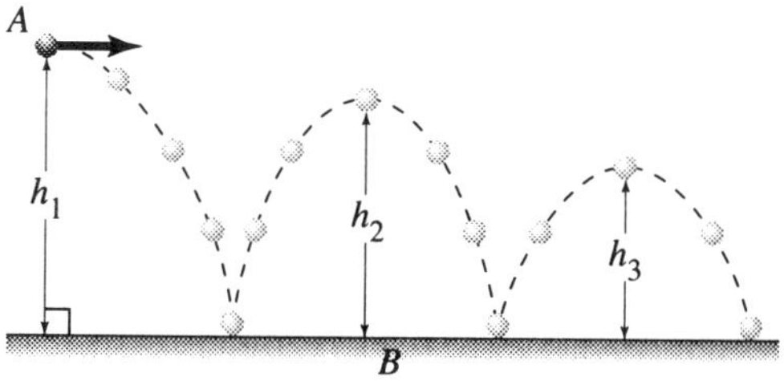

FIGURE P16.53.

16.54 A body weighing 50 N moves with a velocity of 3 m/s directed to the right. It collides with a body weighing 100 N. The impact is central. After impact, the 50 N body moves at a rate of 1 m/s to the right. Determine the initial and final velocities of the 100 N body. Use $e = 0.55$

16.55 Two hockey pucks of equal mass collide as shown in Figure P16.55. Before impact, A has a speed of 100 mph, and B has a speed of 80 mph. Their velocities are perpendicular to each other before impact as indicated in the figure. For $e = 0.96$, find the velocity of each puck after impact.

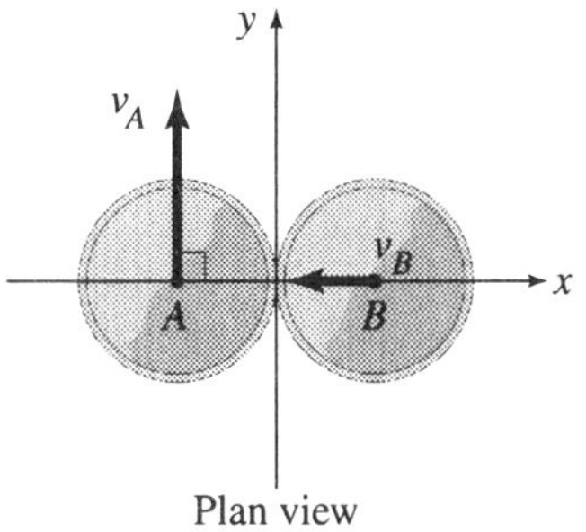

FIGURE P16.55.

16.6 Definition of Angular Momentum and Angular Impulse

The angular momentum $\mathbf{H_O}$ of a particle of mass m with respect to any point O is defined as the moment of the linear momentum vector with respect to point O. Mathematically,

$$\mathbf{H_O} = \mathbf{r} \times \mathbf{L} \tag{16.21}$$

where $\mathbf{H_O}$ is the angular momentum vector, $\mathbf{r}$ is a position vector from O to the particle on its path, and $\mathbf{L}$ is the linear momentum vector for the particle at a given instant. These vectors are shown in Figure 16.13. By Eq. (16.1), $\mathbf{L} = m\mathbf{v}$, and

$$\mathbf{H_O} = \mathbf{r} \times m\mathbf{v} \tag{16.22}$$

From the definition of the cross-product, the angular momentum vector $\mathbf{H_O}$ is perpendicular to the plane defined by the position vector $\mathbf{r}$ and the linear momentum vector $\mathbf{L} = m\mathbf{v}$, as shown in Figure 16.13. The angular momentum vector $\mathbf{H_O}$ may be decomposed into its three rectangular components, as shown in Figure 16.14,

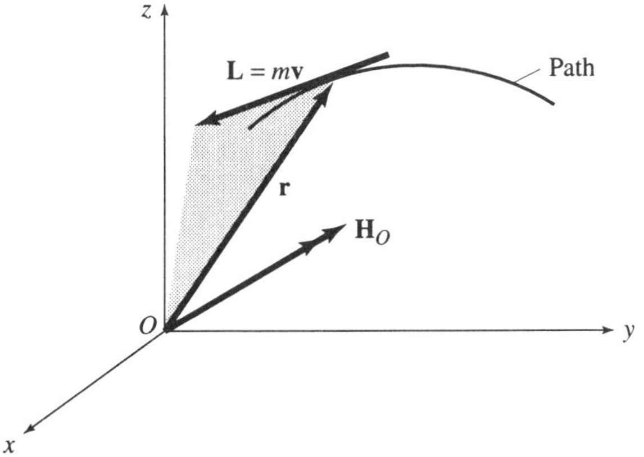

FIGURE 16.13.

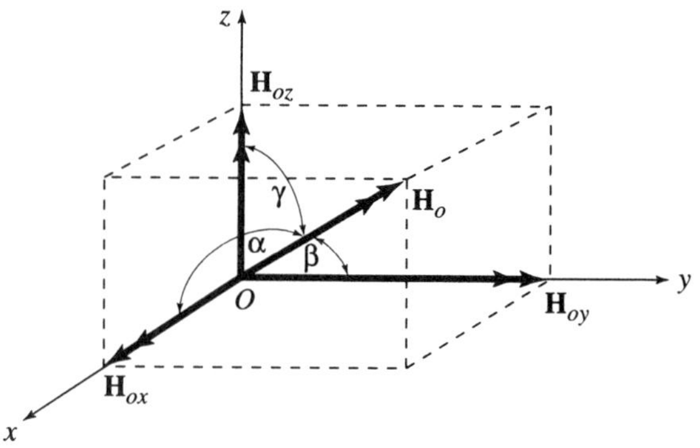

FIGURE 16.14.

$$\left.\begin{aligned} \mathbf{H}_{\mathrm{O}x} &= (H_{\mathrm{O}} \cos\alpha)\mathbf{i} \\ \mathbf{H}_{\mathrm{O}y} &= (H_{\mathrm{O}} \cos\beta)\mathbf{j} \\ \mathbf{H}_{\mathrm{O}z} &= (H_{\mathrm{O}} \cos\gamma)\mathbf{k} \end{aligned}\right\} \tag{16.23}$$

where H_{O} is the magnitude of $\mathbf{H}_{\mathrm{O}}$ and α, β, and γ are the direction angles of $\mathbf{H}_{\mathrm{O}}$. These vector components may also be expressed differently by reference to Eq (16.21). If we let $\mathbf{r} = r_x\mathbf{i} + r_y\mathbf{j} + r_z\mathbf{k}$ and $\mathbf{L} = L_x\mathbf{i} + L_y\mathbf{j} + L_z\mathbf{k}$. Eq. (16.21) may be written in determinant form

$$\mathbf{H}_{\mathrm{O}} = \begin{bmatrix} \mathbf{i} & \mathbf{j} & \mathbf{k} \\ r_x & r_y & r_z \\ L_x & L_y & L_z \end{bmatrix}. \tag{16.24}$$

When expanded, the determinant in Eq. (16.24) yields

$$\left.\begin{aligned} \mathbf{H}_{\mathrm{O}x} &= (r_y L_z - r_z L_y)\mathbf{i} \\ \mathbf{H}_{\mathrm{O}y} &= (r_z L_x - r_x L_z)\mathbf{j} \\ \mathbf{H}_{\mathrm{O}z} &= (r_x L_y - r_y L_x)\mathbf{k} \end{aligned}\right\}, \tag{16.25}$$

where $L_x = mv_x$, $L_y = mv_y$ and $L_z = mv_z$ represent the scalar components of the linear momentum in the x, y and z directions, respectively.

The dimensions of linear momentum have been shown to be the product of force and time. The position vector $\mathbf{r}$ has the dimension of length. Thus, the angular momentum has the dimensions of the product of length times force times time. Typical units are m·N·s and ft·lb·s. Angular momentum, like linear momentum, is evaluated instantaneously.

To define the angular impulse $\mathbf{G}_{\mathrm{O}}$, we begin with the differential linear impulse. Thus,

$$d\mathbf{I} = \mathbf{F}\,dt.$$

The differential angular impulse $d\mathbf{G}_O$ is defined as the moment of the differential linear impulse with respect to O. Thus,

$$d\mathbf{G}_O = \mathbf{r} \times d\mathbf{I} = \mathbf{r} \times \mathbf{F}\,dt. \tag{16.26}$$

Integration over the time interval from t_1 to t_2 enables us to define the angular impulse $\mathbf{G}_O$. Thus,

$$\mathbf{G}_O = \int_{t_1}^{t_2} (\mathbf{r} \times \mathbf{F})\,dt = \int_{t_1}^{t_2} \mathbf{M}_O\,dt. \tag{16.27}$$

The dimensions of the angular impulse are the product of length times force times time, the same as for the angular momentum, $\mathbf{H}_O$. Although the angular momentum is evaluated at a given instant, the angular impulse is an integrated effect taken over the time interval from t_1 to t_2.

16.7 The Principle of Angular Impulse and Momentum

Differentiating Eq. (16.22) with respect to time t yields

$$\frac{d\mathbf{H}_O}{dt} = \frac{d}{dt}(\mathbf{r} \times m\mathbf{v}). \tag{16.28a}$$

The right side of this equation represents the time derivative of the cross-product of two vector functions of time. The vector derivative is obtained in a manner similar to that used to differentiate the product of two scalar functions except that we need to retain the order of the vectors. Thus,

$$\frac{d}{dt}(\mathbf{r} \times m\mathbf{v}) = \mathbf{r} \times m\frac{d\mathbf{v}}{dt} + \frac{d\mathbf{r}}{dt} \times m\mathbf{v}.$$

From Newton's second law, $\mathbf{R} = \sum \mathbf{F} = m\dfrac{d\mathbf{v}}{dt}$. Also, $\dfrac{d\mathbf{r}}{dt} = \mathbf{v}$ and the second term on the right beomes $\mathbf{v} \times m\mathbf{v} = \mathbf{0}$ because the cross-product of a vector multiplied by itself vanishes. Therefore, Eq. (16.28a) reduces to

$$\frac{d\mathbf{H}_O}{dt} = \mathbf{r} \times \mathbf{R} = \mathbf{M}_O. \tag{16.28b}$$

Because the product $\mathbf{r} \times \mathbf{R}$ represents the moment $\mathbf{M}_O$, we conclude that

$$\mathbf{M}_O = \frac{d\mathbf{H}_O}{dt}. \tag{16.29}$$

Equation (16.29) states that the time rate of change of the angular momentum with respect to any point O equals the moment of the resultant force acting on the particle with respect to the same point.

Let us multiply Eq. (16.29) by dt and express it in integral form. Thus,

$$\int_{t_1}^{t_2} \mathbf{M}_O \, dt = \int_{\mathbf{H}_{O1}}^{\mathbf{H}_{O2}} d\mathbf{H}_O.$$

Therefore, using Eq. (16.27), we conclude that

$$\mathbf{G}_O = \mathbf{H}_{O2} - \mathbf{H}_{O1}. \tag{16.30}$$

Equation (16.30) represents the principle of angular impulse and momentum. It states that the angular impulse $\mathbf{G}_O$ (delivered over the time interval from t_1 to t_2) equals the change of angular momentum $(\mathbf{H}_{O2} - \mathbf{H}_{O1})$. Equation (16.30) is conveniently expressed in terms of its three scalar rectangular components. Thus,

$$\left.\begin{aligned} G_{Ox} &= (H_{O2} - H_{O1})_x \\ G_{Oy} &= (H_{O2} - H_{O1})_y \\ G_{Oz} &= (H_{O2} - H_{O1})_z \end{aligned}\right\} \tag{16.31}$$

where G_{Ox} G_{Oy}, and G_{Oz} represent the angular impulses about the x, y, and z axes through point O, respectively, and the quantities $(H_{O2} - H_{O1})_x$, $(H_{O2} - H_{O1})_y$, and $(H_{O2} - H_{O1})_z$ are the changes in the angular momenta about the same axes, respectively.

The following examples illustrate some of the concepts discussed above.

■ Example 16.11

A proposed ride in an amusement park consists of four cars, each with a weight of 322 lb including the passenger, attached to a central shaft, as shown in Figure E16.11. At a given instant (i.e., $t = 0$), the speed of

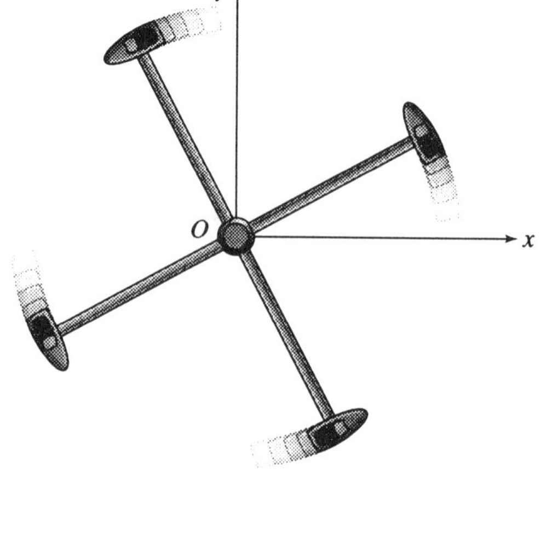

FIGURE E16.11.

the cars is 10 ft/s. For a very short period of time, each car is capable of producing a thrust given by $T = 6t^2$. Also, the central shaft is subjected to a constant frictional moment of 20 lb·ft. Determine the speed of the cars after the thrust is activated. Ignore air resistance and the weight of the connecting rods.

Solution

The principle of angular impulse and momentum given by the third of Eqs.(16.31) is convenient for the solution of this problem. This equation states that $G_{Oz} = (H_{O2} - H_{O2})$ where

$$G_{Oz} = \int_{t_1}^{t_2} M_O\,dt = \int_0^7 [4(6t^2)(10) - 300]\,dt$$

$$= 80t^3 - 300t|_0^7 = 25{,}340 \text{ ft}\cdot\text{lb}\cdot\text{s},$$

$$(H_{O2})_z = 4\left(\frac{322}{32.2}\right)(10)v_2 = 400v_2,$$

$$(H_{O1})_z = 4\left(\frac{322}{32.2}\right)(10)(10) = 4000 \text{ ft}\cdot\text{lb}\cdot\text{s}.$$

Therefore,

$$25{,}340 = 400v_2 - 4000$$

from which

$$v_2 = 73.4 \text{ ft/s}. \qquad \text{ANS.}$$

■ Example 16.12

The position and velocity information for a particle is given below:

$t_1 = 0$: $\quad \mathbf{r}_1 = (2\mathbf{i})$ m $\quad$ and $\quad \mathbf{v}_1 = (4\mathbf{i} + 8\mathbf{k})$ m/s;

$t_2 = 2$ s: $\quad \mathbf{r}_2 = (10\mathbf{i} + 24\mathbf{j} + 16\mathbf{k})$ m $\quad$ and $\quad \mathbf{v}_2 = (4\mathbf{i} + 24\mathbf{j} + 8\mathbf{k})$ m/s.

If the particle has a mass of 2 kg, determine its (a) angular momentum $\mathbf{H}_O$ at t_1 and t_2 and (b) angular impulse $\mathbf{G}_O$ over the time interval from t_1 to t_2. Note that point O is origin of the coordinate system.

Solution

(a) $\mathbf{H}_O = \mathbf{r} \times m\mathbf{v}$,

$$t_1 = 0: \quad (\mathbf{H}_O)_1 = \begin{bmatrix} \mathbf{i} & \mathbf{j} & \mathbf{k} \\ 2 & 0 & 0 \\ 8 & 0 & 16 \end{bmatrix} = -32\mathbf{j} \text{ m}\cdot\text{N}\cdot\text{s}; \qquad \text{ANS.}$$

$$t_2 = 2: \quad (\mathbf{H}_O)_2 = \begin{bmatrix} \mathbf{i} & \mathbf{j} & \mathbf{k} \\ 10 & 24 & 16 \\ 8 & 48 & 16 \end{bmatrix}$$

$$= (-384\mathbf{i} - 32\mathbf{j} + 288\mathbf{k}) \text{ m}\cdot\text{N}\cdot\text{s}. \qquad \text{ANS.}$$

(b) $\mathbf{G}_O = (\mathbf{H}_O)_2 - (\mathbf{H}_O)_1$

$$\mathbf{G}_O = (-384\mathbf{i} - 32\mathbf{j} + 288\mathbf{k}) - (-32\mathbf{j})$$

$$= (-384\mathbf{i} + 288\mathbf{k})\ \text{m}\cdot\text{N}\cdot\text{s}. \qquad \text{ANS.}$$

Example 16.13

A remote-controlled toy airplane with a mass of 0.6 kg flies in space such that its position vector is given by $\mathbf{r} = (3t\mathbf{i} + 2t^2\mathbf{j} + t^3\mathbf{k})$ m where t is measured in seconds. Determine the angular momentum $\mathbf{G}_O$ experienced by the airplane during the time interval from $t_1 = 1$ s to $t_2 = 3$ s. Note that point O is the origin of the coordinate system.

Solution

Because the position vector of the toy airplane is given as a function of time, we can find its velocity vector by differentiating the position vector with respect to time. Thus,

$$\mathbf{v} = \frac{d\mathbf{r}}{dt} = (3\mathbf{i} + 4t\mathbf{j} + 3t^2\mathbf{k})\ \text{m/s}. \qquad \text{ANS.}$$

Therefore,

$$\mathbf{H}_O = \mathbf{r} \times m\mathbf{v},$$

$$\mathbf{H}_O = 0.6\begin{bmatrix} \mathbf{i} & \mathbf{j} & \mathbf{k} \\ 3t & 2t^2 & t^3 \\ 3 & 4t & 3t^2 \end{bmatrix} = (1.2t^4\mathbf{i} - 3.6t^3\mathbf{j} + 3.6t^2\mathbf{k})\ \text{m}\cdot\text{N}\cdot\text{s}. \qquad \text{ANS.}$$

At $t = t_1 = 1$ s,

$$\mathbf{H}_{O1} = (1.2\mathbf{i} - 3.6\mathbf{j} + 3.6\mathbf{k})\ \text{m}\cdot\text{N}\cdot\text{s};$$

at $t = t_2 = 3$ s,

$$\mathbf{H}_{O2} = (97.2\mathbf{i} - 97.2\mathbf{j} + 32.4\mathbf{k})\ \text{m}\cdot\text{N}\cdot\text{s}. \qquad \text{ANS.}$$

By the principle of angular impulse and momentum (Eq. (16.30)),

$$\mathbf{G}_O = \mathbf{H}_{O2} - \mathbf{H}_{O1} = (96.0\mathbf{i} - 93.6\mathbf{j} + 28.8\mathbf{k})\ \text{m}\cdot\text{N}\cdot\text{s}. \qquad \text{ANS.}$$

■

Problems

16.56 A particle of mass $m = 3$ slug is positioned on its path by $\mathbf{r} = t^2\mathbf{i} + 2t\mathbf{j} + (4t + 2)\mathbf{k}$ where t is in seconds and the position vector in feet. (a) When $t = 1$ s and $t = 2$ s, determine the linear and angular momentum vectors. (b) Use the impulse-momentum principles to determine the linear and angular impulses exerted on the particle from $t = 1$ s to $t = 2$ s.

16.57 A particle of mass $m = 2$ kg is positioned on its path by $\mathbf{r} = t^3\mathbf{i} + 4t\mathbf{j} + 8t^2\mathbf{k}$ where t is in seconds and the position vector in meters. (a) When $t = 0$ s and $t = 1$ s, determine the linear and angular momentum vectors. (b) Use the impulse-momentum principles to determine the linear and angular impulses exerted on the particle from $t = 0$ s to $t = 1$ s.

16.58 Refer to Example 16.13, and find the quantities required in parts (b), (c), (e), and (f) for the position vector $\mathbf{r} = 2t^2\mathbf{i} + 8t\mathbf{j} + (t^2 - 4)\mathbf{k}$. Units are ft-s. Let $m = 0.5$ slug, $t_1 = 1$ s, and $t_2 = 4$ s.

16.59 Refer to Example 16.13 and find the quantities required in parts (b), (c), (e) and (f) for the position vector $\mathbf{r} = (2t + 1)\mathbf{i} + (t^2 - 4)\mathbf{j} + (6t - 2)\mathbf{k}$. Units are m-s. Let $m = 0.2$ kg, $t_1 = 0$ s and $t_2 = 3$ s.

16.60 A particle of mass m moves on a circular path such that $\mathbf{r} = r\cos\theta\mathbf{i} + r\sin\theta\mathbf{j}$ (a) Determine the velocity vector of the particle. (b) Find the linear momentum vector of the particle. (c) Determine the angular momentum vector of the particle with respect to the center of the circular path.

16.61 Motion on a parabolic path is defined parametrically by $x = 2t$, $y = 2 + 4t^2$ where t is in seconds and the coordinates are in meters. Determine the vectors **r**, **v**, **L**, and $\mathbf{H}_O$ for a particle of mass m which moves on this path.

16.62 The bob of a spherical pendulum of mass m moves on the surface of a sphere such that $\mathbf{r} = x\mathbf{i} + y\mathbf{j} + z\mathbf{k}$ where $x = a\sin\phi\cos\theta$, $y = a\sin\phi\sin\theta$, and $z = -a\cos\phi$. The radius of the sphere is a which is constant and ϕ and θ are functions of time. Determine the vectors **v**, **L**, and $\mathbf{H}_O$ in terms of a, the angles, and their time derivatives, where point O is the origin of the x-y-z coordinate system.

16.63 The motion of a particle of mass m on a vertical straight line is defined by $\mathbf{r} = b\mathbf{i} + t^2\mathbf{j}$ where b is a constant (ft-s units). Determine the vectors **v**, **a**, **R**, **L**, and $\mathbf{H}_O$ where point O is the origin of the polar coordinate system.

16.64 The motion of a particle of mass m on a horizontal straight line is defined by $\mathbf{r} = t^3\mathbf{i} + b\mathbf{j}$ where b is a constant (m-s units). Determine the vectors **v**, **a**, **R**, **L**, and $\mathbf{H}_O$ where point O is the origin of the polar coordinate system.

16.65 An elliptic path of a particle of mass m is given by $\mathbf{r} = a\cos\omega t\mathbf{i} + b\sin\omega t\mathbf{j}$ where a, b, and ω are constants. Determine the vectors **v**, **a**, **R**, **L**, and $\mathbf{H}_O$. Show that the path is, in fact, elliptic. Note that point O is the origin of the polar coordinate system.

16.66 Initially, a particle of 0.2 slug mass is at point A (1, 1, 1) ft, and, finally, it is at point B (10, 4, −8) ft as shown in Figure P16.66. Determine (a) the initial and final velocities of the particle expressed in terms of the unit vectors **i**, **j**, and **k**. The initial speed is 40 ft/s and the final speed is 60 ft/s. Note that two points are given to direct each of the velocity vectors. (b) the initial and final linear momentum for the particle. (c) the linear impulse exerted on the particle as it moves from point A to point B. (d) the

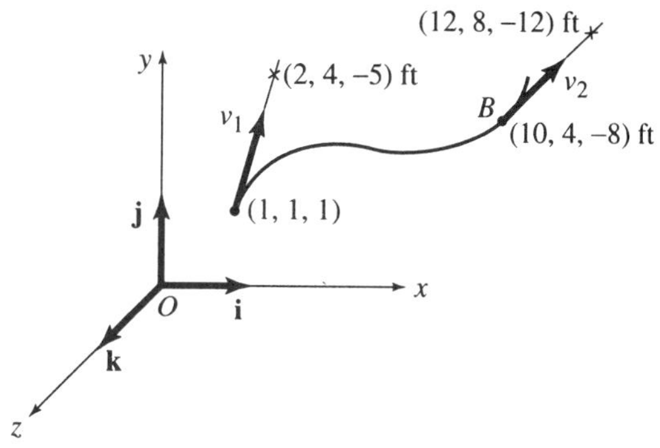

FIGURE P16.66.

initial and final angular momenta with respect to point O. (e) The angular impulse with respect to point O exerted on the particle as it moves from point A to point B.

16.67 Three positions A, B, and C of an auto racer are shown in the plan view of Figure P16.67. The car and driver have a mass of 60 slug and the speeds are 180 mph, 175 mph, and 195 mph for positions A, B and C, respectively. The car moves along a 600-ft radius horizontal circular curve. For each of the three positions, determine (a) the linear momentum vectors and (b) the angular momentum vectors with respect to point O. (c) Find the linear impulse exerted by friction on the racer as it moves from position A to position B. (d) Find the angular impulse with respect to point O exerted on the racer as it moves from position B to position C.

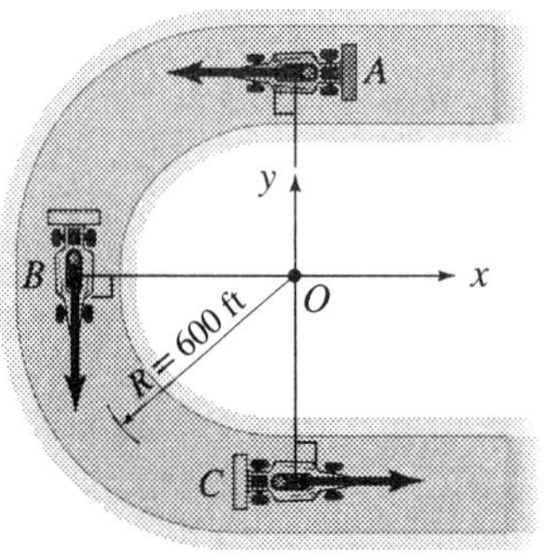

FIGURE P16.67.

16.68 In Figure P16.68 a particle of mass $m = 10$ kg is located at point A at a given instant and has a speed of 5 m/s with a velocity vector directed from A toward B. Determine its angular momentum vector at the given instant with respect to (a) the origin and (b) point C. Point coordinates are expressed in meters.

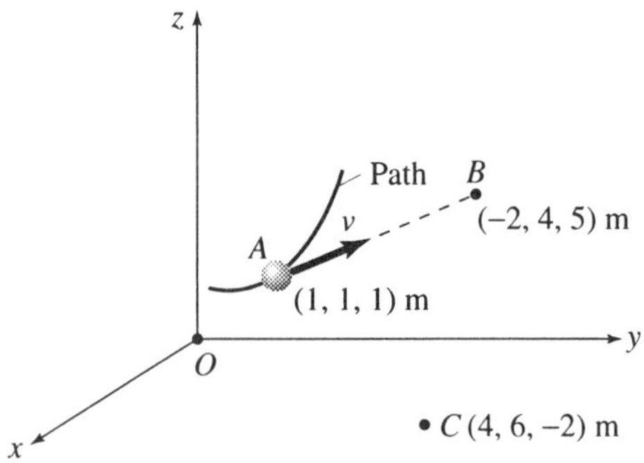

FIGURE P16.68.

16.69 In Figure P16.69 a particle of mass $m = 2$ slug is located at point A at a given instant, and coordinates are expressed in ft. It has a speed of 10 ft/s, and its velocity vector is directed from A toward B. Determine its angular momentum vector at the given instant with respect to (a) the origin and (b) point C.

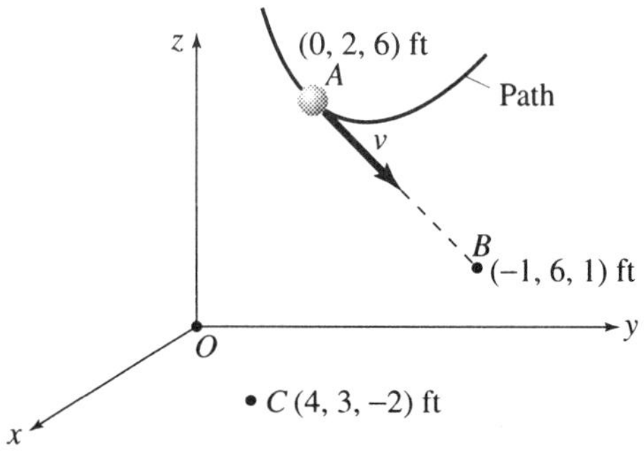

FIGURE P16.69.

16.70 A particle of mass 6 kg shown in Figure P16.70 has an initial velocity $\mathbf{v}_1 = (-4\mathbf{i} + 4\mathbf{j} + 5\mathbf{k})$ m/s and a final velocity $\mathbf{v}_2 = (-2\mathbf{i} + 6\mathbf{j} + 4\mathbf{k})$ m/s. Determine the angular impulse vector exerted on this particle during this time interval, with respect to the origin of the Newtonian reference frame shown.

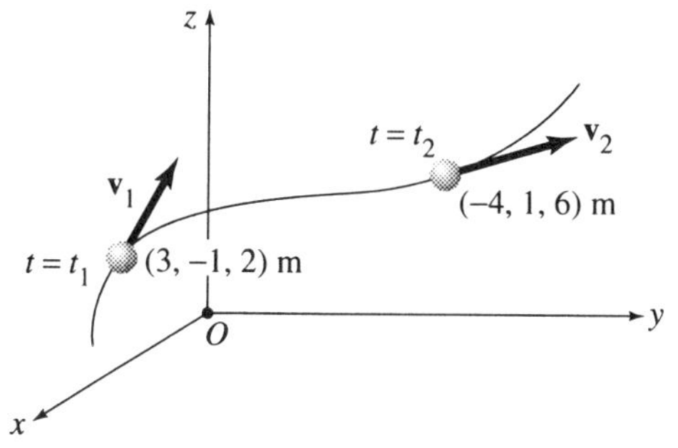

FIGURE P16.70.

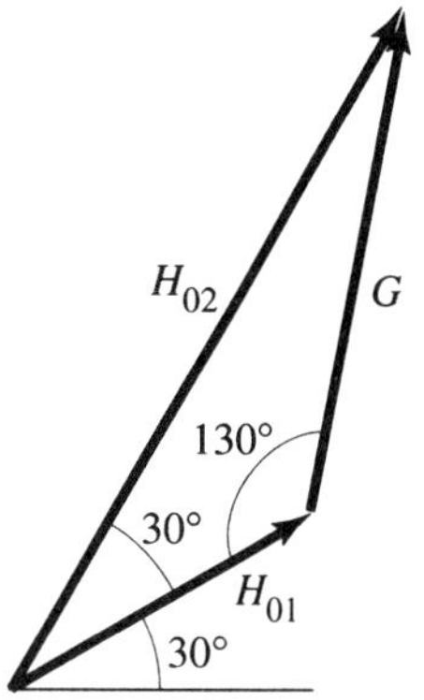

FIGURE P16.71.

16.71 Refer to Figure P16.71 and determine the magnitude of the final angular momentum, given that the magnitude of the initial angular momentum is 342 m·N·s and that the magnitude of the angular impulse is 500 m·N·s. Which of these values are instantaneous values?

16.72 Figure P16.72 is a three-dimensional representation of the angular impulse-momentum equation which you are to assume is scaled. Coordinates are stated in inches, and the scale factor is 1000 in·lb·s/in. Express the initial and final angular momentum vectors in terms of the unit vectors shown.

16.73 Refer to Problem 16.72, and express the angular impulse vector in terms of the unit vectors shown in Figure P16.72.

16.74 Refer to the three component plots of Figure P16.74 showing the variation of moment versus time. Show that the re-

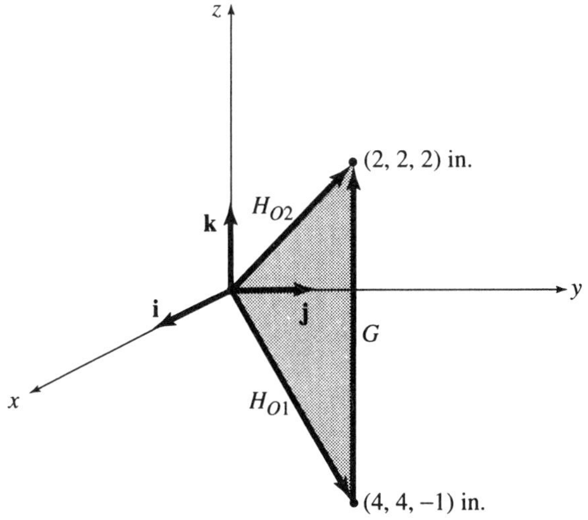

FIGURE P16.72.

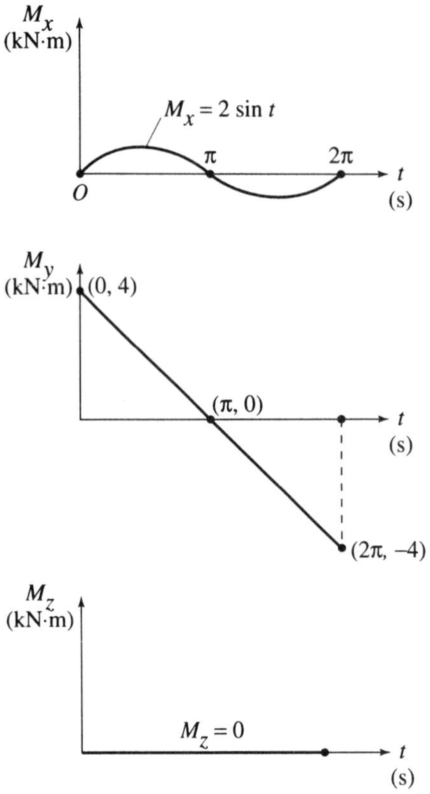

FIGURE P16.74.

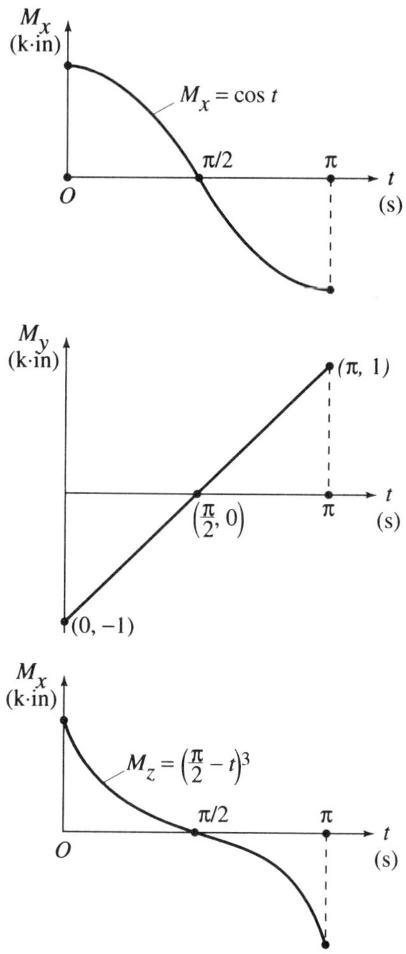

FIGURE P16.76.

sulting angular impulse due to $\mathbf{M} = M_x\mathbf{i} + M_y\mathbf{j} + M_z\mathbf{k}$, exerted over a time interval from $t = 0$ to $t = 2\pi$ s, will vanish. Carry out integrations, and discuss from the graphs, too.

16.75 Use the variations of moment versus time shown in Figure P16.74, and calculate the resulting angular impulse due to $\mathbf{M} = M_x\mathbf{i} + M_y\mathbf{j} + M_z k$, exerted over a time interval from $t = 0$ to $t = \pi$ s.

16.76 Refer to Figure P16.76 where plots are shown for moment components versus time. Determine the angular impulse of the moment given by $\mathbf{M} = M_x\mathbf{i} + M_y\mathbf{j} + M_z\mathbf{k}$ where M_y is a linear function.

16.8 Systems of Particles: The Angular Impulse-Momentum Principle

A system of n particles is shown in Figure 16.15. Consider any particle, the ith particle, of this system where i varies from 1 to n. The resultant external force acting on this particle is $\mathbf{F}_i$, and the resultant internal force acting on it is $\mathbf{f}_i$.

From Eq. (16.28b),

$$\mathbf{M}_{Oi} = (\mathbf{r}_i \times \mathbf{F}_i) + (\mathbf{r}_i \times \mathbf{f}_i) = \frac{d\mathbf{H}_{Oi}}{dt}$$

where $\mathbf{M}_{Oi}$ is the moment of the resultant external and internal forces acting on the ith particle with respect to the origin O of a Newtonian reference frame. Also, by Eq. (16.22), M_{Oi} is equal to the angular momentum of the ith particle with respect to the same origin. If we take the vector sum over all particles of the system, then, for the entire system of particles,

$$\sum \mathbf{M}_{Oi} = \sum \frac{\mathbf{dH}_{Oi}}{dt} = \sum \mathbf{r}_i \times \mathbf{F}_i + \sum \mathbf{r}_i \times \mathbf{f}_i.$$

The summation $\sum \mathbf{r}_i \times \mathbf{f}_i$ will vanish because internal forces occur in equal and opposite pairs and the moment of each collinear pair about the origin will be zero. If we drop the subscript i and understand that the summations extend over the entire number of particles of the system, then,

$$\sum \mathbf{M}_O = \sum \frac{d\mathbf{H}_O}{dt}. \tag{16.32}$$

Equation (16.32) states that the vector sum of the moments of the *external* forces acting on the system of particles, with respect to point O, is equal to the time rate of change of the angular momentum vector for the entire system of particles.

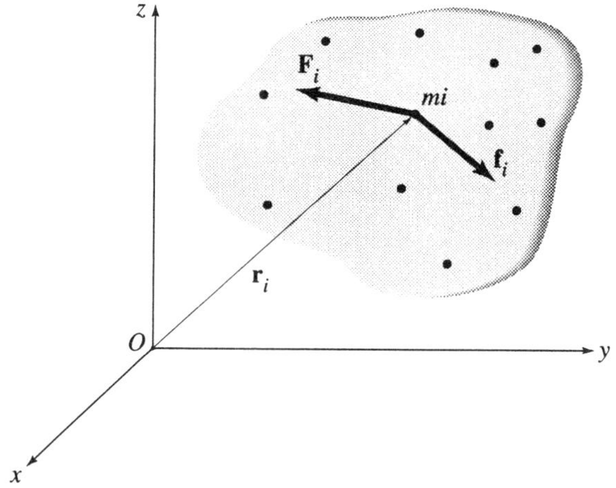

FIGURE 16.15.

Proceeding in the same manner as was done in the development of Eq. (16.30), we conclude that

$$\sum \int_{t_1}^{t_2} \mathbf{M}_O \, dt = \sum \mathbf{H}_{O2} - \sum \mathbf{H}_{O1}. \tag{16.33}$$

Equation (16.33) states that the sum of the angular impulses about point O of *all* external forces acting on the system of particles is equal to the difference between the final angular momenta and the initial angular momenta of the entire system with respect to the same point. Note that Eq. (16.33) for a system of particles has the same form as Eq. (16.30) for a single particle. Note also that point O was selected arbitrarily, and, therefore, Eq. (16.33) is valid with respect to *any* point including the mass center G of the system of particles.

16.9 Conservation of Angular Momentum

The angular impulse-momentum principle for a single particle, Eq. (16.30), states that $\mathbf{G}_O = \mathbf{H}_{O2} - \mathbf{H}_{O1}$. If the angular impulses during the time interval of interest are zero, then, $\mathbf{G}_O$ vanishes and Eq. (16.30) reduces to

$$\mathbf{H}_{O1} = \mathbf{H}_{O2}. \tag{16.34}$$

Equation (16.34) expresses the principle of conservation of angular momentum for a single particle.

Before we may legitimately write the conservation of angular momentum equation, we must show that the angular impulse integral vanishes for the time interval of interest. If either $\mathbf{r}$ or $\mathbf{R}$ in Eq. (16.27) is a null vector for the time interval from t_1 to t_2, then, the angular impulse integral $\mathbf{G}_O$ vanishes, and, thus, the angular momentum is conserved. Even if neither $\mathbf{r}$ or $\mathbf{R}$ is identically a null vector during the time interval of interest, the angular impulses may still vanish about the point of interest, and, thus, the angular momentum of the particle is conserved. For example, an artificial satellite in its orbit around the Earth (see Fig. 16.16) is subjected *only* to the force of attraction $\mathbf{F}$, if it is assumed that there is no atmosphere. This force of attraction is always directed toward the center of Earth, point O. Therefore, its angular impulse about this point vanishes, and we conclude that Eq. (16.34) is applicable because the angular momentum of the satellite about point O is conserved. Further information about the motion of artificial satellites is given in Section 16.13 dealing with space mechanics.

Similarly, in the case of a system of particles, if it can be shown that the angular impulses of all the external forces, about some point O, vanish, then, we conclude that the angular momentum of the entire system, about point O, is conserved, and Eq. (16.33) reduces to

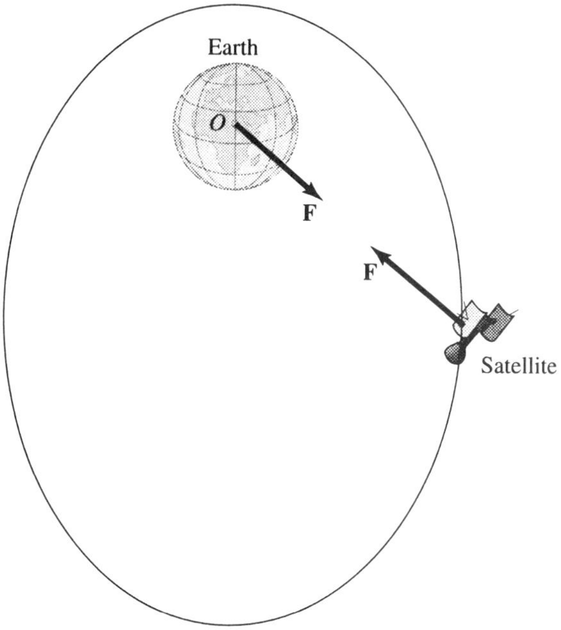

FIGURE 16.16.

$$\sum \mathbf{H}_{O1} = \sum \mathbf{H}_{O2}. \tag{16.35}$$

The following examples illustrate some of the above concepts.

Example 16.14

A block of weight $W = 2$ lb is attached to a cord and rests on a smooth platform as shown in Figure E16.14. Initially when $r = r_1 = 10$ in., the velocity of the block in a direction perpendicular to the cord is $v_1 = 10$ ft/s as shown. The cord is pulled very slowly through the central hole until $r = r_2 = 5$ in. Determine (a) the velocity of the block in a direction perpendicular to the cord when $r = r_2$, (b) the work done by the force F as r changes from r_1 to r_2, and (c) the average value of the force F_{AV} based upon the work done, as r changes from r_1 to r_2.

Solution

(a) The angular momentum of the weight W is conserved with respect to the vertical y axis shown in Figure E16.14. In order to justify this statement, we refer to the forces acting on the free body diagram of the block which are its weight W, the normal force N exerted by the fixed smooth platform, and the cord tension F. The action lines of the weight W and the normal N are both parallel to the vertical y axis and, therefore, produce no moments about this axis. The cord tension F has an action line which intersects the vertical y axis and, thus, produces no moment about this axis. We conclude that the moment about the y

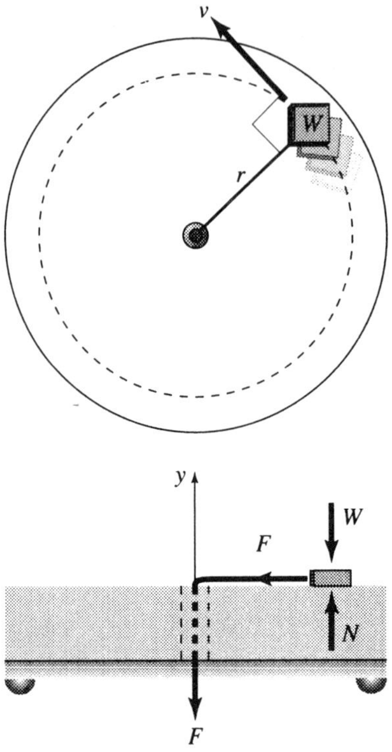

FIGURE E16.14.

axis vanishes for all forces acting on the block. Thus, in the second of Eqs. (16.31), the impulse $G_{Oy} = 0$ and

$$(H_{O1})_y = (H_{O2})_y$$

from which

$$(mv_1 r_1)_y = (mv_2 r_2)_y.$$

We are justified in regarding that the velocity vectors are perpendicular to their respective radii because the cord is pulled slowly such that the radial velocity component is negligible.

Solving for v_2,

$$v_2 = \frac{r_1}{r_2} v_1 = \left(\frac{10}{5}\right)(10) = 20.0 \text{ ft/s}. \qquad \text{ANS.}$$

(b) The work energy principle, $U_{1\to2} = T_2 - T_1$, becomes

$$U_{1\to2} = \frac{1}{2}mv_2^2 - \frac{1}{2}mv_1^2 = \frac{1}{2}\left(\frac{2}{32.2}\right)(20^2) - \frac{1}{2}\left(\frac{2}{32.2}\right)(10^2)$$

$$= 9.32 \text{ lb}\cdot\text{ft}. \qquad \text{ANS.}$$

(c) To find an average value of the force as r changes from r_1 to r_2, we write

$$U_{1\to 2} = F_{AV}(r_1 - r_2).$$

Thus,

$$F_{AV} = \frac{U_{1\to 2}}{(r_1 - r_2)} = \frac{9.32}{\left(\frac{10}{12} - \frac{5}{12}\right)} = 22.4 \text{ lb.} \qquad \text{ANS.}$$

This average force is not to be confused with a time average value of the force. It is the average force, which, exerted at this constant value as r changes from r_1 to r_2, would do the same work as the variable force F associated with the same kinetic energy change.

Example 16.15

A conical pendulum consists of a bob of mass $m = 0.5$ kg attached to a cord of length L passing through a vertical fixed frictionless tube as shown in Figure E16.15(a). The bob is set in circular motion when $L = 0.75$ m and $\theta = 20°$, and the cord is pulled very slowly through the tube by means of the constant force F until the angle θ has changed to 50°. Determine the corresponding length L of the cord for $\theta = 50°$ and the change in kinetic energy as θ changes from 20° to 50°.

Solution

Consider the free-body diagram of the bob, shown in Figure E16.15(b), for any position defined by the angle θ and the corresponding length L. Because the two forces T and W acting on the bob produce no angular impulses about the z axis, the angular momentum of the bob about this axis is conserved as the bob moves from its initial position (position 1) to its final position (position 2). Thus, by the third of Eqs. (16.31)

$$(H_{O1})_z = (H_{O2})_z,$$

$$mv_1 L_1 \sin\theta_1 = mv_2 L_2 \sin 50°,$$

$$v_1(0.75 \sin 20°) = v_2 L_2 \sin 50°,$$

$$v_1 = 2.986 v_2 L_2, \qquad \text{(a)}$$

Also, for position 1 ($\theta_1 = 20°$),

$$\sum F_{z'} = ma_{z'} = 0,$$

$$T_1 \cos\theta_1 - mg = 0,$$

$$T_1 = \frac{mg}{\cos\theta_1} = \frac{9.81 \text{ m}}{\cos 20°} = 10.44 \text{ m}, \qquad \text{(b)}$$

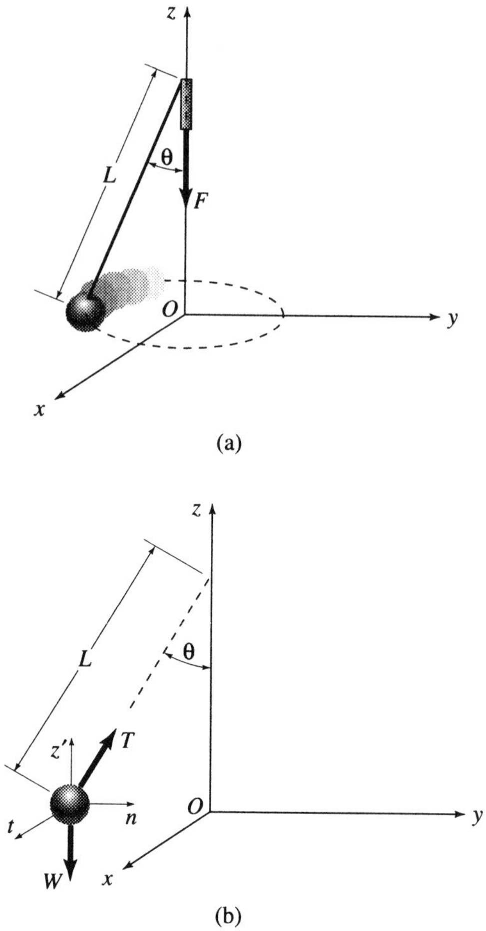

FIGURE E16.15.

$$\sum F_n = ma_n,$$

$$T_1 \sin\theta_1 = m\left(\frac{v_1^2}{L\sin\theta_1}\right),$$

$$T_1 = \frac{mv_1^2}{L_1 \sin^2\theta_1} = \frac{mv_1^2}{(0.75)\sin^2 20^\circ} = 11.40mv_1^2. \tag{c}$$

Substituting Eq. (b) into Eq. (c) and solving for v_1 yields

$$v_1 = 0.957 \text{ m/s}. \tag{d}$$

Similarly, for position 2 ($\theta = 50^\circ$),

$$T_2 = \frac{mg}{\cos\theta_2} = \frac{9.81 \text{ m}}{\cos 50^\circ} = 15.26 \text{ m}, \quad \text{(e)}$$

$$T_2 = \frac{mv_2^2}{L_2 \sin^2\theta_2} = \frac{v_2^2}{L_2 \sin^2 50^\circ} = 1.704\frac{mv_2^2}{L_2}. \quad \text{(f)}$$

Substituting Eq. (e) in Eq. (f) and solving for v_2 leads to

$$v_2 = 2.993\sqrt{L_2}. \quad \text{(g)}$$

Substituting Eqs. (d) and (g) into Eq. (a) yields

$$L_2 = 0.225 \text{ m.} \quad \text{ANS.}$$

From Eq. (g),

$$v_2 = 2.993\sqrt{0.225} = 1.420 \text{ m/s.}$$

Thus, the change in kinetic energy from position 1 to position 2 becomes

$$\Delta T = T_2 - T_1 = \tfrac{1}{2}mv_2^2 - \tfrac{1}{2}mv_1^2 = \tfrac{1}{2}m(v_2^2 - v_1^2),$$

$$= \tfrac{1}{2}(0.5)(1.420^2 - 0.957^2) = 1.101 \text{ N}\cdot\text{m} \quad \text{ANS.}$$

Note that the kinetic energy of the bob increased as it moved from position 1 ($\theta = 20^\circ$) to position 2 ($\theta = 50^\circ$). This increase in kinetic energy is due to part of the work done by the force F pulling the cord. The remainder of the work of this force is used up by the increase in the potential energy of the bob.

■

Problems

16.77 A plan view of a circular table is shown in Figure P16.77. A cord attached to the particle of mass m can be pulled through the central hole at O. The cord is pulled inward at a constant rate of 0.15 m/s as the particle moves on the frictionless table. Initially the particle has a velocity component of 0.8 m/s perpendicular to the radius r where $r = 1.00$ m. Determine the total speed of the particle 2 s later.

16.78 A sphere of weight $W = 3$ lb is attached to one end of a spring of spring constant $k = 4$ lb/ft. The other end of the spring

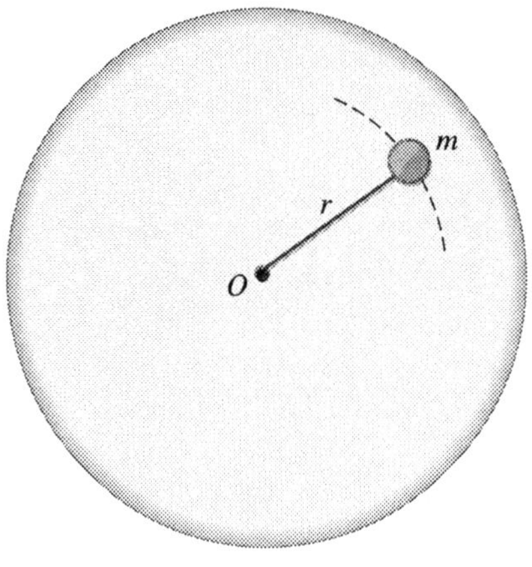

FIGURE P16.77.

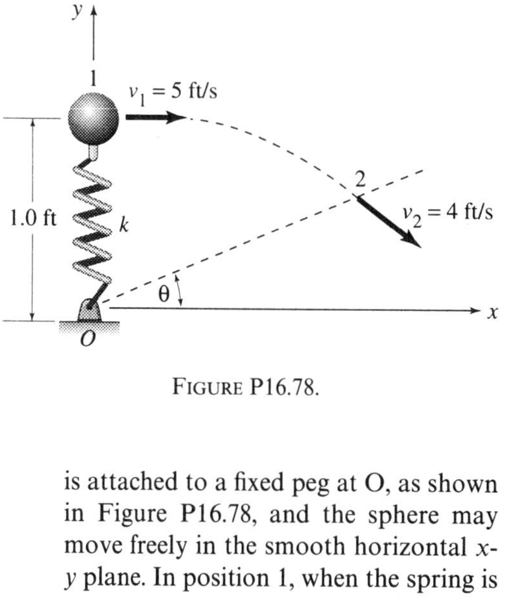

FIGURE P16.78.

is attached to a fixed peg at O, as shown in Figure P16.78, and the sphere may move freely in the smooth horizontal x-y plane. In position 1, when the spring is unstretched, the sphere is given a velocity of $v_1 = 5$ ft/s in a direction perpendicular to the spring. Determine (a) the spring deformation Δs when the sphere is in position 2, at which time its velocity $v_2 = 4$ ft/s and (b) the angle θ defining position 2.

16.79 A top view of a roller-skating rink is shown in Figure P16.79. A skater of mass m attaches a rope to herself and to a stationary post O whose diameter is 0.10 m. As the skater moves around the post in a cw sense, the rope wraps around the post thus decreasing the distance between it and the skater. At the instant shown in Figure P16.79, the distance L between the post and the skater is 20 m and the skater has a speed of 2 m/s perpendicular to the rope. Determine the number of turns the rope has wrapped around the post at the instant the speed of the skater is 4 m/s perpendicular to the rope. Neglect friction and assume that the skater does not exert any effort during the motion.

16.80 As shown in Figure P16.80, a satellite weighing 2000 lb is launched from the surface of the Earth such that its speed at A is 15,000 mph at an angle $\phi = 75°$ where $r_A = 8000$ mi. It proceeds then on a free-flight trajectory executing an elliptic orbit as shown. Determine its speed at both the perigee (point B) and apogee (point D) and the distances from the center of Earth to these two points. Hint: The only force acting on the satellite is the force of attraction of Earth, given by $F = \dfrac{GM_E m_s}{r^2} = \dfrac{gR_E^2 m_s}{r^2}$. Therefore, the angular momentum about Earth's center is conserved. Note that $g = 32.2$ ft/s^2 and $R_E = 3960$ mi.

16.81 Repeat Problem 16.80 for the following input data: the mass of the satellite is 800 kg, the speed at A is 20000 km/h, $\phi = 65°$ and $r_A = 6000$ km.

16.82 A conical pendulum consists of a bob of weight $W = 2$ lb attached to a cord of length L passing through a vertical, fixed, frictionless tube as shown in Figure P16.82. The bob is set in circular motion when $L = 0.75$ ft and $\theta = 40°$, and the cord is released very slowly through the tube by maintaining a constant force F until the angle θ has

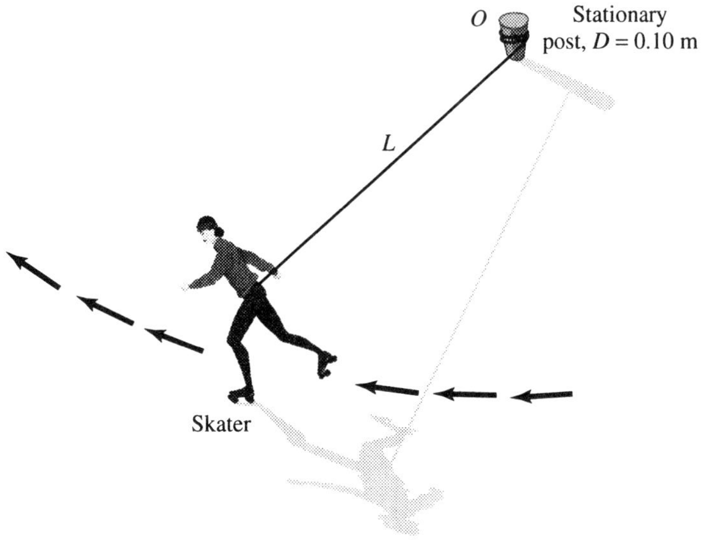

FIGURE P16.79.

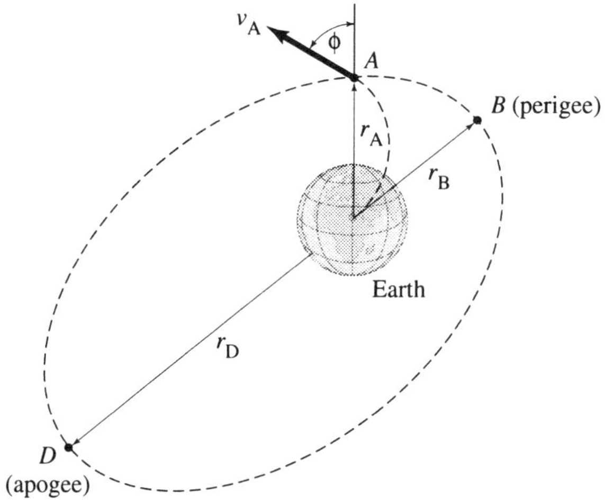

FIGURE P16.80.

changed to 15°. Determine the length L of the cord corresponding to $\theta = 15°$ and the change in kinetic energy as θ changes from 40° to 15°.

16.83 Initially, a particle of unit mass is at point A and finally at point B, as shown in Figure P16.83. The velocity vectors are

$$\mathbf{v}_A = (10\mathbf{i} + 30\mathbf{j} + 20\mathbf{k}) \text{ ft/s},$$

$$\mathbf{v}_B = (-2.5\mathbf{i} - 2\mathbf{j} - 2\mathbf{k}) \text{ ft/s}.$$

(a) Determine the linear impulse vector

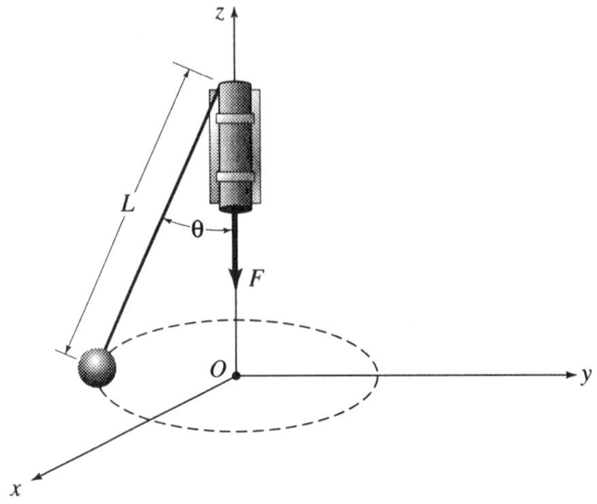

FIGURE P16.82.

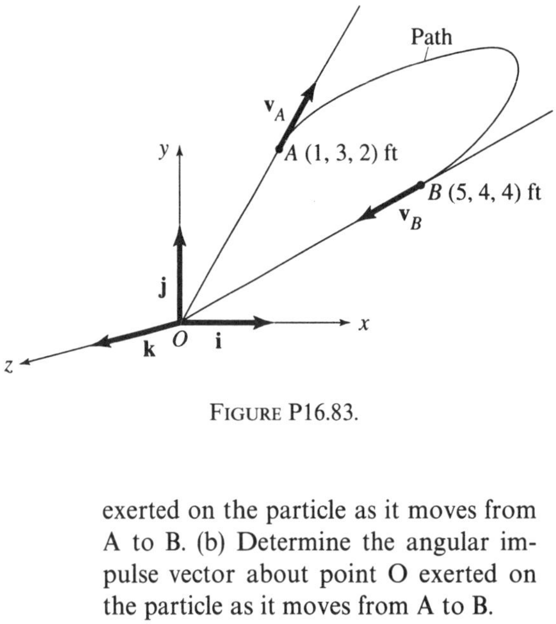

FIGURE P16.83.

exerted on the particle as it moves from A to B. (b) Determine the angular impulse vector about point O exerted on the particle as it moves from A to B.

16.84 The tails of both the initial and final angular momentum vectors for a particle are at the origin and their tips pass through (6, 2, 4)m and (4, 2, 9)m, respectively. The initial angular momentum has a magnitude of 6000 N·m·s, and the final angular momentum vector has a magnitude of 1000 N·m·s. Determine the angular impulse vector exerted on the particle. What is the magnitude of this angular impulse vector about the origin? What are the direction cosines for this vector?

16.85 For a particle, the initial linear momentum vector has a magnitude of 1000 N·s and is directed along the line from point A (0, 2, 4) m toward point B (3, 4, 6) m. The final linear momentum vector has a magnitude of 1500 N·s and is directed along the line from point C (5, 6, 8) m toward point D (9, 7, 10) m. Determine the linear impulse vector exerted on the particle. What is the magnitude of this linear impulse vector? What angles does this vector make with the coordinate axes?

16.86 A variable force given by $\mathbf{R} = 1000 \cdot (\sin t\mathbf{i} + t\mathbf{j} + t^2\mathbf{k})$ N acts on a particle of 2 kg mass during the time interval from $t = 0$ to $t = 1$ s. Its initial linear momentum is given by $\mathbf{L}_1 = 200(\mathbf{i} + 2\mathbf{j} + 3\mathbf{k})$ N·s. Determine its final linear momentum. What is the magnitude of this vector? What angles does it make with each of the three coordinate axes?

16.87 The moment of a force acting on a particle during the time interval from $t = 0$ s to $t = 2$ s, with respect to the origin, is given by $\mathbf{M}_O = 200[(1 - t)\mathbf{i} + (t - 1)\mathbf{j} + (1 - t^3)\mathbf{k}]$ N·m. If the initial angular momentum of the particle is given by $\mathbf{H}_{O1} = (120\mathbf{i} + 150\mathbf{j} + 650\mathbf{k})$ N·m·s, determine its final angular momentum.

16.88 Refer to Figure P16.88 for the time dependence of the three components of the resultant couple $\mathbf{M}_O$ which acts on a particle of mass $m = 200$ kg for a period of 2 s. The initial angular momentum of the particle at $t = 0$ is given by $\mathbf{H}_{O1} = (2000\mathbf{i} + 3000\mathbf{j} + 6500\mathbf{k})$ kN·m·s. Determine the final angular momentum of the particle. In computing the angular impulse it will be most convenient to evaluate the integrals by considering them equal to the area under the appropriate curves.

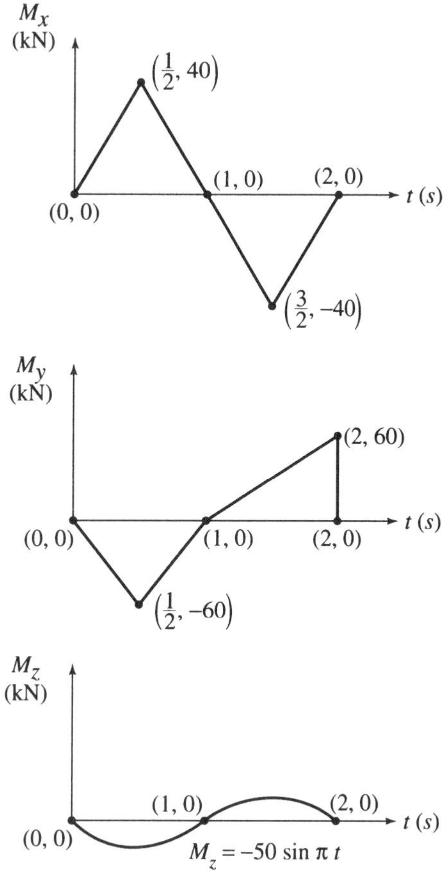

FIGURE P16.88.

16.10 Applications of Energy and Impulse-Momentum Principles

Some problems require a synthesis of a number of concepts for their solution, and we present five examples in this section to illustrate such problems and their solutions. If an impact occurs during the motion, then, the impulse-momentum method and the equation for the coefficient of restitution e will be required. For a given system, if the impulses can be shown to be zero for the time interval of interest, the momentum will be conserved for that system. Energy is lost to sound and heat during impact, except for the perfectly elastic case of $e = 1$. Prior to or after the impact is concluded, the work-energy method is usually the appropriate choice. The conservation of energy equation is applicable unless friction or other nonconservative forces do work on a given system.

Some of the significant features of these example solutions are summarized below:

1. Systems, for which equations are written, are clearly defined.
2. Sign conventions are established and employed in writing equations and interpreting results.
3. Equations are written and solved for the unknown quantities in terms of known quantities. A carefully planned, systematic notation greatly aids in formulation and solution of these problems.
4. Solutions are checked by back substitutions and sometimes by intuitive appraisals of the results.
5. Solutions in symbolic form are presented in dimensionless plots. Such graphic presentations enable us to view the system from a very general standpoint. Dimensionless ratios of quantities are the same in all systems of units which is of considerable advantage during the transition from U.S. Customary to SI units.

Example 16.16

Mass $m_A = 0.5$ slug is held at rest against the spring of constant $k = 200$ lb/ft as shown in Figure E16.16. In this position, the spring which is not attached to the mass, is deformed by an amount $x_O = 6$ in. It is released, moves for a distance $b = 3$ ft along the smooth horizontal plane ($\mu = 0$), and strikes mass $m_B = 0.3$ slug. Following the collision, m_B moves a distance c across the rough plane ($\mu = 0.2$) before it comes to rest. For impact between these masses, the coefficient of restitution is $e = 0.7$. Determine (a) the velocity of mass m_A just before it strikes mass m_B and (b) the velocity of both masses just after impact and the distance c through which m_B slides on the rough horizontal plane before it comes to a complete stop.

Solution

Let us state a sign convention, systematize our notation, and outline the methods to be employed in solving this problem. The positive sense for all velocities will be directed horizontally to the left along the indicated x axis. Carefully establishing a convenient notation quite often provides a great advantage in formulating and solving problems. Thus, the following positions are defined:

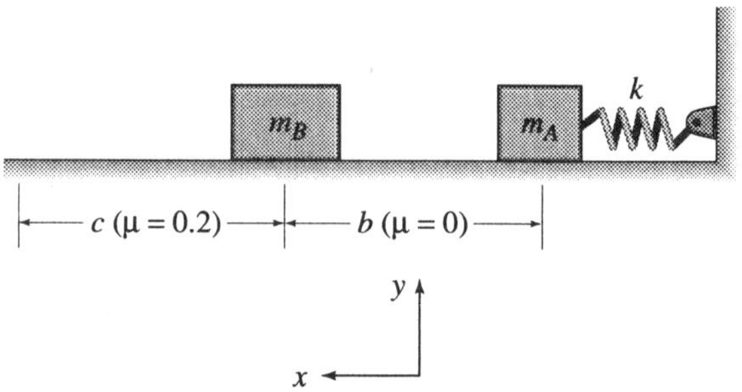

FIGURE E16.16.

1. Given initial position of mass m_A before release.
2. Position of m_A just before impact.
3. Position of both masses m_A and m_B just after impact.
4. Final position of mass m_B.

(a) Consider the motion of mass m_A from position 1 to position 2 where the work-energy method applies. Thus,

$$U_{1\to 2} = T_2 - T_1.$$

Mass m_A is initially at rest and only the spring, which is initially compressed, does work on it. Therefore,

$$\tfrac{1}{2}kx_O^2 = \tfrac{1}{2}m_A v_{A2}^2 - \tfrac{1}{2}m_A v_{A1}^2$$

Because $v_{A1} = 0$, it follows that

$$v_{A2} = \sqrt{\frac{k}{m_A}}x_O = \sqrt{\left(\frac{200}{0.5}\right)}\left(\frac{6}{12}\right) = 10.00 \text{ ft/s.} \qquad \text{ANS.}$$

(b) During impact, the linear momentum is conserved in the x direction. Thus,

$$\sum (L_2)_x = \sum (L_3)_x,$$

$$m_A v_{A2} + m_B v_{B2} = m_A v_{A3} + m_B v_{B3},$$

$$0.5(10) + 0 = 0.5v_{A3} + 0.3v_{B3}.$$

Thus,

$$5v_{A3} + 3v_{B3} = 50. \qquad \text{(a)}$$

Equation (16.20) for the coefficient of restitution yields

$$e = \frac{v_{B3} - v_{A3}}{v_{A2} - v_{B2}},$$

$$0.7 = \frac{v_{B3} - v_{A3}}{10 - 0}, \qquad \text{(b)}$$

$$v_{B3} - v_{A3} = 70.$$

Solve Eqs. (a) and (b) simultaneously to obtain

$$v_{A3} = 3.63 \text{ ft/s } \leftarrow, \qquad \text{ANS.}$$

$$v_{B3} = 10.63 \text{ ft/s } \leftarrow. \qquad \text{ANS.}$$

(c) Consider the motion of mass m_B after impact from position 3 to position 4. Thus, the work energy principle becomes,

$$U_{2\to 3} = T_4 - T_3.$$

Friction is the only force working on m_B as it moves left and reaches a zero velocity in position 4. Thus,

$$-\mu(m_B g)c = \tfrac{1}{2}m_B v_{B4}^2 - \tfrac{1}{2}m_B v_{B3}^2,$$
$$-0.2(0.3)(32.2)c = 0 - \tfrac{1}{2}(0.3)(10.63^2),$$
$$c = 8.77 \text{ ft.} \qquad \text{ANS.}$$

■ Example 16.17

It is desired to drive a pile of weight $W_p = 6000$ lb into the ground by dropping a hammer of weight $W_H = 500$ lb a distance $h = 5$ ft onto the pile as shown in Figure E16.17. Determine the distance that the pile is driven into the ground by a single blow of the hammer, if it is assumed that the ground provides a constant resisting force of 25,000 lb. Assume the impact to be perfectly plastic (i.e., $e = 0$).

Solution

We identify the following positions:

1. Given initial position of the hammer before dropping.
2. Position of the hammer just before impact.
3. Position of both the hammer and the pile just after impact.
4. Final position of the pile.

The work-energy principle is used to determine the velocity of the hammer just before impact. Thus,

$$U_{1\to 2} = T_2 - T_1,$$
$$m_H g(h) = \tfrac{1}{2}m_H v_{H2}^2 - \tfrac{1}{2}m_H v_{H1}^2,$$

Because $v_{H1} = 0$, we conclude that

$$v_{H2} = \sqrt{2gh} = \sqrt{2(32.2)(5)} = 17.94 \text{ ft/s} \downarrow.$$

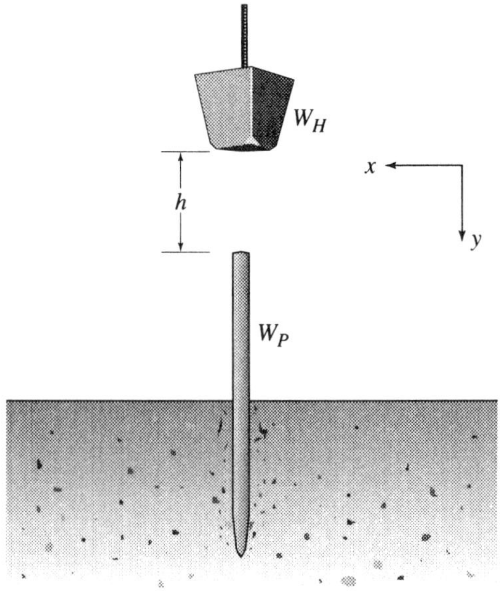

FIGURE E16.17.

During impact, the linear momentum of the system consisting of the hammer and pile is conserved in the y direction. Thus,

$$\sum (L_2)_y = \sum (L_3)_y,$$

$$m_H v_{H2} + m_P v_{P2} = m_H v_{H3} + m_P v_{P3}.$$

Because $v_{P2} = 0$, it follows that

$$\left(\frac{500}{32.2}\right)(17.94) + 0 = \left(\frac{500}{32.2}\right) v_{H3} + \left(\frac{6000}{32.2}\right) v_{P3}. \quad \text{(a)}$$

Also, because the impact is perfectly plastic, we conclude that, immediately after impact,

$$v_{H3} = v_{P3} = v_3. \quad \text{(b)}$$

Substituting Eq. (b) in Eq. (a) and solving,

$$v_{H3} = v_{P3} = v_3 = 1.380 \text{ ft/s} \downarrow.$$

Again, the work-energy principle is used to determine the motion of the system consisting of the hammer and pile, right after impact. Therefore,

$$U_{3\to4} = T_4 - T_3.$$

$$(W_H + W_P)y - 25{,}000y = 0 - \tfrac{1}{2}(m_H + m_P)v_3^2$$

where y represents the distance that the pile is driven into the ground by a single blow. Thus,

$$(500 + 6000)y - 25{,}000y = 0 - \frac{1}{2}\left(\frac{500 + 6000}{32.2}\right)(1.380^2)$$

from which

$$y = 0.0104 \text{ ft} = 0.1245 \text{ in.} \qquad \text{ANS.}$$

■ Example 16.18

The simple pendulum of mass m_A is released from rest in the position shown in Figure E16.18. After moving through the angular displacement θ, it strikes a small block of mass m_B which rests on the edge of a table. This small block becomes a projectile which strikes the floor at a height h below the table. Determine (a) the speed of mass m_A just before it impacts mass m_B, (b) the speed of each mass just after the impact, (c) the condition which assures that mass m_A will move leftward after the impact, and (d) the speed of mass m_B when it strikes the floor. Ignore friction at the pendulum support and on the table, air resistance, and frictional impulses during the impact. Express answers in terms of the known quantities m_A, m_B, L, θ, and the coefficient of resititution e.

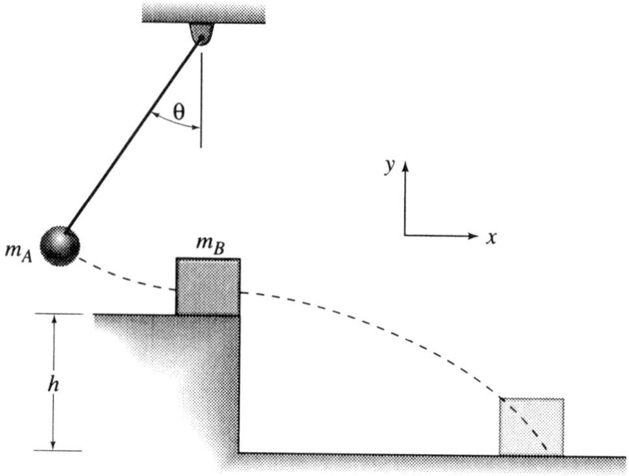

FIGURE E16.18.

Solution

Establish convenient subscripted notation for the speeds with the first subscript referring to the mass and the second subscript referring to the position. These positions are identified as follows:

1. Given initial position of mass m_A before release.
2. Position of m_A just before impact.
3. Position of both masses m_A and m_B just after impact.
4. Position of mass m_B when it hits the floor.

(a) Consider the motion of m_A from position 1 to position 2. The work-energy principle yields

$$U_{1\to 2} = T_2 - T_1.$$

Therefore,

$$m_A g L(1 - \cos\theta) = \tfrac{1}{2} m_A v_{A2}^2 - 0$$

from which

$$v_{A2} = \sqrt{2gL(1 - \cos\theta)}. \qquad \text{(a) ANS.}$$

(b) Now consider the impact of masses m_A and m_B. During impact, the linear momentum of the system in the x direction is conserved. Therefore,

$$\sum (L_2)_x = \sum (L_3)_x.$$

$$m_A v_{A2} + m_B v_{B2} = m_A v_{A3} + m_B v_{B3}$$

where v_{A2} is given by Eq. (a) and $v_{B2} = 0$. Therefore,

$$m_A v_{A2} = m_A v_{A3} + m_B v_{B3} \qquad \text{(b)}$$

The defining equation for the coefficient of restitution e written for the x direction yields

$$e = \frac{v_{B3} - v_{A3}}{v_{A2} - v_{B2}} = \frac{v_{B3} - v_{A3}}{v_{A2} - 0}$$

from which

$$ev_{A2} = v_{B3} - v_{A3} \tag{c}$$

Solve Eqs. (b) and (c) simultaneously to obtain

$$v_{A3} = \left[\frac{m_A - em_B}{m_A + m_B}\right] v_{A2} = \left[\frac{m_A - em_B}{m_A + m_B}\right]\sqrt{2gL(1 - \cos\theta)} \quad \text{ANS.}$$

$$v_{B3} = \left[\frac{m_A(1 + e)}{m_A + m_B}\right] v_{A2} = \left[\frac{m_A(1 + e)}{m_A + m_B}\right]\sqrt{2gL(1 - \cos\theta)} \quad \text{ANS.}$$

(c) Because the positive sense is directed to the right, then v_{A3} will be directed leftward provided

$$em_B > m_A \quad \text{ANS.}$$

(d) Consider block m_B after impact as it moves from position 3 to position 4, and apply the work-energy principle. Thus,

$$U_{3\to4} = T_4 - T_3,$$

$$m_B gh = \tfrac{1}{2}m_B v_{B4}^2 - \tfrac{1}{2}m_B v_{B3}^2.$$

Solve for v_{B4} to obtain

$$v_{B4} = \sqrt{v_{B3}^2 + 2gh} = \sqrt{2g\left[\left(\frac{m_A(1 + e)}{m_A + m_B}\right)^2 (1 - \cos\theta)L + h\right]}. \quad \text{ANS.}$$

■ Example 16.19

Refer to Example 16.18, and let $g = 9.81$ m/s^2, $L = 1$ m, $h = 4$ m, $\theta = 60°$, $m_A = 2$ kg, and $m_B = 3$ kg. (a) For perfectly plastic impact ($e = 0$), determine v_{A2}, v_{B3}, v_{A3} and v_{B4} (b) For perfectly elastic impact ($e = 1$), determine v_{A2}, v_{B3}, v_{A3} and v_{B4} (c) Compute the loss in kinetic energy due to impact for parts (a) and (b).

Solution

While this problem may be solved numerically using the ideas presented in Example 16.18, we will make use of the symbolic answers developed in that example to obtain the needed answers. Thus,

(a) $v_{A2} = \sqrt{2gL(1 - \cos\theta)} = \sqrt{2(9.81)(1)(1 - \cos 60°)} = 3.13$ m/s, ANS.

$$v_{B3} = \frac{m_A(1 + e)}{m_A + m_B}\sqrt{2gL(1 - \cos\theta)},$$

$$v_{B3} = \frac{2(1 + 0)}{2 + 3}\sqrt{2(9.81)(1)(1 - \cos 60°)} = 1.25 \text{ m/s}, \quad \text{ANS.}$$

$$v_{A3} = \frac{m_A - em_B}{m_A + m_B}\sqrt{2gL(1 - \cos\theta)},$$

$$v_{A3} = \frac{2 - 0}{2 + 3}\sqrt{2(9.81)(1)(1 - \cos 60^\circ)} = 1.25 \text{ m/s}, \qquad \text{ANS.}$$

$$v_{B4} = \sqrt{2g\left\{\left[\frac{m_A(1 + e)}{m_A + m_B}\right]^2 (1 - \cos\theta)L + h\right\}},$$

$$v_{B4} = \sqrt{2(9.81)\left\{\left[\frac{2(1 + 0)}{2 + 3}\right]^2 (1 - \cos 60^\circ)(1) + 4\right\}} = 8.95 \text{ m/s}.$$

ANS.

(b) v_{A2} is independent of e and its value is as for part (a). Thus,

$$v_{A2} = 3.13 \text{ m/s}, \qquad \text{ANS.}$$

$$v_{B3} = \frac{2(1 + 1)}{2 + 3}\sqrt{2(9.81)(1)(1 - \cos 60^\circ)} = 2.51 \text{ m/s}, \qquad \text{ANS.}$$

$$v_{A3} = \frac{2 - 1(3)}{2 + 3}\sqrt{2(9.81)(1)(1 - \cos 60^\circ)} = -0.626 \text{ m/s}. \qquad \text{ANS.}$$

The minus sign means that m_A moves leftward after impact.

$$v_{B4} = \sqrt{2(9.81)\left\{\left[\frac{2(1 + 1)}{2 + 3}\right]^2 (1 - \cos 60^\circ)(1) + 4\right\}} = 9.21 \text{ m/s}. \qquad \text{ANS.}$$

(c) Change in kinetic energy $= \Delta T = T_2 - T_3$. For part (a) with $e = 0$,

$$\Delta T = [\tfrac{1}{2}(2)(3.13)^2 + \tfrac{1}{2}(3)(0)^2] - [\tfrac{1}{2}(1.25)^2 + \tfrac{1}{2}(3)(1.25)^2]$$
$$= 6.66 \text{ ft}\cdot\text{lb}. \qquad \text{ANS.}$$

Mechanical energy loss during impact is transformed to sound and heat. For part (b) with $e = 1$,

$$\Delta T = [\tfrac{1}{2}(2)(3.13)^2 + \tfrac{1}{2}(3)(0)^2] - [\tfrac{1}{2}(2)(-0.626)^2 + \tfrac{1}{2}(3)(2.51)^2]$$
$$= 0.045 \text{ ft}\cdot\text{lb} \approx 0. \qquad \text{ANS.}$$

For perfectly elastic impact $e = 1$, therefore, we expect no loss of mechanical energy during impact.

■ Example 16.20

Refer to the solution of Example 16.18 and form the ratios v_{B3}/v_{A2} and v_{A3}/v_{A2}. Express these ratios in terms of the ratio m_B/m_A and plot v_{B3}/v_{A2} and v_{A3}/v_{A2} vs. m_B/m_A. Discuss these plots.

Solution

From Example 16.18,

$$v_{B3} = \left[\frac{m_A(1 + e)}{m_A + m_B}\right] v_{A2}.$$

Therefore,

$$\frac{v_{B3}}{v_{A2}} = \left[\frac{m_A(1+e)}{m_A + m_B}\right] = \frac{1+e}{1+\dfrac{m_B}{m_A}}. \qquad \text{ANS.}$$

Also, from Example 16.18,

$$v_{A3} = \left[\frac{m_A - em_B}{m_A + m_B}\right] v_{A2}$$

Thus,

$$\frac{v_{A3}}{v_{A2}} = \left[\frac{m_A - em_B}{m_A + m_B}\right] = \frac{1 - \left(\dfrac{m_B}{m_A}\right)e}{1+\dfrac{m_B}{m_A}} \qquad \text{ANS.}$$

Figure E16.20(a) shows a plot of v_{B3}/v_{A2} vs. m_B/m_A with the coefficient of restitution e used as a parameter. We observe that the $e = 1$ curve, which corresponds to perfectly elastic impact, has the largest ordinates for v_{B3}/v_{A2} which is to be expected because there are no energy losses during this perfectly elastic impact. For large mass ratios m_B/m_A, the ordinates to the curves become very small, or we say that these curves approach the horizontal axis asymptotically as m_B/m_A become infinite. A large mass ratio means that the block being struck m_B is much more massive than the striking mass m_A. It is not surprising that a small mass striking a much larger one is able to impart only a very small velocity to the larger mass compared to the velocity of the small mass just before impact. Such observations are important because we have confirmed our intuitive assessment by our analytical results, and such critical thinking is often employed by engineers to check their results.

In Figure E16.20(b), a plot is shown of v_{A3}/v_{A2} vs. m_B/m_A with e again used as a parameter. Only the limiting curves for $e = 0$ and $e = 1$ are shown. The curves for intermediate e values would lie between these limiting curves. For a given e value, each curve shows, the variation of the ratio of the speed of mass m_A after impact to the speed of this same mass before impact with respect to the ratio of the struck mass m_B to the striking mass m_A. We observe that, for mass ratios above a unit value (i.e., $m_B/m_A \geq 1$), the ordinates to a number of the curves will be negative which implies that the striking mass m_A moves leftward after impact.

This approach is a very powerful way to explore all possible solutions. When this method is coupled with a computer to calculate values and to plot curves, then, we have utilized modern technology to study a dynamic system. This technology enables us to change our inputs and observe the responses. Engineers of today must be prepared to

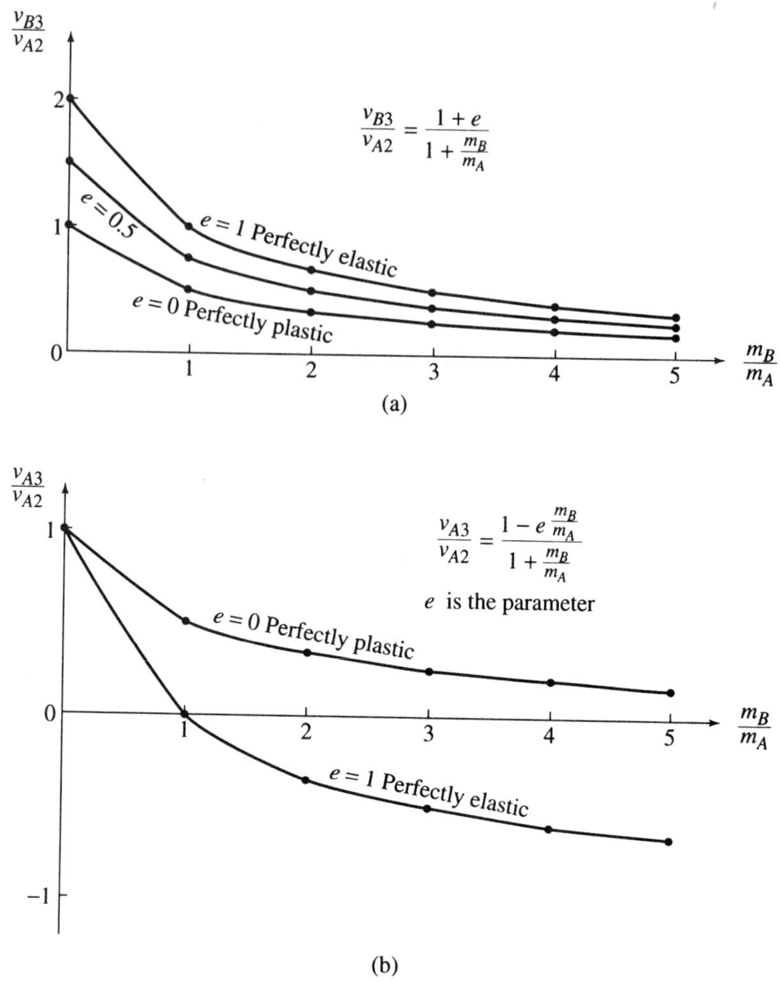

FIGURE E16.20.

employ such techniques to study complex dynamic systems from a very general standpoint.

■

Problems

16.89 Mass m_A is initially at rest in the position shown in Figure P16.89. This mass is released and falls a distance h before striking mass m_B. Neglect air resistance and the impulses of the weights and the spring force during the very small time of impact. The impact is perfectly plastic ($e = 0$). (a) Determine the speed of

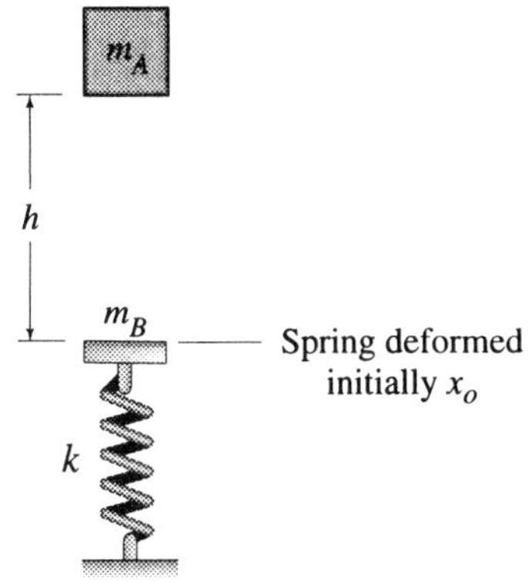

FIGURE P16.89.

m_A just before it impacts m_B, (b) Find the common speed of m_A and m_B just after impact. (c) If $m_A = 2$ slug, $m_B = 4$ slug, $k = 1000$ lb/ft, and $h = 2$ ft, determine the maximum deformation of the spring. Express your answers for parts (a) and (b) in terms of the given quantities m_A, m_B, g, and h.

16.90 In Figure P16.90 mass m_A is released from rest in the position shown and slides down the frictionless inclined plane for a distance s before it impacts m_B. Neglect air resistance and the impulses of the weights and the spring force during the very short time of impact. (a) Determine the speeds of m_A and m_B just after impact in terms of the speed that m_A had just before impact. (b) If $m_A = 5$ kg, $m_B = 2$ kg, $k = 2$ kN/m, $s = 2$ m, $e = 0$ and $\theta = 30°$, determine the maximum spring compression $x_M = x_0 + \Delta x$ where x_0 is its initial deformation.

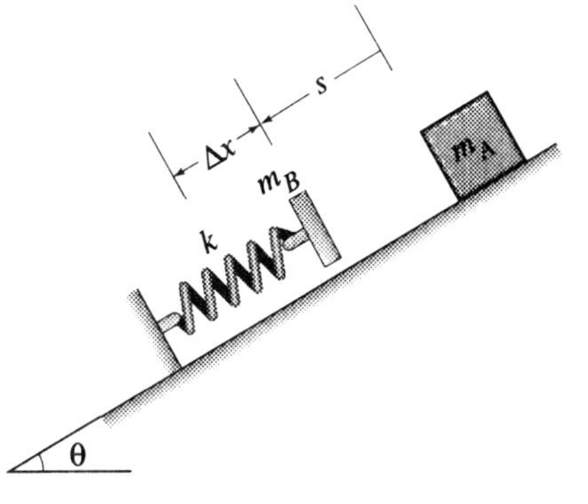

FIGURE P16.90.

16.91 Refer to Figure P16.91, and let $L = 1$ m, $m_A = 2$ kg, $m_B = 4$ kg, $e = 0.6$, $\theta = 30°$ and $\mu = 0.2$. The simple pendulum with bob B is released from rest in the position shown and rotates cw through an angle θ before striking block A, which is at rest. Determine the distance s moved by block A along the horizontal plane before it comes to rest. What is the velocity of bob B immediately after impact? Neglect air resistance, friction at pin O where the pendulum is suspended, and the mass of the light rod OB.

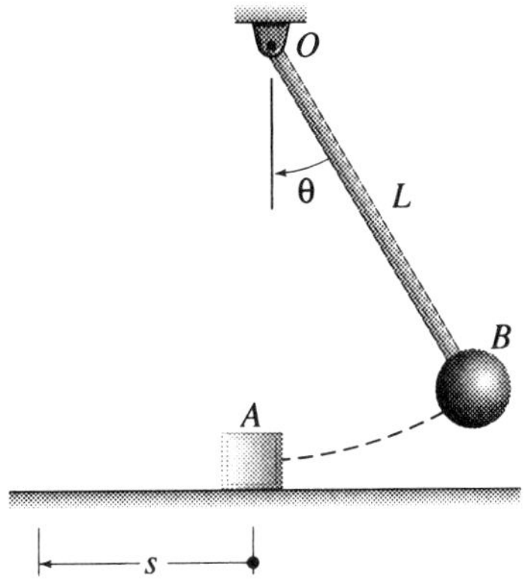

FIGURE P16.91.

16.92 The simple pendulum of Figure P16.91 with bob B is released from rest in the position shown and rotates cw through an angle θ before striking block A, which is at rest. Assign the following numerical values $L = 36$ in, $m_A = 0.1$

slug, $m_B = 0.2$ slug, $e = 0.5$, $\theta = 60°$ and $\mu = 0.2$. Determine the distance s moved by block A along the horizontal plane before it comes to rest. What is the velocity of bob B immediately after impact? Neglect air resistance, friction at pin O where the pendulum is suspended, and the mass of the light rod OB.

16.93 Refer to Figure P16.93 and let $m_A = m_B = m$. The simple pendulum has a bob A which has a negligible velocity directed to the right in the position shown. It rotates cw through a 90° angle before striking mass m_B which is attached to the spring for which the constant is k. Note that, before the bob A strikes m_B, the initial spring deformation x_O is given by mg/k. Assume that the impulses exerted by the weights and the spring force during the very short time of impact are negligible. The coefficient of restitution $e = 0.5$. Determine the maximum deformation of the spring $x_M = x_O + \Delta x$. Assume that, after impact, bob A is stopped and prevented from falling again toward m_B to assure that another impact is avoided. Regard L, m, g, k, and $e = 0.5$ as known quantities. Neglect air resistance, friction at pin O where the pendulum is suspended, and the mass of the light rod OA.

16.94 Refer to Figure P16.94. Regard m_A, m_B, v_{A1}, μ, s, k, e, and g as known quantities, and express your answers in terms of one or more of them. Mass m_A is initially moving to the right with a speed of v_{A1}, and it moves a distance s along a rough plane, for which the coefficient of friction is μ, before striking m_B which is attached to the spring as shown. The spring is initially uncompressed, and m_B does not contact the plane. Determine the velocity of m_A just before it strikes m_B. What condition must be satisfied to assure that m_A does not stop before striking m_B? Neglect frictional and spring impulses during impact. Determine the velocities of m_A and m_B immediately after impact. What condition must be met to assure that mass m_A moves left after impact? Assuming that m_A moves left after impact, determine the maximum spring compression x_M.

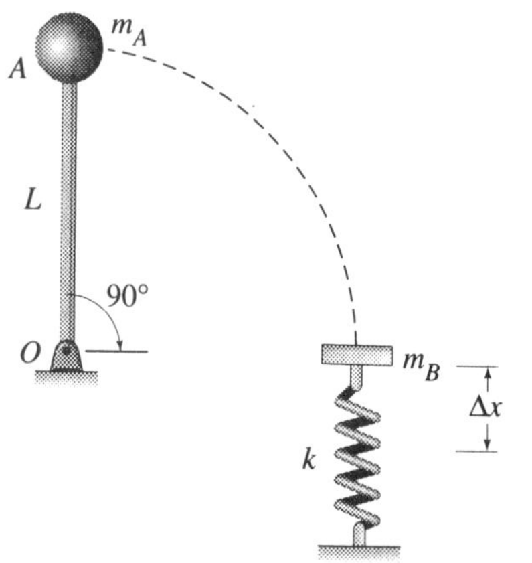

FIGURE P16.93.

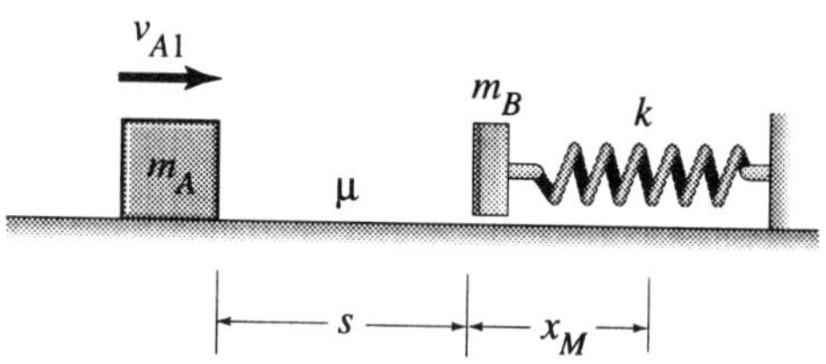

FIGURE P16.94.

16.95 A particle of mass m_A starts from rest in the position indicated and moves down a frictionless path as shown in Figure P16.95 before striking stationary mass m_B. Let $h = 2$ m, $m_A = 2$ kg,

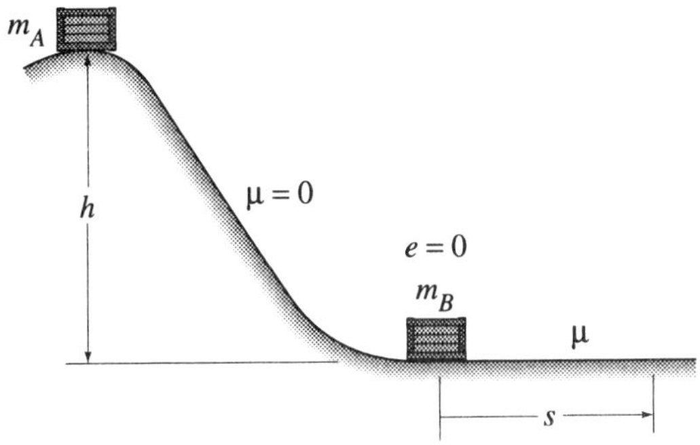

FIGURE P16.95.

$m_B = 3$ kg, $\mu = 0.2$ and $e = 0$ and determine the velocity of m_A just before it impacts m_B. Determine the distance s moved by m_B along the horizontal rough plane before it comes to rest.

16.96 Refer to Figure P16.96 and neglect friction and air resistance. Let $h = 5$ ft, $m_A = 0.2$ slug, $m_B = 0.3$ slug, $k = 100$ lb/ft and $e = 0$. Mass m_A has a negligible velocity directed leftward when it leaves the position shown and travels over the frictionless path. Determine its velocity just before it strikes m_B. During impact, neglect the spring impulse during the very short time of impact. Find the velocities of m_A and m_B just after impact and the maximum deformation of the spring x_M. Prior to impact, the spring is undeformed.

16.97 Mass $m_A = 4$ slug is released from rest in the position shown and slides down the frictionless inclined plane depicted in Figure P16.97 and impacts $m_B = 2$ slug. The coefficient of restitution e is given. Ignore the impulses of the weights and the spring force during the small time of impact. For $L = 10$ ft and $\theta = 30°$ express the loss of energy ΔT during the impact as a function of e. Find this energy loss for $e = 0$, $e = 0.5$, and $e = 1$.

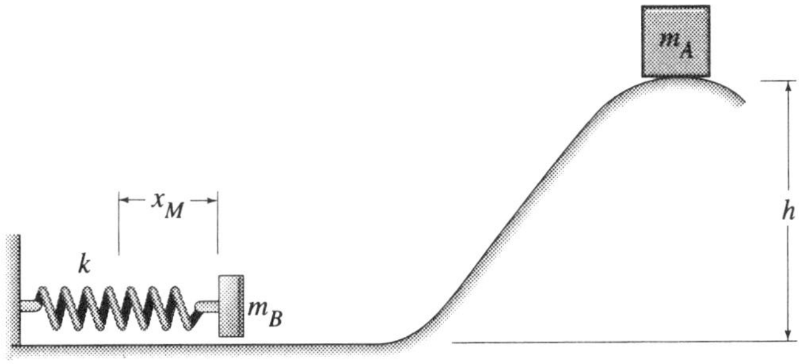

FIGURE P16.96.

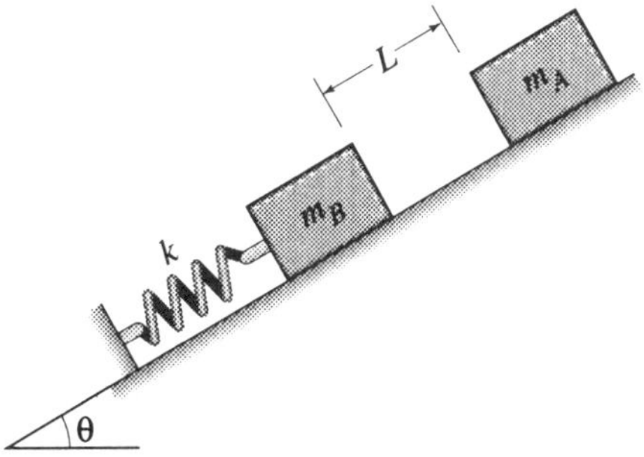

FIGURE P16.97.

16.98 As indicated in Figure P16.98, the horizontal plane between masses m_A and m_B is smooth (i.e., $\mu = 0$) before impact and is rough (i.e. $\mu \neq 0$) after impact.

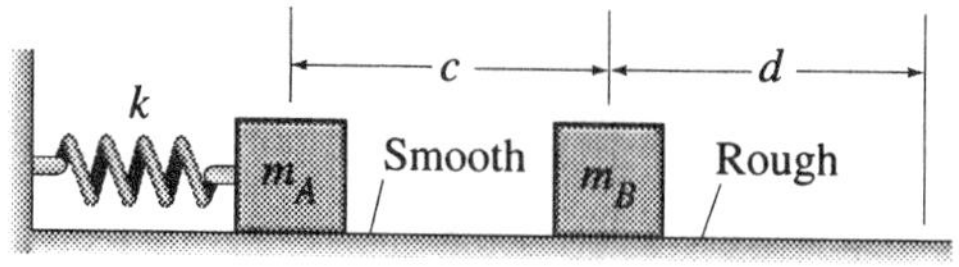

FIGURE P16.98.

The spring constant $k = 6000$ N/m, $m_A = 10$ kg, $m_B = 5$ kg, $c = 0.8$ m, and the spring is initially compressed 0.4 m. Mass m_A is released from rest in the position shown and moves to impact mass m_B, which is at rest. Determine the coefficient of restitution e such that 70% of the energy of mass m_A is lost during the impact. Then, find the distance d through which mass m_B moves before coming to rest if $\mu = 0.6$.

16.99 The pendulum OA of Figure P16.99 has a small nonzero ccw motion in the position shown. It rotates through an angle of 90°, and mass m_A impacts the small platform m_B. This platform is supported by four vertical rods (only two are shown) which are very securely fastened to an overhead massive body. You may regard the platform m_B as rigid and the rods as springs. The spring constant for each rod is k. Neglect the mass of the light rod OA which connects the pendulum bob to the fixed pin O. Determine the velocity of m_A just before it strikes the platform. Find the velocities of m_A and m_B just after impact. Regard k, b, m_A, m_B, and the coefficient of restitution e as given quantities and express your answers in terms of one or more of them. Neglect air resistance and friction at pin O.

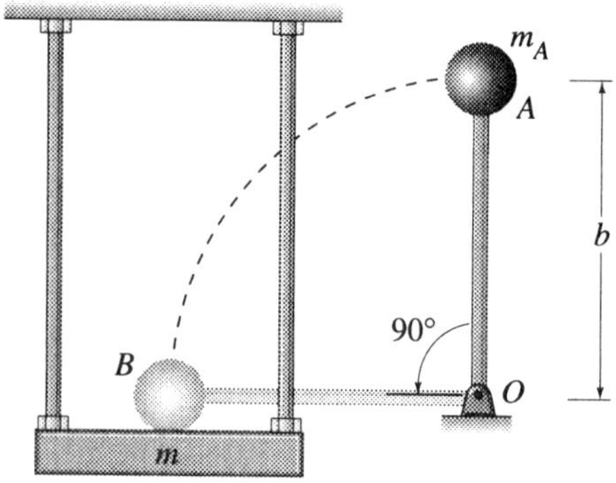

FIGURE P16.99.

16.100 Refer to Figure P16.100, and let $m_A = 0.2$ slug, $m_B = 0.1$ slug, $k = 100$ lb/ft, $\mu = 0.1$, $e = 0.75$, $\theta = 30°$, $L = 30$ ft, and $h = 40$ ft. Initially, the particle m_A is in the position shown and is held there with the spring compressed 6 in.

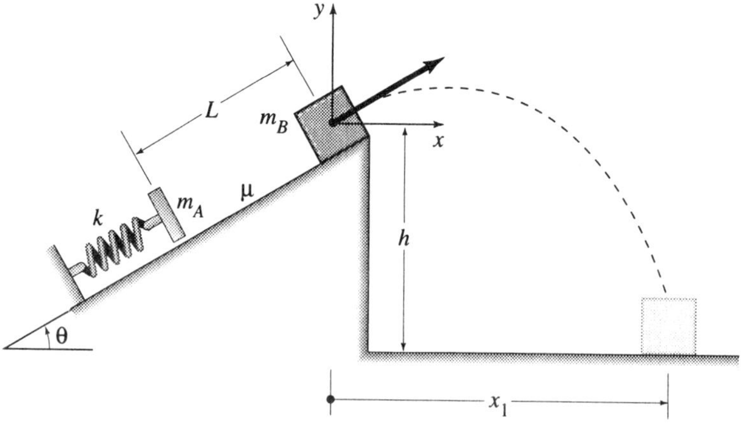

FIGURE P16.100.

It is released from rest moves up the inclined plane, for which the coefficient of friction is μ, and impacts m_B which is at rest. Projectile flight of m_B begins following impact where the coefficient of restitution is e. Finally m_B impacts the horizontal plane which lies at a distance h below. Determine the coordinate x_1 of the impact position and the time of the projectile flight.

16.11 Steady Fluid Flow

The impulse-momentum methods are very useful for studying fluid flow. Steady flow of a fluid is defined as a flow which does not change with time at any given point in the fluid. At any point, the velocity remains constant in magnitude and fixed in direction, but velocity may vary from point to point in a steadily flowing fluid. Thus, consider Figure 16.17 which represents a segment of a pipe of variable cross-sectional area. At a given position where the cross-sectional area is A, we will deal with the average velocity v over that area. Note that a differential volume of fluid flowing across the area in a time Δt may be written as $Av\Delta t$. Multiplication of this volume by the mass density of the fluid ρ will give the differential mass flow $\Delta m = \rho A v \Delta t$. The differential mass of fluid entering a hydraulic device (e.g., vane, pipe) at B equals the differential mass of fluid leaving the same device at C. This conservation statement may be expressed symbolically by

$$\rho_B A_B v_B \Delta t = \rho_C A_C v_C \Delta t$$

Division by Δt yields,

$$\rho_B A_B v_B = \rho_C A_C v_C \tag{16.36}$$

If the fluid mass density is constant then $\rho_B = \rho_C$. This implies that the fluid is incompressible. Therefore,

$$Q = A_B v_B = A_C v_C \tag{16.37}$$

where Q is the volumetric flow rate.

Consider fluid flowing at a steady rate through a hydraulic device as shown in Figure 16.18. Fluid enters at a velocity $\mathbf{v}_B$ at section B and

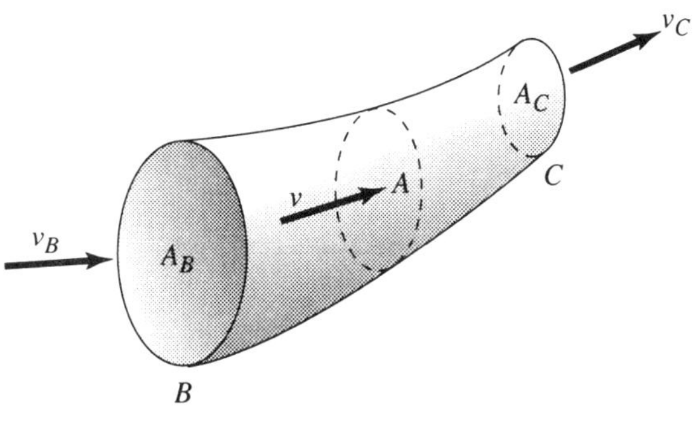

FIGURE 16.17.

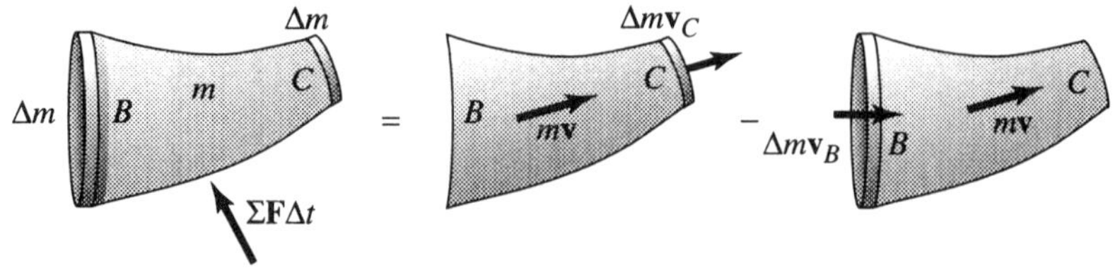

FIGURE 16.18.

leaves the device at a velocity $\mathbf{v}_C$ at section C. The mass of fluid $(m + \Delta m)$ is termed the *control mass* of fluid. It consists of the mass m inside the device at any time t plus the mass increment Δm associated with the time increment Δt. The mass Δm of fluid entering the device is equal to the mass Δm of fluid leaving the device.

Initially the control mass equals the mass m inside the device plus the mass increment Δm which is just about to enter the device. Finally the control mass equals the mass m inside the device plus the mass increment Δm which has just left the device. Either initially or finally, we are referring to a total mass of $(m + \Delta m)$.

Refer to the impulse-momenta diagrams in Figure 16.18, and consider the fluid control mass $(m + \Delta m)$ in a time Δt. External impulses $\sum \mathbf{F}\Delta t$ are exerted on the fluid by the hydraulic device. These impulses change the momentum of the fluid. Fluid contained in the device has a momentum $m\mathbf{v}$ which does not change over the time Δt because the flow is steady. The differential mass Δm has a momentum $\Delta m\mathbf{v}_C$ after it leaves the device and a momentum $\Delta m\mathbf{v}_B$ before it enters the device. The velocities $\mathbf{v}_C$ and $\mathbf{v}_B$ are average values for their respective cross sections of flow A_C and A_B. The linear impulse-momentum equation for this steady fluid flow becomes

$$\sum \mathbf{F}\Delta t = \mathbf{L}_C - \mathbf{L}_B$$

from which

$$\sum \mathbf{F}\Delta t = (m\mathbf{v} + \Delta m\mathbf{v}_C) - (m\mathbf{v} - \Delta m\mathbf{v}_B).$$

If we divide by Δt and let Δt approach zero,

$$\sum \mathbf{F} = \frac{dm}{dt}(\mathbf{v}_C - \mathbf{v}_B). \tag{16.38}$$

This equation states that the vector sum of all the external forces acting on the fluid in the device is equal to the product of the time rate of change of the mass $\frac{dm}{dt}$ and the vector difference of the exit and entrance velocities. For two-dimensional fluid flow in the x-y plane, Eq. (16.38) may be expressed as two scalar equations in the form,

$$\sum F_x = \frac{dm}{dt}(v_{Cx} - v_{Bx}),$$
$$\sum F_y = \frac{dm}{dt}(v_{Cy} - v_{By}). \qquad (16.39)$$

The quantity $\frac{dm}{dt}$ represents the rate of fluid mass flow. As stated earlier, $\Delta m = \rho A v \Delta t$. If we divide by Δt and let Δt approach zero,

$$\frac{dm}{dt} = \rho A v = \rho Q. \qquad (16.40)$$

Another relationship that is useful in the solution of some problems is obtained from the principle of angular impulse and momentum given by Eq. (16.30). Thus, consider the fluid flow, shown in Figure 16.19, occurring at a time interval Δt. Take moments about point O to obtain,

$$\sum M_O \Delta t = \Delta m v_C d_C - \Delta m v_B d_B.$$

Dividing by Δt and letting Δt approach zero yields

$$\sum M_O = \frac{dm}{dt}(v_C d_C - v_B d_B). \qquad (16.41)$$

The following examples illustrate the application of the above concepts in the solution of steady fluid-flow problems.

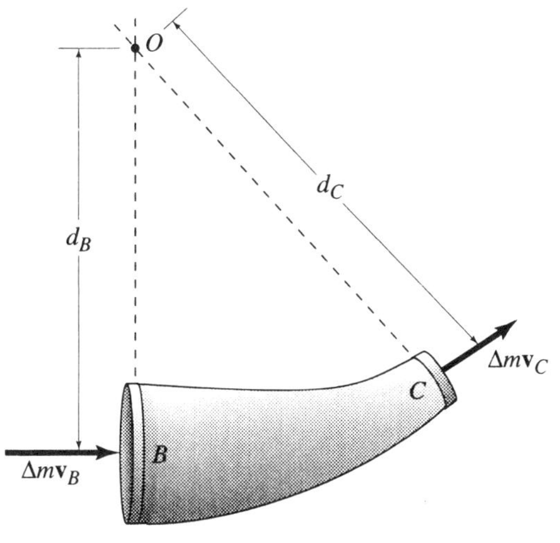

FIGURE 16.19.

■ Example 16.21

A water jet has a diameter of 1.5 in. and moves with a velocity $v_W =$ 110 ft/s before striking a vane which is moving in the same direction with a velocity $v_V = 50$ ft/s as shown in Figure E16.21(a). The deflection angle of the vane is 140° as shown. Find (a) the force components exerted on the vane by the water and (b) the absolute velocity of the water as it leaves the moving vane. Ignore frictional losses as the water moves along the vane. Water weighs 62.4 lb/ft³.

Solution

(a) The impulse-momenta diagrams for a column of water during a time interval Δt are shown in Figure E16.21(b) along with an x-y coordinate system that moves with the vane. Thus,

$$I_x = (L_2 - L_1)_x,$$

$$F_x \Delta t = -\Delta m(v_w - v_v)\cos 40° - \Delta m(v_w - v_v),$$

$$I_y = (L_2 - L_1)_y,$$

$$F_y \Delta t = \Delta m(v_w - v_v)\sin 40° - 0.$$

Divide by Δt, and solve for F_x and F_y. Thus,

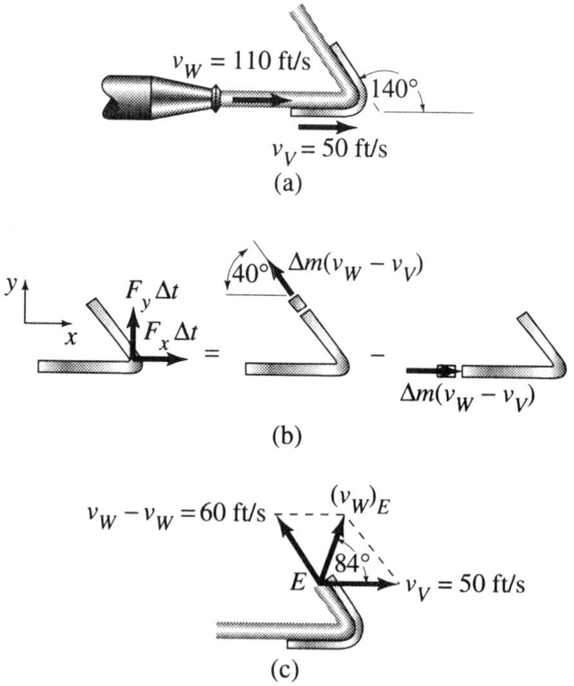

FIGURE E16.21.

$$F_x = \frac{-\Delta m}{\Delta t}(v_w - v_v)(1 + \cos 40°)$$

$$F_y = \frac{\Delta m}{\Delta t}(v_w - v_v)\sin 40°$$

Let $\Delta t \to 0$, and replace $\frac{\Delta m}{\Delta t}$ by $\frac{dm}{dt}$. The mass density of water is given by

$$\rho = \frac{62.4}{32.2} = 1.94\frac{\text{slug}}{\text{ft}^3}.$$

By Eq. (16.40),

$$\frac{dm}{dt} = \rho A(v_w - v_v) = 1.94\left[\frac{\pi\left(\frac{1.5}{12}\right)^2}{4}\right](110 - 50) = 1.428 \text{ slug/s}.$$

Substitution of these values and the given velocities enables us to determine the forces exerted by the vane on the *water*:

$$F_x = -1.428(100 - 50)(1 + \cos 40°) = -151.3 \text{ lb},$$

$$F_y = 1.428(100 - 50)\sin 40° = 55.1 \text{ lb}.$$

Therefore, the force components on the *vane* are

$$F_x = 151.3 \text{ lb} \rightarrow, \qquad \text{ANS.}$$

$$F_y = 55.1 \text{ lb} \downarrow. \qquad \text{ANS.}$$

(b) Consider the water as it exits the vane as shown in Figure E16.21(c) and write

$$v_{wx} = -60\cos 40° + 50 = 4.04 \text{ ft/s} \qquad \text{ANS.}$$

$$v_{wy} = 60\sin 40° + 0 = 38.6 \text{ ft/s} \qquad \text{ANS.}$$

Combine these components to obtain $(v_W)_E$, the absolute velocity of the water as it exits the vane at E. Thus,

$$(v_w)_E = 38.8 \text{ ft/s} \ \angle 84.0° \ . \qquad \text{ANS.}$$

Example 16.22

Water flows at the steady rate of 1 cfs over the fixed vane which is supported by a pin at C and a horizontal link AB as shown in Figure E16.22(a). The velocity of water at D is $v_D = 75$ ft/s, and, if we neglect losses due to friction, then, its velocity at E is $v_E = 75$ ft/s. Neglect the weight of the vane and water on the vane, and determine the force in

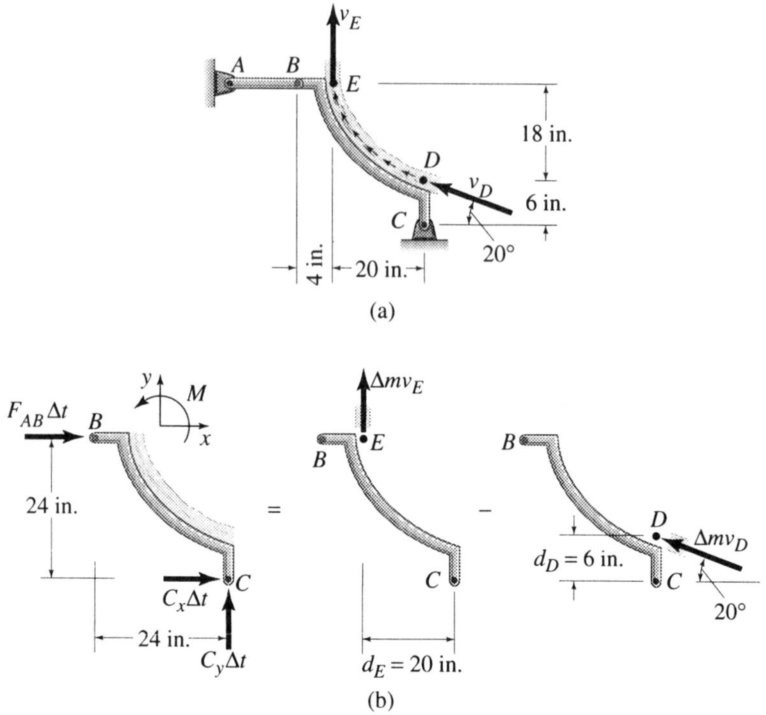

FIGURE E16.22.

link AB and the pin reaction components at C. The unit weight of water is 62.4 lb/ft^3.

Solution

Figure E16.22(b) shows the impulse-momenta diagrams for a column of water during a time interval Δt. An x-y coordinate system is also shown.

Let us choose pin C as a moment center to eliminate C_x and C_y from the following angular impulse-momentum equation. Thus,

$$G_{Cz} = H_{Ez} - H_{Dz},$$

$$\sum M_C \Delta t = -(\Delta m v_E) d_E - (\Delta m v_D \cos 20°) d_D,$$

$$-F_{AB} \Delta t (24) = -\Delta m v_E (20) - \Delta m v_D \cos 20° \; (6).$$

Divide by 24 Δt, substitute $v_E = v_D = 75$, and let $\Delta t \to 0$ to obtain

$$F_{AB} = \frac{dm}{dt}(20 + 6\cos 20°)\frac{75}{24} = 80.12\frac{dm}{dt},$$

$$\frac{dm}{dt} = 1\,\frac{\text{ft}^3}{\text{s}}\left(\frac{62.4 \text{ lb}}{\text{ft}^3}\right)\left(\frac{1 \text{ slug}}{32.2 \text{ lb}}\right) = 1.938\,\frac{\text{slug}}{\text{s}},$$

$$F_{AB} = 80.12\,(1.938) = 124.5 \text{ lb}.$$ ANS.

From the positive sign, we conclude that the force on the water F_{AB} acts to the right and, therefore, by action and reaction, it acts to the left on the vane and link AB is in compression.

Summing components in the x direction gives the following linear impulse-momentum equation. Thus,

$$I_x = (L_E - L_D)_x,$$

$$F_{AB}\Delta t + C_x\Delta t = 0 - (-\Delta m v_D \cos 20°).$$

Divide by Δt, let $\Delta t \to 0$, substitute for $\frac{dm}{dt}$, for $v_E = v_D = 75$ ft/s, and for F_{AB} to give

$$C_x = -F_{AB} + \left(\frac{dm}{dt}\right) v_D \cos 20°$$

$$C_x = -124.5 + 1.938(75)\cos 20° = 12.08 \text{ lb} \rightarrow \text{on water,}$$

$$= 12.08 \text{ lb} \leftarrow \text{on vane.}$$ ANS.

Summing components in the y direction gives the following linear impulse-momentum equation. Thus,

$$I_y = (L_E - L_D)_y,$$

$$C_y\Delta t = \Delta m v_E - \Delta m v_D \sin 20°.$$

Divide by Δt, let $\Delta t \to 0$, substitute for $\frac{dm}{dt}$ and for $v_E = v_D = 75$ ft/s to give

$$C_y = \frac{dm}{dt}(v_E - v_D \sin 20°) = v_D\left(\frac{dm}{dt}\right)(1 - \sin 20°)$$

$$C_y = 1.938\,(75)\,(1 - \sin 20°) = 95.6 \text{ lb} \uparrow \text{ on water.}$$

$$= 95.6 \text{ lb} \downarrow \text{ on vane.}$$ ANS.

■

Problems

16.101 Water flows from a nozzle and strikes a fixed plate as shown in Figure P16.101. At section A-A, the cross sectional area is 3.65 in^2 and the flow rate is 1.20 cfs. Determine the components of the force exerted by the water on the plate. Assume that the volumetric flow splits evenly in the two directions shown. Water has a density of 62.4 lb/ft^3.

16.102 Solve Problem 16.101 if the fluid is oil with a specific gravity of 0.9, and the plate is being moved to the left with a velocity of 15 ft/s.

16.103 Solve Problem 16.101 if the fluid is oil

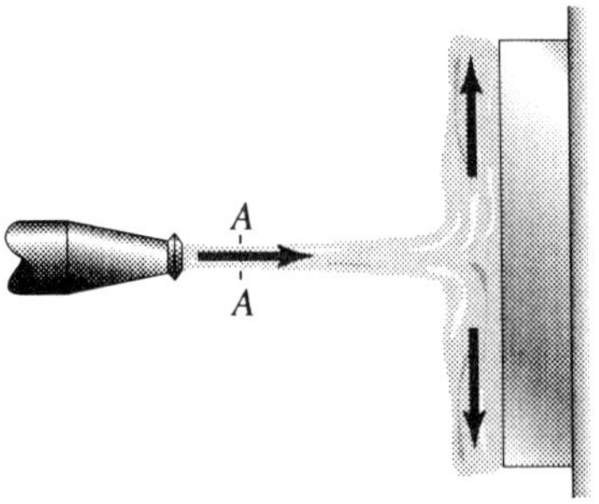

FIGURE P16.101.

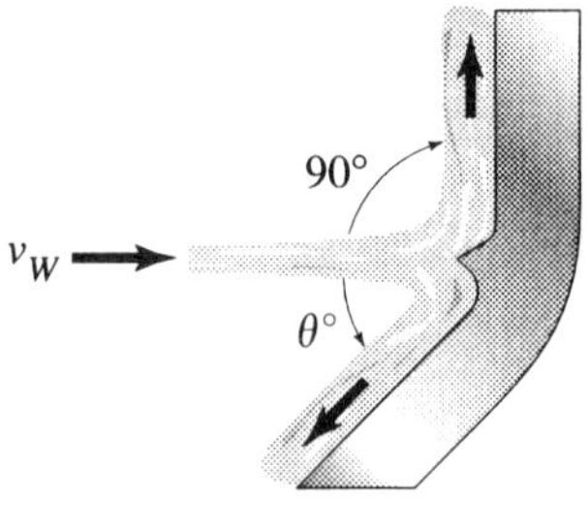

FIGURE P16.106.

with a specific gravity of 0.85, and the plate is being moved to the right with a velocity of 10 ft/s.

16.104 The flow of water from a nozzle is at the rate of 2.5 cfs with a velocity of 60 ft/s. The nozzle is inclined such that the water flows down to the right at 20° to the horizontal. The water jet strikes a fixed vane and is deflected through a 70° ccw angle. It emerges upward to the right at 50° to the horizontal. Determine the horizontal and vertical force components exerted on the fixed vane by the water. Water has a density of 62.4 lb/ft^3.

16.105 A fixed vane deflects a jet of water as shown in Figure P16.105. The flow rate Q = 2.20 cfs with a velocity of 50 ft/s. Determine the horizontal and vertical force components exerted on the fixed vane by the water. Water has a density of 62.4/ft^3.

16.106 Water flows onto a moving vane as shown in Figure P16.106 at a flow rate of $v_W = 0.10$ m^3/s and a velocity of 30 m/s. Half of the water jet is deflected upward and the other half is deflected downward such that $\theta = 45°$. If the vane is moving to the right at $v_V = 10$ m/s, determine the horizontal and vertical force components exerted on the vane. Water has a mass density of 1000 kg/m^3.

16.107 Water flows onto a fixed vane as shown in Figure P16.106 at the rate of 4 cfs and a velocity of 80 ft/s. One third of the water jet is deflected upward and the remainder is deflected downward such that $\theta = 30°$. Determine the horizontal and vertical force components exerted on the vane. Water has a density of 62.4 lb/ft^3.

16.108 Water flows at the steady rate of 0.03 m^3/s over the fixed vane of Figure P16.108 which is supported by a hori-

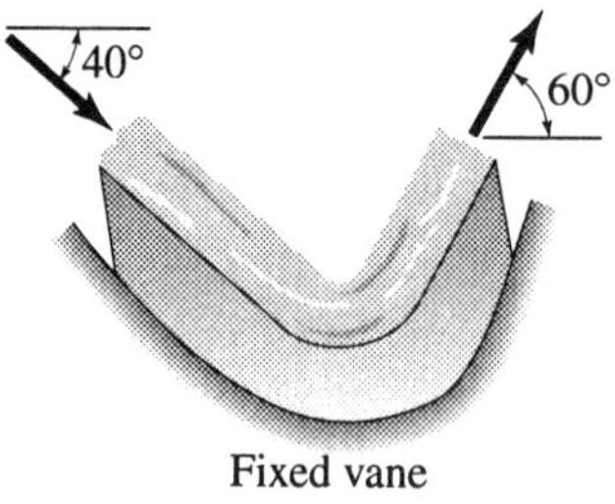

FIGURE P16.105.

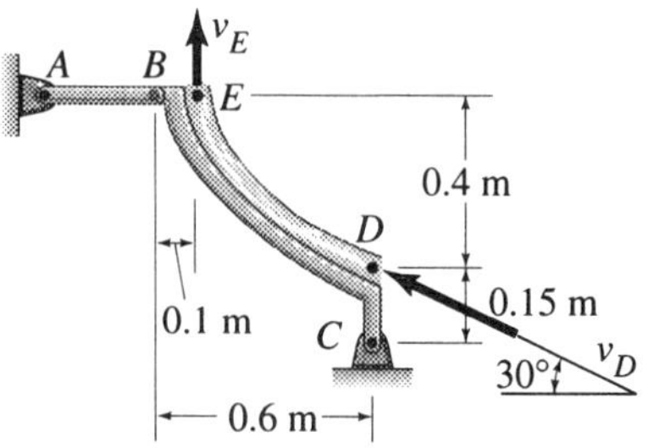

FIGURE P16.108.

zontal link AB and the pin at C. The entering and exit speeds at D and E, respectively, are each equal to 20 m/s. Neglect the weight of the vane and the water on the vane, and determine the force in link AB and the pin reaction components at C. Water has a mass density of 1000 kg/m^3.

16.109 A fixed flat plate contains a *splitter* which divides the incoming flow in half as shown in Figure P16.109. The jet of water has a diameter of 1.75 in and $v = 80$ ft/s. Water has a density of 62.4 lb/ft^3. Determine the magnitude and direction of the force exerted by the water on this plate.

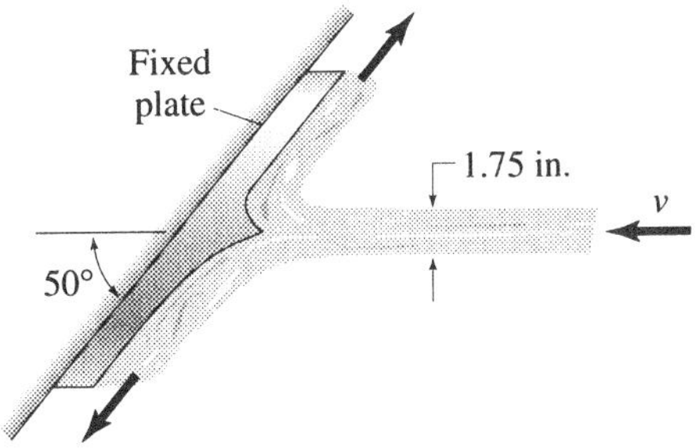

FIGURE P16.109.

16.110 In Figure 16.110 the blade is moving at $v_B = 40$ ft/s horizontally to the left. If the water jet has a velocity $v_W = 90$ ft/s and the water jet diameter is 1.50 in., determine the horizontal and vertical force components exerted by the blade on the water. Water has a density of 62.4 lb/ft^3.

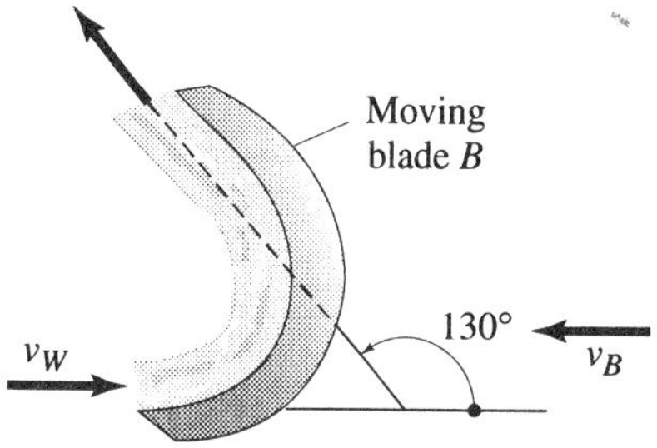

FIGURE P16.110.

16.111 A downward directed jet of water impinges upon the horizontal fixed plate shown in Figure P16.111. If 0.6 of the flow is directed leftward and 0.4 of the flow is directed rightward, determine the force exerted on the water by the plate. State the magnitude and direction of this force. The downward flow is 2.50 cfs and the velocity $v = 75$ ft/s.

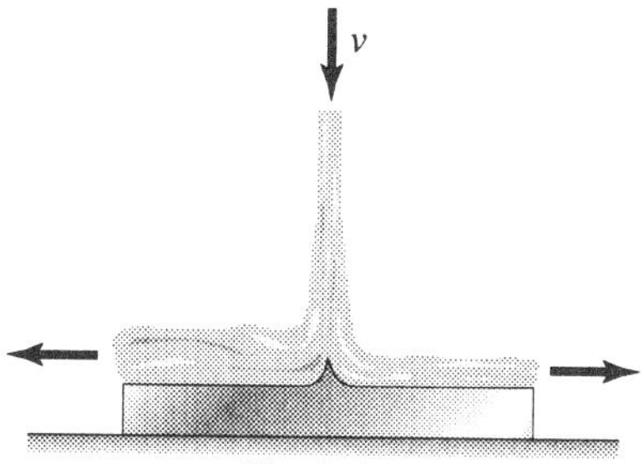

FIGURE P16.111.

16.12 Systems with Variable Mass

Consider the system shown in Figure 16.20(a) which, at a given instant, moves to the right with a velocity v measured from a fixed reference frame. The system is gaining mass by taking in an increment of mass Δm during a time interval Δt from a source that itself moves to the right with a velocity v_s. During this time interval, the velocity of the mass system $(m + \Delta m)$ changes to $v + \Delta v$.

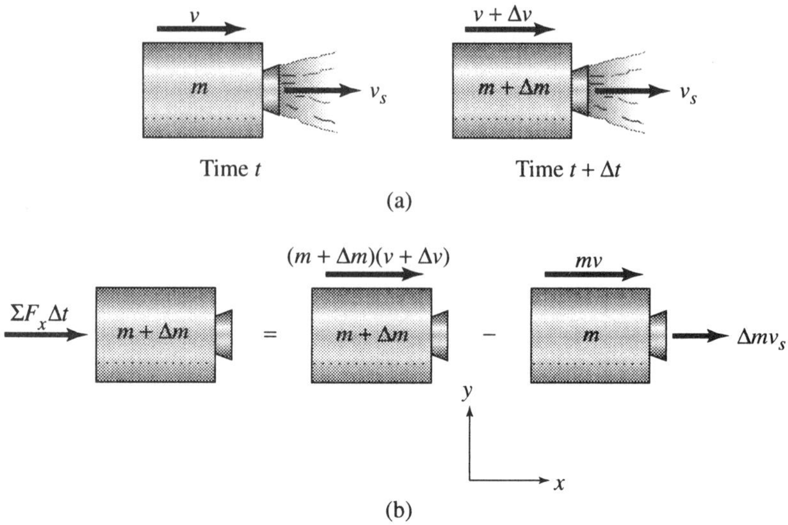

FIGURE 16.20.

The impulse-momenta diagrams for the system during the time interval Δt are shown in Figure 16.20(b). Also, a convenient fixed x-y coordinate system is shown. The quantity $\sum F_x \Delta t$ represents the external impulse delivered, during a time increment Δt, to the system consisting of the mass m and the increment of mass Δm. We observe that any impulses due to forces acting between m and Δm will cancel out. After Δm has become a part of the total mass, the *final* momentum becomes $(m + \Delta m)(v + \Delta v)$ where Δv is the velocity increment developed during a time interval Δt. Initially, the mass m moves with velocity v. The total *initial* momentum consists of two terms mv and $(\Delta m)v_s$ where v_s represents the velocity of the source of the mass increase Δm. All velocities are absolute and, for convenience, we consider motion along the x direction. A more general vector equation will be presented subsequently. The linear impulse-momentum equation in the x direction for this system becomes

$$\sum F_x \Delta t = [(m + \Delta m)(v + \Delta v)] - (mv + \Delta m v_s).$$

Expansion of the final momentum term on the right gives

$$\sum F_x \Delta t = mv + (\Delta m)v + m\Delta v + \Delta m \Delta v - mv - (\Delta m)v_s.$$

We observe that the two mv terms cancel and that $\Delta m \Delta v$ is a second order term compared to the others and may be dropped. Therefore,

$$\sum F_x \Delta t = (\Delta m)v + m(\Delta v) - (\Delta m)v_s.$$

Dividing by Δt and letting Δt approach zero in the limit leads to

$$\sum F_x = \frac{dm}{dt}v + m\frac{dv}{dt} - \frac{dm}{dt}v_s \tag{16.42}$$

where $\frac{dm}{dt}$ represents the rate at which mass is absorbed and $\frac{dv}{dt}$ is the acceleration of the system. Equation (16.42) was developed for a system that gains mass. It may also be used for a system that loses mass in which case $\frac{dm}{dt}$ becomes negative. Note that the velocity v_s should be determined *just before* Δm enters a system that gains mass and *just before* Δm leaves a system that loses mass.

Equation (16.42) may be written in a more general vector form as follows:

$$\sum \mathbf{F} = \frac{dm}{dt}\mathbf{v} + m\frac{d\mathbf{v}}{dt} - \frac{dm}{dt}\mathbf{v}_s. \tag{16.43}$$

In this case, $\sum \mathbf{F}$ represents the vector sum of the external forces acting on the system. The velocities $\mathbf{v}$ and $\mathbf{v_s}$ and the acceleration $\frac{d\mathbf{v}}{dt}$ are, in general, vectors in three dimensions. Recall that the scalar derivative $\frac{dm}{dt}$ is positive for a system, which gains mass, and negative for a system which loses mass.

It is sometimes more convenient to restate Eq. (16.42) in terms of the relative velocity $u = v_s - v$ of the added mass with respect to the system for which the equation is written. Thus, Eq. (16.42) becomes

$$\sum F_x = m\frac{dv}{dt} - \frac{dm}{dt}u. \tag{16.44}$$

If we let $\frac{dv}{dt} = a$ and transpose the second term on the right to the left side of this equation,

$$\sum F_x + \frac{dm}{dt}u = ma. \tag{16.45}$$

The term $\frac{dm}{dt}u$ is referred to as the *thrust* due to the mass being absorbed or expelled and is of considerable interest when we analyze rocket motion or other jet propulsion systems.

Similarly, Eq. (16.43) may be rewritten as

$$\sum \mathbf{F} = m\frac{d\mathbf{v}}{dt} - \frac{dm}{dt}\mathbf{u}. \tag{16.46}$$

Also, Eq. (16.45) may be written in a more general vector form as

$$\sum \mathbf{F} + \frac{dm}{dt}\mathbf{u} = m\mathbf{a}. \tag{16.47}$$

Two methods may be used to solve problems. One method is to use Eq. (16.42) or Eq. (16.43) directly and the other method is to draw the impulse-momentum diagrams and write the equation(s) by reference to these diagrams. The following examples will illustrate both methods.

■ Example 16.23

A perfectly flexible chain of length L rests on a floor. It is to be lifted at a constant velocity v by applying an upward force F to the top of the chain as shown in Figure E16.23(a). If the chain weighs w lb/ft, determine F and the upward reaction R exerted on the chain by the floor. Use Eq. (16.42), and express the answers in terms of w, L, v, and t.

Solution

To determine F, we consider that portion of the chain above the floor as the system and construct its free-body diagram as shown in Figure E16.23(b). A convenient coordinate system is also shown. Thus, by Eq. (16.42),

$$\sum F_y = v\frac{dm}{dt} + m\frac{dv}{dt} - v_s\frac{dm}{dt}.$$

To determine $\frac{dm}{dt}$, we express the mass of chain above the floor as

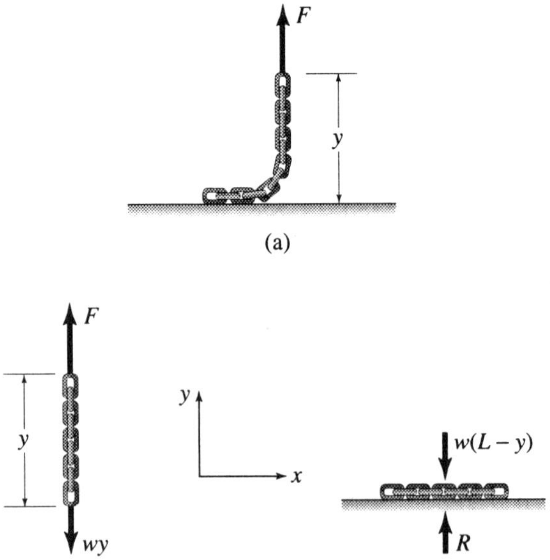

FIGURE E16.23. (b) System gaining mass. (c) System losing mass.

$m = \frac{w}{g}y$. Differentiating with respect to time we obtain $\frac{dm}{dt} = \frac{w}{g}\frac{dy}{dt}$, but $\frac{dy}{dt} = v$ because every point on this moving chain has the same constant velocity v. Therefore, $\frac{dm}{dt} = \frac{w}{g}v$. We observe that $\frac{dm}{dt}$ is positive because that portion of the chain above the floor is gaining mass. In the second term, $m = \frac{w}{g}y$, but $\frac{dv}{dt} = 0$ because v is a constant. The third term contains v_s, which is zero because Δm must be considered just before it enters the system gaining mass and this mass increment is at rest on the floor. Thus, applying Eq. (16.42),

$$F - wy = v\left(\frac{w}{g}v\right) + \frac{w}{g}y(0) - 0\left(\frac{w}{g}v\right)$$

from which

$$F = wy + \frac{w}{g}v^2.$$

Because v is constant and $y = 0$ at $t = 0$, we conclude that $y = vt$. Therefore,

$$F = wvt + \frac{w}{g}v^2 = wv\left(t + \frac{v}{g}\right). \qquad \text{ANS.}$$

We observe that F is a linear function of time and that, when $t = 0$, $F = \frac{wv^2}{g}$.

To determine R, we consider that part of the chain on the floor as the system and construct its free-body diagram as shown in Figure E16.23(c). Again, by Eq. (16.42) and using the same coordinate system as above,

$$\sum F_y = v\frac{dm}{dt} + m\frac{dv}{dt} - v_s\frac{dm}{dt}.$$

To determine $\frac{dm}{dt}$, we express the mass of the chain on the floor as $m = \frac{w}{g}(L - y)$. Differentiating with respect to time, $\frac{dm}{dt} = -\frac{w}{g}\frac{dy}{dt}$, but $\frac{dy}{dt} = v$ because every point on this moving chain has the same constant velocity v. Thus, $\frac{dm}{dt} = -\frac{w}{g}v$. We observe that $\frac{dm}{dt}$ is negative which is consistent with the fact that the chain on the floor is losing mass. However, $v = 0$ because the chain on the floor is at rest. In this case, the mass increment Δm must be considered just before it leaves the system which is losing mass and $v_s = 0$ because this mass increment is at rest on the floor. Thus, Eq. (16.42) yields,

$$R - w(L - y) = 0\left(-\frac{wg}{g}v\right) + \frac{w}{g}(L - y)(0) - 0\left(-\frac{w}{g}\right)v$$

from which

$$R = w(L - y).$$

Again, $y = vt$ and R becomes

$$R = w(L - vt). \qquad \text{ANS.}$$

We note that the reaction R is a linear function of time, and at $t = 0$, $R = wL$. This quantity represents the weight of the entire chain which is initially on the floor.

Example 16.24

Solve Example 16.23 by the use of impulse-momentum diagrams instead of Eq. (16.42).

Solution

The impulse-momentum diagrams for that part of the chain above the floor is shown in Figure E16.24(a) along with a convenient coordinate system. Thus,

$$I_y = (L_2 - L_1)_y,$$

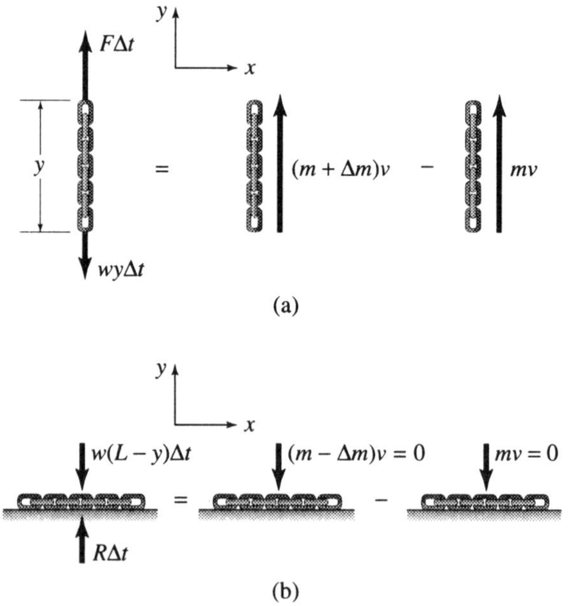

FIGURE E16.24.

$$F\Delta t - wy\Delta t = [(m + \Delta m)v] - mv.$$

Divide by Δt, and let Δt approach zero in the limit to obtain

$$F - wy = v\frac{dm}{dt}.$$

From Example 16.23, $\dfrac{dm}{dt} = \left(\dfrac{w}{g}\right)v$, and $y = vt$. Therefore,

$$F = wvt + \frac{w}{g}v^2 = wv\left(t + \frac{v}{g}\right). \qquad \text{ANS.}$$

Note that the value of F is the same as that obtained by application of Eq. (16.42).

The impulse-momentum diagrams for that part of the chain still on the floor is shown in Figure E16.24(b). Thus, using the same coordinate system as above,

$$I_y = (L_2 - L_1)_y,$$
$$R\Delta t - w(L - y)\Delta t = 0 - 0.$$

Divide by Δt, substitute $y = vt$, and solve for R to obtain

$$R = w(L - vt). \qquad \text{ANS.}$$

Note that the value of R is the same as that obtained by application of Eq. (16.42).

Example 16.25

At any time t, a rocket has a mass $m = m_0 - ct$ where m_0 is the initial mass of the rocket and its fuel and $c = dm/dt$ is the constant rate at which fuel is burned. The rocket is fired vertically upward, and gases are expelled with a downward constant velocity u with respect to the rocket. Use Eq. 16.45 to derive the governing differential equation and to determine the rocket velocity as a function of time. Neglect resistance of the atmosphere, variations of g with altitude, and Coriolis accelerations due to the rotation of Earth.

Solution

Equation (16.45) states that

$$\sum F_x + \frac{dm}{dt} = ma.$$

The free-body diagram of the rocket is shown in Figure E16.25, where T represents the thrust that the expelled gases produce. A coordinate system is also shown. Thus, by Eq. (16.44),

$$-(m_0 - ct)g + cu = (m_0 - ct)\frac{dv}{dt}.$$

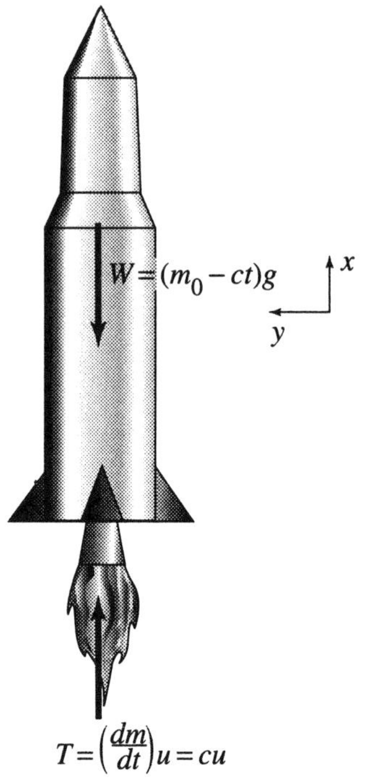

FIGURE E16.25.

Solving for the acceleration $a = \dfrac{dv}{dt}$ enables us to write the governing differential equation of motion. Thus,

$$a = \frac{dv}{dt} = -g + \frac{cu}{m_0 - ct}. \qquad \text{ANS.}$$

This equation states that the rocket acceleration consists of two components: a downward acceleration due to gravity and an upward acceleration due to the rocket thrust, provided that $m_0 > ct$. The rocket thrust T is a force equal to cu. Thus, the second acceleration term on the right hand side is simply the quotient of the thrust and the mass at any time t.

This differential equation is separable. Multiplying by dt and integrating from the initial condition ($t = 0$, $v = 0$) to general upper limits (t, v),

$$\int_0^v dv = -g\int_0^t dt + cu\int_0^t \frac{1}{m_0 - ct}\,dt,$$

$$v = -gt|_0^t - u\ln(m_0 - ct)|_0^t$$

$$v = -gt + u\ln\frac{m_0}{(m_0 - ct)}. \qquad \text{ANS.}$$

■

Problems

16.112 A perfectly flexible chain of length L is held vertically such that its lower end just touches a floor. It is to be lowered onto the floor at a constant velocity v by applying an upward force F at its upper end. If the chain weighs w lb/ft, determine, by applying Eq. (16.42), (a) F as a function of time, (b) the reaction R exerted on the chain by the floor as a function of time, and (c) values of F and R when one half of the chain is on the floor.

16.113 Solve part (a) of Problem 16.112 by reference to appropriate impulse-momentum diagrams.

16.114 A railroad car is being loaded with coal as shown in Figure P16.114. It is initially moving freely to the left with a speed of 8 ft/s and has an initial weight of 10,000 lb. This car has a length of 40 ft and the mass rate of loading is constant $c = dm/dt = 20$ slug/s. Determine (a) the total weight of the car and coal just as the car clears the loader and (b) the speed of the coal car at this same time. (Neglect frictional resistance to motion of the coal car.)

16.115 Water flows from a hose into a 45 lb tank mounted on wheels as shown in Figure P16.115. It flows at the rate of 0.4 slug/s and with a velocity $v_w = 40$ ft/s. Determine the velocity of the tank

FIGURE P16.114.

FIGURE P16.15.

FIGURE P16.16.

at $t = 0.4$ s if the tank is at rest and empty at $t = 0$. Neglect frictional resistance to motion.

16.116 A street-washing truck full of water weighs 24,000 lb. Water flows out at the rate of 3 slug/s with a velocity $v_w = 50$ ft/s relative to the truck as shown in Figure P16.116. If the tractive effort is 2000 lb (exerted by the street on the tires) determine the acceleration of this truck when it is starting a run from rest.

16.117 The rocket and fuel depicted in Figure P16.117 initially weighs 40,000 lb and must reach an upward velocity of 250 ft/s in 8.0 s. The relative velocity of the burning fuel, with respect to the rocket, is 3500 ft/s directed downward. Determine the constant mass rate at which fuel must be burned. What is the thrust at this instant? (At $t = 0$, the rocket has a zero velocity.)

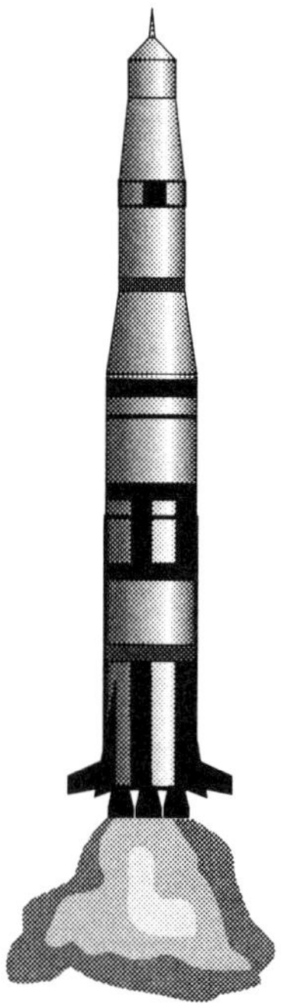

FIGURE P16.117.

16.118 Refer to Problem 16.117, and, at the instant considered there, show the forces acting on the rocket including the thrust. What is the acceleration of the rocket at this instant?

16.119 A rocket to be fired vertically has an initial mass, including fuel, of 1.85×10^6 kg. Fuel is burned at the constant rate of 5.25×10^3 kg/s, and burning gases are expelled at a constant speed of 3.80 km/s relative to the rocket. Express the velocity of this rocket as a function of time. What is the rocket velocity at $t = 20$s?

16.120 Refer to Problem 16.119 and express the acceleration of this rocket as a function of time. Find this acceleration at $t = 10$ s.

16.121 A chain weighing 6 lb/ft is 20 ft long. It is to be lifted from a floor by an up-

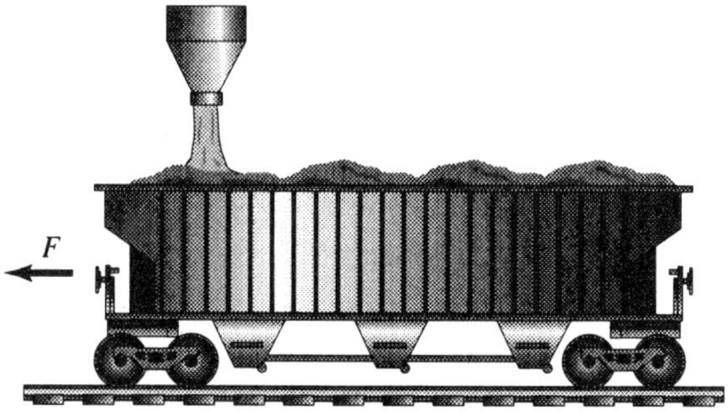

FIGURE P16.122.

ward vertical force F at a constant vertical velocity of 5 ft/s. Find the force F as a function of time. Also find the reaction R exerted on the chain by the floor as a function of time. State the time domain for which these functions are valid.

16.122 In Figure P16.122, pea gravel is being dropped into the railroad car at a rate of 400 lb/s. The empty car weighs 18,000 lb and is subjected to a constant force F which keeps it moving at a constant speed of 25 ft/s. Determine the required magnitude of F. Neglect rolling resistance at the wheels.

16.123 Graded gravel is being spread by the dump truck shown in Figure P16.123. Gravel leaves this truck at 15 ft/s relative to the truck as shown. If the gravel is spread at the rate of 600 lb/s, determine the horizontal tractive force F required at the wheels if the truck is to move to the left at a constant speed of 10 ft/s. Initially, the truck plus the load of gravel weigh 24,000 lb.

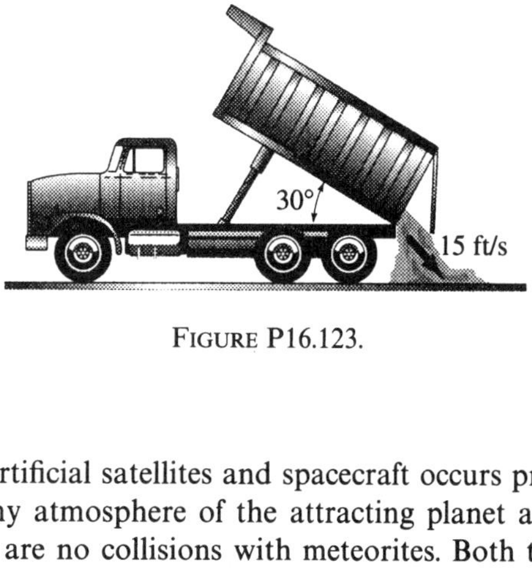

FIGURE P16.123.

16.13 Space Mechanics

The free-flight motion of artificial satellites and spacecraft occurs provided they move above any atmosphere of the attracting planet and provided we assume there are no collisions with meteorites. Both the principle of conservation of angular momentum and that of conservation energy will be shown to apply to this free-flight motion.

In Figure 16.21(a), a satellite is in orbit about an attracting planet, such as the Earth, with a mass center at O. The mass m of the satellite is small compared to the mass M of the attracting planet such that the mass center of the system is for all practical purposes located at point O. Imagine a Newtonian reference frame attached at point O. The satellite is shown in any position by polar coordinates r and θ.

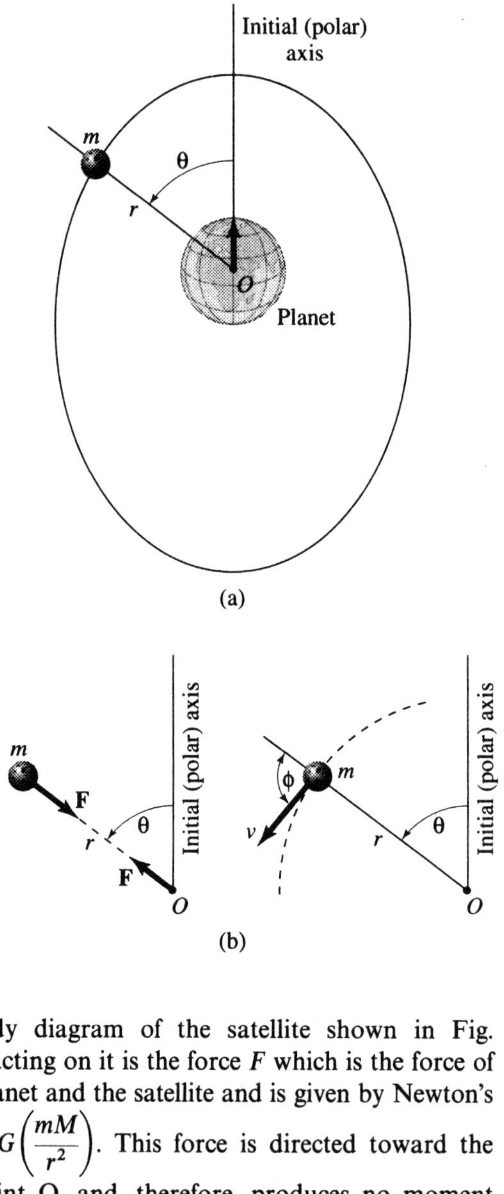

FIGURE 16.21.

Consider the free-body diagram of the satellite shown in Fig. 16.21(b). The only force acting on it is the force F which is the force of attraction between the planet and the satellite and is given by Newton's law of gravitation, $F = G\left(\dfrac{mM}{r^2}\right)$. This force is directed toward the center of the planet, point O, and, therefore, produces no moment about this point. Thus we conclude that the angular momentum of the satellite about point O is conserved and, therefore, for any two positions 1 and 2 along its path

$$H_{O1} = H_{O2}.$$

Consider, for example, the position of the satellite defined by r and θ, as shown in Fig. 16.21(c). The angular momentum H for this position may be written as $H = m(v \sin \phi)r$ where ϕ is the angle between the velocity vector and the polar coordinate r. Thus, the conservation of angular momentum for any two points 1 and 2 along the path becomes

$$m(v_1 \sin \phi_1)r_1 = m(v_2 \sin \phi_2)r_2$$

which reduces to

$$(v_1 \sin \phi_1)r_1 = (v_2 \sin \phi_2)r_2. \tag{16.48}$$

Usually, position 1 is taken as the launch position for which v, ϕ, and r_1 are known quantities.

Because the attractive force of gravity F is conservative, the law of conservation of mechanical energy is applicable to the motion of the satellite. The total energy of the satellite in the launch position 1 and in any other position 2 remains constant with time, which may be expressed as

$$T_1 + V_1 = T_2 + V_2$$

where T and V represent, respectively, the kinetic and potential energies of the satellite. The kinetic energy for any position of the satellite is given by $T = \frac{1}{2}mv^2$. The potential energy may be found by determining the negative of the work done by the force F as the satellite moves from position r to another position r' where $r' > r$. Thus,

$$V = -U = -\int_r^{r'} F\,dr = -\int_r^{r'} \left(\frac{GMm}{r^2}\right) dr = -GMm\left(\frac{1}{r} - \frac{1}{r'}\right).$$

As r' approaches infinity,

$$V = -\frac{GMm}{r}.$$

Substituting for T and V in the above equation for conservation of mechanical energy,

$$\tfrac{1}{2}mv_1^2 - \frac{GMm}{r_1} = \tfrac{1}{2}mv_2^2 - \frac{GMm}{r_2} \tag{16.49a}$$

which, after substituting $GM = gR^2$ from Eq. (1.5) and rearranging terms, becomes

$$v_1^2 - v_2^2 = 2gR^2\left(\frac{1}{r_1} - \frac{1}{r_2}\right). \tag{16.49b}$$

Thus, there are two independent equations available for the solution of a problem in space mechanics. Therefore, we can solve for a maximum of two unknown quantities in a given problem.

The following examples illustrate some of the above concepts.

Example 16.26

A reusable shuttle is fired at point A into an elliptic path AB with angle $\phi_A = 60°$ as shown in Figure E16.26. It is to link up with an orbiting laboratory at point B. Point A is at 50 mi above the earth and the laboratory moves in a circular orbit 540 mi above the surface of the earth. Determine the speeds v_A and v_B. Use the tabular information contained in Chapter 14, p. 181 in the following solution.

Solution

$$r_A = (3960 + 50)(5280) = 2.1173 \times 10^7 \text{ ft},$$

$$r_B = (3960 + 540)(5280) = 2.3760 \times 10^7 \text{ ft}.$$

Also,

$$GM = gR_E^2 = 32.2\,(3960 \times 5280)^2 = 1.4077 \times 10^{16} \text{ ft}^3/\text{s}^2.$$

For the shuttle-laboratory linkup to take place at B, the elliptical and circular paths must have a common tangent at B and $\phi_B = 90°$. The initial angular momentum at point A is equal to the angular momentum at point B. Thus, by Eq. (16.48),

$$(v_A \sin \phi_A) r_A = (v_B \sin \phi_B) r_B,$$

$$v_A = \left(\frac{r_B \sin \phi_B}{r_A \sin \phi_A}\right) v_B = \left(\frac{2.3760 \times 10^7 \sin 90°}{2.1173 \times 10^7 \sin 60°}\right) v_B,$$

$$= 1.296 v_B. \qquad \text{(a)}$$

The total energy is conserved as the shuttle moves from point A to point B along the ellipse. Thus, by Eq. (16.49b),

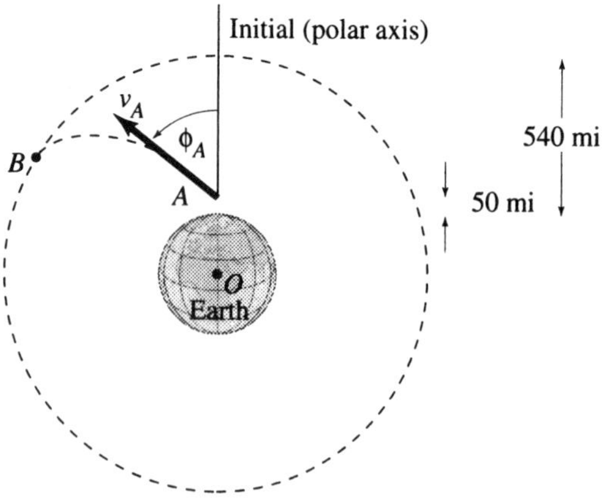

FIGURE E16.26.

$$v_A^2 - v_B^2 = 2gR_E^2\left(\frac{1}{r_A} - \frac{1}{r_B}\right)$$

$$= 2(1.4077 \times 10^{16})\left(\frac{1}{2.1173 \times 10^7} - \frac{1}{2.3760 \times 10^7}\right),$$

$$= 1.4478 \times 10^8. \qquad \text{(b)}$$

Soving Eq. (a) and (b) simultaneously yields

$$v_B = 1.4596 \times 10^4 \text{ ft/s} = 9950 \text{ mph}, \qquad \text{ANS.}$$

$$v_A = 1.8916 \times 10^4 \text{ ft/s} = 12900 \text{ mph}. \qquad \text{ANS.}$$

■ Example 16.27

An artificial satellite enters an elliptic orbit at perigee A as shown in Figure E16.27. The altitude at the perigee A is 5.88×10^5 m and at the apogee B is 3.9405×10^7 m. (a) Determine the speeds at A and B. (b) Determine the radial coordinate and speed at point C on the path at which $\phi_C = 50°$. Note that at A and B the velocity of the satellite is perpendicular to the radial line (i.e., $\phi_A = \phi_B = 90°$). Use the tabular information given in Chapter 14, p. 181 in the following solution.

Solution

(a) $$GM = gR_E^2 = 9.81(6.373 \times 10^6)^2 = 3.9844 \times 10^{14} \text{ m}^3/\text{s}^2,$$

$$r_A = 6.373 \times 10^6 + 5.88 \times 10^5 = 6.961 \times 10^6 \text{ m},$$

$$r_B = 6.373 \times 10^6 + 3.9405 \times 10^7 = 4.5778 \times 10^7 \text{ m}.$$

At the perigee and apogee the angle $\phi = 90°$. The angular momentum

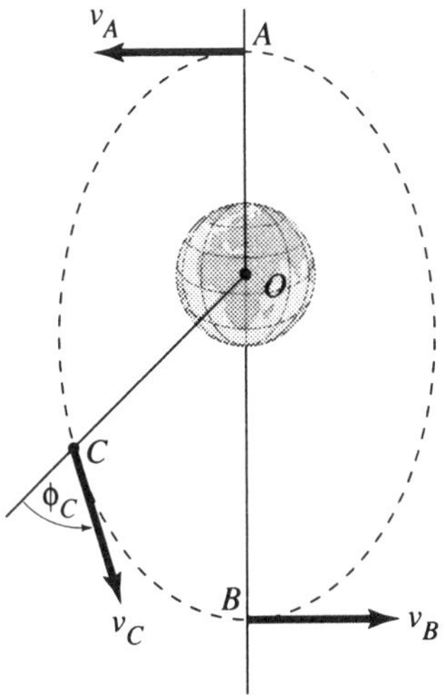

FIGURE E16.27.

of the satellite with respect to O is conserved and Eq. (16.48) yields

$$v_A r_A = v_B r_B$$

$$v_A = \frac{r_B}{r_A} v_B = \frac{4.5778 \times 10^7}{6.961 \times 10^6} v_B = 6.5764 v_B. \qquad \text{(a)}$$

The mechanical energy is also conserved, and Eq. (16.49b) leads to

$$v_A^2 - v_B^2 = 2gR_E^2\left(\frac{1}{r_A} - \frac{1}{r_B}\right)$$

$$= 2(3.9844 \times 10^{14})\left(\frac{1}{6.961 \times 10^6} - \frac{1}{4.5778 \times 10^7}\right),$$

$$= 9.7070 \times 10^7 \qquad \text{(b)}$$

Solve Eqs. (a) and (b) simultaneously to obtain

$$v_B = 1516 \text{ m/s}, \qquad \text{and} \qquad v_A = 9970 \text{ m/s}. \qquad \text{ANS.}$$

(b) Consider points A and C to determine v_C and r_C. The angular momentum with respect to point O is conserved, and Eq. (16.48) reduces to

$$v_C r_C \sin 50^\circ = v_A r_A$$

$$v_C r_C = \frac{v_A r_A}{\sin 50^\circ} = \frac{6.961 \times 10^6 (9970)}{\sin 50^\circ} = 9.06 \times 10^{10} \qquad \text{(c)}$$

Also, the mechanical energy is conserved, and Eq. (16.49b) yields

$$v_C^2 - v_A^2 = 2gR_E^2\left(\frac{1}{r_C} - \frac{1}{r_A}\right) = 2(3.9844 \times 10^{14})\left(\frac{1}{r_C} - \frac{1}{6.961 \times 10^6}\right). \qquad \text{(d)}$$

Substitute $r_C = \dfrac{9.06 \times 10^{10}}{v_C}$ from Eq. (c), and simplify to obtain

$$v_C^2 - 8795.6 v_C + 1.511 \times 10^7 = 0$$

from which

$$v_C = 6460 \text{ m/s} \quad \text{or} \quad 2340 \text{ m/s}. \qquad \text{ANS.}$$

Therefore, Eq. (c) yields,

$$r_C = 1.403 \times 10^7 \quad \text{or} \quad 3.87 \times 10^7 \text{ m}. \qquad \text{ANS.}$$

■

Problems

Use the tabular information given in Chapter 14, p. 181 in the solution of the following problems.

16.124 An artificial satellite enters an elliptic orbit at perigee A as shown in Figure P16.124. The altitude at the perigee A is 5.93×10^5 m and at the apogee B is 3.976×10^7 m. (a) Determine the speeds at A and B. (b) For $\phi = 60°$, determine the radial coordinate and speed.

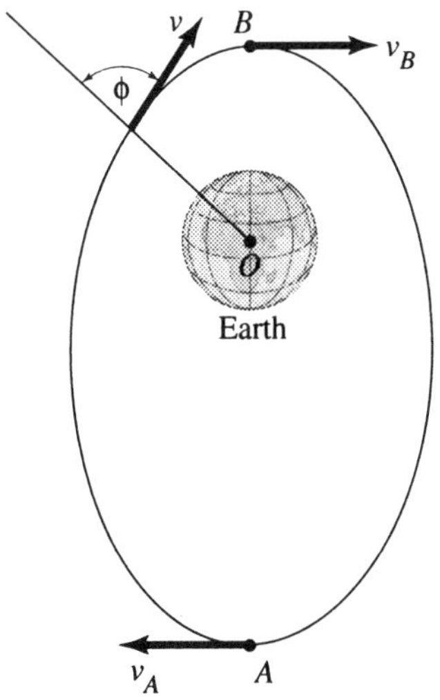

FIGURE P16.124.

16.125 A reusable shuttle is fired at point A into an elliptic path AB with angle $\phi_A = 45°$ as shown in Figure P16.125. It is to link up with an orbiting laboratory at point B. Point A is 60 mi above Earth, and the laboratory moves in a circular orbit 600 mi above the surface of Earth. Determine the speeds v_A and v_B.

16.126 As shown in Figure P16.126, a spacecraft is initially 5 Mm above Earth's surface in circular orbit 1. Retrorockets are fired at A to place it in a transfer elliptic orbit AB. When it reaches point B, the retrorockets are fired again to place the craft in circular orbit 2 at an altitude of 10.5 Mm above Earth. Determine the velocity at A just after firing and that at B just before firing.

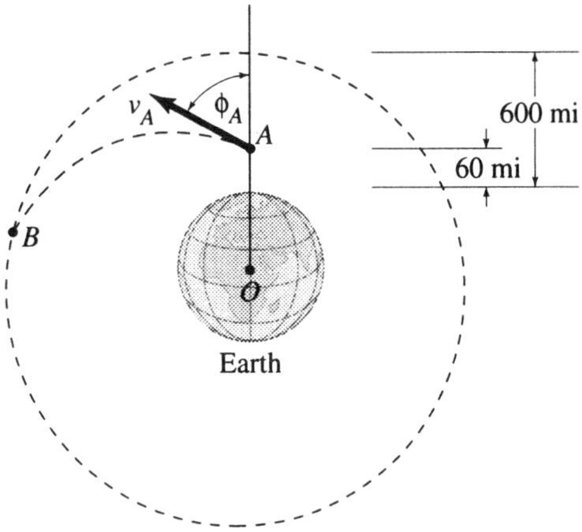

FIGURE P16.125.

16.127 Solve Problem 16.126 for an inner circular orbit altitude of 1.5×10^6 ft and an outer circular orbit altitude of 3.0×10^6 ft.

16.128 Planning for a Martian return mission involves entry into a transfer elliptic orbit AB at point A and into a circular orbit at point B, as shown in Figure P16.128. Point A is at an altitude of 20,000 ft and point B is at an altitude of 100,000 ft above the surface of Mars. If the angle $\phi_A = 40°$, determine the velocity v_A and the velocity change at B required for a change from the elliptic to the circular orbit. (Hint: Use Eq. (14.46) for circular orbits).

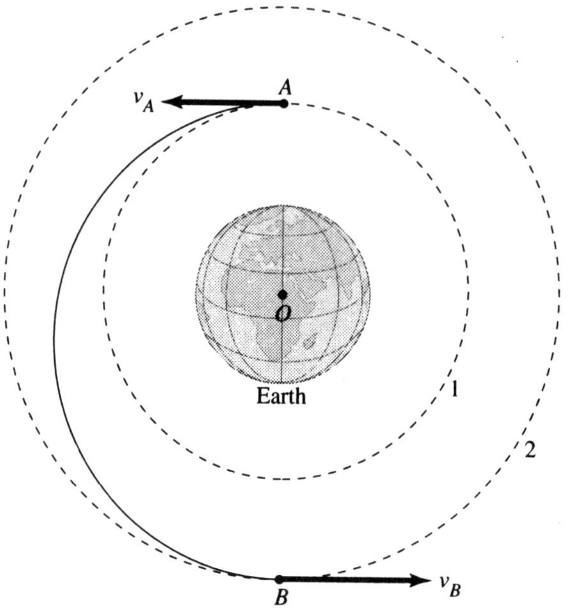

FIGURE P16.126.

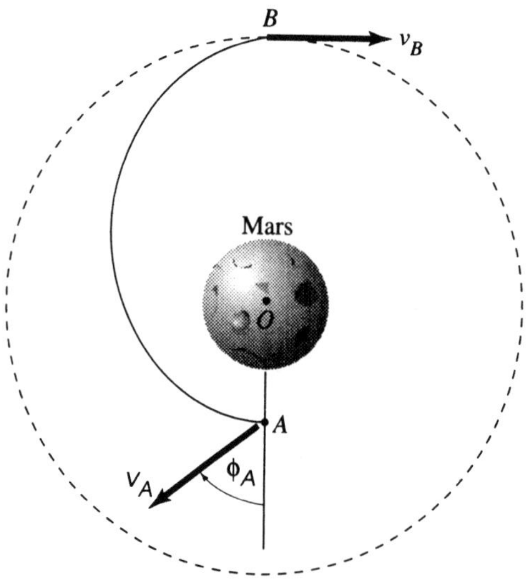

FIGURE P16.128.

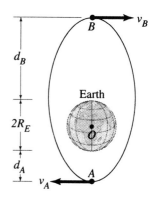

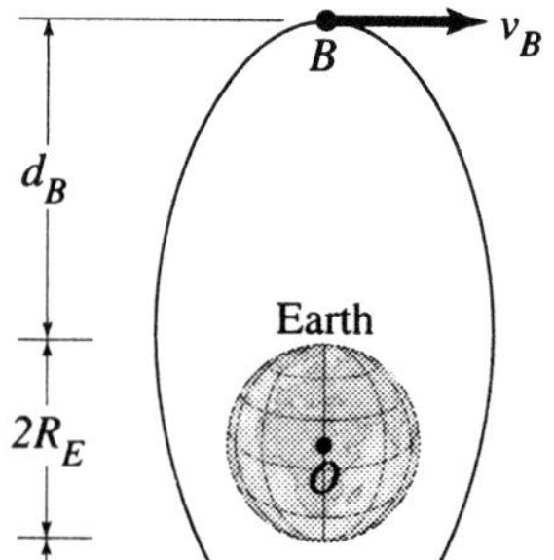

FIGURE P16.130.

16.129 Solve Problem 16.128 for $\phi_A = 50°$, point A at an altitude of 60 km and point B at an altitude of 300 km.

16.130 Express the speeds of an artificial satellite at the perigee A and the apogee B as functions of G, M_E, R_E, d_A, and d_B where d_A and d_B are the altitudes shown in Figure P16.130.

Review Problems

16.131 A remote-controlled toy car moves along the straight-line path shown in Figure P16.131 such that $s = (2t^2 + t + 5)$ ft where t is in seconds. If the car weighs 0.75 lb, determine its momentum vector for $t = 1$ s.

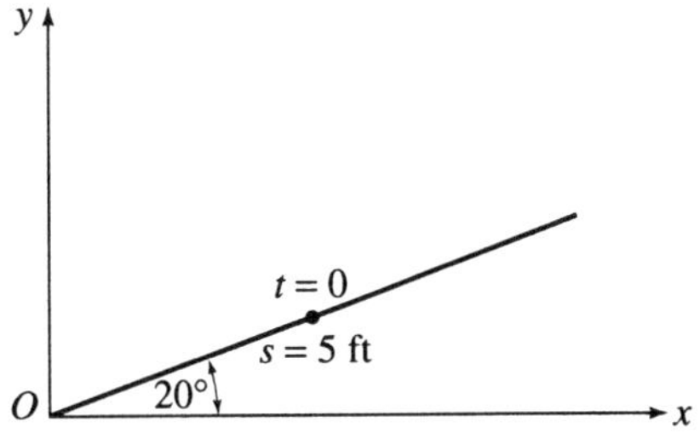

FIGURE P16.131.

16.132 The loading on a building due to a gust of wind is approximated as shown in Figure P16.132. Use the trapezoidal rule to find an approximate value for the impulse created by the wind from $t = 0$ to $t = 0.8$ s. Let $\Delta t = 0.1$ s.

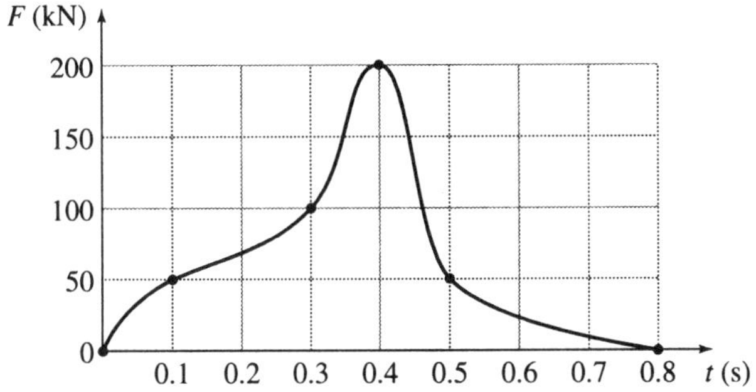

FIGURE P16.132.

16.133 A car weighing 4000 lb is moving on a level stretch of road at a speed of 60 mph when the brakes are suddenly applied locking all four wheels. What is the coefficient of kinetic friction between the tires and the road if it took 4.5 s to bring the car to a complete stop?

16.134 A fighter plane ($W = 20{,}000$ lb) landing on a carrier is brought to a complete stop by an arresting hook which provides a force whose time variation may be approximated by the graph shown in Figure P16.134. The carrier is at rest

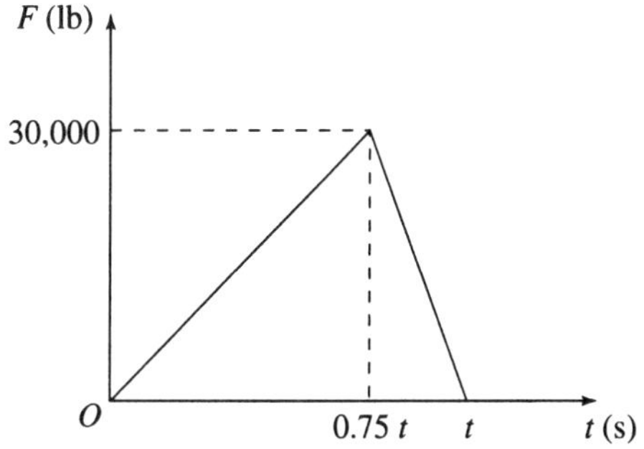

FIGURE P16.134.

when the fighter plane approaches it at a speed of 120 mph. Determine the time t required to bring the plane to a complete stop. Assume frictionless conditions and that the plane's engines are turned off during the stopping operation.

16.135 Two boats A and B tied together by a rope are initially at rest in calm water as shown in Figure P16.135. The boy in boat A pulls on the rope to bring boat B closer to him. If his speed relative to the calm water is 0.5 m/s, determine the speed of boat B relative to boat A. Let $m_A = 60$ kg, $m_B = 30$ kg, and neglect the resisting effects of the water on the boats.

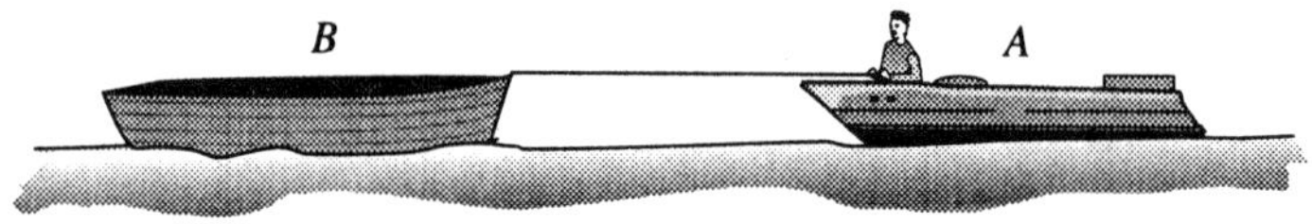

FIGURE P16.135.

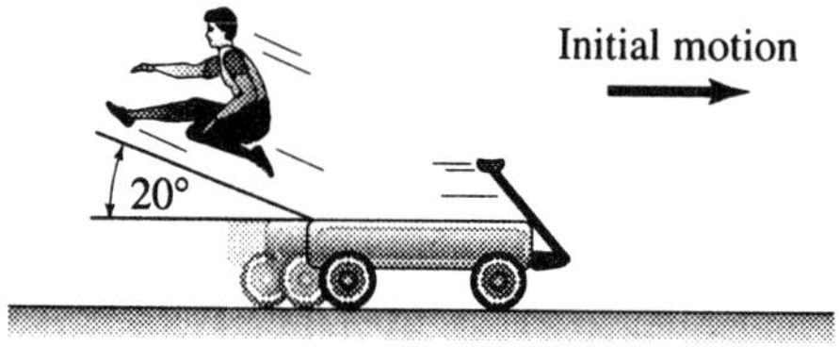

FIGURE P16.136.

16.136 A boy and his wagon are moving on a horizontal surface at a speed of 5 ft/s to the right when suddenly the boy jumps out the back of the wagon with a speed relative to the wagon of 2 ft/s directed to the left at 20° with the horizontal as shown in Figure P16.136. Determine the speed of the wagon immediately after he jumps out. The weight of the boy is 1000 lb and that of the wagon is 25 lb. Ignore friction.

16.137 A hockey puck with a mass $m = 0.2$ kg moving at 140 km/h strikes the boards at an *angle* of 50° in direct central impact. For a coefficient of restitution $e = 0.95$, find the velocity of the puck after impact. Assume frictionless conditions.

16.138 The 5-lb ball A is released from rest when $\theta = 0$ and strikes the 3-lb ball B which is sitting on the ledge as shown in Figure P16.138. If the coefficient of restitution $e = 0.8$, determine (a) the velocity of balls A and B immediately after impact and (b) the distance x_1 at which ball B strikes the ground.

16.139 The driver of car A moving on a horizontal stretch of road at a speed of 100 km/h sees a disabled car B and applies his brakes when he is 40 m away from it as shown in Figure P16.139. Assume that, while moving through the 40-m distance, all four wheels of car A are locked and that the coefficient of kinetic friction $\mu = 0.6$. Let $m_A = 1800$ kg, $m_B = 1350$ kg, and the coefficient of restitution $e = 0.2$, and determine the speed of both cars immediately after impact. Assume frictionless conditions during impact. Find the loss of energy due to impact.

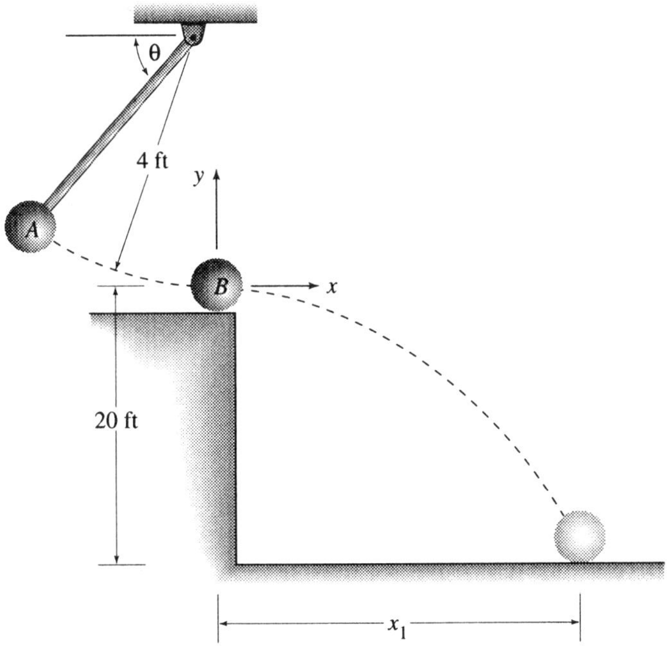

FIGURE P16.138.

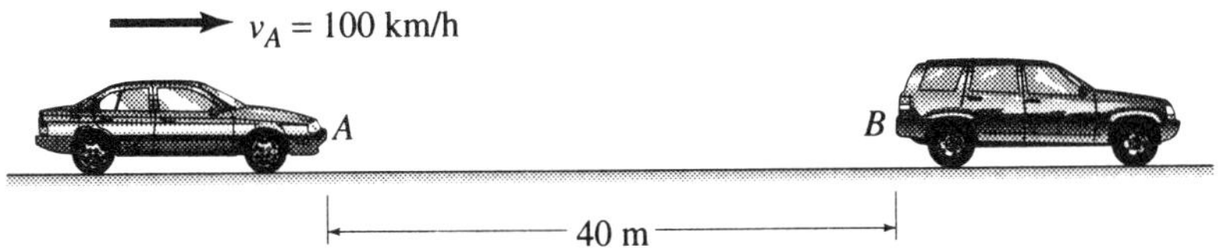

FIGURE P16.139.

16.140 For a short period of time, a remote controlled toy plane of mass $m = 0.75$ kg flies in a vertical plane along a path that may be described parametrically by $x = 20t$ and $y = 120 + 10t^2$ where t is in seconds and x and y are in meters. Find the path expressing it as $y = f(x)$, and determine the angular momentum of the plane about the origin of the coordinate system for $t = 5$ s.

16.141 A children's ride in an amusement park consists of four gondolas (only two are shown) as depicted in Figure P16.141. On average, each gondola and its occupant weigh 120 lb. The system starts from rest and a couple $M = (60 + 8t)$ lb·ft , where t is in seconds, is applied as shown. Determine the linear speed of the gondolas after 10 s. Assume the supporting arms to be weightless.

16.142 A particle of mass $m = 5$ kg has a speed of 10 m/s, when in position A, and a speed of 20 m/s, when in position B, as shown in Figure P16.142. Find

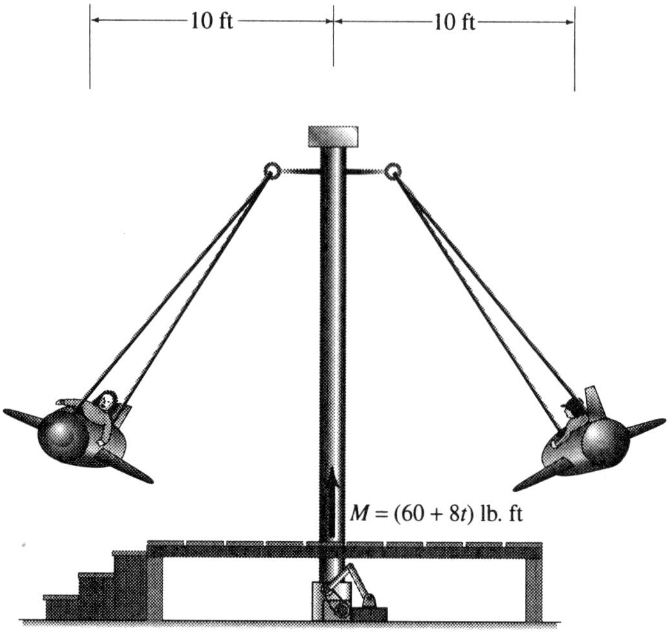

FIGURE P16.141.

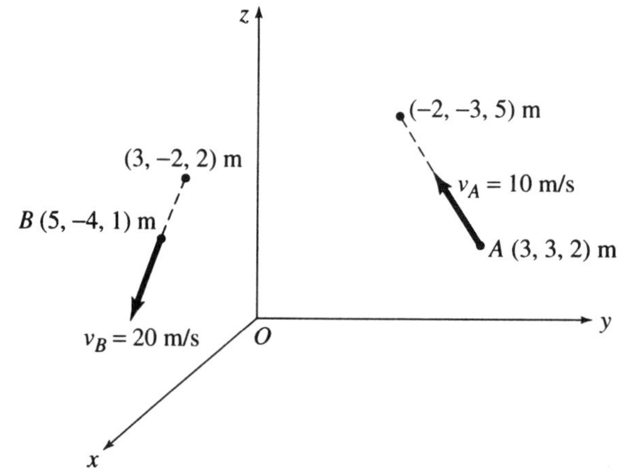

FIGURE P16.142.

the angular impulse vector about point O exerted on the particle as it moves from A to B.

16.143 The system shown in Figure P16.143 consists of four equal weights $W = 10$ lb (only two are shown) attached to a vertical shaft by four identical arms which may be assumed weightless. When the arm length $d = 12$ in, the system rotates about the vertical shaft

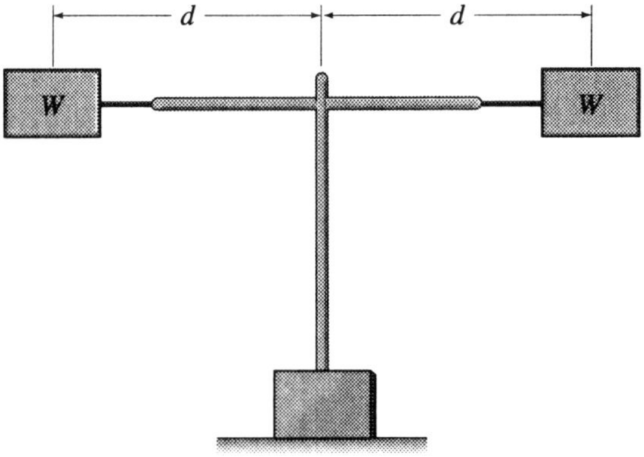

FIGURE P16.143.

axis at an angular speed of 10 rad/s. An internal mechanism requiring negligible energy allows the arms to lengthen so that $d = 24$ in. Determine the linear speed of the weights when $d = 24$ in.

16.144 The 3-kg mass A is released from rest, when $\theta = 0$, and strikes the 4-kg mass B which is at rest on the frictionless horizontal plane, as shown in Figure P16.144. The spring attached to B has a spring constant $k = 2000$ N/m and is initially compressed an amount $x_0 = 0.10$ m by means of the two strings C and D. Determine the maximum additional deformation experienced by the spring.

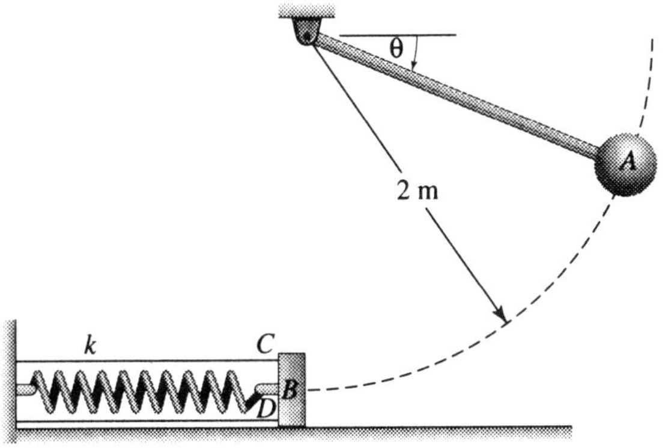

FIGURE P16.144.

16.145 Block A weighing 3 lb is released from the position shown and moves down the smooth circular surface before striking the 4-lb block B as indicated in Figure P16.145. Determine the distance traveled by block B on the horizontal plane where the kinetic coefficient of friction is $\mu = 0.2$. What happens to block A after impact? The coefficient of restitution $e = 0.7$.

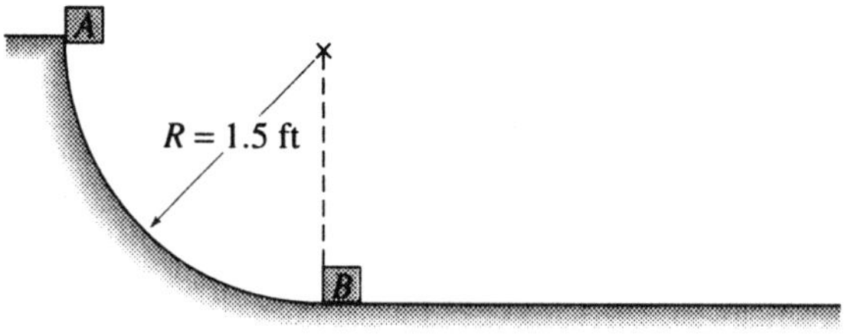

FIGURE P16.145.

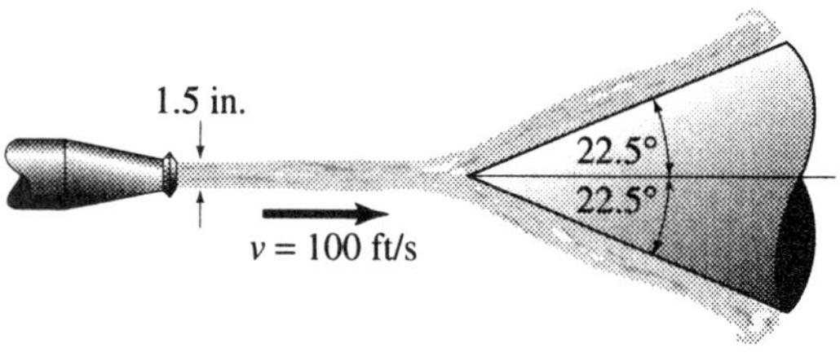

FIGURE P16.146.

16.146 A water jet with a diameter of 1.5 in. strikes a stationary smooth diffuser with a velocity of 100 ft/s as shown in Figure P16.146. Determine the horizontal force that the diffuser exerts on the water jet. Water has a density of 62.4 lb/ft^3.

16.147 A water jet moving at $v_W = 20$ m/s strikes a vane moving in the same direction at $v_V = 4$ m/s relative to the jet as shown in Figure P16.147. The water flows at the rate of 0.025m^3/s and its mass density is 1000 kg/m^3. Find the horizontal and vertical components of the force that the water exerts on the vane.

16.148 The jet plane shown in Figure P16.148 flying at a constant speed v_0, takes in air at the intake port I at the rate of 8 slug/s and discharges it at the exhaust port E at a speed of 2500 ft/s relative to the plane. If the thrust developed by the engine is 13,000 lb, determine the constant speed v_0 at which the plane is flying. Locate the line of action of the thrust in relation to exhaust port E.

16.149 An empty truck weighing 40,000 kN is being loaded with sand as shown in Figure P16.149. The truck is to maintain a constant speed of 5 m/s while being loaded, and the horizontal tractive force exerted by the pavement at

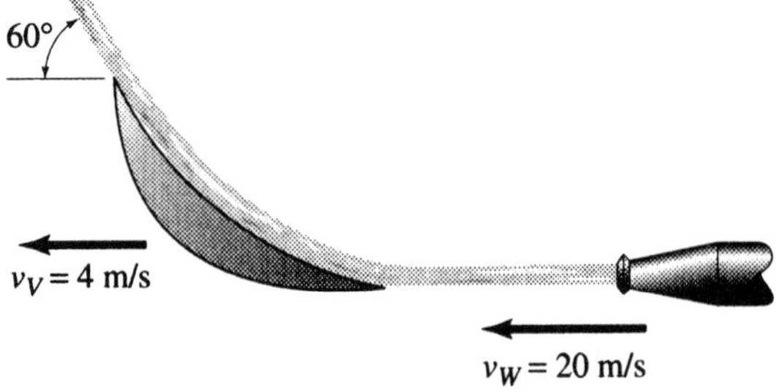

FIGURE P16.147.

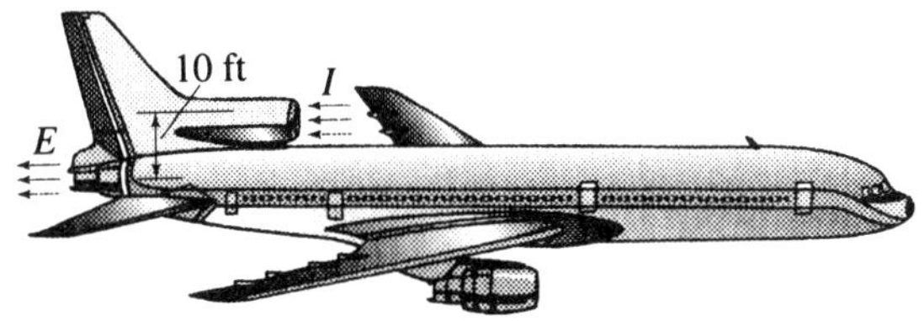

FIGURE P16.148.

FIGURE P16.149.

the front driving wheels is known to be 3.75 kN. Determine the volume of sand that is being loaded onto the truck per second. The mass density of sand is approximately 1500 kg/m^3.

16.150 A reusable shuttle is fired at point A with an initial velocity $v_A = 4860$ m/s into an elliptic path AB as shown in Figure P16.150. It is to link up with an orbiting laboratory at point B. Point A is 80 km above Earth and the laboratory moves in a circular orbit 800 km above the surface of Earth. Determine the launch angle ϕ such that the shuttle reaches point B with a velocity tangent to the circular orbit of the laboratory.

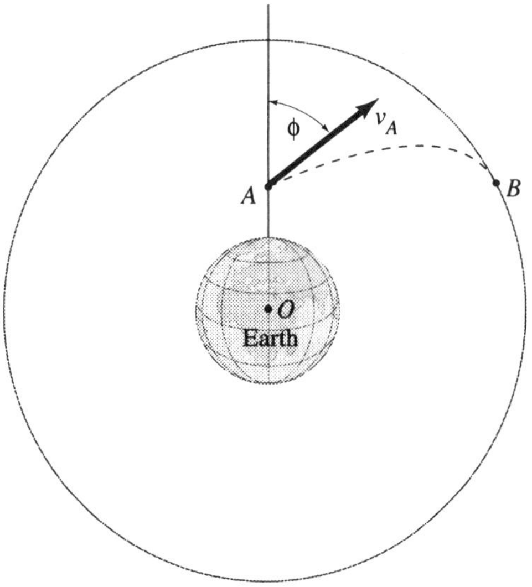

FIGURE P16.150.

17

Two-Dimensional Kinematics of Rigid Bodies

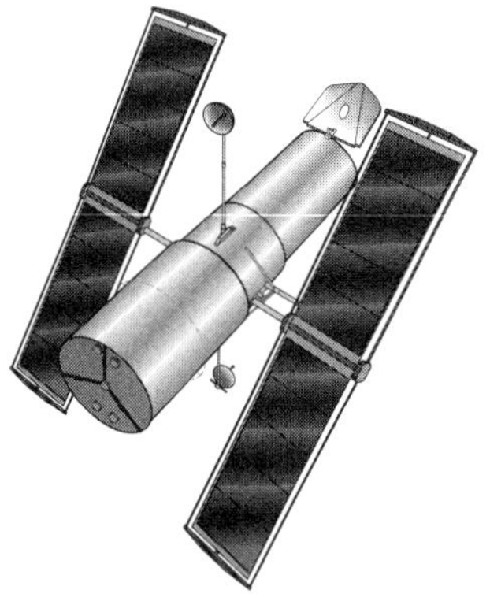

A picture showing the Hubble Telescope. Engineering is a challenging and a rewarding profession in spite of occasional setbacks.

When dealing with the kinematics *of a rigid body, we are concerned with the geometric relationships that exist among displacements, velocities, and accelerations of various particles in a body in motion without regard to the forces causing the motion or caused by it.*

A rigid body is said to be in plane motion *if all of its particles move in parallel planes. Three types of plane motion are identified in this chapter. The first type is known as* translation *which could be either* rectilinear *or* curvilinear. *The second type deals with* rotation about a fixed axis. *The third type is referred to as* general plane motion *which is a combination of a translation and a rotation.*

Outer space provides one of many areas where kinematics plays a primary role. While on Earth, a force is required to maintain a body at a constant velocity because of friction; in outer space, no such force is required. However, basic kinematics applies equally well here on Earth and in space. The implication is that, irrespective of your engineering discipline, the knowledge of basic kinematics of rigid bodies will provide you with the basic foundations upon which dynamic principles are based.

In a society where technology is pervasive, there is a need for understanding basic concepts and training of the individuals who design and organize the production of much of that technology. This chapter should enhance your enthusiasm for and the pleasure you derive from problem solving and reducing concepts to solid expression. The case study accompanying this chapter provides a clear illustration of the joys and frustrations of being an engineer.

The challenges involved in putting the Hubble telescope into orbit and, then, fixing it is a vivid example of where kinematics plays a substantial role. It is hoped that this odyssey through the world of the engineer will convey the patterns of challenges and triumphs that are common to all engineering disciplines. You will learn that a dream fulfilled is often just another challenge to a bigger dream.

17.1 Rectilinear and Curvilinear Translations

A rigid body is said to be in *translation* if a straight line connecting any two particles in the body remains parallel to itself during the motion. *Plane translation* of a rigid body is said to occur when all particles in the body move along paths that maintain constant distances from a fixed plane. Thus, for example, the rigid body shown in Figure 17.1 will have plane translation parallel to the x-y plane if, and only if, particles, such as A, B and C, maintain constant distances d_A, d_B, and d_C, respectively, from the fixed x-y plane during the motion.

Two types of plane translation are identified. The first is known as *rectilinear* translation in which all particles of the rigid body move along parallel *straight* paths; the second is referred to as *curvilinear* translation in which these particles move along *congruent curved* paths. Thus, for example, plate ABCD executes rectilinear translation in the plane of the page to position A′B′C′D′ in Figure 17.2(a) where particles, such as A, B, C and D, all move along parallel straight paths. On the other hand, plate ABCD undergoes curvilinear translation in the plane of the page to position A′B′C′D′ in Figure 17.2(b) where particles, such as A, B, C, D, all move along congruent curves. In both cases, note that a straight line, such as AB, remains parallel to itself throughout the entire motion.

Consider a rigid body in plane motion, as shown in Figure 17.3, which is assumed to be executing curvilinear translation parallel to the x-y plane. The position vectors to particles A and B on this rigid body are $\mathbf{r}_A$ and $\mathbf{r}_B$, respectively, and the position vector of particle B relative to particle A is $\mathbf{r}_{B/A}$. Therefore, the positions of any two particles A and B on a rigid body may be related by the following vector equation:

$$\mathbf{r}_B = \mathbf{r}_A + \mathbf{r}_{B/A}. \tag{17.1}$$

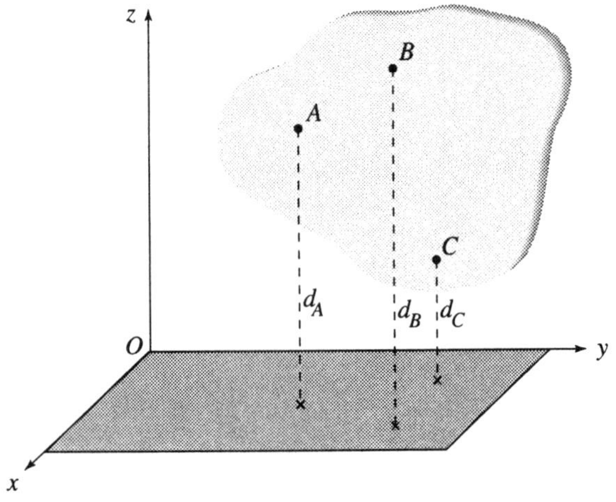

FIGURE 17.1.

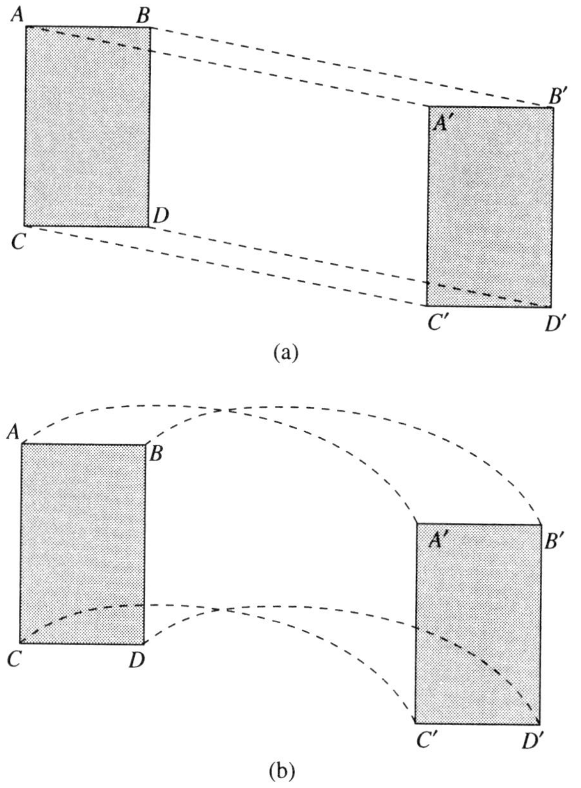

FIGURE 17.2.

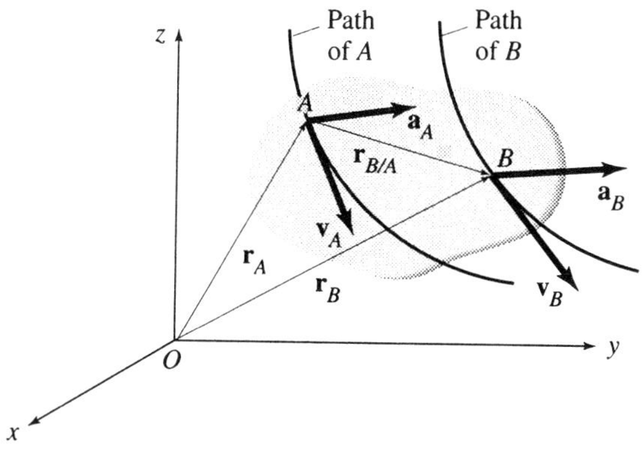

FIGURE 17.3.

Thus, if the position of any one particle on a rigid body, such as A, is known, the position of any other particle, such as B, may be determined by Eq. (17.1) if its position relative to particle A is known.

If we differentiate Eq. (17.1) with respect to time,

$$\frac{d\mathbf{r}_B}{dt} = \frac{d\mathbf{r}_A}{dt} + \frac{d\mathbf{r}_{B/A}}{dt}. \tag{17.2}$$

By definition, because a rigid body implies that the positions of the particles relative to each other do not change, it follows that the term $\frac{d\mathbf{r}_{B/A}}{dt}$ must vanish, and Eq. (17.2) reduces to $\frac{d\mathbf{r}_B}{dt} = \frac{d\mathbf{r}_A}{dt}$. Because $\frac{d\mathbf{r}_B}{dt} = \mathbf{v}_B$ and $\frac{d\mathbf{r}_A}{dt} = \mathbf{v}_A$, it follows that

$$\mathbf{v}_B = \mathbf{v}_A \tag{17.3}$$

where $\mathbf{v}_A$ and $\mathbf{v}_B$ represent the velocities of points B and A, respectively. Thus, according to Eq. (17.3), if the velocity of one particle, such as A, in a rigid body is known, the velocity of any other particle, such as B, may be found. In other words the velocities of all particles in a translating rigid body are the same at a given instant of time. As discussed in Chapter 13, these velocity vectors must be tangent to the paths of motion, as shown in Figure 17.3.

If we now differentiate Eq. 17.3 with respect to time and replace the time derivatives of the velocities by accelerations,

$$\mathbf{a}_B = \mathbf{a}_A \tag{17.4}$$

where $\mathbf{a}_B$ and $\mathbf{a}_A$ represent the acceleration vectors of particles B and A, respectively. Thus, if the acceleration of one particle, such as A, is known, the acceleration of any other particle, such as B, may be determined. In other words, the accelerations of all particles in a translating rigid body are all the same at a give instant of time. The accelerations of the two particles A and B are sketched in Figure 17.3. Note that these accelerations could have any direction depending upon the conditions specified in a given problem.

It is obvious from the above discussion that the translating motion of a rigid body is completely defined if the motion of one of the particles is fully specified. Therefore, the equations developed in Chapter 13 for the motion of a particle may be used equally well to analyze the translating motion of a rigid body.

The concepts discussed above are illustrated in the following examples.

■ Example 17.1

An automobile starts from rest and travels along a straight road. The driver increases the speed to 50 mph at a constant acceleration in a distance of 500 ft and, then, decreases the speed at a constant deceleration of 10 ft/s^2 until he comes to a full stop. Determine (a) the constant acceleration of the automobile during the first 500 ft of travel and (b) the distance and the time needed to stop the automobile from 50 mph.

Solution

(a) The automobile is executing rectilinear translation because all points on it move along parallel straight paths. Because the acceleration is constant, Eq. (13.21) , may be used. Thus,

$$v^2 = v_0^2 + 2a_C(s - s_0)$$

from which

$$a_C = \frac{v^2 - v_0^2}{2(s - s_0)}$$

In this equation, $v_0 = 0$ because the automobile starts from rest, $v = 50$ mph $= 73.3$ ft/s and $(s = s_0) = 500$ ft. Therefore,

$$a_C = \frac{73.3^2 - 0}{2(500 - 0)} = 5.37 \text{ ft/s}^2. \qquad \text{ANS.}$$

(b) Since the deceleration is constant, it follows that

$$v^2 = v_0^2 + 2a_C(s - s_0)$$

and

$$(s - s_0) = \frac{v^2 - v_0^2}{2a_C}$$

In this equation, $v = 0$, $v_0 = 50$ mph $= 73.3$ ft/s and $a_C = -10$ ft/s^2. Thus,

$$\Delta s = (s - s_0) = \frac{0 - 73.3^2}{2(-10)} = 269 \text{ ft.} \qquad \text{ANS.}$$

The time required to stop the automobile at a constant deceleration from a speed of 50 mph may be found by using Eq. (13.19). Thus,

$$v = v_0 + a_C t$$

which leads to

$$t = (v = v_0)/a_C$$

In this equation, $v = 0$, $v_0 = 73.3$ ft/s and $a_C = -10$ ft/s^2. Therefore,

$$t = \frac{0 - 73.3}{-10} = 7.33 \text{ s.} \qquad \text{ANS.}$$

Example 17.2

A rocket is fired vertically upward as shown in Figure E17.2. The acceleration of the rocket may be approximated by the relationship $a = 2y^{1/2}$ where a is the acceleration in m/s^2 and y is the upward displacement in m. (a) Develop expressions for the velocity of the rocket v and for the time of travel t in terms of the displacement y. (b) What are the numerical values of v and t when the rocket has traveled a vertical distance of 200 m?

Solution

(a) The rocket executes rectilinear translation because all of its particles move along parallel straight paths. Since the acceleration of the rocket is expressed in terms of its displacement, it follows from Eq. (13.11) that

$$v\,dv = a\,dy = 2y^{1/2}\,dy. \tag{a}$$

Integrating Eq. (a), using initial conditions for lower limits and general conditions for upper limits,

$$\int_0^v v\,dv = 2\int_0^y y^{1/2}\,dy.$$

Therefore,

$$\left.\frac{v^2}{2}\right|_0^v = 2\left.\left(\frac{y^{3/2}}{3/2}\right)\right|_0^y,$$

and

$$v = \sqrt{\frac{8}{3}}\,y^{3/4}. \qquad \text{(b) ANS.}$$

Using Eq. (b)

$$v = \frac{dy}{dt} = \sqrt{\frac{8}{3}}\,y^{3/4}.$$

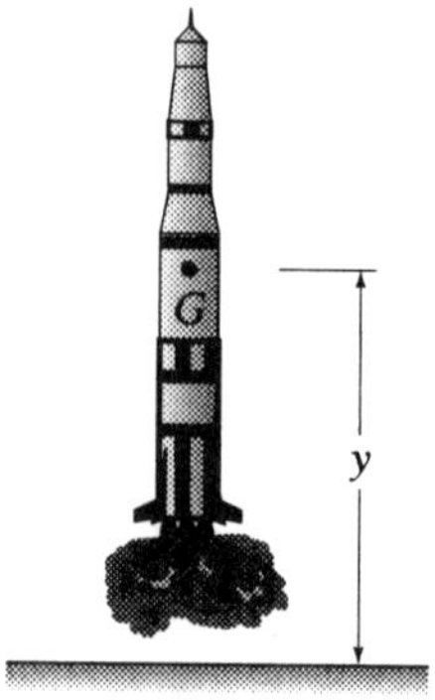

FIGURE E17.2.

Separating variables, integrating, and using initial conditions for lower limits and general conditions for upper limits,

$$\int_0^t dt = \sqrt{\frac{3}{8}} \int_0^y y^{-3/4}\, dy$$

from which

$$t\,\big|_0^t = \sqrt{\frac{3}{8}} \left(\frac{y^{1/4}}{1/4}\right)\Bigg|_0^y$$

and

$$t = \sqrt{6}\, y^{1/4}. \qquad \text{(c) ANS.}$$

(b) Substituting $y = 200$ m into Eqs. (b) and (c) leads to the numerical values of velocity and time. Thus,

$$v = \sqrt{\frac{8}{3}}(200)^{3/4} = 86.9 \text{ m/s}, \qquad \text{ANS.}$$

and

$$t = \sqrt{6}(200)^{1/4} = 9.21 \text{ s}. \qquad \text{ANS.}$$

Example 17.3

A plate is connected to two parallel rods AC and BD by frictionless hinges as shown in Figure E17.3(a). The plate is set in motion in the vertical plane such that $\theta = t^2$ where θ is measured in radians and t is the time in seconds. Determine the velocity and acceleration of the plate expressing them in terms of L and t. Compute numerical values of these quantities for the case when $\theta = 30°$ and $L = 3$ ft.

Solution

Examination of the motion of the plate reveals that it is executing curvilinear translation because a straight line such as AB remains par-

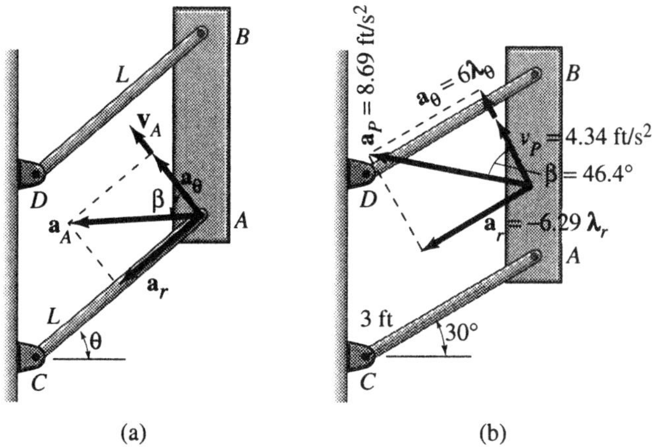

FIGURE E17.3.

allel to itself during the motion, and particles A and B move along two circular congruent curves of radius L. Therefore, the velocity and acceleration of the plate may be completely specified by finding the velocity and acceleration of any one particle, such as A, in the plate.

It is convenient to express the velocity of particle A in the plate in terms of polar coordinates. Thus, by Eq. (13.60),

$$\mathbf{v}_A = \dot{r}\boldsymbol{\lambda}_r + (r\dot{\theta})\boldsymbol{\lambda}_\theta$$

where, for the system under consideration, $r = L$, $\dot{r} = 0$, and $\dot{\theta} = 2t$. Therefore

$$\mathbf{v}_A = 2Lt\boldsymbol{\lambda}_\theta$$

Thus, the magnitude of the velocity of point A is $2Lt$, and its direction is along the transverse direction or along the tangent to the circular path followed by point A. This velocity vector is shown in Figure E17.3(a) and is identical to the plate's velocity $\mathbf{v}_P$. Therefore,

$$\mathbf{v}_P = \mathbf{v}_A = 2Lt\boldsymbol{\lambda}_\theta.$$

Again we express the acceleration of particle A in terms of polar coordinates. Thus, from Eq. (13.61),

$$\mathbf{a}_A = (\ddot{r} - r\dot{\theta}^2)\boldsymbol{\lambda}_r + (r\ddot{\theta} + 2\dot{r}\dot{\theta})\boldsymbol{\lambda}_\theta = a_r\boldsymbol{\lambda}_r + a_\theta\boldsymbol{\lambda}_\theta.$$

For our particular situation $r = L$, $\dot{r} = \ddot{r} = 0$, $\dot{\theta} = 2t$, and $\ddot{\theta} = 2$. Therefore,

$$\mathbf{a}_A = -4Lt^2\boldsymbol{\lambda}_r + 2L\boldsymbol{\lambda}_\theta.$$

This acceleration vector is shown in Figure E17.3(a) and is identical to the acceleration of the plate $\mathbf{a}_P$. Therefore,

$$\mathbf{a}_P = \mathbf{a}_A = -4Lt^2\boldsymbol{\lambda}_r + 2L\boldsymbol{\lambda}_\theta.$$

The magnitude a_P of the plate's acceleration may be written as

$$a_P = \sqrt{(4Lt^2)^2 + (2L)^2} = 2L\sqrt{4t^2 + 1}$$

and its direction is defined by the angle β where

$$\beta = \tan^{-1}\left(\frac{4Lt^2}{2L}\right) = \tan^{-1}(2t^2)$$

For $\theta = 30° = 0.524$ rad, the value of t may be found from the given relation between t and θ. Thus,

$$t = \sqrt{\theta} = \sqrt{0.524} = 0.724 \text{ s}$$

Therefore, because $L = 3$ ft,

$$\mathbf{v}_P = 4.34\boldsymbol{\lambda}_\theta \text{ ft/s},$$ ANS.

and

$$v_P = 4.34 \text{ ft/s.} \qquad \text{ANS.}$$

Also

$$\mathbf{a}_P = -6.29\boldsymbol{\lambda}_r + 6.00\boldsymbol{\lambda}_\theta \text{ ft/s}^2, \qquad \text{ANS.}$$

and

$$a_P = 8.69 \text{ ft/s}^2. \qquad \text{ANS.}$$

These quantities are shown in Figure E17.3(b) where

$$\beta = 46.4^\circ \qquad \text{ANS.}$$

■

Problems

17.1 A motorcyclist starts from rest and travels along a straight road increasing his speed to 100 km/h in a distance of 175 m at a constant acceleration. Then, he decreases his speed at the constant rate of 3m/s^2 until he comes to a complete stop. Determine (a) the constant acceleration of the motorcyclist during the first 175 m of travel and (b) the time needed to stop the motorcycle from 100 km/h.

17.2 As a balloon is rising along a straight path at a speed of 10 f/s, a package is eased overboard as shown in Figure P17.2. Four seconds later the package

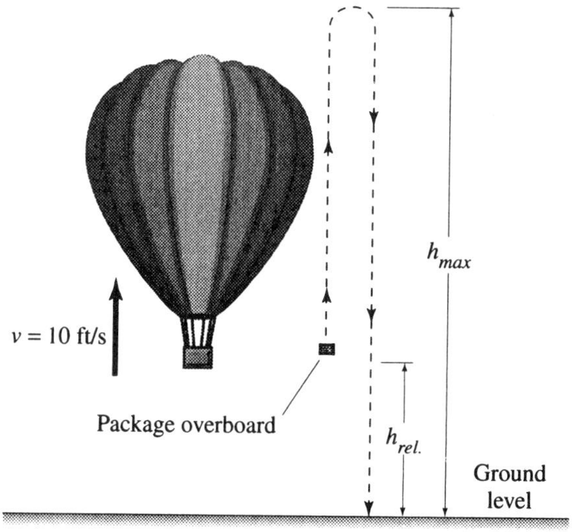

FIGURE P17.2.

hits the ground. Neglect air resistance and find (a) the maximum height above ground level reached by the package and (b) the height above ground level when it was eased overboard.

17.3 An automobile starts at point A with a velocity $v_A = 2.625$ m/s and travels a distance s on a straight level road at constant acceleration to point B, where the velocity is $v_B = 21.375$ m/s, as shown in Figure P17.3. If the time it took to travel from A to B is $t = 25$ s, determine (a) the constant acceleration, (b) the distance s, and (c) the distance traveled in the last 5 s.

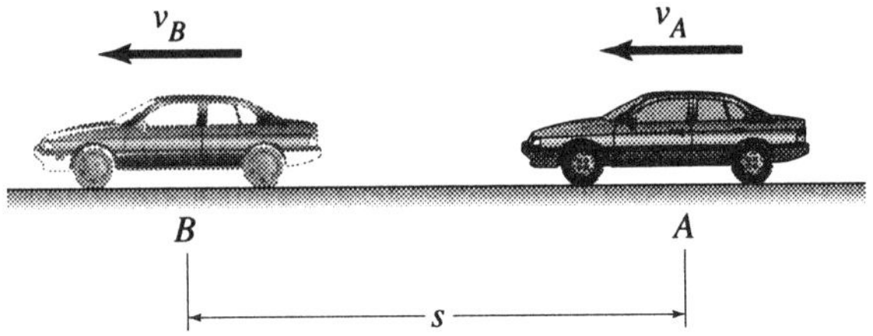

FIGURE P17.3.

17.4 An automobile starts at point A with a velocity v_A and travels a distance s on a straight level road at constant acceleration a to point B, where the velocity is v_B as shown in Figure P17.3. If the time it took to travel the distance from A to B is t, develop expressions for v_A and v_B in terms of s, t, and a. What are the magnitudes of v_A and v_B for $s = 1000$ ft, $t = 20$ s and $a = 5$ ft/s^2.

17.5 The engineer of a locomotive traveling at 60 mph on a 1% straight grade notices a stalled car on the tracks at some distance s away as shown in Figure P17.5. He immediately applies the brakes, thus, reducing the speed of the locomotive at constant deceleration and stopping in time $t = 30$ s just prior to crashing into the stalled car. Determine (a) the constant deceleration and (b) the distance s.

FIGURE P17.5.

17.6 An acrobat is projected vertically upward with an initial velocity v from end A by dropping a weight W at end B of the lever arrangement shown in Figure P17.6. If the time it takes the acrobat to return to her initial position is $t = 2$ s, determine (a) the initial velocity v and (b) the maximum height h reached by the acrobat. Assume the motion of lever to be negligible.

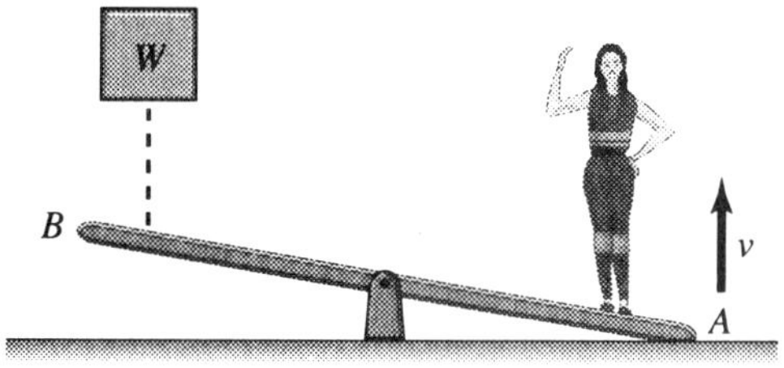

FIGURE P17.6.

17.7 The acceleration a due to Earth's gravity decreases as the altitude y increases above sea level. The relationship may be approximated by the equation $a = \dfrac{D^2 g}{(D + 2y)^2}$ where D is the diameter of Earth and g is the acceleration of gravity at sea level. Develop a relationship between the initial velocity v of a body that is projected vertically upward and the height h reached by the body. Ignore air resistance and express this relationship as $h = h(v)$. If the *escape velocity* v_E is defined as the initial velocity for which h becomes infinite, show that $v_E = \sqrt{gD}$. If $g = 32.2$ ft/s^2 and $D = 41.8 \times 10^6$ ft, determine the escape velocity. What initial velocity should be imparted to a body if it is to reach a height of 4 miles?

17.8 A destroyer releases a depth charge which hits the surface of the water with a vertical velocity which is sufficiently small and may be assumed to be zero. The depth charge moves vertically downward through the water as shown in Figure P17.8. The deceleration due to the resistance of sea water and to buoyancy may be approximated by the expression $a' = gC^2v^2$ where C is a constant, g is the acceleration due to gravity, and v is the velocity of the charge. Develop an expression for the velocity of the depth charge as a function of the constant C and the time t. Use this expression to show that for a sufficiently large value of t, the velocity v approaches a constant value equal to $(1/C)$ which is known as the terminal velocity. (Hint: The acceleration of the depth charge is $a = g - gC^2v^2$).

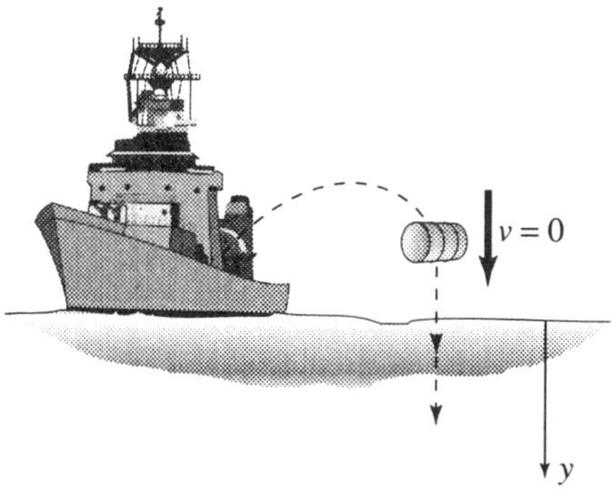

FIGURE P17.8.

17.9 Refer to Problem 17.8 and develop an expression for the velocity of the depth charge as a function of the constant C and the depth y below the surface of the water. Use this expression to show that, for a sufficiently large value of y, the velocity v approaches a constant value equal to $(1/C)$ which is known as the terminal velocity.

17.10 The acceleration of a high speed monorail moving along a straight track, as shown in Figure P17.10, may be approximated by the equation $a = 0.5t^2$ where a is the acceleration in ft/s^2 and t is the time in seconds. When the acceleration is zero, the displacement is zero, but the speed of the monorail is v_1 in

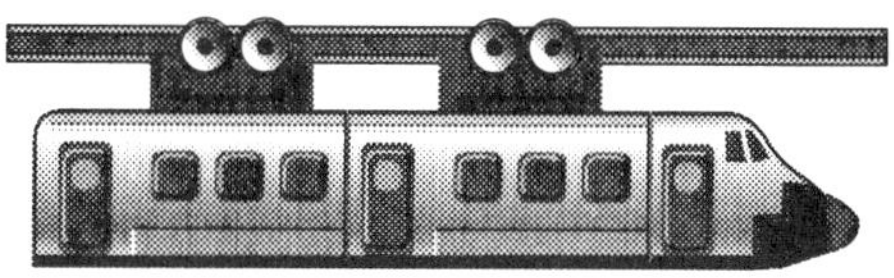

FIGURE P17.10.

ft/s. Derive expressions for the speed and displacement of the monorail in terms of v_1 and t. If $v_1 = 60$ mph, compute the time it takes the monorail to reach a speed of 200 mph and the distance traveled during this period.

17.11 Rod AB is hinged by means of frictionless hinges to two identical wheels as shown in Figure P17.11. The two wheels rotate in a horizontal plane about their respective geometric axes such that $\theta = 3t^2 - t$ where θ is in radians and t is the time in seconds. Develop vector expressions in terms of R and t for the velocity and acceleration of rod AB. What are the magnitudes and directions of these quantities when $\theta = 60°$, $R = 0.8$ m, and $t = 5$ s?

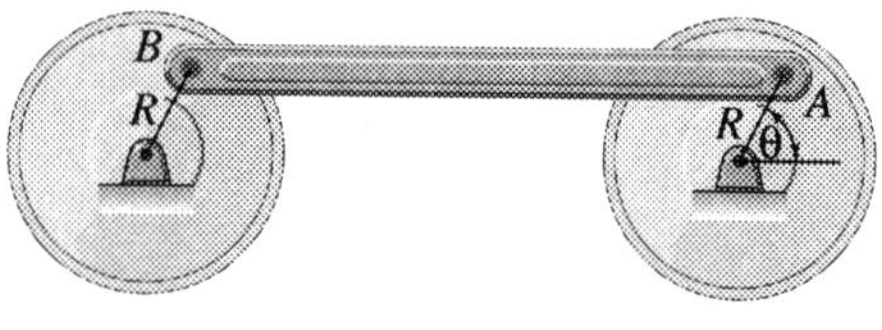

FIGURE P17.11.

17.12 A proposed ride for an amusement park consists of two beams AB and CD hinged to two identical wheels that rotate in a vertical plane as shown in Figure P17.12. The passenger seats are attached rigidly to the two beams. During the accelerating period which takes 20 s, the angle θ may be approximated by the relationship $\theta = 0.03t^2$ where is in radians and t is in seconds. Develop vector expressions in terms of L and t for the velocity and acceleration experienced by the passengers during the accelerating period. What are the magnitudes and directions of these quantities for $t = 20$ s and $L = 5$ ft?

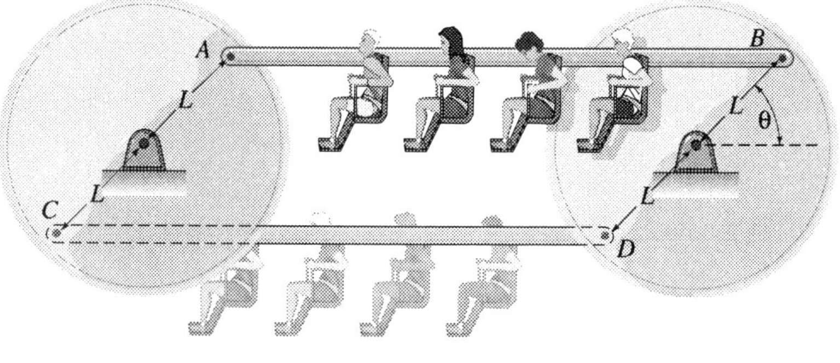

FIGURE P17.12.

17.13 A proposed ride for an amusement park consists of a gondola that is constrained to move along a vertical curve defined by the equation $r = A\cos\theta$ as shown in Figure P17.13. Neglect any rocking action of the gondola during motion and assume $\theta = \frac{\pi}{30}$ t where θ is in radius and t is the time in seconds. Find the velocity and acceleration of the gondola at the instant when $t = 10$ s if A = 10 m.

17.14 Repeat Problem 17.13 for the case where $r = A\sin^2\theta$ as shown in Figure P17.14. In this case $\theta = \frac{\pi}{60}$ t where θ is in radians and t is in seconds.

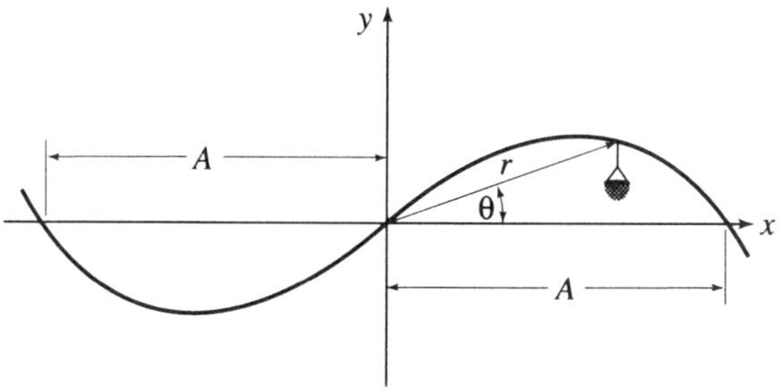

FIGURE P17.13.

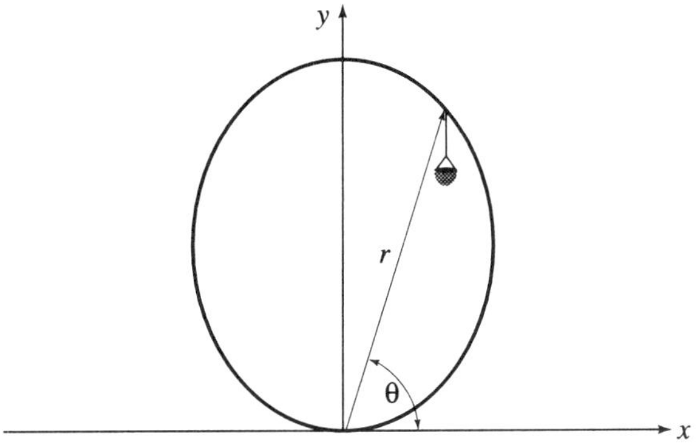

FIGURE P17.14.

17.15 The system shown in Figure P17.15 is used to transport people from a level defined by $\theta = 0$ to other levels for the range $(0 \leq \theta \leq \pi)$. Assume that $\theta = 0.5t^{3/2}$ where θ is in radians and t is in seconds. Develop expressions, in terms of L and t, for the velocity and acceleration of the passengers. If the acceleration of the passengers is to be 6 ft/s^2 when $\theta = \pi/2$, what should be the dimension L? What is the velocity for this position?

FIGURE P17.15.

17.2 Rotation About a Fixed Axis

The rigid body shown in Figure 17.4(a) is assumed to rotate about the fixed x axis. In such a case, any particle, such as A, on a radial line BA describes a circular path whose radius is r. The rotating radial line BA defines a plane perpendicular to the fixed x axis of rotation. By definition, because a rigid body consists of particles that are at fixed distances with respect to each other, it follows that the rotation of a rigid

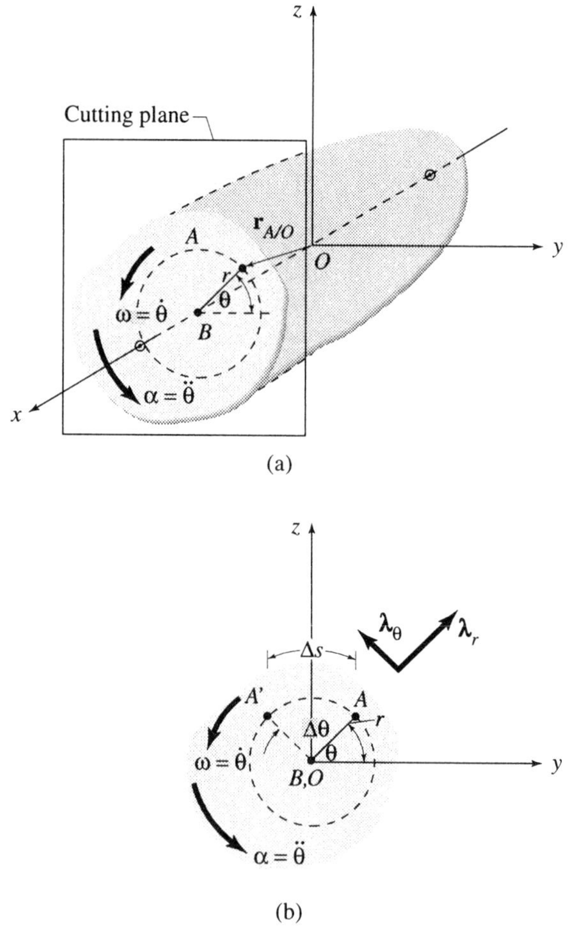

FIGURE 17.4.

body about a fixed axis is fully known if the rotation of one of its particles about this axis is fully defined. Furthermore, as stated earlier, the radial line, on which this particle is located, defines a plane perpendicular to the axis of rotation. It follows, therefore, that we can investigate the rotation of a rigid body about a fixed axis by examining the rotation of a radial line that lies in a thin plate cut from the rigid body by two neighboring cutting planes perpendicular to the axis of rotation. One such cutting plane is shown in Figure 17.4(a) and the resulting thin plate is shown in Figure 17.4(b) which depicts a two-dimensional situation in which the x axis comes out of the page.

Angular Position

The *angular position* of any radial line, such as BA, in the thin plate of Figure 17.4(b) may be specified by giving the angle θ that it makes with some convenient reference line such as the y axis. This angular position

θ is usually measured in radians, but, occasionally, it may be given in angular degrees or in revolutions. During a short interval of time Δt, the radial line BA rotates about the x axis through a small angle $\Delta\theta$ known as the *angular displacement*. Note that, during this displacement, points on the radial line move to new positions. For example, point A moves to A′ through a small arc length Δs.

Angular Velocity

By definition, the average angular velocity ω_{avg} of the radial line is

$$\omega_{avg} = \dot{\theta}_{avg} = \frac{\Delta\theta}{\Delta t} \tag{17.5}$$

The instantaneous angular velocity ω is obtained from Eq. (17.5) by evaluating the ratio $\frac{\Delta\theta}{\Delta t}$ in the limit when Δt approaches zero. Thus,

$$\omega = \lim_{\Delta t \to 0} \left(\frac{\Delta\theta}{\Delta t}\right) = \frac{d\theta}{dt} = \dot{\theta} \tag{17.6}$$

where $\dot{\theta}$ represents the first time derivative of θ.

Angular Acceleration

During the time interval Δt, if the instantaneous angular velocity ω changes by $\Delta\omega$, the *average angular acceleration* α_{avg} is defined by

$$\alpha_{avg} = \ddot{\theta}_{avg} = \frac{\Delta\omega}{\Delta t} \tag{17.7}$$

which, when evaluated in the limit for Δt approaching zero, yields the *instantaneous angular acceleration* α. Thus,

$$\begin{aligned} \alpha &= \lim_{\Delta t \to 0} \left(\frac{\Delta\omega}{\Delta t}\right) = \frac{d\omega}{dt} \\ &= \frac{d\dot{\theta}}{dt} = \frac{d^2\theta}{dt^2} \\ &= \dot{\omega} = \ddot{\theta} \end{aligned} \tag{17.8}$$

where $\dot{\omega}$ is the first time derivative of ω and $\ddot{\theta}$ the second time derivative of θ. The relationship in Eq. (17.8) may be used to develop a useful relationship among the angular acceleration, angular velocity, and angular displacement. Thus,

$$\alpha = \frac{d\omega}{dt} = \left(\frac{d\omega}{d\theta}\right)\left(\frac{d\theta}{dt}\right) = \omega\frac{d\omega}{d\theta} \tag{17.9}$$

A positive angular velocity ω would be directed in the same sense as a positive angular displacement $d\theta$. If ω increases, the angular acceleration is positive and would be measured in the same sense as ω. If, however, ω decreases, the angular acceleration is negative, in which case it is directed in a sense opposite to ω and is referred to as an *angular deceleration.*

Constant Angular Acceleration

When the angular acceleration of a rotating rigid body is constant, special relationships can be developed among the angular displacement, angular velocity, angular acceleration, and time. Thus, from Eq. (17.8), $d\omega = \alpha_C\, dt$ where α_C is the constant angular acceleration. Integration of this equation using the initial conditions $\omega = \omega_0$, when $t = 0$ for lower limits, and general upper limits leads to

$$\int_{\omega_0}^{\omega} d\omega = \alpha_C \int_0^t dt$$

from which

$$\omega = \omega_0 + \alpha_C t. \tag{17.10}$$

Also, from Eq. (17.6), $d\theta = \omega\, dt$ where ω is given by Eq. (17.10). Integrating this equation using the initial conditions $\theta = \theta_0$, when $t = 0$ for lower limits, and general upper limits,

$$\int_{\theta_0}^{\theta} d\omega = \alpha_C \int_0^t (\omega_0 + \alpha_C t)\, dt$$

from which

$$\theta = \theta_0 + \omega_0 t + \tfrac{1}{2}\alpha_C t^2 \tag{17.11}$$

Finally, from Eq. (17.9), $\omega\, d\omega = \alpha_C\, d\theta$. If we integrate this equation using for lower limits the initial conditions $\omega = \omega_0$, when $t = 0$, and general upper limits,

$$\int_{\omega_0}^{\omega} \omega\, d\omega = \alpha_C \int_{\theta_0}^{\theta} d\theta$$

from which

$$\omega^2 = \omega_0^2 + 2\alpha_C(\theta - \theta_0) \tag{17.12}$$

Note the mathematical similarities between the set of equations derived here for the rotation of a rigid body and the set derived in Chapter 13 for the rectilinear motion of a particle.

Motion of a Particle

Position: The position of any particle, such as A, on a rotating line is given by the position vector $\mathbf{r}_{A/O}$ as shown in Figure 17.4(a).

Velocity: The linear velocity of any particle, such as A, on the rotating radial line BA in Figure 17.4(b) may be found from Eq. (13.42a) which expresses the velocity of a particle in polar coordinates. In our particular case, because r is constant, it follows that $\dot{r}$ vanishes and the first term in Eq. (13.42a) is zero. Thus, the linear velocity of any particle, such as A, becomes

$$\mathbf{v} = (r\dot{\theta})\boldsymbol{\lambda}_\theta = (r\omega)\boldsymbol{\lambda}_\theta \tag{17.13a}$$

where the product $r\omega$ is its magnitude and the unit vector $\boldsymbol{\lambda}_\theta$ signifies the fact that its direction is perpendicular to the radius r_A. Thus,

$$v = r\omega. \tag{17.13b}$$

If the sign of $\mathbf{v}$ is positive, its sense is in the direction of $\boldsymbol{\lambda}_\theta$; if negative, its sense is opposite to that of $\boldsymbol{\lambda}_\theta$. Note that, because the path of the particle is circular, we could use the unit vectors $\boldsymbol{\lambda}_n$ (normal) and $\boldsymbol{\lambda}_t$ (tangential), respectively, instead of $\boldsymbol{\lambda}_r$ and $\boldsymbol{\lambda}_\theta$.

Acceleration

The linear acceleration of particle A may be obtained by using Eq. (13.43a) which expresses the acceleration of a particle in polar coordinates. Again, in our particular case, r is constant, and $\dot{r} = \ddot{r} = 0$. Therefore, Eq. (13.43a) reduces to

$$\mathbf{a} = -(r\dot{\theta}^2)\boldsymbol{\lambda}_r + (r\ddot{\theta})\boldsymbol{\lambda}_\theta = -(r\omega^2)\boldsymbol{\lambda}_r + (r\alpha)\boldsymbol{\lambda}_\theta. \tag{17.14a}$$

We recognize the first term as the *radial* or *normal* component of the acceleration whose *magnitude* is $r\omega^2$ and points in the negative direction of the unit vector or toward the center of rotation. Thus,

$$a_r = a_n = r\omega^2 = v\omega = \frac{v^2}{r} \tag{17.14b}$$

The second term in Eq. (17.14a) is the *transverse* or *tangential* component of acceleration whose magnitude is $r\alpha$ which, according to the unit vector $\boldsymbol{\lambda}_\theta$, is directed along the tangent to the circular path of particle A. Therefore,

$$a_\theta = a_t = r\alpha. \tag{17.14c}$$

Vector Representation

Equations (17.13) and (17.14), expressing the velocity and acceleration of any particle in a rotating rigid body, may also be expressed in terms of the vector (cross) product of vectors. This formulation is sometimes convenient and advantageous, particularly when dealing with the three-dimensional motion of a rigid body. Consider the rigid body shown in Figure 17.5 that is rotating about the fixed x axis. The position vector $\mathbf{r}_{A/O}$ locates any particle, such as A, in the rigid body. Because the magnitude of this position vector is $r_{A/O}$, then, the radial distance from the axis of rotation to particle A is $r = (r_{A/O})\sin\psi$ where ψ is the angle between $r_{A/O}$ and the x axis. Therefore, by Eq. (17.13b), $v = \omega(r_{A/O})\sin\psi$. We recognize this quantity as the magnitude of the vector

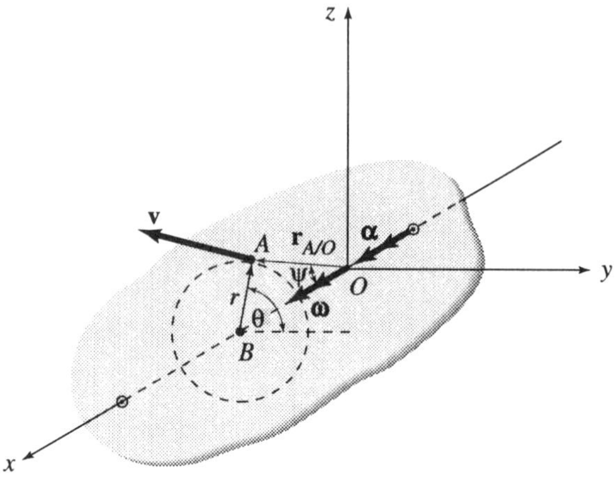

FIGURE 17.5.

product of the two vectors $\boldsymbol{\omega}$ and $\mathbf{r}_{A/O}$. Thus, the velocity of particle A may be expressed in vector form by the equation

$$\mathbf{v} = \boldsymbol{\omega} \times \mathbf{r}_{A/O} \tag{17.15}$$

which yields a vector whose magnitude is $\omega(r_{A/O})\sin\psi = \omega r$, whose direction is perpendicular to the plane formed by $\boldsymbol{\omega}$ and $\mathbf{r}_{A/O}$, and whose sense is given by the right-hand rule. The vectors $\mathbf{v}$ and $\boldsymbol{\omega}$ are shown in Figure 17.5. Note the double-headed arrows used for $\boldsymbol{\omega}$ and $\boldsymbol{\alpha}$ to distinguish them from the linear quantities $\mathbf{v}$ and $\mathbf{a}$ represented by single-headed arrows.

The acceleration of particle A may be obtained by differentiating, with respect to time, the velocity vector given in Eq. (17.15). Thus,

$$\mathbf{a} = \frac{d\mathbf{v}}{dt} = \boldsymbol{\omega} \times \frac{d\mathbf{r}_{A/O}}{dt} + \frac{d\boldsymbol{\omega}}{dt} \times \mathbf{r}_{A/O} \tag{17.16}$$

where $\dfrac{d\mathbf{r}_{A/O}}{dt} = \dot{\mathbf{r}}_{A/O} = \mathbf{v} = \boldsymbol{\omega} \times \mathbf{r}_{A/O}$ and $\dfrac{d\boldsymbol{\omega}}{dt} = \dot{\boldsymbol{\omega}} = \boldsymbol{\alpha}$. Therefore,

$$\mathbf{a} = \boldsymbol{\omega} \times (\boldsymbol{\omega} \times \mathbf{r}_{A/O}) + \boldsymbol{\alpha} \times \mathbf{r}_{A/O} = \boldsymbol{\omega} \times \mathbf{v} + \boldsymbol{\alpha} \times \mathbf{r}_{A/O} \tag{17.17}$$

Since the angle between $\boldsymbol{\omega}$ and $\mathbf{v}$ is 90°, the first term in Eq. (17.17) has a magnitude of ωv which, by Eq. (17.14b), is also equal to $r\omega^2 = \dfrac{v^2}{r}$. Also, by the right-hand rule, the quantity $\boldsymbol{\omega} \times \mathbf{v}$ is directed toward the center of rotation, point **B**. Therefore, the first term in Eq. (17.17) represents $\mathbf{a}_n$, the normal component of acceleration of particle A. The second term in Eq. (17.17) has a magnitude equal to $\alpha r_{A/O}\sin\psi = \alpha r$ and, by the right-hand rule, a direction tangent to the circular path of particle A. It follows, therefore that the second term in Eq. (17.17) represents $\mathbf{a}_t$, the tangential component of acceleration of particle A. Thus, Eq. (17.17) may be written in the form

$$\mathbf{a} = \mathbf{a}_n + \mathbf{a}_t \tag{17.18a}$$

where

$$\left.\begin{aligned} \mathbf{a}_n &= \boldsymbol{\omega} \times \mathbf{v} \\ \mathbf{a}_t &= \boldsymbol{\alpha} \times \mathbf{r}_{A/O} \end{aligned}\right\} \tag{17.18b}$$

Also, the magnitudes of these acceleration components are given by

$$\left.\begin{aligned} a_n &= \omega v = r\omega^2 = \frac{v^2}{r} \\ a_t &= r\alpha \end{aligned}\right\}. \tag{17.18c}$$

The following examples illustrate some of the above concepts.

Example 17.4

The pulley-belt system shown in Figure E17.4 consists of the motorized pulley A which drives pulley B. During the accelerating period, the cw angular acceleration of pulley A may be approximated by the relationship $\alpha_A = \left(\frac{1}{\pi^2}\right)\theta_A$ where α_A is in rad/s^2 and θ_A is in radians. Assume no slippage between the belt and the pulleys and that the system starts from rest (i.e., $\theta = 0$ for $\omega = 0$). Derive expressions for the angular velocity ω_B and angular acceleration α_B of pulley B in terms of those for pulley A.

Solution

Consider point C which is common to both pulley A and the belt. By Eq. (17.13b),

$$v_C = r_A \omega_A. \tag{a}$$

Also consider point D which is common to pulley B and to the belt. By Eq. (17.13b),

$$v_D = r_B \omega_B. \tag{b}$$

If we assume the belt to be inextensible, it follows that $v_C = v_D$. Therefore, $r_A\omega_A = r_B\omega_B$, from which

$$\omega_B = \left(\frac{r_A}{r_B}\right)\omega_A \circlearrowright. \qquad \text{(c) ANS.}$$

Similarly, by Eq. (17.14c), the tangential accelerations of points C and D may be found in terms of the angular accelerations of pulleys A and

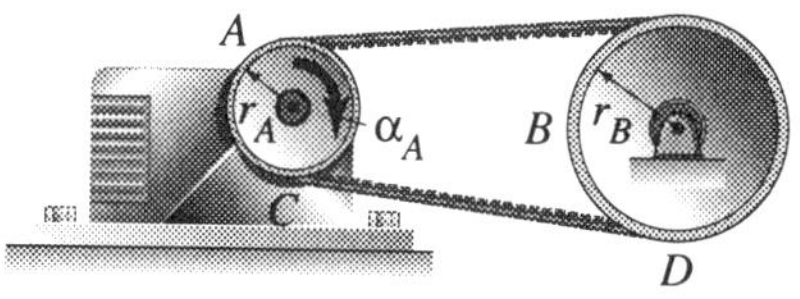

FIGURE E17.4.

B, respectively. Thus,

$$(a_t)_C = r_A\alpha_A, \tag{d}$$

and

$$(a_t)_D = r_B\alpha_B. \tag{e}$$

For the same reason given earlier, $(a_t)_C = (a_t)_D$ which implies that $r_A\alpha_A = r_B\alpha_B$. Hence,

$$\alpha_B = \left(\frac{r_A}{r_B}\right)\alpha_A \circlearrowleft. \qquad \text{(f) ANS.}$$

■ Example 17.5

Refer to Example 17.4 and determine numerical values for ω_B and for α_B for the instant when $\theta_A = 25$ revolutions and for the case where $r_A = \frac{1}{2}r_B$.

Solution

Because the angular acceleration of pulley A is expressed in terms of its angular displacement, Eq. (17.9) may be used to determine the angular velocity of pulley A. Thus, by Eq. (17.9),

$$\int_0^\omega \omega\, d\omega = \int_0^\theta \alpha\, d\theta \tag{g}$$

Substituting the given relationship between α and θ in Eq. (g), after simplification,

$$\omega = \left(\frac{1}{\pi}\right)\theta. \tag{h}$$

Therefore, for $\theta = 25$ rev. $= 50\pi$ rad, from Eq. (h),

$$\omega_A = \left(\frac{1}{\pi}\right)50\pi = 50.0 \text{ rad/s} \circlearrowleft. \qquad \text{ANS.}$$

From Eq. (c),

$$\omega_B = \tfrac{1}{2}(50.0) = 25.0 \text{ rad/s} \circlearrowleft. \qquad \text{ANS.}$$

From the given relationship $\alpha_A = \left(\frac{1}{\pi^2}\right)\theta_A$, we may find the angular acceleration of pulley A when $\theta_A = 25$ rev. $= 50\pi$ rad. Thus,

$$\alpha_A = \left(\frac{1}{\pi^2}\right)50\pi = 15.92 \text{ rad/s}^2 \circlearrowleft. \qquad \text{ANS.}$$

From Eq. (f)

$$\alpha_B = \tfrac{1}{2}(15.92) = 7.96 \text{ rad/s}^2 \circlearrowleft. \qquad \text{ANS.}$$

■ Example 17.6

The pulley arrangement shown in Figure E17.6 is used to move the weight W up the inclined plane. The system starts from rest and the applied load P imparts a constant acceleration a to points on the vertical cable. Determine the velocity and acceleration of the weight W after it has moved a distance s along the inclined plane. Express your answers symbolically in terms of r_1, r_2, s, and a. If $a = 4\ \text{m/s}^2$ and $r_2 = 2r_1$, determine the velocity and acceleration of the weight W after it has moved a distance $s = 10$ m. Assume the belt to be inextensible and that there is no slippage between it and the pulley system.

Solution

When the weight W moves a distance s along the inclined plane, the pulley system rotates through $\theta = \dfrac{s}{r_1}$ rad. Because the acceleration a of points on the vertical cable is constant, point A, which is common to both the vertical cable and to the large pulley, has a constant acceleration equal to a. Insofar as the pulley is concerned, a is the tangential acceleration of point A. Thus, by Eq. (17.14c), the constant angular acceleration of the pulley becomes

$$\alpha_C = \frac{a_t}{r} = \frac{a}{r_2} \circlearrowleft$$

Now, since the angular acceleration of the pulley system is constant, Eq. (17.12) may be used to determine its angular velocity for any angular displacement θ. Therefore, because the system starts from rest, $\theta_0 = \omega_0 = 0$ in Eq. (17.12), and

$$\omega^2 = 0 + 2\alpha_C\theta.$$

If we substitute the values obtained earlier for α_C and for θ,

$$\omega^2 = 2\left(\frac{a}{r_2}\right)\left(\frac{s}{r_1}\right),$$

and

$$\omega = \sqrt{2\left(\frac{a}{r_2}\right)\left(\frac{s}{r_1}\right)}.$$

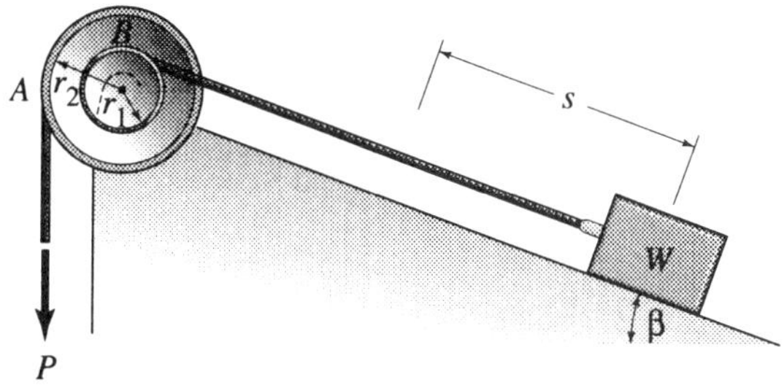

FIGURE E17.6.

The velocity and acceleration of point B, which is common to the small pulley and to the inclined cable, may now be determined from Eqs. (17.13b) and (17.14c) as shown below. Note that the velocity of the weight W is the same as that of point B on the small pulley. Note also that the acceleration of the weight W is the same as the tangential component of acceleration of point B on the small pulley. Thus,

$$v_W = v_B = r\omega = r_1\sqrt{2\left(\frac{a}{r_2}\right)\left(\frac{s}{r_1}\right)} = \sqrt{2\left(\frac{r_1}{r_2}\right)sa}. \qquad \text{ANS.}$$

Also

$$a_W = (a_B)_t = r\alpha = r_1\left(\frac{a}{r_2}\right) = \left(\frac{r_1}{r_2}\right)a \qquad \text{ANS.}$$

For the numerical values provided in the problem,

$$v_W = \sqrt{2(\tfrac{1}{2})(10)(4)} = 6.32 \text{ m/s}, \qquad \text{ANS.}$$

$$a_W = \tfrac{1}{2}(4) = 2.00 \text{ m/s}^2. \qquad \text{ANS.}$$

Example 17.7

At the instant shown the rigid body of Figure E17.7 rotates about axis AD with an angular velocity $\omega = 20$ rad/s and an angular acceleration $\alpha = 10$ rad/s^2. The directions of ω and α are as shown. Determine the velocity and acceleration of point B.

Solution

In view of the three-dimensional nature of this problem, it is advantageous to employ Eqs. (17.15) and (17.17) for the solution. Thus, the unit vector from A to D is

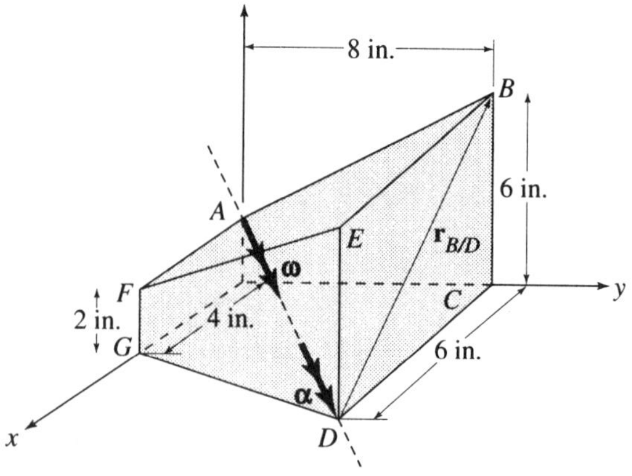

FIGURE E17.7.

$$\boldsymbol{\lambda}_{AD} = \frac{6\mathbf{i} + 8\mathbf{j} - 2\mathbf{k}}{\sqrt{6^2 + 8^2 + 2^2}} = 0.588\mathbf{i} + 0.784\mathbf{j} - 0.196\mathbf{k}.$$

Therefore,

$$\boldsymbol{\omega} = 20\boldsymbol{\lambda}_{AD} = (11.76\mathbf{i} + 15.68\mathbf{j} - 3.92\mathbf{k}) \text{ rad/s},$$

and

$$\boldsymbol{\alpha} = 10\boldsymbol{\lambda}_{AD} = (5.88\mathbf{i} + 7.84\mathbf{j} - 1.96\mathbf{k}) \text{ rad/s}^2.$$

Also, a radius vector to point B may be established from any point on the axis of rotation AD. It is more convenient in this case to use the radius vector $\mathbf{r}_{B/D}$ than $\mathbf{r}_{B/A}$. Thus,

$$\mathbf{r}_{B/D} = (-6\mathbf{i} + 6\mathbf{k}) \text{ in.}$$

Now, by Eq. (17.15)

$$\mathbf{v}_B = \begin{bmatrix} \mathbf{i} & \mathbf{j} & \mathbf{k} \\ 11.76 & 15.68 & -3.92 \\ -6 & 0 & 6 \end{bmatrix} = (94.1\mathbf{i} - 47.0\mathbf{j} + 94.1\mathbf{k}) \text{ in./s},$$

$$= 141.1(0.667\mathbf{i} - 0.333\mathbf{j} + 0.667\mathbf{k}) \text{ in./s.} \qquad \text{ANS.}$$

By Eq. (17.17),

$$\mathbf{a}_B = \boldsymbol{\omega} \times \mathbf{v}_B + \boldsymbol{\alpha} \times \mathbf{r}_{B/D} = \begin{bmatrix} \mathbf{i} & \mathbf{j} & \mathbf{k} \\ 11.76 & 15.68 & -3.92 \\ -6 & 0 & 6 \end{bmatrix}$$

$$+ \begin{bmatrix} \mathbf{i} & \mathbf{j} & \mathbf{k} \\ 5.88 & 7.84 & -1.96 \\ -6 & 0 & 6 \end{bmatrix},$$

$$= 2821(0.458\mathbf{i} - 0.523\mathbf{j} - 0.719\mathbf{k})$$

$$+ 70.6(0.667\mathbf{i} - 0.333\mathbf{j} - 0.667\mathbf{k}) \text{ in./s}^2 \qquad \text{ANS.}$$

where the first term is the normal component of acceleration of point B and points toward the axis of rotation AD and the second term is the tangential component of its acceleration and is tangent to the path of point B. Alternatively, the two components of acceleration may be combined to give the total acceleration of point B as follows

$$\mathbf{a}_B = (1338\mathbf{i} - 1499\mathbf{j} - 198\mathbf{k}) \text{ in./s}^2,$$

$$= 2820(0.474\mathbf{i} - 0.531\mathbf{j} - 0.702\mathbf{k}) \text{ in./s}^2. \qquad \text{ANS.}$$

■

Problems

17.16 After the power is cut off, the rotation of an engine flywheel may be approximated by the equation $\theta = 20t - t^2$ where θ is in radians and t is the time in seconds. Determine (a) the number of revolutions turned in $t = 5$ s as well as the angular velocity and angular acceleration at this instant of time and (b) the total number of revolutions turned and the time elapsed before the flywheel comes to a complete stop.

17.17 The accelerated rotation of an electric motor may be approximated by the equation $\theta = 0.1t^{3/2}$ where θ is the angular displacement in radians and t is the time in seconds. Determine the angular displacement, angular velocity and angular acceleration when $t = 1$ minute.

17.18 The wheel shown in Figure P17.18 starts from rest and rotates with a constant angular acceleration α. Determine the velocity and acceleration of point A after the wheel has turned through θ radians. How much time t does it take for this angular displacement? Express your answers in terms of θ, α, and D. If $\alpha = 5$ rad/s^2, $\theta = 2$ rad, and $D = 4$ ft, determine the magnitudes of t, v_A, and a_A.

17.19 The wheel shown in Figure P17.18 starts from rest and rotates with an angular acceleration a given by the relationship $\alpha = (\frac{1}{100})\theta^2$ where α is in rad/s^2 and θ is in radians. Develop expressions, in terms of D and θ, for the velocity and acceleration of point A. Find the magnitudes of the velocity and acceleration of point A for the case when $D = 1$ m and $\theta = 3$ radians.

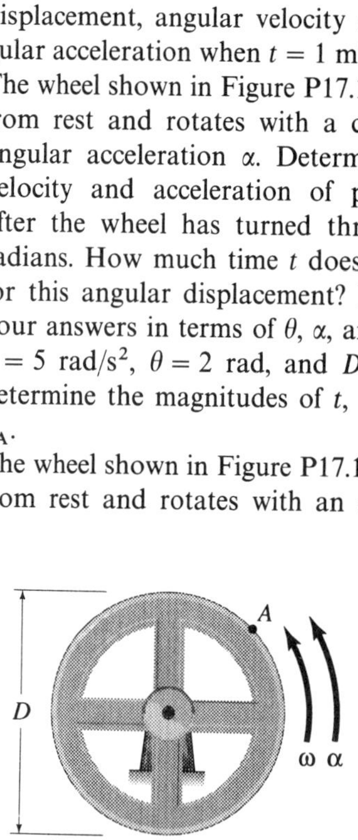

FIGURE P17.18.

17.20 The wheel shown in Figure P17.18 starts from rest and rotates with an angular acceleration α given by the relation $\alpha = 5t^{3/2}$ where α is in rad/s^2 and t is the time in seconds. (a) Develop expressions, in terms of D and t, for the velocity and acceleration of point A. Find the magnitudes of the velocity and acceleration of point A for the case when $D = 24$ in. and $t = 1.5$ s. Show these quantities on a sketch of the wheel. (b) Develop a relation between θ and t.

17.21 The motor-pulley system shown in Figure P17.21 is used to raise the weight W. Assume that the motor accelerates at a constant angular acceleration α_1. Develop a relationship between the acceleration a_W of the weight W, the angular acceleration α_1, and the radii of the pulleys. Also develop a relationship between the velocity v_W of the weight W and the angular velocity ω_1 of the motor and the radii of the pulleys.

17.22 Refer to Figure P 17.21. Assume that the motor starts from rest and accelerates at a constant angular acceleration $\alpha_1 = 5$ rad/s^2. Let $r_1 = 0.05$ m, $r_2 = 0.12$ m, $r_3 = 0.07$ m, $r_4 = 0.12$ m, $r_5 = 0.16$ m and determine the velocity and acceleration of the weight W after a period of 10 s. What is the distance moved by the weight during this time interval?

17.23 Refer to Figure P17.21. Assume that the motor starts from rest and accelerates

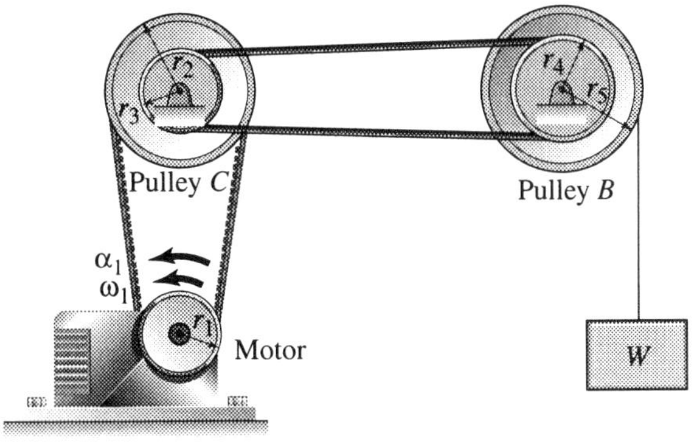

FIGURE P17.21.

with an angular acceleration $\alpha_1 = \frac{1}{30}\theta_1^{3/2}$, where α_1 is in rad/s^2 and θ_1 is the angular displacement of the motor in rad. Assume that when $\theta_1 = 0$, the angular velocity of the motor, $\omega_1 = 0$. Let $r_1 =$ 2.0 in., $r_2 = 5.0$ in., $r_3 = 4.0$ in., $r_4 =$ 6.0 in., $r_5 =$ 7.0 in., and determine the angular velocity and angular acceleration of pulley B when $\theta_1 = 25$ radians.

17.24 The weight W shown in Figure P17.24 is moving upward with a velocity v_W and an acceleration a_W. In terms of v_W, a_W, r_1, r_2, r_3 and r_4, develop expressions for the angular velocity ω_1 and angular acceleration α_1 of gear 1. If $v_W = 0.25$ m/s, $a_W = 0.05$ m/s^2, $r_1 = 0.08$ m, $r_2 = 0.04$ m, $r_3 = 0.03$ m and $r_4 = 0.08$ m, determine ω_1 and α_1.

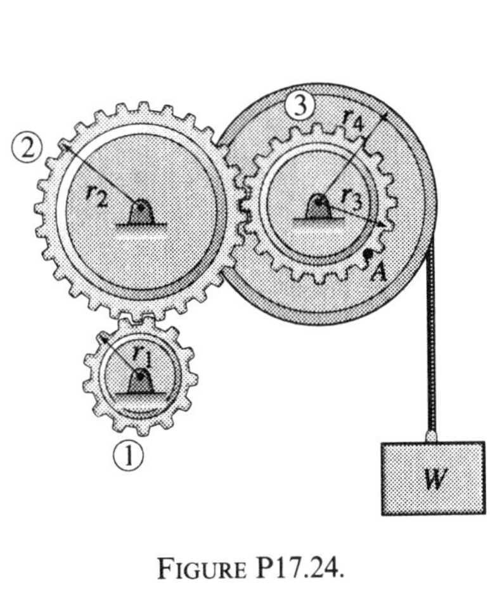

FIGURE P17.24.

17.25 Refer to the gear system shown in Figure P17.24. If the angular velocity and angular acceleration of gear 2, are $\omega_2 =$ 5 rad/s and $\alpha_2 = 2$ rad/s^2, respectively, determine (a) the magnitudes of the velocity and acceleration of point A on gear 3 and (b) the angular velocity and angular acceleration of gear 1. Assume $r_1 = 0.01$ m, $r_2 = 0.05$ m, $r_3 = 0.025$ m, and $r_4 = 0.08$ m.

17.26 Gear 1 in Figure P17.26 has an angular velocity ω_1 and an angular acceleration α_1. In terms of ω_1, α_1, r_1 and r_2, develop expressions for the angular velocity ω_2 and angular acceleration α_2 of gear 2. If $\omega_1 = 3$ rad/s, $\alpha_1 = 4$ rad/s^2, $r_1 = 6$ in. and $r_2 = 4$ in. determine ω_2 and α_2.

17.27 Refer to the gear system shown in Figure P17.26. The rack is moving such that its velocity is 2 ft/s to the right and its acceleration is 3 ft/s^2 to the left. Determine the angular velocities and angular accelerations of gears 1 and 2.

17.28 Refer to the gear system shown in Figure P17.26. Gear 2 starts from rest and accelerates at a constant rate α_2. In terms of α_2, t, r_1 and r_2, develop expres-

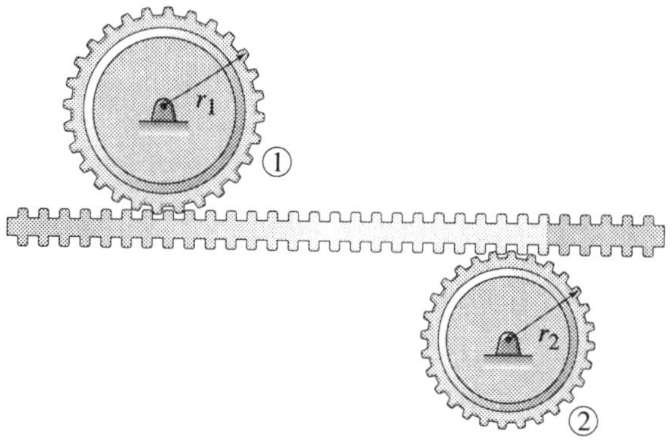

FIGURE P17.26.

sions, for the angular velocity ω_1 and angular acceleration α_1 of gear 1 after a time t has elapsed. If $\alpha_2 = 0.5$ rad/s^2, $t = 5$ s and $r_1 = 1.5\ r_2$, what are the values of ω_1 and α_1?

17.29 A friction drive consists of the two wheels A and B, as shown in Figure P17.29. At a given instant, wheel A has an angular velocity ω_A and is decelerating at a constant rate α_A. At the same instant, wheel B starts with an angular velocity ω_B which is less than ω_A and accelerates at a constant rate a_B. In terms of ω_A, ω_B, α_A, α_B, r_A, and r_B, determine the elapsed time t before wheel B can be brought in contact with wheel A if there is to be no slippage between them. If $\omega_A = 3$ rad/s, $\omega_B = 90$ rad/s, $\alpha_A = \alpha_B = 5$ rad/s^2, and $r_A = 2r_B = 0.20$ m, determine the elapsed time t.

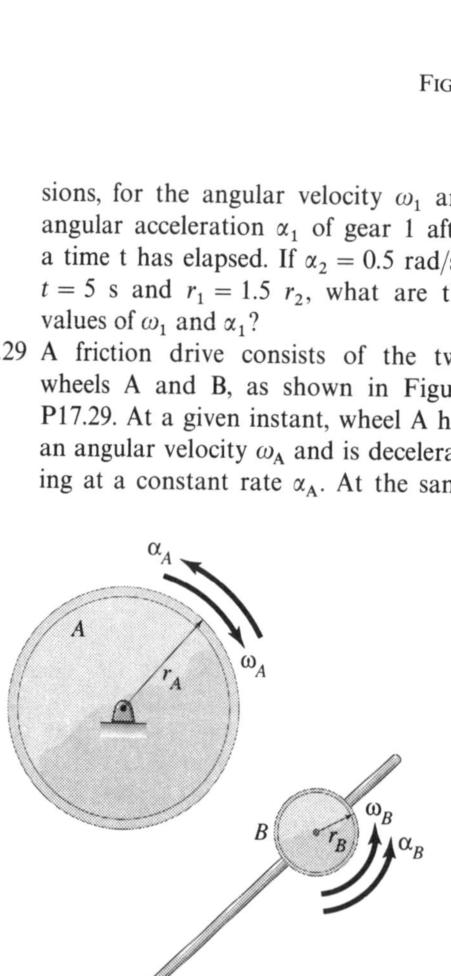

FIGURE P17.29.

17.30 Gear A, which is meshed with gear B in Figure P17.30, rotates at a constant angular velocity ω_A. It is known that the magnitude of the resultant acceleration of point P on gear A is four times that of point P on gear B. Determine the ratio of r_B/r_A.

17.31 Refer to the gear system shown in Figure P17.30. Assume that the system is accelerating and that, at a given instant, $\omega_A = 40$ rad/s and the magnitude of the resultant acceleration of point P on gear A is 10,000 in./s^2. Let $r_A = 5$ in. and $r_B = 16$ in., and determine the magnitude of the resultant acceleration of point P on gear B.

17.32 The gear-pulley system shown in Figure P17.32 is used to raise the weight W. It starts from rest and accelerates at a con-

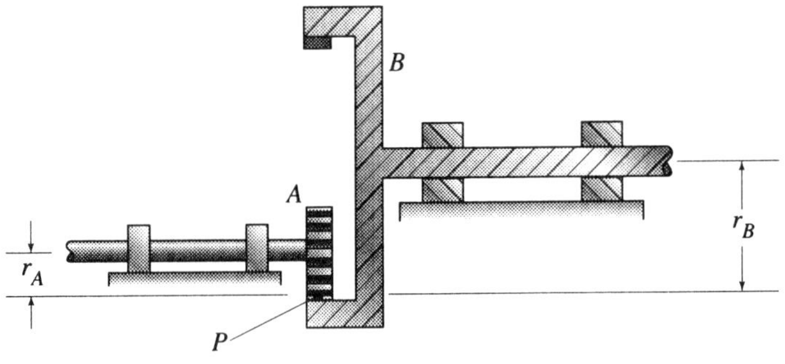

FIGURE P17.30.

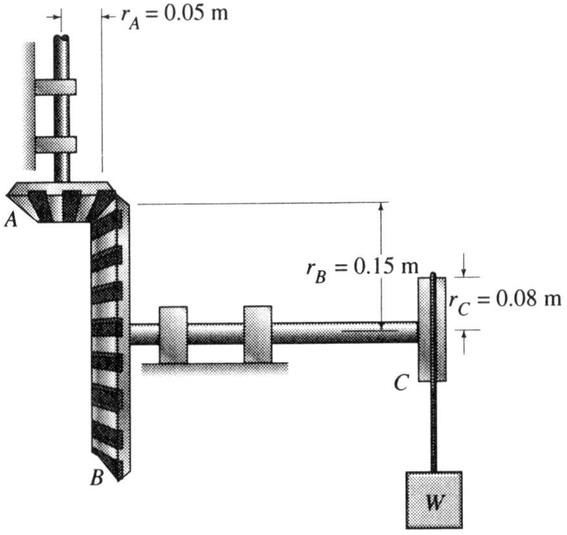

FIGURE P17.32.

stant rate such that, when $t = 3$ s, the speed of the weight is $v = 6$ m/s upward. Determine (a) the angular accelerations of gears B and A and (b) the angular displacements of gears B and A during the first 3 seconds.

17.33 A 6 in. × 8 in. rectangular plate is welded at its center of mass to the midpoint of a 24-in. axis OA as shown in Figure P17.33. In the position shown (when the plate is parallel to the x-z plane), the assembly is rotating about axis OA with a constant angular velocity of 10 rad/s, cw when viewed from O to A. Compute the velocity and acceleration of point C.

17.34 Refer to Problem 17.33. For the position considered, the assembly rotates about axis OA with an angular velocity of 15 rad/s and an angular acceleration of 20 rad/s^2, both ccw when viewed from O to A. Determine the velocity and acceleration of point D.

17.35 At a given instant, the rigid body shown

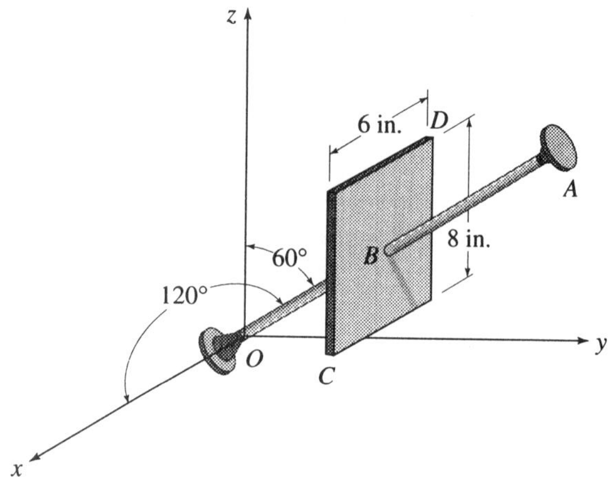

FIGURE P17.33.

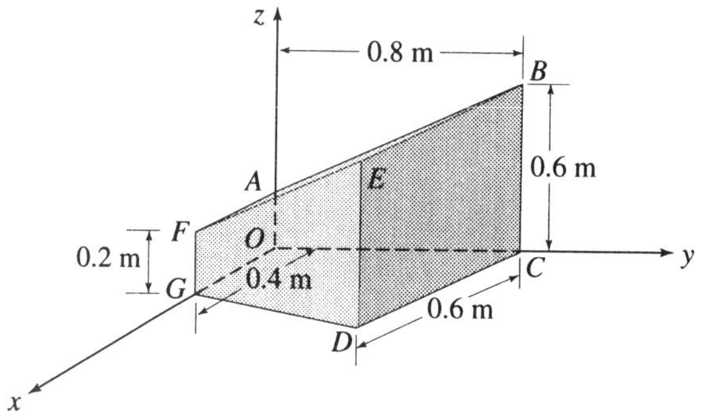

FIGURE P17.35.

in Figure P17.35 rotates about axis GB such that the x component of the velocity of point D is 2 m/s in the positive x direction. At the same instant, the angular acceleration of the rigid body is 8 rad/s^2 cw when viewed from B to G. Determine (a) the angular velocity of the rigid body, (b) the velocity of point C, and (c) the acceleration of point C.

17.36 Earth rotates about its polar axis approximately once every twenty-four hours. Assume the radius of Earth to be 6380 km, and determine the velocity and resultant acceleration of a person on the surface of Earth located at (a) the equator, (b) latitude 50° South and (c) the South Pole.

17.37 The planet Venus has an orbit about the Sun which is approximately circular with a mean radius of 67.2×10^6 miles. It makes one complete revolution, at constant angular velocity, in 224.7 days. Determine the speed and resultant acceleration of the planet.

17.3 Absolute Motion Formulation—General Plane Motion

In the two foregoing sections, special types of the motion of a rigid body in a plane were discussed. These were translation and rotation about a fixed axis. In this section, we will consider the type of motion of a rigid body known as *general plane motion* which does not fit the definition given previously for translation or rotation about a fixed axis. However, a general plane motion may be viewed as a combination of a translation and a rotation about a fixed axis.

For example, consider the rectangular rigid plate shown in Figure 17.6(a) which executes general plane motion from position 1 to position 2 during which time element A_1B_1 moves to its new position A_2B_2. This general plane motion may be viewed in two separate stages. The first stage would translate the plate from position 1 to position 1′, as shown in Figure 17.6(b), during which time element A_1B_1 would remain parallel to itself, assuming location A_2B_1' in position 1′. The second stage of the motion would impart a rotation to the rigid plate about a fixed axis through point A_2 while line AB rotates from position A_2B_1' to position A_2B_2 as shown in Figure 17.6(c). This point of view, in which a general plane motion is considered as a combination of a translation and a rotation, will be utilized in subsequent sections when dealing with relative velocity and relative acceleration formulations discussed in Sections 17.5 and 17.7.

In this section, we will analyze the general plane motion of a rigid body by first expressing, in terms of a convenient parameter, the absolute displacements of selected points in the body measured from a fixed coordinate system. These expressions are then differentiated once with respect to time to obtain the absolute velocities and twice to obtain the absolute accelerations of the points that have been selected. This information may also be used to specify the angular motion of the rigid body.

The following examples illustrate the above concepts.

■ Example 17.8

The rectangular plate ABCD, shown in Figure E17.8, is constrained to move so that corner B slides in a vertical track and corner C in a horizontal track. If corner C moves to the right with a constant velocity $v_C = 0.75$ m/s, determine the angular velocity ω and angular acceleration α of the plate. Express your answers in terms of the angular position θ measured cw from the vertical track.

Solution

From the geometry of Figure 17.8,

$$x_C = a \sin \theta.$$

Thus,

$$v_C = \dot{x}_C = a\dot{\theta} \cos \theta = a\omega \cos \theta \tag{a}$$

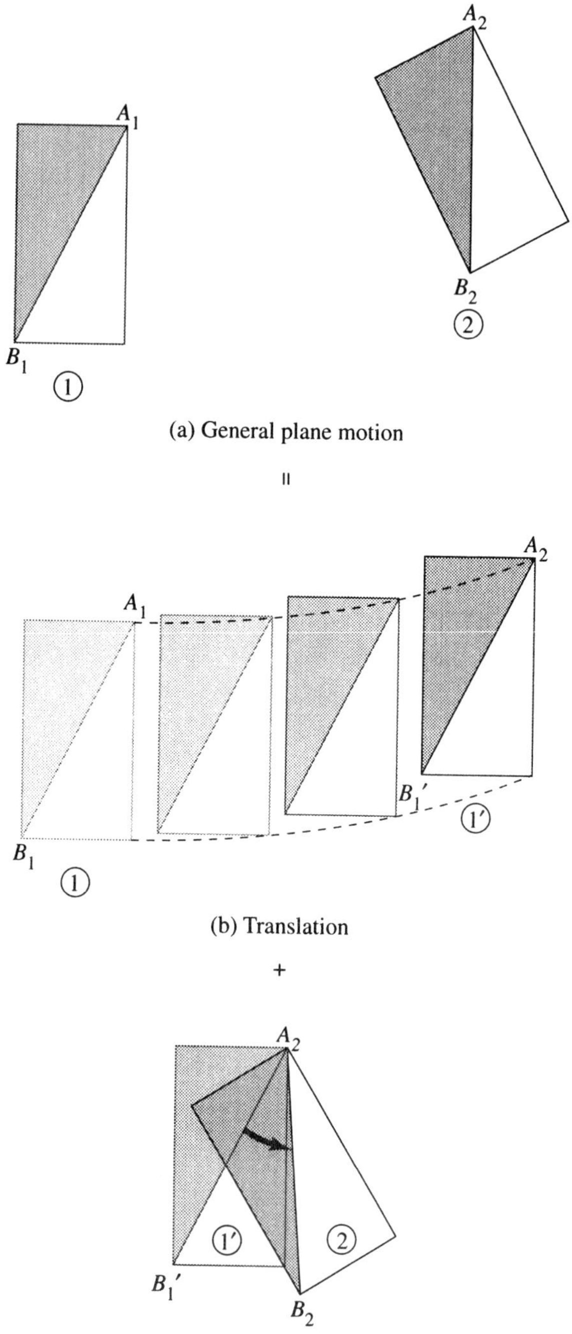

FIGURE 17.6. (c) Rotation about fixed axis A_2

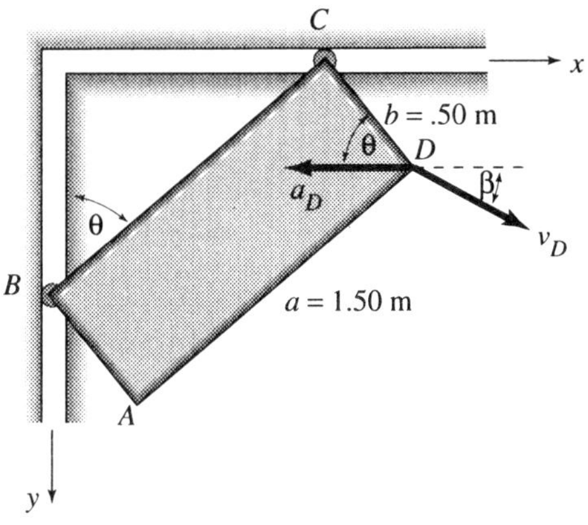

FIGURE E17.8.

where $\omega = \dot{\theta}$ is the angular velocity of the rigid plate. Therefore,

$$\omega = \frac{v_C}{a\cos\theta} = \frac{0.50}{\cos\theta} \text{ rad/s } ↻. \qquad \text{(b) ANS.}$$

Also

$$a_C = \ddot{x}_C.$$

Because v_C is constant, $a_C = 0$ and it follows that

$$a_C = -a\dot{\theta}^2 \sin\theta + a\ddot{\theta}\cos\theta = 0.$$

Thus,

$$-a\omega^2 \sin\theta + a\alpha\cos\theta = 0$$

where $\alpha = \ddot{\theta}$ is the angular acceleration of the rigid plate. Therefore,

$$\alpha = \omega^2 \tan\theta. \qquad \text{(c)}$$

If Eq. (b) is substituted in Eq. (c)

$$\alpha = \frac{v_C^2 \tan\theta}{a^2\cos^2\theta} = \frac{0.25\tan\theta}{\cos^2\theta} \text{ rad/s}^2 ↻. \qquad \text{(d)}$$

■ Example 17.9

Refer to Example 17.8 and determine the velocity v_D and the acceleration a_D of corner D. Express your answers in terms of the angular position θ measured clockwise from the vertical track.

Solution

From the geometry of Figure E17.8,

$$x_D = x_C + b\cos\theta = a\sin\theta + b\cos\theta$$

$$y_D = b\sin\theta$$

Thus,

$$(v_D)_x = \dot{x}_D = a\omega\cos\theta - b\omega\sin\theta \tag{a}$$

and

$$(v_D)_y = \dot{y}_D = b\omega\cos\theta. \tag{b}$$

Using the results given in Eqs. (a) and (b) of Example 17.8 for v_C and ω, respectively,

$$(v_D)_x = v_C - v_C\left(\frac{b}{a}\right)\tan\theta = v_C\left(1 - \frac{b}{a}\tan\theta\right)$$

and

$$(V_D)_y = \frac{b}{a}v_C$$

Because

$$v_D = \sqrt{(v_D)_x^2 + (v_D)_y^2},$$

it follows that

$$v_D = v_C\sqrt{\left(\frac{b}{a}\right)^2 + \left(1 + \frac{b}{a}\tan\theta\right)^2}$$

$$= 0.75\sqrt{\left(\frac{1}{9}\right) + \left(1 - \frac{1}{3}\tan\theta\right)^2}\ \text{m/s} \tag{c}$$

ANS.

The direction of v_D is defined by the angle β (see Figure E17.8) where

$$\beta = \tan^{-1}\left[\frac{(v_D)_y}{(v_D)_x}\right] = \tan^{-1}\left[\frac{b}{a - b\tan\theta}\right] = \tan^{-1}\left[\frac{1.0}{3.0 - \tan\theta}\right] \tag{d}$$

Differentiating Eq. (a),

$$(a_D)_x = \ddot{x}_D = -a\omega^2\sin\theta + a\alpha\cos\theta - b\omega^2\cos\theta - b\alpha\sin\theta$$
$$= -\omega^2(a\sin\theta + b\cos\theta) + \alpha(a\cos\theta - b\sin\theta).$$

Substituting for ω and α from Eqs. (b) and (d) of Example 17.8, respectively, after simplification,

$$(a_D)_x = -\frac{bv_C^2}{a^2\cos^3\theta}$$

Also, a differentiation of Eq. (b) yields

$$(a_D)_y = \ddot{y}_D = -b\omega^2\sin\theta + b\alpha\cos\theta$$

Substituting for ω and α from Eqs. (b) and (d) of Example 17.8, respectively, after simplification

$$(a_D)_y = \frac{bv_C^2}{a^2\cos^3\theta}(-\sin\theta + \sin\theta) = 0.$$

Because

$$a_D = \sqrt{(a_D)_x^2 + (a_D)_y^2},$$

it follows that

$$a_D = (a_D)_x = -\frac{bv_C^2}{a^2 \cos^3 \theta} = \frac{0.125}{\cos^3 \theta} \text{ m/s}^2 \longleftarrow \qquad \text{ANS.}$$

where the negative sign indicates a sense for a_D which is opposite to the positive direction of the x axis. Thus, a_D is pointed to the left as shown in Figure E17.8, for the case when the velocity of corner C is constant and pointed to the right (i.e., in the positive x direction).

■ Example 17.10

The composite wheel shown in Figure E17.10(a) is being pulled up the inclined plane by means of the force F. The large wheel rolls on the inclined plane without slipping. For the position shown, the center C of the composite wheel moves up the inclined plane with a velocity $v_C =$

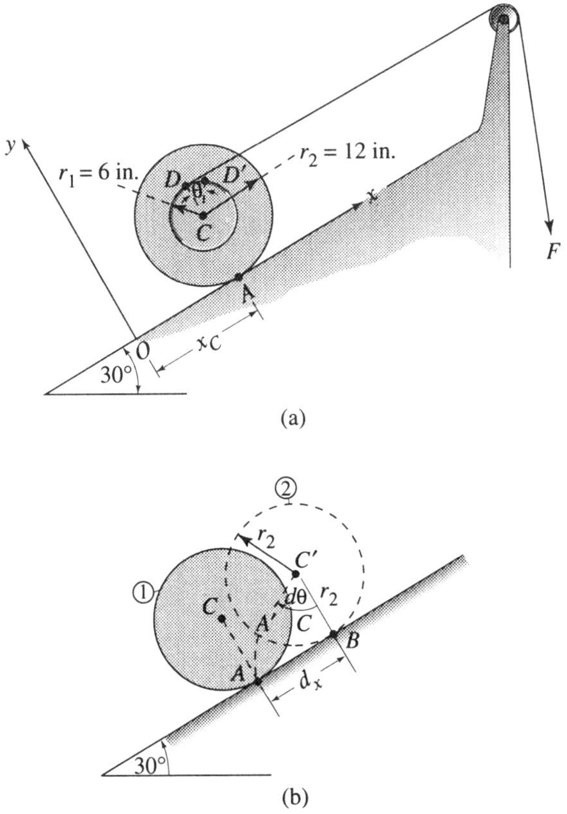

FIGURE E17.10.

10.0 in./s and acceleration $a_C = 5.0$ in./s^2. Determine the angular velocity and angular acceleration of the composite wheel.

Solution

Figure E17.10(b) shows the large wheel rolling up the inclined plane from position 1, represented by the solid circle, to position 2, represented by the dashed circle. During this motion, the center C of the wheel moves up the inclined plane to point C′ through the infinitesimal distance dx, the point of contact A moves to A′, and a new point B on the rim of the wheel comes in contact with the inclined plane. Also, the radial line CA rotates through the infinitesimal angle $d\theta$ to its new position C′A′. Since the wheel rolls without slipping, the length of arc BA′ must be equal to the distance dx measured along the inclined plane. Thus,

$$dx = \mathrm{BA}' = r_2\, d\theta \tag{a}$$

Dividing both sides of this Eq. (a) by dt,

$$\dot{x} = \frac{dx}{dt} = r_2 \frac{d\theta}{dt} \tag{b}$$

or

$$v_C = r_2 \omega \tag{c}$$

where ω is the angular velocity of the composite wheel. Differentiating Eq. (b) with respect to time,

$$\ddot{x} = \frac{d^2x}{dt^2} = r_2 \frac{d^2\theta}{dt^2}$$

or

$$a_C = r_2 \alpha \tag{d}$$

where α is the angular acceleration of the pulley. Equations (c) and (d), relating the angular velocity and angular acceleration of a wheel to the velocity and to the acceleration of its center, are valid as long as the wheel rolls without slipping. Using the given numerical values for v_C, a_C and r_2, from Eq. (c),

$$\omega = \frac{v_C}{r_2} = \frac{10}{12} \text{ rad/s } \circlearrowright, \qquad \text{ANS.}$$

and from Eq. (d),

$$\alpha = \frac{a_C}{r_2} = \frac{5}{12} \text{ rad/s}^2 \circlearrowright. \qquad \text{ANS.}$$

Example 17.11 Refer to Example 17.10 and determine the speed and acceleration with which the cable is being pulled.

Solution

The speed and acceleration with which the cable is being pulled by the force F may be found by determining, respectively, the velocity of point D on the small wheel in Figure E17.10(a), and the tangential component of the acceleration of this point.

Refer to Figure E17.10(a). If the composite wheel rotates cw through the angle θ, point D moves to position D′. The coordinates of point D′ are given by

$$\left.\begin{aligned} x_{D'} &= x_C + r_1 \sin\theta \\ y_{D'} &= r_2 + r_1 \cos\theta \end{aligned}\right\} \qquad \text{(a)}$$

Differentiating Eqs. (a) with respect to time, we obtain the x and y components of the velocity of point D′. Thus

$$(v_{D'})_x = \dot{x}_{D'} = \dot{x}_C + r_1\dot{\theta}\cos\theta = v_C + r_1\omega\cos\theta \qquad \text{(b)}$$

and

$$(v_{D'})_y = \dot{y}_{D'} = -r_1\dot{\theta}\sin\theta = -r_1\omega\sin\theta. \qquad \text{(c)}$$

The components of the velocity of point D are found by setting $\theta = 0$ in Eqs. (b) and (c). Thus,

$$(v_{D'})_x = v_C + r_1\omega$$

and

$$(v_{D'})_y = 0.$$

Therefore, because $v_D = \sqrt{(v_D)_x^2 + (v_D)_y^2}$,

$$v_D = (v_D)_x = v_C + r_1\omega = 15.00 \text{ in./s.} \qquad \text{ANS.}$$

Differentiating Eqs. (b) and (c) with respect to time, we obtain the x and y components of the acceleration of point D′. Thus,

$$(a_{D'})_x = \ddot{x}_{D'} = \ddot{x}_C - r_1\dot{\theta}^2\sin\theta + r_1\ddot{\theta}\cos\theta = a_C - r_1\omega^2\sin\theta + r_1\alpha\cos\theta, \qquad \text{(d)}$$

and

$$(a_{D'})_y = \ddot{y}_{D'} = -r_1\dot{\theta}^2\cos\theta - r_1\ddot{\theta}\sin\theta = -r_1\omega^2\cos\theta - r_1\alpha\sin\theta. \qquad \text{(e)}$$

The components of the acceleration of point D are found by setting $\theta = 0$ in Eqs. (d) and (e). Therefore,

$$(a_D)_x = a_C + r_1\alpha,$$

and

$$(a_D)_y = -r_1\omega^2.$$

If the total acceleration of point D were needed, it would be found from the relationship $a_D = \sqrt{(a_D)_x^2 + (a_D)_y^2}$. However, in this case, only that component of a_D along the cable (the tangential component) is required. This component is $(a_D)_x$. Therefore,

$$a_{CABLE} = (a_D)_x = a_C + r_1\alpha = 7.50 \text{ in./s}^2. \qquad \text{ANS.}$$

■

Problems

17.38 Refer to Example 17.8 (p. 385), and, in terms of v_C, θ, a, and b, develop expressions for the velocity v_B and the acceleration of a_B of corner B. Determine the magnitudes, directions, and senses of v_B and a_B for $\theta = 30°$, $a = 24$ in., $b = 15$ in., and $v_C = 5$ in./s to the left. Assume v_C is constant.

17.39 Refer to Example 17.8 (p. 385), and, in terms of v_C, θ, a, and b, develop expressions for the x and y components of the velocity v_A and the acceleration a_A of corner A. Determine the magnitudes, directions, and senses of v_A and a_A for $\theta = 45°$, $a = 1.00$ m, $b = 0.75$ m, and $v_C = 2.0$ m/s to the right. Assume v_C is constant.

17.40 Member AB of length L executes general plane motion such that end A is constrained to move along a horizontal track while end B moves along a vertical track, as shown in Figure P17.40. If member AB has a constant ccw angular velocity ω, determine in terms of L, ω and θ, the velocity v_A and the acceleration a_A of end A. If $\omega = 4$ rad/s, $\theta = 45°$, and $L = 15$ in., determine v_A and a_A.

17.41 Repeat Problem 17.40 for central point C instead of A.

17.42 Refer to the mechanism shown in Figure

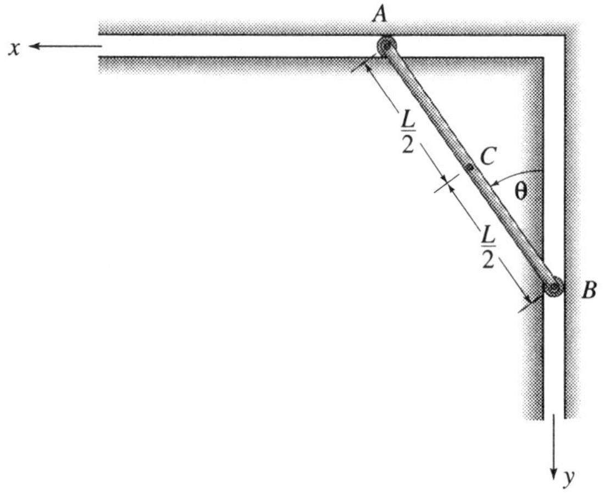

FIGURE P17.40.

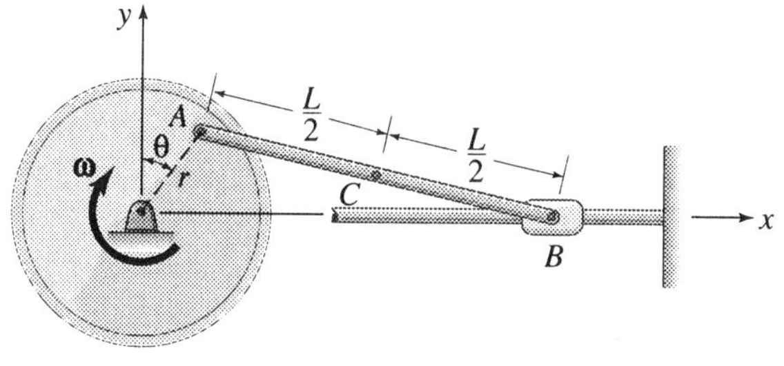

FIGURE P17.42.

P17.42. If the wheel rotates cw at a constant angular velocity ω, determine the velocity v_B and acceleration a_B of point B in terms of ω, θ, r, and L. If $\omega = 500$ rpm, $r = 1.5$ ft, and $L = 5$ ft, determine v_C and a_C for $\theta = 30°$.

17.43 Refer to Problem 17.42. Develop expressions for the x and y components of the velocity v_C and the acceleration a_C in terms of ω, θ, r and L. Note that point C is at the midpoint of the connecting rod AB.

17.44 Member AB in Figure P17.44 starts from rest, when $\theta = 0$, and rotates in a ccw direction at a constant angular acceleration α. In terms of b, θ, and α, develop expressions for the velocity v_B and acceleration a_B of the collar-slider unit B which is constrained to move along a smooth vertical track. If $b = 0.75$ m and $\alpha = 5.0$ rad/s^2, determine v_B and a_B for $\theta = 30°$.

17.45 Member AB moves freely, with roller A maintaining contact with the inclined plane and roller B with the horizontal plane as shown in Figure P17.45. If member AB rotates ccw with a constant angular velocity ω, develop expressions for the velocity v_A and the acceleration a_A of roller A. Express your answers in terms of b, β, ω, and θ. Assume that the rollers at A and B have negligible dimensions.

17.46 Repeat Problem 17.45 for roller B instead of roller A.

17.47 Refer to Figure P17.45 and let $b = 4$ ft

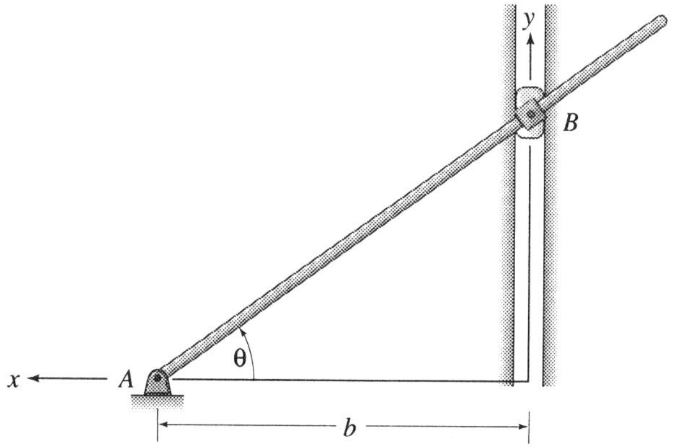

FIGURE P17.44.

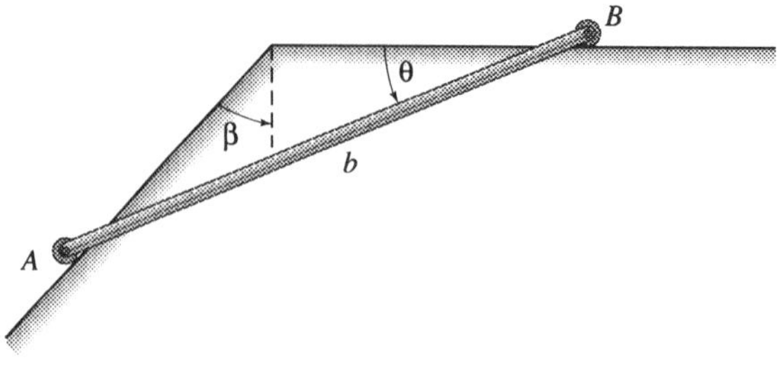

FIGURE P17.45.

and $\beta = 30°$. Roller B starts from rest, when $\theta = 0$, and moves to the left at a constant acceleration $a_B = 2.0$ ft/s^2. Determine the angular velocity ω and angular acceleration α of member AB after roller B has moved a distance of 2.0 ft.

17.48 Arm OA of the gear system, shown in Figure P17.48, rotates cw with a constant angular velocity ω. The small gear rotates on the inner toothed surface of the outer gear which is stationary. In terms of r_0, r, and ω, develop expressions for the angular velocity ω_0 of the small gear and for the velocity v_A of its center.

17.49 The wheel shown in Figure P17.49 rolls without slipping to the right such that, at the instant shown, the velocity and acceleration of its center, point A are $v_A = 5$ m/s and $a_A = 2$ m/s^2. Determine (a) the angular velocity and angular acceleration of the wheel and (b) the velocity and acceleration of point B.

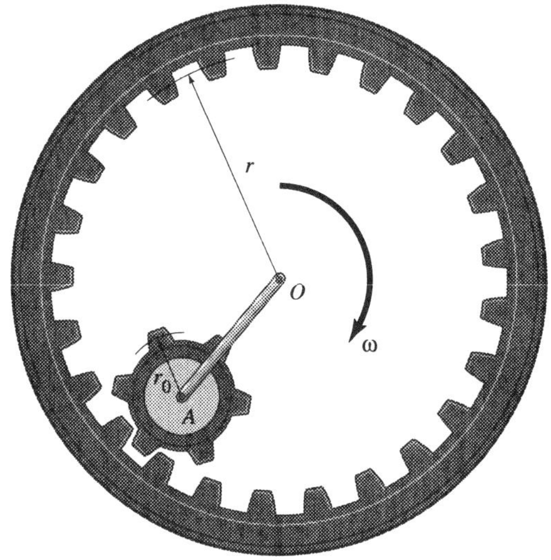

FIGURE P17.48.

17.50 Refer to Problem 17.49 and determine the velocity and acceleration of point C.

17.51 Crank OA of the mechanism shown in

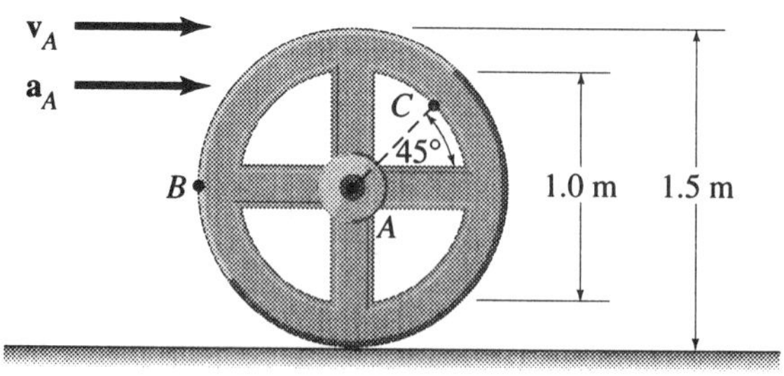

FIGURE P17.49.

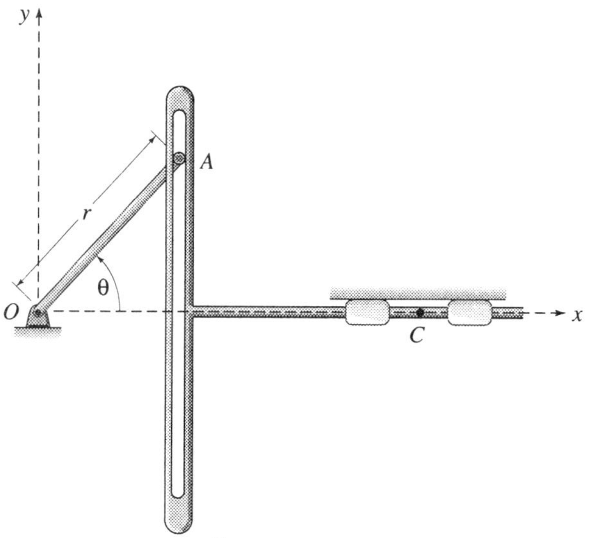

FIGURE P17.51.

Figure P17.51 starts from rest, when $\theta = 0$, and rotates cw with a constant angular acceleration $\alpha = 0.5$ rad/s. Let $r = 0.3$ m, and determine the velocity and acceleration of point C on the horizontal rod after crank OA has rotated for a period of (a) $t = 1$ s and (b) $t = 15$ s.

17.52 The pulley arrangement, shown in Figure P17.52, is used to hoist the weight W. The weight W starts from rest and moves upward with a constant acceleration a_w. After the weight W has moved to height h, determine (a) the angular velocity and angular acceleration of pulley A and (b) the velocity and acceleration at which the force F pulls the horizontal cable. Express your answers in terms of r_1, r_2, h, and a_w.

17.53 The pulley arrangement shown in Figure P17.52 is used to lower the weight W. Points on the horizontal cable move to the left at a constant velocity of 5 ft/s. Determine (a) the velocity and acceleration at which the weight is being lowered and (b) the angular velocity and angular acceleration of pulley A. Let $r_1 = 8$ in and $r_2 = 24$ in.

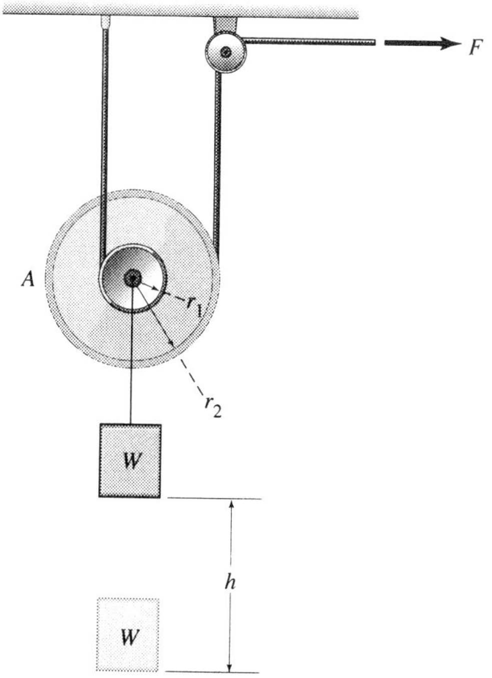

FIGURE P17.52.

17.4 Relative Velocity—Translating Nonrotating Axes

In Section 17.3, it was pointed out that a general plane motion may be viewed as consisting of the combination of a translation and a rotation. As shown in Figure 17.7, a rigid body executes general plane motion from position 1 to position 2 during an infinitesimal time interval dt, during which line BA moves from position B_1A_1 to position B_2A_2. This general plane motion may be looked upon as occurring in two stages. The first stage would result in an infinitesimal translation from position B_1A_1 to parallel position B_2A_1' where the displacement $d\mathbf{r}_B$, for convenience, is shown magnified in Figure 17.7. The second stage would result in a fixed-axis rotation about B_2 through the infinitesimal angle $d\theta$ yielding a further displacement $d\mathbf{r}_{A/B}$ for point A from position A_1' to position A_2. Note that the magnitude of $d\mathbf{r}_{A/B}$ is equal to $(r_{A/B})\,d\theta$ where $r_{A/B}$ represents the fixed distance from point A to point B on the rigid body.

By the geometry shown in Figure 17.7, the following displacement relationship may be written

$$d\mathbf{r}_A = d\mathbf{r}_B + d\mathbf{r}_{A/B} \tag{17.19}$$

where $d\mathbf{r}_A$ and $d\mathbf{r}_B$ are absolute displacements measured from the fixed x-y coordinate system whereas $d\mathbf{r}_{A/B}$ represents a relative displacement of point A as viewed by an observer at reference point B which is the origin of the translating but nonrotating x'-y' coordinate system. Note that, in the development of Eq. (17.19), the translating, nonrotating x'-y' coordinate system was attached to reference point B. This was, in fact, an arbitrary choice, as point A could have been selected as the reference point or as the origin for the translating, nonrotating x'-y' coordinate system. In such a case, instead of the displacement relationship expressed in Eq. (17.19), we would have

$$d\mathbf{r}_B = d\mathbf{r}_A + d\mathbf{r}_{B/A}. \tag{17.20}$$

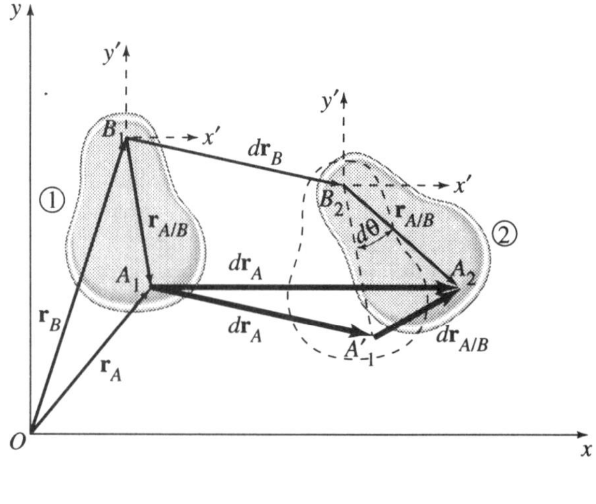

FIGURE 17.7.

The choice between Eq. (17.19) and Eq. (17.20) would depend upon whether we are interested in selecting, as a reference, point A or point B, respectively. Because both of these equations express the same physical concept, we will concentrate on the first, namely, Eq. (17.19).

Dividing Eq. (17.19) by dt,

$$\mathbf{v}_A = \mathbf{v}_B + \mathbf{v}_{A/B} \tag{17.21}$$

where $\mathbf{v}_A$ and $\mathbf{v}_B$ represent absolute velocities and $\mathbf{v}_{A/B}$ represents the relative velocity of point A as viewed by an observer at point B, which is the origin of the translating, nonrotating x'-y' coordinate system. Because we are dealing with a rigid body, the distance $r_{A/B}$, between point A and B is fixed, and, to an observer at point B, point A appears to move along a circular path with center at point B. Thus, the vector $\mathbf{v}_{A/B}$ represents the velocity of point A rotating about an axis through point B, which is momentarily fixed. Therefore, Eq. (17.15), developed in Section 17.2 for rotation about a fixed axis, may be used to express the vector $\mathbf{v}_{A/B}$. Thus,

$$\mathbf{v}_{A/B} = \boldsymbol{\omega}_B \times \mathbf{r}_{A/B} \tag{17.22}$$

whose magnitude is given by

$$v_{A/B} = r_{A/B}\frac{d\theta}{dt} = r_{A/B}\omega \tag{17.23}$$

where ω represents the magnitude of the angular velocity of the rigid body. The direction of vector $\mathbf{v}_{A/B}$ is along a line which is perpendicular to the direction from B to A.

Equation (17.21) shows that the velocity of any point, such as A, on a rigid body may be obtained as the vector sum of the velocity of a second point B and the velocity of A relative to B. Alternatively, Equation (17.21) states that the absolute velocity of any point on a rigid body, executing general plane motion, may be obtained by combining a translation component and a rotation component. This latter point of view is represented in Figure 17.8. Figure 17.8(a) represents a rigid body at a given instant while undergoing general plane motion. Figure 17.8(b) represents the translation component of velocity and Figure 17.8(c) the rotation component of velocity for this general plane motion. Finally, the velocity diagram shown in Figure 17.8(d) shows how these two components are added vectorially by Eq. (17.21) to obtain the velocity $\mathbf{v}_A$.

In the solution of velocity problems, it is recommended that a velocity diagram, similar to that of Figure 17.8(d) be constructed to represent the terms in (17.21). Because Eq. (17.21) is a two-dimensional vector relationship, it represents two independent scalar equations, and, consequently, we can solve for no more than two unknown quantities for every application of the vector equation. The velocity diagram

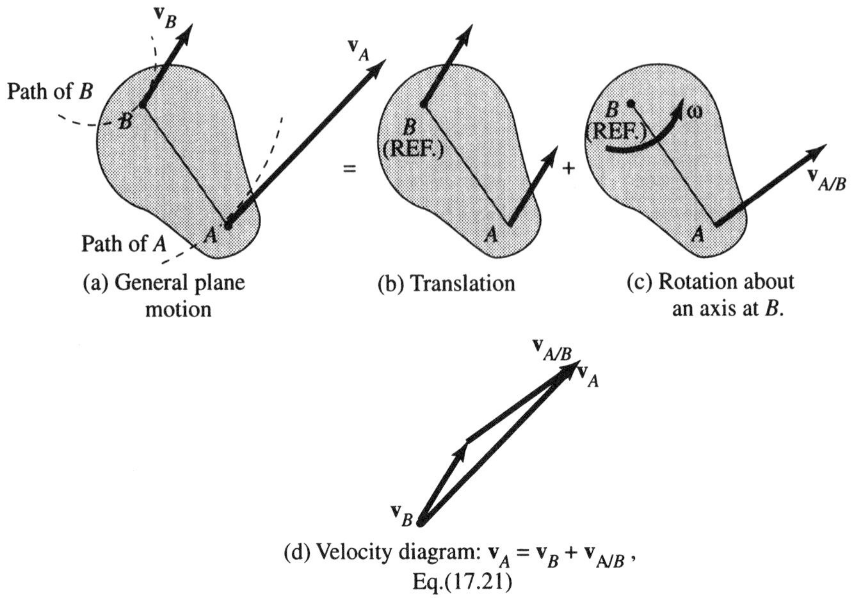

FIGURE 17.8.

may be solved analytically for the unknown quantities using either the scalar trigonometric method or the vector method. There are no special advantages of one method over the other. However, as a rule of thumb, the scalar trigonometric technique is more direct when dealing with right-angle triangles for the velocity diagrams, and the vector method is more suitable when dealing with more complex velocity diagrams. To illustrate the differences and similarities between them, both methods are used in the solution of the examples that follow.

■ Example 17.12

The wheel shown in Figure E17.12(a) rolls without slipping down the 10° inclined plane such that its center C has a velocity of 10.0 ft/s. Determine the magnitudes and directions of the velocities of points (a) A and (b) B. Use Eq. (17.21).

Solution

(a) As discussed in Example 17.10, because the wheel rolls without slippage, its angular velocity is related to the linear velocity v_C of its center C by the equation $v_C = r_2\omega$ where $r_2 = D_2/2 = 2.5$ ft. Thus,

$$\omega = \frac{v_C}{r_2} = 4.0 \text{ rad/s} \circlearrowright .$$

By Eq. (17.21)

$$\mathbf{v}_A = \mathbf{v}_C + \mathbf{v}_{A/C} \tag{a}$$

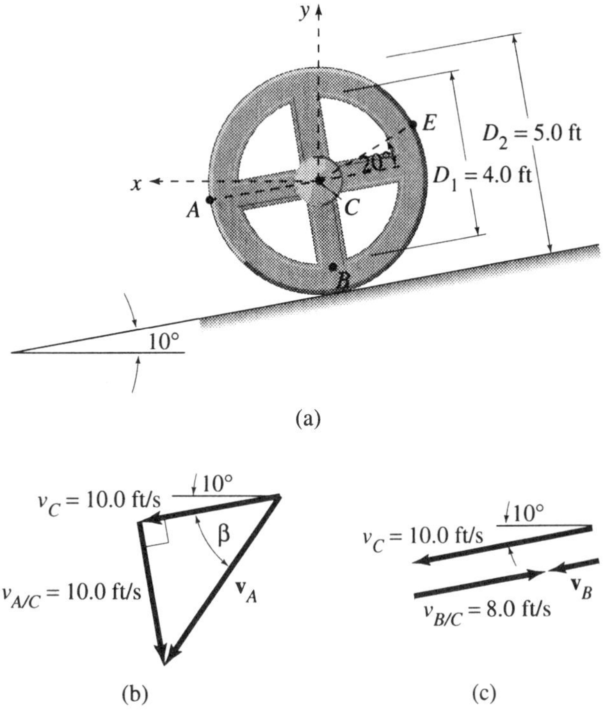

FIGURE E17.12.

The velocity diagram representing Eq. (a) is shown in Figure E17.12(b). Note that the center of the wheel, point C, is the reference point which serves as the origin of the translating, nonrotating x-y coordinate system. This velocity diagram is constructed by first drawing vector $\mathbf{v}_C$ which is known fully in magnitude, direction, and sense. This is followed by drawing the vector $\mathbf{v}_{A/C}$ which is also known in magnitude, direction, and sense. The magnitude of $\mathbf{v}_{A/C}$ is $r_2\omega = 2.5(4.0) = 10.0$ ft/s, its direction is perpendicular to line AC, and its sense is downward and to the right because ω is ccw. The required vector $\mathbf{v}_A$ is found by connecting the tail of $\mathbf{v}_C$ to the head of $\mathbf{v}_{A/C}$. The scalar trigonometric analysis is used because the velocity diagram is a right-angle triangle. Thus,

$$\beta = \tan^{-1}\frac{v_{A/C}}{v_C} = 45.0^\circ,$$

and

$$v_A = \sqrt{(v_{A/C})^2 + v_C^2} = 14.14 \text{ ft/s} \qquad \text{ANS.}$$

(b) By Eq. (17.21)

$$\mathbf{v}_B = \mathbf{v}_C + \mathbf{v}_{B/C}. \qquad \text{(b)}$$

The velocity diagram representing Eq. (b) is shown in Figure E17.12(c). Again, point C is the reference point. This diagram is constructed in a manner similar to that of Figure E17.12(b). Both $\mathbf{v}_C$ and $\mathbf{v}_{B/C}$ are known fully in magnitude, direction, and sense. Note that the magnitude of $\mathbf{v}_{B/C}$ is $r_1\omega$ where $r_1 = D_1/2 = 2$ ft. Therefore, the magnitude $v_{B/C} = 2.0(4.0) = 8.0$ ft/s. Also, its direction is perpendicular to line BC and its sense is upward and to the right because ω is ccw. Thus, $\mathbf{v}_{B/C}$ has the same direction as $\mathbf{v}_C$ but is opposite to it in sense. The vector $\mathbf{v}_B$ is obtained by connecting the tail of $\mathbf{v}_B$ to the head of $\mathbf{v}_{B/C}$. Thus,

$$v_B = 10.0 - 8.0 = 2.00 \text{ ft/s} \quad \overset{10.0^\circ}{\swarrow}. \qquad \text{ANS.}$$

Example 17.13

Refer to Example 17.13 and determine the magnitude and direction of the velocity of point E in Figure E17.12(a). Use Eq. (17.21).

Solution

By Eq. (17.21)

$$\mathbf{v}_E = \mathbf{v}_C + \mathbf{v}_{E/C} \tag{a}$$

The velocity diagram representing Eq. (a) is obtained as explained in Example 17.12 and is shown in Figure E17.13(d). Here again, point C serves as the reference point. The velocity $\mathbf{v}_C$ has a magnitude of 10.0 ft/s and is directed downward and to the left. The required velocity $\mathbf{v}_{E/C}$ has a magnitude of $r_2\omega = 2.5(4.0) = 10.0$ ft/s, a direction perpendicular to radius CE, and is pointed upward and to the left. The required velocity $\mathbf{v}_E$ is obtained by connecting the tail of $\mathbf{v}_C$ to the head of $\mathbf{v}_{E/C}$. It is convenient, in this case, to use the vector method because the velocity diagram is not a right-angle triangle. Using the translating, nonrotating x-y coordinate system attached to point C, Eq. (a) is expressed in terms of $\mathbf{i}$ and $\mathbf{j}$ components. Thus,

$$\begin{aligned}\mathbf{v}_E &= (v_E \cos\gamma)\mathbf{i} + (v_E \sin\gamma)\mathbf{j} \\ &= (10.0 \cos 10.0^\circ)\mathbf{i} - (10.0 \sin 10.0^\circ)\mathbf{j} \\ &= 9.848\mathbf{i} - 1.736\mathbf{j}\end{aligned}$$

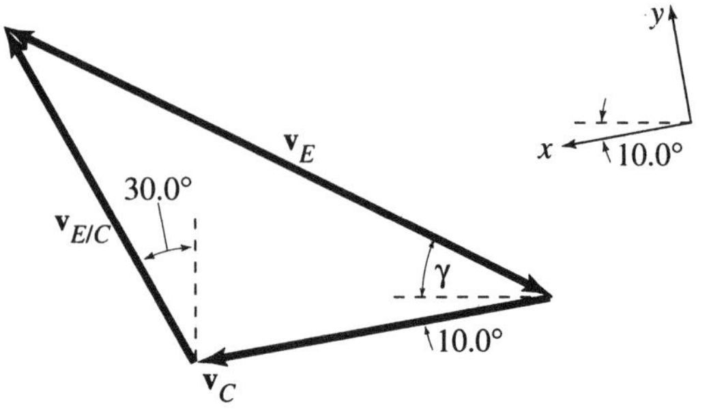

FIGURE E17.13.

$$\mathbf{v}_{E/C} = \boldsymbol{\omega} \times \mathbf{r}_{E/C} = -4.0\mathbf{k} \times [-(2.5\cos 30.0°)\mathbf{i} + (2.5\sin 30.0°)\mathbf{j}]$$
$$= 5.000\mathbf{i} + 8.660\mathbf{j}$$

Therefore, Eq. (a) may be written in the following vector form:

$$(v_E \cos\gamma)\mathbf{i} + (v_E \sin\gamma)\mathbf{j} = 14.848\mathbf{i} + 6.924\mathbf{j}.$$

Equating, separately, the scalar multipliers of the **i** and **j** components we obtain two scalar equations. Thus,

$$v_E \cos\gamma = 14.848, \tag{b}$$

and

$$v_E \sin\gamma = 6.924. \tag{c}$$

Solving Eqs. (b) and (c) simultaneously we obtain the values for γ and v_E as follows:

$$\gamma = 25.0°,$$
$$v_E = 16.38 \text{ ft/s}.$$

Therefore

$$v_E = 16.38 \text{ ft/s}$$ 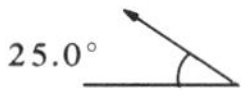

. ANS.

■ Example 17.14

Refer to Example 17.8 (p. 385), and use Eq. (17.21) to solve for (a) the angular velocity ω of the plate and (b) the velocity v_B of corner B. Express answers in terms of θ. (c) Find ω and v_B for $\theta = 30°$.

Solution

(a) For convenience, the rigid plate of Example 17.8 is redrawn in Figure E17.14(a). In addition to corner C, whose velocity is known fully in magnitude and direction, there is only one other point in the rigid plate about whose velocity something is known. This is corner B whose velocity is along the vertical track in the negative y direction. Therefore, using Eq. (17.21) to relate the velocities of these two points on the rigid plate,

$$\mathbf{v}_B = \mathbf{v}_C + \mathbf{v}_{B/C} \tag{a}$$

The velocity diagram representing Eq. (a) is shown in Figure 17.14(b). Note that point C has been chosen as the reference which serves as the origin of the translating, nonrotating x-y coordinate system. This diagram was constructed by first drawing the vector $\mathbf{v}_C$ which is known fully in magnitude, direction, and sense. Because the direction of the relative velocity $\mathbf{v}_{B/C}$ is perpendicular to edge BC and the angular velocity of the rigid plate ω is cw, we conclude that $\mathbf{v}_{B/C}$ must be a vector originating at the head of $\mathbf{v}_C$ and pointing upward and to the left.

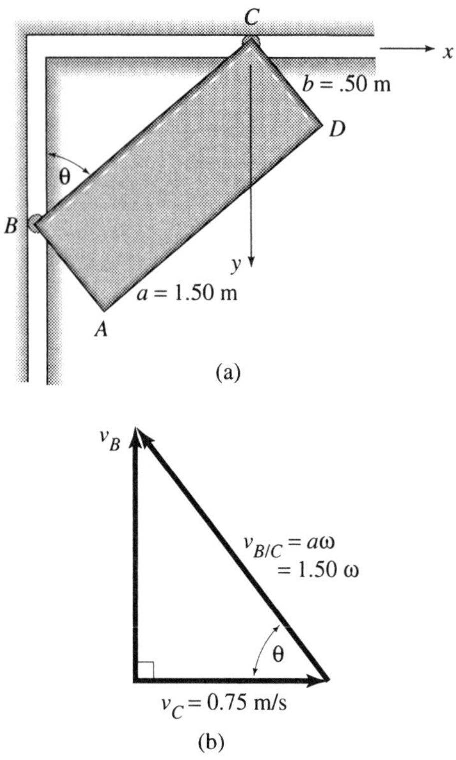

FIGURE E17.14.

Finally, because the direction of $\mathbf{v}_B$ is vertically upward, a line is drawn from the tail of $\mathbf{v}_C$ and extended to intersect the direction of $\mathbf{v}_{B/C}$. Note that the magnitude of the relative velocity $\mathbf{v}_{B/C}$ is equal to $a\omega$ where $a = 1.50$ m is the length of edge BC and ω is the unknown angular velocity of the rigid plate. Thus, $v_{BC} = 1.50\omega$.

Because the velocity diagram in Figure 17.14(b) is a right-angle triangle, we can make use of a scalar trigonometric analysis. Thus,

$$\cos\theta = \frac{v_C}{v_{B/C}} = \frac{v_C}{a\omega}$$

from which

$$\omega = \frac{v_C}{a\cos\theta} = \frac{0.50}{\cos\theta}\ \text{rad/s} \qquad \text{ANS.}$$

This answer is, of course, identical to that found in Example 17.8.

(b) The velocity diagram shown in Figure E17.14(b) may also be used to find the velocity of corner B. Thus,

$$v_B = 0.75\tan\theta. \qquad \text{ANS.}$$

(c) For $\theta = 30°$,

$$\omega = \frac{0.5}{\cos 30°} = 0.577 \text{ rad/s} \circlearrowright, \qquad \text{ANS.}$$

$$\text{and } v_B = 0.75 \tan 30° = 0.433 \text{ m/s.} \uparrow \qquad \text{ANS.}$$

Example 17.15

Wheel OA of the reciprocating mechanism shown in Figure E17.15(a) rotates at a constant ccw angular velocity ω. In terms of r, θ, d, and ω, derive expressions for the velocity v_B of collar B and for the angular velocity ω_{AB} of rod AB.

Solution

Consider the reciprocating mechanism for the position defined by the angle θ. The magnitude of the velocity of point A, which is common to members OA and AB, is given by

$$v_A = r\omega. \tag{a}$$

The direction $\mathbf{v}_A$ is along a line perpendicular to line OA and, because ω is ccw, its sense is upward and to the right.

Equation (17.21) may now be applied to relate the velocity of collar B to that of joint A. Thus,

$$\mathbf{v}_B = \mathbf{v}_A + \mathbf{v}_{B/A}. \tag{b}$$

The velocity diagram representing Eq. (b) is shown in Figure E17.15(b). Note that point A is the one serving as the reference point. The dia-

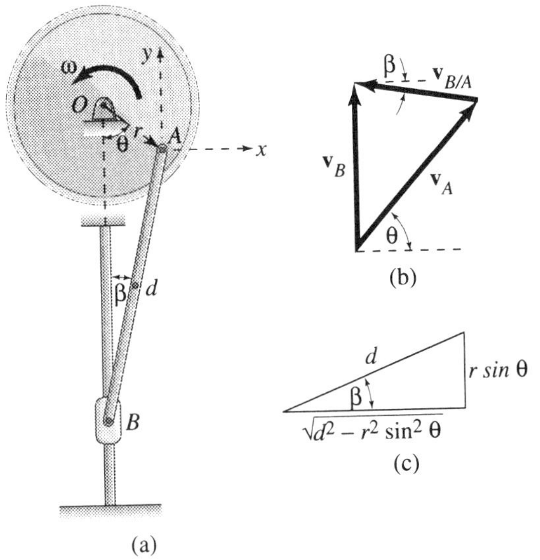

FIGURE E17.15.

gram was obtained by first drawing vector $\mathbf{v}_A$ which is fully known in magnitude, direction, and sense. From the tail of $\mathbf{v}_A$, a vertical line is constructed to represent the velocity of collar B which is constrained to slide along the vertical rod. Finally, a line is drawn from the head of $\mathbf{v}_A$ in a direction perpendicular to member AB to represent the vector $\mathbf{v}_{B/A}$. This line is extended to intersect the previously drawn vertical line. This point of intersection defines the vectors $\mathbf{v}_B$ and $\mathbf{v}_{B/A}$.

Using the coordinate system shown at the reference point A in Figure E17.15(a), we may express Eq. (b) in terms of **i** and **j** components. Thus,

$$(v_B)\mathbf{j} = (v_A \cos\theta)\mathbf{i} + (v_A \sin\theta)\mathbf{j} - (v_{B/A} \sin\beta)\mathbf{i} + (v_{B/A} \sin\beta)\mathbf{j}. \quad \text{(c)}$$

Equating, separately, the coefficients of the **i** and **j** terms, we obtain the following two scalar equations:

$$v_B = v_A \sin\theta - \omega_{AB} d \sin\beta, \quad \text{(d)}$$

and

$$0 = v_A \cos\theta + \omega_{AB} d \cos\beta \quad \text{(e)}$$

where the quantity $\omega_{AB} d$ was substituted for $\mathbf{v}_{B/A}$. A simultaneous solution of Eqs. (d) and (e) yields

$$\omega_{AB} = -\left(\frac{\cos\beta}{d\cos\beta}\right) v_A \quad \text{(f)}$$

and

$$v_B = (\sin\theta + \cos\theta \tan\beta) v_A \quad \text{(g)}$$

From the geometry of Figure E17.15(a), we can show that

$$\sin\beta = \frac{r\sin\theta}{d} \quad \text{(h)}$$

Constructing Figure E17.15(c) with the aid of Eq. (h),

$$\cos\beta = \frac{\sqrt{d^2 - r^2\sin^2\theta}}{d}, \quad \text{(i)}$$

$$\tan\beta = \frac{r\sin\theta}{\sqrt{d^2 - r^2\sin^2\theta}}. \quad \text{(j)}$$

Substituting Eqs. (a), (i), and (j) in Eqs. (f) and (g),

$$\omega_{AB} = -\left(\frac{\cos\theta}{\sqrt{d^2 - r^2\sin^2\theta}}\right) r\omega \qquad \text{ANS.}$$

$$v_B = \left(\sin\theta + \frac{r\sin\theta\cos\theta}{\sqrt{d^2 - r^2\sin^2\theta}}\right) r\omega \qquad \text{ANS.}$$

■

Problems

In the solution of the following Problems, use the scalar method when the velocity diagram is a right-angle triangle and the vector method when it is not.

17.54 Refer to Problem 17.40 (p. 392), and solve for the velocity v_A using Eq. (17.21).

17.55 Refer to Problem 17.45 (p. 393), and solve for the velocity v_A using Eq. (17.21).

17.56 The wheel shown in Figure P17.56 rolls without slipping on the horizontal surface, and its angular velocity is $\omega = 50$ rad/s cw. Determine the magnitude, direction, and sense of the velocity of (a) point C, (b) point A, and (c) point B.

17.57 Refer to Problem 17.56 and determine the magnitude, direction, and sense of the velocity of (a) point D, (b) point E, and (c) point F.

17.58 Member AB executes general plane motion such that its two ends are constrained to move along two perpendicular tracks as shown in Figure P17.58. In the position shown, end B moves upward and to the right with a velocity $v_B = 0.35$ m/s. Determine the angular velocity of member AB and the linear velocity of end A.

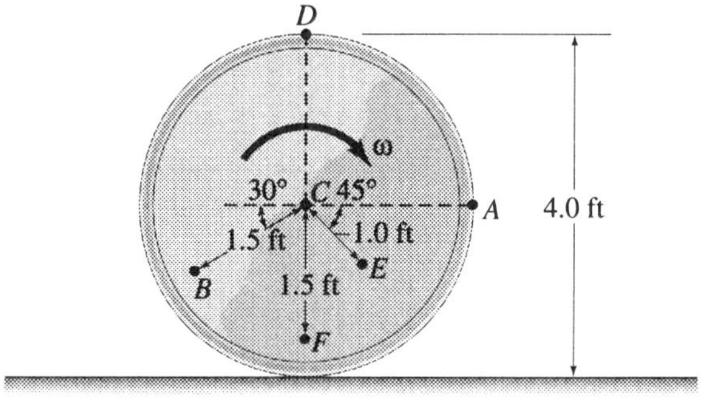

FIGURE P17.56.

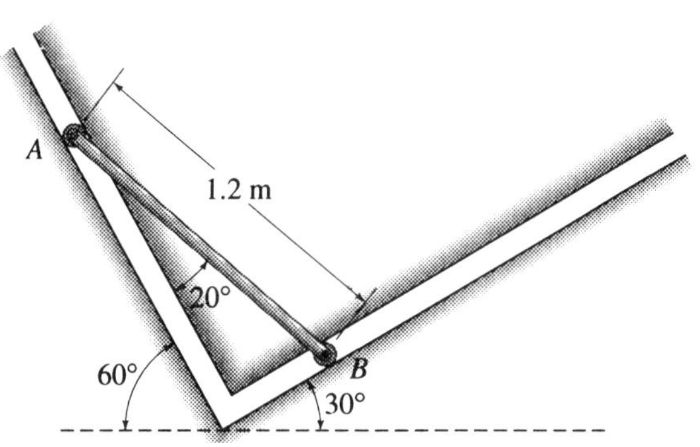

FIGURE P17.58.

17.59 In the reciprocating engine mechanism shown in Figure P17.59, the velocity of the piston B is known to be $v_B = 50$ ft/s downward. If $R = 3$ in. and $L = 10$ in., $\theta = 30°$, determine (a) the angular velocity ω_{AB} of connecting rod AB, (b) the angular velocity ω_{OA} of the crank OA, and (c) the velocity of point A.

17.60 Refer to the reciprocating engine mechanism shown in Figure P17.59. The magnitude of the velocity of point A is known to be $v_A = 20.0$ m/s upwards and to the right. If $R = 0.14$ m, $L = 0.36$ m, and $\theta = 45°$, determine (a) the velocity v_B of piston B, (b) the angular velocity ω_{AB} of connecting rod AB, and (c) the angular velocity ω_{OA} of the crank OA.

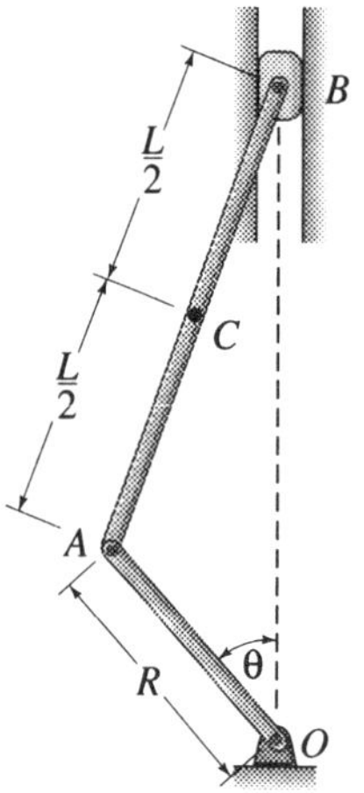

FIGURE P17.59.

17.61 Refer to the reciprocating engine mechanism shown in Figure P17.59. (a) If crank OA rotates ccw at a constant angular velocity ω, develop expressions for the velocity v_B and for the angular velocity ω_{AB} in terms of R, L, ω, and θ. (b) If $R = 0.75$ ft, $L = 2$ ft, $\omega = 60$ rad/s, and $\theta = 60°$, find the velocity v_C where point C is the center point of connecting rod AB.

17.62 Crank OA of the mechanism shown in Figure P17.62 rotates cw at a constant angular velocity $\omega = 1000$ rpm. If $\theta = 45°$, $R = 5.0$ in., $L_1 = 13.0$ in. and $L_2 = 14.0$ in., determine (a) the velocity of point B, (b) the angular velocity of AB and (c) the velocity of point C.

17.63 Refer to the mechanism shown in Figure P17.62. At the instant when $\theta = 30°$, point C is known to have a velocity $v_C = 1.0$ m/s downward. Let $R = 0.5$ m, $L_1 = 1.5$ m and $L_2 = 2.0$ m and deter-

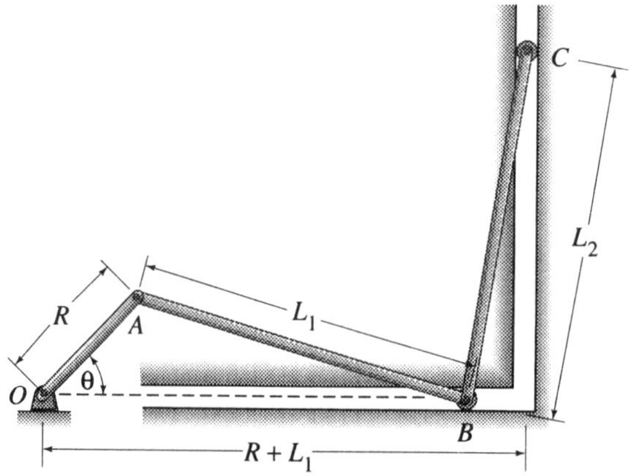

FIGURE P17.62.

mine (a) the velocity of point B, (b) the angular velocity of BC, (c) the velocity of point A and (d) the angular velocity of OA.

17.64 The triangular plate shown in Figure P17.64 is constrained to move so that corner A slides in a horizontal track and corner B in a vertical track. At the instant when $\theta = 45^\circ$, the velocity of point B is known to be 4.0 ft/s downward. Determine (a) the angular velocity of the plate and (b) the velocity of corner A. Let $b = 10.0$ in.

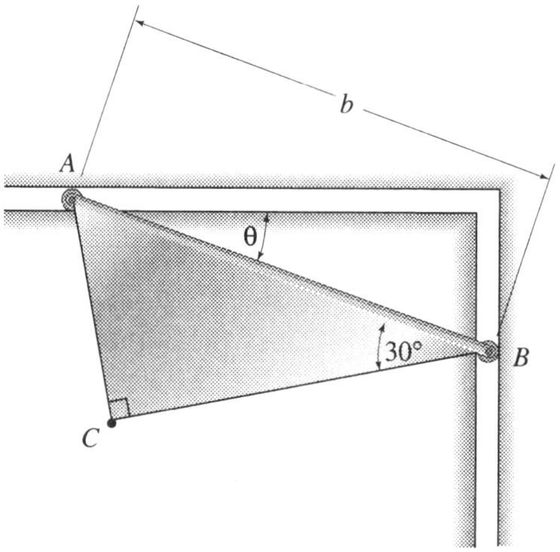

FIGURE P17.64.

17.65 The triangular plate shown in Figure P17.64 is constrained to move so that corner A slides in a horizontal track and corner B in a vertical track. At the instant when $\theta = 20^\circ$, the velocity of point A is 1.5 m/s to the left. Determine (a) the angular velocity of the plate and (b) the velocity of corner C. Let $b = 0.5$ m.

17.66 Member ABCD shown in Figure P17.66 executes general plane motion such that point B slides along a vertical track and point C along a horizontal one. If point B moves downward with a velocity v_B, develop, in terms of v_B and θ, an expression for v_A, the velocity of point A. (Hint: First determine the angular velocity of member ABCD by relating the velocities of point B and C.)

17.67 Repeat Problem 17.66 for point D instead of point A. Note that the same hint given in Problem 17.66 applies.

17.68 Crank AC in Figure P17.68 rotates in a ccw direction at a constant angular velocity ω. In terms of b, ω, and θ, determine an expression for the velocity of slider S which is constrained to move

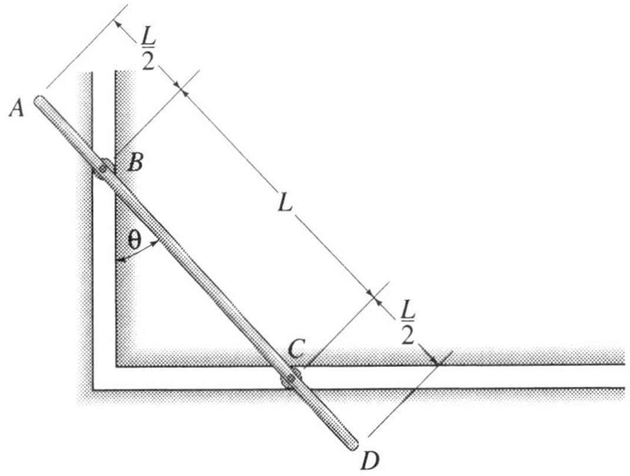

FIGURE P17.66.

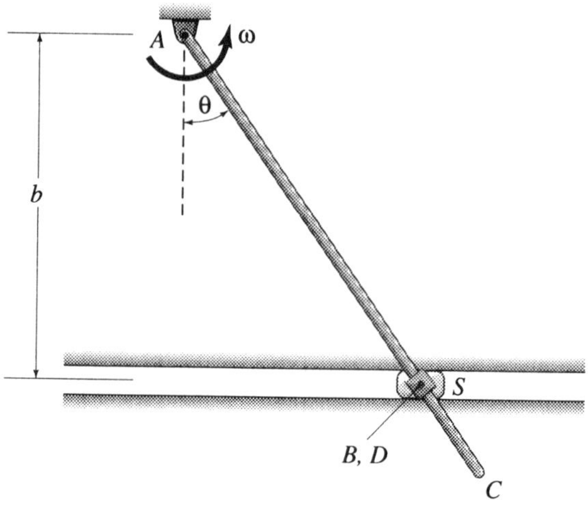

FIGURE P17.68.

along the horizontal track. (Hint: Select point B on crank AC coincident with point D which is on slider S. Relate the velocities of points D and B using Eq. 17.21).

17.69 Crank OA in Figure P17.69 rotates in a ccw direction at a constant angular velocity ω. In terms of r, ω, and θ, determine an expression for the velocity of the vertical member BC. (Hint: Select point D on the horizontal slot coincident with point A which is on the crank

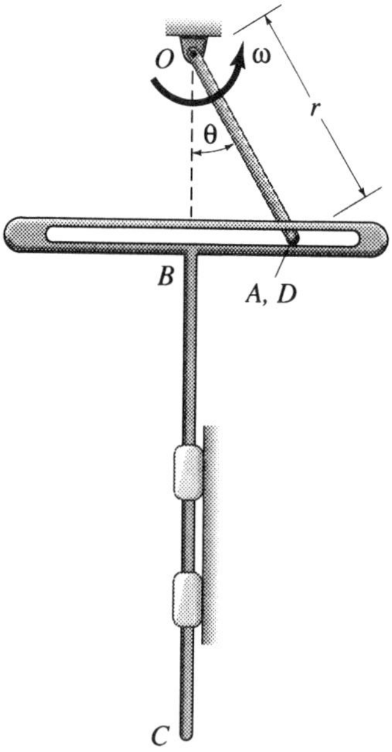

FIGURE P17.69.

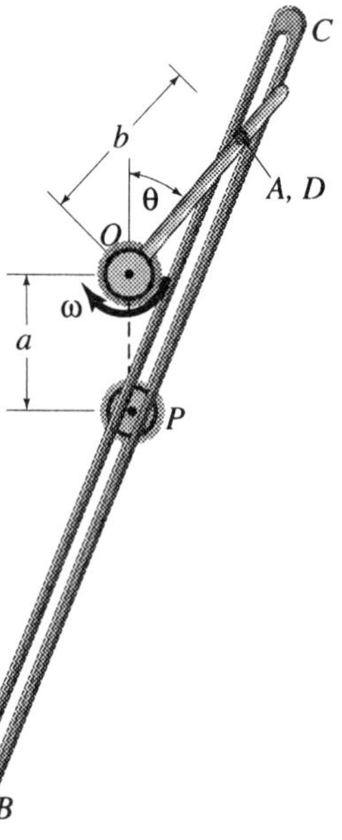

FIGURE P17.70.

OA. Relate the velocities of points D and A by Eq. (17.21).) If $r = 10$ in. and $\omega = 1800$ rpm, determine the velocity of member BC for (a) $\theta = 0°$, (b) $\theta = 30°$, and (c) $\theta = 210°$.

17.70 Crank OA in Figure P17.70 rotates cw about fixed axis O at a constant angular velocity ω. In terms of a, b, ω, and θ, develop an expression for the angular velocity ω_{BC} of member BC which rotates about fixed axis P. (Hint: Select point D on the slot coincident with A which is on crank OA. Relate the velocities of points D and A by Eq. (17.21).) Investigate the angular velocity ω_{BC} for the cases when $b = a$ and $b = 2a$.

17.5 Instantaneous Center of Rotation

The velocities of points on a rigid body, executing general plane motion, may be determined in a relatively simple and direct manner by making use of the concept of the *instantaneous center of rotation*. This concept is based upon the fact that, as the body undergoes general plane motion, for any instant, the body may be considered to rotate about an axis perpendicular to the plane of motion, known as the *instantaneous axis of rotation*. The intersection of this axis with the plane of motion is the instantaneous center of rotation, given the symbol I.

Consider the rigid body shown in Figure 17.9 which is executing general plane motion. Assuming that the instantaneous center of rotation I can be located, the absolute velocity of any point on the rigid body may be expressed in terms of the velocity of I (which is zero) by Eq. (17.21). Thus, for any point A,

$$\mathbf{v}_A = \mathbf{v}_I + \mathbf{v}_{A/I} = \mathbf{0} + \mathbf{v}_{A/I} = \boldsymbol{\omega} \times \mathbf{r}_{A/I}$$

The magnitude of $\mathbf{v}_{A/I}$ is $(r_{A/I})\omega$ where $r_{A/I}$ is the radial distance from point A to point I and ω is the cw angular velocity of the rigid body which has a unique value at a given instant of time. As discussed in Section 17.4, the direction of $\mathbf{v}_{A/I}$ is along a line perpendicular to $\mathbf{r}_{A/I}$ and its sense is determined by the sense of ω. In Figure 17.9, because ω is assumed cw the vector $\mathbf{v}_A$ is pointed upward and to the right. Similarly, for point B,

$$\mathbf{v}_B = \mathbf{0} + \mathbf{v}_{B/I} = \boldsymbol{\omega} \times \mathbf{r}_{B/I}$$

The magnitude of $\mathbf{v}_{B/I}$ is $(r_{B/I})\omega$ where $r_{B/I}$ is the radial distance from point B to point I. The direction of $\mathbf{v}_{B/I}$ is perpendicular to $r_{B/I}$, and its sense, as in the case of $\mathbf{v}_{A/I}$, is upward and to the right. Thus, at the instant depicted in Figure 17.9, the rigid body may be viewed as rotating (but not translating) at a cw angular velocity ω about an axis through point I, the instantaneous center of rotation.

Therefore, if the velocities $\mathbf{v}_A$ and $\mathbf{v}_B$ of any two points A and B on a rigid body, executing general plane motion, are known, the location of the instantaneous center of rotation, point I, may be determined as

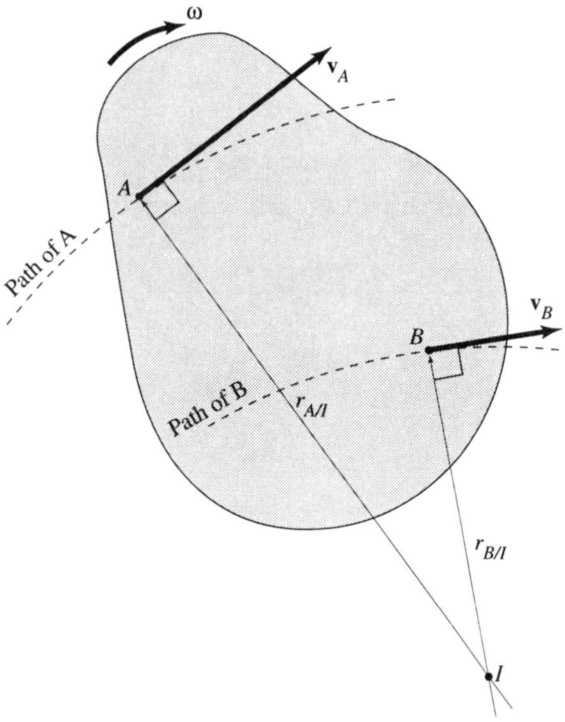

FIGURE 17.9.

indicated in Figure 17.9. Lines perpendicular to $\mathbf{v}_A$ and $\mathbf{v}_B$ are drawn from points A and B, respectively, and extended to intersect at point I, the instantaneous center of rotation. Note that, for this construction, only the directions of the velocities $\mathbf{v}_A$ and $\mathbf{v}_B$ are needed, as long as these two vectors are not parallel to each other. If, however, $\mathbf{v}_A$ and $\mathbf{v}_B$ are parallel to each other, as shown in Figures 17.10(a) and (b), these two vectors must be fully known in magnitude, direction, and sense, to locate the instantaneous center of rotation. As shown in Figures 17.10(a) and (b) the construction consists of first drawing a line connecting the tails of $\mathbf{v}_A$ and $\mathbf{v}_B$, which, of necessity, is perpendicular to these two vectors, and, then, a second line connecting the heads of $\mathbf{v}_A$ and $\mathbf{v}_B$. The intersection of these two lines locates point I, the instantaneous center of rotation. From the geometry in Figures 17.10(a) and (b) and by Eq. (17.23), we observe that the following relationships are valid:

$$\omega = \frac{v_A}{r_{A/I}} = \frac{v_B}{r_{B/I}}$$

where, as stated earlier, ω has a unique value for a rigid body at a given instant. We also conclude, from Figure 17.10(b), that, if $\mathbf{v}_A = \mathbf{v}_B$, point I

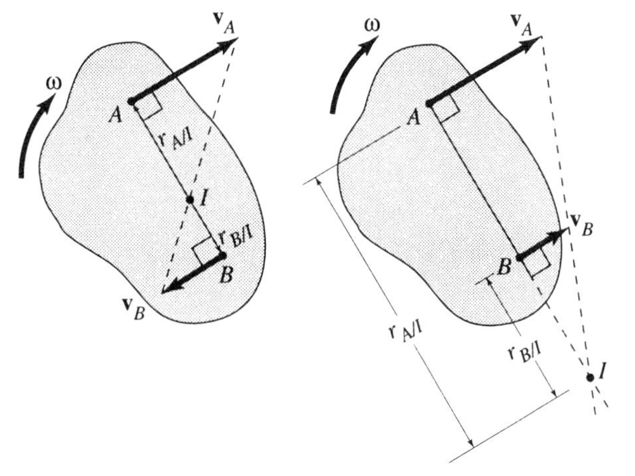

FIGURE 17.10.

will be located at infinity. In such a case, the rigid body does not rotate but undergoes pure translation.

It should be emphasized that the position of point I is not fixed but changes as the rigid body moves. This changing position of point I describes one curve on the rigid body or the rigid body extended, known as the *body centrode*, and a second curve in space, known as the *space centrode*. For any instant, these two curves are tangent to one another at point I. It should also be emphasized that, in general, although point I has zero velocity at a given instant, *it may not* have zero acceleration. Therefore, the student should not make the mistake of using point I as an instantaneous center of zero acceleration in an effort to determine the acceleration of points on the rigid body as though it were actually rotating about a fixed axis through point I.

The examples that follow illustrate the use of the concept of the instantaneous center of rotation in the solution of problems.

■ Example 17.16

Refer to Examples 17.12 and 17.13 (pp. 398, 400), and use the method of instantaneous centers to solve for the velocities of points (a) A, (b) B, and (c) E.

Solution

For convenience, the wheel of Example 17.12 is redrawn in Figure E17.16 where the spokes and other unnecessary details have been omitted.

Because the wheel rolls without slipping down the 10° inclined plane, the point of contact between it and the inclined plane is momentarily at rest (has zero velocity) and, therefore, it is the instantaneous

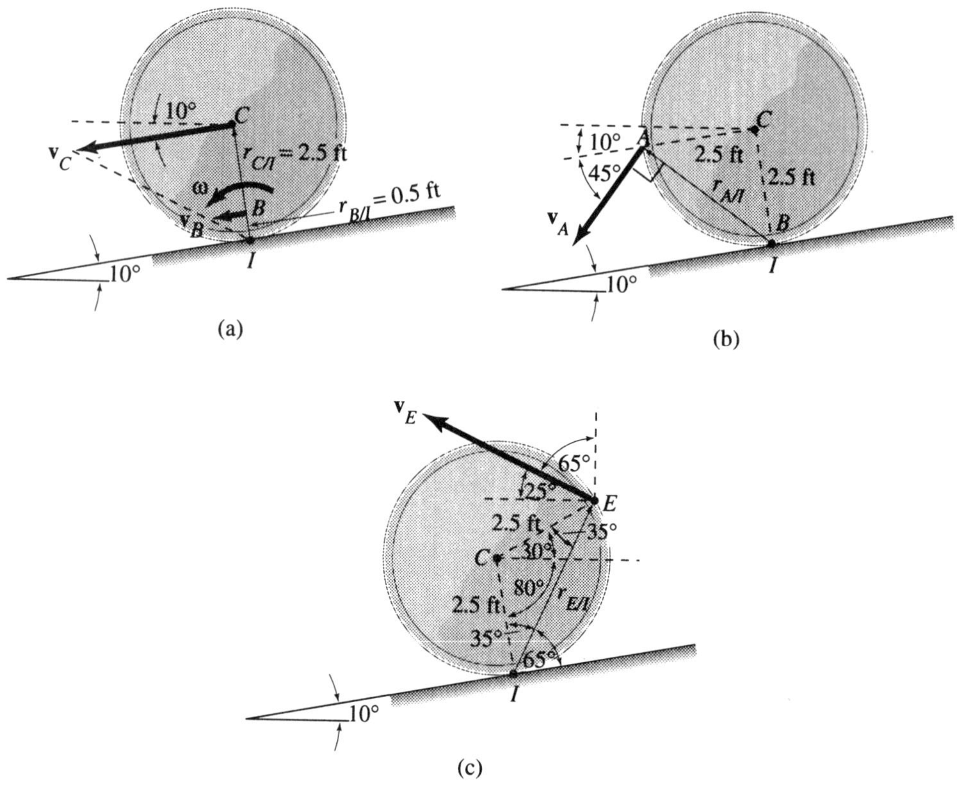

FIGURE E17.16.

center of rotation, point I. Thus, the magnitude of the velocity of its center C may be written as

$$v_C = (r_{C/I})\omega$$

where $r_{C/I} = 2.5$ ft and ω is the angular velocity of the wheel. Thus,

$$\omega = \frac{v_C}{r_{C/I}} = \frac{10.0}{2.5} = 4.0 \text{ rad/s} \circlearrowleft,$$

as shown in Figure E17.16(a). Knowing the position of the instantaneous center I and the angular velocity ω of the wheel, we can now determine the velocity of any point on the wheel as follows:

(a) Refer to Figure E17.16(b).

$$v_A = r_{A/I}\omega = \sqrt{2.5^2 + 2.5^2}(4.0) = 14.14 \text{ ft/s} \quad 55.0^\circ \swarrow. \quad \text{ANS.}$$

(b) Refer to Figure E17.16(a).

$$v_B = r_{B/I}\omega = 0.5(4.0) = 2.00 \text{ ft/s} \quad 10.0^\circ \swarrow \quad \text{ANS.}$$

(c) Refer to Figure E17.16(c).

$$v_E = r_{E/I}\omega = 2.0(2.5\cos 35^\circ)(4.0) = 16.38 \text{ ft/s} \quad 25.0^\circ \qquad \text{ANS.}$$

■ Example 17.17

Refer to Examples 17.8 and 17.9 (pp. 385, 387), and determine by the method of instantaneous centers (a) the angular velocity of the rigid plate ABCD and (b) the linear velocity of corner D. Express these answers in terms of the angular position θ.

Solution

For convenience, the rigid plate used in Examples 17.8 and 17.9 is redrawn in Figure E17.17.

(a) The velocity of corner C is known to have a magnitude of 0.75 m/s, a direction along the horizontal track, and a sense pointing to the right. Therefore, the instantaneous center of rotation, point I, must lie on a vertical line drawn at point C. Also, the velocity of point B, although unknown in magnitude, is known to have a direction along the vertical track and a sense that must be upward. Thus, point I must lie on a horizontal line drawn at point B. Consequently, the intersection of these two lines defines point I as shown in Figure 17.17. Thus, because $v_C = (r_{C/I})\omega$,

$$\omega = \frac{v_C}{r_{C/I}} = \frac{v_C}{a\cos\theta} = \frac{0.50}{\cos\theta} \text{ rad/s} \circlearrowright. \qquad \text{(a) ANS.}$$

(b) From the geometry provided in Figure E17.17, we can determine the radial distance $r_{D/I}$, from point D to point I, by applying the law of cosines to triangle ICD. Thus,

$$r_{D/I}^2 = a^2\cos^2\theta + b^2 - (2ab\cos\theta)\cos(90^\circ - \theta)$$

Because $\cos(90^\circ - \theta) = \sin\theta$,

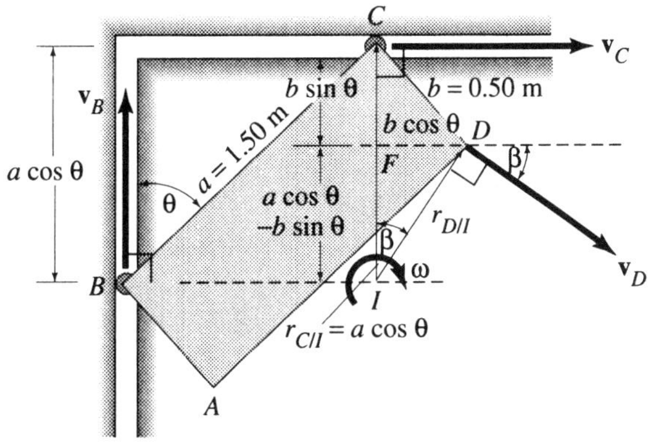

FIGURE E17.17.

$$r_{D/I} = \sqrt{a^2\cos^2\theta + b^2 - 2ab\sin\theta\cos\theta}.$$

Therefore, because $v_D = (r_{D/I})\omega$, after substituting for ω from Eq. (a), we obtain

$$v_D = \left(\frac{\sqrt{a^2\cos^2\theta + b^2 - 2ab\sin\theta\cos\theta}}{a\cos\theta}\right)v_C \qquad \text{(b)}$$

Equation (b) is a perfectly good answer for the velocity of point D. However, it is not in the same form as given in Example 17.9. To transform Eq. (b) to the desired form, we proceed as follows:

$$\begin{aligned} v_D &= \left(\sqrt{1 + \left(\frac{b}{a\cos\theta}\right)^2 - \frac{2b}{a}\tan\theta}\right)v_C \\ &= \left(\sqrt{\left(\frac{b}{a}\right)^2 + 1 - \frac{2b}{a}\tan\theta + \tan^2\theta}\right)v_C \\ &= \left(\sqrt{\left(\frac{b}{a}\right)^2 + \left(1 - \frac{b}{a}\tan\theta\right)^2}\right)v_C \\ &= 0.75\sqrt{\frac{1}{9} + \left(1 - \frac{1}{3}\tan\theta\right)^2} \qquad \text{ANS.} \end{aligned}$$

Also by considering the right-angle triangle FDI, we can determine the angle β defining the direction of v_D. Thus,

$$\begin{aligned} \beta &= \tan^{-1}\left(\frac{b\cos\theta}{a\cos\theta - b\sin\theta}\right) \\ &= \tan^{-1}\left(\frac{b}{a - b\tan\theta}\right) = \tan^{-1}\left(\frac{1.0}{3.0 - \tan\theta}\right). \qquad \text{ANS.} \end{aligned}$$

■

Problems

Use the method of instantaneous centers of rotation to solve all of the following problems:

17.71 Problem 17.56 (p. 405)
17.72 Problem 17.58 (p. 405)
17.73 Problem 17.59 (p. 406)
17.74 Problem 17.60 (p. 406)
17.75 Problem 17.61 (p. 406)
17.76 Problem 17.62 (p. 406)
17.77 Problem 17.63 (p. 406)
17.78 Problem 17.64 (p. 407)
17.79 Problem 17.66 (p. 407)
17.80 Problem 17.67 (p. 407)
17.81 Rod AB of length L is allowed to slide inside a semicircular dish of radius R as shown in Figure P17.81. For the position shown, the velocity of point A is known to have a magnitude v_A. In terms

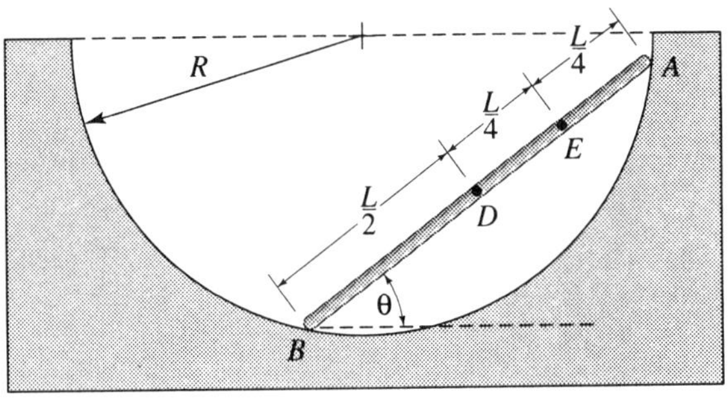

FIGURE P17.81.

of v_A, L, and R, determine (a) the angular velocity of rod AB and (b) the velocity of point B.

17.82 Refer to Problem 17.81 and determine (a) the velocity of point D and (b) the velocity of point E. Express answers in terms of v_A, L, and R.

17.83 Refer to the mechanisms shown in Figure P17.83. For the position shown, the angular velocity of link AB is $\omega = 10$ rad/s cw. Determine (a) the angular velocity of link BC and (b) the angular velocity of link CD.

17.84 Refer to the mechanism shown in Figure 17.84. For the position shown, the angular velocity of member AB is 5.0 rad/s ccw. Determine (a) the angular velocity of member BC and (b) the velocity of the slider S.

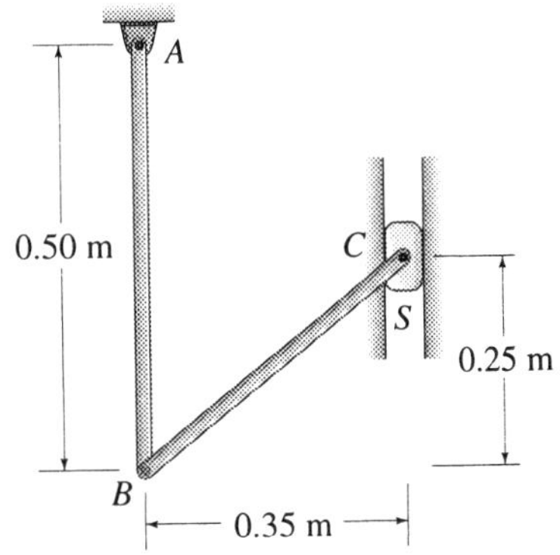

FIGURE P17.84.

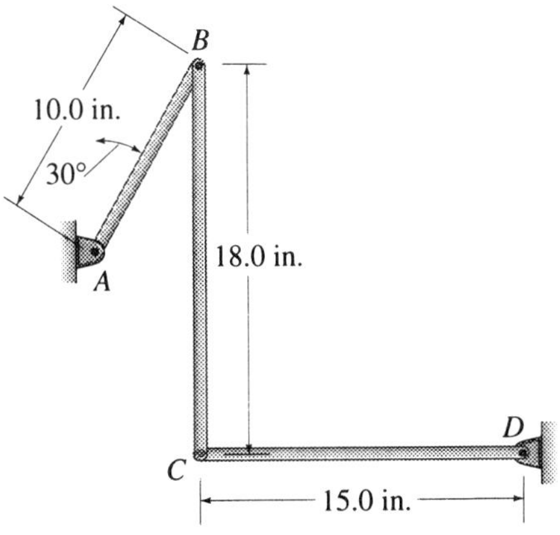

FIGURE P17.83.

17.85 Refer to the mechanism shown in Figure 17.84. For the position shown, the velocity of the slider S is 7.0 m/s downward. Determine (a) the angular velocity of member BC and (b) the angular velocity of member AB.

17.86 Refer to the mechanism shown in Figure P17.86. For the position shown, the angular velocity of member AB is 8.0 rad/s cw. Determine (a) the angular velocity of member CD, (b) the velocity of point E, and (c) the velocity of point F.

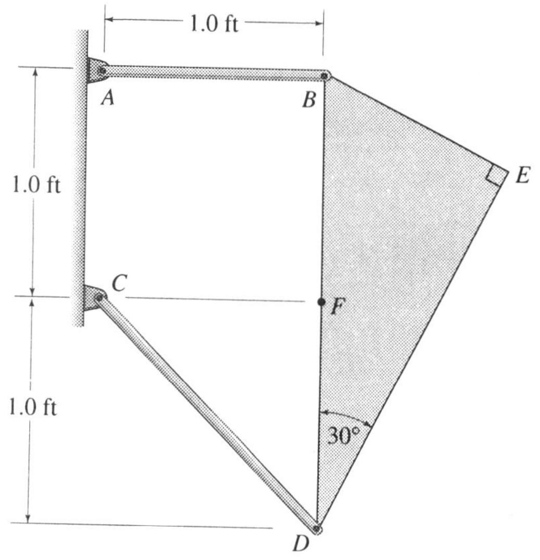

FIGURE P17.86.

17.87 Refer to the mechanism shown in Figure P17.86. For the position shown, the angular velocity of the rigid triangular plate is 3.0 rad/s ccw. Determine (a) the angular velocity of member AB and (b) the angular velocity of member CD.

17.6 Relative Acceleration—Translating Nonrotating Axes

The method developed in Section 17.4 for the determination of absolute velocities from relative velocities in general plane motion is extended here to derive a procedure that we can use to determine absolute accelerations from relative accelerations. Accordingly, if Eq. 17.21 is differentiated with respect to time,

$$\mathbf{a}_A = \mathbf{a}_B + \mathbf{a}_{A/B} \tag{17.24}$$

Physically, Eq. 17.24 means that we can obtain the absolute acceleration of some point A on a rigid body, such as shown in Figure 17.11(a), by adding two quantities vectorially. These two quantities are, first, the absolute acceleration of a second point B on the same rigid body (chosen as a reference), and, second, the acceleration of point A relative to the reference point B. In other words, this latter term, $\mathbf{a}_{A/B}$, represents the acceleration of point A as measured by an observer stationed at point B which serves as the origin of a translating, nonrotating coordinate system. As discussed in Section 17.4, because points A and B lie on the same rigid body, point A to an observer at point B appears to be rotating about an axis at point B. Thus, the vector $\mathbf{a}_{A/B}$ may be conveniently expressed in terms of normal and tangential components, as was done in Section 13.9. Therefore,

$$\mathbf{a}_{A/B} = \boldsymbol{\omega} \times \mathbf{v}_{A/B} + \boldsymbol{\alpha} \times \mathbf{r}_{A/B} = (\mathbf{a}_{A/B})_n + (\mathbf{a}_{A/B})_t \tag{17.25}$$

where

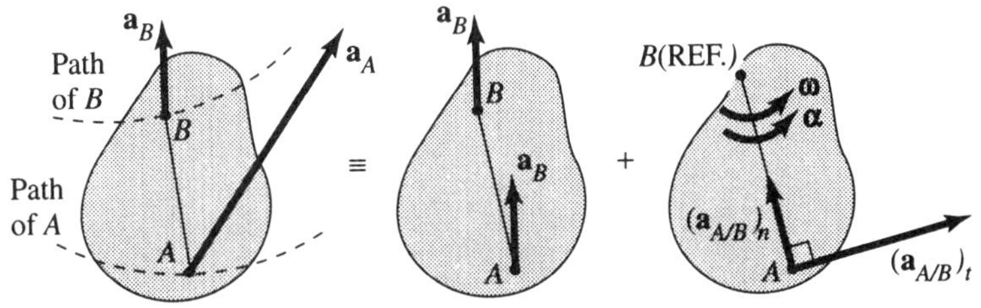

(a) General plane motion
(b) Translation
(c) Rotation about an axis at B.

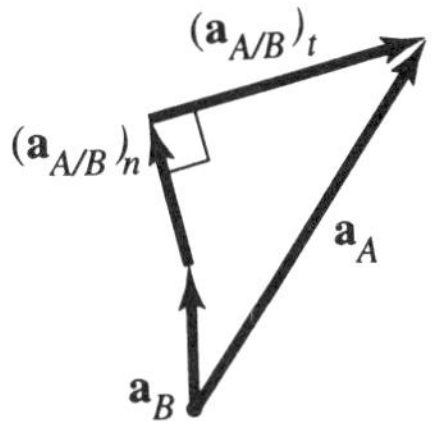

FIGURE 17.11.

$$(\mathbf{a}_{A/B})_n = \boldsymbol{\omega} \times \mathbf{v}_{A/B} = \boldsymbol{\omega} \times (\boldsymbol{\omega} \times \mathbf{r}_{A/B}) = \omega\mathbf{k} \times (\omega\mathbf{k} \times \mathbf{r}_{A/B})$$
$$= -\omega^2\mathbf{r}_{A/B} \tag{17.26a}$$

and

$$(\mathbf{a}_{A/B})_t = \boldsymbol{\alpha} \times \mathbf{r}_{A/B} = \alpha\mathbf{k} \times \mathbf{r}_{A/B} \tag{17.26b}$$

Thus, Eq. (17.24) may be expressed in the following two useful forms.

$$\mathbf{a}_A = \mathbf{a}_B + (\mathbf{a}_{A/B})_n + (\mathbf{a}_{A/B})_t$$
$$= \mathbf{a}_B - \omega^2\mathbf{r}_{A/B} + \alpha\mathbf{k} \times \mathbf{r}_{A/B} \tag{17.27}$$

where ω and α, respectively, are the magnitudes of the angular velocity and angular acceleration of the rigid body that contains particles A and B and $\mathbf{k}$ is a unit vector perpendicular to the page. Note that $\mathbf{r}_{A/B}$ represents the position vector of point A measured from point B which serves as the reference point. Note also that the magnitudes of $(\mathbf{a}_{A/B})_n$ and $(\mathbf{a}_{A/B})_t$ are given by

$$(a_{A/B})_n = (r_{A/B})\omega^2 = (v_{A/B})\omega = \frac{(v_{A/B})^2}{r_{A/B}} \tag{17.28a}$$

and

$$(a_{A/B})_t = (r_{A/B})\alpha. \tag{17.28b}$$

As in the case of velocities discussed in Section 17.4, we may view Eq. 17.27 as yielding the absolute acceleration of any point of a rigid body undergoing general plane motion, by combining a translation component and two rotation components. Such a viewpoint is illustrated in Figure 17.11. Figure 17.11(a) shows a rigid body in general plane motion. Note that, unlike velocities, accelerations of points are not, in general, tangent to the paths of these points. Figure 17.11(b) represents the translation component of acceleration and Figure 17.11(c) represents the two rotational components of acceleration for this general plane motion. The acceleration diagram shown in Figure 17.11(d) illustrates how these three acceleration components are added vectorially by Eq. 17.27 to obtain the acceleration $\mathbf{a}_A$.

The use of this equation in the solution of acceleration problems is illustrated in the following examples. Since Eq. (17.27) is a two-dimensional vector relationship, it represents two independent scalar equations which can be solved to give only two unknown quantities. Note that, as in the case of velocity diagrams, acceleration diagrams may be handled analytically by using either the scalar trigonometric method or the vector method. It is, generally, more direct to use the scalar trigonometric method when dealing with acceleration diagrams that are right-angle triangles. However, the vector method would probably yield a simpler solution for more complex cases.

Example 17.18

Consider the mechanism shown in Figure E17.18(a). For the position shown, crank OA has a ccw angular velocity $\omega = 10.00$ rad/s and a ccw angular acceleration $\alpha = 20.00$ rad/s^2. Determine (a) the angular velocity of rod AB and the linear velocity of the slider at B and (b) the angular acceleration of rod AB and the linear acceleration of the slider at B. Let $r = 0.40$ m and $\theta = 30°$.

Solution

The mechanism shown in Figure E17.18(a) consist of three rigid bodies that execute three different types of motion: member OA rotates about an axis through fixed point O, member AB executes general plane motion, and the slider at B undergoes rectilinear translation.

(a) The angular velocity ω_{AB} of rod AB and the linear velocity of the slider at B are found by the method of the instantaneous center of rotation. The location of the instantaneous center of member AB, I_{AB}, is shown in Figure E17.18(a). Thus,

$$v_A = r\omega = (0.40)(10.00) = 4.00 \text{ m/s} \downarrow .$$

Also,

$$v_A = (r_{A/I})\omega_{AB},$$

and

$$\omega_{AB} = \frac{v_A}{r_{A/I}} = 2.89 \text{ rad/s} \circlearrowleft,$$

$$v_B = (r_{B/I})\omega_{AB} = 2.31 \text{ m/s} \leftarrow .$$

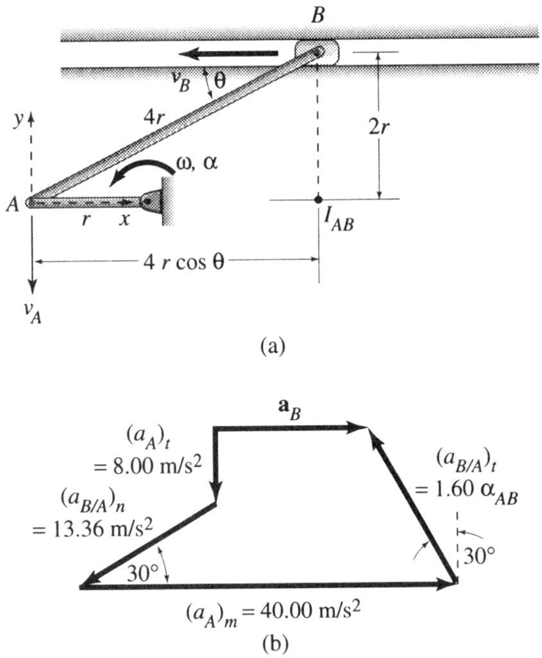

FIGURE E17.18.

(b) The acceleration of the slider at B may be related to the acceleration of point A by the first form of Eq. (17.27). Thus,

$$\mathbf{a}_B = a_A + (\mathbf{a}_{B/A})_n + (\mathbf{a}_{B/A})_t = (\mathbf{a}_A)_n + (\mathbf{a}_A)_t + (\mathbf{a}_{B/A})_n + (\mathbf{a}_{B/A})_t \quad \text{(a)}$$

where the acceleration $\mathbf{a}_A$ has been replaced by its normal and tangential components. Note that A is the reference point for $\mathbf{a}_{B/A}$. The acceleration diagram representing Eq. (a) is shown in Figure E17.18(b). The construction of the acceleration diagram was begun by drawing vector $(\mathbf{a}_A)_t$ whose magnitude is $r\alpha = (0.40)(20) = 8.00 \text{ m/s}^2$, whose direction is perpendicular to OA, and whose sense is downward. This was followed by vector $(\mathbf{a}_{B/A})_n$ which has a magnitude equal to $(r_{B/A})\omega_{AB}^2 = (1.60)(2.89)^2 = 13.26 \text{ m/s}^2$, a direction along BA, and a sense from B to A. Next vector $(\mathbf{a}_A)_n$ was drawn. This vector has a magnitude equal to $r\omega^2 = (0.40)(10)^2 = 40.00 \text{ m/s}^2$, a direction along OA, and a sense from A to 0. The next vector drawn is $(\mathbf{a}_{B/A})_t$ whose magnitude is equal to $(r_{B/A})\omega_{AB} = (1.60)\alpha_{AB}$ where α_{AB} is the unknown angular acceleration of rod AB. Its direction, however, is along a line perpendicular to BA. Thus, from the head of vector $(\mathbf{a}_A)_n$, a line is drawn perpendicular to BA to represent the direction of vector $(\mathbf{a}_{B/A})_t$. Finally, from the tail of the first vector $(\mathbf{a}_A)_t$, a horizontal line is drawn to represent the direction of $\mathbf{a}_B$. This horizontal line is extended to intersect the vector $(\mathbf{a}_{B/A})_t$ at a point that defines the magnitudes of both $(\mathbf{a}_{B/A})_t$ and $\mathbf{a}_B$. Note that other sequences for drawing the component vectors could have been used in the construction of the acceleration diagram.

Using the coordinate system shown at point A in Figure E17.16(a), we may express Eq. (a) in vector form as follows:

$$(a_B)\mathbf{i} + (0)\mathbf{j} = -(8.00)\mathbf{j} - (13.36\cos 30°)\mathbf{i} - (13.36\sin 30°)\mathbf{j}$$
$$+ (40.00)\mathbf{i} - (1.60\alpha_{AB}\sin 30°)\mathbf{i} + (1.60\alpha_{AB}\cos 30°)\mathbf{j}$$

Equating, separately, the coefficients of the **i** and **j** terms we obtain the following two scalar equations:

$$a_B = 28.43 - 0.80\alpha_{AB}, \quad \text{(b)}$$

$$0 = -14.68 + 1.39\alpha_{AB}. \quad \text{(c)}$$

The simultaneous solution of Eqs. (b) and (c) yields

$$\alpha_{AB} = 10.56 \text{ rad/s}^2 \circlearrowleft, \quad \text{ANS.}$$

$$a_B = 19.98 \text{ m/s}^2 \rightarrow. \quad \text{ANS.}$$

Example 17.19

The mechanism ABCD shown in Figure E17.19 moves in the plane of the page. At the instant shown, member AB has a constant cw angular velocity $\omega_{AB} = 2$ rad/s. Determine the angular accelerations of members BC and CD.

Solution

Before the angular acceleration of members BC and CD can be determined, their angular velocities must be found. This is accomplished by the method of instantaneous centers. The instantaneous center I_{BC} for member BC is located in Figure E17.19. Thus,

$$v_B = (r_{B/A})\omega_{AB} = (10)(2) = 20 \text{ in./s} \downarrow.$$

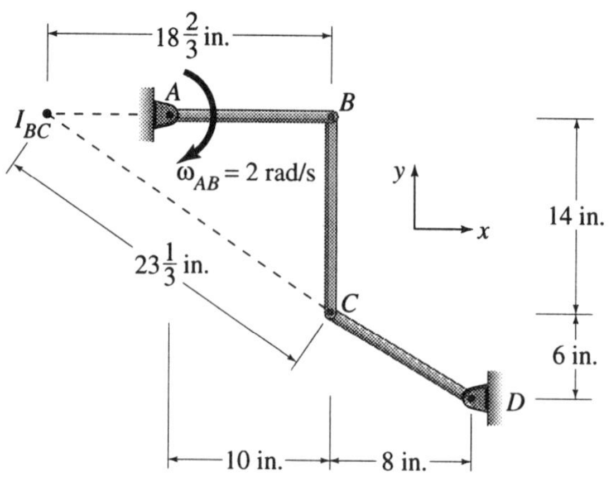

FIGURE E17.19.

Also, because $v_B = (r_{B/I})\omega_{BC}$, it follows that

$$\omega_{BC} = \frac{v_B}{r_{B/I}} = 15/14 \text{ rad/s } \circlearrowleft$$

and

$$v_C = r_{C/I}\omega_{BC} = 25.0 \text{ in./s } \swarrow .$$

Because $v_C = (r_{C/D})\omega_{CD}$, it follows that

$$\omega_{CD} = \frac{v_C}{r_{C/I}} = 2.5 \text{ rad/s } \circlearrowright .$$

Now, using the second form of Eq. (17.27), we may relate the acceleration of point C to that of point B on member CB as follows:

$$\begin{aligned} \mathbf{a}_C &= \mathbf{a}_B - \omega_{BC}^2 \mathbf{r}_{C/B} + \alpha_{BC}\mathbf{k} \times \mathbf{r}_{C/B} \\ &= \mathbf{a}_B - (\tfrac{15}{14})^2(0\mathbf{i} - 14\mathbf{j}) + \alpha_{BC}\mathbf{k} \times (0\mathbf{i} - 14\mathbf{j}) \\ &= \mathbf{a}_B + 16.1\mathbf{j} + 14\alpha_{BC}\mathbf{i} \end{aligned} \tag{a}$$

Again, by the second form of Eq. (17.27), $\mathbf{a}_C$ is related to $\mathbf{a}_D$ on member CD. Thus

$$\begin{aligned} \mathbf{a}_C &= \mathbf{a}_D - \omega_{CD}^2 \mathbf{r}_{C/D} + \alpha_{CD}\mathbf{k} \times \mathbf{r}_{C/D}, \\ &= 0 - (2.5)^2(-8\mathbf{i} + 6\mathbf{j}) + \alpha_{CD}\mathbf{k} \times (-8\mathbf{i} + 6\mathbf{j}), \\ &= (50 - 6\alpha_{CD})\mathbf{i} = (37.5 + 8\alpha_{CD})\mathbf{j}. \end{aligned} \tag{b}$$

Similarly, $\mathbf{a}_B$ is related to $\mathbf{a}_A$ on member BA. Remembering that $\alpha_{AB} = 0$,

$$\begin{aligned} \mathbf{a}_B &= \mathbf{a}_A - \omega_{AB}^2 \mathbf{r}_{B/A} + \alpha_{AB}\mathbf{k} \times \mathbf{r}_{B/A}, \\ &= 0 - (2)^2(10\mathbf{i} + 0\mathbf{j}) = -40\mathbf{j}. \end{aligned} \tag{c}$$

Substitute Eqs. (b) and (c) into Eq. (a), and equate the coefficients of the **i** and **j** terms to obtain the following two simultaneous equations:

$$3\alpha_{CD} + 7\alpha_{BC} - 45 = 0 \tag{d}$$

$$8\alpha_{CD} + 53.6 = 0 \tag{e}$$

Solving Eqs. (d) and (e) simultaneously,

$$\boldsymbol{\alpha}_{CD} = -6.7\mathbf{k} \text{ rad/s}^2 = 6.70 \text{ rad/s}^2 \circlearrowleft, \quad \text{and} \qquad \text{ANS.}$$

$$\boldsymbol{\alpha}_{BC} = 9.3\mathbf{k} \text{ rad/s}^2 = 9.30 \text{ rad/s}^2 \circlearrowright . \qquad \text{ANS.}$$

■

Problems

Use the method of relative accelerations to solve the following problems:

17.88 Problem 17.40 (p. 392)
17.89 Problem 17.41 (p. 392)
17.90 Problem 17.42 (p. 392)
17.91 Problem 17.45 (p. 393)
17.92 Problem 17.51 (p. 394)
17.93 Problem 17.52 (p. 395)
17.94 Crank OA of the mechanism shown in Figure P17.94 rotates ccw at a constant angular velocity ω. At the instant when $\theta = 30°$, the acceleration of the slider at B is measured at 426.5 m/s^2 to the left. If $r = 0.15$ m, $L = 0.45$ m, determine the magnitude of ω.

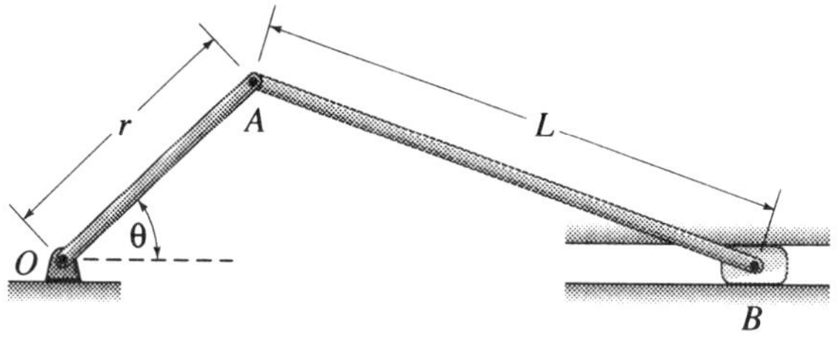

FIGURE P17.94.

17.95 Crank OA of the mechanism shown in Figure P17.94 rotates ccw at a constant angular velocity ω. At the instant that $\theta = 270°$, the acceleration of the slider at B is measured at 18,870.0 in./s^2 to the right. If $r = 6.0$ in. and $L = 20.0$ in., determine (a) the magnitude of ω and (b) the angular acceleration of member AB.

17.96 Refer to the mechanism shown in Figure P17.94. For the position defined by $\theta = 45°$, crank OA rotates ccw at an angular velocity $\omega = 50.0$ rad/s and a ccw angular acceleration $\alpha = 10.0$ rad/s^2. If $r = 6.0$ in. and $L = 15.0$ in., determine the acceleration of the slider at B and the angular acceleration of rod AB.

17.97 Member ABCD shown in Figure P17.97 executes general plane motion such that point B moves along the vertical track and point C along the horizontal track. Point B moves upward at a constant velocity v_B. For any position defined by the angle θ, develop an expression for the acceleration of point C and the angular acceleration of member ABCD in terms of v_B, L, and θ.

17.98 Repeat Problem 17.97 for point A instead of point C.

17.99 At the instant shown, the wheel of Figure P17.99 rolls down the 20° inclined plane with a velocity $v = 5.0$ m/s and an acceleration $a = 10.0$ m/s^2. Determine the angular acceleration of the wheel and the acceleration of point A, the point of contact between the wheel and the inclined plane. Let $r = 0.50$ m.

17.100 At the instant shown, the center of the wheel of Figure P17.99 rolls down the inclined plane with a velocity $v = 12$ ft/s and an acceleration $a = -5$ ft/s^2. Determine the angular acceleration of the wheel and the acceleration of point

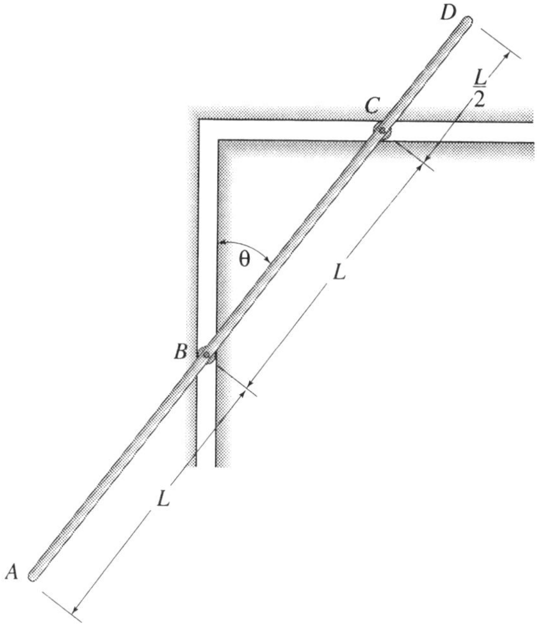

FIGURE P17.97.

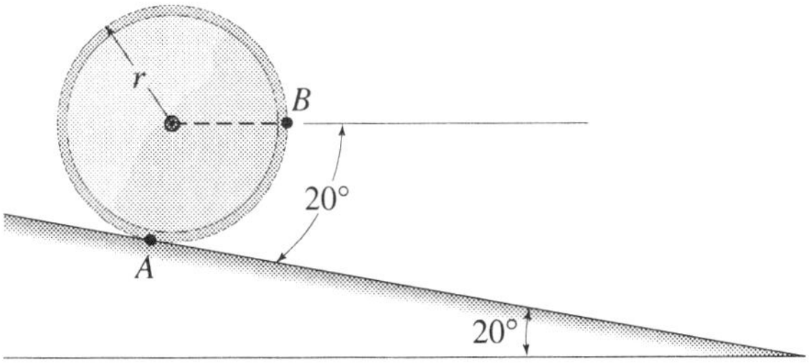

FIGURE P17.99.

B on the rim of the wheel. Let $r = 15$ in.

17.101 Refer to the mechanism shown in Figure P17.101. For the position shown, crank OA has a constant ccw angular velocity $\omega = 10.0$ rad/s. Determine the angular acceleration of links AB and BC. Let $r = 5$ in.

17.102 Refer to the mechanism shown in Figure P17.101. For the position shown, crank OA has a cw angular velocity $\omega = 5.0$ rad/s and a ccw angular acceleration $\alpha = 2.5$ rad/s^2. Determine the angular acceleration of links AB and BC. Let $r = 0.10$ m.

17.103 Refer to the mechanism shown in Figure P17.103. Crank OA rotates at a constant cw angular velocity ω. In terms of R, L, and θ, develop expressions for the acceleration of the collar

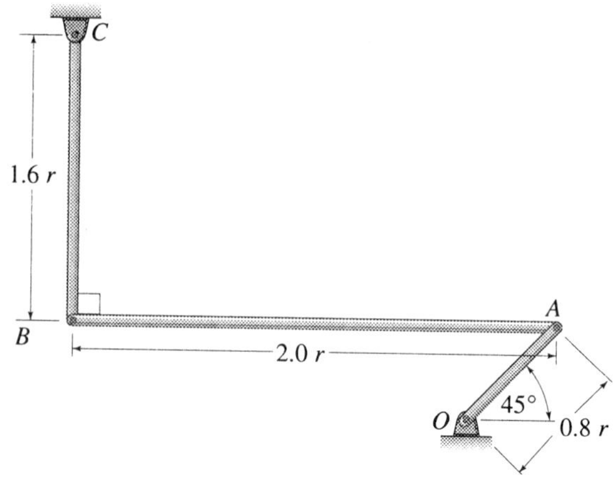

FIGURE P17.101.

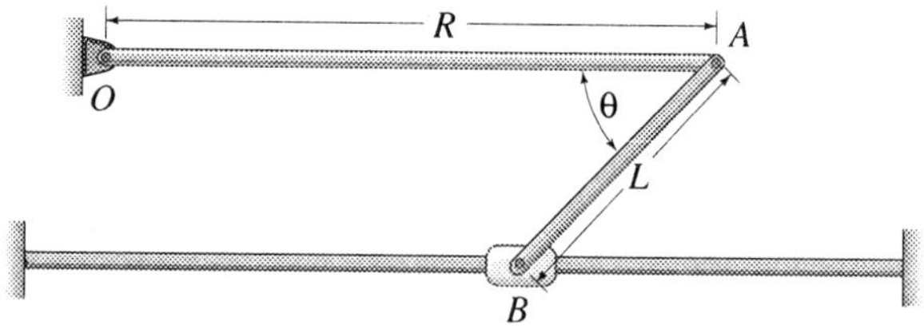

FIGURE P17.103.

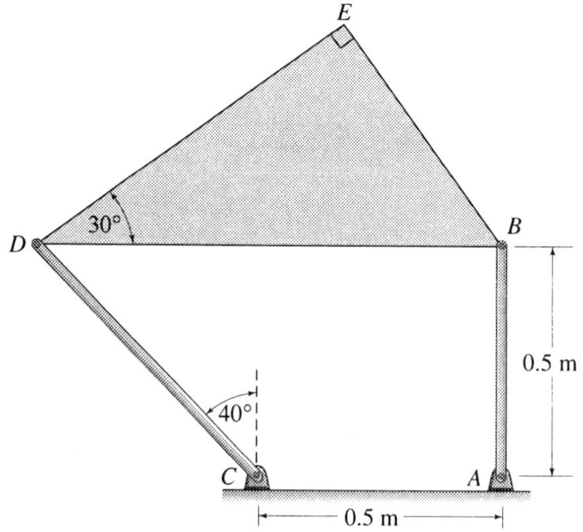

FIGURE P17.105.

at B and for the angular acceleration of link AB, for the position where OA is horizontal.

17.104 Refer to the mechanism shown in Figure P17.103. Let $R = 12.0$ in. and $L = 8.0$ in.. For the position defined by the angle $\theta = 30°$, crank OA has a cw angular velocity $\omega = 7.5$ rad/s and a ccw angular acceleration $\alpha = 5.0$ rad/s^2. Determine the acceleration of the collar at B and the angular acceleration of link AB.

17.105 For the position shown, crank AB, of the mechanism of Figure P17.105, has an angular velocity $\omega = 10$ rad/s and an angular acceleration $\alpha = 6$ rad/s^2, both cw. Determine the acceleration of point D and the angular acceleration of link CD.

17.106 Refer to Problem 17.105 and determine the acceleration of point E and the angular acceleration of the triangular rigid plate BDE.

17.7 Relative Plane Motion—Rotating Axes

Occasionally, we come across a mechanism in which a point on one member moves along a rotating path. This and other similar problems are conveniently formulated by the use of a set of axes that not only translates but also rotates with respect to a fixed frame of reference. The resulting equations relate the velocity and acceleration of one point to those of another point attached to the translating and rotating set of axis. This formulation is sufficiently general to permit the solution of problems in which the two points are either on the same mechanism or moving independently of one another.

Consider the fixed coordinate system X-Y and the two points A and B whose position vectors are $\mathbf{r}_A$ and $\mathbf{r}_B$, respectively, as shown in Figure 17.12. A translating and rotating coordinate system x-y is attached to point B. The angular velocity and angular acceleration of the x-y coordinate system about a Z axis out of the page, as measured

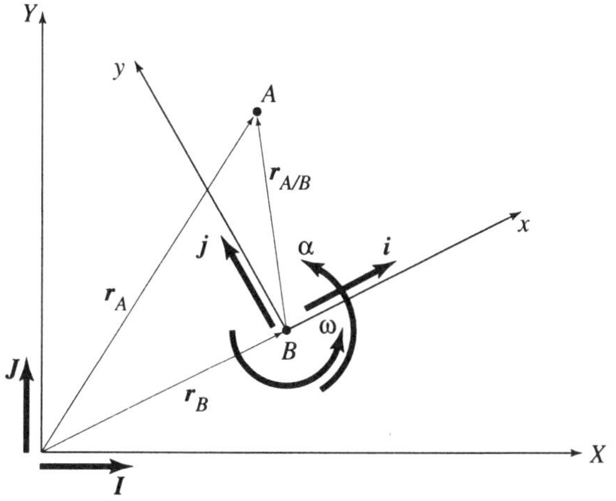

FIGURE 17.12.

from the fixed X-Y coordinate system are, respectively, $\boldsymbol{\omega}$ and $\boldsymbol{\alpha}$. Note that A and B may represent two points moving independently of each other, two points on the same rigid body, or two points on different rigid bodies in the same mechanism.

Using the geometry in Figure 17.12, we can express the absolute position vector of point A, $\mathbf{r}_A$, in terms of the absolute position vector of point B, $\mathbf{r}_B$, and the position of point A relative to point B, $\mathbf{r}_{A/B}$. Thus,

$$\mathbf{r}_A = \mathbf{r}_B + \mathbf{r}_{A/B} \tag{17.29}$$

where

$$\mathbf{r}_{A/B} = x_{A/B}\mathbf{i} + y_{A/B}\mathbf{j} \tag{17.30}$$

In Eq. (17.30), $x_{A/B}$ and $y_{A/B}$ represent the coordinates of point A in the translating-rotating coordinate system, and $\mathbf{i}$ and $\mathbf{j}$ are unit vectors, respectively, along the x and y axes of this system. Taking the time derivative of $\mathbf{r}_A$ in Eq. (17.29) yields the velocity of point A. Accordingly,

$$\frac{d\mathbf{r}_A}{dt} = \frac{d\mathbf{r}_B}{dt} + \frac{d\mathbf{r}_{A/B}}{dt} \tag{17.31}$$

where

$$\frac{d\mathbf{r}_A}{dt} = \mathbf{v}_A, \quad \frac{d\mathbf{r}_B}{dt} = \mathbf{v}_B \tag{17.32a}$$

and

$$\frac{d\mathbf{r}_{A/B}}{dt} = \frac{d}{dt}(x_{A/B}\mathbf{i} + y_{A/B}\mathbf{j}). \tag{17.32b}$$

In Eq. (17.32b), the unit vectors $\mathbf{i}$ and $\mathbf{j}$ are time-dependent because, although their magnitudes are constant, their directions change with the translating-rotating x-y coordinate systems as measured by an observer in the stationary X-Y system. Thus, Eq. (17.32b), may be written as follows

$$\frac{d\mathbf{r}_{A/B}}{dt} = x_{A/B}\frac{d\mathbf{i}}{dt} + \dot{x}_{A/B}\mathbf{i} + y_{A/B}\frac{d\mathbf{j}}{dt} + \dot{y}_{A/B}\mathbf{j}$$

$$= \dot{x}_{A/B}\mathbf{i} + \dot{y}_{A/B}\mathbf{j} + x_{A/B}\frac{d\mathbf{i}}{dt} + y_{A/B}\frac{d\mathbf{j}}{dt} \tag{17.32c}$$

where the quantities $(\dot{x}_{A/B})\mathbf{i}$ and $(\dot{y}_{A/B})\mathbf{j}$ represent the x and y components, respectively, of the velocity of point A as measured from point B. Therefore, the vector sum of the first two terms in Eq. (17.32), represents the velocity $\mathbf{v}_{A/B}$ of point A as measured by a rotating observer stationed at point B. Thus,

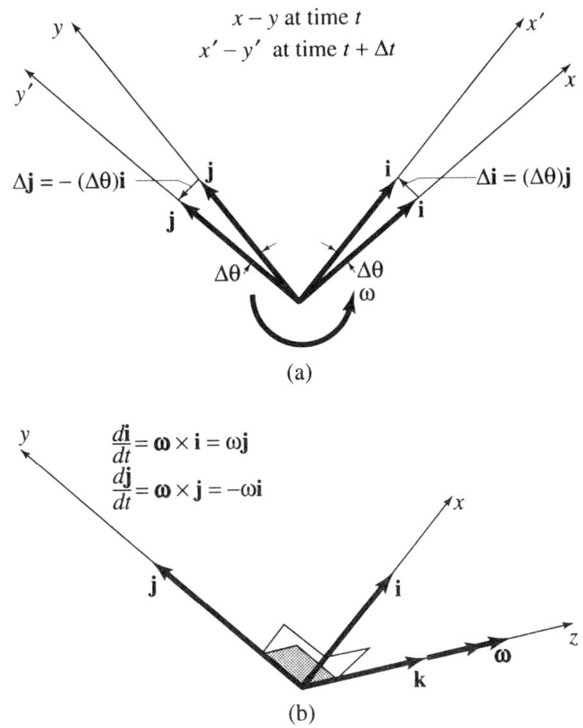

FIGURE 17.13.

$$\dot{x}_{A/B}\mathbf{i} + \dot{y}_{A/B}\mathbf{j} = \mathbf{v}_{A/B}. \tag{17.32d}$$

The time derivatives of the unit vectors in the last two terms of Eq. (17.32c) require interpretation. Consider the translating-rotating x-y coordinate system at time t and at time $t + \Delta t$ as shown in Figure 17.13(a). In the short time interval Δt, the x-y system rotates through the angle $\Delta\theta$ to position x'-y'. During this infinitesimal rotation, the two unit vectors $\mathbf{i}$ and $\mathbf{j}$ retain their magnitudes of unity, but their directions change by the small vectors $\Delta\mathbf{i}$ and $\Delta\mathbf{j}$, respectively, as indicated in Figure 17.13(a). In the limit when Δt approaches zero, the direction of $\Delta\mathbf{i}$ is along the positive y axis and may be expressed by $\mathbf{j}$ and the direction of $\Delta\mathbf{j}$ is along the negative x axis and may be defined by $-\mathbf{i}$. As indicated in Figure 17.13(a), the magnitudes of $\Delta\mathbf{i}$ and $\Delta\mathbf{j}$ are both equal to $\Delta\theta$ which, in the limit when $\Delta t \to 0$, is equal to the arc length representing the product of $\Delta\theta$ and the unit radius. Thus,

$$\frac{d\mathbf{i}}{dt} = \lim_{\Delta t \to 0} \frac{\Delta\mathbf{i}}{\Delta t} = \lim_{\Delta t \to 0} \frac{\Delta\theta}{\Delta t}\mathbf{j} = \frac{d\theta}{dt}\mathbf{j} = \omega\mathbf{j}. \tag{17.32e}$$

Similarly,

$$\frac{d\mathbf{j}}{dt} = -\omega\mathbf{i}. \tag{17.32f}$$

As illustrated in Figure 17.13(b), the quantities $\omega\mathbf{j}$ and $-\omega\mathbf{i}$ may be represented by the cross-product. Thus,

$$\frac{d\mathbf{i}}{dt} = \omega\mathbf{j} = \boldsymbol{\omega} \times \mathbf{i} \tag{17.32g}$$

and

$$\frac{d\mathbf{j}}{dt} = -\omega\mathbf{i} = \boldsymbol{\omega} \times \mathbf{j} \tag{17.32h}$$

Thus, the sum of the last two terms in Eq. (17.32c) becomes

$$x_{A/B}\frac{d\mathbf{i}}{dt} + y_{A/B}\frac{d\mathbf{j}}{dt} = \boldsymbol{\omega} \times [x_{A/B}\mathbf{i} + y_{A/B}\mathbf{j}] = \boldsymbol{\omega} \times \mathbf{r}_{A/B} \tag{17.32i}$$

It follows, therefore, from Eq. (17.32c), after substituting from Eqs. (17.32d) and (17.32i), that

$$\frac{d\mathbf{r}_{A/B}}{dt} = \mathbf{v}_{A/B} + \boldsymbol{\omega} \times \mathbf{r}_{A/B} \tag{17.32j}$$

Substituting the quantities given in Eqs. (17.32a) and (17.32j) in Eq. (17.31) yields

$$\mathbf{v}_A = \mathbf{v}_B + \mathbf{v}_{A/B} + \boldsymbol{\omega} \times \mathbf{r}_{A/B} \tag{17.33}$$

where $\mathbf{v}_A$ and $\mathbf{v}_B$ are the absolute velocities of points A and B, respectively, measured from the fixed X-Y coordinate system. Also, the quantity $\boldsymbol{\omega}$ is the angular velocity of the translating-rotating x-y coordinate system and $\mathbf{r}_{A/B}$ and $\mathbf{v}_{A/B}$ represent the position and velocity of point A, respectively, as measured by an observer stationed at point B, the origin of this coordinate system.

The acceleration of point A is obtained by taking the time derivative of Eq. (17.33). Thus,

$$\frac{d\mathbf{v}_A}{dt} = \frac{d\mathbf{v}_B}{dt} + \frac{d\mathbf{v}_{A/B}}{dt} + \boldsymbol{\omega} \times \frac{d\mathbf{r}_{A/B}}{dt} + \frac{d\boldsymbol{\omega}}{dt} \times \mathbf{r}_{A/B} \tag{17.34}$$

where

$$\frac{d\mathbf{v}_A}{dt} = \mathbf{a}_A, \quad \frac{d\mathbf{v}_B}{dt} = \mathbf{a}_B. \tag{17.35a}$$

Also, by Eq. (17.32b),

$$\frac{d\mathbf{v}_{A/B}}{dt} = \frac{d}{dt}(\dot{x}_{A/B}\mathbf{i} + \dot{y}_{A/B}\mathbf{j}) = \dot{x}_{A/B}\frac{d\mathbf{i}}{dt} + \ddot{x}_{A/B}\mathbf{i} + \dot{y}_{A/B}\frac{d\mathbf{j}}{dt} + \ddot{y}_{A/B}\mathbf{j}$$

$$= \ddot{x}_{A/B}\mathbf{i} + \ddot{y}_{A/B}\mathbf{j} + \dot{x}_{A/B}\frac{d\mathbf{i}}{dt} + \dot{y}_{A/B}\frac{d\mathbf{j}}{dt}. \tag{17.35b}$$

The terms $(\ddot{x}_{A/B})\mathbf{i}$ and $(\ddot{y}_{A/B})\mathbf{j}$ represent the x and y components of the acceleration of point A, respectively, as measured from point B which is the origin of the translating-rotating x-y coordinate system. Therefore, the vector sum of the first two terms in Eq. (17.35b) represents the acceleration $\mathbf{a}_{A/B}$ of point A as measured by a rotating observer stationed at point B. Thus,

$$(\ddot{x}_{A/B})\mathbf{i} + (\ddot{y}_{A/B})\mathbf{j} = \mathbf{a}_{A/B} \tag{17.35c}$$

The last two terms in Eq. (17.35b) may be rewritten with the aid of Eqs. (17.32g) and (17.32h). Thus,

$$\begin{aligned} \dot{x}_{A/B}\frac{d\mathbf{i}}{dt} + \dot{y}_{A/B}\frac{d\mathbf{j}}{dt} &= \dot{x}_{A/B}\boldsymbol{\omega} \times \mathbf{i} + \dot{y}_{A/B}\boldsymbol{\omega} \times \mathbf{j} \\ &= \boldsymbol{\omega} \times [\dot{x}_{A/B}\mathbf{i} + \dot{y}_{A/B}\mathbf{j}] = \boldsymbol{\omega} \times \mathbf{v}_{A/B} \end{aligned} \tag{17.35d}$$

where, by Eq. (17.32d), $\mathbf{v}_{A/B}$ was substituted for the vector sum $(\dot{x}_{A/B})\mathbf{i} + (\dot{y}_{A/B})\mathbf{j}$. Thus,

$$\frac{d\mathbf{v}_{A/B}}{dt} = \mathbf{a}_{A/B} + \boldsymbol{\omega} \times \mathbf{v}_{A/B} \tag{17.35e}$$

Now, using Eq. (17.32j),

$$\begin{aligned} \boldsymbol{\omega} \times \frac{d\mathbf{r}_{A/B}}{dt} &= \boldsymbol{\omega} \times (\mathbf{v}_{A/B} + \boldsymbol{\omega} \times \mathbf{r}_{A/B}) \\ &= \boldsymbol{\omega} \times \mathbf{v}_{A/B} + \boldsymbol{\omega} \times (\boldsymbol{\omega} \times \mathbf{r}_{A/B}) \end{aligned} \tag{17.35f}$$

Also,

$$\frac{d\boldsymbol{\omega}}{\mathrm{dt}} \times \mathbf{r}_{A/B} = \boldsymbol{\alpha} \times \mathbf{r}_{A/B} \tag{17.35g}$$

where $\boldsymbol{\alpha} = \dfrac{d\boldsymbol{\omega}}{dt}$ is the angular acceleration of the translating-rotating x-y coordinate system.

Substituting Eqs. (17.35a), (17.35e), (17.35f), and (17.35g) in Eq. (17.34) yields

$$\mathbf{a}_A = \mathbf{a}_B + \mathbf{a}_{A/B} + 2\boldsymbol{\omega} \times \mathbf{v}_{A/B} + \boldsymbol{\omega} \times (\boldsymbol{\omega} \times \mathbf{r}_{A/B}) + \boldsymbol{\alpha} \times \mathbf{r}_{A/B} \tag{17.36}$$

In Eq. (17.36), $\mathbf{a}_A$ and $\mathbf{a}_B$ are the absolute accelerations of points A and B in Figure 17.12, respectively, as measured from the fixed X-Y coordinate system. The terms $\mathbf{r}_{A/B}$, $\mathbf{v}_{A/B}$, and $\mathbf{a}_{A/B}$ are the position, velocity and acceleration of point A, respectively, measured by a rotating observer stationed at point B, the origin of the translating-rotating coordinate system. Also, as defined previously, the quantities $\boldsymbol{\omega}$ and $\boldsymbol{\alpha}$ are the angular velocity and angular acceleration, respectively, of the translating-rotating x-y coordinate system.

Note that the last three terms in Eq. (17.36) exist because of the angular velocity ω and angular acceleration α of the translating-rotating x-y coordinate system. If the x-y coordinate did not rotate (i.e., $\boldsymbol{\omega} = \boldsymbol{\alpha} = \mathbf{0}$), these three terms will vanish, and Eq. (17.36) reduces to Eq. (17.24) which applies to the case of translating-nonrotating x-y coordinate system. The third term in Eq. (17.36), $2\boldsymbol{\omega} \times \mathbf{v}_{A/B}$, is of special significance in the motion of rockets and artificial satellites that are influenced by Earth's rotation and is given the name *Coriolis acceleration*, after the Frenchman G.C. DeCoriolis (1792–1843) who was the first to describe it. In this book, we will use the symbol a_{COR} to represent this acceleration. As its definition indicates, the Coriolis acceleration of point A represents the combined effect of the velocity of point A relative to point B and the angular velocity of the rotating axes centered at point B. Also, the properties of the cross-product of $\boldsymbol{\omega}$ and $\mathbf{v}_{A/B}$ indicate that Coriolis acceleration is perpendicular to the plane defined by $\boldsymbol{\omega}$ and $\mathbf{v}_{A/B}$ and that its sense is given by the right-hand rule. Furthermore, the Coriolis acceleration vanishes if either ω or $v_{A/B}$ is zero.

Consider, for example, the case of a hydraulic cylinder hinged at B, as shown in Figure 17.14, and rotating at a constant ccw angular velocity $\boldsymbol{\omega}$ while being extended, such that $\mathbf{v}_{A/B}$ is constant. Under these conditions, the absolute acceleration of point A is given by Eq. (17.36) where

$\mathbf{a}_B = \mathbf{0}$ because point B does not translate,

$\mathbf{a}_{A/B} = \mathbf{0}$ because the cylinder is extended at a constant rate,

$2\boldsymbol{\omega} \times \mathbf{v}_{A/B} = 2\omega\mathbf{k} \times v_{A/B}\mathbf{i} = (2\omega v_{A/B})\mathbf{j}$,

$\boldsymbol{\omega} \times (\boldsymbol{\omega} \times \mathbf{v}_{A/B}) = \omega\mathbf{k} \times (\omega\mathbf{k} \times L\mathbf{i}) = -(\omega^2 L)\mathbf{i}$, and

$\boldsymbol{\alpha} \times \mathbf{r}_{A/B} = \mathbf{0}$ because rotation of the cylinder is at a constant rate.

Thus, the absolute acceleration of point A consists of two parts. The first part is the normal component of acceleration $(\mathbf{a}_A)_n$ due to the constant angular rotation of the cylinder whose magnitude is $\omega^2 L$ and is directed toward point B as shown in Figure 17.14. The second part

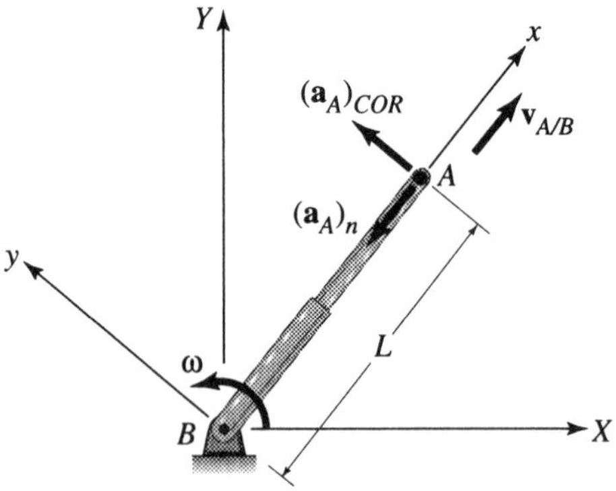

FIGURE 17.14.

represents Coriolis' acceleration $(\mathbf{a}_A)_{COR}$ whose magnitude is $2\omega v_{A/B}$ and is pointed in a direction perpendicular to the cylinder, as shown in Figure 17.14. Note that both of these components would vanish if the angular velocity ω were zero. Thus, the Coriolis acceleration of point A arose as a consequence of the rotation of the x-y coordinate system and because of its velocity relative to the origin of this coordinate system.

The following examples will illustrate the use of Eqs. (17.33) and (17.36) in the solution of problems.

■ Example 17.20

In the position shown ($\theta = 20°$), the boom BA is being rotated at a constant cw angular velocity $\omega = 0.10$ rad/s, and its length is increasing at the constant rate of 0.75 ft/s, as shown in Figure E17.20(a). Determine (a) the absolute velocity of point A, (b) the absolute acceleration of point A, and (c) the Coriolis acceleration of point A.

Solution

As shown in Figure E17.20(a), point B is selected as the origin for both the fixed X-Y and the rotating x-y coordinate systems. Note that, in this case, the x-y coordinate system does not translate.

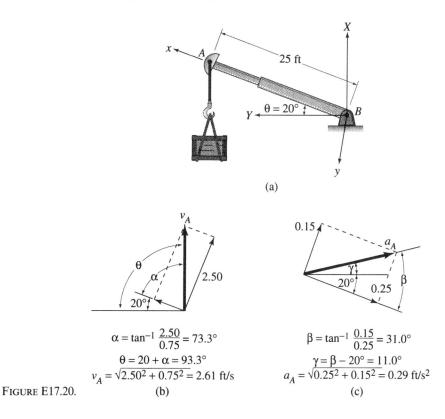

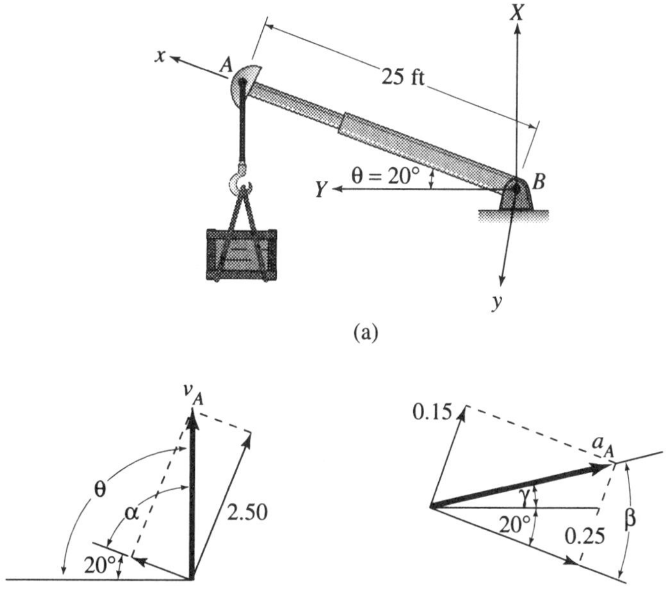

$$\alpha = \tan^{-1}\frac{2.50}{0.75} = 73.3°$$

$$\theta = 20 + \alpha = 93.3°$$

$$v_A = \sqrt{2.50^2 + 0.75^2} = 2.61 \text{ ft/s}$$

(b)

$$\beta = \tan^{-1}\frac{0.15}{0.25} = 31.0°$$

$$\gamma = \beta - 20° = 11.0°$$

$$a_A = \sqrt{0.25^2 + 0.15^2} = 0.29 \text{ ft/s}^2$$

(c)

FIGURE E17.20.

(a) The absolute velocity of point A, $\mathbf{v}_A$, is given by Eq. (17.33). Thus,

$$\mathbf{v}_A = \mathbf{v}_B + \mathbf{v}_{A/B} + \boldsymbol{\omega}_{AB} \times \mathbf{r}_{A/B}$$

where

$\mathbf{v}_B = \mathbf{0}$ because point B does not translate;

$\mathbf{v}_{A/B} = 0.75\mathbf{i}$ ft/s because the length of the boom BA is increasing at the rate of 0.75 ft/s along the rotating x axis;

$\boldsymbol{\omega}_{AB} = -0.10\mathbf{k}$ rad/s as given, the minus sign is introduced because the vector $\boldsymbol{\omega}_{AB}$ is opposite to the chosen z axis;

and

$\mathbf{r}_{A/B} = 25.0\mathbf{i}$ ft as given.

Therefore,

$$\mathbf{v}_A = 0 + 0.75\mathbf{i} + (-0.10\mathbf{k}) \times (25.0\mathbf{i}) = (0.75\mathbf{i} - 2.50\mathbf{j}) \text{ ft/s}$$
$$= (2.61\mathbf{I} - 0.15\mathbf{J}) \text{ ft/s}$$

where $\mathbf{I}$ and $\mathbf{J}$ are unit vectors along the fixed X and Y axes, respectively. Also, as shown in Figure E17.20(b),

$$v_A = 2.61 \text{ ft/s} \quad \angle\, 93.3^\circ.$$ ANS.

(b) The absolute acceleration of point A, $\mathbf{a}_A$, is given by Eq. (17.36). Thus,

$$\mathbf{a}_A = \mathbf{a}_B + \mathbf{a}_{A/B} + 2\boldsymbol{\omega}_{AB} \times \mathbf{v}_{A/B} + \boldsymbol{\omega}_{AB} \times (\boldsymbol{\omega}_{AB} \times \mathbf{r}_{A/B}) + \boldsymbol{\alpha}_{AB} \times \mathbf{r}_{A/B}$$

Some of the terms in the equation above have already been defined in part (a). The remaining terms are defined as follows:

$\mathbf{a}_B = \mathbf{0}$ because point B does not translate;

$\mathbf{a}_{A/B} = \mathbf{0}$ because the length of the boom is changing at a *constant* rate;

$\boldsymbol{\alpha}_{AB} = \mathbf{0}$ because $\boldsymbol{\omega}_{AB}$ is *constant* in magnitude and direction.

Therefore,

$$\mathbf{a}_A = 2(-0.10\mathbf{k}) \times (0.75\mathbf{i}) + (-0.10\mathbf{k}) \times [(-0.10\mathbf{k}) \times (25.0\mathbf{i})]$$
$$= (-0.25\mathbf{i} - 0.15\mathbf{j}) \text{ ft/s}^2$$
$$= (0.06\mathbf{I} - 0.28\mathbf{J}) \text{ ft/s}^2$$

Alternatively, as shown in Figure E17.20(c),

$$\mathbf{a}_A = 0.29 \text{ ft/s}^2 \quad \angle\, 11.0^\circ$$ ANS.

(c) The Coriolis acceleration of point A is given by the third term of Eq. (17.36). Thus,

$$(\mathbf{a}_A)_{COR} = 2\boldsymbol{\omega}_{AB} \times \mathbf{v}_{A/B} = 2(-0.10\mathbf{k}) \times (0.75\mathbf{i}) = -0.1500\mathbf{j} \text{ ft/s}^2 \quad \text{ANS.}$$

■ Example 17.21

In the position shown in Figure E17.21, when member AB is in a vertical position, the constant angular velocity of member AB, $\boldsymbol{\omega}_{AB} =$ 5.0 rad/s ccw. Pin B, which is attached to member AB, slides freely in the slot of member CD. Determine (a) the velocity of the pin relative to the slot and the absolute angular velocity of member CD and (b) the acceleration of pin B relative to the slot and the absolute angular acceleration of member CD.

Solution

(a) Point C is selected as the origin for both the fixed X-Y and the rotating x-y coordinate systems. Note that, in this case, the x-y coordinate system does not translate.

The velocity of the pin relative to the slot, $\mathbf{v}_{B/C}$, and the absolute angular velocity of member CD, $\boldsymbol{\omega}_{CD}$ may be obtained by Eq. (17.33). Thus,

$$\mathbf{v}_B = \mathbf{v}_C + \mathbf{v}_{B/C} + \boldsymbol{\omega}_{CD} \times \mathbf{r}_{B/C}$$

where

$$\mathbf{v}_B = r_{B/A}\omega_{AB} \leftarrow = (0.30)(5.0) \text{ m/s} \leftarrow = 1.50 \text{ m/s} \leftarrow$$

$$= (1.50 \cos 60°)\mathbf{i} + (1.50 \sin 60°)\mathbf{j} = (0.75\mathbf{i} + 1.30\mathbf{j}) \text{ m/s};$$

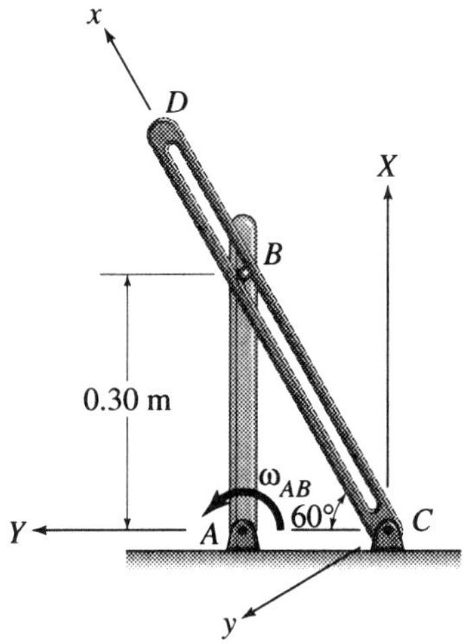

FIGURE E17.21.

$\mathbf{v}_C = \mathbf{0}$ because point C does not translate;

$\mathbf{v}_{B/C} = (v_{B/C})\mathbf{i}$ where the unknown relative velocity $\mathbf{v}_{B/C}$ is assumed along the positive x axis;

$\boldsymbol{\omega}_{CD} = (\omega_{CD})\mathbf{k}$ where the unknown absolute angular velocity $\boldsymbol{\omega}_{CD}$ is assumed along the positive z axis.

$\mathbf{r}_{B/C} = 0.346\mathbf{i}$ as determined from the geometry given.

Therefore,

$$0.75\mathbf{i} + 1.30\mathbf{j} = v_{B/C}\mathbf{i} + (\omega_{CD})\mathbf{k} \times (0.346\mathbf{i}) = v_{B/C}\mathbf{i} + 0.346\omega_{CD}\mathbf{j} \text{ m/s}$$

from which

$$v_{B/C} = 0.750 \text{ m/s} \quad 60.0^\circ \nwarrow, \qquad \text{ANS.}$$

$$\mathbf{v}_{B/C} = 0.750\mathbf{i} \text{ m/s}, \qquad \text{ANS.}$$

$$\omega_{CD} = 3.76 \text{ rad/s} \circlearrowleft, \qquad \text{ANS.}$$

$$\boldsymbol{\omega}_{CD} = 3.76\mathbf{k} \text{ rad/s}, \qquad \text{ANS.}$$

and

(b) The acceleration of the pin relative to the slot, $\mathbf{a}_{B/C}$, and the absolute angular acceleration of member CD, $\boldsymbol{\alpha}_{CD}$, are determined by Eq. (17.36). Thus,

$$\mathbf{a}_B = \mathbf{a}_C + \mathbf{a}_{B/C} + 2\boldsymbol{\omega}_{CD} \times \mathbf{v}_{B/C} + \boldsymbol{\omega}_{CD} \times (\boldsymbol{\omega}_{CD} \times \mathbf{r}_{B/C}) + \mathbf{a}_{CD} \times \mathbf{r}_{B/C}$$

Some of the terms in this equation have already been defined in part (a). The remaining terms are defined as follows:

$$\mathbf{a}_B = (\mathbf{a}_B)_n \downarrow + (\mathbf{a}_B)_t = r_{B/A}\omega_{AB}^2 \downarrow = 7.5 \text{ m/s}^2 \downarrow;$$
$$= (7.5 \sin 60^\circ)\mathbf{i} + (7.5 \cos 60^\circ)\mathbf{j} = -6.495\mathbf{i} + 3.750\mathbf{j}$$

$\mathbf{a}_C = \mathbf{0}$ because point C does not translate.

$\mathbf{a}_{B/C} = (\mathbf{a}_{B/C})\mathbf{i}$ where the unknown relative acceleration $\mathbf{a}_{B/C}$ is assumed along the positive x axis.

$\mathbf{v}_{B/C} = 0.75\text{i}$ as found in part (a);

$\boldsymbol{\omega}_{CD} = 3.76\mathbf{k}$ as found in part (a);

$\boldsymbol{\alpha}_{CD} = (\alpha_{CD})\mathbf{k}$ where the unknown absolute angular acceleration α_{CD} is assumed along the positive z axis.

Therefore,

$$\begin{aligned}-6.495\mathbf{i} + 3.75\mathbf{j} &= (a_{B/C})\mathbf{i} + 2(3.76\mathbf{k}) \times (0.75\mathbf{i}) \\ &\quad + (3.76\mathbf{k}) \times (3.76\mathbf{k} \times 0.346\mathbf{i}) + (\alpha_{CD})\mathbf{k} \times 0.346\mathbf{i} \\ &= (a_{B/C})\mathbf{i} + 5.64\mathbf{j} - 4.888\mathbf{i} + 0.346(a_{CD})\mathbf{j} \\ &= [a_{B/C} - 4.888]\mathbf{i} + [5.64 + 0.346a_{CD}]\mathbf{j}\end{aligned}$$

from which

$$a_{B/C} = 1.607 \text{ m/s}^2 \quad \overset{60.0^\circ}{\searrow}, \qquad \text{ANS.}$$

$$\mathbf{a}_{B/C} = -1.607\mathbf{i} \text{ m/s}^2, \qquad \text{ANS.}$$

and

$$\alpha_{BC} = 5.46 \text{ rad/s}^2 \circlearrowright, \qquad \text{ANS.}$$

$$\boldsymbol{\alpha}_{BC} = -5.46\mathbf{k} \text{ rad/s}^2. \qquad \text{ANS.}$$

■ Example 17.22

Cars A and B travel along their respective circular paths as shown in Figure E17.22 where the velocities and accelerations of both cars are indicated for the instant under consideration. Determine the velocity and acceleration of car B as observed by a passenger in car A.

Solution

Car A is chosen as the origin for both the fixed X-Y and the translating-rotating x-y coordinate systems. The following quantities are needed for the application of Eqs. (17.22) and (17.36):

$$\mathbf{v}_B = -30\mathbf{i} \text{ mph},$$

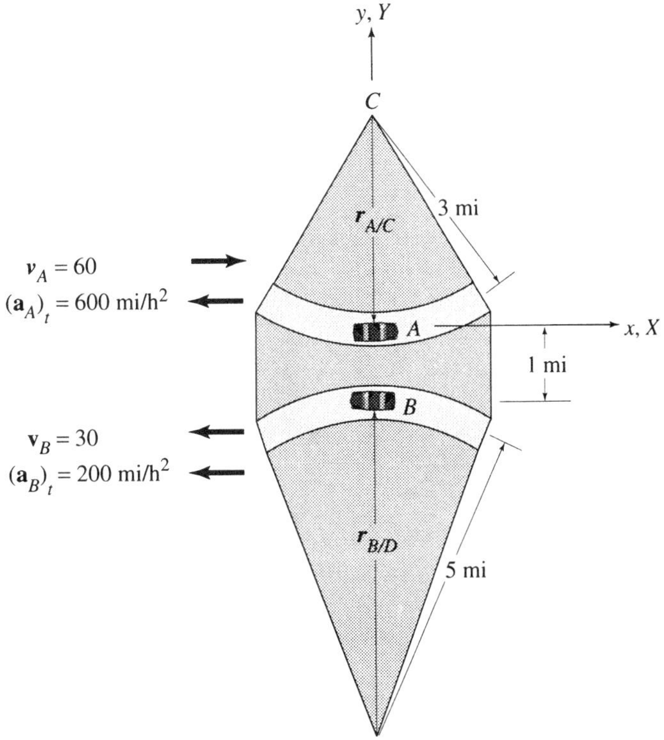

FIGURE E17.22.

$$\mathbf{v}_A = 60\mathbf{i} \text{ mph},$$

$$\boldsymbol{\omega}_{CA} = \left(\frac{v_A}{r_{A/C}}\right)\mathbf{k} = \frac{60}{3}\mathbf{k} = 20\mathbf{k} \text{ rad/h},$$

$$\mathbf{r}_{B/A} = -\mathbf{j} \text{ mi},$$

$$\mathbf{a}_B = (\mathbf{a}_B)_t + (\mathbf{a}_B)_n = -200\mathbf{i} - (30^2/5)\mathbf{j} = (-200\mathbf{i} - 180\mathbf{j}) \text{ mph},$$

$$\mathbf{a}_A = (\mathbf{a}_A)_t + (\mathbf{a}_A)_n = -600\mathbf{i} + (60^2/5)\mathbf{j} = (-600\mathbf{i} + 1200\mathbf{j}) \text{ mph}, \quad \text{and}$$

$$\boldsymbol{\alpha}_{CA} = -\left[\frac{(a_A)_t}{r_{A/C}}\right]\mathbf{k} = -\frac{600}{3}\mathbf{k} = 200\mathbf{k} \text{ rad/h}^2$$

The velocity of B as measured from A, $\mathbf{v}_{B/A}$, is found from Eq. (17.33). Thus,

$$\mathbf{v}_B = \mathbf{v}_A + \mathbf{v}_{B/A} + \boldsymbol{\omega}_{CA} \times \mathbf{r}_{B/A}$$

Therefore,

$$-30\mathbf{i} = 60\mathbf{i} + \mathbf{v}_{B/A} + (20\mathbf{k}) \times (-\mathbf{j})$$

from which

$$\mathbf{v}_{B/A} = -110.0\mathbf{i} \text{ mph}. \qquad \text{ANS.}$$

The acceleration of B as measured from A, $\mathbf{a}_{B/A}$, is found from Eq. (17.36). Thus,

$$\mathbf{a}_B = \mathbf{a}_A + \mathbf{a}_{B/A} + 2\boldsymbol{\omega}_{CA} \times \mathbf{v}_{B/A} + \boldsymbol{\omega}_{CA} \times (\boldsymbol{\omega}_{CA} \times \mathbf{r}_{B/A}) + \boldsymbol{\alpha}_{CA} \times \mathbf{r}_{B/A}$$

$$-200\mathbf{i} - 180\mathbf{j}$$

$$= -600\mathbf{i} + 1200\mathbf{j} + \mathbf{a}_{A/B} + 2(20\mathbf{k}) \times (-110\mathbf{i})$$

$$+ 20\mathbf{k} \times [(20\mathbf{k}) \times (-\mathbf{j})] + (-200\mathbf{k}) \times (-\mathbf{j})$$

which leads to

$$\mathbf{a}_{B/A} = (600\mathbf{i} + 2620\mathbf{j}) \text{ mi/h}^2 \qquad \text{ANS.}$$

■ Example 17.23

A commercial type sprinkler consists of four curved pipes attached to a central stem as shown in Figure E17.23. Each of the four pipes has a constant radius of curvature of 0.500 m as shown. The sprinkler rotates at a constant angular speed $\omega = 200$ rpm and the constant velocity of the water relative to the pipes is 10 m/s. Determine (a) the absolute velocity of a particle of water just prior to leaving the pipe at point A and (b) the absolute acceleration of a particle of water just prior to leaving the pipe at point A.

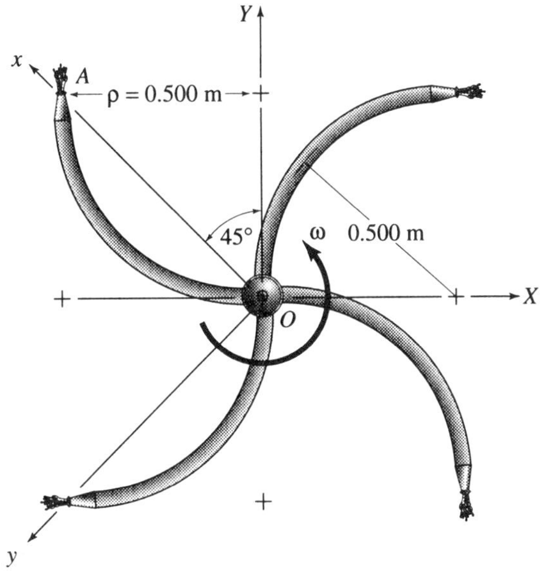

FIGURE E17.23.

Solution

Point O is chosen as the origin of both the fixed X-Y and the rotating x-y coordinate systems. Note that, in this case, the x-y coordinate system does not translate.

(a) The absolute velocity of a water particle just prior to leaving the pipe at point A may be determined by Eq. (17.33). Thus,

$$\mathbf{v}_A = \mathbf{v}_O + \mathbf{v}_{A/O} + \boldsymbol{\omega}_{OA} \times \mathbf{r}_{A/O}$$

where

$\mathbf{v}_O = \mathbf{0}$ since point O does not translate;

$\mathbf{v}_{A/O} = 10\text{ m/s} = (10\cos 45°)\mathbf{i} - (10\sin 45°)\mathbf{j} = (7.071\mathbf{i} - 7.071\mathbf{j})\text{ m/s};$

$\boldsymbol{\omega}_{OA} = \left[\dfrac{200(2\pi)}{60}\right]\mathbf{k} = 20.944\,\mathbf{k}\text{ rad/s};$ and

$\mathbf{r}_{A/O} = 0.707\mathbf{i}\text{ m}.$

Therefore,

$$\begin{aligned}\mathbf{v}_A &= \mathbf{0} + 7.071\mathbf{i} - 7.071\mathbf{j} + 20.944\mathbf{k} \times 0.707\mathbf{i} \\ &= 7.071\mathbf{i} - 7.071\mathbf{j} + 14.807\mathbf{j} \\ &= (7.071\mathbf{i} - 7.736\mathbf{j})\text{ m/s} \\ &= (-10.470\mathbf{I} - 0.475\mathbf{J})\text{ m/s} \\ &= 10.48\text{ m/s } \measuredangle^{2.6°}.\end{aligned}$$ ANS.

(b) The absolute acceleration of a water particle just prior to leaving the pipe at point A is determined by Eq. (17.36). Thus,

$$\mathbf{a}_A = \mathbf{a}_O + \mathbf{a}_{A/O} + 2\boldsymbol{\omega}_{OA} \times \mathbf{v}_{A/O} + \boldsymbol{\omega}_{OA} \times (\boldsymbol{\omega}_{OA} \times \mathbf{r}_{A/O}) + \boldsymbol{\alpha}_{OA} \times \mathbf{r}_{A/O}$$

where

$\mathbf{a}_O = \mathbf{0}$ because point O does not translate;

$$\mathbf{a}_{A/O} = \frac{v^2}{\rho} \to = \frac{10^2}{0.500} \to = 200 \text{ m/s}^2 \to;$$

$$= -(200\cos 45°)\mathbf{i} - (200\sin 45°)\mathbf{j} = -141.421\mathbf{i} - 141.421\mathbf{j};$$

$\boldsymbol{\omega}_{OA} = 20.944\mathbf{k}$ rad/s as in part (a);

$v_{A/O} = (7.071\mathbf{i} - 7.071\mathbf{j})$ m/s as in part (a);

$\mathbf{r}_{A/O} = 0.707\mathbf{i}$ m as in part (a); and

$\boldsymbol{\alpha}_{OA} = \mathbf{0}$ because $\boldsymbol{\omega}_{OA}$ is constant.

Therefore,

$$\begin{aligned}\mathbf{a}_A &= \mathbf{0} - 141.421\mathbf{i} - 141.421\mathbf{j} + 2(20.944\mathbf{k}) \times (7.071\mathbf{i} - 7.071\mathbf{j}) \\ &\quad + 20.944\mathbf{k} \times (20.944\mathbf{k} \times 0.707\mathbf{i}) + \mathbf{0} \\ &= -141.421\mathbf{i} - 141.421\mathbf{j} + 296.903\mathbf{j} + 296.903\mathbf{i} - 310.126\mathbf{i} \\ &= -154.644\mathbf{i} + 155.482\mathbf{j} \\ &= -33.549\mathbf{I} - 219.292\mathbf{J} \\ &= 219 \text{ m/s}^2 \ \overline{89.8°}\!\!\downarrow.\end{aligned}$$

ANS.

■

Problems

17.107 Pin B moves in the slot of member AC as the slotted member rotates about its hinged support at A as shown in Figure P17.107. For the position shown, the angular velocity of member AC is 5 rad/s ccw and its angular acceleration is 1 rad/s² cw. Also, at the same instant, the velocity of the pin relative to the slot is 3 m/s directed toward C and its acceleration is 2 m/s² directed toward A. Determine the absolute velocity and absolute acceleration of the pin.

17.108 A garden sprinkler, consisting of four perpendicular tubes, is shown schematically in Figure P17.108. It rotates with a constant ccw angular velocity $\omega = 20$ rad/s. The velocity of the water relative to the tubes is constant at 15 ft/s. Determine the absolute acceleration of a water particle at point A.

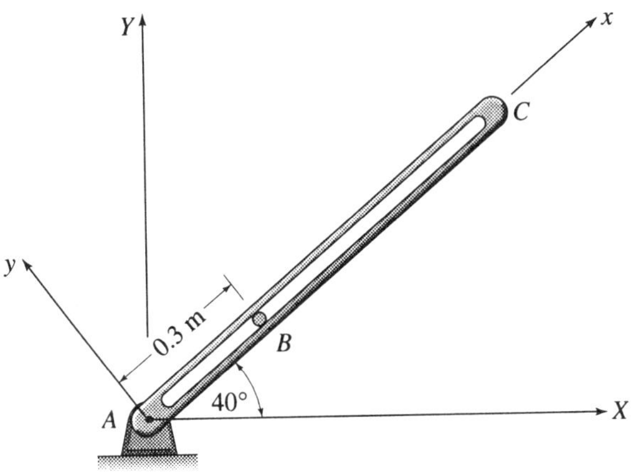

FIGURE P17.107.

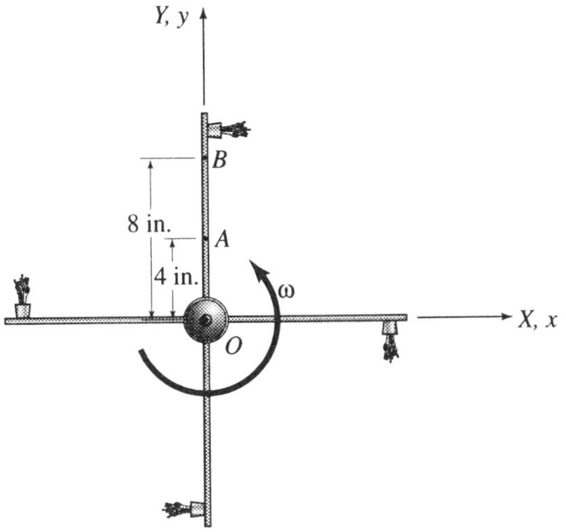

FIGURE P17.108.

17.109 A garden sprinkler, consisting of four perpendicular tubes, is shown schematically in Figure P17.108. At the instant shown, the absolute acceleration of a water particle at B is $\mathbf{a}_B = -300\mathbf{i} - 150\mathbf{j}$. Determine the constant angular velocity of the sprinkler and the constant velocity of the water relative to the tubes.

17.110 The impeller of a centrifugal pump is shown in Figure P17.110. The radius of curvature of the vanes at their exit points is 0.231 m. The impeller rotates at a constant cw speed of 250 rpm, and the speed of the fluid relative to the vanes is 7 m/s. Determine the absolute velocity of a fluid particle just prior to leaving the impeller at A.

17.111 Refer to the impeller of Problem 17.110, and determine the absolute acceleration of a fluid particle just prior to leaving the impeller at A.

17.112 Pin A of Figure P17.112 slides in the straight slot S while the disk rotates at a constant ccw angular velocity $\omega = 12$ rad/s. In the position shown, pin A has a velocity and acceleration relative to the slot given by $v_{A/S} = 10$ in./s and $a_{A/S} = 15$ in./s^2 as shown. Determine the absolute velocity and absolute acceleration of the slider pin A.

17.113 Pin B of Figure P17.112 slides in the circular slots while the disk rotates at a constant ccw angular velocity ω. In the position shown, the pin has an absolute velocity equal to 50 in./s and pointed upward and to the left. Also, in

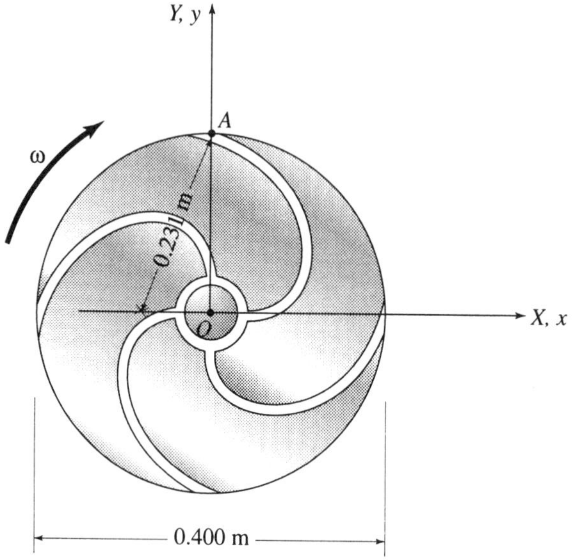

FIGURE P17.110.

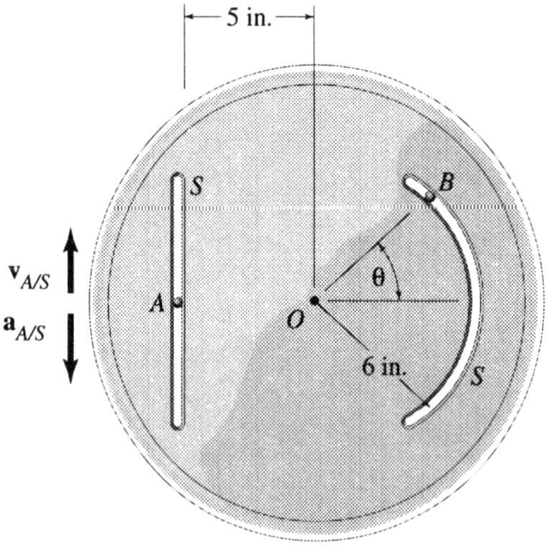

FIGURE P17.112.

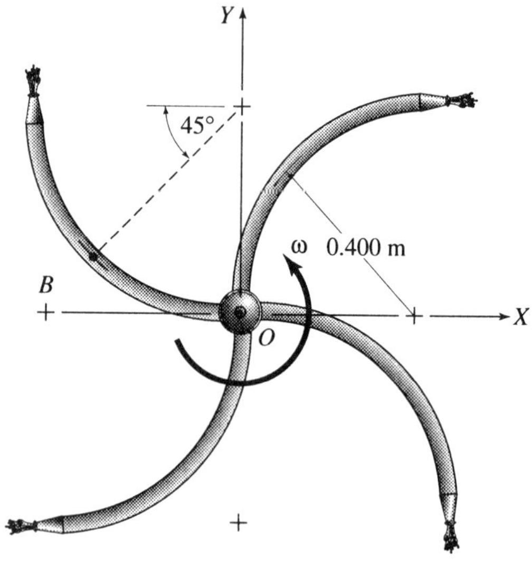

FIGURE P17.114.

this same position, the absolute acceleration of the pin is 1600 in./s^2 and pointed toward the center of the disk. Determine the constant ccw angular velocity ω of the disk and the constant velocity of the pin relative to the slot.

17.114 A commercial type sprinkler consists of four curved pipes attached to a central stem as shown in Figure P17.114. Each of the four pipes has a constant radius of curvature of 0.400 m as shown. The sprinkler rotates at a constant angular speed of 180 rpm, and the constant velocity of the water relative to the pipes is 8 m/s. Determine the absolute velocity of a particle of water at point B.

17.115 Refer to Problem 17.114, and determine the absolute acceleration of a particle of water at Point B.

17.116 Rod CD oscillates about its hinge at D while the collar at C slides freely on member AB as shown in Figure P17.116. In the position shown, rod CD has a constant ccw angular velocity $\omega_{CD} = 6$ rad/s. Determine for this position (a) the absolute angular velocity of member AB and (b) the velocity of the collar relative to member AB.

17.117 Refer to Problem 17.116 and determine (a) the acceleration of the collar relative to member AB and (b) the absolute angular acceleration of member AB.

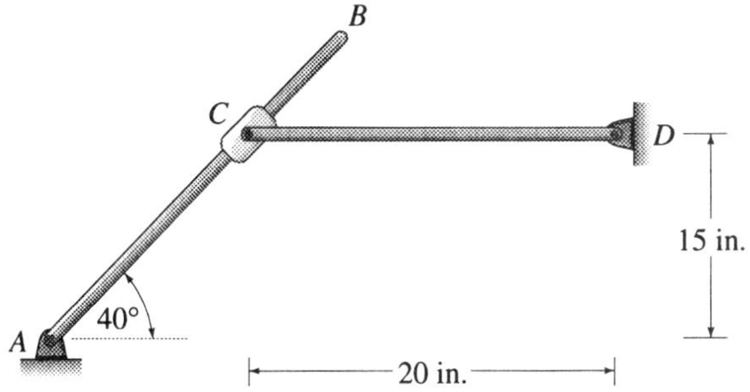

FIGURE P17.116.

17.118 As member OA of Figure P17.118 rotates about its central axis at O, the slotted member BD experiences oscillatory motion about its axis at B. Member OA has a constant cw angular velocity $\omega_{OA} = 10$ rad/s. For the position shown when $\theta = 30°$, determine (a) the absolute angular velocity of slotted member BD and (b) the velocity of pin A relative to the slot.

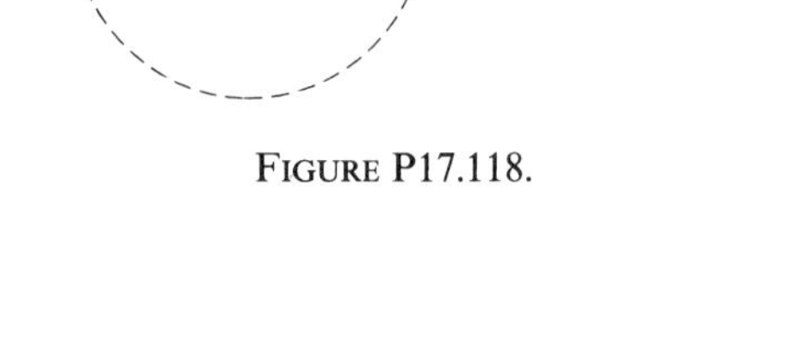

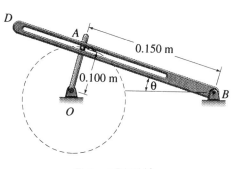

FIGURE P17.118.

17.119 Refer to Problem 17.118 and determine (a) the acceleration of pin A relative to the slot and (b) the absolute angular acceleration of member BD.

17.120 A schematic sketch of a Geneva mechanism is shown in Figure P17.120. Pin P is attached to the driving disk which rotates at a constant cw angular velocity of 8 rad/s. The pin engages one of the four radial slots in the driven disk when $\theta = 45°$ and rotates this disk through a quarter of a revolution before disengaging, resulting in intermittent motion of the driven disk. Determine the angular velocity of the driven disc for (a) $\theta = 0$ and (b) $\theta = 20°$.

17.121 Refer to Problem 17.120 and determine

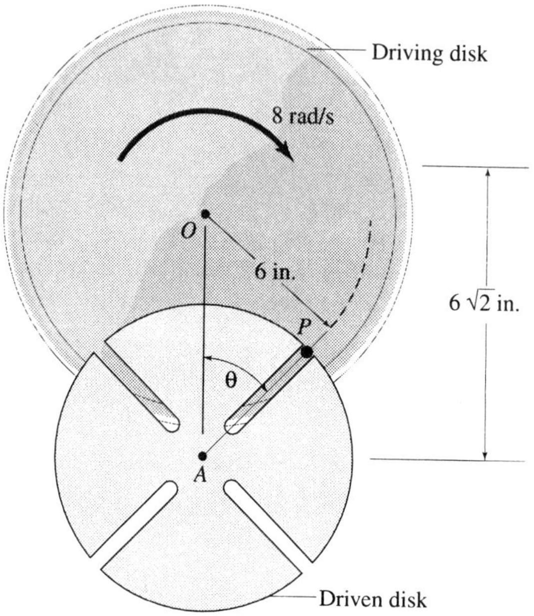

FIGURE P17.120.

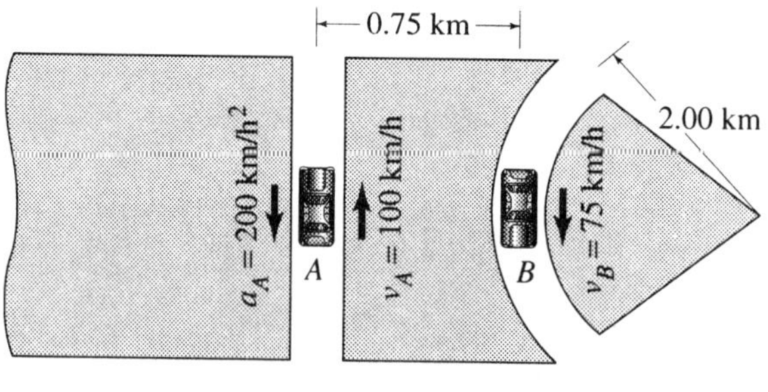

FIGURE P17.122.

the angular acceleration of the driven disk for (a) $\theta = 0$ and (b) $\theta = 20°$.

17.122 At the instant shown in Figure P17.122, car A has a speed of 100 km/h and a deceleration of 200 km/h^2. At the same moment, car B has a constant speed of 75 km/h. Determine the velocity and acceleration of car A as measured by an observer in car B.

17.123 The two planes A and B, shown in Figure P17.123, are in level flight at the same elevation. Plane A has a constant speed of 150 mph along a straight path and, at the instant considered, plane B is on a circular path as shown. In the position shown, an observer in plane B measures the velocity and acceleration of plane A as $\mathbf{v}_{A/B} = (-520\mathbf{i} - 150\mathbf{j})$ mph and $\mathbf{a}_{A/B} = (1927\mathbf{i} - 3467\mathbf{j})$ mph^2, respectively. Determine (a) the velocity

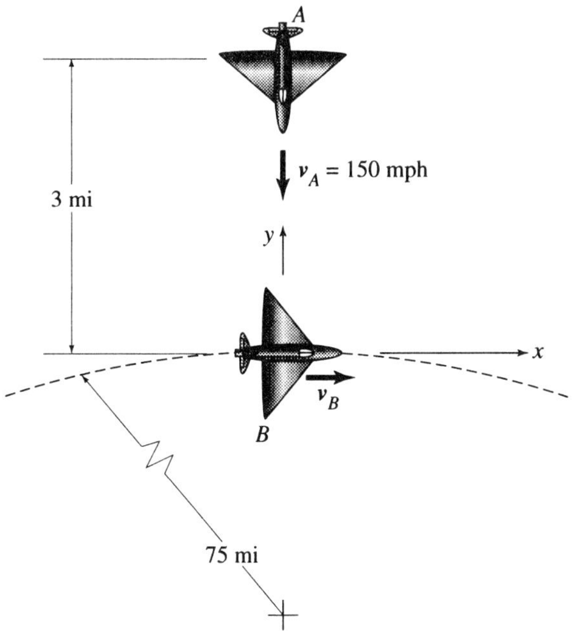

FIGURE P17.123.

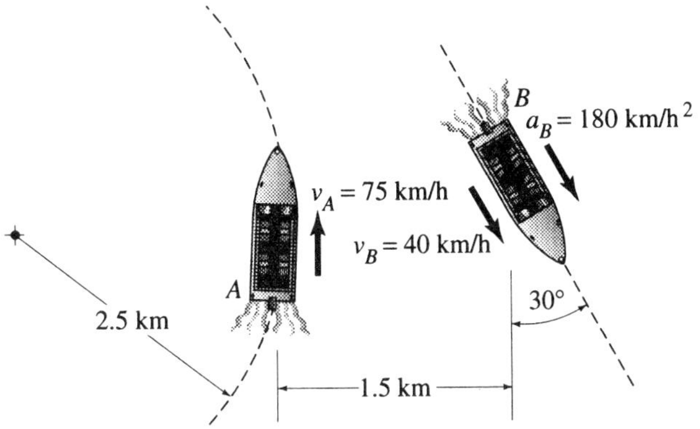

FIGURE P17.124.

of plane B and (b) the acceleration of plane B along the tangent to its path.

17.124 The two speed boats shown in Figure P17.124 are executing the motions indicated. At the instant shown, boat A travels at a constant speed along a curved path of constant radius and boat B moves along a straight path. Determine the velocity and acceleration of boat B relative to boat A. What is the magnitude and direction of the Coriolis acceleration of boat B relative to boat A?

17.125 Two spacecraft are in perfect north-

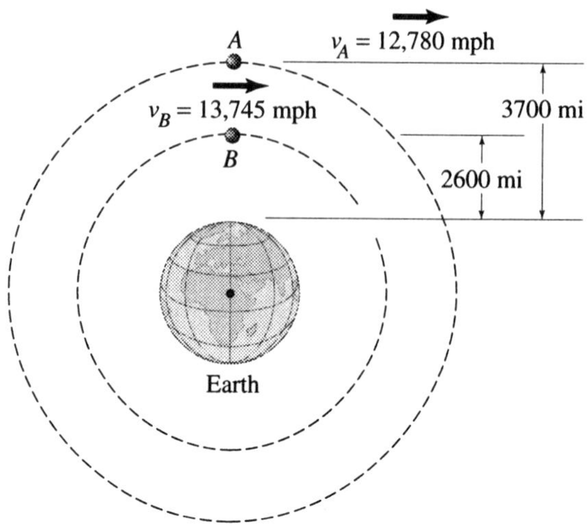

FIGURE P17.125.

south circular orbits about Earth as shown in Figure P17.125. Consider the position shown, and determine the velocity and acceleration of spacecraft A as measured by an observer in spacecraft B. The radius of Earth is 3960 mi.

17.126 Refer to Problem 17.125, and determine the Coriolis accelerations of both spacecraft relative to the center of Earth.

Review Problems

17.127 At the instant shown in Figure P17.127, the boy sitting on the seat of the swing has a speed of 10 ft/s. At this same instant, find the angular velocity of the rods that support the swing. Assume that the boy is still relative to the seat.

17.128 A dead bug lies on a turntable as shown in Figure P17.128. The turntable starts from rest and rotates at a constant angular acceleration $\alpha = 3$ rad/s^2. It is observed that the bug starts to slide on the surface of the turntable after it has turned two complete revolutions. What is the linear speed and total linear acceleration of the bug at the instant it starts to slide?

17.129 The belt-pulley system shown in Figure P17.129 starts from rest, and the driver pulley A accelerates at a constant angular acceleration until it reaches a speed of 1200 rpm. in 5 s. Determine the linear velocity and total linear acceleration of point C on the rim of the driven pulley B, 2 s after the start of the motion. The diameter of pulley A is 0.10 m and that of pulley B is 0.25 m. Assume no slippage.

17.130 The rod shown in Figure P17.130 ro-

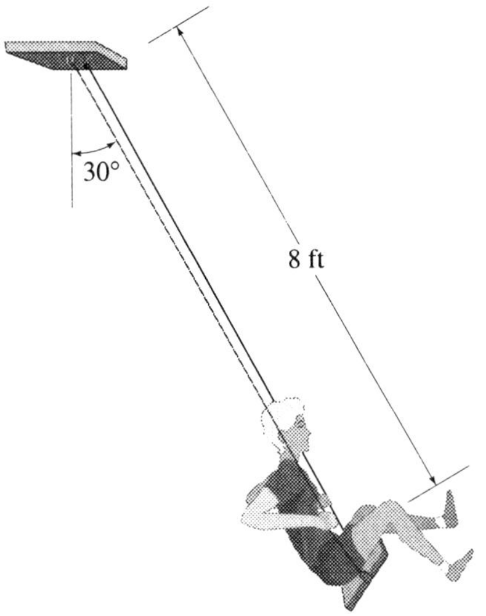

FIGURE P17.127.

tates about axis AB such that $\omega = 50$ rad/s and $\alpha = 200$ rad/s^2 where the senses of ω and α are shown in the sketch. For the instant shown, determine the velocity and acceleration of point C.

17.131 The mechanism shown in Figure P17.131 consists of arm CD that rotates about the hinge at C at a constant ccw angular velocity ω and rod AB that slides freely in cylinder F. The pin P is attached to rod AB and slides in the smooth slot of arm CD. (a) Develop general expressions for the horizontal velocity v_{AB} of rod AB and for the velocity v_E of point E on the rotating arm. Express these in terms of h, ω, L, and θ. (b) For $h = 0.30$ m, $\omega = 20$ rad/s, $L = 0.75$ m, and $\theta = 40°$, find numerical values for v_{AB} and v_E.

17.132 The work platform P shown in Figure P17.132 is being lowered at a constant velocity v_0 by a mechanism on the truck (not shown). (a) Determine an expression for the speed v_A of the roller at A in terms of v_0 and the angle θ. (b)

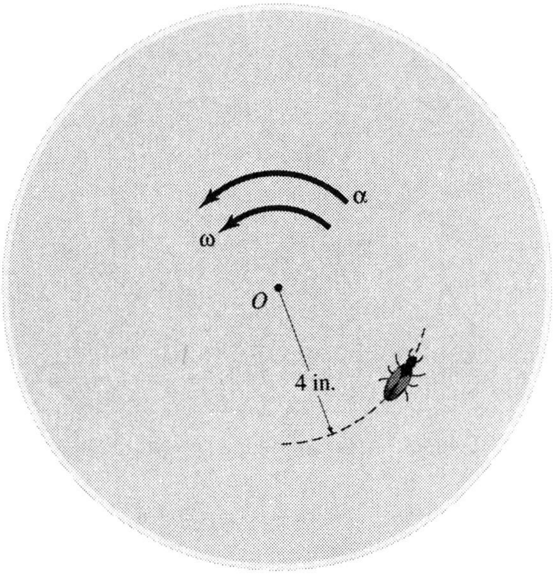

FIGURE P17.128.

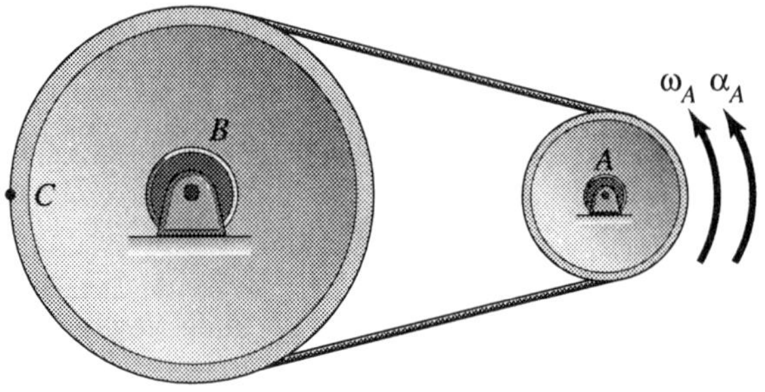

FIGURE P17.129.

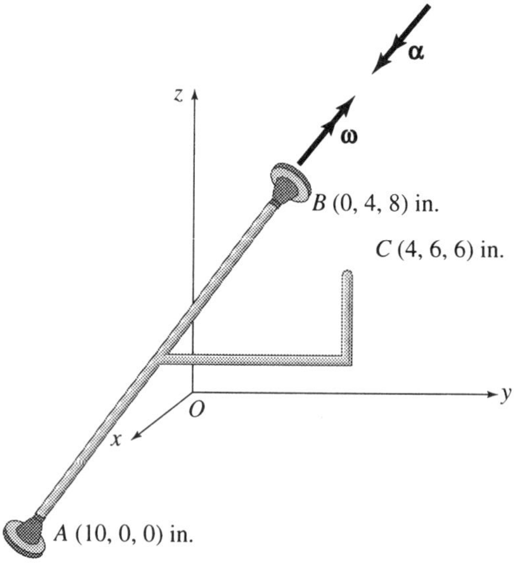

FIGURE P17.130.

For $v_0 = 10$ ft/s and $\theta = 60°$, what is the numerical value of v_A.

17.133 In the planetary gear arrangement shown in Figure P17.133, arm OA rotates at a ccw angular velocity $\omega_{OA} = 10$ rad/s and gear C rotates at a cw angular velocity $\omega_C = 30$ rad/s. Determine the angular velocity of outer gear D.

17.134 In the position shown in Figure P17.134, link AB has a ccw angular velocity of 10 rad/s. Determine the angular velocities of links BC and CD.

17.135 In the reciprocating mechanism shown in Figure P17.135, the wheel rotates at a ccw angular velocity of 800 rpm. For the position defined by $\theta = 90°$, determine the angular velocity of the connecting rod AB and the linear velocity of the collar at B. The length of the connecting rod AB is 15 in., the radius $R = 6$ in., and the distance D = 10 in.

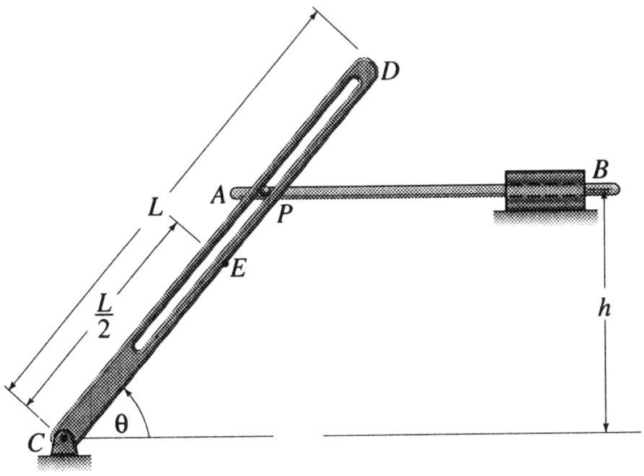

FIGURE P17.131.

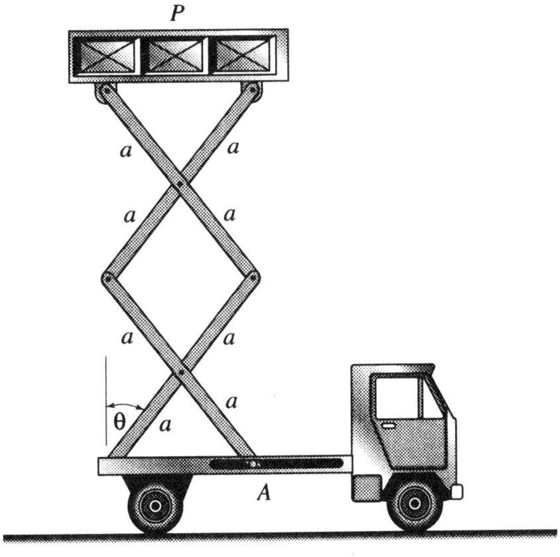

FIGURE P17.132.

17.136 In the reciprocating mechanism shown in Figure P17.135, the wheel rotates at a ccw angular velcoity of 1200 rpm. For the position defined by $\theta = 60°$, determine the angular velocity of the connecting rod AB and the linear velocity of the collar at B. The length of the connecting rod AB is 0.30 m, the radius $R = 0.10$ m, and the distance D = 0.20 m.

17.137 The triangular plate shown in Figure P17.137 moves in the x-y plane and, at the instant shown, the velocity of point A is given by $\mathbf{v}_A = (2.0\mathbf{i} + v_{Ay}\mathbf{j})$ m/s and that of point B by $\mathbf{v}_B = (-1.5\mathbf{i} + 2.5\mathbf{j})$ in/s. Find the component v_{Ay} and the angular velocity of the plate.

17.138 The cord wrapped around the spool-hub system shown in Figure P17.138 is

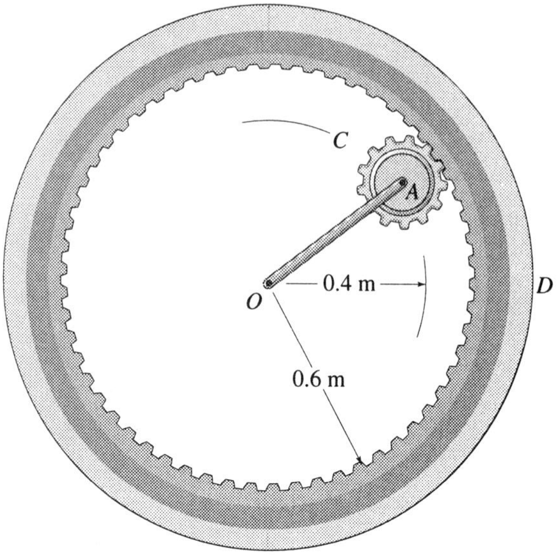

FIGURE P17.133.

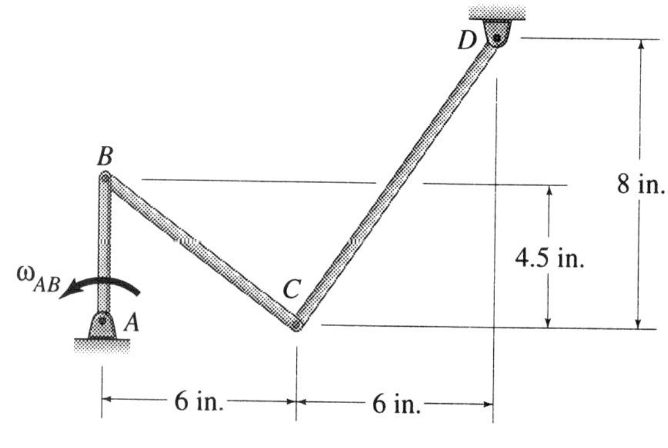

FIGURE P17.134.

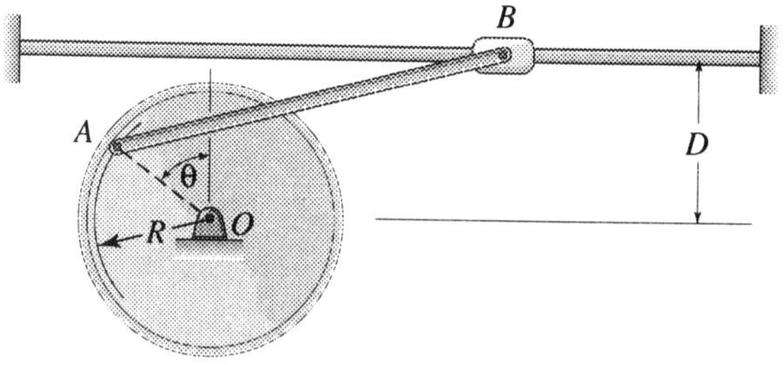

FIGURE P17.135.

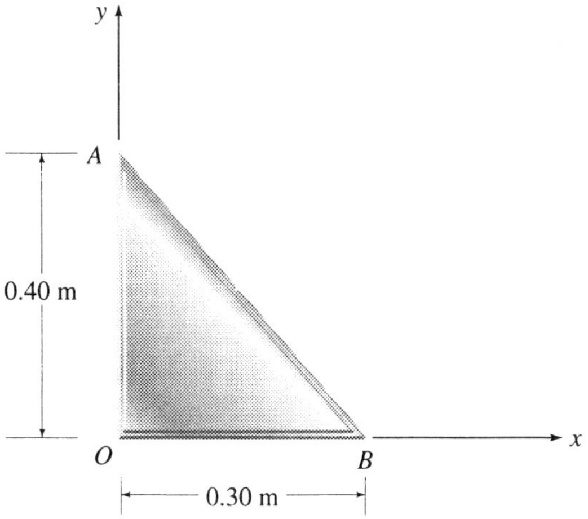

FIGURE P17.137.

pulled so that point A moves to the right at a constant velocity of 10 m/s. As a result, the hub of the system rolls without slipping on the horizontal surface. Determine the angular velocity of the system and the linear velocity of its center, point O.

17.139 Rod AB slides on the smooth inclined planes as shown in Figure P17.139. At the instant shown, point B has a velocity $v_B = 6$ ft/s and an acceleration $a_B = 10$ ft/s^2, both down the plane. Determine the linear acceleration of point A and the angular acceleration of rod AB. The length of rod AB is 4 ft.

17.140 In the position shown in Figure

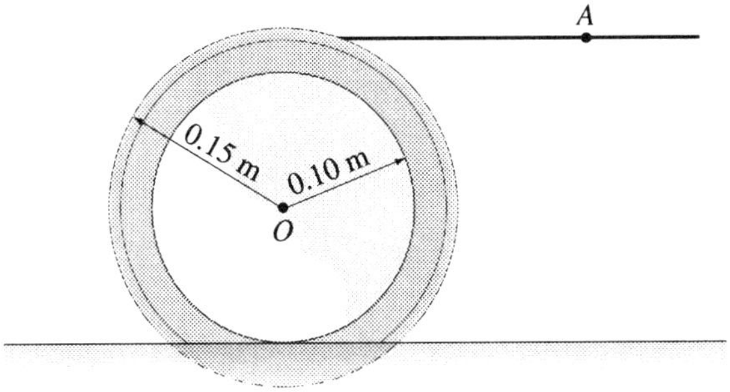

FIGURE P17.138.

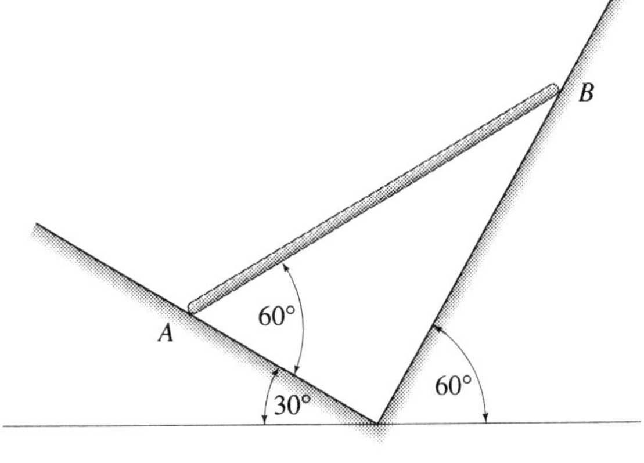

FIGURE P17.139.

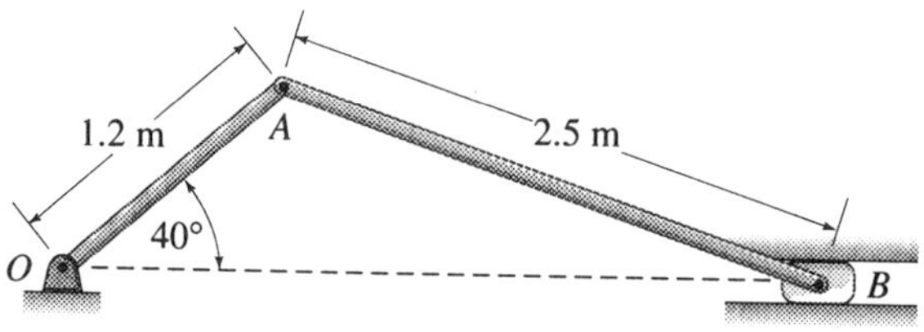

FIGURE P17.140.

P17.140, piston B has a velocity $v_B = 5$ m/s and an acceleration $a_B = 2$ m/s^2 both directed to the left. Determine (a) the velocity of point A on the connecting rod and (b) the angular accelerations of the connecting rod AB and of the crank OA.

17.141 At the instant shown in Figure P17.141, member AB has a cw angular velocity $\omega_{AB} = 4$ rad/s and a ccw angular acceleration $\alpha_{AB} = 2$ rad/s^2. Determine the angular acceleration of member CB and that of member CD.

17.142 Refer to Problem 17.141 and find the acceleration of point E which is the midpoint of member CB.

17.143 Member AB of the mechanisms shown in Figure P17.143 has a constant ccw angular velocity of 10 rad/s. The collar B can slide freely on member CD. For the position shown when member CD is vertical, determine (a) the velocity of the collar relative to member CD and the absolute angular velocity of member CD and (b) the acceleration of the collar relative to member CD and the absolute angular acceleration of member CD.

17.144 Pin P of Figure P17.144 slides in the straight slot S while the disk rotates at a constant ccw angular velocity $\omega = 9$ rad/s. In the position shown, the pin has a velocity and acceleration relative to the slot given by $v_{P/S} = 2$ ft/s and

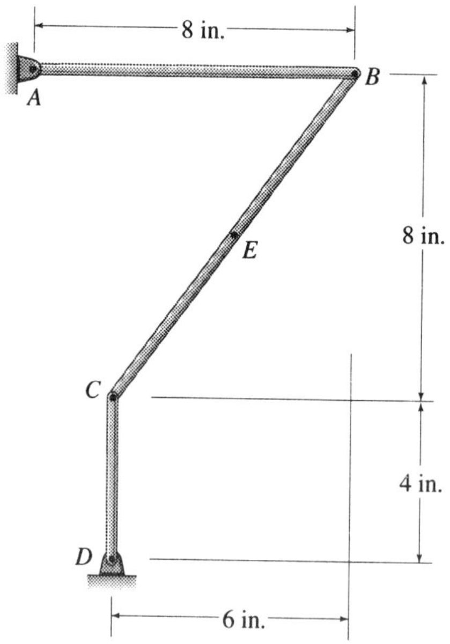

FIGURE P17.141.

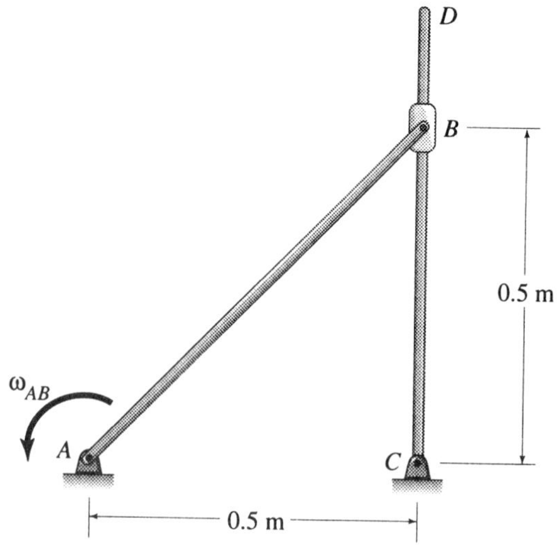

FIGURE P17.143.

$a_{P/S} = 8\ \text{ft/s}^2$ as indicated. Determine the absolute velocity and absolute acceleration of pin P.

17.145 The impeller of a centrifugal pump is shown in Figure P17.145. The radius of curvature of the vanes at their exit points is 9 in. The absolute velocity of the fluid just prior to leaving the impeller at A is known to be $\mathbf{v}_A = (-2.5\mathbf{i} + 10.5\mathbf{j})$ ft/s. Determine (a) the constant clockwise angular speed of the impeller and (b) the absolute acceleration of the fluid just prior to leaving the impeller at A.

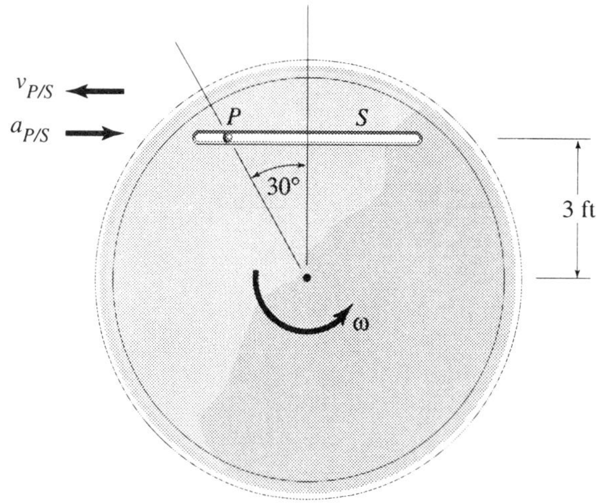

FIGURE P17.144.

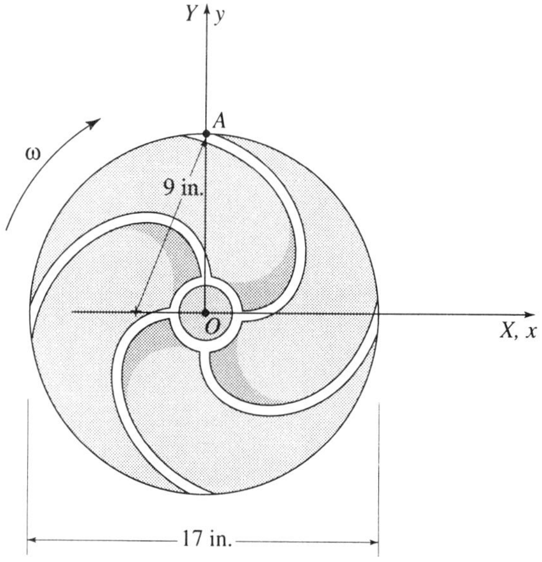

FIGURE P17.145.

18

Two-Dimensional Kinetics of Rigid Bodies: Force and Acceleration

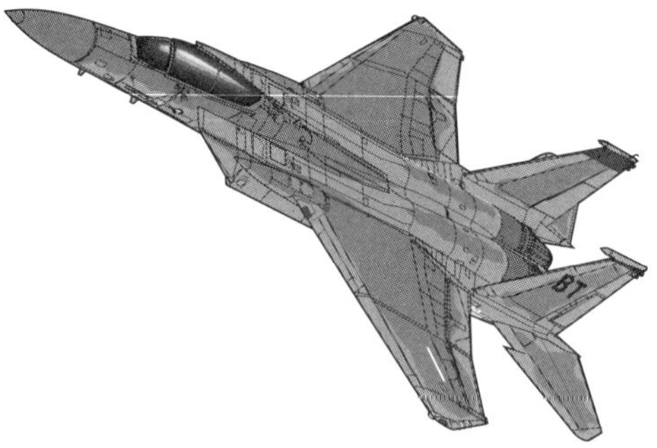
An F-15 Fighter jet thundering into the sky.

Automobile and aircraft manufacturers perform wind tunnel tests to determine resisting forces applied by the air to their models. Recent developments may enable engineers to replace wind tunnels with computer simulation packages similar to the one shown on the cover page of this chapter. The substantial saving in time and money make simulation an attractive alternative to traditional wind tunnel testing. This fact reflects the incredible evolution of the engineering fields and manifests the relentless efforts and genius of engineers from years past. Future engineers need to be exposed to, and learn more about, emerging technologies.

The focus of many new engineering research centers worldwide is graphic simulation of fields, that is, the flow of water around a submarine, air around vehicles, or temperature distribution in an air-conditioned room. Simulation requires the solutions of thousands and even millions of equations. The visual representation of data involves both engineers and artists whose sense of color and composition has helped make new developments possible. Hence, students in art have a chance to learn more about geometry and mechanics and engineering students have something to learn about color and perspective.

Moreover, many of today's consumer products incorporate electronics and computer control. This change means a new role for engineers: they must consider how electronics can be incorporated into the products they design. Consequently, a new discipline called Mechatronics Engineering *was born at the Rensselaer Polytechnic Institute which may pave the way for an overhaul of current engineering curricula everywhere. The emerging disciplines blur the lines among traditional engineering fields for the betterment of mankind.*

The speed with which technology is changing is both challenging and rewarding. The challenge relates to the requirement that an engineer must always be prepared to adapt to new concepts and solve more complex problems. The reward is reflected in the great triumphs in overcoming daunting problems facing our society. In this chapter, you will learn about traditional methods used in solving kinetic problems, including wind tunnels.

18.1 Mass Moments of Inertia

Dynamics is divided into *kinematics* and *kinetics*. Kinematics deals with positions, velocities, and accelerations as functions of time without regard to the forces associated with the motions. In this chapter we will study the kinetics of rigid body motion which is concerned with the forces associated with rigid body motions. It is well to recall what we mean by the idealization which we term a *rigid body*. Such a body is nondeformable no matter how large the forces become during its motion. In reality, there are no such bodies but it is convenient and usually very satisfactory to create mathematical models based upon the rigid body assumption.

To understand the kinetics of rotation of rigid bodies, we need a knowledge of the mass moments of inertia of these bodies. This section provides an introduction to, and a review of this topic, but the reader is well advised to study Chapter 11 of the companion textbook, Statics for Engineers for a more detailed presentation. In studying Chapter 11, we should focus on those parts which deal with mass properties rather than those parts which deal with area properties. Additional information on mass moments of inertia may be found in Appendix B.

Understanding the physical significance of mass moment of inertia requires that we first return to Newton's second law, $\sum F = ma$. (Vector notation is not required because we will focus on magnitudes.) If we solve for the acceleration, we obtain $a = \sum F/m$. From this we note that the acceleration is inversely proportional to the mass. Thus, a given resultant force $\sum F$ applied to a small mass m will produce a larger acceleration than when applied to a larger mass m. Viewing this concept from a different perspective, we know that it is easier to push a child's toy automobile than to push a full sized automobile, and this conveys the concept of inertia intuitively. Now let us consider an equation to be developed in this chapter. It is an equation for the angular motion of a rigid body, $\sum M_O = I_O\alpha$, where I_O is the *moment of inertia* of the rigid body about point O and α is its angular acceleration. Solving for α gives $\alpha = \sum M_O/I_O$. From this we note that the angular acceleration α is inversely proportional to the mass moment of inertia I_O and that the angular acceleration is directly proportional to the applied moment of the external forces $\sum M_O$. Comparing $a = \sum F/m$ to $\alpha = \sum M_O/I_O$, we observe that they are analogous. Thus, a given resultant moment $\sum M_O$ applied to a body with a small moment of inertia I_O will produce a larger angular acceleration than when applied to a body with a larger moment of inertia I_O. We all understand that it is easier to spin the wheel of a toy automobile held in our hand than it is to spin the wheel of a jacked up automobile and this conveys the concept of mass moment of inertia intuitively.

Now, consider a differential element of mass dm located at any position defined by the coordinates x, y and z in a body, as shown in Figure 18.1. The perpendicular distances from the differential element

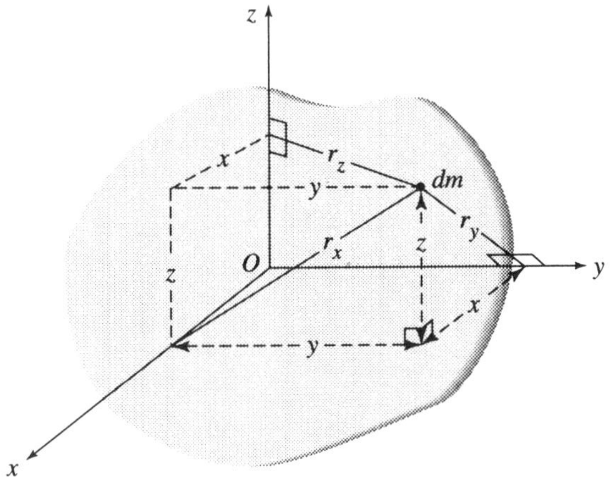

FIGURE 18.1.

of mass dm to the x, y, and z coordinate axes are, respectively, r_x, r_y, and r_z. By definition, the mass moments of inertia of the body with respect to the x, y, and z axes, respectively are given by

$$I_x = \int r_x^2\, dm,$$

$$I_y = \int r_y^2\, dm, \tag{18.1}$$

$$I_z = \int r_z^2\, dm.$$

From the geometry of Figure 18.1, we have $r_x^2 = y^2 + z^2$, $r_y^2 = x^2 + z^2$, and $r_z^2 = x^2 + y^2$. Therefore, Eqs. (18.1) may be written in the form

$$I_x = \int (y^2 + z^2)\, dm,$$

$$I_y = \int (x^2 + z^2)\, dm, \tag{18.2}$$

$$I_z = \int (x^2 + y^2)\, dm.$$

From Eqs. (18.1), we note that the mass moment of inertia has a compound unit equal to a length squared multiplied by the unit of mass. Thus, such units as slug·ft^2 (lb·s^2·ft) are used in the U.S. Customary system and kg·m^2 in the SI system. From their definitions, mass moments of inertia are always positive quantities.

Consider a body of mass m whose moments of inertia with respect to the x, y, and z axes are, respectively I_x, I_y, and I_z. Assume that the

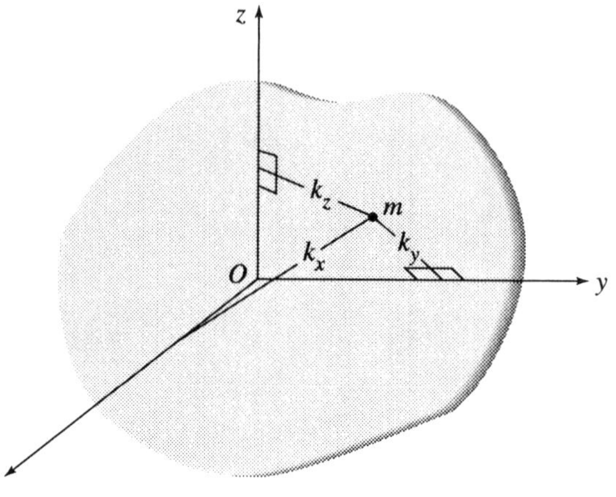

FIGURE 18.2.

entire mass of this body is concentrated at a single point as shown in Figure 18.2. The perpendicular distances of this point, k_x, k_y, and k_z, from the coordinate axes are chosen to satisfy the relationships

$$
\begin{aligned}
I_x &= mk_x^2 \quad \text{or} \quad k_x = \sqrt{\frac{I_x}{m}}, \\
I_y &= mk_y^2 \quad \text{or} \quad k_y = \sqrt{\frac{I_y}{m}}, \\
I_z &= mk_z^2 \quad \text{or} \quad k_z = \sqrt{\frac{I_z}{m}},
\end{aligned}
\tag{18.3}
$$

The quantities k_x, k_y, and k_z in Eq. (18.3) are known as the *radii of gyration of mass* with respect to the x, y, and z axes, respectively.

It should be noted from Eqs. (18.3) that the radii of gyration of mass have units of length. Thus such units as the in. and the ft are used in the U.S. Customary system and the mm and m in the SI system.

The parallel-axis theorem for mass moments of inertia is a very useful tool because it relates moments of inertia with respect to two parallel axes, one of which passes through the center of mass of the body.

Consider the body of mass m shown in Figure 18.3. The X, Y, Z coordinate system has its origin at the center of mass G of the body. A second coordinate system (x, y, z) has its origin at point O such that x is at a distance $(b^2 + c^2)^{1/2}$ from and parallel to X, y is at a distance $(a^2 + c^2)^{1/2}$ from and parallel to Y and z is at a distance $(a^2 + b^2)^{1/2}$ from and parallel to Z. Thus, $x = X + a$, $y = Y + b$ and $z = Z + c$. By the first of Eqs. (18.2) we may write the moment of inertia of mass with respect to the x axis as

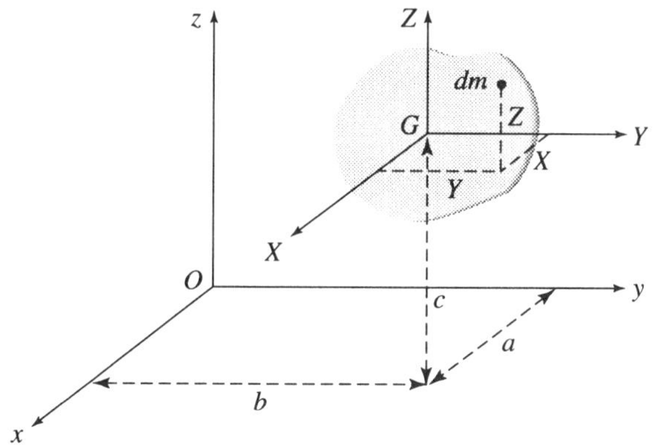

FIGURE 18.3.

$$
\begin{aligned}
I_x &= \int (y^2 + z^2)\, dm, \\
&= \int [(Y + b)^2 + (Z + c)^2]\, dm, \\
&= \int (Y^2 + Z^2)\, dm + (b^2 + c^2) \int dm + 2b \int Y\, dm \\
&\quad + 2c \int Z\, dm. \qquad (18.4)
\end{aligned}
$$

The first integral in Eq. 18.4 represents I_X, the mass moment of inertia with respect to an X axis through the center of mass G. Because the second integral is the mass m, the second term is $(b^2 + c^2)m = d_x^2 m$, where d_x represents the distance between axes x and X. The third and fourth integrals must vanish because they are the first moments of mass with respect to planes passing through the center of mass G of the body. Therefore, Eq. (18.4) may be written as

$$I_x = I_X + m d_x^2. \qquad (18.5)$$

Similarly,

$$
\begin{aligned}
I_y &= I_Y + m d_Y^2 \\
I_z &= I_Z + m d_z^2,
\end{aligned} \qquad (18.6)
$$

where d_y and d_z represent the distances between the y and Y axes and that between the z and Z axes, respectively.

The relationships given in Eqs. (18.5) and (18.6) express the fact that the mass moment of inertia with respect to any axis is equal to the mass moment of inertia with respect to a parallel axis through the center of mass (i.e., centroidal moment of inertia) plus the product of the mass multiplied by the square of the distance between the two axes. This relationship is known as the *Parallel-Axis Theorem* for mass moments of inertia. The theorem may be represented in general form as

$$I = I_G + md^2 \tag{18.7}$$

where I is the mass moment of inertia with respect to any axis, I_G is the mass moment of inertia with respect to a parallel centroidal axis and d is the distance between the two axes. It should be noted that Eq. (18.7) leads to the very important conclusion that mass moments of inertia have minimum values with respect to centroidal axes.

The basic definitions given by Eqs. (18.1) and (18.2) are used to determine the mass moments of inertia for specific masses. A differential element of mass dm is selected and expressed in terms of the chosen coordinate variables and the mass density. Since a differential element of mass is a three-dimensional physical quantity, its mathematical expression could contain as many as three variables which would lead to a triple integration process. However, with a judicious choice of the differential element of mass, it is usually possible to determine the mass moment of inertia of a body by using single or double integrations.

The mass moment of inertia for many bodies of practical interest may be conveniently determined by considering such bodies to consist of an infinite number of thin plates each of which has a constant differential thickness. As shown in the following development, the mass moment of inertia for a differential plate, dI, may be determined in terms of the moment of inertia of the area defined by the contour of the differential plate. The mass moment of inertia for the entire body is, then, obtained by integrating dI. Using this and other techniques, mass moments of inertia for several commonly used bodies have been determined and are listed in Appendix B.

Consider the homogeneous thin plate of constant thickness t shown in Figure 18.4. The mass moments of inertia of this plate is to be determined with respect to the x, y, and z axes. Note that the plate lies in the y-z plane and that the x axis is perpendicular to the plate.

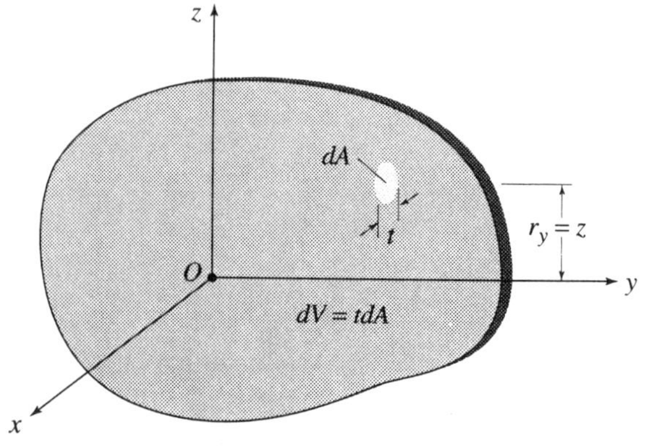

FIGURE 18.4.

The mass moment of inertia of the plate with respect to any in-plane axis, such as the y axis, is given by Eq. (18.1). Thus,

$$I_y = \int r_y^2 \, dm$$

where $r_y = z$, $dm = \rho \, dV = \rho(t \, dA)$ and ρ is the constant mass density. Therefore,

$$I_y = \rho t \int z^2 \, dA$$

in which $\int z^2 \, dA$ is the moment of inertia of the cross-sectional area of the thin plate with respect to the y axis. Consequently, the mass moment of inertia of the thin plate with respect to the centroidal y axis may be written in the form

$$(I_{\text{mass}})_y = \rho t (I_{\text{area}})_y. \tag{18.8a}$$

Similarly,

$$(I_{\text{mass}})_z = \rho t (I_{\text{area}})_z, \tag{18.8b}$$

and

$$(I_{\text{mass}})_x = \rho t (J_{\text{area}})_x. \tag{18.8c}$$

Equations (18.8a) to (18.8c) express the fact that *the mass moment of inertia of a thin homogeneous plate with respect to any axis is equal to the moment of inertia of the cross-sectional area of the plate with respect to the same axis multiplied by the constant thickness of the plate and the constant mass density of the material of the plate.* This important relationship is very useful in the determining mass moments of inertia of bodies by integration.

A composite mass is one that may be decomposed into two or more simple geometric component parts. The moment of inertia of a composite mass, with respect to any axis, is equal to the algebraic sum of the mass moments of inertia of the component parts with respect to the same axis. In finding the mass moments of inertia of the component parts with respect to the desired axis, it becomes necessary to use the parallel-axis theorem expressed in Eq. (18.7), and the moment of inertia for a composite mass becomes

$$I = \sum_{i=1}^{n} (I_G + md^2)_i \tag{18.9}$$

where I represents the moment of inertia of the composite mass with respect to any axis and i is the summation index.

In applying Eq. (18.9), it is desirable to have access to values of mass moments of inertia for simple geometric shapes. A set of values of mass moments of inertia for a selected group of simple geometric shapes is given in Appendix B.

The following examples illustrate the determination of the moments of inertia of masses.

Example 18.1

The body shown in Figure E18.1 is generated by revolving the area bounded by the y axis, the curve $z = (2y)^{1/2}$ and the line $y = h$ about the y axis through one complete revolution. If the mass density of the body ρ is a constant, determine the mass moment of inertia of the body with respect to the y axis.

Solution

Select a thin circular plate of thickness dx for the differential element of mass dm. From Appendix A, the polar moment of inertia of a circular area with respect to a centroidal axis is

$$(J_{\text{area}})_y = \frac{\pi}{32} D^4$$

Substituting this value in Eq. (18.8c),

$$(I_{\text{mass}})_y = \rho t \left(\frac{\pi}{32}\right) D^4 = \rho t \left(\frac{\pi D^2}{4}\right)\left(\frac{D^2}{8}\right) = \left(\frac{D^2}{8}\right) m$$

where the mass of the thin plate m was substituted for the quantity $\rho t \left(\frac{\pi D^2}{4}\right)$. Therefore, the mass moment of inertia of the differential circular plate with respect to the y axis is

$$dI_y = \left(\frac{D^2}{8}\right) dm = \frac{(2z)^2}{8} dm = y\, dm$$

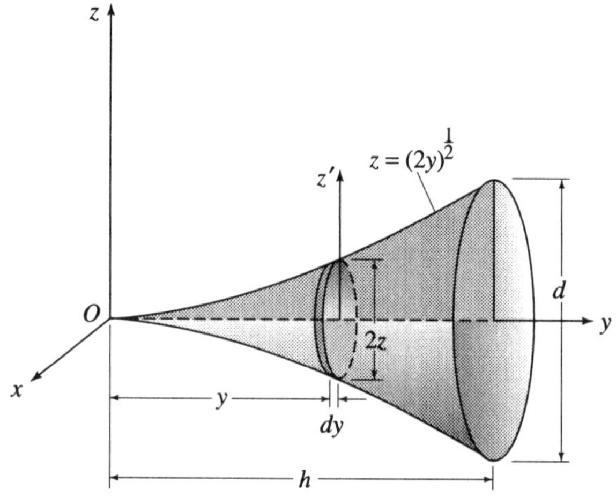

FIGURE E18.1.

where

$$dm = \rho\left(\frac{\pi}{4}\right)(2z)^2\,dy = 2\rho\pi y\,dy$$

and

$$m = \int dm = 2\pi\rho\int_0^h y\,dy = \pi\rho h^2$$

Also,

$$dI_x = 2\pi\rho y^2\,dy$$

and

$$I_x = 2\pi\rho\int_0^h y^2\,dy = \frac{2\rho\pi h^2}{3} = \frac{2}{3}mh = \frac{1}{12}md^2. \qquad \text{ANS.}$$

Example 18.2

Refer to Example 18.1 and determine the mass moment of inertia of the body with respect to the z axis.

Solution

Refer to the thin circular plate selected in Example 18.1 for the differential element dm. From Appendix A, the moment of inertia of a circular area with respect to an in-plane diametrical axis is $(I_{\text{area}})_{y'} = \left(\frac{\pi}{64}\right)D^4$. Substituting this value in Eq. (18.8a),

$$(I_{\text{mass}})_{y'} = \rho t\left(\frac{\pi}{64}\right)D^4 = \rho t\left(\frac{\pi D^2}{4}\right)\left(\frac{D^2}{16}\right) = \left(\frac{D^2}{16}\right)m$$

Thus, the differential mass moment of inertia of the thin plate with respect to an in-plane y' axis is given by

$$dI_{y'} = \left(\frac{D^2}{16}\right)dm = \frac{(2z^2)}{16}dm = \left(\frac{y}{2}\right)dm$$

and with respect to the coordinate y axis, by the parallel-axis theorem Eq. (18.7), it is

$$dI_z = \left(\frac{y}{2}\right)dm + y^2\,dm = \left(\frac{y}{2} + y^2\right)dm.$$

But, from Example 18.1,

$$dm = 2\rho\pi y\,dy$$

and

$$m = \rho\pi h^2.$$

Therefore,

$$dI_z = \left(\frac{y}{2} + y^2\right)(2\rho\pi y\,dy),$$

and

$$I_z = I_x = \rho\pi\left(\int_0^h y^2\,dy + 2\int_0^h y^3\,dy\right) = \rho\pi h^2\left(\frac{h}{3} + \frac{h^2}{2}\right)$$

$$= m\left(\frac{h}{3} + \frac{h^2}{2}\right) = m\left(\frac{d^2}{24} + \frac{d^4}{128}\right) = m\left(\frac{d^2}{24} + \frac{h^4}{128}\right). \qquad \text{ANS.}$$

■ Example 18.3

Determine the mass moment of inertia of the triangular prismatic solid shown in Figure E18.3 with respect to the coordinate y axis. Assume the mass density of the body ρ to be a constant.

Solution

For a differential element of mass dm, select the thin, triangular plate shown in Figure E18.3. From Appendix A, the centroidal moment of inertia of a triangular area is $\frac{1}{36}bh^3$. Therefore, for the chosen differential element of mass with respect to the y' axis,

$$dI_{y'} = \rho\,dx\left(\tfrac{1}{36}ac^3\right).$$

Using the parallel-axis theorem, we can find the mass moment of inertia of this differential element of mass with respect to the coordinate y axis. Thus,

$$dI_y = dI_{y'} + \left[x^2 + \left(\frac{c}{3}\right)^2\right]dm = \rho\,dx\left(\frac{1}{36}\right)ac^3 + \left[x^2 + \left(\frac{c}{3}\right)^2\right]dm.$$

Because $dm = \rho\left(\dfrac{ac}{2}\right)dx$, dI_y becomes

$$dI_y = \frac{1}{36}ac^3\rho\,dx + \frac{1}{2}acx^2\,dx + \frac{1}{18}ac^3\rho\,dx = \frac{ac^3\rho}{12}dx + \frac{ac\rho}{2}x^2\,dx$$

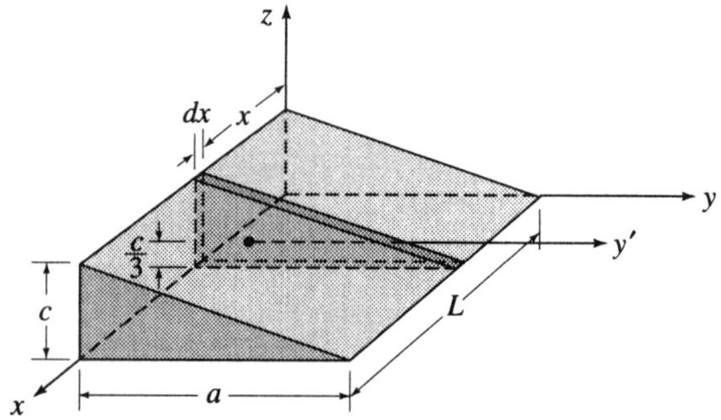

FIGURE E18.3.

and

$$I_y = \frac{ac^3\rho}{12}\int_0^L dx + \frac{ac\rho}{2}\int_0^L x^2\,dx = \frac{ac^3\rho}{12}L + \frac{ac\rho}{6}L^3$$

$$= \rho\left(\frac{ac}{2}\right)L\left(\frac{c^2}{6} + \frac{L^2}{3}\right) = m\left(\frac{c^2}{6} + \frac{L^2}{3}\right) \qquad \text{ANS.}$$

where $m = \rho\left(\frac{ac}{2}\right)L$ is the mass of the triangular prismatic solid.

Example 18.4

Determine the moment of inertia of the composite mass shown in Figure E18.4(a) with respect to the y axis. What is the radius of gyration with respect to this axis? The material has a mass density of $\rho = 500\ \text{kg/m}^3$.

Solution

The composite mass may be viewed as consisting of the four masses m_1, m_2, m_3 and m_4 shown in Figure E18.4(b), that is, the composite mass of Figure E18.4(a) may be obtained from the four component masses in Figure E18.4(b) by first adding the solid conical body of mass m_1 to the solid cylindrical body of mass m_2 and, then, subtracting the solid conical body of mass m_3 and the solid cylindrical body of mass m_4. All of the component masses in Figure E18.4(b) are positioned properly with respect to the desired axis (i.e., the y axis) to show the distances from this axis to the centers of mass of the four component parts. Thus, by Eq. 18.9,

$$I_y = \sum_{i=1}^{n} [(I_G + md^2)_y]_i = [(I_G + md^2)_y]_1 + [(I_G + md^2)_y]_2 - [(I_G + md^2)_y]_3 - [(I_G + md^2)_y]_4$$

Using the values given in Appendix B for mass moments of inertia with respect to centers of mass,

$$\begin{aligned} I_y &= [\tfrac{3}{80}m(D^2 + h^2) + md^2]_1 + [\tfrac{1}{12}m(\tfrac{3}{4}D^2 + h^2) + md^2]_2 \\ &\quad - [\tfrac{3}{80}m(D^2 + h^2) + md^2]_3 - [\tfrac{1}{12}m(\tfrac{3}{4}D^2 + h^2) + md^2]_4 \\ &= [\tfrac{3}{80}m_1(2.4^2 + 3.6^2) + m_1(0.9)^2]_1 \\ &\quad + [\tfrac{1}{12}m_2(0.6^2 + 0.8^2) + m_2(2.8)^2]_2 \\ &\quad - [\tfrac{3}{80}m_3(0.8^2 + 1.2^2) + m_3(2.7)^2]_3 \\ &\quad - [\tfrac{1}{12}m_4(0.225^2 + 3.2^2) + m_4(1.6)^2]_4 \\ &= 1.512m_1 + 7.923m_2 - 7.368m_3 - 3.418m_4. \end{aligned}$$

The values of m_1, m_2, m_3, and m_4 are found as follows:

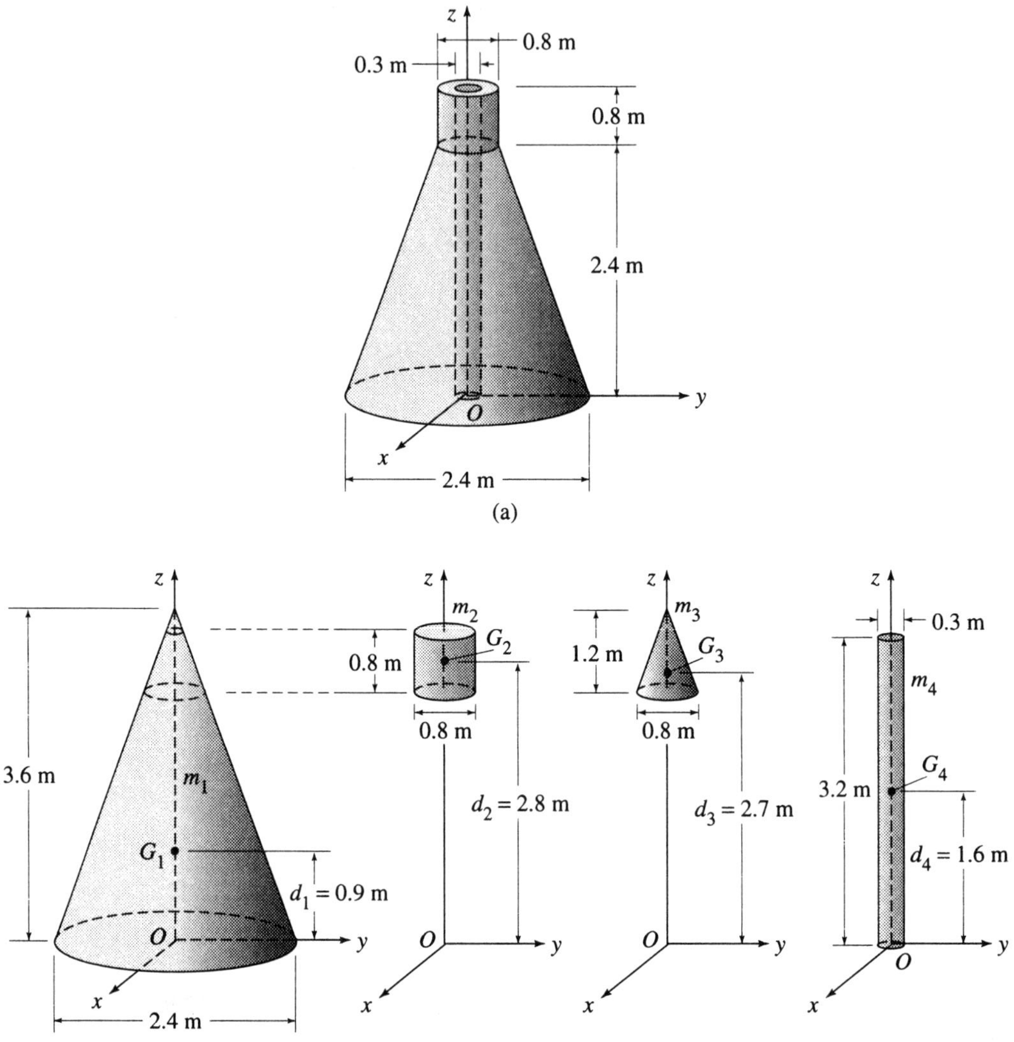

FIGURE E18.4.

$$m_1 = V_1\rho = \left(\frac{\pi D^2}{12}\right)h\rho = \frac{\pi(2.4)^2}{12}(3.6)(5000) = 27143.36 \text{ kg},$$

$$m_2 = V_2\rho = \left(\frac{\pi D^2}{4}\right)h\rho = \frac{\pi(0.8)^2}{4}(0.8)(5000) = 2010.62 \text{ kg},$$

$$m_3 = V_3\rho = \left(\frac{\pi D^2}{12}\right)h\rho = \frac{\pi(0.8)^2}{12}(1.2)(5000) = 1005.31 \text{ kg},$$

and

$$m_4 = V_4\rho = \left(\frac{\pi D^2}{12}\right)h\rho = \frac{\pi(0.3)^2}{12}(3.2)(5000) = 1130.97 \text{ kg}.$$

Therefore

$$I_y = 1.512(27143.36) + 7.923(2010.62) - 7.368(1005.31) - 3.418(1130.97)$$

$$= 4.57 \times 10^4 \text{ kg}\cdot\text{m}^2. \qquad \text{ANS.}$$

By Eq. (18.3), $k_y = \sqrt{\dfrac{I_y}{m}}$ where $m = m_1 + m_2 - m_3 - m_4 = 2.70 \times 10^4$ kg. Therefore,

$$k_y = \sqrt{\frac{4.57 \times 10^4}{2.7 \times 10^4}} = 1.301 \text{ m}. \qquad \text{ANS.}$$

Example 18.5

Four slender rods are attached to a thin disk as shown in Figure E18.5(a). Let the mass of the longest rod be m_r and that for the disk be m_d, and determine the moment of inertia of the composite mass with respect to the x axis. Express the answer in terms of m_r, m_d, and D.

Solution

The composite mass of Figure E18.5(a) may be viewed as consisting of the five masses m_1, m_2, m_3, m_4, and m_5. These masses are labeled in Figure E18.5(a) and properly positioned separately in Figure E18.5(b) with respect to the desired axis (i.e., the x axis), to show clearly the distances to this axis from the mass centers of the five component masses. Therefore, the moment of inertia of the composite mass with respect to the x axis may be obtained by adding the moments of inertia, with respect to the same axis, of the five separate masses. Thus, by Eq. (18.9),

$$I_x = \sum_{i=1}^{n} [(I_G + md^2)_x]_i$$

$$= [(I_G + md^2)_x]_1 + [(I_G + md^2)_x]_2 + [(I_G + md^2)_x]_3 + [(I_G + md^2)_x]_4 + [(I_G + md^2)_x]_5$$

Using the values in Appendix B for the mass moments of inertia with respect to centers of mass and noting that the moments of inertia of m_3 and m_4 are identical,

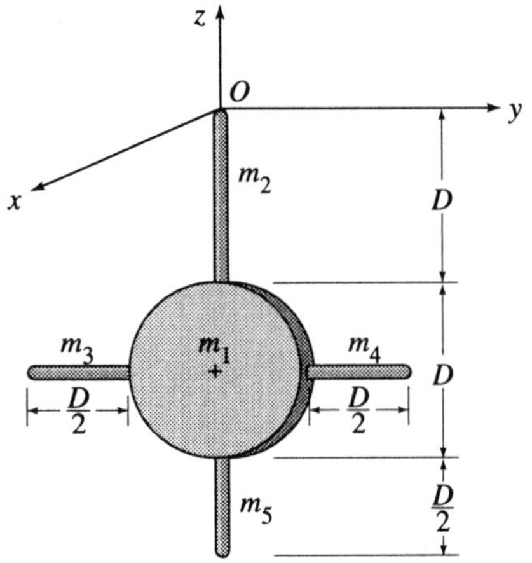

(a)

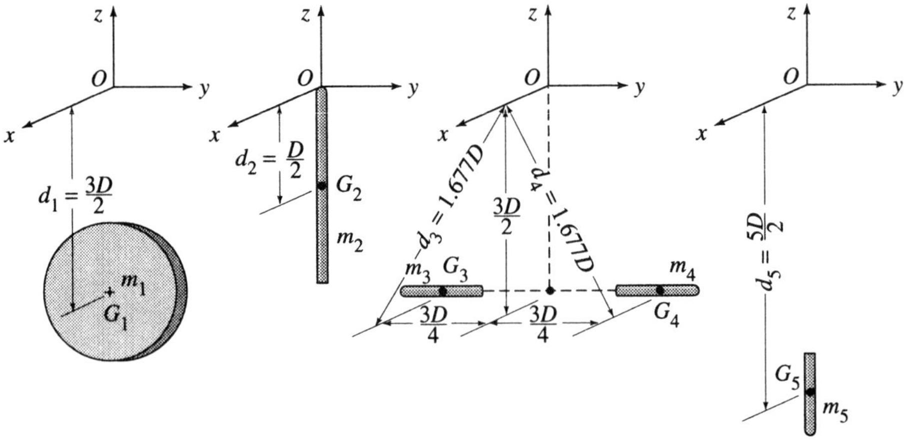

(b)

FIGURE E18.5.

$$I_x = \left[\frac{1}{8}mD^2 + md^2\right]_1 + \left[\frac{1}{12}mL^2 + md^2\right]_2 + 2\left[\frac{1}{12}mL^2 + md^2\right]_3$$

$$+ \left[\frac{1}{12}mL^2 + md^2\right]_5$$

$$= \frac{1}{8}m_\mathrm{d}D^2 + m_\mathrm{d}\left(\frac{3D}{2}\right)^2 + \frac{1}{12}m_\mathrm{r}D^2 + m_\mathrm{r}\left(\frac{D}{2}\right)^2$$

$$+ 2\left[\frac{1}{12}\left(\frac{m_\mathrm{r}}{2}\right)\left(\frac{D}{2}\right)^2 + \left(\frac{m_\mathrm{r}}{2}\right)(1.677D)^2\right]$$

$$+\frac{1}{12}\left(\frac{m_r}{2}\right)\left(\frac{D}{2}\right)^2+\frac{m_r}{2}\left(\frac{5D}{4}\right)^2$$
$$=2.3750m_dD^2+5.7082m_rD^2$$
$$=(2.3750m_d+5.7082m_r)D^2. \qquad \text{ANS.}$$

■

Problems

18.1 The mass moment of inertia of the sphere shown in Figure P18.1 with respect to the y axis is $I_y = (\frac{52}{5})mR^3$ where m is the mass and R is the radius of the sphere. Determine the mass moment of inertia and the mass radius of gyration with respect to the z axis.

18.2 The centroidal mass moment of inertia of the sphere shown in Figure P18.1 is $I_G = 0.08$ kg·m^2. If $R = 0.20$ m and the mass of the sphere is $m = 5$ kg, determine the radius of gyration of the sphere with respect to (a) the x axis and (b) the y axis.

18.3 The mass moment of inertia of the body, shown in Figure P18.3 with respect to the x axis, is $I_x = 5000$ kg·m^2 and, with respect to the z axis, is $I_z = 4000$ kg·m^2. Determine the mass moments of inertia with respect to the X and Z axes passing through the center of mass G if $I_X = 1.5I_Z$. Also determine the mass of the body.

18.4 With respect to the x axis, determine the mass moment of inertia of a thin plate whose cross section is the shaded area shown in Figure P18.4. Assume that both the thickness of the plate and the mass density of the plate material are constants. Express your answers in terms of m, the mass of the plate.

18.5 With respect to the x axis, compute the mass moment of inertia of the prismatic bar shown in Figure P18.5. Assume that

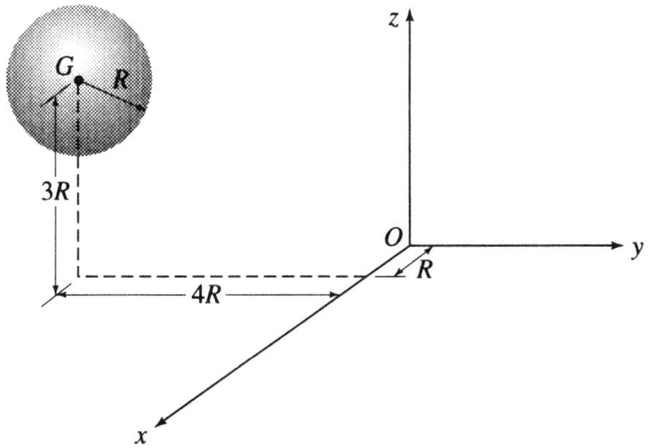

FIGURE P18.1.

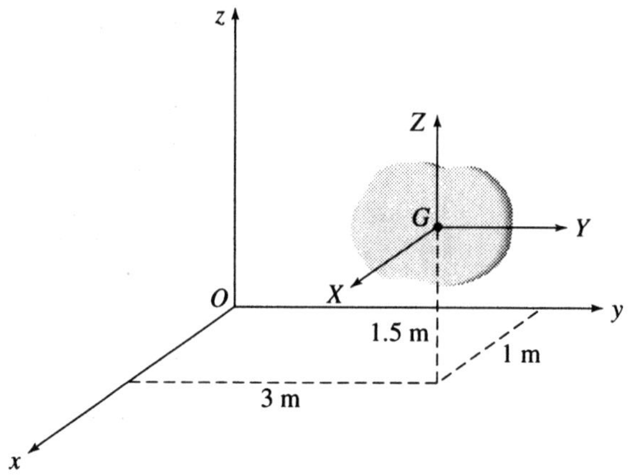

FIGURE P18.3.

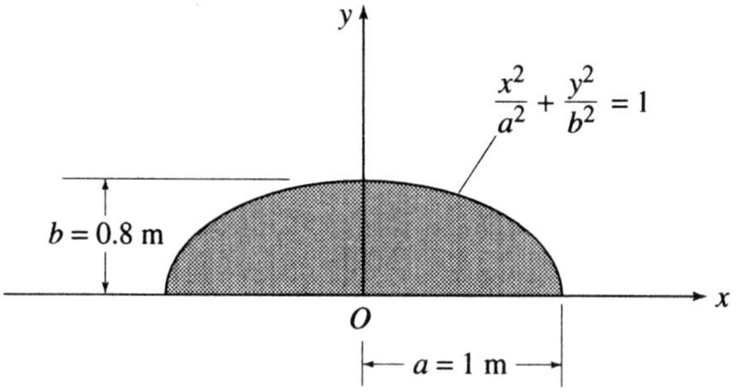

FIGURE P18.4.

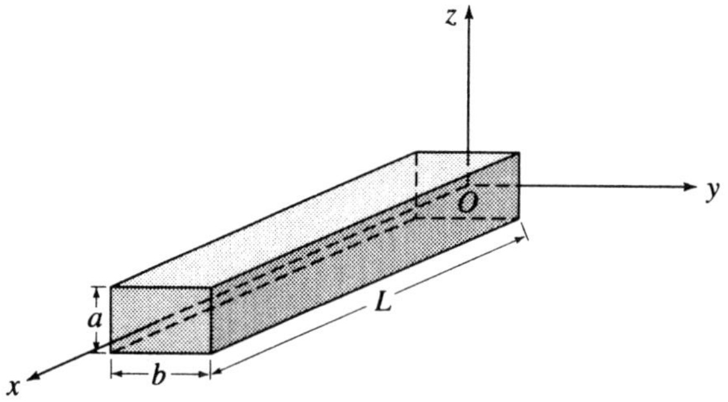

FIGURE P18.5.

the mass density is a constant, and express your answer in terms of the mass m and the dimensions of the bar.

18.6 Find the mass moments of inertia I_X, I_Y, and I_Z of the spherical solid shown in Figure P18.6. Assume that the mass density is a constant, and express your answer in terms of the mass m and the dimensions of the spherical solid.

18.7 With respect to the y axis, determine the mass moment of inertia of the solid shown in Figure P18.7. Assume that the mass density of the solid is a constant, and express your answer in terms of the mass m and the dimensions of the solid.

18.8 With respect to the z axis, find the mass moment of inertia of the solid shown in Figure P18.8. Assume that the mass density of the solid is a constant, and express your answer in terms of the mass m and dimensions of the solid.

18.9 A solid, generated by revolving through one complete revolution about the x axis, is the shaded area shown in Figure P18.9. Compute its mass moment of inertia with respect to the x axis. Let the mass density $\rho = 0.01$ slug/in^3.

18.10 A solid, generated by revolving through one complete revolution about the y axis, is the shaded area shown in Figure P18.10. Compute its mass moment of inertia with respect to the y axis. Let the mass density $\rho = 10$ slug/ft^3.

18.11 Determine the moment of inertia and

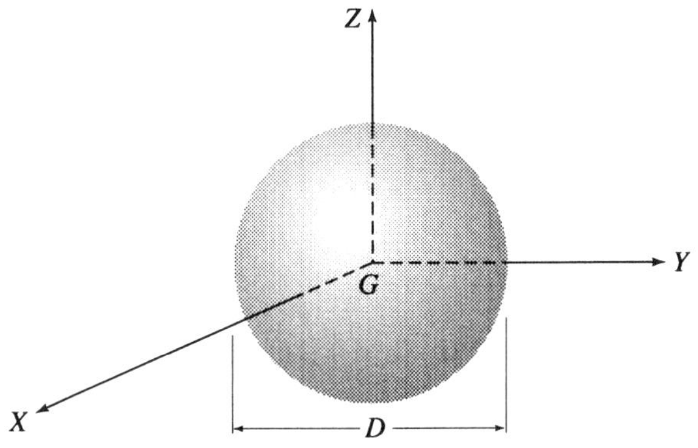

FIGURE P18.6.

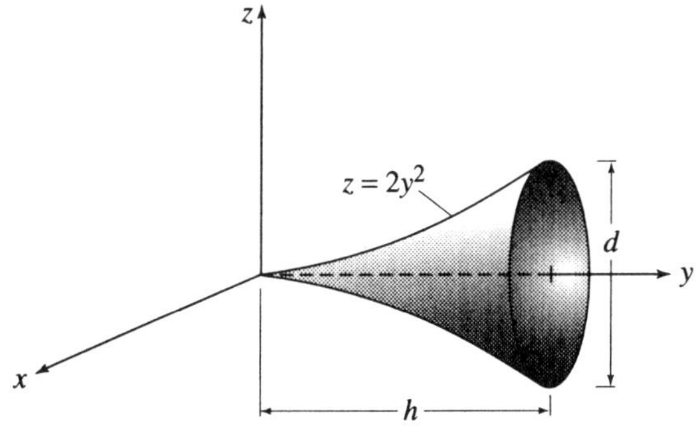

FIGURE P18.7.

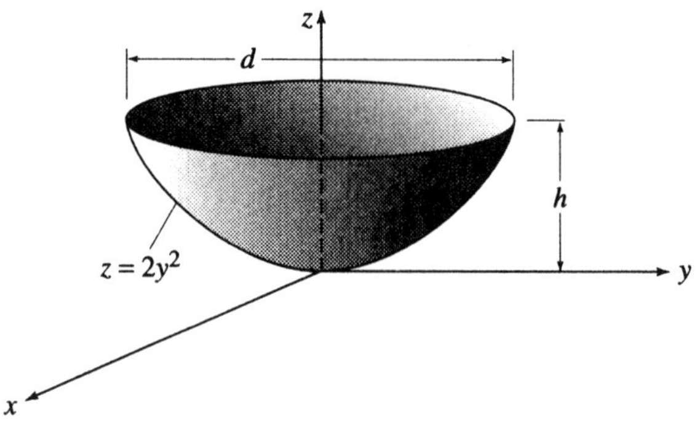

FIGURE P18.8.

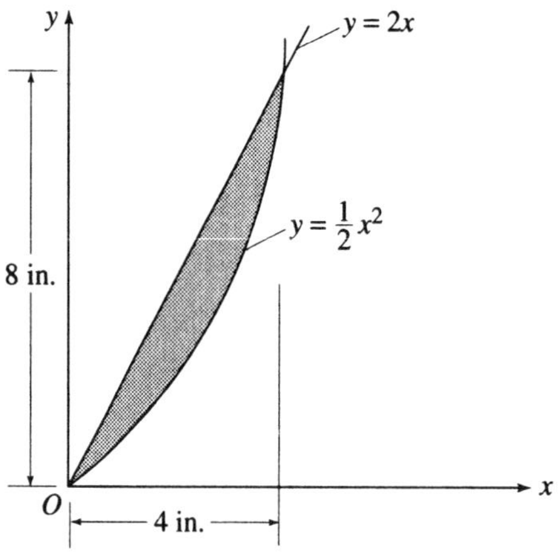

FIGURE P18.9.

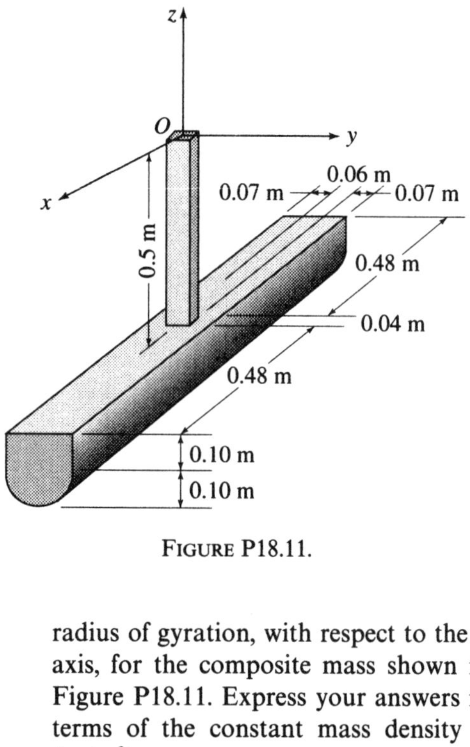

FIGURE P18.11.

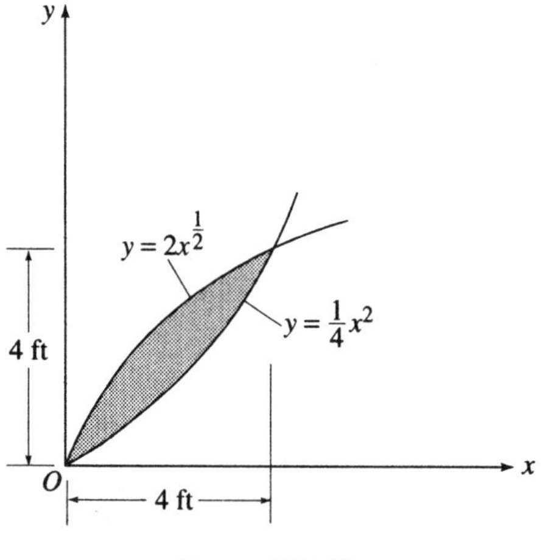

FIGURE P18.10.

radius of gyration, with respect to the x axis, for the composite mass shown in Figure P18.11. Express your answers in terms of the constant mass density ρ (kg/m^3).

18.12 Refer to the composite mass of Figure P18.11, and compute the moment of inertia and radius of gyration with respect to the y axis. Express your answer in terms of the constant mass density ρ (kg/m^3).

18.13 Refer to the composite mass of Figure P18.11, and compute the moment of inertia and radius of gyration with respect to the z axis. Express your answers in terms of the constant mass density ρ (kg/m^3).

18.14 Find the moment of inertia and radius of gyration, with respect to the x and y axes, for the composite mass shown in Figure P18.14. Express your answers in terms of the constant mass density ρ ($slug/in.^3$).

18.15 Refer to the composite mass shown in Figure P18.14, and find the moment of inertia and radius of gyration with respect to the z axis. Express your answers

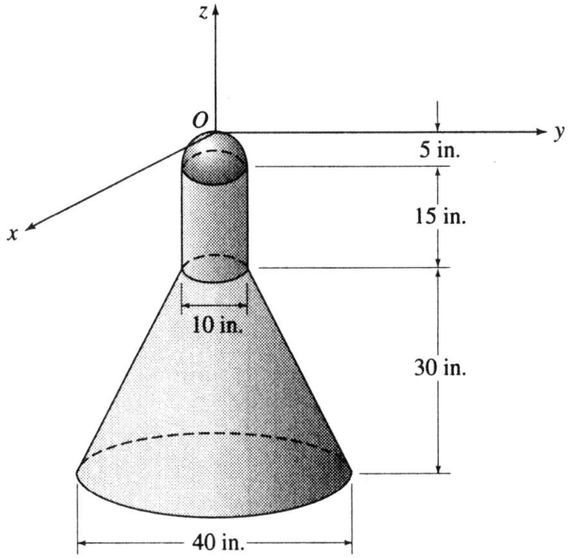

FIGURE P18.14.

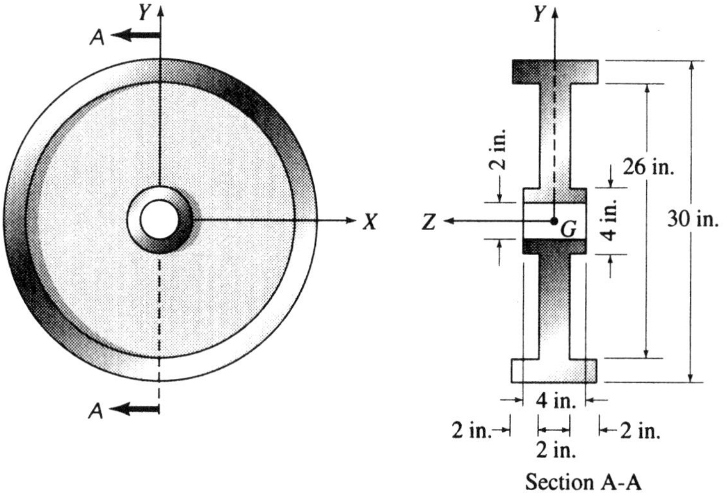

FIGURE P18.16.

in terms of the constant mass density ρ (slug/in.^{3}).

18.16 Compute the moment of inertia and radius of gyration of the aluminum flywheel shown in Figure P18.16 with respect to the centroidal Z axis. The mass density for aluminum is approximately 5.2 slug/ft^3.

18.17 Refer to the flywheel shown in Figure P18.16 and compute the moments of inertia and radii of gyration with respect to the centroidal X and Y axes.

18.18 The composite body shown in Figure P18.18 consists of two thin aluminum plates OABC and OCDE and five steel slender rods DG, BG, GF, EF, AND AF. The mass density for the aluminum plates is 27 kg/m^2 and that for the steel

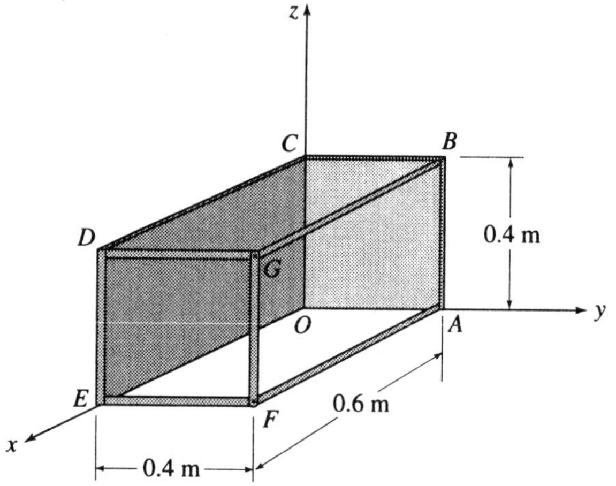

FIGURE P18.18.

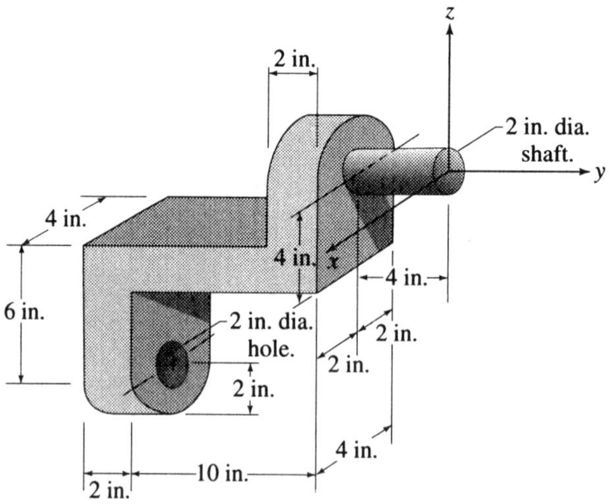

FIGURE P18.20.

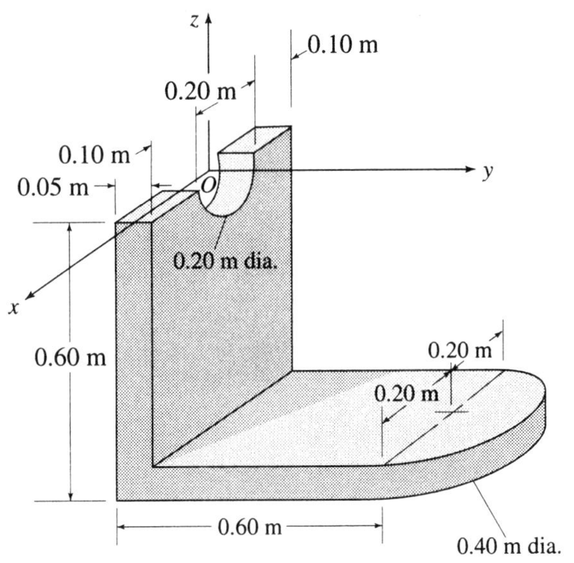

FIGURE P18.21.

rods is 0.6 kg/m. Determine the mass moment of inertia with respect to the x axis.

18.19 Refer to the composite body described in Problem 18.18 and determine the mass moment of inertia and radius of gyration with respect to the y axis. Compute these quantities with respect to the z axis.

18.20 A machine member made of steel is shown in Figure P18.20. The mass density for steel is approximately 15 slug/ft^3. Determine the mass moment of inertia and radius of gyration with respect to the x axis.

18.21 A composite member made of aluminum is shown in Figure P18.21. The mass density for aluminum is approximately 2700 kg/m^3. Determine the mass moment of inertia and radius of gyration with respect to the x axis.

18.22 Determine the moment of inertia of the composite mass, shown in Figure P18.22, with respect the z axis. What is the radius of gyration with respect to this axis? The material has a mass density $\rho = 5000$ kg/m^3.

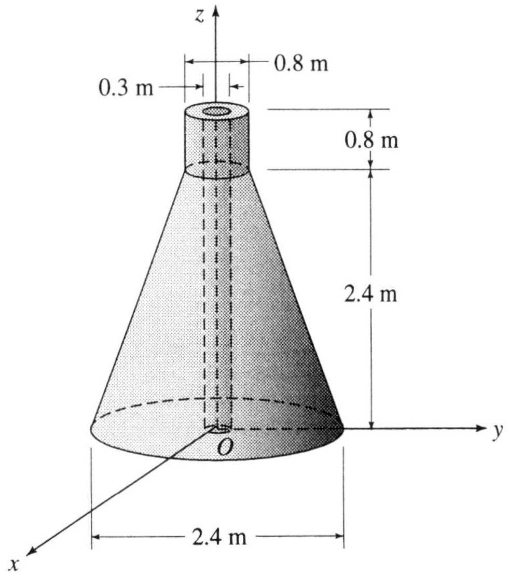

FIGURE P18.22.

18.2 General Equations of Motion

A rigid body is considered to be an infinite collection of particles. This enables us to use results obtained for systems of particles in Sections 14.2 and 16.8 for the motion of rigid bodies.

Consider the translation of the mass center G of a rigid body with respect to a Newtonian reference frame. Newton's second law of motion becomes

$$\sum \mathbf{F} = m\mathbf{a}_G \tag{18.10}$$

This equation states that the vector sum (resultant) of the external forces acting on the body equals the total mass of the body multiplied by the acceleration of the mass center G. This equation is shown pictorially in Figure 18.5. For the motion of the mass center G, both the resultant external force $\sum \mathbf{F}$ and the resultant inertial force $m\mathbf{a}_G$ act through the mass center.

In addition to the translation of the mass center, we must consider the rotation of the rigid body with respect to the mass center G. Equation 16.32, written for a Newtonian reference frame, is also valid for a reference frame attached to the mass center G of a rigid body. Thus,

$$\sum \mathbf{M}_G = \frac{d\mathbf{H}_G}{dt} \tag{18.11}$$

where $\mathbf{H}_G$ now represents the angular momentum of the entire body with respect to the mass center G and the summation on the left hand side extends over all external forces. This equation states that the vector sum of the moment of the external forces (or the resultant external couple) with respect to the mass center G equals the time rate of change of the angular momentum of the body with respect to the same point. This equation is shown pictorially in Figure 18.6. For motion of this rigid body with respect to the mass center G, both the resultant external couple and the resultant inertia couple act through the mass center G.

As discussed in Chapter 17, the general plane motion of a rigid body may be viewed as a combination of a translation and a rotation. Thus, when a rigid body experiences general plane motion, we may consider

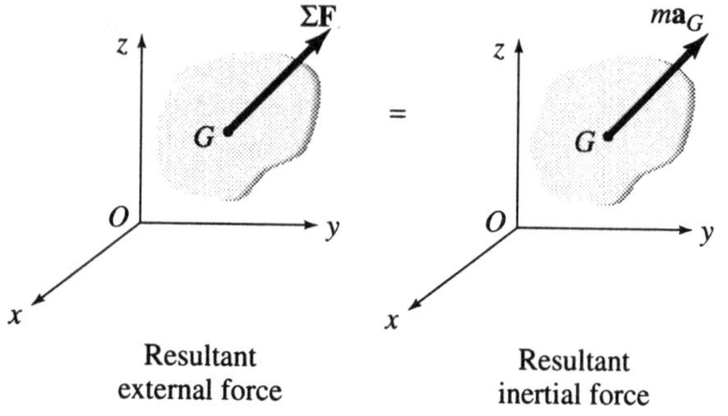

FIGURE 18.5.

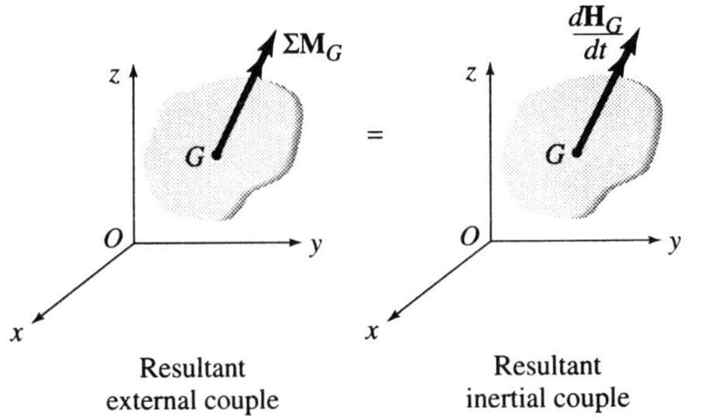

FIGURE 18.6.

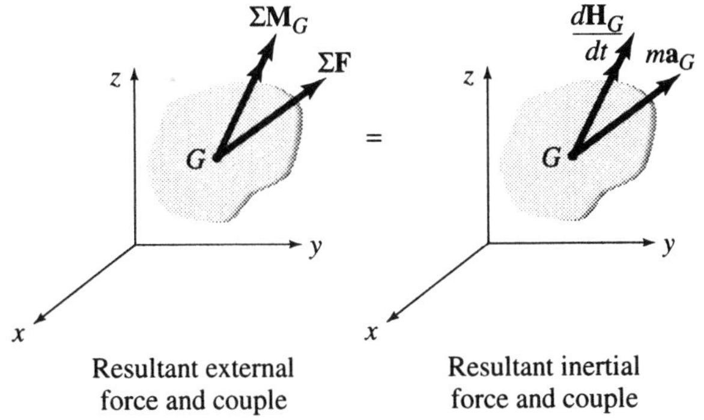

FIGURE 18.7.

this motion as the vector sum of a translation and a rotation. This point of view is illustrated in Figure 18.7.

18.3 Rectilinear and Curvilinear Translation

In Section 17.1, we learned that all points of a translating rigid body have the same velocity at any time. Likewise, all points of a translating rigid body have the same acceleration vector at any time. These instantaneous conditions for the translation of a rigid body are depicted in Figure 18.8 where any two points A and B, in addition to the mass center G, are shown. Using the results obtained in Section 17.1,

$$\mathbf{v}_A = \mathbf{v}_B = \mathbf{v}_G, \quad \text{and} \quad \mathbf{a}_A = \mathbf{a}_B = \mathbf{a}_G. \qquad (18.12)$$

In fact, subscripts are not required, and, for any time, we may write

$$\mathbf{v}_A = \mathbf{v}_B = \mathbf{v}_G = \mathbf{v} \qquad (18.13a)$$

and

$$\mathbf{a}_A = \mathbf{a}_B = \mathbf{a}_G = \mathbf{a}. \qquad (18.13b)$$

If particles of the body travel along parallel straight line paths, the motion is termed *rectilinear, or straight line translation.* In such motion, the magnitudes of the velocity and acceleration vectors may change with time, but their directions are the same along the parallel straight-

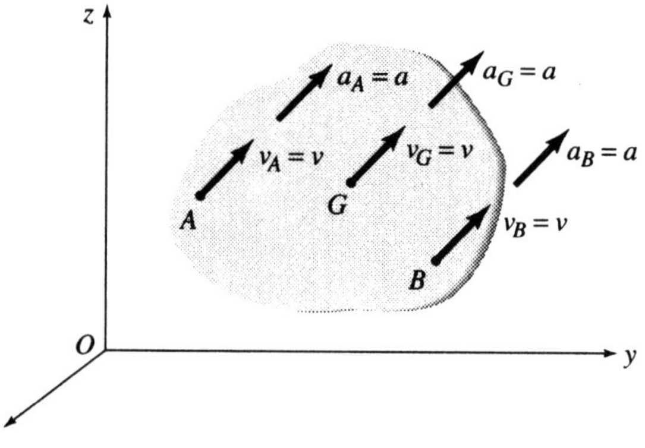

FIGURE 18.8.

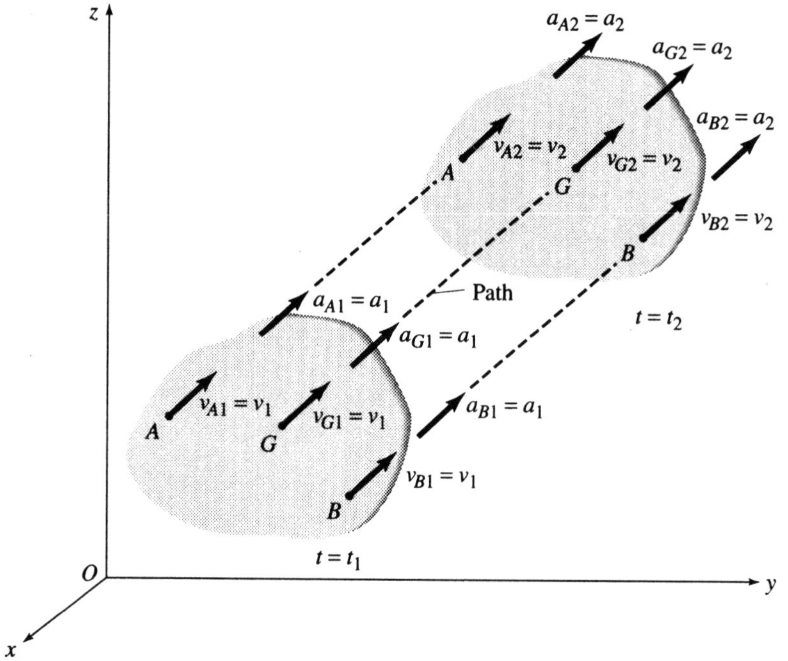

FIGURE 18.9.

line paths. Figure 18.9 shows these kinematic conditions for the two-dimensional rectilinear translation of a rigid body.

If particles of the body travel along curved paths, so that the curvature of all these paths are equal at a given instant, then, the motion is termed *curvilinear translation*. As time passes during a curvilinear translation, in general, the magnitudes and directions of both the velocity and acceleration vectors will change. Figure 18.10 shows these kinematic conditions for the curvilinear translation of a rigid body.

Equations (18.10) and (18.11) are kinetics equations for the general motion of a rigid body and, therefore, apply to the special case of translatory motion. Consider a rigid body in translation with forces

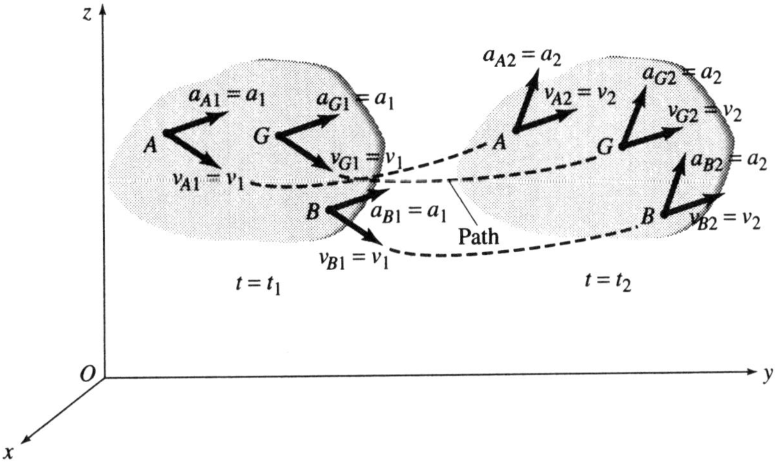

FIGURE 18.10.

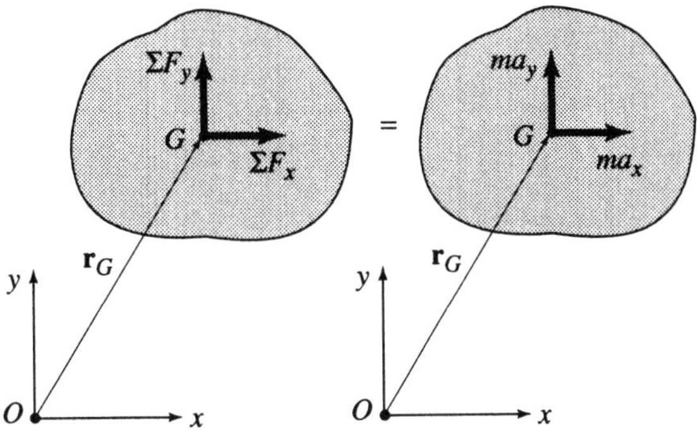

FIGURE 18.11.

applied to it and write Eq. (18.10) in the following form

$$\sum \mathbf{F} = m\mathbf{a}. \tag{18.14}$$

This equation states that the vector sum of the external forces (or the resultant external force) equals the vector sum of the inertial forces (or the resultant inertial, or effective, force). A subscript is not required on the acceleration **a** because, for either rectilinear or curvilinear translation, all points of a rigid body have the same acceleration at a given instant.

A body in translation does not rotate. Thus, we conclude that its angular momentum is a null vector. Furthermore, because $\dot{\mathbf{H}}_G = \dfrac{d\mathbf{H}_G}{dt} = \mathbf{0}$, we conclude, from Eq. (18.11), that

$$\sum \mathbf{M}_G = \mathbf{0}. \tag{18.15}$$

Equation (18.15) states that the vector sum of the moments of the external forces with respect to the mass center will vanish for a body in translation. Equations (18.14) and (18.15) are the fundamental vector equations for the kinetics of translating rigid bodies.

Refer to Figure 18.11 which depicts the special case of translation such that the mass center G of the rigid body moves in the x-y plane. The kinetics equations for this two-dimensional body become

$$\sum F_x = ma_x, \tag{18.16a}$$

$$\sum F_y = ma_y, \tag{18.16b}$$

$$\sum M_G = 0, \tag{18.16c}$$

$$\sum M_O = \sum (M_i)_O. \tag{18.16d}$$

Equation (18.16a) states that the algebraic sum of the external forces in the x direction equals the product of the total mass of the body and the instantaneous acceleration component in the x direction. Likewise, by Eq. (18.16b), the algebraic sum of the external forces in the y direction equals the product of the total mass of the body and the acceleration component in the y direction. If we choose to sum moments of the external forces with respect to an axis perpendicular to the x-y plane through the mass center G, then, by Eq. (18.16c), the algebraic sum of these moments will vanish. Equation (18.16d) is an alternate moment equation in which M_i represents the moment produced by an inertial force and O is a convenient arbitrarily chosen point. It states that the sum of the moments of the external forces with respect to an axis through O, perpendicular to the x-y plane, is equal to the sum of the moments of the inertial forces (i.e., ma_x and ma_y) with respect to this same axis. Equation (18.16d) is sometimes more convenient to use than Eq. (18.16c). Although four algebraic equations have been presented for planar translation, only three independent kinetics equations are available for the solving a given problem.

The following examples illustrate some of the above concepts.

■ Example 18.6

An ore car and its contents, shown in Figure E18.6(a), weigh 15 tons. It is being pulled up a steep inclined plane to charge a furnace. The cable has a force of 28,000 pounds in it which is exerted by an electric motor-pulley system (not shown). Determine the acceleration of the ore car and the forces exerted by each of the 4 wheels on it. Neglect friction and assume the wheels to be massless.

Solution

Because the ore car is in translation along its track, each of its points has an acceleration a directed parallel to the track.

Referring to the free body and inertial force diagrams of Figure E18.6(b), we note that the system of external forces consisting of T, W, N_A, and N_B is equivalent to the single inertial force $(W/g)\,a$ directed parallel to the plane. All particles of the ore car have the same acceleration a at any given time, and this acceleration is constant with time

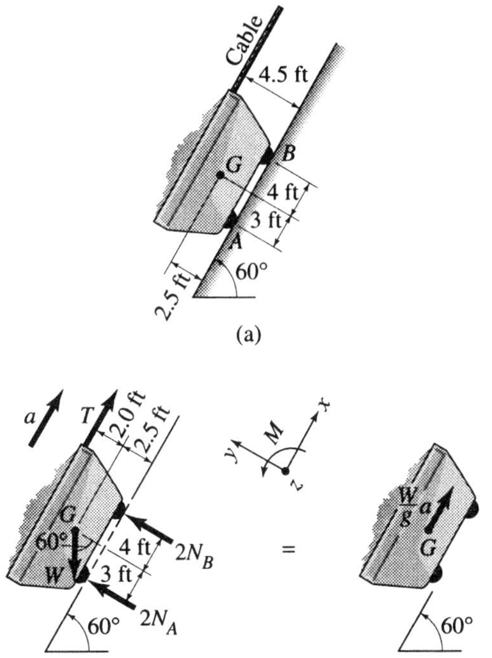

(b) FIGURE E18.6.

because the external forces applied to the ore car do not change with time. The sign convention is based upon assuming that the ore car is accelerated up the plane and the positive x axis is chosen consistent with this assumed acceleration. A coordinate system is shown in Figure E18.6(b). Note that x, y, and the moment sense form a right-handed coordinate system.

Because $W = 15$ tons $= 30{,}000$ lb and the cable tension $T = 28{,}000$ lb there are three unknowns N_A, N_B, and a. To solve this problem, we will write one moment equation and two force summation equations. Thus, using Eq. (18.16a),

$$\sum F_x = ma_x,$$

$$T - W \sin 60^\circ = \frac{W}{g} a.$$

Substitute numerical values for T and W, and, then, solve for a to obtain

$$28000 - 3000 \sin 60^\circ = \frac{3000}{32.2} a,$$

$$a = 2.17 \text{ ft/s}^2 \nearrow. \qquad \text{ANS.}$$

$$\sum F_y = ma_y = 0,$$

$$2N_A + 2N_B - W \cos 60^\circ = 0,$$

$$N_A + N_B = 7500. \qquad \text{(a)}$$

$$\sum M_G = 0,$$
$$2N_B(4) - 2N_A(3) - T(2) = 0,$$
$$-3N_A + 4N_B = T = 28{,}000. \qquad \text{(b)}$$

Solving Eqs. (a) and (b) simultaneously yields

$$N_A = 285 \text{ lb} \nwarrow \qquad \text{ANS.}$$

and

$$N_B = 7215 \text{ lb} \nwarrow. \qquad \text{ANS.}$$

Example 18.7

A sliding door of weight W is subjected to a horizontal force $P = kW$ as shown in Figure E18.7(a). Normal and frictional forces are applied to the door by the massless pegs at A and B. The peg diameters are negligible when compared to other dimensions. Assume W, μ, g, k, b, c, and d as known quantities. Determine the following in terms of these quantities:

(a) the acceleration of the door, (b) the normal forces acting on the door at A and B, and (c) the numerical values for the acceleration and normal forces for $W = 1200$ lb, $\mu = 0.40$, $g = 32.2$ ft/s^2, $k = 0.45$, $b = 2$ ft, $c = 6$ ft, and $d = 3$ ft.

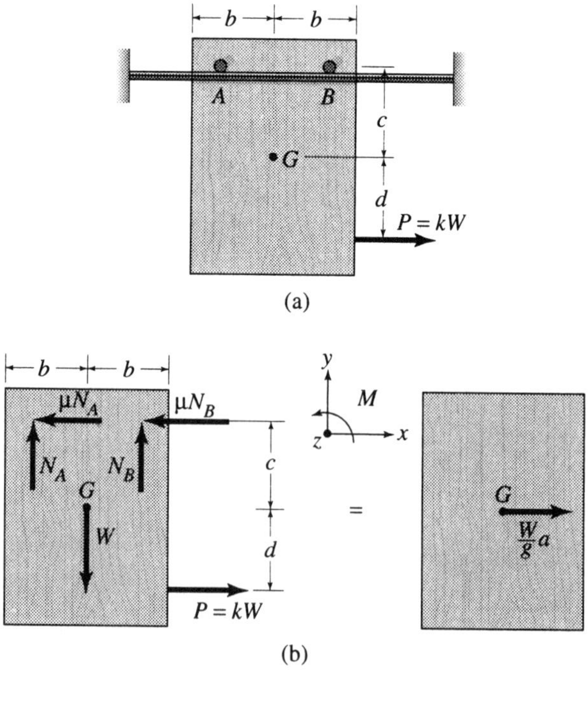

FIGURE E18.7.

Solution

(a) and (b)

Figures E18.7(a) and (b) show the free body and inertial force diagrams along with a right-handed x-y-z coordinate system where the z axis comes out of the page. There are three unknowns a, N_A, and N_B, and three equations available to solve for them. The acceleration of all points of the door are equal because the door translates. Thus, using Eq (18.16a),

$$\sum F_x = ma_x,$$

$$kW - \mu N_A - \mu N_B = \frac{W}{g}a, \tag{a}$$

$$kW - \mu(N_A + N_B) = \frac{W}{g}a.$$

The door is not accelerated vertically, and, therefore, $a_y = 0$. Thus,

$$\sum F_y = ma_y = 0,$$

$$N_A + N_B - W = 0. \tag{b}$$

Since the inertial force acts through the mass center G, the sum of the moments of the external forces with respect to G will equal zero. Therefore,

$$\sum M_G = 0,$$

$$kWd + \mu N_A c + \mu N_B c - N_A b + N_B b = 0. \tag{c}$$

Substitute from Eq. (b) for $N_A + N_B = W$ in Eq. (a) and solve for the acceleration a, noting that the common factor W cancels out. Thus,

$$a = (k - \mu)g \quad \text{(provided } k > \mu\text{)} \qquad \text{(d) ANS.}$$

Substitute from Eq. (b) for $N_B = W - N_A$ in Eq. (c), and solve for N_A to give

$$N_A = \left(1 + k\frac{d}{b} + \mu\frac{c}{b}\right)\frac{W}{2}. \qquad \text{(e) ANS.}$$

Substitute for N_A from (e) into $N_B = W - N_A$ to obtain

$$N_B = \left(1 - k\frac{d}{b} - \mu\frac{c}{b}\right)\frac{W}{2}. \qquad \text{(f) ANS.}$$

(c) Substituting the given numerical values into the equations above yields

$$a = (0.45 - 0.40)32.2 = 1.61 \text{ ft/s}^2 \rightarrow, \qquad \text{ANS.}$$

$$N_A = [1 + 0.45(\tfrac{3}{2}) + 0.40(\tfrac{6}{2})]\tfrac{1200}{2} = 1725 \text{ lb} \uparrow, \qquad \text{ANS.}$$

$$N_B = [1 - 0.45(\tfrac{3}{2}) - 0.40(\tfrac{6}{2})]\tfrac{1200}{2} = -525 \text{ lb}$$

$$= 525 \text{ lb} \downarrow. \qquad \text{ANS.}$$

Example 18.8

A homogeneous plate is supported by two massless rods as shown in Figure E18.8(a). The plate has a mass m centered at G. Frictionless pins at B and E attach the rods to the plate and frictionless pins at A and D attach the rods to a very massive body overhead. The angle θ positions the system initially. The system is released from rest (i.e., $\dot{\theta} = 0$). Immediately after release, determine (a) the angular acceleration $\ddot{\theta}$ of the rods as a function of θ, g, and L and (b) the forces in the rods F_{ED} and F_{BA} as functions of θ and b/d. (c) For $\theta = 40°$, $b = 2$ ft, $d = 3$ ft, $mg = 100$ lb and $L = 5$ ft, determine numerical values for $\ddot{\theta}$, F_{ED} and F_{BA}.

Solution

(a) Point B travels a circular path of radius L centered at A and point E travels a circular path of radius L centered at D. Since the rods and the plate are rigid, then, line BD of the plate remains horizontal during the motion, which means that the plate is in curvilinear translation.

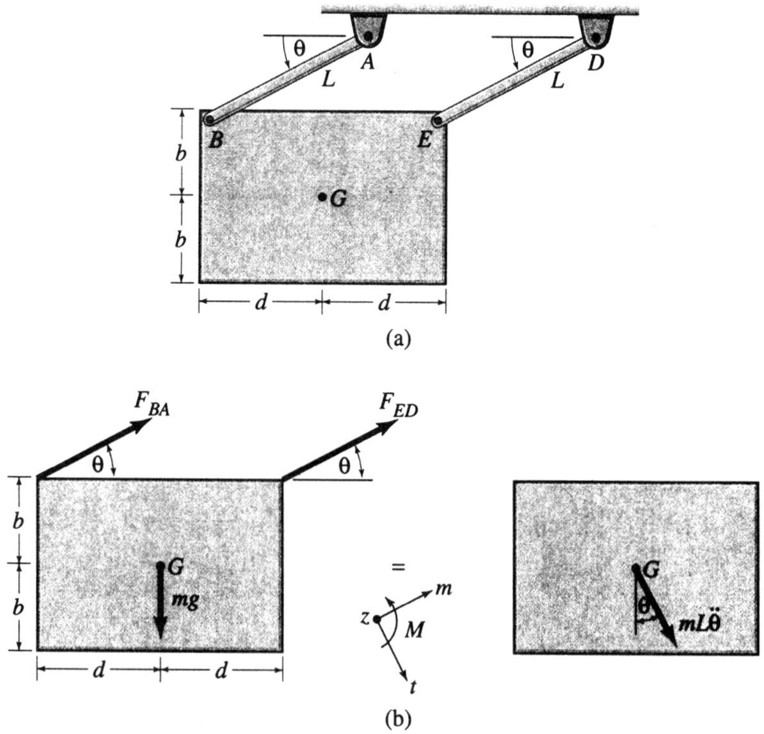

FIGURE E18.8.

The free body and inertial force diagrams are shown in Figure E18.8(b) where a right-handed t-n-z coordinate system is given. Note that the z axis comes out of the page. The tangential axis t is directed tangent to the circular paths of the points and the normal axis n is directed toward the center of these paths. Since the plate is released from rest, the angular velocity $\omega = \dot{\theta}$ vanishes at this instant. The tangential acceleration equals $L\ddot{\theta}$, and this is the total acceleration of any point on the translating plate at the instant of release. Once again we employ Eqs. (18.16) with subscripts x and y replaced by t and n, respectively. Thus,

$$\sum F_t = ma_t, \quad mg\cos\theta = mL\ddot{\theta},$$

$$\ddot{\theta} = \frac{g}{L}\cos\theta. \qquad \text{(a)} \quad \text{ANS.}$$

Provided we know the initial value of θ and values for g and L, then, Eq. (a) enables us to find the angular acceleration $\ddot{\theta}$ at release.

(b) Note that $a_n = 0$ because the initial value of $\dot{\theta}$ is zero. Therefore,

$$\sum F_n = ma_n = 0, \quad F_{BA} + F_{ED} - mg\sin\theta = 0. \qquad \text{(b)}$$

Because the inertial force has an action line passing through the mass center G, then,

$$\sum M_G = 0,$$

$$(F_{ED}\sin\theta)d - (F_{ED}\cos\theta)b - (F_{BA}\sin\theta)d - (F_{BA}\cos\theta)b = 0. \qquad \text{(c)}$$

Solve Eq. (c) for F_{BA} in terms of F_{ED}, substitute in Eq. (b), and solve for F_{ED} to give

$$F_{ED} = \frac{1}{2}mg\left(\sin\theta + \frac{b}{d}\cos\theta\right). \qquad \text{(d)} \quad \text{ANS.}$$

Back substitute for F_{ED} to obtain

$$F_{BA} = \frac{1}{2}mg\left(\sin\theta - \frac{b}{d}\cos\theta\right). \qquad \text{(e)} \quad \text{ANS.}$$

(c) Substituting given numerical values in Eqs. (a), (d), and (e), we obtain

$$\ddot{\theta} = \left(\tfrac{32.2}{5}\right)\cos 40^\circ = 4.93 \text{ rad/s}^2 \circlearrowleft. \qquad \text{ANS.}$$

$$F_{ED} = \tfrac{1}{2}(100)(\sin 40^\circ + \tfrac{2}{3}\cos 40^\circ) = 57.7 \text{ lb} \nearrow. \qquad \text{ANS.}$$

$$F_{BA} = \tfrac{1}{2}(100)(\sin 40^\circ + \tfrac{2}{3}\cos 40^\circ) = 6.60 \text{ lb} \nearrow. \qquad \text{ANS.}$$

■

Problems

18.23 The body depicted in Figure P18.23 translates along the horizontal plane as shown. If $P = 0.2W$, where W is the weight of the body and the coefficient of kinetic friction $\mu_k = 0.1$, determine (a) the acceleration of the body in terms of g, (b) the normal and frictional forces acting on the body, and (c) the horizontal distance from the mass center of this homogeneous body to the vertical action line of the normal force.

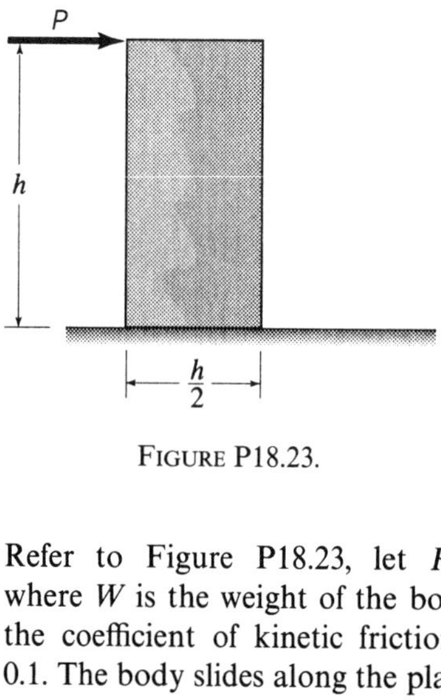

FIGURE P18.23.

18.24 Refer to Figure P18.23, let $P = kW$ where W is the weight of the body and the coefficient of kinetic friction $\mu_k = 0.1$. The body slides along the plane and you are to determine k such that this body is on the verge of tipping about its lower right corner as it slides. What is the acceleration of the body in terms of g?

18.25 An automobile is depicted in Figure P18.25. The rear wheels at A are the driving wheels and friction at these wheels provides the forces which accelerate this automobile. The four wheels have relatively small mass compared to the mass of the vehicle and it is reasonable to neglect their rotary inertia. Also, as a consequence, we consider only normal forces acting at B. Regarding the weight W, μ, g, a, b, and c as known quantities, determine the following in terms of these quantities: (a) the acceleration of the automobile, (b) the friction force at each rear wheel when slip impends, and (c) the normal forces at each of the four wheels.

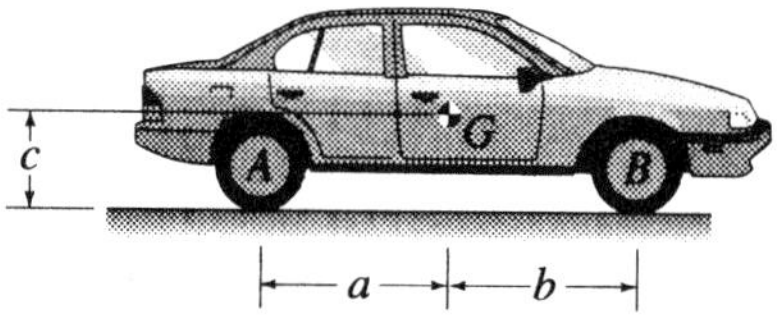

FIGURE P18.25.

18.26 Solve problem 18.25 with $a = 4.5$ ft., $b = 3.5$ ft., $c = 1.6$ ft., $W = 3000$ lb, $g = 32.2$ ft/s^2 and $\mu = 0.6$.

18.27 Solve problem 18.25 with $a = 1.4$ m, $b = 1.0$ m, $c = 0.5$ m, $W = 12$ kN, $g = 9.81$ m/s^2 and $\mu = 0.5$.

18.28 Refer to the system shown in Figure P18.28. The body on the inclined plane is subjected to resultant normal and friction forces and has a weight W which acts through its mass center G. Pulley D has a negligible mass and is to be assumed frictionless. Regard b, c, d and W as known quantities, and express your answer in terms of these quantities. Determine P required for simultaneous sliding and tipping about point A assuming that the acceleration up the plane is $a = 0.2$ g. What is the corresponding coefficient of kinetic friction μ? What condition would have to be satisfied if $\mu = 0$?

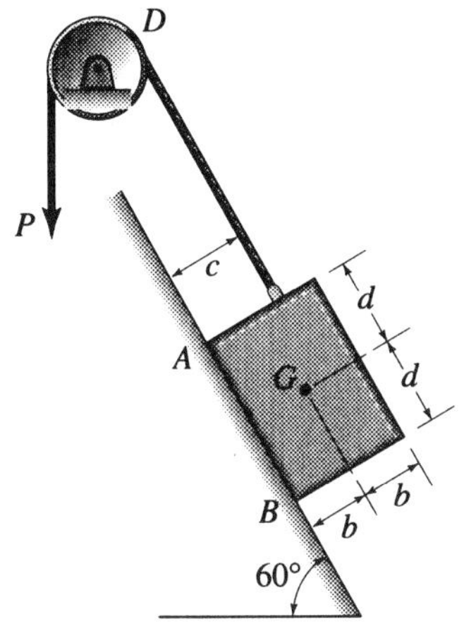

FIGURE P18.28.

18.29 Solve Problem 18.28 for P required for simultaneous sliding and tipping about point B assuming that the acceleration down the plane is $a = 0.1$g. What is the corresponding coefficient of kinetic friction μ? What condition would have to be satisfied if $\mu = 0$?

18.30 Refer to Figure P18.30 and regard W, g, b and $\mu = 0.1$ as known quantities. Body A has a weight equal to W and the pulley at B may be assumed frictionless and massless. Body A is moving to the right. Determine (a) the acceleration of body A, (b) the tension T in the cable connecting the bodies, and (c) the point where

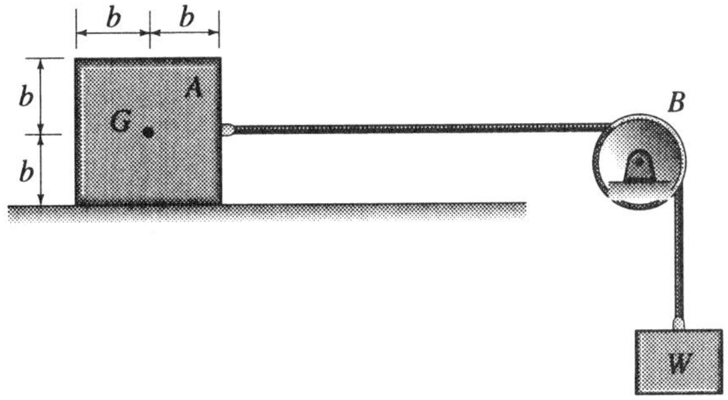

FIGURE P18.30.

the normal force exerted by the horizontal plane acts on the body A with respect to its mass center.

18.31 The system shown in Figure P18.31 has an acceleration of $g/6$ to the right when the body resting on the cart impends rotation about F. The cart has a weight of $0.2W$ where W is the weight of the body on the cart. DE represents two members which are pin connected at their ends. The cart has four wheels and you may ignore their mass and assume frictionless conditions. Determine the tension in each of the two members DE and the reaction components at F. Regard W, b and g as known quantities, and express the answers in terms of these quantities.

18.32 Determine the weight of body E such that the 1600 N sliding door of Figure P18.32 will have an acceleration of 0.5 m/s^2. Determine also the normal reactions at rollers A and B. Neglect the mass of the rollers at A and B and friction acting on them. The pulley D is also assumed massless and frictionless.

18.33 Solve Problem 18.32 except that frictional forces are to be considered acting at rollers A and B. Each of these friction forces is equal to 0.1 the normal forces acting at these rollers.

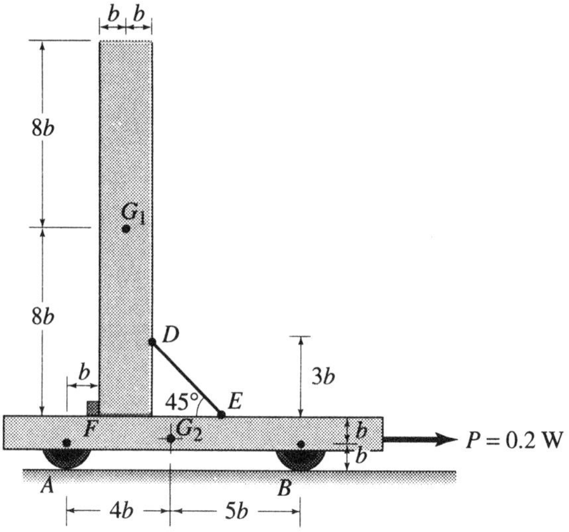

FIGURE P18.31.

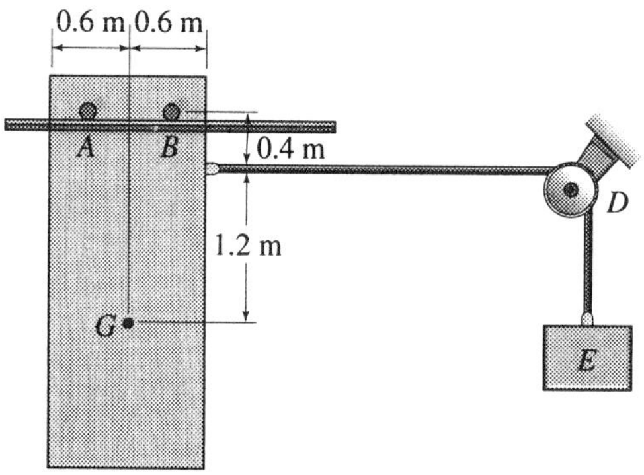

FIGURE P18.32.

18.34 A magnetically powered vehicle is shown in Figure P18.34. During the accelerated phase, equal constant horizontal forces are applied at A and B such that the vehicle accelerates from rest to a velocity of 176 ft/s in a distance of 300 ft. Determine these horizontal forces and the vertical forces at A and B. The vehicle and passengers weigh 20,000 lb. What is the acceleration of this vehicle?

18.35 The small airplane of weight W shown in Figure P18.35 has a velocity v_0 at *touch down* and is to be brought to rest in a distance s by a horizontal force H at each of the two front wheels B. Regard v_0, s, b, c, d, and W as known quantities. Determine the deceleration of this airplane, the force H and the normal forces at the two massless wheels at B and at the massless single rear wheel

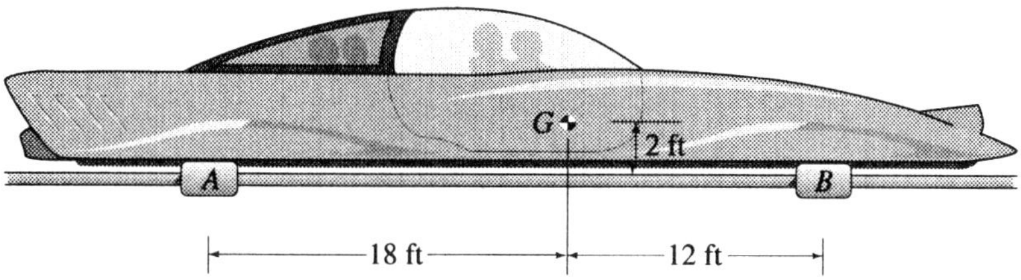

FIGURE P18.34.

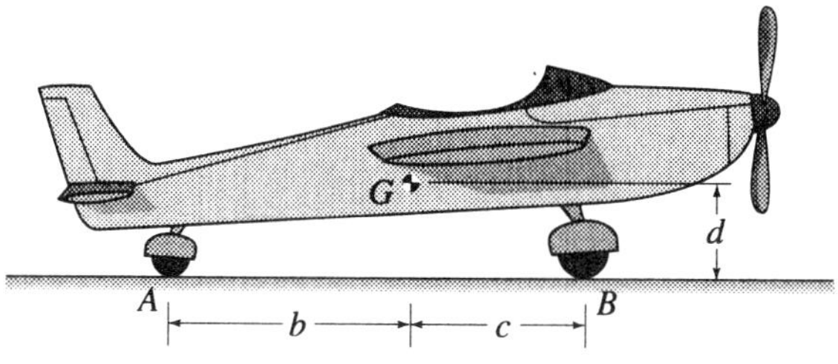

FIGURE P18.35.

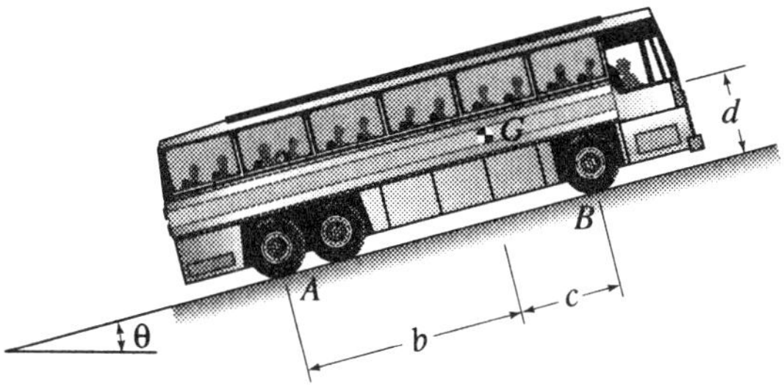

FIGURE P18.36.

at A. Express your answers in terms of the known quantities. Assume that all wheels roll without sliding.

18.36 A bus of weight W is depicted in Figure P18.36. If the maximum friction driving force at each of the rear wheels is denoted by Q and if only normal forces act at the front wheels, determine its maximum uphill acceleration and the normal reactions at A and B. Express your answers in terms of the known quantities b, c, d, Q, θ, W and g. Neglect the mass of the wheels.

18.37 A crate weighing 10,000 lb is being raised by a variable speed motor as shown in Figure P18.37. Rods AB and DE and pulley F are to be assumed massless and frictionless. The cable connecting the crate and the motor is to be assumed inextensible and massless. The position function for the rods is given by $\theta = (\pi/3)t^2$. When $t = 1$ s, find the forces exerted by the rods and the cable on the crate.

18.38 Refer to Figure P18.38 and determine d as a function of the coefficient of kinetic

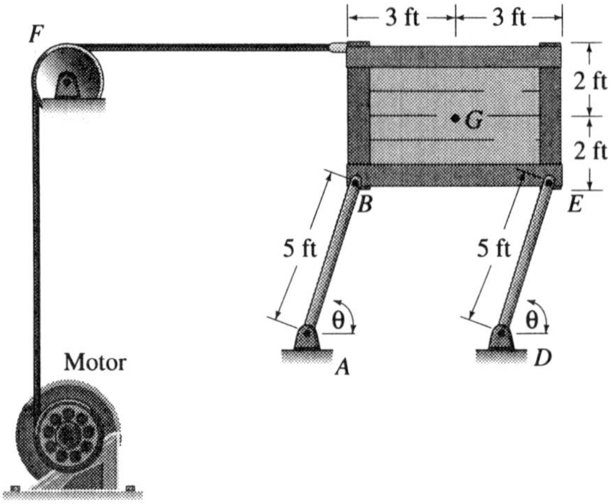

FIGURE P18.37.

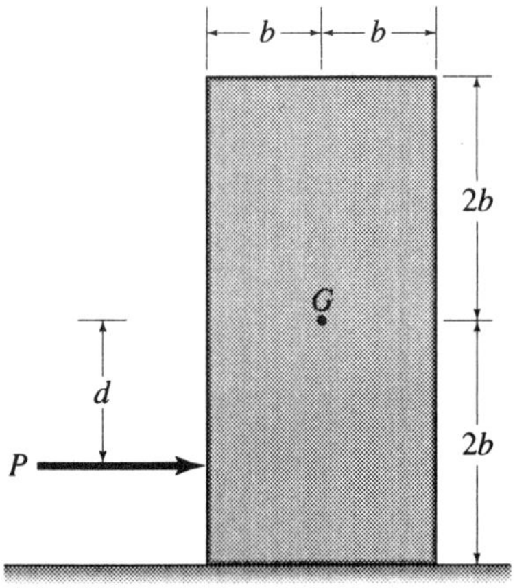

FIGURE P18.38.

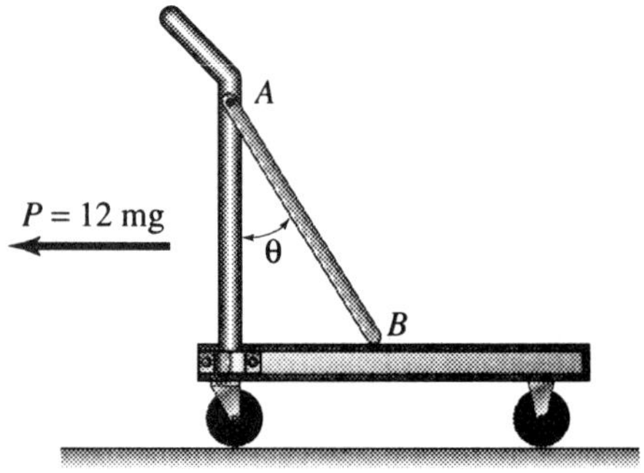

FIGURE P18.39.

friction μ and the dimension b such that the normal force N exerted on the block has an action line through point G. The block has a weight W and an acceleration $a = g$ directed to the right.

18.39 Rod AB has a mass m and length L and is pinned at A to the cart shown in Figure P18.39. The cart has a mass of 5 m and massless wheels whose frictional resistance may be ignored. A force $P = 12mg$ is applied as shown. Determine the angle θ such that the force exerted on the rod at B vanishes. Determine the pin reaction components acting on the rod at A.

18.40 Refer to Problem 18.39, reverse the sense of $P = 12mg$ such that it acts to the right and assign θ a value of 30°. Find the pin reaction components acting on the rod at A and the force exerted on the rod at B. Neglect friction in the system.

18.41 An equilateral triangular plate weighing 1000 N is supported by pin-ended

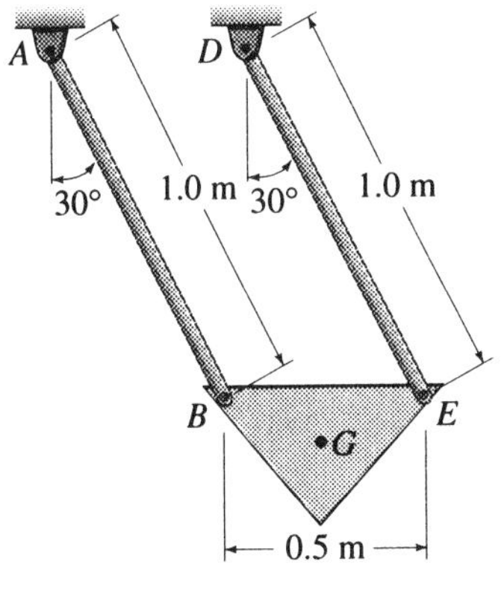

FIGURE P18.41.

weightless rods AB and DE as shown in Figure P18.41. In the position shown, the rods have an angular velocity of 0.2 rad/s cw. Determine the acceleration of the mass center of the plate at this instant and the forces in each of the two rods.

18.42 Two steel cylinders to be drop-forged into armatures, each weighing 1200 lb, are to be transported on a truck as shown in Figure P18.42. On level terrain, the truck acceleration is limited to $\pm g/6$. The cylinders have radii of 1 ft.

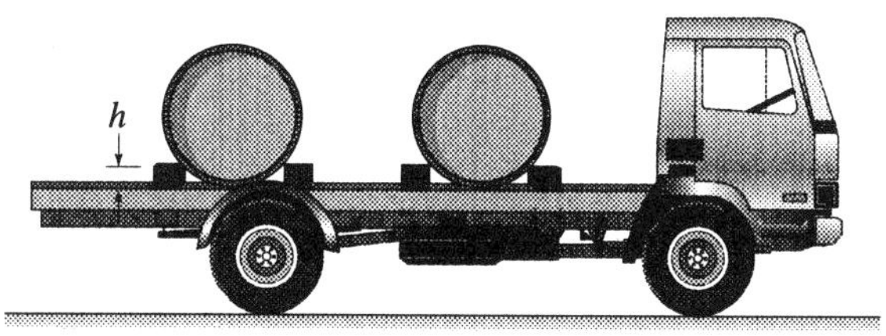

FIGURE P18.42.

Determine the height h of the blocks required to prevent the cylinders from rolling off the truck or rolling forward toward the cab.

18.43 In a highly automated warehouse, a large crate with contents assumed homogeneous, moves a short distance on an electrically powered vehicle, shown in Figure P18.43. The rack is fixed, and there are four gears which propel the vehicle by electric power delivered to motors through a center rail (not shown). The fixed rack delivers a force equal to $mg/8$ acting to the right at the base of each of the four gears. Determine the acceleration of the system and the normal forces at each of the four gears exerted by the rack. Determine also the normal and frictional forces acting on the crate and their position relative to G_1. The vehicle weight is mg and the crate weight is 2 mg. Neglect the mass of the gears.

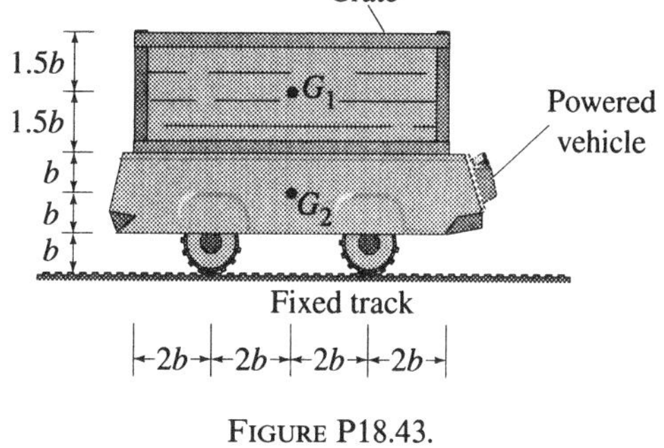

FIGURE P18.43.

18.44 A thin uniform plate of weight W rests on a frictionless platform as shown in

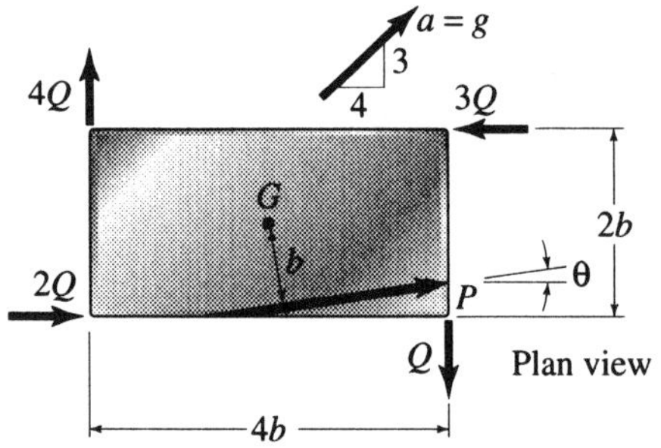

FIGURE P18.44.

plan view in Figure P18.44. An additional force P is to supplement those shown such that the plate will translate with an acceleration $a = g$ directed as shown. Determine the magnitude of the additional force as a multiple of Q and find its action line with respect to the mass center G of the plate.

18.45 A thin uniform triangular plate of weight W rests on a frictionless platform as shown in plan view in Figure P18.45. Determine the magnitude of force P as a multiple of Q such that the plate will translate on the platform. Determine the acceleration of the plate in magnitude and direction, and express this magnitude in terms of W, Q, and g. Find the numerical value of the acceleration for the case where $Q = 0.5$ W and $g = 9.81$ m/s^2.

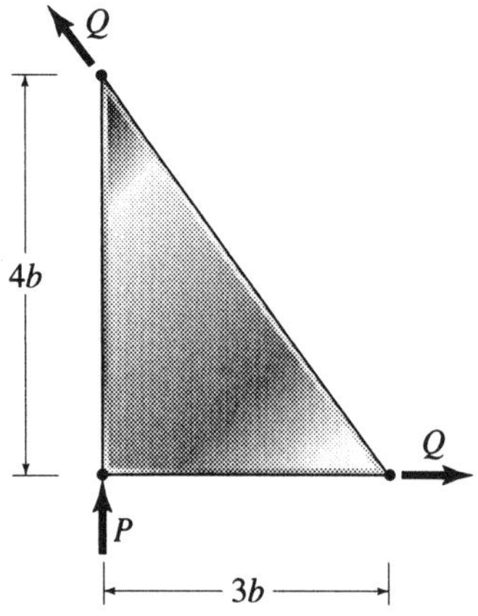

FIGURE P18.45.

18.46 A uniform circular plate of weight W rests on a frictionless table and is subjected to four forces each of magnitude $Q = 2W$ as shown in plan view in Figure P18.46. Determine the magnitude and direction of the acceleration of the plate.

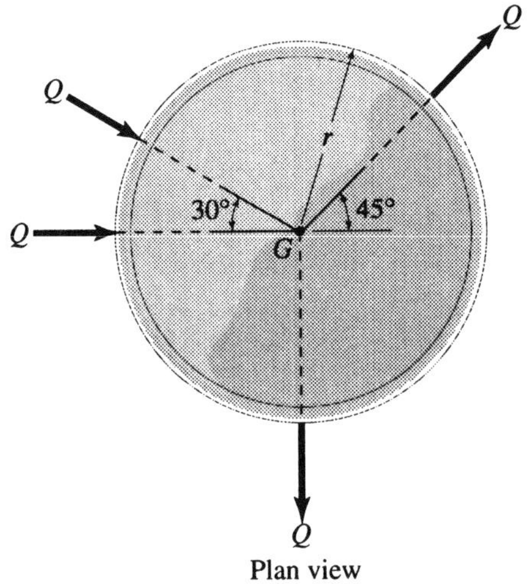

FIGURE P18.46.

18.47 A homogeneous cylinder of radius r and weight W, as shown in Figure P18.47, rests on a plane for which the coefficient of friction is $\mu = 0.1$. If $P = 0.4W$, determine the distance d as a fraction of r such that the cylinder will slide in a translatory motion along the plane. What is the acceleration of the cylinder expressed as a fraction of the acceleration due to gravity g?

18.48 A pendulum of mass m and length L is mounted in a frame of mass $2m$, as shown in Figure P18.48. The assembly

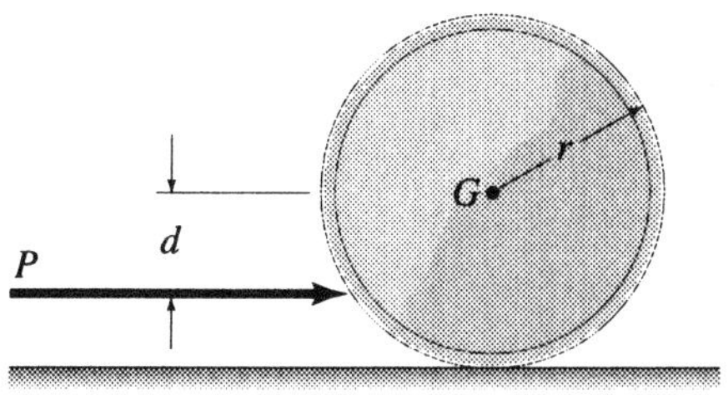

FIGURE P18.47.

moves on a frictionless horizontal plane when subjected to the force $F = 2mg$. Determine the acceleration of the frame and the angle θ as well as the cable tension. Express answers in terms of m and g.

18.49 A semicylindrical body of weight W is accelerated along a frictionless, inclined plane, as shown in Figure P18.49. Determine the distance d as a function of r and θ such that the normal force exerted by the plane on the body passes through point A. What is the acceleration of body expressed in terms of g and θ?

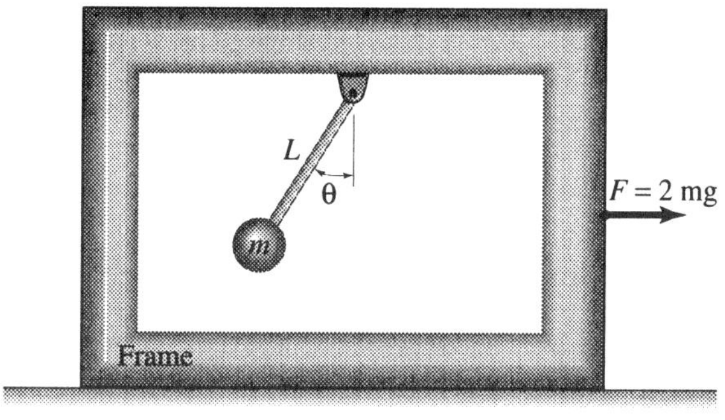

FIGURE P18.48.

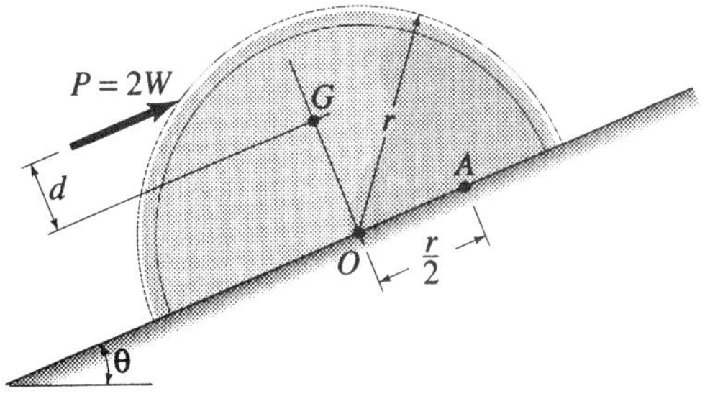

FIGURE P18.49.

18.50 A triangular, homogeneous, prismatic body of mass m translates on a frictionless inclined plane, as shown in Figure P18.50. Determine the force P as a function of m, g and θ such that the normal force exerted on the body by the plane acts through point A. What is the corresponding acceleration a of this body as a function of g and θ? Find P and a for the case when $\theta = 0°$ and $\theta = 90°$.

18.51 A uniform, thin, circular rod of mass m and radius r is attached to a cart of mass $2m$ as shown in Figure P18.51. Ignore friction at the cart's massless wheels, and determine the acceleration of the system due to the force $F = 0.5mg$. Also find the components of the pin reaction at A as well as the force exerted on the rod at B. Ignore friction at the pin and at B. Express the answers in terms of m and g where g is the acceleration due to gravity.

18.52 Refer to Problem 18.51, and determine

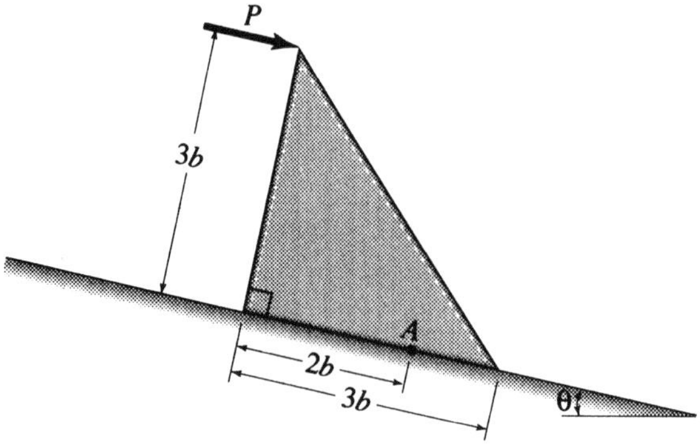

FIGURE P18.50.

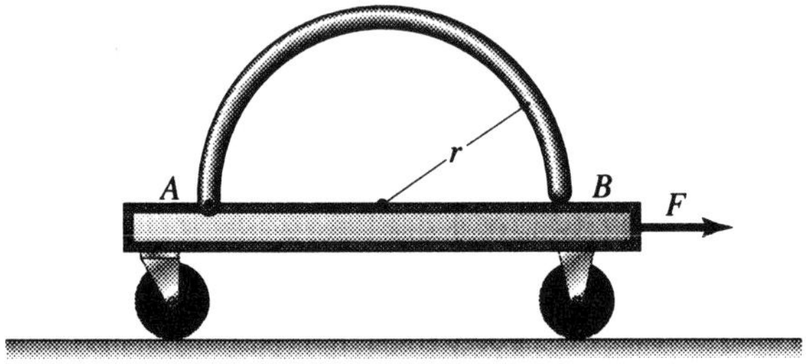

FIGURE P18.51.

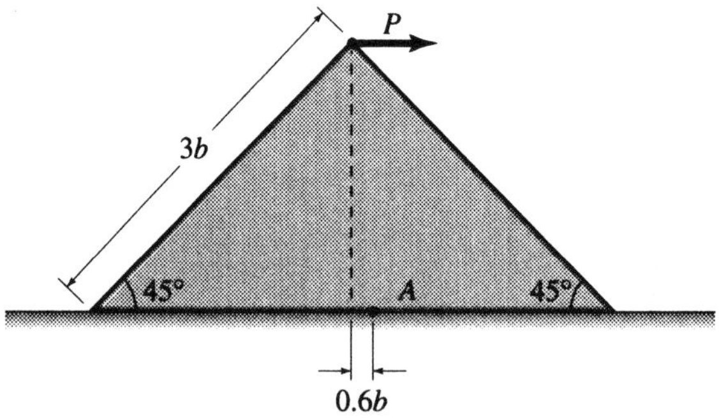

FIGURE P18.53.

the acceleration of the system such that the force acting on the rod at B will vanish. What is the corresponding value of the force F?

18.53 A triangular, prismatic, homogeneous plate of weight W is shown in Figure P18.53. It slides along a frictionless horizontal plane under the action of the force P. Determine the force P and the corresponding acceleration such that the normal force exerted by the plane on the plate acts through point A. Express the answers in terms of W and g where g is the acceleration due to gravity.

18.54 A heavy tractor trailer shown in Figure P18.54 supports crated machinery which weighs 74,000 lb. If the coefficient of friction between the crate and trailer bed is $\mu = 0.9$, determine whether the crate will slide or tip when the truck is accelerated. What is the maximum acceleration? If the gross weight of the tractor and its load is 100,000 lb, what total force must be exerted by friction at the driving wheels for the system to have this maximum acceleration? Neglect the mass of the wheels.

18.55 Automobile manufacturers perform wind tunnel tests to determine resisting forces applied by the air to their vehicles. In Figure P18.55, smoke jets reveal the airflow pattern over an automobile. The basic equation used to determine the drag force is $F_D = \dfrac{C_D A \rho v^2}{2}$ where $F_D =$ drag force in lb, $C_D =$ drag coefficient, $A =$ projected area of the vehicle in ft^2, $\rho =$ mass density of air in slug/ft^3, and $v =$ speed of the air with respect to the vehicle in ft/s. (a) For a given automobile model $C_D = 0.25$ (a very low value, which is desirable), $A = 24.3$ ft^2, $\rho = 0.00237$ slugs/ft^3 (air at 60° F), $v = 60$ mph. Determine the drag force acting on this model. (b) Find the power, measured in hp, required to overcome air resistance at this speed. (Hint $P = F_D v$). (c) If the power capacity of the automobile is rated at 150 hp, determine its maximum acceleration.

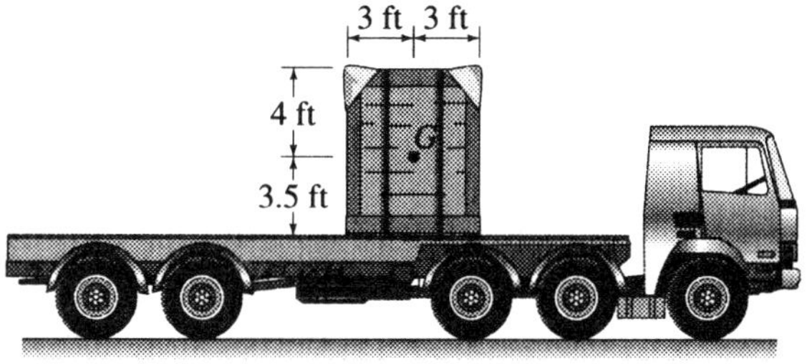

FIGURE P18.54.

FIGURE P18.55.

18.56 The locomotive shown in Figure P18.56 weighs 130 tons. It is traveling at 60 mph up a 4% grade when the brakes are suddenly applied and the locomotive is brought to a complete stop in 15 s. Find the friction force at the massless wheels during the stopping period.

18.57 A crash test is to be conducted on an

FIGURE P18.56.

automobile of mass m, as shown in Figure P18.57. Just before striking the rigid barrier, the automobile has a speed v_0. If the force exerted on the vehicle bumper is given by $F = kx$, where k is the bumper spring constant and x is the vehicle deformation, find (a) the maximum deformation of the front of the vehicle and (b) the maximum deceleration of the automobile. Express answers in terms of m, k, and v_0.

FIGURE P18.57.

18.58 A welded beam assembly shown in Figure P18.58 is to be lifted such that it translates with an upward acceleration of 0.1 g. If the vertical beam weighs 170 lb/ft and the horizontal beam weights 72 lb/ft, determine the lift cable tensions T_1 and T_2.

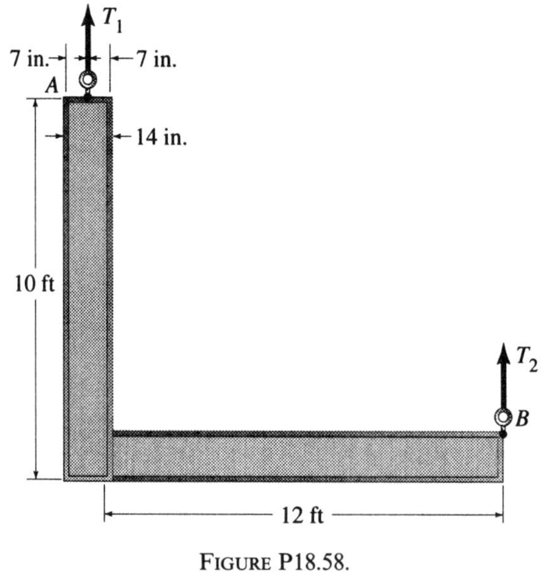

FIGURE P18.58.

18.59 A foundation grillage shown in plan view in Figure P18.59 consists of five beams which are welded together to form a single unit. Special foundation conditions have required the use of beams with weights per foot as given on this plan view. The grillage is to be lifted

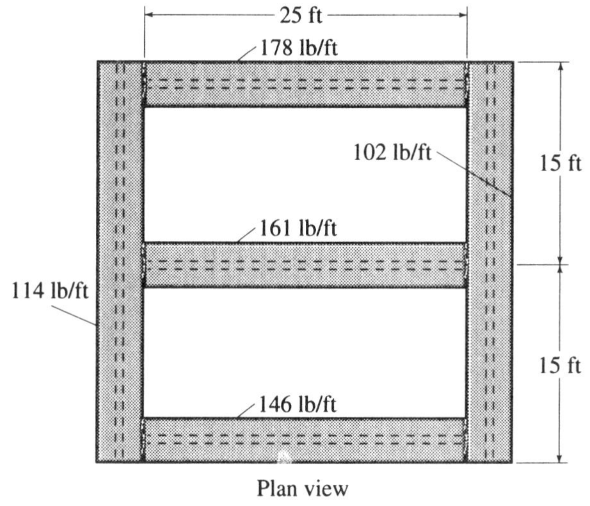

FIGURE P18.59.

by a crane and then lowered into position. Each corner of the grillage is attached to an individual cable. A main lifting cable, which is oriented vertically, is attached to the four individual cables which meet above the mass center of the grillage. (a) If, during lifting, the acceleration is limited to $0.1g$, what is the force in the main cable? (b) If, when the grillage is lowered, the acceleration is to be limited to $0.05g$, what is the force in the main cable?

18.60 A homogeneous block of mass m rests on a horizontal plane and is held against a detached spring which is compressed by an amount x_0, as shown in Figure P18.60. The coefficient of friction between the block and the horizontal plane is μ. (a) Determine an expression for the acceleration of the block, while the spring acts on it. (b) Locate the action line of the normal force N exerted on the block with respect to its mass center. (c) Find the velocity of the block at the instant when the spring reaches zero deformation. (d) Find the distance through which the block moves before coming to rest. Express answers in terms of the mass m, the spring constant k, the coefficient of friction μ, the initial spring deformation x_0, and the acceleration of gravity g. (e) Provide numerical answers for parts (a) through (d) if $\mu = 0.1$, $b = 0.5$ ft, $k = 60$ lb/ft, $m = 1.00$ slug and $x_0 = 1$ ft.

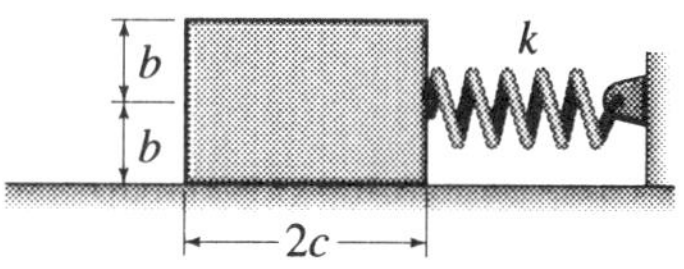

FIGURE P18.60.

18.61 An automobile of weight $W = 4000$ lb has a front-wheel drive. (a) If the coefficient of friction between the pavement and the tires is 0.7, determine the maximum acceleration of the automobile when traveling on a level pavement. (b) Find the stopping distance after the brakes are fully applied, if the automobile is traveling at 60 mph.

18.4 Rotation about a Fixed Axis

A rigid body in rotation about a fixed axis is depicted in Figure 18.12. This body is symmetric with respect to the plane of motion which is the plane in which the mass center G moves. We imagine the plane of motion to be the plane of the page. For every element of mass in front of the plane of motion there is a corresponding element of mass behind the plane of motion. The axis of rotation through point O is perpendicular to the plane of motion and this axis appears as a point in the diagram. The mass center G travels a circular path of radius $OG = r_G$. Figure 18.12 also shows the kinematics of the motion of the mass center G. The velocity of G is $v_G = r_G\omega$ and this vector is directed perpendicular to line OG and consistent with the sense of ω. The acceleration of G is shown in tangential and normal components. The tangential acceleration component of G is $(a_G)_t = r_G\alpha$, and this vector is directed perpendicular to OG and consistent with the sense of α. The normal acceleration component of G is $(a_G)_n = r_G\omega^2$ and is directed from G toward O regardless of the sense of ω. The senses of the acceleration components of G are used to establish positive senses for the tangential and normal axes with origin at G as shown in Figure 18.12.

The general Eqs. (18.10) and (18.11) in two-dimensional form are basic to our development of equations for the special case of rotation about a fixed axis. Consider the right-hand side of Eq. (18.10) which is the inertial force ma_G. The tangential component of this force is the product of the mass of the body and the mass center's tangential acceleration component, namely $mr_G\alpha$. The normal component of this force is the product of the mass of the body and the mass center's normal acceleration component, namely $mr_G\omega^2$.

Consider Eq. (18.11) written for the fixed axis of rotation through point O. Thus,

$$\sum \mathbf{M}_O = \frac{d\mathbf{H}_O}{dt}. \tag{18.17}$$

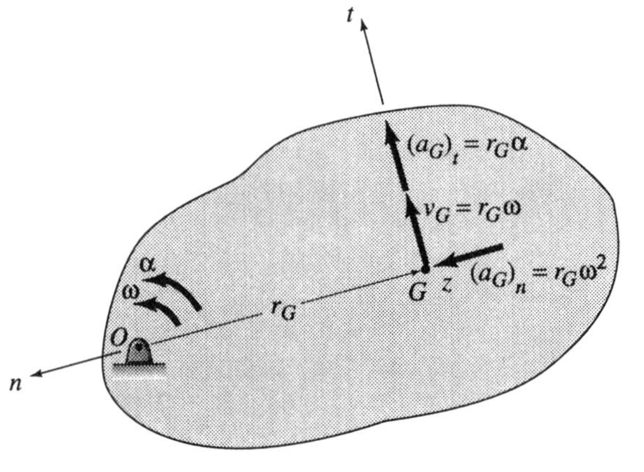

FIGURE 18.12.

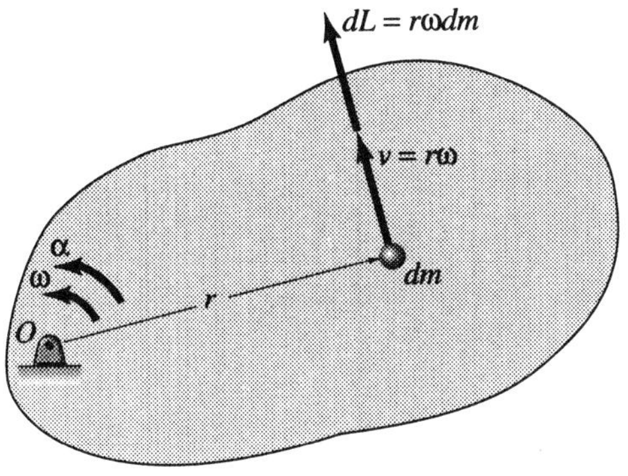

FIGURE 18.13.

To deal with the magnitude of the right-hand side of this equation, we need to refer to Figure 18.13 and develop an equation for H_O. The differential mass dm has a speed $v = r\omega$ where ω is the instantaneous angular velocity of the body and r is measured perpendicular to the axis of rotation through point O. The differential linear momentum dL is given by

$$dL = r\omega\, dm.$$

The differential angular momentum dH_O equals the moment of dL with respect to the axis through point O. Thus,

$$dH_O = r^2\omega\, dm.$$

Integrating both sides of this equation over the mass m and recalling that ω has a single value for a rigid body at a given instant,

$$H_O = \omega \int r^2\, dm.$$

We recognize the integral on the right hand side of this equation as I_O, the mass moment of inertia of the body with respect to the axis of rotation through point O, and write

$$H_O = I_O\omega. \tag{18.18}$$

Differentiating Eq. (18.18) with respect to time and recalling that I_O is time-independent gives

$$\frac{dH_O}{dt} = I_O\frac{d\omega}{dt}.$$

But $\frac{d\omega}{dt} = \alpha$, and using Eq. (18.17), we conclude that

$$\sum M_O = I_O\alpha. \tag{18.19}$$

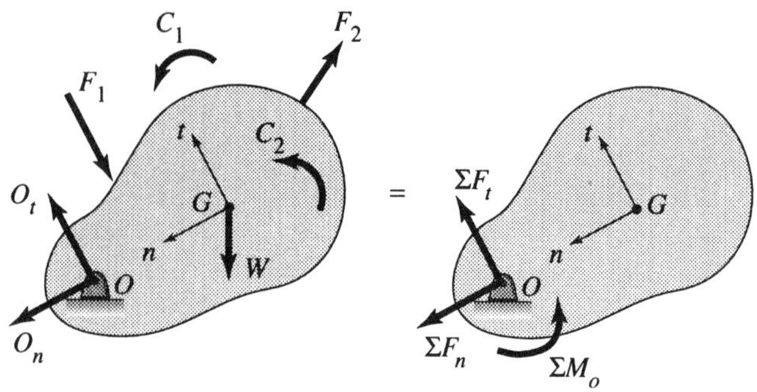

FIGURE 18.14. (a) (b)

Refer to Figure 18.14(a) where the weight W, the reaction components O_n and O_t at the axis of rotation, applied forces, typified by F_1 and F_2, and applied couples, typified by C_1 and C_2, are shown acting on the body. Usually, we would know the applied forces and couples. In Section 5.8 we learned that any given system of forces and couples may be reduced to a single force and a single couple at an arbitrary point. We choose the arbitrary point as the axis of rotation at O and in Figure 18.14(b) show the resultant force in terms of its components $\sum F_n$ and $\sum F_t$ and the resultant couple $\sum M_O$.

The two-dimensional form of the general equations (18.10) and (18.19) are depicted pictorially for a rotating rigid body in Figure 18.15, which shows the free-body diagram on the left and the corresponding inertial force diagram on the right. Also, a right-handed t-n-z coordinate system is shown where the z axis comes out of the page. The sum of the forces in the t and n directions yield, respectively, $\sum F_t = mr_G\alpha$ and $\sum F_n = mr_G\omega^2$. The sum of the moments about the axis of rotation at O yields $\sum M_O = I_G\alpha + (mr_G\alpha)r_G = (I_G + mr_G^2)\alpha$. By the parallel axis theorem, $I_G + mr_G^2 = I_O$. Therefore, $\sum M_O = I_O\alpha$.

Thus, the kinetic equations corresponding to rotation about a fixed axis through an arbitrary point O are

$$\sum F_n = mr_G\omega^2 \tag{18.20a}$$

$$\sum F_t = mr_G\alpha \tag{18.20b}$$

$$\sum M_O = I_O\alpha = \sum (M_i)_O \tag{18.20c}$$

where $\sum (M_i)_O$ represents the sum of the moments of the inertial forces about the axis of rotation.

An alternate formulation of these equations is possible by considering an axis through the mass center G, instead of point O, as shown in Figure 18.16. The corresponding kinetic equations become,

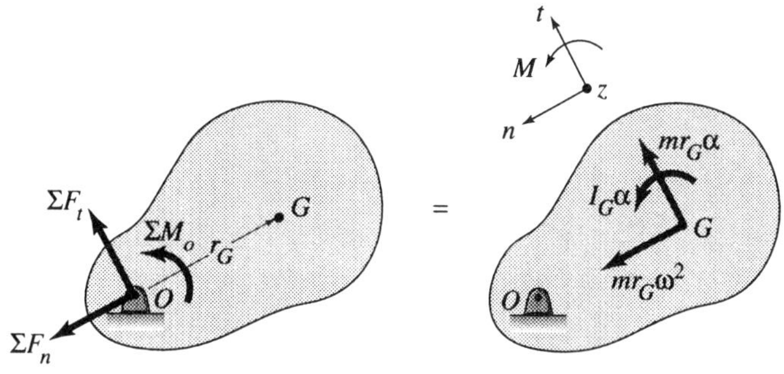

FIGURE 18.15.

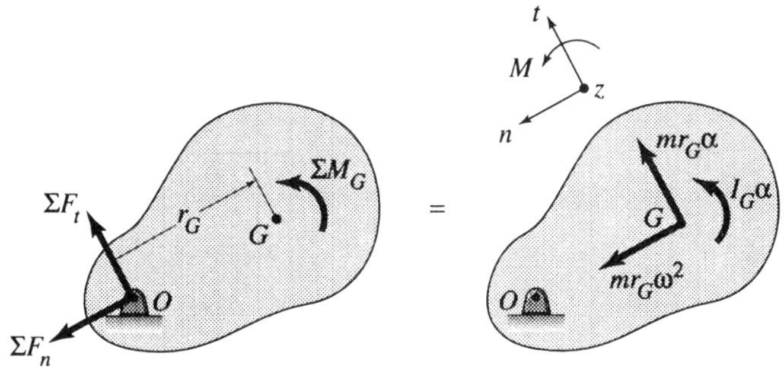

FIGURE 18.16.

$$\sum F_n = mr_G\omega^2 \tag{18.21a}$$

$$\sum F_t = mr_G\alpha \tag{18.21b}$$

$$\sum M_G = I_G\alpha \tag{18.21c}$$

We observe that the Eqs. (18.20a) and (18.20b) are identical to Eqs. (18.21a) and (18.21b) and that only the Eqs. (18.20c) and (18.21c) are different. Either set of equations is valid. Equations (18.20) are usually preferred when compared to Eqs. (18.21) because the pin reaction components do not appear in the former equations.

If a body rotates about an axis through its mass center G, we refer to this as *centroidal rotation.* This special case of motion will be studied by modifying Eqs. (18.21). Because the axis of rotation now coincides with the mass center, then OG $= r_G$ will vanish and the right-hand side of the Eqs. (18.21a) and (18.21b) reduce to zero. In other words, these two equations become statics equations, and we may replace the n and t axes by x and y axes and arbitrarily orient them horizontally and vertically as shown in Figure 18.17. The equations for this special case of centroidal rotation become

$$\sum F_x = 0 \tag{18.22a}$$

$$\sum F_y = 0 \tag{18.22b}$$

$$\sum M_G = I_G\alpha \tag{18.22c}$$

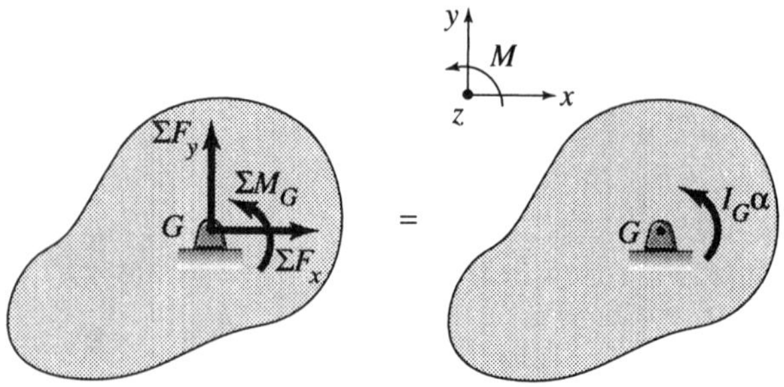

FIGURE 18.17.

These equations are shown pictorially in Figure 18.17 where the free-body diagram is shown on the left and the inertia force diagram on the right of the equal signs. In this case the mass center G is initially at rest and remains at rest which means that it is not accelerated. This in turn implies that the inertia forces reduce to zero but the inertia couple $I_G\alpha$, in general, will have a non-zero value.

The following examples illustrate some of the above concepts.

Example 18.9

A body consists of two homogeneous, slender rods rigidly connected at their centers and is supported by a frictionless pin as shown in Figure E18.9. Each rod is of mass m and length L. This body is subjected to a ccw constant couple $C = 4\,mgL$. Determine the angular acceleration of this body and the horizontal and vertical pin reaction components which support it at the mass center G. Neglect air resistance to the motion.

Solution

We observe that the motion may be classified as centroidal rotation and begin the solution by constructing the free-body diagram and the inertial force diagram of Figure E18.9(b) which also shows a convenient coordinate system.

The forces shown on the free-body diagram consist of the weight $W = 2mg$ and the pin reaction components G_x and G_y. Since the mass center G is pinned and remains at rest, it has zero acceleration components. Therefore, the linear inertial forces $2m(a_G)_x$ and $2m(a_G)_y$ both reduce to zero. The inertial forces reduce to the inertial couple $I_G\alpha$. Because the applied couple C acts in a ccw sense, the angular acceleration and the inertial couple will be directed in a ccw sense also. Thus, using Eqs. (18.22),

$$\sum F_x = m(a_G)_x = 0, \quad G_x = 0, \qquad \text{ANS.}$$

$$\sum F_y = m(a_G)_y = 0, \quad G_y - 2mg = 0,$$

$$G_y = 2mg; \qquad \text{ANS.}$$

$$\sum M_G = I_G\alpha, \quad 4mgL = \tfrac{1}{6}mL^2\alpha.$$

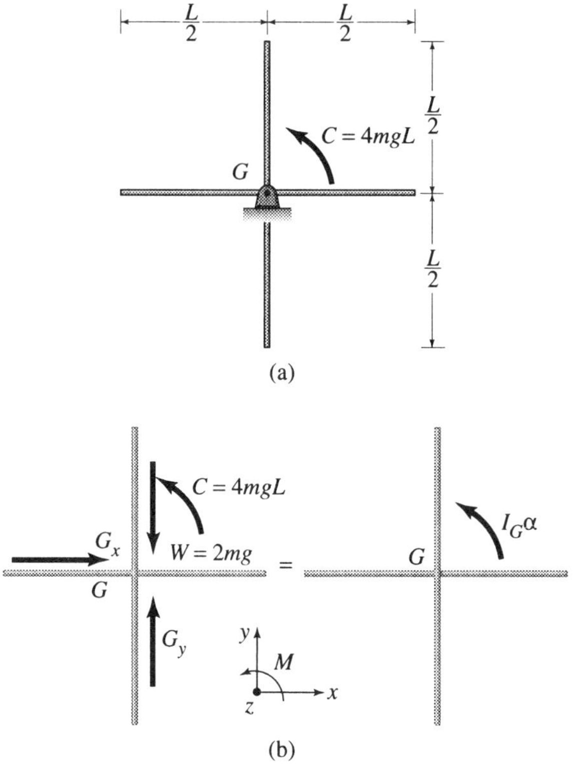

FIGURE E18.9.

where $I_G = \frac{1}{6}mL^2$ was determined using the information in Appendix B. Thus,

$$\alpha = \frac{24g}{L} \circlearrowleft. \qquad \text{ANS.}$$

As long as the applied constant couple acts, the angular acceleration has this constant value, and the pin reactions need support only the weight of the body since its mass center remains at rest.

■ Example 18.10

A rigid body consisting of two homogeneous, slender rods rigidly connected at their centers and supported by a frictionless pin is shown in Figure E18.10(a). This is the same body of Example 18.9 except that the point of suspension has been shifted from the mass center G to point O. Each rod has a mass m and a length L. At the instant shown, the body has an angular velocity ω and an angular acceleration α as indicated. Determine, for the position shown, the angular acceleration of this body and the normal and tangential pin reaction components.

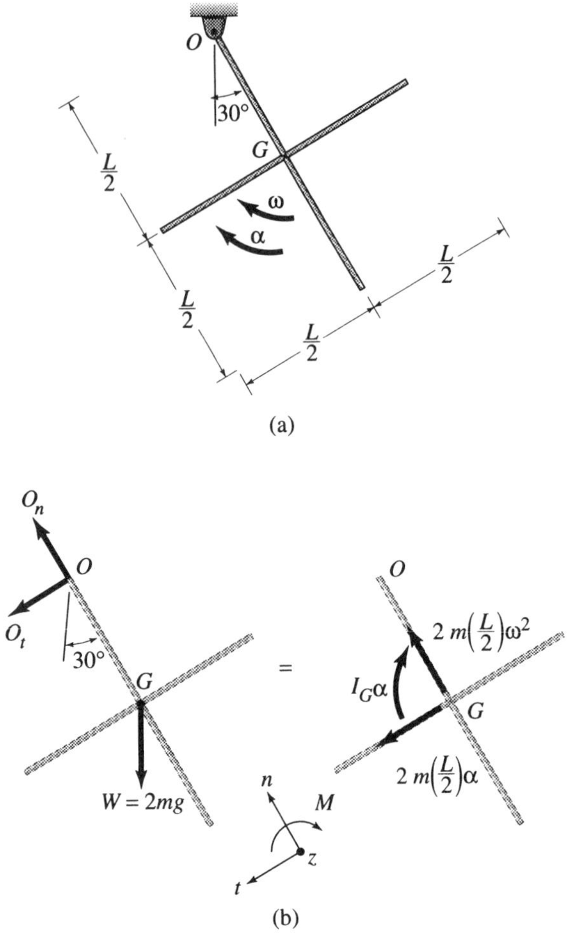

FIGURE E18.10.

Solution

Note that the mass center G moves along a circular path of radius $L/2$ which is centered at O. At the instant shown, the body has a cw angular velocity ω and a cw angular acceleration α. The mass center has acceleration components given by

$$(a_G)_n = \frac{L}{2}\omega^2 \quad \text{and} \quad (a_G)_t = \frac{L}{2}\alpha.$$

The normal component $(a_G)_n$ is always directed toward the center of curvature of the path (i.e. point O) and the tangential component $(a_G)_t$ is directed down to the left which is consistent with the cw sense of α. The sense of these acceleration components establishes the sign convention for forces, and the sense of the inertia forces must agree with this convention.

The free-body and inertial force diagrams are shown in Figure E18.10(b) along with a right-handed t-n-z coordinate system where the z axis goes into the page. Thus, using Eq. (18.20c),

$$\sum M_O = \sum (M_i)_O, \quad (2\text{ mg} \sin 30°)\frac{L}{2} = I_G\alpha + mL\alpha\left(\frac{L}{2}\right),$$

$$= (I_G + mL^2/2)\alpha,$$

$$= I_O\alpha,$$

$$= (\tfrac{2}{3}mL^2)\alpha,$$

where $I_O = \frac{2}{3}mL^2$ was determined using the information in Appendix B. Thus,

$$\alpha = \frac{3}{4}\frac{g}{L} \circlearrowright. \qquad \text{ANS.}$$

Also,

$$\sum F_t = mr_G\alpha, \quad 2mg \sin 30° + O_t = 2m\left(\frac{L}{2}\right)\alpha$$

Substituting $\alpha = \dfrac{3g}{4L}$ and solving for O_t gives

$$O_t = -\frac{mg}{4}. \qquad \text{ANS.}$$

The negative sign means that the tangential component of the pin reaction acts up to the right rather than down to left, as shown in Figure E18.10(b).

$$\sum F_n = 2m\left(\frac{L}{2}\right)\omega^2{:} \quad O_n - 2mg\cos 30° = 2m\left(\frac{L}{2}\right)\omega^2$$

$$O_n = \sqrt{3}mg + mL\omega^2. \qquad \text{ANS.}$$

Note that ω needs to be known before O_n can be fully determined and that the positive sign of O_n implies that the assumed sense shown in Figure E18.10(b) is correct.

■ Example 18.11

Two slender rods of length L and two spheres of radius R are fastened together to form the composite body shown in Figure E18.11(a). All components are homogeneous, each rod has a mass m, and each sphere has a mass 5m. The system is released from rest when $\theta = 30°$. (a) Determine the angular acceleration α and the normal and tangential pin reaction components at the axis of rotation O immediately after

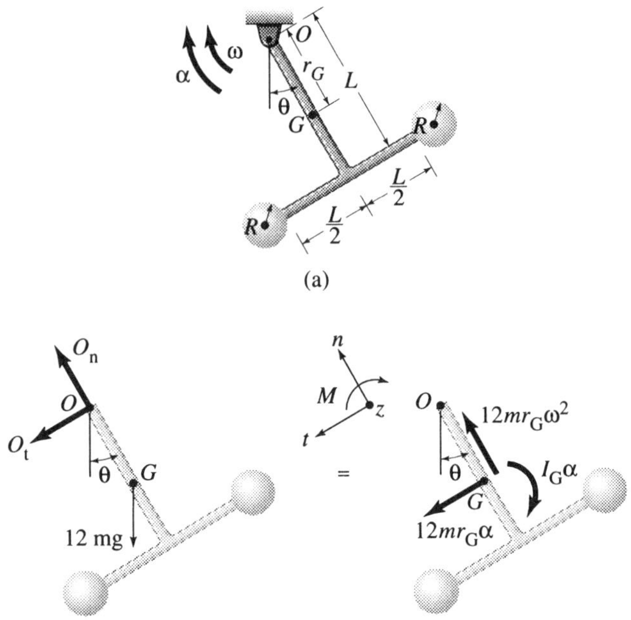

FIGURE E18.11.

release. Regard m, L, and R as known quantities, and use the particular ratio of $R/L = 1/4$. (b) For $m = 2$ kg and $L = 0.80$ m, determine numerical values for α and for the reaction components at O.

Solution

(a) Let us locate the mass center G of this composite body by referring to Figure E18.1(a). By symmetry, we conclude that G will lie as indicated on the center line of the upper rod. Thus,

$$r_G = \frac{m\dfrac{L}{2} + mL + 5mL + 5mL}{m + m + 5m + 5m} = 0.958L.$$

Using the information in Appendix B,

$$I_G = \frac{1}{12}mL^2 + m(0.958L - 0.500L)^2 + \frac{1}{12}mL^2 + m(L - 0.958L)^2$$
$$+ 2\left(\frac{2}{5}\right)(5m)\left(\frac{L}{4}\right)^2 + 2(5m)\left[\left(\frac{L}{2} + \frac{L}{4}\right)^2 + (L - 0.958L)^2\right]$$
$$= 6.27\ mL^2.$$

The free-body and inertial force diagrams are shown in Figure E18.11(b) along with a right-handed t-n-z coordinate system where the

z axis goes into the page. Thus, by Eq. (18.20c),

$$\sum M_O = \sum (M_i)_O, \quad (12\text{ mg} \sin 30^\circ)(0.958L) = I_G\alpha + (12mr_G\alpha)r_G.$$

Substituting for r_G and I_G,

$$(12\text{ mg} \sin 30^\circ)(0.958L) = 6.27\, mL^2\alpha + 12\, m(0.958L)^2\alpha$$

from which

$$\alpha = 0.347\left(\frac{g}{L}\right), \qquad \text{ANS.}$$

$$\sum F_t = mr_G\alpha, \quad O_t + 12\text{ mg} \sin 30^\circ = 12\, mr_G\alpha.$$

Substituting for r_G and for α,

$$O_t = -2.01\, mg = 2.01\, mg\ \nearrow, \qquad \text{ANS.}$$

$$\sum F_n = mr_G\omega^2, \quad O_n - 12\text{ mg} \cos 30^\circ = 0$$

from which

$$O_n = 10.39\text{ mg}\ \nwarrow. \qquad \text{ANS.}$$

(b) Substituting the given numerical values,

$$\alpha = 4.26\text{ rad/s}^2\ \circlearrowleft,$$

$$O_t = 39.4\text{ N}\ \nearrow, \qquad \text{ANS.}$$

and

$$O_n = 204\text{ N}\ \nwarrow.$$

■ Example 18.12

A homogeneous rod of length L and mass m is suspended from a frictionless hinge at O as shown in Figure E18.12(a). If displaced from its vertical equilibrium position through a small ccw angle θ and released, the rod will oscillate about this equilibrium position indefinitely as long as friction and air resistance are ignored. Write the differential equation governing the oscillatory motion of the rod. Solve this differential equation for small oscillations such that $\sin\theta \approx \theta$. Assume that, when $t = 0$, $\theta = \theta_0$ and $\dot{\theta} = \dot{\theta}_0$.

Solution

The free-body and inertial force diagrams are shown in Figure E18.12(b) along with a right-handed t-n-z coordinate system where the z axis comes out of the page. By Eq. (18.20c),

$$\sum M_O = \sum (M_i)_O, \quad -(mg \sin\theta)\left(\frac{L}{2}\right) = I_G\ddot{\theta} + m\left(\frac{L}{2}\right)\ddot{\theta}\left(\frac{L}{2}\right)$$

$$= \left(\frac{1}{12}mL^2\right)\ddot{\theta} + \frac{1}{4}mL^2\ddot{\theta}$$

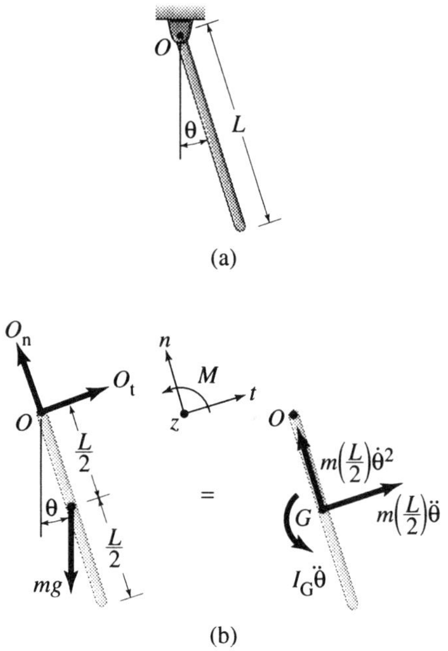

FIGURE E18.12.

where $I_G = \frac{1}{12}mL^2$ was obtained from the information in Appendix B. Simplifying and setting $\sin\theta \approx \theta$,

$$\ddot{\theta} + \frac{3}{2}\left(\frac{g}{L}\right)\theta = 0. \qquad \text{ANS.}$$

For convenience, let $\frac{3}{2}\left(\frac{g}{L}\right) = p^2$. The above differential equation becomes

$$\ddot{\theta} + p^2\theta = 0.$$

The general solution for this type of differential equation is given by

$$\theta = A\sin pt + B\cos pt \tag{a}$$

where A and B are constants that can be determined from the given initial conditions. Thus, for $t = 0$ and $\theta = \theta_O$, and Eq. (a) yields

$$\theta_O = 0 + \mathrm{B} \rightarrow \mathrm{B} = \theta_O.$$

Differentiating Eq. (a) leads to

$$\dot{\theta} = Ap\cos pt - Bp\sin pt. \tag{b}$$

For $t = 0$, $\dot{\theta} = \dot{\theta}_O$. Thus,

$$\dot{\theta}_O = Ap - 0 \Rightarrow A = \frac{\dot{\theta}_O}{p}.$$

Therefore, the solution to the differential equation of motion becomes

$$\theta = \left(\frac{\dot{\theta}_O}{p}\right)\sin pt + \theta_O \cos pt, \qquad \text{ANS.}$$

where $p = \sqrt{\frac{3}{2}\left(\frac{g}{L}\right)}$.

■

Problems

18.62 A uniform homogeneous slender rod of mass m has a length L and is hinged at O as shown in Figure P18.62. It is released from rest from the position shown. Determine the normal and tangential reaction components at O and the angular acceleration of the rod immediately after release. Express your answers in terms of m, L and g.

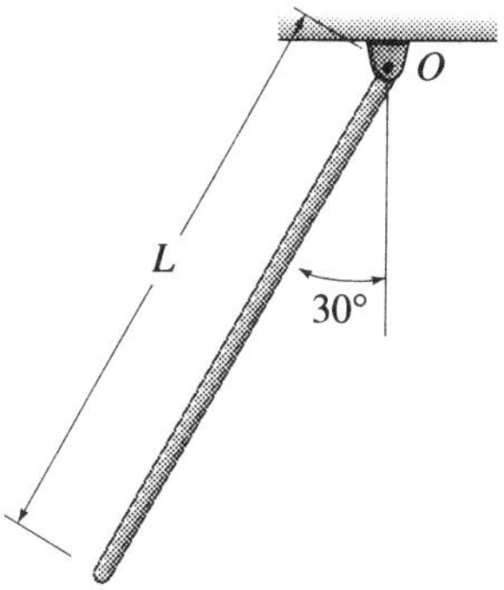

FIGURE P18.62.

18.63 The uniform homogeneous disk shown in Figure P18.63 has a mass m and a radius r. At the instant depicted in Figure P18.63, the disk has a cw angular velocity ω. Determine at this instant (a) the angular acceleration of the disk,

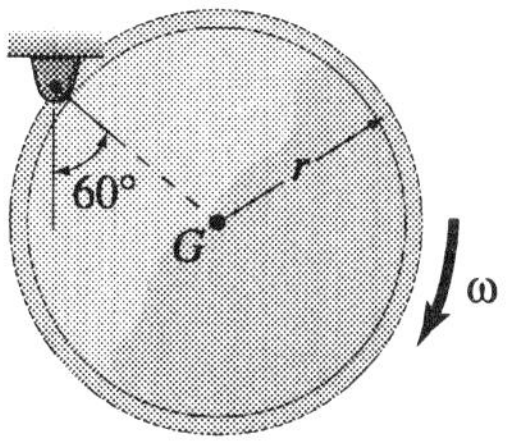

FIGURE P18.63.

and (b) the magnitude of the pin reaction at the axis of rotation O. Express your answers in terms of m r, ω and g.

18.64 The pendulum shown in Figure P18.64 is hinged at O and has properties as follows:

Body	Mass (slug)	Geometry
Sphere	60.0	$r = 1.00$ ft
Slender Rod	4.0	$L = 12.0$ ft

In the position shown, the pendulum has a cw angular velocity $\omega = 1.2$ rad/s. Determine the angular acceleration of this body and the pin reaction at O at this instant.

18.65 The thin plate shown in Figure P18.65 is hinged at O and has a mass m. It is released from rest in the position shown. Determine, immediately after release, its

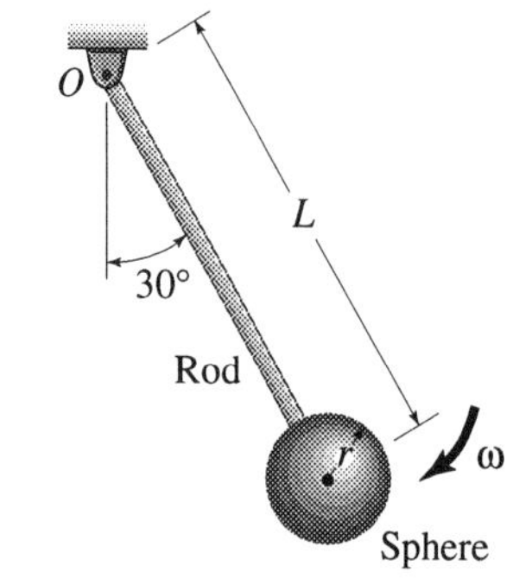

FIGURE P18.64.

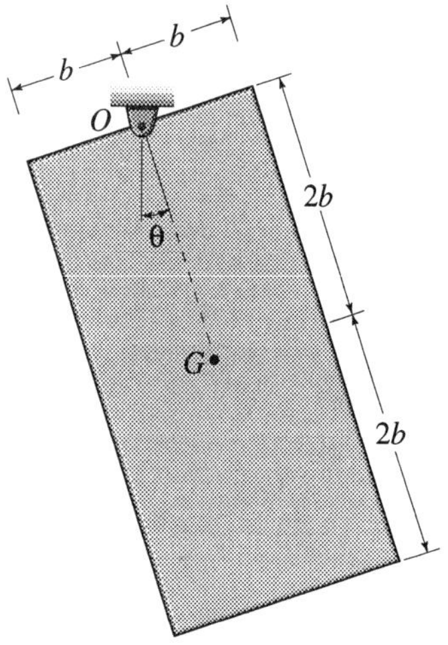

FIGURE P18.65.

angular acceleration and the normal and tangential components of the reaction at O. Express your answers in terms of b, θ and m.

18.66 The disk shown in Figure P18.66 rotates cw at 500 rpm. Determine the force P applied to the brake arm to bring the disk to a complete stop in 10 s. Also find the reaction components at the hinge at O. The coefficient of kinetic friction at the brake shoe is 0.3.

18.67 The plate shown in Figure P18.67 has a uniform thickness and a mass m. In the position shown $\theta = 40°$ and $\omega = 3$ rad/s. Let $b = 0.8$ ft and $m = 1.5$ slug, and determine, for the position shown, the angular acceleration and the normal and

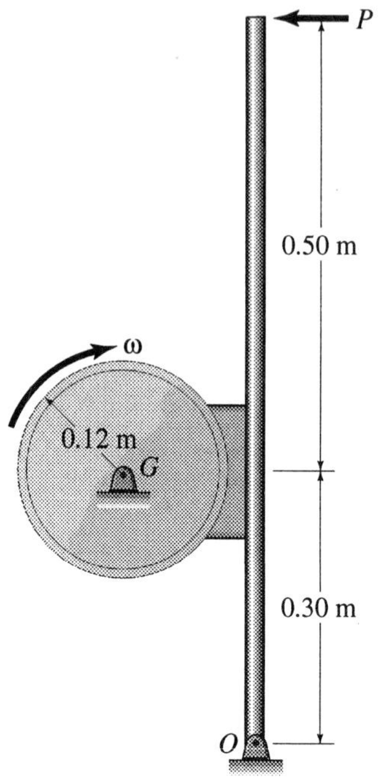

FIGURE P18.66.

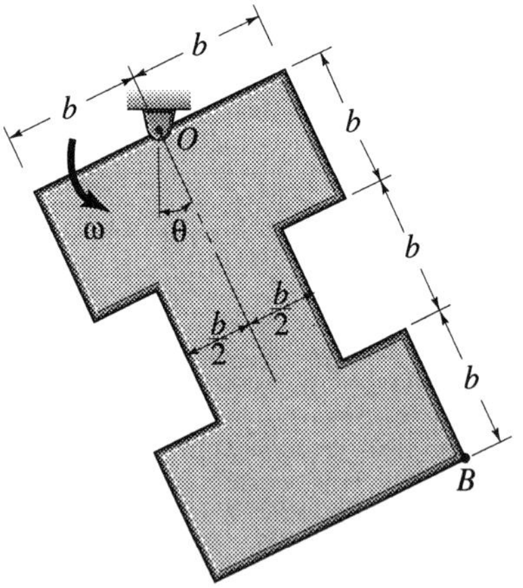

FIGURE P18.67.

tangential components of the reaction at O. Also find the magnitude of the acceleration of point B at the same instant.

18.68 A disk of radius r rotates about an axis through O where a frictionless pin supports it, as shown in Figure P18.68. At the instant shown, the pin reaction components in the normal and tangential directions are 380 lb and 243 lb, respectively, and the ccw angular acceleration is 7.60 rad/s^2. Determine the mass and radius of the disk and its instantaneous angular velocity.

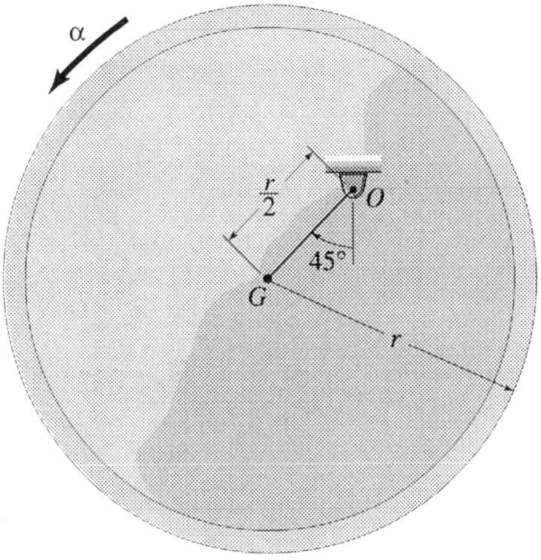

FIGURE P18.68.

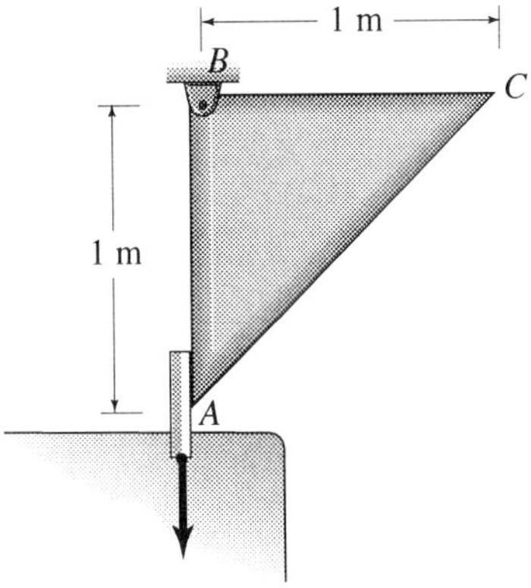

FIGURE P18.69.

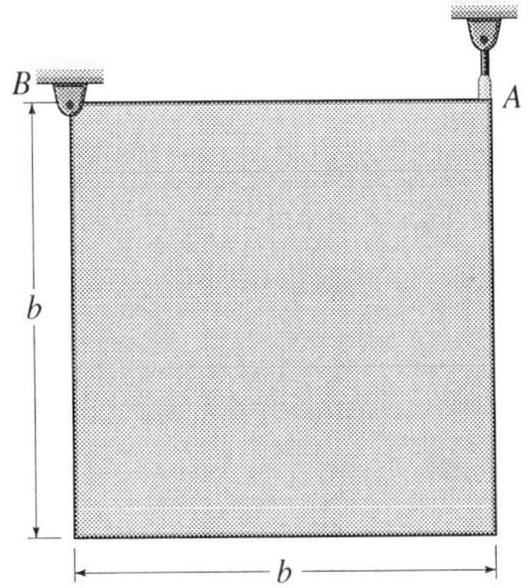

FIGURE P18.70.

18.69 The thin triangular plate shown in Figure P18.69 has a uniform thickness of 0.01 m and weighs 65 kN/m^3. The support at A is suddenly removed. Determine the angular acceleration of the plate and the reaction components at B at the instant support A is removed. The pin B is frictionless.

18.70 A uniform, homogeneous square plate of mass m is supported as shown in Figure P18.70. Determine the angular acceleration of the plate and the reaction components at B immediately after the cable support at A is cut. Find also the acceleration of point C. Express answers in terms of m and b.

18.71 In Figure P18.71, a composite body is depicted which consists of a slender rod of mass m connecting two disks each of mass m. The hinge at O is frictionless, and the rod OA may be considered massless. The system is released from rest when $\theta = 30°$. Determine the angu-

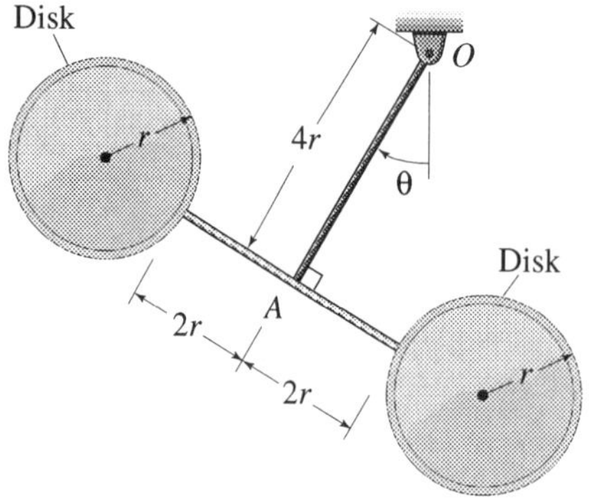

FIGURE P18.71.

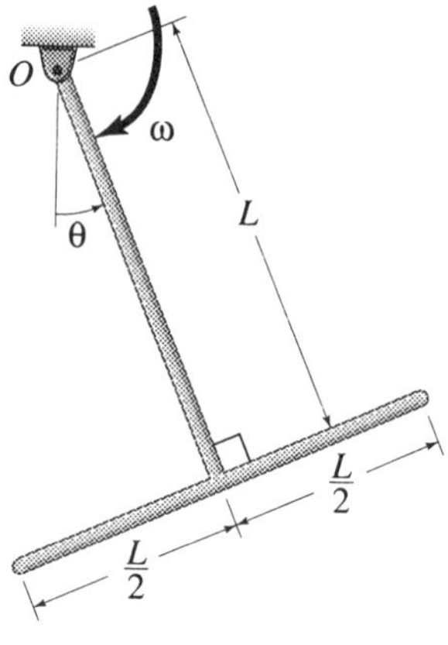

FIGURE P18.73.

lar acceleration α and the components of the pin reaction at O immediately after release. Let $m = 1$ slug and $r = 0.5$ ft.

18.72 A homogeneous, equilateral triangular plate of mass m has a constant thickness and is free to rotate about an axis through its apex O as shown in Figure P18.72. In the position shown, when $\theta = 30°$, $\omega = 5$ rad/s. Let $b = 0.75$ m and $m = 1.3$ kg, and determine the angular acceleration of the plate and the normal and tangential components of the reaction at O.

18.73 Two homogeneous slender rods, each of mass m and length L, are rigidly fastened together as shown in Figure P18.73. In the position shown when $\theta = 50°$, $\omega = 3$ rad/s. Let $L = 3$ ft and $m = 2$ slug, and determine the angular acceleration of the system and the normal and tangential components of the reaction at the frictionless hinge at O.

18.74 Three rods, each of mass m and length L, are welded together to form the triangle shown in Figure P18.74. The triangle is hinged at O and released from rest in the position shown. Determine the angular acceleration of this rigid body and the pin reaction components at the frictionless hinge at O. Express answers in terms of m, L and θ.

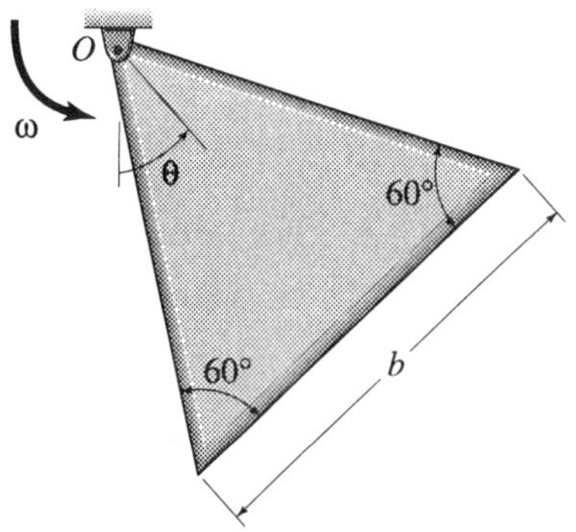

FIGURE P18.72.

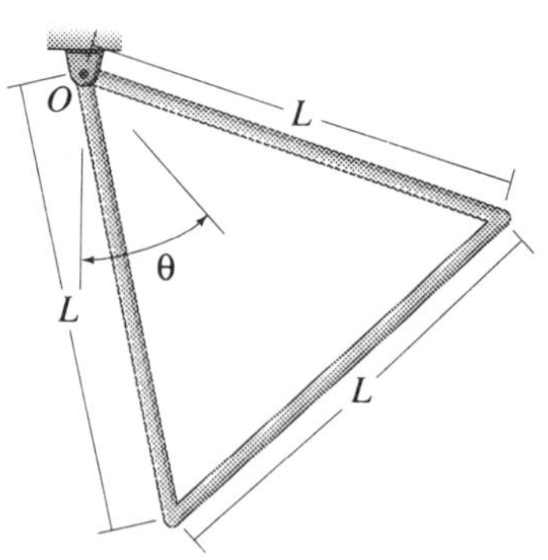

FIGURE P18.74.

18.75 Four rods, each of mass m and length L, are welded together to form a square which is hinged at O as shown in Figure P18.75. The square is displaced through a ccw angle θ and released from rest in this position. Determine the angular acceleration of the square and the magnitude of the linear acceleration of point B immediately after release. Express answers in terms of m, L and θ. Find the numerical values of these quantities if $m = 0.3$ kg, $L = 0.5$ m, and $\theta = 40°$.

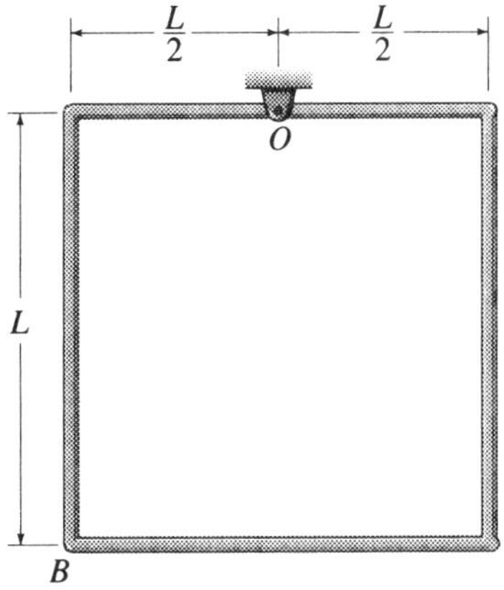

FIGURE P18.75.

18.76 Four slender, homogeneous rods each of mass m and length L are welded to form a square which is hinged at O, as shown in Figure P18.76. It is released from rest in the position shown. In terms of θ, m, and L, determine the angular acceleration of the system immediately after release. What are the reaction components at O at the same instant?

18.77 A system consisting of a square, homogeneous plate of mass 4 m with 4 slender homogeneous rods each of mass m and length L welded to the plate is shown in Figure P18.77. This system is free to rotate about the hinge at the center of mass G. (a) What couple applied in the plane of motion will increase the angular velocity of the system from zero to 23 rad/s ccw? (b) What are the pin reaction components acting at G?

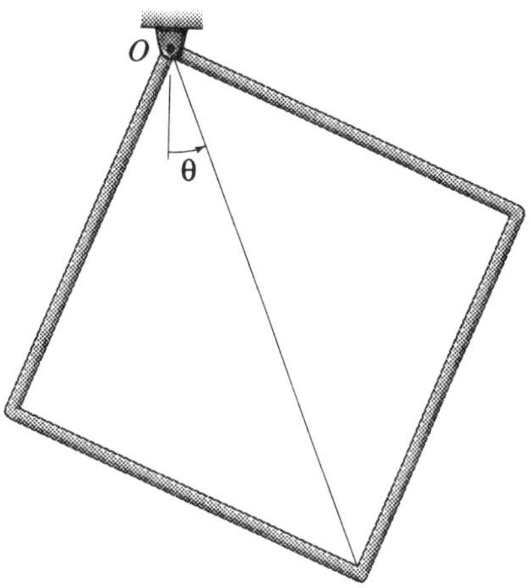

FIGURE P18.76.

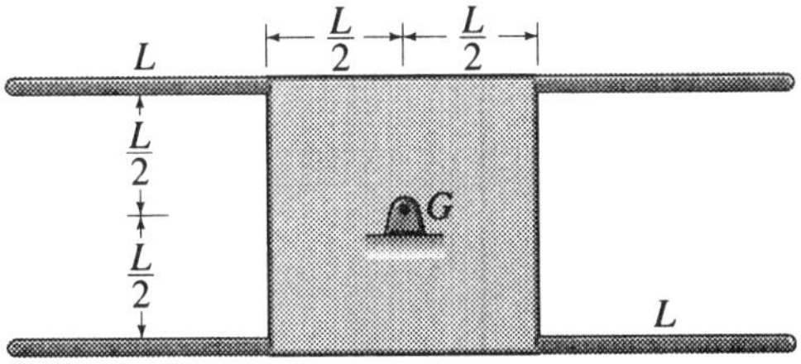

FIGURE P18.77.

18.78 Six slender homogeneous rods are welded together to form the body shown in Figure P18.78. The outside

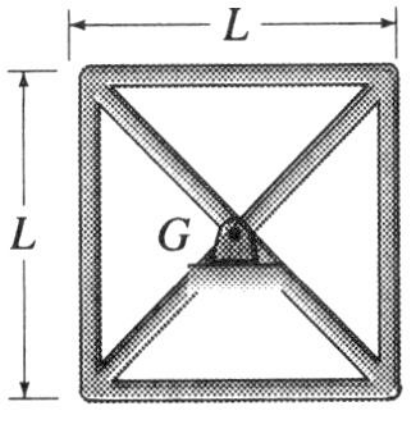

FIGURE P18.78.

rods are each of mass m and length L, and the diagonal rods are each of mass $2m$ and length $\sqrt{2}L$. The body is hinged at its center of mass G, and an external, ccw couple of magnitude 3 mgL is applied. What is the angular velocity after 5 s and the reaction components at G? Let $m = 2$ slug and $L = 3$ ft.

18.79 Refer to the composite body shown in Figure P18.79 and locate its mass center. It is hinged at O and in the position shown, $\theta = 30°$. If $m = 10$ kg and $L = 0.6$ m, determine the angular acceleration α of this body. For the position shown, given that $\omega = 0.50$ rad/s, determine the pin reaction components at the axis of rotation O.

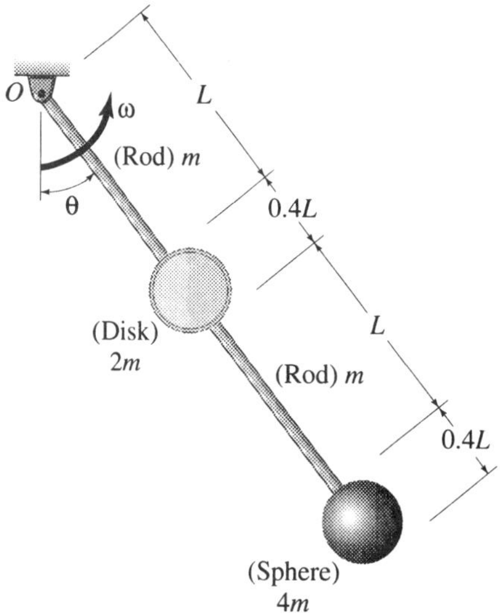

FIGURE P18.79.

18.80 Two spheres are connected by a massless rod as shown in Figure P18.80. Locate the mass center of this composite body, and write the differential equation which describes its oscillatory motion about the frictionless hinge at O. Assume small values of θ and replace $\sin\theta$ by θ in this equation, and solve the linearized equation subject to the initial conditions $t = 0$, $\theta = 0$ and $\dot{\theta} = \dot{\theta}_0$. Hint: See Example 18.12.

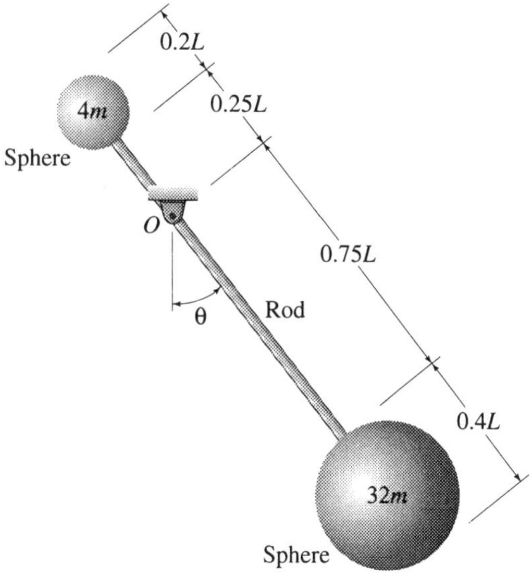

FIGURE P18.80.

18.81 A thin, homogeneous, semicircular plate of mass m and radius r is shown in Figure P18.81. It is released from rest when $\theta = 90°$. At this instant, determine the linear acceleration of point B expressing it in terms of the acceleration of gravity g.

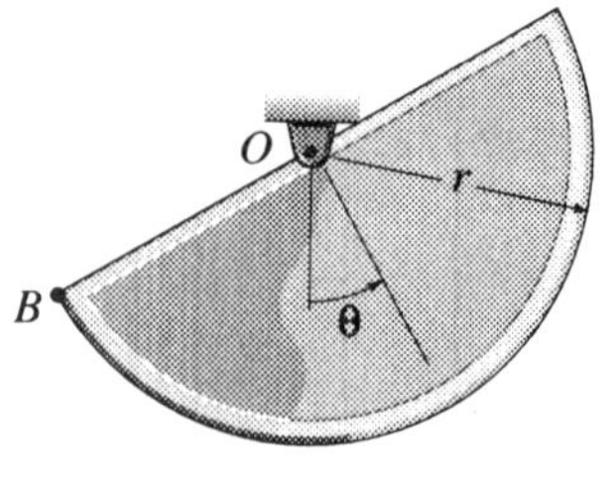

FIGURE P18.81.

18.82 Four spheres are attached to two, mutually perpendicular, massless rods, as shown in Figure P18.82. The center of each sphere is located at a distance b from the supporting frictionless pin at O. All of these spheres are made of the same material and the radii of the smaller ones is $r = b/3$. A ccw couple with a magnitude of 10 mgb is applied in the plane of motion. Determine the angular acceleration α and the corresponding reaction components at the frictionless pin at O. Express answers in terms of m and b.

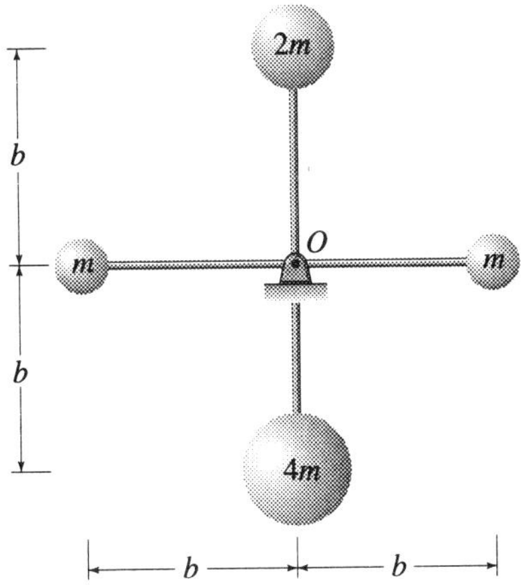

FIGURE P18.82.

18.83 A homogeneous regular hexagonal plate of mass m is supported by a frictionless pin at its mass center G as shown in Figure P18.83. (a) What couple applied to this hexagonal plate will increase its angular velocity from zero to 30 rad/s ccw in 5 s? (b) What pin reaction components support it at the axis of rotation?

18.84 The rectangular plate shown in Figure P18.84 is supported by the frictionless hinge at O. Imagine this plate displaced through a small ccw angle θ and released. Write the differential equation of motion for the plate. Assume small values of θ in this equation of motion, and solve the resulting differential equation for the initial conditions $t = 0$, $\theta = \theta_0$, and $\dot{\theta} = 0$. Hint: See Example 18.12.

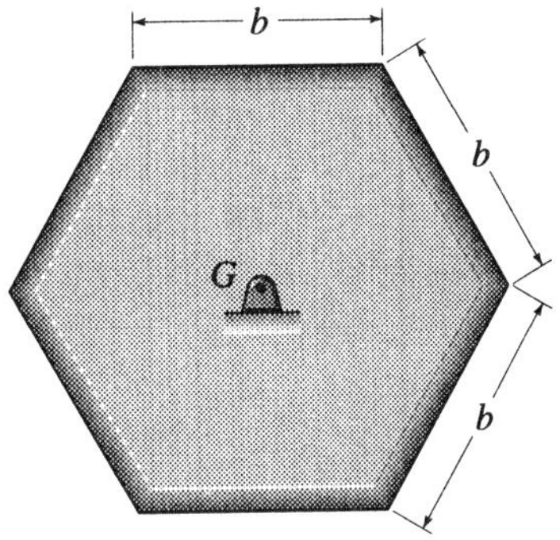

FIGURE P18.83.

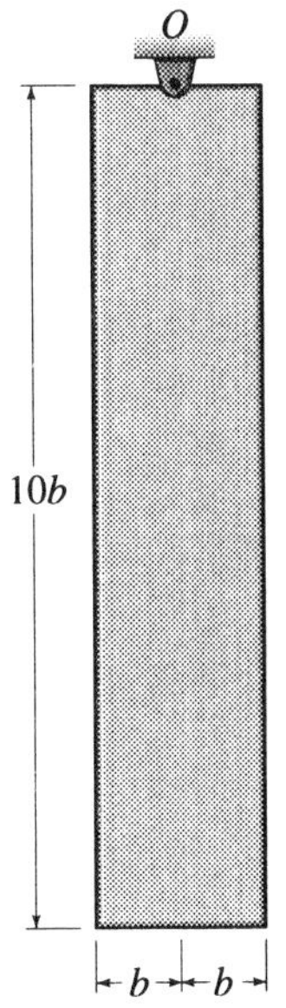

FIGURE P18.84.

18.85 A 15-kg block is attached to a cord which is wrapped around the core of a spool, as shown in Figure P18.85. The

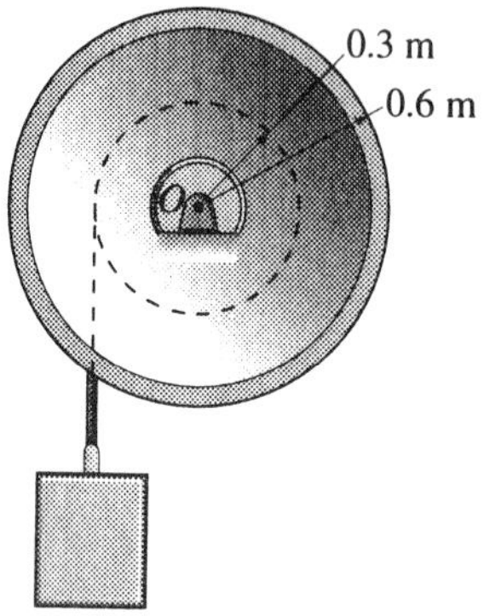

FIGURE P18.85.

spool has a mass of 40 kg and a radius of gyration $k_O = 0.4$ m. The system is released from rest and the block moves down a distance of 5 m. Determine the time required for this motion and the tension in the cord during the motion.

18.86 Two blocks of masses m_1 and m_2 are attached to the compound pulley shown in Figure P18.86. The system is released from rest and the pulley rotates in a ccw sense. The pulley has a weight W and a mass moment of inertia I_O. (a) In terms of m_1, m_2, r_1, r_2 and I_O, determine the angular acceleration of the pulley. (b) If $m_1 = m_2 = m$, $r_1 = 2r$, $r_2 = r$, and $I_O = 4\ mr^2$, what is the angular acceleration of the pulley? (c) Determine the number of revolutions required to increase the angular velocity of the pulley from zero to 30 rad/s.

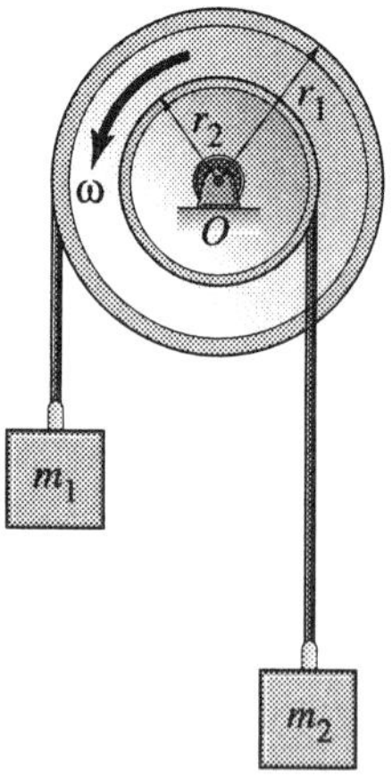

FIGURE P18.86.

18.87 Block B of mass $m = 5$ kg is attached to a cord which is wrapped around pulley P of mass $m_P = 10$ kg and radius of gyration $k_O = 0.5$ m, as shown in Figure P18.87. The coefficient of kinetic friction between the block and the inclined plane is 0.1. (a) If the system is released from rest, determine the speed of the block after it moves a distance of 7 m down the plane. (b) Find the tension in the cord during the motion. Assume that the pulley shaft is frictionless.

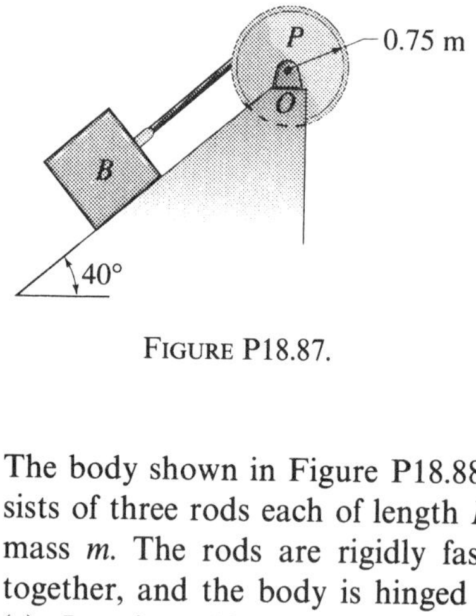

FIGURE P18.87.

18.88 The body shown in Figure P18.88 consists of three rods each of length L and mass m. The rods are rigidly fastened together, and the body is hinged at O. (a) Imagine this body rotated ccw through a small angle θ and released. Write the differential equation of motion. (b) Assume small values of θ, and solve the resulting differential equation subject to the initial conditions $t = 0$, $\theta = \theta_O$, and $\dot{\theta} = \dot{\theta}_O$. Hint: See Example 18.12.

18.89 The body shown in Figure P18.89 con-

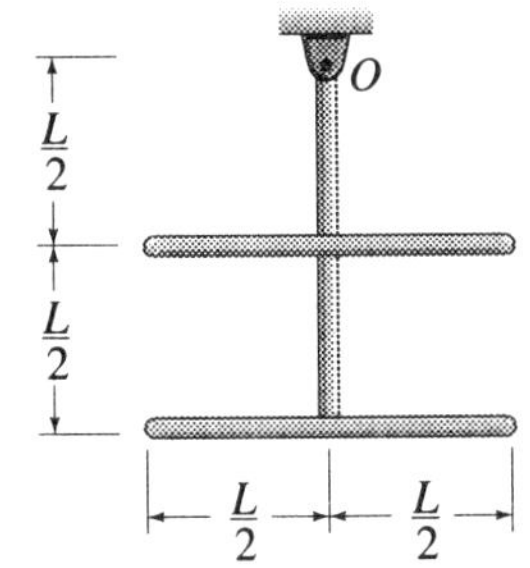

FIGURE P18.88.

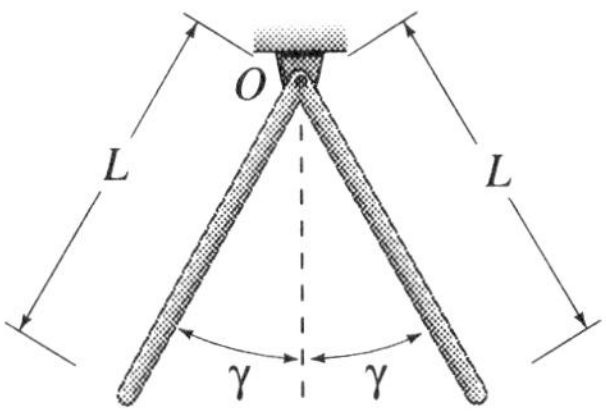

FIGURE P18.89.

sists of two rods each of length L and mass m. Imagine the body rotated ccw through a small angle θ and released. Write the differential equation of motion, and linearize it by assuming small values of θ. Solve the differential equation subject to the initial conditions $t = 0$, $\theta = \theta_O$, and $\dot{\theta} = 0$. Briefly discuss the special cases when $\gamma = 0$ and $\gamma = \pi/2$. Hint: See Example 18.12.

18.90 The system shown in Figure P18.90 consists of a slender rod of mass m connecting two disks each of mass m. The hinge at O is frictionless and rod OA

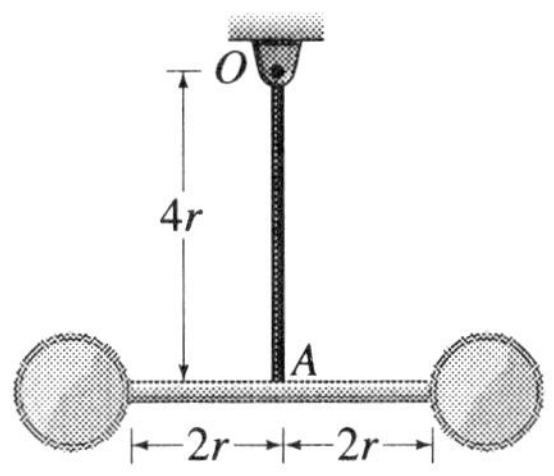

FIGURE P18.90.

may be assumed massless. The system is displaced through a small ccw angle θ and released. Write the differential equation of motion, and linearize it by setting $\sin\theta \approx \theta$. Solve the linearized equation subject to the initial conditions $t = 0$, $\theta = \theta_O$, and $\dot{\theta} = 0$.

18.91 The cylinder shown in Figure P18.91 rotates about the frictionless hinge at a cw angular velocity $\omega_O = 1000$ rpm. A ccw couple $C = 3.75$ lb·ft, applied to the cylinder, brings it to a complete stop in 25 revolutions. If the weight of the cylinders is 250 lb, determine its radius of gyration about point O.

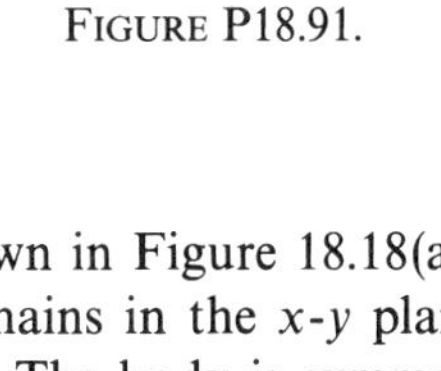

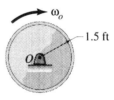

FIGURE P18.91.

18.5 General Plane Motion

A rigid body in plane motion is shown in Figure 18.18(a). This body moves so that its mass center G remains in the x-y plane which we imagine to be the plane of the page. The body is symmetric with respect to this x-y plane which is the plane of motion. General plane motion is a combination of translation and rotation. As shown in Figure 18.18(a), we consider that this motion consists of a translation of the mass center G (Fig. 18.18(b)) and a rotation about the mass

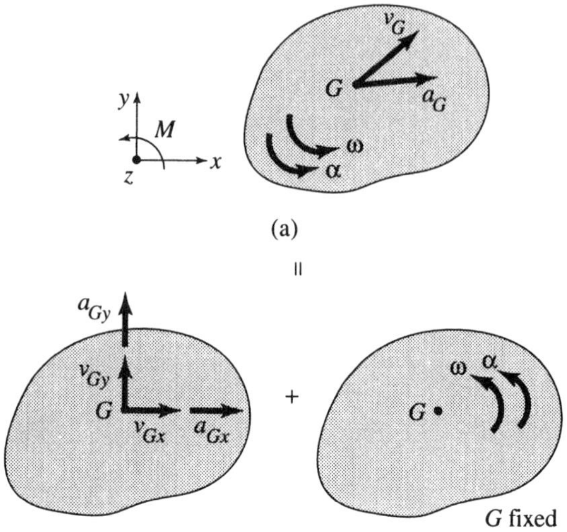

FIGURE 18.18. (b) Translation of mass center G (c) Rotation about mass center G

center G (Fig. 18.18(c)) for which we assume that the mass center is fixed. Of course, the translation and rotation proceed simultaneously.

For translation, the mass center G has instantaneous velocity components v_{Gx} and v_{Gy} and instantaneous acceleration components a_{Gx} and a_{Gy}. For rotation about the mass center, the instantaneous angular velocity of the body is ω, and its instantaneous angular acceleration is α. When we refer to *rotation about the mass center*, we mean that the body rotates about an axis through the mass center perpendicular to the plane of motion. Once we know the angular acceleration α and the linear acceleration components of G (a_{Gx} and a_{Gy}), we can determine the acceleration of any other point of the body by the using kinematic equations. A similar statement can be made for velocities, if we know the angular velocity ω and the linear velocity components v_{Gx} and v_{Gy}.

In Figure 18.19(a), a system of externally applied forces and couples is shown acting on a rigid body executing plane motion. The system consists of applied forces, typified by F_1 and F_2, applied couples, typified by C_1 and C_2, and the weight of the body W. In Section 5.8, we learned that any given system of forces and couples may be reduced to a single force and a single couple acting at an arbitrary point. We choose the mass center G as this arbitrary point. The resultant system acting at the mass center G is shown in Figure 18.19(b). It consists of the resultant force components $\sum F_x$ and $\sum F_y$ and the resultant couple $\sum M_G$. Also shown in Figure 18.19 is a right-handed x-y-z coordinate system where the z axis comes out of the page.

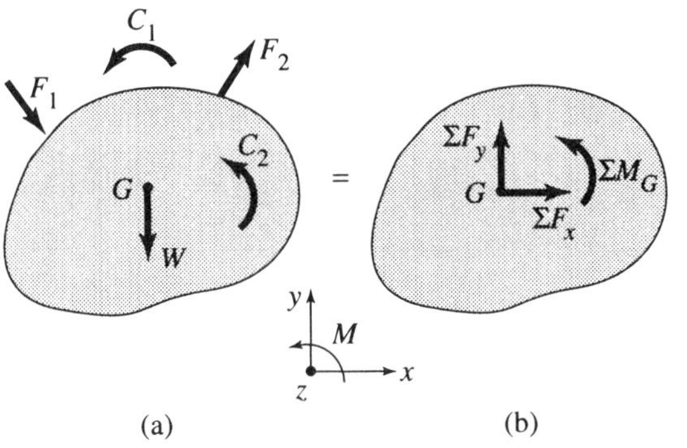

FIGURE 18.19.

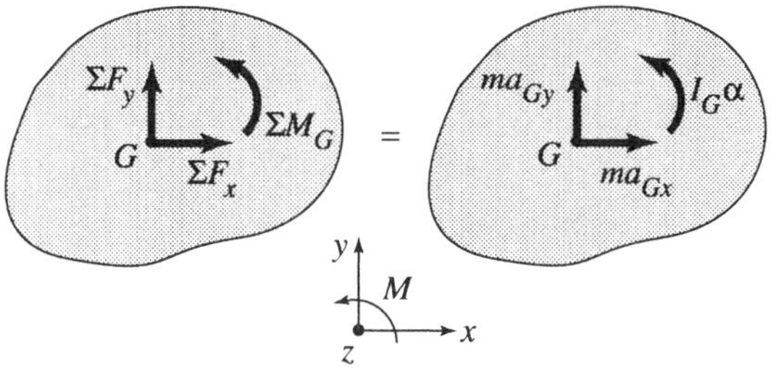

FIGURE 18.20.

Figure 18.20 shows the free-body and inertial force diagrams for a rigid body in general plane motion. A right-handed x-y-z coordinate system is also shown. Therefore, the kinetic equations corresponding to the general plane motion of such a body become

$$\sum F_x = ma_{Gx}, \tag{18.23a}$$

$$\sum F_y = ma_{Gy}, \tag{18.23b}$$

$$\sum M_G = I_G\alpha, \tag{18.23c}$$

These equations are usually supplemented by kinematic equations to solve problems. Equations (18.23a) and (18.23b) state that the components of the resultant external force applied to a rigid body in general plane motion equals the corresponding components of the resultant inertial force. Equation (18.23c) states that the resultant external couple, applied at the mass center G, equals the resultant inertial couple.

Instead of using the mass center G, it is sometimes convenient to use an arbitrary point O. Thus, consider the body shown in Figure 18.21(a) which is executing general plane motion under the action of the given forces and couples. The corresponding inertial force diagram is depicted in Figure 18.21(b). In such a case, Eqs. (18.23) become

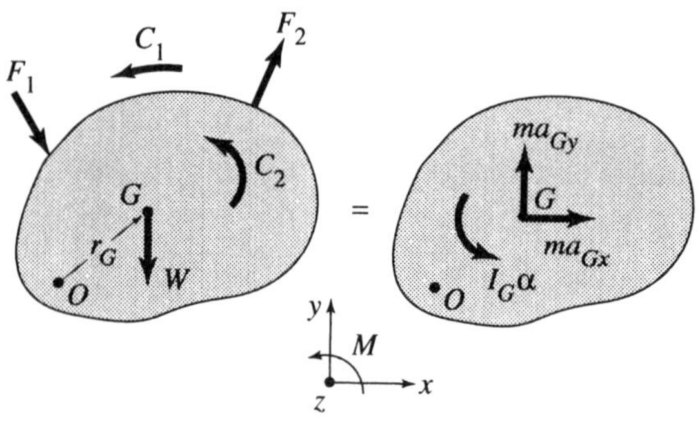

FIGURE 18.21.

$$\sum F_x = ma_{Gx}, \tag{18.24a}$$

$$\sum F_y = ma_{Gy}, \tag{18.24b}$$

$$\sum M_O = \sum (M_i)_O. \tag{18.24c}$$

In Eq. (18.24c), the term $\sum (M_i)_O$ represents the sum of the moments about point O of the inertia forces ma_{Gx} and ma_{Gy} and the inertial couple $I_G\alpha$.

The following examples illustrate some of the concepts discussed above.

Example 18.13

The circular diagram, shown at the top of the inclined plane in Figure E18.13(a), represents three different homogeneous bodies: a sphere, a cylinder, and a hoop. Each has a radius r and a mass m, and all three bodies roll without sliding down the plane. Determine the angular acceleration of each body, and decide the order in which they will arrive at the base of the inclined plane, a distance s from where they were released. Express your answers in terms of the known quantities g, r, s, and θ.

Solution

(a) Each body is in general plane motion as it moves down the inclined plane. Since these bodies roll without sliding, their mass centers, which coincide with their geometric centers, have accelerations equal to $r\alpha$ directed parallel to the plane. The free-body and inertial force diagrams are shown in Figure E18.13(b) along with a right-handed x-y-z coordinate system, where the z axis comes out of the page. Thus, by Eqs. (18.23)

$$\sum F_x = ma_{Gx}, \qquad mg\sin\theta - F = mr\alpha, \tag{a}$$

$$\sum F_y = ma_{Gy} = 0, \quad N - mg\cos\theta = 0, \tag{b}$$

$$\sum M_G = I_G\alpha, \qquad -Fr = -I_G\alpha. \tag{c}$$

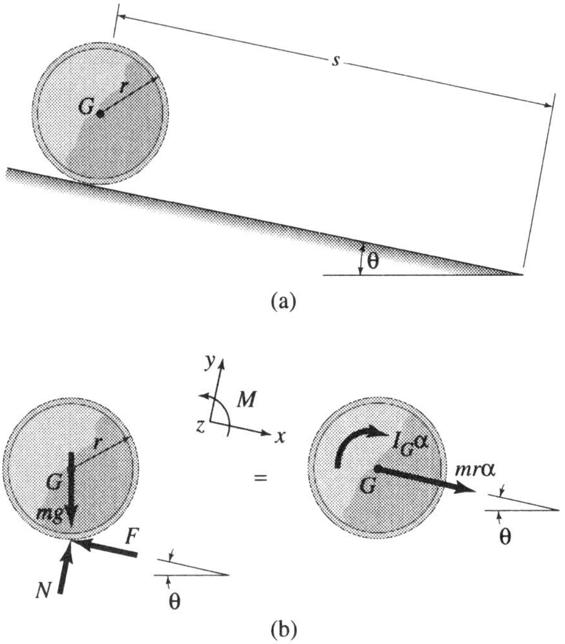

FIGURE E18.13.

Solve Eq. (a) for F to yield

$$F = mg\sin\theta - mr\alpha$$

Substitute this expression for F in Eq. (c) and solve for α to obtain,

$$\alpha = \frac{mgr\sin\theta}{I_G + mr^2}.$$

This value of α depends upon I_G which differs for each of the three bodies. Thus, using the information in Appendix B,

for a sphere,

$$I_G = \frac{2}{5}mr^2 \Rightarrow \alpha = \frac{5}{7}\left(\frac{g}{r}\right)\sin\theta, \qquad \text{ANS.}$$

for a cylinder,

$$I_G = \frac{1}{2}mr^2 \Rightarrow \alpha = \frac{2}{3}\left(\frac{g}{r}\right)\sin\theta, \qquad \text{ANS.}$$

and for a hoop,

$$I_G = mr^2 \Rightarrow \alpha = \frac{1}{2}\left(\frac{g}{r}\right)\sin\theta. \qquad \text{ANS.}$$

Because $5/7 > 2/3 > 1/2$, the sphere arrives first at the base of the inclined plane, followed by the cylinder and hoop in that order.

■ Example 18.14

Refer to Example 18.13 and determine (a) the minimum coefficient of friction required to prevent sliding of each of the three bodies and (b) the time needed for each body to move the distance s along the inclined plane.

Solution

(a) From Eq. (b) of Example 18.13, we solve for N to obtain

$$N = mg\cos\theta.$$

The coefficient of friction μ corresponding to impending slip is equal to the ratio of the friction force F to the normal force N. Thus, using the value of F found in Example 18.13,

$$\mu = \frac{F}{N} = \frac{mg\sin\theta - mr\alpha}{mg\cos\theta}.$$

Substituting for α corresponding to each of the three bodies,

for a sphere,

$$\mu = \tfrac{2}{7}\tan\theta, \qquad \text{ANS.}$$

for a cylinder,

$$\mu = \tfrac{1}{3}\tan\theta, \qquad \text{ANS.}$$

and for a hoop,

$$\mu = \tfrac{1}{2}\tan\theta. \qquad \text{ANS.}$$

Because $2/7 < 1/3 < 1/2$, the sphere is least likely to slip on the plane. Provided $\mu > 1/2\tan\theta$, none of the bodies will slide on the plane.

(b) The mass center of each body moves along a straight line path. For each body, the forces remain constant during the motion which justifies the use of the constant acceleration equations applied to the motion of the mass center of each body. Thus,

$$s = s_0 + v_0 t + \tfrac{1}{2}a_C t^2, \quad s = 0 + 0 + \tfrac{1}{2}a_C t^2.$$

Solving for t and replacing the constant acceleration a_C by $r\alpha$ gives

$$t = \sqrt{\frac{2s}{r\alpha}}.$$

We now substitute the α values for each body, as found in Example 18.13, to determine the corresponding times. Thus, *for a sphere,*

$$t = \sqrt{\frac{2s}{r\left(\frac{5g}{7r}\sin\theta\right)}} = \sqrt{\frac{14s}{5g\sin\theta}}, \qquad \text{ANS.}$$

for a cylinder,

$$t = \sqrt{\frac{3s}{g\sin\theta}}, \qquad \text{ANS.}$$

and for a hoop,

$$t = \sqrt{\frac{4s}{g\sin\theta}}. \qquad \text{ANS.}$$

Because $\sqrt{\frac{14}{5}} < \sqrt{3} < \sqrt{4}$ we note that the sphere requires the least time to roll down the inclined plane, followed by the cylinder and the hoop in order. This is in agreement with our conclusion for part (a) based upon angular accelerations.

■ Example 18.15

A horizontally directed force P is applied to a homogeneous cylinder of weight W and radius r as shown in Figure E18.15(a). Regard W, r, g, P and y as known quantities. (a) If the cylinder rolls without sliding on

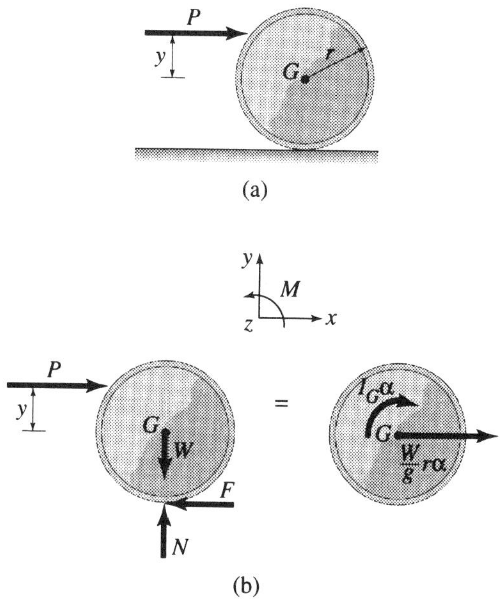

FIGURE E18.15.

the horizontal plane, develop expressions for the angular acceleration α and for the friction and normal forces F and N between the plane and the cylinder. (b) If $P = 20$ lb, $W = 100$ lb, $r = 2.5$ ft and $y = 1.5$ ft, find numerical values for α, F and N.

Solution

(a) Refer to Figure E18.13(b) which shows the free-body and inertial force diagrams for the cylinder along with a right-handed x-y-z coordinate system where the z axis comes out of the page. By Eqs. (18.23),

$$\sum F_x = ma_{Gx}, \quad P - F = \frac{W}{g} r\alpha, \tag{a}$$

$$\sum F_y = ma_{Gy}, \quad N - W = 0, \tag{b}$$

$$\sum M_G = I_G\alpha, \quad -Fr - Py = -I_G\alpha. \tag{c}$$

Eliminate F from Eqs. (a) and (c) by substitution, and solve for α to give

$$\alpha = \frac{P(y + r)}{I_G + \dfrac{W}{g} r^2}$$

From Appendix B, $I_G = \dfrac{1}{2}\dfrac{W}{g} r^2$ for a cylinder. Therefore,

$$\alpha = \frac{2Pg(y + r)}{3Wr^2}. \qquad \text{ANS.}$$

Substitute this value for α in Eq. (a), and solve for F to obtain

$$F = \frac{P}{3}\left(1 - 2\frac{y}{r}\right). \qquad \text{ANS.}$$

Equation (b) yields

$$N = W. \qquad \text{ANS.}$$

(b) Substitute the given numerical values to obtain

$$\alpha = 2.75 \text{ rad/s}^2 \; \circlearrowleft, \qquad \text{ANS.}$$

$$F = -1.333 \text{ lb} = 1.333 \text{ lb} \rightarrow, \qquad \text{ANS.}$$

$$N = 100 \text{ lb} \uparrow. \qquad \text{ANS.}$$

■

Problems

18.92 A homogeneous sphere weighs 322 lb and has a radius of 0.5 ft. It rolls without sliding on a horizontal plane under the action of a horizontal force of 60 lb applied symmetrically to its top. Determine (a) the angular acceleration of the sphere, (b) the acceleration of its mass center, (c) the frictional and normal forces acting on it, and (d) the minimum coefficient of friction required to prevent sliding of the sphere.

18.93 A homogeneous cylinder has a radius of 0.8 ft and weighs 64.4 lb. It rests on a horizontal plane, and a horizontal force of 20 lb is applied symmetrically through the mass center of the cylinder. Determine the angular acceleration, the acceleration of the mass center, and the frictional force acting on the cylinder for (a) $\mu = 0.25$ and (b) $\mu = 0.05$.

18.94 A homogeneous thin hoop weighs 100 N and has a radius of 0.4 m. It rests on a horizontal plane and a horizontal force of 20 N is applied symmetrically through the mass center of the thin hoop. Determine the angular acceleration, the acceleration of the mass center, and the frictional force acting on the hoop for (a) $\mu = 0.25$ and (b) $\mu = 0.05$.

18.95 A homogeneous sphere weighs 100 N and has a radius of 0.5 m. It rolls without sliding on a horizontal plane under the action of a horizontal force of 120 N applied symmetrically through the mass center of the sphere. Determine (a) the angular acceleration of the sphere, (b) the acceleration of the mass center of the sphere, (c) the frictional and normal forces acting on the sphere, and (d) the minimum coefficient of friction required to prevent sliding of the sphere.

18.96 The homogeneous thin hoop of weight W and radius r shown in Figure P18.96 is released from rest and rolls without sliding down the inclined plane under the action of its weight and the applied force P. Determine (a) the angular acceleration of the hoop, (b) the frictional force, (c) the acceleration of the mass center G, and (d) the minimum coefficient of friction required to assure that the thin hoop rolls without sliding. Express answers in terms of W and r.

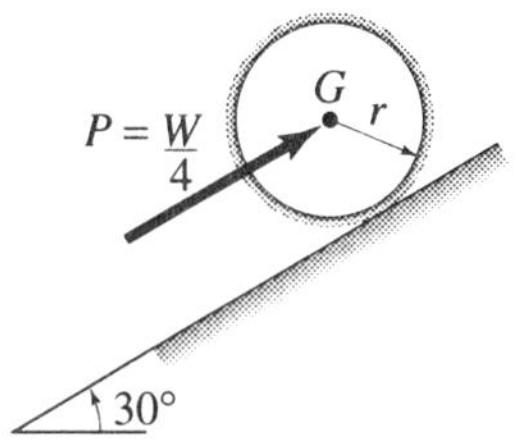

FIGURE P18.96.

18.97 A homogeneous sphere of weight W and radius r shown in Figure P18.97 is released from rest and rolls without sliding down the inclined plane under

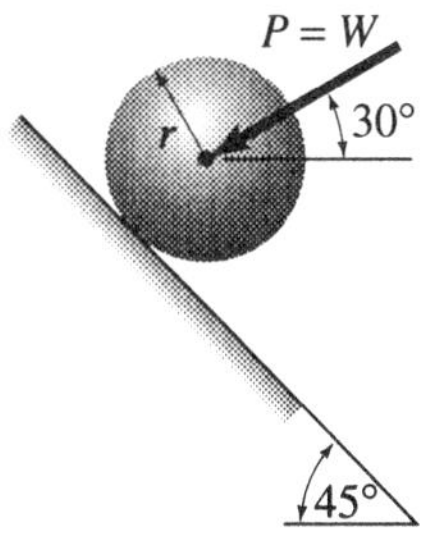

FIGURE P18.97.

the action of its weight and the applied force P. Determine (a) the angular acceleration of the sphere, (b) the frictional force, (c) the acceleration of the mass center of the sphere, and (d) the minimum coefficient of friction required to assure that the sphere rolls without sliding. Express answers in terms of W and r.

18.98 Homogeneous rod AB of length L and mass m is shown in Figure P18.98. It is released from rest in this position, and A moves down the vertical plane while B moves to the right along the horizontal plane. Neglect friction, and consider L and θ to be known quantities. At the instant of release, express the normal forces at A and B and the angular acceleration of the rod in terms of the known quantities.

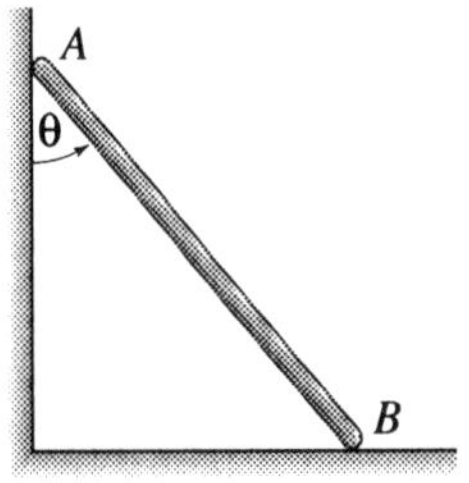

FIGURE P18.98.

18.99 A spoked wheel subjected to a couple C is shown in Figure P18.99. It rolls without sliding on the horizontal plane. The rim has a weight W and may be idealized as a thin hoop. Each spoke has a weight $W/8$ and may be idealized as a slender rod of length r. Regard C, W and r as known quantities. Determine the angular acceleration of the wheel and the frictional and normal forces exerted on the wheel by the horizontal plane. Express your answers in terms of the known quantities.

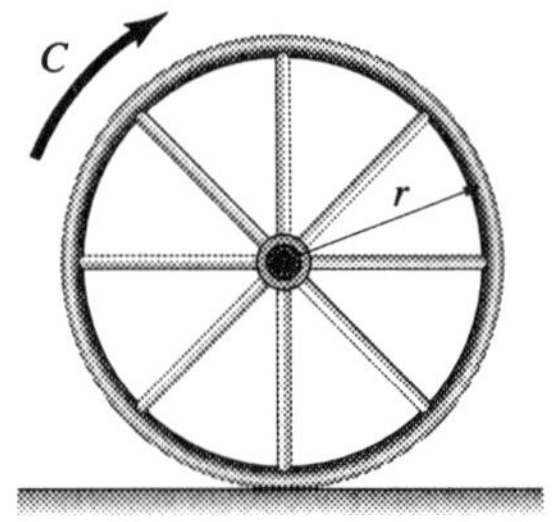

FIGURE P18.99.

18.100 A homogeneous cylinder of weight W and radius r rolls without slipping on the horizontal plane under the action of a ccw couple $C = wr/2$ as shown in Figure P18.100. Determine (a) the angular acceleration of the cylinder, (b) the frictional and normal forces acting on the cylinder, and (c) the minimum coefficient of friction required so that the cylinder does not slip on the plane. Express answers in terms of W and r.

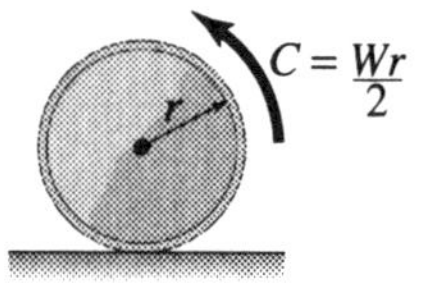

FIGURE P18.100.

18.101 Two cases of a spool subjected to a force $P = W$ are shown in parts A and B of Figure P18.101. The spool has a weight $W = 50$ lb and a radius of gyration $k_G = 1.20r$ where $r = 0.75$ ft. For each case, determine (a) the angular acceleration of the spool, (b) the frictional and normal forces, and (c) the minimum coefficient of friction required for rolling without sliding.

18.102 An upward force $P = W/2$ is applied to the cylinder of weight W shown in Figure P18.102. The cylinder rolls without

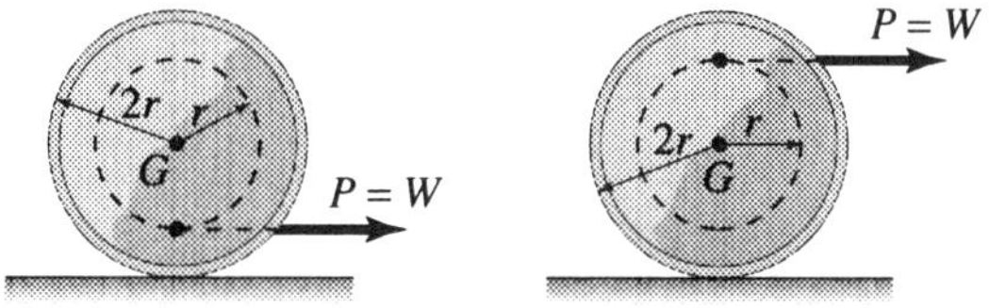

(A) (B)

FIGURE P18.101.

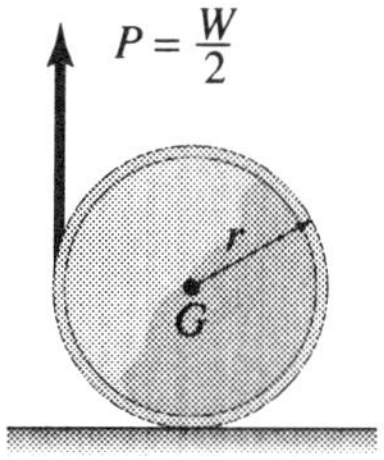

FIGURE P18.102.

slipping on the horizontal plane. In terms of the known quantities W and r, determine (a) the angular acceleration of the cylinder, (b) the frictional force exerted on the cylinder by the plane, and (c) the minimum coefficient of friction required so that the cylinder will roll without sliding.

18.103 The homogeneous cylinder of weight $W = 100$ N and radius $r = 0.25$ m is released from rest as shown in Figure P18.103. Determine the tension in the inextensible cord, the angular acceleration of the cylinder and the acceleration of its mass center G. Assume that the cylinder does not slip on the cord.

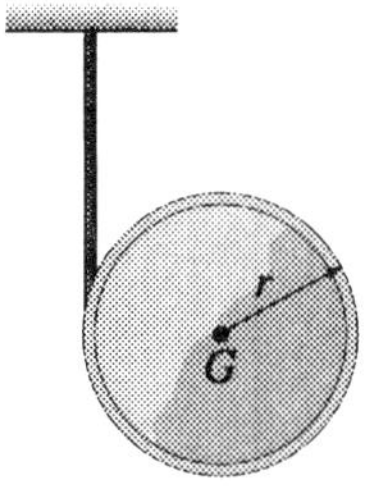

FIGURE P18.103.

18.104 Refer to Example 18.15 (p. 521), and replace the cylinder by a sphere of weight $W = 500$ N, $r = 0.5$ m, $P = 300$ N, and $y = 0.25$ m. Then determine the angular acceleration of the sphere and the minimum coefficient of friction required to prevent sliding.

18.105 Refer to Figure P18.105, which represents a sphere of weight $W = 200$ lb and radius $r = 2$ ft executing plane motion. Does it roll without sliding? What is its angular acceleration? What is the acceleration of its mass center? What frictional force acts on the sphere? Let $K = 3$, $y = 0.6r$, and $\mu = 0.5$.

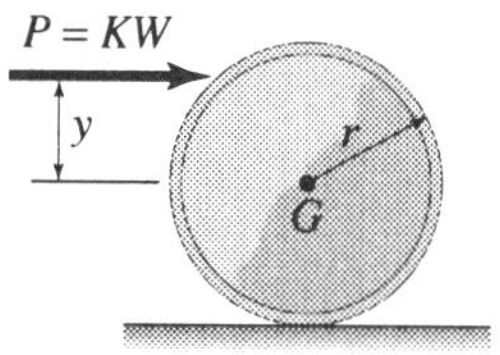

FIGURE P18.105.

18.106 A cylinder of weight $W = 200$ lb and radius $r = 2$ ft is shown in Figure P18.105. Let $K = 1$, $y = 0.4r$ and $\mu = 0.05$. Does this cylinder roll without sliding? What is its acceleration? What is the acceleration of its mass center? What frictional force acts on this cylinder?

18.107 Solve Problem 18.103 assuming the body to be a thin hoop of weight $W = 100$ N and radius $r = 0.25$ m.

18.108 A wheel shown in Figure P18.108 is constructed of a central disk of mass m and radius r connected by massless spokes to an outer hoop of mass m and radius $2r$. A couple $C = 2\ mgr$ is applied to this wheel, and it rolls without sliding on the horizontal plane. In terms of m, g and r, determine (a) the angular acceleration of this wheel, (b) the nor-

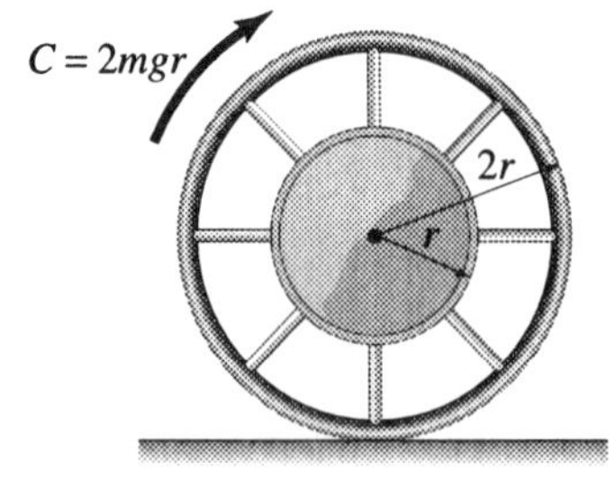

FIGURE P18.108.

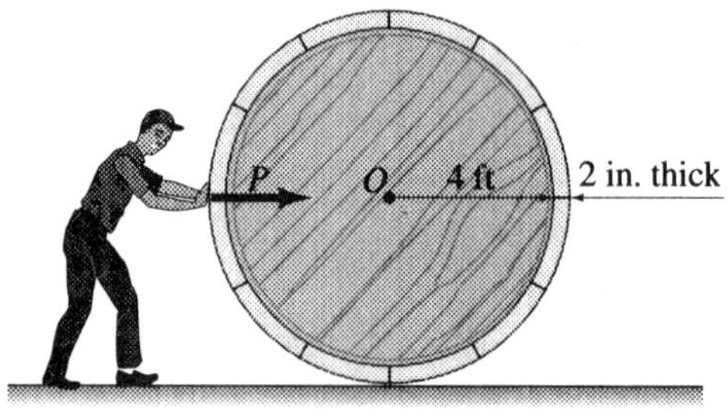

FIGURE P18.110.

mal and frictional force exerted on this wheel by the horizontal plane, and (c) the minimum coefficient of friction required so that it rolls without sliding.

18.109 A small hoop of mass $m/2$ is attached to the inside of a larger hoop of mass m as shown in Figure P18.109. This composite body starts from rest in the position shown and rolls without sliding on the horizontal plane. For the position shown determine (a) the angular acceleration of the body, (b) the acceleration of its mass center, (c) the normal and frictional forces acting on the body, and (d) the minimum coefficient of friction so that it rolls without sliding.

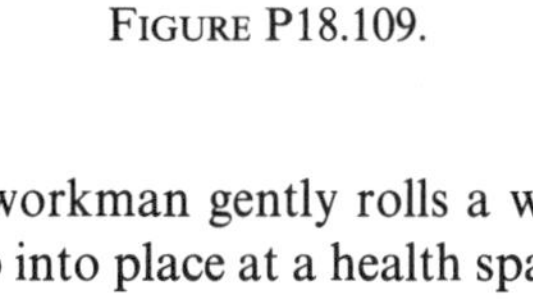

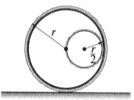

FIGURE P18.109.

18.110 A workman gently rolls a wooden hot tub into place at a health spa, as shown in Figure P18.110. Each stave is 2 in. thick as shown, and, when in use in the proper position, the hot tub is 5 ft. high. Assume that the bottom and the cover (not shown) may each be idealized as a cylinder 8 ft in diameter and 2 in. in thickness. The wood weighs 40 lb/ft^3. Determine the force P so that the angular acceleration is 0.4 rad/s^2.

18.111 The body of radius r and mass m, shown in Figure P18.111, rolls without sliding on the inclined plane under the action of the couple $M = mgd$ where d is a specified length. The centroidal mass moment of inertia of this body with respect to G is mk_G^2 where k_G is the centroidal radius of gyration. (a) Determine the angular acceleration α as a function of r, β, d and k_G. (b) What value of d will reduce α to zero? (c) Find the frictional force F as a function of m, g, β, r, d, and k_G. (d) Find the normal force N as a function of m, g and β. (e) Determine the minimum coefficient of friction required to prevent sliding as a function of β, r, d and k_G.

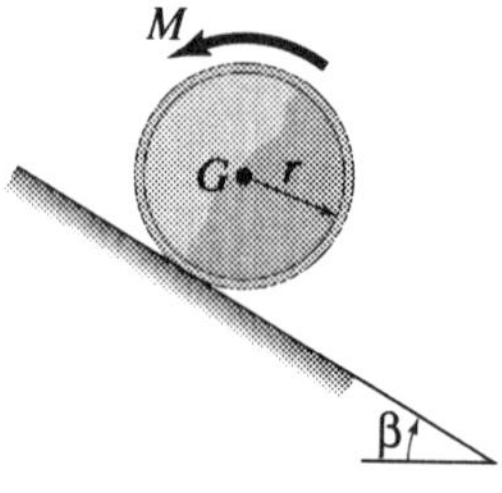

FIGURE P18.111.

18.112 Refer to Problem 18.111. The body is a sphere of mass m and radius r with $d = 0.5r$ and $\beta = 45°$. (a) Find the acceleration of the mass center G. (b) Determine the minimum coefficient of friction required to prevent sliding of the sphere. (c) Express the frictional force as a function of the weight mg.

18.113 Refer to Problem 18.111. The body is a hoop of mass m and radius r with $d = 0.4r$ and $\beta = 30°$. (a) Find the acceleration of the mass center G. (b) Determine the minimum coefficient of friction required to prevent sliding of the hoop. (c) Express the frictional force as a function of the weight mg.

18.114 A cutout in the cylindrical body shown in Figure P18.114 has shifted its mass center from O to G, as shown. The radius of gyration with respect to point O is $0.6r$. In the position shown, the body has a cw angular velocity of 1.0 rad/s. Given that $m = 20$ kg and $r = 0.2$ m, determine, for this position, (a) the cw couple C required for the body to have an angular acceleration of 10.0 rad/s, (b) the corresponding frictional and normal forces, and (c) the minimum coefficient of friction required to prevent slipping.

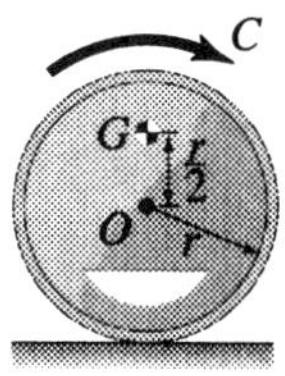

FIGURE P18.114.

18.6 Systems of Rigid Bodies

When dealing with systems of interconnected rigid bodies in general plane motion, the techniques developed in previous sections for analyzing a single rigid body in motion are still applicable. The system of interconnected rigid bodies is separated into the individual rigid bodies composing the system. A free-body diagram and a corresponding inertial force diagram are then, constructed for each of the individual rigid bodies in the system. Equations (18.23) or (18.24) are applied to each rigid body separately, and the resulting relationships are solved for the unknown quantities.

It is possible to construct the free-body and inertial force diagram for the entire system as a whole. Equations (18.23) or (18.24) are, then, applied to the entire system. However, because only three independent equations are available, we cannot solve for more than three unknown quantities.

In the computer age, engineers and scientists must develop their capabilities for viewing problems with a general perspective and for writing equations which permit determining solutions for all unknowns of a system. In this section we consider systems of connected rigid bodies analyzed symbolically to yield general solutions. These general solutions can, then, be used to obtain numerical results for a specific set of conditions or to study the behavior of the system in a general way.

The examples that follow illustrate the concepts and procedures needed for solving this type of problem.

Example 18.16

Consider the frictionless system shown in Figure E18.16(a) where $W_1 = 20$ lb, $W_2 = 40$ lb, $W_3 = 15$ lb, $I_2 = 1.5$ slug ft^2, $r_2 = 0.40$ ft, $R_2 = 0.75$ ft and $\beta = 30°$. Determine the tensions in the two cables and the accelerations of weights W_1 and W_3. Solve the problem symbolically first, and then specialize the general expressions obtained for the specific numerical values given above.

Solution

The free-body and inertial force diagrams for W_1 are shown in Figure E18.16(b) along with a right-handed x-y-z coordinate system. This body is in translation up the inclined plane. Thus,

$$\sum F_x = m_1 a_{1x}, \quad T_1 - W_1 \sin\beta = \left(\frac{W_1}{g}\right) a_1,$$

$$T_1 = W_1(\sin\beta + a_1/g). \tag{a}$$

The free-body and inertial force diagrams for W_2 are shown in Figure E18.16(c) along with a right-handed x-y-z coordinate system. This body is in rotation about the fixed axis at O. Thus,

$$\sum M_O = I_2\alpha_2: T_1 r_2 - T_2 R_2 = -I_2\alpha. \tag{b}$$

The free-body and inertial force diagrams for W_3 are shown in Figure E18.16(d) along with a right-handed x-y-z coordinate system. This body is in translation downward. Thus,

$$\sum F_y = m_3 a_{3y}: W_3 - T_2 = \left(\frac{W_3}{g}\right) a_3$$

$$T_2 = W_3(1 - a_3/g) \tag{c}$$

The equations above must be supplemented by two kinematic relationships to enable us to solve for the five unknown quantities: T_1, T_2, a_1, a_2, and α. These kinematic equations relate a_1 and a_2 to α. Thus,

$$a_1 = r_2\alpha \tag{d}$$

$$a_3 = R_2\alpha \tag{e}$$

Solving Eqs. (a) through (e) simultaneously leads to

$$\alpha = \frac{W_3 R_2 - W_1 r_2 \sin\beta}{(W_1/g)r_2^2 + (W_3/g)R_2^2 + I_2},$$

$$a_1 = \frac{W_3 r_2 R_2 - W_1 r_2^2 \sin\beta}{(W_1/g)r_2^2 + (W_3/g)R_2^2 + I_2}, \qquad \text{ANS.}$$

$$a_3 = \frac{W_3 R_2^2 - W_1 r_2^2 R_2 \sin\beta}{(W_1/g)r_2^2 + (W_3/g)R_2^2 + I_2}, \qquad \text{ANS.}$$

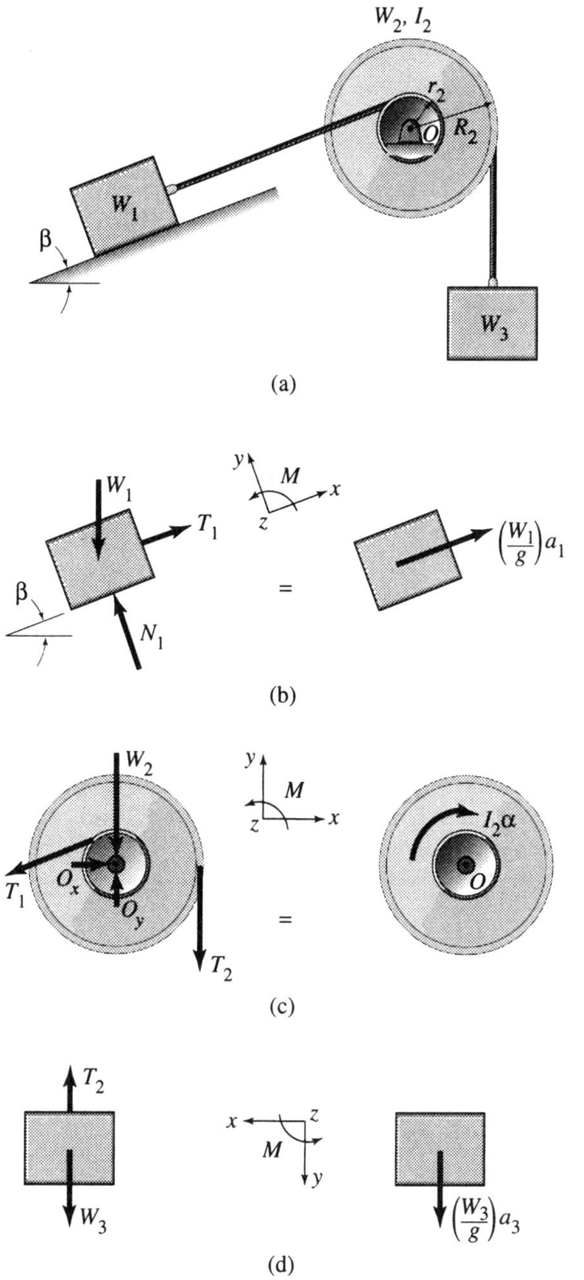

FIGURE E18.16.

$$T_1 = W_1\left[\sin\beta + \frac{W_3 r_2 R_2 - W_1 r_2^2 \sin\beta}{W_1 r_2^2 + W_3 R_2^2 + gI_2}\right], \qquad \text{ANS.}$$

and

$$T_2 = W_3\left[1 - \frac{W_3 R_2^2 - W_1 r_2^2 R_2 \sin\beta}{W_1 r_2^2 + W_3 R_2^2 + gI_2}\right]. \qquad \text{ANS.}$$

Substituting the given numerical values into the above general equations yields

$$a_1 = 1.558 \text{ ft/s}^2, \qquad \text{ANS.}$$

$$a_3 = 2.92 \text{ ft/s}^2, \qquad \text{ANS.}$$

$$T_1 = 10.97 \text{ lb}, \qquad \text{ANS.}$$

$$T_2 = 13.64 \text{ lb}. \qquad \text{ANS.}$$

Example 18.17

Refer to the system shown in Figure E18.17(a) consisting of two homogeneous cylinders of weights W_1 and W_2 and two massless connecting rods, one on each side, and regard W_1, r_1, β, W_2 and r_2 as known quantities. Both cylinders roll without sliding on the horizontal plane. (a) Draw appropriate free-body and inertial force diagrams. Determine the number of unknowns and state the number of equations available to solve this problem. (b) Write these equations and solve for the unknowns in terms of known quantities. (c) Determine numerical values for all of the unknown quantities for the case where $W_1 = 500$ N, $W_2 = 300$ N, $r_1 = 0.75$ m, $r_2 = 0.50$ m and $\beta = 20°$.

Solution

(a) Refer to Figures E18.17 (b) and (c) which represent the free-body and inertial force diagrams for W_1 and W_2, respectively, and list the seven unknowns T, F_1, N_1, α_1, F_2, N_2 and α_2. For each of the two bodies W_1 and W_2, we may write 3 kinetic equations for a total of six equations. These six kinetic equations will be written by referring to Figures E18.16(b) and (c).

Another equation is required, and we turn to kinematics to write this seventh equation. The mass centers G_1 and G_2 of bodies W_1 and W_2 travel straight line paths parallel to the horizontal plane on which the bodies roll. Because the connecting rod is joined by frictionless pins at these mass centers, then, the ends of the connecting rod move along parallel straight-line paths. This leads to the conclusion that the connecting rod executes rectilinear translation and that $r_1\alpha_1 = r_2\alpha_2$. This follows from the facts that the connecting rod translates and the cylinders roll without sliding. Furthermore, we note that the angle β is not time-dependent because the connecting rod translates. The connecting

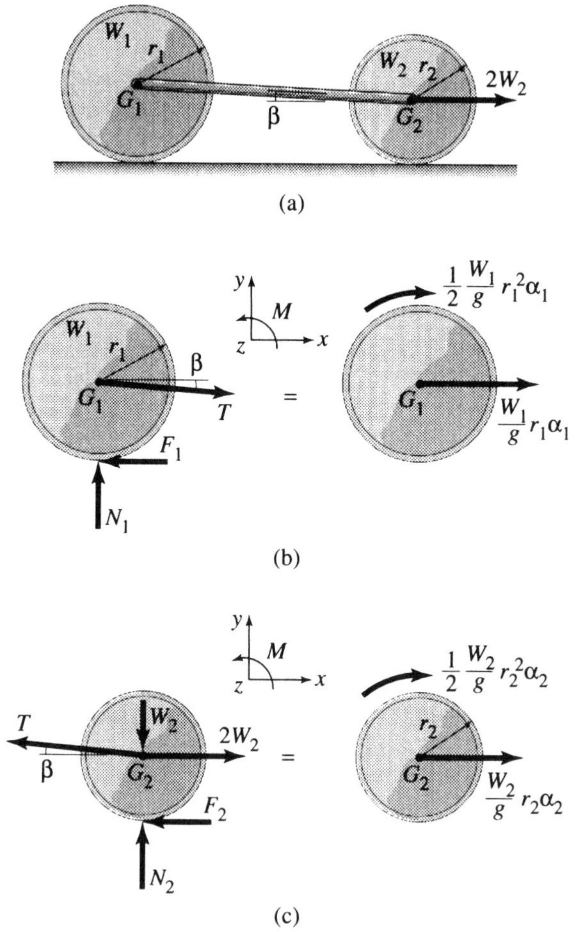

FIGURE E18.17.

rod is a two-force member which we have assumed to carry a tensile force T. Our overall analysis has enabled us to conclude that this problem involves seven unknowns and that we will be able to write seven equations to solve it, in general.

(b) Refer to Figure E18.17(b) for W_1, and use Eqs. (18.23) to obtain

$$\sum F_x = ma_{G1x}, \quad T\cos\beta - F_1 = \frac{W_1}{g} r_1 \alpha_1, \tag{a}$$

$$\sum M_{G1} = I_{G1}\alpha, \quad -F_1 r_1 = -\frac{1}{2}\frac{W_1}{g} r_1^2 \alpha_1, \tag{b}$$

$$\sum F_y = ma_{G1y}, \quad N_1 - T\sin\beta - W_1 = 0. \tag{c}$$

Similarly, for W_2 shown in Figure E18.17(c), Eqs. (18.23) yield

$$\sum F_x = ma_{G2x}, \quad -T\cos\beta - F_2 + 2W_2 = \frac{W_2}{g} r_2 \alpha_2, \tag{d}$$

$$\sum M_G = I_{G2}\alpha, \quad -F_2 r_2 = -\frac{1}{2}\frac{W_2}{g} r_2^2 \alpha_2, \tag{e}$$

$$\sum F_y = ma_{G2y}, \quad N_2 + T\sin\beta - W_2 = 0. \tag{f}$$

Substituting for F_1 from Eq. (b) in Eq. (a), and solving for $T\cos\beta$,

$$T\cos\beta = \frac{3}{2}\frac{W_1}{g} r_1 \alpha_1. \tag{g}$$

Substituting for F_2 from Eq. (e) in Eq. (d), and solving for $T\cos\beta$,

$$T\cos\beta = 2W_2 - \frac{1}{2}\frac{W_2}{g} r_2 \alpha_2. \tag{h}$$

Equating $T\cos\beta$ values of Eqs. (g) and (h), replacing $r_2\alpha_2$ by $r_1\alpha_1$, and solving for α_1,

$$\alpha_1 = \frac{4gW_2}{r_1(3W_1 + W_2)}. \qquad \text{ANS.}$$

Since $\alpha_2 = \dfrac{r_1}{r_2}\alpha_1$,

$$\alpha_2 = \frac{4gW_2}{r_2(3W_1 + W_2)}. \qquad \text{ANS.}$$

Substituting for α_1 in Eq. (g), and solving for T,

$$T = \frac{6W_1 W_2}{(3W_1 + W_2)\cos\beta}.$$

Substituting for α_1 and T in Eq. (a) and solving for F_1,

$$F_1 = \frac{2W_1 W_2}{(3W_1 + W_2)}. \qquad \text{ANS.}$$

Substituting for T in Eq. (c) and solving for N_1,

$$N_1 = W_1\left[1 + \frac{6W_2}{(3W_1 + W_2)}\tan\beta\right].$$

Similar substitutions in Eqs. (e) and (f) enables us to find F_2 and N_2. Thus,

$$F_2 = \frac{2W_2^2}{(3W_1 + W_2)}, \qquad \text{ANS.}$$

$$N_2 = W_2\left[1 - \frac{6W_1}{(3W_1 + W_2)}\tan\beta\right]. \qquad \text{ANS.}$$

(c) Substituting the given numerical values in the above equation yields

$$\alpha_1 = 28.6 \text{ rad/s } \circlearrowright, \quad \alpha_2 = 42.9 \text{ rad/s}^2 \circlearrowright, \qquad \text{ANS.}$$

$$F_1 = 166.7 \text{ N}, \quad F_2 = 100.0 \text{ N}, \qquad \text{ANS.}$$

$$N_1 = 692 \text{ N}, \quad N_2 = 118.0 \text{ N}, \qquad \text{ANS.}$$

$$T = 532 \text{ N}. \qquad \text{ANS.}$$

Example 18.18 The system shown in Figure E18.18(a) consists of two crates weighing $W_1 = W$ and $W_2 = kW$ connected by an inextensible cable which passes over a massless pulley at B. Both planes are rough for which the coefficient of kinetic friction is μ. At the instant shown, W_1 moves down its inclined plane. (a) In terms of W, θ_1, θ_2, k, μ, b, c, and d, determine the tension in the cable, the acceleration of the system, and the normal reaction between each body and its plane giving its location relative to the center of mass of the body. (b) Determine numerical values for

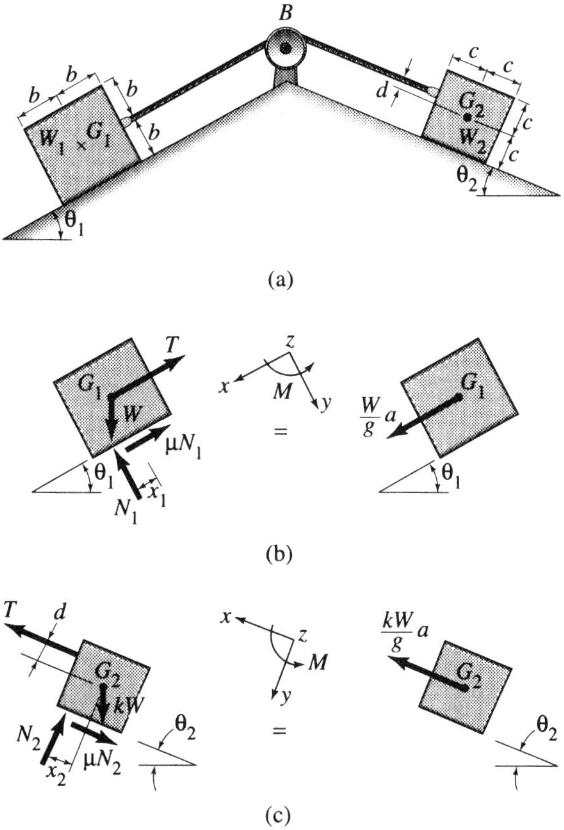

FIGURE E18.18.

these quantities for the case when $W = 10{,}000$ N, $\theta_1 = 60°$, $\theta_2 = 15°$, $k = 0.5$, $\mu = 0.3$, $b = 0.6$ m, $c = 0.4$ m and $d = 0.2$ m.

Solution

(a) The free-body and inertial force diagrams for W_1 are shown in Figure E18.18(b). Note that both crates have the same acceleration a and that the tension T in the cable is the same on either side of the pulley. Thus, using Eqs. (18.23),

$$\sum F_y = m_1(a_{G1})_y = 0, \quad W\cos\theta_1 - N_1 = 0,$$

$$N_1 = W\cos\theta_1, \qquad \text{(a) ANS.}$$

$$\sum M_{G1} = I_{G1}\alpha_1 = 0, \quad \mu N_1 b - N_1 x_1 = 0,$$

$$x_1 = \mu b, \qquad \text{(b) ANS.}$$

$$\sum F_x = m_1(a_{G1})_x, \quad W\sin\theta_1 - \mu N_1 - T = \frac{W}{g}a. \qquad \text{(c)}$$

Substituting for N_1 from Eq. (a) in Eq. (c) and solving for T,

$$T = W\left(\sin\theta_1 - \mu\cos\theta_1 - \frac{a}{g}\right) \qquad \text{(d)}$$

The free-body and inertial force diagrams for W_2 are shown in Figure E18.18(c). Equations (18.23) lead to

$$\sum F_y = m(a_{G2})_y, \quad kW\cos\theta_2 - N_2 = 0,$$

$$N_2 = kW\cos\theta_2. \qquad \text{(e) ANS.}$$

$$\sum M_{G2} = I_{G2}\alpha_2 = 0, \quad Td + \mu N_2 c - N_2 x_2 = 0, \qquad \text{(f)}$$

$$\sum F_x = m(a_{G2})_x\text{:}\ T - \mu N_2 - kW\sin\theta_2 = \left(\frac{kW}{g}\right)a. \qquad \text{(g)}$$

Substituting for N_2 from Eq. (e) in Eq. (g) and solving for T,

$$T = kW\left(\sin\theta_2 + \mu\cos\theta_2 + \frac{a}{g}\right). \qquad \text{(h)}$$

Equating the values of T in Eqs. (d) and (h) and solving for the acceleration a yields

$$a = \frac{(\sin\theta_1 - \mu\cos\theta_1 - k\sin\theta_2 - \mu k\cos\theta_2)g}{1+k}. \qquad \text{(i)}$$

Substituting for a from Eq. (i) in Eq. (d) or (h) yields

$$T = \left(\frac{kW}{1+k}\right)(\sin\theta_1 - \mu\cos\theta_1 + \sin\theta_2 + \mu\cos\theta_2). \qquad \text{(j) ANS.}$$

Substituting for N_2 and for T from Eqs. (e) and (j) in Eq. (f) and solving for x_2 leads to

$$x_2 = \mu c + \left(\frac{d}{1+k}\right)\left[\frac{\sin\theta_1 - \mu\cos\theta_1 + \sin\theta_2 + \mu\cos\theta_2}{\cos\theta_2}\right]. \quad \text{(k)} \quad \text{ANS.}$$

(b) Specializing the general expressions above for the given numerical values yields

$$N_1 = 5000 \text{ N}, x_1 = 0.1800 \text{ m}, a = 2.89 \text{ m/s}^2, \quad \text{ANS.}$$

$$T = 4215 \text{ N}, N_2 = 4830 \text{ N, and } x_2 = 0.295 \text{ m}. \quad \text{ANS.}$$

■

Problems

18.115 The system shown in Figure P18.115 consists of a cylinder of weight W and radius r and two rods AB (one on each side) each of weight $W/4$. The two rods are attached at B with frictionless pins. An externally applied force $P = W$ acts horizontally to the right through B which is the mass center of the cylinder. Ignore friction at contact point A. The cylinder rolls without sliding on the plane. Determine the linear acceleration of the system, the reaction components at B and the normal and frictional forces at C. Express answers in terms of W.

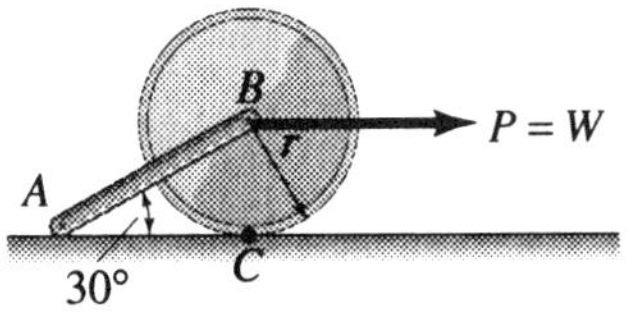

FIGURE P18.115.

18.116 Refer to Figure P18.116 which shows a system consisting of two cylinders, A and B, connected by two light rods attached with frictionless pins at their centers. Each cylinder has a weight W and a radius r. The cylinders roll without slipping on the horizontal plane under the action of the force $P = W$. Assume that the connecting rods are weightless, and determine the angular acceleration of the cylinders, the force in connecting rods, and the normal and frictional forces acting on the two cylinders.

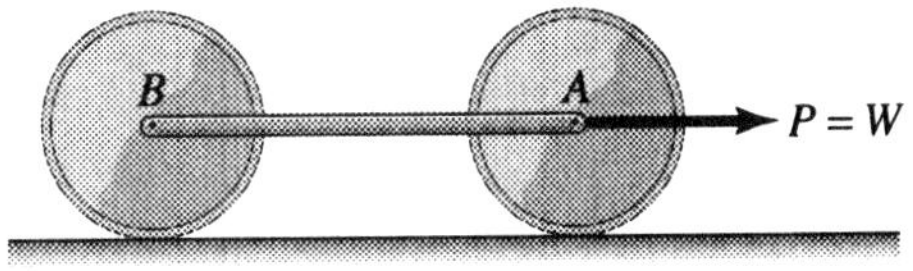

FIGURE P18.116.

18.117 Refer to the system shown in Figure P18.117. Cylinder A rolls without sliding on the horizontal plane, and fixed peg B is frictionless. The cylinder has a weight W and a radius r, and block C has a weight of $W/2$. Determine the acceleration of block C and the tension in the connecting cord. Express answers in terms of the known quantities W and g.

18.118 Consider the system shown in Figure P18.118 consisting of cylinders A and B which are connected by a weightless

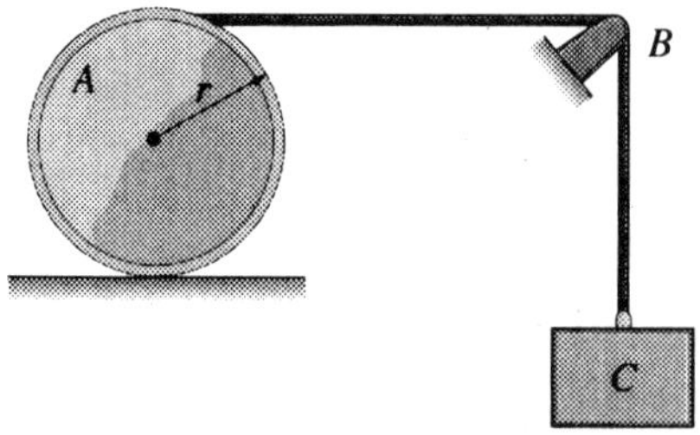

FIGURE P18.117.

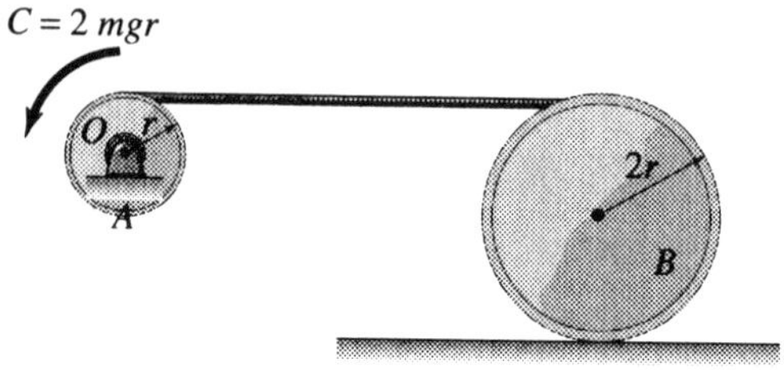

FIGURE P18.118.

cord. The mass of cylinders A is m and that of cylinder B is 2 m. The pin at O is to be regarded as frictionless. When the couple $C = 2\ mgr$ is applied, cylinder B rolls without slipping. Determine the linear acceleration of cylinders B and the tension in the connecting cord. Express answers in terms of m and g.

18.119 Spool A of the system shown in Figure P18.119 has a weight equal to $2W$ and a radius of gyration with respect to its mass center of $1.3r$. Block B has a weight equal to W. Friction between block B and the horizontal plane is negligible. A force $P = 4W$ is applied to block B, and spool A rolls without sliding. Determine the acceleration of block B, the tension in the connecting cord, and the normal and frictional forces acting on the spool at C. Express answers in terms of W and g.

FIGURE P18.119.

18.120 Refer to the system shown in Figure P18.120, consisting of cylinder A of weight W_A and radius r_A, cylinder B of weight W_B and radius r_B, and block C of weight W_C connected by a weightless cord as shown. The system is released from rest, and cylinder B rolls without sliding on the inclined plane. (a) In terms of W_A, W_B, W_C and β, find the linear acceleration of cylinder B and the tensions on both sides of the connecting cord. (b) Find numerical values for these quantities if $W_A = 200$ lb, $W_B = 100$ lb, $W_C = 50$ lb and $\beta = 30°$.

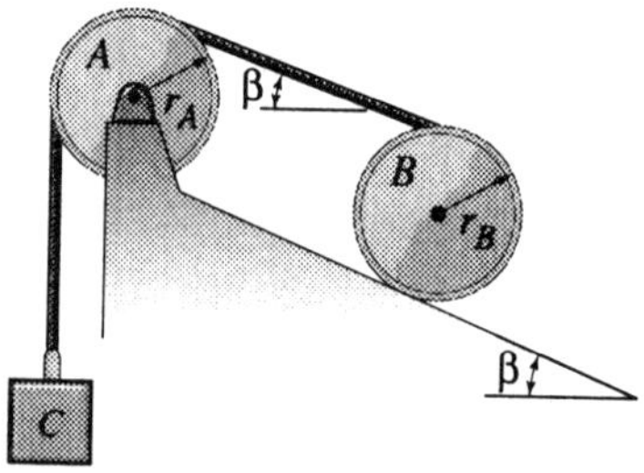

FIGURE P18.120.

18.121 In Figure P18.121 cylinder A has a weight W and radius r and block B has

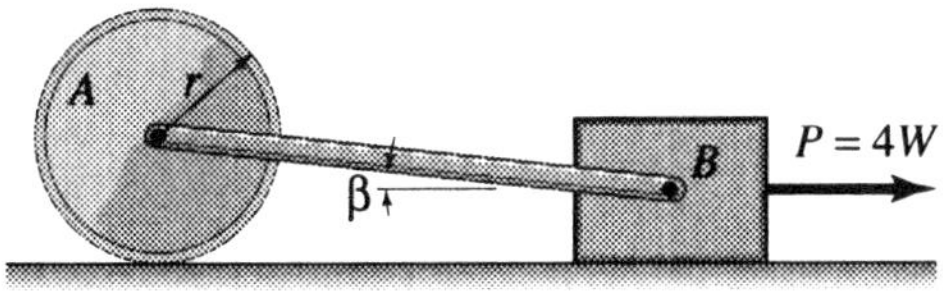

FIGURE P18.121.

a weight $W/2$. A very light strut is connected with frictionless pins to the centers of A and B. When the force $P = 4W$ is applied, the cylinder rolls without sliding on the horizontal plane where the kinetic coefficient of friction is μ. (a) In terms of W, μ, and β, determine the linear acceleration of the system and the force in the connecting strut. (b) Find numerical values for these quantities if $W = 200$ N, $\mu = 0.1$ and $\beta = 10°$.

18.122 The homogeneous member AB shown in Figure P18.122 has a weight of 120 lb and is supported by two identical weightless rods AC and BD each 2 ft long. The hinges at A, B, C, and D are frictionless. In the position shown, the two rods have a ccw angular velocity of 2 rad/s. Determine the forces in rods AC and BD.

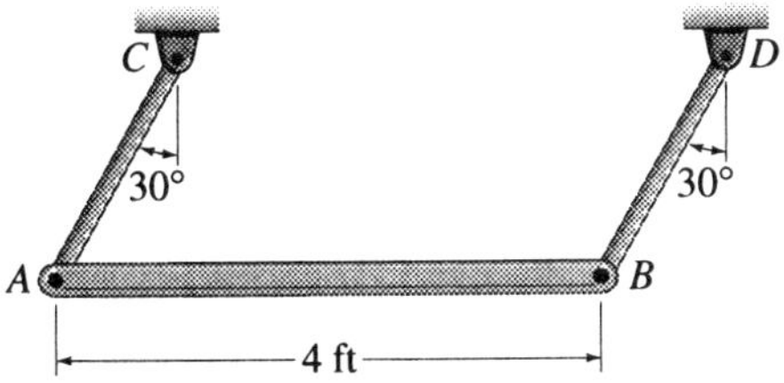

FIGURE P18.122.

18.123 A uniform slender rod of mass m and length $2r$ is attached to a thin hoop of mass m and radius r as shown in Figure P18.123. This body is released from rest in the position shown and rolls without sliding. At the instant of release, determine (a) the angular acceleration of the body, (b) the normal and frictional forces exerted on the body by the plane, (c) the acceleration of the mass center of the body, and (d) the minimum coefficient of friction so that this body will roll without sliding. Express answers in terms of g, r and m. (Hint: First locate the mass center of the body).

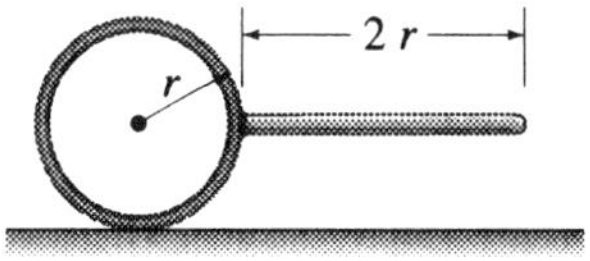

FIGURE P18.123.

18.124 In Figure P18.124, block A has a mass of 6 kg, pulley B has a mass of 2 kg which is concentrated in a thin rim at a distance $r = 0.25$ m from its center O, and cylinder D has a mass of 4 kg and a radius $r = 0.25$ m. If the system is released from rest, determine the acceleration of block A and the tensions in both sides of the connecting cord.

18.125 The system shown in Figure P18.125 consists of two blocks, A and F, two

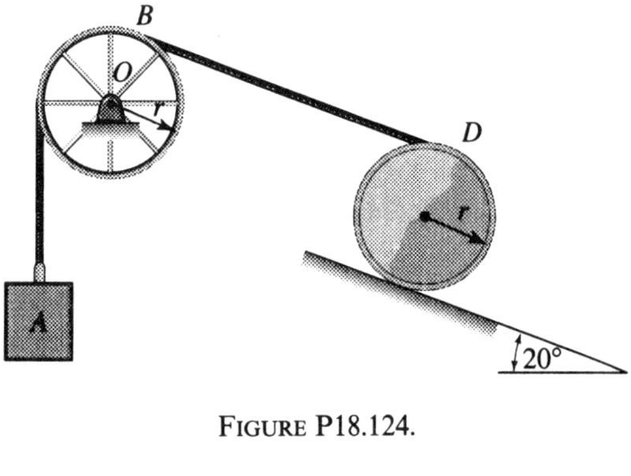

FIGURE P18.124.

fixed, frictionless pegs, B and E, and a spool D which rolls without sliding on the horizontal plane. Block A has a mass m, spool D has a mass $2m$ and radius of gyration $k_G = 1.5r$, and block F has a mass $2m$. If the system is released from rest, determine the distance traveled by block F in 4 s. What are the tensions on both sides of the connecting cord during the motion. Let $m = 2$ slug and $r = 1.2$ ft.

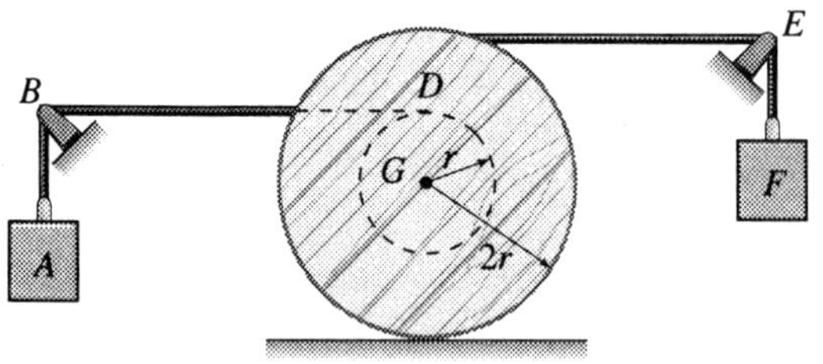

FIGURE P18.125.

18.126 The system shown in Figure P18.126 consists of cylinder A of mass m and radius r, block B of mass m, and block D of mass $2m$ connected together by a weightless cord as shown. Assume frictionless conditions at B and C. If the system is released from rest, determine the velocity of block D after it has moved through a distance s. Express the answer in terms of s.

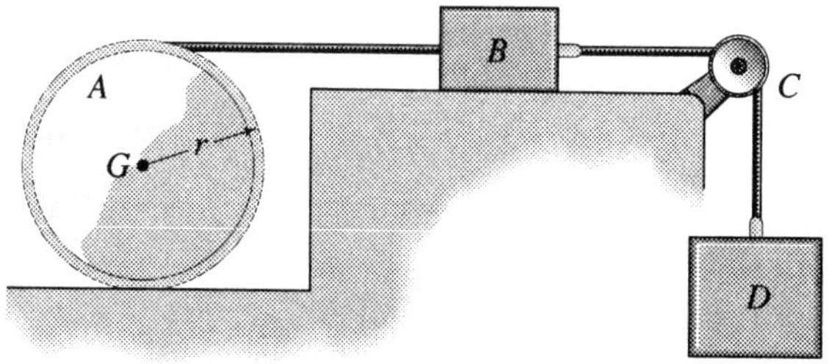

FIGURE P18.126.

18.127 Link BE of the four-bar linkage shown in Figure P18.127 is vertical at the instant shown and it has a mass of 2 slug. Links AB and DE may be assumed massless in your analysis. Link AB has an angular velocity of 2 rad/s cw and zero angular acceleration. (a) Determine the angular velocities of BE and DE. (b) Find the angular accelerations of BE and DE. (c) Determine the horizontal components of the forces at B and E acting on member BE.

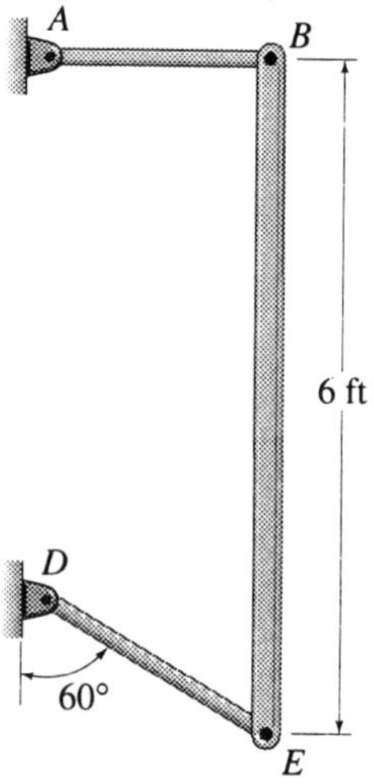

FIGURE P18.127.

18.128 Link DE of the four-bar-linkage shown

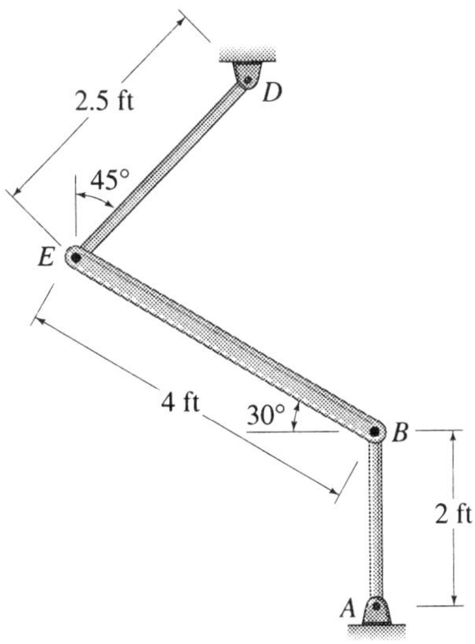

FIGURE P18.128.

in Figure P18.128 has a mass of 3 slug, a ccw angular velocity of 2 rad/s, and zero angular acceleration at the instant depicted. (a) Determine the angular velocities of BE and AB. (b) Find the angular accelerations of BE and AB. (c) Determine the components of the forces normal to EB acting at E and B on member EB. Assume links DE and AB to be massless.

18.129 In Figure P18.129, the cylinder has a mass of 4 slug, the rod has a mass of 2 slug, and the block has a mass of 1 slug. If $r = 0.5$ ft, determine the linear acceleration of the system and the reaction components at A and B. Assume frictionless conditions at the block and that the cylinder rolls without sliding.

18.130 Assume that rods AB and BD of the system, shown in Figure P18.130, are massless. Rod BD has a ccw angular velocity of 1 rad/s and a ccw angular acceleration of 2 rad/s^2. Determine all forces acting on the cylinder of mass m and express them in terms of the known quantities P, m and r. Assume that the cylinder rolls without sliding.

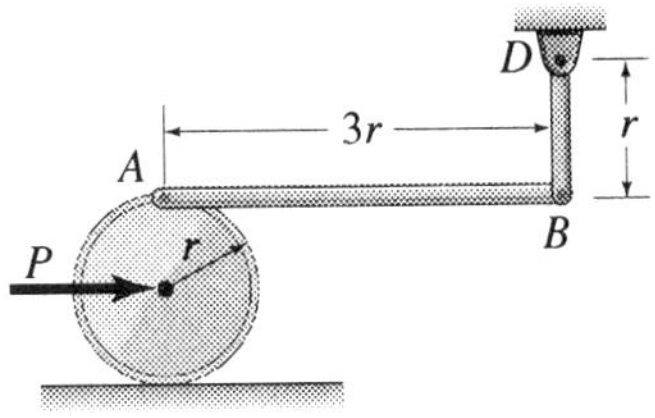

FIGURE P18.130.

18.131 Refer to Figure P18.131, and show that the cylinder cannot roll without sliding, given the following inputs: $M = 2mgb$, mass of cylinder $= 2m$, mass of cube $= 0.5m$, and coefficient of friction $= 0.5$. The small rod BG is massless and connects the mass centers of the cube and cylinder. Hint: Assume initially that the cylinder rolls without sliding.

18.132 Refer to Problem 18.131 and let the coefficient of friction equal 0.90. Show that, in this case, the cylinder rolls

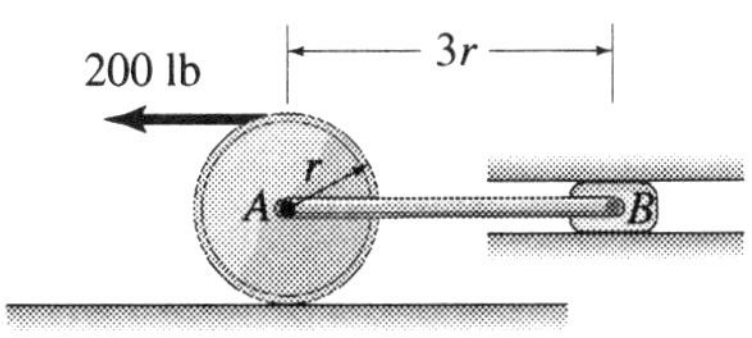

FIGURE P18.129.

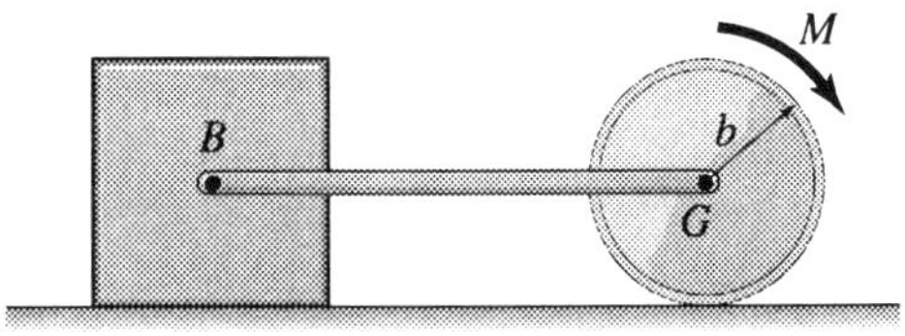

FIGURE P18.131.

without sliding. Determine the force in rod BG and express it in terms of m.

18.133 Refer to Figure P18.133, and determine the acceleration of m_1 and the tensions in both cables after the system is released from rest. Assume that the system is frictionless. Let $m_1 = 2$ slug, $m_2 = 1$ slug, $m_3 = 1$ slug, $r_1 = 0.5$ ft, $r_2 = 1.2$ ft, and $k_2 = 0.85$ ft.

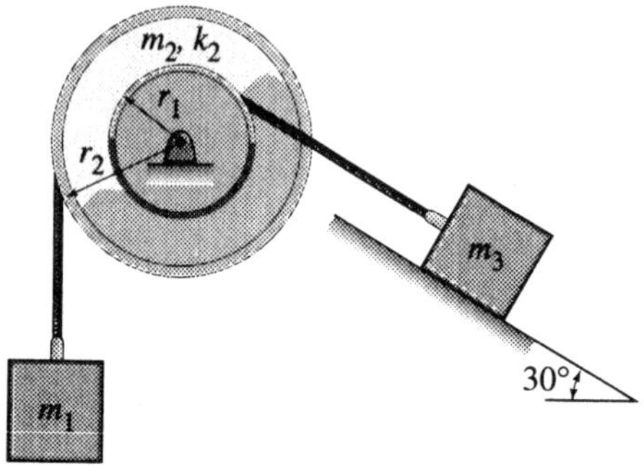

FIGURE P18.133.

18.134 Refer to Figure 18.134, and determine the acceleration of m_3 and the tensions in both cables after the system is released from rest. Assume that the system is frictionless. Let $m_1 = 4$ kg, $m_2 =$

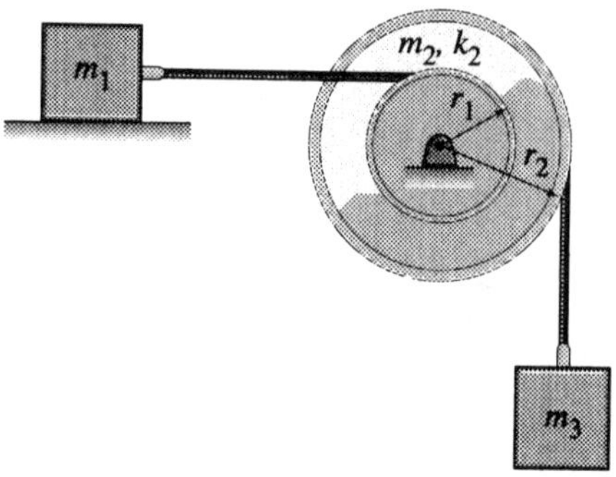

FIGURE P18.134.

6 kg, $m_3 = 8$ kg, $r_1 = 0.4$ m, $r_2 = 0.8$ m, and $k_2 = 0.55$ m.

18.135 Refer to Figure P18.135 and determine the couple C required to give the flywheel B an angular acceleration of 4 rad/s^2. Each flywheel is cylindrical in shape. Assume that the flywheels are the only mass elements of the system and neglect frictional resistance. Input data for the system is given below

Flywheel B	Weight = 6000 lb	Diameter = 4.00 ft
Flywheel E	Weight = 3200 lb	Diameter = 3.00 ft
Gear A		Diameter = 2.80 ft
Gear D		Diameter = 1.20 ft

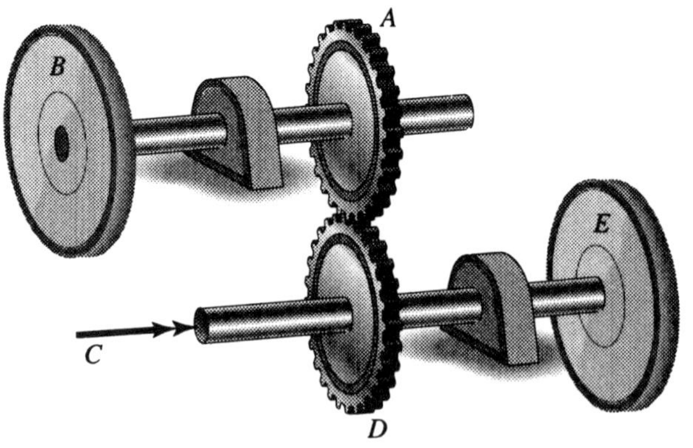

FIGURE P18.135.

18.136 Refer to Figure P18.136 and imagine that the weight W is released from rest. Use the following numerical inputs to find the angular acceleration of gear A and the tooth forces acting on gear B. Assume that A and B are cylinders in

Body	Weight (N)	Radius (m)
A	100	$r_A = 0.1$
B	400	$r_B = 0.3$
C	800	$r_C = 0.4$, $r_D = 0.2$
W	4000	

computing their mass moments of inertia and that the centroidal radius of gyration for C is $k = 0.25$ m.

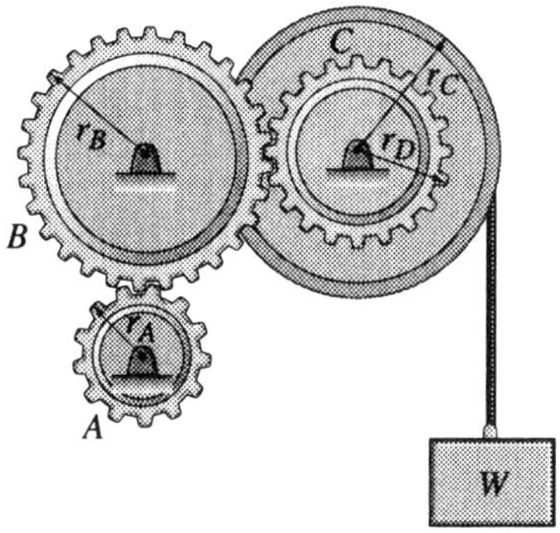

FIGURE P18.136.

18.137 Refer to the gear system shown in Figure P18.137 which is at rest before the force $F = 100$ N is applied. After F is applied, determine the acceleration of rack C and the tooth forces acting on gears A and B. Use the following numerical inputs:

Body	Weight (N)	Radius (m)
A	800	$r_A = 0.20$
B	600	$r_B = 0.15$
C	200	—

Assume that gears A and B are cylinders when computing their mass moments of inertia. Assume that the weight of the rack C is shared equally by body B and the frictionless idler pulley D which has negligible mass.

18.138 Refer to the pulley system shown in Figure P18.138. If the system is released from rest, determine the distance traveled by the weight $W = 100$ lb. Use the following numerical values in the solution.

Body	Weight (lb)	Centroidal Radius of Gyration (ft)	Radius (ft)
A	483	0.95	$r_3 = 0.80$, $r_4 = 1.00$
B	322	1.05	$r_1 = 1.20$, $r_2 = 1.00$

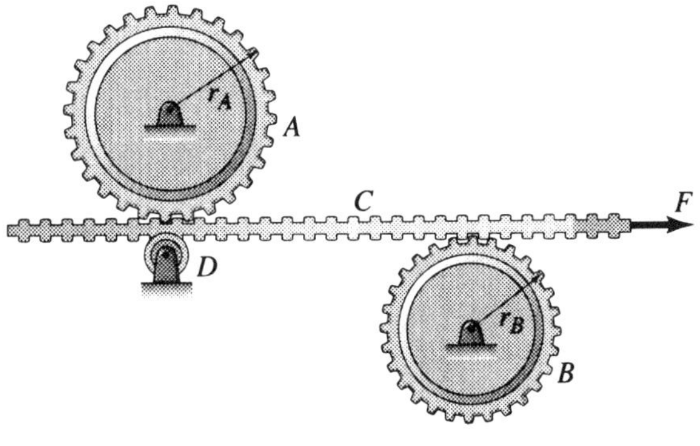

FIGURE P18.137.

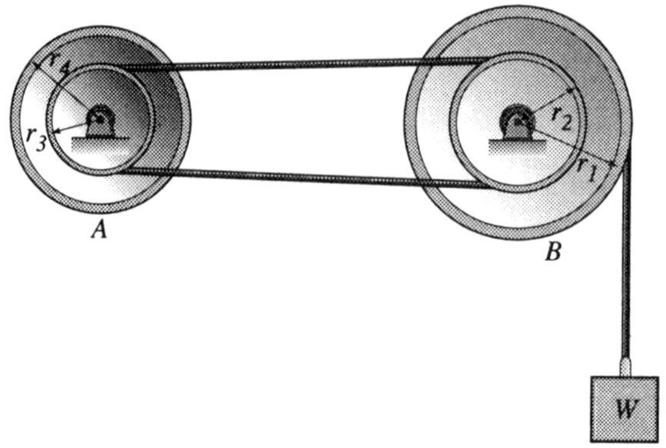

FIGURE P18.138.

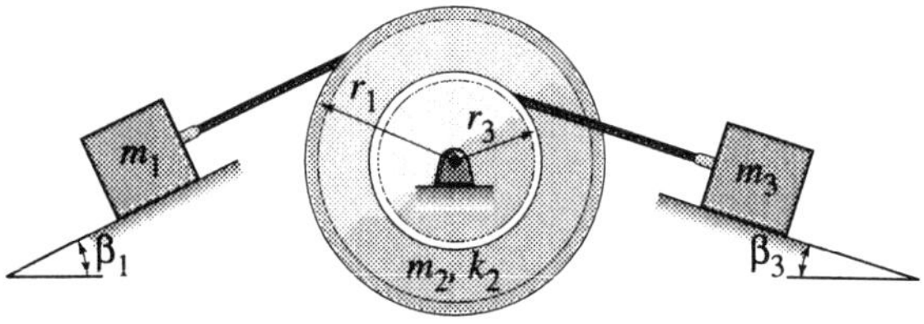

FIGURE P18.139.

18.139 Refer to the frictionless system shown in Figure P18.139 and regard m_1, β_1, m_2, r_1, r_3, k_2, m_3, and β_2 as known quantities. When the system is released from rest, m_1 moves down its inclined plane. Determine the acceleration of m_1 and the tension in both cables. Express answers in terms of the known quantities.

18.140 Refer to the system shown in Figure P18.140. The cylinder of mass m rotates ccw with angular velocity ω_0 when P is applied as shown. The coefficient of friction at the brake shoe is μ, and this is the only friction in the system. If the cylinder is to be brought to rest in two complete revolutions, determine the force P. Let $W = 32.2$ lb, $r = 1.0$ ft, $m = 4$ slug, $\omega_0 = 4$ rad/s, $b = 1.2$ ft and $\mu = 0.1$.

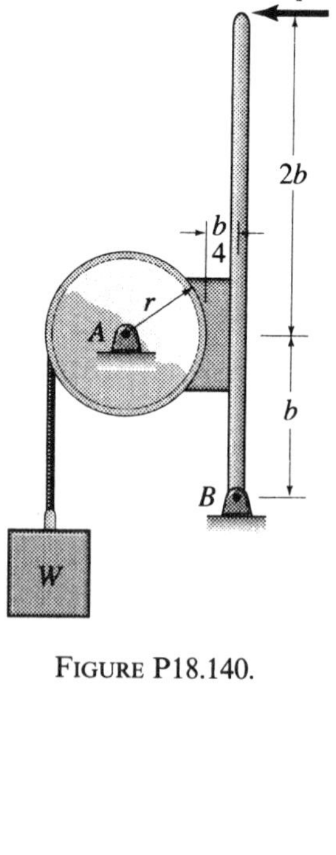

FIGURE P18.140.

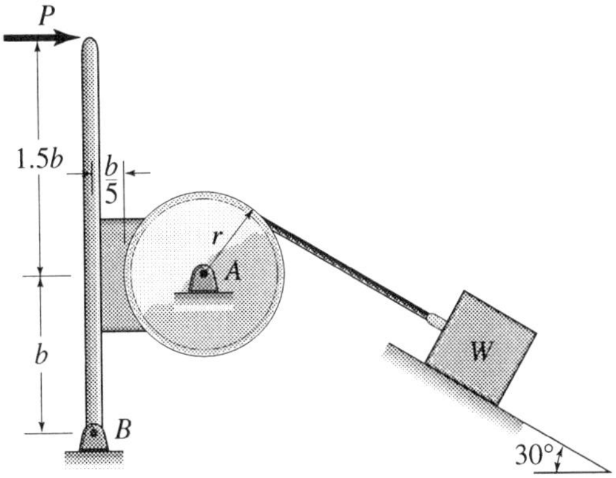

FIGURE P18.141.

18.141 The cylinder of mass m shown in Figure P18.141 is rotating cw with angular velocity ω_0 when P is applied, as shown. The coefficient of friction is μ between the brake shoe and the cylinder, and this is the only friction in the system. Determine the number of revolutions for the cylinder to come to a full stop after the force P is applied. Let $P =$ 200 N, $b = 0.50$ m, $\omega_0 = 4$ rad/s, $r =$ 0.25 m, $m = 50$ kg, $W = 200$ N, and $\mu = 0.3$.

18.142 The essential features of a dump truck mechanism are shown in Figure P18.142. Neglect all masses except that of the truck bed and load which have a mass of 120 slug with a mass center at point G. All pins are to be assumed frictionless. AB is vertical and BE is horizontal at the instant shown, and each of these lengths equals to 2.8 ft. At the given instant, the angular acceleration of the bed and load is 1 rad/s, ccw. Determine the force in the hydraulic cylinder BD. Assume that the bed and load act as a rigid body with a radius of gyration of 6.75 ft with respect to an axis through O.

18.143 A mechanical system is shown in Figure P18.143. Body ABC is the mass element of the system which may be idealized as a 2 ft long slender rod AB of mass 1 slug and a 3 ft long slender

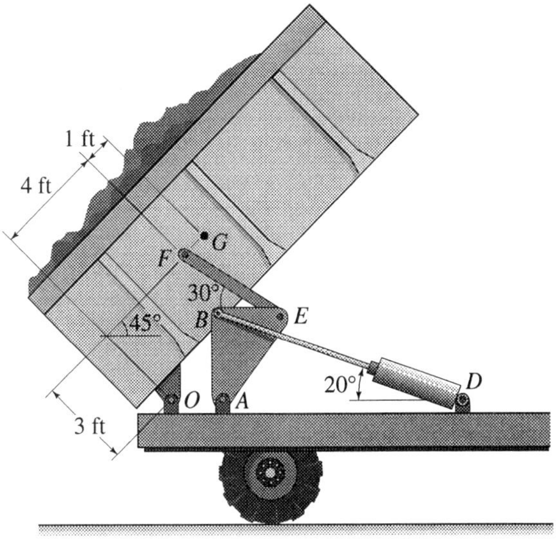

FIGURE P18.142.

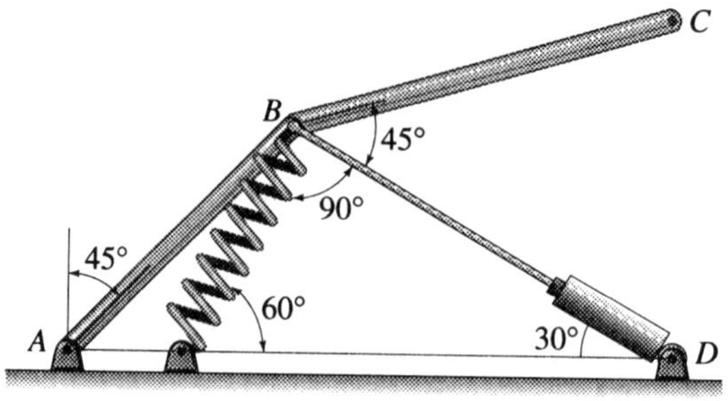

FIGURE P18.143.

rod BC of mass 2.25 slug. At the instant shown, the spring force is 200 lb tensile and the force in the hydraulic cylinder BD is 1000 lb compression. Determine the angular acceleration of the mass element.

Review Problems

18.144 Four slender rods, each of mass m and length L, are welded together as shown in Figure P18.144. In terms of m and L, determine the mass moment of inertia of the composite body with respect to an axis perpendicular to the page through point O.

18.145 The rigid body shown in Figure P18.145 has a mass density of 4.25 slug/ft^3. Use integration to find the mass moment of inertia with respect to the x axis.

18.146 A machine member made of steel is shown in Figure P18.146. The mass density for steel is approximately 15 slug/ft^3. Determine the mass moment of inertia and radius of gyration with respect to the z axis.

18.147 A uniform crate of weight $W = 2000$ lb is placed on the flat bed of a truck as shown in Figure P18.147. If the maximum acceleration of the truck is

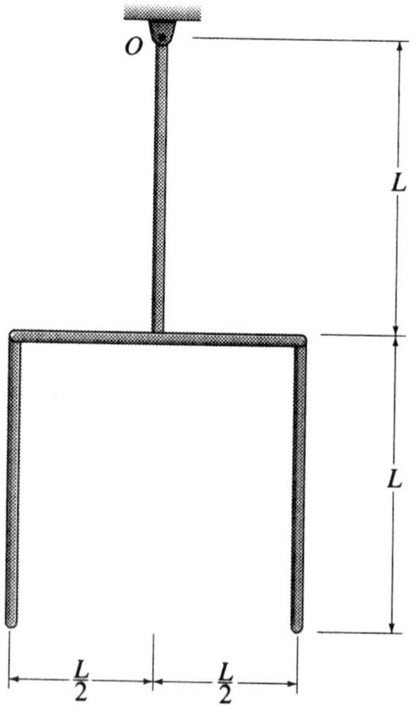

FIGURE P18.144.

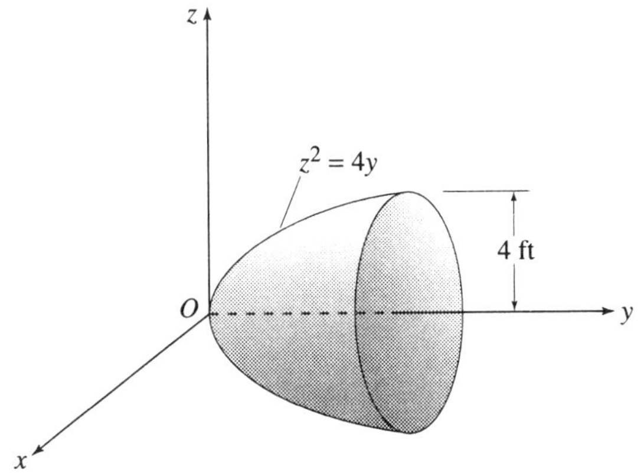

FIGURE P18.145.

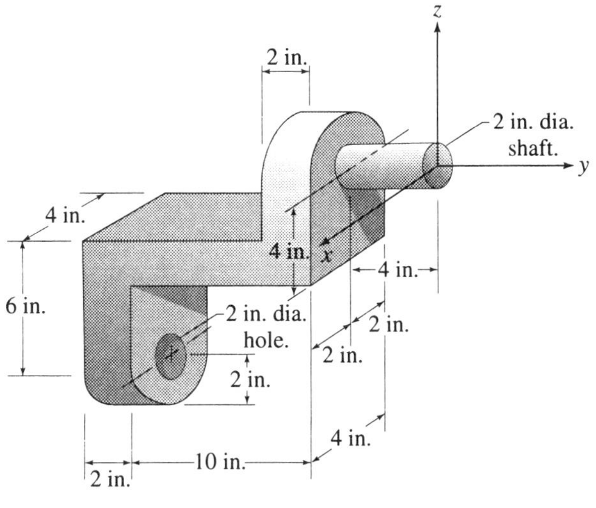

FIGURE P18.146.

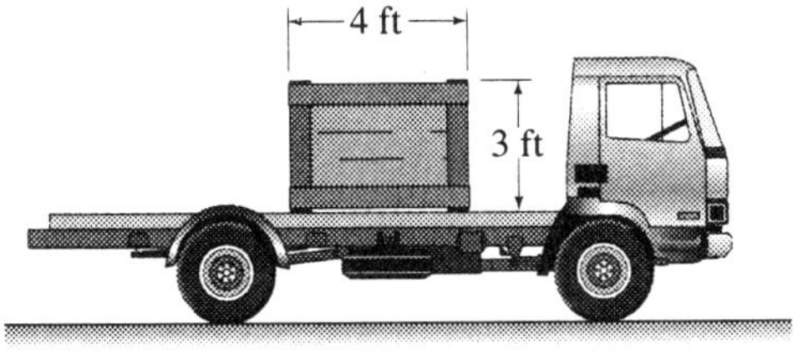

FIGURE P18.147.

10 ft/s^2, determine, for maximum acceleration (a) the minimum coefficient of friction so that the crate does not slide on the truck bed and (b) the position of the normal reaction between the bed and the crate relative to the crate's center of mass.

18.148 The triangular uniform plate of constant thickness shown in Figure P18.148, is constrained to move in the vertical direction. The plate has a weight $W =$

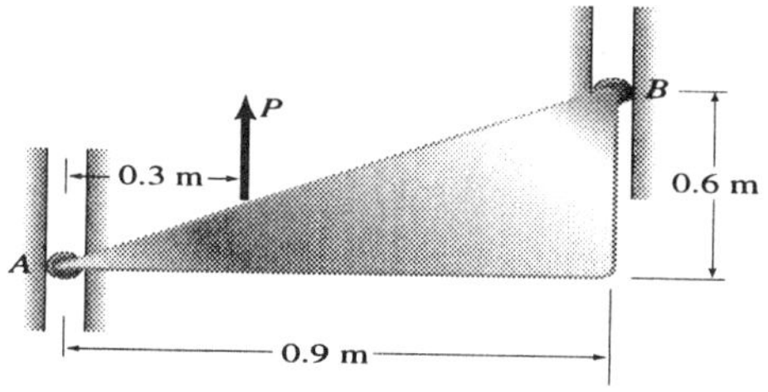

FIGURE P18.148.

400 N and, under the influence of the applied force P, it is accelerated upward at the rate $a = 5$ m/s^2. Determine (a) the magnitude of the applied force P and (b) the reaction components at the two small frictionless rollers at A and B.

18.149 The cable-pulley system shown in Figure P18.149 is used to slide the crate of weight W_1 on the rough horizontal plane for which the coefficient of kinetic friction is μ. (a) Determine the weight W_2 needed to produce an acceleration of the crate equal to a. Express the answer in terms of W_1, a, and μ. (b) Develop an expression in terms of b, h and a for the maximum height H so that the crate will slide without tipping to the left. (c) Find numerical values for W_2 and H for the case when $W_1 = 400$ lb, $b = 2$ ft, $h = 5$ ft and $a = 10$ ft/s^2.

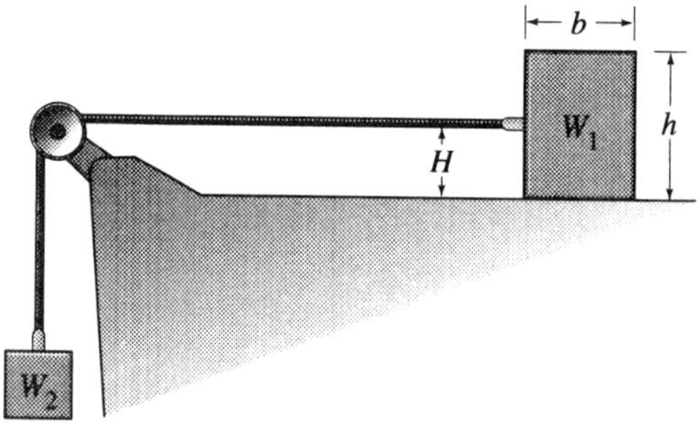

FIGURE P18.149.

18.150 A plate of constant thickness is cut into the shape shown in Figure P18.150 and attached to the two short weightless and parallel links AD and BE. The plate has a mass density of 100 kg/m^2, and the system is released from rest in the position shown. Determine the acceleration of the plate and the forces in the two links immediately after the system is released.

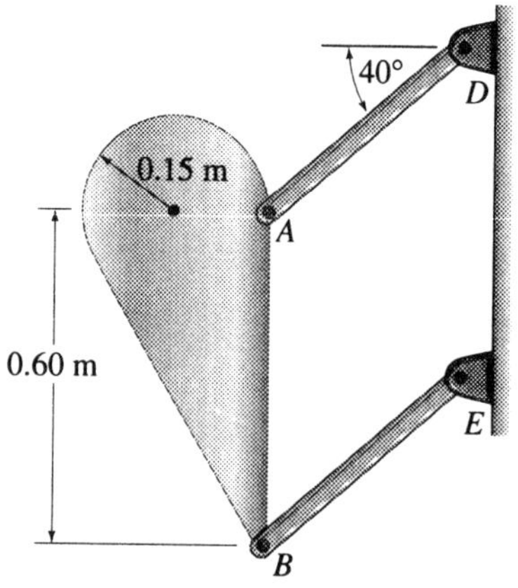

FIGURE P18.150.

18.151 A uniform crate of weight $W = 1500$ lb is placed on the flat bed of a truck as shown in Figure P18.151. The coefficient of friction between the truck bed and the crate is 0.4. The truck applies the brakes while moving forward causing a deceleration a. Determine the range of values for these decelerations so that the crate will neither slide nor tip on the bed of the truck.

18.152 Two homogeneous uniform rods AB and BC, each of mass 5 kg, are welded together and attached to two parallel

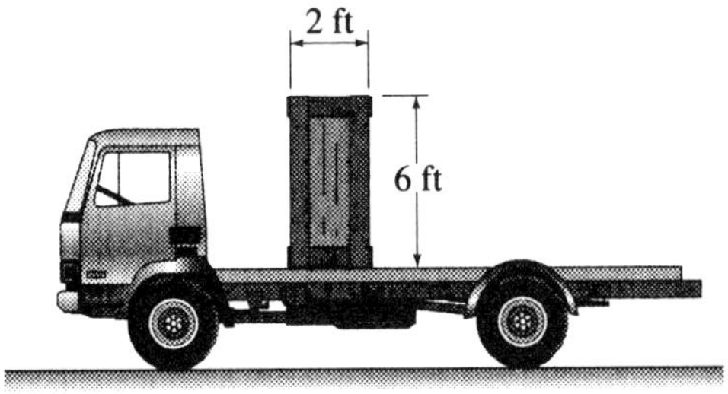

FIGURE P18.151.

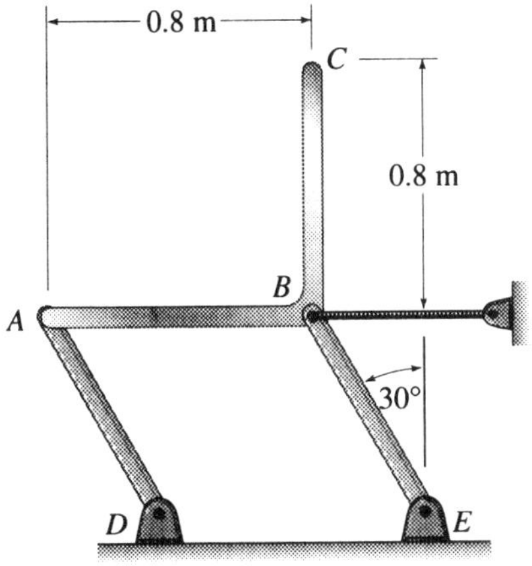

FIGURE P18.152.

and weightless links AD and BE as shown in Figure P18.152. Immediately after the string BF is cut, determine (a) the acceleration of the system consisting of AB and BC and (b) the forces in links AD and BE.

18.153 Four uniform rods, each of mass m and length L, are attached to a cylinder of mass 4 m and radius $L/2$ as shown in Figure P18.153. The system is hinged at its mass center G. One end of a cord is wrapped around the cylinder and the other end to a weight $W = 10$ mg. Assume frictionless conditions at the hinge, and determine the acceleration of W after it is released from rest. What are the reaction components at G?

18.154 The double pulley shown in Figure P18.154 weighs 30 lb and has a centroidal radius of gyration of 10 in. As-

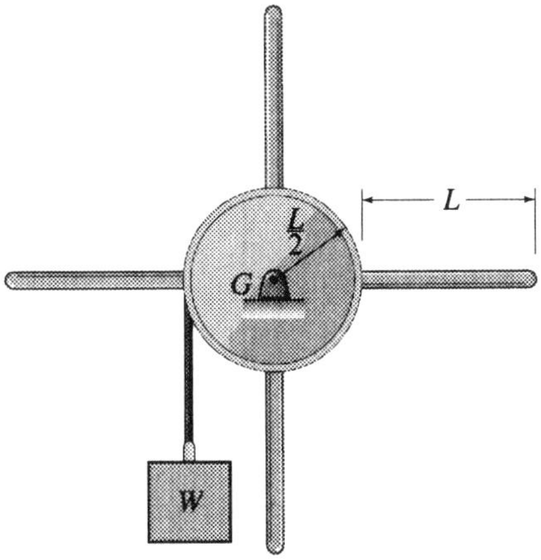

FIGURE P18.153.

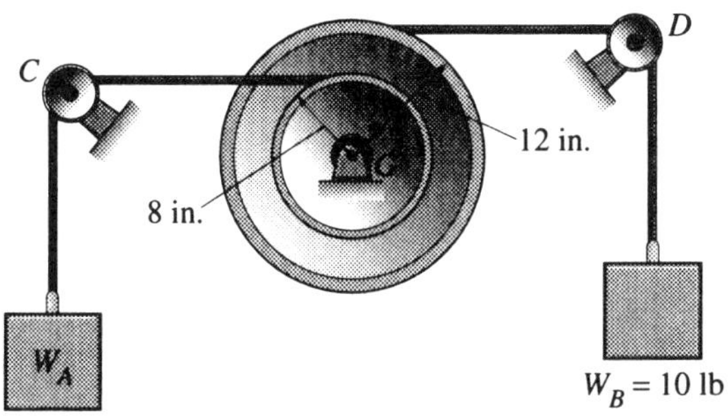

FIGURE P18.154.

sume frictionless conditions at the hinge G and at massless pulleys C and D, and determine (a) the weight of body A needed to produce a downward acceleration of body B equal to 5 ft/s^2 and (b) the reaction components at the hinge G as well as the acceleration of body A.

18.155 The system shown in Figure P18.155 is hinged at O and is displaced through a ccw angle θ and released. Develop the differential equation for the oscillating motion of the system. Assume that the hinge at O is frictionless. The rod has a mass m and a length L, and the L by $L/2$ plate has a mass of 5 m. Assume small values of θ and solve this differential equation for the initial conditions: $t = 0$, $\theta = \theta_O$, and $\dot{\theta} = \dot{\theta}_O$.

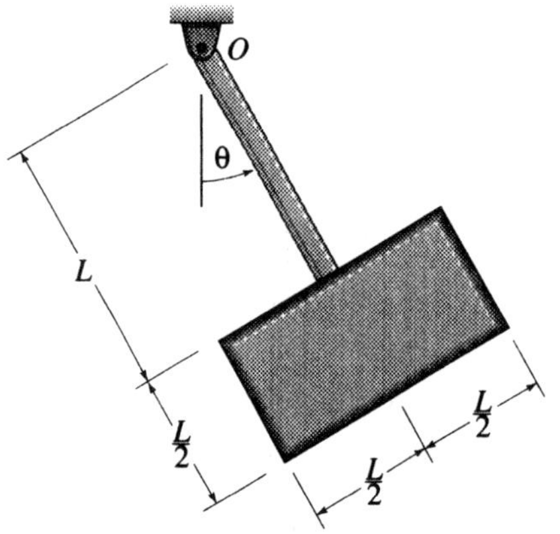

FIGURE P18.155.

18.156 The uniform, homogeneous plate shown in Figure P18.156 has a constant thickness and is supported at A by a frictionless hinge and at B by cord BC. The mass density of the plate is 80 kg/m^2. Determine the angular acceleration of the plate and the reaction components at A immediately after the cord is cut.

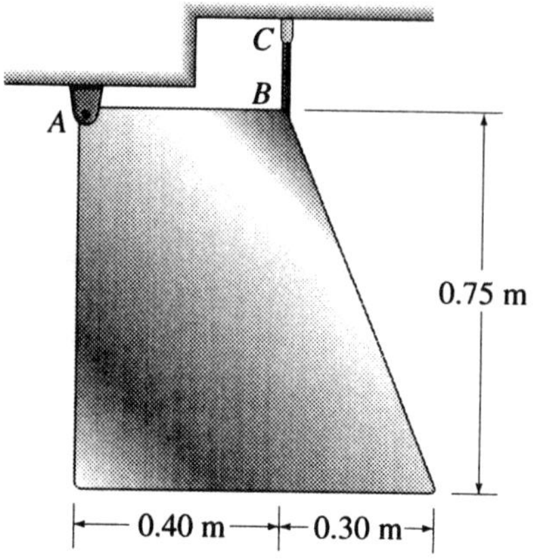

FIGURE P18.156.

18.157 The brake wheel of the band brake system, shown in Figure P18.157, has a centroidal mass moment of inertia $I_G = 5$ slug·ft^2. The wheel is rotating at a cw angular velocity ω_O when a vertical force of 20 lb is applied to the brake lever, as shown. If it takes 30 s for the wheel to come to a complete stop, determine the angular velocity ω_O.

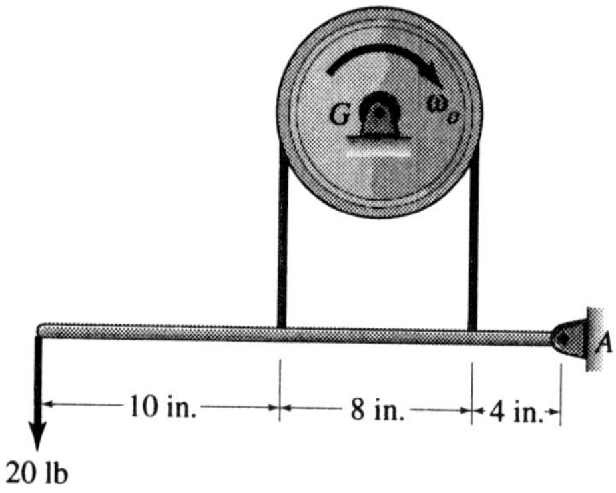

FIGURE P18.157.

18.158 A 3 ft × 4 ft rectangular plate weighing 50 lb is suspended from two cords as shown in Figure P18.158. Immediately

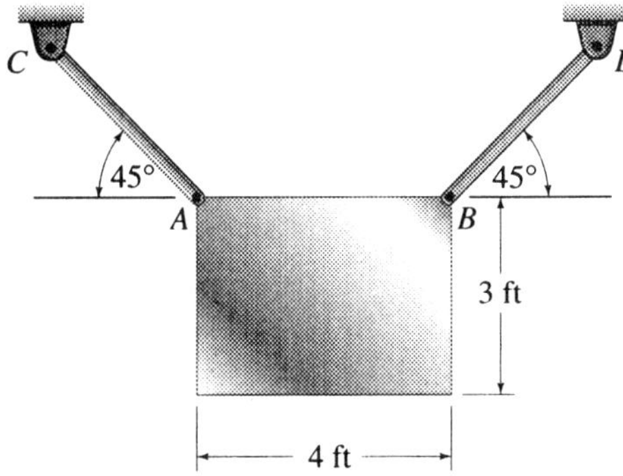

FIGURE P18.158.

after cord AC is cut, determine (a) the angular acceleration of the plate and (b) the linear acceleration of its mass center.

18.159 Cords are wrapped around the inner core and outer rim of a spool as shown in Figure P18.159. The pulley at B may be assumed massless and frictionless. The spool has a mass of 10 kg and a centroidal radius of gyration of 0.3 m. If the system is released from rest, determine (a) the angular acceleration of the spool and the linear acceleration of its mass center G, (b) the tension in cord CD, and (c) the distance traveled by point G in 2 s.

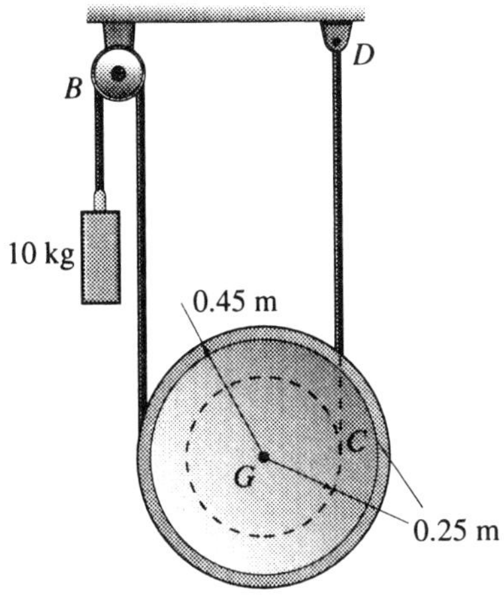

FIGURE P18.159.

18.160 A cylinder of weight $W = 15$ lb and radius $R = 3$ in. is released on a horizontal plane with a linear velocity $v_O = 30$ ft/s and an angular velocity $\omega_O = 20$ rad/s as shown in Figure P18.160. If the coefficient of kinetic friction between the plane and the cylinder is $\mu = 0.3$, determine the angular velocity of the cylinder and the linear speed of its mass center at the instant it starts rolling without sliding.

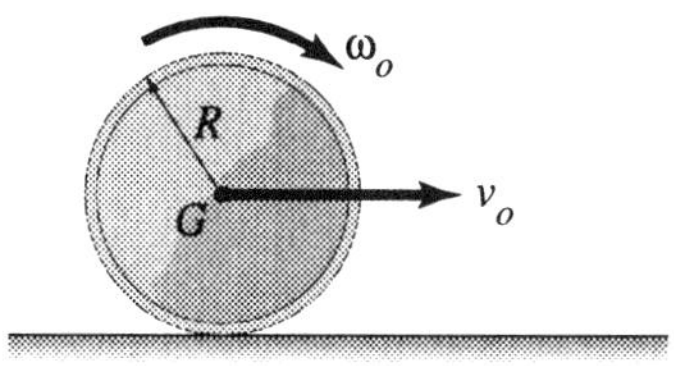

FIGURE P18.160.

18.161 The system shown in Figure P18.161 is released from rest. The spool has a weight $W = 500$ N and a centroidal radius of gyration $k_G = 0.4$m, and the pulley at B may be assumed massless and frictionless. If the spool rolls without slipping, determine (a) the linear and angular velocity of the spool 3 s after release and (b) the minimum coefficient of friction between the inclined plane and the spool so that it rolls without slipping.

18.162 Pulleys A and B are hinged at their respective mass centers as shown in Figure P18.162. Pulley A weighs 20 lb and has a radius of 8 in., and pulley B has a weight of 10 lb and a radius of 5 in. A cw couple $C = 5$ lb. ft is applied to pulley B, as shown. Assume that the pulleys are thin disks, and determine the tension in the cord connecting the two pulleys.

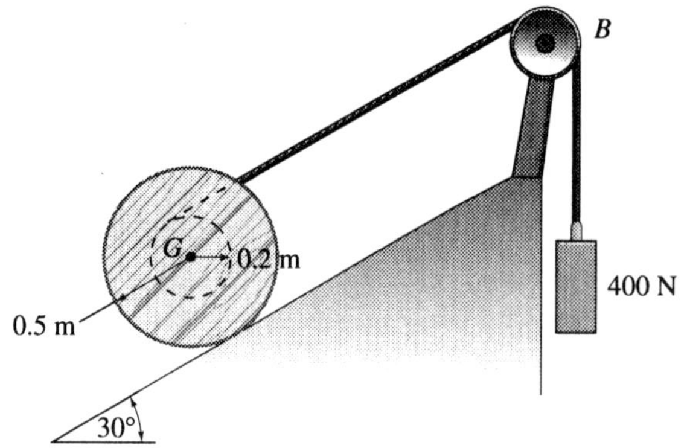

FIGURE P18.161.

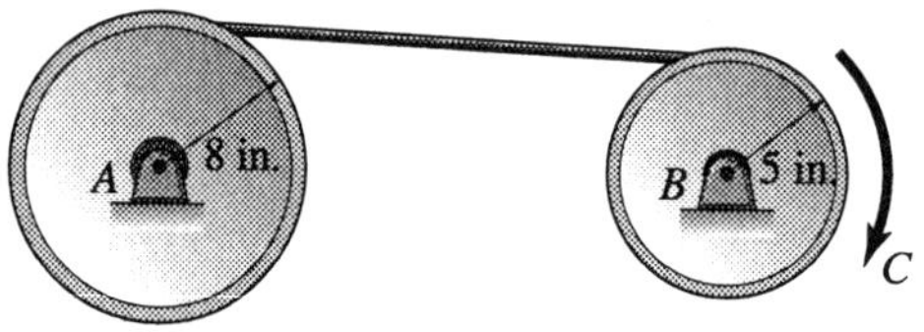

FIGURE P18.162.

18.163 The system shown in Figure P18.163 consists of a cylinder of weight $2W$ and radius R and two rods hinged to the cylinder at its mass center, one on each side. Each rod has a weight W and length $2R$. The system is released from rest on the 30° inclined plane, and the cylinder rolls without sliding. Assume that there is no friction between the rods and the plane at their point of contact. Let $W = 200$ N and $R = 0.5$ m, and determine (a) the angular acceleration of the cylinder and (b) the minimum coefficient of friction between the cylinder and the inclined plane so that it rolls without sliding.

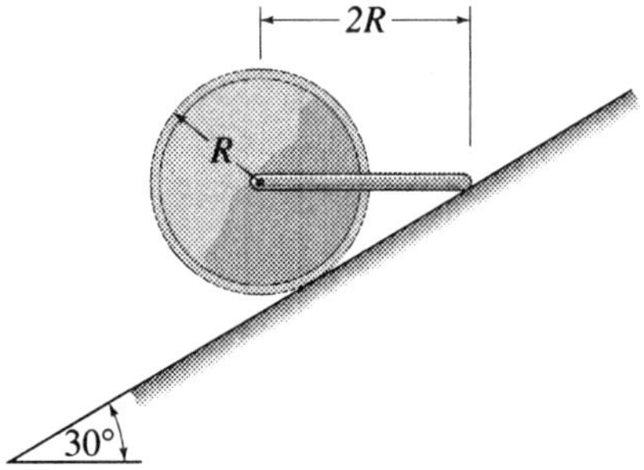

FIGURE P18.163.

19

Two-Dimensional Kinetics of Rigid Bodies: Energy

George W.G. Ferris built his magnificent wheel in 1893. His daring and vision reflect the spirits of a confident and an inspired America.

In 1893, George Washington Ferris was the champion of U.S. technology, the engineer whose vision and relentless pursuit of his dream proved that America could top the Eiffel Tower. That magnificent structure which could carry 2,160 passengers at a time to a height equivalent to 26 story building was a significant engineering triumph in an era when most people had never seen a skyscraper. His reputation quickly grew as the press recounted his struggle to build the machine that other engineers had said could not be built.

The daring of engineers from years past is equaled only by the remarkable giants of today. This continuing journey of triumphs and setbacks illustrates the unyielding commitment of the engineering profession to helping mankind. It is truly phenomenal to note that, when Armstrong set foot on the moon, he commanded a ship with a computer that had 64k of memory! Today, hand held computers can have ten times that capacity. The evolution of engineering technology has inspired and challenged engineers to be more creative and seek alternative solutions to problems.

19.1 Definition of Work

In Chapter 18 we learned how to apply Newton's second law of motion to solving problems dealing with the kinetics of rigid bodies. In this chapter, we will learn how to solve the same type of problems by the principle of work and energy and that of conservation of mechanical energy. The principles of work and energy and of conservation of mechanical energy possess certain advantages over Newton's second law of motion in the solving problems because we can compute changes in the velocity of a given mechanical system without having to compute its acceleration. Also, only those forces that perform work need be considered in analyzing by the work-energy principle. However, it should be pointed out that the principle of conservation of energy is applicable only when dealing with conservative forces.

The concepts of work of a force and that of a couple acting on a rigid body, including the work of conservative forces, were developed in Chapters 12 and 15, and the resulting equations are repeated here for convenience.

Work of a Force

$$U_{1\to 2} = \int_{Q_1}^{Q_2} \mathbf{F}\cdot d\mathbf{r} = \int_{s_1}^{s_2} (F\cos\beta)\,ds. \tag{19.1}$$

The quantity $U_{1\to 2}$ represents the work done by the force **F** as it moves along a path s from position s_1 to position s_2 as shown in Figure 19.1. Note that the quantity $F\cos\beta$ is the component of the force **F** along the tangent to the path since β defines the angle between **F** and this tangent.

Work of a Couple

$$U_{1\to 2} = \int_{\theta_1}^{\theta_2} \mathbf{M}\cdot d\boldsymbol{\theta} = \int_{\theta_1}^{\theta_2} M\,d\theta. \tag{19.2}$$

The quantity **M** is the couple acting on the rigid body and producing the work $U_{1\to 2}$ as it rotates this rigid body from position θ_1 to position

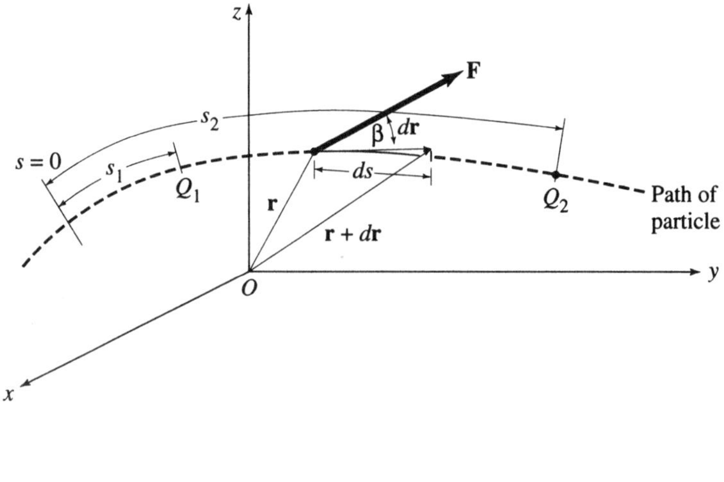

FIGURE 19.1.

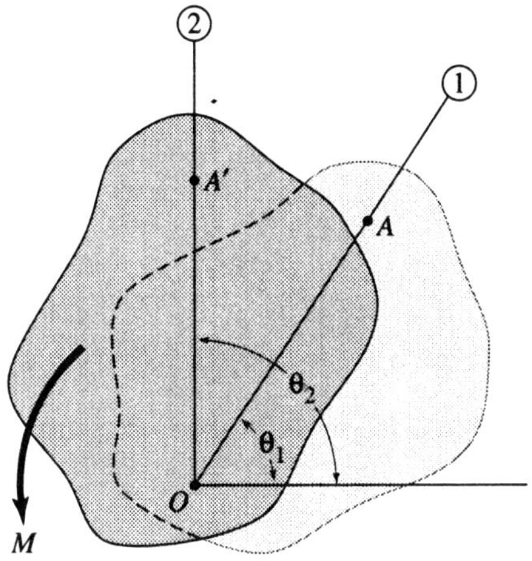

FIGURE 19.2.

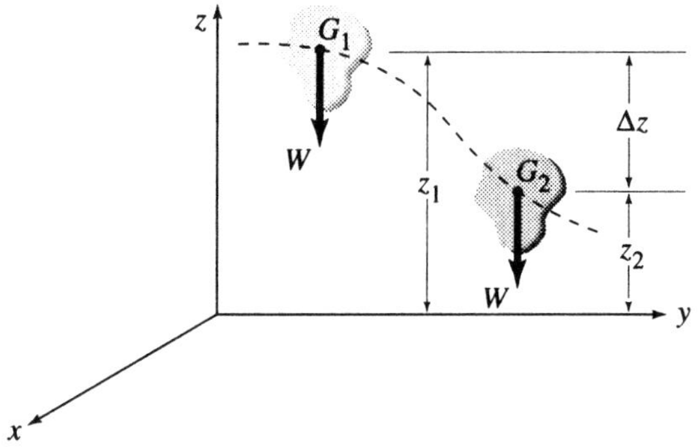

FIGURE 19.3.

θ_2 as shown in Figure 19.2. Note that in Eq. (19.2), the vectors **M** and $d\boldsymbol{\theta}$ are parallel to each other.

Work by the Weight of a Body

$$U_{1\to 2} = W(z_1 - z_2) = W\Delta z. \tag{19.3}$$

The quantity W is the weight of the body and $\Delta z = z_1 - z_2$ represents the change in elevation experienced by the body during its motion as shown in Figure 19.3. Note that $U_{1\to 2}$ is positive if Δz represents a decrease in elevation and negative if Δz represents an increase in elevation.

Work by the Force in a Spring

Linear Spring

$$U_{1\to 2} = \tfrac{1}{2}k(s_1^2 - s_2^2) \tag{19.4a}$$

where k is the linear spring constant and s_1 and s_2 are, respectively, the initial and final linear displacements of the spring, as depicted in Figure 19.4(a). Thus, the work performed by the linear spring force on an attached body is negative, when the system moves away from the

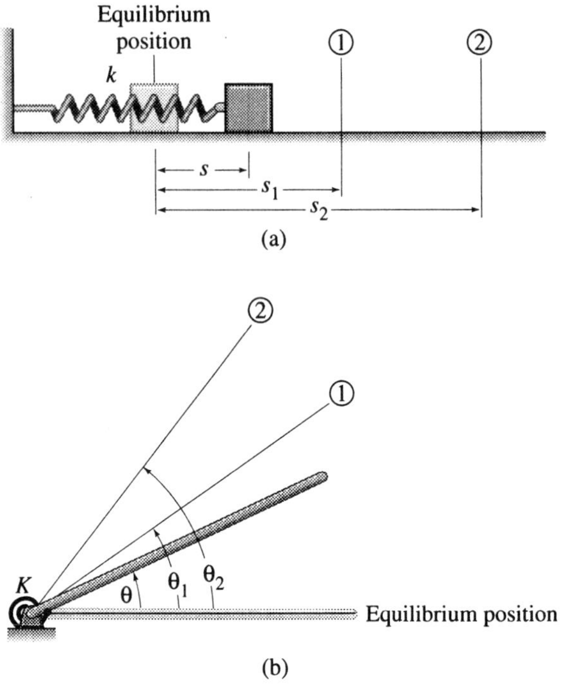

FIGURE 19.4.

equilibrium position, and positive, when the system moves toward the equilibrium position.

Torsional Spring

$$U_{1\to 2} = \tfrac{1}{2}K(\theta_1^2 - \theta_2^2). \qquad (19.4b)$$

where K is the torsional spring constant and θ_1 and θ_2 are, respectively, the initial and final angular displacements of the spring, as shown in Figure 19.4(b). Thus, the work performed by the torsional spring torque on an attached body is negative, when the system moves away from the equilibrium position, and positive, when the system moves toward the equilibrium position.

19.2 Kinetic Energy

The kinetic energy T of a particle of mass m moving at a speed v was derived in Chapter 15 and is given by the equation

$$T = \tfrac{1}{2}mv^2. \qquad (19.5)$$

Using the definition given previously for a rigid body as one consisting of a large number of particles whose positions are fixed in relation to one another, we may use Eq. 19.5 to develop the expression for the kinetic energy of a rigid body in general plane motion.

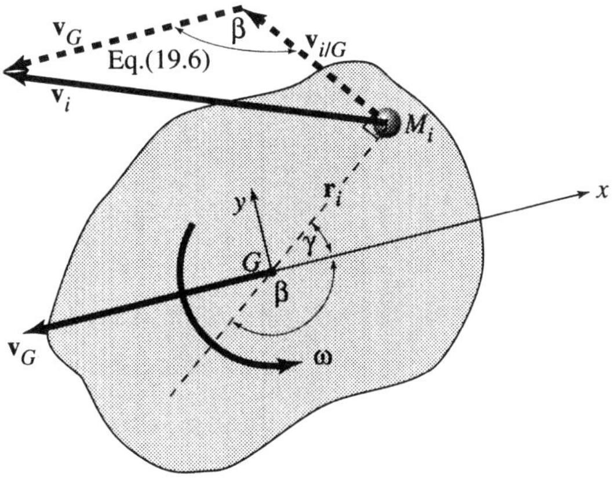

FIGURE 19.5.

The rigid body of mass m shown in Figure 19.5 is undergoing general plane motion in the x-y plane, such that it rotates with an angular velocity $\boldsymbol{\omega}$ and its center of mass G translates with a linear velocity $\mathbf{v}_G$. Using Eq. (17.21), we may relate the velocity $\mathbf{v}_i$ of any particle in the rigid body of mass m_i, located at a distance $\mathbf{r}_i$ from point G, to the velocity $\mathbf{v}_G$ of its center of mass. Thus,

$$\mathbf{v}_i = \mathbf{v}_G + \mathbf{v}_{i/G}. \tag{19.6}$$

The velocity diagram representing Eq. (19.6) is shown in Figure 19.5 where the magnitude v_i of vector $\mathbf{v}_i$ may be found by the law of cosines as follows:

$$v_i^2 = v_G^2 + (r_i\omega)^2 - 2v_G r_i \omega \cos\beta \tag{19.7}$$

where β is the angle between the vectors $\mathbf{v}_G$ and $\mathbf{v}_{i/G}$ as shown in Figure 19.5 and the product $r_i\omega$ represents the magnitude of $\mathbf{v}_{i/G}$. From the geometry in Figure 19.5, we can show that $r_i \cos\beta = -r_i \cos\gamma = -x_i$, so that Eq. (19.7) becomes

$$v_i^2 = v_G^2 + (r_i\omega)^2 + 2v_G \omega x_i. \tag{19.8}$$

Substituting Eq. (19.8) in Eq. (19.5), we obtain the kinetic energy T_i of any particle of mass m_i whose velocity is v_i. Thus,

$$T_i = \tfrac{1}{2}m_i v_i^2 = \tfrac{1}{2}m_i[v_G^2 + (r_i\omega)^2 + 2v_G\omega x_i]. \tag{19.9}$$

The total kinetic energy T of the rigid body is found by summing the kinetic energies of all of its particles. Therefore,

$$\begin{aligned} T = \sum T_i &= \sum \tfrac{1}{2}m_i[v_G^2 + (r_i\omega)^2 + 2v_G\omega x_i] \\ &= \tfrac{1}{2}v_G^2 \sum m_i + \tfrac{1}{2}\omega^2 \sum m_i r_i^2 + v_G\omega \sum m_i x_i. \end{aligned} \tag{19.10}$$

Since $\sum m_i = m$, $\sum m_i r_i^2 = I_G$ and $\sum m_i x_i = m x_G = 0$, Eq. 19.10 becomes

$$T = \tfrac{1}{2}mv_G^2 + \tfrac{1}{2}I_G\omega^2 \tag{19.11}$$

where I_G is the mass moment of inertia of the rigid body about an axis through its mass center, point G, normal to the plane of motion.

Equation 19.11 clearly states that the total kinetic energy T of a rigid body in general plane motion is composed of the two distinct parts, $\frac{1}{2}mv_G^2$ and $\frac{1}{2}I_G\omega^2$. The first part, $\frac{1}{2}mv_G^2$, is the result of the translation of the center of mass G of the rigid body which moves at a speed v_G, and is identical in form (see Eq. (19.5) to the kinetic energy of a particle of mass m. The second part, $\frac{1}{2}I_G\omega^2$, is the result of the rotation of the rigid body at an angular speed ω about an axis through its mass center. This conclusion is in agreement with the kinematic formulation used in Chapter 17, in which the general plane motion of a rigid body was viewed as the sum of a translational component and a rotational component. Note that Eq. (19.11) is very general and applies to any type of plane motion of a rigid body *provided the reference axis passes through the center of mass G.* Three cases of special interest may be derived from Eq. (19.11) as follows:

1. ***Translation***

If the rigid body executes a pure translating motion, its angular speed ω is zero, and Eq. (19.11) reduces to

$$T = \tfrac{1}{2}mv_G^2. \tag{19.12}$$

2. ***Rotation About a Fixed Axis Through the Mass Center***

In the case where the rigid body rotates about a fixed axis through its mass center G and perpendicular to the plane of motion, the linear speed v_G is zero, and Eq. (19.11) reduces to

$$T = \tfrac{1}{2}I_G\omega^2. \tag{19.13}$$

3. ***Rotation About a Fixed Axis Through An Arbitrary Point***

If the rigid body rotates about a fixed axis perpendicular to the plane

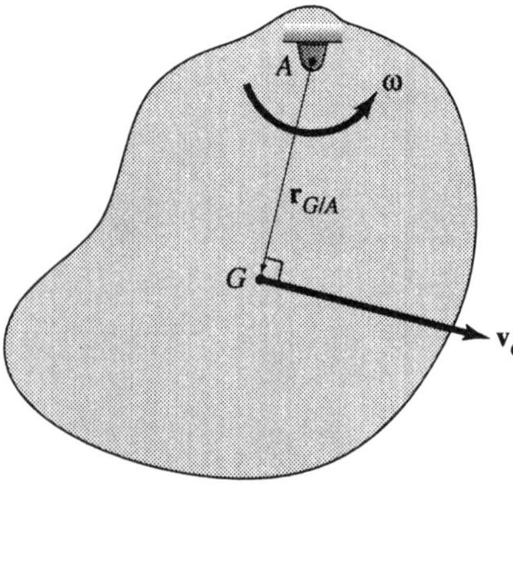

FIGURE 19.6.

of motion through any point other than its mass center, Eq. (19.11) may still be used to find the kinetic energy of the body. However, a special and more compact formulation of this equation may be derived by considering the rigid body shown in Figure 19.6 which is rotating about a fixed axis through any point A other than its center of mass G. Because the magnitude of $\mathbf{v}_G$ is $v_G = r_{G/A}\omega$, we may rewrite Eq. (19.11) as

$$T = \tfrac{1}{2}m(r_{G/A}\omega)^2 + \tfrac{1}{2}I_G\omega^2 = \tfrac{1}{2}(I_G + mr_{G/A}^2)\omega^2 = \tfrac{1}{2}I_A\omega^2 \qquad (19.14)$$

where, according to Eq. (12.18), the quantity $I_G + mr_{G/A}^2$ was replaced by I_A.

19.3 The Work-Energy Principle

The work-energy principle developed in Chapter 15 for the case of a single particle applies equally well to the case of a rigid body. For convenience, this work-energy principle is restated here as Eq. (19.15). Thus,

$$U_{1\to 2} = T_2 - T_1 \qquad (19.15)$$

where $U_{1\to 2}$ is the work performed by *all* of the forces acting on the particle, $T_1 = \frac{1}{2}mv_1^2$, is the initial kinetic energy of the particle and $T_2 = \frac{1}{2}mv_2^2$ is its final kinetic energy. Thus, the quantity $T_2 - T_1$ represents the change in the kinetic energy of the particle and, according to Eq. (19.15), is equal to the work $U_{1\to 2}$ done by all of the forces acting on the particle in displacing it from position 1 to position 2.

For a rigid body in plane motion, the kinetic energies T_1 and T_2 are found by the method developed in Section 19.2. The work $U_{1\to 2}$ performed by forces externally applied to the rigid body is determined as discussed in Section 19.1. Internal forces, however, perform work that reduces to zero as discussed below.

Consider the plane motion of the rigid body shown in Figure 19.7 which may be viewed as a large number of particles whose positions are fixed in relation to each other. In general, as discussed earlier, any particle of mass m_i is subjected to the action of externally applied forces, whose resultant is $\sum \mathbf{F}_i$, and to internally applied forces whose resultant is $\sum \mathbf{f}_i$. Thus, as the rigid body moves from the position defined by the dashed contour to that defined by the solid contour, the total work U_i done on the particle m_i is that due to both $\sum \mathbf{F}_i$ and $\sum \mathbf{f}_i$. When all of the quantities of work U_i, performed by the external and the internal forces acting on all of the particles in the rigid body are added algebraically, we obtain the total work $U_{1\to 2}$ performed on the rigid body. Thus,

$$U_{1\to 2} = \sum (U_i)_{1\to 2}. \qquad (19.16)$$

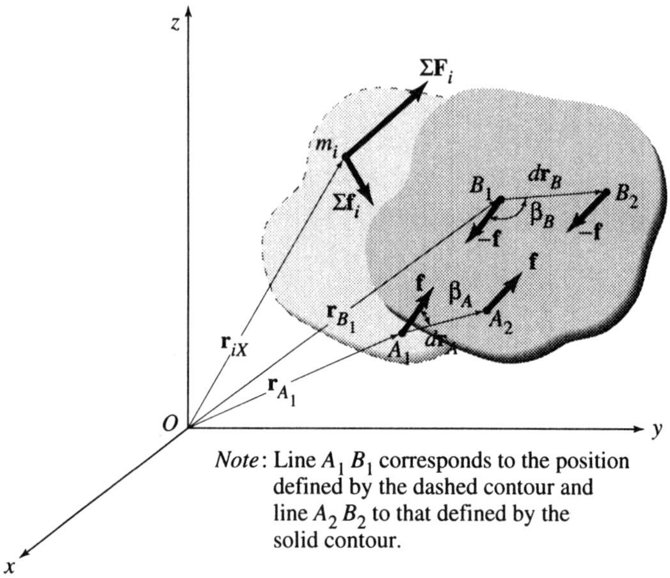

FIGURE 19.7.

Now, consider the two adjacent particles A and B and the two equal and opposite forces of action and reaction $\mathbf{f}$ and $-\mathbf{f}$ as shown in Figure 19.7. As the rigid body executes general plane motion from position A_1B_1 (corresponding to the position defined by the dashed contour) to position A_2B_2, (corresponding to the position defined by the solid contour), particles A and B displace through $d\mathbf{r}_A$ and $d\mathbf{r}_B$, respectively. The work performed by the force $\mathbf{f}$ acting on particle A is $(U_i)_A = \mathbf{f} \cdot d\mathbf{r}_A = f(dr_A \cos \beta_A)$ and that performed by the force $-\mathbf{f}$ acting on particle B is $(U_i)_B = -\mathbf{f} \cdot d\mathbf{r}_B = -f(dr_B \cos \beta_B)$. Note that the quantities $dr_A \cos \beta_A$ and $dr_B \cos \beta_B$ represent, respectively, the components of $d\mathbf{r}_A$ and $d\mathbf{r}_B$ along line AB. Because the body is rigid, the distance between A and B is fixed and the components $dr_A \cos \beta_A$ and $dr_B \cos \beta_B$ must be equal in magnitude and direction. Therefore, the work $(U_i)_A$ must be equal in magnitude but opposite in sign to the work $(U_i)_B$, and, consequently, they cancel each other. Thus, when Eq. (19.16) is used to obtain the work performed on a rigid body during a small displacement from position 1 to position 2, the total work performed by all of the internal forces of action and reaction vanishes, and only the work of external forces remains. In summary, when dealing with a rigid body, the work $U_{1\to 2}$ is *entirely due to external forces* acting on the rigid body. The same conclusion is reached for the case of systems of interconnected rigid bodies in which the internal forces of action and reaction at the connections displace through equal distances, resulting in quantities of work which are equal in magnitude but opposite in sign. Examples of such interconnected systems of rigid bodies include rigid bodies joined together by frictionless hinges and meshed gears. Actually, as illustrated in Examples 19.2, 19.3, and 19.4, one of the primary uses of the work-energy principle is in solving problems dealing with the motion of interconnected rigid bodies. In such a case, Eq. (19.15) may be written in the form

$$\sum U_{1\to 2} = \sum T_2 - \sum T_1. \tag{19.17}$$

The use of the rigid body work-energy principle, as expressed in Eq. (19.15), in solving of problems is illustrated in the following examples. Because the application of the work-energy principle requires computing the work performed by external forces acting on the rigid bodies, it is necessary to construct suitable free-body diagrams to identify the forces that do and those that do not perform work. The free-body diagram used in a given situation may be for an isolated single rigid body or for an isolated system of connected rigid bodies. In either case, care should be exercised in constructing a correct free-body diagram that shows all of the forces external to and acting on the isolated system.

■ Example 19.1

The wheel shown in Figure E19.1(a) is attached to a linear spring whose spring constant is k. The wheel is released from rest in the position shown when the spring is unstretched and rolls without slipping down the inclined plane through a distance d. The weight of the wheel is W, its outside radius is r and its centroidal radius of gyration is k_G. (a) In terms of W, d, k, k_G, r and θ, develop an expression for the velocity v_G of its center of mass after it has moved down the inclined plane through the distance d. (b) Determine v_G and the angular velocity of the wheel if $W = 200$ lb, $d = 2$ ft, $k = 30$ lb/ft, $k_G = 2.0$ ft, $r = 2.5$ ft and $\theta = 20°$. Assume that the spring remains linearly elastic during the motion.

Solution

(a) The solution is accomplished by applying the work-energy principle, Eq. (19.15), to the motion of the wheel. Thus

$$U_{1\to 2} = T_2 - T_1.$$

The free-body diagram of the wheel when it is located somewhere between the initial position 1 and the final position 2 is shown in Figure E19.1(b) where $F_s = ks$ represents the force in the spring, F the friction force and N, the normal force between the wheel and the inclined plane. Of the four forces in Figure E19.1(b) only two, W and F_s, perform work during the motion. Because the wheel does not slide, the point of contact I is the instantaneous center of rotation. The frictional

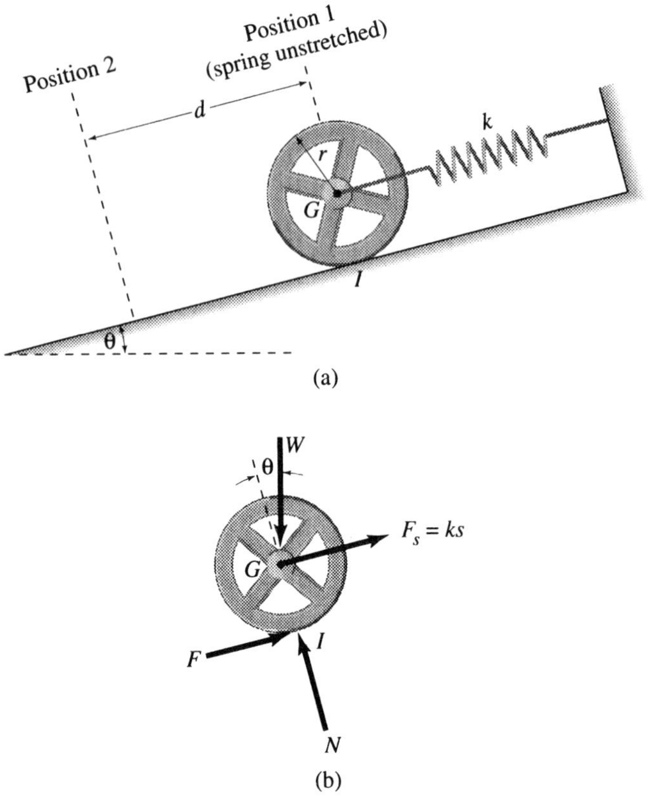

FIGURE E19.1.

force F which acts at point I does not displace and, therefore, its work is zero. The normal force N is perpendicular to the path of motion and does no work. Therefore, in the work-energy principle, the quantity $U_{1\to2}$ represents only the work of W and F_s. Note that in the case of W, only its component along the path of motion performs work equal to $(W \sin\theta)\ d$. Note also that the spring force F_s does negative work because it points opposite to the sense of motion. The magnitude of the work performed by the spring is $\frac{1}{2}kd^2$. Thus,

$$U_{1\to2} = (W \sin\theta)d - \tfrac{1}{2}kd^2$$

Because the wheel is released from rest, its initial kinetic energy $T_1 = 0$. Also, the wheel executes general plane motion as it moves from position 1 to position 2. Therefore, its kinetic energy T_2 is found from Eq. (19.11). Thus,

$$T_2 = \tfrac{1}{2}mv_G^2 + \tfrac{1}{2}I_G\omega^2$$

where $m = \dfrac{W}{g}$, $I_G = mk_G^2$ and $\omega = v_G/r$. Therefore,

$$T_2 = \frac{1}{2}\left(\frac{W}{g}\right)v_G^2 + \frac{1}{2}\left(\frac{W}{g}\right)\left(\frac{k_G}{r}\right)^2 v_G^2$$

Note that the same answer for T_2 may be obtained by using Eq. (19.14) where I_A in this equation represents the moment of inertia of the wheel about its instantaneous center of rotation, point I. Thus,

$$T_2 = \tfrac{1}{2}I_I\omega^2$$

where

$$I_I = \frac{1}{2}mk_I^2 = \frac{W}{g}(k_G^2 + r^2)$$

and

$$\omega = \frac{v_G}{r}.$$

Therefore,

$$T_2 = \frac{1}{2}\left(\frac{W}{g}\right)(k_G^2 + r^2)\left(\frac{v_G}{r}\right)^2 = \frac{1}{2}\left(\frac{W}{g}\right)v_G^2 + \frac{1}{2}\left(\frac{W}{g}\right)\left(\frac{k_G}{r}\right)^2 v_G^2.$$

Applying the work-energy principle (Eq.19.15), solving for v_G, and simplifying,

$$v_G = \sqrt{\left(\frac{2gd}{W}\right)\left[\frac{W\sin\theta - \frac{1}{2}kd}{1 + \left(\frac{k_G}{r}\right)^2}\right]} \qquad \text{ANS.}$$

(b) Substituting the given numerical data,

$$v_G = 3.88 \text{ ft/s} \;\; ⦨ 20^\circ. \qquad \text{ANS.}$$

Because $v_G = r\omega$,

$$\omega = \frac{v_G}{r} = \frac{3.88}{2.5} = 1.552 \text{ rad/s} \;\circlearrowright. \qquad \text{ANS.}$$

Example 19.2

Consider the mechanism shown in Figure E19.2(a). When $\theta = 15^\circ$, the velocity of the slider at C is $v_C = 5.0$ ft/s to the left. The two links AB and BC are homogeneous, weighing 10 lb and 20 lb, respectively. The length of AB is 1.0 ft and that of BC is 1.5 ft. The hinges at A, B, and C and the guide at C are all assumed to be frictionless. Determine the linear velocities v_C and v_B and the angular velocities ω_{AB} and ω_{BC} when $\theta = 90^\circ$. Assume that the weight of the slider at C is negligible.

Solution

Figure E19.2(a) is used to establish the location of the instantaneous center of rotation, point I, for link BC when $\theta = 15^\circ$, and the following

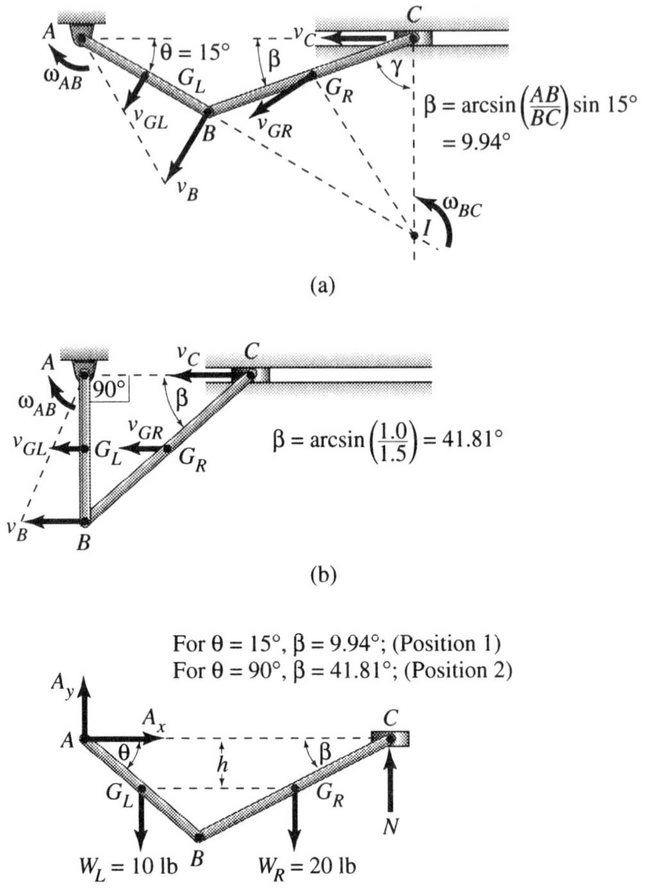

FIGURE E19.2. (c)

geometric relations are found. Note that the mass centers G_L and G_R are located at the geometric centers of homogeneous links AB and BC respectively.

Position 1 ($\theta = 15°$)

$$\beta = \sin^{-1}\left(\frac{AB\sin\theta}{BC}\right) = 9.94°; \quad AC = AB\cos\theta + BC\cos\beta = 2.443 \text{ ft.}$$

$$CI = AC\tan\theta = 0.655 \text{ ft}; \quad \gamma = 90° - \beta = 80.06°$$

$$G_R I = [(CG_R)^2 + (CI)^2 - 2(CG_R)(CI)\cos\gamma]^{1/2} = 0.907 \text{ ft.}$$

$$AI = \frac{AC}{\cos\theta} = 2.529 \text{ ft}; \quad BI = AI - AB = 1.529 \text{ ft.}$$

Therefore, the following velocities are determined.

$$\omega_{BC} = \frac{v_C}{CI} = 7.63 \text{ rad/s} \circlearrowleft; \quad v_{GR} = (G_R I)\omega_{BC} = 6.92 \text{ ft/s}.$$

$$v_B = (BI)\omega_{BC} = 11.67 \text{ ft/s}; \quad v_{GL} = \frac{v_B}{2} = 5.83 \text{ ft/s};$$

$$\omega_{AB} = \frac{v_B}{AB} = 11.67 \text{ rad/s} \circlearrowright.$$

Whereas the mechanism could be dismembered and the work-energy principle applied to each component, it is more convenient to deal with the system as a whole, because the work of the internal forces of action and reaction at the joints is zero.

The kinetic energy $\sum T_1$ for the entire system in position 1 is found from Eq. (19.11) which should account for the kinetic energies of both links AB and BC. Thus,

$$\sum T_1 = (T_{AB})_1 + (T_{BC})_1 = (\tfrac{1}{2}mv_G^2 + \tfrac{1}{2}I_G\omega^2)_{AB} + (\tfrac{1}{2}mv_G^2 + \tfrac{1}{2}I_G\omega^2)_{BC}$$

where $I_G = \frac{1}{12}mL^2$ as found from Appendix B. Therefore,

$$\sum T_1 = \frac{1}{2}\left(\frac{10}{32.2}\right)(5.83)^2 + \frac{1}{2}\left[\frac{1}{12}\left(\frac{10}{32.2}\right)(1.0^2)\right](11.67)^2$$

$$+ \frac{1}{2}\left(\frac{20}{32.2}\right)(6.92)^2 + \frac{1}{2}\left[\frac{1}{12}\left(\frac{20}{32.2}\right)(1.5)^2\right](7.63)^2 = 25.302 \text{ ft·lb}$$

Position 2 ($\theta = 90^\circ$)

When $\theta = 90^\circ$, the configuration of the system is as shown in Figure E19.2(b). Because the velocities v_C and v_B of the two ends of link BC are horizontal, it follows that the instantaneous center of rotation for BC does not exist (i.e., is at infinity) and link BC is undergoing pure translation at this instant. Thus,

$$v_B = v_{GR} = v_C; \quad v_{GL} = \frac{v_B}{2} = \frac{v_C}{2}; \quad \omega_{BC} = 0.$$

and

$$\omega_{AB} = \frac{v_B}{AB} = v_C.$$

Therefore, the kinetic energy $\sum T_2$ for the entire system in position 2 may again be found from Eq. (19.11) as was done for position 1. Note that $\sum T_2$ may be expressed in terms of the single unknown v_C. Thus,

$$\sum T_2 = (T_{AB})_2 + (T_{BC})_2 = \left(\frac{1}{2}mv_G^2 + \frac{1}{2}I_G\omega^2\right)_{AB} + \left(\frac{1}{2}mv_G^2 + \frac{1}{2}I_G\omega^2\right)_{BC}$$

$$= \frac{1}{2}\left(\frac{10}{32.2}\right)\left(\frac{v_C}{2}\right)^2 + \frac{1}{2}\left[\frac{1}{12}\left(\frac{10}{32.2}\right)(1.0^2)\right]v_C^2 + \frac{1}{2}\left(\frac{20}{32.2}\right)v_C^2 + 0$$

$$= 0.3623v_C^2$$

The total work $\sum U_{1\to 2}$ performed by all of the forces acting on the system in moving from position 1 to position 2 may be found by referring to the free-body diagram of the entire system shown in Figure E19.2(c) which represents the system for any position defined by the angle θ. The components A_x and A_y of the pin reaction at A perform no work because the pin does not displace. The normal force N between the slider and the guide does no work because it is normal to the direction of motion of the slider. Note that the internal forces of action and reaction at joints B and C displace through equal distances and produce a net amount of work which is equal to zero. Thus, there are only two forces that perform work, as the system moves from position 1 to position 2. These are the weights of the two links AB and BC. To find the work performed by these forces, we need to determine their displacements as the mechanism moves from position 1 to position 2.

Note that the heights h (Fig. E19.2(c)) of the centers of masses G_L and G_R are always the same, and, therefore, the weights of the two links displace through identical distances as the mechanism moves from position 1 to position 2. In position 1 ($\theta = 15°$), $h_1 = 0.50 \sin 15° = 0.129$ ft. and in position 2 ($\theta = 90°$), $h_2 = 0.50$ feet. Thus, the vertical distance moved by the two weights is $\Delta h = h_2 - h_1 = 0.371$ ft. Therefore,

$$\sum U_{1\to 2} = W_L(\Delta h) + W_R(\Delta h) = \Delta h(W_L + W_R) = 0.371(10 + 20)$$
$$= 11.130 \text{ ft·lb.}$$

The work-energy principle expressed in Eq. (19.17) is now used to find the value of v_C. Thus,

$$\sum U_{1\to 2} = \sum T_2 - \sum T_1$$
$$11.130 = 0.3623 v_C^2 - 25.302$$

from which the final velocities are

$$v_C = 10.03 \text{ ft/s} \leftarrow; \quad v_B = v_C = 10.03 \text{ ft/s} \leftarrow, \qquad \text{ANS.}$$

$$\omega_{AB} = \frac{v_B}{AB} = 10.03 \text{ rad/s} \circlearrowleft; \quad \omega_{BC} = 0 \qquad \text{ANS.}$$

Note that both ω_{AB} and ω_{BC} have decreased in magnitude and, consequently, the angular accelerations α_{AB} and α_{BC} of links AB and BC are both negative during the motion from position 1 to position 2.

Example 19.3

Rework Example 19.2 assuming that friction exists at hinge A and in the guide in which the slider at C moves. Let the friction at A be equivalent to a constant resisting couple $M = 1.5$ ft. lb and that in the guide equivalent to a constant resisting force $F = 2.0$ lb.

Solution

The solution of this problem is again accomplished by the use of the work-energy principle given in Eq. (19.17). Thus,

$$\sum U_{1\to2} = \sum T_2 - \sum T_1$$

Proceeding as in Example 19.2, we find that $\sum T_1$ and $\sum T_2$ have the same values as those found in that solution. Thus,

$$\sum T_1 = 25.302 \text{ ft·lb}$$

and

$$\sum T_2 = 0.3623\, v_C^2.$$

However, the total work $U_{1\to2}$ performed by the forces in moving the system from position 1 ($\theta = 15°$) to position 2 ($\theta = 90°$) is not the same as that found in Example 19.2 because of frictional forces. To enable us to find the work $U_{1\to2}$, we construct the free-body diagram of the entire system as shown in Figure E19.3 which represents the system for any position defined by the angle θ. The forces A_x, A_y and N do not perform work as the system moves from position 1 to position 2. Also, the internal forces of action and reaction at joints B and C displace through equal distances and produce zero net work. Thus, there are only three forces W_L, W_R and F as well as one couple M that contribute to $\sum U_{1\to2}$. Therefore,

$$\sum U_{1\to2} = W_L(\Delta h) + W_R(\Delta h) - F(\Delta s) - M(\Delta\theta)$$
$$= (\Delta h)(W_L + W_R) - F(\Delta s) - M(\Delta\theta)$$

where $\Delta h = 0.371$ ft. was found in the solution of Example 19.2, Δs is the displacement of the slider at C, and $\Delta\theta$ is the angular displacement of joint A. The displacement Δs is found by determining the positions s_1 and s_2 when $\theta = 15°$ and when $\theta = 90°$, respectively, and tak-

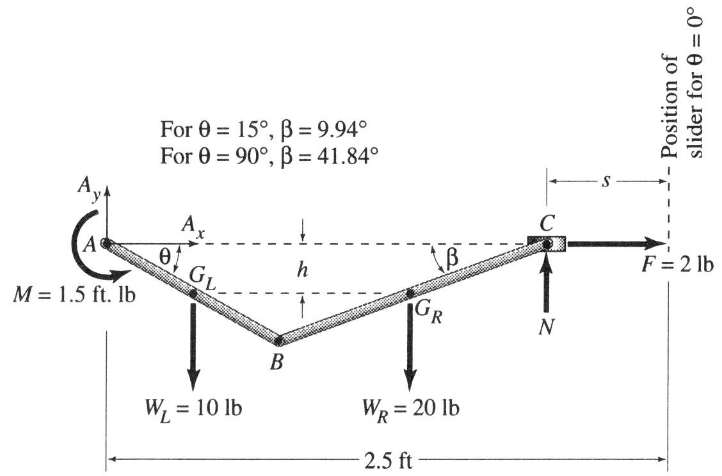

FIGURE E19.3.

ing the difference between them. Utilizing the information developed in Example 19.2, for $\theta = 15°$, $s_1 = 2.5 - (1.0\cos 15° + 1.5\cos 9.94°) = 0.0566$ ft., and, for $\theta = 90°$, $s_2 = 2.5 - 1.5\cos 41.84° = 1.3820$ ft. Thus, $\Delta s = 1.3820 - 0.0566 = 1.325$ ft. The angular displacement $\Delta\theta = 90° - 15° = 75° = 1.309$ rad. Therefore,

$$\sum U_{1\to 2} = 0.371(10 + 20) - 2(1.325) - 1.5(1.309) = 6.517 \text{ ft·lb.}$$

Note that the work of the resisting friction force F and that of the resisting friction couple M are both negative because they are both opposite to the sense of the motion.

The work-energy principle $\sum U_{1\to 2} = \sum T_2 - \sum T_1$ now yields

$$6.517 = 0.3623v_C^2 - 25.302$$

from which the final velocities are

$$v_C = 9.37 \text{ ft/s} \leftarrow, \qquad \text{ANS.}$$

$$v_B = v_C = 9.37 \text{ ft/s} \leftarrow, \qquad \text{ANS.}$$

$$\omega_{AB} = \frac{v_B}{AB} = 9.37 \text{ rad/s} \circlearrowleft, \qquad \text{ANS.}$$

and

$$\omega_{BC} = 0. \qquad \text{ANS.}$$

Example 19.4

The system shown in Figure E19.4(a) consists of two gear-pulley units A and B. The cables connecting weights $W_E = 300$ N and $W_F = 1200$ N to the pulleys are assumed inextensible, and weight W_F is assumed to slide on a frictionless inclined plane. Let the centroidal mass moments of inertia for units A and B be, respectively $I_A = 2.0$ kg·m^2 and $I_B = 1.0$ kg·m^2. The weight W_E is released from rest and allowed to move downward. Determine the velocity v_E of weight W_E after it has moved downward through a vertical distance $h = 2.0$ m.

Solution

The solution to this problem is accomplished by applying the work-energy principle (Eq. 19.17) to the entire system consisting of the two gear-pulley units A and B and the two weights W_E and W_F. The free-body diagram of this system is shown in Figure E19.4(b). Note that the reaction components at A and B, and the weights W_A and W_B do no work because they do not displace. Also, the normal force N_F between the inclined plane and W_F and the components $W_F \cos 30°$ do no work because they are both perpendicular to the path of motion. The internal forces of action and reaction (not shown in Figure E19.4(b)) in the two inextensible cables as well as those at the point of contact in the meshing gears perform a total quantity of work which reduces to zero. For a downward displacement of the weight W_E, the total work $\sum U_{1\to 2}$ performed on the entire system as it displaces from position 1 to position 2, is that due to W_E and W_F. Thus,

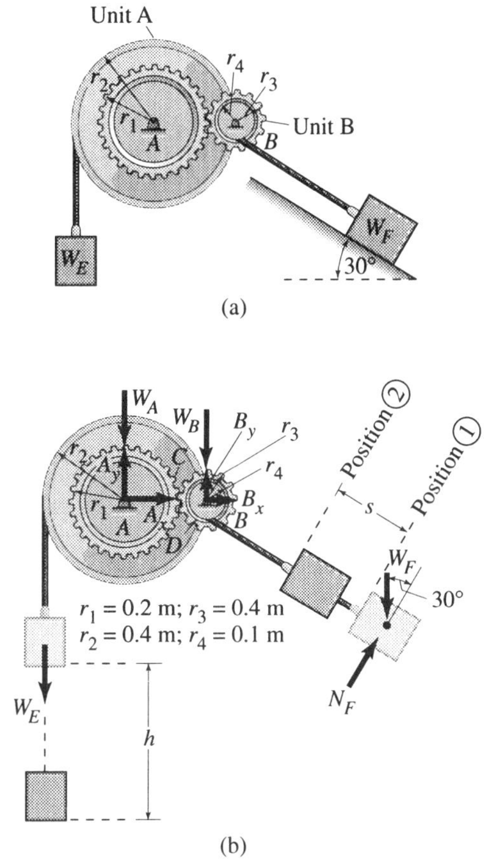

FIGURE E19.4.

$$\sum U_{1\to 2} = W_E h - (W_F \sin 30°)s \tag{a}$$

where s is the displacement of W_F up the inclined plane during the time that W_E displaces through the height h. Note that the work of $W_F \sin 30°$ is negative because its sense is opposite to that of the displacement s. The displacement s may be found in terms of the displacement h as follows.

The angular displacement θ_A of gear-pulley unit A is $\theta_A = h/r_2$. The movement s_C along the arc of radius r_1 of point C which is the point of contact between the two gears is given by $s_C = r_1\theta_A = \left(\frac{r_1}{r_2}\right)h$. Therefore, the angular displacement of gear-pulley unit B is given by $\theta_B = \frac{s_C}{r_3} = \frac{r_1 h}{r_2 r_3}$ and the movement of point D, s_D, on the rim of the small pulley is expressed by $s_D = r_4\theta_B = \frac{r_1 r_4 h}{r_2 r_3}$ which is identical to the displacement s because the cable is inextensible. Therefore

$$s = \frac{r_1 r_4}{r_2 r_3} h$$

and by Eq. (a)

$$\sum U_{1\to 2} = h\left[W_E - (0.5)\left(\frac{r_1 r_4}{r_2 r_3}\right) W_F\right]$$

$$= 2.0\left[300 - (0.5)\left(\frac{0.2(0.1)}{0.4(0.4)}\right)(1200)\right] = 450.0 \text{ N}\cdot\text{m}.$$

Because the system is released from rest, the kinetic energy $\sum T_1$ in position 1 is zero. The kinetic energy $\sum T_2$ in position 2 may be expressed in terms of the velocity v_E of weight W_E after it has moved downward through the height h. This kinetic energy is found by applying Eq. (19.12) to the translating motions of W_E and W_F and Eq. (19.13) to the rotating motions of gear-pulley systems A and B and algebraically adding the results. Thus,

$$\sum T_2 = (T_E)_2 + (T_F)_2 + (T_A)_2 + (T_B)_2$$

$$= \tfrac{1}{2} m_E v_E^2 + \tfrac{1}{2} m_F v_F^2 + \tfrac{1}{2} I_A \omega_A^2 + \tfrac{1}{2} I_B \omega_B^2 \qquad \text{(c)}$$

where v_E and v_F are, respectively, the linear velocities of weights W_E and W_F. Furthermore, ω_A and ω_B are, respectively, the angular velocities of gear-pulley units A and B. The quantities v_F, ω_A, and ω_B may be expressed in terms of the single unknown quantity v_E as follows:

$$\omega_A = \frac{v_E}{r_2} = 2.5 v_E; \qquad v_C = r_1 \omega_A = 0.5 v_E$$

$$\omega_B = \frac{v_C}{r_3} = 1.25 v_E; \quad v_F = v_D = r_4 \omega_B = 0.125 v_E.$$

Therefore, by Eq. (c),

$$\sum T_2 = v_E^2[0.5 m_E + 0.008 m_F + 3.125 I_A + 0.782 I_B]$$

$$= v_E^2\left[0.5\left(\frac{300}{9.81}\right) + 0.008\left(\frac{1200}{9.81}\right) + 3.125(2.0) + 0.782(1.0)\right]$$

$$= 23.3 v_E^2.$$

By the work-energy principle $\sum U_{1\to 2} = \sum T_2 - \sum T_1$,

$$450.0 = 23.3 v_E^2 - 0,$$

from which

$$v_E = 4.39 \text{ m/s}. \qquad \text{ANS.}$$

■

Problems

19.1 The homogeneous wheel of Figure P19.1, weighing 250 lb, rotates cw at a speed of 900 rpm. When the 60-lb force is applied to the brake handle, the wheel comes to a complete stop after 40 revolutions. If the centroidal radius of gyration of the wheel is 8 in., determine the coefficient of kinetic friction between the wheel and the brake shoe.

19.2 A wheel is set in motion at a speed of 1800 rpm about a shaft through its mass center. After the driving power is removed, the wheel rotates 4000 revolutions before coming to a full stop. The average frictional couple between the wheel and its shaft is 40 lb·ft. Determine the centroidal mass moment of inertia of the wheel.

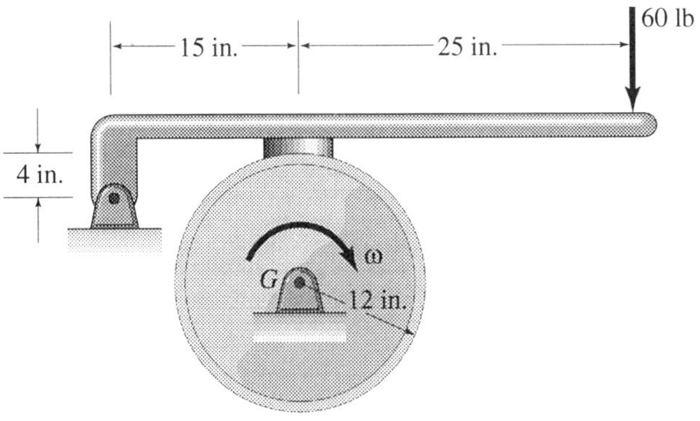

FIGURE P19.1.

19.3 To avoid impact with a stalled car ahead of him, a driver applies his brakes abruptly locking his wheels and skidding to a final stop just prior to hitting the disabled car. If the skid marks were 75 m long on a level pavement and the mass of the driver and his car is 2000 kg, determine his speed just prior to applying the brakes. Let the coefficient of friction between the tires and the pavement be 0.75.

19.4 Refer to the system shown in Figure P19.4. The weight of body A is $W_A =$ 500 lb. The weight of the double wheel unit and its central hub is $W_w = 200$ lb, and the cord which connects the rim of the hub to body A is inextensible. Assume a frictionless pulley at B. Body A is released from rest and moves downward through a vertical distance s. If the wheel rolls up the inclined plane without slipping, determine, when $s = 3$ ft, (a) the velocity v_A of body A, (b) the velocity v_G of the wheel's mass center, point G, and (c) the angular velocity ω of the wheel. Let $r_1 = 12.0$ in. and $r_2 =$ 24.0 in. and the centroidal radius of gyration of the double wheel-hub system is $k = 20.0$ in.

19.5 The flywheel of an engine weighs 2000 N and has a centroidal radius of gyration of 0.50 m. The engine is turned off

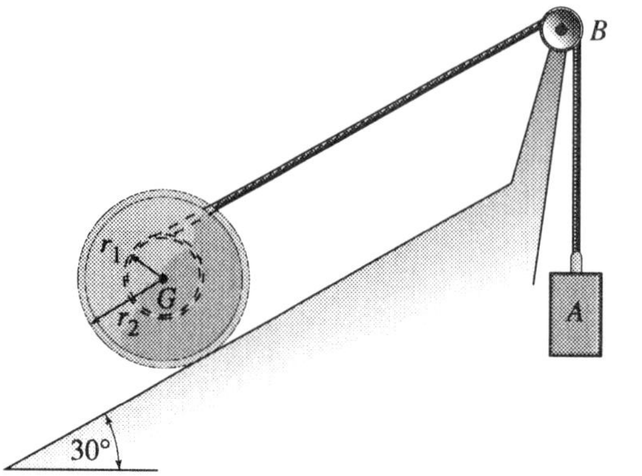

FIGURE P19.4.

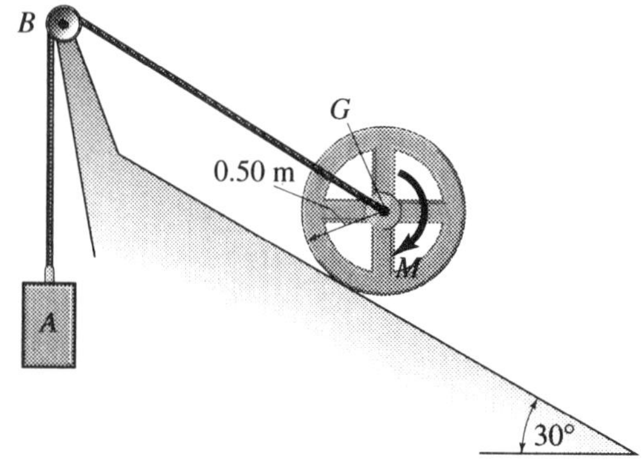

FIGURE P19.7.

when it is running at a speed of 1000 rpm, and the flywheel comes to a complete stop after it has made 5000 revolutions. Assuming that the friction between the flywheel shaft and its bearings develops a constant resisting moment M, determine the magnitude of this moment.

19.6 A method of determining the centroidal radius of gyration k of a flywheel of weight W consists of allowing the flywheel to coast to rest from a given angular velocity ω and counting the number of revolutions N for it to come to a complete stop. If the friction between the shaft of the flywheel and the bearings develops a constant resisting moment M, derive an expression for the radius of gyration k in terms of ω, M, N and W.

19.7 Refer to the system shown in Figure P19.7. The weight of block A is 20 N and that of the wheel is 150 N. The centroidal radius of gyration of the wheel is 0.30 m. The wheel is subjected to a constant cw couple $M = 10$ N·m. In the position shown, the velocity of the wheel's mass center is 2 m/s down the inclined plane. Determine its velocity after block A has moved up a distance of 5 m. Assume that the wheel rolls without slipping and that the peg at B is frictionless.

19.8 The pulley system shown in Figure P19.8 rotates in a ccw direction when released from rest. The centroidal mass moment of inertia of the pulley system is I. (a) Develop an expression for the number of revolutions N that the pulley system must execute in order for the weight W_A to acquire a velocity v_A. Express your answer in terms of W_A, W_B, r_1, r_2 and I. (b) Find a numerical value for N if $W_A =$

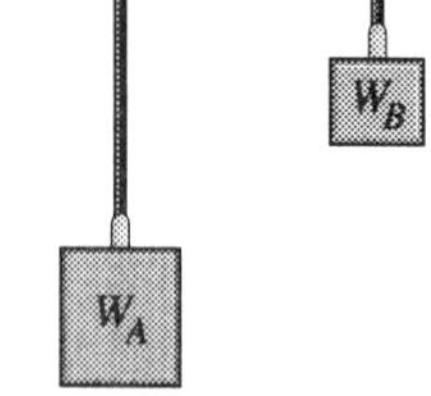

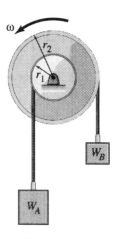

FIGURE P19.8.

100 lb, $W_B = 25$ lb, $r_1 = 6$ in., $r_2 = 10$ in., $I = 0.30\ \text{lb}\cdot\text{s}^2\cdot\text{ft}$ and $v_A = 20$ ft/s.

19.9 A block of weight W is released from rest and slides down the frictionless inclined plane through the distance s before contacting and compressing a spring of spring constant k, as shown in Figure P19.9. (a) Develop an expression in terms of s and θ, for the velocity v of the block just prior to contact with the spring. (b) Develop an expression in terms of s, θ, and k for the maximum deformation of the spring Δ. (c) Relate the spring deformation Δ to the block velocity v just prior to contact with the spring.

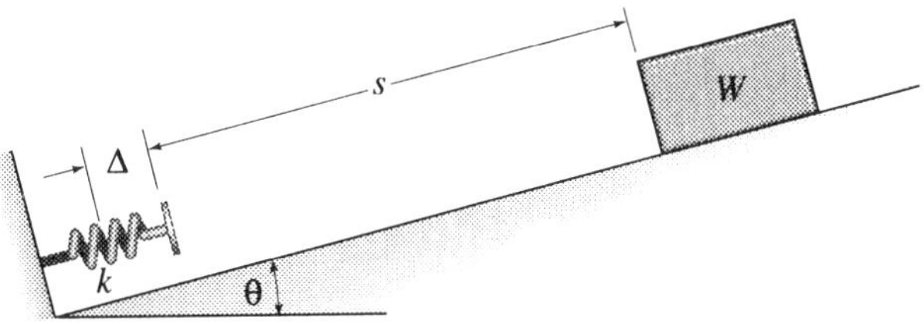

FIGURE P19.9.

19.10 Repeat parts (a) and (b) of Problem 19.9 using a rough inclined plane instead of a smooth one. Assume that the coefficient of kinetic friction is μ.

19.11 The slender homogeneous rod AB shown in Figure P19.11 is hinged at A, and, when in the vertical position, the detached spring is compressed a distance of 2 in. from its equilibrium position. The rod is released from rest in its vertical position and rotates ccw reaching the horizontal position with an angular velocity $\omega = 5$ rad/s. If the spring constant $k = 1000$ lb/in, determine (a) the weight of the homogeneous rod and (b) the reaction components at the hinge when the rod is in the horizontal position.

19.12 The slender homogeneous rod AB shown in Figure P19.12 is hinged at A where a torsional spring is attached. The weight of the homogeneous rod is $W = 10$ N and the torsional spring constant $K = 12\ \text{N}\cdot\text{m/rad}$. When the rod is in the

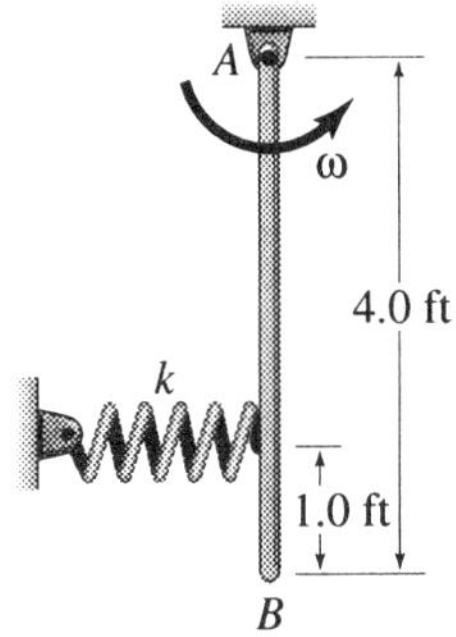

FIGURE P19.11.

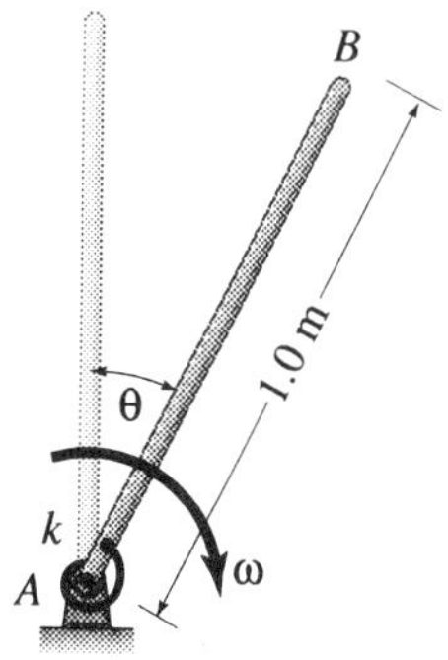

FIGURE P19.12.

vertical position ($\theta = 0°$), the spring is free of any torque. The rod is given an initial cw angular velocity $\omega = 2$ rad/s when in the vertical position. Determine its angular velocity when in the horizontal position ($\theta = 90°$).

19.13 Refer to Problem 19.12 and determine (a) the angle θ at which the rod comes to a complete stop and (b) the angle θ for which the angular velocity of rod AB is maximum.

19.14 The system shown in Figure P19.14 consists of a homogeneous slender rod AB of weight $W = 10$ lb connected at A and B by frictionless hinges to two sliders of weight $W_A = 5$ lb and $W_B = 15$ lb. Slider A is constrained to move in a frictionless horizontal track and slider B in a frictionless vertical track. The system is released from rest when $\theta = 0°$. Determine the velocity of slider A and the angular velocity of rod AB when $\theta = 60°$.

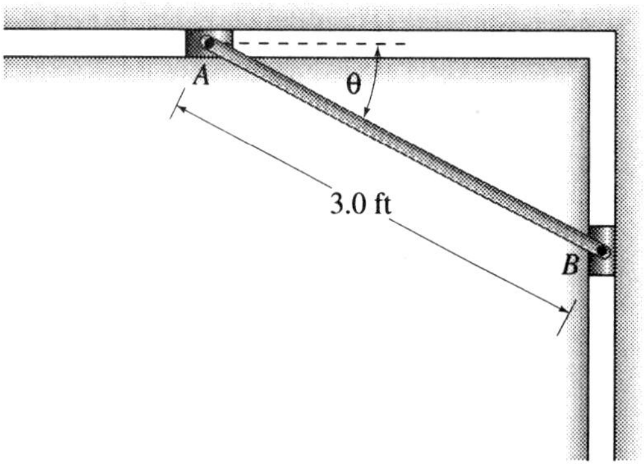

FIGURE P19.14.

19.15 The system shown in Figure 19.15 consists of a homogeneous slender rod AB of weight $W = 50$ N connected at A and B by frictionless hinges, to two sliders of weight $W_A = 30$ N and $W_B = 80$ N. Slider A is constrained to move in a frictionless horizontal track and slider B in a frictionless vertical track. A linear spring of spring constant $k = 500$ N/m is attached to slider B and a cw couple $M = 200$ N·m is applied as shown. The system starts from rest when $\theta = 0°$. Determine the angular velocity of the rod when $\theta = 45°$. There is no stretch in the spring when $\theta = 0°$.

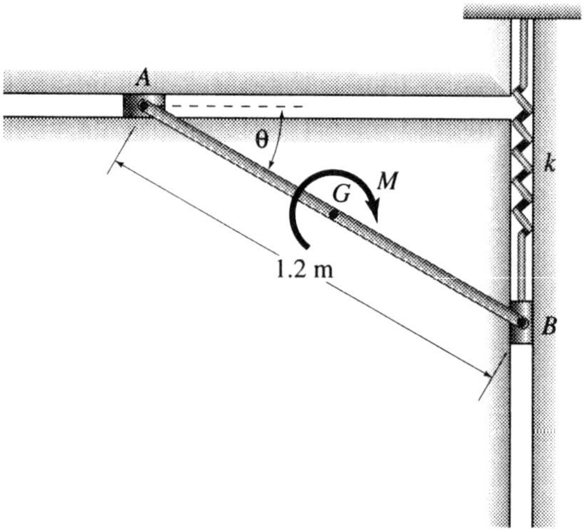

FIGURE P19.15.

19.16 A weight $W_A = 500$ N is attached to one end of an inextensible cord which is wrapped around two identical pulleys each of weight $W = 100$ N as shown in Figure P19.16. The system is released from rest and the weight W_A is allowed to move downward. Determine the linear velocity of point B and the angular velocities of the two pulleys when W_A has moved through a height $h = 2$ m. Assume that there is no slippage between the cord and the pulleys, and treat the two pulleys as though they are thin disks.

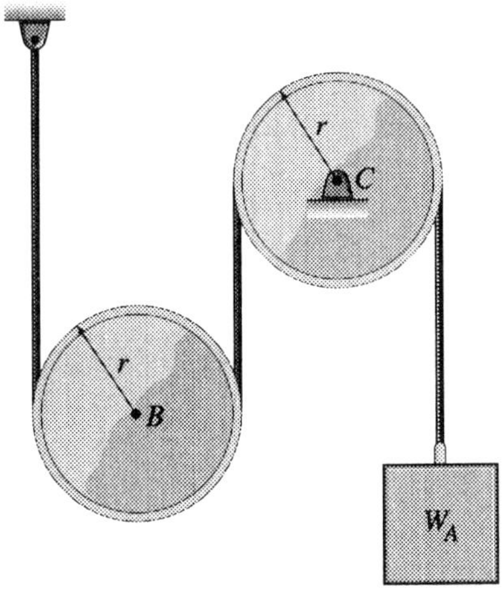

FIGURE P19.16.

19.17 The double-wheel unit and the central hub shown in Figure P19.17 have a combined weight $W = 200$ lb. The cord, which is wrapped around the central hub, is assumed inextensible. In the position shown, the velocity of the center of the unit is $v_C = 10$ ft/s down the inclined plane. Determine the distance down the plane that the wheel unit must move for its center to reach a velocity of 15 ft/s. Assume that the wheel unit rolls without slipping, and let its centroidal radius of gyration $k_C = 12.0$ in.

19.18 The system shown in Figure P19.18 consists of two identical homogeneous slender rods, each of weight W and length L, which are connected together by frictionless hinges. The hinge at A is fixed, and the roller at C has a negligible weight. The system is released from rest when $\theta = 0°$. In terms of W, L, and θ, develop an expression for the velocity v_C for any position of the system defined by the angle θ.

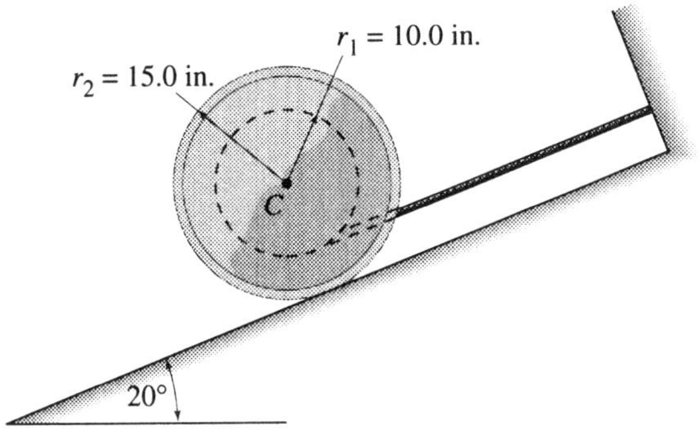

FIGURE P19.17.

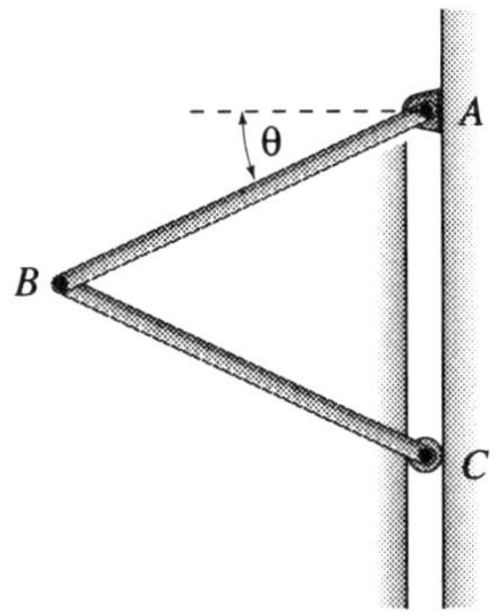

FIGURE P19.18.

19.19 The system shown in Figure P19.19 consists of the two homogeneous slender rods, AB weighing 20 lb and BC weighing 80 lb, which are connected together by frictionless hinges. The hinge

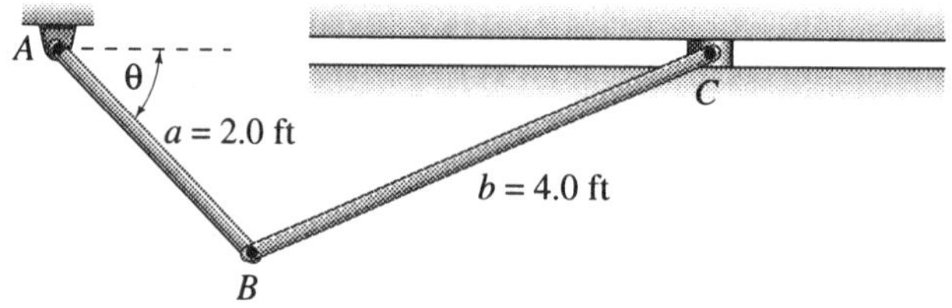

FIGURE P19.19.

at A is fixed, and the slider at C has negligible weight. The friction force between the slider at C and its track is estimated to be constant at 5 lb. The system is released from rest when $\theta = 0°$. Find the velocity v_C of the slider when $\theta = 90°$. What are the angular velocities of links AB and BC in this position?

19.20 Consider the gear-pulley system shown in Figure P19.20. At the instant shown, body A of weight $W_A = 800$ N is moving downward at a speed $v_A = 0.75$ m/s. Determine the speeds of bodies A and B after body A has moved downward a distance of 1.5 m. The weight of body B is $W_B = 200$ N and the weights of gear-pulley systems C and D are, $W_C = 300$ N and $W_D = 150$ N respectively. The centroidal radii of gyration of gear-pulley systems C and D are $k_C = 0.40$ m and $k_D = 0.20$ m, respectively.

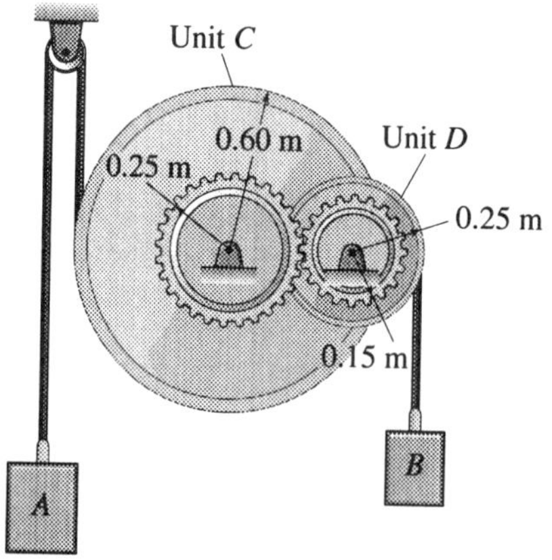

FIGURE P19.20.

19.21 Consider the gear-rack system shown in Figure P19.21. The vertical rack weighing $W_R = 100$ lb is released from rest. Find its velocity after it has moved a distance of 2 ft. The centroidal mass moment of inertia of gear A is $I_A = 4\ \text{lb·s}^2\text{·ft}$ and that of gear B is $I_B = 7\ \text{lb·s}^2\text{·ft}$. Gear B is restrained by a coil spring with a spring constant $K_B = 10$ ft·lb/rad.

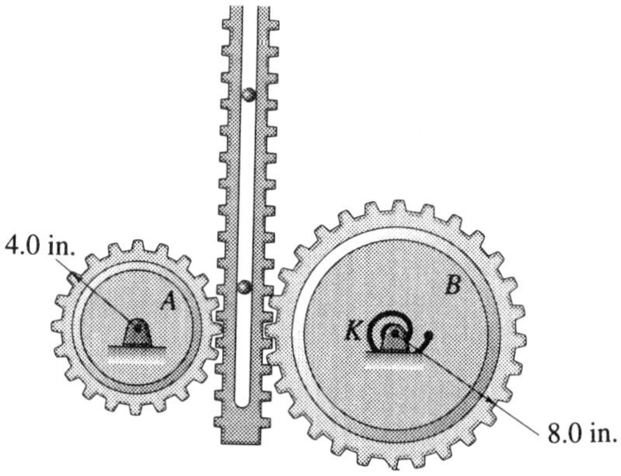

FIGURE P19.21.

19.22 A wheel weighing 20 kg is released from rest in the position shown in Figure P19.22, rolls down the inclined plane without slipping and reaches point A with a linear velocity of 8 m/s. Determine (a) the centroidal mass radius of gyration of the wheel, (b) the linear ve-

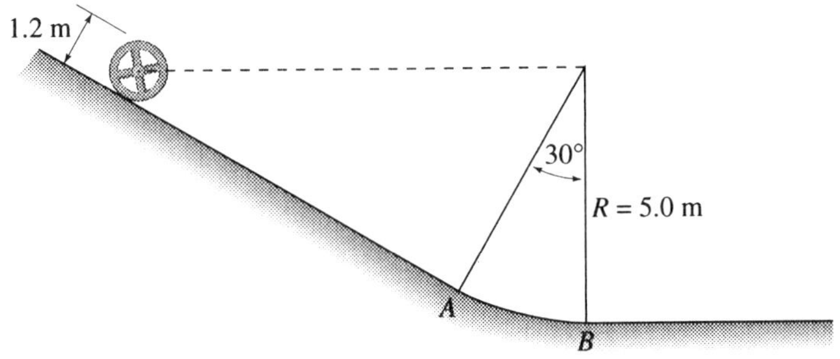

FIGURE P19.22.

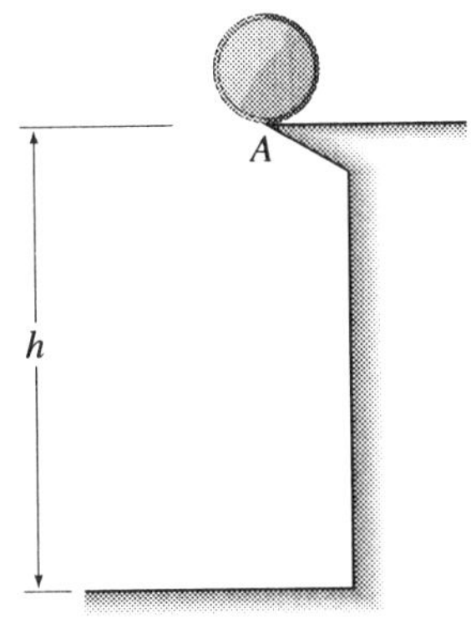

FIGURE P19.23.

locity of the wheel at B, and (c) the normal reaction between the wheel and the contacting surface at B.

19.23 A cylinder of radius $R = 2.5$ ft and weight $W = 200$ lb is released from the position shown in Figure P19.23. It rotates through 90° ccw before separating from its support at A. It, then, falls through a height h before hitting ground. What must be the height h if the cylinder is to make 2 complete revolutions before hitting ground?

19.24 The system shown in Figure 19.24 con-

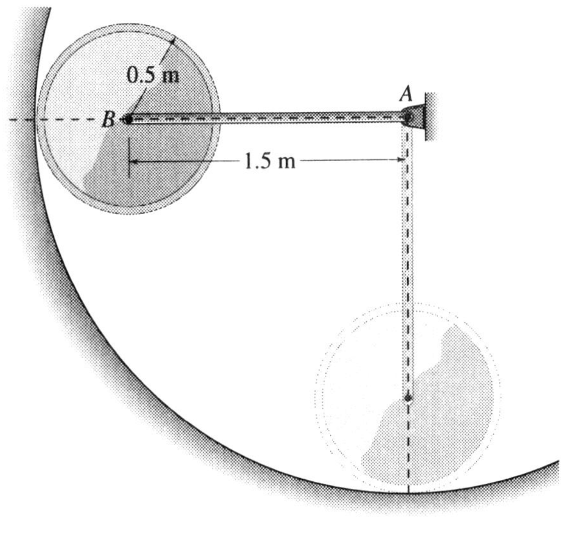

FIGURE P19.24.

sists of a uniform rod AB with a mass of 4 kg hinged at A to a fixed support and at B to a wheel with a mass of 6 kg that rolls without slipping on a fixed circular path. The system is released from rest in the position shown. Determine the centroidal mass moment of inertia of the wheel if the velocity of point B is 3 m/s after rod AB rotates through 90°.

19.25 As shown in Figure P19.25, a plate weighing 20 lb is hinged to two identical homogeneous rods each weighing 3 lb. The system is released from rest when $\theta = 0°$. Determine the velocity of the plate and the angular velocity of the rods when $\theta = 90°$.

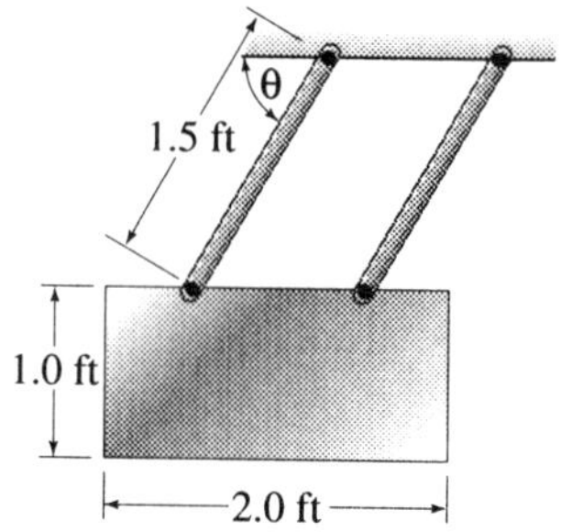

FIGURE P19.25.

19.4 The Conservation of Mechanical Energy Principle

In Chapter 12, we discussed the concepts of gravitational and elastic potential energy functions V_g and V_e, respectively, as they relate to a particle. We concluded that a potential energy function $V = V_g + V_e$ exists only when work is done by conservative forces. We defined a conservative force as one whose work U does not depend upon the path followed by the particle on which it acts. In other words, the work of a conservative force is a function only of the initial and final positions of the particle without regard to the length, shape, or condition of the path. Examples of conservative forces include the weight of a body and the force of a spring. We also concluded that the change in the potential energy function, $V_2 - V_1$, is equal to the negative of the work $U_{1\to 2}$ done by a conservative force in moving the particle from position 1 to position 2. Thus, $V_2 - V_1 = -U_{1\to 2}$, from which

$$U_{1\to 2} = V_1 - V_2. \tag{19.18}$$

In Section 19.3, we concluded that, when dealing with a rigid body, the work of internal forces reduces to zero and only the work of external forces need be considered in determining the quantity $U_{1\to 2}$. Consequently, by Eq. (19.18), the change in the potential energy function of a rigid body, $V_1 - V_2$, is found by determining only the work of the external conservative forces that act on the rigid body. However, we should point out that elastic potential energy functions (i.e., V_e) are entirely due to the work of internal forces in deformable (vs. rigid) bodies, such as linear and torsional springs.

We now direct our attention to the work-energy principle developed in Section 19.3. Substituting Eq. (19.18) in Eq. (19.15),

$$V_1 - V_2 = T_2 - T_1. \tag{19.19}$$

Equation (19.19) states that, when a conservative system of forces acts on a rigid body, the change in its potential energy is identical to the magnitude of the change in its kinetic energy. The equation, however, is usually expressed differently by equating the sum of the potential and kinetic energies of the system in position 1 to their sum in position 2. Thus, rewriting Eq. (19.19) yields,

$$T_1 + V_1 = T_2 + V_2 \tag{19.20}$$

where the sum of T and V represents the total mechanical energy of a system in a given position. Therefore, Eq. (19.20) states that, when a system is acted upon by conservative forces, its total mechanical energy in position 1 is equal to its total mechanical energy in any other position 2. In other words, the total mechanical energy of the system remains a constant or is conserved as the system changes from position to position. Thus, Eq. (19.20) is a mathematical expression of the *principle of conservation of mechanical energy* as it relates to a rigid body.

Application of the principle of conservation of mechanical energy is illustrated in the following examples. As in the case of the principle of work and energy, the selection of the free-body diagram to use in a given case depends upon the problem at hand and upon the choice of the system to be examined. Thus, the free-body diagram may be that for a single rigid body or for a system of connected rigid bodies. In this latter case, the forces of action and reaction at the connections do no work. Also, these forces do not appear as part of the free-body diagram because they are internal to the system of connected rigid bodies. In such a case, Eq. (19.20) may be written in the form

$$\sum T_1 + \sum V_1 = \sum T_2 + \sum V_2. \tag{19.21}$$

Example 19.5

The wheel shown in Figure E19.5(a) is attached to a linear spring whose spring constant is k. The wheel is released from rest in the position shown when the spring is unstretched and rolls without slipping down the inclined plane through a distance d. The weight of the wheel is W, its outside radius is r and its centroidal mass radius of gyrations is k_G. In terms of W, d, k, k_G, r, and θ, develop an expression for the velocity v_G of the center of mass of the wheel after it has moved down the inclined plane through the distance d.

Solution

Note that this is the same problem worked out in Example 19.1(a) by the principle of work and energy.

By the principle of conservation of mechanical energy, Eq.(19.20),

$$T_1 + V_1 = T_2 + V_2.$$

The kinetic energies T_1 and T_2, corresponding to position 1 and 2,

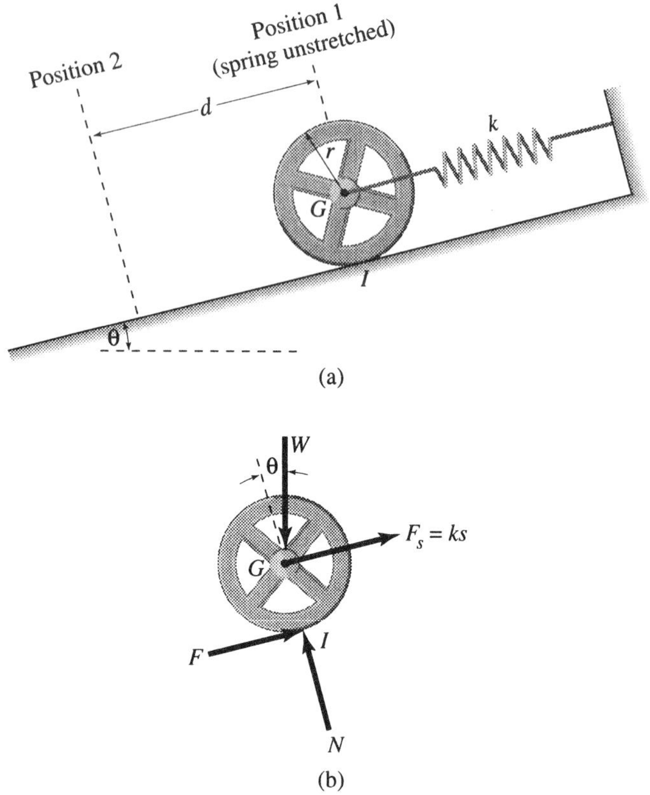

FIGURE E19.5.

respectively, are found as in Example 19.1. These values are simply copied from the solution of Example 19.1 as follows:

$$T_1 = 0$$

and

$$T_2 = \frac{1}{2}\left(\frac{W}{g}\right)v_G^2 + \frac{1}{2}\left(\frac{W}{g}\right)\left(\frac{k_G}{r}\right)^2 v_G^2.$$

Arbitrarily choose position 1 (Fig. 19.5(a)) as the reference or datum for computing the potential energy function. The free-body diagram of the wheel is shown in Figure 19.5(b) for any location between position 1 and 2. As in the solution of Example 19.1, the forces F and N do no work as the wheel displaces from position to position, and, consequently, they do not contribute to the potential energy function. Since position 1 is chosen as the datum, the gravitational potential energy function $(V_g)_1 = 0$. Also because the spring is not stretched in this position, the elastic potential energy function $(V_e)_1 = 0$. Thus,

$$V_1 = (V_g)_1 + (V_e)_1 = 0.$$

When the wheel is in position 2, the center of mass of the wheel is at an elevation of $d \sin\theta$ below the reference position, and the spring is stretched by an amount equal to d. Therefore, $(V_g)_2 = -W(d\sin\theta)$ and $(V_e)_2 = kd^2/2$. The gravitational potential energy $(V_g)_2$ is negative, as a consequence of its lower position, because the capacity of the wheel to do work has been reduced. The elastic potential energy $(V_e)_2$ is positive, as a result of its stretched condition, because elastic strain energy has been stored and the capacity of the spring to do work is increased. Therefore,

$$V_2 = (V_g)_2 + (V_e)_2 = -W(d\sin\theta) + kd^2/2.$$

Using the principle of conservation of energy,

$$0 + 0 = \frac{1}{2}\left(\frac{W}{g}\right)v_G^2 + \frac{1}{2}\left(\frac{W}{g}\right)\left(\frac{k_G}{r}\right)^2 v_G^2 - W(d\sin\theta) + \frac{1}{2}kd^2$$

from which

$$v_G = \sqrt{\left(\frac{2gd}{W}\right)\left[\frac{W\sin\theta - \frac{1}{2}kd}{1 + \left(\frac{k_G}{r}\right)^2}\right]} \qquad \text{ANS.}$$

which, of course, is identical to the answer obtained in the solution of part (a) of Example 19.1

Example 19.6

Consider the mechanism shown in Figure E19.6(a). When $\theta = 15°$, the velocity of the slider at C is $v_C = 5.0$ ft/s to the left. The two links AB and BC are homogeneous, weighing 10 lb and 20 lb, respectively. The length of AB is 1.0 ft and that of BC is 1.5 ft. The hinges at A, B, and C and the guide at C are all assumed to be frictionless. Determine the linear velocities v_C and v_B and the angular velocities ω_{AB} and ω_{BC} when $\theta = 90°$. Assume that the weight of the slider at C is negligible.

Solution

Note that this is the same problem worked out in Example 19.2 by the principle of work and energy.

By the principle of conservation of mechanical energy, Eq. (19.21),

$$\sum T_1 + \sum V_1 = \sum T_2 + \sum V_2.$$

The kinetic energies $\sum T_1$ and $\sum T_2$, corresponding to positions 1 and 2, respectively, are found from the information contained in Figure E19.6(a) and (b), using the same method as in Example 19.2. These values are simply copied from the solution of Example 19.2 as follows:

$$\sum T_1 = 25.302 \text{ ft·lb}$$

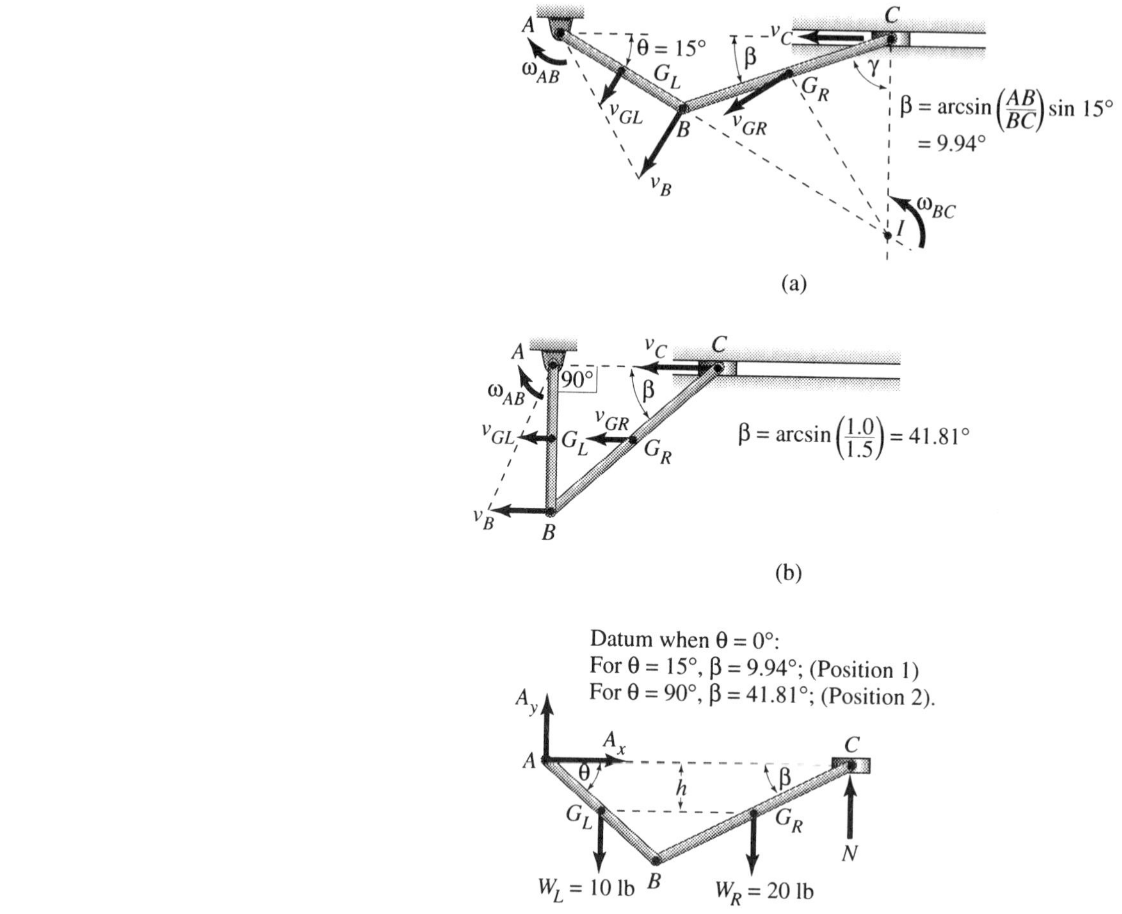

FIGURE E19.6.

and

$$\sum T_2 = 0.3623 v_G^2.$$

The free-body diagram of the mechanism is shown in Figure E19.6(c). Because the choice of reference or datum for measurement of the potential energy function is arbitrary, we will select, as our datum, the position of the mechanism when $\theta = 0°$ where both links AB and BC are horizontal. As discussed in the solution of Example 19.2, the work done by A_x, A_y, and N is zero as the system displaces from position to position. It follows, therefore, that these forces do not contribute to the potential energy function. Also, because there are no deformable components in the system, the elastic potential energy functions, $\sum (V_e)_1 = \sum (V_e)_2 = 0$. As the system displaces to position 1 ($\theta = 15°$), both links undergo a decrease in their capacities to do work because both of their weights have done positive work. Hence, both links acquire negative gravitational potential energies. Thus,

$$\sum (V_g)_1 = -W_L(h_1) - W_R(h_1) = -h_1(W_L + W_R)$$
$$= -0.129(10 + 20) = -3.870 \text{ ft·lb}$$

where $h_1 = 0.129$ ft is found from the geometry in Figure E19.6(c) as was done in Example 19.2. Therefore, because $\sum (V_e)_1 = 0$, the total potential energy function in position 1 is

$$\sum V_1 = \sum (V_g)_1 + \sum (V_e)_1 = -3.870 \text{ ft·lb}.$$

When the system displaces to position 2 ($\theta = 90°$), both links undergo a further decrease in their capacities to do work. Therefore, as in position 1, the gravitational potential energy functions of both links are negative and the total gravitational potential energy function in position 2 becomes

$$\sum (V_g)_2 = -W_L(h_2) - W_R(h_2) = -h_2(W_L + W_R)$$
$$= -0.50(10 + 20) = -15.00 \text{ ft·lb}.$$

As stated earlier, $(V_e)_2 = 0$. It follows, therefore, that the total potential energy function in position 2 is

$$\sum V_2 = \sum (V_g)_2 + \sum (V_e)_2 = -15.00 \text{ ft·lb}.$$

Therefore, applying the principle of conservation of mechanical energy, Eq. (19.21),

$$25.302 - 3.87 = 0.3623 v_C^2 - 15.00$$

from which

$$v_C = 10.03 \text{ ft/s} \leftarrow. \qquad \text{ANS.}$$

As in Example 19.2 and from the geometry contained in Figure E19.6(b), we obtain the values of v_B, ω_{AB}, and ω_{BC} as follows:

$$v_B = v_C = 10.03 \text{ ft/s} \leftarrow, \qquad \text{ANS.}$$

$$\omega_{AB} = \frac{v_B}{AB} = 10.03 \text{ rad/s} \circlearrowleft, \qquad \text{ANS.}$$

and

$$\omega_{AC} = 0. \qquad \text{ANS.}$$

All of the values above are, of course, identical to those found in Example 19.2.

Example 19.7

The system shown in Figure E19.7(a) consists of a slender homogeneous rod AB of length $L = 0.60$ m hinged at A to a fixed support and at B to a wheel whose radius $r = L/2 = 0.30$ m. The weight of rod AB is $W_R = 25.0$ N and that of the wheel is $W_W = 50.0$ N. A torsional spring whose constant is $K = 20.0$ N·m/rad is place at A and provides some restraint to the angular motion of rod AB. The system is released from rest when $\theta = 0°$ and the wheel rolls without slipping on the circular track. (a) Determine the angular velocity of the wheel for any position defined by the angle θ. Treat the wheel as a thin circular disk, and assume that the weight of the torsional spring is negligible and that it is unstressed when $\theta = 0°$. (b) Find the value of the angular velocity of the wheel for $\theta = 45°$ and for $\theta = 90°$.

Solution

(a) Because the wheel W rolls without slipping, the following relationships may be established.

$$\left.\begin{aligned} v_B &= r\omega_W = 0.30\omega_W \\ \omega_R &= \frac{v_B}{L} = 0.50\omega_W \end{aligned}\right\} \qquad \text{(a)}$$

where ω_R and ω_W are, respectively, the angular velocities of the rod and

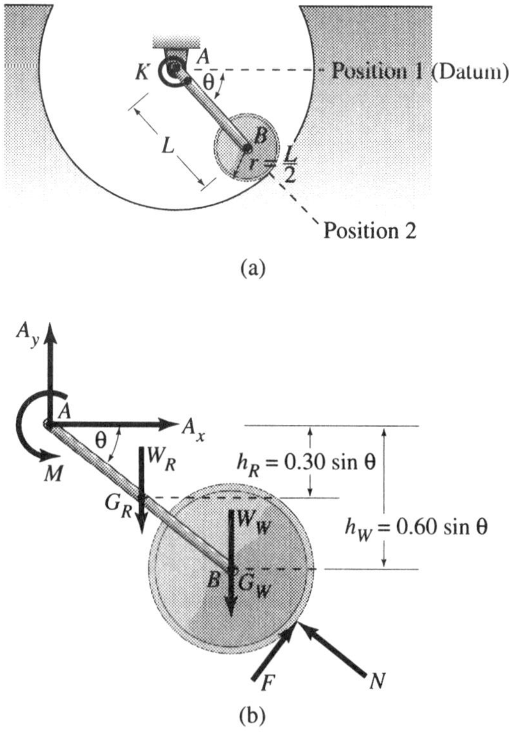

FIGURE E19.7.

the wheel. Because the system is released from rest when $\theta = 0^\circ$ (position 1), it does not have any kinetic energy in this position. Therefore,

$$\sum T_1 = 0. \tag{b}$$

For position 2 defined by any angle θ, the total kinetic energy of the system (consisting of the rod, the wheel, and the spring), is the sum of the kinetic energies of the rod and the wheel because the mass of the torsional spring is negligible. Thus,

$$\sum T_2 = (T_R)_2 + (T_W)_2 = \left(\frac{1}{2} I_A \omega^2\right)_R + \left[\frac{1}{2}\left(\frac{W}{g}\right) v_G^2 + \frac{1}{2} I_G \omega^2\right]_W$$

where $I_A = \dfrac{1}{3}\left(\dfrac{W_R}{g}\right) L^2$ is the mass moment of inertia of the rod about an axis through point A, $v_G = v_B$ is the linear velocity of the wheel's mass center, and $I_G = \dfrac{1}{8}\left(\dfrac{W_W}{g}\right) L^2$ is the mass moment of inertia of the wheel about its mass center. The kinetic energy $\sum T_2$ may be expressed in terms of the single unknown quantity ω_W. Thus, using Eqs. (a),

$$\begin{aligned}\sum T_2 &= \frac{1}{2}\left[\frac{1}{3}\left(\frac{25.0}{9.81}\right)(0.60)^2(0.50\omega_W)^2\right] + \frac{1}{2}\left(\frac{50.0}{9.81}\right)(0.30\omega_W)^2 \\ &\quad + \frac{1}{2}\left[\frac{1}{8}\left(\frac{50.0}{9.81}\right)(0.60)^2\right]\omega_W^2, \\ &= 0.3823\omega_W^2 \text{ N·m}.\end{aligned} \tag{c}$$

The free-body diagram of the system consisting of the rod and the wheel is shown in Figure E19.6(b). The forces A_x and A_y do no work because they do not displace and, therefore, do not contribute to the potential energy function of the system. The force N acting normal to the direction of motion performs no work and contributes nothing to the potential energy function. Because the wheel does not slip, the friction force F, therefore, does no work and its contribution to the potential energy function is zero. Thus, the potential energy function of the system in any position defined by the angle θ is due to the couple M in the torsional spring and to the weights W_R and W_W. Arbitrarily choose position 1 as the datum to measure the potential energy function. In this position both W_R and W_W have zero gravitational potential energy V_g, and, because the torsional spring is not stressed in this position, it has zero elastic potential energy V_e. Consequently, in position 1,

$$\sum V_1 = \sum (V_g)_1 + \sum (V_e)_1 = 0. \tag{d}$$

When the system moves to position 2 defined by any angle θ, the weights W_R and W_W both do positive work and, in so doing, decrease the capacity of the system to do work. Therefore, the gravitational potential energy function due to both weights is negative. Thus,

$$\sum (V_g)_2 = -W_R(h_R) - W_W(h_W) = -25.0(0.30\sin\theta) - 50.0(0.60\sin\theta),$$
$$= -37.5\sin\theta \text{ N·m}.$$

Also, in position 2, the torsional spring is displaced through the angle θ, and, therefore, its capacity to do work has been increased because elastic strain energy has been stored in it. Thus,

$$\sum (V_e)_2 = \tfrac{1}{2}K\theta^2 = \tfrac{1}{2}(20.0)\theta^2 = 10.0\theta^2 \text{ N·m}.$$

Therefore, the total potential energy function in position 2 is

$$\sum V_2 = \sum (V_g)_2 + \sum (V_e)_2 = -37.5\sin\theta + 10.0\theta^2. \qquad \text{(e)}$$

The principle of conservation of mechanical energy states that

$$\sum T_1 + \sum V_1 = \sum T_2 + \sum V_2.$$

Using the results in Eqs. (b), (c), (d), and (e),

$$0 + 0 = 0.3823\omega_W^2 - 37.5\sin\theta + 10.0\theta^2$$

from which

$$\omega_W = \sqrt{98.09\sin\theta - 26.16\theta^2} \qquad \text{(f)} \quad \text{ANS.}$$

(b) For $\theta = 45° = 0.7854$ rad, Eq. (f) yields

$$\omega_W = 7.29 \text{ rad/s} \circlearrowleft. \qquad \text{ANS.}$$

For $\theta = 90° = 1.5708$ rad, Eq. (f) yields

$$\omega_W = 5.79 \text{ rad/s} \circlearrowleft. \qquad \text{ANS.}$$

■

Problems

Note: Do all of the following Problems by the principle of conservation of mechanical energy.

19.26 Problem 19.4 (p. 569)
19.27 Problem 19.7 (p. 570)
19.28 Problem 19.8 (p. 570)
19.29 Problem 19.9 (p. 571)
19.30 Problem 19.11 (p. 571)
19.31 Problem 19.12 (p. 571)
19.32 Problem 19.14 (p. 572)
19.33 Problem 19.15 (p. 572)
19.34 Problem 19.18 (p. 573)
19.35 Problem 19.20 (p. 574)
19.36 Problem 19.21 (p. 574)
19.37 Problem 19.23 (p. 575)
19.38 Problem 19.24 (p. 575)
19.39 The system shown in Figure P19.39 consists of a slender rod AB of weight $W = 25$ lb and length $b = 4$ ft which is hinged at A and attached to a linear

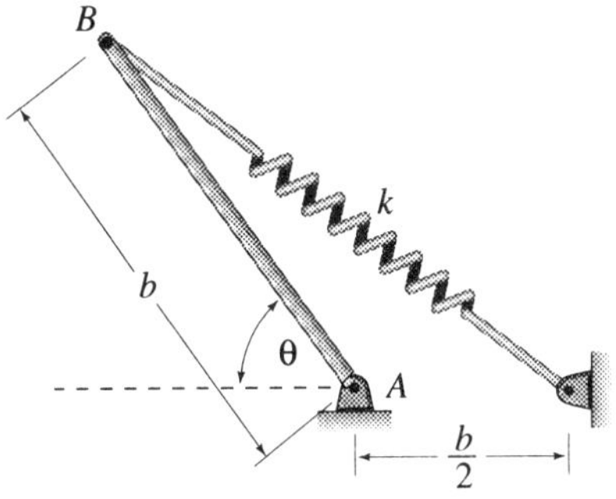

FIGURE P19.39.

spring at B. The linear spring has a spring constant $k = 1000$ lb/ft and is free of any force when the rod is in the vertical position, i.e., when $\theta = 90^\circ$. The rod is released from rest from the near horizontal position when $\theta = 0^\circ$ and allowed to rotate cw. Determine the velocity v_B of point B when (a) $\theta = 45^\circ$ and (b) $\theta = 90^\circ$.

19.40 The system shown in Figure P19.40 consists of a thin disk of radius r and weight W hinged at its center to homogeneous rod AB of length $4r$ and weight $W/2$. End B of the rod is hinged to a fixed support and a linear spring of constant k is attached to the system as shown. In the vertical position shown, the spring is unstretched. Assume frictionless conditions at A and B. The system is released from rest in the almost vertical position and allowed to rotate cw. In terms of W, k, and r, develop an expression for the velocity of point A when the system passes the horizontal position. What is the value of k in terms of W and r, if point A is to reach the horizontal position with zero velocity?

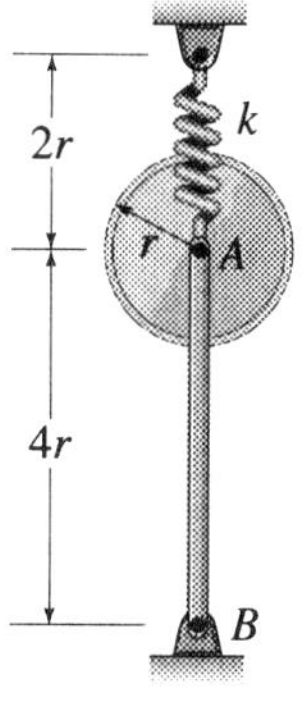

FIGURE P19.40.

19.41 Rework Problem 19.40 if the disk, instead of being hinged to the rod at A, is rigidly fastened to it. All other conditions remain the same.

19.42 The thin disk of weight W and diameter D shown in Figure P19.42 is hinged at point O. In the position shown (i.e., line OA in horizontal position), the angular velocity of the disk is ω, cw. Determine the velocity of point A when line OA is in the vertical position. Express your answer in terms of D and ω.

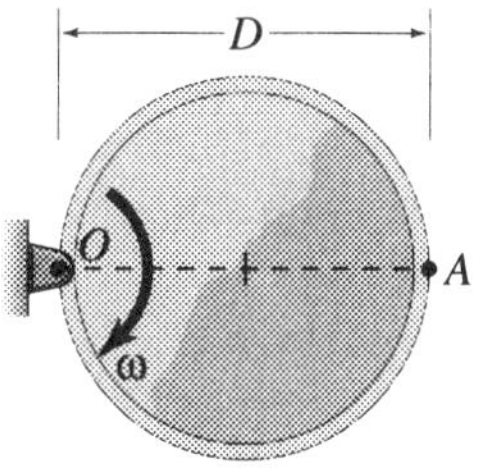

FIGURE P19.42.

19.43 The triangular plate shown in Figure P19.43 is carefully held in the position shown where corner A is attached to a frictionless small hinge. The plate has a uniform thickness $t = 0.05$ m and a weight $W = 250$ N. If the plate is released from rest from the position where edge BC is almost horizontal, determine the angular velocity of the plate and the linear velocity of corner B when edge BC is in a vertical position.

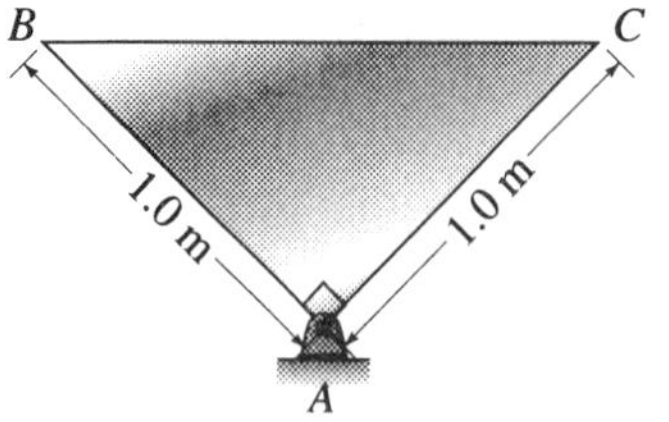

FIGURE P19.43.

19.44 A proposed garage door consists of a uniform 2.4 m × 5.0 m rectangular plate of weight $W = 1000$ N whose side view is shown in Figure P19.44. The structural system, consisting of perpendicular members OA and OB which support the door, has negligible weight, and the door rotates about a horizontal axis through point O. A torsional spring of constant K (N·m/rad) is placed at O to restrain the rotation of the door. Two such structural systems with torsional springs are used, one on each side of the door. When the door is in the horizontal (open) position, the two springs are pretorqued so that their angular displacements are each equal to 0.50 rad. The door is closed by imparting to it an initial ccw angular velocity $\omega_1 = 0.50$ rad/s when in the horizontal position and allowed to rotate into the vertical position. Determine the required value of the spring constant K so that the door reaches the vertical position with an angular velocity (a) $\omega =$ zero and (b) $\omega = 0.2$ rad/s.

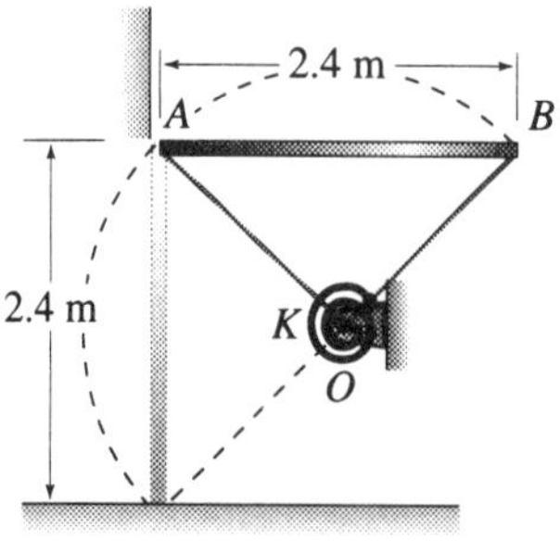

FIGURE P19.44.

19.45 The system shown in Figure P19.45 represents the side view of an 8 ft × 16 ft sectional garage door consisting of four identical sections each having a weight $W = 75$ lb. The door has a torsional spring system attached to each of its two sides with identical spring constants $K = 6.0$ ft·lb/rad each. The torsional springs are mounted on a horizontal shaft through point A. At each of the two ends of the shaft is attached rigidly a pulley of radius r. One end of a cable is wrapped around each pulley and the other end to the bottom section of the door as shown. In the open position shown, the springs are not loaded. Determine the value of the radius of the pulley r so that, when the door is released from rest in the open position, point B will have a linear velocity just before reaching the floor of (a) zero and (b) 0.5 ft/s.

19.46 The system shown in Figure P19.46 is released from rest when $\theta = 0°$. The two torsional springs (not shown) at B and C are free of any torque when $\theta = 90°$ and have spring constants of 10 kN/rad and 20 kN/rad, respectively. The two uniform rods AB and BC have a mass of 10 kg each and the wheel at A, which may be assumed to be a uniform disk, has a mass of 15 kg. If the wheel rolls without slipping, determine the linear speed of point A and the angular speeds of rods AB and BC when $\theta = 90°$.

19.47 Rod AB of length L = 1.5 ft is constrained so that end A moves in a vertical slot and end B in a horizontal one as shown in Figure P19.47. The weight of the uniform rod is $W = 20$ lb. The spring attached at B has a spring constant $k = 5$ lb/in. and is unstretched when $\theta = 0°$. If the system is released from rest when

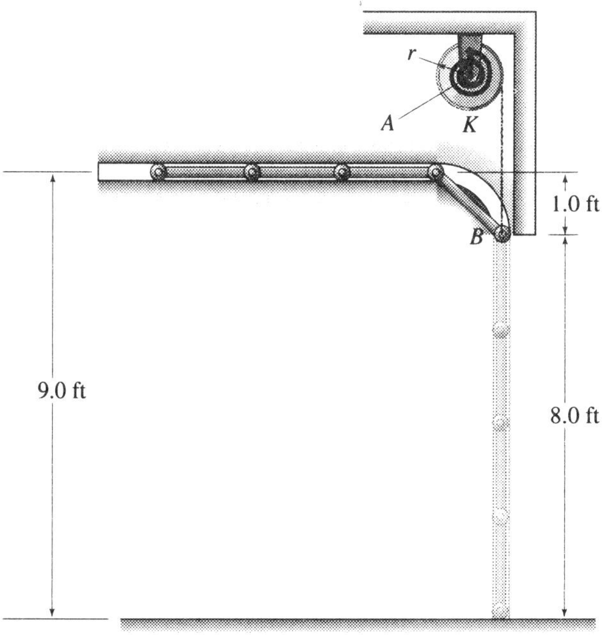

FIGURE P19.45.

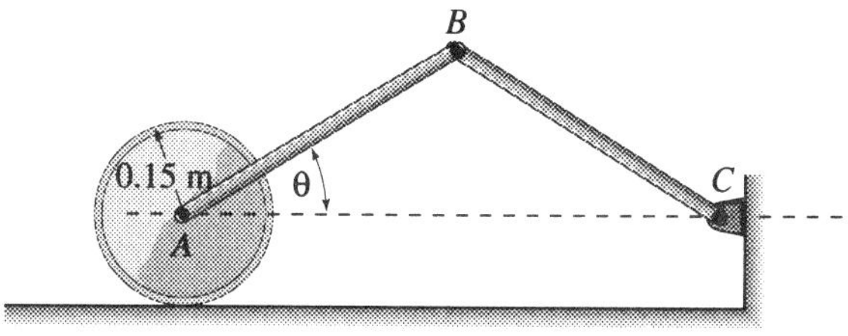

FIGURE P19.46.

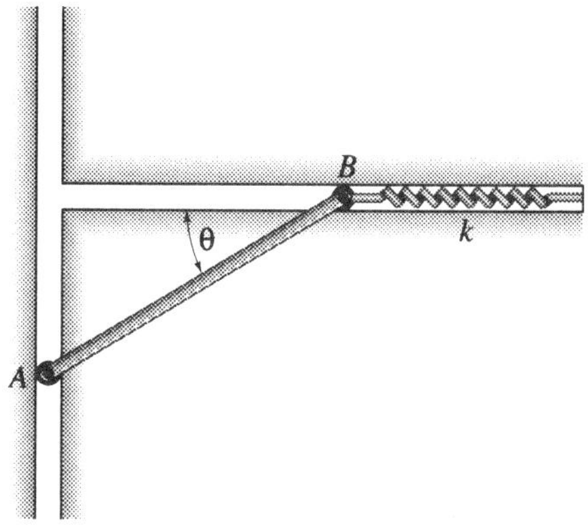

FIGURE P19.47.

$\theta = 90°$, determine the linear speed of points A and B and the angular speed of rod AB when $\theta = 0°$. Assume frictionless conditions.

19.48 A platform of mass $m_P = 100$ kg is supported by four identical wheels each of mass $m_W = 5$ kg and radius $r = 0.15$ m as shown in Figure P19.48. The center of mass for the entire system is located at point G. This system is released from rest when the detached spring is compressed a distance of 0.50 m and moves up the 20% inclined plane. If the wheels roll without slipping, determine (a) the

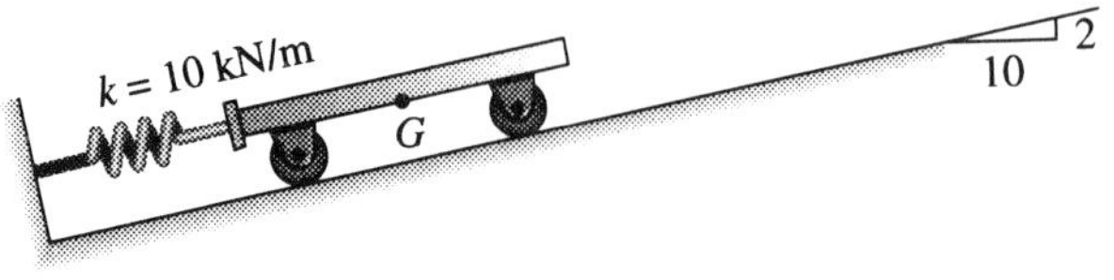

FIGURE P19.48.

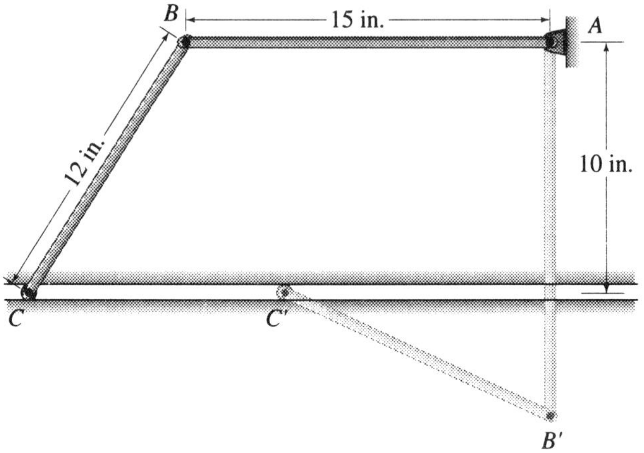

FIGURE P19.49.

speed of the platform after point G has traveled a distance of 3.00 m up the inclined plane and (b) the distance up the inclined plane that the platform will travel before coming to a full stop.

19.49 As shown in Figure P19.49, the uniform rods AB and BC weigh 15 lb and 12 lb, respectively. Point A represents a fixed hinge and point C is constrained to move in the horizontal guide. If the system is released from rest in the position shown, determine the angular velocity of rod AB as it passes the vertical position indicated by the broken lines. What is the corresponding linear velocity of point C? Assume frictionless conditions.

19.50 Refer to Example 19.7 (p. 582) and determine (a) the angle θ defining the position for which the angular velocity of the wheel is zero and (b) the angle θ for which the angular velocity of the wheel is maximum.

19.5 Power and Efficiency

In considering a machine for a given application, it is important to know the speed with which it can do work. Consider, for example, two cars of identical weights, one with a very small engine and the second with a very large engine. Either car can travel the distance between New York City and Los Angeles and perform an equal quantity of work. The only difference between the two would be that the car with the smaller engine would take longer than the car with the large engine to traverse the same distance and to perform the same quantity of work. The concept of *power*, which is defined as the time rate of doing work, is used to enable us to distinguish between two such machines and to provide a scientific means of comparing their performances.

If a machine performs a small quantity of work dU during a small interval of time Δt, the average power P_{avg} generated by the machine is defined by the relationship

$$P_{avg} = \frac{\Delta U}{\Delta t}. \tag{19.22}$$

However, since the time rate at which the machine does work may vary, it is desirable to define the instantaneous power P. This instantaneous power is obtained by evaluating Eq. (19.22) in the limit when Δt approaches zero. Thus,

$$P = \lim_{\Delta t \to 0} \frac{\Delta U}{\Delta t} = \frac{dU}{dt} \tag{19.23}$$

Consider a rigid body in general plane motion acted upon by a force **F** in the plane of motion, passing through the mass center point G, and by a couple **M** perpendicular to the plane of motion as shown in Figure 19.8. Recall from Chapter 17, that general plane motion may be decomposed into translation of the center of mass of the rigid body and rotation about this center of mass. In the case of the force **F**, the work $dU = \mathbf{F}\cdot d\mathbf{r}$ and the power P may be written in the following form:

$$P = \frac{\mathbf{F}\cdot d\mathbf{r}}{dt} = \mathbf{F}\cdot\mathbf{v}_G = (F\cos\beta)v_G \tag{19.24}$$

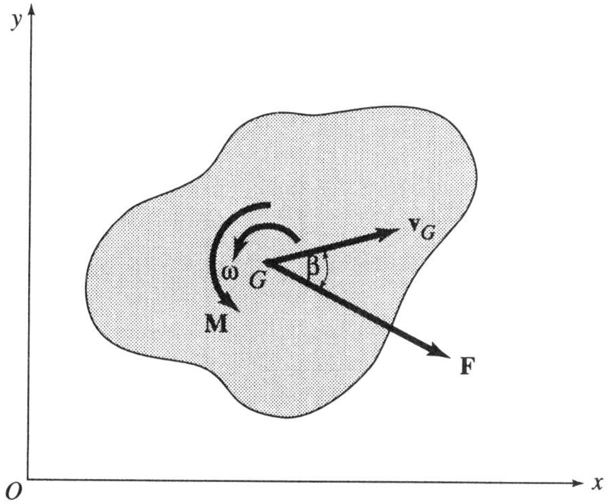

FIGURE 19.8.

In Eq. (19.24) the quantity $d\mathbf{r}/dt$ was replaced by the velocity $\mathbf{v}_G$ of the center of mass of the rigid body. Note that the angle β is the angle between the vectors $\mathbf{F}$ and $\mathbf{v}_G$ and that the quantity $F\cos\beta$ represents the component of $\mathbf{F}$ in the direction of $\mathbf{v}_G$. In the case of the couple $\mathbf{M}$, the work $dU = \mathbf{M}\cdot d\boldsymbol{\theta}$ and the power P may be written in the form

$$P = \frac{\mathbf{M}\cdot d\boldsymbol{\theta}}{dt} = \mathbf{M}\cdot\boldsymbol{\omega} = (M\cos\gamma)\omega. \tag{19.25}$$

In Eq. (19.25) the angle γ is the angle between the vectors $\mathbf{M}$ and $\boldsymbol{\omega}$ and the quantity $d\boldsymbol{\theta}/dt$ was replaced by the angular velocity $\boldsymbol{\omega}$ of the rigid body. Because, in dealing with the plane motion of a rigid body, we generally assume that the directions of $\mathbf{M}$ and $\boldsymbol{\omega}$ coincide, it follows that $\gamma = 0$ and Eq. (19.25) may be written as

$$P = \mathbf{M}\cdot\boldsymbol{\omega} = M\omega. \tag{19.26}$$

It is obvious from Eqs. (19.24) and (19.26) that power is a scalar quantity. By definition, the units used to measure power are those of work per unit of time which must be the same as those obtained when multiplying a force by a linear velocity (Eq. (19.24)) and those obtained when multiplying a moment by an angular velocity (Eq. (19.26)). In the U.S. Customary system of units, if the force is measured in pounds, displacement in feet, and time in seconds, the unit of power would be ft·lb/s. A more commonly used unit of power in the U.S. Customary system is the horsepower (hp) which is defined as

$$1\ \text{hp} = 550\ \text{ft}\cdot\text{lb/s} = 33{,}000\ \text{ft}\cdot\text{lb/min} = 1{,}980{,}000\ \text{ft}\cdot\text{lb/hr}.$$

In the SI units of measure, if the force is in Newtons, the displacement is in meters, and the time in seconds, the unit of power would be N·m/s = J/s which has been given the name *Watt* and the symbol W. A commonly used unit of power in the SI system is the kilowatt (kW) = 1000 Watts. Thus,

$$1\ \text{W} = 1\ \text{N}\cdot\text{m/s} = 1\ \text{J/s},$$

$$1\ \text{kW} = 1000\ \text{W}.$$

A useful relationship between units of power in the U.S. Customary system and those in the SI system is

$$1\ \text{hp} = 746\ \text{W} = 0.746\ \text{kW}$$

A machine is not a perfect device, and it always requires more input power than the output or useful power it is capable of generating. The difference between the input and output powers is lost as a result of one or all of the following three factors: mechanical friction, electrical factors, and thermal effects. The overall *efficiency* η of a machine is defined as the ratio of output power P_{out} to input power P_{in}. Thus,

$$\eta = \frac{P_{out}}{P_{in}} \tag{19.27}$$

where the power loss $(P_{in} - P_{out})$ is due to all of the three factors mentioned above. The term *mechanical efficiency* η_m is used when the entire power loss is due only to mechanical friction effects, the term *electrical efficiency* η_e when the entire power loss is due only to electrical effects, and the term *thermal efficiency* η_t when the entire power loss is due only to thermal effects. Regardless of what factors are responsible for the power loss $(P_{in} - P_{out})$, Eq. (19.27) yields the value of the overall efficiency η. However, in a given system, if the three efficiencies η_m, η_e, and η_t are measured independently, then, the overall efficiency η of the system would be

$$\eta = \eta_m \eta_e \eta_t. \tag{19.28}$$

Example 19.8

The homogeneous rod AB in Figure E19.8(a) is constrained to move so that ends A and B slide freely in the vertical and horizontal tracks,

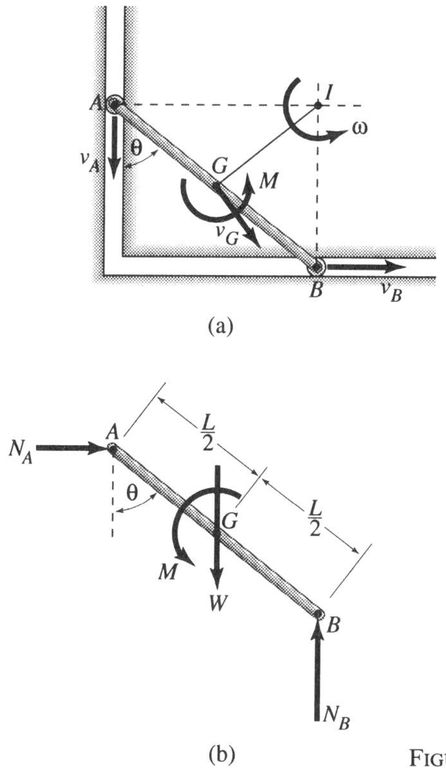

FIGURE E19.8.

respectively. The length of rod AB is $L = 2$ ft and its weight is $W = 50$ lb. A moment $M = 40$ ft·lb is applied as shown, and the system starts from rest when $\theta = 0°$. (a) Determine the power that is generated by W and M for any angle θ $(0 \leq \theta < 90°)$. (b) What is the numerical value of the power for $\theta = 45°$ and $\theta = 85°$.

Solution

(a) Because the system is subjected to the simultaneous action of a force and a couple, the total power at any instant of time would be obtained by adding, algebraically, the results from Eqs. (19.24) and (19.26). To use these equations, we need to determine v_G and ω for any position defined by the angle θ. These are most conveniently found by the work-energy principle and the method of the instantaneous center. The location of this center, point I, is shown in Figure E19.8(a) and the following relations are established:

$$IG = AG = L/2 = 1 \text{ ft},$$

$$v_G = (IG)\omega = \omega$$

where v_G is the linear velocity of the mass center of the rod and ω is its ccw angular velocity. Let position 1 coincide with the position of the rod when $\theta = 0°$ and position 2 with its position for any angle θ $(\theta < 90°)$. By the work-energy principle,

$$U_{1\to 2} = T_2 - T_1.$$

The free-body diagram of the rod is shown in Figure E19.8(b). Note that only W and M do work because N_A and N_B are both normal to the paths of motion of ends A and B, respectively. Thus,

$$U_{1\to 2} = W\left(\frac{L}{2} - \frac{L}{2}\cos\theta\right) + M\theta = \frac{WL}{2}(1 - \cos\theta) + M\theta$$

$$= 50(1 - \cos\theta) + 40\theta \text{ ft·lb}$$

Because the system starts from rest in position 1 $(\theta = 0°)$,

$$T_1 = 0,$$

and, for any position 2 defined by the angle θ $(\theta < 90°)$,

$$T_2 = \frac{1}{2}mv_G^2 + \frac{1}{2}I_G\omega^2 = \frac{1}{2}\left(\frac{W}{g}\right)\omega^2 + \frac{1}{2}\left[\frac{1}{12}\left(\frac{W}{g}\right)L^2\right]\omega^2$$

$$= \frac{1}{2}\left(\frac{50.0}{32.2}\right)\omega^2 + \frac{1}{24}\left(\frac{50.0}{32.2}\right)(2)^2\omega^2$$

$$= 1.0352\omega^2 \text{ ft·lb}.$$

Thus, the work-energy principle yields

$$50(1 - \cos\theta) + 40\theta = 1.0352\omega^2$$

from which

$$\omega = \sqrt{48.30(1 - \cos\theta) + 38.64\theta} \text{ rad/s}$$

and

$$v_G = \omega = \sqrt{48.30(1 - \cos\theta) + 38.64\theta} \quad \searrow^{\theta} \text{ ft/s.}$$

The power may now be determined from Eqs. (19.24) and (19.26). Thus,

$$\begin{aligned} P &= W\cos(90 - \theta)v_G + M\omega \\ &= 50\sin\theta\sqrt{48.30(1 - \cos\theta) + 38.64\theta} \\ &\quad + 40\sqrt{48.30(1 - \cos\theta) + 38.64\theta} \\ &= (50\sin\theta + 40)\sqrt{48.30(1 - \cos\theta) + 38.64\theta} \text{ ft·lb/s.} \end{aligned} \qquad \text{ANS.}$$

(b) For $\theta = 45^\circ = 0.7854$ rad,

$$P = 503 \text{ ft·lb/s} = 0.914 \text{ hp.} \qquad \text{ANS.}$$

For $\theta = 85^\circ = 1.4835$ rad,

$$P = 904 \text{ ft·lb/s} = 1.644 \text{ hp} \qquad \text{ANS.}$$

■ Example 19.9

The motorized pulley system shown in Figure E19.9(a) is used to pull the 500-lb crate up the 30° rough inclined plane. At the instant shown, the crate is moving up the plane at a speed of 10 ft/s and an acceleration of 5 ft/s^2. The coefficient of kinetic friction between the crate and the plane is 0.2 and the efficiency of the motor is 65%. Determine (a) the power output of the motor and (b) the power input to the motor. Assume the pulleys to be massless and frictionless.

Solution

(a) The free body and inertial force diagrams for the crate are shown in Figure E19.9(b) along with a convenient right-handed x-y-z coordinate system. The crate executes pure translation, and, therefore, we may apply the following equations:

$$\sum F_y = ma_y = 0, \quad N - 500\cos 30^\circ = 0,$$

$$N = 433.0 \text{ lb.}$$

$$\sum F_x = ma_x, \qquad F - 500\sin 30^\circ - 0.2(433) = \left(\frac{500}{32.2}\right)(5),$$

$$F = 414.2 \text{ lb.}$$

The free-body diagram of the pulley at A is shown in Figure E19.9(c). Since the pulley is assumed massless and frictionless, Figure E19.9(c) represents a static system, and

$$\sum F_x = ma_x = 0, \quad 2T - 414.2 = 0,$$

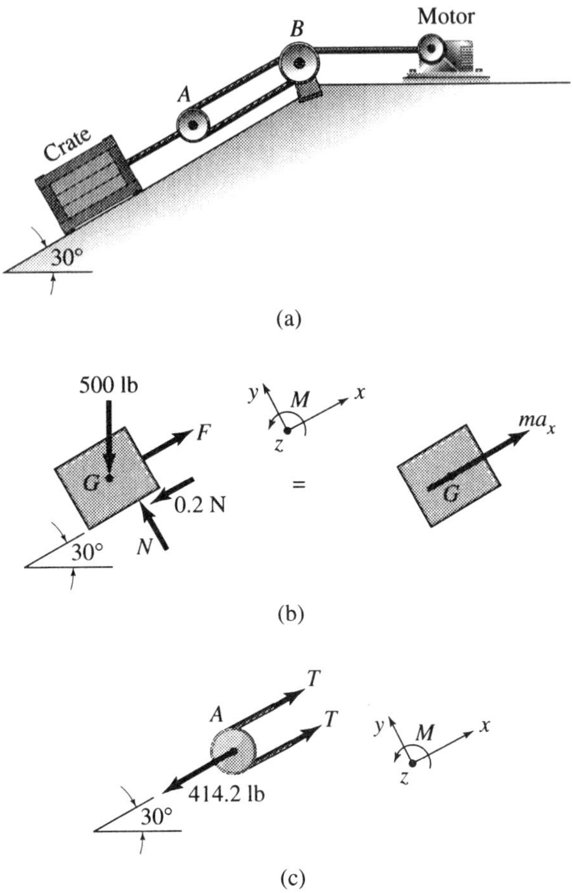

FIGURE E19.9.

$$T = 207.1 \text{ lb.}$$

Because the pulley at B is also massless and frictionless, the tension $T = 207.1$ lb is also the force in the cable wrapped around the motor pulley. Also, by considering the kinematics of the pulley system, we conclude that the speed of the cable wrapped around the motor pulley v_M is twice that of the crate, v_C. Thus,

$$v_M = 2v_C = 2(10) = 20 \text{ ft/s.}$$

Therefore,

$$P_{out} = Tv_M = 207.1(20) = 4142 \text{ ft.lb/s} = 7.53 \text{ hp.} \qquad \text{ANS.}$$

(b) Because the efficiency of the motor is 65%, it follows that

$$P_{in} = \frac{P_{out}}{\eta} = \frac{4142}{0.65} = 6372 \text{ ft.lb/s} = 11.59 \text{ hp.} \qquad \text{ANS.}$$

Example 19.10

An automobile weighing 15000 N travels along a 10% inclined plane as shown in Figure E19.10(a). It starts from rest and accelerates at a constant rate reaching its maximum speed of 150 km/h in 100 s. The automobile has a front-wheel drive, and the resistance to motion due to air drag and rolling friction is estimated at 5% of its weight. Determine (a) the power developed by the engine when reaching the maximum speed, (b) the power capacity of the engine if its efficiency is 0.55, (c) the components of the reactions at the front and rear wheels during the acceleration period, and (d) the power needed to travel at any constant speed v along the 10% inclined plane.

Solution

(a) Assuming that the 10% inclined plane is straight, the automobile undergoes plane translation. The following kinematic relationship is used to determine the linear constant acceleration:

$$v = v_O + a_x t, \qquad \frac{150 \times 10^3}{3600} = 0 + a_x t,$$

and

$$a_x = 0.4167 \text{ m/s}^2.$$

The free-body and inertial force diagrams of the automobile are shown in Figure E19.10(b) along with a convenient right-handed x-y-z coordinate system, where the z axis comes out of the page. The force $R =$

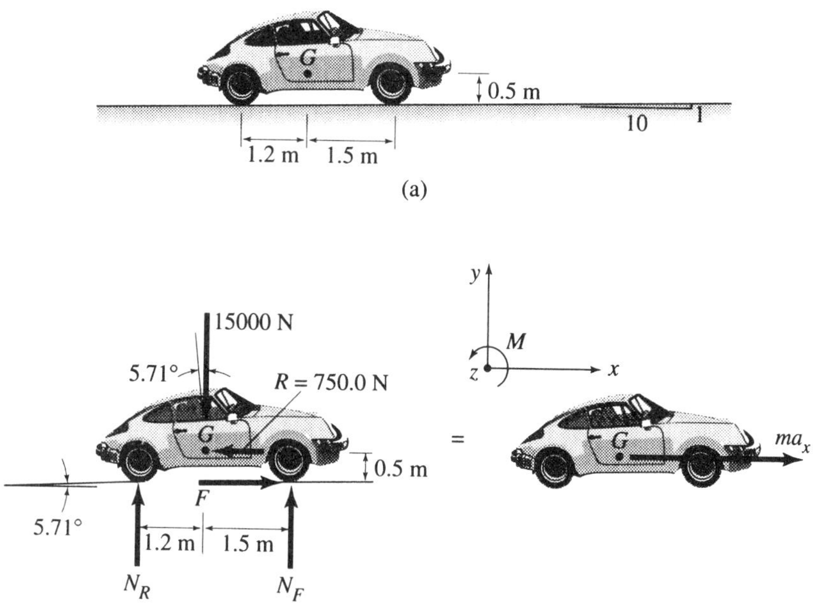

FIGURE E19.10.

0.05 (15000) = 750 N is the resistance to motion due to air drag and rolling friction and, for simplicity, is assumed to act through the mass center of the automobile. The quantities N_R and N_F represent the normal forces at the rear and front wheels, respectively, and the force F is the frictional force that acts on the driving front wheels and propels the automobile forward. Note that no frictional force acts on the rear wheels because they are nondriving wheels.

We apply Newton's equation of motion in the x direction. Thus,

$$\sum F_x = ma_x: \quad F - 750 - 15000 \sin 5.71^\circ = \left(\frac{15000}{9.81}\right)(0.4167)$$

from which

$$F = 2879.56 \text{ N}.$$

By Eq. (19.24),

$$P = Fv_G = 2879.56\left(\frac{150 \times 10^3}{3600}\right) = 119{,}982 \text{ W} = 120.0 \text{ kW}. \qquad \text{ANS.}$$

(b) By Eq. (19.27),

$$P_{in} = \frac{P_{out}}{\eta}$$

where P_{in} represents the engine power capacity, P_{out} represents its output at maximum speed (i.e., 120.0 kW), and η is the stated efficiency. Thus,

$$\text{Power capacity} = P_{in} = \frac{120.0}{0.55} = 218 \text{ kW} \qquad \text{ANS.}$$

(c) Refer to the free-body diagram of Figure E19.10(b), and apply Newton's equations of motions. Thus,

$$\sum F_y = ma_y = 0, \quad N_R + N_F - 15000 \cos 5.71 = 0 \qquad \text{(a)}$$

$$\sum M_G = I_G\alpha = 0, \quad 1.5N_F - 1.2N_R + 0.5(2879.56) = 0 \qquad \text{(b)}$$

The simultaneous solution of Eqs. (a) and (b) yields

$$N_R = 8820 \text{ N} \quad \text{on rear wheels} \qquad \text{ANS.}$$

and

$$N_F = 6100 \text{ N} \quad \text{on front wheels} \qquad \text{ANS.}$$

and, from part (a),

$$F = 2880 \text{ N} \quad \text{on front wheels.} \qquad \text{ANS.}$$

(d) If the automobile travels at any constant speed, $a_x = 0$. Thus

$$\sum F_x = 0, \quad F - 750 - 15000 \sin 5.71 = 0,$$

from which

$$F = 2242.40 \text{ N}$$

and

$$P = Fv = 2242.40v, \quad W = 2.242v \text{ kW} \qquad \text{ANS.}$$

where v is in m/s.

■

Problems

19.51 The wheel shown in Figure P19.51 rolls up the 20° inclined plane without slipping under the action of the force $F =$ 150 lb and the couple $M = 100$ lb·ft. The wheel has a weight $W = 50$ lb and a centroidal radius of gyration $k = 1.2$ ft. Develop an expression for the horsepower generated by F and M in terms of the distance s traveled by the wheel up the inclined plane, assuming that it starts from rest.

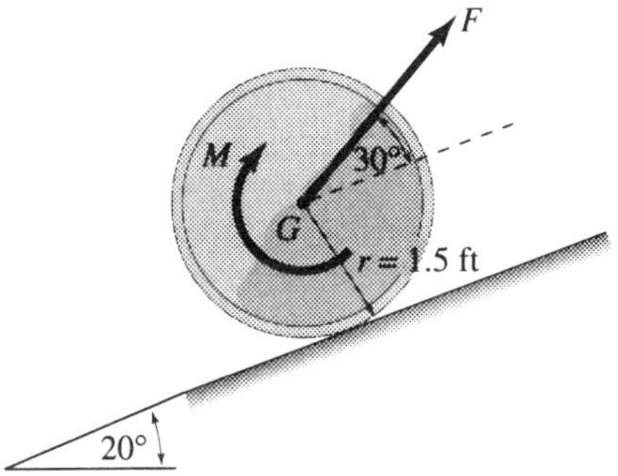

FIGURE P19.51.

19.52 Refer to the system shown in Figure P19.52. The weight of block A is 20 N and that of the wheel is 150 N. The centroidal radius of gyration of the wheel is 0.30 m. The wheel is subjected to a constant cw couple $M = 10$ N·m. In the position shown, the velocity of the wheel's mass center is 2 m/s down the inclined plane. Determine the total power generated by the weight of the wheel and the moment M at the instant when block A has moved up a distance of 5 m. Assume the wheel rolls without slipping and that the peg at B is frictionless.

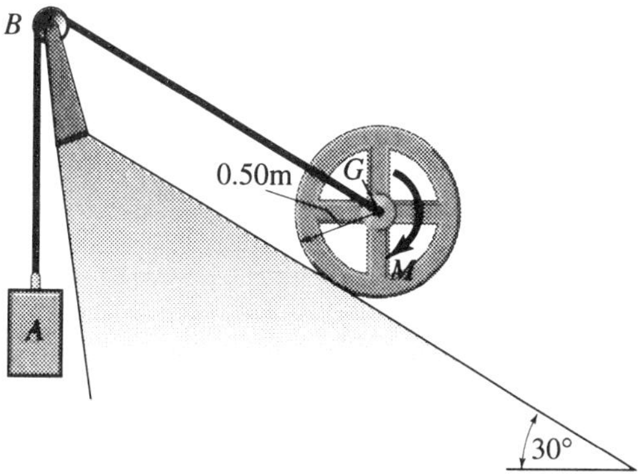

FIGURE P19.52.

19.53 Refer to the system shown in Figure

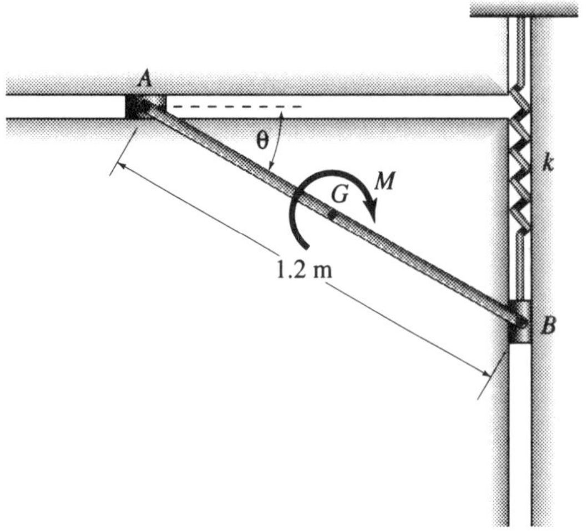

FIGURE P19.53.

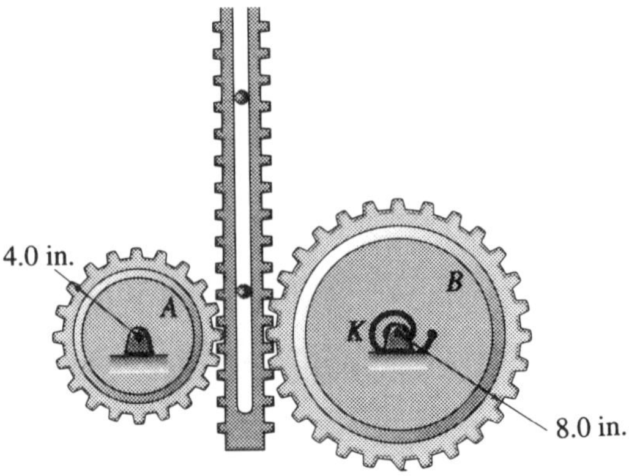

FIGURE P19.54.

P19.53. It consists of a homogeneous slender rod AB of weight $W = 50$ N connected at A and B by frictionless hinges to two sliders of weight $W_A = 30$ N and $W_B = 80$ N. Slider A is constrained to move in a frictionless horizontal track and slider B in a frictionless vertical track. A linear spring of spring constant $k = 500$ N/m is attached to slider B and a cw couple $M = 200$ N·m is applied as shown. The system starts from rest when $\theta = 0°$. Determine the total power generated by the weight of rod AB and by the moment M when $\theta = 45°$. There is no stretch in the spring when $\theta = 0°$.

19.54 Consider the gear-rack system shown in Figure P19.54. The vertical rack weighting $W_R = 100$ lb is released from rest. Find the power generated by W_R after it has moved a distance of 2 ft. The centroidal mass moment of inertia of gear A is $I_A = 4$ lb·s²·ft and that of gear B is $I_B = 7$ lb·s²·ft. Gear B is restrained by a coil spring with a spring constant $K_B = 10$ ft·lb/rad.

19.55 A platform of mass $m_p = 100$ kg is supported by four identical wheels each of mass $m_W = 5$ kg and a radius $r = 0.15$ m as shown in Figure P19.55. The center of mass of the entire system is located at point G. This system is released from rest when the detached spring is compressed a distance of 0.50 m and moves up the 20% inclined plane. If the wheels roll without slipping, determine the power generated by the weight of the platform after point G has traveled a distance of 3.00 m up the inclined plane.

19.56 The pulley system shown in Figure

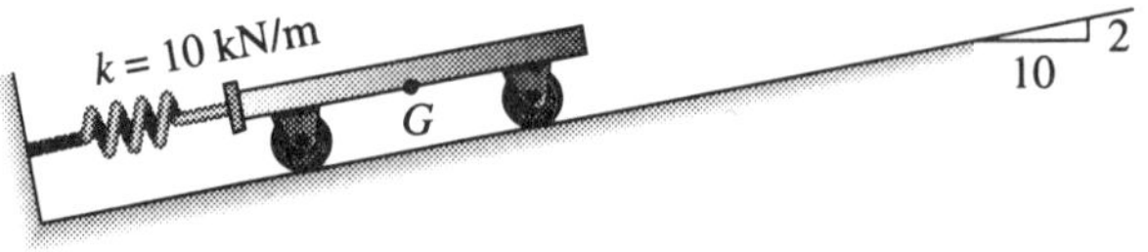

FIGURE P19.55.

P19.56 is used by the worker to raise the weight W. Determine the power generated by the worker (a) if the weight W moves up at a constant velocity v_O and (b) when the weight W has an instantaneous upward velocity v and an upward acceleration a. Express your answers in terms of W, v and a and neglect the weight and friction of the pulleys.

FIGURE P19.56.

19.57 An electric motor that drives a lathe has a rated capacity of 30 kW at 1000 rpm with an efficiency of 80%. What must be the torque carrying capacity of the shaft that transmits the power from the motor to the lathe?

19.58 The elevator system shown in Figure P19.58 is driven by an electric motor with an efficiency of 0.75. The weight of the elevator and its contents are W 5000 lb and that of the counterweight $W_C = 3750$ lb. Determine the power input to the motor (a) if the elevator moves up at a constant speed of 15 ft/s and (b) when the elevator has an instantaneous upward speed of 15 ft/s and an upward acceleration of 4 ft/s^2. Neglect the weight and friction of the pulleys.

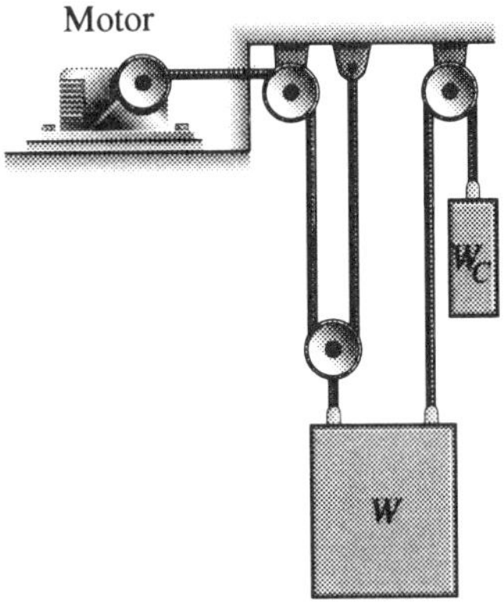

FIGURE P19.58.

19.59 An escalator in a department store is to transport 100 persons per minute from one level to the next, a height of 3 m. The average weight of a person is assumed to be 650 N. Determine (a) the average power needed to do the job and (b) the capacity of the motor required if its efficiency is 55%.

19.60 The automobile shown in Figure P19.60 has a rear-wheel drive and a weight $W =$ 4000 lb. When traveling on a horizontal stretch of road, it increases its speed at a constant rate of acceleration from 20 to 60 mph in a distance of 800 ft. The resistance to motion due to air drag and rolling friction is estimated at 4% of the weight of the automobile. Let $a = 4.5$ ft., $b = 5.5$ ft. and $h = 2.0$ ft, and determine (a) the power developed by the engine when reaching the speed of 60 mph, (b)

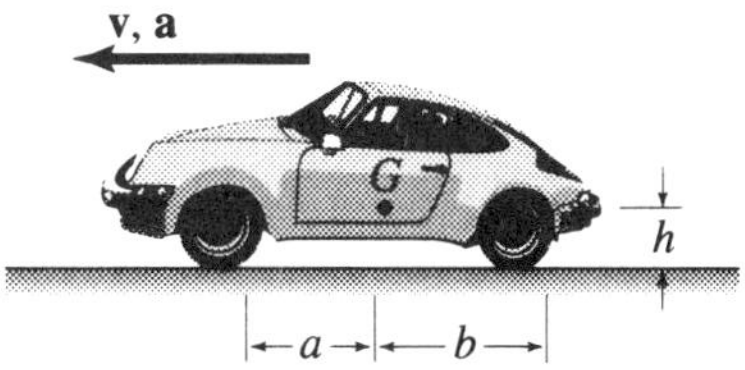

FIGURE P19.60.

the power needed to maintain a speed of 60 mph, and (c) the reaction components at the front and rear wheels when the automobile is accelerating from 20 to 60 mph.

19.61 The engine of the automobile shown in Figure P19.60 is rated at 60 kW and may be assumed to deliver a constant power to the driving wheels of 65% of its rated capacity when traveling on a horizontal road. The weight of the automobile and its contents are $W = 13000$ N, and the resistance to motion due to air drag and rolling friction may be assumed to be negligible. Let $a = 1.50$ m, $b = 1.75$ m and $h = 0.80$ m and determine (a) the time required and (b) the distance traveled for the automobile to increase its speed from 50 to 100 km/h.

Review Problems

19.62 A 20-kg homogenous rod with a length of 1.00 m is hinged at A as shown in Figure P19.62. The hinge at A is frictionless and the rod is attached to a torsional spring at the hinge of spring constant $K = 50$ N·m/rad. In the position shown, the spring is free of any torque. If the rod is released from rest in the horizontal position shown, determine the reaction components at the hinge when the rod reaches the vertical position. Use the work-energy principle.

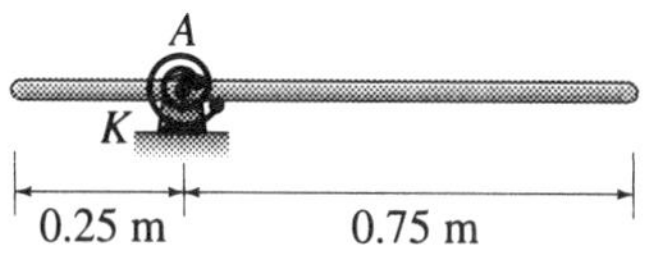

FIGURE P19.62.

19.63 The system shown in Figure P19.63 consists of gear A, having a weight of 15 lb and a radius of 9 in., and gear B which has an outer radius of 6 in. and a 4-in.-radius central pulley welded to it. The composite gear B has a weight of 20 lb and a centroidal radius of gyration of 5 in. One end of a cord is wrapped around the pulley and the other to a weight $W = 10$ lb as shown. The system is released from rest in the position shown. Assume frictionless conditions and determine (a) the speed of the weight W after it has moved down a distance of 2 ft and (b) the magnitude of the tangential force that exists in the gear teeth at the point of contact between the two gears. Use the work-energy principle.

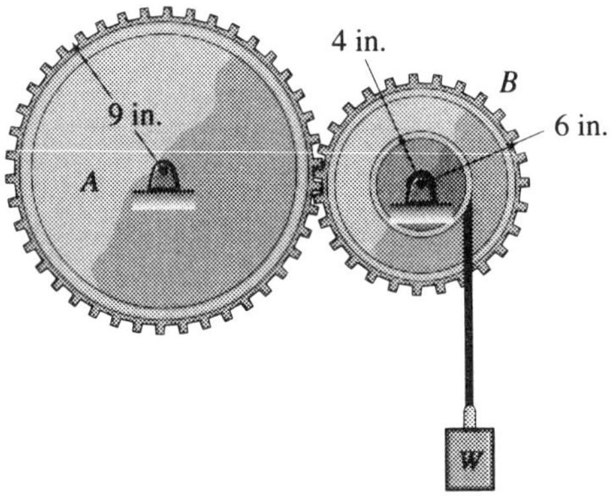

FIGURE P19.63.

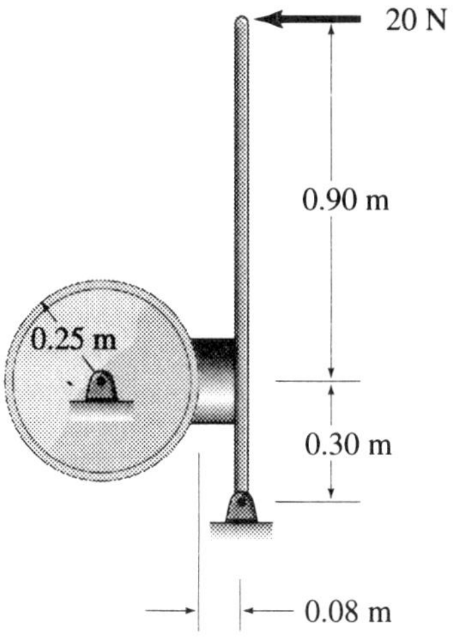

FIGURE P19.64.

19.64 The wheel shown in Figure P19.64 rotates at 1200 rpm. When the 20-N force is applied at the end of the brake lever, the wheel comes to a complete stop in 30 revolutions. Determine the centroidal mass moment of inertia of the wheel.

19.65 The compound pulley shown in Figure P19.65 has a weight of 75 lb and a centroidal radius of gyration of 0.6 ft. The system is released from rest and the 100-lb block slides down the frictionless inclined plane. How far down the plane should the block travel in order to achieve a speed of 10 ft/s?

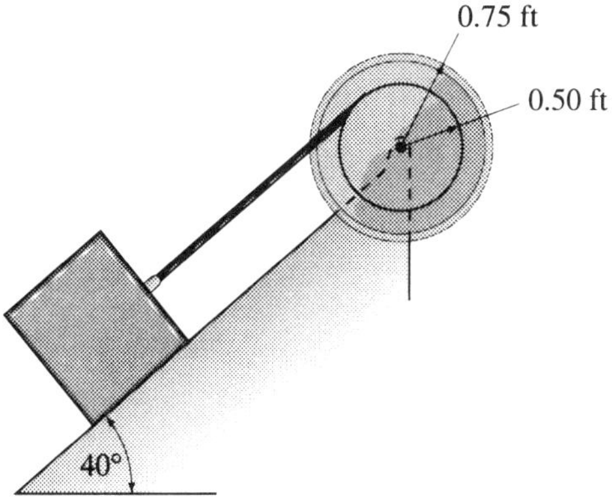

FIGURE P19.65.

19.66 Solve Problem 19.62 by the principle of conservation of mechanical energy.

19.67 Solve Problem 19.63 by the principle of conservation of mechanical energy.

19.68 The 40-kg rectangular plate shown in Figure P19.68 is hinged at corner A where a torsional spring is attached to the support and to the plate. The plate is released from rest in the position shown and allowed to rotate cw. Determine the spring constant K of the torsional spring if the plate is to have zero angular velocity after edge AB has rotated through 180°.

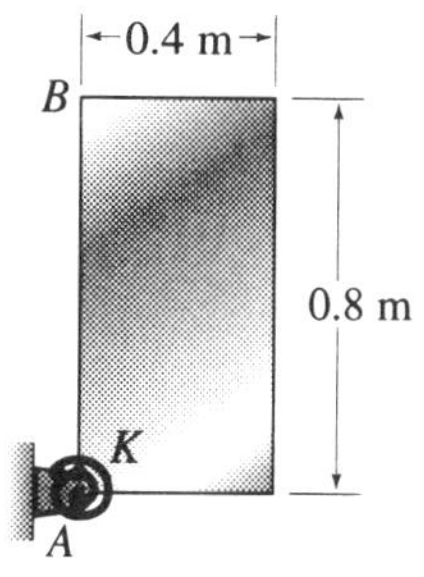

FIGURE P19.68.

19.69 An automobile is traveling along a straight path that is inclined to the horizontal at 10°. The weight of the automobile, excluding the tires, is 2800 lb. Each tire has a weight of 60 lb, a radius of 14 in., and a centroidal radius of gyration of 12 in. If the engine stalls while the automobile is moving at 60 mph, determine the distance that it will travel before coming to a complete stop. Assume the wheels roll without slipping and that the brakes are not applied after the engine stalls.

19.70 The electric motor shown in Figure P19.70 runs at 1800 rpm and operates a lathe connected to the shaft at B. The shaft at B is known to be subjected to a torque $M_B = 100$ N·m. Determine (a) the power generated by the electric motor, (b) the torque acting on the motor shaft at A, and (c) the power input to the electric motor if its efficiency is 85%.

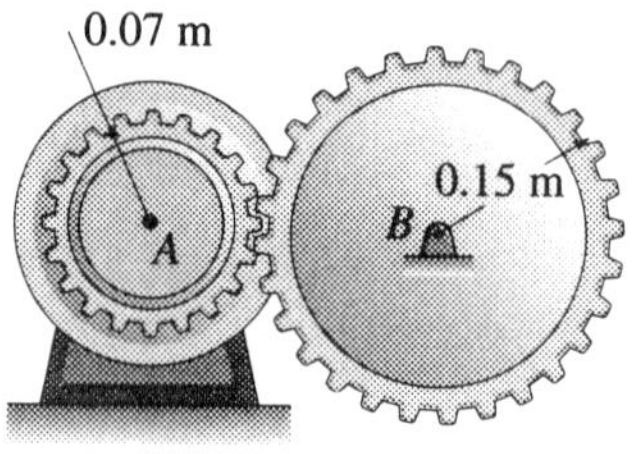

FIGURE P19.70.

20

Two-Dimensional Kinetics of Rigid Bodies: Impulse-Momentum

Mars Observer is the first U.S. probe to visit Mars since Viking 1 and 2 in 1976.

Mars has always been a cause of fascination to scientists and a source of inspiration to poets. The remarkable and vivid pictures sent back to Earth by Vikings 1 and 2 spacecraft revealed that Mars was once a living planet with water flowing freely and possibly a civilization that no longer exists. Despite the preliminary nature of these conclusions, Mars has become a case study of how to prevent this from happening to Earth.

The remarkable engineering feat of sending spacecraft to distant planets and our ability to precisely control their movements demonstrate one of the many contributions engineers have made to reach the unreachable stars and conquer the most distant frontier. Your ability to comprehend the most difficult concepts of kinetics starts with the most elementary principles and understanding the fundamental relationships between force and motion. This chapter provides an alternative technique for solving kinetics problems involving systems encountered here on Earth as well as in outer space.

One common theme running through all of these chapters dealing with kinetics is the uncommon elegance and beauty of the mathematical development of the basic principles relating force to motion. In Chapter 16, you learned how to solve problems dealing with the motion of particles by using the principles of impulse and momentum. In this chapter, you will apply these principles to the planar motion of rigid bodies. Thus, the concept of momentum is developed for a rigid body undergoing three types of plane motion: translation, rotation about a fixed axis through the mass center *and* rotation about a fixed axis through an arbitrary *point. The principles of linear and angular impulse and momentum for rigid bodies and for systems of rigid bodies are also developed.*

The next space ship to set sail to Mars is scheduled to leave Earth in the early part of the next century. We hope that the concepts introduced here will stimulate your hunger to seek the indispensable wisdom to be our first engineer on Mars. You may be the one who will take the next giant leap for mankind. Remember, every journey starts with a small first step and that the burden may be great but the rewards are immense.

20.1 Linear and Angular Momentum

In Chapter 16 we defined the concepts of impulse and momentum and developed the principles of impulse and momentum as they relate to the motion of a particle. In this chapter, we will extend these principles to the general plane motion of a rigid body. Recall from Chapter 17 that this type of motion may be viewed as a combination of translation and rotation.

Translation

In Chapter 16 we defined the linear momentum of a particle by the equation

$$\mathbf{L} = m\mathbf{v}. \tag{20.1}$$

This definition is used as a basis to derive the equation that defines the linear momentum due to the translational component of the general plane motion of a rigid body.

Consider the rigid body of mass m shown in Figure 20.1 which is undergoing general plane motion in the x-y plane, so that it rotates with an angular velocity $\boldsymbol{\omega}$ and its center of mass G translates with a linear velocity $\mathbf{v}_G$. Note that the rigid body is symmetric with respect to the plane of motion. The linear momentum $\mathbf{L}_i$ of any particle of mass m_i in the rigid body moving with a linear velocity $\mathbf{v}_i$, is given by Eq. (20.1). Thus,

$$\mathbf{L}_i = m_i\mathbf{v}_i$$

The linear momentum $\mathbf{L}$ of the entire rigid body is obtained by vectorially adding the linear momenta of all the particles composing the rigid body. Therefore,

$$\mathbf{L} = \sum \mathbf{L}_i = \sum m_i\mathbf{v}_i. \tag{20.2}$$

The position vector $\mathbf{r}_G$ of the center of mass G of the rigid body in Figure 20.1 is given by the relationship (see Eq. 14.6c)

$$m\mathbf{r}_G = \sum m_i\mathbf{r}_i.$$

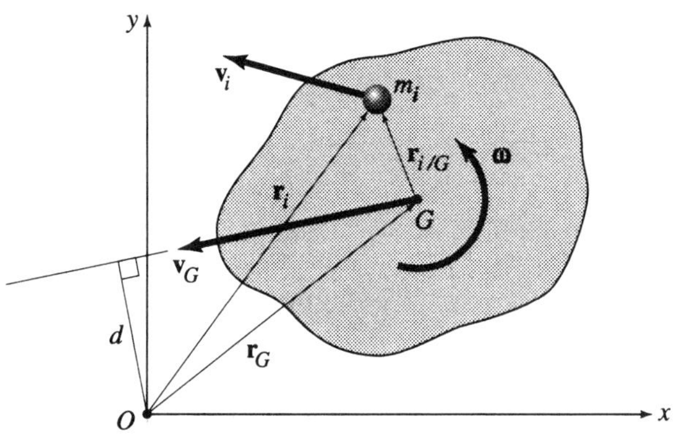

FIGURE 20.1.

Assuming that the masses m and m_i remain constant with time, we may differentiate this equation with respect to time to obtain

$$m\frac{d\mathbf{r}_G}{dt} = \sum m_i \frac{d\mathbf{r}_i}{dt}$$

or

$$m\mathbf{v}_G = \sum m_i \mathbf{v}_i \tag{20.3}$$

where $\mathbf{v}_G$ is the linear velocity of the mass center G of the rigid body. Substituting Eq. (20.3) in Eq. (20.2),

$$\mathbf{L} = m\mathbf{v}_G \tag{20.4}$$

which states that the linear momentum $\mathbf{L}$ due to the translational component of the general plane motion of a rigid body is the product of the mass m of the rigid body and the velocity $\mathbf{v}_G$ of its mass center. The linear momentum of a rigid body is a vector quantity whose magnitude is mv_G and whose direction and sense coincide with those of $\mathbf{v}_G$. The linear momentum has units of mass multiplied by those of velocity. Thus, in the SI System it would have units such as kg·m/s and in the U.S. customary system it would have units such as slug·ft/s. As will be shown later, because momentum is equal to the impulse, these units are reducible, respectively, to N·s and lb·s.

Rotation

In Chapter 16, we defined the angular momentum of a particle about a point O by the equation

$$\mathbf{H}_O = \mathbf{r} \times m\mathbf{v} \tag{20.5}$$

We will use this definition to derive an equation that defines the angular momentum of a rigid body due to the rotational component of its general plane motion. Thus, refer back to the rigid body of Figure 20.1 and recall that it is symmetric with respect to the plane of motion. The angular momentum $\mathbf{H}_{Oi}$ about point O of any particle m_i moving with a linear velocity $\mathbf{v}_i$ is given by Eq. (20.5). Thus,

$$\mathbf{H}_{Oi} = \mathbf{r}_i \times m_i \mathbf{v}_i$$

where $\mathbf{r}_i$ is the position vector of particle m_i. The angular momentum $\mathbf{H}_O$ of the rigid body about point O is obtained by adding vectorially the angular momenta about point O of all of the particles composing the rigid body. Thus,

$$\mathbf{H}_O = \sum \mathbf{H}_{Oi} = \sum \mathbf{r}_i \times m_i \mathbf{v}_i. \tag{20.6}$$

Substituting the relationships $\mathbf{v}_i = \mathbf{v}_G + \mathbf{v}_{i/G}$ and $\mathbf{v}_{i/G} = \boldsymbol{\omega} \times \mathbf{r}_{i/G}$ in Eq. (20.6) and rearranging terms,

$$\mathbf{H}_O = \sum m_i \mathbf{r}_i \times (\mathbf{v}_G + \boldsymbol{\omega} \times \mathbf{r}_{i/G})$$
$$= \sum (m_i \mathbf{r}_i) \times \mathbf{v}_G + \sum [(m_i \mathbf{r}_i) \times (\boldsymbol{\omega} \times \mathbf{r}_{i/G})]. \tag{20.7}$$

In the first term of Eq. (20.7) we replace $\sum m_i \mathbf{r}_i$ by its equivalence $m\mathbf{r}_G$ and in the second term we substitute $\mathbf{r}_i = \mathbf{r}_G + \mathbf{r}_{i/G}$ and rearrange terms to obtain

$$\mathbf{H}_O = \mathbf{r}_G \times m\mathbf{v}_G + \mathbf{r}_G \times (\boldsymbol{\omega} \times \sum m_i \mathbf{r}_{i/G}) + \sum [(m_i \mathbf{r}_{i/G}) \times (\boldsymbol{\omega} \times \mathbf{r}_{i/G})]. \tag{20.8}$$

The first term in Eq. (20.8) represents the angular momentum of the rigid body about point O due to the velocity $\mathbf{v}_G$ of its mass center. It has a magnitude of $(mv_G)d$, where d is defined in Figure 20.1, and a direction and sense that coincide with those of $\boldsymbol{\omega}$. The second term in Eq. (20.8) vanishes because $\sum m_i \mathbf{r}_{i/G} = m\mathbf{r}_{i/G} = \mathbf{0}$ by Varignon's theorem. The vector quantity $\mathbf{r}_{i/G} \times (\boldsymbol{\omega} \times \mathbf{r}_{i/G})$ in the third term of Eq. (20.8) is a vector whose magnitude is $(\mathbf{r}_{i/G})^2\omega$ and whose direction and sense coincide with those of $\boldsymbol{\omega}$, i.e., perpendicular to the plane of motion. Thus, Eq. (20.8) may be written as

$$\mathbf{H}_O = [(mv_G)d]\mathbf{k} + [\sum m_i (r_{i/G})^2 \boldsymbol{\omega}]\mathbf{k} \tag{20.9}$$

where $\sum m_i (\mathbf{r}_{i/G})^2$ represents I_G, the mass moment of inertia of the rigid body about an axis through G parallel to unit vector $\mathbf{k}$. Substituting this value in Eq. (20.9),

$$\mathbf{H}_O = [m\mathbf{v}_G d + I_G \omega]\mathbf{k} \tag{20.10}$$

which is applicable with respect to any fixed point O inside or outside the body. If the rigid body rotates about an axis through fixed point O parallel to unit vector $\mathbf{k}$, then, $v_G = \omega d$, and Eq. (20.10) becomes

$$\mathbf{H}_O = [(I_G + md^2)\omega]\mathbf{k} = (I_G + md^2)\boldsymbol{\omega}. \tag{20.11}$$

By the parallel axis theorem for mass moments of inertia, the quantity $(I_G + md^2)$ is equal to the mass moment of inertia of the rigid body about an axis through point O and parallel to vector $\boldsymbol{\omega}$. Therefore, Eq. (20.11) becomes

$$\mathbf{H}_O = I_O \boldsymbol{\omega} \tag{20.12}$$

which states that the angular momentum of a rigid body about any fixed point O, (i.e., about an axis through point O parallel to vector $\boldsymbol{\omega}$) due to the rotational component of its general plane motion, is obtained by multiplying its angular velocity $\boldsymbol{\omega}$ by its mass moment of inertia I_O.

Since point O was chosen arbitrarily, it may be made to coincide with the mass center, point G, of the rigid body, in which case I_O becomes I_G, $\mathbf{H}_O$ becomes $\mathbf{H}_G$, and Eq. (20.12) becomes

$$\mathbf{H}_G = I_G \boldsymbol{\omega} \tag{20.13}$$

which applies to the rotation of a rigid body about a fixed axis parallel to vector $\boldsymbol{\omega}$ and passing through its mass center, point G. As in the case of the linear momentum **L**, the angular momentum **H** is a vector. The direction and sense of this vector coincides with that of vector $\boldsymbol{\omega}$. As seen from Eqs. (20.12) and (20.13), the angular momentum has units of mass moment of inertia multiplied by those of angular velocity. Thus, in the SI system, it would have such units as $kg \cdot m^2/s$ and in the U.S. Customary system it would have such units as $slug \cdot ft^2/s$. As will be shown later, since the change in angular momentum is equal to the angular impulse, these units are equivalent, respectively, to N·m·s. and lb·ft·s.

In summary, we may identify three special cases of plane motion of a rigid body and their corresponding momenta.

1. ***Translation***
If the rigid body executes a pure translating motion, its angular velocity $\boldsymbol{\omega}$ is a null vector. Therefore, its angular momentum with respect to its mass center G vanishes, and it possesses only linear momentum given by the relation $\mathbf{L} = m\mathbf{v}_G$. A *momentum diagram* for a rigid body in plane translation is shown in Figure 20.2(a).
2. ***Rotation About A Fixed Axis Through the Center of Mass***
When the rigid body rotates about a fixed axis perpendicular to the plane of rotation and passing through its mass center G, its linear velocity $\mathbf{v}_G = 0$. Consequently, its linear momentum vanishes and the rigid body possesses only angular momentum given by the relation $\mathbf{H}_G = I_G\boldsymbol{\omega}$. A *momentum diagram* for this case is shown in Figure 20.2(b).
3. ***Rotation About A Fixed Axis Through An Arbitrary Point***
If the rigid body rotates about a fixed axis perpendicular to the plane of rotation and passing through an arbitrary point O, it possesses both linear and angular momenta given by $\mathbf{L} = m\mathbf{v}_G$ and $\mathbf{H}_G = I_G\boldsymbol{\omega}$, respectively. The *momentum diagram* for this case is shown in Figure 20.2(c). These two types of momenta (i.e., linear and angular) may be replaced by an angular momentum $\mathbf{H}_O = I_O\boldsymbol{\omega}$ (see Eq. (20.12) with respect to point O.

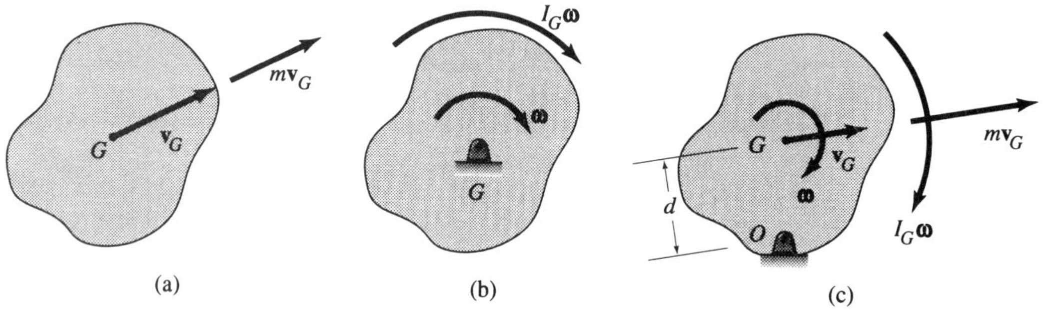

FIGURE 20.2.

20.2 Principles of Impulse and Momentum

Linear Principle

Differentiating Eq. (20.4) with respect to time, assuming m to be constant, yields

$$\frac{d\mathbf{L}}{dt} = \frac{d}{dt}(m\mathbf{v}_G) = m\mathbf{a}_G$$

which, by Newton's second law for rectilinear motion, is equal to the resultant of all the external forces, $\sum \mathbf{F}$, acting on the rigid body. Thus

$$\sum \mathbf{F} = m\mathbf{a}_G = \frac{d}{dt}(m\mathbf{v}_G),$$

and, for constant mass m,

$$\sum \int_{t_1}^{t_2} \mathbf{F}\, dt = m \int_{v_{G1}}^{v_{G2}} d\mathbf{v}_G = m\mathbf{v}_{G2} - m\mathbf{v}_{G1}.$$

Thus, because $m\mathbf{v}_G = \mathbf{L}$, by Eq. (20.4), it follows that

$$\sum \int_{t_1}^{t_2} \mathbf{F}\, dt = \mathbf{L}_2 - \mathbf{L}_1. \tag{20.14}$$

The quantity, $\sum \int_{t_1}^{t_2} \mathbf{F}\, dt$, is the sum of the linear impulses of all external forces acting on the rigid body during the time interval from t_1 to t_2; the quantity $\mathbf{L}_1 = m\mathbf{v}_{G1}$ is the linear momentum of the rigid body at time t_1, and the quantity $\mathbf{L}_2 = m\mathbf{v}_{G2}$ is its linear momentum at time t_2. Thus, Eq. (20.14) expresses the principle of *linear impulse and linear momentum* for a rigid body and states that the change in the linear vector momentum of a rigid body during a time interval from t_1 to t_2, is equal to the linear vector impulses imparted to the rigid body by all external forces during the same time interval.

Angular Principle

Differentiating Eq. (20.12) with respect to time and assuming I_O is constant yields

$$\frac{d\mathbf{H}_O}{dt} = \frac{d}{dt}(I_O\boldsymbol{\omega}) = I_O\boldsymbol{\alpha}$$

which, by Newton's second law for angular motion, is equal to the sum of the moments of all external forces acting on the rigid body about fixed point O. Thus,

$$\sum \mathbf{M}_O = I_O\boldsymbol{\alpha} = \frac{d}{dt}(I_O\boldsymbol{\omega}),$$

and, for I_O = constant,

$$\sum \int_{t_1}^{t_2} \mathbf{M}_O\, dt = I_O \int_{\omega_1}^{\omega_2} d\boldsymbol{\omega} = I_O\boldsymbol{\omega}_2 - I_O\boldsymbol{\omega}_1.$$

Thus, because $I_O \boldsymbol{\omega} = \mathbf{H}_O$ by Eq. (20.12), it follows that

$$\sum \int_{t_1}^{t_2} \mathbf{M}_O \, dt = \mathbf{H}_{O2} - \mathbf{H}_{O1} \tag{20.15}$$

which applies only to the rotation of a rigid body about any fixed point O inside or outside the body which is symmetric with respect to the plane of rotation. The quantity $\sum \int_{t_1}^{t_2} \mathbf{M}_O \, dt$ is the sum of the angular impulses of all the external moments acting on the rigid body about point O during the time interval from t_1 to t_2, the quantity $\mathbf{H}_{O1} = I_O \boldsymbol{\omega}_1$, is the angular momentum of the rigid body about point O at time t_1, and the quantity $\mathbf{H}_{O2} = I_O \boldsymbol{\omega}_2$, is its angular momentum about the same point at time t_2.

Similarly, by differentiating Eq. (20.13), we obtain

$$\sum \int_{t_1}^{t_2} \mathbf{M}_G \, dt = I_G \int_{\omega_1}^{\omega_2} d\boldsymbol{\omega} = I_G \boldsymbol{\omega}_2 - I_G \boldsymbol{\omega}_1.$$

Therefore, because $I_G \boldsymbol{\omega} = \mathbf{H}_G$ by Eq. (20.13), it follows that

$$\sum \int_{t_1}^{t_2} \mathbf{M}_G \, dt = \mathbf{H}_{G2} - \mathbf{H}_{G1} \tag{20.16}$$

which applies only to the rotation of a rigid body about its fixed mass center G.

Equations (20.15) and (20.16) are expressions of the same principle known as the principle of *angular impulse and angular momentum* for a rigid body in plane motion. This principle states that the change in the angular vector momentum of a rigid body about any fixed point during a time interval from t_1 to t_2, is equal to the angular vector impulses about the same point, imparted to the rigid body by all external moments during the same time interval.

Equations (20.14), (20.15) or (20.16) are most conveniently used in scalar forms. Because we are dealing with the two-dimensional motion of a rigid body in the x-y plane, it follows that Eq. (20.14) represents only two useful scalar relations, one in the x direction and a second in the y direction. Also, Eqs. (20.15) or (20.16) each represents only one useful scalar relationship expressing angular impulses and angular momenta about an axis perpendicular to the x-y plane (i.e., about a z axis). Thus, in scalar form, Eqs. (20.14), (20.15), or (20.16) become

$$\sum \int_{t_1}^{t_2} F_x \, dt = (L_2 - L_1)_x, \tag{20.17}$$

and

$$\sum \int_{t_1}^{t_2} F_y \, dt = (L_2 - L_1)_y, \tag{20.18}$$

and

$$\left.\begin{aligned}\sum \int_{t_1}^{t_2} (M_{\text{O}})_z \, dt &= (H_{\text{O}2} - H_{\text{O}1})_z \\ \text{or} \qquad & \\ \sum \int_{t_1}^{t_2} (M_{\text{G}})_z \, dt &= (H_{\text{G}2} - H_{\text{G}1})_z\end{aligned}\right\} \tag{20.19}$$

where the x and y subscripts represent summation of linear impulses and linear momenta in the x and y directions, respectively, and the z subscript represents summation of angular impulses and angular momenta about a z axis through point O or through point G. Thus, when applied to the plane motion of a rigid body, the principles of impulse and momentum represent three independent scalar equations, and, consequently, a maximum of three unknown quantities may be determined with each application of Eqs. (20.17) to (20.19).

In analyzing a system of interconnected rigid bodies in plane motion, it is possible to apply Eqs. (20.17) to (20.19) separately to each one of the interconnected bodies in the system. However, it is convenient, sometimes, to apply these equations to the entire system of interconnected rigid bodies, as long as only three unknown quantities are involved. In such a case, the internal forces of action and reaction at the connections do not contribute to the linear and angular impulses because they occur in pairs of equal and opposite vectors. Also, each of the terms in Eqs. (20.17) to (20.19) must account for *all* the forces, moments, and linear and angular momenta in the entire system. Thus, for a system of interconnected rigid bodies, Eqs. (20.17) to (20.19) may be summarized as follows:

$$\begin{aligned}\left(\sum \text{Syst. Ext. Imp.}\right)_{1\to 2} = \left(\sum \text{Syst. Momenta}\right)_2 \\ - \left(\sum \text{Syst. Momenta}\right)_1.\end{aligned} \tag{20.20}$$

In applying the principles of impulse and momentum to the solution of problems involving the plane motion of rigid bodies, it is essential that Eqs. (20.17) to (20.20) are applied correctly. A convenient way of accomplishing this purpose is to graphically represent the one or more equations that apply in a given situation. Thus, for example, if we were analyzing the plane translation of a rigid body in the x-y plane, the graphical representation of Eqs. (20.17) and (20.18) would be as shown in Figure 20.3. The diagram on the left side of the equal sign is known as the impulse diagram which is essentially a free-body diagram that accounts for the time factor. Thus, it represents the left hand side of Eqs. (20.17) and (20.18). The first diagram on the right of the equal sign is the final momentum diagram (immediately after impulse) and the second is the initial momentum diagram (immediately before impulse).

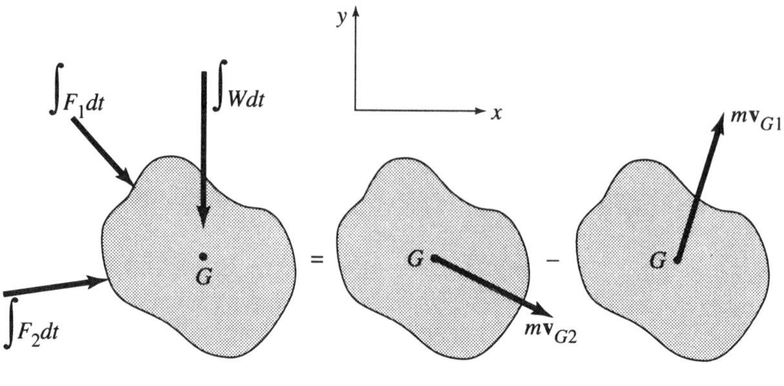

FIGURE 20.3.

Therefore, the two diagrams to the right of the equal sign represent the right hand side of Eqs. (20.17) and (20.18). Similar diagrams may be constructed for Eq. (20.20). Such diagrams, along with the concepts discussed in this section, are illustrated in the solutions of the following examples.

■ Example 20.1

The system shown in Figure E20.1(a) consists of two blocks A and B of weights $W_A = 400$ N and $W_B = 500$ N, connected by an inextensible cord which passes over a frictionless pulley at C. Block A slides on a horizontal rough surface where $\mu_A = 0.3$, and block B slides on a rough inclined surface where $\mu_B = 0.4$. When $t = 0$, block B slides down the inclined plane with a velocity of 0.5 m/s. Determine the velocity of block B and the tension T in the cable when $t = 5$ s.

Solution

Both blocks A and B undergo linear translation and, consequently, possess only linear momenta throughout the motion. Note that since the cord is inextensible $v_A = v_B = v$.

The impulse diagram and the initial and final momenta diagrams for block A are shown in Figure E20.1(b). A coordinate system is also shown to facilitate the application of Eqs. (20.17) and (20.18). Thus,

$$\sum \int_{t_1}^{t_2} F_y \, dt = (L_2 - L_1)_y = 0, \qquad \int_0^5 (N_A - 400) \, dt = 0,$$

from which

$$N_A = 400 \text{ N}. \tag{a}$$

Also

$$\sum \int_{t_1}^{t_2} F_x \, dt = (L_2 - L_1)_x$$

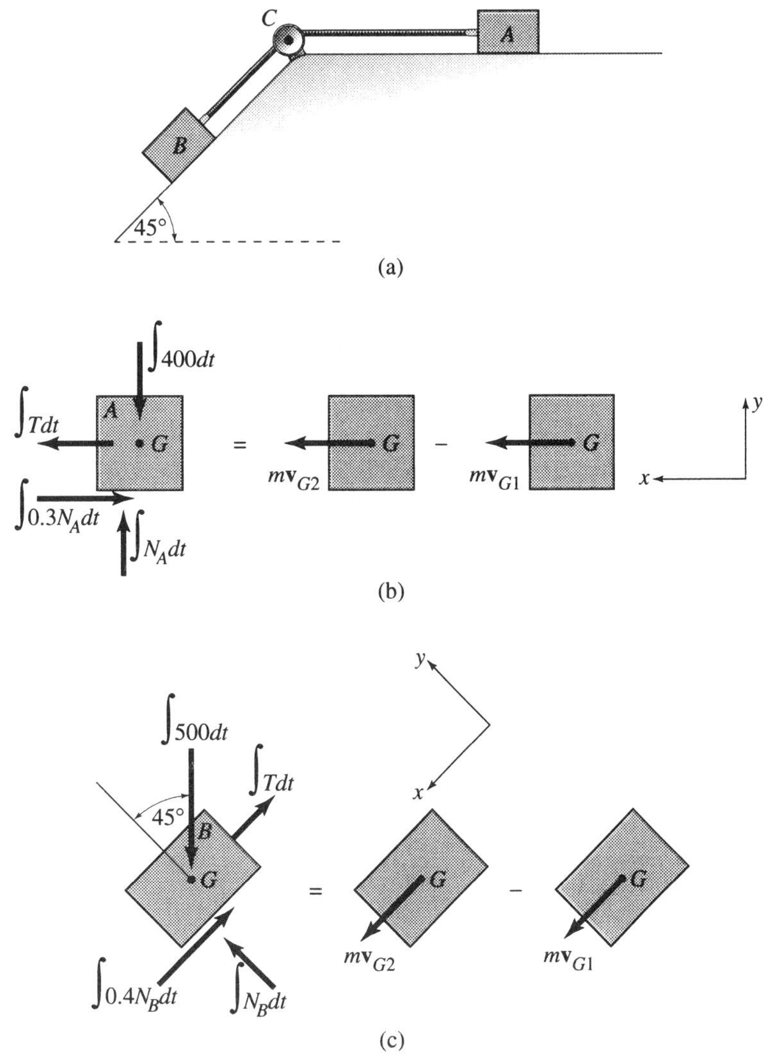

FIGURE E20.1.

where $L_1 = \left(\frac{400}{9.81}\right)(0.5)$ kg m/s and $L_2 = \left(\frac{400}{9.81}\right) v_{G2}$ kg m/s. Therefore,

$$\int_0^5 (T - 0.3(400))\, dt = \left(\frac{400}{9.81}\right)(v_{G2} - 0.5)$$

from which, after simplification, we obtain

$$5T - 40.78 v_{G2} = 579.61. \tag{b}$$

The impulse diagram and the initial and final momenta diagrams for block B are shown in Figure E20.1(c) along with a convenient coordinate system. Therefore,

$$\sum \int_{t_1}^{t_2} F_y\, dt = (L_2 - L_1)_y = 0, \quad \int_0^5 (N_B - 500\cos 45^\circ)\, dt = 0,$$

from which

$$N_B = 353.6 \text{ N}. \tag{c}$$

Also

$$\sum \int_{t_1}^{t_2} F_x\, dt = (L_2 - L_1)_x$$

where $L_1 = \left(\frac{500}{9.81}\right)(0.5)$ kg m/s and $L_2 = \left(\frac{500}{9.81}\right) v_{G2}$ kg m/s. Thus,

$$\int_0^5 [500\sin 45^\circ - 0.4(353.6) - T]\, dt = \left(\frac{500}{9.81}\right)(v_{G2} - 0.5)$$

from which, after simplification, we obtain

$$5T + 50.97v_{G2} = 1086.13. \tag{d}$$

The simultaneous solution of Eqs. (b) and (d) yields

$$v_{G2} = 5.52 \text{ m/s} \qquad \text{ANS.}$$

and

$$T = 160.9 \text{ N}. \qquad \text{ANS.}$$

■ Example 20.2

The wheel shown in Figure E20.2(a) is being pushed up the 30° inclined plane by the horizontal force $Q = (2t + 50)$ lb. The weight of the wheel is $W = 60$ lb and its centroidal radius of gyration $k_G = 1.1$ ft. When $t = 0$, the angular velocity of the wheel is $\omega_1 = 2$ rad/s cw. If the wheel rolls without slipping, develop an expression relating the angular velocity ω of the wheel (in rad/s) and the time t (in seconds).

Solution

The wheel executes general plane motion because it rotates about its center of mass G which is also translating. The impulse diagram and the momenta diagrams immediately before and after impulse are shown in Figure E20.2(b) along with a convenient right-handed x-y-z coordinate system. Note that the z axis comes out of the page. Two alternate methods of solving the problem are presented.

Method A

$$\sum \int_{t_1}^{t_2} F_y\, dt = (L_2 - L_1)_y = 0, \quad \int_0^t (N - 60\cos 30^\circ)\, dt = 0,$$

from which

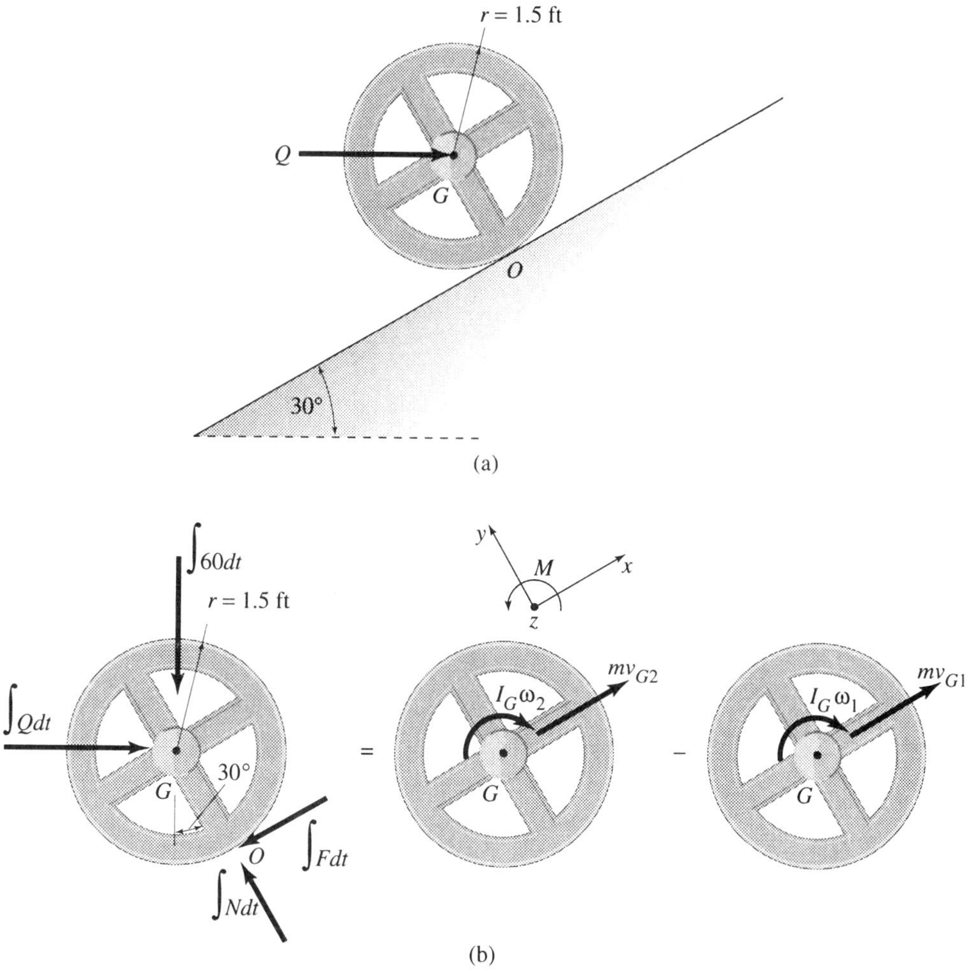

FIGURE E20.2.

$$N = 51.962 \text{ lb.} \tag{a}$$

Also,

$$\sum \int_{t_1}^{t_2} F_x \, dt = (L_2 - L_1)_x.$$

Since the wheel rolls without slipping, we can use the relationship $v_G = r\,\omega = 1.5\omega$ where ω is the angular velocity of the wheel. Thus, because $L_2 = mv_{G2}$ and $L_1 = mv_{G1}$,

$$\int_0^t [(2t + 50)\cos 30^\circ - 60 \sin 30^\circ - F]\, dt = \left(\frac{60}{32.2}\right)[1.5\omega - 1.5(2)]$$

which, after simplification, becomes

$$0.866t^2 + 13.301t - \int_0^t F\,dt = 2.795\omega - 5.590. \tag{b}$$

Note that the force F was kept inside the integral sign because it is a variable. Note also that Eq. (b) contains two unknown quantities, t and F. A second relationship between these quantities is obtained by applying the second of Eqs. (20.19). Thus,

$$\sum \int_{t_1}^{t_2} (M_G)_z\,dt = (H_{G2} - H_{G1})_z$$

where $H_{G1} = I_G\omega_1$, $H_{G2} = I_G\omega_2$ and $I_G = mk_G^2 = 2.255$ slug ft^2. Therefore, referring to the diagrams of Figure E20.2(b) and recalling that $\omega_1 = 2$ rad/s and $\omega_2 = \omega$,

$$\int_0^t (1.5F)\,dt = 2.255(\omega - 2)$$

from which

$$\int_0^t F\,dt = 1.503\omega - 3.007. \tag{c}$$

Substituting Eq. (c) in Eq. (b) and simplifying, we obtain the following quadratic equation in t

$$t^2 + 15.359t + 9.927 - 4.963\omega = 0 \qquad \text{(d)} \quad \text{ANS.}$$

Method B

Note that, in the preceding method, two simultaneous equations had to be obtained to reach a solution. This may be avoided by summing angular impulses and angular momenta about point O, which is the point of contact of the wheel with the inclined plane, i.e., the instantaneous center. The use of the instantaneous center of rotation as a moment center is justified because the wheel rolls without slipping. This procedure requires the use of the first of Eqs. (20.19) and would eliminate the unknown quantity $\int_0^t F\,dt$. Thus,

$$\sum \int_{t_1}^{t_2} M_O\,dt = (H_{O2} - H_{O1})_z$$

where $H_{O1} = I_O\omega_1$, $H_{O2} = I_O\omega_2$ and $I_O = I_G + mr^2 = 6.448$ slug ft^2. Therefore, referring to the diagrams of Figure E20.2(b),

$$\int_0^t [1.5(60)\sin 30^\circ - 1.5(2t + 50)\cos 60^\circ]\,dt = -6.448(\omega - 2).$$

After simplification, this equation reduces to

$$1.299t^2 + 19.952t - 6.448\omega + 12.896 = 0$$

from which

$$t^2 + 15.359t + 9.928 - 4.964\omega = 0 \qquad \text{(e)} \quad \text{ANS.}$$

which is almost identical to Eq. (d) obtained by Method A. The slight differences that appear between Eqs. (d) and (e) are due to roundoff during the computation.

Example 20.3

The system shown in Figure E20.3(a) consists of two pulleys connected together by an inextensible cord. Pulley A rotates freely about a fixed axis at O which coincides with its mass center, and pulley B unwraps itself from the cord as it moves downward. Let $r_A = 2r_B$ and $I_A = 4I_B$ where r_A and I_A are the radius and centroidal moment of inertia, respectively, of pulley A and r_B and I_B, those of pulley B. The weights of both pulleys are identical, i.e., $W_A = W_B$. (a) Develop expressions for the angular velocities ω_A and ω_B of pulleys A and B after time t has elapsed, if the system is released from rest. Express your answers in terms of r_B, W_B, I_B and t. (b) Find numerical values for ω_A and ω_B for the case when $W_B = 50$ N, $r_B = 0.25$ m, $I_B = 0.5$ kg·m^2 and $t = 5$ s.

Solution

(a) Pulley A executes rotation about a fixed axis at O whereas pulley B executes general plane motion.

The impulse diagram and the initial and final momenta diagrams for pulley A are shown in Figure E20.3(b) along with a suitable right-handed x-y-z coordinate system. We now apply the angular impulse-momentum relationship about a z axis through point O. Thus, letting T be the tension in the cord,

$$\sum \int_{t_1}^{t_2} (M_O)_z \, dt = (H_{O2} - H_{O1})_z$$

where $H_{O1} = I_O\omega_1 = 0$ and $H_{O2} = I_O\omega_2 = I_A\omega_A$. Thus,

$$\int_0^t (Tr_A)\, dt = I_A\omega_A - 0$$

from which

$$T = \left(\frac{I_A}{tr_A}\right)\omega_A = \left(\frac{2I_B}{tr_B}\right)\omega_A. \qquad \text{(a)}$$

The impulse diagram and the initial and final momenta diagrams for pulley B are shown in Figure E20.3(c) along with a suitable coordinate system. Application of the linear and angular impulse-momentum relationships follows:

$$\sum \int_{t_1}^{t_2} F_y \, dt = (L_2 - L_1)_y$$

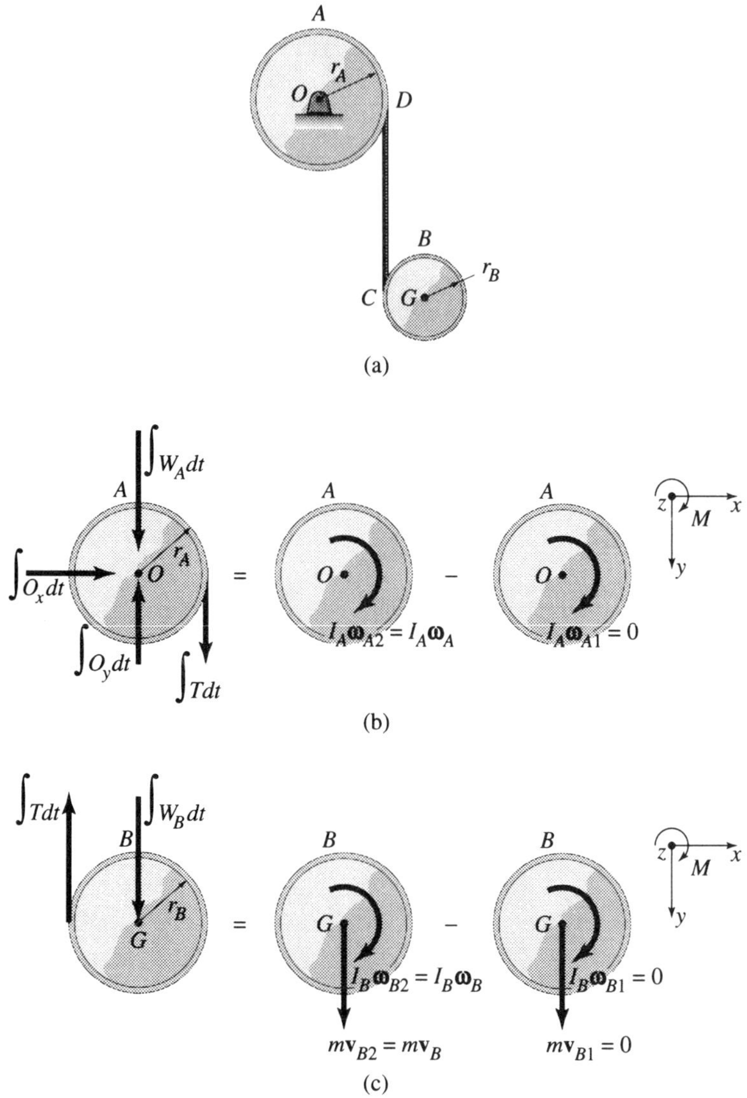

FIGURE E20.3.

where $L_1 = mv_{G1} = 0$ and $L_2 = mv_{G2} = \left(\frac{W_B}{g}\right)v_B$. Therefore,

$$\int_0^t (W_B - T)\,dt = \left(\frac{W_B}{g}\right)v_B - 0$$

from which

$$(W_B - T)t = \left(\frac{W_B}{g}\right)v_B. \tag{b}$$

Also,

$$\sum \int_{t_1}^{t_2} (M_G)_z \, dt = (H_{G2} - H_{G1})_z$$

where $H_{G1} = I_G \omega_1 = 0$ and $H_{G2} = I_G \omega_2 = I_B \omega_B$. Thus,

$$\int_0^t (Tr_B) \, dt = I_B \omega_B - 0$$

from which

$$T = \left(\frac{I_B}{tr_B} \right) \omega_B. \tag{c}$$

Equations (a), (b), and (c) contain the four unknown quantities, T, ω_A, ω_B, and v_B. A fourth relationship among the unknown quantities is, therefore, needed. This is obtained from the relative velocity relationship $\mathbf{v}_G = \mathbf{v}_C + \mathbf{v}_{G/C}$. Because all of these vectors are in the y direction and have the same sense, we need deal only with the magnitudes of these quantities. Thus, because $v_C = v_D = r_A \omega_A$ and $v_{G/C} = r_B \omega_B$,

$$v_G = r_A \omega_A + r_B \omega_B = 2r_B \omega_A + r_B \omega_B \tag{d}$$

which is the fourth needed relationship.

Substituting Eqs. (c) and (d) in Eq. (b) and solving for ω_A in terms of ω_B,

$$\omega_A = \frac{gt}{2r_B} - \left[\left(\frac{gI_B}{2r_B^2 W_B} \right) + \frac{1}{2} \right] \omega_B \ \circlearrowright. \tag{e}$$

Equating Eqs. (a) and (c) and solving for ω_A in terms of ω_B yields

$$\omega_A = \tfrac{1}{2} \omega_B \ \circlearrowright. \tag{f}$$

Solving Eqs. (e) and (f) simultaneously, after simplification, we obtain

$$\omega_A = \frac{gtr_B W_B}{4r_B^2 W_B + 2gI_B} \qquad \text{ANS.}$$

and

$$\omega_B = 2\omega_A. \qquad \text{ANS.}$$

(b) Substituting the given numerical values into the general relationships obtained in part (a),

$$\omega_A = 27.5 \text{ rad/s} \ \circlearrowright,$$

and

$$\omega_B = 2\omega_A = 55.0 \text{ rad/s} \ \circlearrowright. \qquad \text{ANS.}$$

■

Problems

20.1 The block shown in Figure P20.1 is moving to the right with a velocity of 5 ft/s when the force $Q = (5t + 20)$, where Q is in lb and t is in s, is applied. The block weighs 100 lb. Determine the time it takes the force Q to bring the block to a complete stop if the coefficient of kinetic friction between the block and the floor is (a) $\mu = 0$ and (b) $\mu = 0.3$.

20.2 The system shown in Figure P20.2 consists of two blocks connected together by an inextensible cord. When $t = 0$, the 200-N block is moving to the right with a velocity of 0.75 m/s and the force $Q = (3t^2 + 10)$, where Q is in N and t is in s, is applied as shown. Determine the time it takes the force Q to bring the system to a complete stop if the coefficient of kinetic friction between the 200-N block and the horizontal plane is $\mu = 0$.

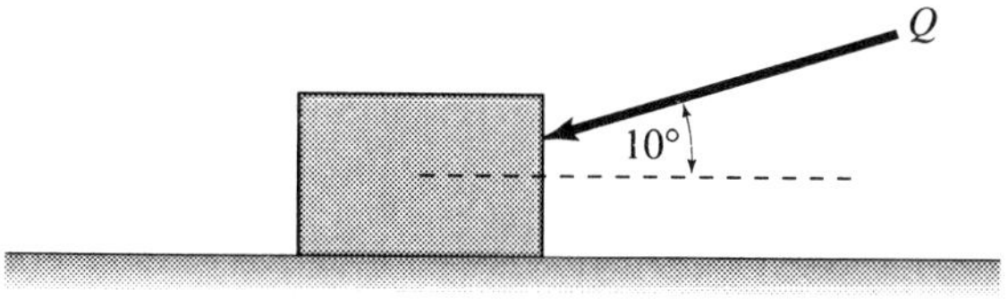

FIGURE P20.1.

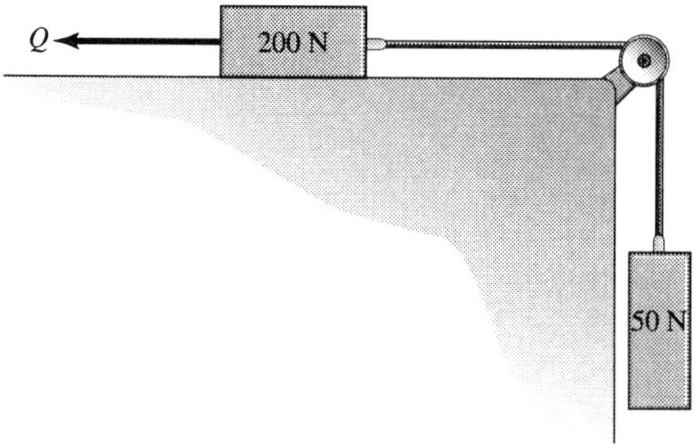

FIGURE P20.2.

20.3 Solve Problem 20.2 if the coefficient of kinetic friction between the 200-N block and the horizontal plane is $\mu = 0.15$.

20.4 A locomotive weighing 30 tons is pulling two coal cars each weighing 15 tons up a 10° inclined plane as shown in Figure P20.4. At $t = 0$, when the velocity of the system is 75 mph, the brakes are suddenly applied locking all wheels of the locomotive and bringing the system to a full stop in 15 s. Assume that the wheels of the locomotive and those of the freight cars have negligible weights and that the wheels of the freight cars

FIGURE P20.4.

roll freely without slipping during the stopping period. Determine (a) the constant frictional force developed between the locked wheels of the locomotive and the tracks, (b) the force in the coupling between the locomotive and the first freight car and (c) the force in the coupling between the first and second freight cars.

20.5 The locomotive-freight train system shown in Figure P20.4 accelerates uniformly from 30 mph to 60 mph in 20 s. Assume that the wheels of the freight cars roll freely without slipping. Determine (a) the propelling frictional force between the locomotive wheels and the tracks (b) the force in the coupling between the locomotive and the first freight car and (c) the force in the coupling between the first and second freight cars.

20.6 The wheel shown in Figure P20.6 weighs 150 N and has a centroidal radius of gyration of 0.45 m. When $t = 0$, its linear velocity is 2 m/s down the inclined plane. Determine its linear and angular

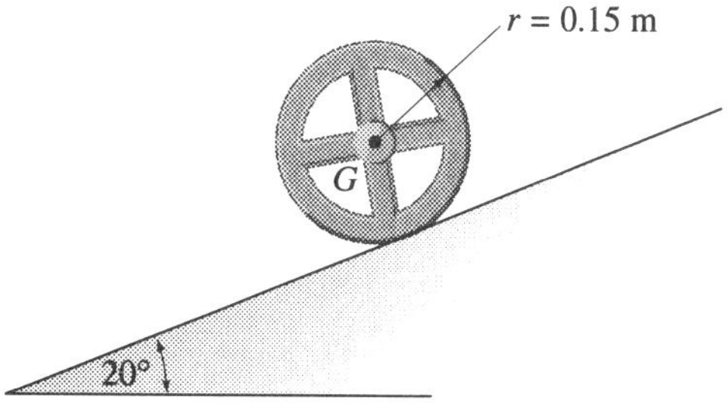

FIGURE P20.6.

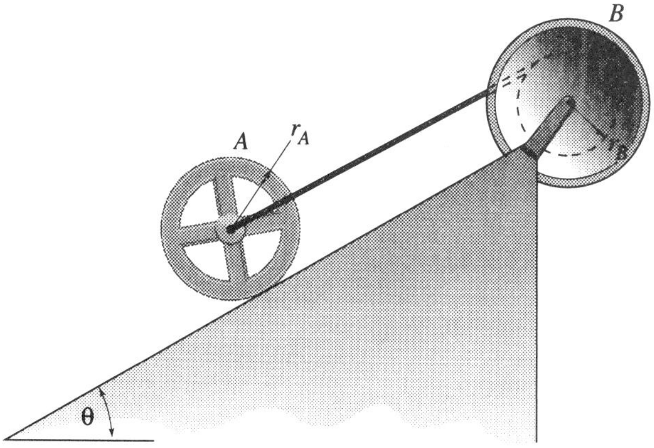

FIGURE P20.7.

velocities when $t = 10$ s. What is the magnitude of the constant frictional force during the motion? Assume that the wheel rolls without slipping.

20.7 The system shown in Figure P20.7 consists of wheel A and the spool B connected together by an inextensible cord. Wheel A has a weight W_A and a centroidal radius of gyration k_A. The spool B has a weight W_B and a centroidal radius of gyration k_B. The system is released from rest and wheel A rolls down the inclined plane without slipping. (a) In terms of W_A, k_A, r_A, W_B, k_B, r_B, θ, and the time t, develop expressions for the linear velocity of the wheel A and the angular velocity of the spool B. (b) Find numerical values for the quantities in (a) and (b) for the special case where $W_A = 50$ lb, $k_A = 0.9$ ft, $r_A = 1.2$ ft, $W_B = 20$ lb, $k_B = 0.6$ ft, $r_B = 0.8$ ft, $\theta = 30°$, and $t = 3$ s.

20.8 The two disks A and B are connected to a rigid and weightless rod as shown in Figure P20.8 by means of frictionless hinges that are located at their respective mass centers. Disk A has a weight $W_A = 100$ lb and disk B a weight $W_B = 400$ lb. The system has a linear velocity of 5 ft/s to the right when $t = 0$, at which time the force $Q = 50$ lb is applied as shown. Determine the time needed for the system to reach a linear velocity (a) of zero and (b) of 5 ft/s to the left. Assume both disks roll on the horizontal surface without sliding.

20.9 The system shown in Figure P20.9 consists of a spool A of weight W_A connected to a block B of weight W_B by an inextensible cord. The system is released

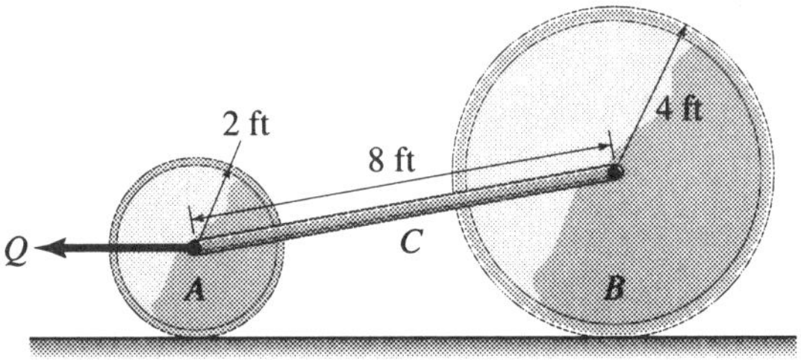

FIGURE P20.8.

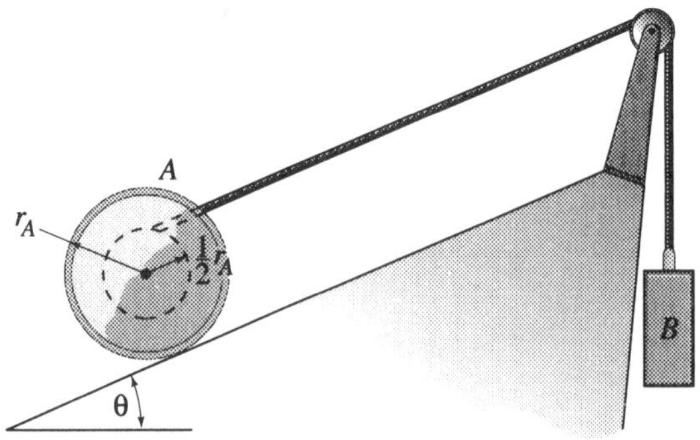

FIGURE P20.9.

from rest, and block B moves downward. (a) Develop expressions for the linear and angular velocities of the spool and the linear velocity of the block after an elapsed time t. The centroidal radius of gyration of the spool is k_A. Assume that the spool rolls without sliding, and express your answers in terms of W_A, r_A, k_A, W_B, θ and t. (b) Find numerical values for the quantities in part (a) for the special case where $W_A = 50$ N, $r_A = 0.5$ m, $k_A = 0.7$ m, $W_B = 40$ N, $\theta = 25°$, and $t = 4$ s.

20.10 A 100-N block is attached to an inextensible cord which passes over a pulley as shown in Figure P20.10. The weight of the pulley is 200 N, and its centroidal radius of gyration is 0.35 m. When $t = 0$, a constant cw couple $C = 60$ N m is applied to the pulley, at which time the block is moving downward with a velocity of 5 m/s. Determine the time required for the block to reach a velocity of (a) 2 m/s downward (b) zero and (c) 5 m/s upward.

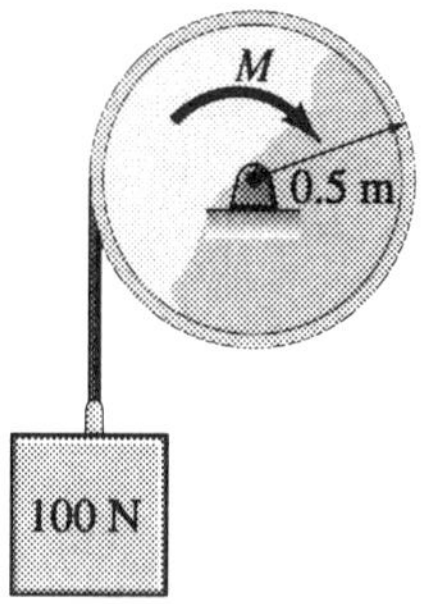

FIGURE P20.10.

20.11 Blocks A and B of weights $W_A = 75$ lb and $W_B = 300$ lb are connected to a pulley system C as shown in Figure P20.11 by means of inextensible cords. The pulley system C has a weight $W_C = 100$ lb and a centroidal radius of gyration $k_C = 0.75$ ft. At $t = 0$, block A is moving downward with a velocity of 10 ft/s. Determine its velocity after 10 s. Let $r_1 = 0.5$ ft and $r_2 = 1.2$ ft, and assume frictionless conditions at the hinge at D and between block B and the inclined plane.

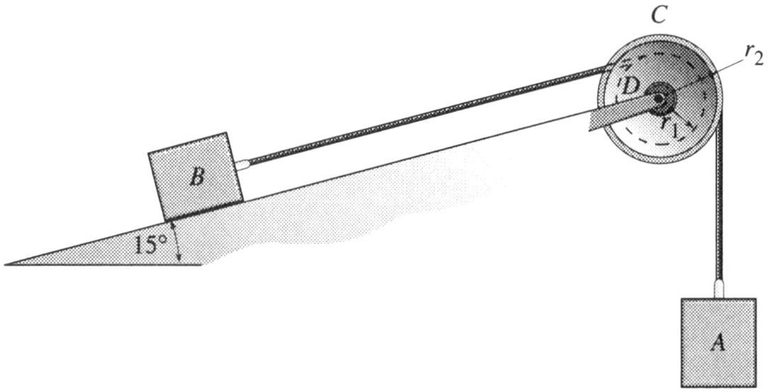

FIGURE P20.11.

20.12 Repeat Problem 20.11 if the coefficient of kinetic friction between B and the inclined plane is $\mu = 0.05$ and the kinetic friction at the hinge at D is equivalent to a constant couple $M = 10.0$ lb ft.

20.13 The arrangement shown in Figure P20.13 consists of two identical disks weighing 150 N each, connected by means of an inextensible cord to block B weighing 300 N. The friction at the

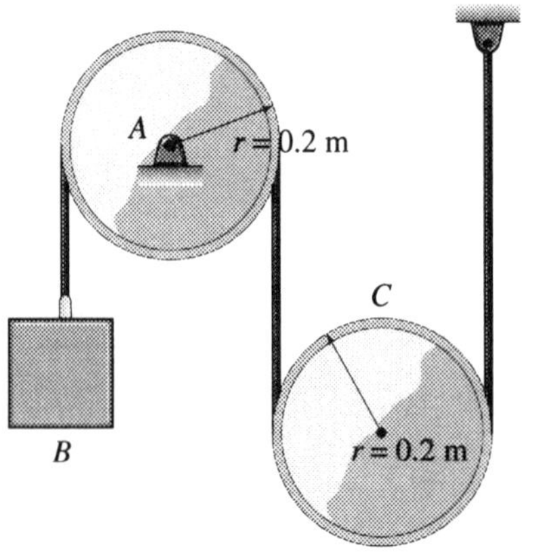

FIGURE P20.13.

fixed hinge at A is equivalent to a constant couple $C = 50$ N m. At $t = 0$, the block is moving downward with an initial velocity of 3 m/s. Determine the time needed for the velocity of the block to become (a) 1 m/s downward and (b) zero.

20.14 The system shown in Figure P20.14 consists of a double wheel which rolls without slipping on a horizontal surface under the influence of the tensions induced in the cords by the weights $W_A = 250$ lb and $W_B = 200$ lb. Find the angular velocity of the wheel and the linear velocity of its mass center 10 s after the system is released from rest. Let the weight of the double wheel be 500 lb and its centroidal radius of gyration be 1.5 ft.

20.15 The arrangement shown in Figure P20.15 consists of two identical disks attached to two identical parallel rods by frictionless hinges. The two disks rotate about stationary frictionless hinges at A and B. The weight of each disk is W_D and the rods may be assumed weightless. The disks are rotating at a constant cw angular velocity ω when a ccw couple M is applied at A, as shown. Develop an expression for the time required for the disks to have an angular velocity of (a) zero and (b) ω ccw. Express answers in terms of W_D, M, r, and ω.

20.16 Consider the gear-rack arrangement

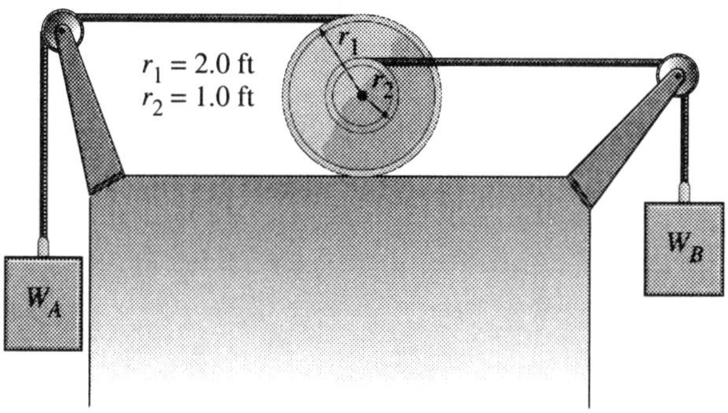

FIGURE P20.14.

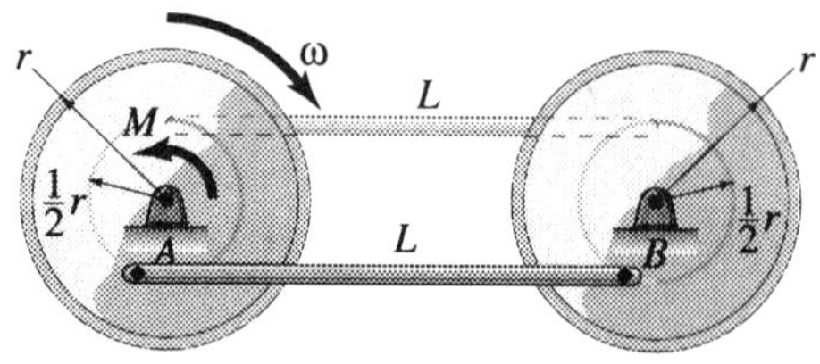

FIGURE P20.15.

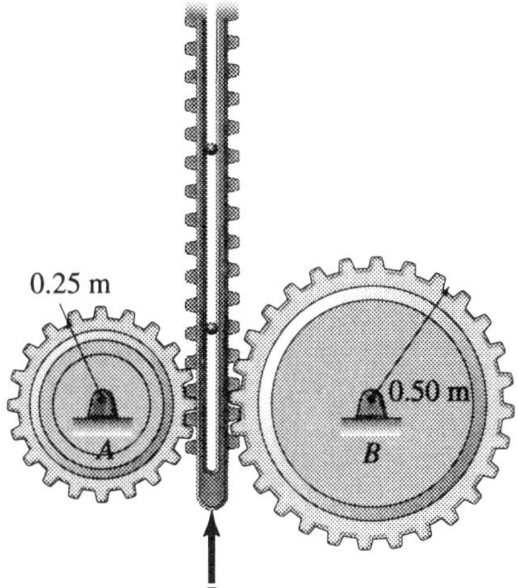

FIGURE P20.16.

shown in Figure P20.16. The vertical rack has a weight $W_R = 30$ N. The centroidal radius of gyration of gear A is $I_A = 0.050$ kg m^2 and that of gear B is $I_B = 0.100$ kg m^2. At $t = 0$, the vertical rack has a downward velocity of 5.0 m/s when the force $P = (2t + 10)$ N, where t is in s, is applied. Determine the time required for the rack to reach a velocity of zero. Assume that the width of the rack is negligible.

20.17 Refer to Problem 20.16 and determine the time required for the rack to reach a velocity of 2.0 m/s upward.

20.18 Consider the brake system shown in Figure P20.18. The homogeneous drum has a weight of 150 lb and rotates at a constant ccw angular speed of 200 rpm before the force $Q = 3t + 5$ is applied. The force Q is in lb and t is in seconds. The coefficient of kinetic friction between the drum and the brake shoe is 0.30. Determine the time t required for the angular speed of the drum to become zero. Neglect the weight of lever ABC.

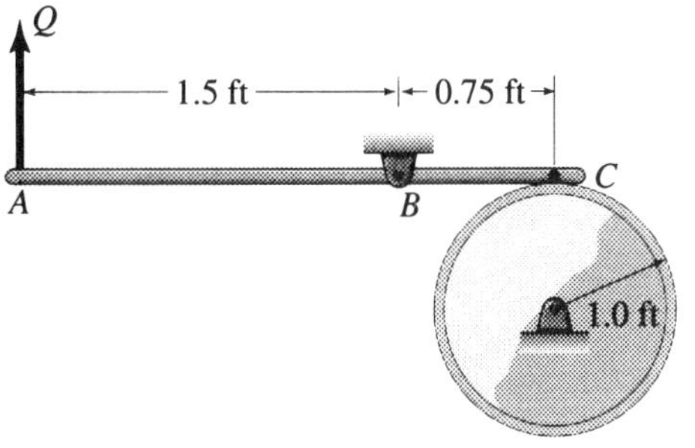

FIGURE P20.18.

20.19 Refer to Problem 20.18 and determine the time required for the drum to reach an angular speed of 100 rpm ccw.

20.20 The brake system shown in Figure P20.20 consists of the thin disk A of weight W_A and radius r and the brake arm OB of negligible weight and of length $6r$. The coefficient of kinetic friction between the disk and the brake arm is μ. The disk rotates at a constant cw angular velocity ω before the constant force P is applied. Develop an expression for the time t needed to reduce the angular velocity of the disk to zero.

20.21 Refer to Problem 20.20 and develop an expression for the time t needed to reduce the angular velocity of the disk to $(1/4)\omega$ cw.

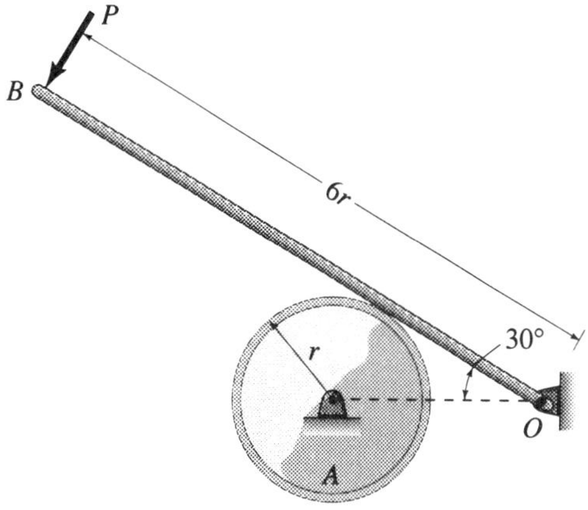

FIGURE P20.20.

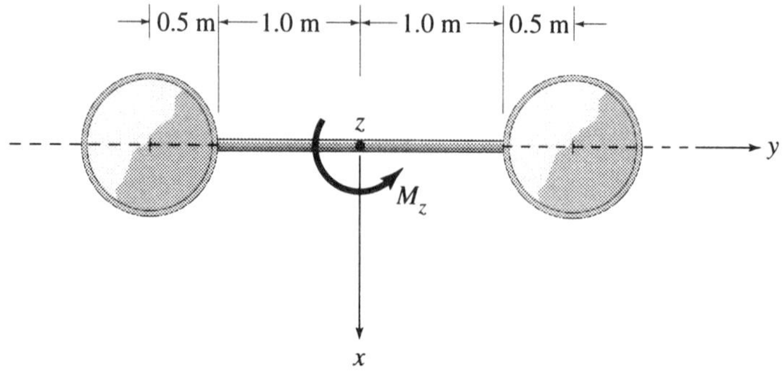

FIGURE P20.22.

20.22 The system shown in Figure P20.22 consists of two identical thin disks each of weight $W_D = 10$ N and of radius $r = 0.5$ m mounted at the ends of a slender rod of weight $W_R = 20$ N and of length $L = 2.0$ m. This arrangement is mounted on a vertical shaft which is in the z direction, about which it can rotate under the action of the couple $M_z = 20$ N m. The two disks may be oriented in any desired position with respect to the horizontal x-y plane. When $t = 0$, the system is at rest at which time the couple M_z is applied. Determine the angular velocity of the system when $t = 10$ s if the two disks are fixed to the rod so that their planes remain horizontal as shown. Ignore air resistance.

20.23 Refer to Problem 20.22. Determine the angular velocity of the system when $t = 10$ s if the two disks are fixed to the rod so that their planes remain vertical, that is, the two discs are rotated 90° about the y axis in Figure 20.22. Ignore air resistance.

20.24 A homogeneous cylinder is attached symmetrically to four identical, thin square plates as shown in Figure P20.24. The system is made to rotate under the action of a constant couple M_z, about a vertical z axis through the center of mass of the cylinder. The weight of the cylinder is 400 lb and that of each plate is 50 lb. Ignore the weight and dimension of the attachments between the cylinder and the plates. These plates may be oriented in any desired position relative to the horizontal x-y plane. Determine the magnitude of M_z needed to increase the angular speed of the system from zero to 50 rad/s in 25 s if the four plates are fixed so that their planes remain horizontal. Ignore air resistance.

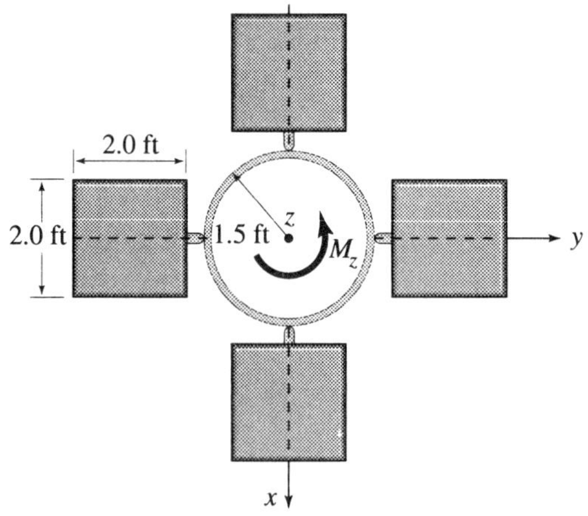

FIGURE P20.24.

20.25 Refer to Problem 20.24. Determine the magnitude of M_z needed to increase the angular speed of the system from zero to 50 rad/s in 25 s if the four plates are fixed so that their planes remain vertical, that is, two of the four plates are rotated 90° about the y axis and the other two 90° about the x axis in Figure P20.24. Ignore air resistance.

20.26 The spacecraft shown in Figure P20.26 has a mass $m = 2\,500$ kg and a centroidal mass moment of inertia $I_x = 40\,000$ kg m^2. The craft is moving in a straight path at a constant speed of 25 000 km/h when the engine at A is turned on providing a thrust $T = 250(1 - e^{-0.5t})$ kN where t is in seconds. Determine (a) the speed of the spacecraft 0.3 s later and (b) its angular velocity. Assume that the effects of gravity are negligible.

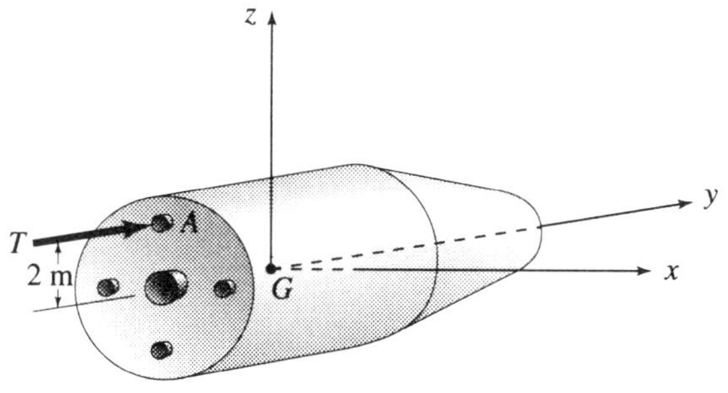

FIGURE P20.26.

20.3 Principles of Conservation of Momentum

There are numerous situations in which the sum of the external forces (resultant force) acting on a rigid body in a given direction during a specified time interval is either equal to zero or is sufficiently small and may be ignored. In such cases, the resultant linear impulse in the chosen direction is either identically zero or sufficiently close to zero and may be neglected. Under these conditions, Eqs. (20.17) and (20.18) reduce to

$$L_{2x} = L_{1x}, \qquad L_{2y} = L_{1y}. \tag{20.21}$$

Equations (20.21) are mathematical expressions of the principle of *conservation of linear momentum.* This principle states that, in the absence of a resultant linear impulse, the change in the linear momentum is zero. In other words, the linear momentum of the rigid body remains the same or is *conserved.* If we assume that the mass of the rigid body is constant, as a consequence of Eqs. (20.21), it follows that the velocity v_G of the mass center of the rigid body remains unchanged, i.e., $(v_{G2})_x = (v_{G1})_x$ and $(v_{G2})_y = (v_{G1})_y$.

Similarly, if the sum of the external moments (resultant moment) acting on a rigid body about an axis through fixed point O or through the center of mass point G of the rigid body during a given time interval is either equal to zero or is negligibly small, then, the angular impulses about these axes are either identically zero or sufficiently small and may be neglected. Under such conditions, Eqs. (20.19) become

$$
\begin{aligned}
(H_{O2})_z &= (H_{O1})_z, \\
(H_{G2})_z &= (H_{G1})_z.
\end{aligned}
\tag{20.22}
$$

Equations (20.22) are mathematical expressions of the principle of *conservation of angular momentum.* This principle states that, in the absence of a resultant angular impulse about an axis through fixed point O or through the center of mass G, the corresponding angular momentum remains the same or is *conserved.* If we assume that the mass moment of inertia I_O or I_G is constant, as a consequence of Eqs. (20.22), it follows that the angular velocity ω of the rigid body remains unchanged, i.e., $\omega_2 = \omega_1$. On the other hand, if the mass moment of inertia changes, the angular velocity also changes so that their product remains constant. A good illustration of this phenomenon may be had by observing the spinning of a figure skater. When the skater has her arms extended, she spins at a given angular velocity about a vertical axis through her mass center. When the skater brings her arms down and places them close to her body, she decreases her centroidal mass moment of inertia and, consequently, her angular velocity increases.

It should be pointed out that there are cases in which the linear momentum is not conserved even though the angular momentum about a given axis is conserved. Such is the case when the lines of action of the external forces, producing the linear impulses, all pass through the axis about which the angular momenta are obtained. In such a case, the resultant angular impulse about this axis vanishes, and the angular momentum is conserved. However, the resultant linear impulse may not be zero and, consequently, the linear momentum may not be conserved.

The principles of conservation of momentum derived above for a single rigid body are also applicable in the case of a system of interconnected rigid bodies. This is so because, in dealing with an entire system of rigid bodies, its internal forces and moments at the connections occur in pairs of equal and opposite vectors which produce impulses that cancel each other and contribute nothing to the external impulses. However, each of the terms in Eqs. (20.21) and (20.22) must account for all of the momenta in the entire system. Thus, for a system of interconnected rigid bodies, Eqs. (20.21) and (20.22) may be summarized as

$$(\textstyle\sum \text{Syst. Momenta})_2 = (\textstyle\sum \text{Syst. Momenta})_1. \tag{20.23}$$

The use of the principles of conservation of momentum is illustrated in the following examples. As in the case of the principles of impulse and momentum, the student is strongly urged to draw impulse diagrams and initial and final momenta diagrams representing Eqs. (20.21), (20.22), and (20.23).

■ Example 20.4

A young person B weighing 400 N starts from rest at the top of a smooth slide and leaves the bottom of the slide with a horizontal velocity parallel to the x axis. At the same time, a friend imparts a constant velocity $v_C = 10$ m/s to a cart C weighing 200 N which moves in the smooth x-y plane. The velocity of the cart makes an angle of 45° with the x axis. All of this information is depicted in the plan and elevation views shown in Figure E20.4(a). The young person and the cart reach the bottom of the slide simultaneously enabling him to land on the cart without moving relative to it, and the two of them continue

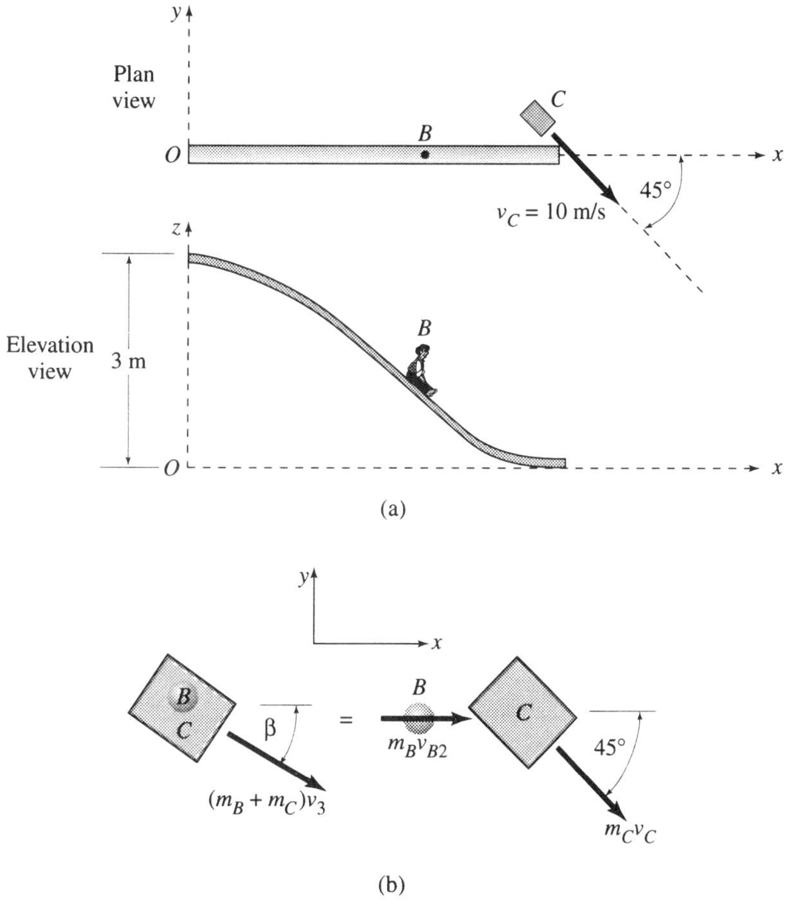

FIGURE E20.4.

to move as a single entity in the x-y plane. Determine their combined velocity immediately after the young person lands on the cart.

Solution

The velocity of the young person B just prior to leaving the slide may be determined by using the work-energy principle. Let position 1 be the top of the slide, position 2 the bottom of the slide just prior to landing and position 3 immediately after landing on the cart. Thus,

$$U_{1\to 2} = T_2 - T_1, \qquad 400(3) = \frac{1}{2}\left(\frac{400}{9.81}\right)v_{\mathrm{B2}}^2 - 0$$

from which

$$v_{\mathrm{B2}} = 7.67 \text{ m/s}$$

where v_{B2} is the velocity of B in position 2 and is directed along the positive x axis.

Since positions 2 and 3 define motion in the x-y plane, we will confine our discussion to this plane. Because both the slide and the x-y plane are smooth, the only forces acting on B and the cart are their weights and the normal reactions from the slide and the x-y plane, respectively. All of these forces are in the z direction and produce no impulses in the x or the y direction. Thus, the linear momentum is conserved in the x-y plane and Eq. (20.23) applies.

Figure E20.4(b) shows the momenta diagrams in positions 2 and 3 representing Eq. (20.23). Thus, by Eq. (20.23) applied in the x direction to the system consisting of the young person and the cart,

$$[(m_{\mathrm{B}} + m_{\mathrm{C}})v_3]_x = (m_{\mathrm{B}}v_{\mathrm{B2}} + m_{\mathrm{C}}v_{\mathrm{C}})_x,$$

$$\left(\frac{400 + 200}{9.81}\right)(v_3)_x = \left[\left(\frac{400}{9.81}\right)(7.67) + \left(\frac{200}{9.81}\right)(10\cos 45^\circ)\right]$$

from which

$$(v_3)_x = 7.47 \text{ m/s} \rightarrow$$

where $(v_3)_x$ is the velocity in the x direction of the young person and the cart combined immediately after landing. Also, by Eq. (20.23) applied in the y direction

$$[(m_{\mathrm{B}} + m_{\mathrm{C}})v_3]_y = (m_{\mathrm{B}}v_{\mathrm{B2}} + m_{\mathrm{C}}v_{\mathrm{C}})_y,$$

$$\left(\frac{400 + 200}{9.81}\right)(v_3)_y = \left[0 - \left(\frac{200}{9.81}\right)(10\sin 45^\circ)\right]$$

from which

$$(v_3)_y = 2.36 \text{ m/s} \downarrow$$

where $(v_3)_y$ is the velocity in the y direction of the system immediately after landing. The resultant velocity $v_3 = \sqrt{(v_3)_x^2 + (v_3)_y^2}$. Thus,

$$v_3 = 7.83 \text{ m/s} \qquad \text{ANS.}$$

The direction of v_3 is defined by the angle β measured from the positive x axis as shown in Figure E20.4(b). Thus,

$$\beta = \tan^{-1}\left(\frac{2.36}{7.47}\right) = 17.53°. \qquad \text{ANS.}$$

Example 20.5

The uniform rectangular crate shown in Figure E20.5 is released from point A and slides a distance L down the frictionless 20° inclined plane to point B where it impacts a small obstruction (not shown) with negligible rebound. The crate then pivots about the obstruction and, if it has sufficient velocity when it reaches point B, it will rotate about the obstruction and finally reach the position indicated by the letters

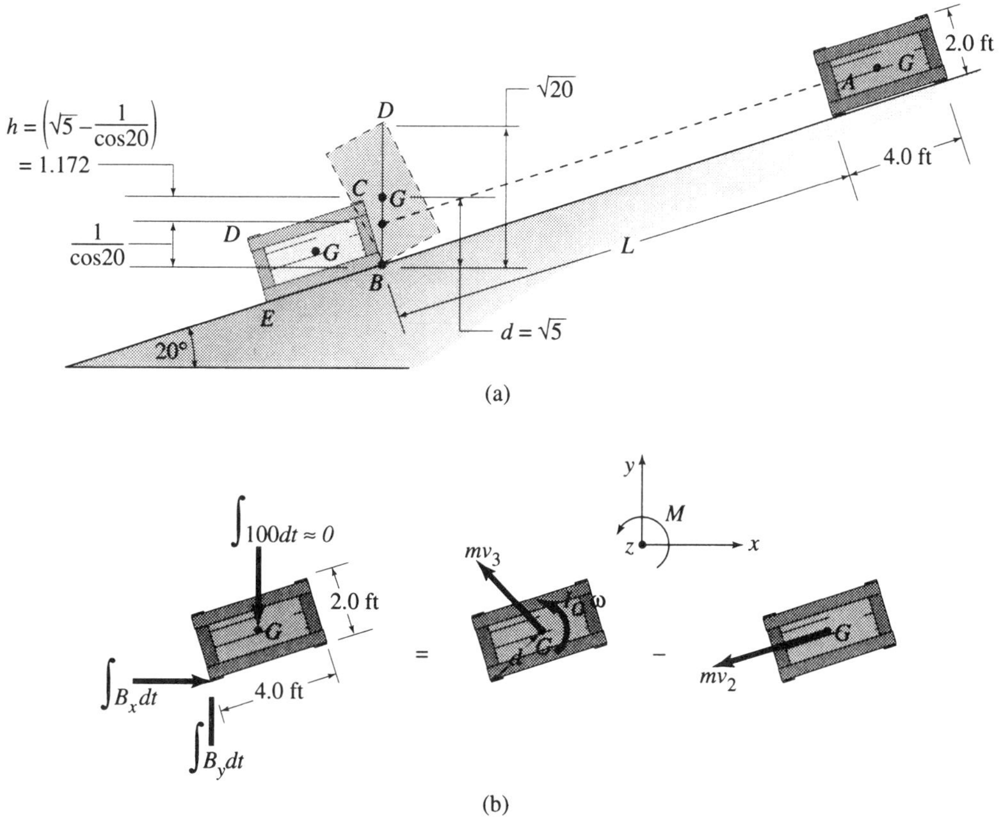

FIGURE E20.5.

BCDE. (a) Determine the minimum distance L between points A and B on the inclined plane so that the crate is finally able to reach position BCDE. The weight of the crate is 100 lb. (b) Find the percent loss of kinetic energy due to the impact between the crate and the obstruction.

Solution

(a) The crate will be able to finally reach position BCDE if its linear velocity, when it reaches point B, is sufficiently large to allow it to rotate about the obstruction at B and reach a position just past the one where its diagonal BD is vertical. Note that when it reaches this position, the crate would have no linear or angular velocity. Thus, the minimum distance L is determined consistent with the critical linear velocity of the crate at point B.

Let position 1 be that of the crate just before it is released from rest at A, position 2 that of the crate just before hitting the obstruction, position 3 that of the crate just after hitting the obstruction, and position 4 that of the crate when its diagonal BD is vertical where the crate has no linear or angular momenta. The principle of work and energy may be used to relate the critical velocity v_2 to the distance L. Thus,

$$U_{1\to 2} = T_2 - T_1, \quad 100L \sin 20^\circ = \frac{1}{2}\left(\frac{100}{32.2}\right)v_2^2 - 0$$

from which

$$v_2 = 4.693\ L^{1/2}\ \text{ft/s} \ \swarrow . \qquad \text{(a)}$$

The impulse diagram for the crate as well as the initial (position 2) and final (position 3) momenta of the crate are shown in Figure E20.5(b) along with a right-handed x-y-z coordinate system. If impulses and momenta are summed about a z axis through point B, the only force producing an angular impulse is the weight of the crate. However, this angular impulse may be ignored because the time of impact is extremely short. Therefore, the angular momenta about a z axis through point B is conserved. Thus, by the first of Eqs. (20.22),

$$(H_{B3})_z = (H_{B2})_z \qquad \text{(b)}$$

where

$$(H_B)_3 = I_B\omega_3 = (I_G + md^2)\left(\frac{v_3}{d}\right)$$

$$= \left[\frac{1}{12}\left(\frac{100}{32.2}\right)(2^2 + 4^2) + \left(\frac{100}{32.2}\right)(5)\right]\frac{v_3}{\sqrt{5}} = 9.259v_3$$

$$(H_B)_2 = (mv_2)(1.0) = \left(\frac{100}{32.2}\right)(4.693L^{1/2})14.575L^{1/2}$$

Therefore, using Eq. (b),

$$v_3 = 1.574L^{1/2} \text{ ft/s}. \tag{c}$$

Also, because $d = \sqrt{5}$,

$$\omega_3 = \frac{v_3}{d} = 0.704L^{1/2} \text{ rad/s} \circlearrowleft. \tag{d}$$

The work-energy principle may again be used to investigate the transition from position 3 to position 4 where diagonal BD is vertical and the crate possesses no kinetic energy. Thus,

$$U_{3\to4} = T_4 - T_3, \quad -100h = 0 - (\tfrac{1}{2}I_G\omega_3^2 + \tfrac{1}{2}mv_3^2) \tag{e}$$

where

$$I_G = \frac{1}{12}\left(\frac{100}{32.2}\right)(2^2 + 4^2) = 5.176 \text{ slug ft}^2,$$

$$h = \sqrt{5} - \frac{1}{\cos 20°} = 1.172 \text{ ft (see Fig. E20.5(a))}$$

Therefore, Eq. (e) becomes

$$100(1.172) = \frac{1}{2}(5.176)(0.704L^{1/2})^2 + \frac{1}{2}\left(\frac{100}{32.2}\right)(1.574L^{1/2})^2$$

from which

$$L = 22.8 \text{ ft.} \qquad \text{ANS.}$$

(b) The loss of kinetic energy is found by determining the kinetic energies immediately before and immediately after impact and obtaining the difference between them. Thus,

$$T_2 = \tfrac{1}{2}mv_2^2 = 779.74 \text{ lb ft},$$

and

$$T_3 = \tfrac{1}{2}I_B\omega_3^2 = 116.98 \text{ lb ft}.$$

Therefore, the percent loss of energy is

$$100\left(\frac{779.74 - 116.98}{779.74}\right) = 85.0\%.$$

Example 20.6

A homogeneous cylinder of weight W_C is attached symmetrically to four identical plates each of weight W_P. The plates may be positioned horizontally as in Figure E20.6(a) or vertically as in Figure E20.6(b). Ignore the weight and size of the connections between the plates and the cylinders. The system is capable of rotating about a vertical axis through the geometric center of the cylinder, point O. When the four

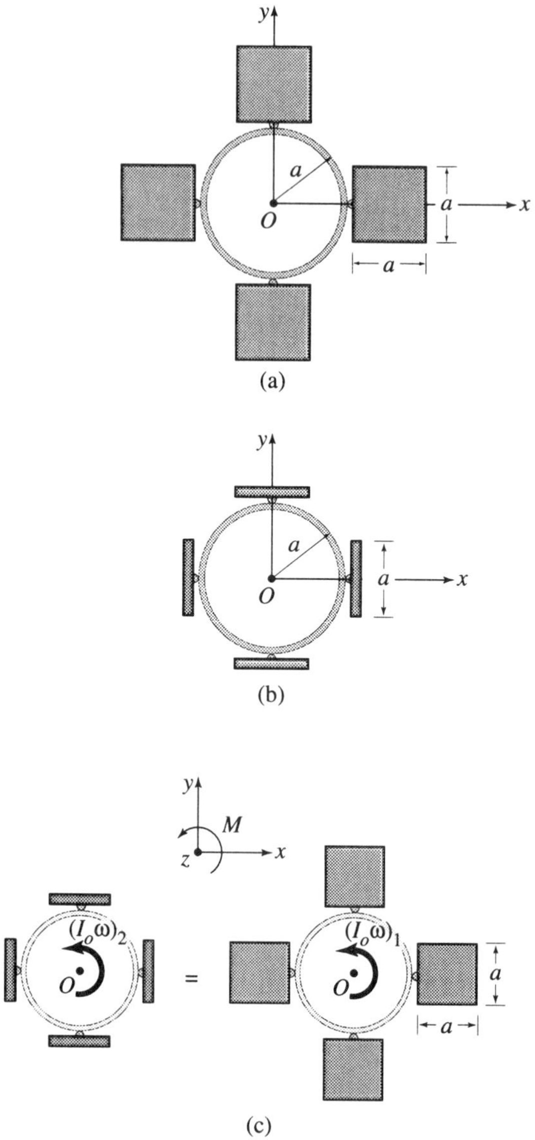

FIGURE E20.6.

plates are as shown in Figure E20.6(a), the system rotates with an angular velocity ω_1. An internal mechanism requiring negligible energy changes the position of the four plates to that shown in Figure E20.6(b) while the system is rotating. In terms of W_C, W_P and ω_1, develop an expression for the angular velocity ω_2 of the system immediately after the four plates reach the position shown in Figure E20.6(b).

Solution

If we consider the angular impulses of all forces acting on the system about a vertical axis through point O, we conclude that their resultant is zero and, consequently, the angular momentum about this axis is conserved. This situation is depicted in Figure E20.6(c) which shows the initial and final angular momenta diagrams for the system representing the first of Eqs. (20.22). Also, a right-handed x-y-z coordinate system is shown. Thus, by the first of Eqs. (20.22),

$$(H_{O2})_z = (H_{O1})_z, \quad [(I_O\omega)_2]_z = [(I_O\omega)_1]_z$$

where (see Appendix B)

$$(I_O)_1 = (\textstyle\sum I_O)_1 = \frac{1}{2}\left(\frac{W_C}{g}\right)a^2 + 4\left[\frac{1}{12}\left(\frac{W_P}{g}\right)(a^2 + a^2) + \left(\frac{W_P}{g}\right)\left(\frac{2}{3}a^2\right)\right]$$

$$= \frac{a^2}{g}\left(\frac{1}{2}W_C + \frac{13}{3}W_P\right)$$

$$(I_O)_2 = (\textstyle\sum I_O)_2 = \frac{1}{2}\left(\frac{W_C}{g}\right)a^2 + 4\left[\frac{1}{12}\left(\frac{W_P}{g}\right)a^2 + \frac{W_P}{g}a^2\right]$$

$$= \frac{a^2}{g}\left(\frac{1}{2}W_C + \frac{13}{3}W_P\right)$$

Therefore,

$$\frac{a^2}{g}\left(\frac{1}{2}W_C + \frac{13}{3}W_P\right)\omega_2 = \frac{a^2}{g}\left(\frac{1}{2}W_C + \frac{10}{3}W_P\right)\omega_1$$

from which

$$\omega_2 = \left[\frac{\frac{1}{2}W_C + \frac{10}{3}W_P}{\frac{1}{2}W_C + \frac{13}{3}W_P}\right]\omega_1 = \left[\frac{3 + 20\frac{W_P}{W_C}}{3 + 26\frac{W_P}{W_C}}\right]\omega_1. \qquad \text{ANS.}$$

Example 20.7

A bullet having a mass $m_B = 0.005$ kg is fired with a velocity of 1000 m/s into a thin triangular plate of mass $m_P = 5$ kg. The plate is suspended in a vertical position from the frictionless hinge O–O as shown in Figure E20.7(a). The bullet is imbedded in the plate which is initially at rest and the combination of the bullet and the plate swings about the axis O–O through an angle θ. Determine (a) the angle θ and (b) the percent loss of kinetic energy due to the impact of the bullet and the plate.

Solution

(a) When the plate is in the vertical position just before and just after impact, the forces acting on it (i.e., its weight and the reaction compo-

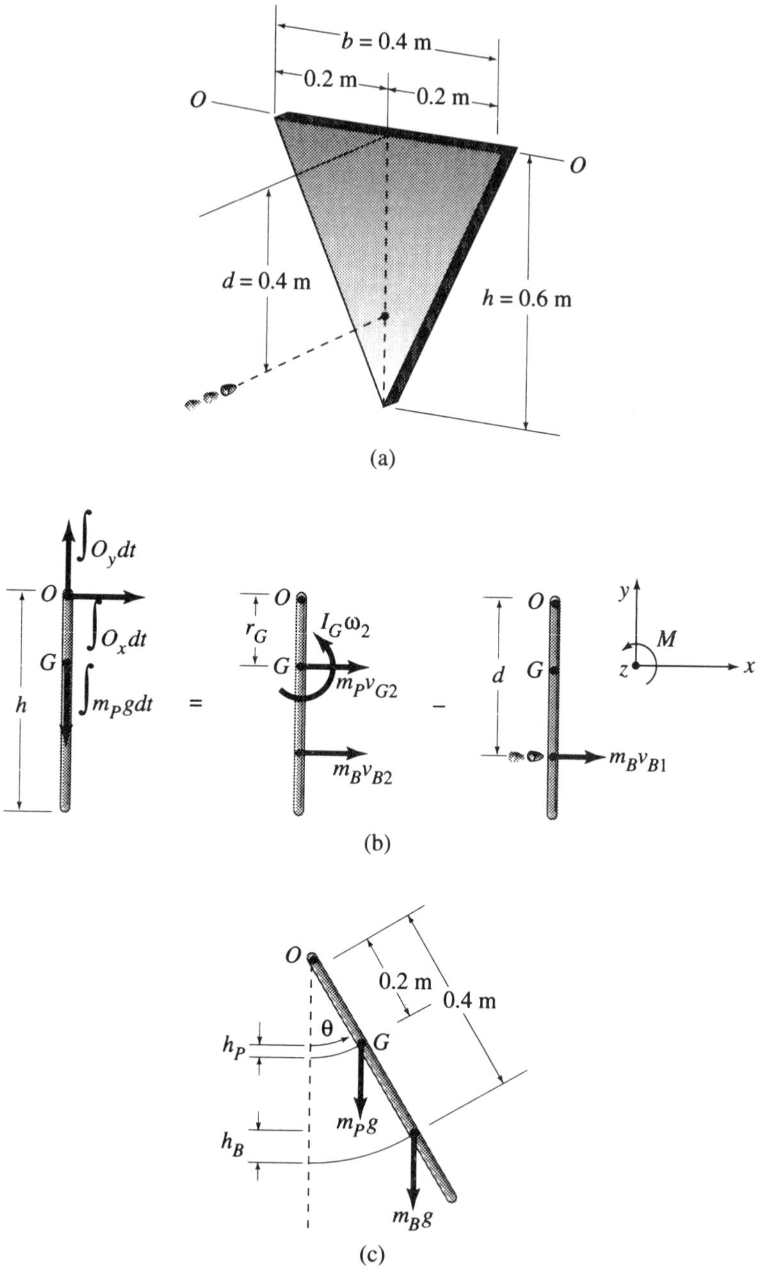
b = 0.4 m
0.2 m
0.2 m
O
O
d = 0.4 m
h = 0.6 m
(a)
∫O_y dt
∫O_x dt
∫m_P g dt
O
G
h
=
O
r_G
G
I_G ω_2
m_P v_G2
m_B v_B2
−
O
d
G
m_B v_B1
y
x
z
M
(b)
O
0.2 m
0.4 m
θ
G
h_P
m_P g
h_B
m_B g
(c)

FIGURE E20.7.

nents at the hinge) produce no angular impulses about hinge O–O as shown, in edge view, in Figure E20.7(b). Thus, the angular momentum about hinge O–O is conserved. Also shown in Figure E20.7(b) are the initial and final angular momenta diagrams and a right-handed x-y-z coordinate system. Thus, by the first of Eqs. (20.22),

$$H_{O2} = H_{O1} \tag{a}$$

where 1 and 2 indicate positions just before and just after impact, respectively.

Now, by reference to Figure E20.7(b),

$$H_{O1} = (m_B v_{B1})d = 0.005(1000)(0.4) = 2 \text{ kg m}^2/\text{s}.$$

Also,

$$\begin{aligned} H_{O2} &= I_G\omega_2 + (m_p v_{G2})r_G + (m_B v_{B2})d \\ &= I_G\omega_2 + (m_p r_G\omega_2)r_G + (m_B d\omega_2)d \\ &= (I_G + m_p r_G^2)\omega_2 + m_B d^2\omega_2 \\ &= (I_O)_P\omega_2 + (I_O)_B\omega_2 \\ &= [(I_O)_P + (I_O)_B]\omega_2. \end{aligned} \tag{b}$$

Now

$$\begin{aligned} (I_O)_B &= m_B d^2 = 0.005(0.4^2) = 0.0008 \text{ kg m}^2, \\ (I_O)_P &= \rho t[(I_O)_P]_{AREA} \qquad \text{(Eq. (11.20))} \\ &= \left(\frac{m}{V}\right)t\left(\frac{1}{12}bh^3\right) = \left(\frac{m}{At}\right)t\left(\frac{1}{12}bh^3\right) \\ &= \left(\frac{m}{A}\right)\left(\frac{1}{12}bh^3\right) = \left[\frac{5}{0.5(0.4)(0.6)}\right]\left[\frac{1}{12}(0.4)(0.6^3)\right] \\ &= 0.300 \text{ kg m}^2. \end{aligned}$$

Thus, by Eq. (b),

$$H_{O2} = 0.3008\omega_2 \text{ kg m}^2/\text{s},$$

and by Eq. (a),

$$0.3008\omega_2 = 2$$

from which

$$\omega_2 = 6.65 \text{ rad/s} \circlearrowleft.$$

Note that the contribution of the bullet to the moment of inertia, $(I_O)_B$, is, in this case, sufficiently small and could have been ignored without significantly affecting the magnitude of ω_2.

The work-energy method is now used to find the angle θ. Thus,

$$U_{2\to 3} = T_3 - T_2$$

where position 3 indicates the final position of the plate. Now, by reference to the geometry in Figure E20.7(c), we conclude that

$$\begin{aligned} U_{2\to 3} &= -m_{\mathrm{P}}gh_{\mathrm{P}} - m_{\mathrm{B}}gh_{\mathrm{B}} \\ &= -5(9.81)(0.2)(1 - \cos\theta) - 0.005(9.81)(0.4)(1 - \cos\theta) \\ &= -9.8296(1 - \cos\theta). \end{aligned}$$

$$T_3 = 0.$$

$$T_2 = \tfrac{1}{2}(I_{\mathrm{O}})\omega_2^2 = \tfrac{1}{2}(0.3008)(6.65^2) = 6.6511.$$

Therefore,

$$9.8296(1 - \cos\theta) = 6.6511$$

from which

$$\theta = 71.1^\circ \circlearrowleft. \qquad \text{ANS.}$$

Again, the contribution of the bullet to the energy of the system could have been neglected.

(b) The loss of kinetic energy due to impact is found by determining the difference between the energy of the system before and after impact. Thus,

$$T_1 = \tfrac{1}{2}m_{\mathrm{B}}v_{\mathrm{B}}^2 = \tfrac{1}{2}(0.005)(1000^2) = 2500 \text{ Nm}$$

$$T_2 = \tfrac{1}{2}(I_{\mathrm{O}})\omega_2^2 = \tfrac{1}{2}(0.3008)(6.65^2) = 6.6511 \text{ Nm}$$

Therefore,

$$\text{Loss of energy} = (2500 - 6.6511)/2500 = 99.7\%. \qquad \text{ANS.}$$

■ Example 20.8

The homogenous rectangular prism of weight $W = 5$ lb is equipped with two weightless pins at A as shown in Figure E20.8(a). The prism is released from rest and translates (without rotating) a distance $h = 5$ ft before engaging the two hooks B in the fixed support as shown, forcing the prism to rotate cw. Assume that there is no rebound when engagement takes place. Find (a) the angular velocity of the prism immediately after engagement and (b) the velocity of point P after the prism has rotated through 90°.

Solution

(a) Consider the impulse-momenta diagrams shown in side view in Figure E20.8(b) along with a right-handed x-y-z coordinate system. The only force producing an angular impulse about point A is the

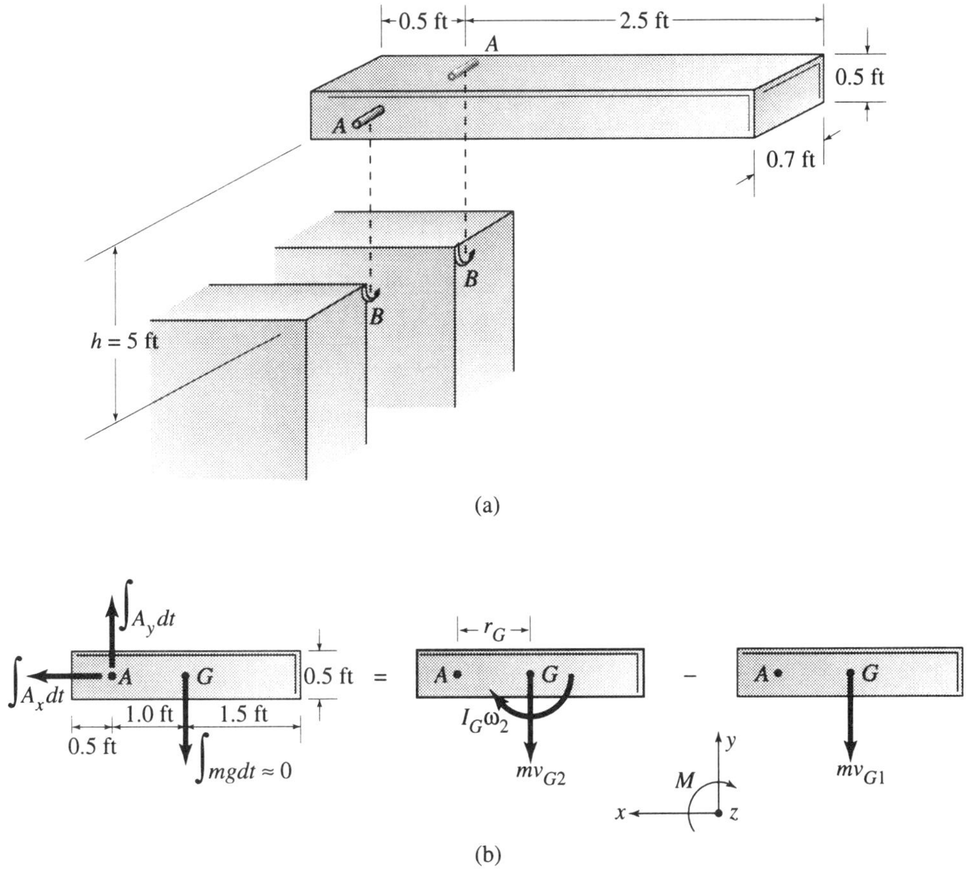

FIGURE E20.8.

weight of the prism. This, however, may be ignored because the weight is small and the time of engagement is extremely short. Thus, we conclude that the angular momentum of the prism about A during engagement is conserved. Thus, by the first of Eqs. (20.22),

$$H_{A2} = H_{A1} \tag{a}$$

where 1 and 2 represent positions just before and after engagement. To determine H_{A1}, we need to find the speed of the prism just prior to engagement. This is accomplished by the work-energy method. Thus,

$$U_{O\to 1} = T_1 - T_O, \quad Wh = \frac{1}{2}\left(\frac{W}{g}\right)v_{G1}^2 - 0$$

where O represents the position of the prism at release. Therefore,

$$v_{G1} = \sqrt{2gh} = \sqrt{2(32.2)5} = 17.94 \text{ ft/s} \downarrow.$$

Hence,

$$H_{A1} = \left(\frac{W}{g}\right) v_{G1}(r_G) = \left(\frac{5}{32.2}\right)(17.94)(1.0) = 2.786 \text{ slug ft}^2\text{/s}.$$

Also,

$$H_{A2} = I_G\omega_2 + (mv_{G2})r_G = I_G\omega_2 + (mr_G\omega_2)r_G$$
$$= (I_G + mr_G^2)\omega_2 = I_A\omega_2 \qquad \text{(b)}$$

where I_A is the mass moment of inertia of the prism with respect to A. Thus, using the information in Appendix B,

$$I_A = \frac{1}{2}\left(\frac{W}{g}\right)(3^2 + 0.5^2) + \frac{W}{g}(1.0^2) = 5.625\left(\frac{W}{g}\right) = 0.8734 \text{ slug ft}^2,$$

and, by Eq. (b),

$$(H_A)_2 = I_A\omega_2 = 0.8734\omega_2 \text{ slug ft}^2\text{/s}.$$

Therefore, using Eq. (a),

$$0.8734\omega_2 = 2.786$$

from which

$$\omega_2 = 3.19 \text{ rad/s} \circlearrowright. \qquad \text{ANS.}$$

(b) The work-energy method may again be used to find the angular velocity of the prism after it has rotated through 90°. Thus,

$$U_{2\to3} = T_3 - T_2$$

where 3 designates the position of the prism after rotating through 90°. Thus,

$$U_{2\to3} = W(1.0) = 5.0 \text{ lb ft},$$

$$T_3 = \tfrac{1}{2}I_A\omega_3^2 = \tfrac{1}{2}(0.8734)\omega_3^2 = 0.4367\omega_3^2 \text{ lb ft},$$

and

$$T_2 = \tfrac{1}{2}I_A\omega_2^2 = \tfrac{1}{2}(0.8734)(3.19^2) = 4.444 \text{ lb ft}.$$

Therefore,

$$5.0 = 0.4367\omega_3^2 - 4.444$$

from which

$$\omega_3 = 4.65 \text{ rad/s} \circlearrowright.$$

Thus,

$$v_P = r_P\omega_3 = (3.5)(4.65) = 16.28 \text{ ft/s} \leftarrow. \qquad \text{ANS.}$$

■

20.4 Eccentric Impact

The concepts of *direct* and *oblique central* impact of particles were discussed in Section 16.5. These concepts are expanded here to cover the *eccentric impact* of rigid bodies.

Two rigid bodies are said to experience *eccentric impact* if the mass centers of the two bodies do not lie on the *line of impact*. As in the case of particle impact, the impact of rigid bodies may be described as occurring in two stages. The first stage consists of a period of *deformation** at the end of which the points of contact of the two bodies reach the same velocity. The second stage consists of a period of restitution at the end of which the points of contact, in general, possess different velocities.

Consider the unconstrained motion of the two rigid bodies shown in Figure 20.4 which are experiencing eccentric impact. The symbols A and B are used to signify not only the two rigid bodies but also the two specific points on the bodies that come in contact during the collision. The n axis represents the line of impact, and the t axis (perpendicular to n) is the common tangent plane, sometimes known as the plane of contact. Figure 20.4(a) shows the two rigid bodies just prior to impact, both moving to the right, where the velocities of points A and B are, $\mathbf{v}_{A1}$ and $\mathbf{v}_{B1}$, respectively, and their components along the line of impact are $(v_{A1})_n$ and $(v_{B1})_n$ such that $(v_{A1})_n > (v_{B1})_n$. Also, the angular velocities of bodies A and B just before impact are $\boldsymbol{\omega}_{A1}$ and $\boldsymbol{\omega}_{B1}$, respectively, both ccw. Figure 20.4(b) shows the two rigid bodies during the period of deformation at the end of which the velocities of points A and B are, $\mathbf{v}_A$ and $\mathbf{v}_B$, respectively, and their components along the line of impact are $(v_A)_n$ and $(v_B)_n$ so that $(v_A)_n = (v_B)_n$. Also, the angular velocities of bodies A and B at the end of this period are $\boldsymbol{\omega}_A$ and $\boldsymbol{\omega}_B$, respectively. Finally, Figure 20.4(c) shows the two rigid bodies during the period of *restitution* at the end of which the velocities of points A and B are $\mathbf{v}_{A2}$ and $\mathbf{v}_{B2}$, respectively, and their components along the line of impact are $(v_{A2})_n$ and $(v_{B2})_n$ so that $(v_{A2})_n < (v_{B2})_n$. The angular velocities of the two bodies at the end of the period of restitution are $\boldsymbol{\omega}_{A2}$ and $\boldsymbol{\omega}_{B2}$, respectively, both ccw.

If it is assumed that the two bodies are frictionless, then, the impact forces that they exert on each other must be directed along the line of impact. Let F_D be the force developed during the period of deformation and F_R that during the period of restitution. The coefficient of restitution e may now be stated as

*It is contradictory to speak of *deformation* when discussing the behavior of *rigid* bodies. However, to obtain a clear understanding of the physical behavior of impacting bodies, we must momentarily abandon the concept of rigidity during impact.

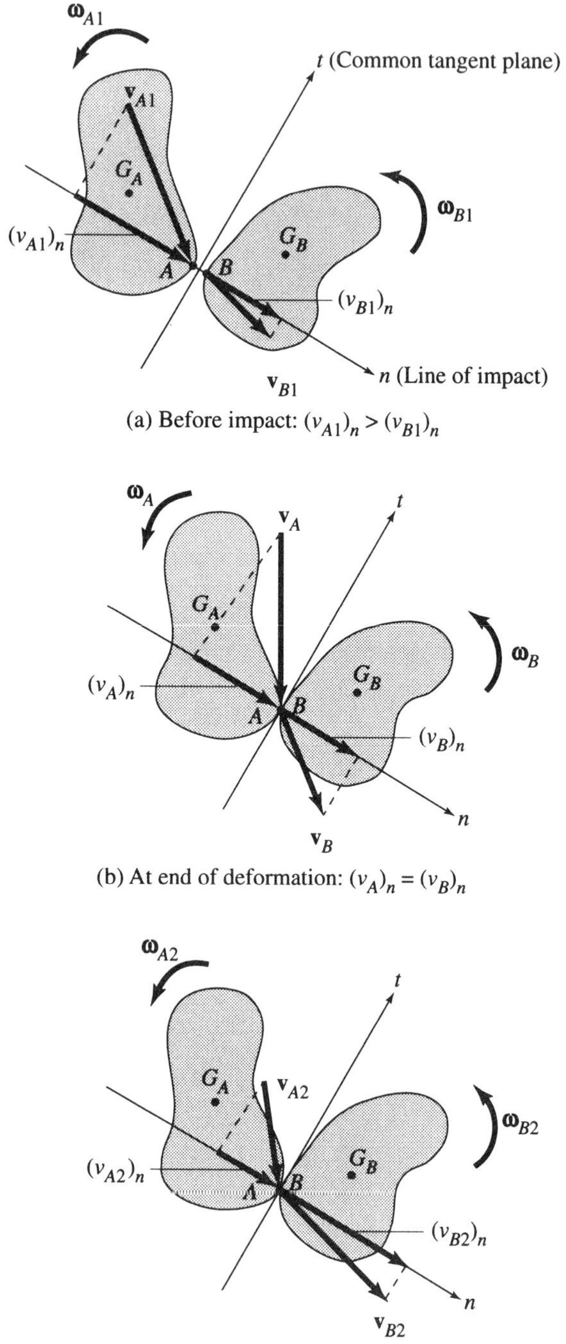

FIGURE 20.4. (c) At end of restitution: $(v_{A2})_n < (v_{B2})_n$

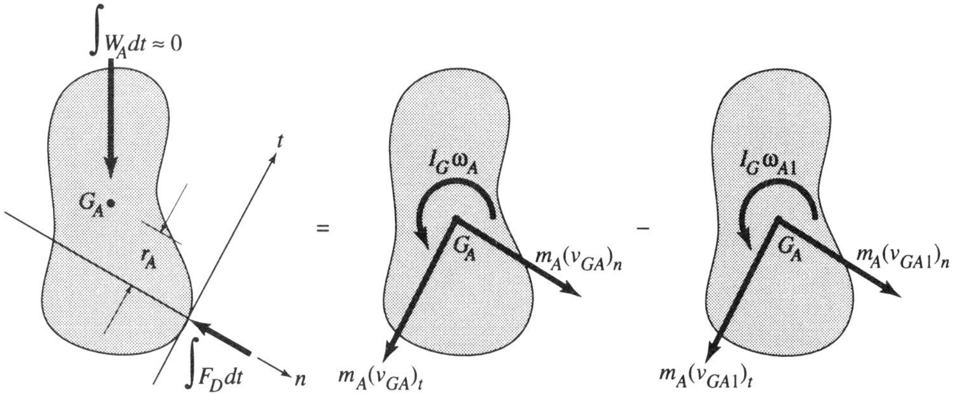

FIGURE 20.5.

$$e = \frac{\int F_R \, dt}{\int F_D \, dt}. \tag{20.24}$$

The impulse and momenta diagrams for body A during the period of deformation are shown in Figure 20.5. Note that, because the motions of the rigid bodies prior to impact are unconstrained and if it is assumed that their weights are nonimpulsive, the only impulsive force acting on body A during this period is F_D. Thus, applying Eq. (20.17) along the n axis we obtain

$$-\int F_D \, dt = m_A(v_{GA})_n - m_A(v_{GA1})_n. \tag{20.25a}$$

Also, the second of Eqs. (20.19) yields

$$-r_A \int F_D \, dt = I_G \omega_A - I_G \omega_{A1}. \tag{20.25b}$$

In a similar manner, we consider the period of restitution for body A for which the impulse and momenta diagrams are shown in Figure 20.6. Thus,

$$-\int F_R \, dt = m(v_{GA2})_n - m_A(v_{GA})_n \tag{20.25c}$$

and

$$-r_A \int F_R \, dt = I_G \omega_{A2} - I_G \omega_A. \tag{20.25d}$$

If Eqs. (20.25a) and (20.25c) are substituted in Eq. (20.24) we obtain

$$e = \frac{(v_{GA})_n - (v_{GA2})_n}{(v_{GA1})_n - (v_{GA})_n}. \tag{20.25e}$$

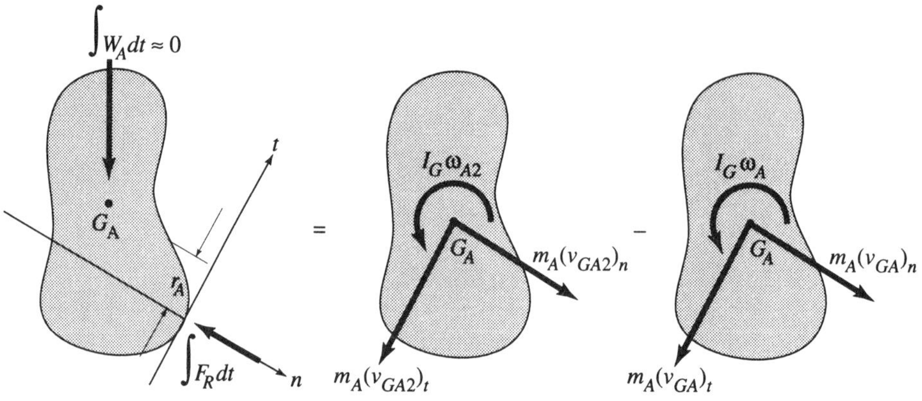

FIGURE 20.6.

Also, when Eqs. (20.25b) and (20.25d) are substituted in Eq. (20.24) the result is

$$e = \frac{\omega_{\mathrm{A}} - \omega_{\mathrm{A2}}}{\omega_{\mathrm{A1}} - \omega_{\mathrm{A}}}. \tag{20.25f}$$

Multiplying the numerator and denominator of Eq. (20.25f) by r_{A},

$$e = \frac{r_{\mathrm{A}}\omega_{\mathrm{A}} - r_{\mathrm{A}}\omega_{\mathrm{A2}}}{r_{\mathrm{A}}\omega_{\mathrm{A1}} - r_{\mathrm{A}}\omega_{\mathrm{A}}} \tag{20.25g}$$

Adding the numerators of Eqs. (20.25e) and (20.25g) and dividing the result by the sum of the denominators of the same two equations, we obtain

$$e = \frac{[(v_{\mathrm{GA}})_{\mathrm{n}} + r_{\mathrm{A}}\omega_{\mathrm{A}}] - [(v_{\mathrm{GA2}})_{\mathrm{n}} + r_{\mathrm{A}}\omega_{\mathrm{A2}}]}{[(v_{\mathrm{GA1}})_{\mathrm{n}} + r_{\mathrm{A}}\omega_{\mathrm{A1}}] - [(v_{\mathrm{GA}})_{\mathrm{n}} + r_{\mathrm{A}}\omega_{\mathrm{A}}]}. \tag{20.25h}$$

Noting that $(v_{\mathrm{GA}})_{\mathrm{n}} + r_{\mathrm{A}}\omega_{\mathrm{A}} = (v_{\mathrm{A}})_{\mathrm{n}}$, $(v_{\mathrm{GA2}})_{\mathrm{n}} + r_{\mathrm{A}}\omega_{\mathrm{A2}} = (v_{\mathrm{A2}})_{\mathrm{n}}$, and $(v_{\mathrm{GA1}})_{\mathrm{n}} + r_{\mathrm{A}}\omega_{\mathrm{A1}} = (v_{\mathrm{A1}})_{\mathrm{n}}$, Eq. (20.25h) becomes

$$e = \frac{(v_{\mathrm{A}})_{\mathrm{n}} - (v_{\mathrm{A2}})_{\mathrm{n}}}{(v_{\mathrm{A1}})_{\mathrm{n}} - (v_{\mathrm{A}})_{\mathrm{n}}}. \tag{20.26}$$

Similarly, by considering the periods of deformation and restitution for body B, we obtain

$$e = \frac{(v_{\mathrm{B2}})_{\mathrm{n}} - (v_{\mathrm{B}})_{\mathrm{n}}}{(v_{\mathrm{B}})_{\mathrm{n}} - (v_{\mathrm{B1}})_{\mathrm{n}}} \tag{20.27}$$

As stated earlier, at the end of the period of deformation (beginning of restitution), $(v_{\mathrm{B}})_{\mathrm{n}} = (v_{\mathrm{A}})_{\mathrm{n}}$. Therefore, Eq. (20.27) becomes

$$e = \frac{(v_{B2})_n - (v_A)_n}{(v_A)_n - (v_{B1})_n} \tag{20.28}$$

Adding the numerators of Eqs. (20.26) and (20.28) and dividing the result by the sum of the denominators of the same two equations,

$$e = \frac{(v_{B2})_n - (v_{A2})_n}{(v_{A1})_n - (v_{B1})_n} \tag{20.29}$$

Equation (20.29) states that the coefficient of restitution for the eccentric impact of two bodies, is obtained as the ratio of the relative velocity along the line of impact of points A and B, just after impact, to the relative velocity along the line of impact of these two points just before impact. Thus, the relation expressed in Eq. (20.29) is essentially the same as Eq. (16.20) obtained previously for the collision of two particles. Also, whereas Eq. (20.29) was derived for the case where the motions of both rigid bodies are unconstrained, it may be shown that it remains valid for the case in which either one or both of the colliding rigid bodies is constrained to rotate about an axis through a fixed point.

The use of Eq. (20.29) in the solution of problems dealing with colliding rigid bodies is illustrated in Example 20.9.

■ Example 20.9

The slender rod OA whose mass is 10 kg is suspended from a frictionless hinge at O and released from rest in a horizontal position as shown in Figure E20.9(a). Swinging in a ccw direction, the rod reaches a vertical position at which time its end A strikes a small stationary ball on a frictionless plane whose mass is 3 kg. If the coefficient of restitution is $e = 0.6$, determine, just after impact, the angular velocity of the rod and the linear velocity of the ball.

Solution

The angular velocity of the rod ω_1, just prior to impact, may be determined from the principle of conservation of energy. Thus,

$$T_O + V_O = T_1 + V_1$$

Using the horizontal position of the rod as the datum for the potential energy,

$$\begin{aligned} 0 + 0 &= \tfrac{1}{2} I_O \omega_1^2 - mgh \\ &= \tfrac{1}{2}\left[\tfrac{1}{3}(10)(0.8)^2\right]\omega_1^2 - 10(9.81)(0.4) \end{aligned}$$

from which

$$\omega_1 = 6.065 \text{ rad/s} \circlearrowleft.$$

Using Eq. (20.29),

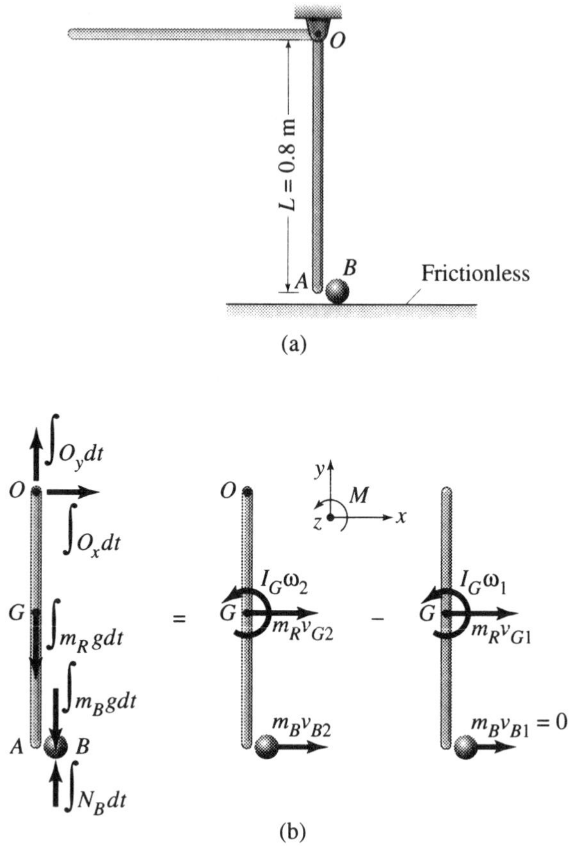

FIGURE E20.9.

$$0.60 = \frac{v_{B2} - v_{A2}}{v_{A1} - v_{B1}} = \frac{v_{B2} - v_{A2}}{0.8(6.065) - 0}$$

from which

$$v_{B2} - v_{A2} = 2.9112 \tag{a}$$

where v_{B2} and v_{A2} represent the linear velocity of the ball and that of point A on the rod, respectively, immediately after impact.

To solve for v_{B2} and v_{A2}, we need a second relationship between these two unknown quantities. This second relationship is obtained from the conservation of angular momenta about the hinge at O for the system consisting of the rod and the ball. Note that the momenta of the system are conserved because the impulsive forces at O and the weights of the rod and the ball, as well as the normal reaction under the ball, produce no angular impulses about point O (see the impulse and momenta diagrams shown in Figure E20.9(b)). Also, the impulsive forces between the rod and the ball during impact are equal and opposite and cancel each other. Thus, referring to Figure E20.9(b) and using the first of Eqs. (20.22),

$$(H_O)_1 = (H_O)_2$$

from which

$$m_B v_{B2}(0.8) + m_R v_{G2}(0.4) + I_G\omega_2 = m_R v_{G1}(0.4) + I_G\omega_1$$

where ω_2 is the angular velocity of the rod just after impact. Substituting $m_B = 3$ kg, $m_R = 10$ kg, $I_G = \frac{1}{12}m_R L^2 = 0.533$ kg m^2, $\omega_2 = \frac{v_{A2}}{0.8}$, $v_{G1} = \frac{L}{2}\omega_1 = 2.426$ m/s, and $\omega_1 = 6.065$ rad/s and, after simplification,

$$v_{B2} + 1.111v_{A2} = 5.3913 \qquad \text{(b)}$$

The simultaneous solution of Eqs. (a) and (b) yields

$$v_{B2} = 4.09 \text{ m/s} \rightarrow \qquad \text{ANS.}$$

and

$$v_{A2} = 1.175 \text{ m/s} \rightarrow. \qquad \text{ANS.}$$

Since $v_{A2} = 0.8\omega_2$,

$$\omega_2 = 1.469 \text{ rad/s } \circlearrowleft. \qquad \text{ANS.}$$

■

Problems

20.27 A young person of weight W_P starts from rest at point A and slides down the frictionless plane at the bottom of which there is a toboggan of weight W_T waiting as shown in Figure P20.27. The young person lands on the toboggan and does not move with respect to it. The combined unit slides on a horizontal plane

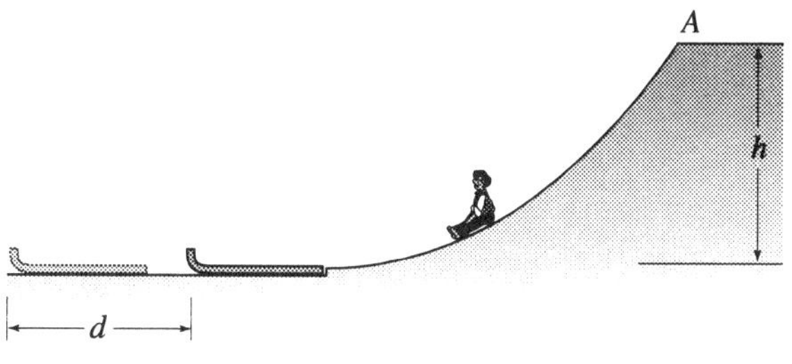

FIGURE P20.27.

with a coefficient of kinetic friction μ. Develop an expression for the distance d traveled by the combined unit of the young person and the toboggan before it comes to rest. If $W_p = 550$ N, $W_T = 50$ N, $h = 4$ m and $\mu = 0.15$, find a numerical value for d.

20.28 A 0.15 lb bullet is fired horizontally with a velocity v. The bullet strikes and imbeds itself in a 10 lb block which is at rest at the bottom of a frictionless circular path whose radius is 12 ft as shown in Figure P20.28. If the unit consisting of the bullet and the block slides up the path and comes to rest when $\theta = 10°$, determine (a) the velocity v of the bullet just before striking the block and (b) the loss of kinetic energy as a result of the bullet impacting the block and imbedding itself in it.

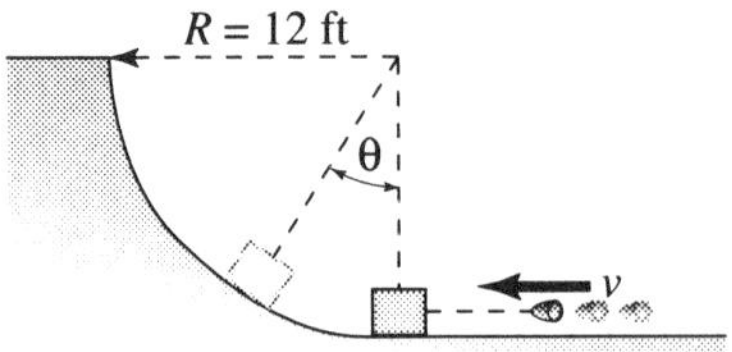

FIGURE P20.28.

20.29 Freight car A of weight W_A moves on horizontal tracks with a velocity v_A and overtakes freight car B of weight W_B which is moving on the same tracks with a velocity v_B as shown in Figure P20.29. (a) If the two cars become coupled and move together as a unit, develop an expression for their combined velocity immediately after coupling. (b) If $W_A = W_B$, determine the combined velocity in terms of v_A and v_B.

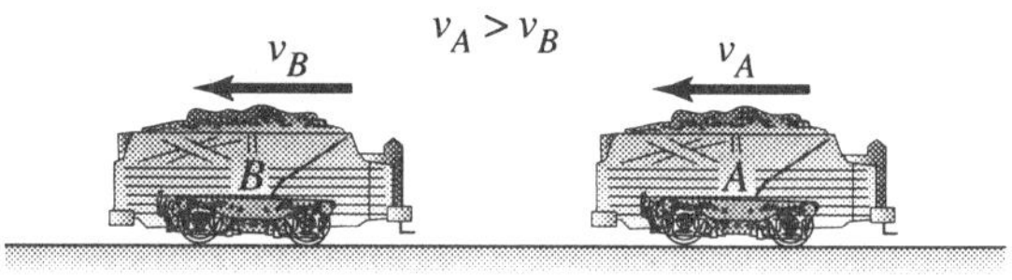

FIGURE P20.29.

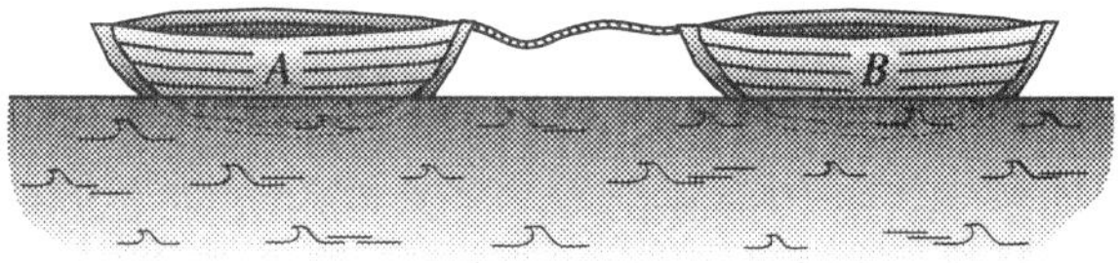

FIGURE P20.30.

20.30 Two identical boats each of weight $W =$ 2000 N are tied together by a cord which is not taut as shown in Figure P20.30. The two boats are at rest in still water. When boat A is given a push which imparts to it a velocity of 2 m/s to the left, the cord is momentarily tightened and imparts a velocity to boat B and decreases the velocity of boat A. If one half of the kinetic energy of boat A is lost as a result of the tightening of the cord, find the common velocity of the two boats immediately after the cord is tightened. Assume that water resistance is negligible.

20.31 A bullet A of weight $W_A = 0.2$ lb is fired

horizontally into the box of sand B of weight $W_B = 12$ lb which is suspended by means of a rigid rod of length $L = 3$ ft and weight $W_R = 2$ lb as shown in Figure P20.31. The bullet impacts the box with a velocity $v_A = 500$ ft/s and becomes imbedded in it and the unit consisting of the box, the rod and the bullet swings about point O through the angle θ. Determine the angle θ.

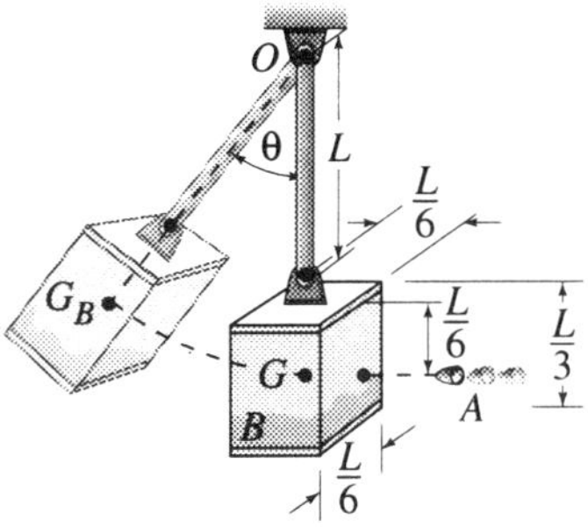

FIGURE P20.31.

20.32 Pendulum A whose bob weighs 4 lb and whose length is $L = 15$ in. is released from the position shown and swings in the y-z plane about a hinge at point C on the z axis as shown in Figure P20.32. Pendulum B whose bob weighs 2 lb and whose length is $L = 15$ in. is released from the position shown and swings in the x-z plane about the hinge at C. The two bobs strike, and stick to each other at the bottom of their swing (point O). Treat the two bobs as particles and determine (a) the magnitude and direction of the combined velocity immediately after impact and (b) the height above the x-y plane to which the combined weights will rise after impact. Assume that the two cords of length L are weightless.

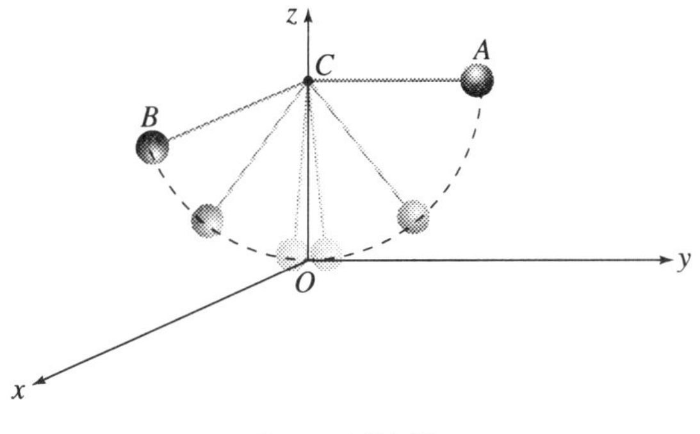

FIGURE P20.32.

20.33 Solve Example 20.6 (p. 621) if the internal mechanism changes the position of *only* two plates diametrically opposite to that shown in Figure E20.6(b) while the other two diametrically opposite plates remain as shown in Figure E20.6(a).

20.34 A figure skater goes into a spin rotating at an angular velocity of 10 rad/s with her arms fully extended in a horizontal plane. Determine her angular velocity if she brings her arms down and places them vertically close to her body. For purposes of computations, assume the body of the skater to be a homogeneous cylinder 0.30 m in diameter, 1.70 m long and weighing 500 N, and her arms to be two slender rods each having a length of 0.70 m and a weight of 40 N. Neglect friction between the skates and the ice floor.

20.35 A bullet weighing 0.20 lb is fired with a velocity of 3000 ft/s into a thin rectangular plate weighing 150 lb. The plate is suspended in a vertical position by the frictionless hinge O–O as shown in Figure P20.35. The bullet is imbedded in the plate which is initially at rest. The combination of the bullet and the plate swings about the axis O–O through an angle θ. Determine the value of the angle θ. What is the loss of kinetic energy due to the impact of the bullet with the plate?

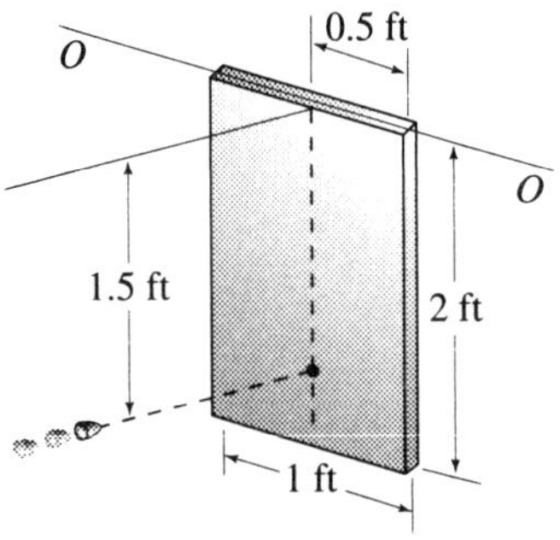

FIGURE P20.35.

20.36 Refer to Problem 20.35. Assume that the bullet passes through the plate and emerges with a velocity of 500 ft/s. Determine the value of the angle θ. What is the loss of kinetic energy due to the impact and penetration of the bullet?

20.37 A bullet of weight W_B is fired at a homogeneous rectangular plate of weigh W_P with a velocity v_0 as shown in Figure P20.37. The plate is hinged at point O and can rotate freely about a horizontal axis through this point. The bullet becomes imbedded in the plate and the combination swings about the axis at point O. (a) What is the angular velocity of the plate immediately after the bullet becomes imbedded in it? Express your answer in terms of W_P, W_B, b, h and v_0. (b) Find a numerical value for the angular velocity for the case where $W_B = 0.06$ N, $W_P = 40$ N, $h = 0.8$, $b = 0.4$ m, and $v_0 = 800$ m/s.

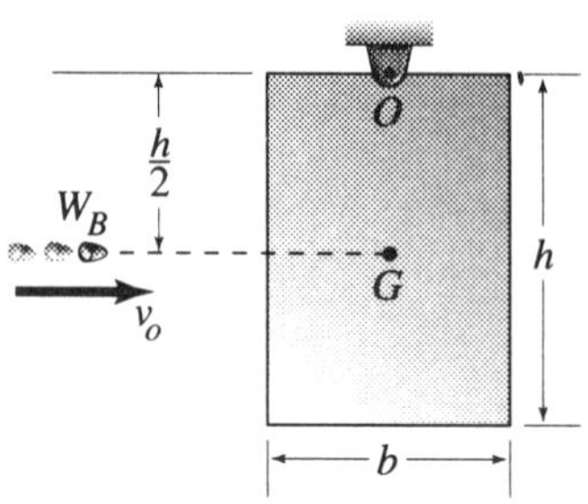

FIGURE P20.37.

20.38 A bullet B of weight $W_B = 0.50$ N is fired at a homogeneous disk D of weight $W_D = 200$ N with a velocity $v_0 = 800$ m/s as shown in Figure P20.38. The bullet becomes imbedded in the disk which is suspended from a flexible weightless cord as shown. Determine the velocities of points E and H on the disk immediately after the bullet becomes imbedded in it. Let $r = 0.30$ m.

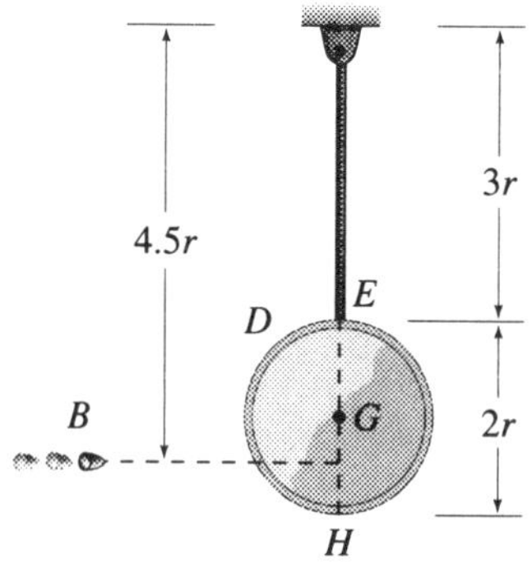

FIGURE P20.38.

20.39 The system shown in Figure P20.39 consists of a central homogeneous cylinder of radius $L/2$ and weight W attached symmetrically to four homogeneous slender rods each of length L and weight $W/8$. The four rods are held in the verti-

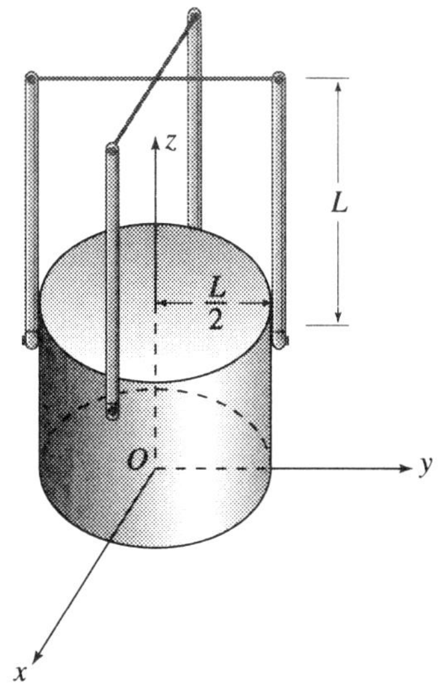

FIGURE P20.39.

cal position (parallel to the z axis) by two weightless strings and the system is rotated about the z axis at an angular velocity ω_1. If the two strings are suddenly cut, while the system is rotating at this angular velocity, the four rods swing outward and assume positions parallel to the x-y plane. Determine the angular velocity ω_2 of the system when the four rods reach the horizontal position.

20.40 Repeat Problem 20.39 if only one of the two strings is cut while the other continues to hold two of the rods in their original positions.

20.41 A 2 ft × 4 ft homogeneous rectangular plate (shown in Figure P20.41) weighing 100 lb is moving downward with a velocity v_0 (no angular velocity) when the two supporting cords become taut simultaneously without causing a rebound of the plate. If the angular velocity of the plate immediately after the cords become taut is 2 rad/s, determine the magnitude of v_0.

20.42 A uniform cylinder of weight W and radius r is released from rest and rolls without slipping down the inclined plane as shown in Figure P20.42. It travels a

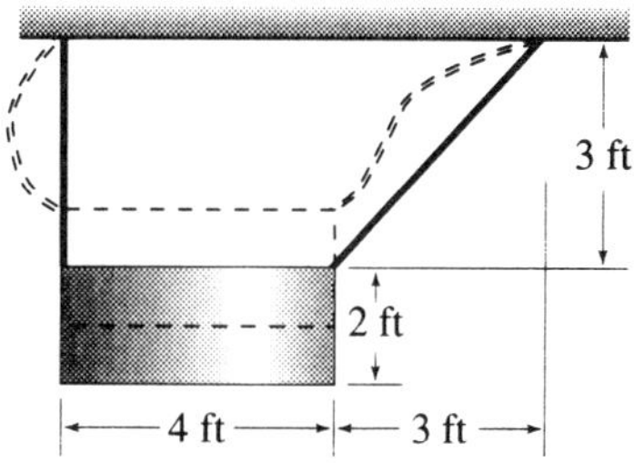

FIGURE P20.41.

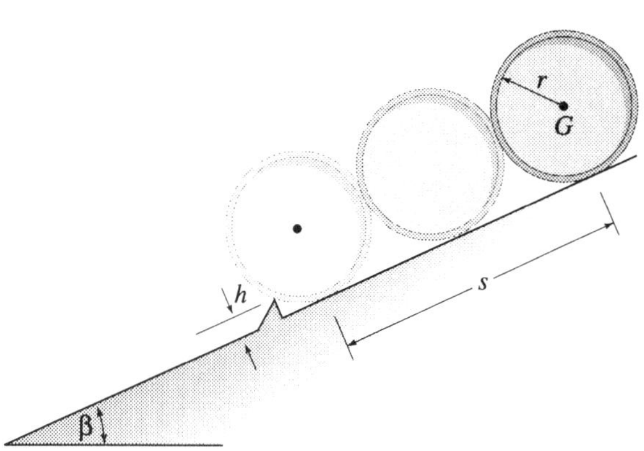

FIGURE P20.42.

distance s down the plane when it impacts an obstruction of height h without rebounding. Derive an expression in terms of r, h and β for the minimum distance s required for the wheel to just roll over the obstruction.

20.43 Plan and elevation views of a mechanism are shown in Figure P20.43. This mechanism consists of a disk A that rotates freely about a vertical axis through its mass center, point O, and two identical disks B mounted on vertical axes that pass through their respective mass centers, points G, and are fixed to and rotate with disk A. The weight of disk A is 200 N and that of each of the two disks B is 150 N. Ignore the weights of the three vertical shafts in the analysis. The two identical disks are retained in their slightly elevated positions and are set in motion about their respective axes at angular cw velocities of 50 rad/s. An internal mechanism requiring negligible energy releases both disks B from their elevated positions and they slide down their respective frictionless shafts, through a very small negligible height, and come to rest in contact with disk A simultaneously. Determine the angular velocity of the entire system immediately after the two disks B come to rest on disk A. How much kinetic energy is lost as a result of disks B coming to rest on disk A? Assume that disk A is initially at rest.

20.44 The homogeneous rod shown in Figure P20.44 has a weight W and is equipped with two pins at A (one on each side) which may be assumed weightless. The rod is dropping with a linear velocity v_0 (no angular velocity) when fixed hooks at B engage the pins at A forcing the rod to rotate cw. Assume that there is no rebound when engagement occurs. (a) In terms of v_0 and L, determine the angular velocity ω of the rod immediately after engagement with the hooks at B. How much kinetic energy is lost as a result of the engagement? (b) Find a numerical value for ω if $v_0 = 10$ ft/s and $L = 2$ ft.

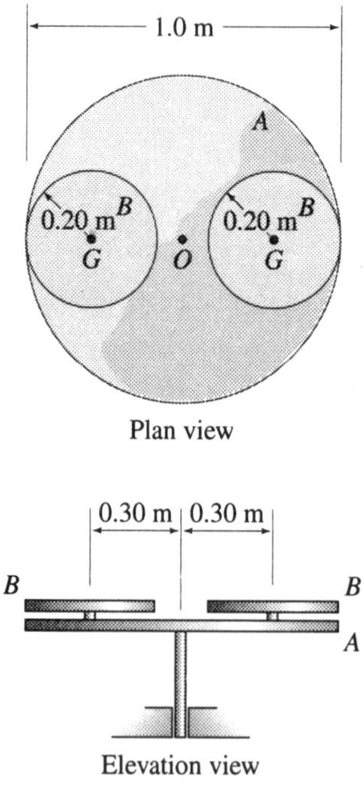

FIGURE P20.43.

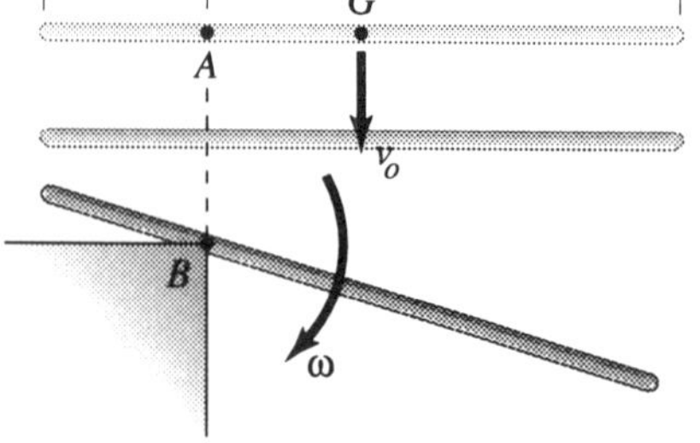

FIGURE P20.44.

20.45 A homogeneous rod of weight $W_R = 5$ lb is suspended from a frictionless hinge and is at rest in the vertical position when struck by a small sphere of weight $W_S = 0.5$ lb with a velocity of 25 ft/s as shown in Figure P20.45. Let the coefficient of restitution $e = 0.75$ and determine (a) the linear velocity of the sphere and the angular velocity of the rod immediately after impact and (b) the maximum angular rotation of the rod after impact.

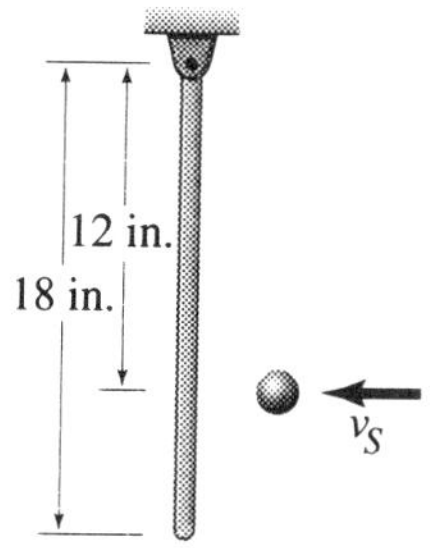

FIGURE P20.45.

20.46 A homogeneous rod of mass $m_R = 12$ kg is released from rest when $\theta = 0°$ and strikes a small ball of mass $m_B = 2$ kg when $\theta = 90°$ as shown in Figure P20.46. At the moment of impact, the ball is moving to the left at $v_B = 7$ m/s. Assume the coefficient of restitution $e = 0.5$, and determine the linear velocity of the ball and the angular velocity of the rod immediately after impact.

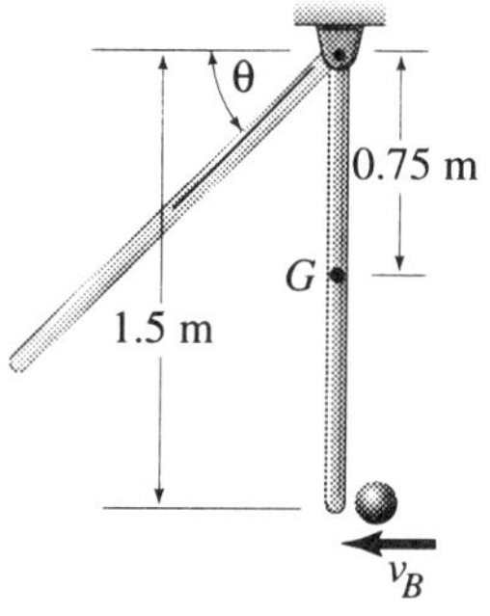

FIGURE P20.46.

20.47 The pendulum shown in Figure P20.47 consists of a 6-lb disk attached to a 1-lb rod. It is released from rest when $\theta = 0°$ and rotates cw until the disk strikes the smooth inclined plane at which time $\theta = 60°$. Determine the maximum angle of rebound measured from the inclined plane. Let $e = 0.75$.

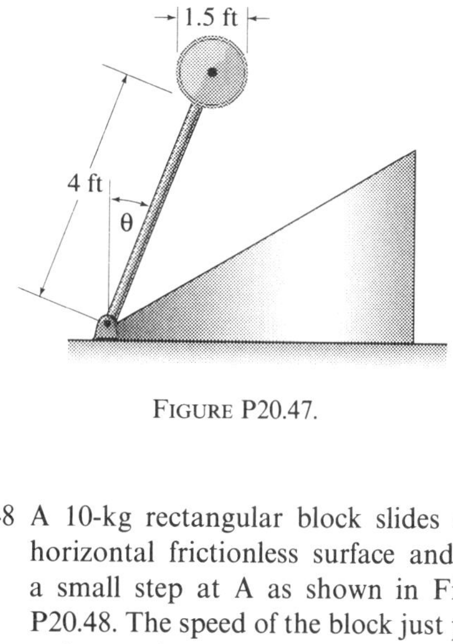

FIGURE P20.47.

20.48 A 10-kg rectangular block slides on a horizontal frictionless surface and hits a small step at A as shown in Figure P20.48. The speed of the block just prior to impact is 3 m/s to the left. Let $e = 0$

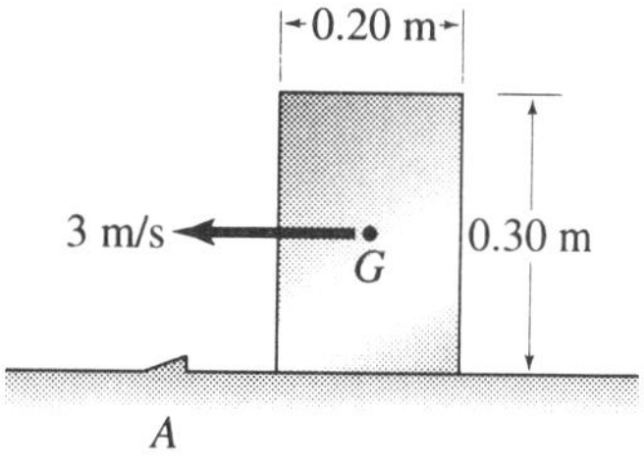

FIGURE P20.48.

(perfectly plastic impact). Determine the linear speed of point G and the angular velocity of the block immediately after impact.

20.49 Solve Problem 20.48 if the impact between the block and the step at A is perfectly elastic ($e = 1.0$).

20.50 A thin rectangular plate A weighing 10 lb is suspended from a frictionless hinge at O as shown in Figure P20.50. It is at rest in a vertical position when struck by block B weighing 4 lb sliding on a frictionless plane with a velocity of $v =$ 10 ft/s to the left. If the impact is perfectly plastic ($e = 0$), determine the angular velocity of the plate and the linear velocity of its mass center immediately after impact. What is the linear velocity of block B immediately after impact?

20.51 Repeat Problem 20.50 if the impact is perfectly elastic ($e = 1.0$).

20.52 The 5-kg homogeneous square plate shown in Figure P20.52 is dropping with a linear velocity of 8 m/s (without any angular velocity) when the supporting string becomes taut. If the impact is perfectly plastic ($e = 0$), determine the angular velocity of the plate and the linear velocity of its mass center immediately after impact.

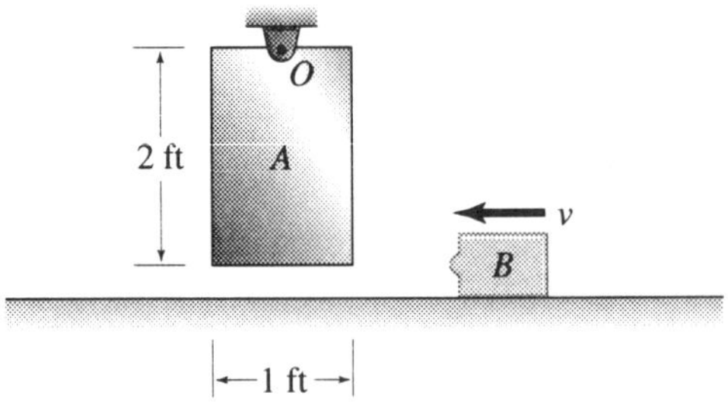

FIGURE P20.50.

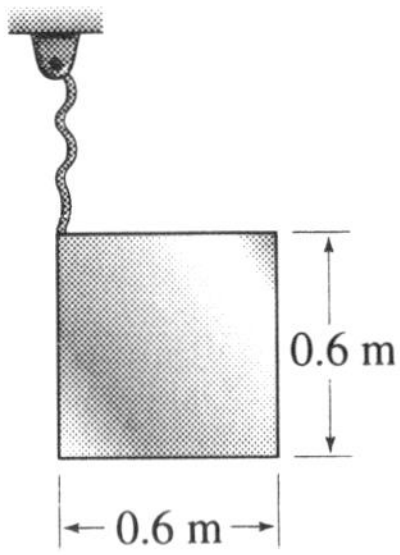

FIGURE P20.52.

Review Problems

20.53 A cylinder of weight $W = 20$ lb and radius $R = 3$ in., rotating with a cw angular velocity $\omega = 50$ rad/s, is carefully placed in the corner between two walls as shown in Figure P20.53. If the time it took the cylinder to come to a complete stop is $t = 1.5$ s, determine the coefficient of kinetic friction between the cylinder and the walls.

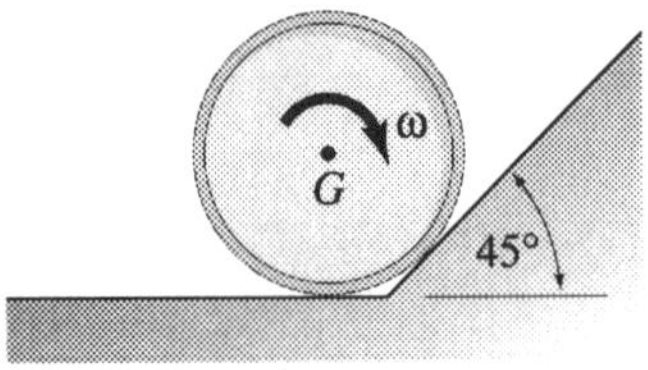

FIGURE P20.53.

20.54 The system shown in Figure P20.54 consists of gear A with an attached central pulley having a combined mass $m_A = 12$ kg and radius of gyration $k_A =$ 0.125 m and gear B with a mass $m_B = 8$ kg and radius of gyration $k_B = 0.200$ m. One end of a cord is wrapped around the pulley of gear A and the other to a weight $W = 60$ N. If the system is released from rest, determine (a) the angu-

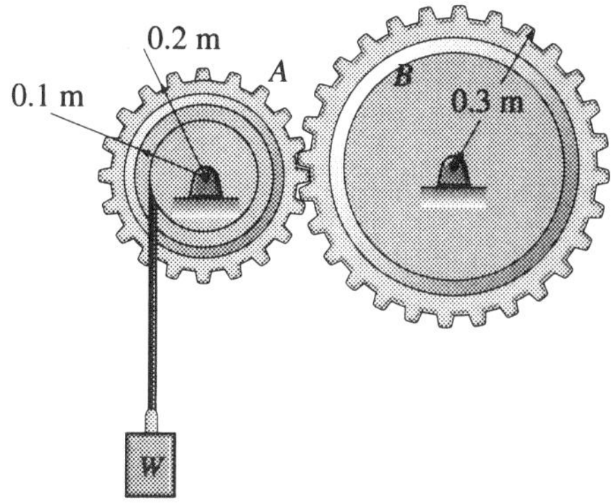

FIGURE P20.54.

lar speed of gear B after 2 s and (b) the magnitude of the tangential force that exists in the gear teeth at the point of contact between the two gears.

20.55 The brake system indicated in Figure P20.55 consists of drum A of weight W_A and radius b and the brake arm BCD of negligible weight and dimensions as shown. The coefficient of kinetic friction between the drum and the brake shoe is μ. The drum rotates at a constant cw angular velocity ω before the constant force P is applied. (a) In terms of W_A, P, μ, and ω, derive an expression for the dimension b needed to stop the drum in time t. (b) Find a numerical value for b for the specific case when $W_A = 40$ lb, $P = 10$ lb, $\mu = 0.4$, $\omega = 50$ rad/s and $t = 0.5$ s.

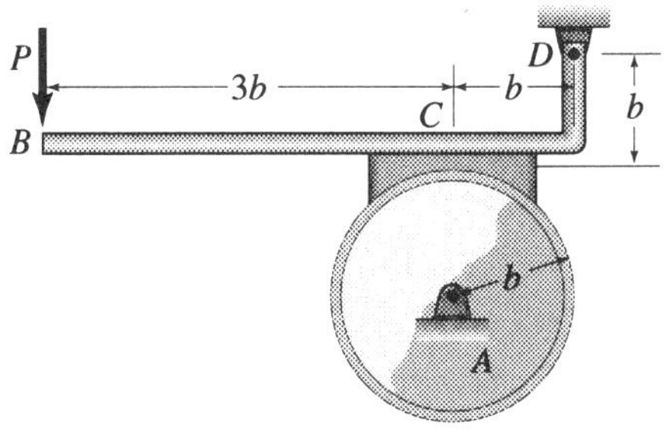

FIGURE P20.55.

20.56 The system shown in Figure P20.56 consists of two identical disks each of mass $m = 10$ kg and radius $R = 0.15$ m. If the system is released from rest, determine (a) the angular velocity of each disk 2 s later and (b) the tension in the inextensible cord connecting the two disks.

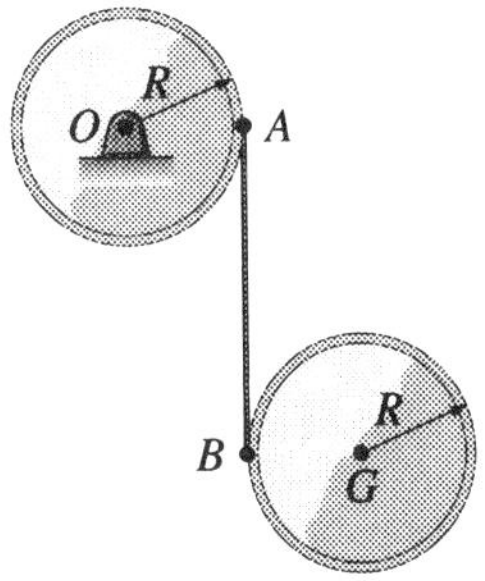

FIGURE P20.56.

20.57 A 2 ft × 4 ft rectangular plate weighing 10 lb is released from rest and allowed to translate vertically (without rotation) through a distance of 3 ft as shown in Figure P20.57, when the supporting string becomes taut. If the impact is perfectly plastic ($e = 0$), determine the angular velocity of the plate and the linear velocity of its mass center immediately after impact.

20.58 A homogeneous cylinder C of mass

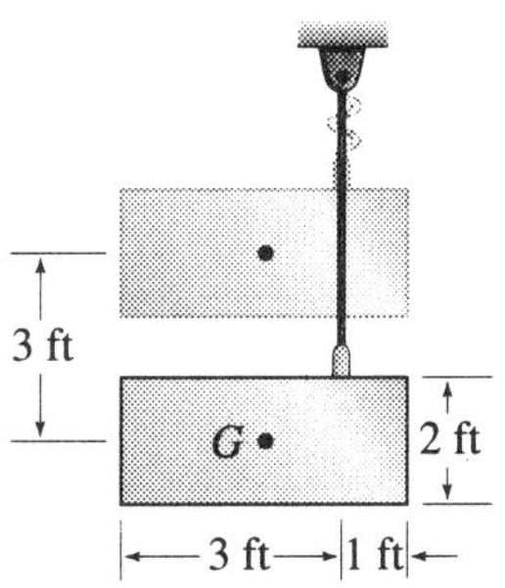

FIGURE P20.57.

$m_C = 5$ kg, radius $r = 0.15$ m and length $L = 1.80$ m is suspended from a frictionless hinge at O as shown in Figure P20.58. A bullet B of mass $m_B = 0.04$ kg is fired as shown at the cylinder which is at rest in the vertical position. The initial velocity of the bullet, which strikes and imbeds itself in the cylinder, is 500 m/s. Assume that the penetration of the bullet into the cylinder is negligibly small. Determine (a) the angular velocity of the cylinder immediately after the bullet is imbedded, (b) the corresponding velocity of point P, and (c) the reaction components at the hinge during impact if it takes 0.01 s for the bullet to imbed itself.

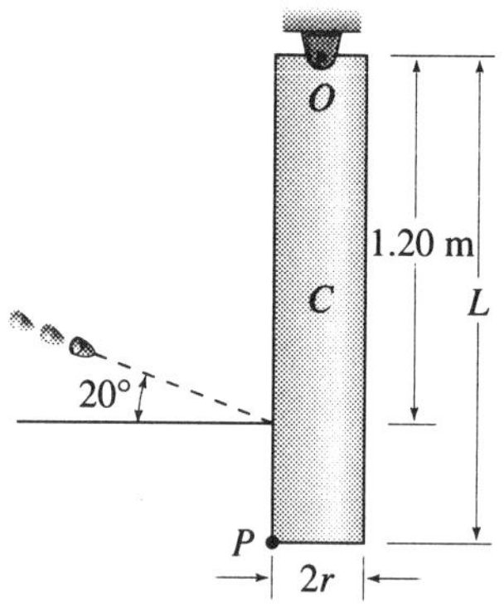

FIGURE P20.58.

20.59 Four uniform rods each of length $R = 0.75$ ft and mass m are attached symmetrically to a homogeneous circular plate of radius R and mass 4 m as shown in Figure P20.59. The system is set in motion on a frictionless horizontal plane so that it has a ccw angular velocity $\omega_1 = 20$ rad/s and its mass center has a velocity $v_1 = 10$ ft/s to the left. Determine its angular velocity and the velocity of its mass center immediately after one of the rods strikes the fixed peg at O without rebounding.

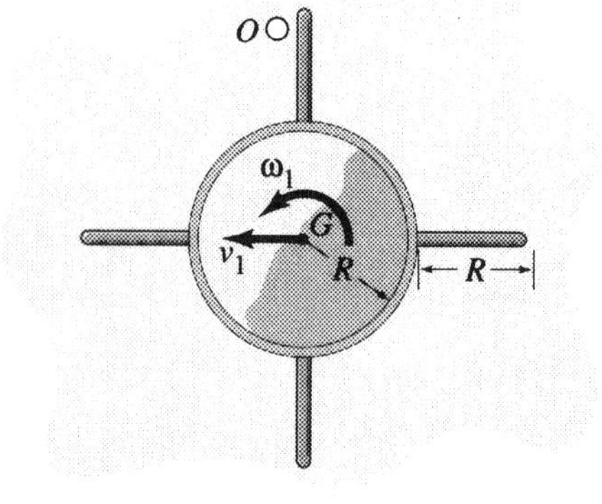

FIGURE P20.59.

20.60 Rod AB of mass $m_R = 3$ kg and length $L = 1.0$ m is released from rest, when in the horizontal position, and swings about the frictionless axis at A. As shown in Figure P20.60, when rod AB reaches the vertical position, its end B strikes point A of the 0.4 m × 1.20 m plate of mass $m_P = 6$ kg which is free to rotate about the smooth vertical shaft O–O. Let the coefficient of restitution be $e = 0.6$ and determine the angular velocities of the plate and the rod immediately after impact.

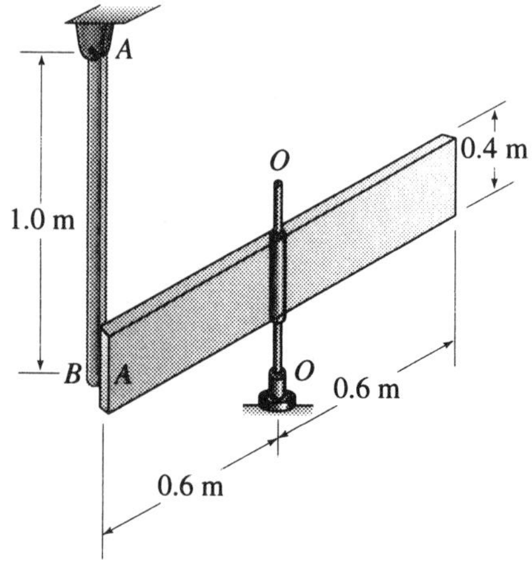

FIGURE P20.60.

21

Three-Dimensional Kinematics of Rigid Bodies

Picture of a radar station

Engineers are increasingly challenged to develop the technology necessary for faster, smaller, lighter, cheaper, and more efficient systems. Modern engineering systems, such as cars, airplanes, buildings, bridges, space stations, and others, require expertise offered by different disciplines and especially by mechanical, civil, and electrical engineering. With so many technical disciplines involved and shrinking resources, the solution is partnering for success. *Consequently, future engineers are faced with the daunting but exciting task of having to work as teams rather than individuals.*

As student, you are exposed to the latest technological advances and taught to compete as an individual. As an engineer, you must learn to compete in a new environment as a member of a team. These contrasting themes continue to embroil the engineering profession worldwide. The new climate reflects the sophistication and evolution of engineering technology. The need to excel and effectively compete in a new world economy makes it imperative that future engineers are properly equipped in the areas of analysis, design, and manufacturing. Furthermore, the engineer must possess the proper awareness of the need for teamwork and have the appropriate technical background needed to meet the challenges facing our society and others around the world.

The radar station shown on the cover page is one of several sophisticated engineering systems examined in this chapter. Although the basic theory is fundamentally understandable, you will learn that the applications are quite varied and arduous. Because of their increased uses and applications, the presentations of numerous examples and problems dealing with robots should be useful to engineers in all disciplines. We believe that the treatment of the principles of kinematics of rigid bodies will expand your horizon in Dynamics profoundly and substantively. This chapter will help you acquire the necessary knowledge needed to refine your interests and focus your energies in the engineering mechanics domain.

21.1 Motion about a Fixed Point

A first course in dynamics deals primarily with one-and two-dimensional motion because three-dimensional motion of rigid bodies is rather complex. In three-dimensional motion, points on the rigid body move in three-dimensional space. Also, both the magnitudes and directions of the angular velocity and angular acceleration vectors may change with time. Dealing with such changes by the scalar method becomes extremely difficult, if not impossible and, therefore, the vector method is generally used.

Before discussing the motion of a rigid body about a fixed point, let us examine some of the special characteristics of the motion of a rigid body in three-dimensional space. These special characteristics include a knowledge that finite rotations *are not vectors*, a familiarity with Euler's Theorem, and an understanding that infinitesimal rotations are indeed vectors despite the fact that finite rotations are not.

Finite Rotations are not Vectors

To show that finite rotations of a rigid body are not vector quantities, we consider the rectangular prism shown in Figure 21.1. The reader may wish to use a book to follow the steps and visualize the operations described here. For convenience, we assume that each component rotation is $\theta = 90^\circ$ and consistent with the right-hand rule. In Figure 21.1(a), the prism is first rotated 90° about the x axis which is then followed by a 90° rotation about the y axis. In Figure 21.1(b), the prism is first

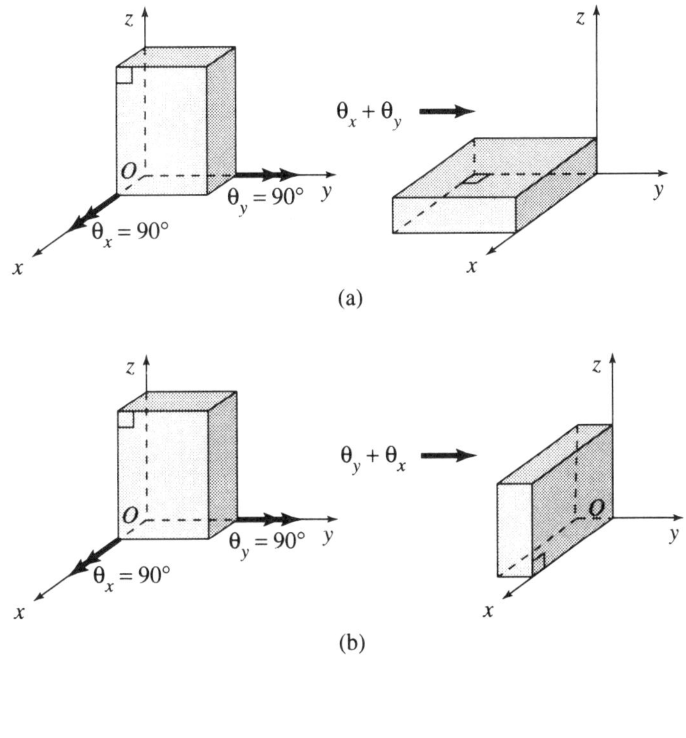

FIGURE 21.1.

rotated 90° about the y axis and then 90° about the x axis. The final positions in Figures 21.1(a) and 21.1(b) are obviously not the same. Therefore, we conclude that finite rotations *do not obey* the commutative law for vector addition (i.e., $\theta_x + \theta_y \neq \theta_y + \theta_x$) and thus, although they possess a magnitude and direction, finite rotations cannot be added by the law of the parallelogram and, consequently, *cannot* be classified as vectors. However, as will be shown later, infinitesimal rotations are in fact vector quantities unlike finite rotations.

Euler's theorem

Leonard Euler (1707–1783), a Swiss mathematician, made many contributions to mathematics, science, and engineering. He holds the all-time record for mathematical productivity, as he wrote 80 volumes of mathematics, many of enduring interest and usefulness. One of his lasting developments is a theorem that carries his name (*Euler's theorem*) which states that: *The most general displacement of a rigid body, one of whose points is fixed in space, is equivalent to a rotation of the body about an axis through this fixed point.* The proof of this theorem is developed as follows:

If a rigid body moves about a fixed point O, any particle, such as A, in the rigid body maintains a constant distance $r_{A/O}$ from the fixed point, and therefore, as shown in Figure 21.2, the path of particle A lies on a spherical surface of radius $r_{A/O}$ and center at O. The position of a rigid body is completely defined by specifying any three points in it. Thus, points O, A, and B define the initial position of the rigid body and points O, A′, and B′ define its final position. However, because the body is rigid, arc AB on the great circle of the spherical surface is equal to arc A′B′ of the same great circle. The next step in the development is

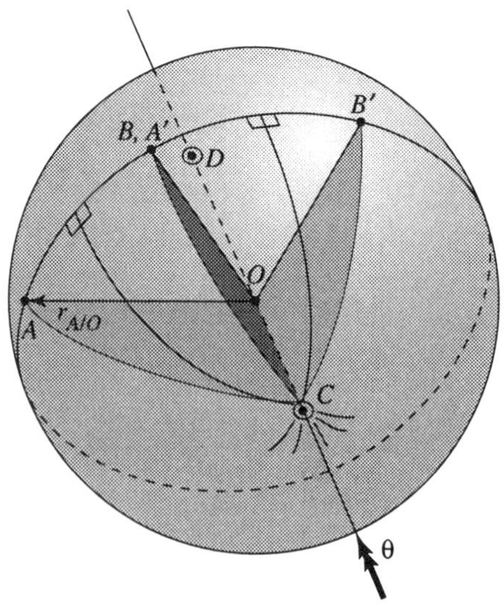

FIGURE 21.2.

to construct great-circle perpendicular bisectors to arcs AB and A′B′ which, if extended, will intersect at the two points C and D, where C is in front of the sphere and D is behind it, as shown in Figure 21.2. Let us focus our attention on point C, which we can clearly see, and construct the great circles AC, B(A′)C and B′C as shown. Since point C is equidistant from points A, B(A′) and B′ by construction, it follows that the two spherical triangles ACB(A′) and B(A′)CB′ are congruent and the spherical angles ACB(A′) and B(A′)CB′ are identical. Thus, if we designate these identical angles by the symbol θ, we conclude that the rigid body is moved from its initial position AB to its final position A′B′ by a rotation of magnitude θ about axis COD, as expressed by Euler's theorem.

Infinitesimal Rotations are Vectors

As stated earlier, although finite rotations of a rigid body cannot be classified as vectors, we will show that infinitesimal rotations are indeed vectors because they possess not only a magnitude and direction but may be added by the law of parallelograms. To this end, consider Figure 21.3. When a rigid body moves about a fixed point O, any particle, such as A, in the rigid body maintains a constant distance $r_{A/O}$ from the fixed point and, consequently, the path of the particle lies on a spherical surface of radius $r_{A/O}$. Consider two infinitesimal rotations $d\theta_1$ and $d\theta_2$ occurring about axes OD and OE, respectively. Let us assume that $d\theta_1$ takes place first followed by $d\theta_2$ in which case particle A would move through $d\mathbf{r}_1$ to position B_1 where $d\mathbf{r}_1 = d\theta_1 \times \mathbf{r}_{A/O}$, then through $d\mathbf{r}_2$ to position C_1 where $d\mathbf{r}_2 = d\theta_2 \times \mathbf{r}_{A/O}$. On the other hand,

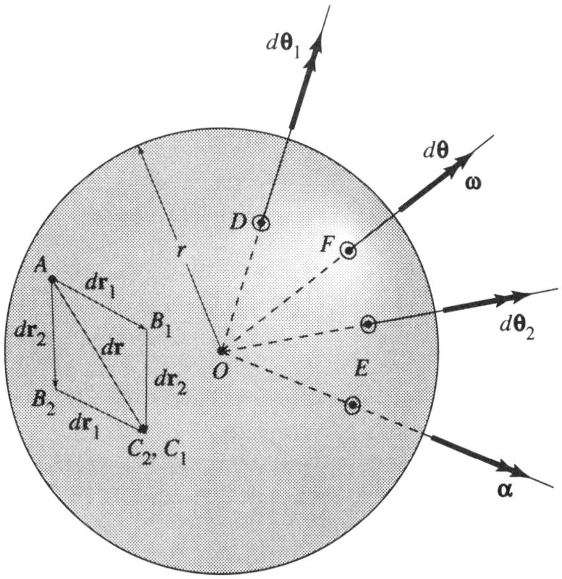

FIGURE 21.3.

if we assume that $d\theta_2$ occurs first, followed by $d\theta_1$, then, particle A would move through $d\mathbf{r}_2 = d\theta_2 \times \mathbf{r}_{A/O}$ to position B_2 and, then, through $d\mathbf{r}_1 = d\theta_1 \times \mathbf{r}_{A/O}$ to position C_2. However, because we are dealing with infinitesimal rotations, the curvature of the spherical surface may be ignored. Thus, the infinitesimal vectors $d\mathbf{r}_1$ and $d\mathbf{r}_2$ may be assumed to occur on a flat surface, in which case C_2 coincides with C_1 as shown in Figure 21.3. Therefore, the sequence (i.e., $d\theta_1 + d\theta_2$ vs. $d\theta_2 + d\theta_1$) is inconsequential and the resultant displacement $d\mathbf{r}$ may be obtained in one of two identical ways. Thus,

$$\begin{aligned} d\mathbf{r} &= d\mathbf{r}_1 + d\mathbf{r}_2 = d\theta_1 \times \mathbf{r}_{A/O} + d\theta_2 \times \mathbf{r}_{A/O} \\ &= (d\theta_1 + d\theta_2) \times \mathbf{r}_{A/O} = d\theta \times \mathbf{r}_{A/O}, \end{aligned}$$

or

$$\begin{aligned} d\mathbf{r} &= d\mathbf{r}_2 + d\mathbf{r}_1 = d\theta_2 \times \mathbf{r}_{A/O} + d\theta_1 \times \mathbf{r}_{A/O} \\ &= (d\theta_2 + d\theta_1) \times \mathbf{r}_{A/O} = d\theta \times \mathbf{r}_{A/O} \end{aligned}$$

where

$$d\theta_1 + d\theta_2 = d\theta_2 + d\theta_1 = d\theta.$$

Thus, we conclude that the two separate rotations ($d\theta_1 + d\theta_2$ or $d\theta_2 + d\theta_1$) may be replaced by a single rotation $d\theta$ about axis OF as shown in Figure 21.3 and that infinitesimal rotations are vector quantities because they have magnitudes and directions and they obey the distributive law of the parallelogram.

Angular Velocity

When a rigid body moving about a fixed point experiences an infinitesimal angular displacement $d\theta$ as shown in Figure 21.3, we may obtain its angular velocity $\boldsymbol{\omega}$ by differentiating the vector $d\theta$ with respect to time. Thus,

$$\boldsymbol{\omega} = \frac{d\theta}{dt} = \dot{\theta}. \tag{21.1}$$

Because $d\theta$ was shown to be a vector quantity, it follows that $\boldsymbol{\omega}$ is also a vector quantity. Obviously, then, the direction of $\boldsymbol{\omega}$ is the same as the direction of $d\theta$ and in Figure 21.3, it is along the axis OF. This axis is referred to as the *instantaneous axis of rotation.* In general, $\boldsymbol{\omega}$ varies with time not only in magnitude but also in direction, and, therefore, the instantaneous axis of rotation is not fixed in space but changes direction with time.

Angular Acceleration

The angular acceleration of a rigid body moving about a fixed point may be obtained by differentiating its angular velocity $\boldsymbol{\omega}$ with respect to time. Thus,

$$\boldsymbol{\alpha} = \frac{d\boldsymbol{\omega}}{dt} = \ddot{\theta}. \tag{21.2}$$

In general, since $\boldsymbol{\omega}$ changes in magnitude and direction with time, it follows that $\boldsymbol{\alpha}$ will also change with time in magnitude and direction. Thus, in general, as shown in Figure 21.3, the direction of $\boldsymbol{\alpha}$ will also change in magnitude and direction with time. Thus, in general, as shown in Figure 21.3, the direction of $\boldsymbol{\alpha}$ does not coincide with the direction of $\boldsymbol{\omega}$.

As the instantaneous axis of rotation (or the vector $\boldsymbol{\omega}$) moves with the passage of time, it generates a conical surface. When the motion of the instantaneous axis of rotation is considered by an observer on the rigid body, the conical surface it generates is known as the *body cone*, and, when its motion is considered by an observer outside the body (in space), the generated conical surface is known as the *space cone* as shown in Figure 21.4. For a given position of the space observer, the space cone is fixed in space. As shown in Figure 21.4, at a given instant, the two conical surfaces share the instantaneous axis of rotation (or the vector $\boldsymbol{\omega}$) as a common tangent and the body cone appears to rotate either on the outside surface of the space cone (Fig. 21.4(a)) or on its inside surface (Fig. 21.4(b)). Examples of this phenomenon will be presented in Chapter 22 where gyroscopic motion of a rigid body is discussed. Because $\boldsymbol{\alpha}$ is the time derivative of $\boldsymbol{\omega}$, it also follows that the vector $\boldsymbol{\alpha}$ is tangent to the contour traced by the head of the vector $\boldsymbol{\omega}$ as shown in Figure 21.4.

Linear Velocity

If the displacement $d\mathbf{r}$ of any point A in the rigid body of Figure 21.3 takes place during a time interval dt, then, as was shown in Section 17.2, the linear velocity $\mathbf{v}$ of any point A becomes

$$\mathbf{v} = \frac{d\mathbf{r}}{dt} = \boldsymbol{\omega} \times \mathbf{r}_{A/O}. \tag{21.3}$$

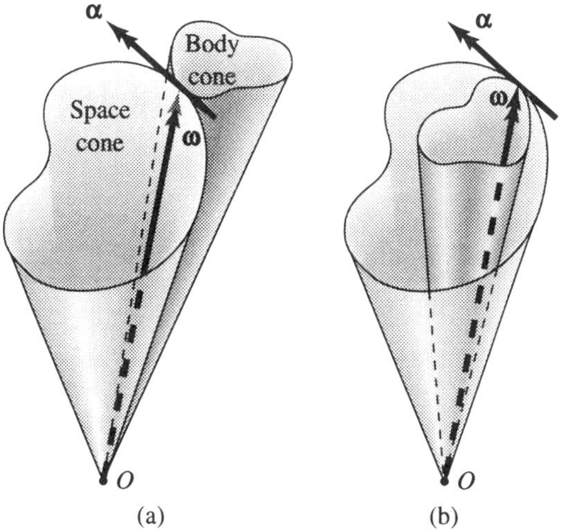

FIGURE 21.4.

where $\mathbf{r}_{A/O}$ is the position vector locating point A relative to the fixed point O on the rigid body.

Linear Acceleration

The linear acceleration $\mathbf{a}$ of any point A in the rigid body of Figure 21.3 may now be obtained by differentiating Eq. (21.3) with respect to time. Thus, as in Section 17.2,

$$\mathbf{a} = \frac{d\mathbf{v}}{dt} = \boldsymbol{\omega} \times \mathbf{v} + \boldsymbol{\alpha} \times \mathbf{r}_{A/O} \tag{21.4}$$

where $\boldsymbol{\omega}$, $\boldsymbol{\alpha}$ and $\mathbf{v}$ are defined by Eqs. (21.1), (21.2), and (21.3), respectively.

21.2 General 3-D Motion (Translating, Nonrotating Axes)

In Chapter 17, we concluded that the general two-dimensional motion of a rigid body may be looked upon as consisting of a translation added vectorially to a rotation. This same point of view may be adopted in discussing the general three-dimensional motion of a rigid body. Thus, consider the rigid body shown in Figure 21.5 which is executing general motion in space with an angular velocity $\boldsymbol{\omega}$ and an angular acceleration $\boldsymbol{\alpha}$. Any two points A and B are selected in the rigid body. The X-Y-Z coordinate system is fixed in space and point B is chosen as the origin for the translating, nonrotating x-y-z coordinate system.* As shown in Figure 21.5, $\mathbf{r}_A$ and $\mathbf{r}_B$ are the position vectors of points A and B with respect to the fixed origin, point O, and $\mathbf{r}_{A/B}$ represents the position vector of point A relative to the translating origin, point B. Thus,

$$\mathbf{r}_A = \mathbf{r}_B + \mathbf{r}_{A/B}. \tag{21.5}$$

Differentiation of Eq. (21.5) with respect to time yields

$$\mathbf{v}_A = \mathbf{v}_B + \mathbf{v}_{A/B} \tag{21.6a}$$

where $\mathbf{v}_A$ and $\mathbf{v}_B$ are the velocities of points A and B with respect to the fixed X-Y-Z system and, therefore, represent absolute velocities of these two points. *However, the quantity* $\mathbf{v}_{A/B}$ *represents the velocity* of point A relative to the translating $\mathbf{x}$-y-z system. Because points A and B are two points on the same rigid body, the magnitude of $\mathbf{r}_{A/B}$ is constant and, to an observer stationed at B, point A appears to be executing motion about a fixed point and to move on a spherical surface whose center is at B and whose radius $\mathbf{r}_{A/B}$ is constant. Thus, by

---

*The solution of some problems is simplified by considering a translating, rotating coordinate system instead of a translating, nonrotating system as is done in this section. The case of a translating and rotating coordinate system is dealt with in Section 21.4.

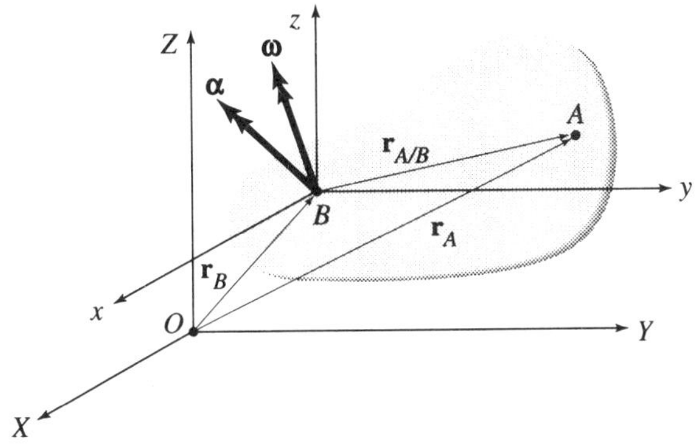

FIGURE 21.5.

Eq. (21.3), $\mathbf{v}_{A/B} = \boldsymbol{\omega} \times \mathbf{r}_{A/B}$ and

$$\mathbf{v}_A = \mathbf{v}_B + \boldsymbol{\omega} \times \mathbf{r}_{A/B}. \tag{21.6b}$$

Therefore, we conclude that the velocity of any point A in a rigid body executing general motion in space consists of a translation component $\mathbf{v}_B$ plus a rotation component $\boldsymbol{\omega} \times \mathbf{r}_{A/B}$.

Similarly, by taking the time derivative of Eqs. (21.6a), we conclude that

$$\mathbf{a}_A = \mathbf{a}_B + \mathbf{a}_{A/B}. \tag{21.7a}$$

By Eqs. (21.4) and (21.3), $\mathbf{a}_{A/B} = \boldsymbol{\omega} \times \mathbf{v}_{A/B} + \boldsymbol{\alpha} \times \mathbf{r}_{A/B} = \boldsymbol{\omega} \times (\boldsymbol{\omega} \times \mathbf{r}_{A/B}) + \boldsymbol{\alpha} \times \mathbf{r}_{A/B}$. It follows, therefore, that

$$\mathbf{a}_A = \mathbf{a}_B + \boldsymbol{\omega} \times (\boldsymbol{\omega} \times \mathbf{r}_{A/B}) + \boldsymbol{\alpha} \times \mathbf{r}_{A/B}. \tag{21.7b}$$

Equations (21.7) show that the acceleration of any point A in a rigid body undergoing general motion in space consists of a translation component $\mathbf{a}_B$ plus a rotation component $\boldsymbol{\omega} \times (\boldsymbol{\omega} \times \mathbf{r}_{A/B}) + \boldsymbol{\alpha} \times \mathbf{r}_{A/B}$.

The following examples illustrate some of the concepts discussed above.

Example 21.1

The robotics assembly shown in Figure E21.1 rotates about the vertical z axis at a constant angular velocity $\omega_1 = 5$ rad/s while the arm OP rotates about the horizontal x axis at a constant angular velocity $\omega_2 = 8$ rad/s. At the instant depicted, arm OP is positioned by the direction angles $\theta_x = 75^\circ$ and $\theta_z = 60^\circ$. The length of arm OP is $L = 20$ in. Determine the linear velocity of point P at the instant depicted in Figure E21.1.

Solution

The resultant angular velocity $\boldsymbol{\omega}$ of arm OP is given by

$$\boldsymbol{\omega} = \omega_x\mathbf{i} + \omega_y\mathbf{j} + \omega_z\mathbf{k} = 8\mathbf{i} + 5\mathbf{k}.$$

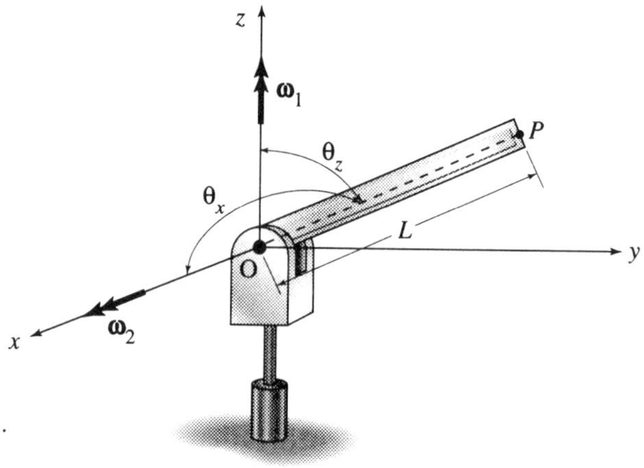

FIGURE E21.1.

Now

$$\cos^2\theta_x + \cos^2\theta_y + \cos^2\theta = 1$$

$$\cos^2 75^\circ + \cos^2\theta_y + \cos^2 60^\circ = 1.$$

Therefore

$$\theta_y = 34.3^\circ.$$

Thus,

$$\mathbf{r}_{P/O} = (20\cos 75^\circ)\mathbf{i} + (20\cos 34.3^\circ)\mathbf{j} + (20\cos 60^\circ)\mathbf{k}$$
$$= 5.176\mathbf{i} + 16.522\mathbf{j} + 10.000\mathbf{k}.$$

$$\mathbf{v}_P = \mathbf{v}_O + \boldsymbol{\omega}\times\mathbf{r}_{P/O} = \mathbf{0} + \begin{bmatrix} \mathbf{i} & \mathbf{j} & \mathbf{k} \\ 8 & 0 & 5 \\ 5.176 & 16.522 & 10.000 \end{bmatrix},$$

$$= (-82.6\mathbf{i} - 54.1\mathbf{j} + 132.2\mathbf{k})\ \text{in./s.} \qquad \text{ANS.}$$

■ Example 21.2

The mechanism shown in Figure E21.2 consists of a disk rotating in the y-z plane at a constant angular velocity $\omega_D = 30$ rad/s about a shaft parallel to the x axis, a collar A that slides freely on a rod that lies along the x axis and rod AB of length $L = 1.4$ m that is connected to the disk and collar by ball-and-socket joints. Determine, for the position shown, (a) the absolute velocity of the collar and (b) the angular velocity of rod AB which, at the instant considered, is spinning about its own axis at the constant rate $\omega_S = 10$ rad/s as shown.

Solution

(a)

$$\mathbf{v}_B = \boldsymbol{\omega}_D \times \mathbf{r}_{B/C} = 30\mathbf{i}\times 0.4\mathbf{k} = -12\mathbf{j}\ \text{m/s},$$

$$\mathbf{v}_A = v_A\mathbf{i}.$$

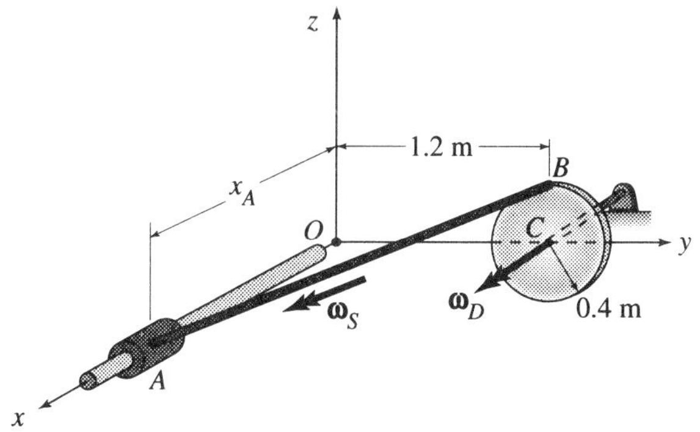

FIGURE E21.2.

From the given geometry and the length of rod AB,

$$(1.4)^2 = x_A^2 + (1.2)^2 + (0.4)^2$$

from which

$$x_A = 0.6 \text{ m}.$$

The position of A relative to B is given by

$$\mathbf{r}_{A/B} = 0.6\mathbf{i} - 1.2\mathbf{j} - 0.4\mathbf{k}.$$

The angular velocity $\boldsymbol{\omega}$ of rod AB may be expressed in terms of its x, y, and z components by

$$\boldsymbol{\omega} = \omega_x\mathbf{i} + \omega_y\mathbf{j} + \omega_z\mathbf{k}.$$

Now, by Eq. (21.6b),

$$\mathbf{v}_A = \mathbf{v}_B + \boldsymbol{\omega} \times \mathbf{r}_{A/B}.$$

Therefore,

$$\begin{aligned} v_A\mathbf{i} &= -12\mathbf{j} + \begin{bmatrix} \mathbf{i} & \mathbf{j} & \mathbf{k} \\ \omega_x & \omega_y & \omega_z \\ 0.6 & -1.2 & -0.4 \end{bmatrix}, \\ &= (1.2\omega_z - 0.4\omega_y)\mathbf{i} - (12 - 0.4\omega_x - 0.6\omega_z)\mathbf{j} - (1.2\omega_x + 0.6\omega_y)\mathbf{k} \end{aligned}$$

from which, by equating the coefficients of the **i**, **j**, and **k** terms on both sides of the equation, we obtain

$$\mathbf{i}: \ 1.2\omega_z - 0.4\omega_y = v_A, \qquad \text{(a)}$$

$$\mathbf{j}: \ 0.4\omega_x + 0.6\omega_z = 12, \qquad \text{(b)}$$

and

$$\mathbf{k}: \ 1.2\omega_x + 0.6\omega_y = 0. \qquad \text{(c)}$$

Multiply Eq. (b) by -2, divide Eq. (c) by 1.5, and add the three equations to obtain

$$v_A - 24 = 0 \quad \text{which implies that} \quad v_A = 24 \text{ m/s}.$$

Therefore,

$$\mathbf{v}_A = (24.0\mathbf{i}) \text{ m/s}. \qquad \text{ANS.}$$

(b) Equations (a), (b), and (c) cannot be solved for the components ω_x, ω_y, and ω_z because the determinant of their coefficients vanishes. Thus, we need to supplement them with the given information relative to the spin of rod AB which provides an additional equation that allows the determination of ω_x, ω_y, and ω_z. This information indicates that the component of $\boldsymbol{\omega}$ along axis BA is 10 rad/s. Thus,

$$\begin{aligned} 10 &= \boldsymbol{\omega}\cdot\boldsymbol{\lambda}_{BA} = \boldsymbol{\omega}\cdot\left(\frac{\mathbf{r}_{B/A}}{r_{B/A}}\right) \\ &= (\omega_x\mathbf{i} + \omega_y\mathbf{j} + \omega_z\mathbf{k})\cdot\left(\frac{-0.6\mathbf{i} + 1.2\mathbf{j} + 0.4\mathbf{k}}{1.4}\right) \\ &= -0.429\omega_x + 0.857\omega_y + 0.286\omega_z. \qquad \text{(d)} \end{aligned}$$

Any two of Eqs. (a), (b), and (c) plus Eq. (d) may be solved simultaneously to obtain

$$\omega_x = -1.834 \text{ rad/s}, \omega_y = 3.67 \text{ rad/s}, \quad \text{and} \quad \omega_z = 21.2 \text{ rad/s}.$$

Therefore,

$$\boldsymbol{\omega} = (-1.834\mathbf{i} + 3.67\mathbf{j} + 21.2\mathbf{k}) \text{ rad/s}. \qquad \text{ANS.}$$

Example 21.3

Refer to the mechanism of Example 21.2 and, for the position shown in Figure E21.3, determine the linear acceleration of collar A and (b) the

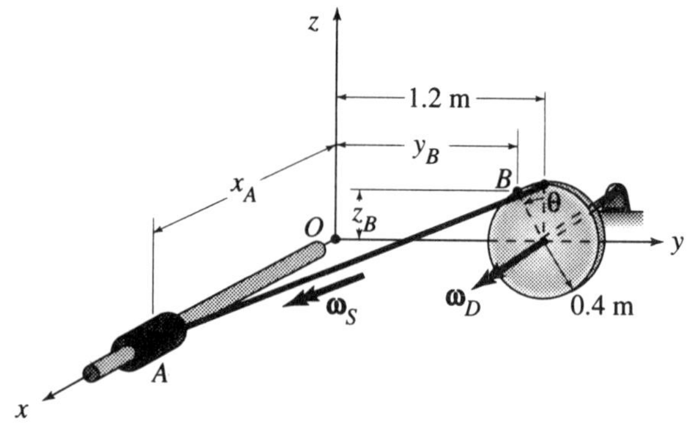

FIGURE E21.3.

angular acceleration of rod AB. Recall that rod AB is spinning about its own axis at the constant rate $\omega_S = 10$ rad/s as shown.

Solution

(a) Consider Figure E21.3 depicting any position of joint B where $y_B = 1.2 - 0.4\sin\theta$ and $z_B = 0.4 - 0.4\cos\theta$. Thus, because the fixed length of rod AB is 1.4 m, it follows that

$$\begin{aligned} 1.4^2 &= x_A^2 + y_B^2 + z_B^2 \\ &= x_A^2 + (1.2 - 0.4\sin\theta)^2 + (0.4 - 0.4\cos\theta)^2. \end{aligned} \tag{a}$$

Differentiating Eq. (a) with respect to time yields

$$0 = 2x_A\dot{x}_A - 0.96\dot{\theta}\cos\theta + 0.32\dot{\theta}\sin\theta. \tag{b}$$

In Eq. (b), $x_A = 0.6$ m (from Example 21.2), $\dot{\theta} = \omega_D = 30$ rad/s, and $\theta = 0$. Solving for $\dot{x}_A$, we obtain

$$v_A = \dot{x}_A = 24 \text{ m/s}$$

which agrees with the answer obtained in Example 21.2.

Differentiating Eq. (b) with respect to time yields

$$\begin{aligned} 0 &= 2x_A\ddot{x}_A + 2\dot{x}_A^2 + 0.96\dot{\theta}^2\sin\theta + 0.32\dot{\theta}^2\cos\theta \\ &\quad - 0.96\ddot{\theta}\cos\theta + 0.32\ddot{\theta}\sin\theta. \end{aligned} \tag{c}$$

In Eq. (c), $x_A = 0.6$ m, $\dot{x}_A = 24$ m/s, $\dot{\theta} = 30$ rad/s, $\theta = 0$, and $\ddot{\theta} = 0$ (because $\dot{\theta} = \omega_D =$ constant.). Substituting and solving for $\ddot{x}_A$ yields

$$a_A = \ddot{x}_A = -1200 \text{ m/s}^2.$$

Therefore,

$$\mathbf{a}_A = -(1200\mathbf{i}) \text{ m/s}^2. \qquad \text{ANS.}$$

(b) By Eq. (21.7b),

$$\mathbf{a}_A = \mathbf{a}_B + \boldsymbol{\omega} \times (\boldsymbol{\omega} \times \mathbf{r}_{A/B}) + \boldsymbol{\alpha} \times \mathbf{r}_{A/B}$$

where

$$\mathbf{a}_A = -1200\mathbf{i},$$

$$\begin{aligned} \mathbf{a}_B &= \mathbf{a}_C + \boldsymbol{\omega}_D \times (\boldsymbol{\omega}_D \times \mathbf{r}_{B/C}) + \boldsymbol{\alpha}_D \times \mathbf{r}_{B/C}, \\ &= \mathbf{0} + 30\mathbf{i} \times (30\mathbf{i} \times 0.4\mathbf{k}) + \mathbf{0} \\ &= -360\mathbf{k}. \end{aligned}$$

$$\boldsymbol{\omega} = -1.834\mathbf{i} + 3.67\mathbf{j} + 21.2\mathbf{k} \quad \text{(from Example 21.2)},$$

$$\mathbf{r}_{A/B} = 0.6\mathbf{i} - 1.2\mathbf{j} - 0.4\mathbf{k} \quad \text{(from Example 21.2)},$$

$$\boldsymbol{\omega} \times \mathbf{r}_{A/B} = \mathbf{v}_{A/B} = \mathbf{v}_A - \mathbf{v}_B = 24\mathbf{i} + 12\mathbf{j},$$

and

$$\boldsymbol{\alpha} = \alpha_x\mathbf{i} + \alpha_y\mathbf{j} + \alpha_z\mathbf{k}.$$

Therefore,

$$-1200\mathbf{i} = -360\mathbf{k} + \begin{bmatrix} \mathbf{i} & \mathbf{j} & \mathbf{k} \\ -1.834 & 3.67 & 21.2 \\ 24 & 12 & 0 \end{bmatrix} + \begin{bmatrix} \mathbf{i} & \mathbf{j} & \mathbf{k} \\ \alpha_x & \alpha_y & \alpha_z \\ 0.6 & -1.2 & -0.4 \end{bmatrix},$$

$$= (-254.4 - 0.4\alpha_y + 1.2\alpha_z)\mathbf{i} - (-508.8 - 0.4\alpha_x - 0.6\alpha_z)\mathbf{j}$$
$$+ (-110.1 - 1.2\alpha_x - 0.6\alpha_y)\mathbf{k}. \quad \text{(d)}$$

Equating the coefficients of the **i**, **j**, and **k** terms on both sides of Eq. (d), we obtain

$$\mathbf{i}: \quad -0.4\alpha_y + 1.2\alpha_z = -945.6, \quad \text{(e)}$$

$$\mathbf{j}: \quad 0.4\alpha_x + 0.6\alpha_z = -508.8, \quad \text{(f)}$$

and

$$\mathbf{k}: \quad 1.2\alpha_x + 0.6\alpha_y = -470.1. \quad \text{(g)}$$

Equations (e), (f), and (g) cannot be solved for the components α_x, α_y and α_z because the determinant of their coefficients vanishes. Thus, we need to supplement these equations with the given information relative to the spin of rod AB. Because the spin of rod AB is constant, the component of $\boldsymbol{\alpha}$ along rod AB must vanish. Thus,

$$0 = \boldsymbol{\alpha} \cdot \boldsymbol{\lambda}_{AB} = \boldsymbol{\alpha} \cdot \left(\frac{\mathbf{r}_{A/B}}{r_{A/B}}\right)$$

$$= (\alpha_x\mathbf{i} + \alpha_y\mathbf{j} + \alpha_z\mathbf{k}) \cdot \left(\frac{0.6\mathbf{i} - 1.2\mathbf{j} - 0.4\mathbf{k}}{1.4}\right)$$

$$= 0.429\alpha_x - 0.857\alpha_y - 0.286\alpha_z. \quad \text{(h)}$$

Any two of Eqs. (e)–(g) plus Eq. (h) may be solved simultaneously to obtain

$\alpha_x = -734.6\ \text{rad/s}^2$, $\alpha_y = 160.1\ \text{rad/s}^2$, and $\alpha_z = -170.0\ \text{rad/s}^2$.

Therefore,

$$\boldsymbol{\alpha} = (-735\mathbf{i} + 160.1\mathbf{j} - 170.0\mathbf{k})\ \text{rad/s}^2. \quad \text{ANS.}$$

21.3 Time Derivative of a Vector with Respect to Rotating Axes

In dealing with the three-dimensional motion of a rigid body, it is convenient to attach to the body an x-y-z coordinate system which rotates with the body. Under these conditions, the angular velocity $\boldsymbol{\omega}$ of the body would be expressed in terms of its three components ω_x, ω_y, and ω_z, and the angular acceleration $\boldsymbol{\alpha}$ of the body would be obtained as the time derivative of the angular velocity $\boldsymbol{\omega}$. To accomplish this task, we need to be able to compute the time derivative of a vector, such as $\boldsymbol{\omega}$, with respect to a rotating set of axis.

Consider, for example, the rigid body shown in Figure 21.6 which is executing general motion in space. Two coordinate systems are attached to some arbitrary point O in the body. The first system X-Y-Z

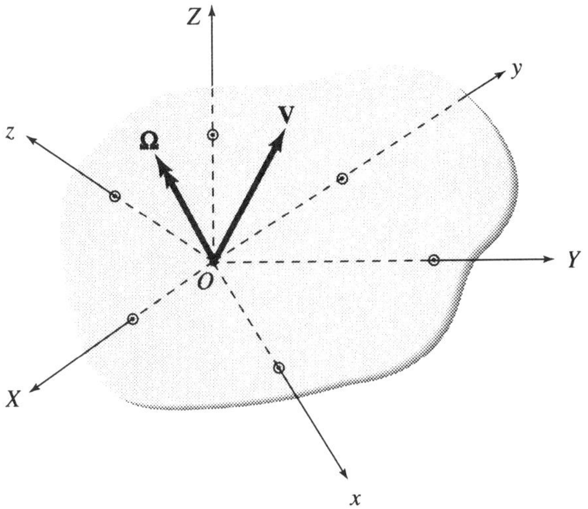

FIGURE 21.6.

is fixed in orientation but may translate (without rotation) with the rigid body. The second system x-y-z not only may translate with the rigid body but also rotates with an angular velocity $\mathbf{\Omega}$ relative to the X-Y-Z coordinate system. Note that the angular velocity $\mathbf{\Omega}$ may or may not be the same as the angular velocity $\boldsymbol{\omega}$ of the rigid body. Note also that the unit vectors corresponding to the rotating x-y-z coordinate system are $\mathbf{i}$, $\mathbf{j}$, and $\mathbf{k}$, respectively, whereas those corresponding to the X-Y-Z system are $\mathbf{I}$, $\mathbf{J}$, and $\mathbf{K}$.

Let us now consider the time derivative of any vector $\mathbf{V} = V_x\mathbf{i} + V_y\mathbf{j} + V_z\mathbf{k}$ with respect to the rotating x-y-z coordinate system. Because the unit vectors $\mathbf{i}$, $\mathbf{j}$, and $\mathbf{k}$ rotate with the x-y-z system and have constant magnitudes, to an observer rotating with this coordinate system, these unit vectors do not change in magnitude or direction. Thus,

$$(\dot{\mathbf{V}})_{xyz} = \dot{V}_x\mathbf{i} + \dot{V}_y\mathbf{j} + \dot{V}_z\mathbf{k}.$$

However, to an observer using the nonrotating X-Y-Z system, the unit vectors $\mathbf{i}$, $\mathbf{j}$, and $\mathbf{k}$ will change direction with time, and

$$\begin{aligned}(\dot{\mathbf{V}})_{XYZ} &= \dot{V}_x\mathbf{i} + \dot{V}_y\mathbf{j} + \dot{V}_z\mathbf{k} + V_x\frac{d\mathbf{i}}{dt} + V_y\frac{d\mathbf{j}}{dt} + V_z\frac{d\mathbf{k}}{dt}\\ &= (\dot{\mathbf{V}})_{xyz} + V_x\frac{d\mathbf{i}}{dt} + V_y\frac{d\mathbf{j}}{dt} + V_z\frac{d\mathbf{k}}{dt}.\end{aligned}$$

Using the procedure followed in Section 17.7 (see Eqs. 17.32), we can show that

$$\frac{d\mathbf{i}}{dt} = \mathbf{\Omega} \times \mathbf{i}, \quad \frac{d\mathbf{j}}{dt} = \mathbf{\Omega} \times \mathbf{j}, \quad \text{and} \quad \frac{d\mathbf{k}}{dt} = \mathbf{\Omega} \times \mathbf{k}.$$

Therefore, the last three terms in the equation for $(\dot{\mathbf{V}})_{XYZ}$ may be replaced by $\mathbf{\Omega} \times \mathbf{V}$. Thus,

$$(\dot{\mathbf{V}})_{XYZ} = (\dot{\mathbf{V}})_{xyz} + \mathbf{\Omega} \times \mathbf{V}. \tag{21.8}$$

Thus, Eq. (21.8) shows that the time derivative of any vector **V** with respect to a nonrotating, translating (or fixed) X-Y-Z coordinate system consists of two parts. The first part $(\dot{\mathbf{V}})_{xyz}$ is the time derivative of **V** with respect to the rotating x-y-z system and represents the change in the magnitude of **V**. The second, $\mathbf{\Omega} \times \mathbf{V}$, is due to the rotation of the x-y-z coordinate system and represents the change in the direction of **V**.

The following example illustrates the use of Eq. (21.8) in solving problems.

■ Example 21.4

Refer to the robotics assembly of Example 21.1 and, for the position depicted in Figure E21.1, determine (a) the angular acceleration of arm OP and (b) the linear acceleration of point P.

Solution

(a) From Example 21.1, we concluded that the angular velocity $\boldsymbol{\omega}$ of arm OP is

$$\boldsymbol{\omega} = \boldsymbol{\omega}_x + \boldsymbol{\omega}_z = 8\mathbf{i} + 5\mathbf{k}.$$

The angular acceleration $\boldsymbol{\alpha}$ of arm OP is the time derivative of $\boldsymbol{\omega}$.

To determine the time derivative of $\boldsymbol{\omega}$ we use Eq. (21.8) which is based upon a rotating x-y-z coordinate system as well as a fixed X-Y-Z system. Figure E21.4 is identical to Figure E21.1 except that a fixed X-Y-Z system has been added which, at the instant considered, coincides with the rotating x-y-z system. This latter system rotates about the z (or Z) axis at the angular velocity $\boldsymbol{\omega}_1$ which is constant in both magnitude and direction. Thus,

$$\boldsymbol{\alpha} = \dot{\boldsymbol{\omega}} = \dot{\boldsymbol{\omega}}_x + \dot{\boldsymbol{\omega}}_z = \dot{\boldsymbol{\omega}}_2 + \dot{\boldsymbol{\omega}}_1 = \dot{\boldsymbol{\omega}}_2 + \mathbf{0}.$$

Using Eq. (21.8) with $\mathbf{V} = \boldsymbol{\omega}_2$ and $\mathbf{\Omega} = \boldsymbol{\omega}_1$,

$$\boldsymbol{\alpha} = (\dot{\boldsymbol{\omega}}_x)_{XYZ} = (\dot{\boldsymbol{\omega}}_x)_{xyz} + \boldsymbol{\omega}_2 \times \boldsymbol{\omega}_1.$$

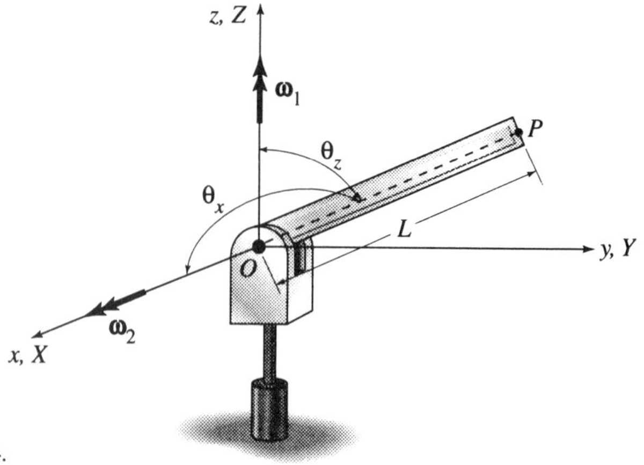

FIGURE E21.4.

Because $\boldsymbol{\omega}_1$ rotates with the x-y-z system and is constant in magnitude, its time derivative with respect to this system vanishes. Therefore,

$$\boldsymbol{\alpha} = \mathbf{0} + \boldsymbol{\omega}_2 \times \boldsymbol{\omega}_1,$$

$$\boldsymbol{\alpha} = 5\mathbf{k} \times 8\mathbf{i} = 40.0\mathbf{j} \text{ rad/s}^2. \qquad \text{ANS.}$$

(b) By Eq. (21.7b),

$$\mathbf{a}_A = \mathbf{a}_B + \boldsymbol{\omega} \times (\boldsymbol{\omega} \times \mathbf{r}_{P/O}) + \boldsymbol{\alpha} \times \mathbf{r}_{P/O} = \mathbf{0} + \boldsymbol{\omega} \times \mathbf{v}_P + \boldsymbol{\alpha} \times \mathbf{r}_{P/O},$$

$$= \begin{bmatrix} \mathbf{i} & \mathbf{j} & \mathbf{k} \\ 8 & 0 & 5 \\ -82.6 & -54.1 & 132.2 \end{bmatrix} + \begin{bmatrix} \mathbf{i} & \mathbf{j} & \mathbf{k} \\ 0 & 40.0 & 0 \\ 5.176 & 16.522 & 10.000 \end{bmatrix}$$

$$= (671\mathbf{i} - 1471\mathbf{j} - 640\mathbf{k}) \text{ in./s}^2. \qquad \text{ANS.}$$

■

Problems

21.1 Assume that each component rotation $\theta = 90°$ and is consistent with the right-hand rule. On this basis, the rectangular prism of Figure P21.1(a) assumes the final position shown in Figure P21.1(b) after the sequence of rotations $-\theta_x - \theta_y$. Find two other sequences, consisting of four rotations each, that would take the prism from its initial to its final position.

21.2 Assume that each component rotation $\theta = 90°$ and is consistent with the right-hand rule. On this basis, the rectangular prism of Figure P21.2(a) assumes the final position shown in Figure P21.2(b) after the sequence of rotations $\theta_y + \theta_x$. Find two other sequences, consisting of four rotations each, that would take the prism from its initial to its final position.

21.3 The mechanism shown in Figure P21.3 consists of a wheel of radius $R = 0.15$ m that rotates about a horizontal shaft parallel to the x axis at a constant angular velocity $\omega_1 = 60$ rad/s. This shaft is held by a vertical bracket-shaft arrangement that rotates about the z axis at a

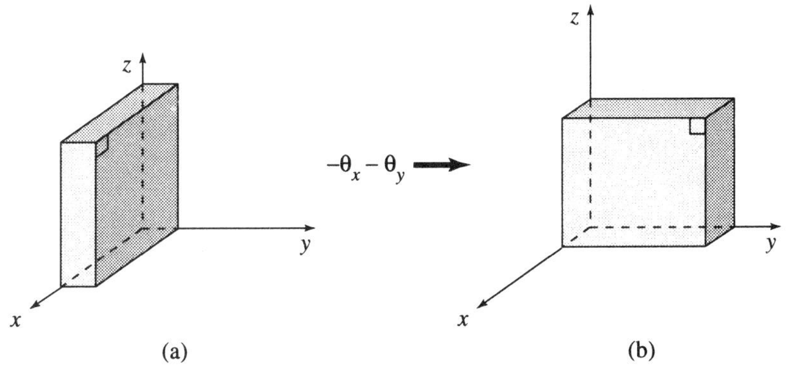

FIGURE P21.1.

constant angular velocity $\omega_2 = 20$ rad/s. Determine (a) the resultant angular velocity of the wheel and (b) the resultant linear velocity of point P at the instant shown in Figure P21.3.

21.4 Refer to Problem 21.3, and compute (a) the resultant angular acceleration of the wheel and (b) the resultant linear acceleration of point P.

21.5 In the position shown in Figure P21.5, the blades of the helicopter rotate at a constant angular velocity $\boldsymbol{\omega}_1 = \omega_1 \mathbf{k}$, relative to the frame, while the helicopter makes a vertical turn (i.e., in the y-z plane) at a constant angular velocity $\boldsymbol{\omega}_2 = \omega_2 \mathbf{i}$, as shown. If the length of each blade is L, determine (a) the resultant angular velocity of the blades and (b) the resultant angular acceleration of the blades.

21.6 The propeller of the plane shown in Figure P21.6 has a constant angular ve-

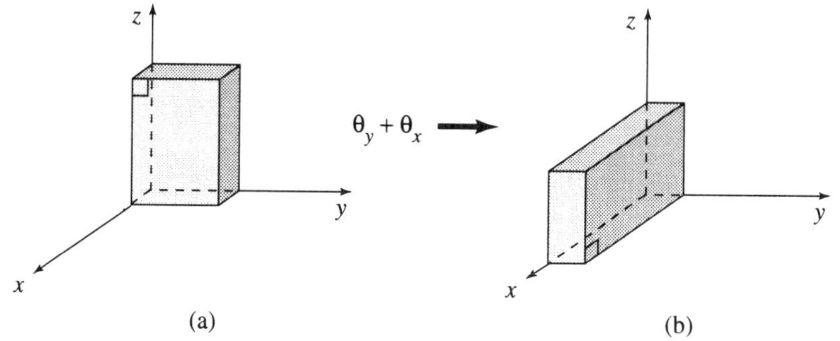

FIGURE P21.2.

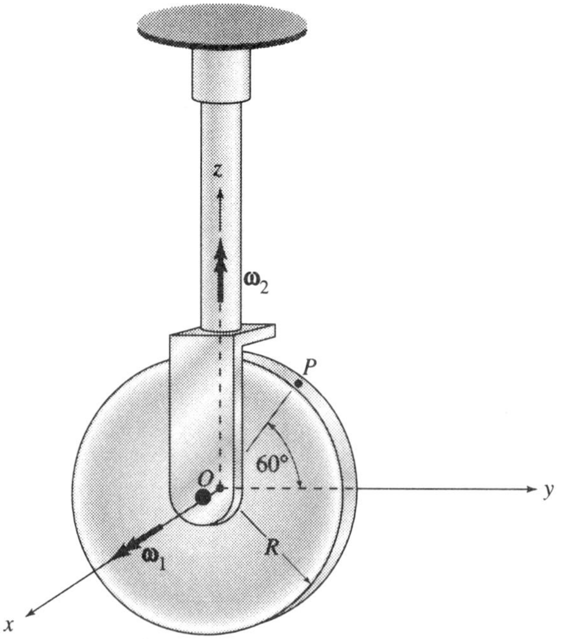

FIGURE P21.3.

locity relative to the frame of $\omega_1 = 300$ rad/s. At the instant shown, the plane is making a left turn at a constant angular velocity $\omega_2 = 20$ rad/s. Determine (a) the resultant angular velocity of the blade and (b) the resultant angular acceleration of the blade.

21.7 The assembly shown in Figure P21.7 rotates about the ball and socket at O such that, at the instant shown, $\omega_z = 7$ rad/s and $\mathbf{v}_P = [5\mathbf{i} + (v_P)_y\mathbf{j} - 3\mathbf{k}]$ in./s. Determine $(v_P)_y$ and the angular velocity of the assembly.

21.8 Refer to Problem 21.7, and determine the angular acceleration of the assembly and the linear acceleration of P if at the

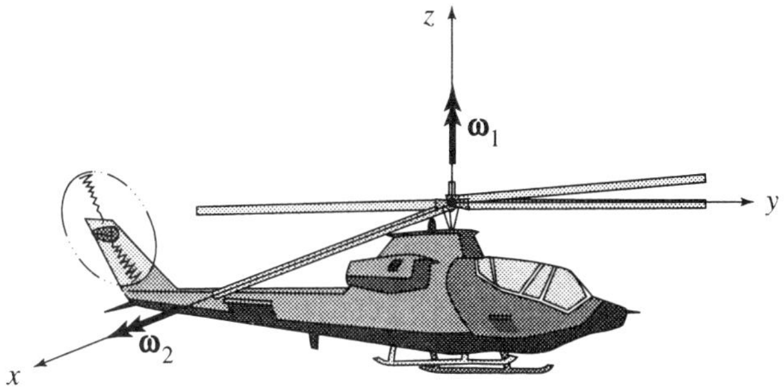

FIGURE P21.5.

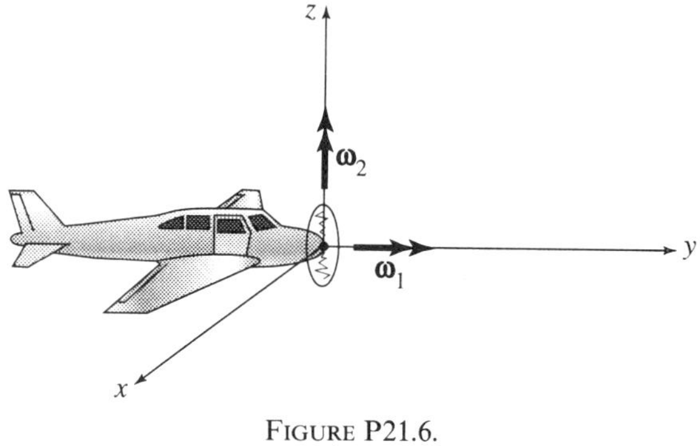

FIGURE P21.6.

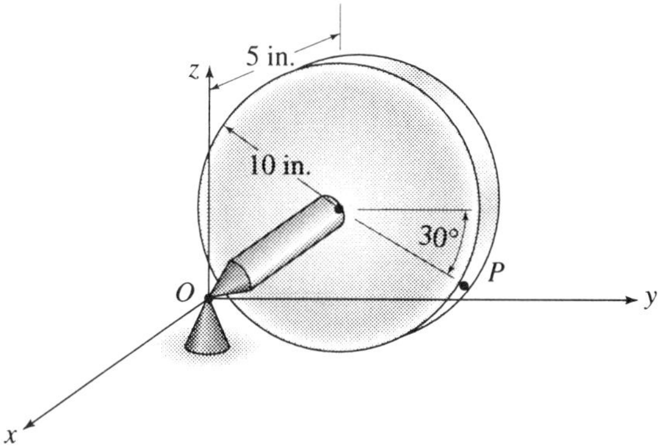

FIGURE P21.7.

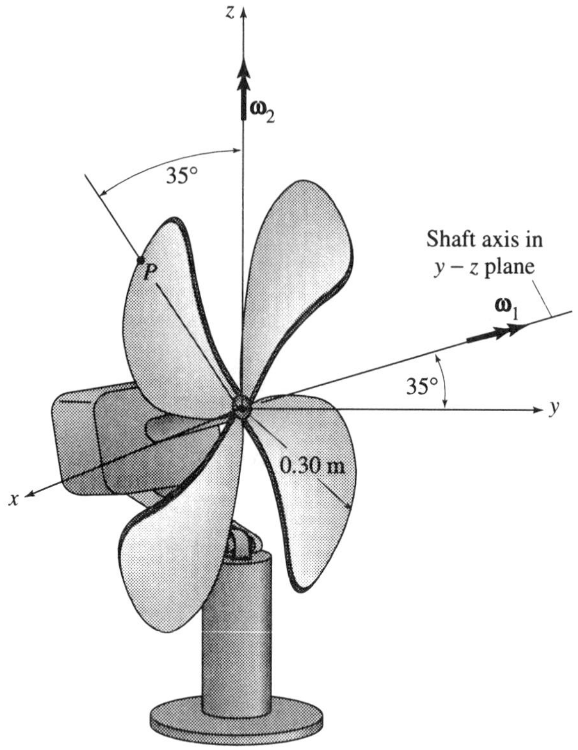

FIGURE P21.9.

instant shown $\alpha_x = 20$ rad/s^2 and $\mathbf{a}_P = [100\mathbf{i} + 75\mathbf{j} + (\mathbf{a}_P)_z\mathbf{k}]$ in./s^2.

21.9 The fan blades shown in Figure P21.9 rotate about the motor shaft at a constant angular velocity $\omega_1 = 40$ rad/s. Simultaneously, the entire motor assembly rotates about the vertical z axis at the constant rate $\omega_2 = 0.6$ rad/s. For the position indicated in Figure P21.9, determine (a) the resultant angular velocity of the blades and (b) the resultant linear velocity of point P which lies in the y-z plane.

21.10 Refer to Problem 21.9, and compute (a) the resultant angular acceleration of the blades and (b) the resultant linear acceleration of point P.

21.11 The wheel shown in Figure P21.11 rolls without slipping on the horizontal plane. The shaft on which the wheel is mounted is parallel to the x-y plane and rotates about the vertical z axis at the constant rate $\omega_1 = 5$ rad/s. Determine (a) the resultant angular velocity of the wheel and (b) the resultant linear velocity of point P.

21.12 Refer to Problem 21.11, and determine (a) the resultant angular acceleration of the wheel and (b) the resultant linear acceleration of point P.

21.13 The robotics assembly shown in Figure P21.13 rotates about the vertical z axis at the constant rate $\omega_1 = 0.25$ rad/s and, simultaneously, about the horizontal y axis at the constant rate $\omega_2 = 0.35$ rad/s. If the length of arm OP is $L = 0.85$ m, determine (a) the resultant angular velocity of arm OP and (b) the resultant linear velocity of point P.

21.14 Refer to Problem 21.13, and compute (a)

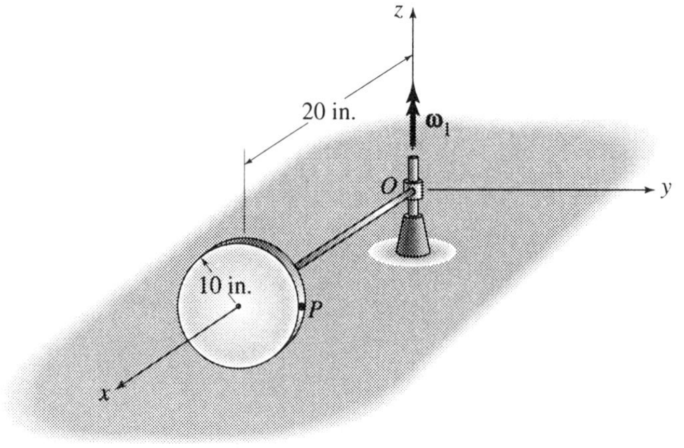

FIGURE P21.11.

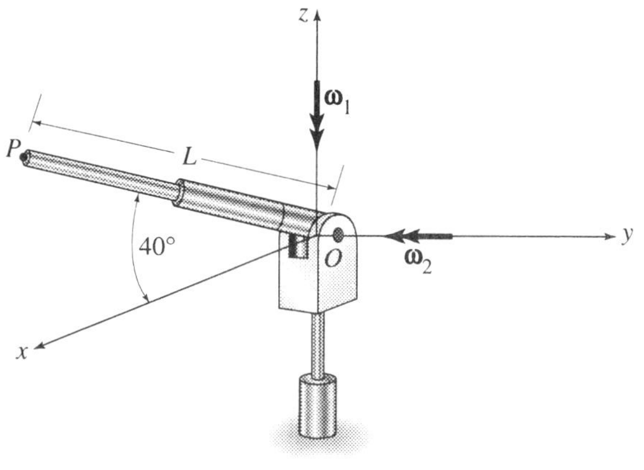

FIGURE P21.13.

the resultant angular acceleration of arm OP and (b) the resultant linear acceleration of point P.

21.15 In Figure P21.15, the horizontal gear is fixed and the inclined bevel gear rotates about the z axis at a constant angular velocity $\omega_1 = 20$ rad/s as shown. Find (a) the resultant angular velocity of the bevel gear, (b) the rate of spin of the bevel gear, and (c) the resultant linear velocity of point P.

21.16 Refer to Problem 21.15, and determine (a) the resultant angular acceleration of the bevel gear and (b) the resultant linear acceleration of point P.

21.17 At the instant shown in Figure P21.17, arm OP has a length $L = 4.5$ m and is being manipulated to bring the worker close to the elevated power and telephone lines. This is accomplished by simultaneously rotating the arm about the vertical z axis at the constant rate $\omega_1 = 0.30$ rad/s and about the horizontal x axis at the constant rate $\omega_2 = 0.50$ rad/s. For the position shown, determine (a) the resultant angular velocity of arm

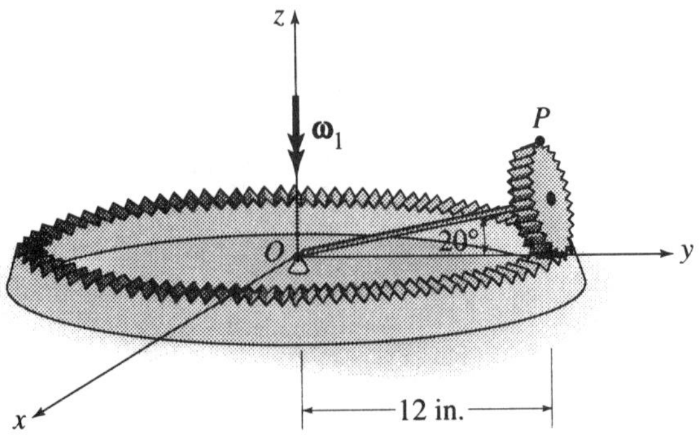

FIGURE P21.15.

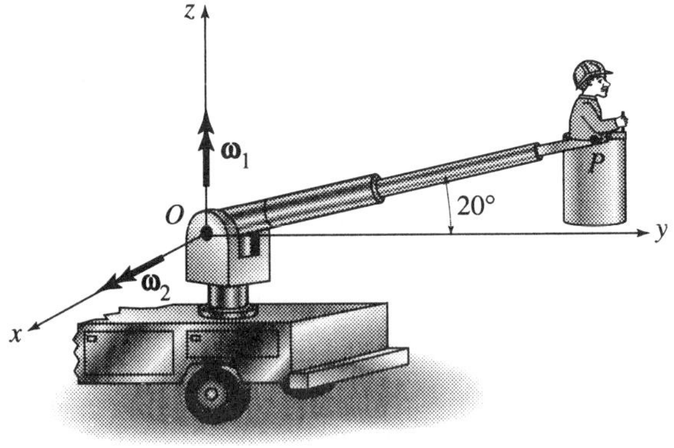

FIGURE P21.17.

OP and (b) the resultant linear velocity of the worker.

21.18 Refer to Problem 21.17, and determine (a) the resultant angular acceleration of arm OP and (b) the resultant linear acceleration of the worker.

21.19 At the instant shown in Figure P21.19, the turret of a tank is rotating about the vertical z axis at the constant rate $\omega_1 =$ 3 rad/s. Simultaneously, the gun barrel, which lies in the y-z plane and whose length $L = 12$ ft, is rotating about the horizontal x axis at the constant rate $\omega_2 = 0.75$ rad/s. For the position shown, determine (a) the resultant angular velocity of the barrel and (b) the resultant linear velocity of point P.

21.20 Refer to Problem 21.19, and find (a) the resultant angular acceleration of the barrel and (b) the resultant linear acceleration of point P.

21.21 The radar antenna shown in Figure P21.21 is tracking the path of a fighter plane. At the instant shown, the antenna is rotating about the vertical z axis at the constant rate $\omega_1 = 2.5$ rad/s and simultaneously about the horizontal x axis at the constant rate $\omega_2 = 1.5$ rad/s. Arm OP has a length of 1.2 in. and, at the instant shown, it lies in the y-z plane. For the position shown, determine (a) the resultant angular velocity of the antenna and (b) the resultant linear velocity of point P.

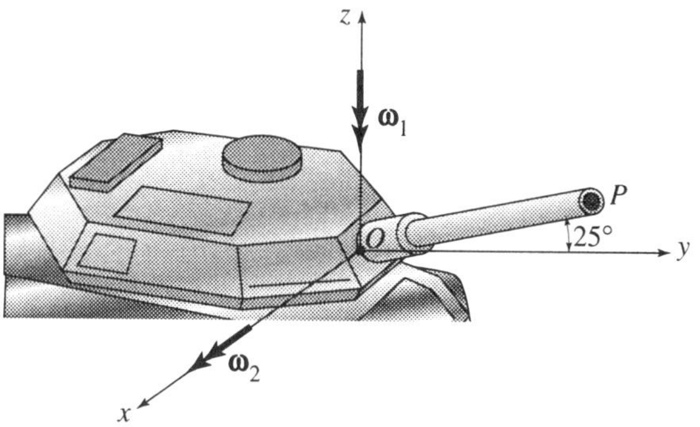

FIGURE P21.19.

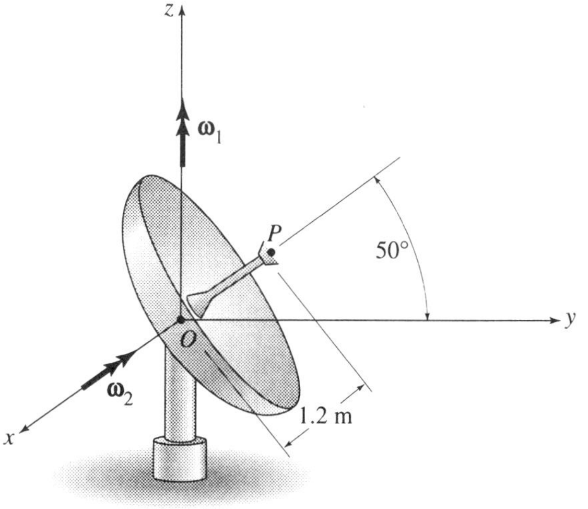

FIGURE P21.21.

21.22 Refer to Problem 21.21, and compute (a) the resultant angular acceleration of the antenna and (b) the resultant linear acceleration of point P.

21.23 The mechanism shown in Figure P21.23 consists of arm OA rotating in the x-z plane about the y axis at a constant angular velocity $\omega_1 = 20$ rad/s, a collar B that slides freely on a rod that lies along the y axis, and rod AB of length $L =$ 1.8 m that is connected to arm OA and collar B by ball-and-socket joints. For the position shown, determine the velocity of the collar.

21.24 Refer to Problem 21.23, and determine the angular velocity of rod AB which, at the instant considered, is spinning about its own axis at the constant rate $\omega_S = 7$ rad/s, as shown in Figure P21.23.

21.25 Collar A is constrained to move on a rod parallel to the z axis, and collar B is constrained to move on a rod along the y axis, as shown in Figure P21.25. Rod AB of length $L = 22$ in., is attached to the collars by ball-and-socket connections. In the position shown, collar B has a constant velocity $\mathbf{v}_B = (75\mathbf{j})$ in./s. Determine the velocity of collar A at the same instant.

21.26 Refer to Problem 21.25, and determine

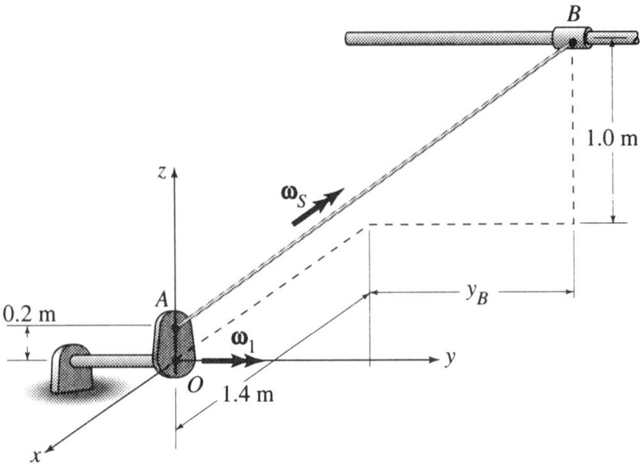

FIGURE P21.23.

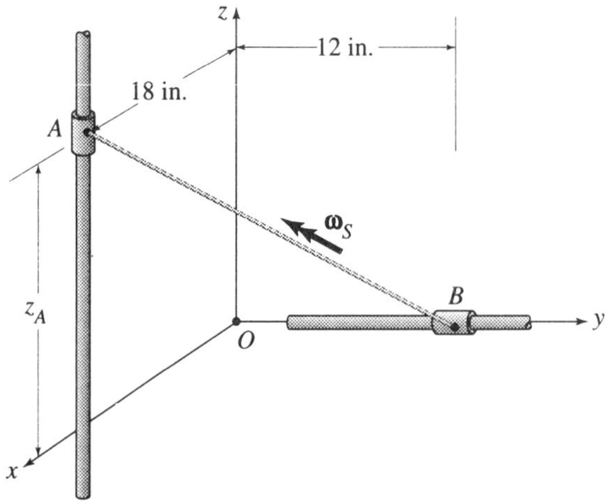

FIGURE P21.25.

the angular velocity of rod AB which, for the position shown, is spinning about its own axis at the constant rate $\omega_S = 25$ rad/s as indicated in Figure P21.25.

21.27 Collar A slides on a rod that is parallel to the x axis, and collar B slides on a rod that lies in the y-z plane, as shown in Figure P21.27. Rod AB of length $L = 3$ m, is attached to the collars by ball-and-socket joints. In the position shown, collar A has a constant velocity $\mathbf{v}_A = -(2\mathbf{i})$ m/s. Determine the velocity of collar B at the same instant.

21.28 Refer to Problem 21.27, and determine the rate of spin of rod AB about its own axis, for the position shown in Figure P21.27, if the component of its angular velocity about the x axis $\omega_x = 1.5$ rad/s.

21.29 The mechanism shown in Figure P21.29 consists of a wheel of radius $R = 0.15$ m that, at the instant shown, rotates about a horizontal shaft that lies along the x

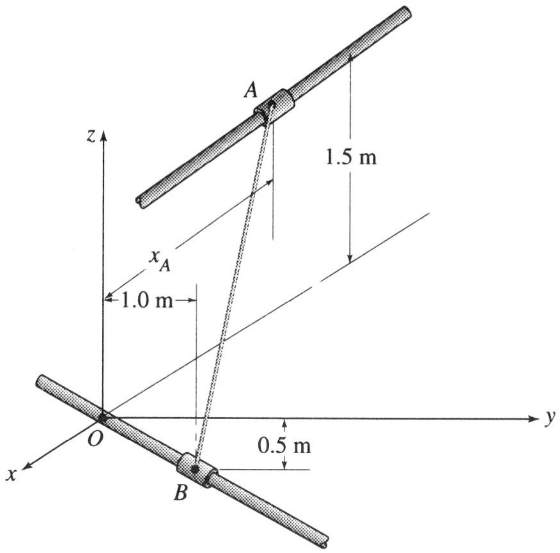

FIGURE P21.27.

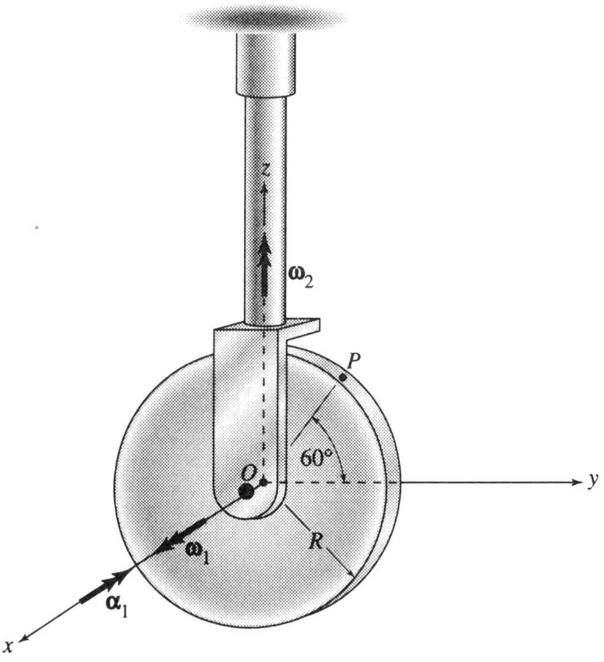

FIGURE P21.29.

axis at an angular velocity $\omega_1 = 60$ rad/s and an angular acceleration $\alpha_1 = 300$ rad/s^2. This shaft is held by a vertical bracket-shaft arrangement that rotates about the z axis at a constant angular velocity $\omega_2 = 20$ rad/s. Determine (a) the resultant angular acceleration of the wheel and (b) the resultant linear acceleration of point P.

21.30 The fan blades rotate about the motor shaft at a constant angular velocity $\omega_1 = 40$ rad/s as shown in Figure P21.30. Simultaneously, the entire motor assembly rotates about the vertical z axis and, at the instant shown, $\omega_2 = 0.6$ rad/s and $\alpha_2 = 1.5$ rad/s as indicated. Determine (a) the resultant angular acceleration of the blades and (b) the resultant linear acceleration of point P which lies in the y-z plane.

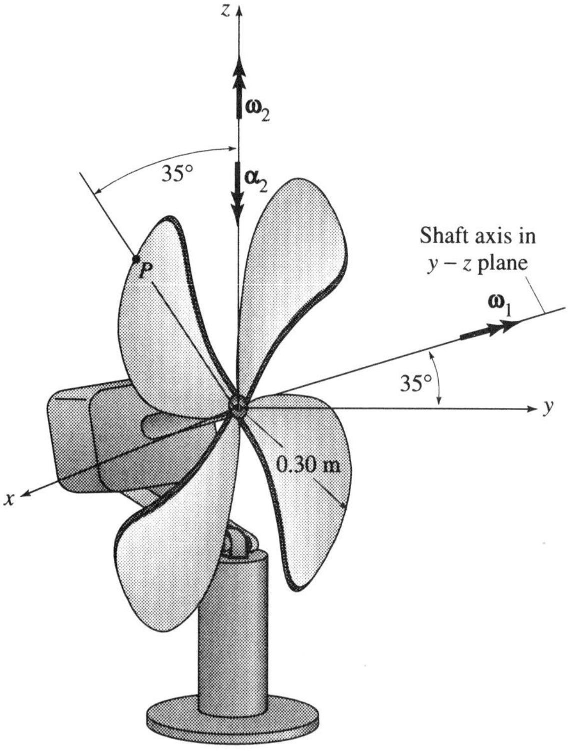

FIGURE P21.30.

21.31 The robotics assembly shown in Figure P21.31 rotates about the vertical z axis and, at the instant shown, $\omega_1 = 0.40$ rad/s and $\alpha_1 = 1.75$ rad/s^2 as indicated. Simultaneously, arm OP rotates about the horizontal y axis and, at the instant shown, $\omega_2 = 0.60$ rad/s and $\alpha_2 = 0.75$ rad/s^2 as indicated. For the position shown $\theta_x = 50°$ and $\theta_z = 60°$. If the length of arm OP is $L = 1.80$ ft, determine (a) the resultant angular acceleration of arm OP and (b) the resultant linear acceleration of point P.

21.32 Refer to Problem 21.23, and, for the position shown in Figure P21.23, determine the linear acceleration of collar B and the angular acceleration of rod AB which, at the instant indicated, is spin-

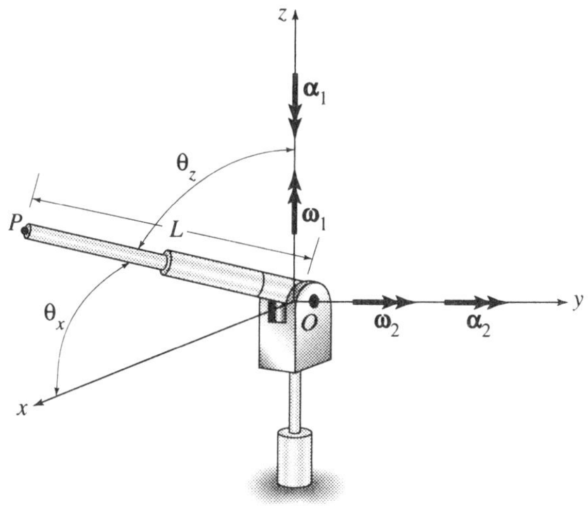

FIGURE P21.31.

ning about its own axis at the constant rate $\omega_S = 7$ rad/s, as shown.

21.33 Refer to Problem 21.25, and determine, for the position shown in Figure P21.25, the linear acceleration of collar A and the angular acceleration of rod AB which, at the instant indicated, is spinning about its own axis at the constant rate $\omega_S = 25$ rad/s, as shown.

21.34 Refer to Problem 21.27, and for the position shown, determine the linear acceleration of collar B and the angular acceleration of rod AB which, at the instant considered, is spinning about its own axis at the constant rate $\omega_S = 2$ rad/s, cw when viewed from A to B.

21.4 General 3-D Motion (Translating and Rotating Axes)

In Section 21.2, we referred the three-dimensional motion of a rigid body to a translating, nonrotating coordinate system. In the solution of a certain type of problem, it is convenient to refer the three-dimensional motion of a rigid body to a coordinate system that not only translates, but also rotates at some angular velocity $\mathbf{\Omega}$ and an angular acceleration $\dot{\mathbf{\Omega}}$.

To this end, consider two points A and B, as shown in Figure 21.7 which represent two particles on the same rigid body or two particles on two different rigid bodies. The X-Y-Z coordinate system is fixed in space and the x-y-z coordinate system attached to point B is translating and rotating at an angular velocity $\mathbf{\Omega}$ relative to the fixed X-Y-Z system. The vectors $\mathbf{r}_A$ and $\mathbf{r}_B$ represent the positions of points A and B, respectively, relative to the fixed X-Y-Z system whereas the vector $\mathbf{r}_{A/B}$ represents the position of point A as measured by an observer in the translating and rotating x-y-z system. Thus,

$$\mathbf{r}_A = \mathbf{r}_B + \mathbf{r}_{A/B}. \tag{21.9}$$

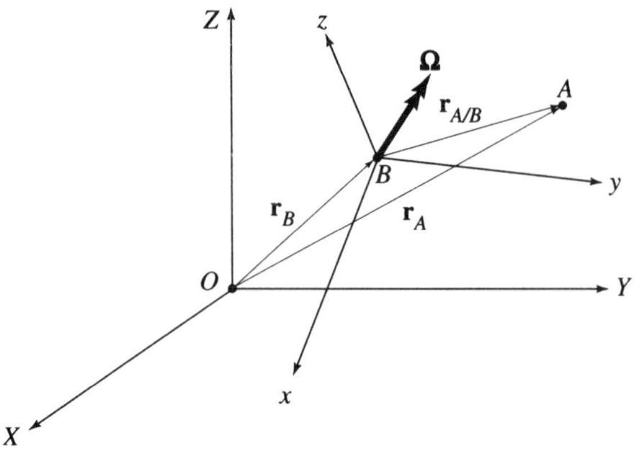

FIGURE 21.7.

Differentiating Eq. (21.9) with respect to time yields

$$\dot{\mathbf{r}}_A = \dot{\mathbf{r}}_B + \dot{\mathbf{r}}_{A/B} \tag{21.10a}$$

where $\dot{\mathbf{r}}_A = \mathbf{v}_A$, $\dot{\mathbf{r}}_B = \mathbf{v}_B$ and $\dot{\mathbf{r}}_{A/B} = \mathbf{v}_{A/B}$. Because the time derivatives are defined in terms of the fixed X-Y-Z system, the term $\mathbf{v}_{A/B} = \dot{\mathbf{r}}_{A/B}$ has to be determined by Eq. (21.8). Thus,

$$\mathbf{v}_{A/B} = \dot{\mathbf{r}}_{A/B} = (\dot{\mathbf{r}}_{A/B})_{xyz} + \boldsymbol{\Omega} \times \mathbf{r}_{A/B} = (\mathbf{v}_{A/B})_{xyz} + \boldsymbol{\Omega} \times \mathbf{r}_{A/B}. \tag{21.10b}$$

Thus, on the basis of Eqs. (21.10a) and (21.10b), we conclude that

$$\mathbf{v}_A = \mathbf{v}_B + (\mathbf{v}_{A/B})_{xyz} + \boldsymbol{\Omega} \times (\mathbf{r}_{A/B})_{xyz}. \tag{21.11}$$

In Eq. (21.11), $\mathbf{v}_A$ and $\mathbf{v}_B$ represent the absolute velocities (i.e., measured from the fixed X-Y-Z system) of points A and B, respectively. The quantity $(\mathbf{v}_{A/B})_{xyz}$ is the velocity of point A as measured by an observer stationed at point B. Finally, $\boldsymbol{\Omega}$ represents the angular velocity of the x-y-z system and, as stated earlier, $(\mathbf{r}_{A/B})_{xyz}$ is the position of A relative to B as measured by an observer in the x-y-z system.

Similarly, the absolute acceleration of point A is obtained by differentiating Eq. (21.11) with respect to time. Thus,

$$\dot{\mathbf{v}}_A = \dot{\mathbf{v}}_B + \dot{\mathbf{v}}_{A/B} + \boldsymbol{\Omega} \times \dot{\mathbf{r}}_{A/B} + \dot{\boldsymbol{\Omega}} \times \mathbf{r}_{A/B}. \tag{21.12a}$$

Because the time derivatives are defined with respect to the fixed X-Y-Z system, $\dot{\mathbf{v}}_A = \mathbf{a}_A$, $\dot{\mathbf{v}}_B = \mathbf{a}_B$, and, by Eq. (21.10b), $\dot{\mathbf{r}}_{A/B} = (\mathbf{v}_{A/B})_{xyz} + \boldsymbol{\Omega} \times \mathbf{r}_{A/B}$. Also, $\dot{\mathbf{v}}_{A/B}$ is evaluated by Eq. (21.8), yielding

$$\dot{\mathbf{v}}_{A/B} = \mathbf{a}_{A/B} = (\mathbf{a})_{xyz} + \boldsymbol{\Omega} \times (\mathbf{v}_{A/B})_{xyz}. \tag{21.12b}$$

Therefore, on the basis of Eqs. (21.12a) and (21.12b), we conclude that

$$\mathbf{a}_A = \mathbf{a}_B + (\mathbf{a}_{A/B})_{xyz} + 2\boldsymbol{\Omega} \times (\mathbf{v}_{A/B})_{xyz} + \boldsymbol{\Omega} \times [\boldsymbol{\Omega} \times (\mathbf{r}_{A/B})_{xyz}] + \dot{\boldsymbol{\Omega}} \times (\mathbf{r}_{A/B})_{xyz}. \tag{21.13}$$

In Eq. (21.13), $\mathbf{a}_A$ and $\mathbf{a}_B$ are the absolute accelerations (i.e., measured relative to the fixed X-Y-Z system), respectively, of points A and B; the quantities $(\mathbf{v}_{A/B})_{xyz}$ and $(\mathbf{a}_{A/B})_{xyz}$ represent the relative velocity and relative acceleration, respectively, of point A as measured by an observer in the x-y-z system; the quantities $\mathbf{\Omega}$ and $\dot{\mathbf{\Omega}}$ are the angular velocity and angular acceleration of the x-y-z system, respectively, and, as stated earlier, the quantity $(\mathbf{r}_{A/B})_{xyz}$ is the position of point A relative to point B as measured by an observer moving with the x-y-z system. Note that the quantity $2\mathbf{\Omega} \times (\mathbf{v}_{A/B})_{xyz}$ represents Coriolis acceleration which was discussed in some detail in Section 17.7 dealing with the two-dimensional motion of a rigid body. Note also that Eq. (21.13) is identical to Eq. (17.36) developed in Chapter 17 for the two-dimensional motion of a rigid body. It should be pointed out, however, that the application of Eq. (21.13) is a little more complex than Eq. (17.36) because of the three-dimensional nature of the problems for which Eq. (21.13) is used.

The following examples illustrate some of the above concepts.

Example 21.5

The robotics assembly shown in Figure E21.5 rotates about the vertical z axis at a constant angular velocity $\omega_1 = 5$ rad/s while arm OP rotates about the horizontal x axis at a constant angular velocity $\omega_2 = 8$ rad/s. At the instant depicted, arm OP of length $L = 20$ in. is positioned by the direction angles $\theta_x = 75°$ and $\theta_z = 60°$. Use the method of Section 21.4 to determine (a) the linear velocity of point P and (b) the linear acceleration of point P.

Solution

(a) By Eq. (21.11),

$$\mathbf{v}_P = \mathbf{v}_O + (\mathbf{v}_{P/O})_{xyz} + \mathbf{\Omega} \times (\mathbf{r}_{P/O})_{xyz}$$

where the x-y-z coordinate system rotates at the constant angular velocity $\mathbf{\Omega} = \boldsymbol{\omega}_1 = 5\mathbf{k}$ rad/s. The various terms in the above equation are determined as follows:

$\mathbf{v}_O = \mathbf{0}$ because point O does not translate.

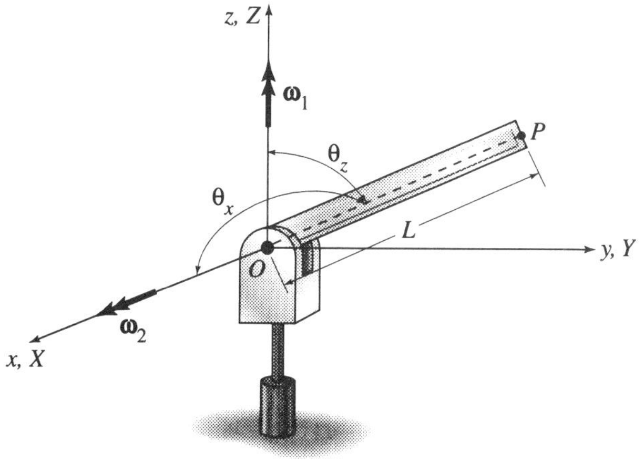

FIGURE E21.5.

Now,

$$\cos^2\theta_x + \cos^2\theta_y + \cos^2\theta_z = 1$$

$$\cos^2 75^\circ + \cos^2\theta_y + \cos^2 60^\circ = 1$$

Therefore

$$\theta_y = 34.3^\circ,$$

$$(\mathbf{r}_{P/O})_{xyz} = (20\cos 75^\circ)\mathbf{i} + (20\cos 34.3^\circ)\mathbf{j} + (20\cos 60^\circ)\mathbf{k}$$

$$= 5.176\mathbf{i} + 16.522\mathbf{j} + 10.000\mathbf{k},$$

$$(\mathbf{v}_{P/O})_{xyz} = \boldsymbol{\omega}_2 \times (\mathbf{r}_{P/O})_{xyz} = \begin{bmatrix} \mathbf{i} & \mathbf{j} & \mathbf{k} \\ 8 & 0 & 0 \\ 5.176 & 16.522 & 10.000 \end{bmatrix}$$

$$= -80.0\mathbf{j} + 132.176\mathbf{k},$$

and

$$\boldsymbol{\Omega} \times (\mathbf{r}_{P/O})_{xyz} = \begin{bmatrix} \mathbf{i} & \mathbf{j} & \mathbf{k} \\ 0 & 0 & 5 \\ 5.176 & 16.522 & 10.000 \end{bmatrix}$$

$$= -82.61\mathbf{i} + 25.88\mathbf{j}.$$

Substituting in the equation for $\mathbf{v}_P$ yields

$$\mathbf{v}_P = (-82.6\mathrm{i} - 54.1\mathbf{j} + 132.2\mathbf{k})\ \text{in./s.} \qquad \text{ANS.}$$

(b) By Eq. (21.13),

$$\mathbf{a}_P = \mathbf{a}_O + (\mathbf{a}_{P/O})_{xyz} + 2\boldsymbol{\Omega} \times (\mathbf{v}_{P/O})_{xyz}$$

$$+ \boldsymbol{\Omega} \times [\boldsymbol{\Omega} \times (\mathbf{r}_{P/O})_{xyz}] + \dot{\boldsymbol{\Omega}} \times (\mathbf{r}_{P/O})_{xyz}$$

The various terms in the above equation are found as follows:

$\mathbf{a}_O = \mathbf{0}$ because point O does not translate.

$$(\mathbf{a}_{P/O})_{xyz} = \boldsymbol{\omega}_2 \times [\boldsymbol{\omega}_2 \times (\mathbf{r}_{P/O})_{xyz}] = \boldsymbol{\omega}_2 \times (\mathbf{v}_{P/O})_{xyz}$$

$$= \begin{bmatrix} \mathbf{i} & \mathbf{j} & \mathbf{k} \\ 8 & 0 & 0 \\ 0 & -80.0 & 132.176 \end{bmatrix}$$

$$= -1057.408\mathbf{j} - 640.0\mathbf{k},$$

$$2\boldsymbol{\Omega} \times (\mathbf{v}_{P/O})_{xyz} = 2\begin{bmatrix} \mathbf{i} & \mathbf{j} & \mathbf{k} \\ 0 & 0 & 8 \\ 0 & -80.0 & 132.176 \end{bmatrix}$$

$$= -800.0\mathbf{i},$$

$$\boldsymbol{\Omega} \times [\boldsymbol{\Omega} \times (\mathbf{r}_{P/O})_{xyz}] = \begin{bmatrix} \mathbf{i} & \mathbf{j} & \mathbf{k} \\ 0 & 0 & 5 \\ -82.61 & 25.88 & 0 \end{bmatrix}$$
$$= -129.4\mathbf{i} - 413.05\mathbf{j},$$

and

$$\dot{\boldsymbol{\Omega}} \times (\mathbf{r}_{P/O})_{xyz} = \mathbf{0} \quad \text{because } \boldsymbol{\Omega} \text{ is constant in both magnitude and direction}$$

Substituting in the equation for $\mathbf{a}_P$, we conclude that

$$\mathbf{a}_P = (671\mathbf{i} - 1471\mathbf{j} - 640\mathbf{k}) \text{ in./s}^2. \qquad \text{ANS.}$$

■ Example 21.6

The mechanism shown in Figure E21.6 consists of a wheel that rotates at a constant angular velocity $\omega_1 = 15$ rad/s about a shaft at B which is parallel to the X axis. This shaft is attached to arm BAO which rotates at a constant angular velocity $\omega_2 = 5$ rad/s about a vertical shaft that lies along the Z axis. For the position indicated (i.e., the wheel is parallel to the Y-Z plane), determine (a) the linear velocity of point P and (b) the linear acceleration of point P.

Solution

(a) By Eq. (21.11),

$$\mathbf{v}_P = \mathbf{v}_B + (\mathbf{v}_{P/B})_{xyz} + \boldsymbol{\Omega} \times (\mathbf{r}_{P/B})_{xyz}$$

where the x-y-z coordinate system rotates at the constant angular velocity $\boldsymbol{\Omega} = \boldsymbol{\omega}_2 = 5\mathbf{k}$ rad/s. The various terms in the above equation are found as follows:

$$\mathbf{v}_B = \mathbf{v}_O + \boldsymbol{\omega}_2 \times (\mathbf{r}_{B/O})_{xyz}$$
$$= \mathbf{0} + 5\mathbf{k} \times (0.75\mathbf{i} + 0.50\mathbf{j})$$
$$= -2.50\mathbf{i} + 3.75\mathbf{j},$$

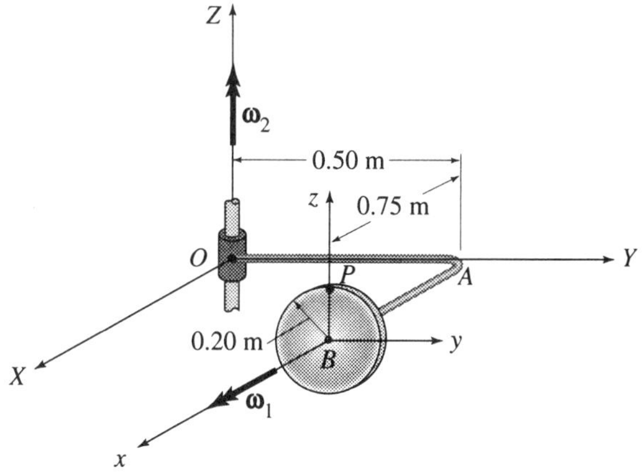

FIGURE E21.6.

$$(\mathbf{v}_{P/B})_{xyz} = \boldsymbol{\omega}_1 \times (\mathbf{r}_{P/B})_{xyz}$$
$$= 15\mathbf{i} \times 0.20\mathbf{k}$$
$$= -3.00\mathbf{j},$$

and

$$\boldsymbol{\Omega} \times (\mathbf{r}_{P/B})_{xyz} = 5\mathbf{k} \times 0.20\mathbf{k} = \mathbf{0}.$$

Therefore, substituting these values in the equation for $\mathbf{v}_P$ yields

$$\mathbf{v}_P = (-2.50\mathbf{i} + 0.75\mathbf{j})\ \text{m/s}. \qquad \text{ANS.}$$

(b) By Eq. (21.13),

$$\mathbf{a}_P = \mathbf{a}_B + (\mathbf{a}_{P/B})_{xyz} + 2\boldsymbol{\Omega} \times (\mathbf{v}_{P/B})_{xyz}$$
$$+ \boldsymbol{\Omega} \times [\boldsymbol{\Omega} \times (\mathbf{r}_{P/B})_{xyz}] + \dot{\boldsymbol{\Omega}} \times (\mathbf{r}_{P/B})_{xyz}.$$

The various terms in the above equation are determined as follows:

$$\mathbf{a}_B = \mathbf{a}_O + \mathbf{a}_{B/O} = \mathbf{0} + \boldsymbol{\omega}_2 \times (\boldsymbol{\omega}_2 \times \mathbf{r}_{B/O})$$
$$= 5\mathbf{k} \times (-2.50\mathbf{i} + 3.75\mathbf{j})$$
$$= -18.75\mathbf{i} - 12.50\mathbf{j},$$

$$(\mathbf{a}_{P/B})_{xyz} = \boldsymbol{\omega}_1 \times (\boldsymbol{\omega}_1 \times \mathbf{r}_{P/B})$$
$$= 15\mathbf{i} \times (15\mathbf{i} \times 0.20\mathbf{k})$$
$$= -45.0\mathbf{k},$$

$$2\boldsymbol{\Omega} \times (\mathbf{v}_{P/B})_{xyz} = 2(5\mathbf{k}) \times (-3.00\mathbf{j}) = 30.0\mathbf{i},$$

$$\boldsymbol{\Omega} \times [\boldsymbol{\Omega} \times (\mathbf{r}_{P/B})_{xyz}] = \mathbf{0} \quad \text{because, from part (a),} \quad \boldsymbol{\Omega} \times (\mathbf{r}_{P/B})_{xyz} = \mathbf{0},$$

and

$$\dot{\boldsymbol{\Omega}} \times (\mathbf{r}_{P/B})_{xyz} = \mathbf{0} \quad \text{because} \quad \dot{\boldsymbol{\Omega}} = \mathbf{0}.$$

Substituting in the equation for $\mathbf{a}_P$, we obtain

$$\mathbf{a}_P = (11.25\mathbf{i} - 12.50\mathbf{j} - 45.0\mathbf{k})\ \text{m/s}^2. \qquad \text{ANS.}$$

Example 21.7

The assembly shown in Figure E21.7 rotates about the vertical shaft AB which lies along the Z axis so that, at the instant depicted, the angular velocity is $\omega = 10$ rad/s and the angular acceleration is a $= 75$ rad/s^2 as shown. The collar C slides freely on arm OD which is rigidly attached to shaft AB and, in the position shown, the relative linear velocity $\mathbf{v}_R$ and relative linear acceleration $\mathbf{a}_R$ of the collar with respect to arm OD have magnitudes of 30 in./s and 200 in./s^2, respectively. For the position shown, determine (a) the absolute linear velocity of the collar and (b) the absolute linear acceleration of the collar.

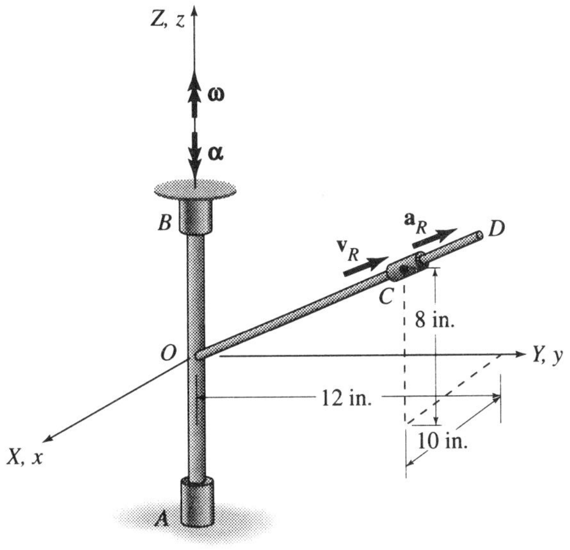

FIGURE E21.7.

Solution

(a) From the geometry given in Figure E21.7, we compute the direction cosines of arm OD to be

$$\theta_x = 55.264°,\ \theta_y = 46.862°, \quad \text{and} \quad \theta_z = 62.881°.$$

By Eq. (21.11),

$$\mathbf{v}_C = \mathbf{v}_O + (\mathbf{v}_{C/O})_{xyz} + \mathbf{\Omega} \times (\mathbf{r}_{C/O})_{xyz}$$

where, at the instant considered, the x-y-z coordinate system rotates at the angular velocity $\mathbf{\Omega} = \boldsymbol{\omega} = 10\mathbf{k}$ rad/s and at an angular acceleration $\dot{\mathbf{\Omega}} = \boldsymbol{\alpha} = -75\mathbf{k}$ rad/s^2. The various terms in the above equation are computed as follows:

$$\mathbf{v}_O = \mathbf{0} \quad \text{because point 0 does not translate.}$$

$$(\mathbf{v}_{C/O})_{xyz} = (30\cos 55.264°)\mathbf{i} + (30\cos 46.862°)\mathbf{j} + (30\cos 62.881°)\mathbf{k}$$
$$= 17.094\mathbf{i} + 20.513\mathbf{j} + 13.675\mathbf{k},$$
$$\mathbf{\Omega} \times (\mathbf{r}_{C/O})_{xyz} = 10\mathbf{k} \times (10\mathbf{i} + 12\mathbf{j} + 8\mathbf{k}),$$
$$= -120\mathbf{i} + 100\mathbf{j}.$$

Substituting in the equation for $\mathbf{v}_C$ yields

$$\mathbf{v}_C = (-102.9\mathbf{i} + 120.5\mathbf{j} + 13.68\mathbf{k}) \text{ in./s.} \qquad \text{ANS.}$$

(b) By Eq. (21.13)

$$\mathbf{a}_C = \mathbf{a}_O + (\mathbf{a}_{C/O})_{xyz} + 2\mathbf{\Omega} \times (\mathbf{v}_{C/O})_{xyz} + \mathbf{\Omega} \times [\mathbf{\Omega} \times (\mathbf{r}_{C/O})_{xyz}] + \dot{\mathbf{\Omega}} \times (\mathbf{r}_{C/O})_{xyz}.$$

The various terms in this equation are determined as follows:

$$\mathbf{a}_{\mathrm{O}} = \mathbf{0} \quad \text{because point O does not translate.}$$

$$(\mathbf{a}_{\mathrm{C/O}})_{xyz} = (200\cos 55.264°)\mathbf{i} + (200\cos 46.862°)\mathbf{j} + (200\cos 62.881°)\mathbf{k}$$
$$= 113.959\mathbf{i} + 136.752\mathbf{j} + 91.168\mathbf{k},$$

$$2\boldsymbol{\Omega} \times (\mathbf{v}_{\mathrm{C/O}})_{xyz} = 20\mathbf{k} \times (17.094\mathbf{i} + 20.513\mathbf{j} + 13.675\mathbf{k})$$
$$= -410.26\mathbf{i} + 341.88\mathbf{j},$$

$$\boldsymbol{\Omega} \times [\boldsymbol{\Omega} \times (\mathbf{r}_{\mathrm{C/O}})_{xyz}] = 10\mathbf{k} \times (-120\mathbf{i} + 100\mathbf{j})$$
$$= -1000\mathbf{i} - 1200\mathbf{j},$$

and

$$\dot{\boldsymbol{\Omega}} \times (\mathbf{r}_{\mathrm{C/O}})_{xyz} = -75\mathbf{k} \times (10\mathbf{i} + 12\mathbf{j} + 8\mathbf{k})$$
$$= 900\mathbf{i} - 750\mathbf{j}.$$

Substituting in the equation for $\mathbf{a}_{\mathrm{C}}$ we obtain

$$\mathbf{a}_{\mathrm{C}} = (-396\mathbf{i} - 1471\mathbf{j} + 91.2\mathbf{k}) \text{ in./s}^2. \qquad \text{ANS.}$$

■

Problems

21.35 The mechanism shown in Figure P21.35 consists of a wheel of radius $R = 0.15$ m that rotates at a constant angular velocity $\omega_1 = 60$ rad/s about a horizontal shaft that lies along the X axis. This shaft is held by a vertical bracket-shaft arrangement that rotates about the Z axis at a constant angular velocity $\omega_2 = 20$ rad/s. For the position shown, use the method of Section 21.4 to find (a) the absolute velocity of point P and (b) the absolute acceleration of point P. (Note that this problem consists of Problems 21.3(b) and 21.4(b)).

21.36 The fan blades rotate about the motor shaft at a constant angular velocity $\omega_1 = 40$ rad/s as shown in Figure P21.36. Simultaneously, the entire motor assembly rotates about the vertical Z axis at the constant rate $\omega_2 = 0.6$ rad/s. For the position indicated in Figure P21.36, determine, by the method of Section 21.4, (a) the absolute velocity of point P and (b) the absolute acceleration of point P. (Note that this problem consists of problems 21.9(b) and 21.10(b)).

21.37 At the instant shown in Figure P21.37, arm OP has a length $L = 4.5$ m and is being manipulated to bring the worker close to the elevated power and telephone lines. This is accomplished by simultaneously rotating the arm about the vertical Z axis at the constant rate $\omega_1 = 0.30$ rad/s and about the horizontal X axis at the constant rate $\omega_2 = 0.50$ rad/s. For the position shown, use the method of Section 1.4 to determine, (a) the absolute velocity of the worker and (b) the absolute acceleration of the worker. (Note that this problem consists of problems 21.17(b) and 21.18(b)).

21.38 The robotics assembly shown in Figure

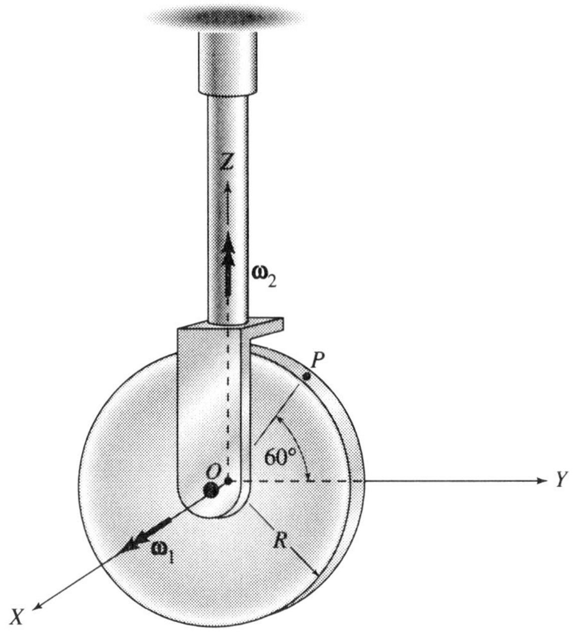

FIGURE P21.35.

P21.38 rotates about the vertical Z axis and, at the instant shown, $\omega_1 = 0.40$ rad/s and $\alpha_1 = 1.75$ rad/s^2 as indicated. Simultaneously, arm OP of length $L = 1.80$ ft rotates about the horizontal Y axis and, at the instant shown, $\omega_2 = 0.60$ rad/s and $\alpha_2 = 0.75$ rad/s^2 as indicated. For the position shown, where $\theta_x = 50°$ and $\theta_z = 60°$, determine, by the method of Section 21.4, (a) the absolute velocity of point P and (b) the absolute acceleration of point P. (Note that this problem is based upon problem 21.31).

21.39 The mechanism shown in Figure P21.39 consists of two wheels each of radius $R = 0.15$ m that rotate at a constant angular velocity $\omega_1 = 60$ rad/s about a shaft that lies along the X axis. This shaft is attached to another vertical shaft that rotates about the Z axis and, at the instant considered, $\omega_2 = 6$ rad/s and $\alpha_2 = 50$ rad/s^2. For the position indicated, determine (a) the absolute velocity of point A and (b) the absolute acceleration of point A.

21.40 Refer to Problem 21.39, and for the position indicated in Figure P21.39, determine (a) the absolute velocity of point B and (b) the absolute acceleration of point B.

21.41 The fan blades shown in Figure P21.41 have a diameter of 2.5 ft and rotate at a constant angular velocity $\omega_1 = 50$ rad/s about a horizontal shaft at C that is parallel to the X axis. This shaft is attached to frame OABC which simultaneously rotates at a constant angular velocity $\omega_2 = 10$ rad/s about a vertical shaft at O that lies along the Z axis. For the position indicated, determine (a) the absolute velocity of point D and (b) the absolute acceleration of point D.

21.42 Repeat Problem 21.41 for point E instead of point D.

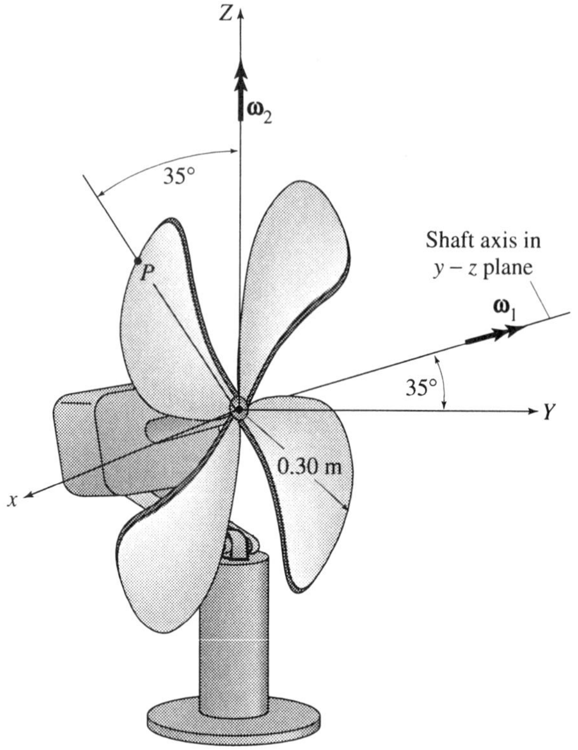

FIGURE P21.36.

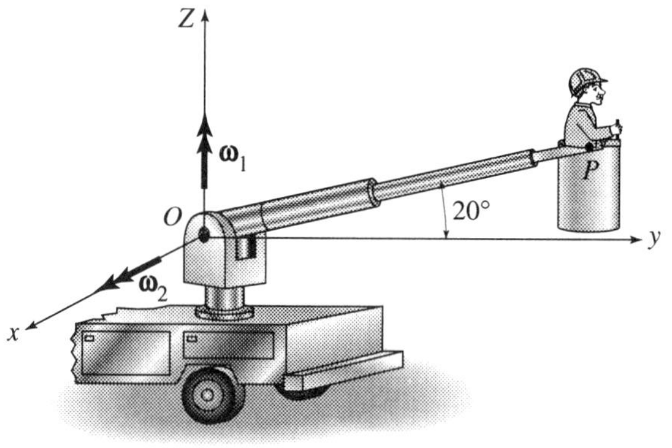

FIGURE P21.37.

21.43 Repeat Problem 21.41 if, at the instant shown in Figure 21.41, $\omega_2 = 10$ rad/s and is increasing at the rate $\alpha_2 = 200$ rad/s^2. All other conditions in Problem 21.41 remain unchanged.

21.44 The robotics assembly shown in Figure

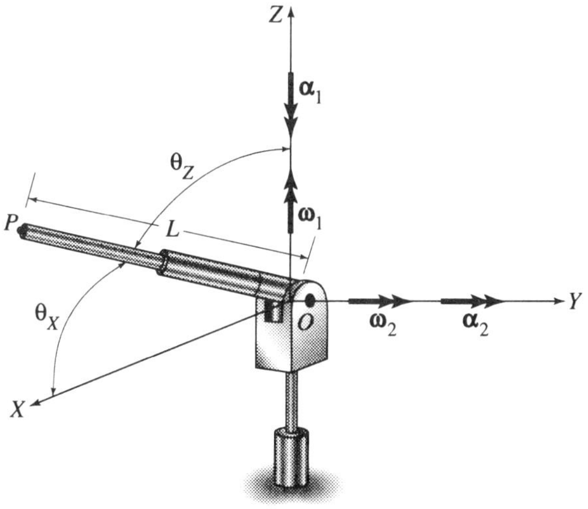

FIGURE P21.38.

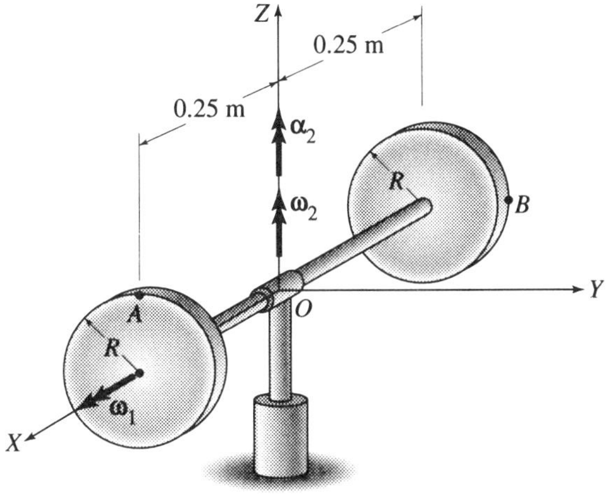

FIGURE P21.39.

P21.44 lies in the X-Z plane and rotates about the vertical Z axis at O at the rate $\omega_1 = 3$ rad/s which is increasing at the rate $\alpha_1 = 20$ rad/s^2. Simultaneously, arm AB rotates about the hinge at A which is parallel to the Y axis at the rate $\omega_2 = \dot{\theta} = 5$ rad/s which is increasing at the rate $\alpha_2 = \ddot{\theta} = 40$ rad/s^2. The length of arm OA is 1.5 m and that of arm BA is 1.0 m. For the position shown, determine (a) the absolute velocity of point B and (b) the absolute acceleration of point B. Note that $\beta = 50° =$ constant and that $\dot{\beta} = \ddot{\beta} = 0$.

21.45 The essential features of a lawn sprinkler are shown in Figure P 21.45. The two tubes OA and OB are each 4 in. long and make angles of 30° with the

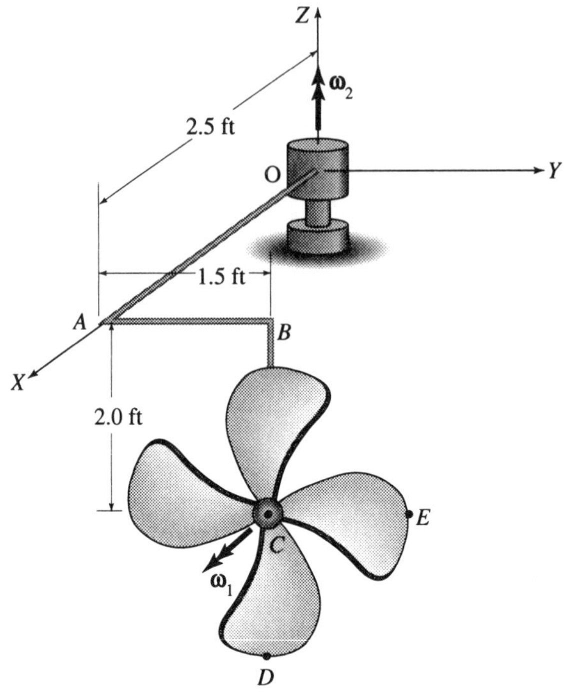

FIGURE P21.41.

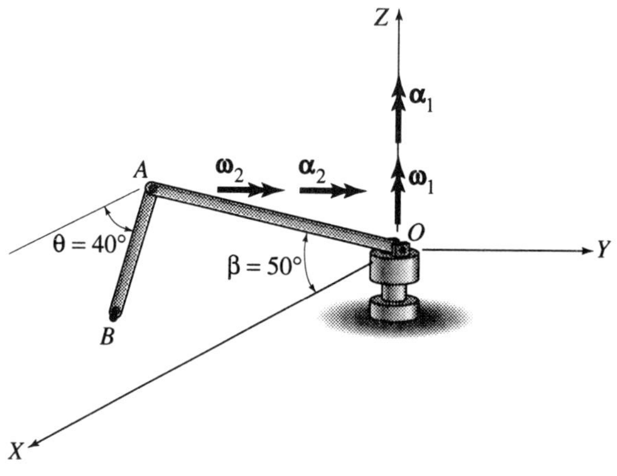

FIGURE P21.44.

horizontal and, in the position shown, they both lie in the Y-Z plane. The velocity of water relative to the tube is constant at a magnitude of 40 ft/s. If the sprinkler rotates about the Z axis at a constant rate of 240 rpm, for the position shown, determine the absolute velocity and absolute acceleration of a particle of water just before exiting at point A.

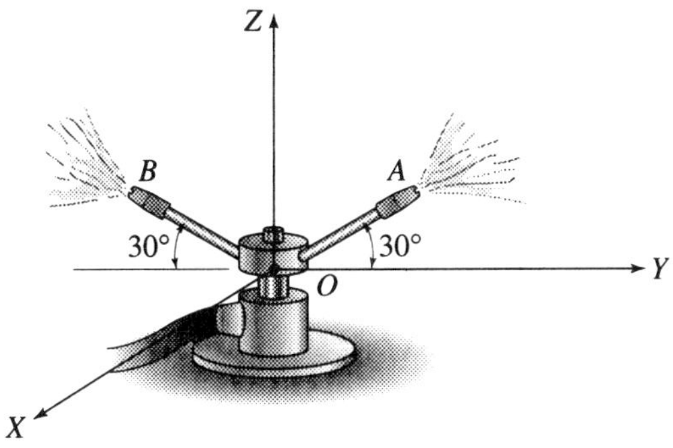

FIGURE P21.45.

21.46 The square frame shown in Figure P21.46 rotates at the constant rate $\omega = 10$ rad/s about the axle AB which lies along the Z axis. Simultaneously, the collar C slides freely along the diagonal rod EF and, at the instant depicted, s = 0.30 m, and $\dot{s} = 5$ m/s. For the position shown, determine (a) the absolute velocity of the collar and (b) the absolute acceleration of the collar.

21.47 Solve Problem 21.46 if, at the instant shown, $\omega = 10$ rad/s which is increasing at the rate $\alpha = 50$ rad/s^2, s = 0.30 m, $\dot{s} = 5$ m/s, and $\ddot{s} = -20$ m/s^2.

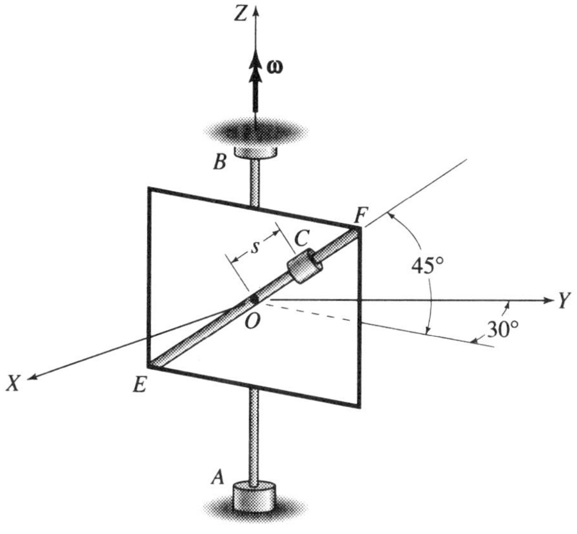

FIGURE P21.46.

Review Problems

21.48 The essential features of a ship's turbine rotor are shown in Figure P21.48. The rotor rotates about its shaft AB at a constant angular velocity $\boldsymbol{\omega}_1 = \omega_1 \mathbf{i}$. At the instant shown, if the ship makes a turn at a constant angular velocity $\boldsymbol{\omega}_2 = \omega_2 \mathbf{k}$, determine the resultant acceleration of point P on the rim of the rotor.

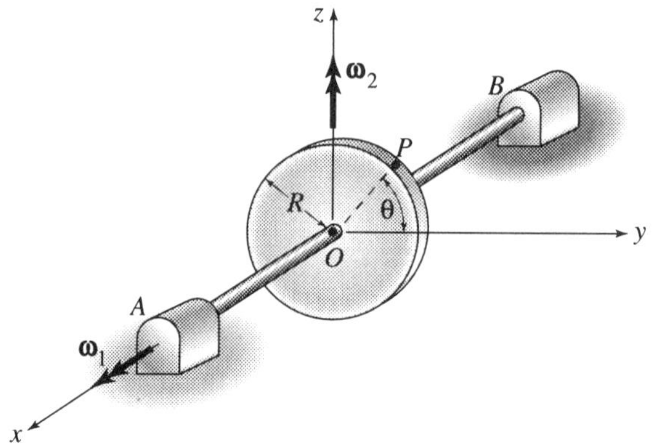

FIGURE P21.48.

21.49 Solve Problem 21.48 if, at the instant considered, $\boldsymbol{\omega}_2$ is changing at the rate $\boldsymbol{\alpha}_2 = -\alpha_2\mathbf{k}$.

21.50 A thin disk of radius $R = 6$ in. is attached to end A of rod OA whose other end is connected to a ball-and-socket at point O as shown in Figure P21.50. The disk rolls, without slipping, on a horizontal surface that lies in the x-y plane so that, at the instant shown, rod OA rotates about the z axis at an angular velocity $\omega_1 = 15$ rad/s which is decreasing at the rate $\alpha_1 = 30$ rad/s^2, as shown. For the position shown, determine the absolute acceleration of point P on the rim of the disk. The length of rod OA is $L = 12$ in.

21.51 Collar A moves on a rod that is parallel to the x axis while collar B moves on a rod that lies in the x-y plane as shown in Figure P21.51. Rod AB of length $L =$ 3 m is attached to the collars by ball-and-socket connections. In the position shown, collar A has a constant velocity $\mathbf{v}_A = (3\mathbf{i})$ m/s. Determine the velocity of collar B at the same instant.

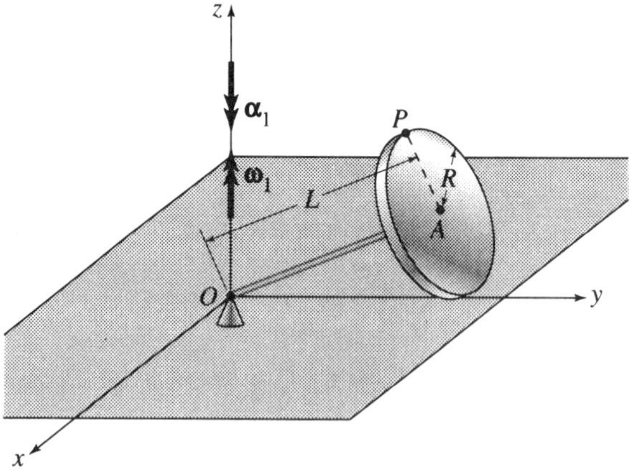

FIGURE P21.50.

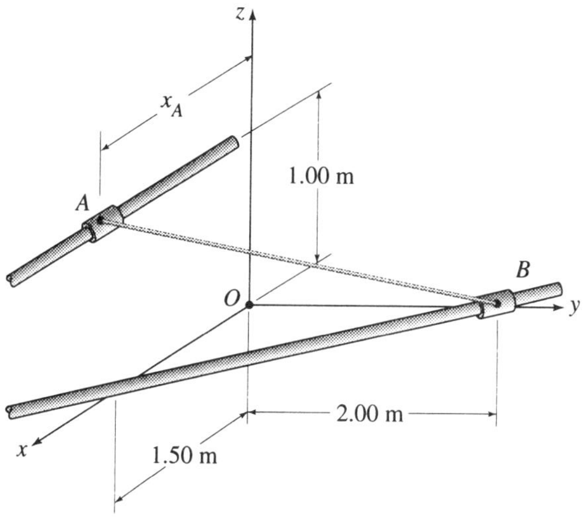

FIGURE P21.51.

21.52 Refer to Problem 21.51, and determine the angular velocity of rod AB which may be assumed *not* to spin about its own axis.

21.53 Frame OAB rotates about the vertical Z axis at an angular velocity $\omega_1 = 50$ rad/s which is increasing at the rate $\alpha_1 = 200$ rad/s as shown in Figure P21.53. Collar C slides freely on rod AB of the frame and, in the position shown, has a velocity relative to the rod $v_R = 40$ in./s which is decreasing at the relative rate $a_R = 800$ in./s^2. For the position shown, determine (a) the absolute velocity of the collar and (b) the absolute acceleration of the collar.

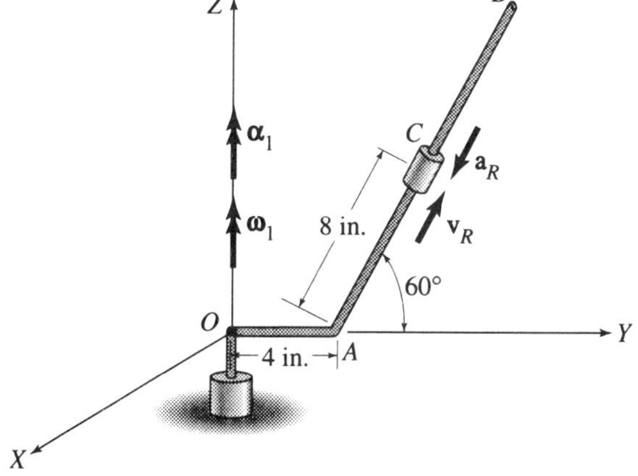

FIGURE P21.53.

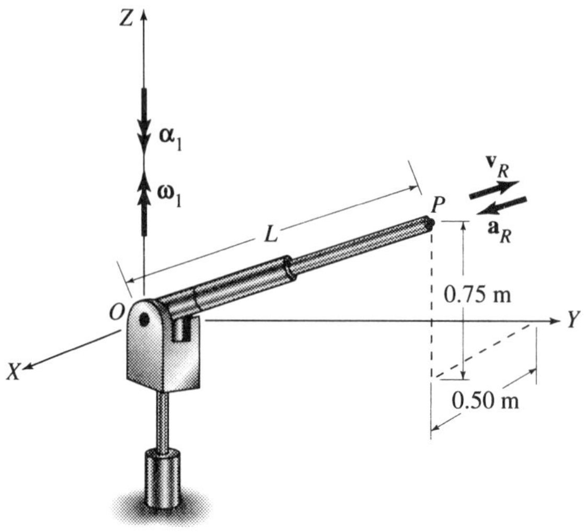

FIGURE P21.54.

21.54 The robotics arm OP, shown in Figure P21.54, rotates about the vertical Z axis at an angular velocity $\omega_1 = 4$ rad/s which is decreasing at the rate $\alpha_1 = 25$ rad/s^2, as shown. In the position shown, the arm has a length $L = 1.20$ m which is increasing at the rate $v_R = 0.25$ m/s. At the instant considered, this rate of length increase is slowing down at the rate $a_R = 1.50$ m/s^2. For the position shown, determine (a) the absolute velocity of point P and (b) the absolute acceleration of point P.

22

Three-Dimensional Kinetics of Rigid Bodies

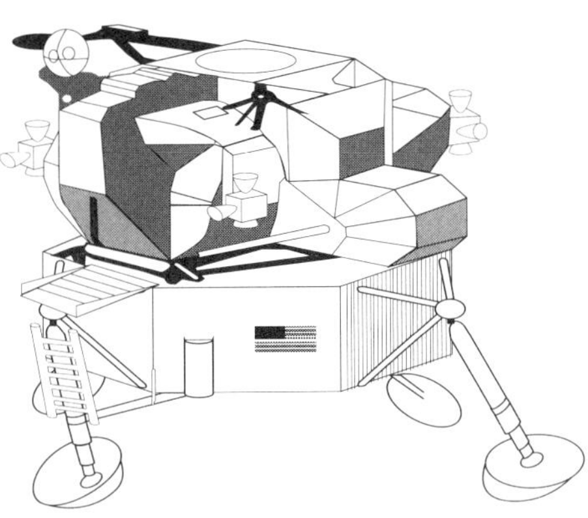

The Lunar Lander the Apollo 11 Spacecraft heads for a landing on the Moon.

On the occasion of its 25th anniversary, the National Academy of Engineering selected the Moon landing as the single most impressive engineering achievement in the past 25 years. That event, which was witnessed on television in 1969 by six hundred million people marked the pinnacle of engineering vision and daring. When Neil Armstrong set the first earthling foot onto that celestial body, human consciousness leaped into space to a degree unequaled since the discovery of the New World. That small step by Armstrong was a triumph for humankind and a fulfillment of the hopes and aspirations of people everywhere without regard to race and national origin. The Moon landing illustrated the vitality and importance of teamwork while eloquently emphasizing the essence of a great nation.

The landing of astronauts on the Moon and returning them safely to Earth ranks with the Egyptian pyramids, the Panama Canal, and the Hanging Gardens of Babylon as the outstanding engineering achievements of all time. It stands alone for the advances and technological sophistication over prior technology. Apollo 11 required a rocket 15 times more powerful, a unique Lunar Lander, and special electronic gear not available at the time the Apollo program was conceived. Engineers were required to break different practical and theoretical barriers and develop modern technologies to meet these challenges. The Apollo program demanded a modern approach to ensure that thousands of government and industrial organizations worked together effectively so that millions of parts would fit their intended function. This was accomplished with utmost precision and unequalled productivity.

In this chapter, you will learn about the interrelationships between forces and the resulting motion they produce in rigid bodies. You will also learn the basic theoretical principles involved in the analysis of spaceships, satellites, and other complex engineering systems. We hope that the topics covered herein will inspire you and ignite the fire that will consume the uncertainty that resides within. The world yearns for another Neil Armstrong and a new Apollo program to meet the challenges facing humankind in the depth of space and here on Earth.

22.1 Moments of Inertia of Composite Masses

A composite mass is one that may be decomposed into two or more simple geometric component parts. The same type of development that was used for the moment of inertia of a composite area in Chapter 11 may be used to show that the moment of inertia of a composite mass, with respect to any axis, is equal to the *algebraic* sum of the mass moments of inertia of the component parts with respect to the same axis. In finding the mass moments of inertia of the component parts with respect to the desired axis, it becomes necessary to use the parallel-axis theorem expressed in Eq. (11.17) and an expression similar to that in Eq. (11.22) may be written for mass moments of inertia. Thus,

$$I = \sum_{i=1}^{n} (I_G + md^2)_i \tag{22.1}$$

where I represents the moment of inertia of the composite mass with respect to any axis and i is the summation index.

In applying Eq. (22.1), it is desirable to have access to values of mass moments of inertia for simple geometric shapes. A set of values of mass moments of inertia for a selected group of simple geometric shapes is given in Appendix B.

The following example illustrates the use of Eq. (22.1) in solving problems.

■ Example 22.1

Four slender rods are attached to a thin disk as shown in Figure E22.1(a). Let the mass of the longest rod be m_r and, for the disk of diameter D, be m_d, and determine the moment of inertia of the composite mass with respect to the x axis. Express the answer in terms of m_r, m_d, and D.

Solution

The composite mass of Figure E22.1(a) may be viewed as consisting of the five masses m_1, m_2, m_3, m_4, and m_5. These masses are labeled in Figure E22.1(a) and properly positioned separately in Figure E22.1(b) with respect to the desired axis (i.e., the x axis), to show clearly the distances to this axis from the mass centers of the five component masses. Therefore, the moment of inertia of the composite mass, with respect to the x, axis may be obtained by adding the moments of inertia, with respect to the same axis, of the five separate masses. Thus, by Eq. (22.1),

$$\begin{aligned} I_x &= \sum_{i=1}^{n} [(I_G + md^2)_x]_i, \\ &= [(I_G + md^2)_x]_1 + [(I_G + md^2)_x]_2 + [(I_G + md^2)_x]_3, \\ &\quad + [(I_G + md^2)_x]_4 + [(I_G + md^2)_x]_5. \end{aligned}$$

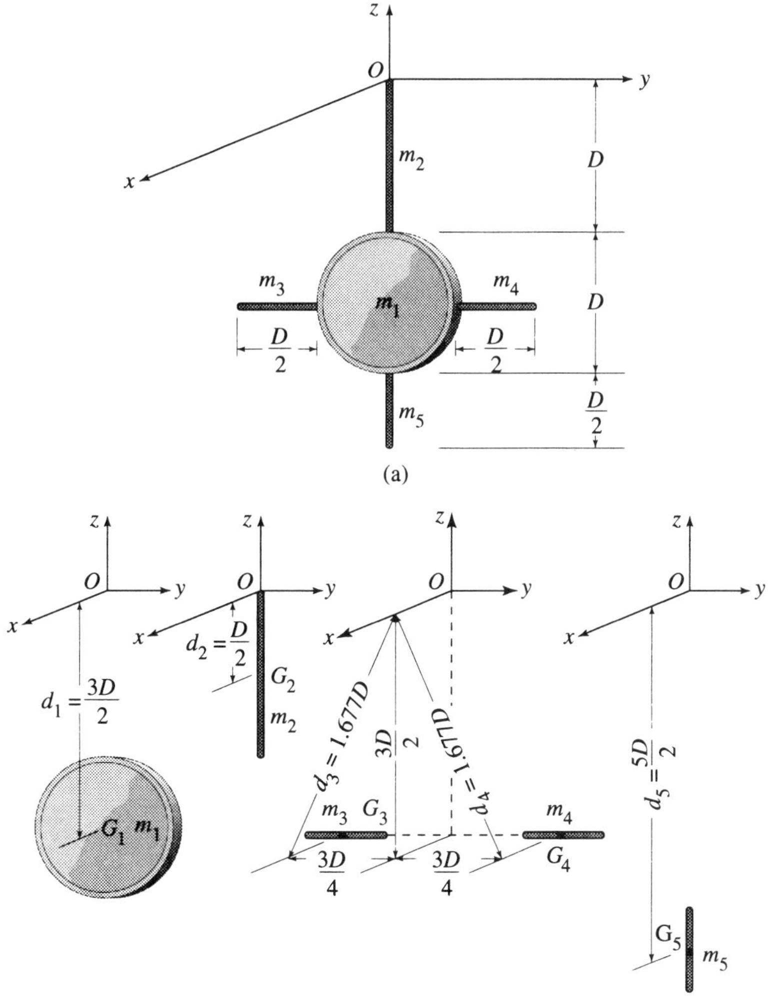

FIGURE E22.1.

Using the values given in Appendix B for the mass moments of inertia with respect to centers of mass and noting that the moments of inertia of m_3 and m_4 are identical, we obtain

$$I_x = \left[\frac{1}{8}mD^2 + md^2\right]_1 + \left[\frac{1}{12}mL^2 + md^2\right]_2$$

$$+ 2\left[\frac{1}{12}mL^2 + md^2\right]_3 + \left[\frac{1}{12}mL^2 + md^2\right]_5.$$

$$I_x = \frac{1}{8}m_d D^2 + m_d\left(\frac{3D}{2}\right)^2 + \frac{1}{12}m_r D^2 + m_r\left(\frac{D}{2}\right)^2$$
$$+ 2\left[\frac{1}{12}\left(\frac{m_r}{2}\right)\left(\frac{D}{2}\right)^2 + \left(\frac{m_r}{2}\right)(1.677D)^2\right]$$
$$+ \frac{1}{12}\left(\frac{m_r}{2}\right)\left(\frac{D}{2}\right)^2 + \left(\frac{m_r}{2}\right)\left(\frac{9D}{4}\right)^2,$$
$$= 2.3750 m_d D^2 + 0.3333 m_r D^2 + 2.8332 m_r D^2 + 2.5417 m_r D^2,$$
$$= (2.3750 m_d + 5.7082 m_r)D^2. \qquad \text{ANS.}$$

■

22.2 Mass Principal Axes and Principal Moments of Inertia

In this section we will consider the moment of inertia of a mass, with respect to an arbitrary axis q passing through some point O, as shown in Figure 22.1. From the fundamental definition of mass moment of inertia,* we may write the mass moment of inertia, I_q, with respect to the arbitrary q axis as

$$I_q = \int s^2\, dm \tag{22.2}$$

where s represents the perpendicular distance from the element of mass dm to the q axis, as shown in Figure 22.1. In view of the three-dimensional nature of the problem, it is convenient to express the distance s using vector algebra as the magnitude of the cross product of two vectors. Thus, denoting by **r** the position vector of the differential

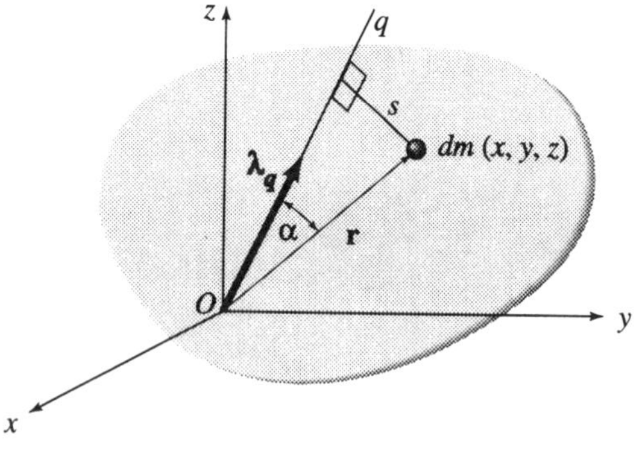

FIGURE 22.1.

* A more complete discussion of moments of inertia (areas and masses) is presented in Chapter 11.

mass dm from point O and by α the angle between $\mathbf{r}$ and the arbitrary q axis, we conclude that the perpendicular distance is $s = r \sin \alpha$ where r is the magnitude of $\mathbf{r}$. By the definition of the vector product of two vectors we may express the vector $\mathbf{s}$ as

$$\mathbf{s} = \boldsymbol{\lambda}_q \times \mathbf{r} \tag{22.3}$$

where $\boldsymbol{\lambda}_q$ is a unit vector along the q axis. Thus,

$$\mathbf{s} = \begin{vmatrix} \mathbf{i} & \mathbf{j} & \mathbf{k} \\ \lambda_x & \lambda_y & \lambda_z \\ x & y & z \end{vmatrix} = (\lambda_y z - \lambda_z y)\mathbf{i} + (\lambda_z x - \lambda_x z)\mathbf{j} + (\lambda_x y - \lambda_y x)\mathbf{k} \tag{22.4}$$

where λ_x, λ_y, and λ_z are the x, y, and z components of the unit vector $\boldsymbol{\lambda}_q$ (i.e., the direction cosines of the q axis) and x, y, and z the scalar components of the position vector $\mathbf{r}$ (i.e., the coordinates of the differential mass dm). Therefore,

$$s^2 = \mathbf{s}\cdot\mathbf{s} = (\lambda_y z - \lambda_z y)^2 + (\lambda_z x - \lambda_x z)^2 + (\lambda_x y - \lambda_y x)^2. \tag{22.5}$$

Thus, Eq. (22.2) may be written as

$$I_q = \int [(\lambda_y z - \lambda_z y)^2 + (\lambda_z x - \lambda_x z)^2 + (\lambda_x y - \lambda_y x)^2]\, dm \tag{22.6}$$

Squaring the quantities in Eq. (22.6) and rearranging terms, we obtain

$$\begin{aligned} I_q &= \lambda_x^2 \int (y^2 + z^2)\, dm + \lambda_y^2 \int (x^2 + z^2)\, dm + \lambda_z^2 \int (x^2 + y^2)\, dm \\ &\quad - 2\lambda_x\lambda_y \int xy\, dm - 2\lambda_x\lambda_z \int xz\, dm - 2\lambda_y\lambda_z \int yz\, dm \\ &= \lambda_x^2 I_x + \lambda_y^2 I_y + \lambda_z^2 I_z - 2\lambda_x\lambda_y \int xy\, dm \\ &\quad - 2\lambda_x\lambda_z \int xz\, dm - 2\lambda_y\lambda_z \int yz\, dm \end{aligned} \tag{22.7}$$

where the integrals $\int (y^2 + z^2)\, dm$, $\int (x^2 + z^2)\, dm$, and $\int (x^2 + y^2)\, dm$ were replaced, respectively, by their equivalent mass moments of inertia I_x, I_y, and I_z. The mixed integrals $\int xy\, dm$, $\int xz\, dm$, and $\int yz\, dm$ in Eq. (22.7) are known as the *mass products of inertia* with respect to the x-y-z coordinate system. In keeping with the notation for area products of inertia of Chapter 11, we define the following three mass products of inertia,

$$I_{xy} = \int xy\,dm,$$

$$I_{xz} = \int xz\,dm, \tag{22.8}$$

$$I_{yz} = \int yz\,dm,$$

where, by referring to Figure 22.2, we may interpret $I_{xy} = \int xy\,dm$ as the mass product of inertia with respect to the z-x and z-y planes, $I_{xz} = \int xz\,dm$ as the mass product of inertia with respect to the y-x and y-z planes, and $I_{yz} = \int yz\,dm$ as the mass product of inertia with respect to the x-y and x-z planes. Thus, for example, I_{xy} is zero if either z-x or z-y is a plane of symmetry. Similarly, I_{xz} is zero if either y-x and y-z is a plane of symmetry, and I_{yz} is zero if either x-y or x-z is a plane of symmetry. As in the case of area products of inertia, these conditions of symmetry are useful in determining mass products of inertia. Note that planes of symmetry are not the only planes for which the mass product of inertia is zero. The mass product of inertia may vanish with respect to planes which may not be planes of symmetry. Whether due to symmetry or not, if the mass product of inertia vanishes, the corresponding planes are known as principal planes.

Substituting Eq. (22.8) in Eq. (22.7), we obtain

$$I_q = \lambda_x^2 I_x + \lambda_y^2 I_y + \lambda_z^2 I_z - 2\lambda_x\lambda_y I_{xy} - 2\lambda_x\lambda_z I_{xz} - 2\lambda_y\lambda_z I_{yz}. \tag{22.9}$$

As in the case of area products of inertia, a parallel axis theorem may be developed for mass products of inertia. Thus, consider the body of mass m shown in Figure 22.3 where the X, Y, Z coordinate system originates at the mass center G and the x, y, and z axes are parallel, respectively, to the X, Y, and Z axes, and have an origin at an arbitrary point O. Because $x = X + a$, $y = Y + b$ and $z = Z + c$, we may write

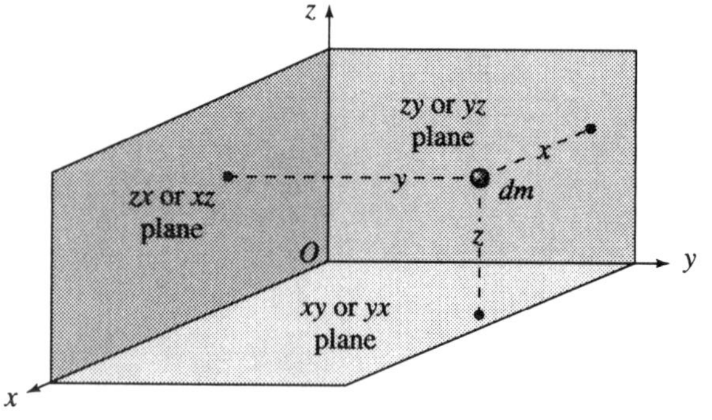

FIGURE 22.2.

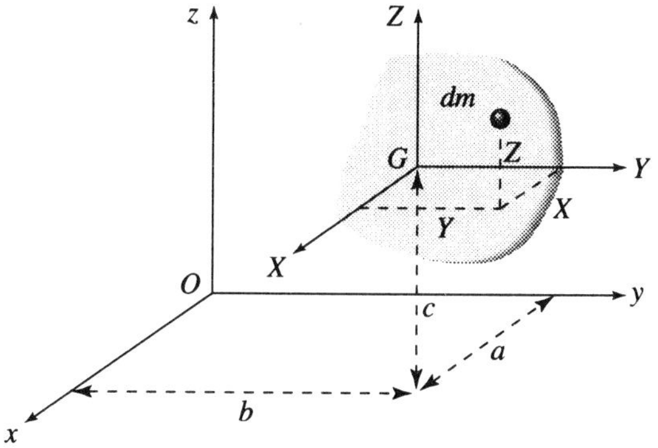

FIGURE 22.3.

$$I_{xy} = \int xy\,dm = \int (X+a)(Y+b)\,dm$$
$$= \int XY\,dm + ab\int dm + a\int Y\,dm + b\int X\,dm. \qquad (22.10)$$

The first integral is I_{XY}, the mass product of inertia with respect to the centroidal Z-X and Z-Y planes. Because $m = \int dm$, the second term may be written as $(ab)m$. The third and fourth integrals must vanish because they represent the first moments of mass with respect to planes through the center of mass, point G. Thus,

$$I_{xy} = I_{XY} + (ab)m. \qquad (22.11)$$

Similarly,

$$I_{xz} = I_{XZ} + (ac)m, \qquad (22.12)$$

and

$$I_{yz} = I_{YZ} + (bc)m. \qquad (22.13)$$

Let us return now to Eq. (22.9). This equation defines the mass moment of inertia I_q, with respect to an arbitrary q axis through some point O, in terms of the direction cosines of the axis q and the mass moments and products of inertia with respect to the x, y, z coordinate system originating at the same point O, as shown in Figure 22.4. By varying the direction cosines, we can obtain the mass moment of inertia I_q with respect to different q axes. As shown in Figure 22.4, if we locate a point K on axis q at a distance $\mathrm{OK} = d = \dfrac{1}{\sqrt{I_q}}$, we conclude that,

$$I_q = \frac{1}{d^2} \qquad (22.14a)$$

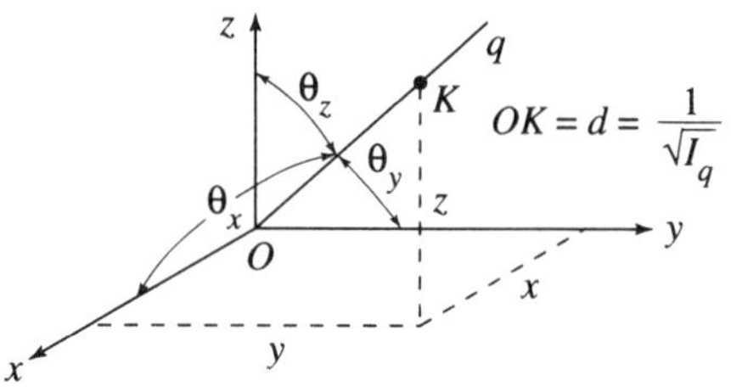

FIGURE 22.4.

and

$$\left.\begin{aligned} \lambda_x &= \cos\theta_x = \frac{x}{d} \\ \lambda_y &= \cos\theta_y = \frac{y}{d} \\ \lambda_z &= \cos\theta_z = \frac{z}{d} \end{aligned}\right\}. \tag{22.14b}$$

Substituting Eqs. (22.14) in Eq. (22.9) and simplifying yields

$$1 = I_x x^2 + I_y y^2 + I_z z^2 - 2I_{xy}xy - 2I_{xz}xz - 2I_{yz}yz. \tag{22.15}$$

As will be shown below, Eq. (22.15) defines an ellipsoid with center at point O as in Figure 22.5(a). Such an ellipsoid is known as the *ellipsoid of inertia* for the body at the specific point O. The ellipsoid of inertia is unique for a specific point O on the body and the coordinates of points on its surface are proportional to the mass moments and mass products of inertia with respect to the coordinate axes originating at point O. If the coordinate axes through point O are changed, the mass moments and mass products of inertia would also change, although the ellipsoid of inertia remains the same. If the x, y, and z coordinate axes through point O are rotated to bring them into coincidence with the principal axes, u, v, w of the ellipsoid, the mass products of inertia I_{uv}, I_{uw}, and I_{vw} vanish because, under these conditions, uv, uw, and vw are principal planes. If the axis rotation is such that x coincides with u, y with v, and z with w, Eq. (22.15) yields,

$$1 = I_u u^2 + I_v v^2 + I_w w^2. \tag{22.16}$$

Because $I_u = 1/d_u^2$, $I_v = 1/d_v^2$ and $I_w = 1/d_w^2$ by Eq. (22.14a) it follows that

$$1 = \frac{u^2}{d_u^2} + \frac{v^2}{d_v^2} + \frac{w^2}{d_w^2} \tag{22.17}$$

which is the equation of an ellipsoid in terms of the principal u, v, and w axes as shown in Figure 22.5(b). The quantities I_u, I_v, and I_w in

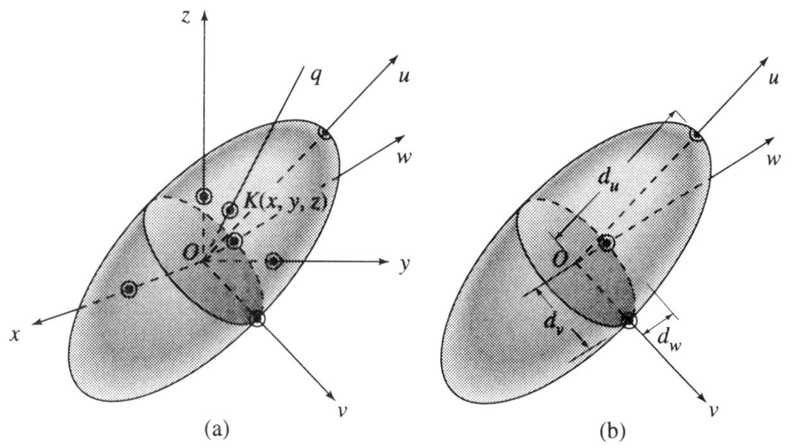

FIGURE 22.5.

Eq. (22.16) are known as the *mass principal moments of inertia*, and the corresponding u, v, and w axes are the *mass principal axes of inertia*.

Computing the three principal mass moments of inertia of a body from knowledge of the mass moments and mass products of inertia with respect to an arbitrary x-y-z coordinate system requires solving a cubic equation. Thus, as shown in Section 22.4, Eqs. (22.37), the x, y, and z components of the angular momentum $\mathbf{H}$ of a body are given by

$$\begin{aligned} H_x &= I_x\omega_x - I_{xy}\omega_y - I_{xz}\omega_z, \\ H_y &= -I_{yx}\omega_x + I_y\omega_y - I_{yz}\omega_z, \\ H_z &= -I_{zx}\omega_x - I_{zy}\omega_y + I_z\omega_z. \end{aligned} \tag{22.18}$$

In general, the angular momentum vector $\mathbf{H}$ does not coincide with the angular velocity vector $\boldsymbol{\omega}$ except when the rigid body rotates about one of its three principal axes of inertia. If the body rotates at an angular velocity $\boldsymbol{\omega}$ about one of its three principal axes of inertia, with respect to which the moment of inertia is I, the corresponding ellipsoid of inertia will be as shown in Figure 22.6. By Eq. (22.38) of Section 2.4, the angular momentum $\mathbf{H}$ is a vector that *coincides* with the principal axis of rotation as shown in Figure 22.6. It follows, therefore, that $H_x = I\omega_x$, $H_y = I\omega_y$ and $H_z = I\omega_z$ and Eqs. (22.18) become

$$\begin{aligned} (I_x - I)\omega_x - I_{xy}\omega_y - I_{xz}\omega_z &= 0, \\ -I_{yx}\omega_x + (I_y - I)\omega_y - I_{yz}\omega_z &= 0, \\ -I_{zx}\omega_x - I_{zy}\omega_y + (I_z - I)\omega_z &= 0. \end{aligned} \tag{22.19}$$

According to the theory of algebraic equations* a nontrivial solution

* See for example Numerical Methods in Engineering Practice by Amir W. Al-Khafaji and John R. Tooley, Holt, Reinhart & Winston, 1 ed., 1986.

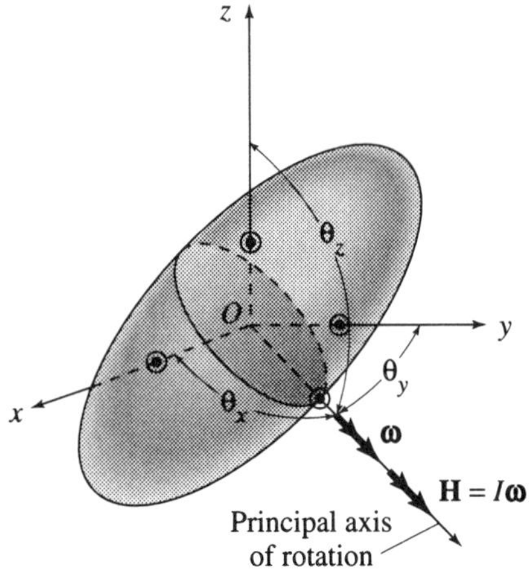

FIGURE 22.6.

of Eqs. (22.19) is possible only if the determinant of the coefficients of ω_x, ω_y, and ω_z vanishes. Thus,

$$\begin{vmatrix} (I_x - I) & -I_{xy} & -I_{xz} \\ -I_{yx} & (I_y - I) & -I_{yz} \\ -I_{zx} & -I_{zy} & (I_z - I) \end{vmatrix} = 0 \tag{22.20}$$

which is a determinant expressing a 3×3 matrix known as the inertial tensor. Upon expanding this determinant and knowing that $I_{yx} = I_{xy}$, $I_{zx} = I_{xz}$, and $I_{zy} = I_{yz}$, we may rewrite Eq. (22.20) in the form,

$$\begin{aligned} (I_x - I)(I_y - I)(I_z - I) - I_{xy}^2(I_z - I) - I_{xz}^2(I_y - I) \\ - I_{yz}^2(I_x - I) - 2I_{xy}I_{xz}I_{yz} = 0. \end{aligned} \tag{22.21}$$

Equation (22.21) is a cubic equation in I whose three roots provide the mass principal moments of inertia I_u, I_v, and I_w.

By reference to Figure 22.6 we conclude that $\omega_x = \omega \cos\theta_x = \omega\lambda_x$, $\omega_y = \omega\cos\theta_y = \omega\lambda_y$ and $\omega_z = \omega\cos\theta_z = \omega\lambda_z$. Substituting in Eqs. (22.19) leads to

$$\begin{aligned} (I_x - I)\lambda_x - I_{xy}\lambda_y - I_{xz}\lambda_z &= 0, \\ -I_{yx}\lambda_x + (I_y - I)\lambda_y - I_{yz}\lambda_z &= 0, \\ -I_{zx}\lambda_x - I_{zy}\lambda_y + (I_z - I)\lambda_z &= 0. \end{aligned} \tag{22.22}$$

In Eq. (22.22), the quantity I represents one of the three principal moments of inertia, say I_u, and the quantities λ_x, λ_y, and λ_z will then be the direction cosines for the corresponding u axis.

It should be pointed out, however, that because the determinant of the coefficients of λ_x, λ_y, and λ_z vanishes (see Eq. (22.20)), Eqs. (22.22) cannot be solved independently for these quantities unless they are supplemented by the relationship that exists among the λs. Thus, any two of Eqs. (22.22) plus the relationship $\lambda_x^2 + \lambda_y^2 + \lambda_z^2 = 1$ are used to obtain the values of λ_x, λ_y, and λ_z. This type of problem is referred to as an *Eigenproblem* and, for more detailed information, the reader is referred to the text: Numerical Methods in Engineering Practice by Amir W. Al-Khafaji and John R. Tooley, Holt, Reinhart & Winston, 1st ed., 1986, Chapter 7.

The three principal mass moments of inertia I_u, I_v, and I_w represent the maximum, minimum, and intermediate values of the mass moments of inertia through a point O inside or outside of a body. For the sake of consistency with area principal moments of inertia (see Chapter 11), we will reserve I_u for the maximum, I_v for the minimum, and I_w for the intermediate value.

The following examples illustrate some of the concepts discussed above. Additional examples are presented in Chapter 11.

Example 22.2

Use integration to determine the mass products of inertia I_{xy}, I_{xz}, and I_{yz} for the body shown in Figure E22.2. Assume that the mass density ρ is a constant. Express the answers in terms of the mass m of the body.

Solution

Select a rectangular element of mass dm as shown in Figure E22.2 where $dm = \rho\, dv = \rho\,(xz\, dy)$. The width x and the height z of the rectangular element vary according to the relationships

$$x = 4a - \tfrac{1}{3}y$$

and

$$z = a + \tfrac{1}{3}y.$$

Thus,

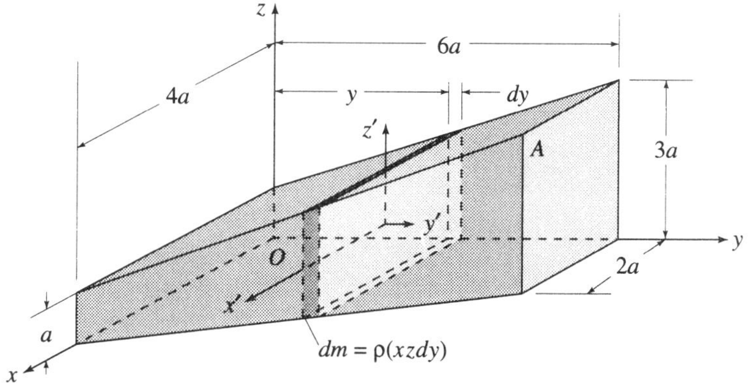

FIGURE E22.2.

$$dm = \rho(4a - \tfrac{1}{3}y)(a + \tfrac{1}{3}y)\,dy = \rho(4a^2 + ay - \tfrac{1}{9}y^2)\,dy,$$

and

$$m = \int dm = \rho \int_0^{6a} \left(4a^2 + ay - \frac{1}{9}y^2\right) dy = \rho\left[4a^2 y + \frac{a}{2}y^2 - \frac{1}{27}y^3\right]\Bigg|_0^{6a}$$
$$= 34a^3\rho.$$

Observing that the centroidal $z'x'$, $z'y'$, and $x'y'$ planes for the element dm are planes of symmetry and using the parallel axis theorem, Eq. (22.11), we obtain

$$dI_{xy} = 0 + \left(\frac{x}{2}\right)(y)\,dm = \frac{1}{2}\left(4a - \frac{1}{3}y\right)(y)\rho\left(4a^2 + ay - \frac{1}{9}y^2\right)dy$$
$$= \frac{1}{2}\rho\left(16a^3 y + \frac{8}{3}a^2y^2 - \frac{7}{9}ay^3 + \frac{1}{27}y^4\right)dy.$$

Therefore,

$$I_{xy} = \int dI_{xy},$$
$$I_{xy} = \frac{1}{2}\rho \int_0^{6a} \left(16a^3 y + \frac{8}{3}a^2y^2 - \frac{7}{9}ay^3 + \frac{1}{27}y^4\right)dy = 142.8\rho a^5$$
$$= 4.2a^2(34a^3\rho) = 4.2ma^2. \qquad \text{ANS.}$$

Similarly, by Eq. (22.12),

$$I_{xz} = \frac{1}{4}\rho \int_0^{6a} \left(16a^4 + 8a^3 y + \frac{1}{9}a^2y^2 - \frac{2}{9}ay^3 + \frac{1}{81}y^4\right)dy$$
$$= 48.8\rho a^5 = 1.435a^2(34a^3\rho) = 1.435ma^2, \qquad \text{ANS.}$$

and, by Eq. (22.13), we obtain

$$I_{yz} = \frac{1}{2}\rho \int_0^{6a} \left(4a^3 y + \frac{7}{3}a^2y^2 + \frac{2}{9}ay^3 - \frac{1}{27}y^4\right)dy$$
$$= 127.2\rho a^5 = 3.741a^2(34a^3\rho) = 3.741ma^2. \qquad \text{ANS.}$$

■ Example 22.3

Use integration to find the mass moment of inertia with respect to the line OA for the body shown in Figure E22.3. Express the answer in terms of the mass m of the body. Assume that the mass density ρ is a constant. Note that this is the same body used in Example 22.2.

Solution

Determining the mass moment of inertia I_{OA} may be accomplished by using Eq. (22.9). This equation requires knowledge of the mass

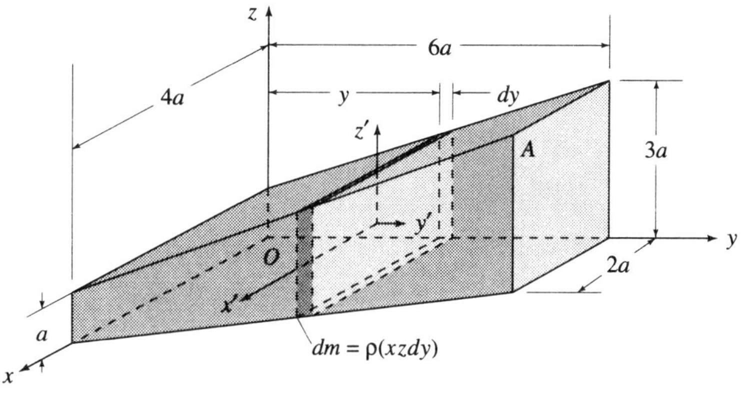

FIGURE E22.3.

moments of inertia I_x, I_y, and I_z as well as the direction cosines of line OA. Using the parallel axis theorem for mass moments of inertia and the information given in Appendix B,

$$dI_x = \tfrac{1}{12}z^2\,dm + [y^2 + (\tfrac{1}{2}z)^2]\,dm = (y^2 + \tfrac{1}{3}z^2)\,dm.$$

From Example 22.2, we have $z = a + y/3$ and $dm = \rho(4a^2 + ay - \frac{1}{9}y^2)dy$. Therefore,

$$dI_x = \left[y^2 + \frac{1}{3}\left(a + \frac{1}{3}y\right)^2\right]\rho\left(4a^2 + ay - \frac{1}{9}y^2\right)dy$$

$$= \rho\left(\frac{4}{3}a^4 + \frac{11}{9}a^3y + \frac{117}{27}a^2y^2 + \frac{82}{81}ay^3 - \frac{28}{243}y^4\right)dy,$$

and

$$I_x = \int_0^{6a} dI_x = 490.8\rho a^5 = 14.435a^2(34a^3\rho) = 14.435ma^2 \qquad \text{ANS.}$$

Similarly,

$$dI_y = \frac{1}{12}(x^2 + z^2)\,dm + \left[\left(\frac{1}{2}x\right)^2 + \left(\frac{1}{2}z\right)^2\right]dm = \left(\frac{1}{3}x^2 + \frac{1}{3}z^2\right)dm$$

$$= \frac{1}{3}\rho\left(68a^4 + 9a^3y - 3a^2y^2 + \frac{4}{9}ay^3 - \frac{2}{81}y^4\right)dy,$$

and

$$I_y = \int_0^{6a} dI_y = 153.2\rho a^5 = 4.506a^2(34a^3\rho) = 4.506ma^2. \qquad \text{ANS.}$$

Also,

$$dI_z = \frac{1}{12}x^2\,dm + \left[\left(\frac{1}{2}x\right)^2 + y^2\right]dm = \left(\frac{1}{3}x^2 + y^2\right)dm$$

$$= \frac{4}{3}\rho\left(16a^4 + \frac{4}{3}a^3y + 2a^2y^2 + \frac{23}{27}ay^3 - \frac{7}{81}y^4\right)dy,$$

and

$$I_z = \int_0^{6a} dI_z = 540.8\rho a^5 = 15.906a^2(34a^3\rho) = 15.906ma^2 \qquad \text{ANS.}$$

Find a unit vector $\boldsymbol{\lambda}_{\mathrm{OA}}$ along line OA. Thus,

$$\boldsymbol{\lambda}_{\mathrm{OA}} = \frac{2a\mathbf{i} + 6a\mathbf{j} + 3a\mathbf{k}}{\sqrt{(2a)^2 + (6a)^2 + (3a)^2}} = \left(\frac{2}{7}\right)\mathbf{i} + \left(\frac{6}{7}\right)\mathbf{j} + \left(\frac{3}{7}\right)\mathbf{k},$$

and

$$(\lambda_{\mathrm{OA}})_x = \frac{2}{7},\ (\lambda_{\mathrm{OA}})_y = \frac{6}{7},\quad \text{and} \quad (\lambda_{\mathrm{OA}})_z = \frac{3}{7}.$$

Also, from the solution of Example 22.2

$$I_{xy} = 4.200ma^2,\ I_{xy} = 1.435ma^2,\quad \text{and} \quad I_{xy} = 3.741ma^2.$$

Thus, by Eq. (22.9),

$$I_{\mathrm{OA}} = \left(\frac{2}{7}\right)^2(14.435ma^2) + \left(\frac{6}{7}\right)^2(4.506ma^2) + \left(\frac{3}{7}\right)^2(15.906ma^2)$$

$$- 2\left(\frac{2}{7}\right)\left(\frac{6}{7}\right)(4.200ma^2) - 2\left(\frac{2}{7}\right)\left(\frac{3}{7}\right)(1.435ma^2)$$

$$- 2\left(\frac{6}{7}\right)\left(\frac{3}{7}\right)(3.471ma^2),$$

$$I_{\mathrm{OA}} = 2.253ma^2. \qquad \text{ANS.}$$

■ Example 22.4

Refer to the body shown in Figure E22.4. Determine the mass principal moments of inertia I_u, I_v, and I_w. Also determine the direction cosines corresponding to each of the three mass principal axes of inertia u, v, and w through point O. The following values are obtained from the solutions of Examples 22.2 and 22.3.

$$I_x = 14.435ma^2,\ I_y = 4.506ma^2,\ I_z = 15.906ma^2,$$

$$I_{xy} = 4.200ma^2,\ I_{xz} = 1.435ma^2,\quad \text{and} \quad I_{yz} = 3.741ma^2.$$

Solution

Substituting the given mass moments and products of inertia in Eq. (22.21) and simplifying yields

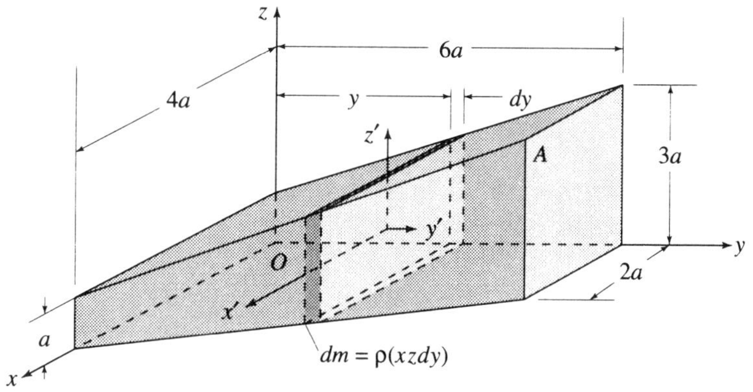

FIGURE E22.4.

$$I^3 - 34.847I^2 + 332.625I - 497.618 = 0.$$

Any iterative technique, such as Newton's method, may be used to obtain

$$I_u = 17.049ma^2, I_w = 15.970ma^2, \quad \text{and} \quad I_v = 1.828ma^2. \qquad \text{ANS.}$$

Substituting $I_u = 17.049ma^2$ for I in Eq. (22.22) and simplifying yields the following relationships:

$$2.614\lambda_x + 4.200\lambda_y + 1.435\lambda_z = 0,$$

$$4.200\lambda_x + 12.543\lambda_y + 3.741\lambda_z = 0,$$

and

$$1.435\lambda_x + 3.741\lambda_y + 1.143\lambda_z = 0.$$

Any two of the three equations above plus the relationship $\lambda_x^2 + \lambda_y^2 + \lambda_z^2 = 1$ may be solved simultaneously to obtain the direction cosines for the principal u axis. Thus,

$$\lambda_x = -0.145, \lambda_y = -0.238, \quad \text{and} \quad \lambda_z = 0.960. \qquad \text{ANS.}$$

Proceeding in the same manner for I_w, we obtain the following direction cosines for the principal w axis. Thus,

$$\lambda_x = 0.932, \lambda_y = -0.358, \quad \text{and} \quad \lambda_z = 0.052. \qquad \text{ANS.}$$

Similarly, the direction cosines for the principal v axis are

$$\lambda_x = 0.332, \lambda_y = 0.903, \quad \text{and} \quad \lambda_z = 0.274. \qquad \text{ANS.}$$

■

Problems

22.1 Determine the moment of inertia and radius of gyration, with respect to the x axis, for the composite mass shown in Figure P22.1. Express your answers in terms of the constant mass density ρ (kg/m^3).

22.2 A machine member made of steel is shown in Figure P22.2. The mass density for steel is approximately 15 slug/ft^3. Determine the mass moment of inertia and radius of gyration with respect to the x axis.

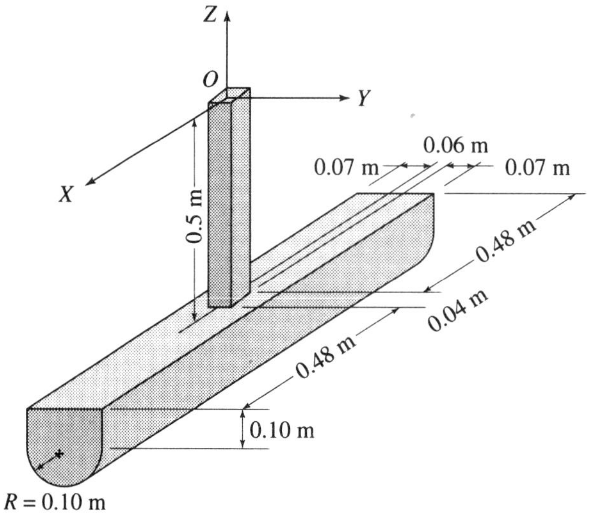

FIGURE P22.1.

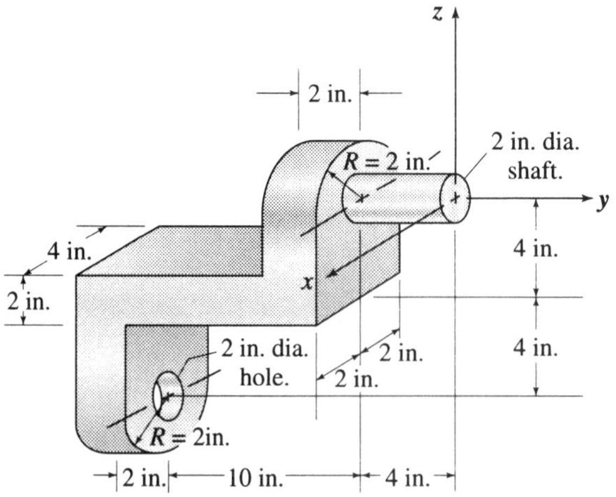

FIGURE P22.2.

22.3 A composite member made of aluminum is shown in Figure P22.3. The mass density for aluminum is approximately 2700 kg/m^3. Determine the mass moment of inertia and radius of gyration with respect to the x axis.

22.4 Four slender rods are attached to a thin disk of diameter D as shown in Figure P22.4. Let the mass of the longest rod be m_r and that for the disk be m_d, and determine the moment of inertia of the composite mass with respect to the y axis. Express your answer in terms of m_r, m_d, and D.

22.5 Use integration to compute the mass products of inertia I_{xy}, I_{xz}, and I_{yz} for the body shown in Figure P22.5. Assume that the mass density ρ is a con-

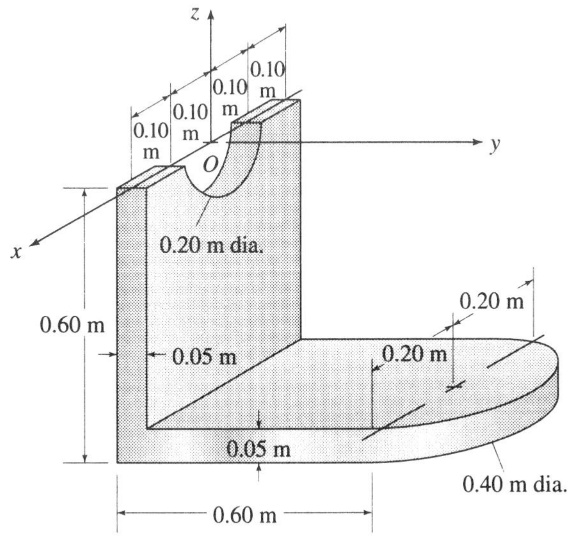

FIGURE P22.3.

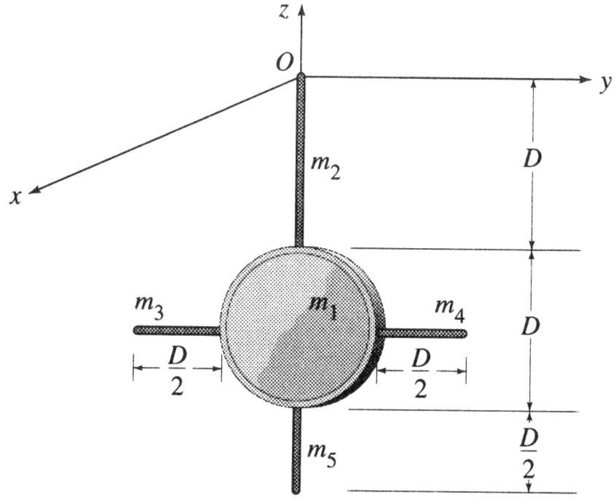

FIGURE P22.4.

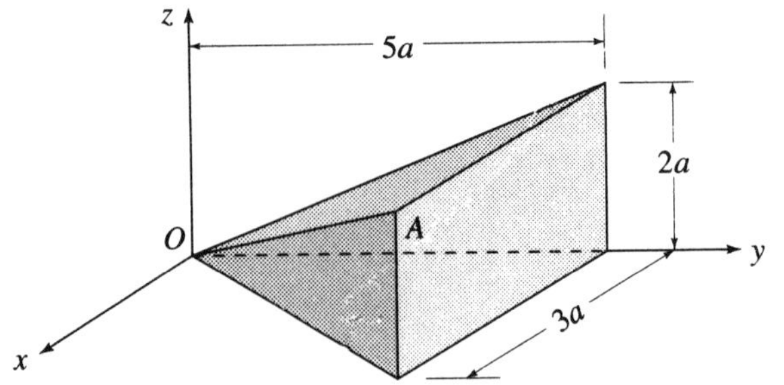

FIGURE P22.5.

stant. Express the answers in terms of the mass m of the body.

22.6 Refer to Problem 22.5. Determine the mass moment of inertia with respect to line OA. Express the answers in terms of the mass m of the body.

22.7 Use integration to compute the mass products of inertia I_{xy}, I_{xz}, and I_{yz} for the body shown in Figure P22.7. The density of the material $\rho = 490$ lb/ft^3.

22.8 Determine the products of inertia I_{xy}, I_{xz}, and I_{yz} for the homogeneous bent rod shown in Figure P22.8. Express the answers in terms of the mass m of the entire rod.

22.9 Refer to Problem 22.8. Compute the

22.10 Find the products of inertia I_{xy}, I_{xz}, and I_{yz} for the homogeneous bent plate shown in Figure P22.10. The thickness t of the plate is uniform, and $t = 0.03$ m. The material is steel whose mass density $\rho = 7850$ kg/m^3.

22.11 Refer to Problem 22.10. Determine the mass moment of inertia with respect to line OA.

22.12 Find the products of inertia I_{xy} and I_{xz} for the composite body shown in Figure P22.12. Assume that the mass density ρ is a constant. Express the answer in terms of the mass m of the composite body.

22.13 Use the results of Problems 22.5 and 22.6 to compute the mass principal moments of inertia with respect to axes through point O in Figure P22.5. Also compute the direction cosines for the u

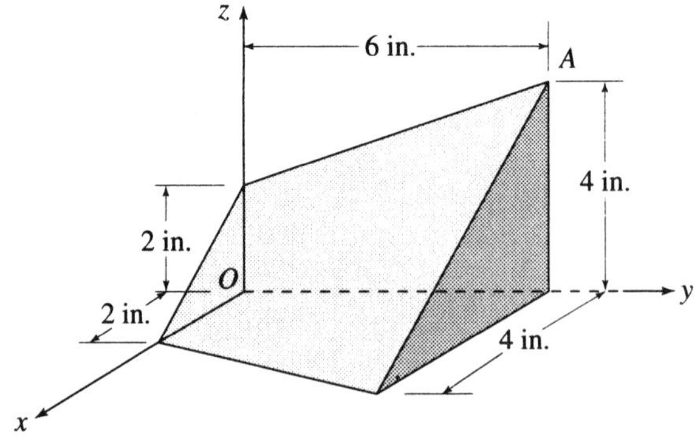

FIGURE P22.7.

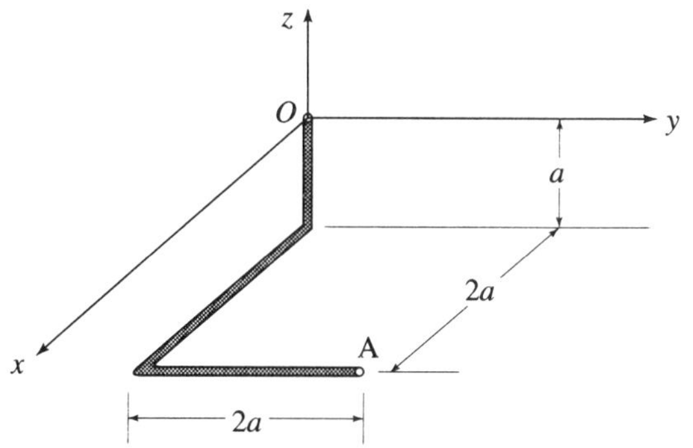

FIGURE P22.8.

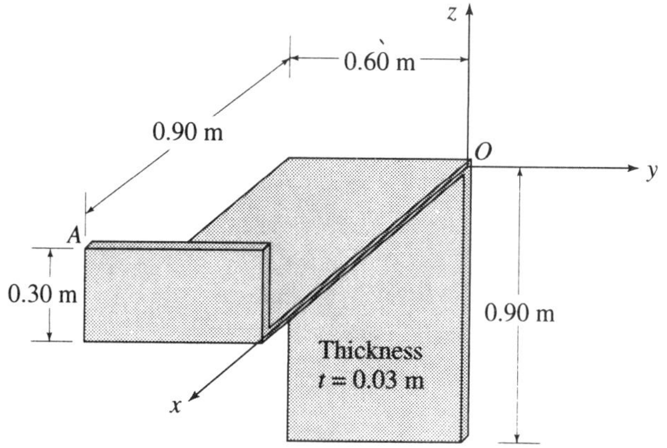

FIGURE P22.10.

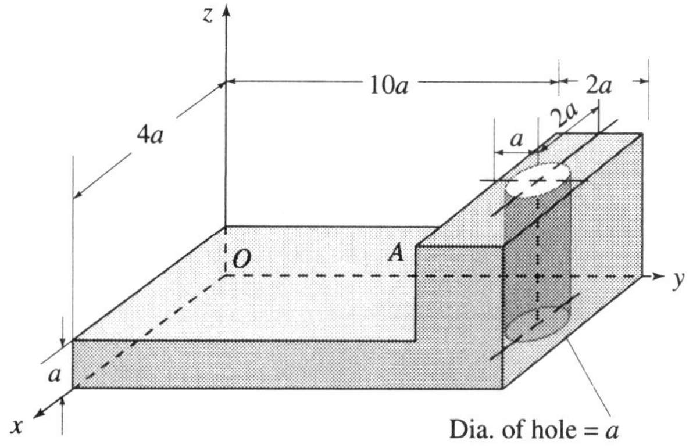

FIGURE P22.12.

axis with respect to which the mass moment of inertia is maximum.

22.14 Use the result of Problems 22.8 and 22.9 to compute the mass principal moments of inertia with respect to axes through point O in Figure P22.8. Also find the direction cosines for the v axis with respect to which the mass moment of inertia is minimum.

22.15 Use the results of Problems 22.10 and 22.11 to determine the mass principal moments of inertia with respect to axes through point O in Figure P22.10. Also compute the direction consines for the w axis with respect to which the mass moment of inertia is intermediate between the maximum and the minimum.

22.3 The Work-Energy Principle

The work-energy principle was developed in Chapter 19 for the case of rigid bodies executing plane motion. This principle is equally applicable to the case of rigid bodies in three-dimensional motion. For convenience, an expression of this principle is given below as Eq. (22.23).

$$\sum U_{1\to 2} = \sum T_2 - \sum T_1 \tag{22.23}$$

The work performed on a given system, $\sum U_{1\to 2}$ by the forces and moments acting on the system is determined in the same manner discussed in Chapter 19 and elsewhere in the text. However, the kinetic energy of the system, at any instant during its three-dimensional motion represented by the quantity $\sum T$, requires formulation. Thus, consider the rigid body shown in Figure 22.7, which is executing motion in three-dimensional space, characterized at a given instant by the linear velocity of its mass center $\mathbf{v}_G$ and its angular velocity $\boldsymbol{\omega}$. The X-Y-Z coordinate system is fixed in space whereas the x-y-z coordinate system, attached to the body's mass center, is arbitrarily oriented with respect to the X-Y-Z coordinate system.

On the basis of the relationships developed in Chapter 15 for the kinetic energy of a system of particles, the kinetic energy T for the rigid body of Figure 22.7, composed of an infinite number of particles, may be written in the form

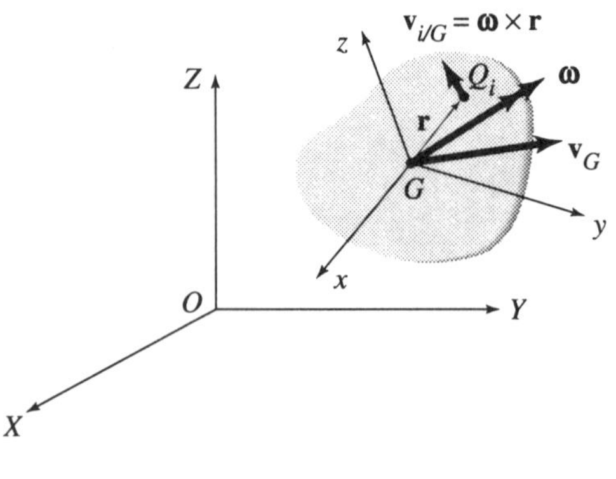

FIGURE 22.7.

$$T = \frac{1}{2} m v_G^2 + \frac{1}{2} \int v_{i/G}^2 \, dm. \tag{22.24}$$

In Eq. (22.24), m is the total mass of the rigid body, $v_{i/G}$ is the relative velocity of any particle Q_i of mass dm having a position relative to the x-y-z coordinate system defined by the position vector $\mathbf{r}$. Because $\mathbf{v}_{i/G} = |\boldsymbol{\omega} \times \mathbf{r}|$, $\boldsymbol{\omega} = \omega_x \mathbf{i} + \omega_y \mathbf{j} + \omega_z \mathbf{k}$, and $\mathbf{r} = x\mathbf{i} + y\mathbf{j} + z\mathbf{k}$, the second term in Eq. (22.24) may be evaluated as follows:

$$\begin{aligned}
\frac{1}{2} \int v_{i/G}^2 \, dm &= \frac{1}{2} \int |\boldsymbol{\omega} \times \mathbf{r}|^2 \, dm, \\
&= \frac{1}{2} \int |(\omega_x \mathbf{i} + \omega_y \mathbf{j} + \omega_z \mathbf{k}) \times (x\mathbf{i} + y\mathbf{j} + z\mathbf{k})|^2 \, dm, \\
&= \frac{1}{2} \int [(\omega_y z - \omega_z y)^2 + (\omega_z x - \omega_x z)^2 + (\omega_x y - \omega_y x)^2] \, dm, \\
&= \frac{1}{2} \omega_x^2 \int (y^2 + z^2) \, dm + \frac{1}{2} \omega_y^2 \int (x^2 + z^2) \, dm \\
&\quad + \frac{1}{2} \omega_z^2 \int (x^2 + y^2) \, dm - \omega_x \omega_y \int xy \, dm \\
&\quad - \omega_x \omega_z \int xz \, dm - \omega_y \omega_z \int yz \, dm.
\end{aligned} \tag{22.25}$$

Using the relationships $I_x = \int (y^2 + z^2) \, dm$, $I_y = \int (x^2 + z^2) \, dm$, $I_z = \int (x^2 + y^2) \, dm$, $I_{xy} = \int xy \, dm$, $I_{xz} = \int xz \, dm$, and $I_{yz} = \int yz \, dm$, Eq. (22.25) may be stated in terms of the mass moments and products of inertia as follows:

$$\begin{aligned}
\frac{1}{2} \int v_{i/G}^2 \, dm = \frac{1}{2} (\omega_x^2 I_x + \omega_y^2 I_y + \omega_z^2 I_z) - \omega_x \omega_y I_{xy} - \omega_x \omega_z I_{xz} \\
- \omega_y \omega_z I_{yz}.
\end{aligned} \tag{22.26}$$

Therefore, Eq. (22.24), expressing the kinetic energy of a rigid body in three-dimensional motion, may be written in the following form:

$$\begin{aligned}
T = \tfrac{1}{2} m v_G^2 + \tfrac{1}{2} (\omega_x^2 I_x + \omega_y^2 I_y + \omega_z^2 I_z) - \omega_x \omega_y I_{xy} - \omega_x \omega_z I_{xz} \\
- \omega_y \omega_z I_{yz}.
\end{aligned} \tag{22.27}$$

Kinetic Energy Using Principal Axes of Inertia

If the x, y, and z axes coincide with the u, v and w principal centroidal axes of inertia, respectively, then, $I_{xy} = I_{uv} = 0$, $I_{xz} = I_{uw} = 0$, and $I_{yz} = I_{vw} = 0$, and Eq. (22.27) becomes

$$T = \tfrac{1}{2} m v_G^2 + \tfrac{1}{2} (\omega_u^2 I_u + \omega_v^2 I_v + \omega_w^2 I_w). \tag{22.28}$$

Kinetic Energy When Body Rotates about a Fixed Point

Equation (22.27) may be specialized to the case when the rigid body rotates at an angular velocity $\boldsymbol{\omega}$ about a fixed point O as shown in Figure 22.8. In such a case, $\mathbf{v}_G = \boldsymbol{\omega} \times \mathbf{R}$ where $\mathbf{R}$ represents the position vector of the mass center, point G, relative to the fixed point O. Thus, the first term in Eq. (22.27) may be evaluated in the following manner:

$$\begin{aligned}\tfrac{1}{2}mv_G^2 &= \tfrac{1}{2}m|\boldsymbol{\omega} \times \mathbf{R}|^2 \\ &= \tfrac{1}{2}m|[(\omega_X \mathbf{I} + \omega_Y \mathbf{J} + \omega_Z \mathbf{K}) \times (X\mathbf{I} + Y\mathbf{J} + Z\mathbf{K})]|^2\end{aligned}$$

where X, Y, and Z are the coordinates of G measured with respect to the fixed X-Y-Z coordinate system and $\mathbf{I}$, $\mathbf{J}$, and $\mathbf{K}$ represent unit vectors along the X, Y, and Z axes, respectively. Evaluation of the above vector product yields

$$\begin{aligned}\tfrac{1}{2}mv_G^2 &= \tfrac{1}{2}m[(\omega_Y Z - \omega_Z Y)^2 + (\omega_Z X - \omega_X Z)^2 + (\omega_X Y - \omega_Y X)^2] \\ &= \tfrac{1}{2}m[\omega_X^2(Y^2 + Z^2) + \omega_Y^2(X^2 + Z^2) + \omega_Z^2(X^2 + Y^2)] \\ &\quad - 2\omega_X\omega_Y XY - 2\omega_X\omega_Z XZ - 2\omega_Y\omega_Z YZ.\end{aligned}$$

Substituting this expression for $\frac{1}{2}mv_G^2$ in Eq. (22.27) and combining terms yields

$$\begin{aligned}T &= \tfrac{1}{2}\omega_X^2[I_x + m(Y^2 + Z^2)] + \tfrac{1}{2}\omega_Y^2[I_y + m(X^2 + Z^2)] \\ &\quad + \tfrac{1}{2}\omega_Z^2[I_z + m(X^2 + Y^2)] - \omega_X\omega_Y[I_{xy} + mXY] \\ &\quad - \omega_X\omega_Z[I_{xz} + mXZ] - \omega_Y\omega_Z[I_{yz} + mXY].\end{aligned} \tag{22.29}$$

Utilizing the parallel axis theorems for mass moments and products of inertia, Eq. (22.29) may be written in the form

$$\begin{aligned}T &= \tfrac{1}{2}(\omega_X^2 I_X + \omega_Y^2 I_Y + \omega_Z^2 I_Z) - \omega_X\omega_Y I_{XY} - \omega_X\omega_Z I_{XZ} \\ &\quad - \omega_Y\omega_Z I_{YZ}\end{aligned} \tag{22.30}$$

where the X, Y, and Z axes are defined with respect to the fixed point O as shown in Figure 22.8. Alternatively, the kinetic energy of the body may be expressed in terms of principal axes of inertia U, V, and W at

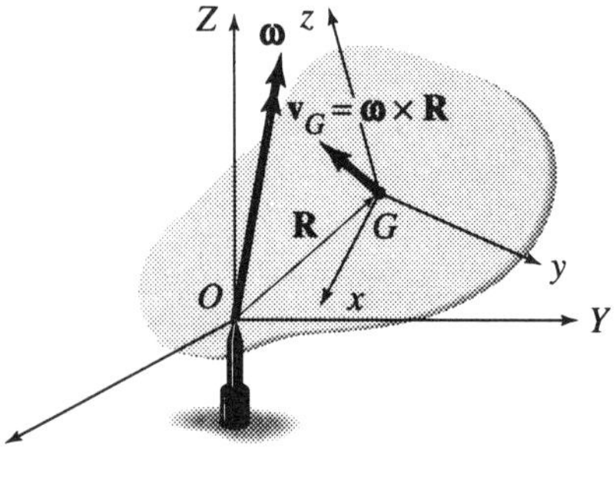

FIGURE 22.8.

fixed point O, in which case $I_{XY} = I_{XZ} = I_{YZ} = I_{UV} = I_{UW} = I_{VW} = 0$. This leads to

$$T = \tfrac{1}{2}(\omega_U^2 I_U + \omega_V^2 I_V + \omega_W^2 I_W). \tag{22.31}$$

Example 22.5 illustrates the use of the work-energy method in the solution of a three-dimensional problem.

Example 22.5

The thin homogeneous rectangular plate shown in Figure E22.5(a) is supported at A and B by two bearings and is made to rotate about axis AB by an agent not shown. The driving agent is suddenly removed when the plate is in the position shown for which the X axis is perpendicular to the plate and its angular velocity is $\omega = 8$ rad/s, ccw when viewed along the Y axis toward the origin. If the plate coasts to a complete stop after 12.5 revolutions, determine the average friction couple developed in the two bearings. Let the weight of the plate $W = 30$ lb.

Solution

The work-energy principle states that

$$\sum U_{1\to 2} = \sum T_2 - \sum T_1$$

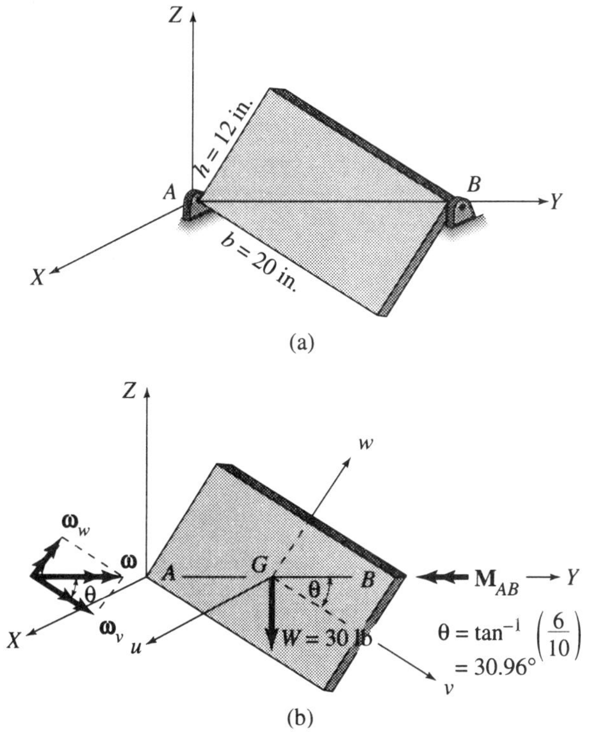

FIGURE E22.5.

where $\sum T_2 = 0$ because, in its final position, the plate is at a stand-still and has no kinetic energy. Also, from the free-body diagram shown in Figure E22.5(b), the reactions at bearings A and B (not shown) and the weight W of the plate perform no work because they do not displace. Thus, the only work performed during the slowdown period is that of the friction couple M_{AB} which is shown as a single vector in Figure E22.5(b) but actually is due to the friction in both bearings. Thus,

$$\sum U_{1\to 2} = -M_{AB}\theta = -M_{AB}(12.5)(2\pi) = -78.54M_{AB}.$$

Employing the u, v, and w principal axes of inertia shown in Figure E22.5(b), we may use Eq. (22.28), with $v_G = 0$, to find the kinetic energy of the plate in the initial position (when the driving agent was removed). The use of Eq. (22.28) requires determining the principal centroidal moments of inertia I_u, I_v, and I_w as well as the scalar components ω_u, ω_v and ω_w of the angular velocity of the plate about the u, v, and w axes, respectively. Thus,

$$I_u = \frac{1}{12}m(h^2 + b^2) = \frac{1}{12}\left(\frac{30}{32.2}\right)\left[\left(\frac{12}{12}\right)^2 + \left(\frac{20}{12}\right)^2\right] = 0.2933 \text{ ft·lb·s}^2,$$

$$I_v = \frac{1}{12}mh^2 = \frac{1}{12}\left(\frac{30}{32.2}\right)\left(\frac{12}{12}\right)^2 = 0.0776 \text{ ft·lb·s}^2,$$

$$I_w = \frac{1}{12}mb^2 = \frac{1}{12}\left(\frac{30}{32.2}\right)\left(\frac{20}{12}\right)^2 = 0.2157 \text{ ft·lb·s}^2,$$

$$\omega_u = 0,$$

$$\omega_v = 8\cos 30.96^\circ = 6.860 \text{ rad/s},$$

and

$$\omega_w = 8\sin 30.96^\circ = 4.116 \text{ rad/s}.$$

Therefore, by Eq. (22.28),

$$\begin{aligned}\sum T_1 &= \tfrac{1}{2}mv_G^2 + \tfrac{1}{2}(\omega_u^2 I_u + \omega_v^2 I_v + \omega_w^2 I_w),\\ &= 0 + \tfrac{1}{2}[0 + (6.860)^2(0.0776) + (4.116)^2(0.2157)]\\ &= 3.653 \text{ ft·lb}.\end{aligned}$$

Thus, the work-energy principle leads to

$$-78.54M_{AB} = 0 - 3.653$$

from which the average friction couple is

$$M_{AB} = 4.65 \times 10^{-2} \text{ ft·lb}. \qquad \text{ANS.}$$

■

22.4 Principles of Linear and Angular Momentum

The equations developed in Chapter 20, describing the principles of linear and angular impulse and momentum for a rigid body in plane motion, are equally valid for its three-dimensional motion. These equations are repeated here for convenience as Eqs. (22.32).

$$\sum \int_{t_1}^{t_2} \mathbf{F}\, dt = \mathbf{L}_2 - \mathbf{L}_1, \tag{22.32a}$$

$$\sum \int_{t_1}^{t_2} \mathbf{M}_\mathrm{O}\, dt = \mathbf{H}_{\mathrm{O}2} - \mathbf{H}_{\mathrm{O}1}, \tag{22.32b}$$

$$\sum \int_{t_1}^{t_2} \mathbf{M}_\mathrm{G}\, dt = \mathbf{H}_{\mathrm{G}2} - \mathbf{H}_{\mathrm{G}1}. \tag{22.32c}$$

The linear momentum $\mathbf{L}$ defined in Chapter 20 as $\mathbf{L} = m\mathbf{v}_\mathrm{G}$ for a rigid body in plane motion is still valid for the three-dimensional motion of this body except that, in general, the velocity vector contains three components instead of only two. The angular momentum $\mathbf{H}$ of a rigid body executing three-dimensional motion, however, requires determination. To this end, consider the rigid body shown in Figure 22.9 which is executing three-dimensional motion. The X-Y-Z coordinate system is fixed in space whereas the centroidal x-y-z coordinate system, attached to the body's mass center, point G, is arbitrarily oriented with respect to the X-Y-Z coordinate system. The angular momentum for a particle was developed in Chapter 16. On the basis of that development, the angular momentum $\mathbf{H}_\mathrm{G}$ of the rigid body of Figure 22.9 about its mass center G may be written in form

$$\mathbf{H}_\mathrm{G} = \sum_{i=1}^{n} \mathbf{r}_i \mathbf{x} m_i \mathbf{v}_i \tag{22.33}$$

In Eq. (22.33), $\mathbf{r}_i$ and $\mathbf{v}_i$ are the position and velocity vectors, respectively, for any particle Q_i of mass m_i as measured from the translating, nonrotating x-y-z coordinate system. Because a rigid body is com-

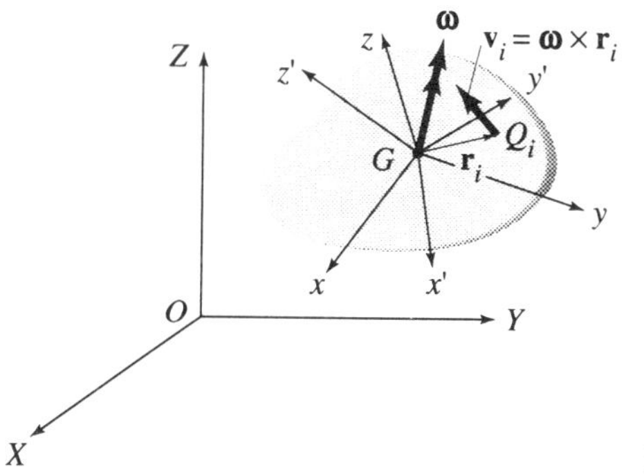

FIGURE 22.9.

posed of an infinite number of particles, we may replace the summation symbol by an integral sign and the particle mass m_i by the differential dm_i. Also, $\mathbf{v}_i = \boldsymbol{\omega} \times \mathbf{r}_i$ where $\boldsymbol{\omega}$ is the instantaneous angular velocity of the rigid body. And, for the sake of notation simplification, if we ignore the subscript i, Eq. (22.33) may be written as

$$\mathbf{H}_G = \int \mathbf{r} \times (\boldsymbol{\omega} \times \mathbf{r})\, dm. \tag{22.34}$$

Performing the indicated vector operations, Eq. (22.34) may be expressed in terms of the x, y and z rectangular components of $\mathbf{r}$ and $\boldsymbol{\omega}$. Thus,

$$\begin{aligned}\mathbf{H}_G = &\left[\omega_x \int (y^2 + z^2)\, dm - \omega_y \int xy\, dm - \omega_z \int xz\, dm\right]\mathbf{i} \\ &+ \left[-\omega_x \int xy\, dm + \omega_y \int (x^2 + z^2)\, dm - \omega_z \int yz\, dm\right]\mathbf{j} \\ &+ \left[-\omega_x \int xz\, dm - \omega_y \int yz\, dm + \omega_z \int (x^2 + y^2)\, dm\right]\mathbf{k}.\end{aligned} \tag{22.35}$$

Because $I_x = \int (y^2 + z^2)\, dm$, $I_y = \int (x^2 + z^2)\, dm$, $I_z = \int (x^2 + y^2)\, dm$, $I_{yx} = I_{xy} = \int xy\, dm$, $I_{zx} = I_{xz} = \int xz\, dm$, and $I_{zy} = I_{yz} = \int yz\, dm$, it follows that

$$\begin{aligned}\mathbf{H}_G = &(I_x\omega_x - I_{xy}\omega_y - I_{xz}\omega_z)\mathbf{i} + (-I_{yx}\omega_x + I_y\omega_y - I_{yz}\omega_z)\mathbf{j} \\ &+ (-I_{zx}\omega_x - I_{zy}\omega_y + I_z\omega_z)\mathbf{k}.\end{aligned} \tag{22.36}$$

Therefore, the three rectangular components of the angular momentum vector $\mathbf{H}_G$ with respect to the x-y-z centroidal coordinate system are

$$\begin{aligned}H_x &= I_x\omega_x - I_{xy}\omega_y - I_{xz}\omega_z, \\ H_y &= -I_{yx}\omega_x + I_y\omega_y - I_{yz}\omega_z,\end{aligned} \tag{22.37}$$

and

$$H_z = -I_{zx}\omega_x - I_{zy}\omega_y + I_z\omega_z.$$

Angular Momentum Using Principal Axes of Inertia

If the x, y, and z axes coincide, respectively, with the u, v and w principal centroidal axes of inertia, then, $I_{xy} = I_{uv} = 0$, $I_{xz} = I_{uw} = 0$, and $I_{yz} = I_{vw} = 0$, and Eq. (22.36) reduces to

$$\mathbf{H}_G = (I_u\omega_u)\boldsymbol{\lambda}_u + (I_v\omega_v)\boldsymbol{\lambda}_v + (I_w\omega_w)\boldsymbol{\lambda}_w \tag{22.38}$$

where $\boldsymbol{\lambda}_u$, $\boldsymbol{\lambda}_v$, and $\boldsymbol{\lambda}_w$ are unit vectors along the u, v and w principal centroidal axes, respectively.

Angular Momentum When Body Rotates about a Fixed Point

If a rigid body rotates about a fixed point O, as shown in Figure 22.10, its angular momentum $\mathbf{H}_O$ with respect to the fixed point is determined in a manner comparable to that used in developing Eq. (22.36). Thus, on the basis of the angular momentum of a particle developed in Chapter 16, the angular momentum H_O may be stated as

$$\mathbf{H}_O = \sum_{i=1}^{n} \mathbf{r}_i \times m_i \mathbf{v}_i. \tag{22.39}$$

Because a rigid body is composed of an infinite number of particles, we may replace the summation symbol by an integral sign and the particle mass m_i by the differential quantity dm_i. Replacing $\mathbf{v}_i$ by $\boldsymbol{\omega} \times \mathbf{r}_i$ and dropping the subscript i, we may write Eq. (22.39) in the form

$$\mathbf{H}_O = \int \mathbf{r} \times (\boldsymbol{\omega} \times \mathbf{r})\, dm \tag{22.40}$$

where $\mathbf{r}$ is the position vector from fixed point O to any particle Q_i, as shown in Figure 22.10, and $\boldsymbol{\omega}$ is the instantaneous angular velocity of the rigid body. Employing the same procedure used in the transformation of Eq. (22.34), we can transform Eq. (22.40) and express it in terms of the X, Y, and Z rectangular components of $\boldsymbol{\omega}$ and the moments and products of inertia of the rigid body relative to point O. Thus,

$$\mathbf{H}_O = (I_X\omega_X - I_{XY}\omega_Y - I_{XZ}\omega_Z)\mathbf{I} + (-I_{YX}\omega_X + I_Y\omega_Y - I_{YZ}\omega_Z)\mathbf{J} + (-I_{ZX}\omega_X - I_{ZY}\omega_Y + I_Z\omega_Z)\mathbf{K}. \tag{22.41}$$

Therefore, the three rectangular components of the angular momentum vector $\mathbf{H}_O$, with respect to the X-Y-Z coordinate system with origin at fixed point O, are

$$H_X = I_X\omega_X - I_{XY}\omega_Y - I_{XZ}\omega_Z,$$
$$H_Y = -I_{YX}\omega_X + I_Y\omega_Y - I_{YZ}\omega_Z, \tag{22.42}$$

and

$$H_Z = -I_{ZX}\omega_X - I_{ZY}\omega_Y + I_Z\omega_Z.$$

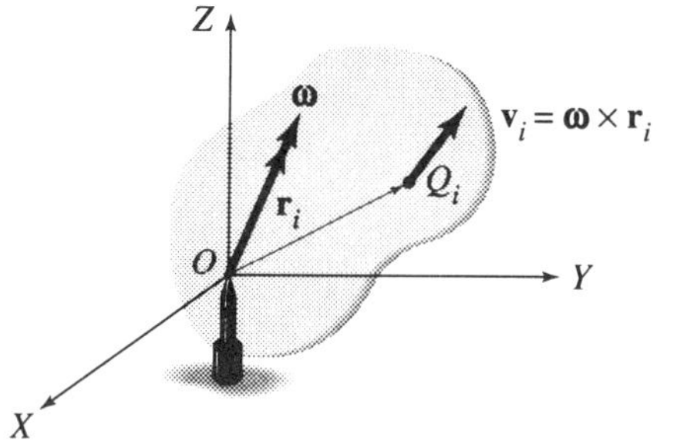

FIGURE 22.10.

Equation (22.41) may also be expressed in terms of the principal moments of inertia about principal axes U, V, and W at fixed point O, in which case $I_{XY} = I_{XZ} = I_{YZ} = I_{UV} = I_{UW} = I_{VW} = 0$. Thus,

$$\mathbf{H}_O = (I_U\omega_U)\boldsymbol{\lambda}_U + (I_V\omega_V)\boldsymbol{\lambda}_V + (I_W\omega_W)\boldsymbol{\lambda}_W \tag{22.43}$$

where $\boldsymbol{\lambda}_U$, $\boldsymbol{\lambda}_V$, and $\boldsymbol{\lambda}_W$ are unit vectors, respectively, along the U, V, and W principal axes of inertia at O.

It should be observed that, for a given angular velocity $\boldsymbol{\omega}$, a rigid body possesses a unique value for its angular momentum $\mathbf{H}_G$ or $\mathbf{H}_O$ regardless of the choice of the coordinate axes. Thus, for example, in Figure 22.9, the angular momentum $\mathbf{H}_G$, as determined with respect to the arbitrary x-y-z coordinate system, would be the same as that found with respect to any other x'-y'-z' coordinate system.

Example 22.6 is designed to illustrate some of the concepts discussed in this section.

Example 22.6

Three thin homogeneous rods are attached rigidly together to form the assembly shown in Figure E22.6(a). The assembly is suspended from point A by a ball and socket as shown and it is standing still when impacted at point C by a force $\mathbf{F}$ in the positive X direction causing an impulse $\mathbf{F}\Delta t = (10\mathbf{I})$ N·s. If the mass of the rods is 1.0 kg/m, determine, immediately after impact, (a) the instantaneous angular velocity of the assembly and (b) its instantaneous axis of rotation.

Solution

(a) By Eq. (22.32), we may write

$$\sum \int_{t_1}^{t_2} \mathbf{M}_A \, dt = \mathbf{H}_{A2} - \mathbf{H}_{A1}. \tag{a}$$

Because the assembly is initially at rest, it follows that

$$\mathbf{H}_{A1} = \mathbf{0}, \tag{b}$$

and the angular impulse-angular momenta diagram reduces to that shown in Figure E22.6(b). Also,

$$\sum \int_{t_1}^{t_2} \mathbf{M}_A \, dt = \mathbf{r}_{C/A} \times (F\Delta t)\mathbf{I} = (-0.75\mathbf{J} - 1.50\mathbf{K}) \times 10\mathbf{I}$$

$$= 7.5\mathbf{K} - 15.0\mathbf{J}. \tag{c}$$

To determine $\mathbf{H}_{A2}$ immediately after impact, we need to find the moments and products of inertia of the assembly relative to the X', Y', and Z' axes which are attached to the body and rotate with it. Note that for the instant considered, the fixed X, Y, and Z axes coincide, respectively, with the rotating X', Y', and Z' axes. Thus,

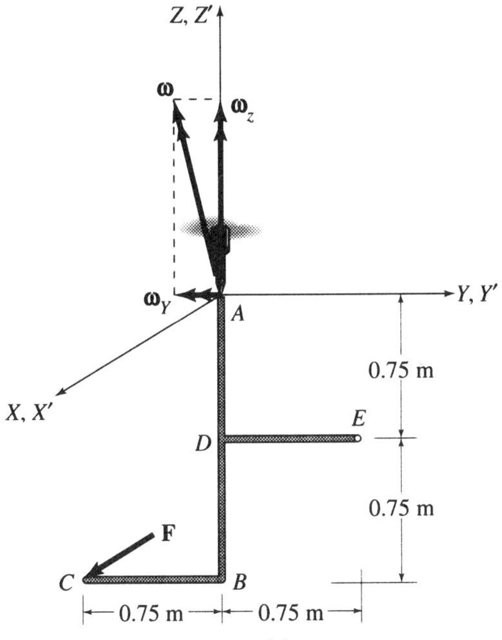

(a)

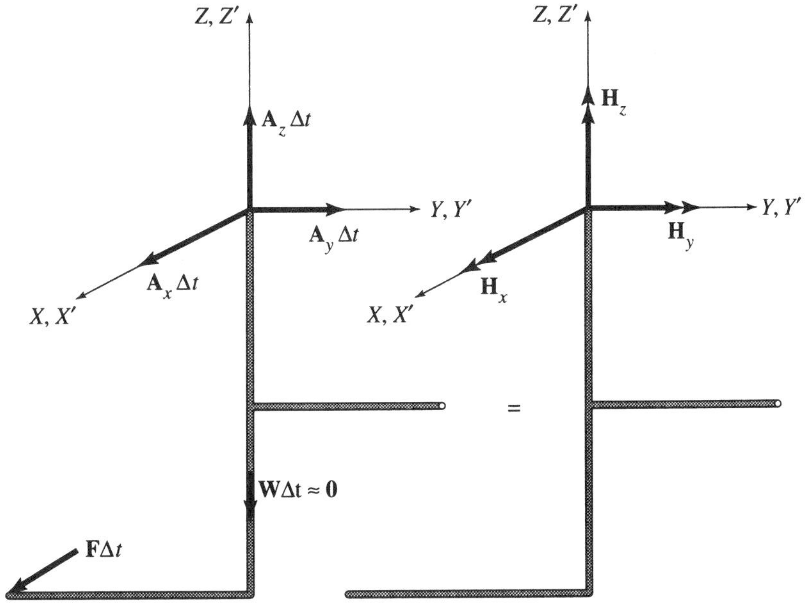

FIGURE E22.6. (b)

$$I_{X'} = (I_{X'})_{AB} + (I_{X'})_{CB} + (I_{X'})_{DE},$$
$$= [\tfrac{1}{3}(1.5 \times 1.0)(1.5^2)]$$
$$+ [\tfrac{1}{12}(0.75 \times 1.0)(0.75^2) + (0.75 \times 1.0)(0.375^2 + 1.50^2)]$$
$$= 3.516 \text{ kg·m}^2.$$
$$I_{Y'} = (I_{Y'})_{AB} + (I_{Y'})_{CB} + (I_{Y'})_{DE},$$
$$= [\tfrac{1}{3}(1.5 \times 1.0)(1.5^2)]$$
$$+ [0 + (0.75 \times 1.0)(1.5^2)] + [0 + (0.75 \times 1.0)(0.75^2)]$$
$$= 3.234 \text{ kg·m}^2.$$
$$I_{Z'} = (I_{Z'})_{AB} + (I_{Z'})_{CB} + (I_{Z'})_{DE},$$
$$= [0] + [\tfrac{1}{12}(0.75 \times 1.0)(0.75^2) + (0.75 \times 1.0)(0.375^2)]$$
$$+ [\tfrac{1}{12}(0.75 \times 1.0)(0.75^2) + (0.75 \times 1.0)(0.375^2)]$$
$$= 0.281 \text{ kg·m}^2.$$
$$I_{X'Y'} = (I_{X'Y'})_{AB} + (I_{X'Y'})_{CB} + (I_{X'Y'})_{DE} = 0,$$
$$I_{X'Z'} = (I_{X'Z'})_{AB} + (I_{X'Z'})_{CB} + (I_{X'Z'})_{DE} = 0,$$
$$I_{Y'Z'} = (I_{Y'Z'})_{AB} + (I_{Y'Z'})_{CB} + (I_{Y'Z'})_{DE}$$
$$= [0] + [(0.75 \times 1.0)(0.375)(1.5)] + [(0.75 \times 1.0)(-0.375)(0.75)]$$
$$= 0.211 \text{ kg·m}^2.$$

Thus, Eq. (22.41) reduces to

$$\mathbf{H}_{A2} = (I_{X'}\omega_X)\mathbf{I} + (I_{Y'}\omega_Y - I_{Y'Z'}\omega_Z)\mathbf{J} + (-I_{Y'Z'}\omega_Y + I_{Z'}\omega_Z)\mathbf{K}$$

which, after substituting the values for moments and products of inertia found above, yields

$$\mathbf{H}_{A2} = (3.516\omega_X)\mathbf{I} + (3.234\omega_Y - 0.211\omega_Z)\mathbf{J}$$
$$+ (-0.211\omega_Y + 0.281\omega_Z)\mathbf{K}. \tag{d}$$

Substituting Eqs. (b), (c) and (d) in Eq. (a) yields

$$7.5\mathbf{K} - 15.0\mathbf{J} = (3.516\omega_X)\mathbf{I} + (3.234\omega_Y - 0.211\omega_Z)\mathbf{J}$$
$$+ (-0.211\omega_Y + 0.281\omega_Z)\mathbf{K} - \mathbf{0}.$$

Equating the coefficients of the **I**, **J** and **K** terms, we obtain

$$\omega_X = 0$$
$$0.211\omega_Z - 3.234\omega_Y = 15.0 \tag{e}$$
$$0.281\omega_Z - 0.211\omega_Y = 7.5 \tag{f}$$

Solving Eqs. (e) and (f) simultaneously leads to

$$\omega_Y = -3.05 \text{ rad/s}$$

and

$$\omega_Z = 24.40 \text{ rad/s}.$$

Therefore,

$$\boldsymbol{\omega} = (-3.05\mathbf{J} + 24.4\mathbf{K}) \text{ rad/s}. \qquad \text{ANS.}$$

Thus, the instantaneous angular velocity of the assembly immediately after impact lies entirely in the *Y-Z* plane because it has no component along the *X* direction as shown in Figure E22.6(a).

(b) The instantaneous axis of rotation for the assembly must pass through fixed point A and is directed along the instantaneous angular velocity $\boldsymbol{\omega}$. Thus, its direction may be expressed by the unit vector $\boldsymbol{\lambda}$ given by

$$\boldsymbol{\lambda} = \frac{\boldsymbol{\omega}}{\omega} = -0.1240\mathbf{J} + 0.992\mathbf{K}. \qquad \text{ANS.}$$

■

Problems

22.16 Two homogeneous thin rectangular plates are welded to a horizontal shaft AB, as shown in Figure P22.16. The shaft is rotated by a motor (not shown) at a constant angular velocity of 200 rpm. Determine the kinetic energy of the system . Each plate weighs 10 lb and the shaft may be assumed to be a massless slender rod.

22.17 Refer to Problem 22.16. If the motor is suddenly turned off when the angular velocity is 200 rpm, determine the number of revolutions that the shaft will make before coming to a complete stop if the total friction in the bearings at both A and B is assumed to be a constant couple $M = 0.5$ lb. ft.

22.18 A thin homogeneous rectangular plate

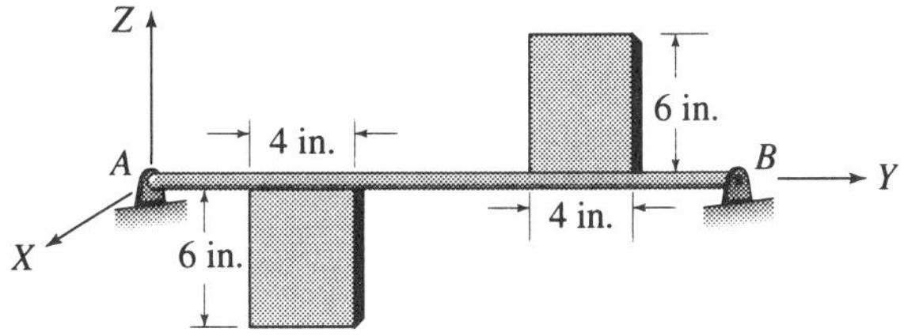

FIGURE P22.16.

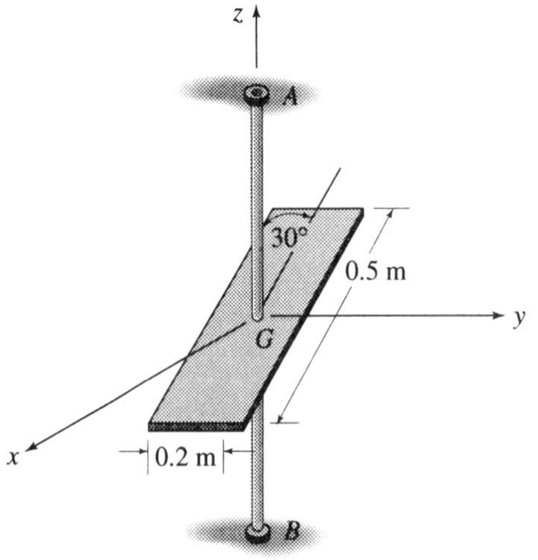

FIGURE P22.18.

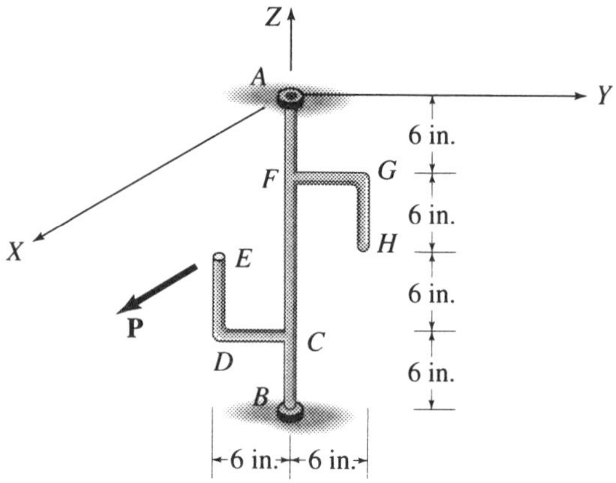

FIGURE P22.20.

of mass $m = 10$ kg is attached rigidly to a weightless shaft AB so that one of the principal centroidal axes of the plate makes a 30° angle with the shaft, as shown in Figure P22.18. The assembly is made to rotate about axis AB at an angular velocity $\omega = 10$ rad/s. Determine the kinetic energy of the assembly at this speed. Assume that the shaft is a weightless slender rod.

22.19 Refer to Problem 22.18. The driving agent is suddenly removed when the angular velocity $\omega = 10$ rad/s, and the assembly is allowed to coast to a complete stop. If it takes 30 revolutions to come to a complete stop, determine the total friction couple at bearings A and B if it is assumed to be constant.

22.20 Three slender rods are welded together at F and C and the assembly is supported by bearings at A and B as shown in Figure P22.20. A force $P = 2e^{0.2t}$ lb where t is in seconds, is applied at E when the assembly has an angular velocity $\boldsymbol{\omega} = (4\mathbf{K})$ rad/s and remains perpendicular to the plane of EDC. Determine the angular velocity of the assembly after the force P has acted for 1 second. All rods weigh 2 lb/in.

22.21 A slender homogeneous rod having a mass of 2 kg/m is bent into the shape OABC shown in Figure P22.21. In the

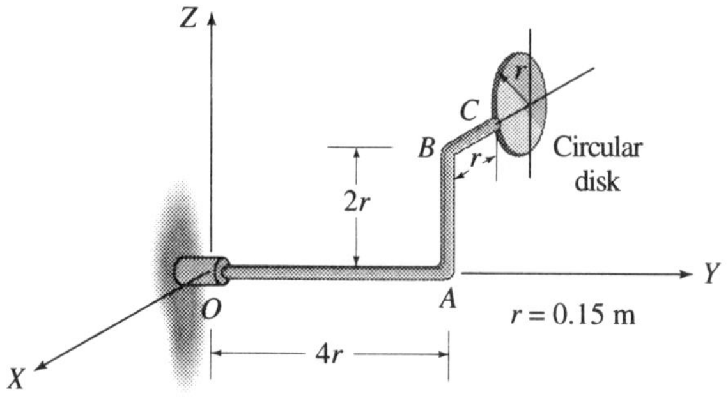

FIGURE P22.21.

position shown, BC is parallel to the X axis. A thin disk having a mass of 2 kg is attached rigidly at C so that its plane is always parallel to the X-Z plane. End O of the system is supported by a frictionless bearing that contains a torsional spring (not shown) of spring constant $K = 3$ N·m/rad that tends to inhibit rotation of the system about the Y axis. If the system is released from rest from the position shown and allowed to rotate about the Y axis, determine its angular velocity after segment AB rotates through 90°. Let $r = 0.15$ m.

22.22 Refer to Problem 22.21. If the system is released from rest from the position shown in Figure P22.21, determine the maximum rotation experienced by the system as measured from the vertical position of segment AB.

22.23 A disk weighing 15 lb rotates at a constant angular velocity $\omega_1 = 20$ rad/s relative to the supporting frame OABC, which itself rotates about the Z axis at O, at a constant angular velocity $\omega_2 = 5$ rad/s, as shown in Figure P22.23. Determine the kinetic energy of the disk.

22.24 Refer to Problem 22.16, and determine the angular momentum of the system about point A. Find the magnitude and direction of this angular momentum.

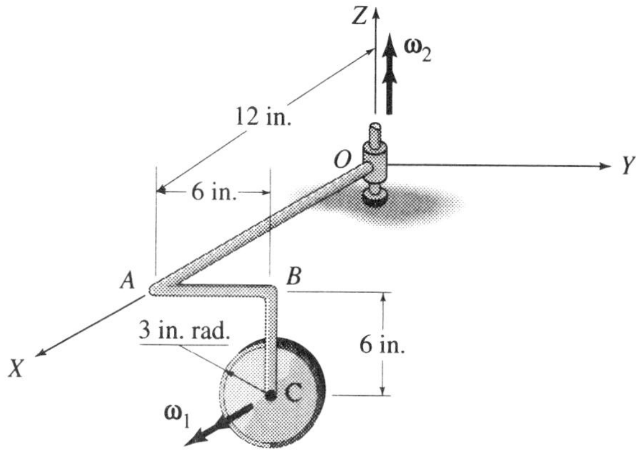

FIGURE P22.23.

22.25 Refer to Problem 22.18, and determine the angular momentum of the assembly about point G. Find the magnitude and direction of this angular momentum.

22.26 Refer to Problem 22.20, and determine the angular momentum of the system about point A prior to the application of P, when the assembly has an angular velocity $\boldsymbol{\omega} = (4 \text{ rad/s})\,\mathbf{K}$. Find the magnitude and direction of this angular momentum.

22.27 Refer to Problem 22.23, and determine the angular momentum of the disk about its center point C. Find the magnitude and direction of this angular momentum.

22.28 Determine the angular momentum of the disk in Problem 22.23 about point O. Find the magnitude and direction of this angular momentum.

22.29 Two thin homogeneous plates are welded to a slender rod and the assem-

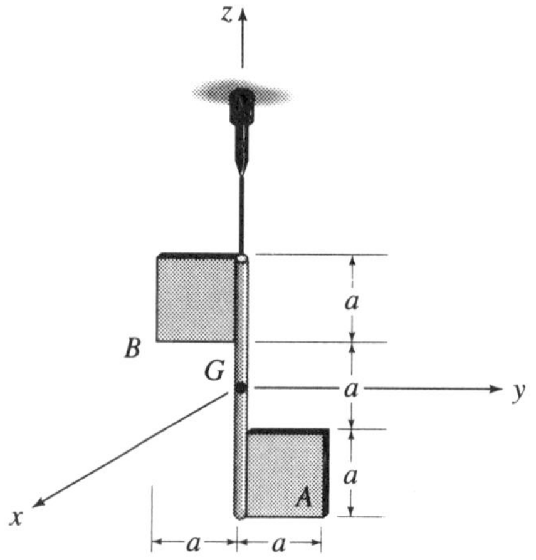

FIGURE P22.29.

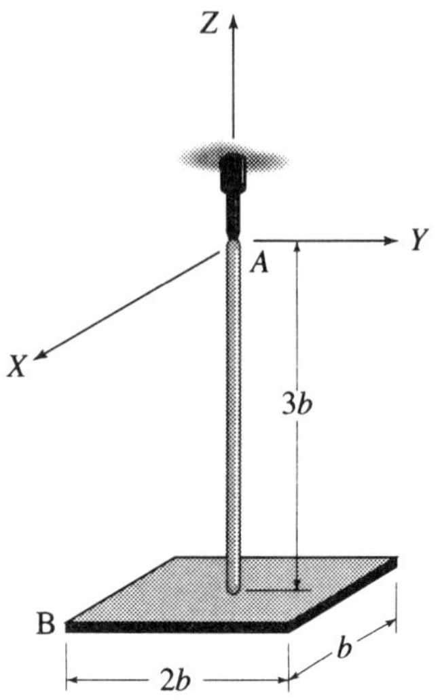

FIGURE P22.31.

bly is suspended, as shown in Figure P22.29. The mass of each plate is 8 kg and the rod may be assumed massless. The assembly is at rest when force **F** is applied at point A causing an impulse $\mathbf{F}\Delta t = (20\ \text{N·s})\mathbf{i}$. If $a = 0.4$ m, determine, immediately after impact, (a) the angular velocity of the assembly and (b) the velocity of its mass center.

22.30 Two thin homogeneous square plates are welded to a slender rod and the assembly is suspended, as shown in Figure P22.29. The weight of each plate is 10 lb and the rod may be assumed weightless. The assembly is at rest when a force **F** is applied at point B causing an impulse $\mathbf{F}\Delta t = (-5\mathbf{i} + 10\mathbf{j})$ lb·s. If $a = 6$ in., determine, immediately after impact, (a) the angular velocity of the assembly and (b) the velocity of its mass center.

22.31 A thin homogeneous rectangular plate of mass m is welded to the end of a slender homogeneous rod of mass $m/3$, as shown in Figure P22.31. The assembly is suspended from a ball and socket at A and it is subjected to a force **F** at B causing an impulse $\mathbf{F}\Delta t = -(2Q)\mathbf{I} + (Q)\mathbf{J}$ where Q is a constant expressed in units of force multiplied by units of time. If the assembly is initially at rest when the force **F** is applied, determine, immediately after impact, (a) the angular velocity of the assembly and (b) the instantaneous axis of rotation.

22.32 Two identical circular homogeneous disks each of mass 4 kg are welded to a slender uniform rod of mass 1 kg, as shown in Figure P22.32. The assembly is suspended from a ball and socket at A and hit at B by a force **F** causing an impulse $\mathbf{F}\Delta t = (-15\mathbf{I})$ N·s. Assume that the assembly is at rest when hit, and determine, just after impact, (a) the angular velocity of the assembly, and (b) the instantaneous axis of rotation. Let $r = 0.12$ m.

22.33 Two identical circular homogeneous disks each weighing 5 lb are welded to a slender uniform rod weighing 2 lb as shown in Figure P22.32. The assembly is suspended from a ball and socket at A and hit at C by a force **F** causing an impulse $\mathbf{F}\Delta t = (0.5\mathbf{I} + 1.0\mathbf{J} - 1.5\mathbf{K})$ lb·s. Assume that the assembly is at rest when hit and determine, immediately after impact, (a) the angular velocity of the

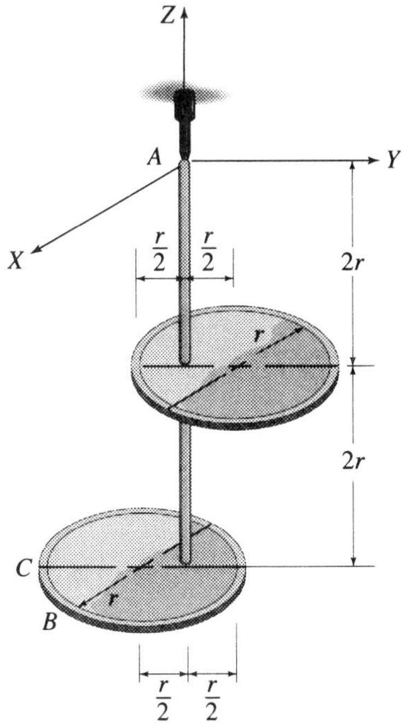

FIGURE P22.32.

assembly and (b) the instantaneous axis of rotation. Let $r = 4$ in.

22.34 A uniform rod of mass m is bent into the shape shown in Figure P22.34 and suspended from its mass center G by a thin wire of negligible mass. The bent rod is hit at A by a force **F** causing an impulse $\mathbf{F}\Delta t = -(Q)\mathbf{k}$ where Q is a constant expressed in units of force multiplied by units of time. If the bent rod is initially at rest when hit, determine, just after impact, (a) the angular velocity of the rod, (b) the linear velocity of its mass center, and (c) the impulse developed in the wire.

22.35 A uniform rod weighing 8 lb is bent into the shape shown in Figure P22.34 and suspended from its mass center G by a thin wire of negligible weight. The bent rod is hit at B by a force **F** causing an impulse $\mathbf{F}\Delta t = (0.2\mathbf{i} - 0.4\mathbf{j} - 0.3\mathbf{k})$ lb·s. If the bent rod is initially at rest when hit, determine, immediately after impact, (a) the angular velocity of the bent rod and (b) the linear velocity of its mass center. Let $d = 4$ in.

22.36 A space satellite of mass 1200 kg is moving at a constant linear speed but with no angular velocity when it is hit at A, as shown in Figure P22.36, by a meteoroid of mass 0.75 kg at a velocity $\mathbf{v}_m = (0.55\mathbf{i} - 0.70\mathbf{j} - 0.95\mathbf{k})$ km/s. The radii of gyration for the satellite are $k_y = k_z = 0.75$ m and $k_x = 0.50$ m and the coordinates of point A are -0.60 m, 0.45 m, 0.40 m. Determine the angular velocity of the satellite immediately after the meteoroid becomes imbedded in it. Assume that the given x, y, and z axes are principal axes of inertia.

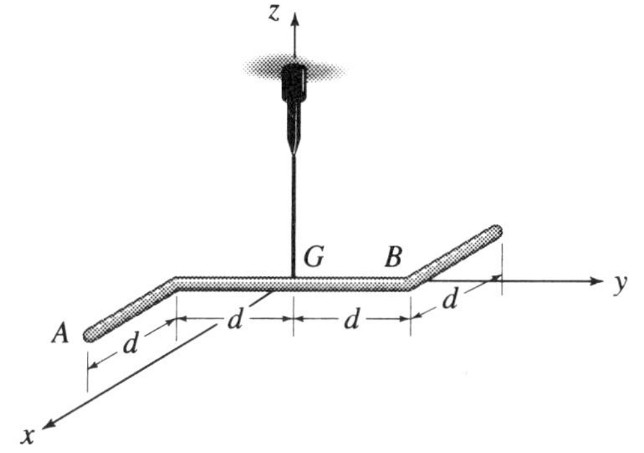

FIGURE P22.34.

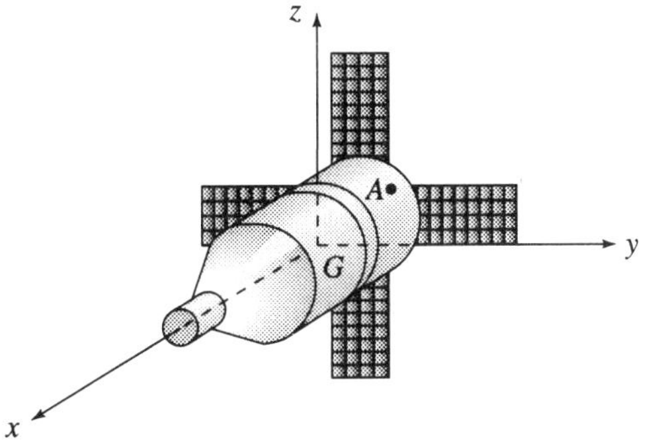

FIGURE P22.36.

22.37 A space satellite weighing 2500 lb is moving at a constant linear speed with a constant angular velocity $\boldsymbol{\omega} = (2.5 \text{ rad/s})\mathbf{i}$ when it is hit at A, as shown in Figure P22.36, by a meteoroid weighing 5 lb and moving at a speed of velocity $v_m = 5000$ ft/s. The meteoroid becomes imbedded in the satellite and the angular velocity of the latter, immediately after imbedment, becomes $\boldsymbol{\omega} = (-0.4\mathbf{i} - 0.7\mathbf{i} + 0.3\mathbf{i})$ rad/s. Determine the components of v_m at impact, if the radii of gyration of the satellite are $k_y = k_z = 2.5$ ft and $k_x = 2.0$ ft. Let the coordinates of point A be -2.0 ft, 1.5 ft, 1.8 ft. Assume that the given x, y, and z axes are principal axes of inertia.

22.38 Two slender homogeneous rods AB and CD are welded together to form a cross as shown in Figure P22.38. The cross is suspended from a ball and socket at A and is made to rotate about the vertical Z axis at a constant angular velocity $\boldsymbol{\omega}_1$ when impacted by an obstruction introduced suddenly at C. Assume that the impact at C is perfectly plastic, and determine, just after impact, (a) the angular velocity of the cross and (b) the linear velocity of its mass center G. Let the mass of CD be m and that of AB be $1.5m$.

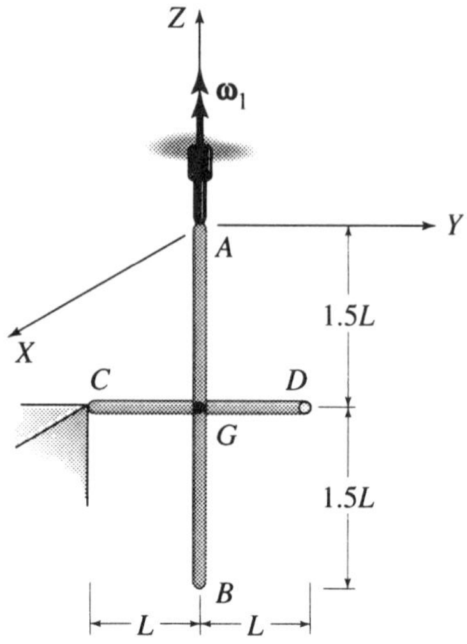

FIGURE P22.38.

22.39 Four identical homogeneous slender rods, each with a mass of 2 kg and a length of 0.2m, are attached rigidly to a thin homogeneous disk of mass 10 kg and radius $r = 0.2$ m as shown in Figure P22.39. The assembly is held horizontally and released from rest. It travels vertically downward through a height

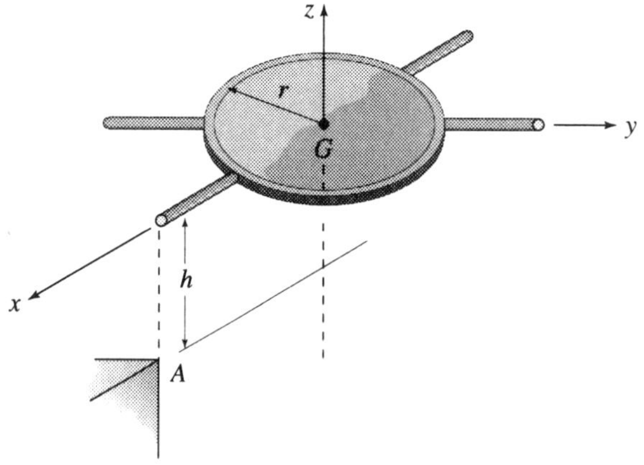

FIGURE P22.39.

$h = 0.75$ m when one of the four rods hits an obstruction at A as shown. Assume that the impact at A is perfectly plastic, and determine, just after the impact, (a) the angular velocity of the assembly and (b) the linear velocity of its mass center G.

22.5 General Equations of Motion

Translation

The equation of motion of the mass center of a system of particles was developed in Chapter 14 and utilized in Chapter 18 to describe the translational component of the two-dimensional motion of a rigid body. For convenience this equation is repeated here as Eq. (22.44). Thus,

$$\sum \mathbf{F} = m\mathbf{a}_G \tag{22.44}$$

where $\sum \mathbf{F}$ represents the sum of all external forces acting on the rigid body, m its mass, and $\mathbf{a}_G$ the acceleration of its mass center. Equation (22.44) forms the basis for the translational component of the three-dimensional motion of a rigid body.

Generally speaking, when dealing with the three-dimensional motion of a rigid body, Eq. (22.44) is used in its vector form. However, there are occasions when Eq. (22.44) is conveniently used in terms of rectangular components. The specific choice of the rectangular components would depend upon the problem under consideration. Thus, for example,

$$\sum \mathbf{F}_X = m(\mathbf{a}_G)_X,$$

$$\sum \mathbf{F}_Y = m(\mathbf{a}_G)_Y, \tag{22.45}$$

and

$$\sum \mathbf{F}_Z = m(\mathbf{a}_G)_Z$$

where the X-Y-Z coordinate system is fixed in space as shown in Figure 22.11.

Rotation

The fundamental relationship governing the rotational component of the general motion of a system of particles was developed in Chapter

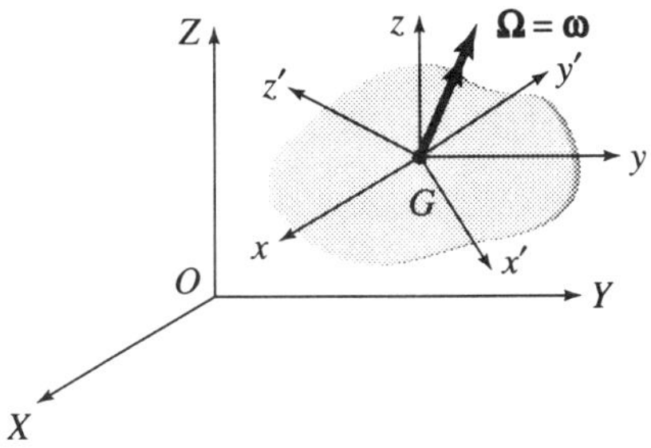

FIGURE 22.11.

16 and utilized in two alternate forms in Chapter 18 to describe the rotational component of the two-dimensional motion of a rigid body. For convenience, the two alternate forms of this equation are repeated here as Eqs. (24.46). Thus,

$$\sum \mathbf{M}_{\mathrm{G}} = \dot{\mathbf{H}}_{\mathrm{G}}$$

and (22.46)

$$\sum \mathbf{M}_{\mathrm{O}} = \dot{\mathbf{H}}_{\mathrm{O}}$$

where $\sum \mathbf{M}_{\mathrm{G}}$ and $\dot{\mathbf{H}}_{\mathrm{G}}$ represent the sum of moments of all external forces and the time derivative of the angular momentum, respectively, both taken about the body's mass center, point G. Similarly, for $\sum \mathbf{M}_{\mathrm{O}}$ and $\dot{\mathbf{H}}_{\mathrm{O}}$ about point O.

As indicated in Section 22.3, the angular momentum of a rigid body, with respect to any point, is independent of the coordinate system chosen through that point. Thus, in analyzing the three-dimensional motion of a rigid body, it is convenient to determine the angular momentum with respect to axes attached to the body and rotating with it, because the moments of inertia with respect to these axes are constant. Let the centroidal x', y', and z' axes shown in Figure 22.11 represent such a rotating coordinate system whose angular velocity is $\boldsymbol{\Omega}$ relative to the translating x-y-z coordinate system or to the fixed X-Y-Z coordinate system. Thus, $\boldsymbol{\Omega}$ is identical to the angular velocity $\boldsymbol{\omega}$ of the rigid body relative to the x-y-z or X-Y-Z coordinate system. In Chapter 21, it was shown that the time derivative of any vector, such as $\mathbf{H}_{\mathrm{G}}$, relative to a nonrotating x-y-z system is equal to its time derivative relative to a rotating x'-y'-z' system plus the cross product $\boldsymbol{\Omega} \times \mathbf{H}_{\mathrm{G}}$. In other words,

$$(\dot{\mathbf{H}}_{\mathrm{G}})_{xyz} = (\dot{\mathbf{H}}_{\mathrm{G}})_{x'y'z'} + \boldsymbol{\Omega} \times \mathbf{H}_{\mathrm{G}} \tag{22.47}$$

where

$\boldsymbol{\Omega}$ = the angular velocity of the rotating x'-y'-z' system relative to the nonrotating x-y-z system.

$\mathbf{H}_{\mathrm{G}}$ = the angular momentum of the rigid body. Note that $\mathbf{H}_{\mathrm{G}}$ may be determined using the rotating x'-y'-z' system or the nonrotating x-y-z system, whichever is more convenient.

$(\dot{\mathbf{H}}_{\mathrm{G}})_{x'y'z'}$ = time derivative of $\mathbf{H}_{\mathrm{G}}$ with respect to the rotating x'-y'-z' system.

It should be emphasized that the first term in Eq. (22.47) is due to the change in magnitude of $\mathbf{H}_{\mathrm{G}}$ whereas the second term $(\boldsymbol{\Omega} \times \mathbf{H}_{\mathrm{G}})$ is due to the change in orientation of vector $\mathbf{H}_{\mathrm{G}}$.

Now, using the first of Eqs. (22.46), we conclude that

$$\sum \mathbf{M}_{\mathrm{G}} = (\dot{\mathbf{H}}_{\mathrm{G}})_{x'y'z'} + \boldsymbol{\Omega} \times \mathbf{H}_{\mathrm{G}}. \tag{22.48}$$

Similarly, using the second of Eqs. (22.46) it may be shown that

$$\sum \mathbf{M}_O = (\dot{\mathbf{H}}_O)_{X'Y'Z'} + \boldsymbol{\Omega} \times \mathbf{H}_O \tag{22.49}$$

where the X'-Y'-Z' coordinate system is attached to the body at an arbitrary point O and rotates with it at an angular velocity $\boldsymbol{\Omega} = \boldsymbol{\omega}$.

Note that, throughout our discussion, the rotating coordinate system x'-y'-z' or X'-Y'-Z' was assumed to be attached to the rigid body. Under these conditions, the angular velocity $\boldsymbol{\Omega}$ of the coordinate system is the same as the angular velocity $\boldsymbol{\omega}$ of the rigid body. There are cases, as, for example, in the motion of a gyroscope discussed in Section 22.5, where it is more convenient to select a coordinate system which rotates differently from the rigid body and with respect to which the moments and products of inertia remain constant throughout the motion. Although resulting in $\boldsymbol{\Omega} \neq \boldsymbol{\omega}$, this approach leads to simplifications in the computations for $\boldsymbol{\omega}$, $\mathbf{H}$, and their derivatives.

Euler's equations

If the rotating x', y', and z' axes coincide, respectively, with the u, v, and w principal centroidal axes of inertia and if $\boldsymbol{\Omega} = \boldsymbol{\omega}$, then, Eq. (22.48) becomes

$$\sum \mathbf{M}_G = (\dot{\mathbf{H}}_G)_{uvw} + \boldsymbol{\omega} \times \mathbf{H}_G. \tag{22.50}$$

By Eq. (22.38),

$$\mathbf{H}_G = (I_u\omega_u)\boldsymbol{\lambda}_u + (I_v\omega_v)\boldsymbol{\lambda}_v + (I_w\omega_w)\boldsymbol{\lambda}_w \tag{22.51}$$

where $\boldsymbol{\lambda}_u$, $\boldsymbol{\lambda}_v$, and $\boldsymbol{\lambda}_w$ represent, respectively, unit vectors along the u, v, and w principal centroidal axes of inertia. Because I_u, I_v, and I_w are constant, it follows that

$$\dot{\mathbf{H}}_G = (I_u\dot{\omega}_u)\boldsymbol{\lambda}_u + (I_v\dot{\omega}_v)\boldsymbol{\lambda}_v + (I_w\dot{\omega}_w)\boldsymbol{\lambda}_w. \tag{22.52}$$

Also,

$$\boldsymbol{\omega} = \omega_u\boldsymbol{\lambda}_u + \omega_v\boldsymbol{\lambda}_v + \omega_w\boldsymbol{\lambda}_w. \tag{22.53}$$

Substituting Eqs. (22.51), (22.52) and (22.53) in Eq. (22.50), carrying out the indicated operations, and simplifying,

$$\sum \mathbf{M}_G = [I_u\dot{\omega}_u) - (I_v - I_w)\omega_v\omega_w]\boldsymbol{\lambda}_u + [I_v\dot{\omega}_v) - (I_w - I_u)\omega_w\omega_u]\boldsymbol{\lambda}_v + [I_w\dot{\omega}_w) - (I_u - I_v)\omega_u\omega_v]\boldsymbol{\lambda}_w \tag{22.54}$$

which represents *Euler's equations of motion* expressed in vector form, named after the Swiss mathematician L. Euler (1707–1783). The three scalar components of Eq. (22.54) represent Euler's three scalar equations of motion as follows:

$$\begin{aligned} \sum \mathbf{M}_u &= I_u\alpha_u - (I_v - I_w)\omega_v\omega_w \\ \sum \mathbf{M}_v &= I_v\alpha_v - (I_w - I_u)\omega_w\omega_u \\ \sum \mathbf{M}_w &= I_w\alpha_w - (I_u - I_v)\omega_u\omega_v \end{aligned} \tag{22.55}$$

where $\alpha_u = \dot{\omega}_u$, $\alpha_v = \dot{\omega}_u$ and $\alpha_w = \dot{\omega}_w$ are the u, v, and w components, respectively, of the body's angular acceleration.

Example 22.7 and 22.8 illustrate some of the concepts discussed above.

■ Example 22.7

A thin homogeneous rectangular plate is welded to a horizontal shaft AB as shown in Figure E22.7(a). The weight of the plate is 20 lb and that of the shaft is 4 lb. The bearings at A and at B can resist only radial forces. The assembly rotates at a constant angular velocity of 250 rpm. For the position shown, determine the components of the reaction (a) at the bearing at A and (b) at the bearing at B.

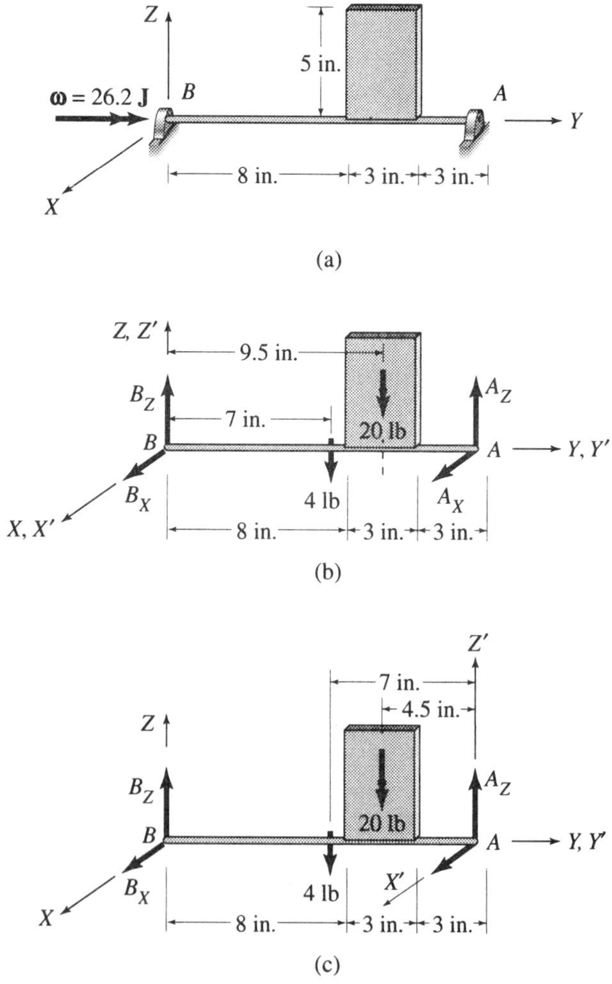

FIGURE E22.7.

Solution

(a) The solution to this problem may be obtained by using Eq. (22.49) which states that

$$\sum \mathbf{M}_B = (\dot{\mathbf{H}}_B)_{X'Y'Z'} + \mathbf{\Omega} \times (\mathbf{H}_B)_{X'Y'Z'} \tag{a}$$

where the X'-Y'-Z' coordinate system is attached to the assembly at B and rotates with it. Note that, for the position shown in Figure E22.7(b), the rotating X'-Y'-Z' coordinate system coincides with the fixed X-Y-Z coordinate system. Thus, $\mathbf{\Omega}$ is equal to $\boldsymbol{\omega}$, the angular velocity of the assembly, where

$$\omega = 250 \text{ rpm} = 26.2 \text{ rad/s}.$$

Thus, $\omega_{X'} = 0$, $\omega_{Y'} = 26.2$ rad/s, $\omega_{Z'} = 0$, and

$$\boldsymbol{\omega} = (0)\mathbf{I} + (26.2)\mathbf{J} + (0)\mathbf{K} \tag{b}$$

where the **I**, **J**, and **K** unit vectors remain fixed in magnitude and direction. The angular momentum vector $(\mathbf{H}_B)_{X'Y'Z'}$ is determined from Eq. (22.41) which reduces to

$$(\mathbf{H}_B)_{X'Y'Z'} = (-I_{X'Y'}\omega_{Y'})\mathbf{I} + (I_{Y'}\omega_{Y'})\mathbf{J} + (-I_{Y'Z'}\omega_{Y'})\mathbf{K}. \tag{c}$$

The necessary moment and products of inertia are determined as follows:

$$I_{Y'} = (I_{Y'})_{\text{SHAFT}} + (I_{Y'})_{\text{PLATE}},$$

$$= 0 + \frac{1}{12}\left(\frac{20}{32.2}\right)\left(\frac{5}{12}\right)^2 + \left(\frac{20}{32.2}\right)\left(\frac{2.5}{12}\right)^2$$

$$= 0.03494 \text{ lb}\cdot\text{ft}\cdot\text{s}^2.$$

$$I_{X'Y'} = (I_{X'Y'})_{\text{SHAFT}} + (I_{X'Y'})_{\text{PLATE}} = 0$$

$$I_{Y'Z'} = (I_{Y'Z'})_{\text{SHAFT}} + (I_{Y'Z'})_{\text{PLATE}},$$

$$= 0 + 0 + \left(\frac{20}{32.2}\right)\left(-\frac{9.5}{12}\right)\left(-\frac{9.5}{12}\right)$$

$$= 0.10244 \text{ lb}\cdot\text{ft}\cdot\text{s}^2.$$

Therefore, Eq. (c) becomes

$$(\mathbf{H}_B)_{X'Y'Z'} = (0)\mathbf{I} + (0.942)\mathbf{J} - (2.684)\mathbf{K}. \tag{d}$$

Because $\mathbf{\Omega}$ is constant and because $I_{Y'}$, $I_{X'Y'}$, and $I_{Y'Z'}$ remain unchanged with time, it follows that

$$(\dot{\mathbf{H}}_B)_{X'Y'Z'} = \mathbf{0}. \tag{e}$$

Also, the quantity, $\sum \mathbf{M}_B$ in Eq. (a), is determined from the free-body diagram shown in Figure E22.6(b). Thus,

$$\sum \mathbf{M}_B = \left(\frac{7}{12}\right)\mathbf{J} \times (-4\mathbf{K}) + \left(\frac{9.5}{12}\right)\mathbf{J} \times (-20\mathbf{K})$$

$$+ \left(\frac{14}{12}\right)\mathbf{J} \times [(A_X)\mathbf{I} + (0)\mathbf{J} + (A_Z)\mathbf{K}],$$

$$= (1.17A_Z - 18.16)\mathbf{I} + (0)\mathbf{J} - (1.17A_X)\mathbf{K}. \quad \text{(f)}$$

Substituting Eqs. (b), (d), (e) and (f) in Eq. (a),

$$(1.17A_Z - 18.16)\mathbf{I} + (0)\mathbf{J} - (1.17A_X)\mathbf{K}$$

$$= \mathbf{0} + [(0)\mathbf{I} + (26.2)\mathbf{J} + (0)\mathbf{K}] \times [(0)\mathbf{I} + (0.942)\mathbf{J} - (2.684)\mathbf{K}]$$

$$= -(70.32)\mathbf{I} + (0)\mathbf{J} + (0)\mathbf{K}.$$

Equating the coefficients of the **I**, **J**, and **K** unit vectors, we conclude that

$$1.17A_Z - 18.16 = -70.32,$$

$$-1.17A_X = 0,$$

from which

$$A_X = 0 \qquad \text{ANS.}$$

$$A_Z = -44.6 \text{ lb.} \qquad \text{ANS.}$$

(b) The components of the reaction at B are found similarly by using Eq. (22.49) applied about point A. Thus,

$$\sum \mathbf{M}_A = (\dot{\mathbf{H}}_A)_{X'Y'Z'} + \boldsymbol{\Omega} \times (\mathbf{H}_A)_{X'Y'Z'} \quad \text{(g)}$$

where the rotating X', Y' and Z' axes are attached to the assembly at point A as shown in Figure E22.6(c). Thus, as in part (a),

$$\boldsymbol{\Omega} = \boldsymbol{\omega} \times (0)\mathbf{I} + (26.2)\mathbf{J} + (0)\mathbf{K}, \quad \text{(h)}$$

$$(\mathbf{H}_A)_{X'Y'Z'} = (0)\mathbf{I} + (0.942)\mathbf{J} + (1.271)\mathbf{K}, \quad \text{(i)}$$

$$(\dot{\mathbf{H}}_A)_{X'Y'Z'} = \mathbf{0}, \quad \text{(j)}$$

$$\sum \mathbf{M}_A = (9.83 - 1.17B_Z)\mathbf{I} + (0)\mathbf{J} - (1.17B_X)\mathbf{K}. \quad \text{(k)}$$

Substituting Eqs. (h) to (k) in Eq. (g), combining terms, and equating the coefficients of the **I**, **J**, and **K** unit vectors, we obtain the following two scalar equations:

$$9.83 - 1.17B_Z = 33.30,$$

and

$$1.17B_X = 0.$$

Therefore,

$$B_X = 0, \qquad \text{ANS.}$$

and

$$B_Z = -20.1 \text{ lb.} \qquad \text{ANS.}$$

The solution above may be checked in part by applying the equation $\sum F_Z = m(a_G)_Z = mr\omega^2$. To apply this equation and insure that the left-hand side is, in fact, equal to the right-hand side, we need to locate the Z coordinate for the center of mass of the system in the position shown. Thus,

$$Z_G = \frac{\sum m_i z_i}{\sum m_i} = \frac{\left(\frac{20}{32.2}\right)\left(\frac{5}{2}\right) + 0}{\left(\frac{20 + 4}{32.2}\right)} = 2.083 \text{ in.} = 0.1736 \text{ ft.}$$

Therefore,

$$\sum F_Z = -44.6 - 20.1 - 4 - 20 = -88.7 \text{ lb},$$

and

$$mr\omega^2 = -\left(\frac{24}{32.2}\right)(0.1736)(26.2^2) = -88.8 \text{ lb}$$

which indicates that both sides of the equation $\sum F_Z = mr\omega^2$ are equal within the limits of the approximations used during the solution.

■ Example 22.8

A thin homogeneous rectangular plate having a mass of 12 kg is attached rigidly to the midpoint of shaft AB such that one of the principal centroidal axes of the plate makes a 25° angle with the shaft as shown in Figure E22.8(a). Starting from rest, a couple $\mathbf{M} = (50\mathbf{k})$ N·m is applied, causing the assembly to accelerate. Determine the components of the reactions at the bearings A and B assuming that, when in the position shown, the assembly has reached an angular velocity $\boldsymbol{\omega} = (40\mathbf{k})$ rad/s. Also determine the angular acceleration of the assembly. Assume that the shaft is weightless. Also, the bearing at A can resist only radial forces whereas the bearing at B can resist both radial and axial forces.

Solution

The free-body diagram of the assembly at the instant it reaches an angular velocity $\boldsymbol{\omega} = 40\mathbf{k}$ rad/s is shown in Figure E22.8(b). Because the mass center of the assembly, point G, does not translate, it follows that $\mathbf{a}_G = \mathbf{0}$. Application of Eq. (22.44), $\sum \mathbf{F} = m\mathbf{a}_G$ leads to

$$(A_x\mathbf{i} + A_y\mathbf{j} + 0\mathbf{k}) + (B_x\mathbf{i} + B_y\mathbf{j} + B_z\mathbf{k}) - 117.7\mathbf{k} = \mathbf{0}. \qquad \text{(a)}$$

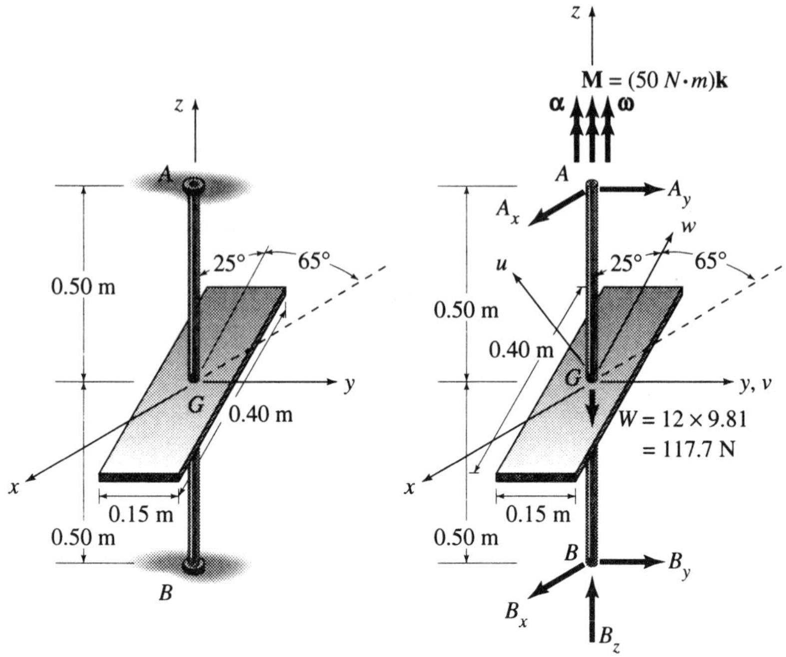

FIGURE E22.8. (a) (b)

After combining terms and equating the coefficients of the **i**, **j**, and **k** unit vectors, we obtain the following three scalar equations

$$A_x + B_x = 0, \tag{b}$$

$$A_y + B_y = 0, \tag{c}$$

and

$$B_z = 117.7 \text{ N}. \tag{d}$$

The next step in the solution may be accomplished by using Euler's equations expressed in Eqs. (22.55), where the u, v, and w principal axes of the plate are as indicated in Figure E22.7(b). The use of Eqs. (22.52) requires determining the following quantities:

$$I_u = \tfrac{1}{12}(12)(0.15^2 + 0.40^2) = 0.1825 \text{ kg·m}^2,$$

$$I_v = \tfrac{1}{12}(12)(0.40^2) = 0.1600 \text{ kg·m}^2,$$

$$I_w = \tfrac{1}{12}(12)(0.15^2) = 0.0225 \text{ kg·m}^2,$$

$$\omega_u = 40 \sin 25^\circ = 16.905 \text{ rad/s},$$

$$\omega_v = 0,$$

$$\omega_w = 40 \cos 25^\circ = 36.252 \text{ rad/s},$$

$$\alpha_u = \alpha \sin 25^\circ = 0.4226\alpha \text{ rad/s}^2,$$

$$\alpha_v = 0,$$

$$\alpha_w = \alpha \cos 25^\circ = 0.9063\alpha \text{ rad/s}^2.$$

Thus, using Eqs. (22.55) along with the free-body diagram of Figure E22.8(b)

$$\sum M_u = I_u\alpha_u - (I_v - I_w)\omega_v\omega_w$$

$$-A_y(0.5\cos 25^\circ) + B_y(0.5\cos 25^\circ) + 50\sin 25^\circ = (0.1825)(0.4226a) - 0$$

from which

$$B_y - A_y = 0.1702\alpha - 46.63. \tag{e}$$

Also,

$$\sum M_v = I_v\alpha_v - (I_w - I_u)\omega_w\omega_u,$$

$$A_x(0.5) - B_x(0.5) = 0 - (0.0225 - 0.1825)(36.252)(16.905),$$

which leads to

$$A_x - B_x = 196.11, \tag{f}$$

and

$$\sum M_w = I_w\alpha_w - (I_u - I_v)\omega_u\omega_v,$$

$$A_y(0.5\cos 25^\circ) - B_y(0.5\sin 25^\circ) + 50\cos 25^\circ = (0.0225)(0.9063\alpha) - 0,$$

which reduces to

$$A_y - B_y = 0.0965\alpha - 214.45. \tag{g}$$

Solving Eqs. (b) through (g) simultaneously,

$$\alpha = 979 \text{ rad/s}^2, \qquad \text{ANS.}$$

$$A_x = 98.1 \text{ N}, \qquad \text{ANS.}$$

$$A_y = -60.0 \text{ N}, \qquad \text{ANS.}$$

$$A_z = 117.7 \text{ N}, \qquad \text{ANS.}$$

$$B_x = -98.1 \text{ N}, \qquad \text{ANS.}$$

and

$$B_y = 60.0 \text{ N}. \qquad \text{ANS.}$$

Note that the moments $\sum M_u$, $\sum M_v$, and $\sum M_w$ were obtained from the scalar equations $\sum M_u = \boldsymbol{\lambda}_u \cdot \mathbf{M}_G$, $\sum M_v = \boldsymbol{\lambda}_v \cdot \mathbf{M}_G$ and $\sum M_w = \boldsymbol{\lambda}_w \cdot \mathbf{M}_G$.

■

Problems

22.40 A rectangular plate weighing 25 lb is fastened rigidly to a weightless shaft as shown in Figure P22.40. The assembly is supported at A and B by bearings that can resist only radial forces and made to rotate at a constant angular velocity $\boldsymbol{\omega} = (15 \text{ rad/s})\mathbf{j}$. Determine the reaction components at A and B for the position shown. Let $a = 4$ in. and $b = 8$ in.

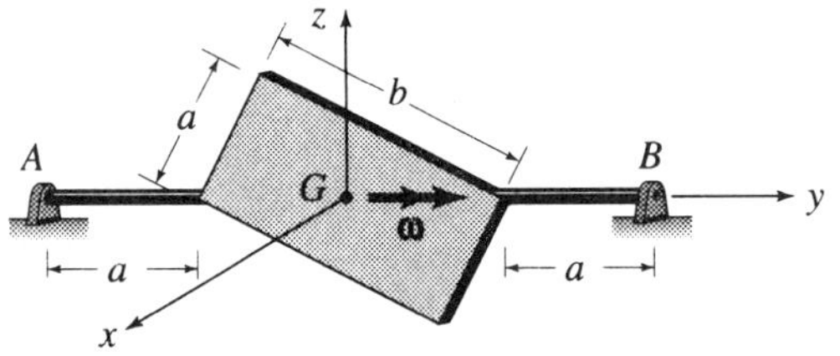

FIGURE P22.40.

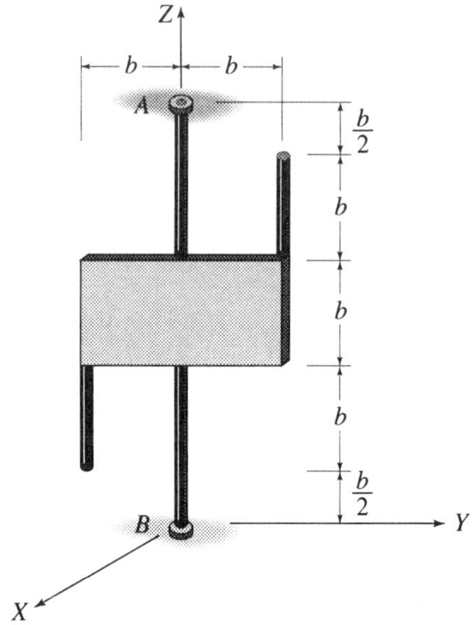

FIGURE P22.42.

22.41 A rectangular plate having a mass of 15 kg is fastened rigidly to a weightless shaft as shown in Figure P22.40. The assembly is supported at A and B by bearings that can resist only radial forces. Starting from rest, a couple $\mathbf{M} = (30\mathbf{j})$ N·m is applied, causing the assembly to accelerate. Determine the components of the reactions at A and B assuming that, when in the position shown, the assembly has reached an angular velocity $\boldsymbol{\omega} = (25\mathbf{j})$ rad/s. Also determine the angular acceleration α of the assembly. Let $a = 0.15$ m and $b = 0.3$ m.

22.42 A thin homogeneous plate of mass m is attached rigidly to two identical slender rods, each of mass $1/4m$, and to a weightless shaft AB as shown in Figure P22.42. The bearing at A can resist only radial forces whereas that at B can resist both radial and axial forces. If the assembly rotates at a constant angular velocity of $\boldsymbol{\omega} = \omega\mathbf{K}$, determine the reaction components at A and B for the position shown.

22.43 A thin homogeneous plate of mass m is attached rigidly to two identical slender rods, each of mass 1/8m, and to a weightless shaft AB as shown in Figure P22.42. The bearing at A can resist only radial forces whereas that at B can resist both radial and axial forces. Starting from rest, a couple $\mathbf{M} = M\mathbf{K}$ is applied, causing the assembly to accelerate. Determine the components of the reactions at A and B assuming that, when in the position shown, the assembly has reached an angular velocity $\boldsymbol{\omega} = \omega\mathbf{K}$. Also determine the angular acceleration of the assembly.

22.44 Two identical thin homogeneous plates,

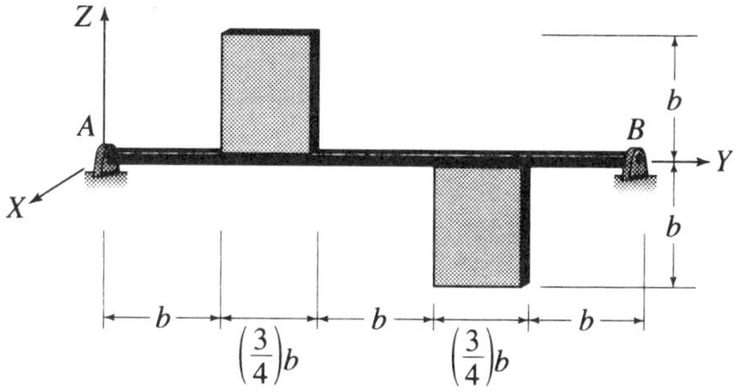

FIGURE P22.44.

each weighing 15 lb, are fastened rigidly to shaft AB weighing 5 lb, and the assembly is supported by bearings at A and B, as shown in Figure P22.44. The bearings at A and B are capable of resisting only radial forces. If the assembly rotates at a constant speed of 500 rpm, determine the reaction components at A and B for the position shown. Let $b = 4$ in.

22.45 Two identical thin homogeneous plates, each with a mass $m = 10$ kg, are fastened rigidly to shaft AB with a mass of 3 kg, and the assembly is supported by bearings at A and B, as shown in Figure P22.44. The bearings at A and B are capable of resisting only radial forces. Starting from rest, a couple $\mathbf{M} = M\mathbf{J}$ is applied to the assembly, causing it to develop an acceleration $\boldsymbol{\alpha} = (10\mathbf{J})$ rad/s^2. Determine (a) the couple **M** and (b) the reaction components at A and B, for the position shown, which corresponds to 2 seconds after the start of the motion. Let $b = 0.12$ m.

22.46 A slender homogeneous rod weighing 1 lb/in. is bent into the shape shown in Figure P22.46, and supported by bearings at A and B which can support only radial forces. If the rod rotates about axis AB at a constant angular velocity $\boldsymbol{\omega} = (80\mathbf{J})$ rad/s, determine the reaction components at support B for the position shown. Let $b = 5$ in.

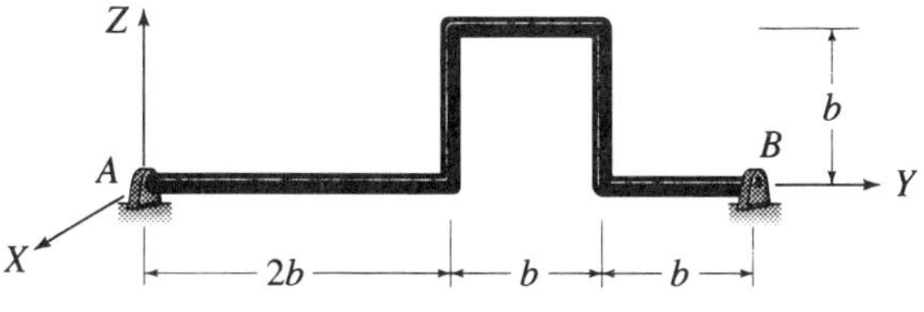

FIGURE P22.46.

22.47 Refer to Problem 22.46, and determine the reaction components at support A.

22.48 A slender homogeneous rod, with a mass of 20 kg/m, is bent into the shape shown in Figure P22.46 and supported by bearings at A and B which can resist only radial forces. Starting from rest, a couple $\mathbf{M} = (80\mathbf{J})$ N·m is applied, causing the rod to accelerate. Determine the components of the reaction support B assuming that, when in the position shown, the rod has reached an angular velocity $\boldsymbol{\omega} = (40\mathbf{J})$ rad/s. Also determine the angular acceleration $\boldsymbol{\alpha}$ of the rod. Let $b = 0.10$ m.

22.49 Refer to Problem 22.48, and determine the reaction components at support A.

22.50 A thin homogeneous disk of mass m and radius R is welded to a vertical weightless shaft AB so that the plane of the disk makes an angle γ with the vertical, as shown in Figure P22.50. The bearing at A can resist only radial forces, and that at B can resist radial and axial forces. The assembly rotates at a constant angular velocity $\boldsymbol{\omega} = \omega\mathbf{k}$. Determine the reaction components at bearings A and B.

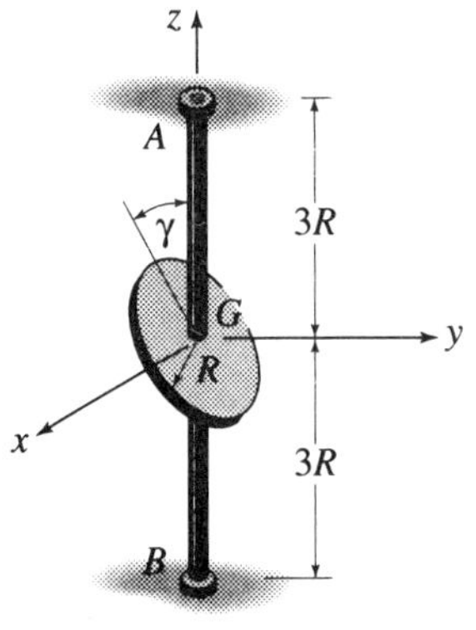

FIGURE P22.50.

22.51 A thin homogeneous disk weighing 180 lb and having a radius $R = 6$ in. is welded to a vertical weightless shaft AB so that the plane of the disk makes an angle $\gamma = 30°$ with the vertical, as shown in Figure P22.50. The bearing at A can resist only radial forces, and that at B can resist radial and axial forces. Starting from rest, a couple $\mathbf{M} = M\mathbf{k}$ is applied to the assembly, causing it to develop an acceleration $\boldsymbol{\alpha} = (15\mathbf{k})$ rad/s^2. Determine (a) the couple M and (b) the reaction components at A and B, for the position shown, which corresponds to 3 seconds after the start of the motion.

22.52 A thin homogeneous disk of mass m and radius R is attached rigidly to a weightless vertical shaft so that the plane of the disk makes an angle θ with the horizontal, as shown in Figure P22.52. The bearing at A can resist only radial forces whereas B can resist radial and axial forces. Starting from rest, a couple $\mathbf{M} = M\mathbf{k}$ is applied, causing the assembly to develop an acceleration $\boldsymbol{\alpha} = \alpha\mathbf{k}$. Determine (a) the magnitude of the constant couple M and (b) the reaction components at supports A and B, for the position shown, which corresponds to t seconds after the start of the motion. Express your answers in terms of the known quantities m, R, α, θ, and t.

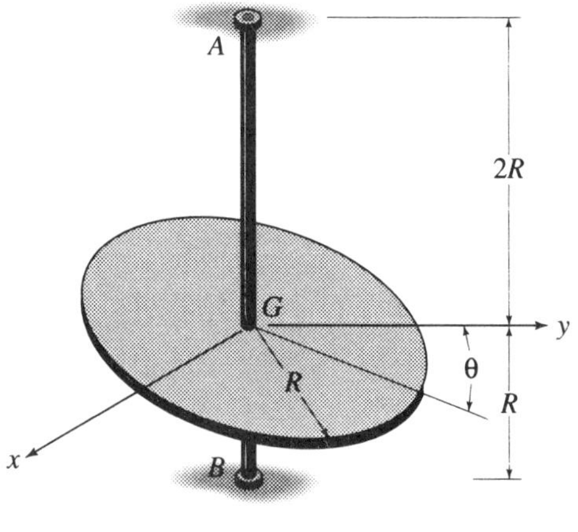

FIGURE P22.52.

22.53 A homogeneous cylinder of weight $W = 75$ lb, radius $R = 6$ in., and length $L =$

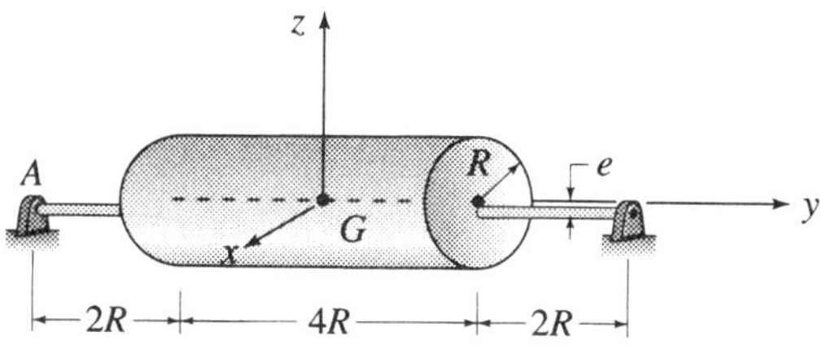

FIGURE P22.53.

$4R$ is mounted eccentrically on shaft AB with an eccentricity $e = 3/16$ in. as shown in Figure P22.53. The bearings at A and B are capable of resisting only radial forces. If the assembly rotates at a constant speed of 500 rpm, determine the reaction components at the two bearings for the position shown.

22.54 A homogeneous cylinder of mass $m = 30$ kg, radius $R = 0.12$ m, and length $L = 4R$, is mounted eccentrically on shaft AB with an eccentricity $e = 0.005$ m, as shown in Figure P22.53. The bearings at A and B are capable of resisting only radial forces. Starting from rest, a couple $\mathbf{M} = (100\mathbf{j})$ N·m is applied, causing the assembly to accelerate. Determine the components of the reaction at supports A and B assuming that, when in the position shown, the assembly has reached an angular velocity $\boldsymbol{\omega} = (25\mathbf{j})$ rad/s. Also determine the magnitude of the angular acceleration $\boldsymbol{\alpha}$ of the assembly.

22.55 One end of a homogeneous rod of mass m and length $3r$ is attached rigidly to a homogeneous sphere of mass $4m$ and radius r. The other end of the rod is attached to a frictionless hinge as shown in Figure P22.55. The assembly rotates about the vertical Z axis at a constant angular velocity $\boldsymbol{\omega} = \omega\mathbf{K}$. Determine the angle θ that the rod-sphere system makes with the axis of rotation and the reaction components at the pin at A. Express the answers in terms of m, r, ω, and g.

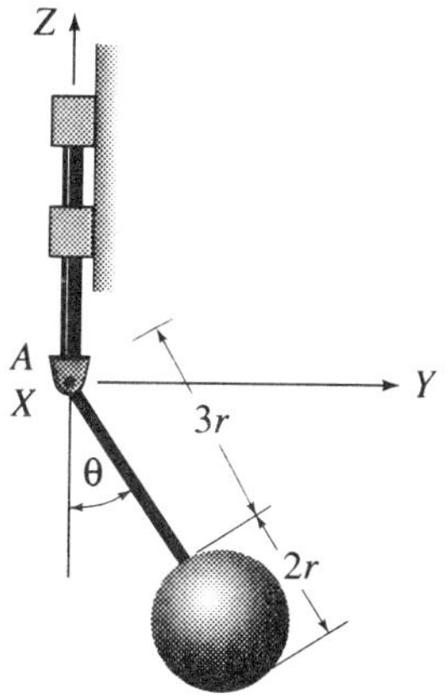

FIGURE P22.55.

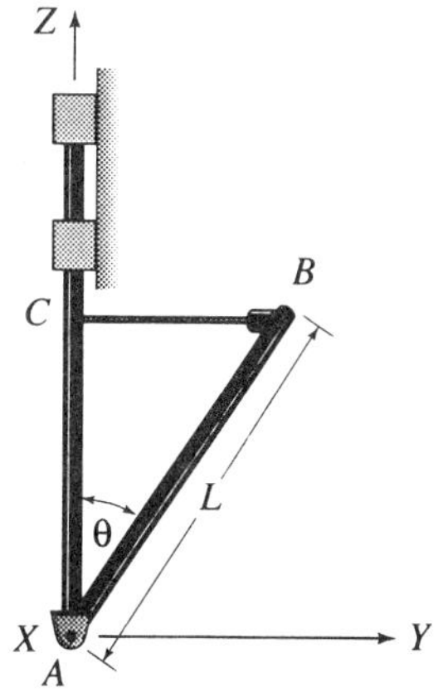

FIGURE P22.56.

22.56 The homogeneous rod AB of length L and weight W, shown in Figure P22.56, is hinged at A to a vertical shaft that rotates at the constant angular velocity $\boldsymbol{\omega} = \omega\mathbf{K}$. It is maintained in the position dened by the angle θ by the horizontal wire BC. Determine the tension in the wire and the reaction components at hinge A. Express the answers in terms of W, L, ω, and θ.

22.57 A homogeneous disk of radius $r = 0.15$ m is hinged along a diameter AB to a clevis that is supported by a bearing at C, as shown in Figure P22.57. The mass of the disk $m = 20$ kg. The mass of the clevis may be assumed negligible. The disk spins about its diametrical axis at a constant angular velocity $\boldsymbol{\omega}_1 = (10\mathbf{J})$ rad/s relative to the clevis while the entire assembly rotates about the bearing at C at a constant angular velocity $\boldsymbol{\omega}_2 = (5\mathbf{I})$ rad/s. For the position shown, determine the force-couple system developed at the bearing C.

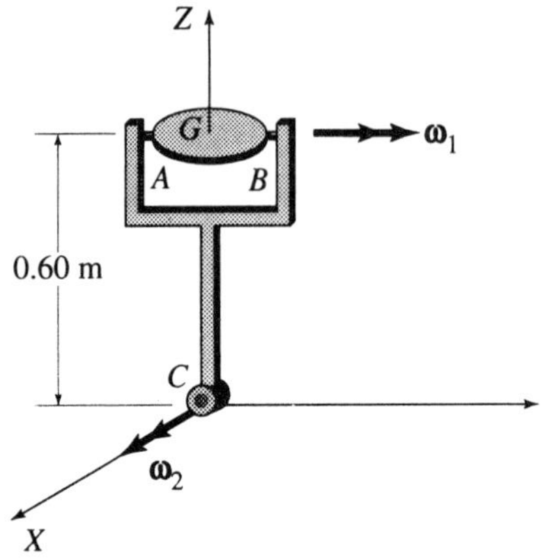

FIGURE P22.57.

22.58 A disk weighing 25 lb rotates at a constant angular velocity $\omega_1 = 20$ rad/s relative to the supporting frame OABG, which itself rotates at a constant angular velocity $\omega_2 = 10$ rad/s about the Z axis at O, as shown in Figure P22.58. Assume that the frame OABG is weightless, and determine the force-couple system developed at the bearing at O.

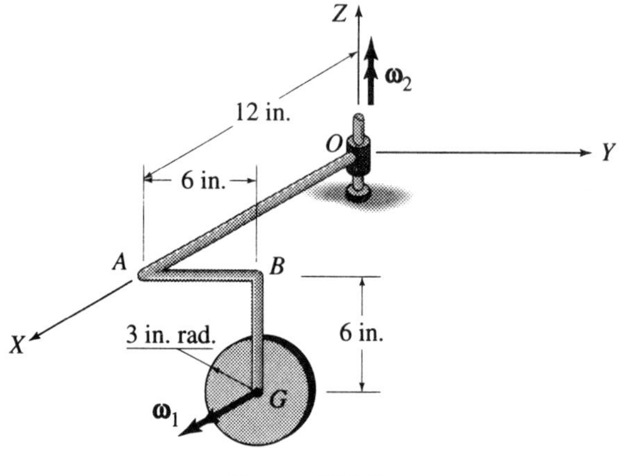

FIGURE P22.58.

22.59 As shown in Figure P22.59, a homogeneous disk of mass $m = 30$ kg and radius $r = 0.25$ m rotates at a constant angular velocity $\boldsymbol{\omega}_1 = (50\mathbf{i})$ rad/s about its horizontal axle at G while the forked shaft ABG rotates at a constant angular velocity $\boldsymbol{\omega}_2 = (10\mathbf{j})$ rad/s. Find the couple **M** that the shaft ABG exerts on the disk.

22.60 In the position shown, the blades of the

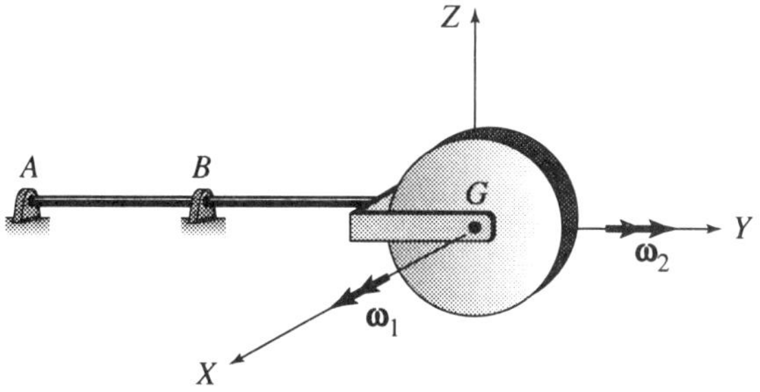

FIGURE P22.59.

helicopter rotate at a constant angular velocity $\boldsymbol{\omega}_1 = \omega_1 \mathbf{K}$ relative to its frame while the frame itself makes a vertical turn (i.e., in the Y-Z plane) at a constant angular velocity $\boldsymbol{\omega}_2 = \omega_2 \mathbf{I}$, as shown in Figure P22.60. Each blade has a mass m and may be assumed to be a slender rod of length L. At the instant shown, determine the couple $\mathbf{M}$ exerted by the blades on the attached shaft.

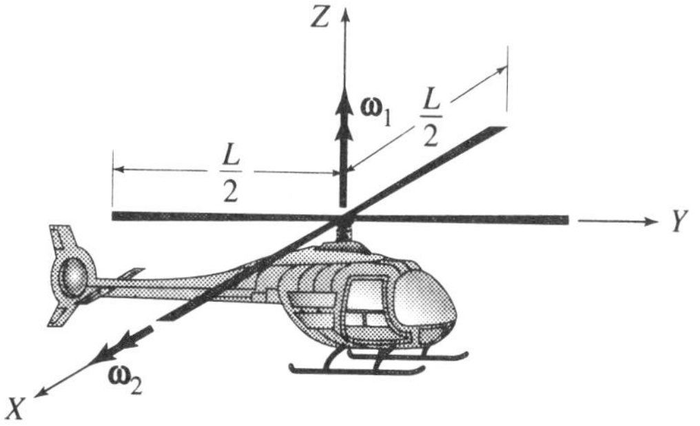

FIGURE P22.60.

22.6 General Gyroscopic Motion

By definition, a gyroscope consists of a body of revolution (axisymmetric body) that rotates (spins) about its own axis of symmetry which in turn rotates (precesses) about some other axis. Consider, for example, the circular cone in Figure 22.12, spinning about the z axis at an angular velocity $\frac{d\psi}{dt} = \dot{\psi}$ known as the *rate of spin.* Point O and the coordinate axes X, Y, and Z are fixed in space. Plane 1 is perpendicular to the Z axis (thus fixed in space) whereas plane 2 is perpendicular to the z axis. In addition to the angle ψ, which defines the rotation experienced by the cone measured from some reference axis, as in Figure 22.12, complete definition of the position of the cone in space, at any instant, requires that we know the angles θ and ϕ. The angle θ defines the position of the axis of spin (the z axis) relative to the fixed Z axis (the axis of precession) and its derivative with respect to time, $\frac{d\theta}{dt} = \dot{\theta}$ is known as the *rate of nutation.* When $\theta = 0$, plane 2 coincides with plane 1 and when $\theta \neq 0$, plane 2 intersects plane 1 along the x axis, as shown in Figure 22.12. The angle ϕ defines the position of the x axis

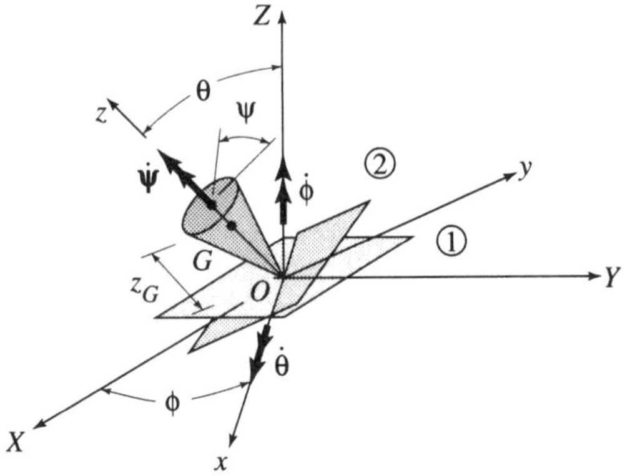

FIGURE 22.12.

relative to the fixed X axis, and its derivative with respect to time $\frac{d\phi}{dt} = \dot{\phi}$ is known as the *rate of precession*. The y axis is perpendicular to both the x and z axes and, obviously, lies in plane 2 and, also, in the plane defined by the z and Z axes. Note that both the X-Y-Z and the x-y-z coordinate systems are right-handed and that the x-y-z coordinate systems are principal axes of inertia for the cone. Note also that, although the x-y-z coordinate system nutates and precesses, it does not spin with the rotating cone. The angles ψ, θ, and ϕ define the position of the cone at any instant during its motion and are known as Euler's angles.

The equations defining the motion of a gyroscope will be derived with the help of Figure 22.12. The angular velocity $\boldsymbol{\omega}$ of the rotating cone is expressed in terms of the rates of spin, nutation, and precession as follows:

$$\boldsymbol{\omega} = \dot{\theta}\mathbf{i} + \dot{\psi}\mathbf{j} + \dot{\phi}\mathbf{K} \tag{22.56}$$

where, from the geometry in Figure 22.12,

$$\mathbf{K} = (\sin\theta)\mathbf{j} + (\cos\theta)\mathbf{k}. \tag{22.57}$$

Substituting Eq. (22.57) in Eq. (22.56) yields

$$\boldsymbol{\omega} = \dot{\theta}\mathbf{i} + (\dot{\phi}\sin\theta)\mathbf{j} + (\dot{\psi} + \dot{\phi}\cos\theta)\mathbf{k}. \tag{22.58}$$

Thus, the x, y, and z components of the cone's angular velocity are

$$\left.\begin{aligned} \omega_x &= \dot{\theta}, \\ \omega_y &= \dot{\phi}\sin\theta, \\ \text{and} \quad \omega_z &= \dot{\psi} + \dot{\phi}\cos\theta \end{aligned}\right\}. \tag{22.59}$$

Recall that the x, y, and z axes are principal axes of inertia, and let $I_x = I_y = I_1$ and $I_z = I_2$ where I_1, and I_2 remain constant with time. Therefore, the angular momentum $\mathbf{H}_O$ of the rotating cone may be found using Eq. (22.43). Thus,

$$\mathbf{H}_O = (I_1 \dot{\theta})\mathbf{i} + (I_1 \dot{\phi} \sin\theta)\mathbf{j} + I_2(\dot{\psi} + \dot{\phi}\cos\theta)\mathbf{k}, \tag{22.60}$$

and

$$\dot{\mathbf{H}}_O = (I_1 \ddot{\theta})\mathbf{i} + I_1(\dot{\theta}\dot{\phi}\cos\theta + \ddot{\phi}\sin\theta)\mathbf{j} + I_2 \frac{d}{dt}(\dot{\psi} + \dot{\phi}\cos\theta)\mathbf{k}. \tag{22.61}$$

The angular velocity $\boldsymbol{\Omega}$ of the rotating x-y-z coordinate system is

$$\boldsymbol{\Omega} = \dot{\theta}\mathbf{i} + \dot{\phi}\mathbf{K}. \tag{22.62}$$

Using Eq. (22.57), Eq. (22.62) becomes

$$\boldsymbol{\Omega} = \dot{\theta}\mathbf{i} + (\dot{\phi}\sin\theta)\mathbf{j} + (\dot{\phi}\cos\theta)\mathbf{k} \tag{22.63}$$

By comparing Eqs. (22.58) and (22.63), we conclude that $\boldsymbol{\omega} \neq \boldsymbol{\Omega}$ in this particular case.

Substituting Eqs. (22.60), (22.61), and (22.63) in Eq. (22.49), we obtain, after simplification, the following vector equation describing the gyroscopic motion of the cone in Figure 22.12:

$$\begin{aligned}\sum \mathbf{M}_O = {} & [I_1(\ddot{\theta} - \dot{\phi}^2 \sin\theta\cos\theta) + I_2\dot{\phi}\sin\theta(\dot{\psi} + \dot{\phi}\cos\theta)]\mathbf{i} \\ & + [I_1(\ddot{\phi}\sin\theta + 2\dot{\theta}\dot{\phi}\cos\theta) - I_2\dot{\theta}(\dot{\psi} + \dot{\phi}\cos\theta)]\mathbf{j} \\ & + \left[I_2 \frac{d}{dt}(\dot{\psi} + \dot{\phi}\cos\theta)\right]\mathbf{k}.\end{aligned} \tag{22.64}$$

Equation (22.64) may be expressed in terms of its x, y, and z scalar components as follows:

$$\left.\begin{aligned}\sum M_X &= I_1(\ddot{\theta} - \dot{\phi}^2\sin\theta\cos\theta) + I_2\dot{\phi}\sin\theta(\dot{\psi} + \dot{\phi}\cos\theta) \\ \sum M_y &= I_1(\ddot{\phi}\sin\theta + 2\dot{\theta}\dot{\phi}\cos\theta) - I_2\dot{\theta}(\dot{\psi} + \dot{\phi}\cos\theta) \\ \sum M_Z &= I_2\frac{d}{dt}(\dot{\psi} + \dot{\phi}\cos\theta)\end{aligned}\right\}. \tag{22.65}$$

Note that Eqs. (22.65) were derived for the gyroscopic motion of an axisymmetric body about a fixed arbitrary point O on its axis of symmetry. It may be shown, however, that these same equations apply equally well to the motion of an axisymmetric body about its mass center. Also note that the relations expressed in Eqs. (22.65) represent a set of nonlinear, second-order differential equations. In general, therefore, it is difficult to obtain closed-form solutions and resort is made to less exact numerical methods. There are, however, two special cases of much practical interest for which closed-form solutions may be obtained. These two special cases are discussed in the following two sections.

22.6 Gyroscopic Motion with Steady Precession

The *steady precession* of a gyroscope is a special type of gyroscopic motion which results when the rate of spin $\dot{\psi}$, the nutation angle θ, and the rate of precession $\dot{\phi}$ are all constant during the motion. Under these conditions, $\ddot{\psi} = \dot{\theta} = \ddot{\phi} = 0$. Substitution of these values in Eqs. (22.65) lead, after reorganization, to

$$\left.\begin{aligned} \sum M_x &= I_2\dot{\phi}\dot{\psi}\sin\theta + (I_2 - I_1)\dot{\phi}^2\sin\theta\cos\theta = \text{constant} \\ \sum M_y &= 0 \\ \sum M_z &= 0 \end{aligned}\right\}. \qquad (22.66)$$

Equations (22.66) show that, to maintain gyroscopic motion with steady precession, it is necessary to apply a constant moment about the x axis given by the first of these equations and shown in Figure 22.13. Note that, in such a case, the moment axis is perpendicular to the plane formed by the axis of spin (the z axis) and the axis of precession (the Z axis). For the spinning cone of Figure 22.13, this moment becomes $\sum M_x = Wz_G \sin\theta$ where z_G locates point G along the axis of spin and W is the weight of the cone. Substituting this value of $\sum M_x$ in the first of Eqs. (22.66) and simplifying yields

$$Wz_G = I_2\dot{\phi}\dot{\psi} + (I_2 - I_1)\dot{\phi}^2\cos\theta. \qquad (22.67)$$

For very small values of the rate of precession $\dot{\phi}$, the second term on the right-hand side of Eq. (22.67) becomes very small in comparison to the first and may be neglected, leading to

$$Wz_G = I_2\dot{\phi}\dot{\psi}. \qquad (22.68)$$

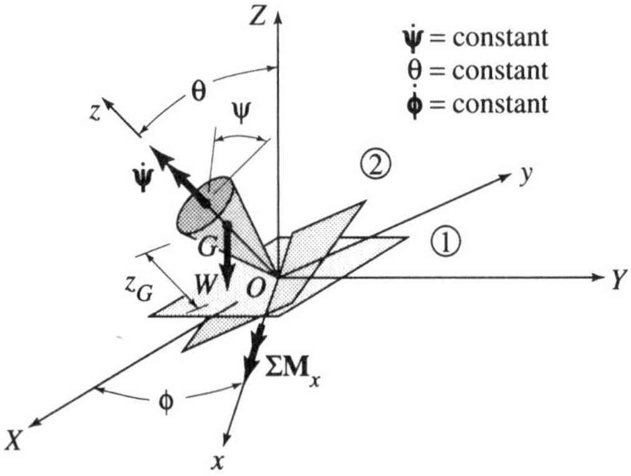

FIGURE 22.13.

Thus, under these conditions, the rate of spin becomes

$$\dot{\psi} = \frac{Wz_G}{I_2\dot{\phi}}. \tag{22.69}$$

It should be noted here that Eqs. (22.66) could also be derived by applying Eq. (22.49) which states that $\sum \mathbf{M}_O = \dot{\mathbf{H}}_O + \mathbf{\Omega} \times \mathbf{H}_O$. Thus, from Eq. (22.58), realizing that $\dot{\theta} = 0$, we obtain

$$\boldsymbol{\omega} = (\dot{\phi} \sin\theta)\mathbf{j} + (\dot{\psi} + \dot{\phi}\cos\theta)\mathbf{k}. \tag{22.70}$$

Because the x, y, and z axes are principal axes of inertia, we can use Eq. (22.43) to get

$$\mathbf{H}_O = (I_1\dot{\phi}\sin\theta)\mathbf{j} + \mathbf{I}_2(\dot{\psi} + \dot{\phi}\cos\theta)\mathbf{k} \tag{22.71}$$

from which we conclude that

$$\dot{\mathbf{H}}_O = \mathbf{0}. \tag{22.72}$$

Also, from Eq. (22.63) with $\dot{\theta} = 0$,

$$\mathbf{\Omega} = (\dot{\phi}\sin\theta)\mathbf{j} + (\dot{\phi}\cos\theta)\mathbf{k}. \tag{22.73}$$

Thus, Eq. (22.49) reduces to

$$\begin{aligned} \sum \mathbf{M}_O &= \mathbf{0} + [(\dot{\phi}\sin\theta)\mathbf{j} + (\dot{\phi}\cos\theta)\mathbf{k}] \\ &\quad \times [(I_1\dot{\phi}\sin\theta)\mathbf{j} + I_2(\dot{\psi} + \dot{\phi}\cos\theta)\mathbf{k}] \\ &= [I_2\dot{\phi}\dot{\psi}\sin\theta + (I_2 - I_1)\dot{\phi}^2\sin\theta\cos\theta]\mathbf{i} \end{aligned} \tag{22.74}$$

which, of course, agrees with Eq. (22.66).

A case of extreme practical interest is the case of gyroscopic motion with steady precession when $\theta = \pi/2$. Substituting this value of θ in the first of Eqs. (22.66), (or into Eq. (22.74), leads to

$$\sum M_x = I_2\dot{\phi}\dot{\psi}. \tag{22.75}$$

Also, as shown earlier for steady gyroscopic motion $\sum M_y = \sum M_z = 0$. Thus, to maintain this type of gyroscopic motion requires a constant moment about the x axis obtained by multiplying the moment of inertia about the axis of spin by the rate of spin and by the rate of precession. The directional relationship that exists among the axes of spin, moment, and precession is depicted in Figure 22.14. Note that, because the axis of spin (z axis) lies in the fixed X-Y plane, the y axis, then, coincides with the fixed Z axis of precession. Also, the x axis and, thus, the vector $\sum M_x$, lies in the fixed X-Y plane and is always normal to the plane defined by the z and Z axes. It is evident that, under these conditions, the axis of spin (z axis), the moment axis (x axis) and the axis of precession (Z axis) form a right-handed coordinate system. In this configuration, the gyroscope seems to defy gravity and "floats" horizontally.

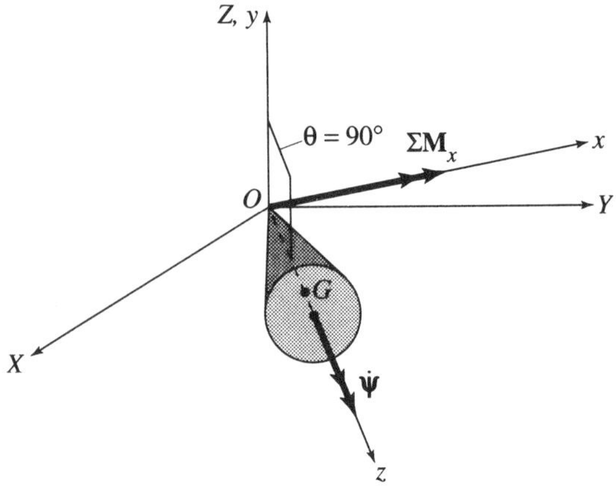

FIGURE 22.14.

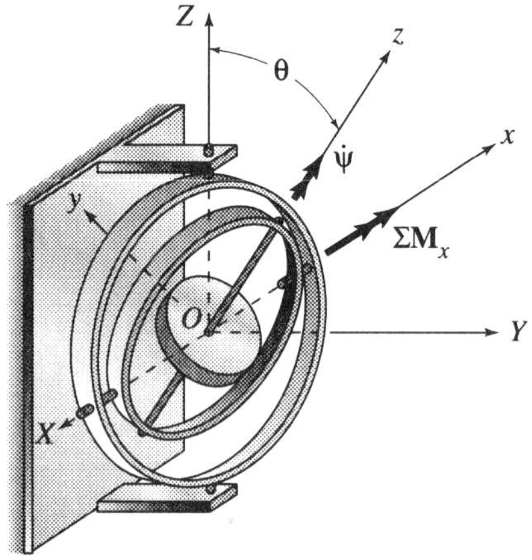

FIGURE 22.15.

Gyroscopic effects are utilized in several practical applications. The large gyroscopic moment developed by the motion of a gyroscope, for example, is used to stabilize ships against rolling motion. When mounted in gimbals rings, as shown in Figure 22.15, the gyroscope serves as a gyrocompass with its axis of spin maintaining a fixed north–south orientation. Also, because of the fact that its axis of spin maintains a fixed direction relative to the gimbal frame to which it is mounted, a gyroscope is very useful in the design of navigational and guidance devices. However, there are situations where gyroscopic effects are harmful and must be taken into account for proper design. Such is the case for spinning rotor shafts that, during operation, are subject to precession because of the gyroscopic moment that develops. Thus, for example, the bearings supporting the shaft of a ship's turbine must be designed to account for the additional reactive forces developed while the ship makes its turns.

22.7 Gyroscopic Motion with Zero Centroidal Moment

Another special case of gyroscopic motion with steady precession is encountered when a body of revolution moves under the influence of zero moment about its mass center. Such is the case with symmetrical space craft, artificial satellites, and missiles, if air resistance is ignored. After the powered portion of its flight, such a body is subjected only to the action of its weight which, obviously, produces no moment about the center of mass. Thus, from the first of Eqs. (22.46), we conclude that $\dot{\mathbf{H}}_G = \mathbf{0}$ which leads to the conclusion that $\mathbf{H}_G$ is constant in both magnitude and direction. As shown in Figure 22.16, the axis of precession (the Z axis) is chosen to coincide with the constant direction of vector $\mathbf{H}_G$. Because the rotating x-y-z coordinate system is chosen as described in Section 22.5, it follows that $\mathbf{H}_G$ lies entirely in the y-z plane with no component along the x axis. Also, because the x, y, and z axes are principal axes of inertia, we conclude that

$$\left.\begin{aligned}(H_G)_x &= I_1\omega_x = 0,\\ (H_G)_y &= I_1\omega_y = H_G\sin\theta,\\ (H_G)_z &= I_2\omega_z = H_G\cos\theta\end{aligned}\right\}. \tag{22.76}$$

Recalling that $\omega_x = \dot{\theta}$, $\omega_y = \dot{\phi}\sin\theta$, and $\omega_z = \dot{\psi} + \dot{\phi}\cos\theta$ (see Eq. (22.59)) and utilizing Eqs. (22.76), we obtain

$$\left.\begin{aligned}\theta &= \text{constant},\\ \dot{\phi} &= \frac{H_G}{I_1} = \text{constant},\\ \dot{\psi} &= H_G\cos\theta\left(\frac{I_1 - I_2}{I_1 I_2}\right) = \text{constant}\end{aligned}\right\}. \tag{22.77}$$

Equations (22.77), therefore, indicate that the motion under consider-

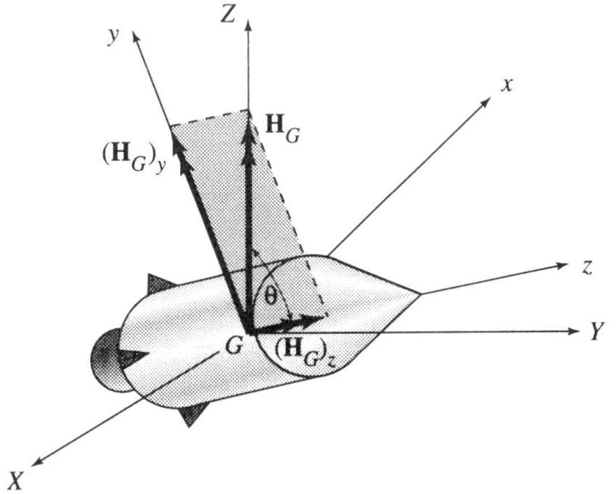

FIGURE 22.16.

ation is gyroscopic motion with steady precession. The rate of precession $\dot{\phi}$ may be expressed in terms of the rate of spin $\dot{\psi}$ and the angle of nutation by eliminating the quantity H_G from the second and third of Eqs. (22.77). This leads to

$$\dot{\phi} = \frac{I_2\dot{\psi}}{(I_1 - I_2)\cos\theta}. \tag{22.78}$$

Note that Eq. (22.78) may also be obtained by setting $\sum M_x = 0$ in the first of Eqs. (22.66).

Let us now direct our attention to the angular velocity vector $\boldsymbol{\omega}$ whose rectangular components may be obtained from Eqs, (22.76). Thus,

$$\left.\begin{aligned} \omega_x &= 0, \\ \omega_y &= \left(\frac{H_G}{I_1}\right)\sin\theta, \\ \omega_z &= \left(\frac{H_G}{I_2}\right)\cos\theta \end{aligned}\right\}. \tag{22.79}$$

Obviously this velocity vector lies in the y-z plane, as shown in Figure 22.17, making the angle β with the z direction so that

$$\beta = \tan^{-1}\left(\frac{I_2}{I_1}\tan\theta\right). \tag{22.80}$$

Therefore, because θ, I_1, and I_2 are all constants, it follows that the angle β is also a constant during the motion. The following two cases of practical* interest are identified:

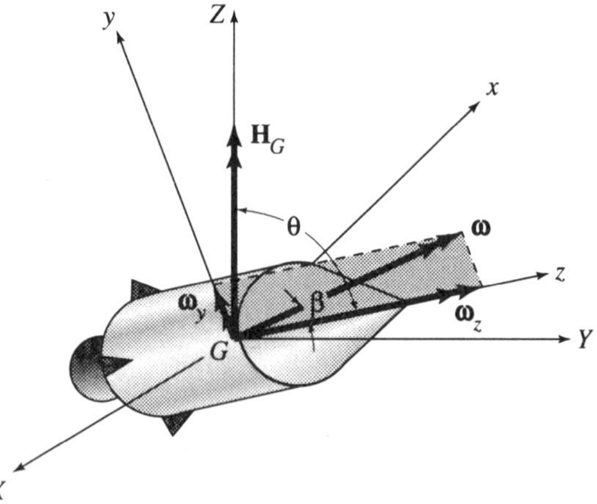

FIGURE 22.17.

* The case for which $I_2 = I_1$ is not practically feasible because, by Eq. (22.78), the rate of precession $\dot{\phi}$ becomes infinite under this condition.

(a) When $I_2 < I_1$, the axisymmetric body is relatively long compared to its cross-sectional dimensions. Thus, Eq. (22.80) shows that $\beta < \theta$ whereas Eq. (22.78) indicates that the rate of spin $\dot{\phi}$ is positive as depicted in Figure 22.18. In such a case, the precession is said to be *direct*. Also, under these conditions, the body cone rolls on the outside surface of the space cone as discussed in Chapter 21.

(b) When $I_2 > I_1$, the axisymmetric body is relatively short compared to its cross-sectional dimensions. Therefore, from Eq. (22.80), we conclude that $\beta > \theta$ and, from Eq. (22.78), that the rate of spin $\dot{\phi}$ is negative as shown in Figure 22.19. This type of precession is said to be *retrograde*. Under such conditions, the body cone rolls on the inside surface of the space cone as discussed in Chapter 21.

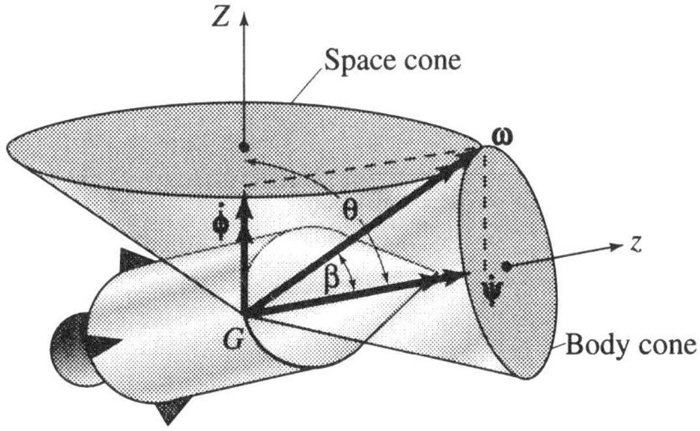

FIGURE 22.18.

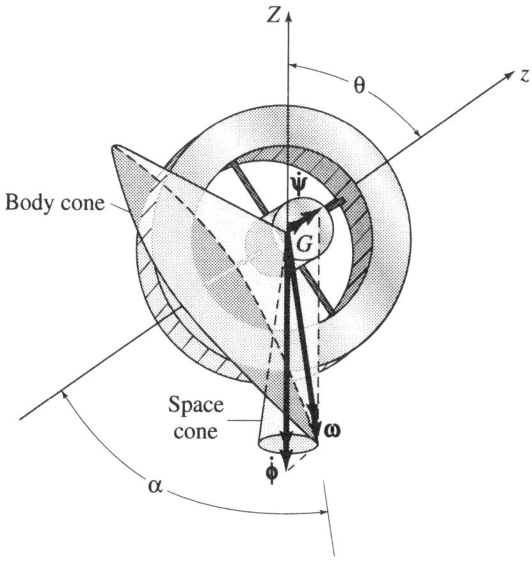

FIGURE 22.19.

Example 22.9

The gyroscope shown in Figure E22.9(a) consists of a thin circular disk of radius $r = 0.10$ m and mass $m = 5$ kg which is free to spin about the weightless axle OA. The axle OA has a length of 0.20 m and is held at O by a ball and socket which is attached to a vertical shaft as shown. If the disk spins about axle OA at the constant rate of 2000 rpm and if the system precesses about the Z axis at the constant rate of 2 rad/s, determine the constant angle of nutation θ. Also, find the resultant angular velocity of the disk.

Solution

Because the motion described here represents gyroscopic motion with steady precession, it follows that Eqs. (22.66) apply. These equations state that $\sum M_y = \sum M_z = 0$ and

$$\sum M_x = I_2 \dot{\phi}\dot{\psi} \sin\theta + (I_2 - I_1)\dot{\phi}^2 \sin\theta \cos\theta$$

where

$$\dot{\phi} = 2 \text{ rad/s as in Figure E22.8(b)},$$

$$\dot{\psi} = 2000 \text{ rpm} = 209.44 \text{ rad/s as in Figure E22.8(b)},$$

$$I_1 = I_x = I_y = \tfrac{1}{4}(5)(0.10^2) + 5(0.20^2) = 0.2125 \text{ kg·m}^2,$$

and

$$I_2 = I_z = \tfrac{1}{2}(5)(0.10^2) = 0.025 \text{ kg·m}^2.$$

Because $\mathbf{W} = (-5)(9.81)\mathbf{K} = -49.05\mathbf{K}$, it follows that

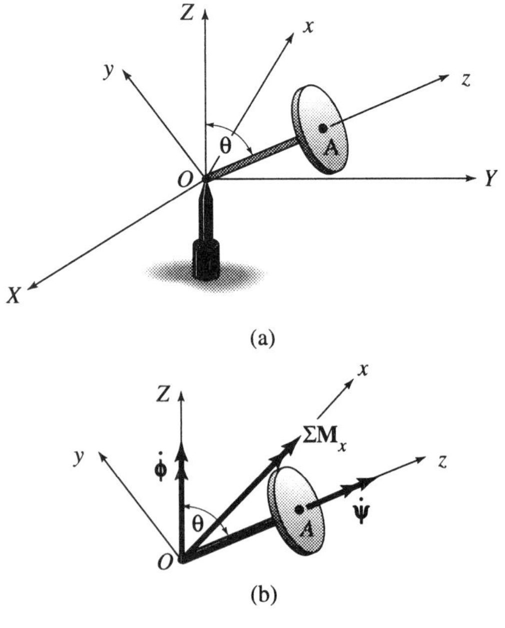

FIGURE E22.9.

$$\sum \mathbf{M}_O = 0.2\mathbf{k} \times (-49.05\mathbf{K}) = -9.81\mathbf{k} \times \mathbf{K}.$$

Also, by Eq. (22.57), $\mathbf{K} = (\sin\theta)\mathbf{j} + (\cos\theta)\mathbf{k}$. Therefore,

$$\sum \mathbf{M}_O = -9.81\mathbf{k} \times [(\sin\theta)\mathbf{j} + (\cos\theta)\mathbf{k}] = (9.81\sin\theta)\mathbf{i}$$

Thus,

$$\sum \mathbf{M}_x = \sum \mathbf{M}_O \cdot \mathbf{i} = (9.81\sin\theta)\mathbf{i}\cdot\mathbf{i} = 9.81\sin\theta.$$

Therefore, using the first of Eqs. (22.66),

$$9.81\sin\theta = 0.025(2)(209.44)\sin\theta + (0.025 - 0.2125)(2^2)\sin\theta\cos\theta$$

from which

$$\cos\theta = 0.8826,$$

and

$$\theta = 28.0^\circ. \qquad \text{ANS.}$$

The resultant angular velocity $\boldsymbol{\omega}$ of the disk may be obtained from Eq. (22.70). Thus,

$$\begin{aligned}\boldsymbol{\omega} &= (\dot\phi\sin\theta)\mathbf{j} + (\dot\psi + \dot\phi\cos\theta)\mathbf{k} = (2\sin 28^\circ)\mathbf{j} + (209.44 + 2\cos 28^\circ)\mathbf{k} \\ &= (0.934\mathbf{j} + 210\mathbf{k})\ \text{rad/s}. \qquad \text{ANS.}\end{aligned}$$

Example 22.10

The space craft shown schematically in Figure E22.10(a) has a weight $W = 10{,}000$ lb, a radius of gyration $k_z = 1.5$ ft, and radii of gyration $k_x = k_y = 2.5$ ft. At the instant shown, if the rate of spin $\dot\psi = 2$ rad/s and $\theta = 50^\circ$, determine the period of precession and sketch the space and body cones. Is the motion direct or retrograde?

Solution

The spacecraft moves with zero centroidal moment. Therefore, the rate of precession may be obtained from Eq. (22.78). Thus,

$$\dot\phi = \frac{I_2\dot\psi}{(I_1 - I_2)\cos\theta}$$

where

$$I_1 = I_x = I_y = \left(\frac{10{,}000}{32.2}\right)(2.5^2) = 1941.0\ \text{slug·ft}^2,$$

$$I_2 = I_z = \left(\frac{10{,}000}{32.2}\right)(1.5^2) = 698.8\ \text{slug·ft}^2,$$

and

$$\dot\psi = 2\ \text{rad/s and}\ \theta = 50^\circ.$$

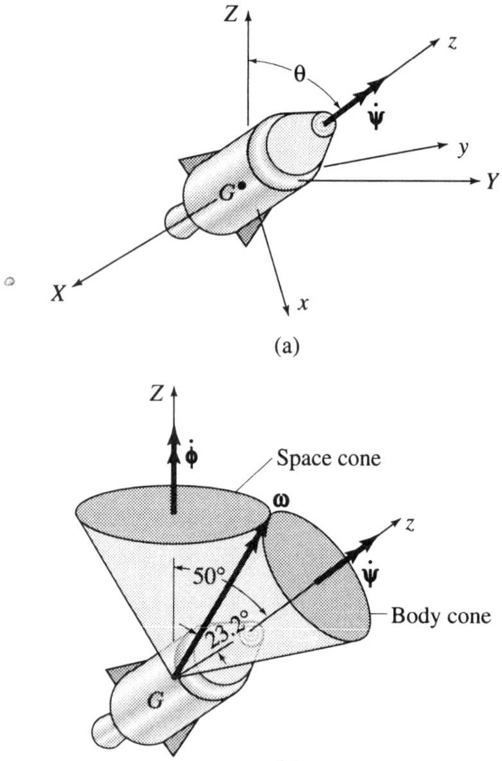

FIGURE E22.10.

Therefore,

$$\dot{\phi} = \frac{698.8(2)}{(1941.0 - 698.8)\cos 50^\circ} = 1.75 \text{ rad/s},$$

and the period of precession becomes

$$\tau = \frac{2\pi}{\dot{\phi}} = \frac{2\pi}{1.75} = 3.59 \text{ s}. \qquad \text{ANS.}$$

To construct the space and body cones, we need the angle β, defining the orientation of the angular velocity vector relative to the z axis. Thus, by Eq. (22.80),

$$\beta = \tan^{-1}\left(\frac{I_2}{I_1}\tan\theta\right) = \tan^{-1}\left(\frac{698.8}{1941.0}\tan 50^\circ\right) = 23.2^\circ.$$

The space and body cones are constructed as shown in Figure E22.9(b). Note that, because $\beta < \theta$, the precession is direct, and the body cone appears to roll on the outside of the space cone. ANS.

■

Problems

22.61 The right circular cone shown in Figure P22.61 is supported at O by a frictionless ball and socket. At the instant shown, $\theta = 40°$ and the constant rate of precession about the Z axis is $\dot{\phi} = (3\mathbf{K})$ rad/s. If $R = 5$ in., $H = 12$ in. and the weight of the cone $W = 3$ lb, determine its rate of spin.

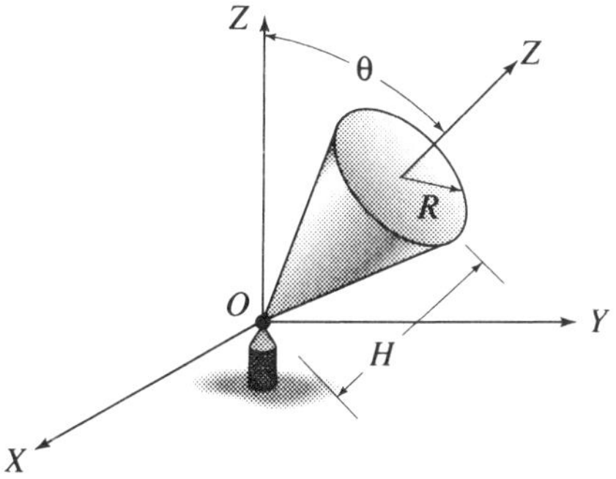

FIGURE P22.61.

22.62 The right circular cone shown in Figure P22.61 is supported at O by a frictionless ball and socket. At the instant shown, $\theta = 20°$ and the constant rate of spin about the z axis $\dot{\psi} = (300\mathbf{k})$ rad/s. If $R = 0.15$ m, $H = 0.25$ m, and the mass of the cone $m = 1.5$ kg, determine the rate of precession about the Z axis.

22.63 Consider the gyroscope shown in Figure P22.63. The rotor has a weight $W =$ 25 lb and the frame may be assumed weightless. The constant rate of precession $\dot{\phi} = (-2\mathbf{K})$ rad/s. Determine the rate of spin. Assume that the rotor is a thin circular disk with $R = 4$ in.

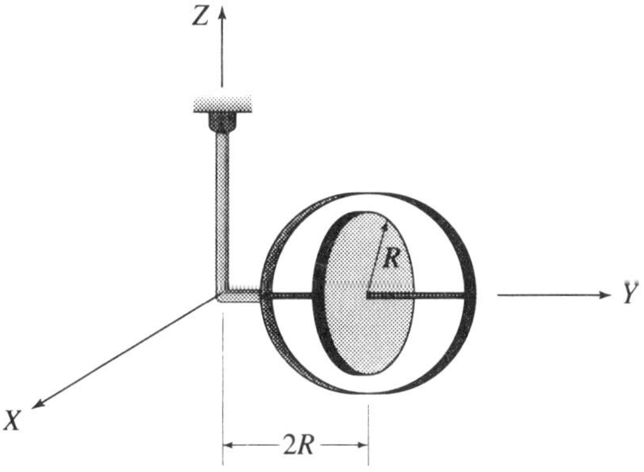

FIGURE P22.63.

22.64 The system shown in Figure P22.64 consist of a rotor of mass $m = 4$ kg and the attached arm OA which may be assumed weightless. The attachment at O to vertical shaft OB is such that the angle AOB is always 90°. The shaft OB spins about its axis at the constant rate of 100 rpm, cw when viewed from O to B whereas the rotor spins about its own axis at the constant rate of 3000 rpm, cw when viewed from O to A. (a) Determine the magnitude and direction of the gyroscopic moment developed during the motion. The radius of gyration of the rotor about its axis of rotation is 0.05 m. (b) Find the length of arm OA.

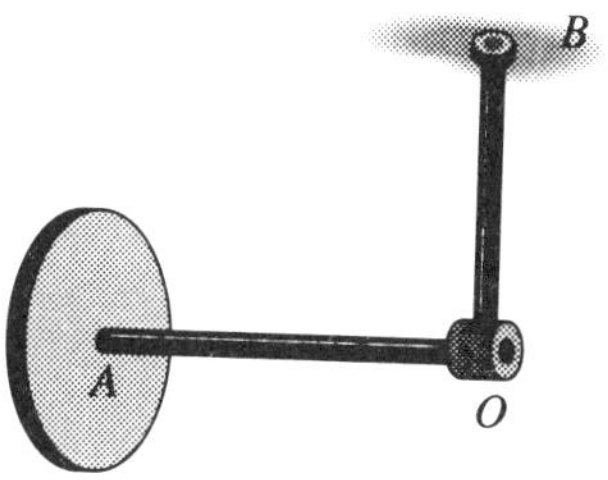

FIGURE P22.64.

22.65 The system shown in Figure P22.65 consists of a rotor of weight $W = 10$ lb and radius $r = 5$ in. which is free to spin

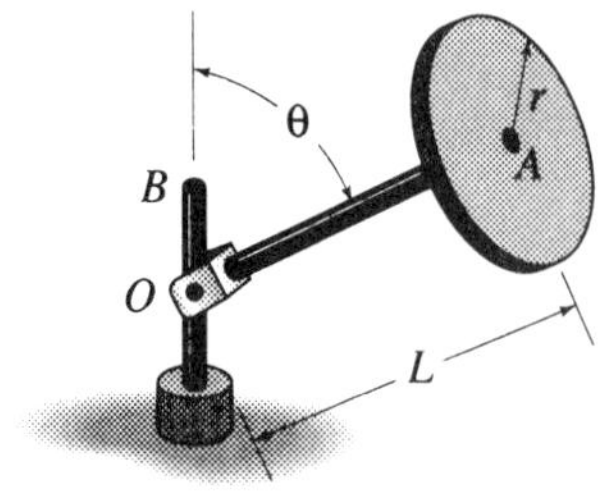

FIGURE P22.65.

about the weightless shaft OA. The shaft OA has a length $L = 12$ in. and is held at O by a hinge which is attached to a vertical axle OB as shown. If the rate of precession of the system about the vertical axle is 1.5 rad/s, cw, when viewed from O to B, determine the rate of spin of the rotor about shaft OA. Assume that the rotor is a thin circular disk, and let $\theta = 40°$.

22.66 The system shown in Figure P22.65 consists of a rotor of weight W and radius r which is free to spin about the weightless shaft OA. The shaft OA has a length $L = 2r$ and is held at O by a hinge which is attached to the vertical axle OB as shown. If the rotor spins about shaft OA at a constant rate $\dot{\psi} = \omega_O$, cw, when viewed from O to A, and if $\theta = \theta_O$ is constant, find the rate of precession about the vertical axle. Express the answer in terms of ω_O, θ_O, r and g.

22.67 A solid circular cylinder of radius $r = 0.10$ m and length $L = 0.30$ m is mounted on axle AB as shown in Figure P22.67. Axle AB is attached to a weightless frame which can rotate freely about the vertical shaft OC. The mass of the rotor is 20 kg, and it spins about axle AB at the constant rate of 4000 rpm, cw when viewed from B to A, while the frame rotates about its vertical shaft at the constant rate of 400 rpm, ccw when viewed from C to O. If the angle θ between axle

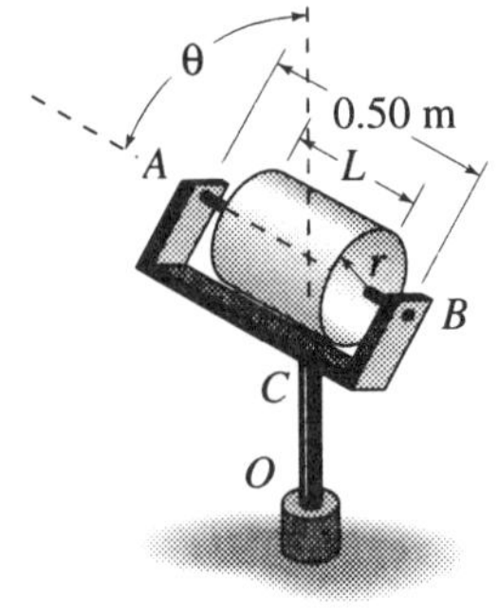

FIGURE P22.67.

AB and shaft OC is 45°, determine the reactions at bearings A and B developed during the motion. Assume that the bearing at A can resist both radial and axial forces whereas the bearing at B can resist only radial forces.

22.68 The 40-lb wheel of radius $R = 5$ in. rolls without slipping on the inside surface of a cone as shown in Figure P22.68. The weightless axle of the wheel is attached to the vertical shaft OA which rotates at the constant rate of 200 rpm. Find the normal reaction between the wheel and the surface of the cone. Assume that the wheel is a thin circular disk.

22.69 A 35-kg disk of radius $R = 0.50$ m shown in Figure P22.69 rotates about

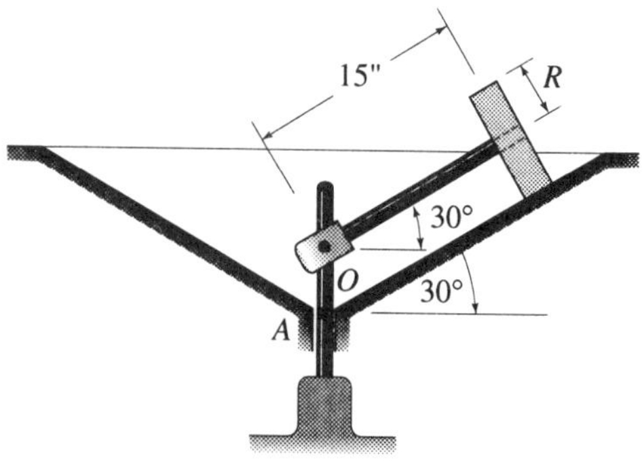

FIGURE P22.68.

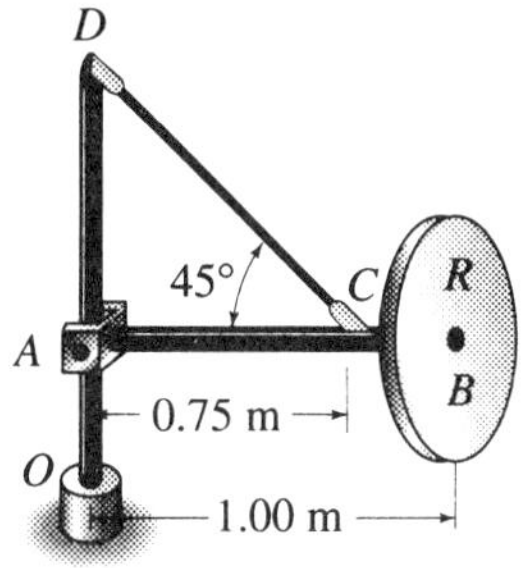

FIGURE P22.69.

axle AB at the constant rate of 15 rad/s, cw when viewed from B to A, while the system rotates about vertical axis OA at the constant rate of 8 rad/s, cw when viewed from O to A. Determine the tension in cable DC which holds the axle AB in the horizontal position.

22.70 The basic elements of a ship's stabilizer are shown in Figure P22.70. The rotor of the gyroscope has a weight W and rotates about the vertical z axis at a constant angular velocity $\dot{\psi} = \omega_1 \mathbf{k}$. The attached precession wheel A rotates about the horizontal y axis at the rate of $\dot{\phi} = -\omega_2 \mathbf{j}$. The system is attached to the hull of the ship at supports B and C. Determine (a) the magnitude and direction of the gyroscopic couple exerted on the ship and (b) the vertical reaction components at bearing supports B and C. The radius of gyration of the rotor about its axis of rotation is k.

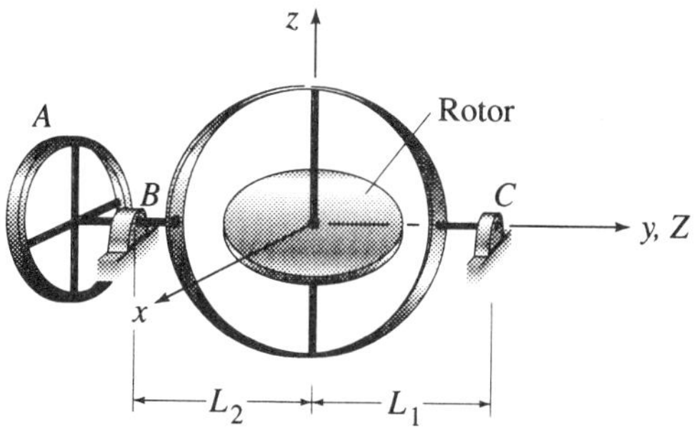

FIGURE P22.70.

22.71 In the position shown in Figure P22.71, the blades of the helicopter rotate at a constant angular velocity 10 rad/s, ccw when viewed from the top, and the helicopter itself is making a vertical turn of radius $R = 100$ m at a speed of 120 km/h. The blades have a combined mass of 110 kg and a radius of gyration about the axis of rotation $k = 1.40$ m. Determine, at the instant considered, when the helicopter is at lowest point of its vertical turn, the magnitude of the gyroscopic moment exerted on the frame.

FIGURE P22.71.

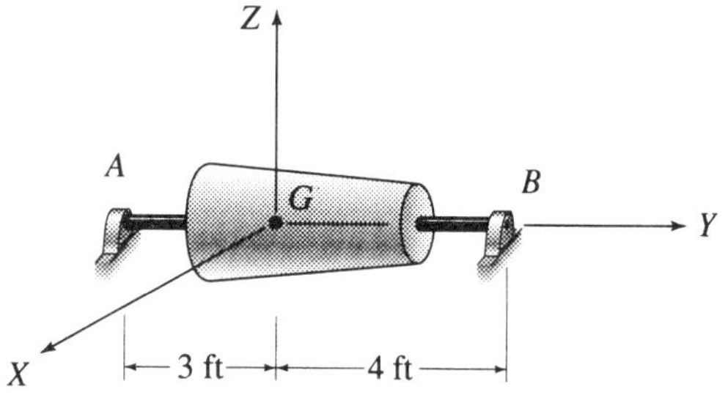

FIGURE P22.72.

22.72 The rotor of a ship's turbine is shown schematically in Figure P22.72. It has a weight $W = 500$ lb and a radius of gyration about its axis of rotation $k = 1.2$ ft. It is supported by bearings at A and B and rotates at a constant rate of 4000 rpm, ccw when viewed from A to B which is the forward direction on the ship. Find the vertical reactions at supports A and B when the ship makes a right turn (in the X-Y plane) of radius $R = 1000$ ft at a speed $v = 30$ ft/s.

22.73 The blades and rotor of the fan shown in Figure P22.73 are supported by bearings at A and B and have a mass $m = 5$ kg and a radius of gyration about the axis of rotation $k = 0.20$ m. The constant speed of the fan is 1800 rpm, ccw when viewed from A to B. The assembly is mounted on bracket CD which rotates about the vertical axis OC at the constant rate of 60 rpm, cw when viewed from O to C. Determine the vertical reactions at bearing supports A and B.

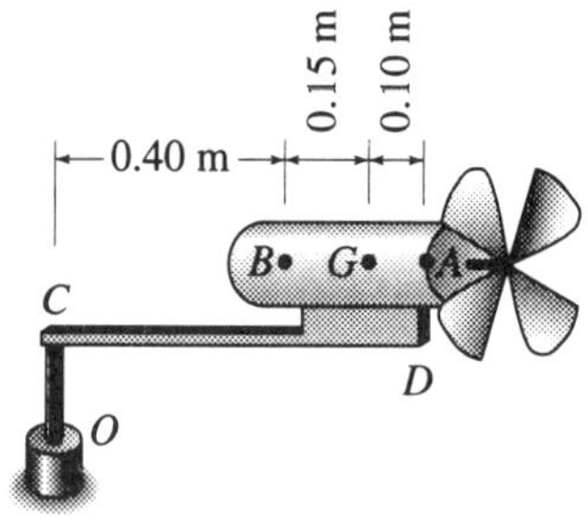

FIGURE P22.73.

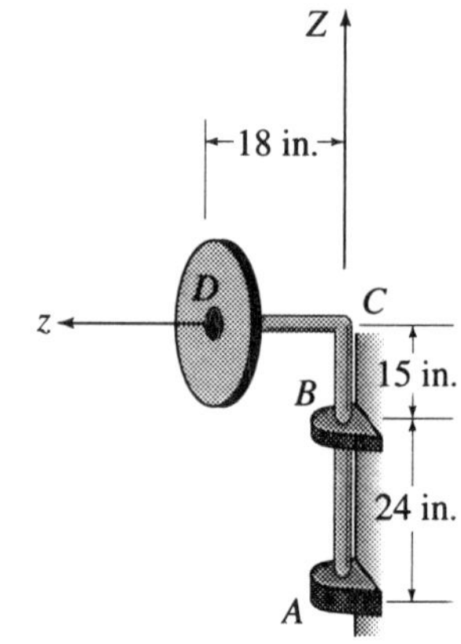

FIGURE P22.74.

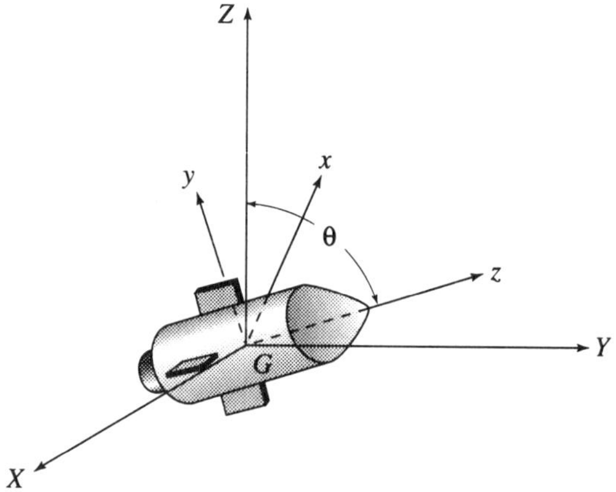

FIGURE P22.75.

22.74 The disk shown in Figure P22.74 with a radius of 10 in. and weighing 25 lb rotates about horizontal axle CD at a constant rate $\boldsymbol{\omega}_1 = (100\mathbf{k})$ rad/s. Axle CD is attached rigidly to the vertical shaft AB which rotates at the constant rate $\boldsymbol{\omega}_2 = (10\mathbf{K})$ rad/s. Determine the z component of the reactions at bearings A and B and the axial thrust at support A.

22.75 The space vehicle shown in Figure P22.75 precesses about the Z axis at the constant rate $\dot{\phi} = (1.5\mathbf{K})$ rad/s. The mass of the vehicle and its contents $m = 3000$ kg and its radii of gyration are $k_z = 2.0$ m and $k_x = k_y = 3.0$ m. If the angle $\theta = 60°$, determine the rate of spin of the vehicle. Also, determine the magnitude and direction of the angular velocity vector.

22.76 The space vehicle shown in Figure P22.75 has a weight $W = 5000$ lb and radii of gyration $k_z = 4$ ft and $k_x = k_y = 6$ ft. If the rate of spin is $\dot{\psi} = (5\mathbf{k})$ rad/s and $\theta = 50°$, determine the rate of precession about the Z axis. Is the motion direct or retrograde?

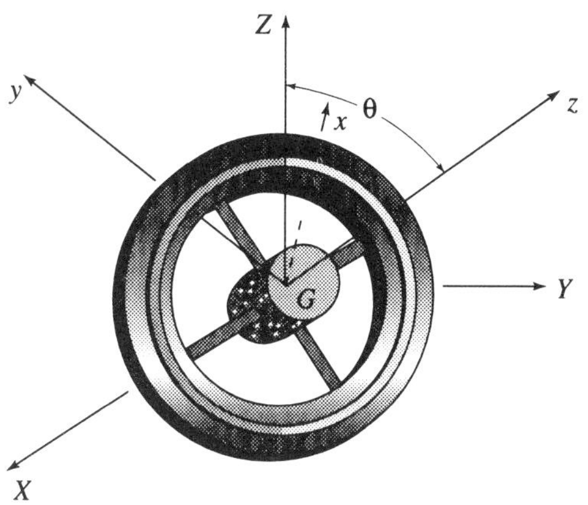

FIGURE P22.77.

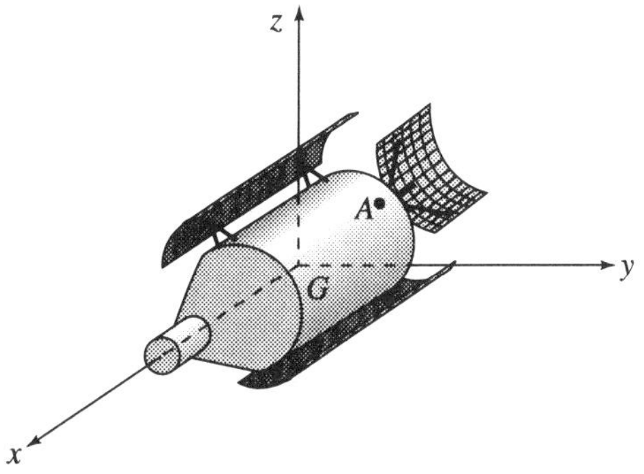

FIGURE P22.78.

22.77 The space station shown in Figure P22.77 of mass $m = 10 \times 10^3$ kg precesses about the Z axis at the constant rate $\dot{\phi} = (-20\mathbf{K})$ rev./h. If the centroidal radii of gyration are $k_z = 10.5$ m and $k_x = k_y = 3.0$ m and if $\theta = 65°$, find the rate of spin about the z axis. Also, find the magnitude and direction of the angular velocity vector.

22.78 A space satellite of mass 1200 kg is moving at a constant linear speed when it is hit at A, as shown in Figure P22.78, by a meteoroid of mass 0.75 kg at a velocity $\mathbf{v}_m = (0.55\mathbf{i} - 0.70\mathbf{j} - 0.95\mathbf{k})$ km/s relative to the satellite. The centroidal radii of gyration are $k_y = k_z = 0.75$ m and $k_x = 0.5$ m, and the coordinates of point A are -0.60 m, 0.45 m, and 0.40 m. Determine the rates of precession and spin of the satellite immediately after the meteoroid becomes imbedded in it. The satellite has no angular velocity prior to impact.

Review Problems

22.79 The composite body shown in Figure P22.79 consists of two thin aluminum plates OABC and OCDE and five steel slender rods DG, BG, GF, EF, and AF. The mass density for the aluminum plates is 27 kg/m^2 and that for the steel rods is 0.6 kg/m. Determine the mass moments of inertia with respect to the x, y, and z axes.

22.80 Refer to Problem 22.79 and determine I_{xy}, I_{xz}, and I_{yz}.

22.81 Use the results obtained in Problems 22.79 and 22.80 to find the mass principal moments of inertia I_u, I_v, and I_w at O. Also, find the direction cosines corresponding to the u mass principal axis of inertia with respect to which the mass moment of inertia is maximum.

22.82 A thin disk of radius $R = 10$ in. and weight $W = 5$ lb is mounted on a shaft of length $L = 20$ in. as shown in Figure P22.82. The shaft is parallel to the X-Y plane and rotates about the vertical Z axis at the constant rate $\omega = 5$ rad/s as

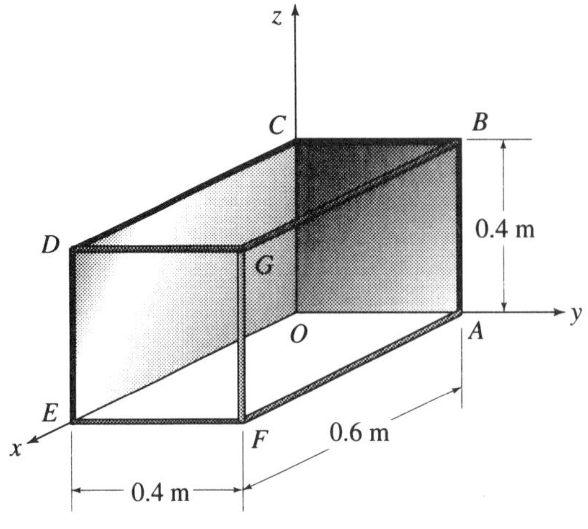

FIGURE P22.79.

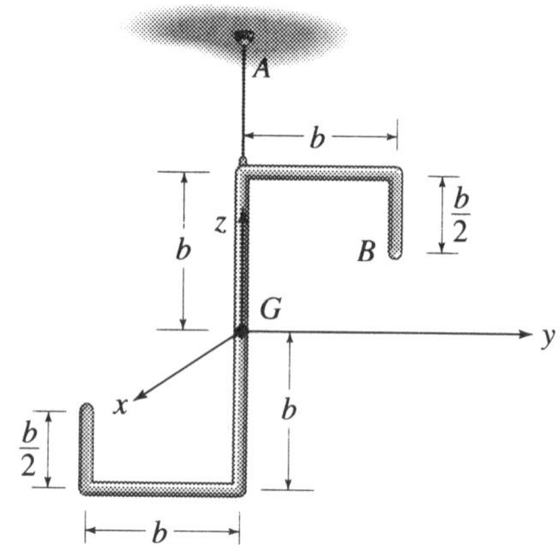

FIGURE P22.84.

shown. For the position shown, determine the kinetic energy of the disk. Assume that the shaft is weightless and that the disk rolls without slipping.

22.83 Refer to Problem 22.82, and determine the angular momentum of the disk about point G. Find the magnitude and direction of this angular momentum.

22.84 A uniform rod length $5b$ and total mass of $5m$ is bent into the shape shown in Figure P22.84. It is suspended from a ball and socket at A and hit at B by a force **F** causing an impulse $\mathbf{F}\Delta t = -(a)\mathbf{i} + (1.5a)\mathbf{j} + (0.5a)\mathbf{k}$ where a has units of force multiplied by units of time. Assume that the rod is at rest when hit, and determine, immediately after impact, (a) the angular velocity of the rod and (b) the velocity of its mass center. Express answers in terms of a, b, and m.

22.85 After being struck by a meteoroid, the space vehicle shown in Figure P22.85 experiences an angular velocity $\boldsymbol{\omega} = (0.03\mathbf{i} - 0.06\mathbf{j})$ rad/s. To correct this and bring the angular velocity back to zero, A and D of the four available jets are brought into action producing a thrust of 12 lb each, parallel to the z axis. Note that, while jet A is directed in the positive z direction, jet D is in the negative z direction. The space vehicle has a mass of 100 slug and radii of gyration $k_x =$

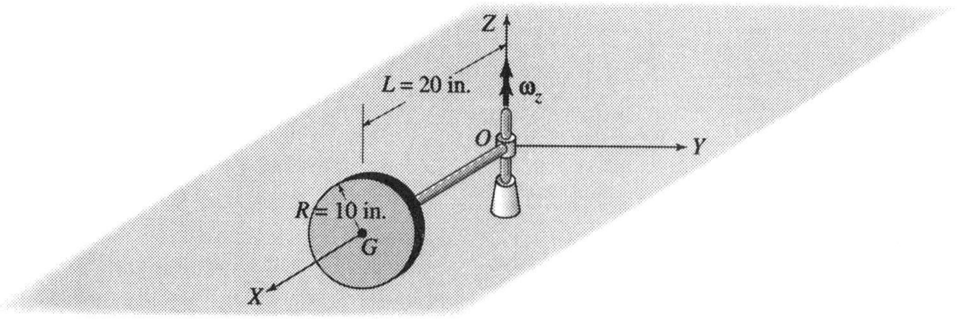

FIGURE P22.82.

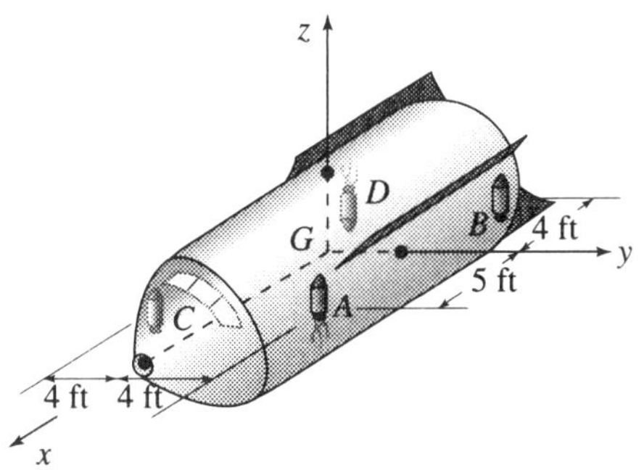

FIGURE P22.85.

3.5 ft and $k_y = k_z = 4.3$ ft where the x, y, and z axes are principal axes of inertia. Determine the time of action, for each of the two jets, needed to accomplish the correction.

22.86 Three thin rectangular plates are welded together to form the shape shown in Figure P22.86 whose total mass is $5m$. The composite plate is connected to weightless shafts which are supported by bearings at A and B. The bearing at A can resist only radial forces whereas that at B can resist radial and axial forces. The assembly rotates at a constant angular velocity $\boldsymbol{\omega} = (\omega \mathbf{J})$ rad/s. For the position shown, determine the components of the reactions at A and B for the case where the $m = 0.12$ slug, $b = 6$ in. and $\omega = 20$ rad/s.

22.87 Three thin rectangular plates are welded together to form the shape shown in Figure P22.86 whose total mass is $5m$. The composite plate is connected to weightless shafts which are supported by bearings at A and B. The bearing at A can resist only radial forces whereas that at B can resist radial and axial forces. Starting from rest, a couple $\mathbf{M} = (5\mathbf{J})$ N·m is applied, causing the assembly to accelerate. Determine the components of the reactions at A and B assuming that, in the position shown, the assembly has reached an angular velocity $\boldsymbol{\omega} = (15\mathbf{J})$ rad/s. Also, determine the angular acceleration $\boldsymbol{\alpha}$ of the assembly. Let $m = 0.20$ kg and $b = 0.25$ m.

22.88 A homogeneous sphere of radius r is hinged along a diameter AB to a clevis that is supported by a bearing at C, as shown in Figure P22.88. The mass of the sphere is m and that of the clevis may be assumed negligible. The sphere spins about its diametral axis at a constant angular velocity $\boldsymbol{\omega}_1 = (\omega_1 \mathbf{K})$ rad/s relative to the clevis while the entire assembly rotates about the bearing at C at a constant angular velocity $\boldsymbol{\omega}_2 = (\omega_2 \mathbf{I})$ rad/s. For the position shown, determine the reaction components at C.

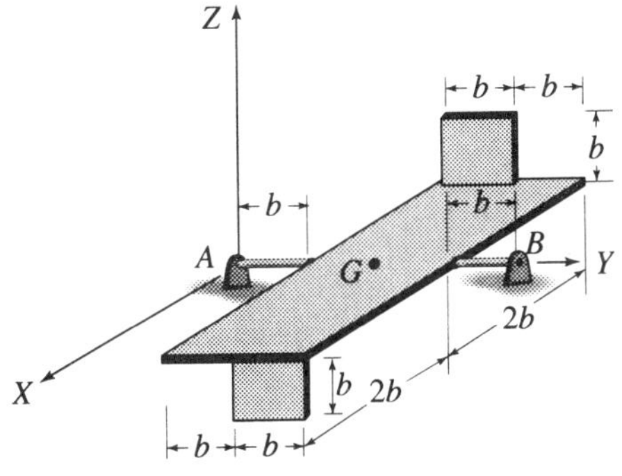

FIGURE P22.86.

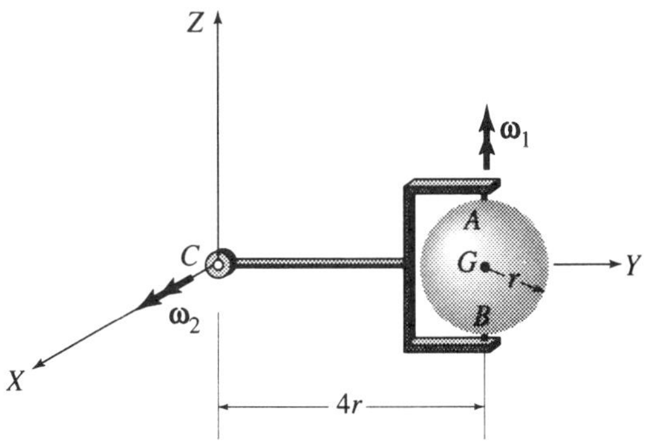

FIGURE P22.88.

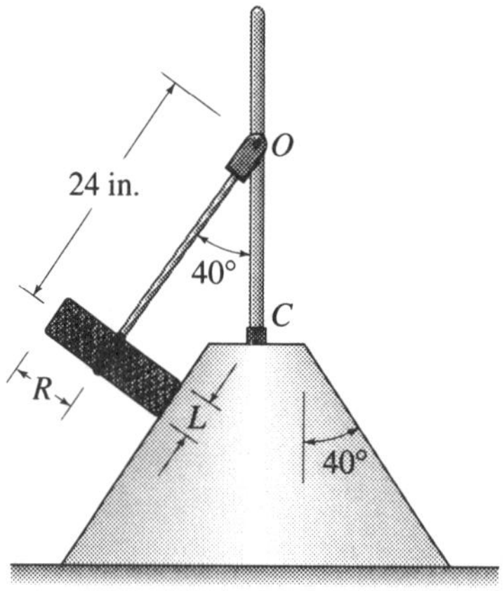

FIGURE P22.89.

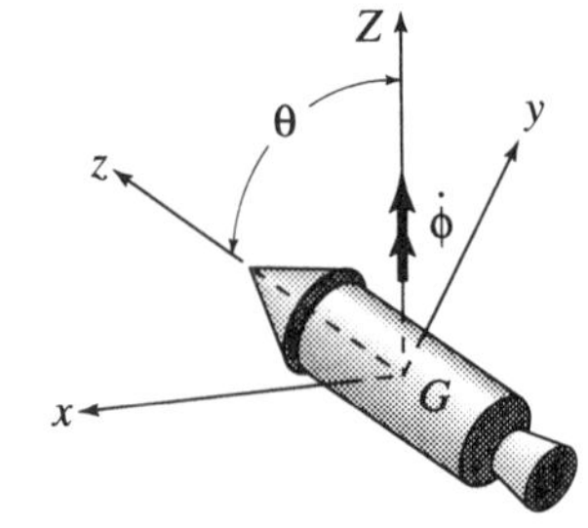

FIGURE P22.90.

22.89 The 100-lb wheel of radius $R = 6$ in. and length $L = 3$ in. rolls without slipping on the outside surface of a truncated cone as shown in Figure P22.89. The weightless axle of the wheel is attached to the vertical shaft OC which rotates at a constant angular velocity ω rad/s. Determine the range of values for ω so that the normal reaction between the wheel and surface of the cone does not exceed 20 lb. Assume that the wheel is a circular cylinder.

22.90 A schematic sketch of a missile in free flight is shown in Figure P22.90 before reentering Earth's atmosphere. At the instant shown, the missile precesses about the Z axis at the rate $\dot{\phi} = 8.50$ rad/s. The mass of the missile is 30 kg, its radius of gyration about the axis of spin $k_z = 0.12$ m and those about the x and y axis $k_x = k_y = 0.35$ m. If $\theta = 40°$, determine the rate at which the missile spins about its axis of symmetry, and sketch the space and body cones.

23

Vibrations

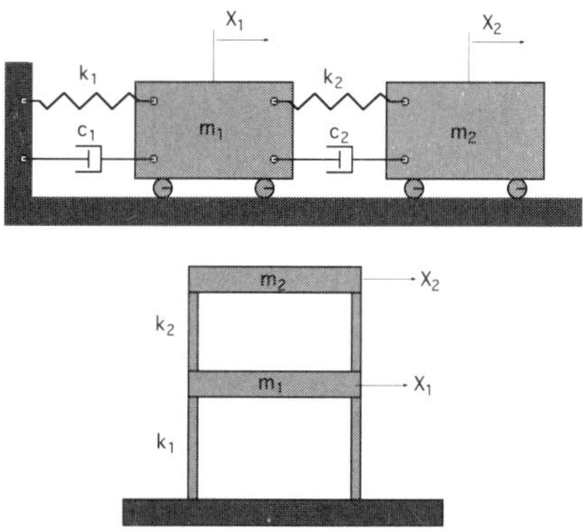

Schematic view of an idealized model corresponding to a two story building. The resulting mathematical model helps engineers to assess structural behavior when such structures are subjected to earthquakes, hurricanes, and other forces.

Natural and man-made vibrations are widespread on Earth and in the depths of space. Vibrations affect our lives in a variety of ways ranging from the destructive forces of hurricanes and earthquakes to the modest forces involving machines with rotating parts. We usually think of vibratory motions as harmful but they can be used constructively to screen substances and to drive piles. Mechanical, aerospace and structural engineers are often concerned with high frequency vibrations of metallic parts of aircraft that could lead to premature fatigue failures with catastrophic consequences. Civil engineers are required to design buildings, dams, and bridges capable of withstanding both man-made and natural forces. These challenging tasks are just a few of the many examples where vibrational principles play a vital role. Frequently, engineers are required to creatively idealize and predict the response of engineering systems irrespective of the forces involved and before the systems are built.

Mathematical models can often be developed to study the dynamic behavior of mechanical systems using mass, shock absorber damping, and spring stiffness. Such models may be more difficult to formulate when dealing with a multistory building because the required physical parameters can not be adequately assessed. Furthermore, the forces involved are unknown or may not be well defined as is the case with earthquakes and hurricanes. This introductory chapter on vibration deals primarily with systems with a single degree of freedom where mass, damping, and stiffness are combined in one location. Time-Dependent coordinate axes are required to properly define the resulting displacement, velocity, and acceleration of single and two degree-of-freedom systems. Both undamped and damped vibration models are covered.

The model shown on the cover page is an illustration of a practical application of vibrational principles and how engineering models are developed. As a student you learn how to solve problems, but as an engineer you will often be required to formulate as well as solve them. We hope that your interest in this important area of Dynamics is enhanced and that the desire to seek more knowledge is invigorated.

23.1 Free Vibrations of Particles—Force and Acceleration

Rectilinear Motion—Governing Differential Equation

If a particle of mass m is suspended by a spring* from an overhead support as shown in Figure 23.1(a), then, the force of gravity, $W = mg$, will stretch the spring by an amount δ_{ST}. If k is the spring constant the linearly elastic spring will resist the pull of gravity with a force equal to $k\delta_{ST}$. The particle is in equilibrium and

$$\sum F_x = 0, \quad mg - k\delta_{ST} = 0. \tag{23.1}$$

The corresponding position of the particle is termed the equilibrium position (E.P.).

Imagine that the particle is displaced below the equilibrium position and released from rest. The particle, then, moves along a vertical line and its position at any time t is given by a single coordinate x. Because a single coordinate, that is x, suffices to position the particle, we term the system a single-degree-of-freedom system. Refer to the free-body and inertial force diagrams of Figure 23.1(b) and write

$$\sum F_x = ma_x, \quad mg - k(x + \delta_{ST}) = m\ddot{x}.$$

Transposing to the right side of this equation gives

$$m\ddot{x} + kx - mg + k\delta_{ST} = 0.$$

The third and fourth terms vanish by Eq. (23.1). This is always true if the displacement x is measured from the static equilibrium position. Thus,

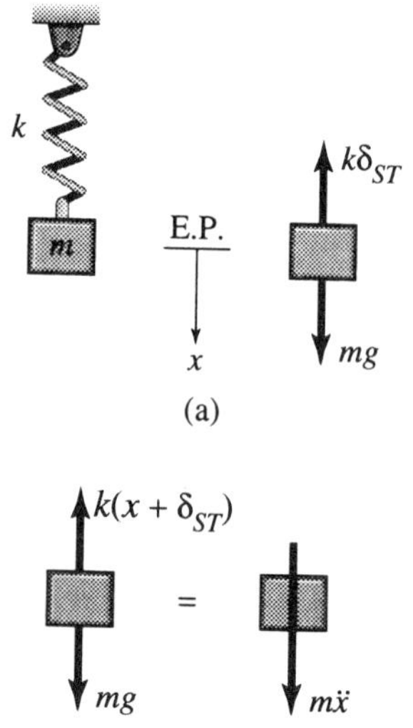

FIGURE 23.1.

* A detailed discussion of springs was given in Chapter 2 and further elaborated on in Chapters 12 and 15.

$$m\ddot{x} + kx = 0. \tag{23.2}$$

This linear, second-order, ordinary differential equation has constant coefficients m and k. It is the governing differential equation for free vibration of single-degree-of-freedom systems which is referred to as *simple harmonic motion.*

Solutions of the Equation of Motion

For the case where there is no external force acting on the mass, the solution is obtained by dividing Eq. (23.2) by m and letting $k/m = p^2$. Thus,

$$\ddot{x} + p^2 x = 0. \tag{23.3}$$

The general solution of this differential equation can be written as

$$x = A \sin pt + B \cos pt \tag{23.4}$$

where $p = \sqrt{\frac{k}{m}}$ is termed the *natural circular frequency* expressed in rad/s. The constants A and B depend upon the initial conditions of the motion. Because the differential equation is of the second order, two integrations are required to express $\ddot{x}$ as a function of time. Each integration gives a constant of integration. Consider the following general initial conditions $t = 0$, $x = x_0$, and $\dot{x} = \dot{x}_0$ where x_0 and $\dot{x}_0$ are known initial values of position and velocity. Substituting $t = 0$, $x = x_0$ in Eq. (23.4) yields

$$x_0 = A \sin 0 + B \cos 0 \Rightarrow B = x_0$$

Differentiating the general solution given by Eq. (23.4) with respect to time gives

$$\dot{x} = Ap \cos pt - Bp \sin pt.$$

Substituting $t = 0$, $\dot{x} = \dot{x}_0$ gives

$$\dot{x}_0 = Ap \sin 0 - Bp \sin 0 \Rightarrow A = \frac{\dot{x}_0}{p}$$

Substituting for A and B in Eq. (23.3) yields the desired solution to Eq. (23.2). Thus,

$$x = \frac{\dot{x}_0}{p} \sin pt + x_0 \cos pt \tag{23.5}$$

If k and m are specified, then, the natural circular frequency can be found from the relationship $p = \sqrt{\frac{k}{m}}$. We regard k and m as natural properties of the system. The initial position x_0 and the initial velocity $\dot{x}_0$ together with p enable us to express x as a function of time by Eq. (23.5).

The velocity v and acceleration a of the vibrating mass, for any time t, may now be obtained by differentiating Eq. (23.5). Thus,

$$v = \dot{x} = \dot{x}_0 \cos pt - x_0 p \sin pt, \tag{23.6}$$

and

$$a = \ddot{x} = -\dot{x}_0 p \sin pt - x_0 p^2 \cos pt. \tag{23.7}$$

The general solution to the governing differential equation (Eq. (23.4)) can be expressed in the alternate form

$$x = x_M \sin(pt + \phi) \tag{23.8}$$

where x_M the amplitude and ϕ the phase angle (see Fig. 23.2) are constants to be determined from the initial conditions. As will be shown shortly, Eq. (23.8) follows directly from Eq. (23.4), but Eq. (23.8) lends itself more readily to physical interpretation. The amplitude x_M is defined as the maximum displacement of the mass from its equilibrium position. If we let $t = 0$ in Eq. (23.8), we see that the initial value of x is equal to $x_M \sin\phi$ which is the product of the amplitude and the sine of the phase angle. This defines mathematically what we mean by the phase angle $\phi = \sin^{-1}\left(\dfrac{x_0}{x_M}\right)$. Again, consider the following general initial conditions $t = 0$, $x = x_0$ and $\dot{x} = \dot{x}_0$ where x_0 and $\dot{x}_0$ are known initial values of displacement and velocity. Substituting $t = 0$, $x = x_0$ in the general solution of Eq. (23.8) yields

$$x_0 = x_M \sin\phi. \tag{23.9}$$

Differentiate the general solution with respect to time to obtain

$$\dot{x} = x_M p \cos(pt + \phi)$$

Substituting $t = 0$, $\dot{x} = \dot{x}_0$ in the general solution yields

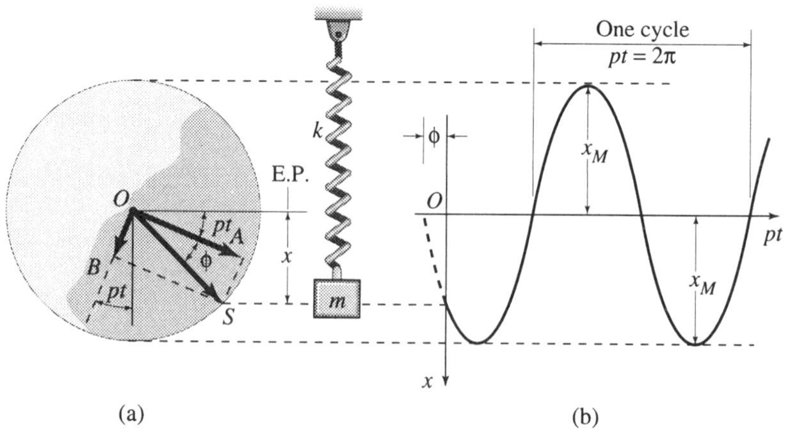

FIGURE 23.2.

$$\dot{x}_0 = x_M p \cos\phi. \tag{23.10}$$

Divide Eq. (23.9) by Eq. (23.10) to obtain

$$\tan\phi = \frac{px_0}{\dot{x}_0}. \tag{23.11}$$

Square and add Eqs. (23.9) and (23.10) to give

$$x_M = \frac{1}{p}\sqrt{(px_0)^2 + \dot{x}_0^2}, \tag{23.12a}$$

and, by using Eqs. (23.7) and (23.8), Eq. (23.12a) becomes

$$x_M = \sqrt{(x_M \sin\phi)^2 + (x_M \cos\phi)^2}. \tag{23.12b}$$

Equations (23.11) and (23.12) enable us to find the phase angle ϕ and the amplitude x_M from the known values of p, x_0, and $\dot{x}_0$ and the general solution (Eq. (23.5)) expresses x as a function of time.

The Auxiliary Circle

By Eq. (23.4), the displacement x is the sum of the two components $A \sin pt$ and $B \cos pt$. Because the sine and cosine functions of an angle are 90° out of phase, we can obtain the displacement x by considering two rotating vectors of magnitudes A and B positioned as shown in Figure 23.2(a) for any time t. Thus, for any time t, the geometry of Figure 23.2(a) yields

$$x = A \sin pt + B \cos pt$$
$$= \text{OS} \sin(pt + \phi).$$

If we let OS, the radius of the circle, be equal to the amplitude x_M, we are led to Eq. (23.8), namely $x = x_M \sin(pt + \phi)$. We note, from Figure 23.2(a), that the phase angle ϕ is the angle by which the sine component of the motion (OA) lags behind the resultant amplitude (OS).

The displacement x is a sine function of time and may be represented graphically as shown in Figure 23.2(b). Also, the constant angular velocity p of the radius OS is the natural circular frequency introduced earlier. The radius OS completes one revolution (one cycle) in $t = \tau$ seconds, after rotating through 2π radians. Thus, $pt = 2\pi$, from which

$$\tau = 2\pi/p \tag{23.13}$$

where τ is referred to as the *period* of vibration of the spring-mass system. The number of cycles per second (cps), termed the *frequency* and given the symbol f, is the reciprocal of the period τ. Thus,

$$f = \frac{1}{\tau} = \frac{p}{2\pi}. \tag{23.14}$$

In Eq. (23.14), the frequency f is measured in *hertz* (Hz) where 1 Hz = 1 cps.

The velocity v and acceleration a of the mass for any time t may be obtained by differentiating Eq. (23.8). Thus,

$$v = \dot{x} = px_M \cos(pt + \phi) \tag{23.15}$$

and

$$a = \ddot{x} = -p^2 x_M \sin(pt + \phi) \tag{23.16}$$

Because the sine and cosine functions have maximum and minimum values of $+1$ and -1, respectively, it follows that the maximum magnitudes of the velocity and acceleration of the vibrating mass are

$$v_{MAX} = px_M,$$

and

$$a_{MAX} = p^2 x_M. \tag{23.17}$$

Equivalent Spring Constants

In many applications, several structural components exist that may be treated as elastic springs. In such applications, it becomes desirable to reduce these springs to a single equivalent spring to simplify solving the vibration problem. Two types of spring arrangements, *springs in parallel* and *springs in series*, are considered in the following paragraphs:

Consider the three springs arranged in *parallel*, as shown in Figure 23.3(a), and determine the equivalent spring constant k_e for this arrangement. This means that we are to find a single spring as shown in Figure 23.3(b) with a spring constant so that the force-deformation

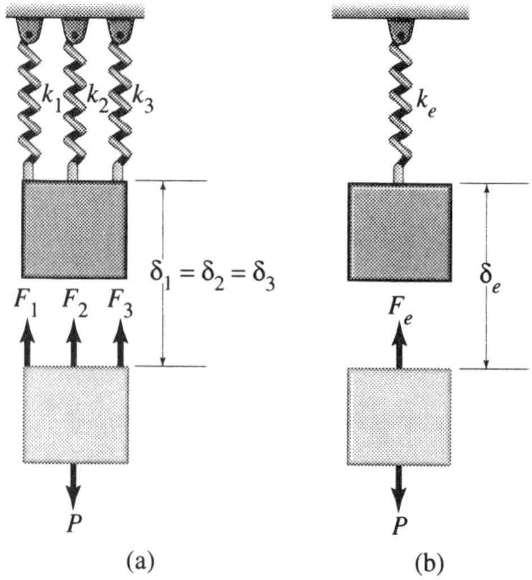

FIGURE 23.3.

relationship will be the same for the original system and for the single spring system.

We apply a single force P to stretch the three springs equally so that the supported body moves in translation without rotation. An equal force P is applied to the equivalent system which deforms by the amount δ_e. Equilibrium equations for the given and equivalent systems give

$$P = F_1 + F_2 + F_3 \quad \text{and} \quad P = F_e.$$

Because the same force P is applied to each system, we conclude that

$$F_1 + F_2 + F_3 = F_e,$$

but $F_1 = k_1\delta_1$, $F_2 = k_2\delta_2$, and $F_3 = k_3\delta_3$. Also, $F_e = k_e\delta_e$. Substitution gives

$$k_1\delta_1 + k_2\delta_2 + k_3\delta_3 = k_e\delta_e.$$

The deformations are all equal. That is, $\delta_1 = \delta_2 = \delta_3 = \delta_e$ which enables us to cancel the deformations and write

$$k_e = k_1 + k_2 + k_3.$$

Thus, for a system of springs in *parallel*, the equivalent spring constant equals the sum of the individual spring constants. Obviously, this relationship would hold for any number of springs provided the force P is applied so that there is no rotation. We conclude, therefore, that the equivalent spring constant k_e for any number of spring arranged in parallel becomes

$$k_e = \sum_{i=1}^{n} k_i \tag{23.18}$$

where n represents the total number of springs in the given parallel system.

Consider the three springs arranged in *series* as shown in Figure 23.4(a), and determine the equivalent spring constant for this arrangement. This means that we are to find a single spring, as shown in Figure 23.4(b) with a spring constant k_e so that the force-deformation relationship will be the same for both systems.

In this case, we begin by writing the deformation relationship as

$$\delta_1 + \delta_2 + \delta_3 = \delta_e$$

For each individual spring, we recall that the deformation equals the force divided by spring constant. Thus,

$$\frac{F_1}{k_1} + \frac{F_2}{k_2} + \frac{F_3}{k_3} = \frac{F_e}{k_e}$$

Equilibrium dictates that

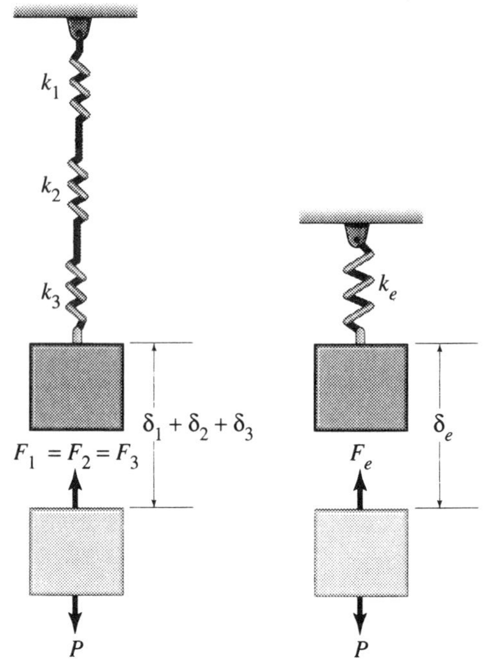

FIGURE 23.4. (a) (b)

$$F_1 = F_2 = F_3 = P \quad \text{and} \quad F_e = P.$$

Because each spring force equals the applied force P, substitution in the deformation equation enables us to cancel the forces to give

$$\frac{1}{k_e} = \frac{1}{k_1} + \frac{1}{k_2} + \frac{1}{k_3}.$$

Solving for k_e yields

$$k_e = \frac{1}{\dfrac{1}{k_1} + \dfrac{1}{k_2} + \dfrac{1}{k_3}}$$

Thus, for a system of springs in *series*, the equivalent spring constant k_e equals the reciprocal of the sum of the reciprocals of the given spring constants. Obviously this relationship holds for any number of springs arranged in series and

$$k_e = \frac{1}{\sum_{i=1}^{n} 1/k_i} \tag{23.19}$$

The following examples illustrate some of the above concepts.

Example 23.1

Replace the six given springs shown in Figure E23.1(a) with a single equivalent spring with the same force-deformation response as the given system.

Solution

Top three springs: These three springs are arranged in parallel, and Eq. (23.18) applies. Thus,

$$(k_e)_{TOP} = 100 + 50 + 100 = 250 \text{ lb/in.}$$

Bottom three springs: These three springs are arranged in series, and Eq. (23.19) applies. Thus,

$$(k_e)_{BOTTOM} = \frac{1}{\frac{1}{60} + \frac{1}{80} + \frac{1}{120}} = 26.7 \text{ lb/in.}$$

The intermediate equivalent system is shown in Figure E23.1(b). Let us use a fundamental approach for finding the single spring to replace these two springs by applying a downward arbitrary force P to this system. Note that the top spring would lengthen by an amount δ and the bottom spring would shorten by the same amount δ. The top spring resists the applied force in tension and the bottom spring resists the applied force in compression. Both spring forces act upward on the supported body. In each case the spring force equals the product of the spring constant k and the deformation δ. Referring to the free body diagram of Figure E23.1(b),

$$\sum F_x = 0, \quad P - 250\delta - 26.7\delta = 0,$$

$$P = 276.7\delta.$$

Divide this equation by δ to obtain the ratio P/δ which represents the final equivalent spring constant k_e for the given system. Thus

$$k_e = \frac{P}{\delta} = 276.7 = 277 \text{ lb/in.} \qquad \text{ANS.}$$

This final equivalent system which has the same force-deformation response as the given system is shown in Figure E23.1(c).

Example 23.2

Consider the beam shown in Figure E23.2 carrying a concentrated weight W at its center. Then, (a) Determine the natural circular frequency for vertical motion assuming $L = 180$ in., $EI = 10^9$ lb·in^2, and $W = 25{,}000$ lb. (b) If the initial displacement and the initial velocity of the weight are $y_0 = 0.5$ in. and $\dot{y}_0 = 15$ in./s, respectively, determine the displacement, velocity, and acceleration of W when time $t = 2$ s. Neglect the mass of the beam and assume that the beam is fixed at both ends.

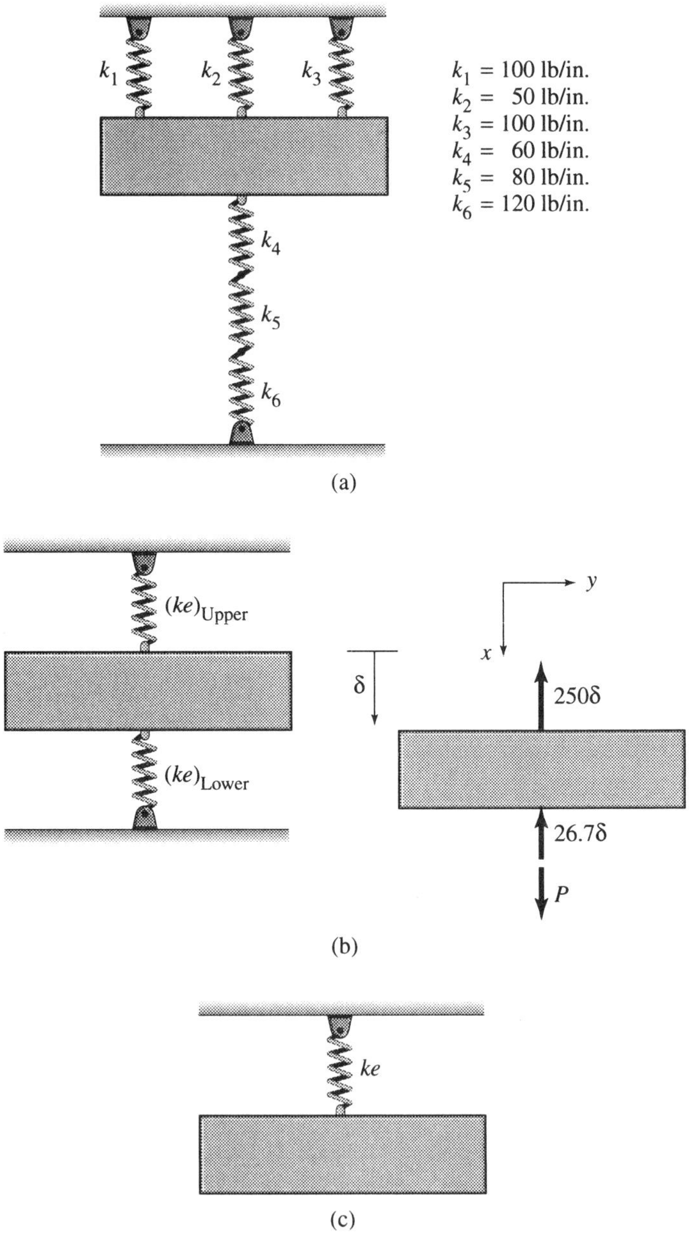
k_1
k_2
k_3
$k_1 = 100$ lb/in.
$k_2 = 50$ lb/in.
$k_3 = 100$ lb/in.
$k_4 = 60$ lb/in.
$k_5 = 80$ lb/in.
$k_6 = 120$ lb/in.
k_4
k_5
k_6
(a)
$(ke)_{\text{Upper}}$
$(ke)_{\text{Lower}}$
δ
y
x
250δ
26.7δ
P
(b)
ke
(c)

FIGURE E23.1.

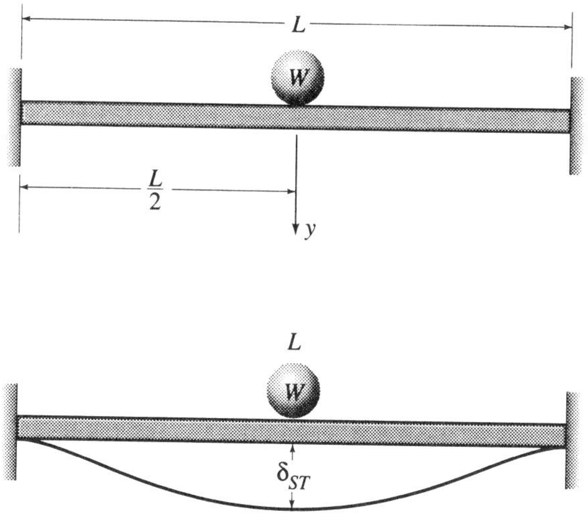

FIGURE E23.2.

Solution

(a) The natural circular frequency of the beam is given by

$$p = \sqrt{\frac{k}{m}} = \sqrt{\frac{kg}{W}}$$

The stiffness is not as readily obvious as was the case in Example E23.1. However, it can be computed by considering that the beam at mid-span behaves like a spring, that is, if the weight W is applied under static conditions, then, a displacement δ_{ST} will result:

$$k = \frac{W}{\delta_{\text{ST}}} = \frac{W}{\dfrac{WL^3}{192EI}} = \frac{192EI}{L^3}$$

Note that δ_{ST} is found in any textbook in structural analysis. Hence, substituting the equivalent spring constant for the beam in the frequency equation gives

$$p = \sqrt{\frac{192EIg}{WL^3}} = \sqrt{\frac{(192)(10^9)(386.4)}{25000(180^3)}} = 22.557 \text{ rad/s.} \qquad \text{ANS.}$$

(b) For undamped free vibration, the displacement is given by

$$y = y_0 \cos pt + \frac{\dot{y}_0}{p} \sin pt \tag{a}$$

Therefore, the velocity and acceleration are computed by taking the first and second derivatives, respectively. Thus,

$$\dot{y} = -py_0 \sin pt + \dot{y}_0 \cos pt \tag{b}$$

and

$$\ddot{y} = -p^2 y_0 \cos pt - p\dot{y}_0 \sin pt. \tag{c}$$

Substituting $p = 22.557$, $y_0 = 0.5$ in., $\dot{y}_0 = 15$ in./s, and $t = 2$ s. in equations (a), (b), and (c) gives the displacement, velocity, and acceleration of the weight W at $t = 2$ s. Thus,

$$y = 0.5\cos[22.557(2)] + \frac{15}{22.557}\sin[22.557(2)] = 0.815 \text{ in.,}$$

ANS.

$$\dot{y} = 22.557(0.5)\sin[22.557(2)] + 15\cos[22.557(2)] = -3.83 \text{ in./s,}$$

ANS.

and

$$\ddot{y} = -22.557^2(0.5)\cos[22.557(2)] - 22.557(15)\sin[22.557(2)]$$
$$= -414 \text{ in./s}^2.$$

ANS.

The student should verify that using Eq. 23.8 will give exactly the same answers given above.

■

Rotational Motion—Governing Differential Equation

A particle of mass m is attached to a light rod of negligible mass which is pinned at its upper end A as shown in Figure 23.5(a), for any position defined by the angle θ. Because the angle θ suffices to position the pendulum, it is described as a single-degree-of-freedom system.

To write the governing differential equation of motion for this pendulum we consider the free-body and inertial force diagrams for the mass m as shown in Figure 23.5(b) along with a right-handed t-n coordinate system. Applying Newton's second law of motion along the t axis,

$$\sum F_t = ma_t, \quad -mg\sin\theta = mL\ddot{\theta}.$$

Transposing and dividing by mL,

$$\ddot{\theta} + \frac{g}{L}\sin\theta = 0. \tag{23.20}$$

This equation is the governing differential equation of motion of a simple pendulum for large or small values of θ. It can be solved with elliptic functions, but modern computer solutions employ numerical methods.

If we restrict θ to small values, then, $\sin\theta \approx \theta$, and, for small oscillations, Eq. (23.20) becomes

$$\ddot{\theta} + \frac{g}{L}\theta = 0.$$

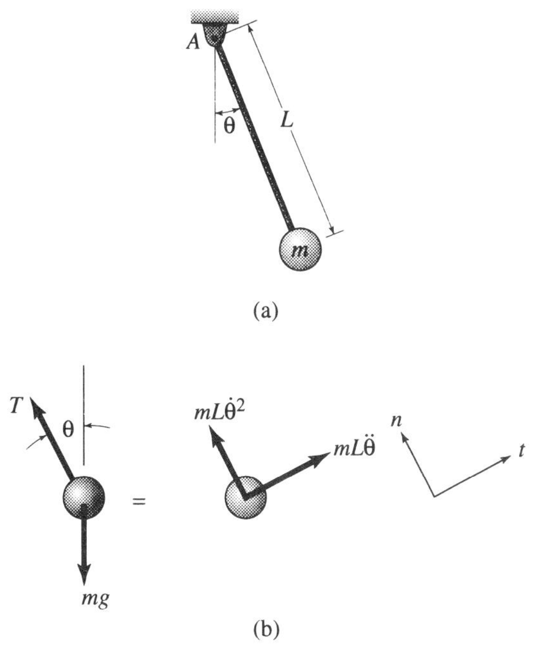

FIGURE 23.5.

Let $p^2 = \dfrac{g}{L}$, and

$$\ddot{\theta} + p^2\theta = 0. \tag{23.21}$$

For comparison, Eq. (23.3) states that

$$\ddot{x} + p^2 x = 0,$$

We observe that the equations for rotational (small rotations) and rectilinear motion have the same mathematical form. The dependent variables are θ and x, and the independent variable is time t. The angle θ positions the pendulum whereas the coordinate x positions a particle at any time t.

The circular frequency of the simple pendulum is given by p where $p = \sqrt{\dfrac{g}{L}}$ is expressed in rad/s. The period of this motion is given by

$$\tau = \frac{2\pi}{p} = 2\pi\sqrt{\frac{L}{g}}$$

where τ is expressed in seconds. We observe that the period of this vibration is independent of the amplitude. Galileo was first to note this fact. He observed swinging chandeliers in church and measured time with his pulse. The development of dynamics was retarded by the lack of devices to measure time accurately.

The frequency f becomes

$$f = \frac{1}{\tau} = \frac{1}{2\pi}\sqrt{\frac{g}{L}}$$

where f is expressed in cycles per second (Hertz).

The general solution of this governing differential equation is given by

$$\theta = A \sin pt + B \cos pt$$

where, as in the case of rectilinear motion, A and B are determined from the initial conditions.

The following example illustrates some of the above concepts.

Example 23.3

A simple pendulum is shown in Figure E23.3. It has a length $L =$ 4.00 ft and a mass $m = 1.00$ slug. (a) Determine the natural circular frequency, period, and frequency of this motion. (b) Express θ as a function of time for the initial conditions $t = 0$, $\theta = 25°$, and $\dot{\theta} = 0$. (c) Express θ as a function of time for initial conditions $t = 0$, $\theta = 0°$ and $\dot{\theta} = 1.00$ rad/s. Assume small oscillations.

Solution

(a) The circular frequency p is given by

$$p = \sqrt{\frac{g}{L}} = \sqrt{\frac{32.2}{4.00}} = 2.84 \text{ rad/s.} \qquad \text{ANS.}$$

The period is computed next as

$$\tau = \frac{2\pi}{p} = \frac{2\pi}{2.84} = 2.21 \text{ s.} \qquad \text{ANS.}$$

The frequency is given by

$$f = \frac{1}{\tau} = \frac{1}{2.21} = 0.452 \text{ Hz.} \qquad \text{ANS.}$$

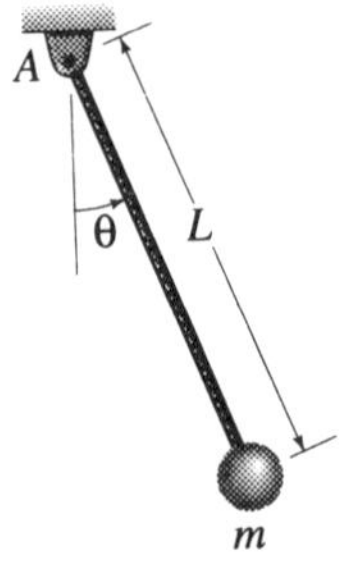

FIGURE E23.3.

(b) These initial conditions state that the pendulum is displaced 25° ccw and released from rest. The general solution is given by $\theta = A \sin pt + B \cos pt$. Using the condition at $t = 0$, $\theta = 25°\left(\frac{\pi}{180°}\right) = 0.436$ rad,

$$0.436 = A \sin 0 + B \cos 0.$$

$$B = 0.436 \text{ rad.}$$

Using the condition at $t = 0$, $\dot{\theta} = 0°$ gives

$$0 = Ap \cos 0° - Bp \sin 0°,$$

$$A = 0.$$

Substitution of these values for A and B into the general solution gives the position θ as a function of time. Thus,

$$\theta = (0.436 \cos 2.84t) \text{ rad.} \qquad \text{ANS.}$$

(c) These initial conditions state that, when the pendulum is vertical, it is given an angular velocity of 1.00 rad/s ccw. Using the condition at $t = 0$, $\theta = 0°$,

$$0 = A \sin 0° + B \cos 0°,$$

$$B = 0.$$

Using the condition at $t = 0$, $\dot{\theta} = 1.00$ rad/s yields

$$1.00 = A \cos 0° - Bp \sin 0°,$$

$$A = 1/p = 0.352 \text{ rad.}$$

Substitution of these numerical values for A and B in the general solution yields

$$\theta = (0.352 \sin 2.84t) \text{ rad.} \qquad \text{ANS.}$$

■

Problems

23.1 The spring supported mass, shown in Figure P23.1, is displaced downward 1 in. from its equilibrium position and released. Determine the circular frequency, period, and frequency of this motion. What is the amplitude, maximum velocity, and maximum acceleration of this oscillatory motion. Let $k = 1200$ lb/ft and $m = 3$ slug.

23.2 Solve problem 23.1 for $k = 6$ kN/m and $m = 8$ kg.

23.3 The mass m shown in Figure P23.3 moves horizontally on a frictionless plane. Given $k_1 = 30$ lb/in., $k_2 = 40$ lb/

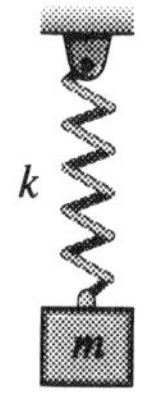

FIGURE P23.1.

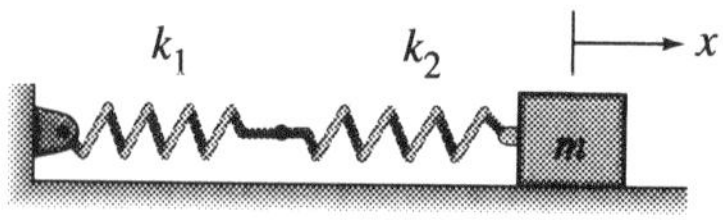

FIGURE P23.3.

in., $m = 0.5$ slug, and the initial conditions $t = 0$, $x = 0.5$ in., and $\dot{x} = 0$, (a) Determine the equivalent spring constant and the period of this vibratory motion. (b) Determine the position as a function of time and the amplitude of this motion. (c) Find the maximum velocity and acceleration of this motion.

23.4 Solve problem 23.3 for $k_1 = 25$ lb/in., $k_2 = 50$ lb/in., and $m = 0.40$ slug. Initial conditions: $t = 0$, $x = 0$, and $\dot{x} = 4$ ft/s.

23.5 The position of a simple pendulum is given as a function of time by $\theta = 0.5 \cos 3t$. Assume that θ is in radians and t is in seconds. (a) What is the amplitude of this motion? (b) What is the maximum angular velocity of this pendulum? (c) What is the maximum angular acceleration of this pendulum? (d) If $g = 32.2$ ft/s^2, what is the length of this pendulum? (e) What is the period of this motion? (f) What is the frequency of this motion? (g) State and describe the initial conditions.

23.6 Refer to Problem 23.5 and answer the same questions for the function of time given by $\theta = 0.6 \sin 4t$. Assume θ is in radians and t is in seconds. Let $g = 9.81$ m/s^2.

23.7 Draw the free-body and inertial force diagrams for a pendulum in the position $\theta = 0°$ which means that it hangs vertically. (a) Show that, in this position, the angular acceleration vanishes. (b) Show that the tension in the rod attached to the bob is given at this time by $T = mg + mL\dot{\theta}^2$.

23.8 The general solution for rectilinear vibratory motion may be expressed in either of two forms as follows:

$$x = A \sin pt + B \cos pt$$

$$x = x_M \sin(pt + \phi)$$

Given the initial conditions $t = 0$, $x = 1$ in., and $\dot{x} = 0$, determine (a) A and B directly, (b) x_M and ϕ directly, and (c) x_M and ϕ from A and B using the equation expressing x_M and ϕ in terms of A and B.

23.9 Show that both forms of the general solution given in problem 23.8 satisfy the governing differential equation of motion for single-degree-of-freedom oscillatory motion.

23.10 Imagine that a sign error is made and the governing differential equation is written $\ddot{x} - p^2 x = 0$. (a) Show that the correct general solution to this incorrect equation is given by

$$x = \text{A} \sinh pt + B \cosh pt$$

(Hint: Differentiate this general solution twice, and substitute for acceleration and position functions in the differential equation.) (b) Explain why a general solution composed of hyperbolic sines and hyperbolic cosines is characterized as nonoscillatory, and, thus, cannot be accepted as a general solution of a vibration problem.

23.11 Write the governing differential equation of motion for the system of Figure P23.11. Determine the circular frequency, period, and frequency of this

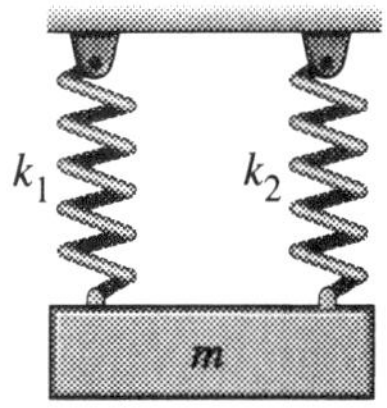

FIGURE P23.11.

motion. Let $k_1 = k_2 = 6$ kN/m and $m =$ 12 kg.

23.12 A simple pendulum is shown in Figure P23.12. It is shown displaced cw at any instant t. The sign convention should be chosen positive in the cw sense in this case. (a) Show that the governing differential equation is $\ddot{\theta} + p^2\theta = 0$ by constructing the free-body and inertial force diagrams. (b) Show that the tension T in the rod supporting the bob is given at any time t by $T = mg\cos\theta + mL\dot{\theta}^2$.

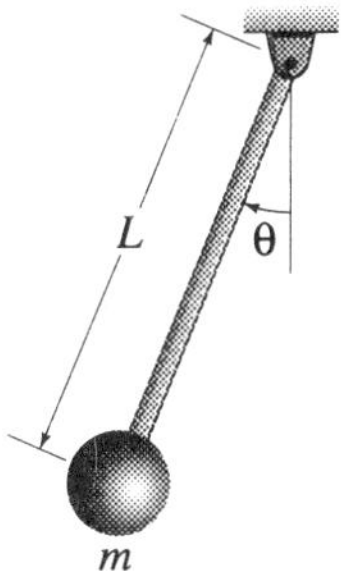

FIGURE P23.12.

23.13 Refer to Figure P23.12, and express the pendulum's position, angular velocity, and angular acceleration as functions of time. The initial conditions of the motion are given by $t = 0$, $\theta = 0.15$ rad cw, and $\dot{\theta} = 0°$.

23.14 Solve problem 23.13 for initial conditions given by $t = 0$, $\theta = 0°$ and $\dot{\theta} =$ 0.2 rad/s cw.

23.15 (a) Determine the equivalent spring constant for the system shown in Figure P23.15 given $k_1 = 10$ lb/in., $k_2 = 20$ lb/in., and $k_3 = 40$ lb/in. (b) If $m = 0.5$ slug, express the position, velocity, and acceleration of the particle m as functions of time given that it is displaced downward 1 in. and released from rest. Displacements are measured from the equilibrium position.

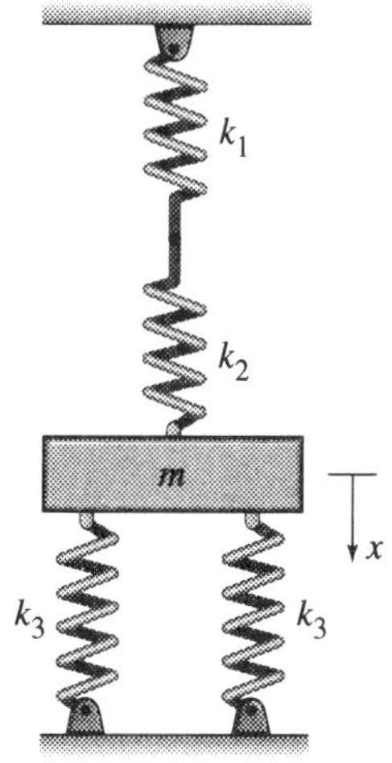

FIGURE P23.15.

23.16 Solve problem 23.15 for $k_1 = 500$ N/m, $k_2 = 1000$ N/m, and $k_3 = 2000$ N/m. Let $m = 20$ kg, and use the initial conditions $t = 0$, $x = 0.02$ m, and $\dot{x} = 0$.

23.17 The springs shown in Figure P23.17 are arranged symmetrically with respect to the point of suspension of mass m. The mass of rod AB is negligible compared to $m = 20$ kg. Determine the period and frequency of motion for the mass m. Let $k_1 = 20$ kN/m and $k_2 = 40$ kN/m.

23.18 Refer to problem 23.17 and express the position, velocity, and acceleration of the mass m as functions of time. What is the position of the mass at $t = 1$ s? Use

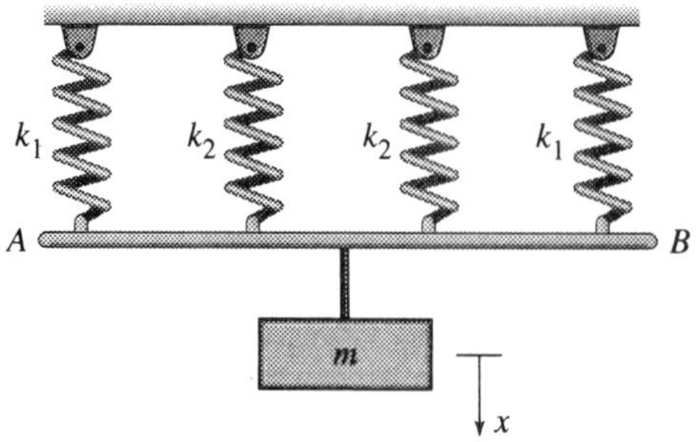

FIGURE P23.17.

initial conditions as follows: $t = 0$, $x = 0.1$ m, and $\dot{x} = 0$. Is the mass above or below its equilibrium position at $t = 1$ s? Why? Assume x is measured from the equilibrium position.

23.19 In all three spring arrangements of Figure P23.19, the mass $m = 1$ slug, and the period of the motion equals 0.2 s. Find the value of k for each of the three arrangements.

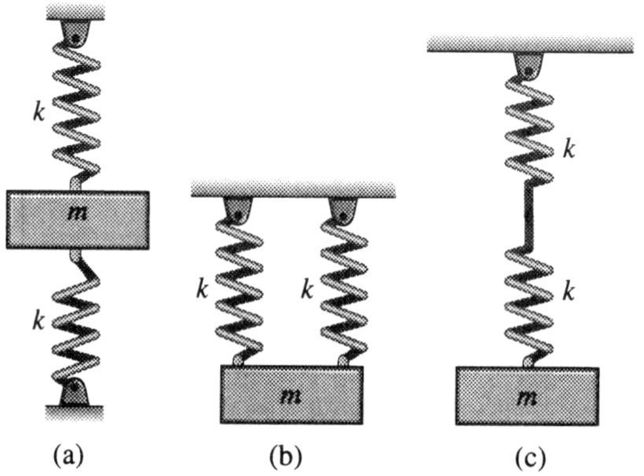

FIGURE P23.19.

23.20 A mass m_1 is suspended from a spring of constant k_1 and a mass m_2 is suspended from a spring of constant k_2. (a) If these masses are to execute simple harmonic motion of the same circular frequency, what relationship must hold between m_1, k_1, m_2, and k_2? (b) If these masses have equal circular frequencies, will they have equal periods and frequencies? Justify your answers. (c) If $p_1 = p_2$ for these masses and their motions start with the same initial conditions, will their motions be identical? Carefully explain your answer based upon general solutions for their motions.

23.21 The general solution for rectilinear vibratory motion of a mass is given by

$$x = 5 \sin 4t + 12 \cos 4t$$

where x is in inches and t is in seconds. (a) Determine the amplitude of this motion. (b) Determine the phase angle ϕ. (c) Express the velocity of this mass at any time t. (d) Express the acceleration of this mass at any time t.

23.22 (a) A given simple harmonic motion has an amplitude $x_M = 1.00$ ft. and a phase angle $\phi = 30°$. (a) Determine A and B in the general solution of Eq. (23.4). (b) For $p = 2.00$ rad/s, determine the initial conditions of this motion.

23.23 Consider two simple pendulums of equal length L and equal bob mass m. One is executing simple harmonic motion on Earth and the other on the Moon. The acceleration due to gravity on the Moon is about one-sixth of that on Earth. For each listed quantity, form the ratio of the value for the pendulum on Earth to the value for the pendulum on the Moon: (a) periods, (b) frequencies, (c) maximum angular velocities, and (d) maximum angular accelerations. Assume both pendulums begin their motions with identical initial conditions. If measurements were taken to compare with theoretical predictions would you expect the correlation to be better on Earth or on the Moon? Why?

23.24 A pendulum is rotated ccw through an

angle of 10° and released from rest. It has a length of 1.00 m. Determine (a) the circular frequency, period, and frequency of small oscillations, (b) the general solution for rotation angle θ as a function of time, (c) the maximum angular velocity and the corresponding position, and (d) the maximum angular acceleration and the corresponding position.

23.25 The collar of mass m shown in Figure P23.25 is attached to the spring of constant k. The collar is displaced 1.00 in downward from its equilibrium position and released from rest. Determine the position, velocity, and acceleration of the collar at $t = 0.50$ s. Let $m = 0.20$ slug and $k = 10$ lb/in. Ignore friction between the collar and the rod.

23.26 The collar of mass m shown in Figure P23.25 is attached to the spring of constant k. The collar is displaced 0.020 m downward from its equilibrium position and released from rest. Determine the position, velocity, and acceleration of the collar at $t = 1.00$ s. Let $m = 3$ kg and $k = 2000$ N/m. Ignore friction between the collar and the rod.

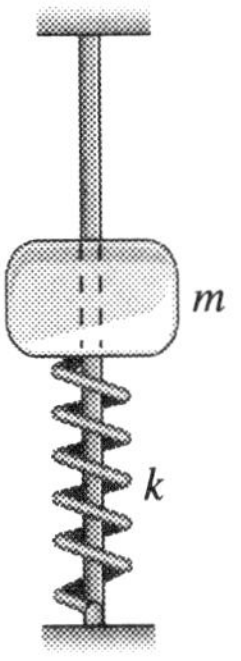

FIGURE P23.25.

23.2 Free Vibrations of Rigid Bodies —Force and Acceleration

If the position of a rigid body during its vibratory motion can be described fully by a single variable for any instant, the body is said to possess a single degree of freedom and its motion is described as a single-degree-of-freedom motion. Such motion is the only type of motion considered in this section and may be, therefore, analyzed similarly to that for a particle described in Section 23.1.

Consider, for example, the composite rigid body shown in Figure 23.6(a) which is free to rotate about the frictionless hinge at A and which is shown in any position defined by the single variable θ. Before we can proceed to apply the methods of Section 23.1, we need to locate the center of mass G for this composite body consisting of two slender rods, each of length L and mass m, welded together as shown. Thus, using the methods of Chapter 10, we can show that $y_G = \frac{3}{4}L$ as indicated in Figure 23.6.

The free-body and inertial force diagrams for any position θ are shown in Figure 23.6(b) along with a convenient t-n-z coordinate system. Applying the moment equation of motion about point A,

$$\sum M_A = \sum (M_i)_A, \quad -2mg(\tfrac{3}{4}L\sin\theta) = I_G\ddot{\theta} + m(a_G)_t(\tfrac{3}{4}L).$$

Because the mass center G travels along a circular path of radius $\frac{3}{4}L$, it follows that $(a_G)_t = (\frac{3}{4}L)\ddot{\theta}$. Substituting in the moment equation above

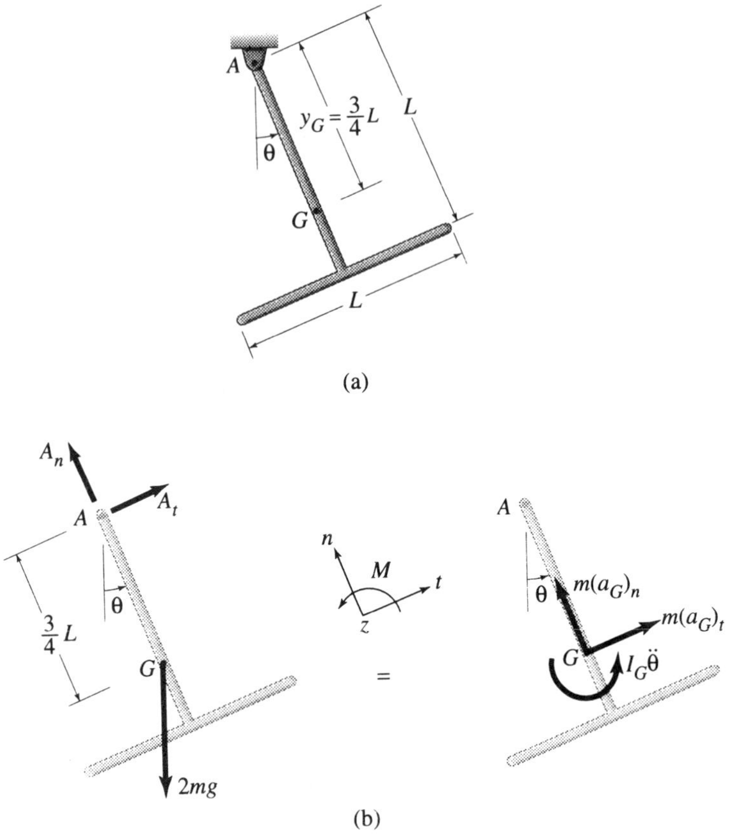

FIGURE 23.6.

and rearranging terms yields

$$[I_G + m(\tfrac{3}{4}L)^2]\ddot{\theta} + (\tfrac{3}{2}mgL)\sin\theta = 0.$$

Because $I_G + m(\frac{3}{4}L)^2 = I_A$ and, if we assume small oscillations for which $\sin\theta \approx \theta$, the equation above reduces to

$$I_A\ddot{\theta} + (\tfrac{3}{2}mgL)\theta = 0$$

from which

$$\ddot{\theta} + \left(\frac{\frac{3}{2}mgL}{I_A}\right)\theta = 0. \tag{23.22}$$

Equation (23.22) represents the differential equation of motion for the rigid body of Figure 23.6(a). When we compare Eq. (23.22) to Eq. (23.21), we conclude that they are mathematically similar and become identical if we let $p^2 = (\frac{3}{2}mgL)/I_A$. Therefore, Eq. (23.22) represents a simple harmonic motion, and the natural circular frequency of the rigid body of Figure 23.6(a) becomes

$$p = \sqrt{\left(\frac{\frac{3}{2}mgL}{I_A}\right)}.$$

The period of oscillation τ and the frequency f are $\tau = \dfrac{2\pi}{p}$ and $f = \dfrac{1}{\tau}$, respectively. Also, the general solution of Eq. (23.22) may be expressed in one of the following two forms:

$$\theta = A \sin pt + B \cos pt, \tag{23.23a}$$

or

$$\theta = \theta_M \sin(pt + \phi) \tag{23.23b}$$

where A, B, θ_M and ϕ are constants of the motion that can be determined from the initial conditions. In the second of Eqs. (23.23), the quantity θ_M represents the amplitude of vibration and ϕ is the phase angle as discussed earlier for the case of a particle.

The following examples illustrate some of the concepts above.

Example 23.4

A circular homogeneous disk is suspended by a pin at A as shown in Figure E23.4(a). It weighs 32.2 lb and has a radius $R = 9.00$ in. It is released from rest at $t = 0$ when $\theta = 10°$. Determine (a) the governing differential equation of motion for *small* oscillations, (b) the circular frequency, period, and frequency of this motion, and (c) the solution of the equation of motion which satisfies the initial conditions.

Solution

(a) Refer to the free-body and inertial force diagrams of Figure E23.4(b) and sum the moments with respect to the axis of rotation through A. Thus,

$$\sum M_A = \sum (M_i)_A, \quad -mgR \sin\theta = I_G\ddot{\theta} + mR\ddot{\theta}(R).$$

Therefore

$$\ddot{\theta} + \left(\frac{mgR}{I_A}\right)\sin\theta = 0.$$

For small oscillations, we can replace $\sin\theta$ by θ to give

$$\ddot{\theta} + \left(\frac{mgR}{I_A}\right)\theta = 0$$

where

$$I_A = \tfrac{1}{2}mR^2 + mR^2 = \tfrac{3}{2}mR^2.$$

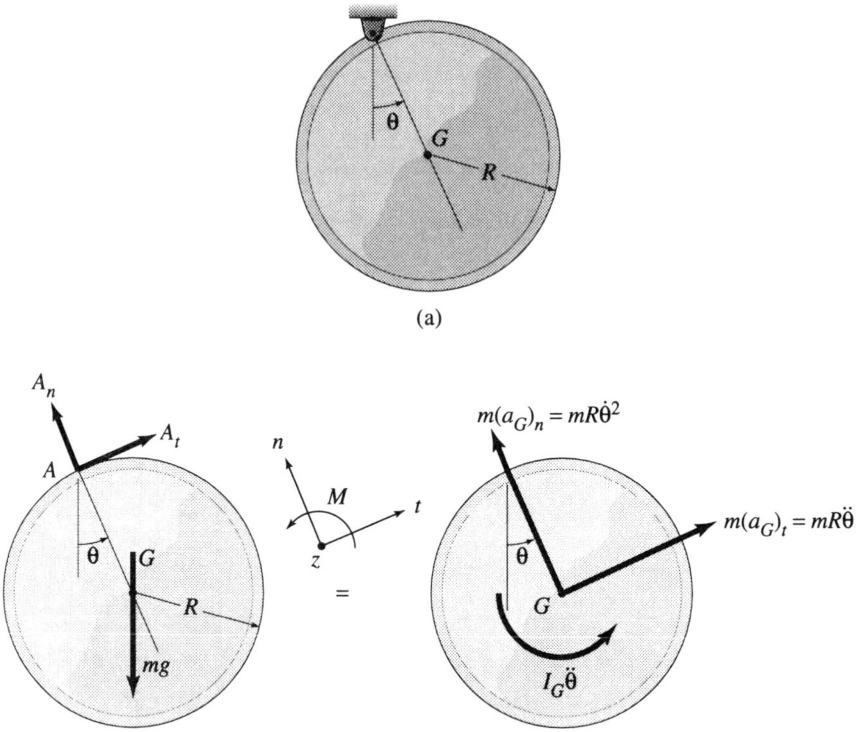

FIGURE E23.4. (b)

Thus,

$$\ddot{\theta} + \left(\frac{2g}{3R}\right)\theta = 0 \qquad \text{ANS.}$$

(b) The coefficient of θ in this equation is the circular frequency squared. Thus,

$$p^2 = \frac{2g}{3R},$$

$$p = \sqrt{\frac{2(32.2)}{3\left(\frac{9.00}{12}\right)}} = 5.35 \text{ rad/s.} \qquad \text{ANS.}$$

The period is given by

$$\tau = \frac{2\pi}{p} = \frac{2\pi}{5.35} = 1.174 \text{ s.} \qquad \text{ANS.}$$

The frequency becomes

$$f = \frac{1}{\tau} = \frac{1}{1.174} = 0.852 \text{ Hz.} \qquad \text{ANS.}$$

(c) The general solution to the equation of motion is given by $\theta = A \sin pt + B \cos pt$. To apply the initial conditions, we need a general equation for the angular velocity $\dot{\theta}$ which we obtain by differentiating this general solution with respect to time. Thus,

$$\dot{\theta} = Ap \cos pt - Bp \sin pt.$$

The initial conditions may be stated as $t = 0$, $\theta = 10^\circ = 0.175$ rad, and $\dot{\theta} = 0$. Using the second condition in the equation for $\dot{\theta}$,

$$0 = Ap(1) - Bp(0) \Rightarrow A = 0.$$

Substitute the first condition in the equation for θ to yield

$$0.175 = A(0) + B(1) \Rightarrow B = 0.175 \text{ rad}.$$

The general solution, thus, becomes

$$\theta = (0.175 \cos 5.35t) \text{ rad}. \qquad \text{ANS.}$$

This equation predicts where the disk will be at any time during its motion.

Example 23.5

A slender rod is supported by a frictionless pin at A and is attached to a spring at B as shown in Figure E23.5(a). It is released from rest when $\theta = 5^\circ$. The rod is horizontal when it is in static equilibrium. Given that $m = 15$ kg, $L = 1$ m, and $k = 10$ kN/m, determine (a) the governing differential equation of motion for *small* oscillation, (b) the circular frequency, period, and frequency of this motion, (c) the solution of the equation of motion which satisfies the initial conditions, and (d) the position, angular velocity, and angular acceleration of the rod at $t = 0.10$ s.

Solution

(a) Consider the static position of the rod shown in Figure 23.5(b), and sum the moments with respect to A. The rod is in equilibrium in this position, and we will measure its position θ at any time t with respect to this equilibrium position. Thus,

$$\sum M_A = 0, \quad -mg(0.3L) + k\delta_{ST}(0.4L) = 0.$$

$$k\delta_{ST} = 0.75 \text{ mg} \qquad \text{(a)}$$

Refer to the free-body and inertial force diagrams of Figure E23.5(c) and sum the moments with respect to A. Thus,

$$\sum M_A = \sum (M_i)_A,$$

$$[-mg(0.3L) + k(\delta_{ST} + 0.4L\theta)(0.4L)] \cos\theta = I_G\ddot{\theta} + m(a_G)_t(0.3L)$$

where $(a_G)_t = 0.3L\ddot{\theta}$ and $I_G = \frac{1}{12}mL^2$. Also, for small angles $\cos\theta \approx 1$. Hence,

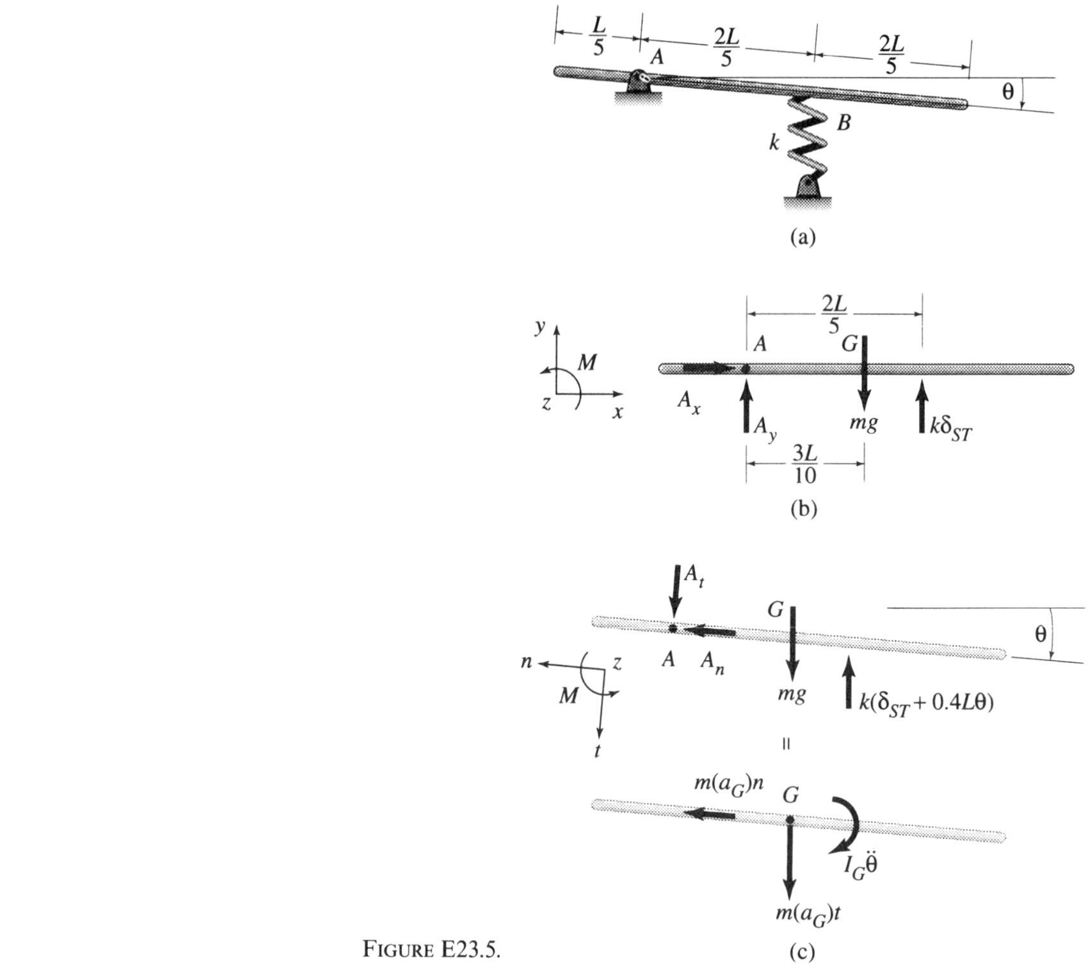

FIGURE E23.5.

$$[\tfrac{1}{12}mL^2 + m(0.3L)^2]\ddot{\theta} + k(0.4L)^2\theta + k\delta_{\mathrm{ST}}(0.4L) - mg(0.3L) = 0.$$

Because of Eq. (a), the sum of the last two terms on the left side of this equation vanishes. Simplifying gives the governing differential equation of motion. Thus,

$$\ddot{\theta} + 0.923\left(\frac{k}{m}\right)\theta = 0. \qquad \text{ANS.}$$

(b) The circular frequency squared equals the coefficient of θ in this equation. Thus,

$$p^2 = 0.923\left(\frac{k}{m}\right) = \frac{0.923(10{,}000)}{15}$$

from which

$$p = 24.8 \text{ rad/s}. \qquad \text{ANS.}$$

The period is computed as

$$\tau = \frac{2\pi}{p} = \frac{2\pi}{24.81} = 0.253 \text{ s}. \qquad \text{ANS.}$$

The frequency is determined to be

$$f = \frac{1}{\tau} = \frac{1}{0.253} = 3.95 \text{ Hz}. \qquad \text{ANS.}$$

(c) The general solution of the equation of motion is given by

$$\theta = A \sin pt + B \cos pt.$$

The constants A and B are determined from the given initial conditions $t = 0$, $\theta = 5^\circ = 0.0873$ rad, and $\dot{\theta} = 0$. Hence, substituting the first condition in the general solution gives

$$0.0873 = A(0) + B(1) \Rightarrow B = 0.0873.$$

The angular velocity $\dot{\theta}$ is obtained by differentiating the general solution with respect to time. Thus,

$$\dot{\theta} = Ap \cos pt - Bp \sin pt.$$

Substituting the second condition gives

$$0 = Ap(1) - Bp(0) \Rightarrow A = 0.$$

The general solution becomes

$$\theta = (0.0873 \cos 24.8t) \text{ rad} \qquad \text{ANS.}$$

This function gives the position of the rod at any time t during its oscillatory motion.

(d) To determine the angular velocity of this rod at any time, we differentiate the position function. Thus,

$$\dot{\theta} = 0.0873(-\sin 24.8t)(24.8) = -2.17 \sin 24.8t.$$

To determine the angular acceleration of this rod at any time, we differentiate the angular velocity function with respect to time to give

$$\ddot{\theta}\, 53.8 \cos 24.8t.$$

Evaluation of these functions at $t = 0.10$ s gives the desired values of θ, $\dot{\theta}$, and $\ddot{\theta}$. Thus,

$$\theta = 0.0873 \cos 2.48 = -0.0690 \text{ rad} = -3.95^\circ, \qquad \text{ANS.}$$

$$\dot{\theta} = -2.17 \sin 2.48 = -1.333 \text{ rad/s}, \qquad \text{ANS.}$$

and

$$\ddot{\theta} = -53.8 \cos 2.48 = 42.4 \text{ rad/s}^2. \qquad \text{ANS.}$$

The negative sign for θ and $\dot{\theta}$ is to be interpreted as ccw because the positive sense for θ is shown cw. At this time $t = 0.10$ s, the rod is 3.95° above the horizontal equilibrium position and is moving ccw while it is being angularly accelerated cw due to the moment of the restoring spring force about the axis of rotation A.

■

Problems

23.27 The homogeneous slender rod of mass m and length L, shown in Figure P23.27, is free to vibrate in a vertical plane about the frictionless hinge at A. (a) Write the governing differential equation of motion for small oscillations of this rod. (b) Express the circular frequency, period, and frequency as functions of g, L, m, and k. The rod is vertical when static.

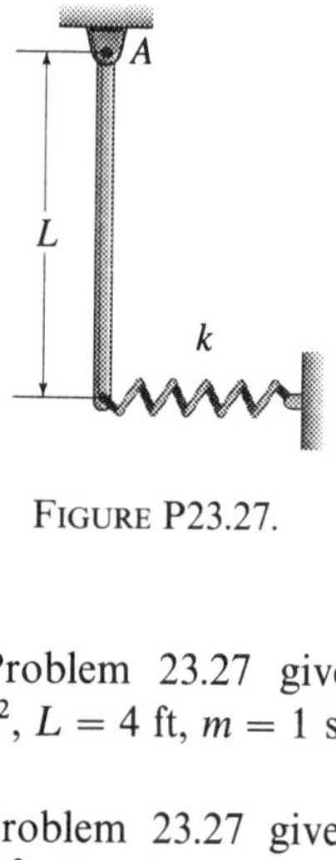

FIGURE P23.27.

23.28 Solve Problem 23.27 given that $g = 32.2$ ft/s^2, $L = 4$ ft, $m = 1$ slug, and $k = 20$ lb/ft.

23.29 Solve Problem 23.27 given that $g = 9.81$ m/s^2, $L = 1.0$ m, $m = 12$ kg, and $k = 400$ N/m.

23.30 Two uniform slender rods are welded together as shown in Figure P23.30. The vertical rod has a mass m and the horizontal rod has a mass $m/2$. The composite body is free to oscillate in a vertical plane about the frictionless hinge at A. Assume small oscillations, and develop equations for the circular frequency, period, and frequency as functions of g and L.

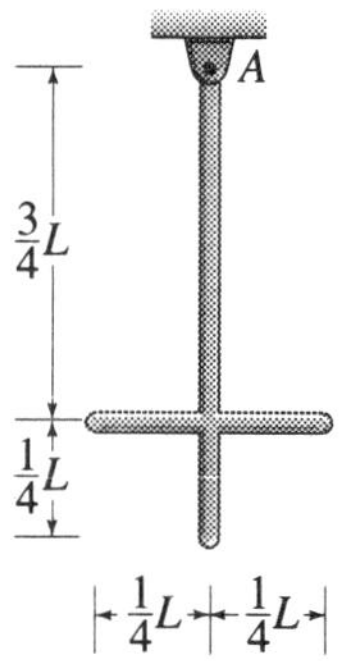

FIGURE P23.30.

23.31 Solve Problem 23.30 for $g = 9.81$ m/s^2 and $L = 0.5$ m. If this body is released from rest when $\theta = 8°$ express θ, $\dot{\theta}$, and $\ddot{\theta}$ as functions of time.

23.32 A uniform rectangular plate is suspended at A and is free to oscillate in a vertical plane as shown in Figure P23.32. Assume small oscillations, and develop equations for the circular frequency, period, and frequency as functions of g, a, and b. Specialize these results for the case $a = b$.

23.33 Solve problem 23.32 for the plate sus-

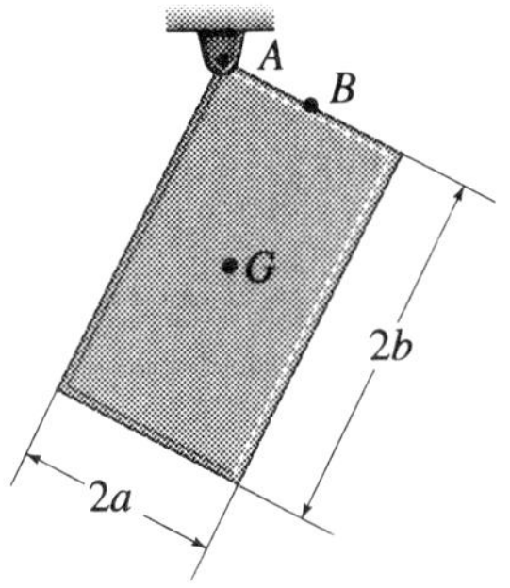

FIGURE P23.32.

pended at point B, the midpoint of the side of length $2a$.

23.34 A steel plate has a thickness of 0.020 ft and is suspended from A as shown in Figure P23.34. It has a parabolic shape given by $y = 0.5x^2$. Determine the circular frequency, period, and frequency of small oscillations of this plate. Steel weighs 490 lb/ft^3, and the coordinates are expressed in ft.

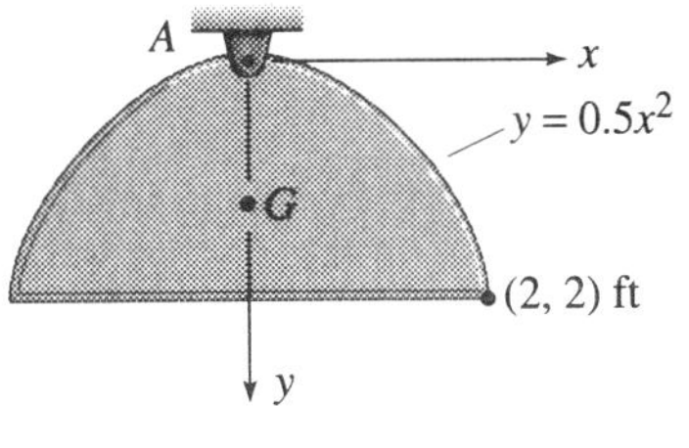

FIGURE P23.34.

23.35 A slender rod of mass m and length L is shown in its horizontal equilibrium position in Figure P23.35. It is pinned at A and attached to a spring at B. Develop equations for small oscillations of this rod which give the circular frequency, period, and frequency as functions of k and m. The horizontal position of the beam represents its equilibrium position.

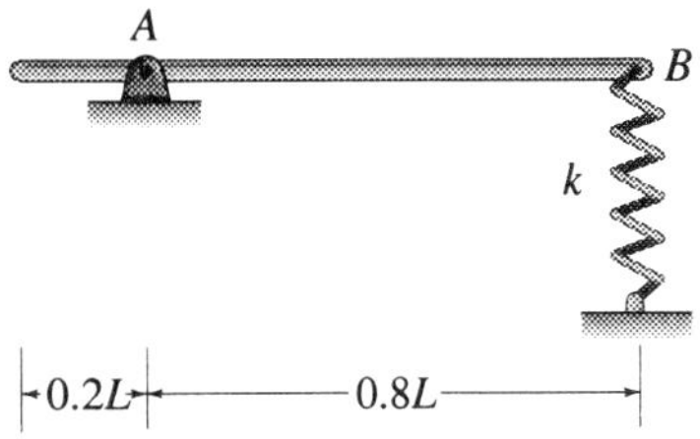

FIGURE P23.35.

23.36 Solve Problem 23.35 for $k = 2000$ N/m and $m = 25$ kg.

23.37 Solve Problem 23.35 for $k = 200$ lb/ft and $m = 3$ slug.

23.38 A half-circular disk of mass m and radius R is shown in Figure P23.38. It is suspended by a frictionless pin at A. Determine equations for the circular frequency, period, and frequency of small oscillations as functions of g and R.

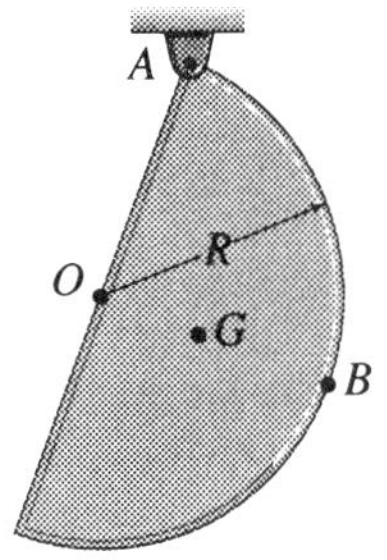

FIGURE P23.38.

23.39 Solve Problem 23.38 for a suspension point at B which lies on the line joining O to G.

23.40 A slender rod of mass m and length $L = 4R$ is welded to a disk of mass $6m$ and radius R, as shown in Figure P23.40. The composite body is suspended by a frictionless hinge at A. Express the circular frequency, period, and frequency

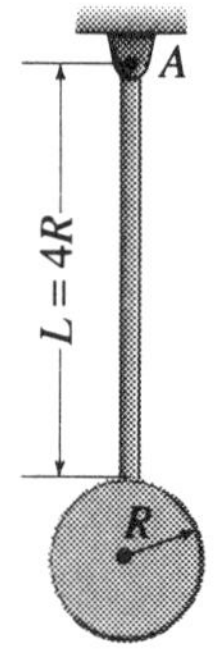

FIGURE P23.40.

of small oscillations as functions of g and R.

23.41 Solve Problem 23.40 for $R = 0.20$ m.

23.42 Solve Problem 23.40 for $R = 1.00$ ft.

23.43 A half disk of radius R is attached to a rectangular plate ($2R \times 2H$), as shown in Figure P23.43. The composite body, suspended by a frictionless hinge at A, has a constant mass density and a constant thickness. Determine the circular frequency of small oscillations as a function of g, H, and R.

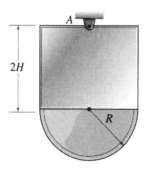

FIGURE P23.43.

23.44 Solve Problem 23.43 for $H = R = 0.5$ ft. Also find the corresponding period and frequency of this motion.

23.45 A cylinder of mass m and radius R is attached to a spring, as shown in Figure P23.45. This body is in equilibrium in the position shown and will roll without sliding when displaced cw 0.10 rad and released from rest. Derive the equation of motion for small oscillations of this cylinder. Express the period of this motion as a function of k and m. If $k = 600$ lb/ft and $m = 5$ slug, find the period and express θ, $\dot{\theta}$, and $\ddot{\theta}$ as functions of time.

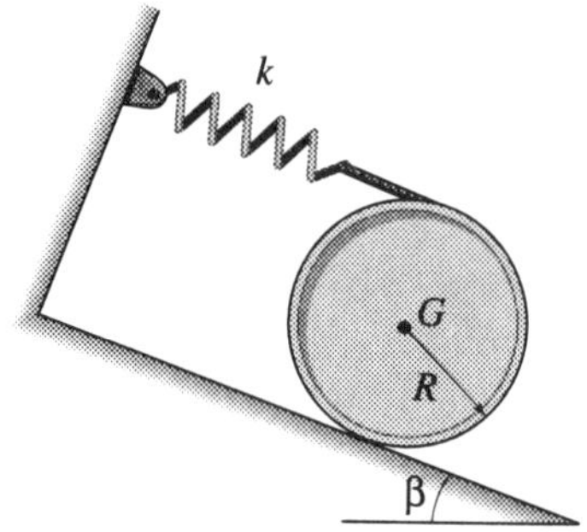

FIGURE P23.45.

23.46 Solve Problem 23.45 for $k = 4000$ N/m and $m = 40$ kg.

23.47 In Figure P23.47, the cylinder of mass m and radius R is at rest, and the spring is unstretched. Imagine that the cylinder is rotated through a cw angle of 12° and released from rest. If the cylinder rolls without sliding on the horizontal plane, express θ, $\dot{\theta}$, and $\ddot{\theta}$ as functions of time given that $k = 300$ lb/ft and $m = 3.50$ slug.

23.48 Four slender rods are welded together to form a square as shown in Figure P23.48. Each rod has a mass m and a

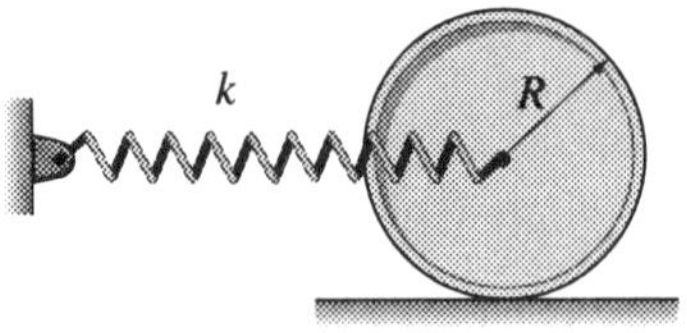

FIGURE P23.47.

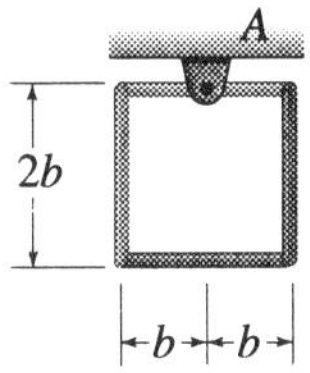

FIGURE P23.48.

length 2b. The square is suspended by a frictionless hinge at A. Express the period of small oscillations of this body in terms of g and b.

23.49 Solve Problem 23.48 if the body is suspended from one corner of the square.

23.50 A slender rod of mass m is suspended from a spring on the left and is pinned at A on the right, as shown in Figure P23.50. It is in equilibrium in the horizontal position shown. In terms of k and m, determine the period of small oscillations of this rod about its equilibrium position.

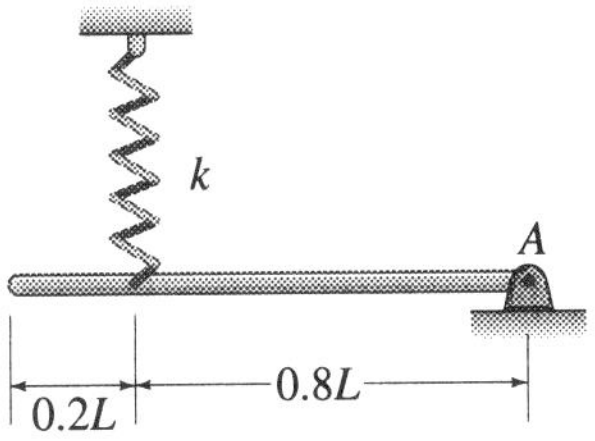

FIGURE P23.50.

23.51 In Figure P23.51, the square plate has a hole of radius $R = b/2$ cut from its center. If the mass density and thickness of the plate are constant, determine the frequency of small oscillations of this plate about the hinge at A in terms of g and b.

23.52 A composite body of constant mass density consisting of 2 spheres joined by a slender rod is shown in Figure P23.52. The top sphere has a mass m with a diameter $d = 0.2L$ whereas the bottom sphere has a mass equal to $8m$ with a diameter $D = 0.4L$. The slender rod of length L has a mass m_R with a diameter of $0.05L$. Determine the period of small oscillations about the hinge at A in terms of g and L.

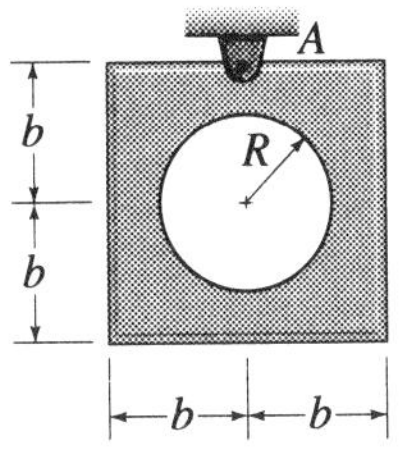

FIGURE P23.51.

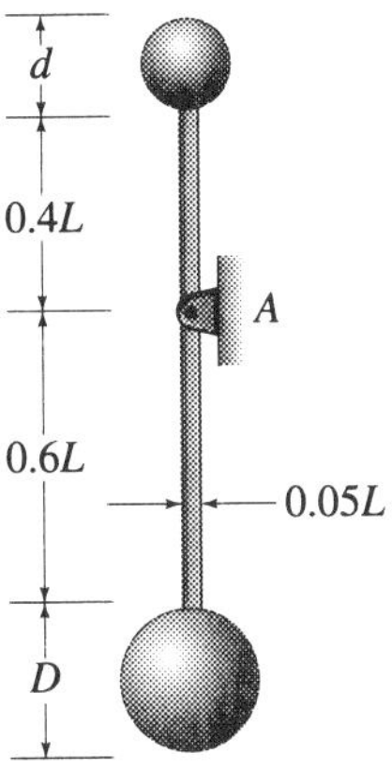

FIGURE P23.52.

23.53 The slender rod, which is spring supported, as shown in Figure P23.53, is in equilibrium when horizontal. Determine the frequency of small oscillations of this rod about the hinge at A as a function of k and m.

23.54 Three slender rods, of mass m each, are welded together to form an equilateral triangle as shown in Figure P23.54. Find the period of small oscillations of this body, pinned at A, in terms of g and b.

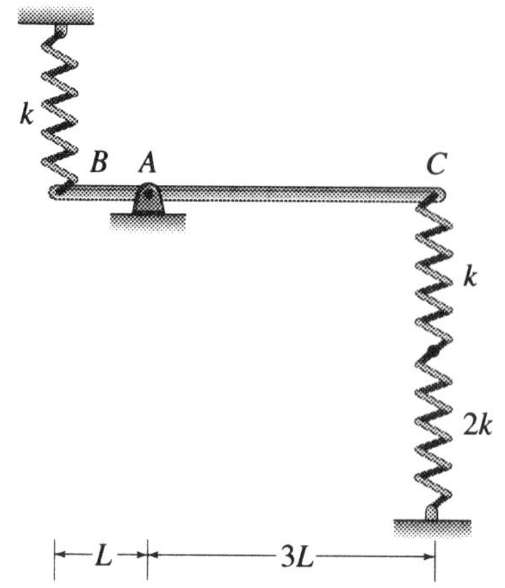

FIGURE P23.53.

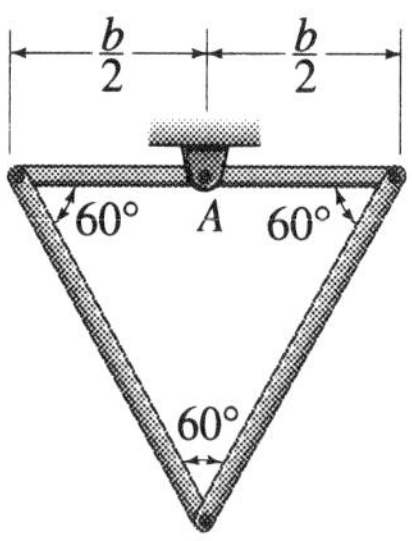

FIGURE P23.54.

23.55 Solve Problem 23.54 for a supporting pin placed at one of the vertices of the equilateral triangle.

23.56 A slender rod of length L and mass $4m$ is attached to a particle of mass m as shown in Figure P23.56. Determine the circular frequency, period, and frequency of small oscillations of this body pinned at A and spring supported at B. This body is horizontal when in equilibrium.

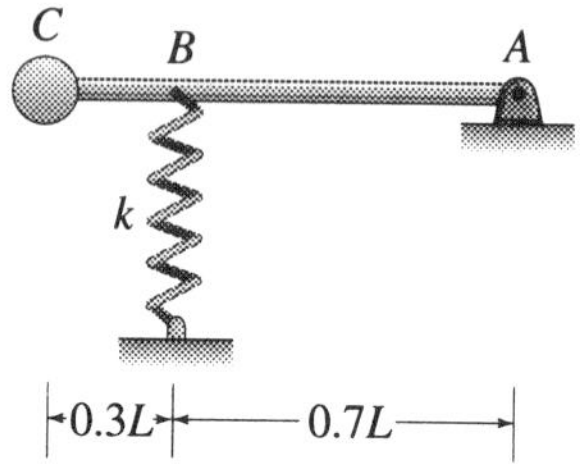

FIGURE P23.56.

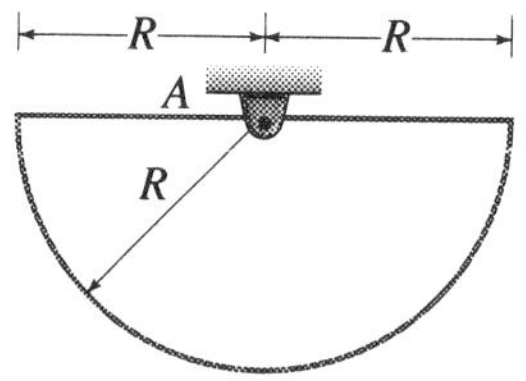

FIGURE P23.57.

23.57 A small diameter wire of uniform mass density is bent in the shape shown in Figure P23.57. If this body is rotated cw through an angle of 0.20 radians and released from rest, determine θ, $\dot{\theta}$, and $\ddot{\theta}$ as functions of g, R, and time t.

23.3 Free Vibrations of Rigid Bodies —Energy

The principle of conservation of energy may be used to write differential equations of motion for rigid bodies because we are neglecting air resistance and frictional forces which, in general, are nonconservative forces. Consider, for example, the cylinder shown in Figure 23.7(a) which rolls without sliding on the inclined plane. The free-body diagram of the cylinder is shown in Figure 23.7(b). Because the cylinder rolls without sliding, the frictional force F acting at A does no work and will not enter the energy equation. The normal force N acting at A is perpendicular to the plane and does no work. As before, if the stable

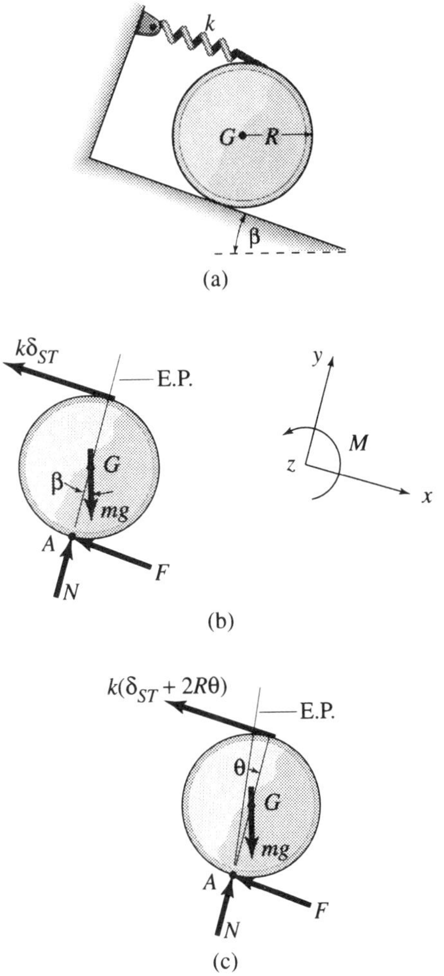

FIGURE 23.7.

equilibrium position is chosen as the reference for dynamic displacements, then, the effect of the force of gravity is canceled by the static spring force. Referring to Figure 23.7(b),

$$\sum M_A = 0, \quad k\delta_{ST}(2R) - mgR\sin\beta = 0. \tag{23.24}$$

This equation will be used below to show that we may neglect the gravitational potential energy changes if we measure dynamic displacements from a stable equilibrium position. Consistent with this, we ignore the static spring force.

Refer to Figure 23.7(c) which shows the free-body diagram of the cylinder for any small angle θ measured from the static equilibrium position. Here again, F and N perform no work. The principle of conservation of energy, developed in Chapter 19 for a rigid body, may be written in the form

$$T + V = E$$

where T is the kinetic energy, V is the potential energy, and E is the total energy which remains constant. In this case,

$$T = \tfrac{1}{2}mv_G^2 + \tfrac{1}{2}I_G\omega^2 = \tfrac{1}{2}m(R\dot{\theta}) + \tfrac{1}{2}I_G\dot{\theta}^2,$$

and

$$V = V_e + V_g = \tfrac{1}{2}k(2R\theta + \delta_{ST})^2 - mgR(\sin\beta)\theta.$$

Therefore,

$$\tfrac{1}{2}m(R\dot{\theta})^2 + \tfrac{1}{2}I_G\dot{\theta}^2 + \tfrac{1}{2}k(2R\theta + \delta_{ST})^2 - mgR(\sin\beta)\theta = E$$

If we differentiate this energy equation with respect to time, then $\dfrac{dE}{dt} = 0$ because the total energy E does not vary with time. Thus,

$$mR^2\dot{\theta}\ddot{\theta} + I_G\dot{\theta}\ddot{\theta} + k(2R\theta + \delta_{ST})2R\dot{\theta} - mg(\sin\beta)R\dot{\theta} = 0.$$

In general, $\dot{\theta}$ is not zero, even though it may vanish for certain times. We are considering the body at any time t and may factor $\dot{\theta}$ from all terms of this equation. By Eq. 23.24, $k\delta_{ST}(2R) - mg\sin\beta R = 0$, and the governing differential equation of motion becomes

$$(I_G + mR^2)\ddot{\theta} + (4kR^2)\theta = 0.$$

It should be pointed out that Eq. (23.24) is always valid as long as displacements are measured from the static equilibrium position. Thus, if we measure displacements from the static equilibrium position, we can ignore the effect of the weight and that of the static deformation. Recalling that $I_G = \tfrac{1}{2}mR^2$ for a cylinder, the coefficient of $\ddot{\theta}$ becomes $\tfrac{3}{2}mR^2$ which also equals I_A, where A is the instantaneous center of rotation of the cylinder as shown in Figure 23.7(c). The above differential equation, therefore, becomes

$$(\tfrac{3}{2}mR^2)\ddot{\theta} + (4kR^2)\theta = 0$$

from which

$$\ddot{\theta} + \left(\frac{8k}{3m}\right)\theta = 0.$$

Comparing with Eq. (23.22), we conclude that

$$p = \sqrt{\frac{8k}{3m}}.$$

The period is given by

$$\tau = \frac{2\pi}{p} = 2\pi\sqrt{\frac{3m}{8k}}.$$

The frequency is given by

$$f = \frac{1}{\tau} = \frac{1}{2\pi}\sqrt{\frac{8k}{3m}}.$$

The general solution of this differential equation is given by $\theta = A\sin pt + B\cos pt$ where constants A and B are determined from the initial conditions of the motion.

The energy method provides an alternate method for deriving the differential equation. Application of the force, mass, and acceleration method of Sec. 23.2 would result in the same governing differential equation.

The following examples illustrate the use of the principle of conservation of energy in solving vibration problems.

Example 23.6

Use the principle of conservation of energy to derive the governing differential equation of motion for small oscillations for the disk shown in Figure E23.6(a). The disk has a mass m and a radius R. Note that this is the same disk analyzed in Example 23.4.

Solution

The free-body diagram of the disk for any small ccw angular displacement θ is shown in Figure E23.6(b). Note that the reaction components at A do no work because they do not displace. Thus, from the principle of conservation of energy ($T + V = E$),

$$T = \tfrac{1}{2}mv_G^2 + \tfrac{1}{2}I_G\omega^2 = \tfrac{1}{2}m(R\dot{\theta})^2 + \tfrac{1}{2}I_G\dot{\theta}^2,$$

and

$$V = V_e + V_g = 0 + mgR(1 - \cos\theta).$$

Therefore, the principle of conservation of energy yields

$$\tfrac{1}{2}m(R\dot{\theta})^2 + \tfrac{1}{2}I_G\dot{\theta}^2 + mgR(1 - \cos\theta) = E.$$

Because E remains constant with time, time differentiation of the above equation yields

$$mR^2\dot{\theta}\ddot{\theta} + I_G\dot{\theta}\ddot{\theta} + mgR(\sin\theta)\dot{\theta} = 0.$$

Dividing by $\dot{\theta}$, replacing $\sin\theta$ by θ and I_G by $\frac{1}{2}mR^2$ and simplifying,

$$(\tfrac{3}{2}mR^2)\ddot{\theta} + (mgR)\theta = 0.$$

Division by $\frac{3}{2}mR^2$ yields

$$\ddot{\theta} + \left(\frac{2}{3}\frac{g}{R}\right)\theta = 0.$$

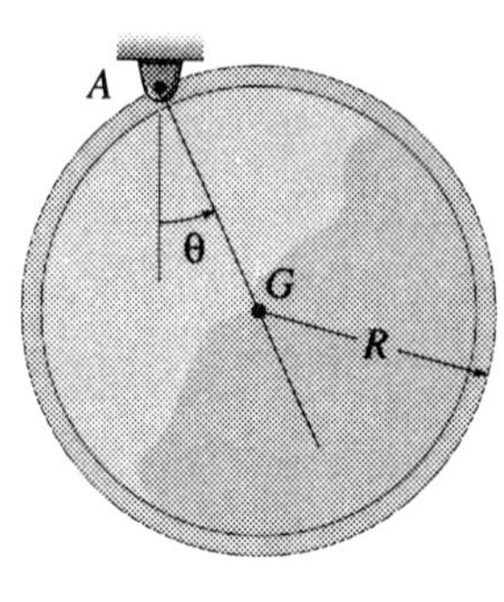

(a)

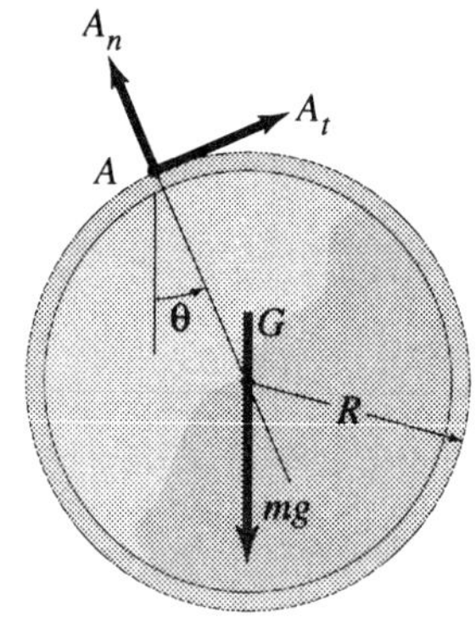

FIGURE E23.6. (b)

This result is obviously the same as that obtained in the solution of Example 23.4.

■ Example 23.7

Use the principle of conservation of energy to derive the governing differential equation of motion for small oscillations for the slender rod shown in Figure E23.7(a). The rod has a mass $m = 15$ kg and a length $L = 1$ m, and the spring has a spring constant $k = 10$ kN/m. Note that this is the same rod analyzed in Example 23.5.

Solution

The free-body diagram of the rod for any small cw angular displacement θ is shown in Figure E23.7(b). Note that the reaction components at A do no work because they do not displace. Also, note that, since displacements are measured from the horizontal static equilibrium position, we can ignore the effects of gravity and static deformation. Thus, the free-body diagram of Figure E23.7(b) does not include the weight of the rod mg and the static deflection δ_{ST}. Therefore, from the principle

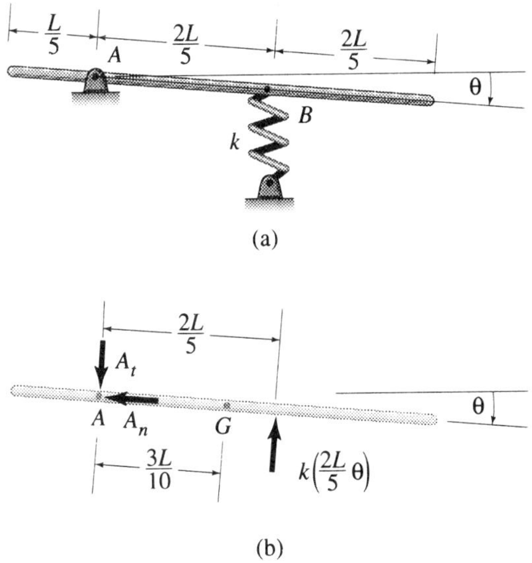

FIGURE E23.7.

of conservation of energy ($T + V = E$),

$$\begin{aligned} T &= \tfrac{1}{2}mv_G^2 + \tfrac{1}{2}I_G\omega^2 \\ &= \tfrac{1}{2}m(0.3L\dot{\theta})^2 + \tfrac{1}{2}(\tfrac{1}{3}mL^2)\dot{\theta} = \tfrac{1}{2}I_A\dot{\theta}^2 \\ &= 0.0867mL^2\dot{\theta}^2, \end{aligned}$$

$$\begin{aligned} V &= V_e + V_g \\ &= \tfrac{1}{2}k(\tfrac{2}{5}L\theta)^2 \\ &= 0.08kL^2\theta^2. \end{aligned}$$

Therefore, the principle of conservation of energy ($T + V = E$) yields

$$0.0867mL^2\dot{\theta}^2 + 0.08kL^2\theta^2 = E.$$

Because E remains constant with time, the time derivative of the above relationship leads to

$$0.1734mL^2\dot{\theta}\ddot{\theta} + 0.16kL^2\theta\dot{\theta} = 0.$$

Dividing by $0.1734mL^2\dot{\theta}$,

$$\ddot{\theta} + 0.923\left(\frac{k}{m}\right)\theta = 0$$

which is the same differential equation of motion derived in Example 2.5.

■

Problems

Use the principle of conservation of mechanical energy to derive the governing differential equation of motion in each case.

23.58 Refer to Figure P23.58, and derive the differential equation of motion for small oscillations of the system shown. The equilibrium position is shown and the body has a mass m. Consider θ positive cw. Use $k_G = 1.4R$, and assume rolling without sliding.

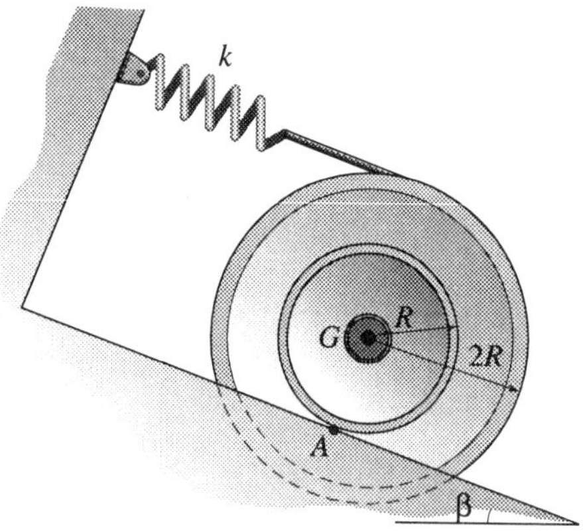

FIGURE P23.58.

23.59 The slender rod of mass m and length L shown in Figure P23.59 is in equilibrium when horizontal. Consider θ positive cw, and derive the differential equation of motion for small oscillations.

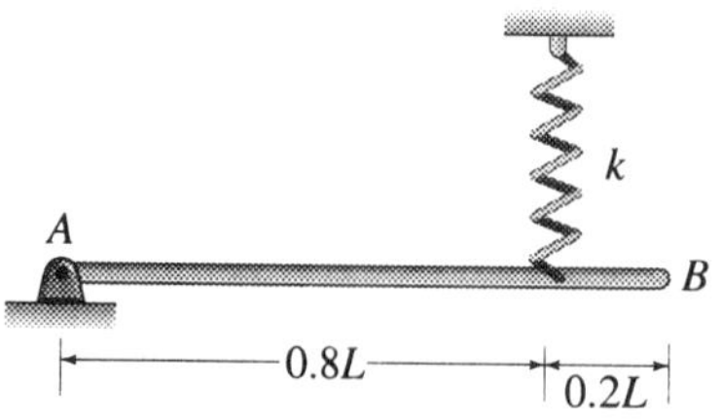

FIGURE P23.59.

23.60 Three slender rods are fastened together, as shown in Figure P23.60. Each has mass m and length L. Derive the differential equation of motion for small oscillations about A. Express the period as a function of g and L.

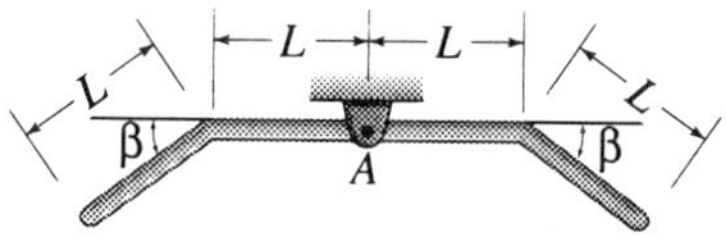

FIGURE P23.60.

23.61 Two particles each of mass m are attached to a rod of negligible mass as depicted in Figure P23.61. Derive the differential equation of motion for small oscillations about A.

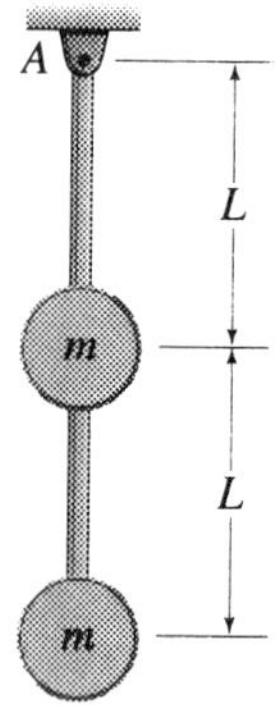

FIGURE P23.61.

23.62 A thin plate of constant thickness and of mass and constant mass density is shown in Figure P23.62. It has the shape of an isosceles triangle. Derive the differential equation of motion for small

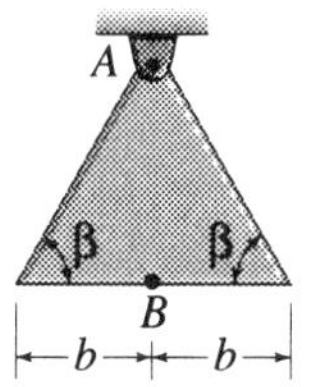

FIGURE P23.62.

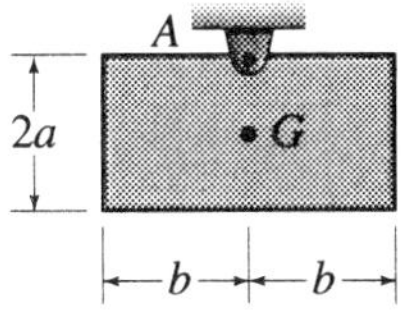

FIGURE P23.65.

oscillations of this plate. Express p as a function of g, b, and β.

23.63 Solve Problem 23.62 for a pin support placed at point B instead of point A.

23.64 A rod hangs vertically as shown in Figure P23.64. The spring is unstretched in this position. Derive the differential equation of motion for this system for small oscillations. The rod has mass m and length L. Express the period of this motion as a function of k, m, g, and L.

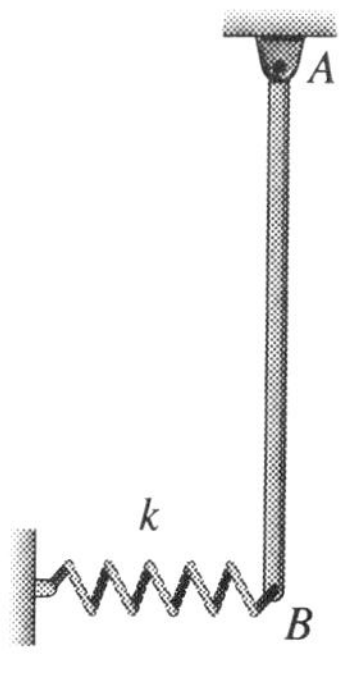

FIGURE P23.64.

23.65 A thin homogeneous plate is supported by a pin at A as shown in Figure P23.65. Derive the differential equation of motion for small oscillations of this plate. Express the circular frequency of this motion as a function of g, a, and b. Suppose $b \ll a$, and describe the body and the frequency for this special case.

23.66 In the horizontal equilibrium position shown, the slender rod of mass m in Figure P23.66 is supported by the pin at A. The springs are free of forces. Write the differential equation of motion of this rod for small oscillations. If the rod is rotated 0.08 radians cw and released from rest, determine θ as a function of time. Let $m = 1$ slug and $k = 144$ lb/ft.

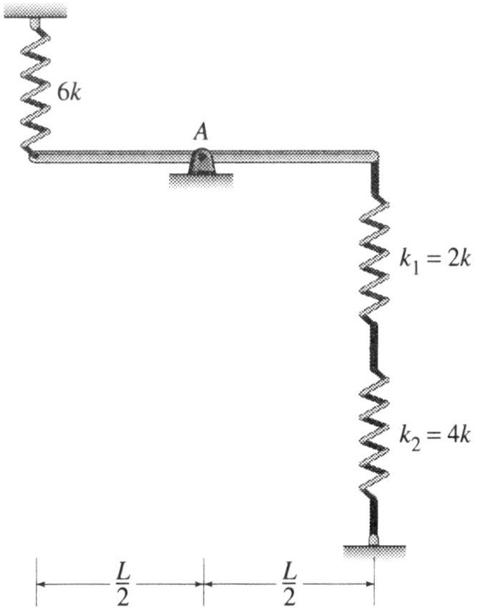

FIGURE P23.66.

23.67 Solve Problem 23.66 for $m = 10$ kg and $k = 1500$ N/m.

23.68 The body of mass m, shown in Figure P23.68, has a radius of gyration $k = 1.5R$ and rolls without sliding on the

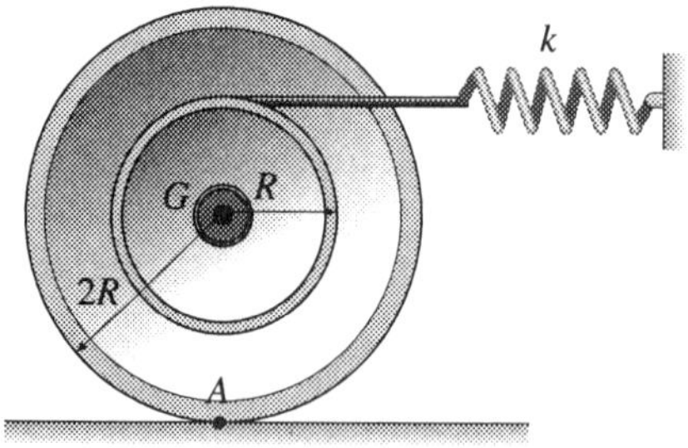

FIGURE P23.68.

horizontal plane. Derive the differential equation of motion for small oscillations. Express the frequency f as a function of k and m.

23.69 Solve Problem 23.68 for $m = 4$ slug and $k = 800$ lb/ft.

23.70 A square is formed by four slender rods welded together as shown in Figure P23.70. Each rod has mass m and length $2b$. Derive the differential equation of motion for small oscillations of this body pinned at A. If it is rotated ccw through an angle of 0.1 radian and released from rest, express θ as a function of time t.

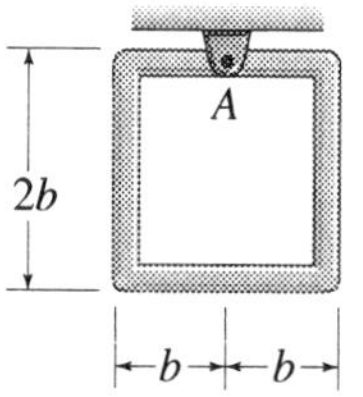

FIGURE P23.70.

23.71 A quarter circular plate of mass m has a constant thickness and mass density. Refer to Figure P23.71, and write the differential equation of motion for small oscillations of this plate. Express the period of this motion as a function of g and R.

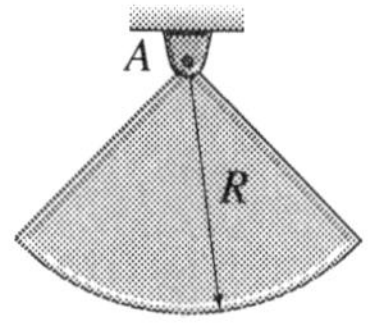

FIGURE P23.71.

23.72 Solve Problem 23.71 for $g = 32.2$ ft/s^2 and $R = 2$ ft.

23.73 Slender rods of total mass $4m$ form the rigid body shown in Figure P23.73. Write the differential equation of motion for small oscillations of this body, and express the period of motion as a function of g and L.

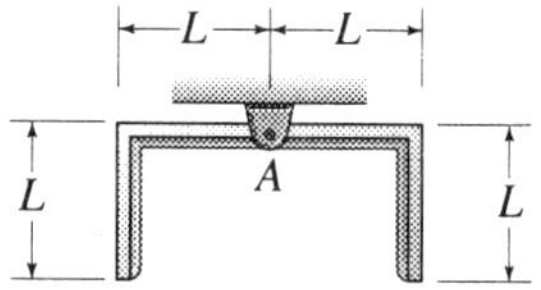

FIGURE P23.73.

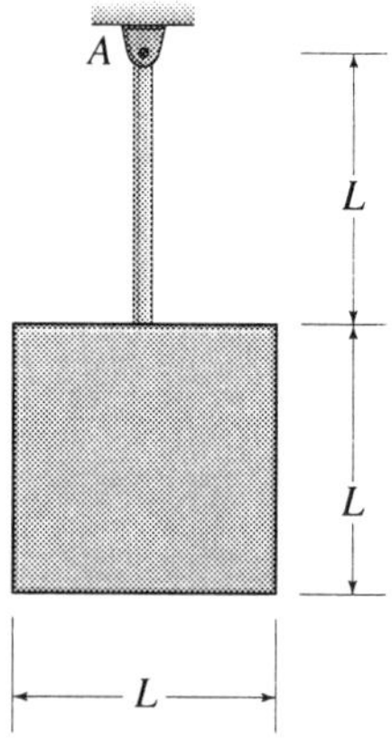

FIGURE P23.74.

23.74 A square plate of mass $10m$ is attached to the lower end of a slender rod of mass m as shown in Figure P23.74. Write the differential equation of motion for small oscillations of this body. Determine the circular frequency, period, and frequency of this vibratory motion in terms of g and L.

23.75 A slender rod of mass m is in equilibrium in the position shown in Figure 23.75. Imagine that it is rotated cw through a small angle and released from rest. Write the differential equation of motion for small oscillations of this rod. Show that the period of this motion is a function of k and m.

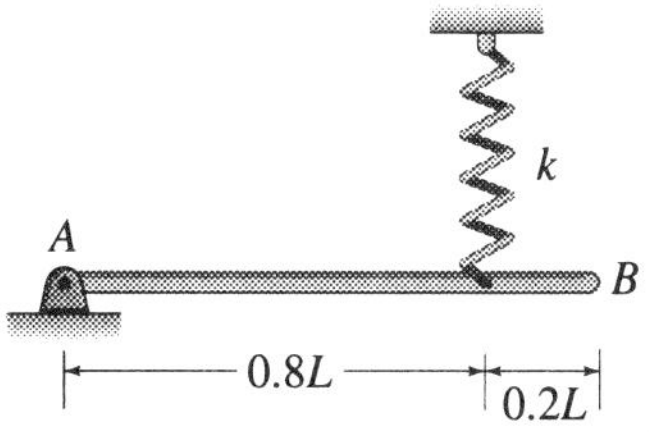

FIGURE P23.75.

23.76 Solve Problem 23.75 for the spring attached to the mass center of the rod.

23.77 Three uniform slender rods form an isosceles triangle as shown in Figure P23.77. Let $\beta = 50°$ and the total mass of this body be m. Derive the differential equation of motion for small oscillations. Determine the period of this motion as a function of g and b.

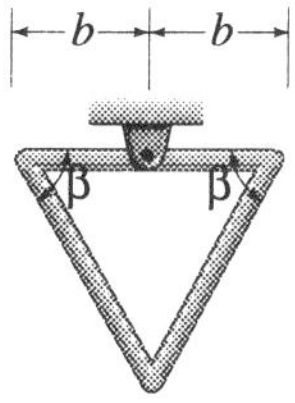

FIGURE P23.77.

23.78 Solve Problem 23.77 for $\beta = 45°$, and again let the total mass be m.

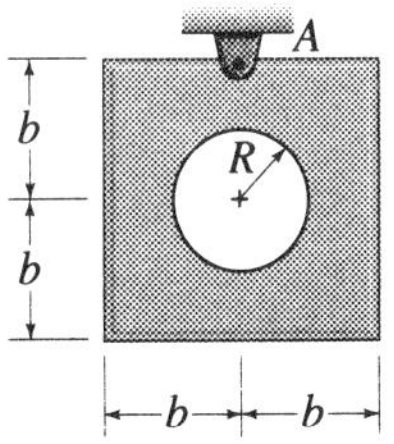

FIGURE P23.79.

23.79 A square plate has a center hole of radius $R = b/2$ as shown in Figure P23.79. It has a constant thickness and mass density. Derive the differential equation of motion for small oscillations of this plate. Repeat the derivation for a square plate without a hole. Form the ratio of the periods of these two plates.

23.80 The parabolic shaped steel plate shown in Figure P23.80 has a contour defined by $y = 0.5x^2$ where x and y are given in ft. It has a constant thickness of 0.020 ft and a specific weight of 490 lb/ft^3. Determine the frequency of small oscillations of this plate which is pinned at A.

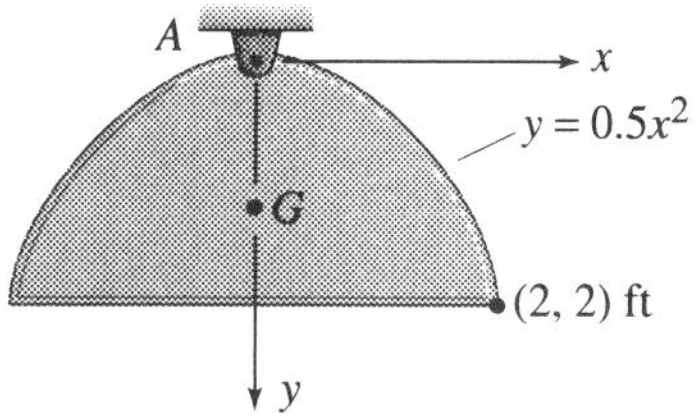

FIGURE P23.80.

23.81 Two slender rods each of mass $2m$ and length $2L$ are fastened together, as shown in Figure P23.81. Write the governing differential equation of motion for small oscillations of this body. De-

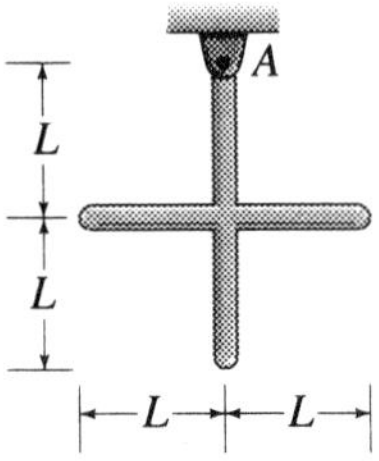

FIGURE P23.81.

termine the period of this motion in terms of g and L.

23.82 A slender rod of mass $4m$ and length $4L$ is attached to two square plates each of mass $12m$ and measuring $L \times L$, as shown in Figure P23.82. Determine the period of small oscillations of this composite body in terms of g and L.

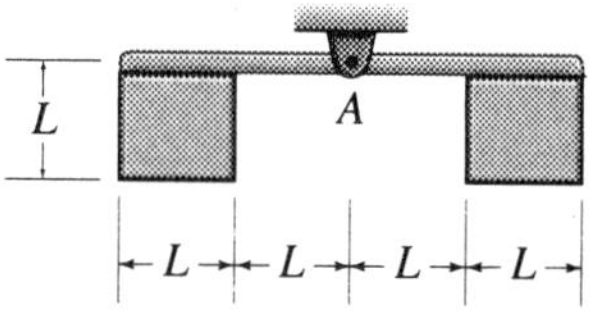

FIGURE P23.82.

23.4 Lagrange's Method—Conservative Forces

A duality exists between Lagrange's equation and Newton's equations of motion. If we assume Newton's laws, then, we can prove Lagrange's equation, and, if we assume Lagrange's equation, then, we can prove Newton's laws. The equation presented by J.L. Lagrange (1736–1813) is very general, but we limit the discussion in this Section to a single degree of freedom and to conservative forces. Lagrange's method applies both to particles and rigid bodies.

Let us define the *Lagrangian* l as the difference between the kinetic and potential energies. Thus,

$$l = T - V.$$

Then for a one-degree-of-freedom system, Lagrange's equation* is given by

$$\frac{d}{dt}\left(\frac{\partial l}{\partial \dot{x}}\right) - \frac{\partial l}{\partial x} = 0 \tag{23.25a}$$

where $\dfrac{\partial l}{\partial \dot{x}}$ and $\dfrac{\partial l}{\partial x}$ are the first partial derivatives of the Lagrangian with respect to the velocity $\dot{x}$ and the position coordinate x, respectively. The operator $\dfrac{d}{dt}$ is the first time derivative performed on the partial derivative $\dfrac{\partial l}{\partial \dot{x}}$ after it has been found.

* Derivation of Lagrange's equation is beyond the scope of this text. Interested readers may refer to Section 8.4, pp. 268–282, of Advanced Dynamics for Engineers by Bruce J. Torby, Holt, Reinhart & Winston, New York, 1984.

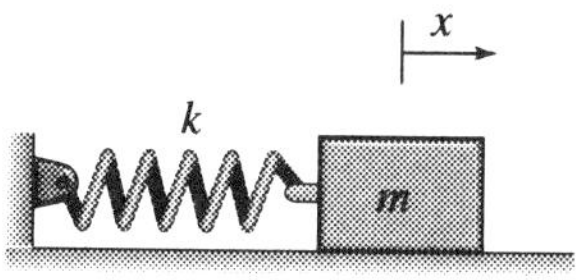

FIGURE 23.8.

As an illustration, consider the mass m of Figure 23.8 which slides on a frictionless platform and is connected to a spring of constant k. For this system, $T = \frac{1}{2}m\dot{x}^2$ and $V = \frac{1}{2}kx^2$. Note that the gravitational potential energy may be taken as zero because the particle does not change elevation during its motion. The Lagrangian becomes

$$l = \tfrac{1}{2}m\dot{x}^2 - \tfrac{1}{2}kx^2.$$

The Lagrangian is a function of the velocity $\dot{x}$ and the position coordinate x. In other words, the dependent variable l is a function of two independent variables x and $\dot{x}$. Thus,

$$\frac{\partial l}{\partial \dot{x}} = m\dot{x}; \quad \frac{d}{dt}\left(\frac{\partial l}{\partial \dot{x}}\right) = \frac{d}{dt}(m\dot{x}) = m\ddot{x}; \quad \frac{\partial l}{\partial x} = -kx.$$

Note that $\dfrac{\partial l}{\partial \dot{x}} = m\dot{x}$ represents the linear momentum of the system. Substituting in Eq. (23.25a) yields

$$m\ddot{x} + kx = 0$$

which agrees with Eq. (23.2), which was derived by applying Newton's second law. We have derived the equation of motion by starting with a function l which is a scalar function because it is expressible in terms of energies which are scalars. This is an interesting feature of Lagrange's method. A single scalar function l contains all the information about the system even though this system may have many degrees of freedom. The equation of motion was derived by performing mathematical operations on this scalar function l.

As a second illustration, consider the simple pendulum of Figure 23.9, and treat the angle θ as a coordinate. In this case, Lagrange's equation becomes

$$\frac{d}{dt}\left(\frac{\partial l}{\partial \dot{\theta}}\right) - \frac{\partial l}{\partial \theta} = 0. \tag{23.25b}$$

For this system $T = \frac{1}{2}m(L\dot{\theta})^2$ and $V = mgL(1 - \cos\theta)$. Thus,

$$l = T - V = \tfrac{1}{2}m(L\dot{\theta})^2 - mgL(1 - \cos\theta),$$

$$\frac{\partial l}{\partial \dot{\theta}} = mL^2\dot{\theta},$$

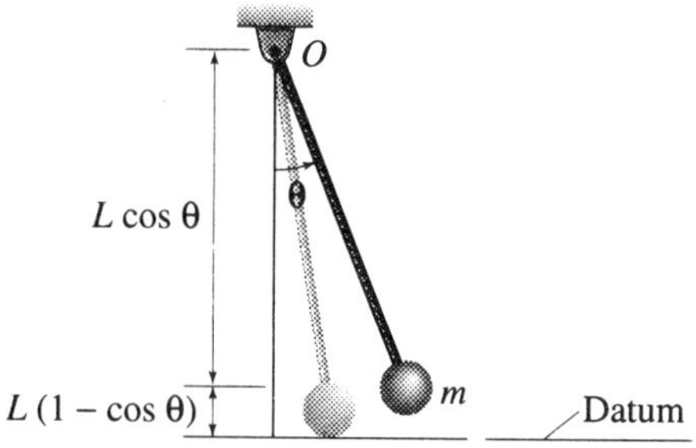

FIGURE 23.9.

$$\frac{d}{dt}\left(\frac{\partial l}{\partial \dot{\theta}}\right) = \frac{d}{dt}(mL^2\dot{\theta}) = mL^2\ddot{\theta},$$

and

$$\frac{\partial l}{\partial \theta} = -mgL\sin\theta.$$

Note that $\frac{\partial l}{\partial \dot{\theta}} = mL^2\dot{\theta}$ represents the angular momentum with respect to O. Substituting these derivatives in Eq. (23.25b) gives

$$mL^2\ddot{\theta} + mgL\sin\theta = 0.$$

Dividing by mL^2,

$$\ddot{\theta} + \frac{g}{L}\sin\theta = 0,$$

agreeing with Eq. (23.20), which was derived by applying Newton's second law.

Lagrange's method deals with writing the equations of motion and not with the problem of solving these equations. The computer revolution has changed the engineer's focus to formulating governing equations, because numerical solutions of acceptable accuracy can almost always be found with the aid of a computer.

The following example illustrates further the use of Lagrange's method in deriving the differential equation of motion.

■ Example 23.8

A cylinder of mass m and radius r is attached to a spring of constant k as shown in Figure E23.8. It rolls without sliding on the inclined plane. Use the Lagrangian to derive the equation of vibratory motion of this cylinder. Express the period of this motion as a function of k and m.

Solution

If displacements are measured from the equilibrium position, then, we can ignore changes in the gravitational potential energy and need not

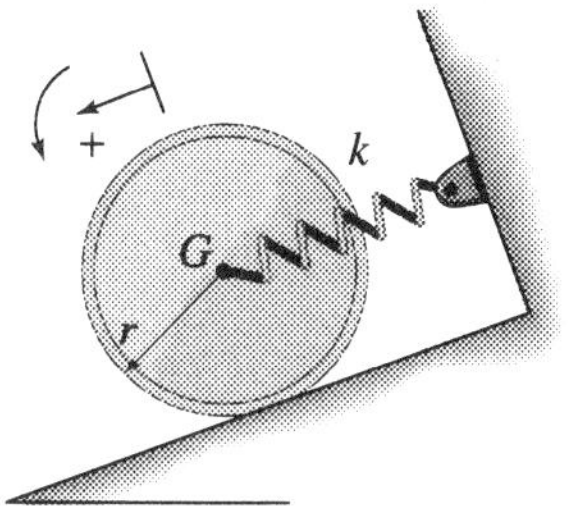

FIGURE E23.8.

consider static deformation of the spring. Thus,

$$T = \tfrac{1}{2}I_G\dot{\theta}^2 + \tfrac{1}{2}mv_G^2 = \tfrac{1}{2}(\tfrac{1}{2}mr^2)\dot{\theta}^2 + \tfrac{1}{2}m(r\dot{\theta})^2 = \tfrac{3}{4}mr^2\dot{\theta}^2,$$

and

$$V = \tfrac{1}{2}kx^2 = \tfrac{1}{2}k(r\theta)^2 = \tfrac{1}{2}kr^2\theta^2.$$

Therefore, the *Lagrangian* becomes

$$l = \tfrac{3}{4}mr^2\dot{\theta}^2 - \tfrac{1}{2}kr^2\theta^2.$$

Taking the appropriate derivatives of the Lagrangian gives

$$\frac{\partial l}{\partial \dot{\theta}} = \frac{3}{2}mr^2\dot{\theta},$$

$$\frac{d}{dt}\left(\frac{\partial l}{\partial \dot{\theta}}\right) = \frac{3}{2}mr^2\ddot{\theta},$$

and

$$\frac{\partial l}{\partial \theta} = -kr^2\theta.$$

Substituting the derivatives in Lagrange's Eq. (23.25b) yields

$$\tfrac{3}{2}mr^2\ddot{\theta} + kr^2\theta = 0.$$

Division by $\frac{3}{2}mr^2$ yields the differential equation of motion. Thus,

$$\ddot{\theta} + \left(\frac{2}{3}\frac{k}{m}\right)\theta = 0. \qquad \text{ANS.}$$

Because $p = \sqrt{\dfrac{2k}{3m}}$, the period is determined as

$$\tau = \frac{2\pi}{p} = 2\pi\sqrt{\frac{3m}{2k}}. \qquad \text{ANS.}$$

■

Problems

23.83 Use Lagrange's method to write the governing differential equation of motion for the spring supported mass shown in Figure P23.83. Measure x from the equilibrium position.

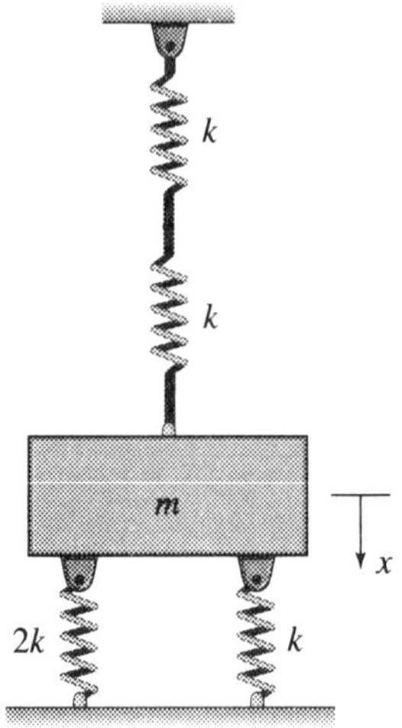

FIGURE P23.83.

23.84 Two uniform slender rods are welded together to form the body shown in Figure P23.84. Consider small oscillations of this rigid body about an axis through O. Use the Lagrangian to write the differential equation of motion. Express the period of this motion as a function of g and L.

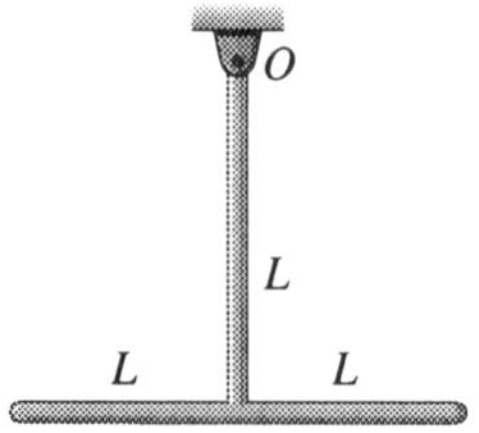

FIGURE P23.84.

23.85 A thin plate is pinned at O as shown in Figure P23.85. Write the Lagrangian for this plate and derive the differential equation of motion. Consider small oscillations, and determine the solution for θ as a function of time if the plate is released from rest when $\theta = 0.4$ rad cw.

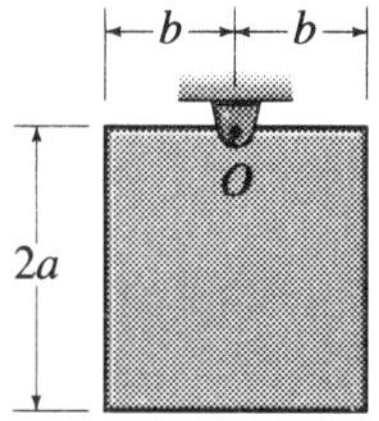

FIGURE P23.85.

23.86 A disk of mass $36m$ is welded to a slender rod of mass $12m$, as shown in Figure P23.86. Use Lagrange's method to write the differential equation of motion for this body for rotation about O. Express the frequency f for small oscillations as a function of g and L.

23.87 If the block shown in Figure P23.87 is displaced to the right or left and released

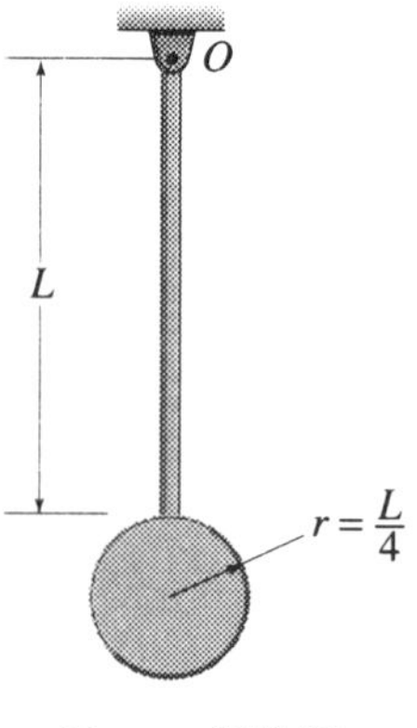

FIGURE P23.86.

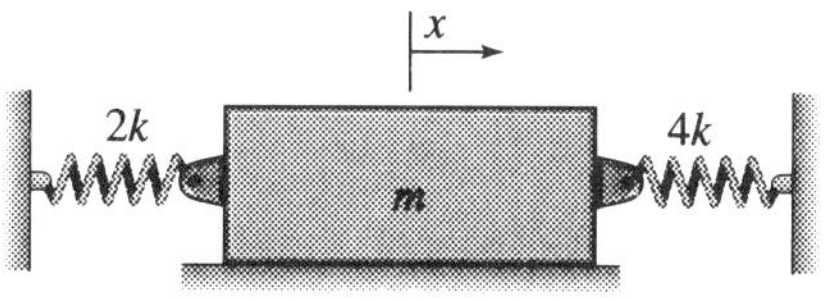

FIGURE P23.87.

it will oscillate. Assume that the horizontal plane on which the block slides is frictionless and that the springs are initially unstressed. Use the Lagrangian to write the governing differential equation of motion. Express the circular frequency, the period, and the frequency as functions of k and m.

23.88 Two slender rods each of mass $12m$ and length $12L$ are welded tangent to a disk of mass $40m$ and radius $5L$ as shown in Figure P23.88. Use Lagrange's method to write the differential equation of motion for rotation of this body about O. Consider small oscillations, and express θ as a function of time if the body is released from rest after being rotated 0.2 rad ccw.

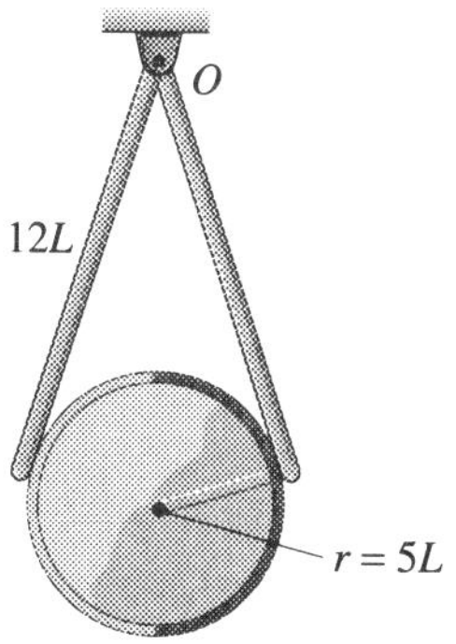

FIGURE P23.88.

23.89 Four slender rods, each of mass m, are joined to form a square as shown in Figure P23.89. Use the Lagrangian to determine the period and frequency of small oscillations of this body in rotation about an axis through point O.

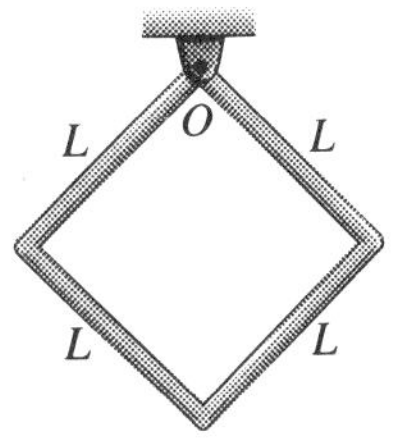

FIGURE P23.89.

23.90 A cylinder of mass m and radius r is attached to a spring of constant k, as shown in Figure P23.90. When displaced from its equilibrium position and released, it will vibrate. Use Lagrange's method to write the governing differential equation of motion for this cylinder which rolls without sliding on the inclined plane. Express its period of oscillatory motion as a function of k and m.

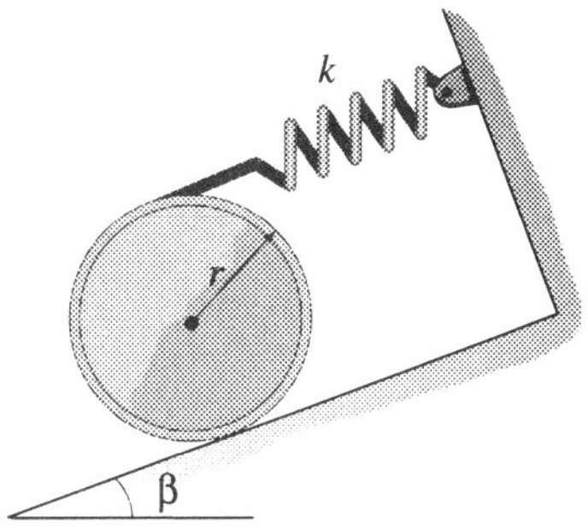

FIGURE P23.90.

23.91 In Figure P23.91 a slender rod of mass m and length $12b$ is attached at its left end A to a spring of constant k and to a frictionless pin at point O. When the rod is horizontal, it is in static equilibrium. Use the Lagrangian to determine the differential equation of motion, and express the period of small oscillations as a function of k and m.

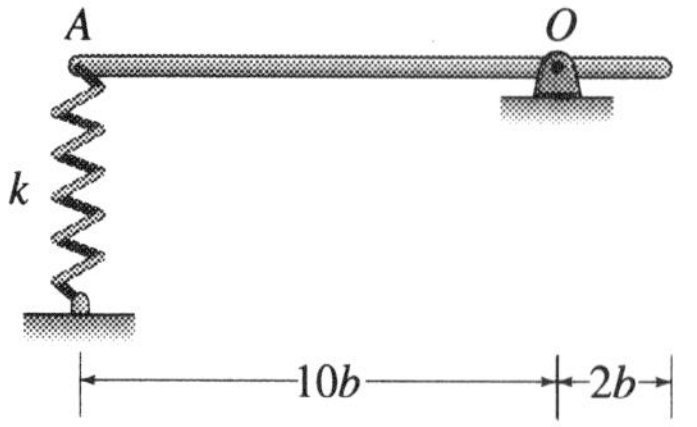

FIGURE P23.91.

23.92 The plate of constant thickness and mass density, shown in Figure P23.92, executes small oscillations about an axis through O. Use Lagrange's method to write the differential equation of motion. Express the circular frequency, period, and frequency as functions of g and b.

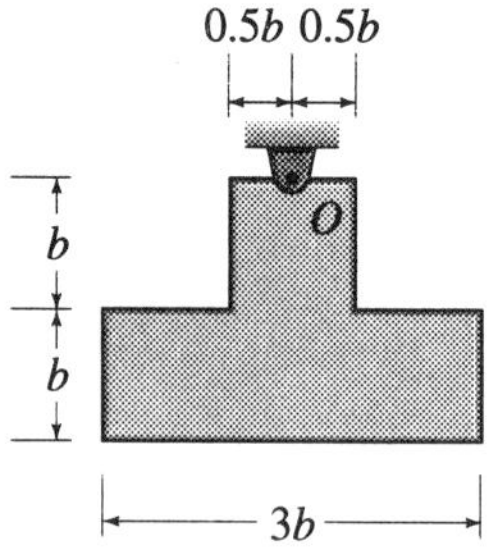

FIGURE P23.92.

23.93 A plate of constant thickness and mass density, supported at a pin O, has a hole cut from it, as shown in Figure P23.93. The plate has a mass of $12m$ before the hole is cut from it. Use Lagrange's method to write the differential equation of motion for this plate. Determine the period of this vibration for small oscillations if $b = 0.20$ m.

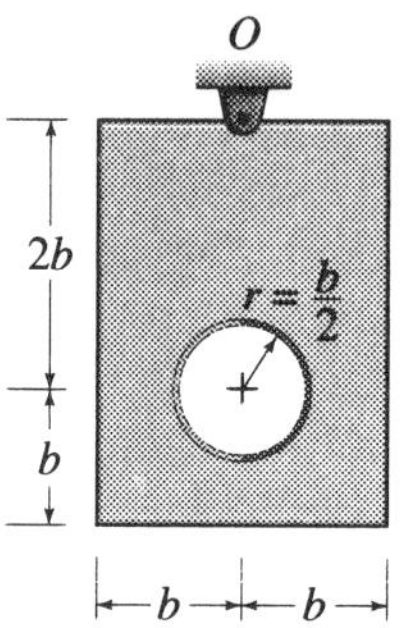

FIGURE P23.93.

23.94 Solve Problem 23.93 for $b = 6$ in. Also, find the frequency f and express θ as a function of time if this plate is rotated 0.1 rad ccw and released from rest.

23.5 Forced Vibrations—Force and Acceleration

If displaced from the equilibrium position and released from rest, spring-supported masses and pendulums will vibrate naturally. These oscillations are defined by natural characteristics of the system, such as spring constants, masses, and pendulum lengths. In contrast with these natural or free vibrations, we now consider *forced vibrations*. They are termed forced vibrations because a time-dependent force is applied externally to the system. This applied force is termed a *forcing function*.

In Figure 23.10(a), the spring supported mass is in equilibrium and as before,

$$\sum F_x = 0, \quad mg - k\delta_{ST} = 0.$$

This equilibrium position will be used as a reference from which the position $x = f(t)$ will be measured. In Figure 23.10(b), the system is

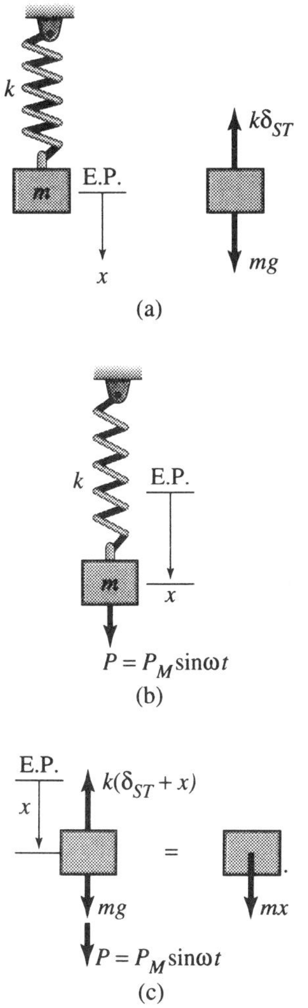

FIGURE 23.10.

subjected to a forcing function $P = P_{\mathrm{M}} \sin \omega t$. This function varies sinusoidally with time, has an amplitude P_{M} which is the maximum force, and a *forcing frequency* ω. It varies from a maximum tension of P_{M} to a maximum compression of equal magnitude P_{M}.

Refer to the free-body and inertial force diagrams of Figure 23.10(c) and write

$$\sum F_x = ma_x, \quad -k(\delta_{\mathrm{ST}} + x) + mg + P_{\mathrm{M}} \sin \omega t = m\ddot{x}.$$

Recall that $mg - k\delta_{\mathrm{ST}} = 0$, and transpose to obtain

$$m\ddot{x} + kx = P_{\mathrm{M}} \sin \omega t.$$

Dividing through by m and recalling that $k/m = p^2$ gives

$$\ddot{x} + p^2 x = \left(\frac{P_M}{m}\right) \sin \omega t. \tag{23.26}$$

This is the governing nonhomogeneous differential equation for forced vibrations of the one-degree-of-freedom system. Its general solution x consists of a linear combination of the *homogeneous* solution x_H and the *particular* solution x_P. Thus,

$$x = x_H + x_P.$$

The homogeneous solution is also referred to as the *complementary solution.* If we equate the forcing function to zero, then, the differential equation becomes

$$\ddot{x} + p^2 x = 0.$$

As for free vibrations, the solution termed the *homogeneous solution* is given by

$$x_H = A \sin pt + B \cos pt.$$

The particular solution must satisfy the entire differential equation and, because the term on the right side is sinusoidal, we assume a particular solution of the form

$$x_P = x_M \sin \omega t.$$

The general solution may, then, be written as

$$x = A \sin pt + B \cos pt + x_M \sin \omega t.$$

Thus, the displacement function consists of two components. The first component (the homogeneous or complementary component given by $x_H = A \sin pt + B \cos pt$) is very quickly eliminated by frictional and other damping forces in the system and is, therefore, known as the *transient* component. The effects of damping will be discussed in Sections 23.6 and 23.7. The second component (the particular component given by $x_P = x_M \sin \omega t$) is the only component that remains after the transient component is damped out and is referred to as the *steady-state* component. In other words, we have solved the forced vibration problem once we find a particular solution x_P which satisfies the entire differential equation. We differentiate this particular solution twice to obtain

$$\ddot{x}_P = -\omega^2 x_M \sin \omega t.$$

Substituting in the entire differential equation (Eq. (23.26)) gives

$$-\omega^2 x_M \sin \omega t + p^2 x_M \sin \omega t = \frac{P_M}{m} \sin \omega t.$$

Dividing by ωt, because, in general, it is not zero, gives

$$-\omega^2 x_M + p^2 x_M = \frac{P_M}{m}.$$

Solving for x_M yields

$$x_M = \frac{\left(\dfrac{P_M}{m}\right)}{p^2 - \omega^2}.$$

Dividing the numerator and denominator on the right by p^2,

$$x_M = \frac{\left(\dfrac{P_M}{mp^2}\right)}{1 - \left(\dfrac{\omega}{p}\right)^2}.$$

Recall that $p^2 = k/m$ or $mp^2 = k$, and divide by P_M/k to obtain

$$\frac{x_M}{(P_M/k)} = \frac{x_M}{(\delta_M)} = \frac{1}{1 - \left(\dfrac{\omega}{p}\right)^2}. \tag{23.27}$$

The dimensionless ratio given by Eq. (23.27) is known as the *amplification factor* and is a function of the dimensionless ratio of the circular frequency ω of the forcing function to the circular natural frequency p. This function is plotted in Figure 23.11. The ordinates are values of the amplification factor which may be interpreted as the ratio of the maximum dynamic displacement x_M to the static maximum displacement δ_M. Because $P_M/k = \delta_M$ is a displacement calculated from the amplitude of the forcing function P_M, as though it were applied statically, we refer to it as the static maximum displacement. If $\omega = p$ or $\omega/p = 1$, then, the amplification factor becomes infinite, and this condition is termed *resonance*. In other words, if the forcing function frequency ω equals

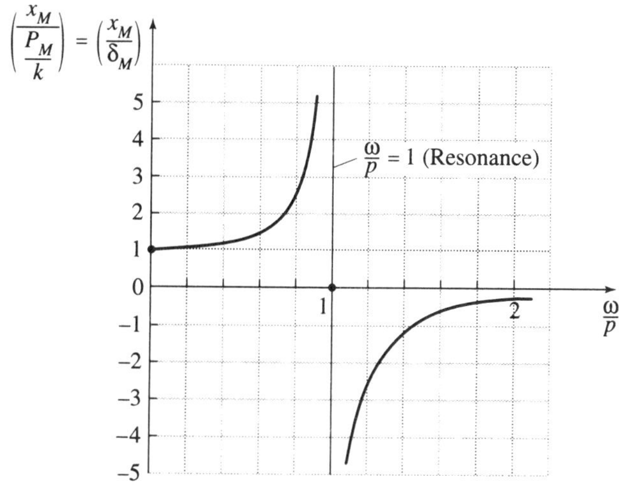

FIGURE 23.11.

the natural circular frequency p, then, the maximum dynamic displacement x_M becomes infinite. Of course, no real system can withstand infinite displacements, but knowledge of the resonance condition is valuable. In many cases, we may select system properties so that p will not equal expected input ω values to avoid resonance. In other cases, it may be desirable to have p approach the value of an expected input ω to usefully employ the large resulting dynamic displacements associated with near resonance conditions.

From Figure 23.11, we note that, for circular frequency ratios ω/p between zero and one, the amplification factor x_M/δ_M is positive, which means that the forcing function is *in phase* with the natural vibration of the system. For ω/p greater than one, the amplification factor is negative, which means that the forcing function is 180° *out of phase* with the natural vibration of the system.

As an example of forced vibration, consider the reciprocating mechanism shown in Figure 23.12(a) which illustrates the effect of support movement on the vibration of a spring-mass system. As the crank OB of length δ_M rotates about the hinge at O, it forces the piston P to oscillate horizontally within the smooth guides. One end of a spring with spring constant k is attached to the piston and the other end to a mass m that can slide freely on the smooth horizontal plane as shown.

The horizontal movement of the spring support P, given by $\delta_M \sin \omega t$, results in the forced vibration of the mass m. This movement is to the

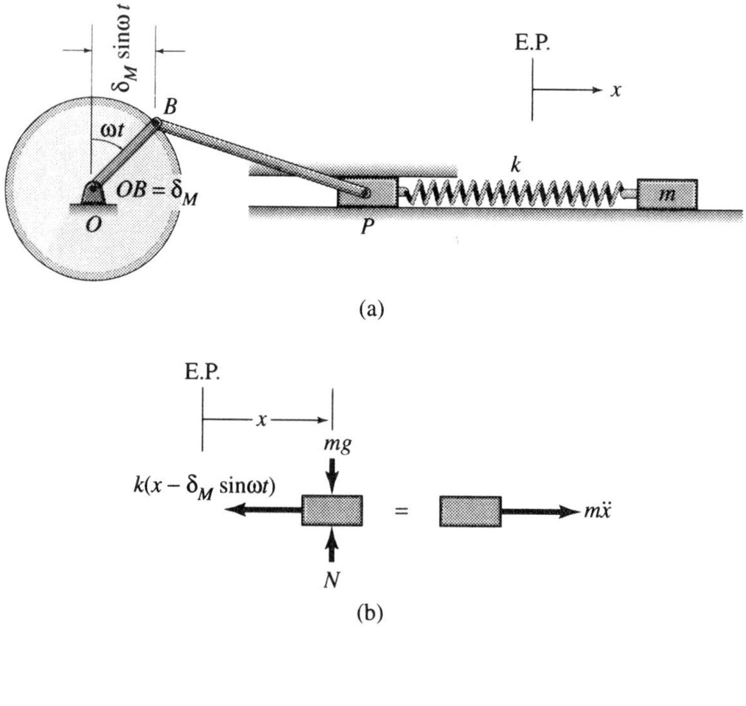

FIGURE 23.12.

right in the positive sense and subtracts from the displacement x of the mass m.

Refer to the free-body and inertial force diagrams shown in Figure 23.12(b), and apply Newton's equation of motion. Thus,

$$\sum F_x = ma_x, \quad -k(x - \delta_M \sin \omega t) = m\ddot{x}$$

from which

$$m\ddot{x} + kx = k\delta_M \sin \omega t.$$

Dividing by m and recalling that $k/m = p^2$ gives

$$\ddot{x} + p^2 x = \frac{k\delta_M}{m} \sin \omega t. \tag{23.28}$$

This is the governing differential equation of motion for forced vibrations due to support movement of this one-degree-of-freedom system.

If Eqs. (23.26) and (23.28) are compared, we note that they are identical if we let $P_M = k\delta_M$. Once again, the homogeneous or transient solution will be damped out with time, and only the particular or steady-state solution is of interest. If we assume a particular solution of the form $x_P = x_M \sin \omega t$, then, the amplification factor will be given by Eq. (23.27) derived earlier and plotted in Figure 23.11.

The following examples illustrate some of the above concepts.

■ Example 23.9

A spring-supported mass m is shown in Figure E23.9. The springs are symmetrically placed with respect to a vertical axis along which the forcing function $P = P_M \sin \omega t$ acts. (a) Determine an expression for the equivalent spring constant k_e in terms of k_1 and k_2, (b) let $k_1 = 80$ lb/ft, $k_2 = 120$ lb/ft, $m = 4$ slug, and $P = 50 \sin 20t$ (lb), then find the static maximum displacement, the amplification factor, and the maximum dynamic displacement.

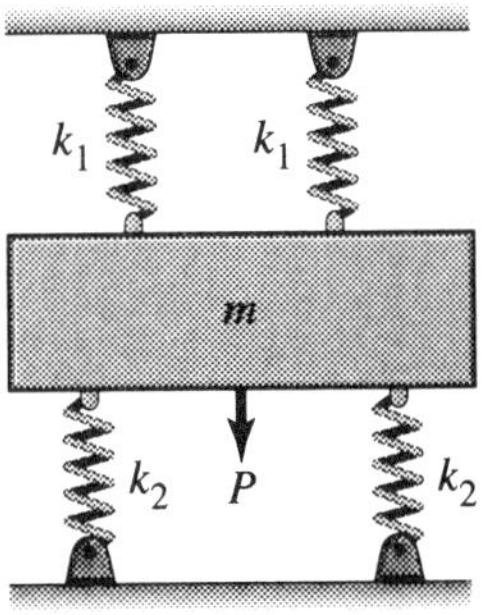

FIGURE E23.9.

Solution

(a) Because both the top and bottom springs are in parallel, Eq. (23.18) may be used to find k_e. Thus,

$$k_e = \sum k = 2k_1 + 2k_2 \qquad \text{ANS.}$$

(b) Use the given data to compute a numerical value for k_e. Thus,

$$k_e = 2(80) + 2(120) = 400 \text{ lb/ft}.$$

The maximum static displacement is given by

$$\delta_M = \frac{P_M}{k_e} = \frac{50}{400} = 0.125 \text{ ft} = 1.500 \text{ in.} \qquad \text{ANS.}$$

The natural circular frequency p is given by

$$p = \sqrt{\frac{k_e}{m}} = \sqrt{\frac{400}{4}} = 10 \text{ rad/s}.$$

From the given forcing function $P = 50 \sin 20t$, we see that the forcing circular frequency $\omega = 20$ rad/s and the ratio $\omega/p = 20/10 = 2$. The amplification factor is given by Eq. (23.27). Thus,

$$\frac{x_M}{\delta_M} = \frac{1}{1 - \left(\dfrac{\omega}{p}\right)^2} = \frac{1}{1 - 2^2} = -0.333. \qquad \text{ANS.}$$

The maximum dynamic displacement is given by

$$x_M = -0.333\delta_M = -0.500 \text{ in.} \qquad \text{ANS.}$$

■ Example 23.10

The roadway over which the camper of Figure E23.10 is being pulled is assumed to be sinusoidal with a wave length of 25 ft and an amplitude of 0.40 in. The camper and contents weigh 850 lb and its equivalent spring constant is 1600 lb/ft. Find (a) the maximum dynamic displacement when the camper is being pulled at a constant speed of 55 mph and (b) the constant speed of the camper corresponding to resonance.

Solution

(a) This is the case of a forced displacement function, and we can imagine the camper held fixed while a *demon* pulls the roadway to the

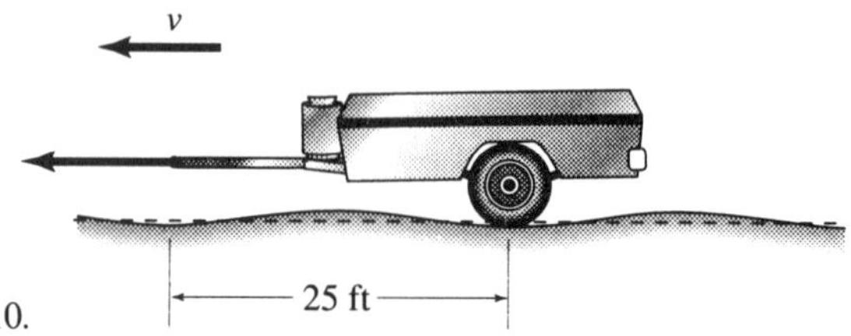

FIGURE E23.10.

right at constant speed. Therefore, since the camper moves at constant speed we have

$$s = vt$$

For $s = 25$ ft, $v = 55$ mph $= 80.7$ ft/s and $t = \tau$ for one period,

$$25 = 80.7\tau,$$

and

$$\tau = 0.310 \text{ s}.$$

Because $\omega\tau = 2\pi$ and $\tau = 0.310$ s, we conclude that $\omega = 20.3$ rad/s. Also, the natural circular frequency is given as

$$p = \sqrt{\frac{k}{m}} = \sqrt{\frac{1600}{850/32.2}} = 7.79 \text{ rad/s}.$$

The ratio of forcing frequency to natural circular frequency is given by

$$\frac{\omega}{p} = \frac{20.3}{7.79} = 2.61.$$

Using Eq. (23.27),

$$\frac{x_M}{\delta_M} = \frac{1}{1 - \left(\frac{\omega}{p}\right)^2} = \frac{1}{1 - (2.61)^2} = -0.172.$$

The maximum dynamic displacement becomes

$$x_M = -0.172\delta_M = -0.172(0.4) = -0.0688 \text{ in.} \qquad \text{ANS.}$$

(b) For resonance,

$$\omega = p = 7.79 \text{ rad/s}.$$

Thus, because $\omega t = 2\pi$, we conclude that

$$\tau = \frac{2\pi}{\omega} = \frac{2\pi}{7.79} = 0.807 \text{ s}.$$

For constant speed of the camper, $s = vt$ from which

$$v = \frac{s}{t} = \frac{s}{\tau} = \frac{25}{0.807} = 30.979 \text{ ft/s}$$

$$= 21.1 \text{ mph}. \qquad \text{ANS.}$$

■

Problems

23.95 A-spring supported mass is subjected to a sinusoidal forcing function so that the ratio of the maximum dynamic deflection to P_M/k equals (a) 4 and (b) -4. Determine the ratio of ω/p for each of these cases. Sketch the amplification factor versus ω/p and show these two points on your sketch.

23.96 Determine the amplification factor for each of the following ω/p ratios: 0.9, 0.999, 1.1, 1.01, and 1.001. If $\delta_M = P_M/k = 1.00$ in., calculate the corresponding maximum dynamic displacements for each of these (ω/p) ratios. Explain why some of these displacements are physically unrealistic.

23.97 A spring-supported mass is subjected to the forcing function $P = 50 \sin 4t$ where P is given in lb and t in s. If $m = 6$ slug and $k = 1350$ lb/ft, determine (a) the amplification factor and (b) the maximum dynamic displacement.

23.98 Determine the spring constant k given the following information about a spring-supported mass. The spring support moves according to $\delta = 0.8 \sin 2t$ where δ is expressed in in. and t in s. The body weighs 100 lb and the maximum dynamic displacement is to be 1.60 in.

23.99 Refer to the frictionless system of Figure P23.99, and let $k_1 = 200$ lb/in. and $k_2 = 400$ lb/in., $m = 2$ slug, and $P_M = 300$ lb. (a) Determine ω corresponding to resonance. (b) If $\omega = 6$ rad/s, determine the maximum dynamic displacement.

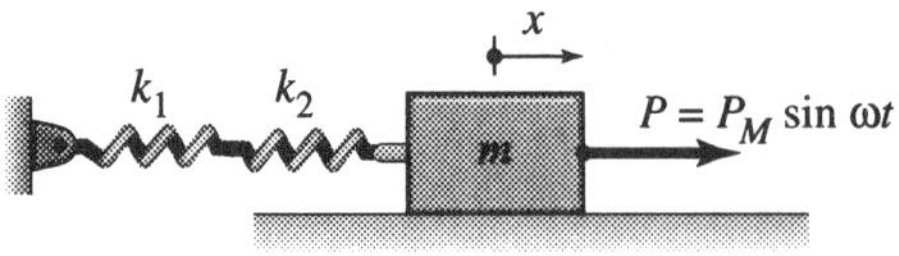

FIGURE P23.99.

23.100 The mass $m = 20$ kg, shown in Figure P23.99, is subjected to a forcing function so that $P_M = 1600$ N and $\omega = 10$ rad/s. In this case $k_1 = k_2 = 30{,}000$ N/m. For this frictionless system, determine (a) the natural circular frequency of vibration, (b) the amplification factor, and (c) the maximum dynamic displacement.

23.101 (a) Determine the equivalent spring constant for the system shown in Figure P23.101 given that $k_1 = 3$ kN/m, $k_2 = 4$ kN/m and $k_3 = 6$ kN/m. (b) For $P_M = 2$ kN, $\omega = 5$ rad/s, and $m = 12$ kg, determine the maximum dynamic displacement.

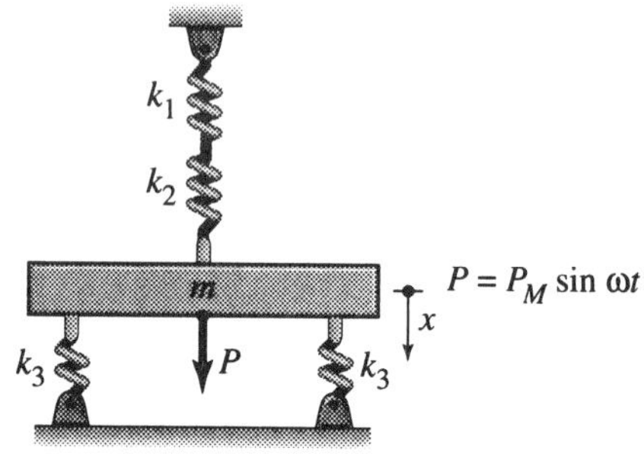

FIGURE P23.101.

23.102 Refer to Figure P23.101, and assign values as follows: $P_M = 50$ lb, $\omega = 2$ rad/s, $m = 0.80$ slug, $k_1 = 100$ lb/ft, $k_2 = 120$ lb/ft, and $k_3 = 80$ lb/ft. (a) What is the natural circular frequency of vibration of this system? (b) What is the maximum dynamic displacement?

23.103 Derive the governing differential equation for the frictionless system shown in Figure P23.103. If $\omega = 6$ rad/s and

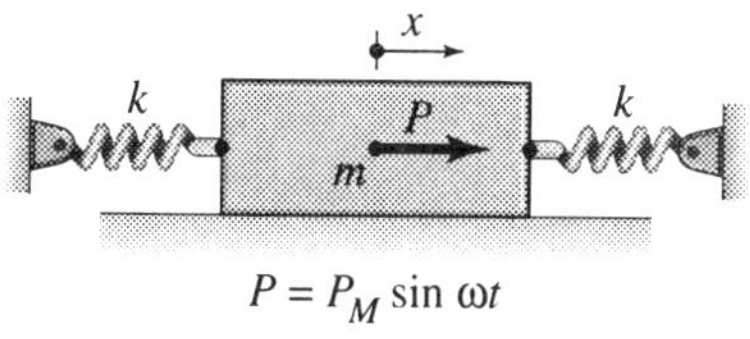

FIGURE P23.103.

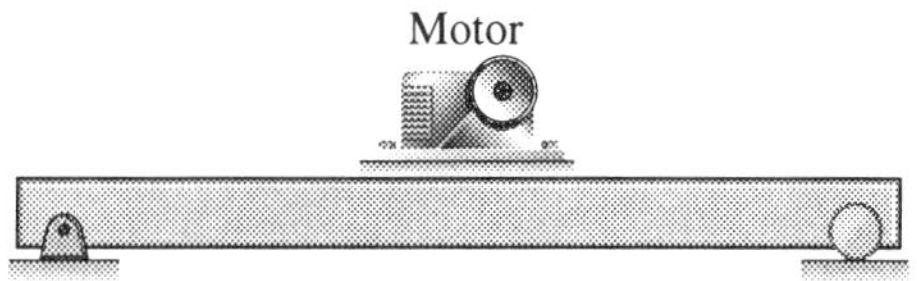

FIGURE P23.107.

$m = 4$ slug determine k corresponding to the resonance of the system.

23.104 Refer to Figure P23.103, and assign data as follows: $m = 1$ slug, $k = 80$ lb/ft, $P_M = 40$ lb, and $\omega = 8$ rad/s. Find (a) the natural circular frequency of the system and (b) the maximum dynamic displacement.

23.105 In Figure P23.105 the bars AB and CD are rigid but of negligible mass. (a) Express the equivalent spring constant in terms of k. (b) Let $k = 8$ kN/m, $m = 10$ kg, $P_M = 4$ kN, and $\omega = 5$ rad/s. Determine the natural circular frequency and the maximum dynamic displacement.

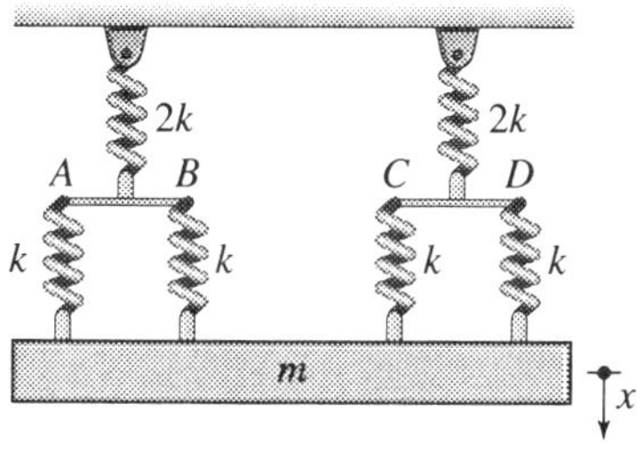

FIGURE P23.105.

23.106 Refer to Figure P23.105, and derive the governing differential equation of motion for this system if the forcing function is $P = P_M \sin \omega t$.

23.107 An electric motor is mounted on a beam as depicted in Figure P23.107. Flooring attached to the beam restricts the motion to the vertical direction. The axis of rotation of the motor is parallel to the longitudinal axis of the beam. The forcing function arises due to a small imperfection in the rotor. A small mass m at a distance r from the motor axis represents this imperfection. A centrifugal force $mr\omega^2$, where ω is the angular velocity of the motor, yields a vertical forcing function

$$P = mr\omega^2 \sin \omega t.$$

To limit the maximum dynamic displacement to a very small value, a choice is to be made between two suggested values of the equivalent spring constant of the beam. These values are 8500 lb/in and 10,000 lb/in. Which one should be chosen? Justify your answer with calculations. Remaining inputs: Motor weight plus an allowance for beam weight is 1200 lb. Angular velocity of the motor is 500 rpm. Weight corresponding to the small mass m is 0.3 lb and $r = 0.75$ in.

23.108 In Figure P23.108 is shown a mathematical model for a small trailer moving at a constant velocity of 50 mph along a roadway. The roadway is assumed to be represented by a sinusoidal function given by $\delta = 0.8 \sin \frac{\pi x}{40}$ where δ and x are expressed in in. If the trailer weighs 440 lb and $k = 1200$ lb/ft, determine its maximum dynamic displacement.

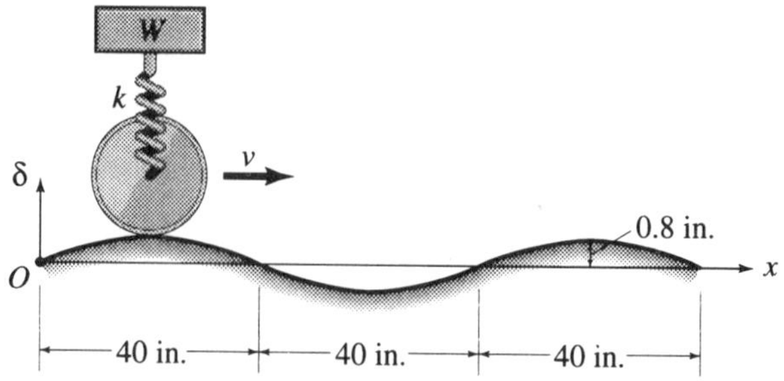

FIGURE P23.108.

23.109 Refer to Problem 23.108 and determine a new value of the spring constant k which would correspond to resonance.

23.110 Solve Problem 23.108 for the following data. Constant velocity = 60 km/hr. The roadway surface is given by $\delta = 0.015 \sin(\pi x/40)$ where δ and x are expressed in meter. Trailer weight = 2 kN, and spring constant $k = 20$ kN/m.

23.111 A spring-supported mass is subjected to a forcing function $P = P_M \cos \omega t$. (a) Derive the differential equation of motion. (b) Determine the particular solution of this differential equation. (Assume that displacements are measured from the equilibrium position.)

23.112 A mathematical model for a machine weighing 5 kN and rotating at 1750 rpm is shown in Figure P23.112. Four springs support the machine for which each $k = 30$ kN/m. The maximum centrifugal force is 1 kN for the unbalanced rotor. Determine the amplification factor and the maximum dynamic displacement of this machine. Note that the machine is prevented from moving horizontally, and, as in problem 23.107, the forcing function is $P = mr\omega^2 \sin \omega t$.

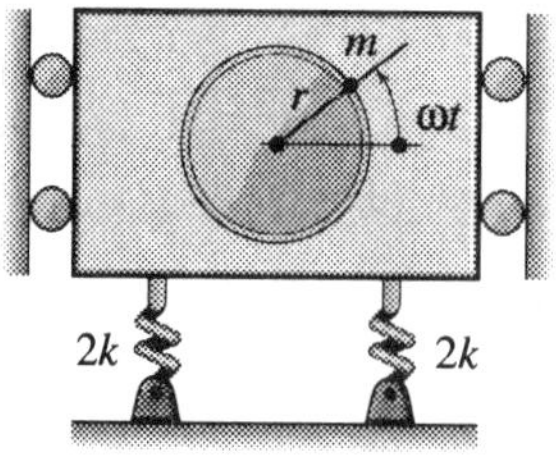

FIGURE P23.112.

23.113 Refer to Problem 23.112, and determine a new value of k for each spring which would correspond to resonance.

23.114 Determine the equivalent spring constant for the mathematical model shown in Figure P23.114. Then, derive the differential equation of motion, and find the steady-state solution of this equation.

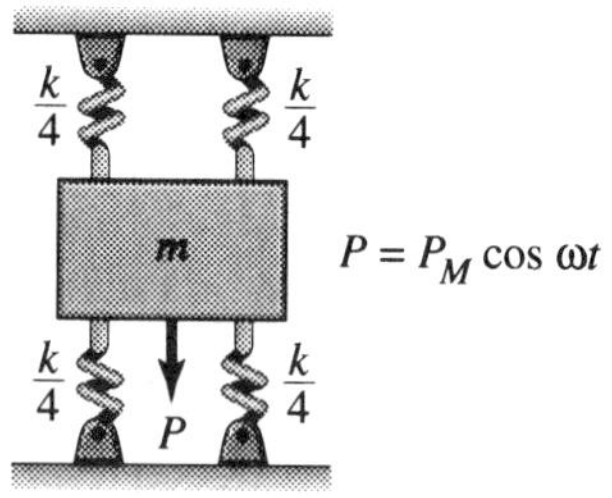

FIGURE P23.114.

23.115 The shaft shown in Figure P23.115 rotates with a constant angular velocity

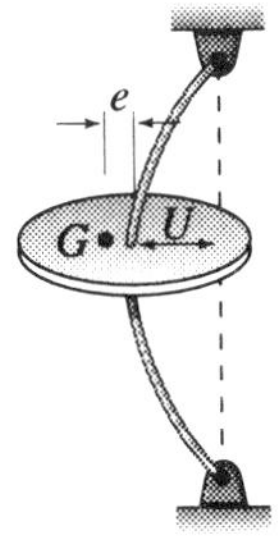

FIGURE P23.115.

ω about the vertical axis. At the midpoint of this shaft a disk of mass m is attached so that the mass center G of the disk is eccentric a distance e from the shaft center. Show that the lateral shaft deflection U is given by

$$U = \frac{e}{\left(\frac{p}{\omega}\right)^2 - 1}$$

where $p^2 = k/m$ and k is the spring constant for the shaft for lateral deflection.

23.116 Refer to Example 23.10 and solve it for the following data: Wave length of sinusoidal roadway = 10 m, amplitude of roadway = 0.0120 m, weight of camper and contents = 4000 N, spring constant = 24 kN/m, and constant speed of camper = 70 km/hr.

23.117 Refer to Example 23.9, and solve it for k_1 = 1.5 kN/m, k_2 = 3.0 kN/m, m = 35 kg and, $P = 400 \sin 15t$ where P is in N and t is in s.

23.6 Damped Free Vibrations—Force and Acceleration

When a physical system is set into motion by any means, the resulting displacement is reduced with time because of damping. Damping is a physical property of physical systems which reflects the inherent potential of the system to dissipate energy. The coefficient of damping can be readily evaluated when dealing with the shock absorbers of a car, but it is more difficult to assess for a building. All physical systems have damping which may be classified as follows:

1. ***Coulomb or dry friction damping.*** *Surfaces in contact which move relative to each other dissipate mechanical energy in the form of heat.*
2. ***Internal or molecular damping.*** *Materials are imperfectly elastic, and mechanical energy is converted to heat inside bodies due to molecular friction.*
3. ***Viscous or fluid friction damping.*** *Bodies which move in fluids are subjected to resisting forces. All real fluids have viscosity, and they dissipate mechanical energy in the form of heat when bodies move in them.*

Of these three types of damping, we will consider only viscous damping. In Figure 23.13(a), the dashpot has a plunger attached to mass m and the piston or lower end of the plunger moves in a fluid when the system vibrates. Resistance to this motion is proportional to the velocity $\dot{x}$ of the plunger. The damping force F_D which acts to oppose motion is given by $F_D = c\dot{x}$ where c is the *damping coefficient*

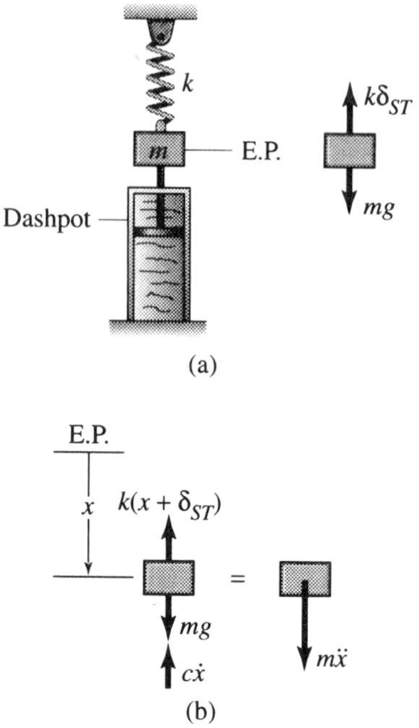

FIGURE 23.13.

which has dimensions of force divided by dimensions of velocity. The units of the damping coefficient c, for SI and for U.S. Customary, are N·s/m and lb·s/ft, respectively.

The mass m is supported by a spring of constant k and is in equilibrium, as shown in Figure 23.13(a). The dashpot plunger attached to the mass is at rest, $\dot{x} = 0$, and, therefore there is no damping force exerted on the mass. Thus,

$$\sum F_x = 0, \quad mg - k\delta_{\text{ST}} = 0.$$

The free-body and inertial force diagrams for any displacement x are shown in Figure 23.13(b). The force of gravity mg acts downward and the spring and damping forces act to oppose the downward positive displacement x and the downward positive velocity $\dot{x}$.

Newton's second law is readily written by reference to Figure 23.13(b). Thus,

$$\sum F_x = m\ddot{x}, \quad -k(x + \delta_{\text{ST}}) - c\dot{x} + mg = m\ddot{x}$$

Recall the equilibrium condition $mg - k\delta_{\text{ST}} = 0$, and transpose all terms to the right side:

$$m\ddot{x} + c\dot{x} + kx = 0. \tag{23.29}$$

This governing differential equation for one-degree-of-freedom, damped, free vibration expresses the fact that the sum of the forces equals zero, that is,

$$(\text{Inertial force}) + (\text{Damping force}) + (\text{Spring force}) = 0$$

It is a second-order, linear, ordinary differential equation with constant coefficients. Let us assume a particular solution of this equation of the form $x = e^{\beta t}$. Differentiation gives $\dot{x} = \beta e^{\beta t}$ and $\ddot{x} = \beta^2 e^{\beta t}$. Substitution in Eq. (23.29) yields

$$m\beta^2 e^{\beta t} + c\beta e^{\beta t} + ke^{\beta t} = 0.$$

Because $e^{\beta t} \neq 0$, we divide by it to obtain

$$m\beta^2 + c\beta + k = 0.$$

This quadratic equation is termed the *characteristic* equation and it has roots given by

$$\beta = -\frac{c}{2m} \pm \sqrt{\left(\frac{c}{2m}\right)^2 - \frac{k}{m}}. \tag{23.30}$$

Three types of damping are identified and discussed in the following paragraphs.

Critical damping is defined by the value of $c = c_c$ which will make the radical in Eq. (23.30) vanish. Thus,

$$\left(\frac{c_c}{2m}\right)^2 - \frac{k}{m} = 0$$

from which

$$c_c = \sqrt{\frac{4m^2k}{m}} = 2m\sqrt{\frac{k}{m}} = 2mp. \tag{23.31}$$

Because the radical vanishes for critical damping, the roots of the characteristic equation are negative, real, repeated and given by $\beta = \frac{c_c}{2m}$. The differential equation has a general solution given by

$$x = (A + Bt)e^{-pt}. \tag{23.32}$$

This solution is not oscillatory and is shown qualitatively in Figure 23.14. It is of interest in the design of recoil mechanisms in the sense

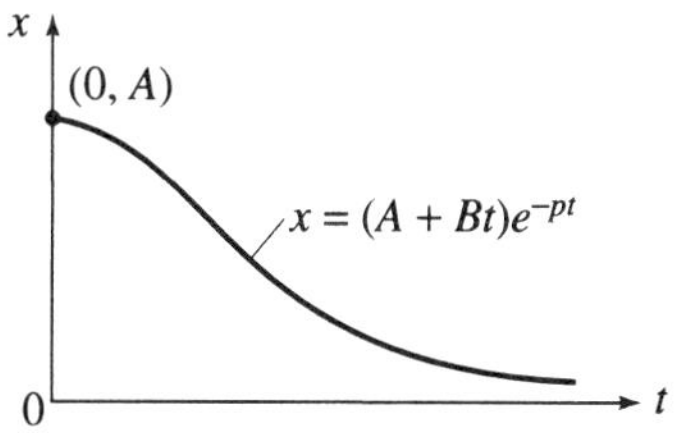

FIGURE 23.14.

that, if a system is critically damped and released from rest, then, it will return to equilibrium in the smallest possible time.

Consider the case of *heavy* damping for which the damping coefficient is greater than the critical value $c > c_c$. In this case, the two roots of the characteristic equation are negative, real and unequal. The general solution of the differential equation is given by

$$x = Ae^{\beta_1 t} + Be^{\beta_2 t} \tag{23.33}$$

where β_1 and β_2 are the roots of the characteristic equation. This solution is nonoscillatory and only of minor interest to engineers who study vibrations.

The case of *light* damping is of major interest. The coefficient of damping is less than its critical value or $c < c_c$ where $c_c = 2mp$. To consider this case, return to the solutions of the characteristic equation given by Eq. (23.30), and define the negative of the quantity under the radical as

$$q^2 = \frac{k}{m} - \left(\frac{c}{2m}\right)^2. \tag{23.34a}$$

Equation (23.34a) may also be written in dimensionless form as

$$q = p\sqrt{1 - (c/c_c)^2} \tag{23.34b}$$

where p was defined earlier and is the undamped circular natural frequency whereas q, defined by Eq. (23.34b), is known as the damped circular frequency. It is left as an exercise for the student (see Problem 23.118) to derive Eq. (23.34b) from Eq. (23.34a). Then, the roots of the characteristic equation become

$$\beta = -\frac{c}{2m} \pm qi \quad \text{where } i^2 = -1.$$

These roots are a complex conjugate pair, and the general solution of the governing differential equation is in the form

$$x = Ae^{(-c/2m+qi)t} + Be^{(-c/2m-qi)t} \tag{23.35}$$

Factoring $e^{-(c/2m)t}$ gives

$$x = e^{(-c/2m)t}(Ae^{qit} + Be^{-qit}) \tag{23.36}$$

where A and B are constants which may be determined from the initial conditions. An identity discovered by Euler will enable us to show that this solution is oscillatory. With θ as an argument, this identity is

$$e^{i\theta} = \cos\theta + i\sin\theta.$$

Substituting these quantities in Eq. (23.36) gives

$$x = e^{(-c/2m)t}[A(\cos qt + i\sin qt) + B(\cos qt - i\sin qt)].$$

Collecting terms,

$$x = e^{(-c/2m)t}[(A - B)i \sin qt + (A + B)\cos qt].$$

Both $(A - B)i$ and $(A + B)$ are real constants, say C_1 and C_2, because the resulting displacement x is real. This enables us to write the general solution for lightly damped free vibrations for a one-degree-of-freedom system as

$$x = e^{(-c/2m)t}[C_1 \sin qt + C_2 \cos qt] \tag{23.37}$$

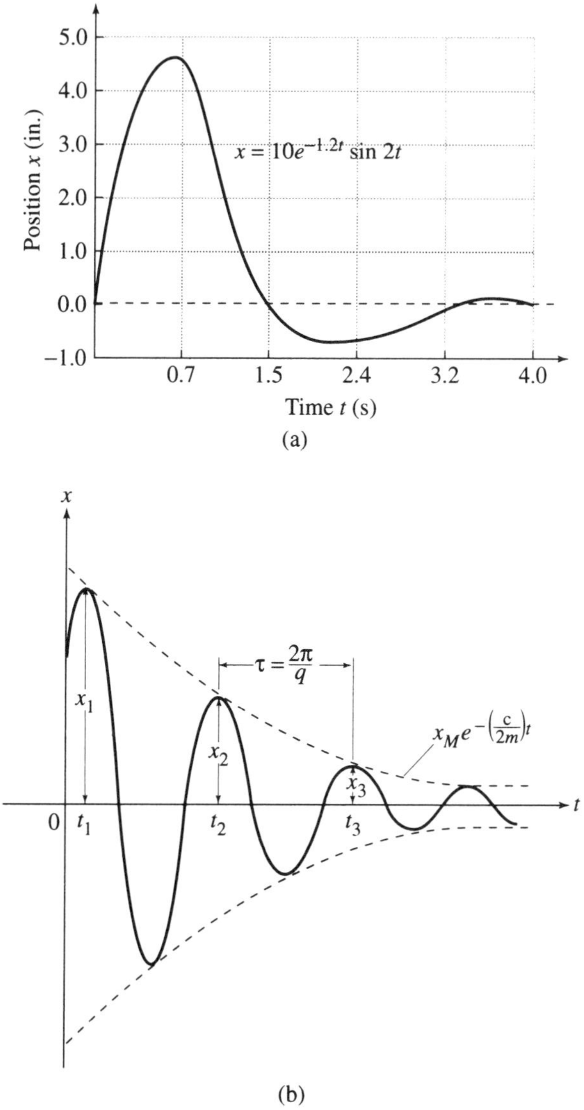

FIGURE 23.15.

where the constants C_1 and C_2 depend on the initial conditions of the motion.

Equation (23.37) represents an oscillatory motion which decreases in amplitude with time. Consider the special case where $C_2 = 0$ and write

$$x = C_1 e^{(-c/2m)t} \sin qt. \tag{23.38}$$

A typical, damped, vibratory motion curve of x versus t is shown in Figure 23.15(a). This is a plot of Eq. (23.38) for $C_1 = 10$, $q = 2$, and $\dfrac{c}{2m} = 1.2$.

The general solution expressed in Eq. (23.37) may also be written in terms of amplitude x_M and a phase angle ϕ. Thus,

$$x = x_M e^{(-c/2m)t} \sin(qt + \phi) \tag{23.39}$$

where x_M and ϕ depend on the initial conditions of the motion. Clearly Eq. (23.39) represents a damped oscillatory motion as indicated schematically in Figure 23.15(b). Note that, because of the damping, the amplitudes of vibration (i.e., x_1, x_2, etc.) decrease progressively with time. However, the period τ for damped vibrations remains constant.

The following example illustrates some of the concepts discussed above.

Example 23.11

A spring-supported body is connected to a damper as shown in Figure E23.11. The body weighs 16.1 lb, the spring constant $k = 600$ lb/ft, and two cases of damping are to be considered: (a) critical damping c_c and (b) $c = 20$ lb·s/ft. In each case, the initial conditions are given by $t = 0$, $x = 0.1$ ft, and $\dot{x} = 5$ ft/s. For each value of the damping coefficient, express x as a function of time, and describe the motion.

Solution

(a) The critical damping coefficient is given by

$$p = \sqrt{\frac{k}{m}} = \sqrt{\frac{500}{\left(\dfrac{16.1}{32.2}\right)}} = 34.64 \text{ Hz},$$

$$c_c = 2mp = 2\left(\frac{16.1}{32.2}\right)(34.64) = 34.64 \text{ lb·s/ft}.$$

The general solution for critically damped motion is given by Eq. (23.32) as

$$x = (A + Bt)e^{-pt}.$$

Use the initial conditions to determine the constants A and B. Thus, substituting $t = 0$ and $x = 0.1$ ft into this equation for x yields

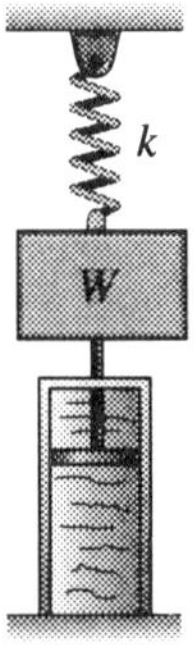

FIGURE E23.11.

$$0.1 = (A + 0)e^0 \Rightarrow A = 0.10.$$

Differentiate Eq. (23.31) to express $\dot{x}$ as a function of time. Thus,

$$\dot{x} = -Ape^{-pt} + B(e^{-pt} - pte^{-pt}).$$

Substitute $t = 0$, $\dot{x} = 5$ ft/s, and the found values of A and p into this equation to obtain B. Thus,

$$5 = -(0.1)(34.64) + B(1 - 0) \Rightarrow B = 8.46.$$

Therefore,

$$x = (0.10 + 8.46t)e^{-34.64t} \text{ ft.} \qquad \text{ANS.}$$

This damped motion is nonoscillatory. The displaced mass returns to its equilibrium position in minimum time.

(b) When $c = 20$ lb·s/ft, $c < c_c$, and the motion is lightly damped, the general solution for x is given by Eq. (23.37) which states that

$$x = e^{-(c/2m)t}[C_1 \sin qt + C_2 \cos qt]$$

where, by Eq. (23.34),

$$q = \sqrt{\frac{k}{m} - \left(\frac{c}{2m}\right)^2} = \sqrt{\frac{600}{(16.1/32.2)} - \left(\frac{20}{2(16.1)/32.2}\right)^2} = 28.28.$$

Use the initial conditions to determine the constants C_1 and C_2. Substitute $t = 0$ and $x = 0.1$ ft in Eq. (23.37). Thus,

$$0.1 = e^0(C_1 \cdot 0 + C_2 \cdot 1) \Rightarrow C_2 = 0.10.$$

Differentiating Eq. (23.37) to express $\dot{x}$ as a function of time gives

$$\dot{x} = e^{-(c/2m)t}(C_1 q \cos qt - C_2 q \sin qt)$$
$$- \frac{c}{2m} e^{-(c/2m)t}(C_1 \sin qt + C_2 \cos qt).$$

Substituting $t = 0$ and $\dot{x} = 5$ ft/s in this equation yields

$$5 = e^0[C_1 q(1) - C_2 q(0)] - \frac{c}{2m} e^0[C_1(0) + C_2(1)].$$

Substituting $c = 20$, $m = 16.1/32.2 = 0.5$, $q = 28.28$, and $C_2 = 0.10$ permits the determination of C_1. Thus,

$$5 = 28.28C_1 - \frac{20}{2(0.5)}(0.1) \Rightarrow C_1 = 0.248.$$

Substitution in Eq. (23.37) enables us to express x as a function of time. Thus,

$$x = e^{-20t}(0.248 \sin 28.28t + 0.100 \cos 28.28t) \text{ ft.} \qquad \text{ANS.}$$

This lightly damped, free vibration is oscillatory with an amplitude which decreases with time.

23.7 Damped Forced Vibrations—Force and Acceleration

A spring-supported mass attached to a dashpot and subjected to an externally applied forcing function $P = P_M \sin \omega t$ is shown in Figure 23.16(a). Referring to the free-body and inertial force diagrams of Figure 23.16(b),

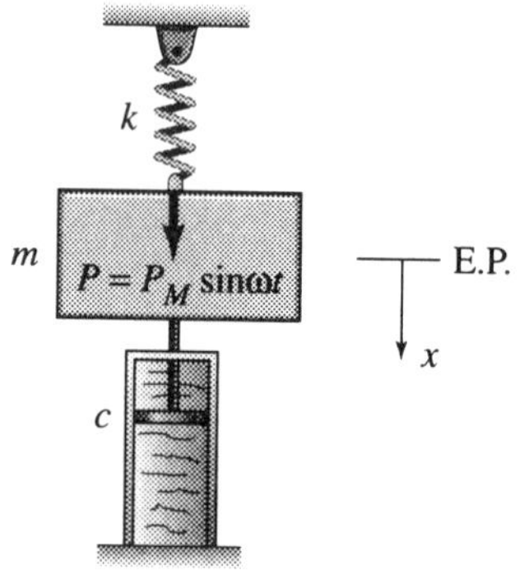

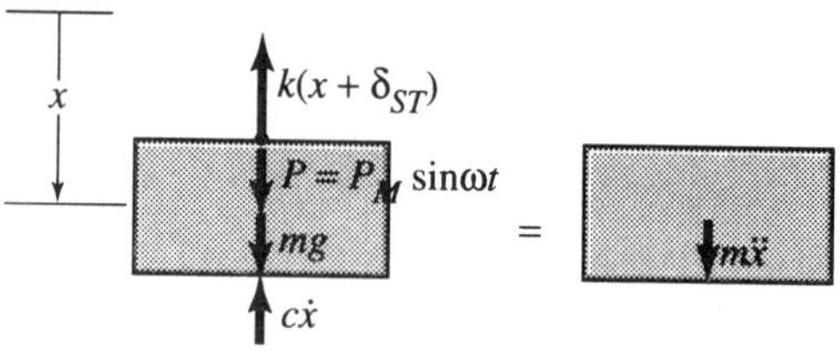

FIGURE 23.16.

$$\sum F_x = m\ddot{x}, \quad -k(x + \delta_{ST}) + P_M \sin \omega t + mg - c\dot{x} = m\ddot{x}.$$

Because x is measured from the equilibrium position, $-k\delta_{ST} + mg = 0$. Rearranging terms,

$$m\ddot{x} + c\dot{x} + kx = P_M \sin \omega t. \tag{23.40}$$

This is the governing differential equation for damped, forced vibrations of a single-degree-of-freedom system. It is a mathematical statement that the inertial force plus the damping force plus the spring force equals the applied forcing function at any time t during the motion. As in Sec. 23.5, the homogeneous or transient solution will be damped out and will not be considered. The particular or steady-state solution is the only one of practical interest and will be dealt with as follows:

Assume a particular solution of the form

$$x_P = x_M \sin(\omega t - \phi) \tag{23.41}$$

where x_M is the amplitude and ϕ is the phase angle. Differentiating this equation twice with respect to time gives

$$\dot{x}_P = \omega x_M \cos(\omega t - \phi)$$

and

$$\ddot{x}_P = -\omega^2 x_M \sin(\omega t - \phi).$$

Substituting in Eq. (23.40) gives

$$-m\omega^2 x_M \sin(\omega t - \phi) + c\omega x_M \cos(\omega t - \phi) + kx_M \sin(\omega t - \phi)$$
$$= P_M \sin \omega t. \tag{23.42}$$

Let $\omega t - \phi = 0$ in this equation to obtain

$$c\omega x_M = P_M \sin \phi. \tag{23.43}$$

Let $\omega t - \phi = \pi/2$, and recall that $\sin(\phi + \pi/2) = \cos \phi$. Then, substitute in Eq. (23.42) to obtain

$$(k - m\omega^2)x_M = P_M \cos \phi. \tag{23.44}$$

Equations (23.43) and (23.44) contain two unknowns x_M and ϕ. Therefore, dividing Eq. (23.43) by Eq. (23.44) gives

$$\tan \phi = \frac{c\omega}{k - m\omega^2}. \tag{23.45}$$

Squaring and adding Eqs. (23.43) and (23.44) yields

$$[(k - m\omega^2)^2 + (c\omega)^2]x_M^2 = P_M^2$$

Solving for x_M yields

$$x_M = \frac{P_M}{\sqrt{[(k - m\omega^2)^2 + (c\omega)^2]}}. \tag{23.46}$$

Equations (23.45) and (23.46) enable us to determine the phase angle ϕ and the amplitude x_M given the physical quantities of the system m, k, c, P_M, and ω. Once ϕ and x_M are known, then, the particular solution, $x_P = x_M \sin(\omega t - \phi)$, gives the position of the mass as a function of time. The derivatives of Eq. (23.40) provide the velocity and acceleration as functions of time.

Equation (23.46) and (23.45) can be written in dimensionless form by using $p^2 = k/m$ and $c_c = 2mp$. The amplification factor for damped, forced vibration becomes

$$\frac{x_M}{\frac{P_M}{k}} = \frac{x_M}{\delta_M} = \frac{1}{\sqrt{\left[1 - \left(\frac{\omega}{p}\right)^2\right]^2 + \left[2\left(\frac{c}{c_2}\right)\left(\frac{\omega}{p}\right)\right]^2}}. \tag{23.47}$$

The tangent of the phase angle ϕ is given by

$$\tan\phi = \frac{2\left(\frac{c}{c_c}\right)\left(\frac{\omega}{p}\right)}{1 - \left(\frac{\omega}{p}\right)^2}. \tag{23.48}$$

Both the amplification factor and the tangent of the phase angle are functions of two dimensionless ratios: ω/p, the ratio of the forcing frequency to the natural frequency and c/c_c, the ratio of the damping coefficient to the critical damping coefficient. As for the undamped case, these equations apply to support displacements given by $\delta = \delta_M \sin \omega t$ where $\delta_M = P_M/k$.

Figure 23.17 is a plot of Eq. (23.47) for $c/c_c = 0.2$. This represents the variation of the amplification factor x_M/δ_M versus the frequency ratio ω/p for light damping because $c < c_c$ or $c/c_c < 1$. If $c/c_c = 0$ (the system

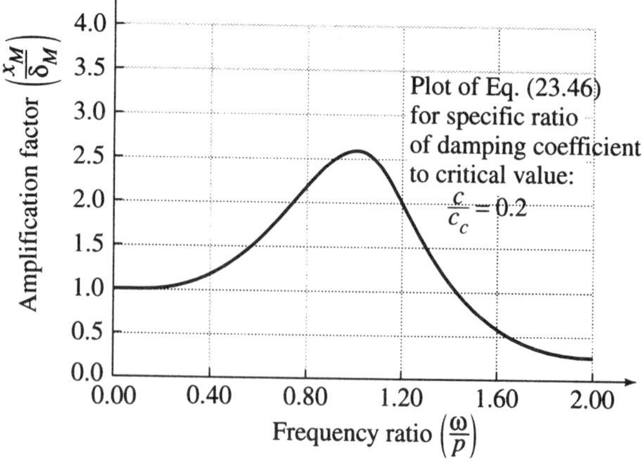

FIGURE 23.17.

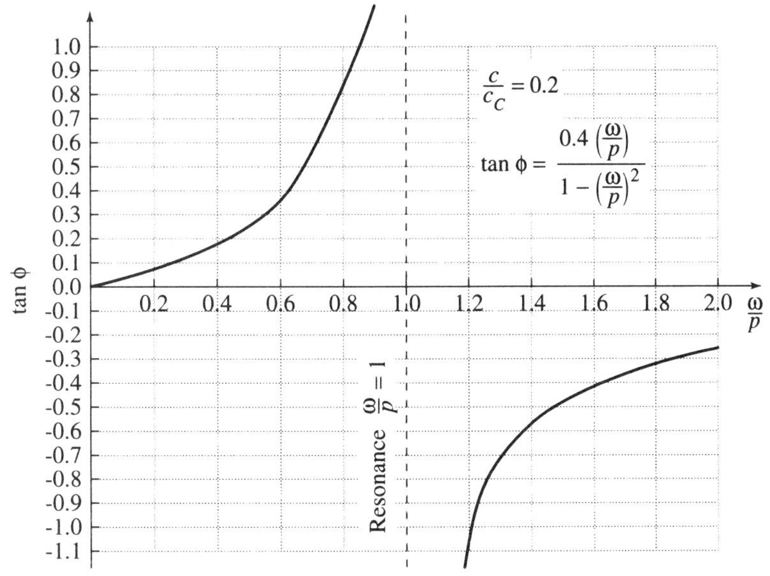

FIGURE 23.18.

is undamped), then, $(x_M/\delta_M) \to \infty$ for $\omega/p = 1$. Because damping is present, we note that $x_M/\delta_M = 2.5$ for $\omega/p = 1$ at resonance. In other words, the maximum dynamic amplitude is greatly reduced by the presence of damping in the system. Figure 23.18 is a plot of Eq. (23.48) for $c/c_c = 0.2$. This represents the variation of the tangent of the phase angle ($\tan\phi$) versus the frequency ratio ω/p for light damping. At resonance $\omega/p = 1$, $\tan\phi \to \infty$ or $\phi = 90°$. Additional plots of Eqs. (23.47) and (23.48) for other c/c_c ratios are readily obtained by the use of a pocket calculator, a personal computer, or a mainframe computer. Their general shapes would be similar to those shown in Figures 23.17 and 23.18.

The following examples illustrate some of the concepts discussed in this section.

■ Example 23.12

A forcing function $P = P_M \sin\omega t$ is externally applied to the system depicted in Figure E23.12. The mass $m = 0.5$ slug, the damping coefficient $c = 20$ lb·s/ft, the spring constant $k = 600$ lb/ft, the amplitude of the forcing function $P_M = 60$ lb, and the forcing frequency $\omega = 25$ rad/s. Determine the amplification factor x_M/δ_M, the maximum dynamic displacement, the phase angle ϕ, and express x as a function of time after the transient solution has been damped out.

Solution

Equations (23.47) and (23.48) apply, and they are functions of two dimensionless rations ω/p and c/c_c. These are found as follows:

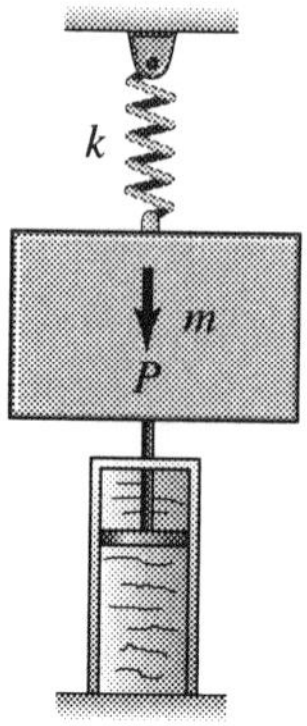

FIGURE E23.12.

$$\grave{p}\; X\frac{k}{m} = \sqrt{\frac{600}{0.5}} = 34.64 \text{ rad/s.},$$

$$\frac{\omega}{p} = \frac{25}{34.64} = 0.722,$$

$$c_c = 2mp = 2(0.5)(34.64) = 34.64 \text{ lb·s/ft},$$

and

$$\frac{c}{c_c} = \frac{20}{34.64} = 0.577.$$

Using Eq. (23.47) for the amplification factor, we obtain

$$\frac{x_M}{\frac{P_M}{k}} = \frac{1}{\sqrt{[1 - (0.722)^2]^2 + [2(0.577)(0.722)]^2}} = 1.041, \qquad \text{ANS.}$$

$$\delta_M = \frac{P_M}{k} = \frac{60}{600} = 0.1000 \text{ ft}.$$

The maximum dynamic displacement x_M is computed by

$$x_M = 1.041\left(\frac{P_M}{k}\right) = 1.041\left(\frac{60}{600}\right) = 0.1041 \text{ ft}. \qquad \text{ANS.}$$

We observe for the given inputs that this value for the dynamic amplitude is only 4% more than the static deflection δ_M.

The tangent of the phase angle ϕ is given by Eq. (23.48). Thus,

$$\tan\phi = \frac{2(0.577)(0.722)}{1 - (0.722)^2} = 1.741$$

$$\phi = \tan^{-1} 1.741 = 60.1^\circ = 1.049 \text{ rad} \qquad \text{ANS.}$$

The steady-state solution is given by Eq. (23.41). Substituting for x_M, ω, and ϕ, we obtain

$$x = 0.1041 \sin(25t - 1.049) \text{ ft.} \qquad \text{ANS.}$$

■ Example 23.13

A force $P(t) = P_M \sin 8\pi t$ is externally applied to the single story building for 0.25 second as depicted in Figure E23.13. Determine the displacement and the velocity of the mass m at the time the external force is removed. Assume that $P_M = 35{,}000$ lb and that the structure has a mass $m = 253.3$ lb·in²/s, a damping coefficient $c = 200$ lb·s/ft, and a stiffness $k = 10{,}000$ lb/in. Note that the columns in the building are assumed massless.

Solution

From the given information, the natural circular frequency is calculated as

$$p = \sqrt{\frac{10{,}000}{253.3}} = 6.283 \text{ rad/s.}$$

Therefore, because the frequency of excitation is $\omega = 8\pi$, the frequency ratio becomes

$$\frac{\omega}{p} = \frac{8\pi}{6.283} = 4.0.$$

Now,

$$c_c = 2mp = 2(253.3)(6.283) = 3186.97 \text{ lb·s/in}$$

and the damping ratio becomes

$$\frac{c}{c_c} = \frac{200}{3186.97} = 0.0628.$$

Note that the damping ratio for most buildings is on the order of 5% and seldom exceeds 20%. The maximum dynamic displacement when the external force is acting on the structure is computed by Eq. (23.46). Thus,

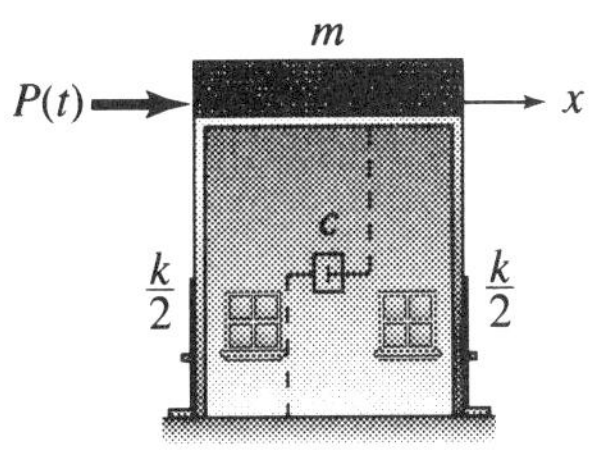

FIGURE E23.13.

$$x_M = \frac{35000}{\sqrt{[10{,}000 - 253.3(8\pi)^2]^2 + [200(8\pi)]^2}} = 0.233 \text{ in.}$$

Also, by Eq. (23.48), the phase becomes

$$\phi = \tan^{-1}\left(\frac{200(8\pi)}{10{,}000 - (253.3)(8\pi)^2}\right) = 0.9448 \text{ rad.}$$

Therefore, when the load is acting on the structure, the displacement is given by Eq. (23.41). Thus,

$$x = 0.233 \sin(8\pi t - 0.9448) \text{ in.}$$

Differentiating gives the velocity, when the load is acting on the structure, as

$$\dot{x} = 0.233(8\pi)\cos(8\pi t - 0.9448) \text{ in./s.}$$

Thus, at time $t = 0.25$ s,

$$x(0.25) = 0.233 \sin[8\pi(0.25) - 0.9448] = 0.189 \text{ in.} \qquad \text{ANS.}$$

and

$$\dot{x}(0.25) = 0.233(8\pi)\cos[8\pi(0.25) - 0.9448] = 3.43 \text{ in./s.} \qquad \text{ANS.}$$

Note that, at $t > 0.25$, the displacement of the structure corresponds to the free-vibration response due to the initial conditions at $t = 0.25$ s.

23.8 Lagrange's Method—Nonconservative Forces and MDOF

In this section, we extend the method of Lagrange of Sec. 23.4 to include nonconservative forces and multidegrees of freedom (MDOF). (Conservative and nonconservative forces are discussed in Sec. 12.5.). Consider the single-degree-of-freedom system shown in Figure 23.19(a) which includes two nonconservative forces, the damping force of the form $c\dot{x}$ and the externally applied forcing function $P = P_M \sin \omega t$.

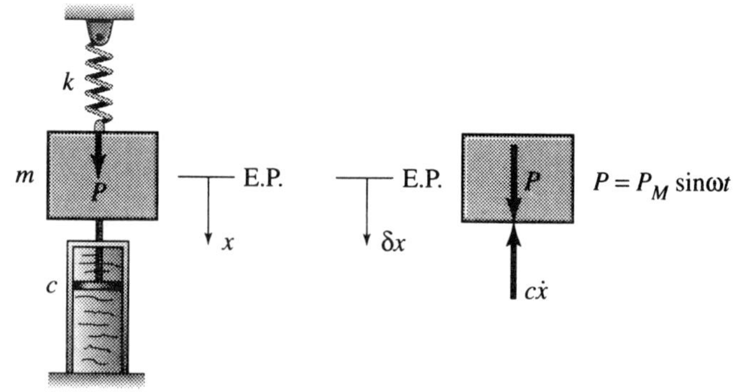

FIGURE 23.19. (a) (b)

Lagrange's equation for this single-degree-of-freedom system is given by

$$\frac{d}{dt}\left(\frac{\partial l}{\partial \dot{x}}\right) - \frac{\partial l}{\partial x} = Q_x \tag{23.49}$$

where $l = T - V$ and $\frac{\partial l}{\partial \dot{x}}$ and $\frac{\partial l}{\partial x}$ are the first partial derivatives, respectively, of the Lagrangian with respect to the velocity $\dot{x}$ and the position coordinate x. The operator d/dt is the first time derivative performed on the partial derivative $\frac{\partial l}{\partial \dot{x}}$ after it has been found. The Lagrangian l will include all conservative forces of the system in the potential energy V. In this case, the conservative force is the spring force for which the potential energy is given by $V = \frac{1}{2}kx^2$. The gravitational force mg is also a conservative force, but we need not include it in V because we are measuring the coordinate x from the equilibrium position. The term Q_x on the right side of Lagrange's equation is the generalized force associated with the nonconservative forces (the damping force and the forcing function) of the system. This term vanishes if there are only conservative forces in the system. To find Q_x, we refer to Figure 23.19(b) and imagine that the nonconservative forces perform virtual work during a downward virtual displacement δ_x. (Virtual work and virtual displacements are discussed in Chapter 12 of *Statics for Engineers* by Muvdi et. al.). Thus,

$$Q_x \delta_x = P_M \sin \omega t \delta_x - c\dot{x}\delta_x$$

Because the virtual displacement δ_x is not zero, we may divide by it to give $Q_x = P_M \sin \omega t - c\dot{x}$. Now,

$$l = T - V = \tfrac{1}{2}m\dot{x}^2 - \tfrac{1}{2}kx^2,$$

$$\frac{\partial l}{\partial \dot{x}} = m\dot{x},$$

$$\frac{\partial l}{\partial x} = -kx,$$

and

$$\frac{d}{dt}\left(\frac{\partial l}{\partial \dot{x}}\right) = m\ddot{x}.$$

Substituting the appropriate derivatives and Q_x in Eq. (23.49) gives

$$m\ddot{x} + kx = P_M \sin \omega t - c\dot{x}.$$

Transposing the damping force $c\dot{x}$ to the left side gives

$$m\ddot{x} + c\dot{x} + kx = P_M \sin \omega t. \tag{23.50}$$

This equation derived by Lagrange's method is identical to Eq. (23.40) derived from Newton's second law.

If we consider a system of n degrees of freedom (i.e., one that requires n coordinates for complete definition), then, Lagrange's equations may be written compactly as

$$\frac{d}{dt}\left(\frac{\partial l}{\partial \dot{\alpha}_r}\right) - \frac{\partial l}{\partial \alpha_r} = Q_r \tag{23.51}$$

where the subscript $r = 1, 2, 3 \ldots n$. The symbols α_r and $\dot{\alpha}_r$ represent the generalized coordinates and the generalized velocities. Other symbols have been defined earlier.

Application to a two-degree-of-freedom system is given in Example 23.14.

Example 23.14

A two-degree-of-freedom system is shown in Figure E23.14(a). Use Lagrange's method to write the two differential equations of motion for this system. Measure x_1 and x_2 from the equilibrium positions of the masses as shown.

Solution

Two coordinates x_1 and x_2 are required to position the two masses m_1 and m_2 at any time t, and we, therefore, refer to the system as a two-degree-of-freedom system. We assume, arbitrarily, $x_2 > x_1$ and $\dot{x}_2 > \dot{x}_1$. The opposite assumption will lead to the same equations of motion.

The kinetic energy

$$T = \tfrac{1}{2}m_1\dot{x}_1^2 + \tfrac{1}{2}m_2\dot{x}_2^2.$$

The potential energy

$$V = 2(\tfrac{1}{2}kx_1^2) + 2(\tfrac{1}{2}k_1)(x_2 - x_1)^2 = kx_1^2 + k_1(x_2 - x_1)^2.$$

The Lagrangian

$$l = \tfrac{1}{2}m_1\dot{x}_1^2 + \tfrac{1}{2}m_2\dot{x}_2^2 - kx_1^2 - k_1(x_2 - x_1)^2.$$

Determine the appropriate derivatives of the Lagrangian l, which is a function of $\dot{x}_1$, $\dot{x}_2$, x_1, and x_2, that is,

$$\frac{\partial l}{\partial \dot{x}_1} = m_1\dot{x}_1,$$

$$\frac{d}{dt}\left(\frac{\partial l}{\partial \dot{x}_1}\right) = m_1\ddot{x}_1,$$

$$\frac{\partial l}{\partial \dot{x}_2} = m_2\dot{x}_2,$$

$$\frac{d}{dt}\left(\frac{\partial l}{\partial \dot{x}_2}\right) = m_2\ddot{x}_2,$$

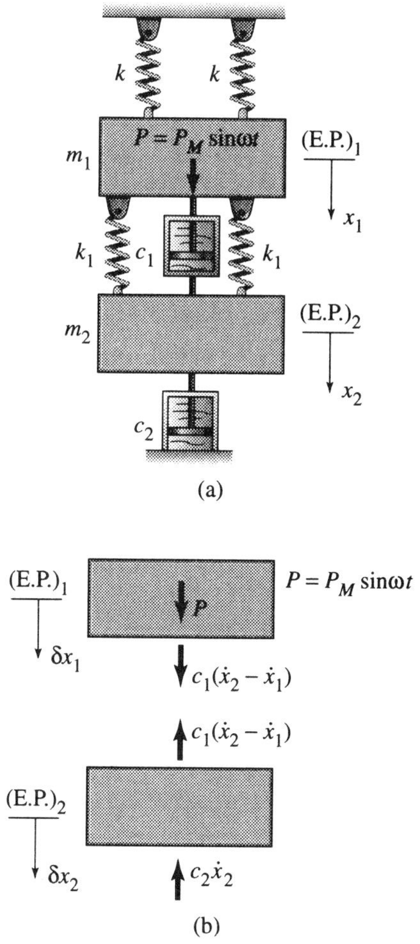

FIGURE 23.14.

$$\frac{\partial l}{\partial x_1} = -2kx_2 + 2k_1(x_2 - x_1)$$

and

$$\frac{\partial l}{\partial x_2} = -2k_1(x_2 - x_1).$$

The nonconservative forces of this system are shown in Figure E23.14(b). For this system there are two generalized forces, Q_1 and Q_2 which we determine from the virtual work equation with the positive sense chosen downward. Thus,

$$Q_1\delta_{x1} + Q_{x2}\delta_2 = [P_{\rm M}\sin\omega t + c_1(\dot{x}_2 - \dot{x}_1)]\delta_{x1}$$
$$+ [-c_1(\dot{x}_2 - \dot{x}_1) - c_2\dot{x}_2]\delta_{x2}$$

The virtual displacements, δ_{x1} and δ_{x2} are independent of each other. Thus,

$$\delta_{x2} \to 0 \qquad Q_1 = P_M \sin \omega t + C_1(\dot{x}_2 - \dot{x}_1)$$

and

$$\delta_{x1} \to 0 \qquad Q_2 = -C_1(\dot{x}_2 - \dot{x}_1) - C_2\dot{x}_2.$$

Refer to Eq. (23.51) with $r = 1, 2$ (i.e., $n = 2$) and let $\alpha_1 = x_1$, $\alpha_2 = x_2$, $\dot{\alpha}_1 = \dot{x}_1$, and $\dot{\alpha}_2 = \dot{x}_2$. Lagrange's equations become

$$\frac{d}{dt}\left(\frac{\partial l}{\partial \dot{x}_1}\right) - \frac{\partial l}{\partial x_1} = Q_1$$

and

$$\frac{d}{dt}\left(\frac{\partial l}{\partial \dot{x}_2}\right) - \frac{\partial l}{\partial x_2} = Q_2.$$

Substituting for appropriate derivatives and the generalized forces Q_1 and Q_2 from above gives

$$m_1\ddot{x}_1 + 2kx_1 - 2k_1(x_2 - x_1) = P_M \sin \omega t + c_1(\dot{x}_2 - \dot{x}_1)$$

and

$$m_2\ddot{x}_2 + 2k_1(x_2 - x_1) = -c_1(\dot{x}_2 - \dot{x}_1) - c_2\dot{x}_2.$$

Rearranging terms yields the two differential equations of motion for this system. Thus,

$$m_1\ddot{x}_1 + 2(k + k_1)x_1 - 2k_1x_2 + c_1(\dot{x}_2 - \dot{x}_1) = P_M \sin \omega t \qquad \text{ANS.}$$

and

$$m_2\ddot{x}_2 + 2k_1x_2 - 2k_1x_1 + (c_1 + c_2)\dot{x}_2 - c_1\dot{x}_1 = 0 \qquad \text{ANS.}$$

■

Problems

23.118 Refer to Eq. (23.34) and use $p^2 = k/m$ and $c_c = 2mp$ to rewrite this equation in dimensionless form:

$$\frac{q}{p} = \sqrt{1 - \left(\frac{c}{c_c}\right)^2}$$

Clearly state each algebraic step taken to achieve this. Plot q/p as a function of c/c_c. Assign c/c_c the values 0, 0.2, 0.4, 0.6, 0.8 and 1.0.

23.119 Refer to Eq. (23.37) and use the following initial conditions $t = 0$, $x = 0$ and $t = 0$, $\dot{x} = 0.30$ ft/s to determine values for C_1 and C_2. Let $q = 6$ rad/s., and state the units of these constants.

23.120 Data for a damped forced vibration

system is: $m = 10$ kg, $k = 12$ kN/m, $c = 300$ N·s/m, $P_M = 2$ kN and $\omega = 6$ rad/s. Determine the phase angle ϕ and the amplitude x_M, and write the steady-state solution for this motion. Write the corresponding displacment, velocity, and acceleration equations for this motion. Let $P = P_M \sin \omega t$.

23.121 For a damped forced vibration of a one-degree-of-freedom-system, the ratio of the forced circular frequency to the natural circular frequency is 0.6, and the ratio of the coefficient of damping to the coefficient of critical damping is 0.6. (a) Determine the ratio of x_M/δ_M. (b) Determine the maximum dynamic displacement if $\delta_M = 1.00$ in.

23.122 The solution for a damped, free vibration of a single-degree-of-freedom system can be expressed in the form

$$x = e^{-(c/2m)t}[C_1 \sin qt + C_2 \cos qt].$$

Data for a given system is $k = 12{,}000$ N/m, $m = 50$ kg, and $c = 600$ N·s/m. (a) Show that this system is lightly damped. (b) Given the initial conditions $t = 0$, $x = 0$, and $\dot{x} = 14.28$ m/s, determine the constants C_1 and C_2, and write the solution for x as a function of t.

23.123 A free vibration of a single-degree-of-freedom system is critically damped. Initial conditions for this motion are $t = 0$, $x = 0$, and $\dot{x} = 5$ m/s. Numerical inputs are $k = 15{,}000$ N/m and $m = 60$ kg. (a) Determine the critical damping factor. (b) Determine the constants of integration, and write the position x as a function of time. (c) Sketch x vs t. Is the motion vibratory?

23.124 Data for a damped, free-vibration system of a one-degree-of-freedom system is $k = 12{,}000$ N/m, $m = 50$ kg, and $c = 2000$ N·s/m. (a) Show that the system is heavily damped. (b) Determine the roots of the characteristic equation. (c) Determine the constants A and B in the general solution of the differential equation for the initial conditions $t = 0$, $x = 0.1$ m, and $\dot{x} = 0$.

23.125 A single-degree-of-freedom system is damped and sinusoidally forced. It is characterized by the following data: $m = 1$ slug, $k = 1400$ lb/ft, $c = 200$ lb·s/ft, $P_M = 500$ lb, and $\omega = 4$ rad/s. Determine the phase angle ϕ and the amplitude x_M, and write the steady-state solution for this motion. Write the corresponding displacment, velocity, and acceleration equations for this motion.

23.126 The spring-supported mass of Figure P23.126 is subjected to damping. Use the method of Lagrange to write the differential equation of motion. Check your resulting equation by writing Newton's second law.

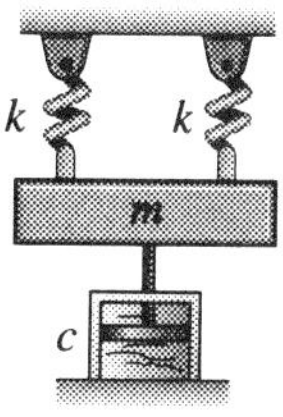

FIGURE P23.126.

23.127 If the plate of Figure P23.127 is rotated ccw about an axis through O, then, gravity provides the only restoring force tending to return it to the stable equilibrium position shown. Use Lagrange's method to write the differential equation describing this oscillatory motion. Note that $Q = 0$ in this case.

23.128 Use Lagrange's method to write the differential equation of motion for the system shown in Figure P23.128. Derive the characteristic equation and

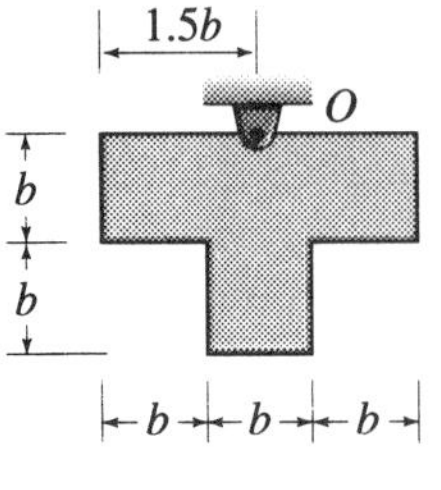

FIGURE P23.127.

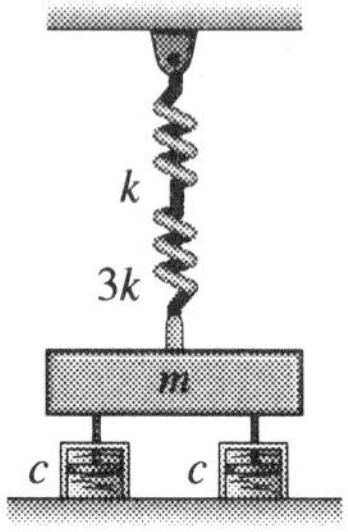

FIGURE P23.128.

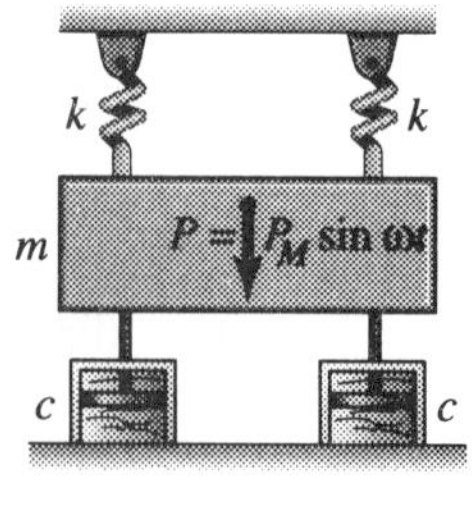

FIGURE P23.130.

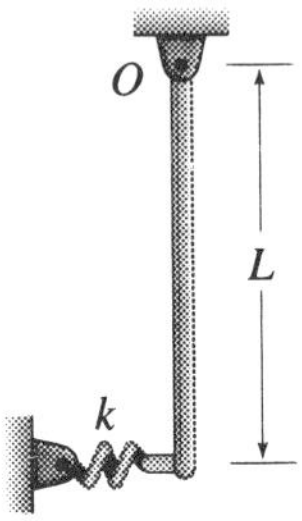

FIGURE P23.131.

express the critical damping coefficient c_c in terms of k and m.

23.129 The platform on which the block of Figure P23.129 slides is frictionless. Derive the differential equation of motion using Lagrange's method.

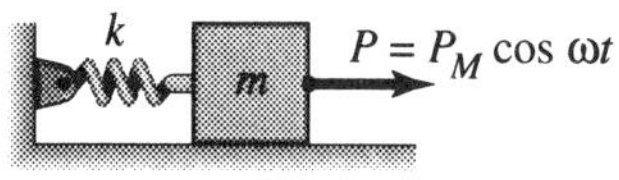

FIGURE P23.129.

23.130 Refer to Figure P23.130 and measure the displacement of the block from the equilibrium position. Use Lagrange's method to write the governing differential equation for this oscillatory motion.

23.131 Imagine displacement of the slender rod of mass m in Figure P23.131 through a small ccw angle θ. Both the spring and gravity will tend to restore the rod of mass m to the vertical equilibrium position. Because these forces are conservative, the generalized force Q will vanish. Use Lagrange's method to write the governing differential equation of motion of this rod.

23.132 If the slender rod of Figure 23.132 is rotated cw through a small angle θ and released from rest it will vibrate about the horizontal equilibrium position. Use Lagrange's method to write the differential equation describing this vibratory motion.

23.133 A two-degree-of-freedom system is shown in Figure P23.133. The platform

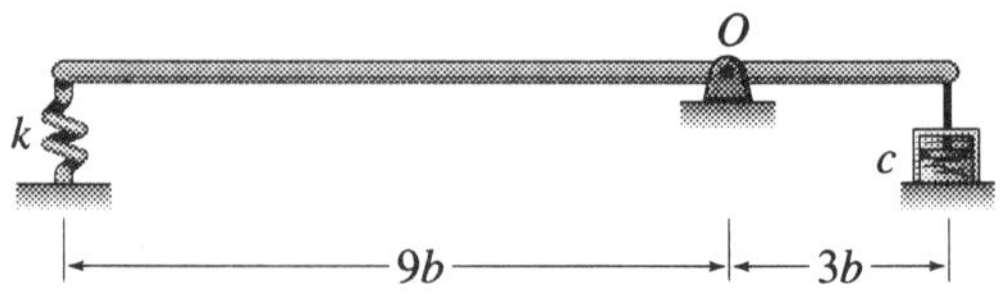

FIGURE P23.132.

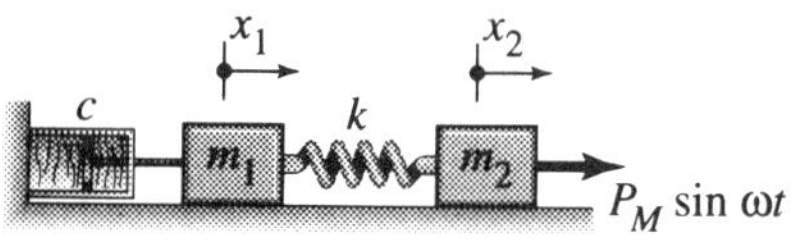

FIGURE P23.133.

on which the masses move is frictionless. Use the method of Lagrange to write the two differential equations of motion.

23.134 Displacements x_1 and x_2 of Figure P23.134 are measured from the equilibrium positions of the masses. Derive the two differential equations of motion for this two-degree-of-freedom system using Lagrange's method.

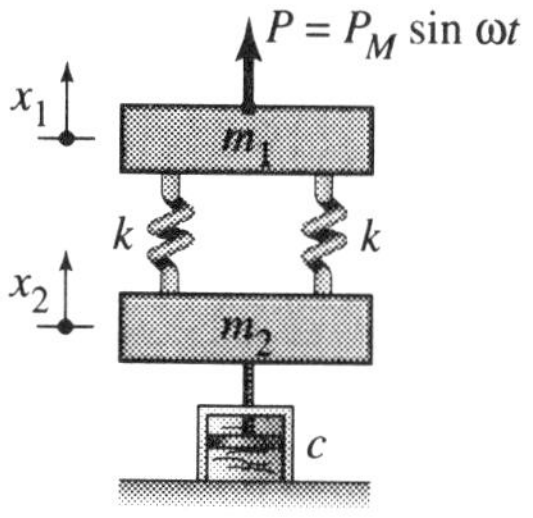

FIGURE P23.134.

23.135 Refer to the system shown in Figure P23.135. Use the method of Lagrangian to write the two differential equations of motion for this system.

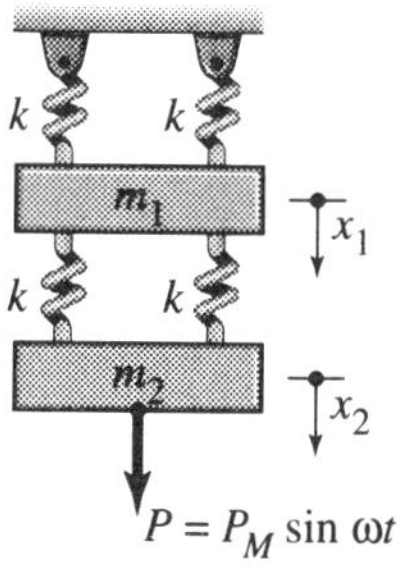

FIGURE P23.135.

23.136 A two-degree-of-freedom system is shown in Figure P23.136. Write the two differential equations which describe the vibratory motion of this system. Use the method of Lagrange. Assume that the platform on which the masses move is frictionless.

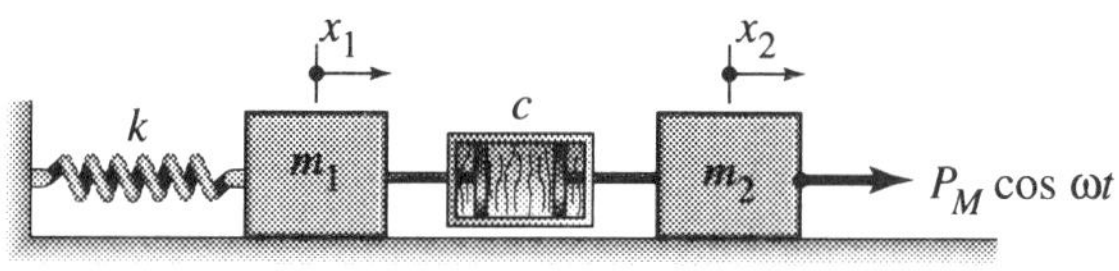

FIGURE P23.136.

Review Problems

23.137 The mass m shown in Figure P23.137 can slide freely on the frictionless horizontal plane. If $m = 2$ kg, $k = 1.5$ kN/m, and the mass is displaced from its equilibrium position by 0.05 m and released from rest, determine the position, velocity, and acceleration of the mass 2 s after the start of the motion.

23.138 A water storage tower is modeled mathematically by a concentrated weight W on top of a weightless cantilever beam of spring constant k as shown in Figure

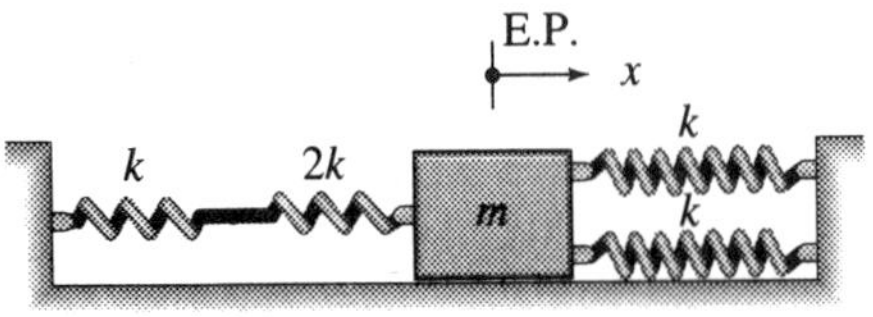

FIGURE P23.137.

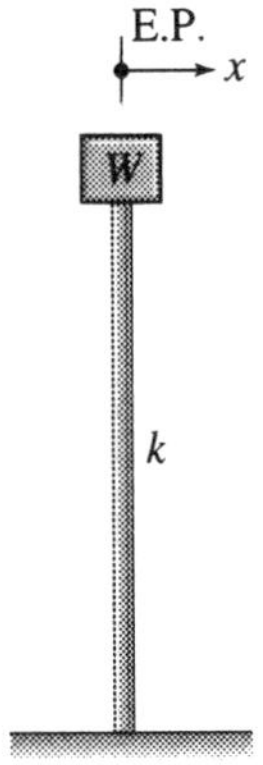

FIGURE P23.138.

P23.138. If $W = 1 \times 10^5$ lb and the spring constant for the beam is $k = 3 \times 10^6$ lb/ft, determine (a) the circular natural frequency and the period of oscillations, (b) the maximum velocity and maximum acceleration of the weight W if the weight is displaced a distance of 2 in. from its equilibrium position and released from rest and (c) the displacement, velocity, and acceleration 3 s after the start of the motion.

23.139 The simple pendulum shown in Figure P23.139 is provided with a torsional spring at the hinge at O. The torsional spring constant is K whose units are those of moment per rad. Thus, the restoring moment due to an angular displacement θ is $K\theta$. Determine the period for small oscillations of the pendulum. Express the answer in terms of K, L, and m where L is the length of the weighless rod and m is the mass of the bob.

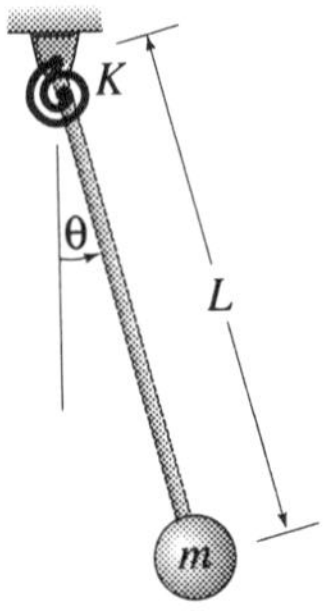

FIGURE P23.139.

23.140 The period of vibration for the weightless beam and attached blocks A and B shown in Figure P23.140 is measured at 0.4 s. After the 4-kg block A is removed, the period of oscillation becomes 0.3 s. Determine (a) the mass of block B and (b) the effective spring constant k of the beam.

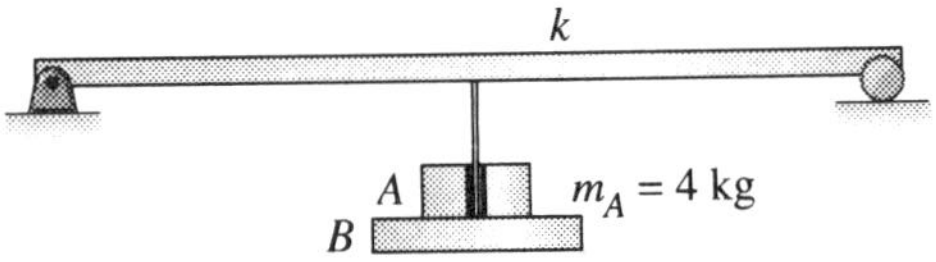

FIGURE P23.140.

23.141 The blades of a helicopter of weight W_B are suspended from a rigid shaft a shown in Figure P23.141. The upper end of the shaft is connected to a torsional spring (not shown) which is also attached to the bearing at A. The torsional spring has a spring constant K whose units are those of moment per rad. Thus, the restoring moment due to an angular displacement θ about the z axis is $K\theta$. The period of small vibrations of the blades about the z axis is

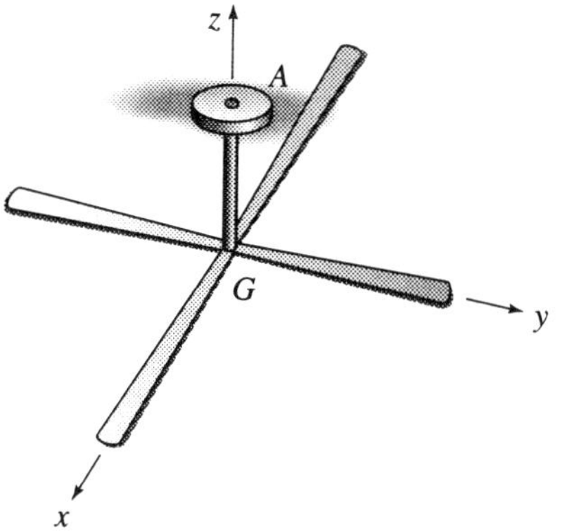

FIGURE P23.141.

τ_B. When the blades are removed and a cylinder of diameter D and weight equal to $4W_B$ is suspended such that its axis of symmetry is along the z axis, its period of vibration $\tau_C = 0.2\tau_B$. Use Newton's laws of motion to determine the centroidal radius of gyration of the blades in terms of the diameter D of the cylinder.

23.142 Four slender rods, each of mass m and length b, are welded together to form a square as shown in Figure P23.142. The square is hinged at A and suspended at B by a spring whose constant is k. Use Newton's laws of motion to determine the period of small oscillations about the hinge at A. Express the answer in terms of m and k.

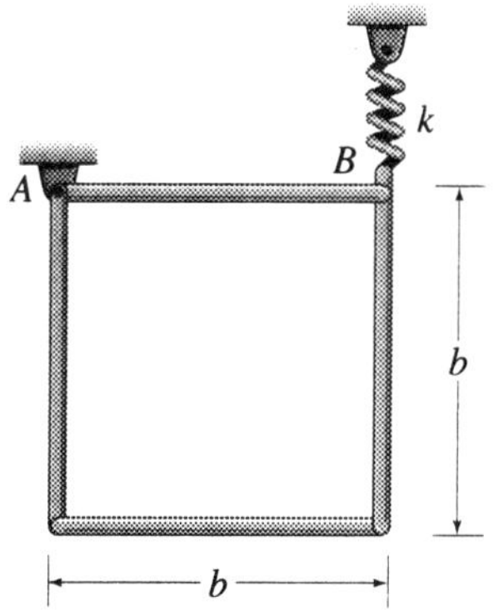

FIGURE P23.142.

23.143 Show that, for a pendulum of any shape, as shown in Figure P23.143, vibrating about a hinge at O, the centroidal radius of gyration k_G may be found from knowledge of the period τ for small oscillations from the relation $k_G = \sqrt{\left(\frac{gb}{4\pi^2}\right)\tau^2 - b^2}$ where b defines the distance from the point of suspension O to the center of mass G. If the period for small oscillations of a thin slender rod suspended at one end is measured at 2 s, what is the length of the rod in ft?

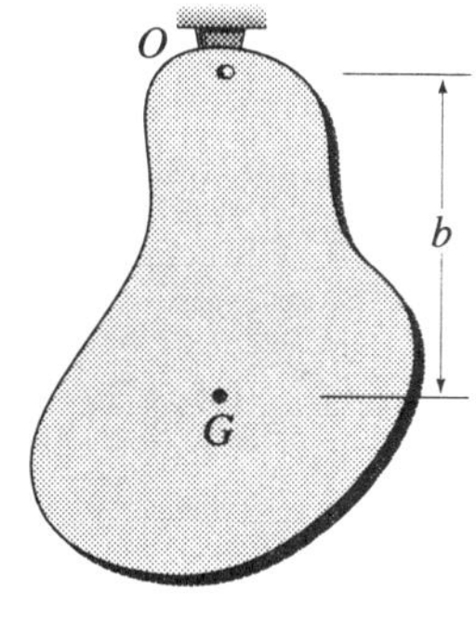

FIGURE P23.143.

23.144 Use the conservation of energy method to solve Problem 23.141.

23.145 Use the conservation of energy method to solve Problem 23.142.

23.146 Use the method of conservation of energy to find the period of small oscillations for the homogeneous square plate hinged at corner A as shown in Figure P23.146. Express the answer in terms of the dimension b. Find a numerical value for the period if b is (a) 2 ft and (b) 0.50 m.

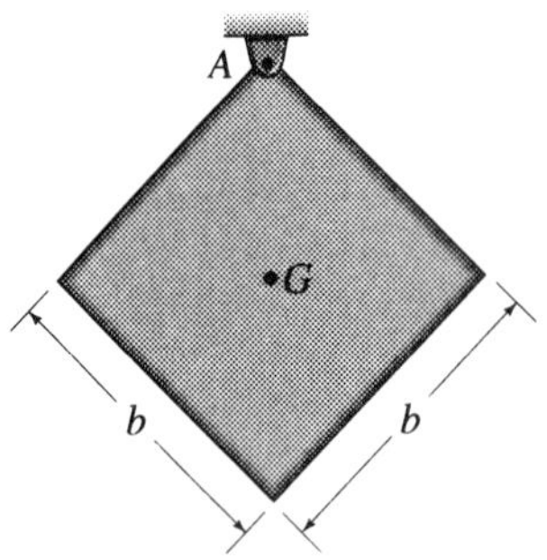

FIGURE P23.146.

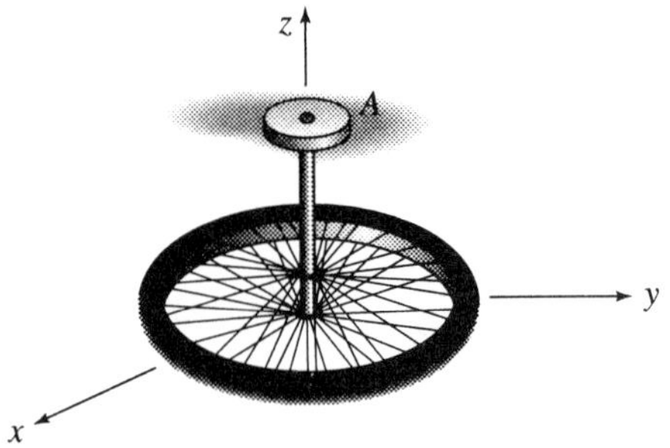

FIGURE P23.148.

23.147 Use the method of conservation of energy to find the circular natural frequency for small oscillations of the homogeneous square plate hinged at O and supported by two identical springs as shown in Figure P23.147. Assume that the springs are undeformed in the position shown, and express the answer in terms of the mass m, the dimension b, and the spring constant k.

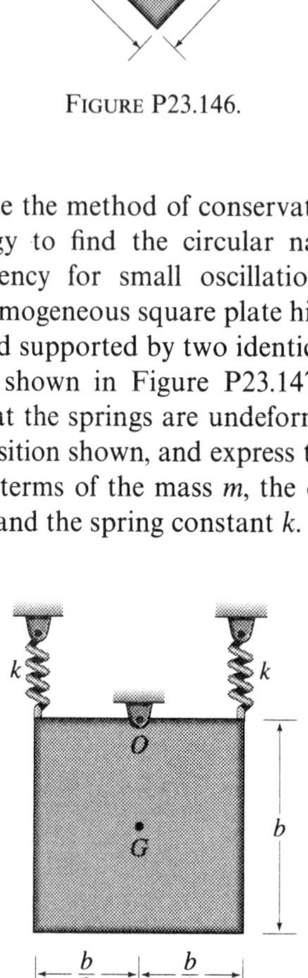

FIGURE P23.147.

23.148 A bicycle wheel is suspended from a rigid shaft as shown in Figure P23.148. The upper end of the shaft is connected to a torsional spring (not shown) which is also attached to the bearing at A. The torsional spring has a spring constant K whose units are those of moment per rad. Thus, the restoring moment due to an angular displacement θ about the z axis is $K\theta$. (a) If the mass of the wheel is m and its centroidal radius of gyration is k_G, find its period for small vibrations about the z axis. Express your answer in terms of m, k_G, and K. (b) Find the period for small vibrations for the case when $m = 1$ kg, $k_G = 0.35$ m, and $K = 2$ N·m/rad.

23.149 (a) Determine the period for small oscillations of the rod shown in Figure P23.149 which is hinged at O and supported at B by a spring whose constant is k. The rod has a mass m and a length of $4b$. Express the answer in terms of m and k. Use Lagrange's method. (b) If the weight of the rod is 10 lb and the spring constant $k = 20$ lb/ft, find a numerical value for the period.

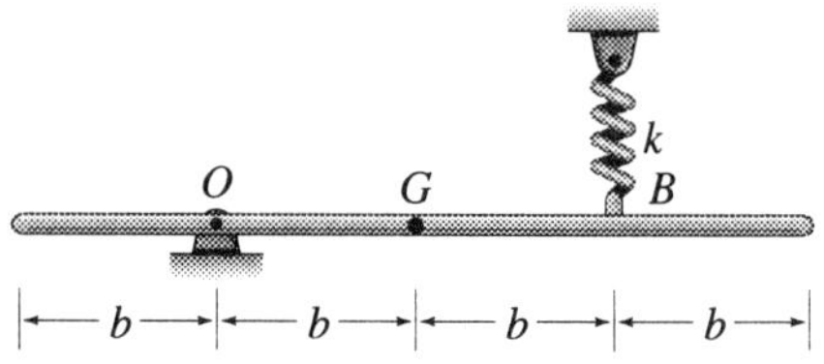

FIGURE P23.149.

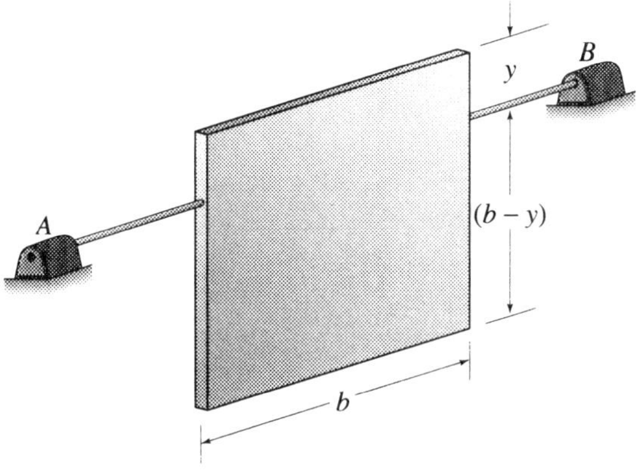

FIGURE P23.150.

23.150 (a) Use Lagrange's method to find the period of small oscillations about the weightless shaft AB for the system shown in Figure P23.150. Express the answer in terms of the side b of the square homogeneous plate and the distance y locating the shaft AB. (b) Specialize this answer for the case where $y = 1/4b$. (c) Is there another value of y for which the period is the same as that found in part (b)? If so, find it.

23.151 Because of equipment that is not perfectly balanced, the floor of an industrial building vibrates such that $\delta = 0.25 \sin 10t$ where δ is in in. and t is in s.

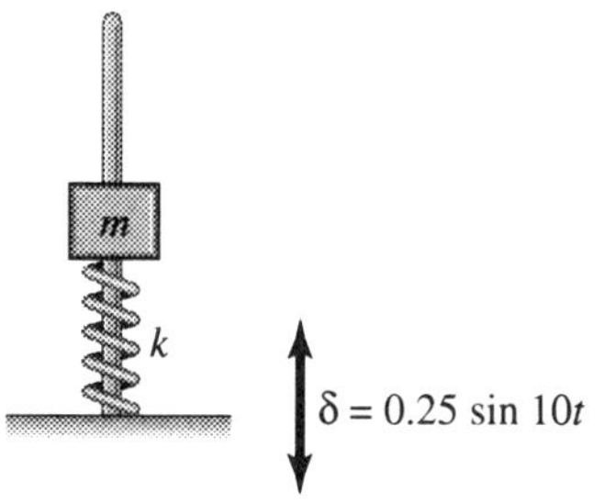

FIGURE P23.151.

A collar of mass $m = 0.15$ slug, that slides freely on a rod fastened to this floor, is attached to a spring constant k, as shown in Figure P23.151. Determine the spring constant k if (a) the maximum dynamic displacement of the mass is $+0.50$ in., (b) the maximum dynamic displacement of the mass is -0.50 in., and (c) the maximum dynamic displacement of the mass is theoretically infinite.

23.152 A trailer is being pulled over a road that may be assumed to be sinusoidal with a wave length of 5 m and amplitude of 0.10 m as shown in Figure P23.152. If the total mass of the trailer and the boat is 250 kg, determine (a) the equivalent spring constant for the springs in the trailer if the maximum dynamic displacement of the boat is measured at -0.018 m (out of phase) at a speed of 80 km/h and (b) the speed of the trailer corresponding to resonance.

23.153 A motor weighing 300 lb is placed at the free end of a cantilever beam as shown in Figure P23.153. The imbalance in the motor may be approxi-

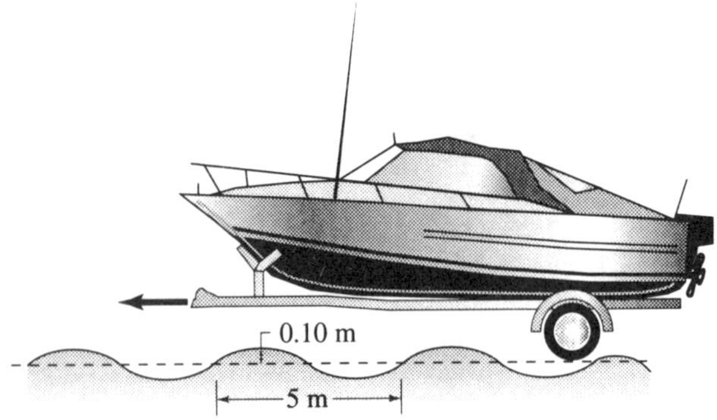

FIGURE P23.152.

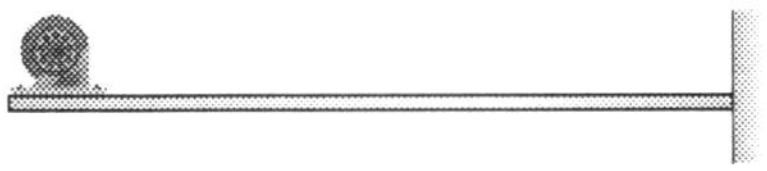

FIGURE P23.153.

mated by a 1 oz mass at a distance of 0.50 in. from the motor axis of rotation. When the motor runs at a speed of 360 rpm, the maximum dynamic displacement is measured at $x_M = 0.010$ in. Determine the equivalent spring constant for the beam if the amplification factor is 10.

23.154 Refer to Figure 23.15(b), and assume that the displacements x_1, x_2, etc. correspond to the maximum displacements given by the envelope equation $x = x_M e^{-(c/2m)t}$. (a) Show that the natural log of the ratio of any two consecutive amplitudes, say x_n and x_{n+1}, is given by

$$\ln\left(\frac{x_n}{x_{n+1}}\right) = \frac{2\pi(c/c_c)}{\sqrt{1 - (c/c_c)^2}}$$

where $\ln\left(\frac{x_n}{x_{n+1}}\right)$ is referred to as the *logarithmic decrement*. (b) Show that, for very small values of the damping coefficient c, the logarithmic decrement may be approximated by

$$\ln\left(\frac{x_n}{x_{n+1}}\right) = 2\pi(c/c_c).$$

23.155 A machine component may be approximately described by the mathematical model shown in Figure P23.155. If $k = 80$ kN/m, $c = 1000$ N·s/m, $m = 400$ kg, and the forcing function $P = 200 \sin 5t$ where P is in Newtons and t in s, determine the amplitude of the steady-state vibration of the machine component.

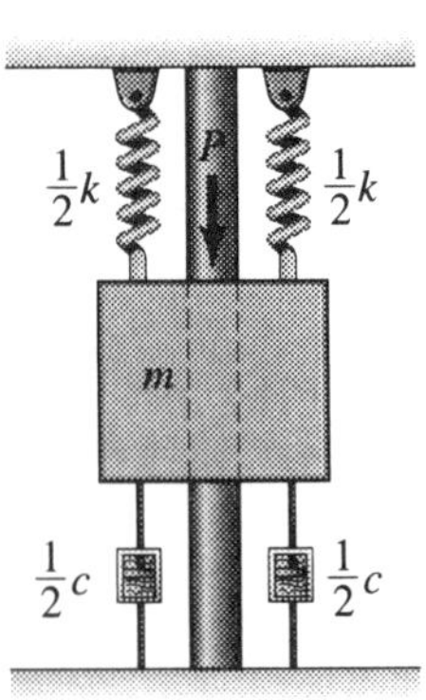

FIGURE P23.155.

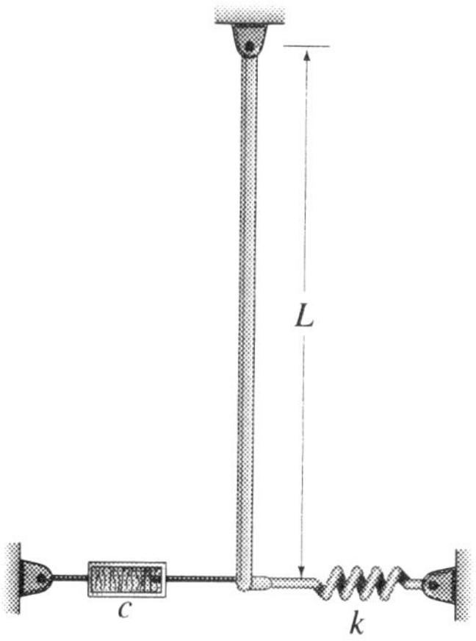

FIGURE P23.156.

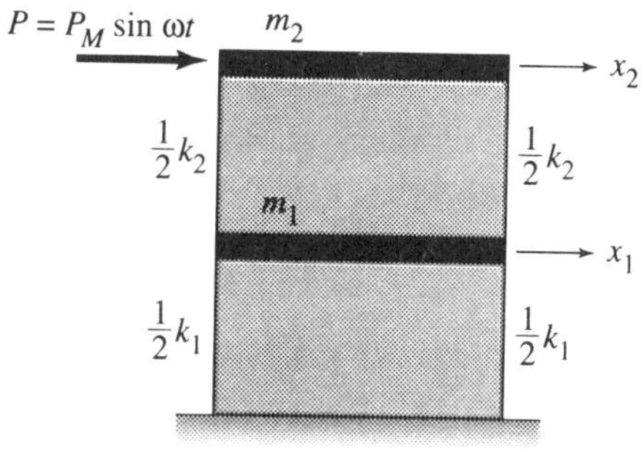

FIGURE P23.157.

23.156 Use Lagrange's method to derive the differential equation for small oscillations of the rod about the hinge at O as shown in Figure P23.156.

23.157 The structure of Figure P23.157 is subjected to a sinusoidal loading as shown. Determine the differential equations whose solutions give the displacement functions x_1 and x_2. Assume that the structure is undamped, and solve this problem using Newton's second law and by summing forces for each of the two masses.

Appendix A Properties of Selected Lines and Areas

Shape	Length or Area	Centroid Location	Centroidal Moments of Inertia	Centroidal Radii of Gyration
Arc of a circle 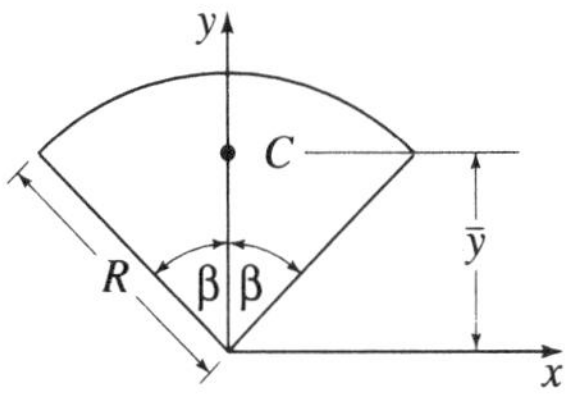	$L = 2\beta R$	$\bar{x} = 0$ $\bar{y} = \dfrac{R \sin \beta}{\beta}$		
Arc of a quarter circle 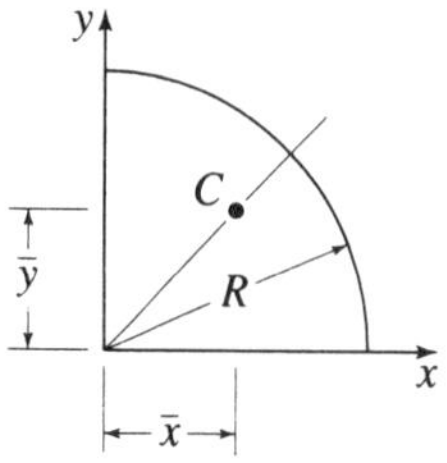	$L = \dfrac{\pi R}{2}$	$\bar{x} = \bar{y} = \dfrac{2R}{\pi}$		

Shape	Length or Area	Centroid Location	Centroidal Moments of Inertia	Centroidal Radii of Gyration
Rectangular Area 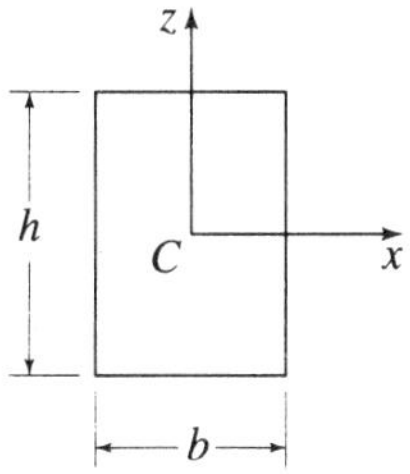	$A = bh$	$\bar{x} = \bar{y} = 0$	$I_x = \frac{1}{12} bh^3$ $I_y = \frac{1}{12} hb^3$	$r_x = \frac{h}{\sqrt{12}}$ $r_y = \frac{b}{\sqrt{12}}$
Triangular Area 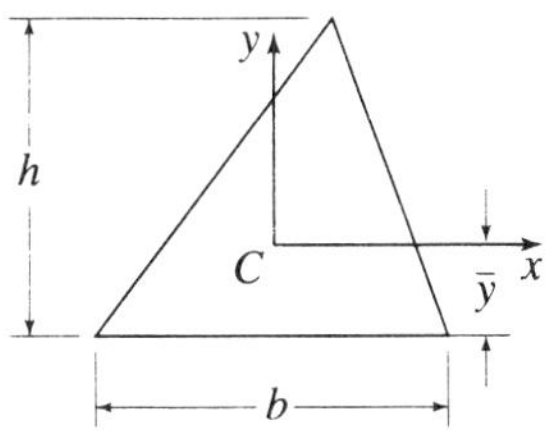	$A = \frac{1}{2} bh$	$\bar{y} = \frac{1}{3} h$	$I_x = \frac{1}{36} bh^2$	$r_x = \frac{h}{\sqrt{18}}$
Circular sector area 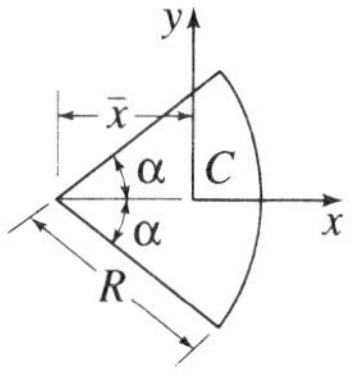	$A = \alpha R^2$	$\bar{x} = \frac{2}{3} \frac{R \sin\alpha}{\alpha}$ $\bar{y} = 0$	$I_x = \frac{R^4}{4}\left(\alpha - \frac{1}{2}\sin 2\alpha\right)$	
α = π leads to a circular area 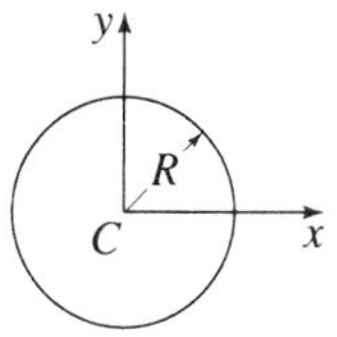	$A = \pi R^2$	$\bar{x} = 0$ $\bar{y} = 0$	$I_x = I_y = \frac{\pi R^4}{4}$ $J_c = I_x + I_y = \frac{\pi R^4}{2}$	$r_x = r_y = \frac{R}{2}$

Shape	Length or Area	Centroid Location	Centroidal Moments of Inertia	Centroidal Radii of Gyration
Semicircular area 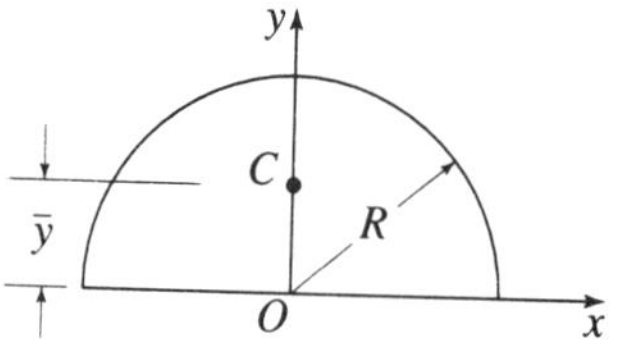	$A = \pi R^2/2$	$\bar{x} = 0$ $\bar{y} = \dfrac{4R}{3\pi}$	$I_x = I_y = \dfrac{\pi R^4}{8}$ $J_O = \dfrac{\pi R^4}{4}$	
Elliptical area 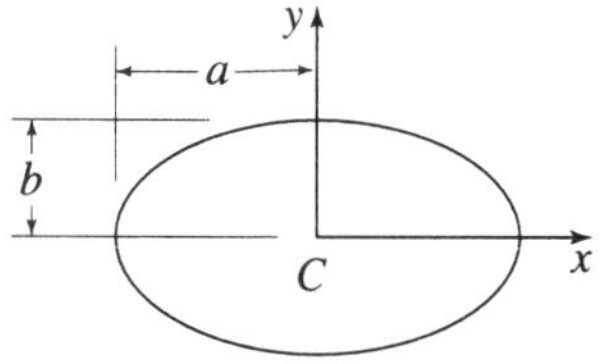	$A = \pi ab$	$\bar{x} = 0$ $\bar{y} = 0$	$I_x = \dfrac{\pi ab^3}{4}$ $I_y = \dfrac{\pi ba^3}{4}$	$r_x = \dfrac{b}{2}$ $r_y = \dfrac{a}{2}$
nth-degree Parabolic quadrant 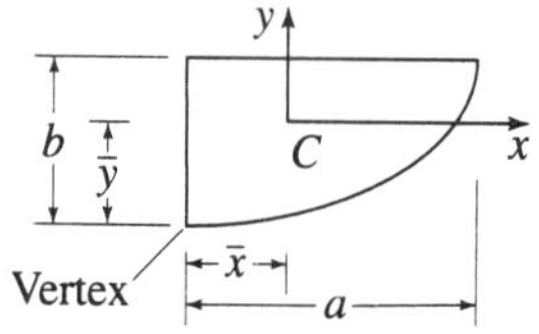	$A = \dfrac{nab}{n+1}$	$\bar{x} = \dfrac{(n+1)a}{2(n+2)}$ $\bar{y} = \dfrac{(n+1)b}{2n+1}$		
nth-degree Parabolic spandrel 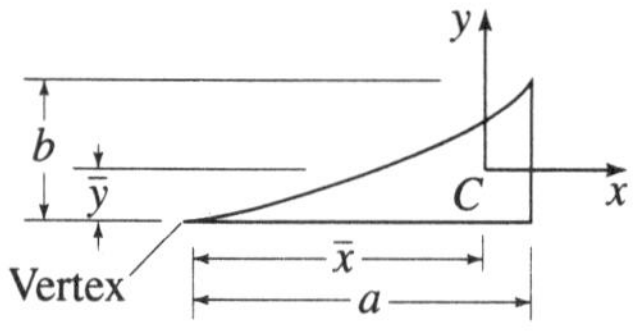	$A = \dfrac{ab}{n+1}$	$\bar{x} = \dfrac{(n+1)a}{n+2}$ $\bar{y} = \dfrac{(n+1)b}{2(2n+1)}$		

Appendix B
Properties of Selected Masses

Body and dimensions	Volume	Center of mass	Centroidal Moments of Inertia
Solid circular cylinder 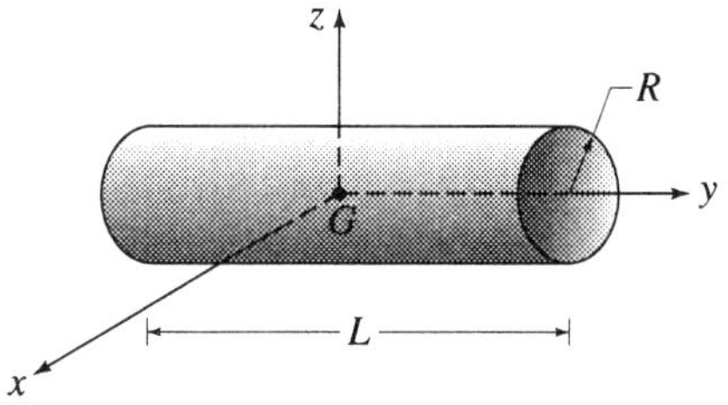	$V = \pi R^2 L$	$\bar{x} = 0$ $\bar{y} = 0$ $\bar{z} = 0$	$I_x = I_z = \frac{1}{12} m(3R^2 + L^2)$ $I_y = \frac{1}{2} mR^2$
Thin cylindrical shell 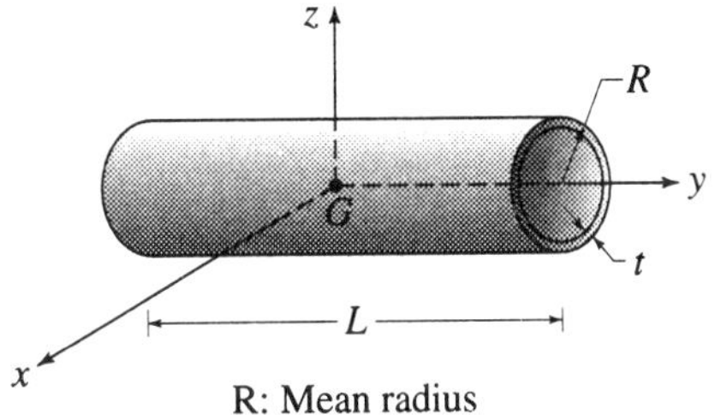 R: Mean radius	$V = 2\pi RtL$	$\bar{x} = 0$ $\bar{y} = 0$ $\bar{z} = 0$	$I_x = I_z = \frac{1}{4} m\left(2R^2 + \frac{1}{3} L^2\right)$ $I_y = mR^2$

<table>
<tr><th>Body and dimensions</th><th>Volume</th><th>Center of mass</th><th>Centroidal Moments of Inertia</th></tr>
<tr><td>Long slender rod
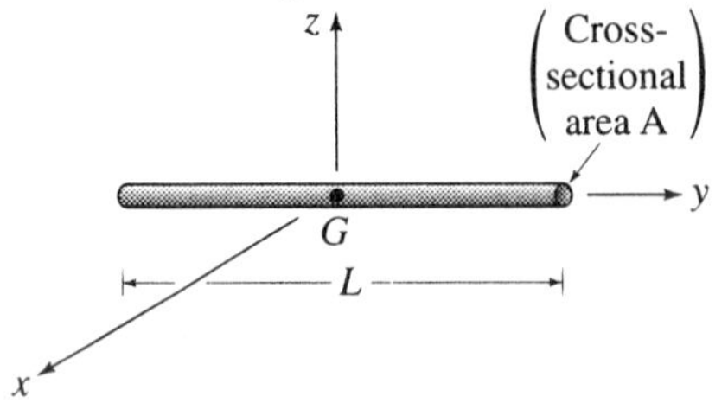</td><td>$V = AL$</td><td>$\bar{x} = 0$
$\bar{y} = 0$
$\bar{z} = 0$</td><td>$I_x = I_z = \frac{1}{12} mL^2$</td></tr>
<tr><td>Thin rectangular plate
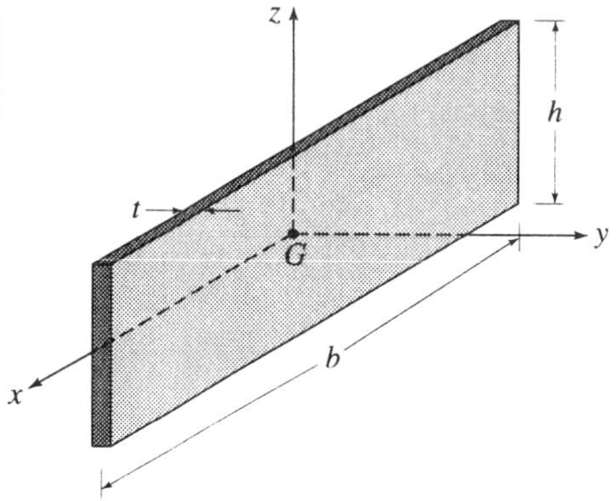</td><td>$V = bht$</td><td>$\bar{x} = 0$
$\bar{y} = 0$
$\bar{z} = 0$</td><td>$I_x = \frac{1}{12} mh^2$
$I_y = \frac{1}{12} m(h^2 + b^2)$
$I_z = \frac{1}{12} mb^2$</td></tr>
<tr><td>Rectangular prism
(or parallelepiped)
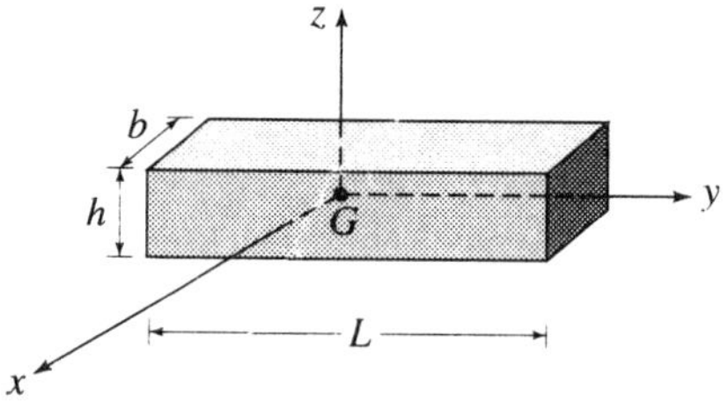</td><td>$V = bhL$</td><td>$\bar{x} = 0$
$\bar{y} = 0$
$\bar{z} = 0$</td><td>$I_x = \frac{1}{12} m(L^2 + h^2)$
$I_y = \frac{1}{12} m(h^2 + b^2)$
$I_z = \frac{1}{12} m(L^2 + b^2)$</td></tr>
</table>

Body and dimensions	Volume	Center of mass	Centroidal Moments of Inertia
Solid right circular cone 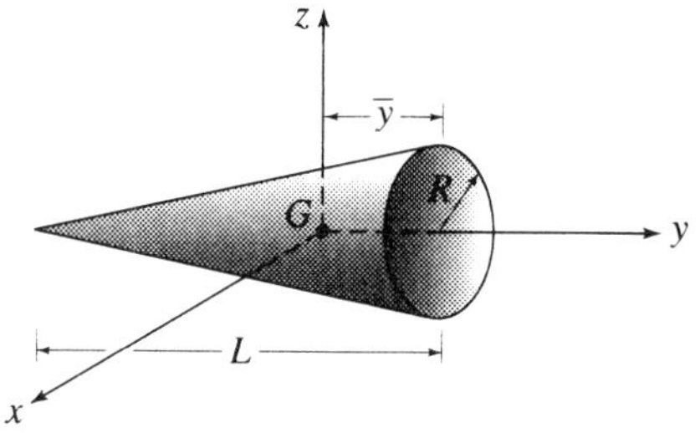	$V = \frac{1}{12}\pi D^2 L$	$\bar{x} = 0$ $\bar{z} = 0$ $\bar{y} = \frac{1}{4}L$	$I_x = I_z = \frac{3}{80}m\left(4R^2 + \frac{1}{3}L^2\right)$ $I_y = \frac{3}{30}mR^2$
Thin circular conical shell 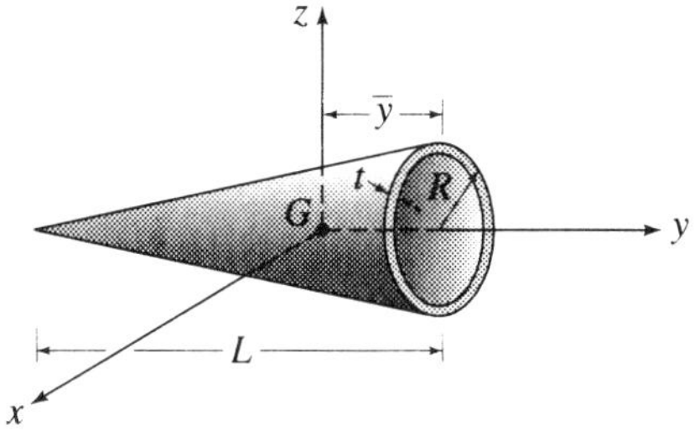 R: Mean radius	$V = \pi R t L$	$\bar{x} = 0$ $\bar{z} = 0$ $\bar{y} = \frac{1}{3}L$	$I_x = I_z = \frac{1}{12}m\left(\frac{1}{2}R^2 + \frac{1}{9}L^2\right)$ $I_y = \frac{1}{12}mR^2$
Solid sphere 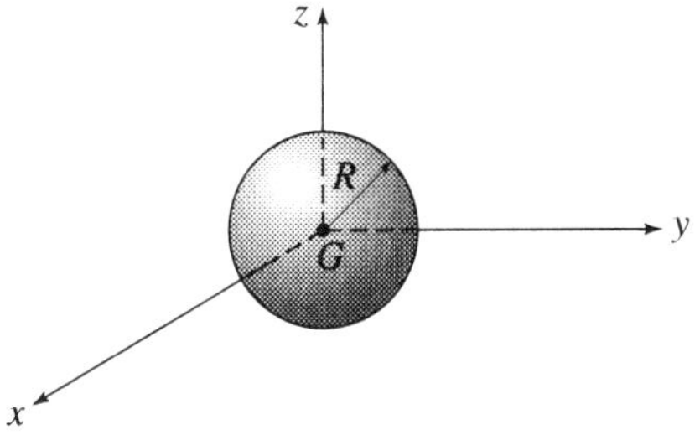	$V = \frac{4}{3}\pi R^3$	$\bar{x} = 0$ $\bar{y} = 0$ $\bar{z} = 0$	$I_x = I_y = I_z = \frac{2}{5}mR^2$
Thin spherical shell 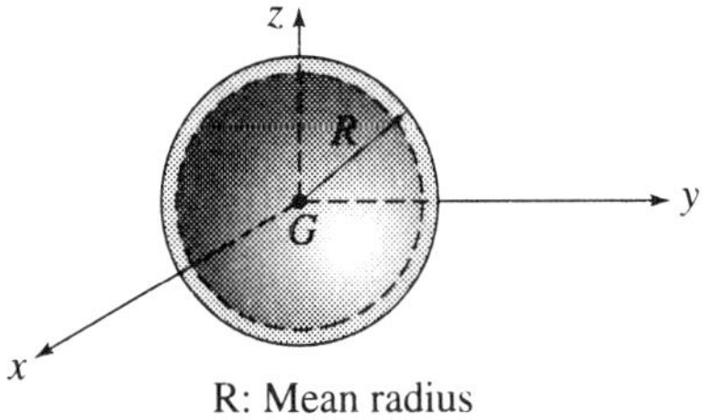 R: Mean radius	$V = 4\pi R t$	$\bar{x} = 0$ $\bar{y} = 0$ $\bar{z} = 0$	$I_x = I_y = I_z = \frac{2}{3}mR^2$

Body and dimensions	Volume	Center of mass	Centroidal Moments of Inertia
Thin circular disk 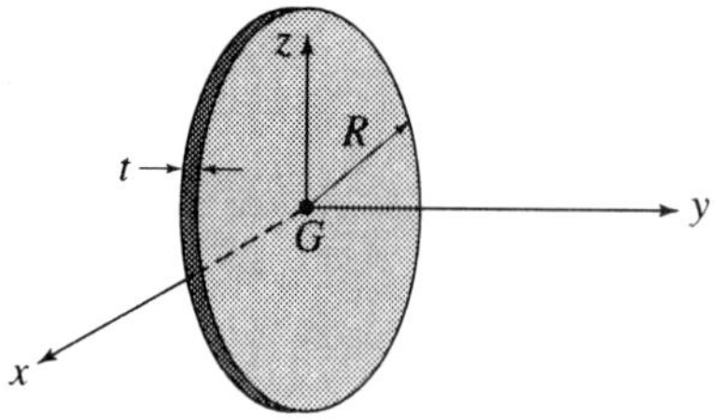	$V = \pi R^2 t$	$\bar{x} = 0$ $\bar{y} = 0$ $\bar{z} = 0$	$I_x = I_z = \frac{1}{4} mR^2$ $I_y = \frac{1}{2} mR^2$
Hemisphere 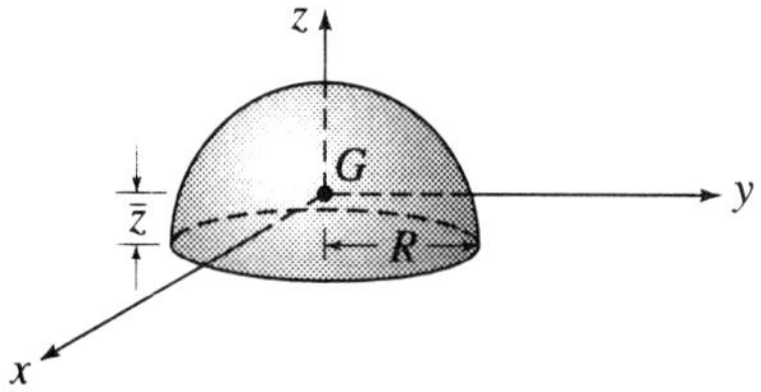	$V = \frac{2}{3}\pi R^3$	$\bar{x} = 0$ $\bar{y} = 0$ $\bar{z} = \frac{3}{8} R$	$I_x = I_y = \frac{83}{640} mR^2$ $I_z = \frac{2}{5} mR^2$
Circular paraboloid 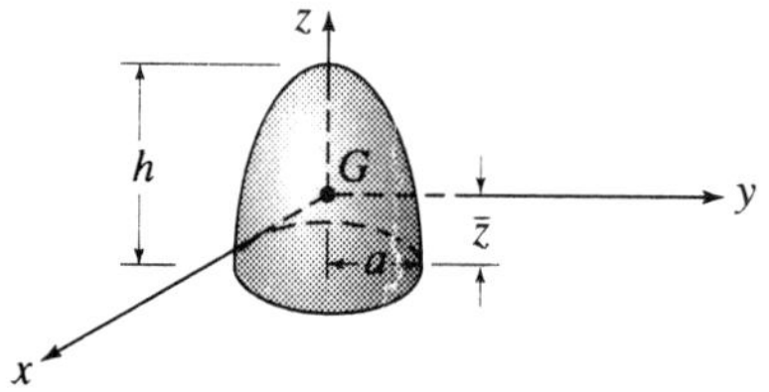	$V = \frac{1}{2}\pi h a^2$	$\bar{x} = 0$ $\bar{y} = 0$ $\bar{z} = \frac{1}{3} h$	$I_x = I_y = \frac{1}{15} m(3a^2 + h^2)$ $I_z = \frac{1}{3} ma^2$
Semicylinder 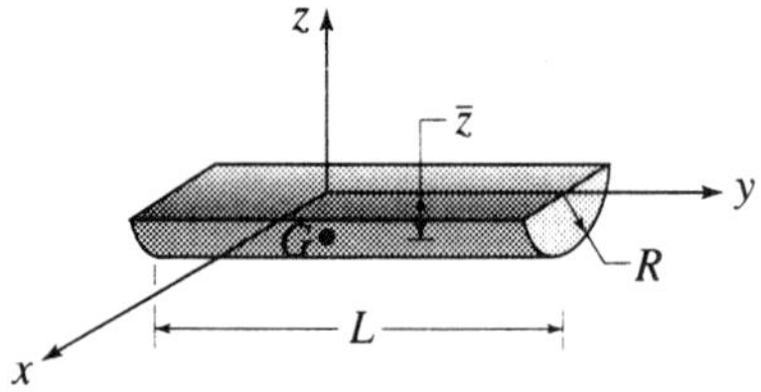	$V = \frac{1}{2}\pi R^2 L$	$\bar{x} = 0$ $\bar{y} = 0$ $\bar{z} = \frac{4R}{3\pi}$	$I_x = I_z = \frac{1}{4} m\left(R^2 + \frac{L^2}{3}\right)$ $I_y = \frac{1}{2} mR^2$

Appendix C Useful Mathematical Relations

Approximations of Areas

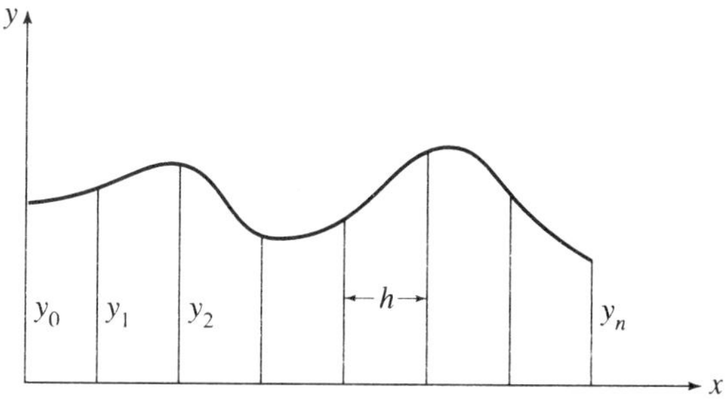

Trapezoidal Rule

$$A = h\left(\frac{1}{2}y_0 + y_1 + y_2 + \cdots + y_{n-1} + \frac{1}{2}y_n\right)$$

Simpson's Rule (Note that n must be even)

$$A = \frac{h}{3}(y_0 + 4y_1 + 2y_2 + 4y_3 + 2y_4 + \cdots + 2y_{n-2} + 4y_{n-1} + y_n)$$

Trigonometric Functions

$$\sin\theta = \frac{b}{c}; \quad \csc\theta = \frac{1}{\sin\theta}$$

$$\cos\theta = \frac{a}{c}; \quad \sec\theta = \frac{1}{\cos\theta}$$

$$\tan\theta = \frac{b}{a}; \quad \cot\theta = \frac{1}{\tan\theta}$$

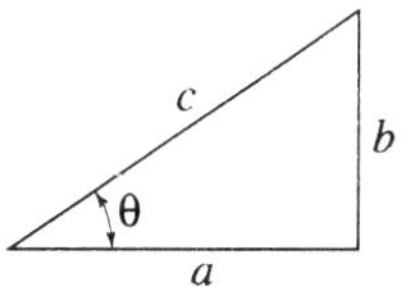

Selected Trigonometric Identities

$$\sin(\theta_1 \pm \theta_2) = \sin\theta_1 \cos\theta_2 \pm \cos\theta_1 \sin\theta_2$$

$$\cos(\theta_1 \pm \theta_2) = \cos\theta_1 \cos\theta_2 \mp \sin\theta_1 \sin\theta_2$$

$$\sin 2\theta = 2\sin\theta\cos\theta$$

$$\cos 2\theta = \cos^2\theta - \sin^2\theta$$

$$\sin^2\theta + \cos^2\theta = 1$$

$$2\cos^2\theta = 1 + \cos 2\theta$$

$$2\sin^2\theta = 1 - \cos 2\theta$$

$$\sec^2\theta = 1 + \tan^2\theta$$

$$\csc^2\theta = 1 + \cot^2\theta$$

Hyperbolic Functions

$$\sinh u = \frac{e^u - e^{-u}}{2}$$

$$\cosh u = \frac{e^u + e^{-u}}{2}$$

$$\tanh u = \frac{e^u - e^{-u}}{e^u + e^{-u}}$$

Series Expansions

$$\sin u = u - \frac{u^3}{3!} + \frac{u^5}{5!} - \frac{u^7}{7!} + \cdots.$$

$$\cos u = 1 - \frac{u^2}{2!} + \frac{u^4}{4!} - \frac{u^6}{6!} + \cdots.$$

$$\sinh u = u + \frac{u^3}{3!} + \frac{u^5}{5!} + \frac{u^7}{7!} + \cdots.$$

$$\cosh u = 1 + \frac{u^2}{2!} + \frac{u^4}{4!} + \frac{u^6}{6!} + \cdots.$$

Quadratic Equations

$$Ax^2 + Bx + C = 0$$

$$x = \frac{-B \pm \sqrt{B^2 - 4AC}}{2A}$$

Appendix D
Selected Derivatives

$$\frac{d}{du}(au) = a\frac{du}{dx}$$

$$\frac{d}{dx}(uv) = u\frac{dv}{dx} + v\frac{du}{dx}$$

$$\frac{d}{dx}\left(\frac{u}{v}\right) = \frac{1}{v^2}\left(v\frac{du}{dx} - u\frac{dv}{dx}\right)$$

$$\frac{d}{dx}(u^n) = nu^{n-1}\left(\frac{du}{dx}\right)$$

$$\frac{d}{dx}[f(u)] = \frac{d}{du}[f(u)]\left(\frac{du}{dx}\right)$$

$$\frac{d}{dx}(\ln u) = \frac{1}{u}\left(\frac{du}{dx}\right)$$

$$\frac{d}{dx}(e^u) = e^u\left(\frac{du}{dx}\right)$$

$$\frac{d}{dx}(\sin u) = \cos u\left(\frac{du}{dx}\right)$$

$$\frac{d}{dx}(\cos u) = -\sin u\left(\frac{du}{dx}\right)$$

$$\frac{d}{dx}(\tan u) = \sec^2 u\left(\frac{du}{dx}\right)$$

$$\frac{d}{dx}(\csc) = -(\csc u)(\cot u)\left(\frac{du}{dx}\right)$$

$$\frac{d}{dx}(\sec) = (\sec u)(\tan u)\left(\frac{du}{dx}\right)$$

$$\frac{d}{dx}(\cot u) = -(\csc^2 u)\left(\frac{du}{dx}\right)$$

$$\frac{d}{dx}(\sinh u) = \cosh u\left(\frac{du}{dx}\right)$$

$$\frac{d}{dx}(\cosh u) = \sinh u \left(\frac{du}{dx}\right)$$

$$\frac{d}{dx}(\tanh u) = \operatorname{sech}^2 u \left(\frac{du}{dx}\right)$$

$$\frac{d}{dx}(\operatorname{csch} u) = -(\operatorname{csch} u)(\coth u) \left(\frac{du}{dx}\right)$$

$$\frac{d}{dx}(\operatorname{sech} u) = -(\operatorname{sech} u)(\tanh u) \left(\frac{du}{dx}\right)$$

$$\frac{d}{dx}(\coth u) = -\operatorname{csch}^2 u \left(\frac{du}{dx}\right)$$

Appendix E
Selected Integrals

$$\int \sin x\,dx = -\cos x + c$$

$$\int \cos x\,dx = \sin x + c$$

$$\int \tan x\,dx = \ln \sec x + c$$

$$\int \csc x\,dx = \ln \tan\left(\frac{x}{2}\right) + c$$

$$\int \sec x\,dx = \ln \tan\left(\frac{\pi}{4} + \frac{x}{2}\right) + c$$

$$\int \cot x\,dx = \ln \sin x + c$$

$$\int x \sin(ax)\,dx = \frac{1}{a^2}\sin(ax) - \frac{x}{a}\cos(ax) + c$$

$$\int x^2 \sin(ax)\,dx = \frac{2x}{a^2}\sin(ax) - \left(\frac{a^2x^2 - 2}{a^3}\right)\cos(ax) + c$$

$$\int x \cos(ax)\,dx = \frac{1}{a^2}\cos(ax) + \frac{x}{a}\sin(ax) + c$$

$$\int x^2 \cos(ax)\,dx = \frac{2x\cos(ax)}{a^2} + \left(\frac{a^2x^2 - 2}{a^3}\right)\sin(ax) + c$$

$$\int \sinh x\,dx = \cosh x + c$$

$$\int \cosh x\,dx = \sinh x + c$$

$$\int \tanh x\,dx = \ln(\cosh x) + c$$

$$\int \operatorname{csch} x\,dx = \ln\left[\tanh\left(\frac{x}{2}\right)\right] + c$$

$$\int \operatorname{sech} x\, dx = \tan^{-1}(\sinh x) + c$$

$$\int \coth x\, dx = \ln(\sinh x) + c$$

$$\int a\, dx = ax + c$$

$$\int u\, dv = uv - \int v\, du + c$$

$$\int x^n\, dx = \frac{x^{n+1}}{n+1} + c \quad \text{except for } n = -1$$

$$\int e^{ax}\, dx = \frac{e^{ax}}{a} + c$$

$$\int \ln x\, dx = x \ln x - x + c$$

$$\int \frac{dx}{x(a+bx)} = -\frac{1}{a}\ln\left(\frac{a+bx}{x}\right) + c$$

$$\int \frac{dx}{a+bx^2} = \frac{1}{\sqrt{ab}}\tan^{-1}\left(x\sqrt{\frac{b}{a}}\right) + c \quad \text{for } a > 0 \text{ and } b > 0$$

$$\int \frac{dx}{a^2 - b^2x^2} = \frac{1}{2ab}\ln\left(\frac{a+bx}{a-bx}\right) + c$$

$$\int \frac{dx}{a+bx} = \frac{1}{b}\ln(a+bx) + c$$

$$\int \frac{x\, dx}{a+bx} = \frac{x}{b} - \frac{a}{b^2}\ln(a+bx) + c$$

$$\int \sqrt{a+bx}\, dx = \frac{2}{3b}\sqrt{(a+bx)^3} + c$$

$$\int x\sqrt{a+bx}\, dx = -\frac{2(2a-3bx)\sqrt{(a+bx)^3}}{15b^2} + c$$

$$\int x^2\sqrt{a+bx}\, dx = \frac{2(8a^2 - 12abx + 15b^2x^2)\sqrt{(a+bx)^3}}{105b^3} + c$$

$$\int \frac{dx}{\sqrt{a+bx}} = \frac{2\sqrt{(a+bx)}}{b} + c$$

$$\int \frac{x\, dx}{\sqrt{a+bx}} = -\frac{2(2a-bx)\sqrt{(a+bx)}}{3b^2} + c$$

$$\int \frac{x^2\,dx}{\sqrt{a+bx}} = \frac{2(8a^2 - 4abx + 3b^2x^2)\sqrt{(a+bx)}}{15b^3} + c$$

$$\int \frac{dx}{x\sqrt{a+bx}} = \frac{1}{\sqrt{a}} \ln\left(\frac{\sqrt{(a+bx)} - \sqrt{a}}{\sqrt{(a+bx)} + \sqrt{a}}\right) + c \quad a > 0$$

$$\int \frac{dx}{x\sqrt{a+bx}} = \frac{2}{\sqrt{-a}} \tan^{-1} \sqrt{\frac{(a+bx)}{a}} + c \quad a < 0$$

$$\int \frac{dx}{x^2\sqrt{a+bx}} = -\frac{\sqrt{(a+bx)}}{ax} - \frac{b}{2a} \int \frac{dx}{x\sqrt{a+bx}} + c$$

$$\int \sqrt{a^2 - x^2}\,dx = \frac{1}{2}\left(x\sqrt{a^2 - x^2} + a^2 \sin^{-1}\frac{x}{|a|}\right) + c$$

$$\int x\sqrt{a^2 - x^2}\,dx = -\frac{1}{3}\sqrt{(a^2 - x^2)^3} + c$$

$$\int x^2\sqrt{a^2 - x^2}\,dx = -\frac{x}{4}\sqrt{(a^2 - x^2)^3} + \frac{a^2}{8}\left(x\sqrt{(a^2 - x^2)} + a^2 \sin^{-1}\frac{x}{|a|}\right) + c \quad (a > 0)$$

$$\int \sqrt{a^2 \pm x^2}\,dx = \frac{1}{2}[x\sqrt{x^2 \pm a^2} \pm a^2 \ln(x\sqrt{x^2 \pm a^2})] + c$$

$$\int x\sqrt{a^2 \pm x^2}\,dx = \frac{1}{3}\sqrt{(x^2 \pm a^2)^3} + c$$

$$\int x^2\sqrt{a^2 \pm x^2}\,dx = \frac{x}{4}\sqrt{(x^2 \pm a^2)^3} \mp \left(\frac{a^2}{8}\right)x\sqrt{x^2 \pm a^2} - \left(\frac{a^4}{8}\right)\ln(x + \sqrt{x^2 \pm a^2}) + c$$

$$\int \frac{dx}{\sqrt{x^2 \pm x^2}} = \ln(x + \sqrt{x^2 \pm a^2}) + c$$

$$\int \frac{x\,dx}{\sqrt{x^2 \pm x^2}} = \sqrt{x^2 \pm a^2} + c$$

$$\int \frac{x^2\,dx}{\sqrt{x^2 \pm a^2}} = \frac{x}{2}\sqrt{x^2 \pm a^2} \mp \left(\frac{a^2}{2}\right)\ln(x + \sqrt{x^2 \pm a^2}) + c$$

$$\int x \ln x\,dx = \left(\frac{x^2}{2}\right)\ln x - \frac{x^2}{4} + c$$

Appendix F
Supports and Connections

<table>
<tr><th>Type of Connection or Support</th><th>Reactive Force Components</th><th>Special Features</th></tr>
<tr><td>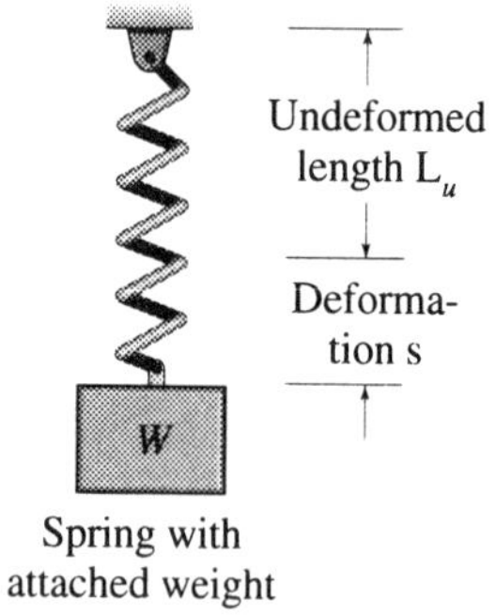

1</td><td>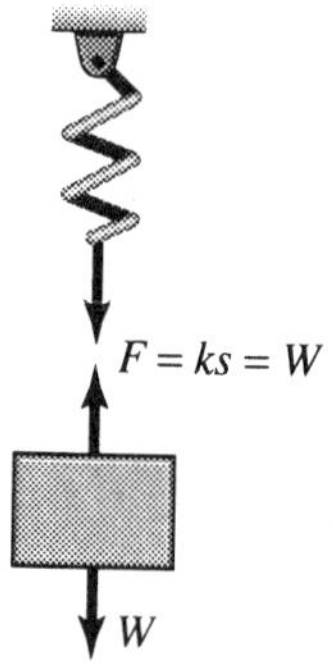
</td><td>The force F in a deformed spring is directed along its axis. The sense of this force is such that it is tension if the spring is stretched and compression if it is shortened. Also, $F = ks$ where k is known as the spring constant, equal to the force needed to deform the spring a unit distance and s is the total deformation.</td></tr>
<tr><td>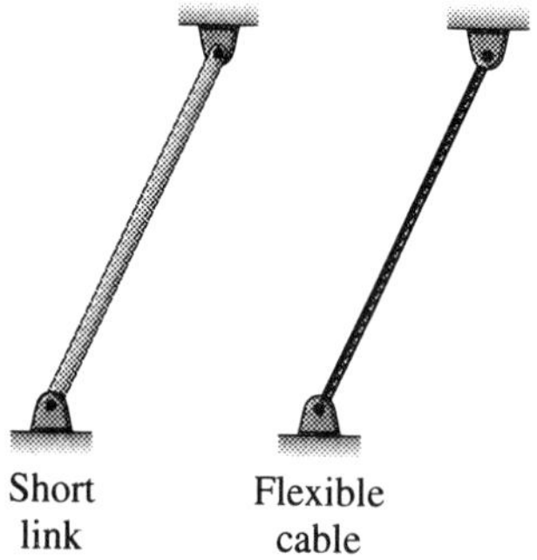

2</td><td>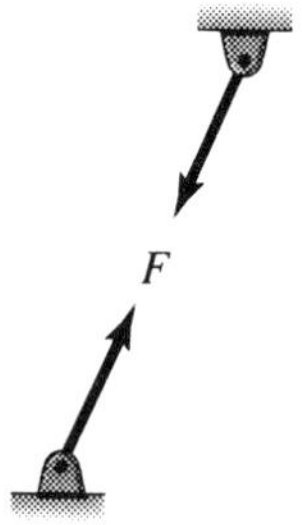
</td><td>One reactive force component F of known direction since it must act along the axis of the link or cable. This force F is unknown only in magnitude.</td></tr>
</table>

	Type of Connection or Support	Reactive Force Components	Special Features
3	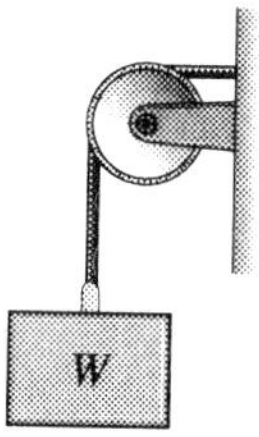 Flexible cable or belt around frictionless pulley. 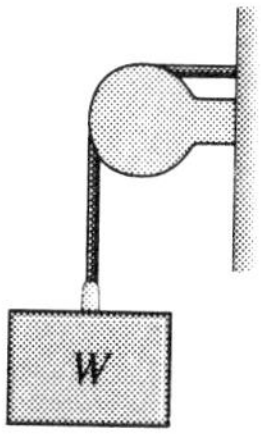 Flexible cable or belt around frictionless peg.	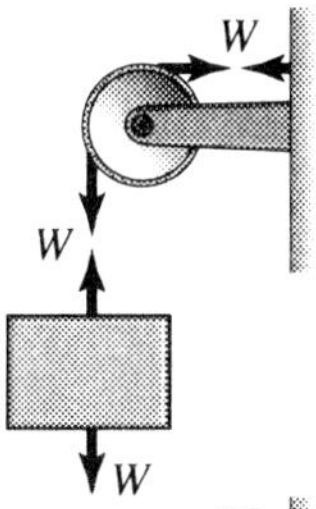 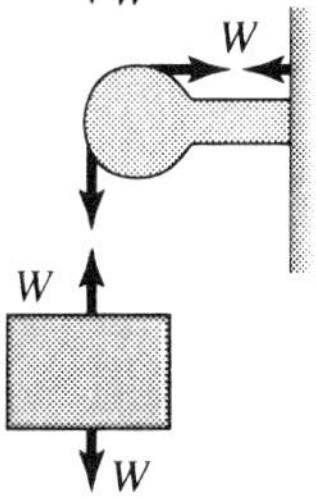	A frictionless pulley or frictionless peg changes the direction but not the magnitude of the force in the flexible cable or flexible belt. Thus, the force in the flexible cable or flexible belt is a constant along the entire length and must always be tension.
4	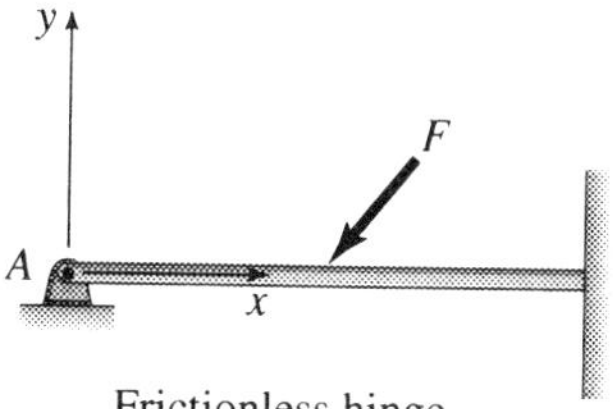 Frictionless hinge or pin.	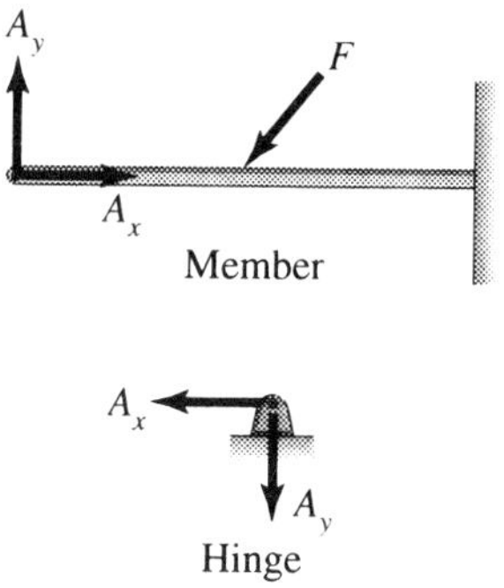	A frictionless hinge prevents any translation but allows rotation of the member about the pin axis. The reaction at the hinge is usually expressed in terms of its x and y components. Thus, the reaction at a frictionless hinge contains two unknowns.

Type of Connection or Support	Reactive Force Components	Special Features
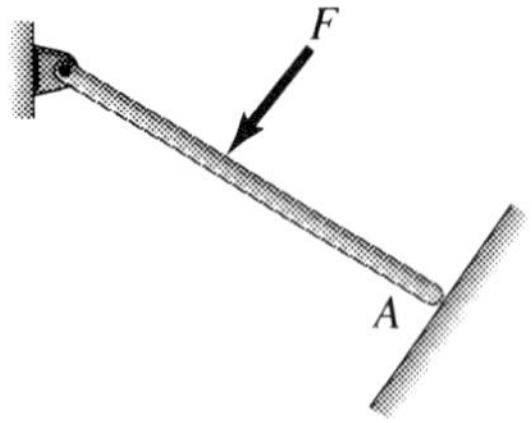 Frictionless plane 5	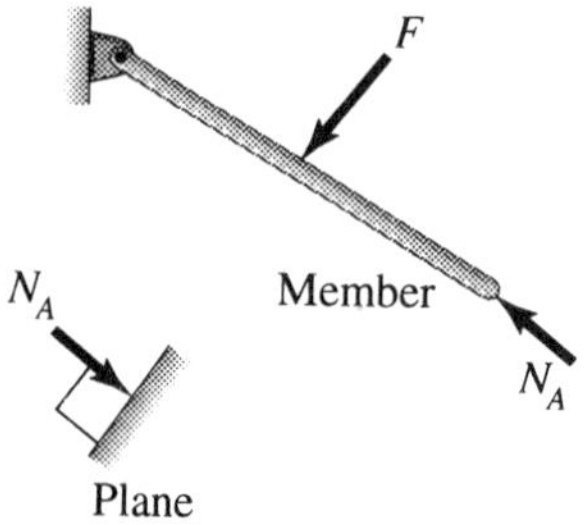	A frictionless plane prevents translation in a direction perpendicular to the plane but allows rotation and translation along the plane. Thus, the support reaction consists of only one component N perpendicular to the frictionless plane.
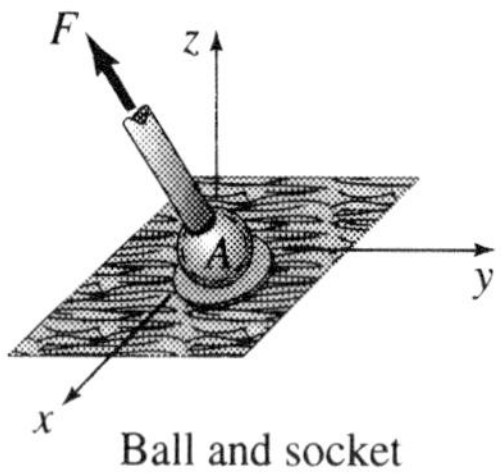 Ball and socket 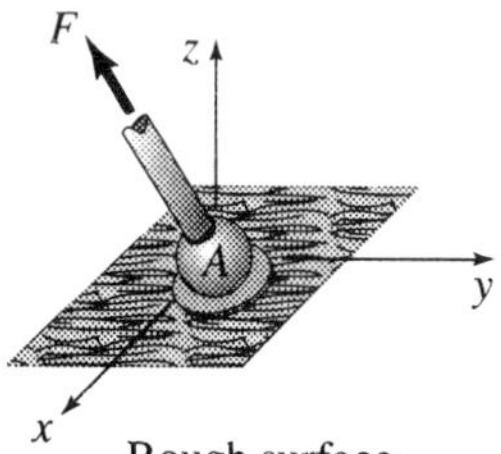 Rough surface 6	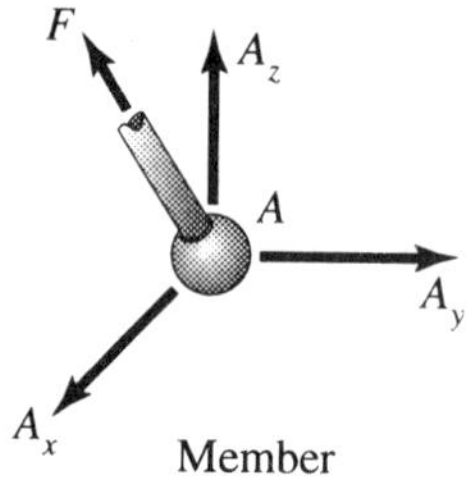 Member	These supports prevent translation in any direction but permit rotation about any axis. The reaction at the support is generally expressed in terms of its x, y, and z components. Thus, the reaction at the support contains three unknowns.

Type of Connection or Support	Reactive Force Components	Special Features
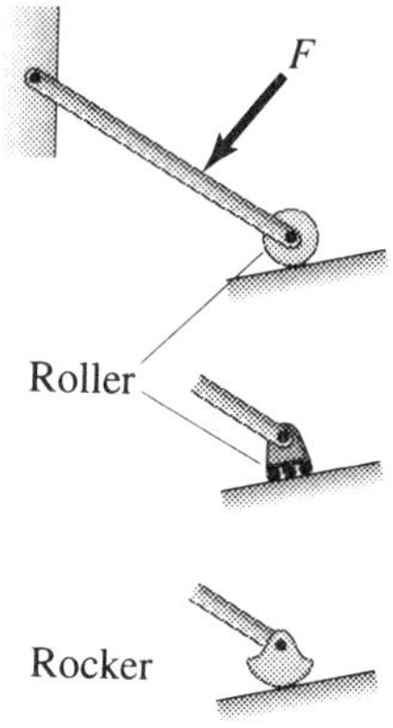 7	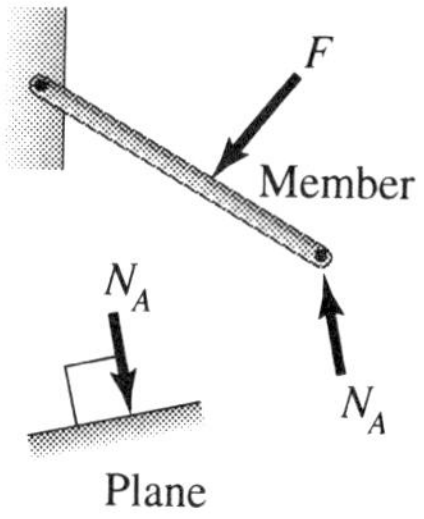	A roller or rocker support is capable of resisting a force only in a direction perpendicular to the plane supporting the roller or rocker. Both allow rotation and translation parallel to the plane and cannot transmit a moment or a force parallel to the plane. Thus, the support reaction consists of only one component N perpendicular to the plane.
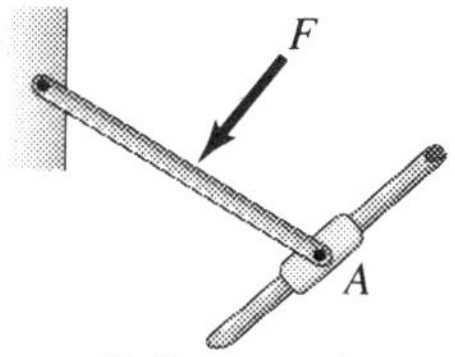 Collar on rod 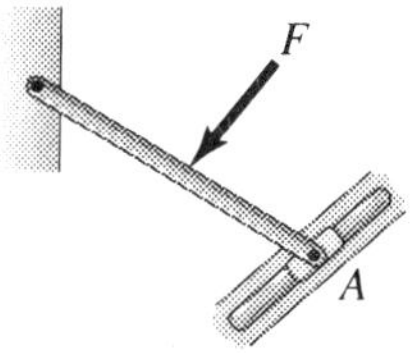 Slider in slot 8	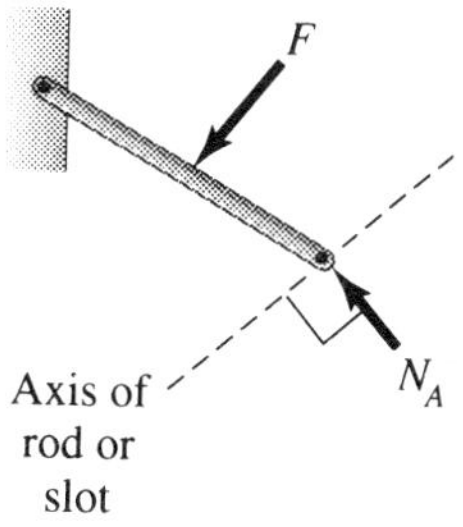	As in cases 5 and 7, the support reaction consists of only one component N perpendicular to the axis of the rod or the slot.
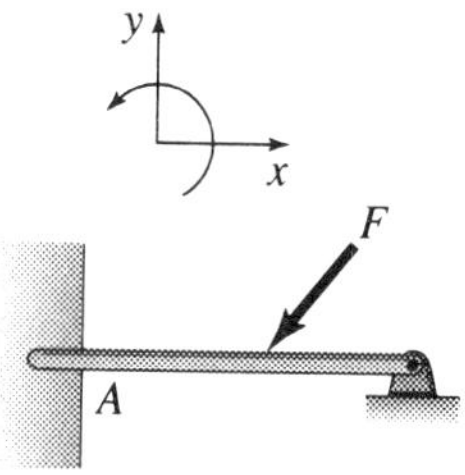 Fixed support 9	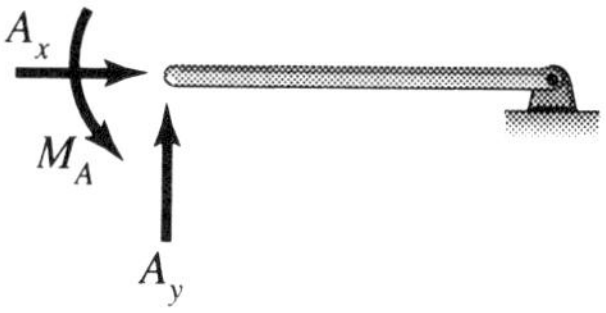	A fixed support prevents translation in any direction and rotation about a z axis at the support. Thus, the support reaction consists of x and y components as well as a moment about the z axis.

Type of Connection or Support	Reactive Force Components	Special Features
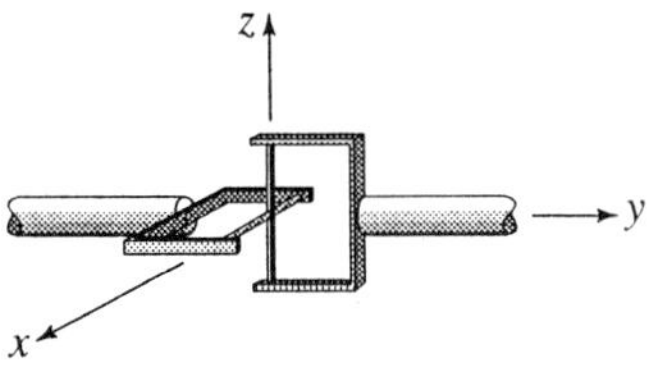 Universal joint 10	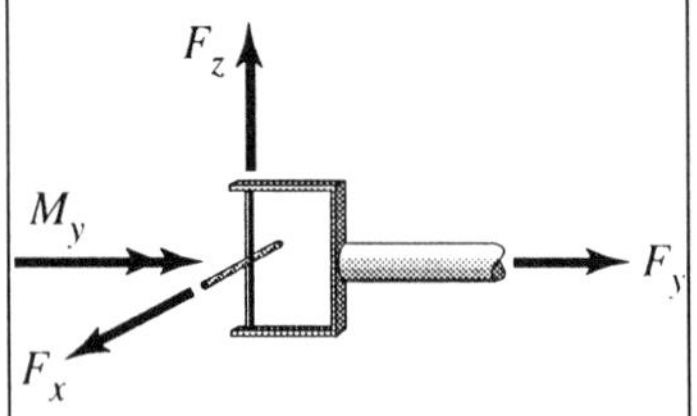	This universal joint allows only relative rotations about the y and z axis but no relative translation along any axis. Thus, there is only one reactive moment component and three reactive force components.
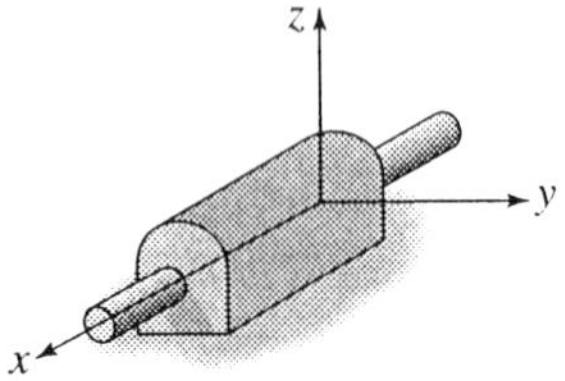 Bearing resisting no axial thrust. 11	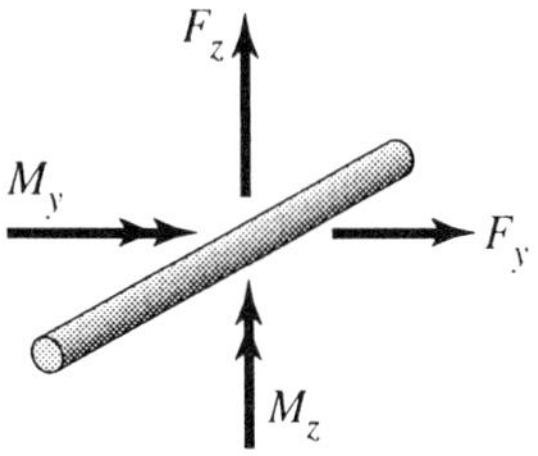	This bearing allows the shaft to translate only in the x direction and to rotate only about the x axis. Therefore, there are two force components and two moment components.
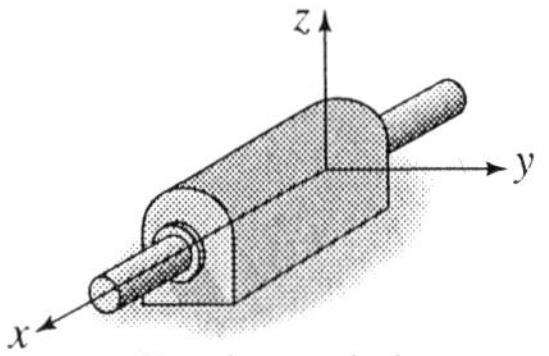 Bearing resisting axial thrust 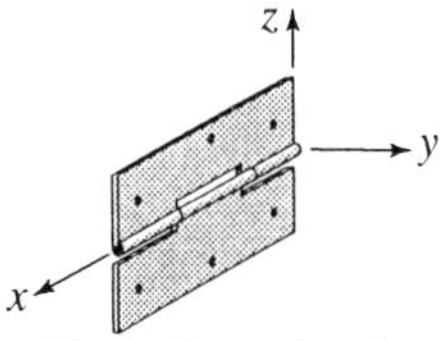 Three-dimensional hinge 12	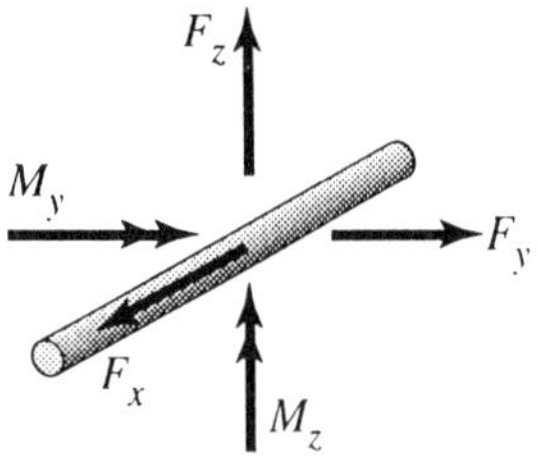 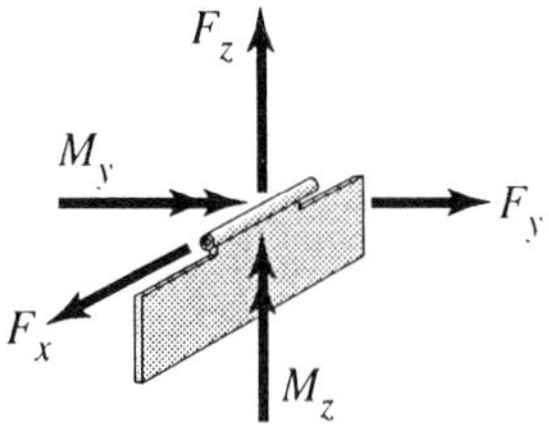	The bearing resisting axial thrust and the hinge are identical in action in that they allow only rotation about the axis of the shaft. Thus, there are three force components and two moment components.

Type of Connection or Support	Reactive Force Components	Special Features
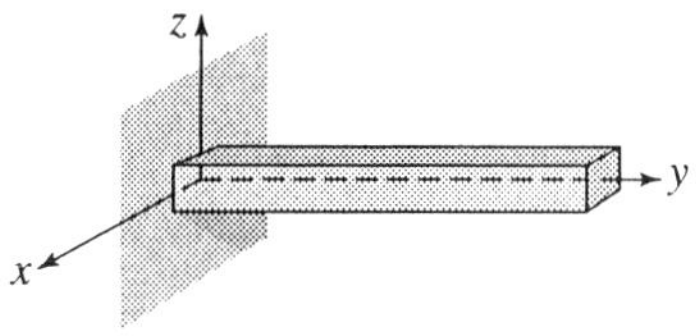 Three-dimensional fixed support 13	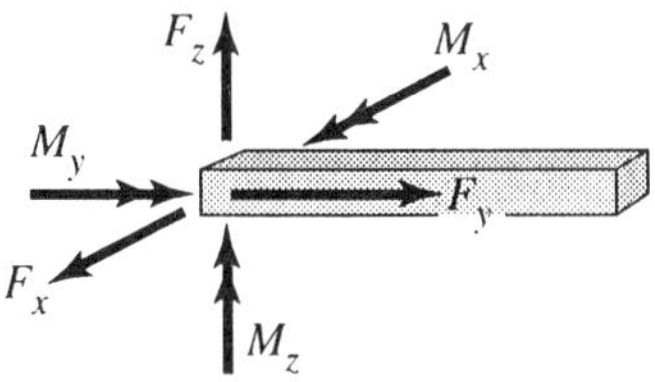	The fixed support prevents translation along and rotation about any axis. Therefore, this support develops three force and three moment reactive components.

Answers

Chapter 13

13.1 $v = 6t^2$, $a = 12t$.

13.3 $v = 2t - 4$, $t = 2$ s.

13.5 (a) $s = -6.00$ m; (b) $s = 0$; (c) $\Delta s = 6.00$ m; (d) $t = 1.423$ s; (e) DT $= 6.77$ m.

13.7 (a) $t = 12.00$ s; (b) $t = 8.00$ s; (c) $\Delta s = 0$; (d) DT $= 32.0$ ft.

13.9 (a) $v = 2{,}000 \cos 10t$; (b) $a = -20{,}000 \sin 10t$; (c) Amp. $= 200$ mm; (d) $s = 119.7$ mm, $v = -1602$ mm/s, $a = -11{,}970$ mm/s^2.

13.11 (a) $v = 2.00$ in./s; (b) $a = 0$; (c) $s = 273$ in.; (d) $v = 7.52$ in./s.

13.13 (a) $t = 0$ and 2.00 s; (b) $v = 3t^2 - 8t + 4$; (c) $t = 0.667$ and 2.00 s; (d) $\Delta s = 0$; (e) DT $= 5.60$ m.

13.15 (a) $a = 12.00$ m/s^2; (b) $s = 6t + 4$.

13.17 (a) $v = 2.00$ ft/s; (b) $a = 3.00$ ft/s^2; (c) $\Delta s = 4.67$ ft; (d) $v_{\text{AV.}} = 3.00$ ft/s; (e) $a_{\text{AV.}} = 3.00$ ft/s^2.

13.19 (a) $a = 8.00$ m/s^2; (b) $s = 4t^2 - 8$; (c) $v = 8t$.

13.21 (a) $s = -4.00$ m; (b) $s = 5.00$ m; (c) $\Delta s = 9.00$ m; (d) $t = 0.417$ s.

13.23 (a) $v = 10 \cos 5t$; (b) $a = -50 \sin 5t$; (c) Amp. $= 2.00$ ft; (d) $s = 1.683$ ft; $v = 5.40$ ft/s, $a = -42.1$ ft/s^2.

13.25 (a) $v = 0.500$ in./s; (b) $a = 0$; (c) $s = 5.88$ in.; (d) $v = 0.772$ in./s.

13.27 (a) $t = 0$, $s = -6.00$ in., $t = 3$s, $s = 0$; (b) $v = 3t^2 - 12t + 11$; (c) $t = 1.423$ and 2.58 s; (d) DT $= 6.77$ in.

13.29 (a) $a = 0.500$ in./s^2; (b) $s = \frac{1}{4}t^2 + t$.

13.31 (a) $a = 2t - 25$; (b) $s = \frac{1}{3}t^3 - \frac{25}{2}t^2 + 150t$; (c) $t = 10.00$ and 15.00 s; (d) $\Delta s = 479$ ft; (e) DT $= 571$ ft.

13.33 (a) $v = \left(\dfrac{10}{\cosh k - 1}\right)(\cosh kt + \cosh k - 2)$;
(b) $s = \left(\dfrac{10}{\cosh k - 1}\right)\left[\dfrac{1}{k}\sinh kt + (\cosh k - 2)t\right]$.

13.35 (a) $v = 0.25t^4 - 2t^3 + 5.5t^2 - 6t$; (b) $s = 0.05t^5 - 0.5t^4 + 1.833t^3 - 3t^2$.

13.37 (a) $v = -A\pi \sin \pi t$; (b) $s = A(\cos \pi t - 1) + 1$.

13.39 (a) $v = At - \dfrac{B}{\pi}\cos \pi t$; (b) $s = t^2 - \dfrac{1}{\pi^2}\sin \pi t$ for $A = 2$ and $B = 1$.

13.41 $t = 3$ s, $v = 16.00$ ft/s; $t = 10$ s, $v = 30.0$ ft/s.

13.43 (a) $v = 3t^2 - 16t + 17$; (b) $s = t^3 - 8t^2 + 17t - 10$;
(c) $\Delta s = 10.00$ m.

13.45 $a = -2t$, $v = \frac{1}{2}t^2 - 6t + 8$.

13.47 $s = -45.0$ m, $v = -51.0$ m/s, $a = -32.5$ m/s^2.

13.49 (a) $v = 2(\cos 2t + 1)$; (b) $s = \sin 2t + 2t + 2$.

13.51 $t = 0$, $s = 0$, $v = 0$, $a = 1.000$ m/s^2; $t = 2$ s, $s = 2.00$ m,
$v = 2.00$ m/s,
$a = \pm 1.000$ m/s^2; $t = 4$ s, $s = 4.00$ m, $v = 0$, $a = -1.000$ m/s^2.

13.53 $t = 0$, $s = 5.00$ m/s, $v = 10.00$ m/s, $a = 0$; $t = 10$ s, $s = 105.0$ m,
$v = 10.00$ m/s, $a = 0$; $t = 20$ s, $s = 372$ m, $v = 60.0$ m/s,
$a = 10.00$ m/s^2; $t = 24$ s, $s = 692$ m, $v = 100.0$ m/s,
$a = 10.00$ m/s^2.

13.55 $s = 76.0$ m, $v = 66.0$ m/s, $a = 48.0$ m/s^2.

13.57 $v = 4 - ks$, $k = 1.000 \dfrac{1}{s}$.

13.59 (a) $v = \sqrt{2s^2 + 112}$; (b) $t = \dfrac{1}{\sqrt{2}} \ln\left(\dfrac{s + \sqrt{s^2 + 56}}{12.49}\right)$.

13.61 (a) $v = \pm\sqrt{200s - (\frac{10}{3})s^3}$; (b) $v = \pm 22.6$ m/s.

13.63 (a) $v = v_0 + s - s_0$; (b) $s = v_0 e^t + s_0 - v_0$.

13.65 (a) $v = 10 - 0.2s$; (b) $s = 50.0$ m, (c) $t = \infty$, (d) $t = 19.56$ s.

13.67 $v = -10\sqrt{2}s^{-1/2}$, $v = -2.00$ in./s.

13.69 $\Delta s = 217$ ft, $t = 5.38$ s.

13.71 (a) $v = 40.0$ ft/s^2; (b) $\Delta s = 80.0$ ft.

13.73 $a_c = -8.00$ ft/s^2.

13.75 $\Delta s = 318$ m, $t = 9.53$ s.

13.77 $a_c = 25.3$ ft/s^2, $t = 1.779$ s.

13.79 $a_c = 22.0$ ft/s^2.

13.81 $t = 0.434$ s, $v_{AV.} = 25.4$ ft/s.

13.83 (a) $s_{A/B} = -15.00$ ft; (b) $s_{B/A} = 15.00$ ft; (c) $v_{A/B} = -35.0$ f/s;
(d) $a_{A/B} = -5.50$ ft/s^2.

13.85 $v_{A/B} = 35.0$ ft/s, $a_{A/B} = 6.50$ ft/s^2.

13.87 $v_{1/2} = 45.0$ mph.

13.89 $v_B = 0.200$ m/s $\downarrow$, $a_B = 0.050$ m/s^2 $\uparrow$.

13.91 $v_B = 0.375$ ft/s $\downarrow$, $v_C = 0.750$ ft/s $\downarrow$, $a_B = 0.1000$ ft/s^2 $\uparrow$,
$a_C = 0.200$ ft/s^2 $\uparrow$, $v_{B/A} = 1.875$ ft/s $\downarrow$, $a_{B/A} = 0.500$ ft/s^2 $\uparrow$.

13.93 $v_{B/A} = 15.00$ ft/s $\swarrow$, $a_{B/A} = 8.00$ ft/s^2 $\swarrow$.

13.95 $y = 2x$, $\mathbf{v} = \mathbf{i} + 2\mathbf{j}$, $\mathbf{a} = \mathbf{0}$.

13.97 (a) $y = 4 + x^2$, $\mathbf{v} = 4.00\mathbf{i} + 16.00\mathbf{j}$, $\mathbf{a} = 4.00\mathbf{i} + 48.0\mathbf{j}$.

13.99 $\mathbf{v} = (a\omega \cos \omega t)\mathbf{i} - (a\omega \sin \omega t)\mathbf{j}$,
$\mathbf{r} = (a \sin \omega t + c)\mathbf{i} + (a \cos \omega t + d)\mathbf{j}$, $(x - c)^2 + (y - d)^2 = a^2$.

13.101 $\mathbf{v} = 4.00\mathbf{i} + 8.00\mathbf{j}$, $\mathbf{r} = 0.500\mathbf{i} + 2.50\mathbf{j}$.

13.103 $(x/a)^2 + (y/b)^2 = 1$, $\mathbf{v} = (a\omega \cos \omega t)\mathbf{i} - (b\omega \sin \omega t)\mathbf{j}$,
$\mathbf{a} = -[(a\omega^2 \sin \omega t)\mathbf{i} + (b\omega^2 \cos \omega t)\mathbf{j}]$.

13.105 $y = 6 + 24t$, $\mathbf{r} = (4t)\mathbf{i} + (6 + 24t)\mathbf{j}$, $\mathbf{v} = 4.00\mathbf{i} + 24.0\mathbf{j}$, $\mathbf{a} = \mathbf{0}$.

13.107 Trajectory 3 by 26.4 ft.

13.109 (a) $x = 200 + 383t$, $y = 400 + 321t - 16.1t^2$; (b) $x = 8400$ ft.
13.111 $6.99° < \theta < 7.97°$.
13.113 $v = 31.0$ in./s, $a_t = -26.0$ in./s^2, $a_n = 96.1$ in./s^2.
13.115 $a = 1.000$ m/s^2 $\uparrow$.
13.117 $\mathbf{v} = (2t - 8)(0.707\mathbf{i} - 0.707\mathbf{j})$, $\mathbf{a} = 2(0.707\mathbf{i} - 0.707\mathbf{j})$.
13.119 $\mathbf{v} = 0$, $a_t = 1.000$ ft/s^2, $a_n = 0$.
13.121 $a = 37.1$ m/s^2 24.9° .
13.123 $a = 14.28$ m/s^2 65.2° .
13.125 $a = 32.2$ ft/s^2 directed toward the axis of rotation.
13.127 (a) $\mathbf{v} = \mathbf{i} - (\sin t)\mathbf{j}$, $\mathbf{a} = -(\cos t)\mathbf{j}$, (b) $\mathbf{v} = \mathbf{i}$, $\mathbf{a} = \mathbf{j}$.
13.129 The path is a straight line at $\pi/4$ ccw from the x axis, $v = 6.00$ in./s, $a = 2.00$ in./s^2.
13.131 (a) $\mathbf{v} = 9.00\boldsymbol{\lambda}_t$, $\mathbf{a} = 2.00\boldsymbol{\lambda}_t + 8.10\boldsymbol{\lambda}_n$; (b) $\mathbf{v} = 9.00\boldsymbol{\lambda}_\theta$, $\mathbf{a} = 2.00\boldsymbol{\lambda}_\theta - 8.10\boldsymbol{\lambda}_r$.
13.133 $\mathbf{v} = -2.00\boldsymbol{\lambda}_r + 4.00\boldsymbol{\lambda}_\theta$, $\mathbf{a} = -4.00\boldsymbol{\lambda}_r - 8.00\boldsymbol{\lambda}_\theta$.
13.135 (a) $v = 6.85$ ft/s $\uparrow$; (b) $v = 6.85$ ft/s $\uparrow$.
13.137 $a = 42.6$ ft/s^2 $\uparrow$.
13.139 (a) $\mathbf{r} = 6.62\boldsymbol{\lambda}_r$; (b) $\mathbf{v} = -0.734\boldsymbol{\lambda}_r + 6.62\boldsymbol{\lambda}_\theta$.
13.141 $\mathbf{v} = 28{,}105\boldsymbol{\lambda}_r + 4{,}220\boldsymbol{\lambda}_\theta$, $\mathbf{a} = 1{,}425{,}000\boldsymbol{\lambda}_r + 224{,}840\boldsymbol{\lambda}_\theta$.
13.143 $v = \sqrt{2}b\omega e^{\omega t}$, $a = 2b\omega^2 e^{\omega t}$.
13.145 $a = \sqrt{(\ddot{r} - r\dot{\theta}^2)^2 + (r\ddot{\theta} + 2\dot{r}\dot{\theta})^2 + \ddot{z}^2}$.
13.147 $\mathbf{a} = (\ddot{r} - r\dot{\theta}^2)\boldsymbol{\lambda}_r + (r\ddot{\theta} + 2\dot{r}\dot{\theta})\boldsymbol{\lambda}_\theta + \ddot{z}\boldsymbol{\lambda}_z$.
13.149 $\mathbf{v}_{B/A} = 30.0\mathbf{i} + 40.0\mathbf{j}$, $\mathbf{a}_{B/A} = -40.0\mathbf{i} + 18.00\mathbf{j}$.
13.151 $v_{B/A} = 43.6$ mph 36.6° , $a_{B/A} = 7200$ mph^2 60.0° .
13.153 (a) $v_{A/B} = 12.21$ m/s 35.0° , $r_{A/B} = 154.0$ m 13.13° .
13.155 $v_R = 18.54$ ft/s, $v_{B/R} = 25.7$ ft/s 51.2° .
13.157 $\mathbf{r}_{B/A} = 0.173\mathbf{i} - 0.379\mathbf{j}$, $\mathbf{v}_{B/A} = 0.757\mathbf{i} - 0.653\mathbf{j}$.
13.159 $v = 15.38$ m/s, $s = 12.82$ m.
13.161 (a) $v = 4t^2 + t + 60$; (b) $s = \frac{4}{3}t^3 + \frac{1}{2}t^2 + 60t + 200$.
13.163 (a) $v = 100 - 2s$; (b) $v = 100e^{-2t}$.
13.165 $v_B = 12.37$ m/s.
13.167 (a) $\alpha = 377$ rad/s^2; (b) $\Delta\theta = 7.50$ rev.
13.169 $v_A = 12.00$ m/s 30° , $a_A = 6.00$ m/s^2 30° .
13.171 $x_1 = 1.219$ km, $v_1 = 487$ km/h. 37.6° .
13.173 $a = 10.73$ m/s^2 21.3° .
13.175 $a = 6.75$ m/s^2 directed toward point B.
13.177 $\mathbf{r}_{A/B} = -148.7\mathbf{i} + 100.0\mathbf{j}$, $\mathbf{v}_{A/B} = -6.21\mathbf{i} - 10.00\mathbf{j}$, $\mathbf{a}_{A/B} = -1.732\mathbf{i} - 1.000\mathbf{j}$.

Chapter 14

14.1 $v = 8.44$ ft/s.
14.3 $s = 2.49$ m.
14.5 $\mu = 0.267$.

14.7 (a) $a = 3.51\ \text{m/s}^2$  ; (b) $s = 43.9$ m;
(c) $v = 3.75$ m/s  .

14.9 $W = 112.0$ lb.

14.11 $P = 569$ N, $v = 6.00$ m/s.

14.13 (a) $a = (g/W)(k\delta - W)\uparrow$; (b) $a = (g/W)(W - k\delta)\downarrow$.

14.15 $F_A = 35.7$ lb, $F_B = 72.4$ lb.

14.17 $a = 0.095\ \text{m/s}^2 \uparrow$, $t = 10.53$ s.

14.19 (a) $F = 5{,}280$ lb; (b) $Q = 622$ lb (C).

14.21 $F_{AB} = 11{,}380$ lb (C), $F_{AC} = 3{,}950$ lb (T).

14.23 (a) $v = 38.7$ ft/s, $s = 96.7$ ft; (b) $t = 4.50$ s.

14.25 $h = 9{,}680$ ft.

14.27 $h = (W/g)\left[\dfrac{v_1 - v_2}{C} + \dfrac{W}{C^2}\ln\left(\dfrac{W - Cv_1}{W - Cv_2}\right)\right]$.

14.29 $v = 60.8$ ft/s.

14.31 $F = 210$ lb.

14.33 (a) $T = 0.085$ lb; (b) $T = 0.335$ lb.

14.35 (a) $v = \sqrt{gL\tan\theta\sin\theta}$; (b) $\omega = 3.62$ rad/s.

14.37 (a) $F_n = 4{,}010$ lb; (b) $F_n = 7{,}130$ lb.

14.39 (a) $v = \sqrt{2Rg\sin\theta}$, $N = 3W\sin\theta$;
(b) $v = 17.94$ ft/s, $N = 7.50$ lb;
(c)$v_{MAX} = 25.4$ ft/s, $N_{max} = 15.00$ lb both for $\theta = 90°$.

14.41 $\cos\theta = 1 + \dfrac{v_2^2 - v_1^2}{2gR}$, $N = (W/2gR)(2gR - v_1^2 + 3v_2^2)$, $\theta = 86.5°$,
$N = 181.7$ lb both for $v_2 = 0$.

14.43 (a) $R = 962$ ft; (b) $N = 320$ lb.

14.45 $T_{AC} = 8.66$ N, $v = 1.763$ m/s.

14.47 (a) $v = \sqrt{gR\left(\dfrac{\tan\theta + \mu}{1 - \mu\tan\theta}\right)}$; (b) $v = \sqrt{gR\tan\theta}$, $\theta = 33.8°$.

14.49 $N_A = 31.3$ lb, $N_B = 21.7$ lb.

14.51 $t = \sqrt{\dfrac{T_0 + \mu mg}{mR\alpha^2}}$, $t = 2.74$ s.

14.53 $v_B = 47.9$ mph.

14.55 (a) $R_1/R_2 = 0.500$; (b) $\alpha = 48.2°$.

14.57 $\omega_{MIN} = 3.28$ rad/s.

14.59 (a) $v_{MAX} = 17.16$ m/s; (b) $N = 3.14$ kN.

14.61 $k = 20.3$ lb/ft.

14.63 See answers to Problem 14.35.

14.65 See answers to Problem 14.39.

14.67 See answers to Problem 14.43.

14.69 (a) $F_r = 76.4$ N, $F_\theta = -9.60$ N, $F_z = 40.0$ N;
(b) $F_r = 58.4$ N, $F_\theta = 14.40$ N, $F_z = 41.2$ N.

14.71 $F_r = -0.1379$ lb, $F_\theta = -0.412$ lb, $F_z = 1.969$ lb.

14.73 $\mu = 0.1510$.

14.75 (a) $F_r = mb\omega_0^2(2\cos 2\theta - \sin^2\theta)$, $F_\theta = 2mb\omega_0^2 \sin 2\theta$;
(b) $(F_r)_{MAX} = 2mb\omega_0^2$ at $\theta = 0°$, $(F_\theta)_{MAX} = 2mb\omega_0^2$ at $\theta = 45°$.

14.77 $F_r = -\dfrac{m(R^2\omega_0^2)(\sin\theta\tan^2\theta)}{h(1 - \sin^2\theta - \mu\sin\theta\cos\theta)}$.

14.79 $F_A/W = -113.0$, $F_S/W = 84.9$.

14.81 (a) $v = 17{,}260$ mph, $e = 0$;
(b) $17{,}260 < v < 24{,}400$ mph, $0 < e < 1.000$;
(c) $v = 24{,}400$ mph, $e = 1.000$; (d) $v > 24{,}400$ mph, $e > 1.000$.

14.83 (a) $e = 0.935$; (b) $h_D = 73{,}500$ mi, $v_D = 338$ mph; (c) $\tau = 132.2$ hr.

14.85 (a) $\Delta v_D = 3{,}240$ mph; (b) $\tau_{ss} = 10.63$ hr, $\tau_{cs} = 24.0$ hr.

14.87 (a) $\tau = 9.67$ h; (b) $\Delta v_B = 1496$ m/s.

14.89 (a) $\Delta v_D = -82.0$ m/s; (b) $\Delta v_B = -85.0$ m/s.

14.91 $\theta = 133.1°$.

14.93 (a) $\Delta v_B = 1{,}300$ m/s; (b) $\Delta v_B = 2{,}020$ m/s.

14.95 (a) $v'_B = 12{,}970$ ft/s; (b) $\Delta v_{D'} = 1{,}223$ ft/s.

14.97 $r = (1.7152 \times 10^{11})/(1 + 0.558\cos\theta)$.

14.99 (a) $v = 4.68$ m/s $\uparrow$; (b) $t = 2.14$ s.

14.101 $a = 5.10$ m/s^2.

14.103 $m = 1{,}043$ kg, $F_{AB} = 24{,}500$ N(C).

14.105 (a) $v = 131.0$ km/h; (b) $N = 1.600W$.

14.107 (a) $N = 1{,}415$ lb; (b) $N = 4{,}580$ lb.

14.109 (a) $F = 2.20$ kN $\nearrow$; (b) $F = 1.286$ kN $\uparrow$.

14.111 $r = 2R\sin\theta$, (a) $F_R = 4R\omega^2\left(\dfrac{W}{g}\right) \leftarrow$; (b) $F_R = 4R\omega^2\left(\dfrac{W}{g}\right) \downarrow$.

14.113 (a) $\Delta v_B = 529$ km/h; (b) $\Delta v_D = 500$ km/h.

Chapter 15

15.1 $U_{1\to2} = -900$ lb·ft.

15.3 $U_{1\to2} = 6{,}200$ lb·ft.

15.5 $U_{1\to2} = 3{,}600$ lb·ft.

15.7 $U_{1\to2} = -5{,}100$ lb·ft.

15.9 $U_{1\to2} = \left(\dfrac{1}{k}\right)W^2$.

15.11 $U_{1\to2} = \pi C - 8b^2k$.

15.13 $U_{1\to2} = \pi C - 8b^2k$.

15.15 $v_2 = 25.6$ ft/s, $a = 1.932$ ft/s^2.

15.17 $v_2 = 1.774b\sqrt{\dfrac{k}{m}}$.

15.19 $a_1 = (2.598kb/m) - 0.5g$, $a_2 = -0.5g$.

15.21 $v_2 = 10.95$ ft/s 45° $\swarrow$.

15.23 $v_2 = \sqrt{4gr\sin\theta}$, $v_3 = \sqrt{2gr(1 + \sin\theta)}$; for $\theta = 90°$,
$v_2 = v_3 = 2\sqrt{gr}$.

15.25 $v_2 = \sqrt{2gy + \left(\frac{k}{m}\right)(\sqrt{y^2 + b^2} - b^2)}.$
15.27 $(v_A)_2 = (v_B)_2 = 5.63$ m/s.
15.29 $s = 4.97$ ft to the left.
15.31 $(v_A)_2 = (v_B)_2 = 6.62$ m/s.
15.33 $Q = 13.70$ lb.
15.35 $v_A = 8.03$ ft/s $\downarrow$, $v_B = 19.26$ ft/s $\leftarrow$.
15.37 $s = (2m_B g \sin\beta)/k$; for $\beta = 0°$, system does not move; for $\beta = 90°$, $s = (2m_B g)/k$.
15.39 (a) $U_{OB} = 60.0$ lb·ft; (b) $U_{OAB} = 40.0$ lb·ft.
15.41 $U_{OAO} = 0$.
15.43 **F** is nonconservative.
15.45 $k = 2{,}030$ lb/ft.
15.47 $k = 2mg/(h - 2h^{1/2} + 1)$.
15.49 $x = 1.843$ ft.
15.51 $h = 2.00$ ft, $a = 225$ ft/s^2 $\uparrow$.
15.53 $v_2 = \sqrt{2g(h_1 - h_2) + v_1^2}$, $N = m\left(g + \frac{v_2^2}{\rho}\right)$.
15.55 $s = \frac{W\sin\beta}{k} + \sqrt{\left(\frac{W\sin\beta}{k}\right)^2 + \frac{2Wd\sin\beta}{k}}$.
15.57 $(v_A)_2 = 4.25$ ft/s ∠45°, $(v_B)^2 = 4.25$ ft/s $\downarrow$.
15.59 $(v_C)_2 = 2.01$ ft/s ∠30° (downward to the right).
15.61 $U_{1\to2} = 6.42$ N·m.
15.63 $h = 5.27$ m.
15.65 $v_A = 9.81$ m/s $\rightarrow$, $v_B = 3.27$ m/s $\downarrow$.
15.67 $\rho_B = 16.00$ m.

Chapter 16

16.1 $\mathbf{L}_1 = (50.3 \text{ lb·s})\mathbf{j}$, $\mathbf{L}_2 = -(51.8 \text{ lb·s})\mathbf{i}$.
16.3 (a) $\mathbf{L}_1 = (2.00 \text{ lb·s})\mathbf{i}$, (b) $y = x^2 - 4x + 8$, (c) $\mathbf{a} = (8.00 \text{ ft/s}^2)\mathbf{j}$.
16.5 $r = \frac{2}{\cos\theta}$, $\mathbf{v} = (23.1\boldsymbol{\lambda}_r + 14.80\boldsymbol{\lambda}_\theta)$ m/s, $\mathbf{L} = (46.2\boldsymbol{\lambda}_r + 29.6\boldsymbol{\lambda}_\theta)$ N·s.
16.7 $\mathbf{L} = (-0.390\mathbf{i} + 0.553\mathbf{j})$ N·s.
16.9 $I_x = 200$ kN·s.
16.11 $I_z = 4.00$ kN·s.
16.13 $I_y = 5.88$ kN·s.
16.15 (a) $I_z = 29.3$ N·s; (b) $I_z = 30.0$ N·s; (c) % Error = 2.39.
16.17 $F_x = 10.00$ N $\leftarrow$, $a_x = -10.00$ m/s^2.
16.19 $v_2 = 23.2$ m/s ∠40°.
16.21 (a) $v_2 = v_1 + \frac{1}{m}[(t_2^4 - t_1^4) + 5(t_2^2 - t_1^2)]$; (b) $v_2 = 1165$ ft/s.

16.23 $\mathbf{I} = (40.6\mathbf{i} + 56.6\mathbf{j})$ lb·s.

16.25 $T = \left(\dfrac{m_A m_B}{m_A + m_B}\right)(1 + \sin\theta)g,$

$$v_2 = v_1 + \left[\left(\frac{m_B}{m_A + m_B}\right) - \left(\frac{m_A}{m_A + m_B}\right)\sin\theta\right](t_2 - t_1)g.$$

16.27 (a) $t = 3.00$ s; (b) $t = 7.46$ s.

16.29 (a) $\mathbf{I} = (8\ \text{mt})\mathbf{j}$; (b) $\mathbf{I} = (8\ \text{mt})\mathbf{j}$.

16.31 (a) $v_{2x} = \left(\dfrac{m}{m + M}\right)v_0\cos\theta;$

(b) $F_x\Delta t = \left(\dfrac{mM}{m + M}\right)v_0\cos\theta \leftarrow,\ F_y\Delta t = mv_0\sin\theta \uparrow;$

(c) % loss of energy $= \left[1 - \left(\dfrac{m}{m + M}\right)\right](100).$

16.33 (a) $v_{2x} = \left(\dfrac{M_A}{M_A + M_B}\right)v_0 \leftarrow$; (b) $F_x\Delta t = \left(\dfrac{M_A M_B}{M_A + M_B}\right)v_0 \leftarrow.$

16.35 (a) $v_{B2} = 3.35$ ft/s $\leftarrow$; (b) $v_{B2} = 3.35$ ft/s $\leftarrow$;
(c) $v_{B2} = 3.35$ ft/s $\leftarrow$.

16.37 (a) $v_{C2} = 4.57$ m/s $\leftarrow$; (b) $Q\Delta t = 3{,}050$ N $\uparrow$.

16.39 $v_{B2} = -\left(\dfrac{m_A}{m_B}\right)\sqrt{2gL\sin\theta\cos^2\theta}.$

16.41 (a) $v_f = -\left(\dfrac{m_B}{m_A + m_B}\right)v_0$; (b) $F_{AV.} = \left(\dfrac{m_A m_B}{m_A + m_B}\right)\left(\dfrac{v_0}{t_c}\right).$

16.43 $\mathbf{L} = (800\mathbf{i} + 400\mathbf{j} - 1{,}600\mathbf{k})$ N·s, $\alpha = 64.1°$, $\beta = 77.4°$,
$\gamma = 150.8°$.

16.45 (a) $v_{A2} = 5.58$ m/s $\leftarrow$, $v_{B2} = 4.38$ m/s; (b) $\Delta T = 119.2$ N·m.

16.47 $v_{A2} = 0.1250$ m/s along the positive x axis,
$v_{B2} = 6.10$ m/s along an axis 88.8° ccw from the positive x axis,
$\Delta T = 2.95$ N·m.

16.49 $e = \sqrt{h'/h}.$

16.51 $v_{A2} = 4.50$ m/s at 117.4° ccw from the positive x axis,
$v_{B2} = 2.88$ m/s at 79.0° ccw from the positive x axis.

16.53 $h_2 = 0.0625$ m, $h_3 = 0.00391$ m.

16.55 $v_{A2} = 127.1$ mph at 128.1° ccw from the positive x axis.

16.57 (a) for $t = 0$ s, $\mathbf{L} = (8.00\mathbf{j})$ N·s, $\mathbf{H}_O = \mathbf{0}$ for $t = 1$ s,
$L = (6.00\mathbf{i} + 8.00\mathbf{j} + 32.0\mathbf{k})$ N·s;
(b) $\mathbf{I} = (6.00\mathbf{i} + 32.0\mathbf{k})$ N·s,
$\mathbf{G} = (64.0\mathbf{i} + 16.00\mathbf{j} - 16.00\mathbf{k})$ m·N·s.

16.59 (b) $\mathbf{L} = (0.400\mathbf{i} + 0.4t\mathbf{j} + 1.200\mathbf{k})$ N·s; (c) $\mathbf{I} = (1.200\mathbf{j})$ N·s;
(e) $\mathbf{H}_O = [(0.8t - 1.2t^2 - 4.80)\mathbf{i} - 2.00\mathbf{j}$
$+ (0.4t^2 + 0.4t + 1.600)\mathbf{k}]$ m·N·s;
(f) $\mathbf{G} = (-8.40\mathbf{i} + 4.80\mathbf{k})$ m·N·s.

16.61 $\mathbf{r} = [(2t)\mathbf{i} + (2 + 4t^2)\mathbf{j}]$ m., $\mathbf{v} = (2\mathbf{i} + 8t\mathbf{j})$ m/s,
$\mathbf{L} = m(2\mathbf{i} + 8t\mathbf{j})$ N·s, $\mathbf{H}_O = [m(8t^2 - 4)\mathbf{k}]$ m·N·s.

16.63 $\mathbf{v} = (2t\mathbf{j})$ ft/s, $\mathbf{a} = (2\mathbf{j})$ ft/s^2, $\mathbf{R} = (2m\mathbf{j})$ lb, $\mathbf{L} = (2mt)\mathbf{j}$ lb·s, $\mathbf{H}_O = [(2bmt)\mathbf{k}]$ ft·lb·s.

16.65 $\mathbf{v} = -(a\omega \sin \omega t)\mathbf{i} + (b\omega \cos \omega t)\mathbf{j}$, $\mathbf{a} = -\omega^2\mathbf{r}$, $\mathbf{R} = -m\omega^2\mathbf{r}$, $\mathbf{L} = m\omega[(-a \sin \omega t)\mathbf{i} + (b \cos \omega t)\mathbf{j}]$, $\mathbf{H}_O = (mab\omega \cos 2\omega t)\mathbf{k}$.

16.67 (a) $\mathbf{L}_A = (-15.84\mathbf{i})$ k·s, $\mathbf{L}_B = (-15.40\mathbf{j})$ k·s, $\mathbf{L}_C = (17.16\mathbf{i})$ k·s; (b) $\mathbf{H}_{OA} = (9{,}500\mathbf{k})$ ft·k·s, $\mathbf{H}_{OB} = (9{,}240\mathbf{k})$ ft·k·s, $\mathbf{H}_{OC} = (10{,}300\mathbf{k})$ ft·k·s; (c) $\mathbf{I}_{AB} = (15.84\mathbf{i} - 15.40\mathbf{j})$ k·s; (d) $\mathbf{G}_{BC} = (1{,}060\mathbf{k})$ ft·k·s.

16.69 (a) $\mathbf{H}_O = (-10.49\mathbf{i} - 18.54\mathbf{j} + 6.18\mathbf{k})$ ft·lb·s; (b) $\mathbf{H}_C = (-83.3\mathbf{i} - 86.5\mathbf{j} - 52.5\mathbf{k})$ ft·lb·s.

16.71 $H_{O2} = 776$ m·N·s.

16.73 $\mathbf{G} = (-2{,}000\mathbf{i} - 2{,}000\mathbf{j} + 3{,}000\mathbf{k})$ in.·lb·s.

16.75 $\mathbf{G} = (2.00\mathbf{i} + 2\pi\mathbf{j})$ m·kN·s.

16.77 $v_2 = 1.153$ m/s.

16.79 $N = 31.8$ turns.

16.81 $v_B = 6{,}600$ km/hr, $r_B = 1{,}818$ km, $v_D = 20{,}000$ km/hr, $r_D = 6{,}000$ km.

16.83 (a) $\mathbf{I} = (-12.50\mathbf{i} - 32.0\mathbf{j} - 22.0\mathbf{k})$ lb·s; (b) $\mathbf{G}_O = \mathbf{0}$.

16.85 $\mathbf{I} = (582\mathbf{i} - 158.0\mathbf{j} + 169.0\mathbf{k})$ N·s, $I = 626$ N·s, $\alpha = 21.6°$, $\beta = 104.6°$, $\gamma = 74.3°$.

16.87 $\mathbf{H}_{O2} = (120.0\mathbf{i} + 150.0\mathbf{j} - 550\mathbf{k})$ m·N·s.

16.89 (a) $v_{A2} = \sqrt{2gh}$; (b) $v_{A3} = v_{B3} = \left(\dfrac{m_A}{m_A + m_B}\right)\sqrt{2gh}$; (c) $x_M = 5.92$ in.

16.91 $s = 0.763$ m, $v_B = 0.756$ m/s.

16.93 $x_M = \dfrac{mg}{k} + \sqrt{\left(\dfrac{9}{8}\right)\dfrac{mgL}{k}}$.

16.95 $v_A = 6.26$ m/s, $s = 1.593$ m.

16.97 For $e = 0$, $\Delta T = 215$ lb·ft; for $e = 0.5$, $\Delta T = 161.3$ lb·f; for $e = 1.0$, $\Delta T = 0$.

16.99 Before impact, $v_A = \sqrt{2gb}$;

after impact, $v_A = \left[\dfrac{1 - (m_B/m_A)e}{1 + (m_B/m_A)}\right]\sqrt{2gb}$

$v_B = \left[\dfrac{1 + e}{1 + (m_B/m_A)}\right]\sqrt{2gb}$.

16.101 $F = 110.2$ lb →.

16.103 $F = 73.7$ lb →.

16.105 $F_x = 56.7$ lb →, $F_y = 321$ lb ↑.

16.107 $F_x = 978$ lb →, $F_y = 0$.

16.109 $R = 150.7$ lb ←.

16.111 $R = 370$ lb 78.7° ↖.

16.113 $F = w(L - vt)$.

16.115 $v = 4.11$ ft/s ←.

16.117 $c = 21.0$ slug/s, Thrust $= 73{,}500$ lb ↑.

16.119 $v = -9.81t + 3.80 \times 10^3 \ln\left(\dfrac{1.85 \times 10^6}{1.85 \times 10^6 - 5.25 \times 10^3 t}\right)$;
for $t = 20$ s, $v = 17.60$ mph ↑.
16.121 $F = (30t + 4.66)$ lb, $R = (120 - 30t)$ lb for $0 \le t \le 4$ s.
16.123 $F = 242$ lb.
16.125 $v_A = 16{,}030$ ft/s, $v_B = 9{,}990$ ft/s.
16.127 Just before firing, $v_B = 23{,}900$ ft/s;
just after firing, $v_A = 25{,}500$ ft/s.
16.129 $v_A = 4{,}760$ m/s, $\Delta v_B = 2{,}110$ m/s.
16.131 $\mathbf{L} = (0.048\mathbf{i} + 0.1062\mathbf{j})$ lb·s.
16.133 $\mu = 0.607$.
16.135 $v_{B/A} = 1.500$ m/s →.
16.137 $v = 37.8$ m/s 48.5° .
16.139 $v_A = 8.44$ m/s →, $v_B = 11.92$ m/s →, $\Delta T = 111.8$ kN·m.
16.141 $v = 6.71$ ft/s.
16.143 $v = 5.00$ ft/s.
16.145 $x_A = 0.006$ ft.
16.147 $F_x = 100.0$ N ←, $F_y = 347$ N ↑.
16.149 $dV/dt = 0.500$ m³/s.

Chapter 17

17.1 (a) $a = 2.20$ m/s²; (b) $t = 9.26$ s.
17.3 (a) $a = 0.75$ m/s²; (b) $s = 300$ m; (c) $s = 97.5$ m.
17.5 (a) $a = -2.93$ ft/s²; (b) $s = 1{,}322$ ft.
17.7 $h = \dfrac{v^2D^2}{2(D^2g - v^2D)}$, $v_E = 25{,}000$ mph, $v = 795$ mph.
17.9 $v = \sqrt{\dfrac{1}{C^2}[1 - e^{-2C^2}gy]}$.
17.11 $\mathbf{v} = R(6t - 1)\boldsymbol{\lambda}_\theta$, $\mathbf{a} = -R(6t - 1)^2\boldsymbol{\lambda}_r + 6R\boldsymbol{\lambda}_\theta$,
$v = 23.2$ m/s 30°, $a = 672.8$ m/s² 59.6°.
17.13 $v = 1.407$ m/s 30°, $a = 0.219$ m/s² 60°.
17.15 $v = [(0.75L)t^{1/2}]\boldsymbol{\lambda}_\theta$, $a = -[(0.5625L)t]\boldsymbol{\lambda}_r + [(0.375L)t^{-1/2}]\boldsymbol{\lambda}_\theta$,
$L = 4.86$ ft, $v = 5.34$ ft/s ↓.
17.17 $\theta = 46.5$ rad, $\omega = 1.162$ rad/s, $\alpha = 0.00968$ rad/s².
17.19 $v_A = 0.0408D\theta^{3/2}$, $a_A = 3.162 \times 10^{-3}D\theta^2\sqrt{2.5 + 1.111\theta^2}$;
$v_A = 0.212$ m/s, $a_A = 0.1006$ m/s².
17.21 $a_W = \left(\dfrac{r_1r_3r_5}{r_2r_4}\right)\alpha_1$, $v_W = \left(\dfrac{r_1r_3r_5}{r_2r_4}\right)\omega_1$.
17.23 $\omega_B = 2.43$ rad/s, $\alpha_B = 1.111$ rad/s².
17.25 (a) $v_A = 0.250$ m/s, $a_A = 2.50$ m/s²;
(b) $\omega_1 = 25.0$ rad/s, $\alpha_1 = 10.00$ rad/s².
17.27 $\omega_1 = 4.00$ rad/s ↺, $\alpha_1 = 6.00$ rad/s² ↻;
$\omega_2 = 6.00$ rad/s ↻, $\alpha_2 = 9.00$ rad/s² ↺.

17.29 $t = \dfrac{r_A \omega_A - r_B \omega_B}{r_A \alpha_A + r_B \alpha_B}$, $t = 10.00$ s.

17.31 $(a_P)_B = 6{,}500$ in./s^2.

17.33 $\mathbf{v}_C = -35.7(0.792\mathbf{i} + 0.140\mathbf{j} + 0.594\mathbf{k})$ in./s,
$\mathbf{a}_C = 357(0.350\mathbf{i} + 0.693\mathbf{j} + 0.630\mathbf{k})$ in./s^2.

17.35 (a) $\omega = -4.49(-0.371\mathbf{i} + 0.743\mathbf{j} + 0.557\mathbf{k})$ rad/s;
(b) $\mathbf{v}_C = 2.24(0.894\mathbf{i} + 0.447\mathbf{j})$ m/s;
(c) $\mathbf{a}_C = 10.80(0.562\mathbf{i} - 0.298\mathbf{j} + 0.772\mathbf{k})$ m/s^2.

17.37 $v = 78{,}300$ mph, $a = 91.2$ mph^2.

17.39 $(v_A)_x = -\left(\dfrac{b}{a}\right) v_C \tan\theta$, $(v_A)_y = v_C\left(\dfrac{b}{a} - \tan\theta\right)$,
$(a_A)_x = -\dfrac{b v_C^2}{a^2 \cos^3\theta}$,
$(a_A)_y = \left(\dfrac{v_C^2}{a^2\cos^2\theta}\right)[2b\sin\theta + a(\cos\theta - \sin\theta\tan\theta)]$,
$v_A = 1.581$ m/s 18.43°, $a_A = 12.00$ m/s^2 45.0°.

17.41 $v_C = \left(\dfrac{L}{2}\right)\omega$ θ, $a_C = \left(\dfrac{L}{2}\right)\omega^2$ θ,
$v_C = 30.0$ in./s 45.0°, $a_C = 120.0$ in./s^2 45.0°.

17.43 $(v_C)_x = \left[r\cos\theta + \dfrac{r^2\sin\theta\cos\theta}{2(L^2 - r^2\cos^2\theta)^{1/2}}\right]\omega$, $(v_C)_y = -(\tfrac{1}{2}r\sin\theta)\omega$,
$(a_C)_x = \left[\dfrac{r^2(\cos^2\theta - \sin^2\theta)}{2(L^2 - r^2\cos^2\theta)^{1/2}} - \dfrac{r^4\sin^2\theta\cos^2\theta}{2(L^2 - r^2\cos^2\theta)^{3/2}}\right]\omega^2$,
$(a_C)_y = -(\tfrac{1}{2}r\cos\theta)\omega^2$.

17.45 $v_A = \left(\dfrac{b\omega}{\cos\beta}\right)\cos\theta$ β, $a_A = \left(\dfrac{b\omega^2}{\cos\beta}\right)\sin\theta$ β.

17.47 $\omega = 0.679$ rad/s ↺, $\alpha = 0.258$ rad/s^2 ↺.

17.49 (a) $\omega = 6.67$ rad/s ↻, $\alpha = 2.67$ rad/s^2 ↻; (b) $v_B = 5.00$ m/s →,
$a_B = 35.5$ m/s^2 3.23°.

17.51 (a) $v_C = 0.0371$ m/s ←, $a_C = 0.1098$ m/s^2 ←;
(b) $v_C = 0.662$ m/s →, $a_C = 16.08$ m/s^2 ←.

17.53 (a) $v_W = 1.250$ ft/s ↓, $a_W = 0$; (b) $\omega = 1.875$ rad/s ↻, $\alpha = 0$.

17.55 $v_A = \left(\dfrac{\cos\theta}{\cos\beta}\right)b\omega$ β.

17.57 (a) $v_D = 200$ ft/s →; (b) $v_E = 73.7$ ft/s 28.7°;
(c) $v_F = 25.0$ ft/s →.

17.59 (a) $\omega_{AB} = 83.2$ rad/s ↻; (b) $\omega_{OA} = 317$ rad/s ↺;
(c) $v_A = 79.2$ ft/s 30.0°.

17.61 (a) $v_B = R\omega\sin\theta\left[1 + \dfrac{R\cos\theta}{\sqrt{L^2 - R^2\sin^2\theta}}\right]$ ↓,
$\omega_{AB} = \dfrac{R\omega\cos\theta}{\sqrt{L^2 - R^2\sin^2\theta}}$ ↻; (b) $v_C = 44.3$ ft/s 75.2°.

17.63 (a) $v_B = 22.7$ m/s $\leftarrow$; (b) $\omega_{BC} = 11.37$ rad/s ↻;
(c) $v_A = 35.1$ m/s $60°$; (d) $\omega_{OA} = 70.2$ rad/s ↺.

17.65 (a) $\omega = 8.77$ rad/s ↺; (b) $v_C = 0.759$ m/s $30°$.

17.67 $v_D = \left(\dfrac{\sqrt{1 + 8\cos^2\theta}}{2\sin\theta}\right) v_B$ β , where $\beta = \tan^{-1}(\frac{1}{3}\tan\theta)$.

17.69 $v_{BC} = r\omega\sin\theta \uparrow$.

17.71 (a) $v_C = 100.0$ ft/s $\rightarrow$; (b) $v_A = 141.4$ ft/s $45°$;
(c) $v_B = 90.1$ ft/s $46.1°$.

17.73 See answers to Problem 17.59.

17.75 See answers to Problem 17.61.

17.77 See answers to Problem 17.63.

17.79 $v_A = \left(\dfrac{\sqrt{1 + 8\cos^2\theta}}{2\sin\theta}\right) v_B$ β ,
where $\beta = \sin^{-1}\left(\dfrac{3\tan\theta}{\sqrt{1 + 9\tan^2\theta}}\right)$.

17.81 (a) $\omega_{AB} = v_A/R$ ↻; (b) $v_B = v_A$ γ ,
where $\gamma = \dfrac{\pi}{2} - \theta - \cos^{-1}\left(\dfrac{L}{2R}\right)$.

17.83 (a) $\omega_{BC} = 4.81$ rad/s ↻; (b) $\omega_{CD} = 3.33$ rad/s ↺.

17.85 (a) $\omega_{BC} = 20.0$ rad/s ↻; (b) $\omega_{AB} = 10.00$ rad/s ↻.

17.87 (a) $\omega_{AB} = 6.00$ rad/s ↺; (b) $\omega_{CD} = 6.00$ rad/s ↺.

17.89 See answers to Problem 17.41.

17.91 See answers to Problem 17.45.

17.93 (a) $\omega = \sqrt{2a_W h/r_1}$, $\alpha = a_W/r_1$;
(b) $v_F = \left(\dfrac{r_1 + r_2}{r_1}\right)\sqrt{2a_W h}$, $a_F = \left(\dfrac{r_1 + r_2}{r_1}\right) a_W$.

17.95 (a) $\omega = 959$ rpm ↺; (b) $\alpha_{AB} = 989$ rad/s^2 ↺.

17.97 $a_C = \dfrac{v_B^2}{L\sin\theta}(1 + \text{ctn}^2\theta) \leftarrow$, $\alpha_{ABCD} = \dfrac{v_B^2\cos\theta}{L^2\sin^3\theta}$ ↻.

17.99 $\alpha = 20.0$ rad/s^2 ↻, $a_A = 50.0$ m/s^2 $20°$.

17.101 $\alpha_{AB} = 38.4$ rad/s^2 ↻, $\alpha_{BC} = 25.4$ rad/s^2 ↻.

17.103 $a_B = \left[\dfrac{R}{L\cos^3\theta} - 1\right] R\omega^2 \rightarrow$, $\alpha_{AB} = \left(\dfrac{R^2\tan\theta}{L^2\cos^2\theta}\right)\omega^2$ ↺.

17.105 $\mathbf{a}_D = (22.1\mathbf{i} - 66.7\mathbf{j})$ m/s^2, $\alpha_{CD} = 39.7$ rad/s ↺.

17.107 $\mathbf{v}_B = (3.00\mathbf{i} + 1.500\mathbf{j})$ m/s, $\mathbf{a}_B = (-9.50\mathbf{i} + 29.7\mathbf{j})$ m/s^2.

17.109 $\omega = (15.00\mathbf{k})$ rad/s, $v_{B/O} = (10.00\mathbf{j})$ ft/s.

17.111 $\mathbf{a}_A = (77.2\mathbf{i} - 3.37\mathbf{j})$ m/s^2.

17.113 $\omega = 11.34$ rad/s ↺, $v_{B/S} = 18.01$ in./s θ .

17.115 $\mathbf{a}_B = (-141.2\mathbf{i} + 217\mathbf{j})$ m/s^2 i $22.5°$ j.

17.117 (a) $a_{C/A} = 914$ in./s^2 ∠40°; (b) $\alpha_{AB} = 45.9$ rad/s^2 ↻.
17.119 (a) $a_{A/B} = 2.24$ m/s^2 30°; (b) $\alpha_{BD} = 82.6$ rad/s^2 ↺.
17.121 (a) $\alpha = 0$; (b) $\alpha = 330$ rad/s^2 ↺.
17.123 (a) $v_B = 500$ mph →; (b) $(a_B)_t = 70.2$ mph^2 →.
17.125 $v_{A/B} = 3{,}270$ mph ←, $a_{A/B} = 1{,}395$ mph^2 ←.
17.127 $\omega = 1.250$ rad/s ↻.
17.129 $v_C = 2.51$ m/s ↓, $a_C = 50.5$ m/s^2 1.424°.

17.131 (a) $v_{AB} = \dfrac{h\omega \sec^2\theta}{\tan^2\theta}$ ←, $v_E = \left(\dfrac{L}{2}\right)\omega$ θ;

(b) $v_{AB} = 14.52$ m/s ←, $v_E = 7.50$ m/s 40°.

17.133 $\omega_D = 3.33$ rad/s ↻.
17.135 $\omega_{AB} = 45.0$ rad/s ↺, $v_B = 450$ in./s ←.
17.137 $v_{Ay} = 5.13$ m/s ↑, $\omega = 8.75$ rad/s ↻.
17.139 $a_A = 54.7$ ft/s^2 30°, $\alpha_{AB} = 10.59$ rad/s^2 ↺.
17.141 $\alpha_{CB} = 116.6$ rad/s^2 ↺, $\alpha_{CD} = 244$ rad/s^2 ↻.
17.143 (a) $v_{B/C} = 5.00$ m/s ↑; $\omega_{CD} = 10.00$ rad/s ↺.
17.145 (a) $\omega = 39.0$ rad/s ↻; (b) $\mathbf{a}_A = (-706\mathbf{i} + 1{,}069\mathbf{j})$ ft/s^2.

Chapter 18

18.1 $I_z = \left(\dfrac{87}{5}\right)mR^2$, $k_z = \left(\sqrt{\dfrac{87}{5}}\right)R$.
18.3 $I_X = 2{,}000$ kg·m^2, $I_Z = 1{,}333$ kg·m^2, $m = 267$ kg.
18.5 $I_x = (\frac{1}{12})m(a^2 + b^2)$.
18.7 $I_y = (\frac{10}{9})mh^4$.
18.9 $I_x = 22.9$ slug·in.^{2}
18.11 $I_x = 0.01277\rho$ kg·m^2, $k_x = 0.588$ m.
18.13 $I_z = 0.00308\rho$ kg·m^2, $k_z = 0.289$ m.
18.15 $I_z = 2.026 \times 10^6\rho$ slug·in.^{2}, $k_z = 10.63$ in.
18.17 $I_X = I_Y = 3.19$ slug·ft^2, $k_X = k_Y = 0.705$ ft $= 8.46$ in.
18.19 $I_y = 1.808$ kg·m^2, $k_y = 0.384$ m;
$I_z = 1.533$ kg·m^2, $k_z = 0.354$ m.
18.21 $I_x = 23.7$ kg·m^2, $k_x = 0.588$ m.
18.23 (a) $a = 0.1g$ →; (b) $N = W$ ↑, $F_f = 0.1W$; (c) $x = 0.15h$.
18.25 (a) $a = \left(\dfrac{\mu b}{a + b - \mu c}\right)g$ →; (b) $F_f = \dfrac{\mu bW}{2(a + b - \mu c)}$ →;
(c) $N_A = \dfrac{bW}{2(a + b - \mu c)}$ ↑, $N_B = \dfrac{(a - \mu c)W}{2(a + b - \mu c)}$ ↑.
18.27 (a) $a = 2.28$ m/s^2 →; (b) $F_f = 1{,}395$ N →;
(c) $N_A = 2{,}790$ N ↑, $N_B = 3{,}210$ N ↑.

18.29 $P = \left(\frac{W}{c}\right)(0.766b - 0.500d)$, $\mu = 1.532\left(1 - \frac{b}{c}\right) + \frac{d}{c}$,
$1.532\left(\frac{b}{c}\right) - \left(\frac{d}{c}\right) = 1.532$.

18.31 $T = 0.0471W$, $F_x = 0.050W$, $F_y = 0.533W$.

18.33 $W_E = 255$ N, $N_A = 771$ N $\uparrow$, $N_B = 829$ N $\uparrow$.

18.35 $a = v_0^2/2s \leftarrow$, $H = \left(\frac{W}{4gs}\right)v_0^2 \leftarrow$, $N_A = \left(\frac{W}{b+c}\right)\left(c - \frac{dv_0^2}{2gs}\right) \uparrow$,
$N_B = \left(\frac{W}{2(b+c)}\right)\left(b + \frac{dv_0^2}{2gs}\right)$.

18.37 $F_B = 8{,}250$ lb ∠60°, $F_E = 1{,}634$ lb 60°,
$T = 9{,}530$ lb $\leftarrow$.

18.39 $\theta = 63.4°$, $A_x = 2mg \leftarrow$, $A_y = mg \uparrow$.

18.41 $a = 5.11$ m/s^2 30°, $T_B = 675$ N 60°,
$T_E = 338$ N 60°.

18.43 $a = g/6$, $N_R = 0.979mg \uparrow$, $N_F = 0.521mg \uparrow$, $N = 2mg \uparrow$,
$F_f = mg/3 \rightarrow$, $x = b/4$.

18.45 $a = 3.75\left(\frac{Q}{W}\right)g$ ∠83.9°, $a = 18.39$ m/s^2 ∠83.9°.

18.47 $d = 0.25r$, $a = 0.3g \rightarrow$.

18.49 $d = (\frac{1}{4})r\cos\theta$, $a = (2 - \sin\theta)g$.

18.51 $a = (\frac{1}{6})g \rightarrow$, $A_x = (\frac{1}{6})mg \rightarrow$, $A_y = 0.553mg \uparrow$,
$B_y = 0.447mg \uparrow$.

18.53 $P = 0.424W$, $a = 0.424g \rightarrow$.

18.55 (a) $F_D = 55.7$ lb; (b) $P = 8.91$ hp; (c) $a_{MAX} = 9.46$ ft/s^2.

18.57 (a) $x_{MAX} = \sqrt{\frac{mv_0^2}{k}}$; (b) $a_{MAX} = \sqrt{\frac{kv_0^2}{m}}$.

18.59 (a) $T = 20{,}500$ lb; (b) $T = 17{,}670$ lb.

18.61 (a) $a_{MAX} = 12.95$ ft/s^2; (b) $x = 299$ ft.

18.63 (a) $\alpha = \left(\frac{\sqrt{3}}{3}\right)\left(\frac{g}{r}\right)$ ↻;
(b) $O_t = \left(\frac{\sqrt{3}}{6}\right)mg$ ∠60°, $O_n = \frac{mg}{2} + mr\omega^2$ 30°.

18.65 $\alpha = \left(\frac{6g}{17b}\right)\sin\theta$ ↻, $O_t = 0.294mg\sin\theta$ ∠θ, $O_n = mg\cos\theta$ θ.

18.67 $\alpha = 11.36$ rad/s^2 ↻, $O_t = 10.60$ lb ∠40°,
$O_n = 53.2$ lb 50°, $a_B = 36.7$ ft/s^2.

18.69 $\alpha = 9.82$ rad/s^2 ↻, $B_t = 0.0765$ N ∠45°,
$B_n = 0.230$ N 45°.

18.71 $\alpha = 10.63$ rad/s^2 ↺, $O_t = 27.0$ lb 30°, $O_n = 83.7$ lb ∠60°.

18.73 $\alpha = 8.70$ rad/s^2 ↻, $O_t = 59.5$ lb $\angle 50°$, $O_n = 123.3$ lb $40°\angle$.

18.75 $\alpha = 0.857\left(\dfrac{g}{L}\right)\sin\theta$ ↻, $a_B = 0.958g\sin\theta$, $\alpha = 10.81$ rad/s^2 ↻,
$a_B = 6.04$ m/s^2.

18.77 (a) $C = 3.16$ N·m ↺; (b) $G_x = 0$, $G_y = 78.5$ N ↑.

18.79 $r_G = 1.140$ m, $\alpha = 3.62$ rad/s^2 ↻, $O_t = 62.3$ N $\angle 30°$,
$O_n = 702$ N $60°\angle$.

18.81 $a_B = \left(\dfrac{8}{3\pi}\right)g$ ←.

18.83 (a) $C = 4.50$ N·m ↺; (b) $G_x = 0$, $G_y = 49.1$ N ↑.

18.85 $t = 2.42$ s, $T = 121.5$ N.

18.87 (a) $v = 6.42$ m/s; (b) $T = 13.07$ N.

18.89 $\theta = \theta_0 \cos pt$, $p = 1.225\sqrt{\left(\dfrac{g}{L}\right)\cos\gamma}$; for $\gamma = 0$, $p = 1.225\sqrt{\dfrac{g}{L}}$;
for $\gamma = \pi/2$, $p = 0$.

18.91 $k_0 = 1.204$ ft.

18.93 (a) $\alpha = 8.33$ rad/s^2 ↻, $a_G = 6.66$ ft/s^2 →, $F = 6.67$ lb;
(b) $\alpha = 1.006$ rad/s^2 ↻, $a_G = 9.60$ ft/s^2 →, $F = 0.805$ lb ←.

18.95 (a) $\alpha = 16.82$ rad/s^2 ↻; (b) $a_G = 8.41$ m/s^2 →;
(c) $F = 34.3$ N ←, $N = 100.0$ N ↑; (d) $\mu_{MIN} = 0.343$.

18.97 (a) $\alpha = 0.320g/r$ ↻; (b) $F = 0.1283W$ $45°\angle$;
(c) $N = 1.673W$ $\angle 45°$; (d) $\mu_{MIN} = 0.0767$.

18.99 $\alpha = (3gC)/(10Wr^2)$ ↻, $F = (6C)/(10r)$ →, $N = 2W$ ↑.

18.101 Case A: (a) $\alpha = 7.89$ rad/s^2 ↻;
(b) $F = 31.6$ lb ←, $N = 50.0$ lb ↑, (c) $\mu_{MIN} = 0.632$;
Case B: (a) $\alpha = 23.7$ rad/s^2 ↻,
(b) $F = 5.21$ lb →, $N = 50.0$ lb ↑, (c) $\mu_{MIN} = 0.1042$.

18.103 $T = \dfrac{W}{3}$, $\alpha = \dfrac{2g}{3r}$ ↻, $a_G = \dfrac{2g}{3}$ ↓.

18.105 The sphere rolls without sliding, $\alpha = 55.2$ rad/s^2 ↻,
$a_G = 110.4$ ft/s^2 →, $F = 85.8$ lb →.

18.107 $T = 50.0$ N, $\alpha = 19.62$ rad/s^2 ↻, $a_G = 6.54$ m/s^2 ↓.

18.109 (a) $\alpha = (0.0909g)/r$ ↻; (b) $a_G = 0.09215g$ $\angle 9.46°$;
(c) $F = 0.1364mg$ →, $N = 1.477mg$ ↑; (d) $\mu_{MIN} = 0.0923$.

18.111 (a) $\alpha = \dfrac{(d - r\sin\beta)g}{k_G^2 + r^2}$ ↺; (b) $d = r\sin\beta$;
(c) $F = mg\left[\sin\beta + \left(\dfrac{rd - r^2\sin\beta}{k_G^2 + r^2}\right)\right]$; (d) $N = mg\cos\beta$;
(e) $\mu_{MIN} = \tan\beta + \dfrac{rd - r^2\sin\beta}{(k_G^2 + r^2)\cos\beta}$.

18.113 (a) $a_G = 0.05g$ ↘30°; (b) $\mu_{MIN} = 0.520$;
(c) $F = 0.45mg$ 30°↖.

18.115 $a = g/2 \rightarrow$, $B_x = 0.125W \rightarrow$, $B_y = 0.1611W \uparrow$, $N_A = 0.0889W \uparrow$, $N_C = 1.322W \uparrow$, $F = 0.25W \leftarrow$.

18.117 $a = 4g/7 \downarrow$, $T = 3W/14$.

18.119 $a_B = 1.767g \rightarrow$, $T = 2.233W$, $N_C = 2W \uparrow$, $F_C = 0.123W \rightarrow$.

18.121 (a) $a = \left(\dfrac{8-\mu}{4-3\mu\tan\beta}\right)g \rightarrow$, $F = \left(\dfrac{8-\mu}{4-3\mu\tan\beta}\right)\left(\dfrac{3W}{2\cos\beta}\right)(T)$;
(b) $a = 19.63$ m/s $\rightarrow$, $F = 610$ N (C).

18.123 (a) $\alpha = \left(\dfrac{3}{11}\right)\left(\dfrac{g}{r}\right)$ ↻; (b) $N = \left(\dfrac{16}{11}\right)mg \uparrow$, $F = \left(\dfrac{6}{11}\right)mg \rightarrow$;
(c) $a_G = \sqrt{2}r\alpha$ ↘45°; (d) $\mu_{MIN} = \dfrac{6}{16}$.

18.125 $s = 96.3$ ft, $T_{Upper} = 80.6$ lb, $T_{Lower} = 82.5$ lb.

18.127 (a) $\omega_{BE} = 0.385$ rad/s ↺, $\omega_{DE} = 2.00$ rad/s ↻;
(b) $\alpha_{BE} = 0.359$ rad/s^2 ↻, $\alpha_{DE} = 1.864$ rad/s^2 ↻;
(c) $B_x = 8.72$ lb $\leftarrow$, $E_x = 9.43$ lb $\leftarrow$.

18.129 $a = 44.4$ ft/s^2 $\leftarrow$, $A_x = 133.3$ lb $\leftarrow$, $A_y = 32.2$ lb $\uparrow$, $B_x = 44.4$ lb $\rightarrow$, $B_y = 32.2$ lb $\uparrow$.

18.131 Cylinder slides and rolls.

18.133 $a_1 = 21.6$ ft/s^2 $\downarrow$, $T_{Vertical} = 21.2$ lb, $T_{Inclined} = 25.1$ lb.

18.135 $C = 1{,}682$ lb·ft ↻.

18.137 $a = 1.090$ m/s^2 $\rightarrow$, $F_A = 44.4$ N $\leftarrow$; $F_B = 33.3$ N $\leftarrow$.

18.139 $a_1 = \left(\dfrac{r_1 m_1 \sin\beta_1 - r_3 m_3 \sin\beta_3}{m_1 r_1^2 + m_2 k_2^2 + m_3 r_3^2}\right) r_1 g$,

$$T_1 = \left[g\sin\beta_1 - \left(\frac{r_1 m_1 \sin\beta_1 - r_3 m_3 \sin\beta_3}{m_1 r_1^2 + m_2 k_2^2 + m_3 r_3^2}\right) r_1 g\right] m_1,$$

$$T_3 = \left[g\sin\beta_3 + \left(\frac{r_1 m_1 \sin\beta_1 - r_3 m_3 \sin\beta_3}{m_1 r_1^2 + m_2 k_2^2 + m_3 r_3^2}\right) r_2 g\right] m_3.$$

18.141 $\theta = 2.18$ rev.

18.143 $\alpha = 45.7$ rad/s^2 ↺.

18.145 $I_x = 4{,}560$ slug·ft^2.

18.147 (a) $\mu_{MIN} = 0.311$; (b) $x = 0.466$ ft.

18.149 (a) $W_2 = \left(\mu + \dfrac{a}{g}\right)W_1$; (b) $H = \dfrac{bg + ha}{2(\mu g + a)}$;
(c) $W_2 = 284$ lb, $H = 2.50$ ft.

18.151 $a < 10.73$ ft/s^2.

18.153 $a_W = (\frac{15}{29})g \downarrow$, $G_x = 0$, $G_y = 18mg \uparrow$.

18.155 $\theta = (\dot{\theta}_0/p)\sin pt + \theta_0 \cos pt$, where $p = \sqrt{0.7343\left(\dfrac{g}{L}\right)}$.

18.157 $\omega_0 = 47.9$ rad/s ↻.

18.159 (a) $\alpha = 28.9$ rad/s^2 ↻, $a_G = 7.24$ m/s^2 $\uparrow$; (b) $T = 72.4$ N;
(c) $y = 14.48$ m.

18.161 (a) $a = 3.71\ \text{m/s}^2$ ∠30°; $\alpha = 7.42\ \text{rad/s}^2$ ↻; (b) $\mu_{\text{MIN}} = 0.090$.
18.163 (a) $\alpha = 7.36\ \text{rad/s}^2$ ↺; (b) $\mu_{\text{MIN}} = 0.1760$.

Chapter 19

19.1 $\mu_k = 0.346$.
19.3 $v = 119.6$ km/hr.
19.5 $M = 8.90$ N·m.
19.7 $v_G = 6.07$ m/s ∠30°.
19.9 (a) $v = \sqrt{2sg\sin\theta}$; (b) $\Delta = \sqrt{\left(\dfrac{2W\sin\theta}{k}\right)s}$; (c) $\Delta = \left(\sqrt{\dfrac{W}{kg}}\right)v$.
19.11 (a) $W = 40.9$ lb; (b) $A_x = 63.5$ lb ←; $A_y = 10.23$ lb ↑.
19.13 (a) $\theta = 170.4°$; (b) $\theta = 111.2°$.
19.15 $\omega = 3.55$ rad/s ↻.
19.17 $s = 9.31$ ft.
19.19 $v_C = 8.06$ ft/s ←, $\omega_{BC} = 0$, $\omega_{AB} = 4.03$ rad/s ↻.
19.21 $v_R = 2.38$ ft/s ↓.
19.23 $h = 182.0$ ft.
19.25 $v_P = 10.05$ ft/s →, $\omega_R = 6.70$ rad/s ↺.
19.27 See answer to Problem 19.7.
19.29 See answers to Problem 19.9.
19.31 $\omega = 4.00$ rad/s ↻.
19.33 See answer to Problem 19.15.
19.35 $v_A = 4.32$ m/s ↓, $v_B = 3.00$ m/s ↑.
19.37 See answer to Problem 19.23.
19.39 (a) $v_B = 62.2$ ft/s ∠45°, (b) $v_B = 92.9$ ft/s →.
19.41 $v_A = \sqrt{\left(\dfrac{192}{115W}\right)(5Wr - 13.5778kr^2)}$, $k = 0.36825\left(\dfrac{W}{r}\right)$.
19.43 $\omega = 5.27$ rad/s ↻; $v_B = 5.27$ m/s ∠45°.
19.45 (a) $r = 3.55$ in.; (b) $r = 3.55$ in.
19.47 $v_A = 22.5$ ft/s ↑, $v_B = 0$, $\omega = 15.01$ rad/s ↻.
19.49 $\omega_{AB} = 6.40$ rad/s ↺, $v_C = 8.00$ ft/s →.
19.51 $P = 4.243\sqrt{s}$ hp.
19.53 $P = 0.785\ kW$.
19.55 $P = -0.676\ kW$.
19.57 $M = 229$ N·m.
19.59 (a) $P_{\text{Ave.}} = 3.25kW$; (b) $P_{\text{Cap.}} = 5.91kW$.
19.61 (a) $t = 9.83$ s; (b) $s = 212$ m.
19.63 (a) $v_W = 4.71$ ft/s ↓; (b) $F = 1.938$ lb.
19.65 $s = 5.02$ ft.
19.67 See answers to Problem 19.63.
19.69 $s = 683$ ft.

Chapter 20

20.1 (a) $t = 0.723$ s; (b) $t = 0.308$ s.

20.3 $t = 3.87$ s.

20.5 (a) $F = 29.0\ k$; (b) $T_1 = 14.48\ k(\text{T})$; (c) $T_2 = 7.22\ k(T)$.

20.7 (a) $v_A = \dfrac{(g \sin\theta)t}{(k_A/r_A)^2 + (W_B/W_A)(k_B/r_B)^2 + 1}$,

$\omega_A = \dfrac{(g \sin\theta)t/r_A}{(k_A/r_A)^2 + (W_B/W_A)(k_B/r_B)^2 + 1}$;

(b) $v_A = 27.0$ ft/s 30°, $\omega_A = 22.5$ rad/s ↺.

20.9 (a) $v_B = \dfrac{9gr_A^2(W_B - \frac{2}{3}W_A \sin\theta)t}{4W_A(k_A^2 + r_A^2) + 9W_B r_A^2}$, $v_S = \dfrac{2}{3}v_B$, $\omega_S = \dfrac{2}{3}\left(\dfrac{v_B}{r_A}\right)$;

(b) $v_B = 9.61$ m/s ↓, $v_S = 6.41$ m/s 25°, $\omega_S = 12.82$ rad/s ↻.

20.11 $v_A = 92.7$ ft/s ↓.

20.13 (a) $t = 3.52$ s; (b) $t = 5.28$ s.

20.15 (a) $t = \left(\dfrac{W_D}{gm}\right)r^2\omega$; (b) $t = \left(\dfrac{2W_D}{gM}\right)r^2\omega$.

20.17 $t = 21.4$ s.

20.19 $t = 3.80$ s.

20.21 $t = 0.1083\left(\dfrac{W_A r\omega}{\mu_P g}\right)$.

20.23 $\omega = 37.1$ rad/s ↺.

20.25 $M_z = 109.7$ lb·ft.

20.27 $d = \left(\dfrac{h}{\mu}\right)\left(\dfrac{W_P}{W_P + W_T}\right)^2$, $d = 22.4$ m.

20.29 (a) $v = \dfrac{W_A v_A + W_B v_B}{W_A + W_B}$; (b) $v = \frac{1}{2}(v_A + v_B)$.

20.31 $\theta = 42.8°$.

20.33 $\omega = \dfrac{2}{3}\left(\dfrac{3 + 58(W_P/W_C)}{2 + 22(W_P/W_C)}\right)$.

20.35 $\theta = 54.4°$, loss of kinetic energy = 27,900 lb·ft.

20.37 (a) $\omega = \dfrac{W_B h v_0}{(W_B/2)h^2 + (W_P/6)(b^2 + 4h^2)}$; (b) $\omega = 2.12$ rad/s ↺.

20.39 $\omega_2 = (\frac{3}{8})\omega_1$.

20.41 $v_0 = 14.67$ ft/s ↓.

20.43 $\omega = 5.17$ rad/s ↻, loss of kinetic energy = 1,450 N·m.

20.45 (a) $v = 13.60$ ft/s →, $\omega = 5.15$ rad/s ↻; (b) $\theta = 54.0°$.

20.47 $\gamma = 21.4°$.

20.49 $v_G = 2.08$ m/s 86.8°, $\omega = 20.8$ rad/s ↺.

20.51 $\omega = 5.30$ rad/s ↻, $v_G = 5.30$ ft/s ←, $v_B = 0.608$ ft/s ←.

20.53 $\mu = 0.1250$.

20.55 (a) $b = \dfrac{8\mu g P t}{(1 - \mu)W_A\omega}$; (b) $b = 0.429ft$.

20.57 $\omega = 5.21$ rad/s ↺, $v_G = 5.21$ ft/s ↓.
20.59 $\omega = 0.645$ rad/s ↻, $v_G = 0.968$ ft/s ←.

Chapter 21

21.1 $\theta_z + \theta_y + \theta_x + \theta_z$ and $-\theta_z - \theta_z - \theta_x + \theta_y$.
21.3 (a) $\boldsymbol{\omega} = (60.0\mathbf{i} + 20.0\mathbf{k})$ rad/s;
(b) $\mathbf{v}_P \equiv (-1.500\mathbf{i} - 7.80\mathbf{j} + 4.50\mathbf{k})$ m/s.
21.5 (a) $\boldsymbol{\omega} = \omega_2\mathbf{i} + \omega_1\mathbf{k}$; (b) $\boldsymbol{\alpha} = -(\omega_1\omega_2)\mathbf{j}$.
21.7 $(v_P)_y = 1.155$ in./s, $\boldsymbol{\omega} = (7.23\mathbf{i} - 13.12\mathbf{j} + 7.00\mathbf{k})$ rad/s.
21.9 (a) $\boldsymbol{\omega} = (32.8\mathbf{j} + 23.5\mathbf{k})$ rad/s; (b) $\mathbf{v}_P = (12.10\mathbf{i})$ m/s.
21.11 (a) $\boldsymbol{\omega} = (-10.00\mathbf{i} + 5.00\mathbf{k})$ rad/s;
(b) $\mathbf{v}_P = (-50.0\mathbf{i} + 100.0\mathbf{j} - 100.0\mathbf{k})$ in./s.
21.13 (a) $\boldsymbol{\omega} = (-0.350\mathbf{j} - 0.250\mathbf{k})$ rad/s;
(b) $\mathbf{v}_P = (-0.1911\mathbf{i} - 0.1628\mathbf{j} + 0.228\mathbf{k})$ m/s.
21.15 (a) $\omega = 54.9$ rad/s; (b) $\omega_S = 58.5$ rad/s, (c) $\mathbf{v}_P = (423\mathbf{i})$ in./s.
21.17 (a) $\boldsymbol{\omega} = (0.500\mathbf{i} + 0.300\mathbf{k})$ rad/s;
(b) $\mathbf{v} = (-1.269\mathbf{i} - 0.770\mathbf{j} + 2.11\mathbf{k})$ m/s.
21.19 (a) $\boldsymbol{\omega} = (0.750\mathbf{i} - 3.00\mathbf{k})$ rad/s;
(b) $\mathbf{v}_P = (32.6\mathbf{i} - 3.80\mathbf{j} + 8.16\mathbf{k})$ ft/s.
21.21 (a) $\boldsymbol{\omega} = (-1.500\mathbf{i} + 2.50\mathbf{k})$ rad/s;
(b) $\mathbf{v}_P = (-1.928\mathbf{i} + 1.379\mathbf{j} - 1.157\mathbf{k})$ m/s.
21.23 $\mathbf{v}_B = (-7\mathbf{k})$ m/s.
21.25 $\mathbf{v}_A = (-225\mathbf{k})$ in./s.
21.27 $\mathbf{v}_B = (-2.00\mathbf{j} + 1.000\mathbf{k})$ m/s.
21.29 (a) $\boldsymbol{\alpha} = (-300\mathbf{i} + 1200\mathbf{j})$ rad/s^2;
(b) $\mathbf{a}_P = (312\mathbf{i} - 261\mathbf{j} - 491\mathbf{k})$ m/s^2.
21.31 (a) $\boldsymbol{\alpha} = (-0.240\mathbf{i} + 0.750\mathbf{j} - 1.750\mathbf{k})$ rad/s^2;
(b) $\mathbf{a}_P = (1.902\mathbf{i} - 1.760\mathbf{j} - 1.192\mathbf{k})$ ft/s^2.
21.33 $\mathbf{a}_A = (-14{,}060\mathbf{k})$ in./s^2, $\boldsymbol{\alpha} = (349\mathbf{i} + 599\mathbf{j} + 227\mathbf{k})$ rad/s^2.
21.35 (a) $\mathbf{v}_P = (-1.500\mathbf{i} - 7.80\mathbf{j} + 4.50\mathbf{k})$ m/s;
(b) $\mathbf{a}_P = (312\mathbf{i} - 300\mathbf{j} - 468\mathbf{k})$ m/s^2.
21.37 (a) $\mathbf{v}_P = (-1.269\mathbf{i} - 0.770\mathbf{j} + 2.11\mathbf{k})$ m/s;
(b) $\mathbf{a}_P = (0.462\mathbf{i} - 1.436\mathbf{j} - 0.385\mathbf{k})$ m/s^2.
21.39 (a) $\mathbf{v}_A = (-7.50\mathbf{j})$ m/s; (b) $\mathbf{a}_A = (99.0\mathbf{i} + 12.50\mathbf{j} - 540\mathbf{k})$ m/s^2.
21.41 (a) $\mathbf{v}_D = (-15.00\mathbf{i} + 87.5\mathbf{j})$ ft/s;
(b) $\mathbf{a}_D = (-1{,}500\mathbf{i} - 150.0\mathbf{j} + 3{,}310\mathbf{k})$ ft/s^2.
21.43 (a) $\mathbf{v}_D = (-15.00\mathbf{i} + 87.5\mathbf{j})$ ft/s;
(b) $\mathbf{a}_D = (-1{,}800\mathbf{i} + 350\mathbf{j} + 3{,}130\mathbf{k})$ ft/s^2.
21.45 $\mathbf{v}_A = (-7.26\mathbf{i} + 34.6\mathbf{j} + 20.0\mathbf{k})$ ft/s, $\mathbf{a}_A = (-1{,}741\mathbf{i} - 182.5\mathbf{j})$ ft/s^2.
21.47 (a) $\mathbf{v}_C = (-0.069\mathbf{i} + 4.12\mathbf{j} + 3.54\mathbf{k})$ m/s;
(b) $\mathbf{a}_C = (-88.1\mathbf{i} + 10.05\mathbf{j} - 14.14\mathbf{k})$ m/s^2.
21.49 $\mathbf{a}_P = (2\omega_1\omega_2 R\sin\theta + \alpha_2 R\cos\theta)\mathbf{i} - [(\omega_1^2 + \omega_2^2)R\cos\theta]\mathbf{j} - (\omega_1^2 R\sin\theta)\mathbf{k}$.
21.51 $\mathbf{v}_B = (1.714\mathbf{i} - 1.286\mathbf{j})$ m/s.

21.53 (a) $\mathbf{v}_C = (-420\mathbf{i} + 34.6\mathbf{k})$ in./s;
(b) $\mathbf{a}_C = (-1{,}600\mathbf{i} - 18{,}400\mathbf{j} - 693\mathbf{k})$ in./s^2.

Chapter 22

22.1 $I_x = 0.01277$ kg·m^2, $k_x = 0.588$ m.
22.3 $I_x = 23.7$ kg·m^2, $k_x = 0.588$ m.
22.5 $I_{xy} = 4.5a^2m$, $I_{xz} = 0.9a^2m$, $I_{yz} = 3.0a^2m$.
22.7 $I_{xy} = 1.036$ slug·in.^{2}, $I_{xz} = I_{yz} = 1.036$ slug·in.^{2}.
22.9 $I_{OA} = 0.830a^2m$.
22.11 $I_{OA} = 60.4$ kg·m^2.
22.13 $I_u = 17.41ma^2$, $I_w = 17.19ma^2$, $I_v = 0.606ma^2$, $\lambda_x = 0.047$, $\lambda_y = -0.212$, $\lambda_z = 0.976$, for u axis.
22.15 $I_u = 130.6$ kg·m^2, $I_w = 94.6$ kg·m^2, $I_v = 54.5$ kg·m^2, $\lambda_x = 0.310$, $\lambda_y = 0.424$, $\lambda_z = -0.851$, for the w axis.
22.17 $N = 3.62$ rev.
22.19 $M_{AB} = 0.057$ N·m.
22.21 $\omega = 6.25$ rad/s.
22.23 $T = 10.28$ ft·lb.
22.25 $\mathbf{H}_G = 1.209\boldsymbol{\lambda}_u + 1.804\boldsymbol{\lambda}_v$, $H_G = 2.17$ kg·m^2/s, $\boldsymbol{\lambda}_{HG} = 0.557\boldsymbol{\lambda}_u + 0.831\boldsymbol{\lambda}_v$.
22.27 $\mathbf{H}_C = 0.2912\boldsymbol{\lambda}_u + 0.0364\boldsymbol{\lambda}_w$, $H_C = 0.293$ slug·ft^2/s, $\boldsymbol{\lambda}_{HC} = 0.994\boldsymbol{\lambda}_u + 0.124\boldsymbol{\lambda}_w$.
22.29 (a) $\boldsymbol{\omega} = (1.437\mathbf{j} - 9.37\mathbf{k})$ rad/s, (b) $v_G = (1.250\mathbf{i})$ m/s.
22.31 (a) $\boldsymbol{\omega} = \left(\dfrac{Q}{mb}\right)(0.290\mathbf{I} + 0.595\mathbf{J} - 3.560\mathbf{K})$;
(b) $\boldsymbol{\lambda} = 0.080\mathbf{I} + 0.164\mathbf{J} - 0.983\mathbf{K}$.
22.33 (a) $\boldsymbol{\omega} = (5.22\mathbf{I} - 1.320\mathbf{J} + 8.77\mathbf{K})$ rad/s;
(b) $\boldsymbol{\lambda} = 0.507\mathbf{I} - 0.128\mathbf{J} + 0.852\mathbf{K}$.
22.35 (a) $\boldsymbol{\omega} = (43.5\mathbf{i} - 130.5\mathbf{j} - 30.5\mathbf{k})$ rad/s;
(b) $\mathbf{v}_G = (0.805\mathbf{i} - 1.610\mathbf{j})$ ft/s.
22.37 $v_x = 3{,}120$ ft/s, $v_y = -30.0$ ft/s, $v_z = -3{,}910$ ft/s.
22.39 (a) $\boldsymbol{\omega} = (-4.32\mathbf{j})$ rad/s; (b) $\mathbf{v}_G = (-1.728\mathbf{k})$ m/s.
22.41 $A_x = 15.97$ N, $A_z = 58.7$ N, $B_x = -15.97$ N, $B_z = 88.5$ N, $\alpha = 667$ rad/s^2.
22.43 $A_X = -(\frac{3}{20})(M/b)$, $A_Y = -(\frac{1}{8})mb\omega^2$, $B_X = (\frac{3}{20})(M/b)$, $B_Y = (\frac{1}{8})mb\omega^2$, $B_Z = (\frac{3}{2})mg$, $\alpha = (\frac{6}{5})(M/mb^2)$.
22.45 (a) $M = 0.96$ N·m, (b) $A_X = 2.33$ N, $A_Z = 19.48$ N, $B_X = -2.33$ N, $B_Z = 206$ N.
22.47 $A_X = 0$, $A_Z = -294$ lb.
22.49 $A_X = 360$ N, $A_Z = -186.0$ N, $\alpha = 2400$ rad/s^2.
22.51 (a) $M = 6.55$ lb·ft, (b) $A_x = 0.758$ lb, $A_y = 102.0$ lb, $B_x = -0.758$ lb, $B_y = -102.0$ lb, $B_z = 180.0$ lb.
22.53 $A_x = B_x = 0$, $A_z = -12.50$ lb, $B_z = -12.50$ lb.

22.55 $\theta = \cos^{-1}\left(\frac{0.261g}{r\omega^2}\right)$, $A_X = 0$, $A_Y = -17.5m\sqrt{(r\omega^2)^2 - (0.261g)^2}$,
$A_Z = 5mg$.
22.57 $\mathbf{C} = -(103.8\mathbf{k})\ N$, $\mathbf{M}_C = (5.625\mathbf{k})\ \text{N}\cdot\text{m}$.
22.59 $\mathbf{M} = -(469\mathbf{k})\ \text{N}\cdot\text{m}$.
22.61 $\dot{\psi} = 179.0$ rad/s ↗↗z.
22.63 $\dot{\psi} = 193.4$ rad/s ↞ →Y.
22.65 $\dot{\psi} = 256$ rad/s ↗↗z.
22.67 $A_x = -B_x$ ↙↗x, $A_y = 2{,}380$ N ↑y, $A_z = 138.7$ N ← →z,
$B_y = 2{,}240$ N ↗.
22.69 $T = 1{,}638$ N.
22.71 $M_x = 718\ \text{N}\cdot\text{m}$.
22.73 $A_x = -B_x$ ↗↗x, $A_y = 918$ N y↑ ↓, $B_y = 967$ N ↑.
22.75 $\dot{\psi} = 0.938$ rad/s ↗↗z, $\beta = 37.6°$, $\omega = 2.13$ rad/s 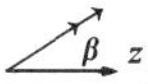 z.
22.77 $\dot{\psi} = 0.0135$ rad/s ↗↗z, $\beta = 87.8°$, $\omega = 0.0339$ rad/s 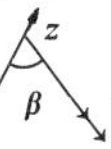 .
22.79 $I_x = 1.094\ \text{kg}\cdot\text{m}^2$, $I_y = 1.808\ \text{kg}\cdot\text{m}^2$, $I_z = 1.533\ \text{kg}\cdot\text{m}^2$.
22.81 $I_u = 1.996\ \text{kg}\cdot\text{m}^2$, $I_w = 1.777\ \text{kg}\cdot\text{m}^2$, $I_v = 0.663\ \text{kg}\cdot\text{m}^2$.
22.83 $\mathbf{H}_G = (-0.539\mathbf{I} + 0.135\mathbf{K})\ \text{lb}\cdot\text{ft}\cdot\text{s}$, $H_G = 0.556\ \text{lb}\cdot\text{ft}\cdot\text{s}$,
$\boldsymbol{\lambda}_{HG} = -0.969\mathbf{I} + 0.243\mathbf{K}$.
22.85 $\Delta t_A = 1.683$ s, $\Delta t_D = 0.208$ s.
22.87 $A_X = 0.889$ N ↗, $A_Z = 2.98$ N ↑, $B_X = 0.889$ N ↙, $B_Y = 0$,
$B_Z = 6.83$ N ↑, $\alpha = 41.4\ \text{rad/s}^2$ ↠.
22.89 $3.98 \le \omega \le 4.80$ rad/s.

Chapter 23

23.1 $p = 20.0$ rad/s, $\tau = 0.314$ s, $f = 3.18$ cps, $x_M = 0.0833$ ft,
$v_{MAX} = 1.667$ ft/s, $a_{MAX} = 33.3\ \text{ft/s}^2$.
23.3 (a) $k_e = 206$ lb/ft, $\tau = 0.310$ s; (b) $x = 0.0417\cos(20.3t)$,
$x_M = 0.0417$ ft; (c) $v_{MAX} = 0.847$ ft/s, $a_{MAX} = 17.18\ \text{ft/s}^2$.
23.5 (a) $\theta_M = 0.500$ rad; (b) $\dot{\theta}_{MAX} = 1.500$ rad/s;
(c) $\ddot{\theta}_{MAX} = 4.50\ \text{rad/s}^2$; (d) $L = 3.58$ ft; (e) $\tau = 2.09$ s;
(f) $f = 0.478$ cps; (g) At $t = 0$, $\theta = 0.500$ rad and $\dot{\theta} = 0$.
23.7 (a) $\ddot{\theta} = 0$; (b) $T = mg + mL\dot{\theta}^2$.
23.9 In both cases, $\ddot{x} + p^2x = 0$ is satisfied.
23.11 $\ddot{x} + 1000x = 0$, $p = 31.6$ rad/s, $\tau = 0.1988$ s, $f = 5.03$ cps.
23.13 $\theta = 0.15\cos pt$, $\dot{\theta} = -0.15p\sin pt$, $\ddot{\theta} = -0.15p^2\cos pt$.
23.15 (a) $k_e = 86.7$ lb/in.; (b) $x = 0.0833\cos(45.6t)$,
$\dot{x} = -3.8\sin(45.6t)$, $\ddot{x} = -173.3\cos(45.6t)$.
23.17 $\tau = 0.0811$ s, $f = 12.33$ cps.
23.19 (a) $k = 493$ lb/ft; (b) $k = 493$ lb/ft; (c) $k = 1{,}974$ lb/ft.

23.21 (a) $x_M = 13.00$ in.; (b) $\phi = 1.176$ rad;
(c) $\dot{x} = 20\cos(4t) - 48\sin(4t)$, (d) $\ddot{x} = -80\sin(4t) - 192\cos(4t)$.

23.23 (a) $\tau_1/\tau_2 = 1/\sqrt{6}$; (b) $f_1/f_2 = \sqrt{6}$; (c) $\dot{\theta}_{M1}/\dot{\theta}_{M2} = \sqrt{6}$;
(d) $\ddot{\theta}_{M1}/\ddot{\theta}_{M2} = 6$.

23.25 $x = 0.0792$ ft, $\dot{x} = 0.635$ ft/s, $\ddot{x} = -47.5$ ft/s^2.

23.27 (a) $\ddot{\theta} + [(3g/2L) + (3k/m)]\theta = 0$; (b) $p = \sqrt{(3g/2L) + (3k/m)}$,
$\tau = 2\pi/\sqrt{(3g/2L) + (3k/m)}$, $f = \left(\dfrac{1}{2\pi}\right)\sqrt{(3g/2L) + (3k/m)}$.

23.29 (a) $\ddot{\theta} + 114.715\,\theta = 0$;
(b) $p = 10.71$ rad/s, $\tau = 0.587$ s, $f = 1.705$ cps.

23.31 $\theta = 0.1396\cos(5.24t)$, $\dot{\theta} = -0.7315\sin(5.24t)$,
$\ddot{\theta} = -3.8331\cos(5.24t)$.

23.33 $p = \sqrt{3gb/(a^2 + 4b^2)}$, $\tau = 2\pi\sqrt{(a^2 + 4b^2)/3gb}$,
$f = \dfrac{1}{2\pi}\sqrt{3gb/(a^2 + 4b^2)}$; for $a = b$, $p = 0.775\sqrt{g/b}$,
$\tau = 8.112\sqrt{b/g}$, $f = 0.1233\sqrt{g/b}$.

23.35 $p = 1.922\sqrt{k/m}$, $\tau = 3.27\sqrt{m/k}$, $f = 0.306\sqrt{k/m}$.

23.37 $p = 15.69$ rad/s, $\tau = 0.400$ s, $f = 2.50$ cps.

23.39 $p = 0.576\sqrt{g/R}$, $\tau = 10.91\sqrt{R/g}$, $f = 0.0917\sqrt{g/R}$.

23.41 $p = 3.15$ rad/s, $\tau = 1.996$ s, $f = 0.500$ cps.

23.43 $p = \sqrt{\dfrac{(H + \pi R/4 + R^2/4H)g}{\frac{\pi}{16}(R^3/H) + (\frac{4}{3})R^2 + (\frac{4}{3})H^2 + (\frac{\pi}{2})HR}}$.

23.45 $\ddot{\theta} + (4k/3m)\theta = 0$, $\tau = 2\pi\sqrt{3m/4k}$. For $k = 600$ lb/ft and
$m = 5$ slug, $\tau = 0.497$ s, $\theta = 0.10\cos(12.65t)$,
$\dot{\theta} = -1.265\sin(12.65t)$, $\ddot{\theta} = -16.00\cos(12.65t)$.

23.47 $\theta = 0.209\cos(7.56t)$, $\dot{\theta} = -1.580\sin(7.56t)$,
$\ddot{\theta} = -11.95\cos(7.56t)$.

23.49 $\tau = 9.65\sqrt{b/g}$.

23.51 $f = 0.201\sqrt{g/b}$.

23.53 $f = 0.404\sqrt{k/m}$.

23.55 $\tau = 5.85\sqrt{b/g}$.

23.57 $\theta = 0.20\cos(0.725\sqrt{g/R}t)$,
$\dot{\theta} = -(0.145\sqrt{g/R})\sin(0.725\sqrt{g/R}t)$,
$\ddot{\theta} = -(0.1051\sqrt{g/R})\cos(0.725\sqrt{g/R}t)$.

23.59 $\ddot{\theta} + (mgL\sin\beta/I_A)\theta = 0$, where $I_A = (\frac{10}{3} + 2\cos\beta)mL^2$.

23.61 $\ddot{\theta} + (3g/5L)\theta = 0$.

23.63 $\ddot{\theta} + (g/b\tan\beta)\theta = 0$, $p = \sqrt{g/b\tan\beta}$.

23.65 $\ddot{\theta} + [3ga/(b^2 + 4a^2)]\theta = 0$, $p = \sqrt{3ga/(b^2 + 4a^2)}$.

23.67 $\theta = 0.08\cos(56.3t)$.

23.69 $f = 4.50$ cps.

23.71 $\ddot{\theta} + [(8\sqrt{2}/3\pi)(g/R)]\theta = 0$, $\tau = 5.73\sqrt{R/g}$.

23.73 $\ddot{\theta} + \left(\frac{3}{10}\right)\left(\frac{g}{L}\right)\theta = 0, \tau = 11.47\sqrt{L/g}.$

23.75 $\ddot{\theta} + (1.92k/m)\theta = 0, \tau = 4.53\sqrt{m/k}.$

23.77 $\ddot{\theta} + 0.583(g/b)\theta = 0, \tau = 8.23\sqrt{b/g}.$

23.79 Plate with hole: $\ddot{\theta}_1 + (0.6g/b)\theta_1 = 0$; plate without hole:

$\ddot{\theta}_2 + (0.556g/b)\theta_2 = 0, \frac{\tau_1}{\tau_2} = 0.963.$

23.81 $\ddot{\theta} + 0.75\left(\frac{g}{L}\right)\theta = 0, \tau = 7.26\sqrt{L/g}.$

23.83 $\ddot{x} + (8k/3m)x = 0.$

23.85 $\ddot{\theta} + [3ga/(4a^2 + b^2)]\theta = 0, \theta = 0.4 \cos pt$, where

$p = \sqrt{\frac{3ga}{4a^2 + b^2}}.$

23.87 $\ddot{x} + (6k/m)x = 0, p = 2.45\sqrt{k/m}, \tau = 2.57\sqrt{m/k},$
$f = 0.390\sqrt{k/m}.$

23.89 $\tau = 6.82\sqrt{L/g}, f = 0.1466\sqrt{g/L}.$

23.91 $\ddot{\theta} + (3.57k/m)\theta = 0, \tau = 3.32\sqrt{m/k}.$

23.93 $\ddot{\theta} + (0.443g/b)\theta = 0, \tau = 1.346$ s.

23.95 (a) $\frac{\omega}{p} = \frac{\sqrt{3}}{2}$; (b) $\frac{\omega}{p} = \frac{\sqrt{5}}{2}$.

23.97 (a) A.F. = 1.077; (b) $x_M = 0.0399$ ft.

23.99 (a) $\omega = 28.3$ rad/s; (b) $x_M = 0.1964$ ft.

23.101 (a) $k_e = 13.71$ kN/m; (b) $x_M = 0.1491$ m.

23.103 $\ddot{x} + (2k/m) = (P_M/m) \sin \omega t, k = 72.0$ lb/ft.

23.105 (a) $k_e = 2k$, (b) $p = 40.0$ rad/s, $x_M = 0.254$ m.

23.107 Use $k = 10{,}000$ lb/in.

23.109 $k = 453$ lb/ft.

23.111 (a) $\ddot{x} + (k/m)x = (P_M/m) \cos \omega t$; (b) $x_P = \left[\frac{(P_M/m)}{(k/m) - \omega^2}\right] \cos \omega t.$

23.113 $k = 4{,}280$ kN/m.

23.115 $U = e/[(p/\omega)^2 - 1].$

23.117 (a) $k_e = 2(k_1 + k_2)$;
(b) $k_e = 9$ kN/m, $\delta_M = 0.0444$ m, $x_M = 0.354$ m.

23.119 $C_1 = 0.05$ ft, $C_2 = 0$.

23.121 (a) $x_M/\delta_M = 1.038$; (b) $x_M = 1.038$ in.

23.123 (a) $c_c = 1{,}897$ N·s/m; (b) $x = 5te^{-(15.81t)}$.

23.125 $\phi = 0.524$ rad, $x_M = 0.313$ ft, $x_P = 0.313 \sin(4t - 0.524)$,
$\dot{x}_P = 1.251 \cos(4t - 0.524), \ddot{x}_P = -5.00 \sin(4t - 0.524).$

23.127 $\ddot{\theta} + 0.5294(g/b)\theta = 0.$

23.129 $\ddot{x} + (k/m)x = (P_M/m) \cos \omega t.$

23.131 $\ddot{\theta} + \left(\frac{3k}{m} + \frac{3g}{2L}\right)\theta = 0.$

23.133 $m_1\ddot{x}_1 + c_1\dot{x}_1 - k(x_2 - x_1) = 0,$
$m_2\ddot{x}_2 + k(x_2 - x_1) = P_M \sin \omega t.$
23.135 $m_1\ddot{x}_1 + 2kx_1 - 2k(x_2 - x_1) = 0,$
$m_2\ddot{x}_2 + 2k(x_2 - x_1) = P_M \sin \omega t.$
23.137 $x = 0.00675$ m, $\dot{x} = -2.21$ m/s, $\ddot{x} = -13.48$ m/s^2.
23.139 $\tau = 2\pi\sqrt{\left(\frac{mL^2}{K + mgL}\right)}.$
23.141 $k_G = (\sqrt{\frac{25}{8}})D.$
23.143 $L = 6.01$ ft.
23.145 $\tau = 2\pi\sqrt{\frac{10m}{3k}}.$
23.147 $p = \sqrt{6(kb^2 + mgb)/(5mb^2)}.$
23.149 (a) $\tau = 2\pi\sqrt{7m/12k}$; (b) $\tau = 0.598$ s.
23.151 (a) $k = 30.0$ lb/ft; (b) $k = 10.00$ lb/ft; (c) $k = 15.00$ lb/ft.
23.153 $k = 1{,}226$ lb/in.
23.155 $x_M = 0.00385$ m.
23.157 $m_1\ddot{x}_1 + k_1x_1 - k_2(x_2 - x_1) = 0,$
$m_2\ddot{x}_2 + k_2(x_2 - x_1) = P_M \sin \omega t.$

Index